	No.				
QUADRILATERALS	67	a ☐ a	Square	A	
	68	b ▭ a	Rectangle	Area $= ab$	
	69	a ⬓ h b	Parallelogram: Diagonals bisect each other	Area $= bh$	
	70	a ⬓ h a	Rhombus: Diagonals intersect at right angles	Area $= ah$	
	71	a ⬡ h b	Trapezoid	Area $= \dfrac{(a+b)h}{2}$	
POLYGON	72	n sides	Sum of Angles $= (n-2)\,180°$		189
CIRCLE	73	Definition of π	$\pi = \dfrac{\text{circumference}}{\text{diameter}} \cong 3.1416$		191
	74		Circumference $= 2\pi r = \pi d$		191
	75		Area $= \pi r^2 = \dfrac{\pi d^2}{4}$		47, 191
	76		Central angle θ (radians) $= \dfrac{s}{r}$		413
	77		Area of sector $= \dfrac{rs}{2} = \dfrac{r^2\theta}{2}$		410
	78		1 revolution $= 2\pi$ radians $= 360°$ $1° = 60$ minute 1 minute $= 60$ seconds		408
SOLIDS	87		Cube	Volume $= a^3$	198
	88		Cube	Surface area $= 6a^2$	198
	89		Rectangular Parallelepiped	Volume $= lwh$	197
	90		Rectangular Parallelepiped	Surface area $= 2(lw + hw + lh)$	197
	91		Any Cylinder or Prism	Volume $= (\text{area of base})(\text{altitude})$	197, 201
	92	Base	Right Cylinder or Prism	Lateral area (not incl. bases) $= (\text{perimeter of base})(\text{altitude})$	197, 201
	93		Sphere	Volume $= \frac{4}{3}\pi r^3$	203
	94		Sphere	Surface area $= 4\pi r^2$	203
	95		Any Cone or Pyramid	Volume $= \frac{1}{3}(\text{area of base})(\text{altitude})$	198, 202
	96		Right Circular Cone or Regular Pyramid	Lateral area $= \frac{1}{2}(\text{perimeter of base}) \times (\text{slant height})$ Does not include the base.	198, 202
	97	A_1	Any Cone or Pyramid	Volume $= \dfrac{h}{3}\left(A_1 + A_2 + \sqrt{A_1 A_2}\right)$	198, 202
	98	Frustum A_2	Right Circular Cone or Regular Pyramid	Lateral area $= \dfrac{s}{2}(\text{sum of base perimeters}) = \dfrac{s}{2}(P_1 + P_2)$	198, 202

www.wileyplus.com

ALL THE HELP, RESOURCES, AND PERSONAL SUPPORT YOU AND YOUR STUDENTS NEED!

www.wileyplus.com/resources

2-Minute Tutorials and all of the resources you & your students need to get started.

Student support from an experienced student user.

Collaborate with your colleagues, find a mentor, attend virtual and live events, and view resources.
www.WhereFacultyConnect.com

Pre-loaded, ready-to-use assignments and presentations. Created by subject matter experts.

Technical Support 24/7 FAQs, online chat, and phone support.
www.wileyplus.com/support

Your *WileyPLUS* Account Manager. Personal training and implementation support.

TECHNICAL MATHEMATICS
WITH CALCULUS

PAUL A. CALTER, MSME, MFA

Professor Emeritus
Vermont Technical College

MICHAEL A. CALTER, PH.D.

Associate Professor
Wesleyan University

WILEY   JOHN WILEY & SONS, INC.

◆◆◆

To Rachel, Christopher, and Kaitlin

VP & PUBLISHER	Laurie Rosatone
EDITOR	Jennifer Brady
MARKETING MANAGER	Debi Doyle
MEDIA EDITOR	Melissa Edwards
PRODUCTION MANAGER	Dorothy Sinclair
PRODUCTION EDITOR	Sandra Dumas
DESIGNER	Wendy Lai
PHOTO DEPARTMENT MANAGER	Hilary Newman
PRODUCTION MANAGEMENT SERVICES	Preparé, Inc.
COVER PHOTO	© Tetra Images/Getty Images, Inc.

This book was typeset in 10/12 Times at Preparé and printed and bound by Courier (Kendallville). The cover was printed by Courier (Kendallville).

Founded in 1807, John Wiley & Sons, Inc. has been a valued source of knowledge and understanding for more than 200 years, helping people around the world meet their needs and fulfill their aspirations. Our company is built on a foundation of principles that include responsibility to the communities we serve and where we live and work. In 2008, we launched a Corporate Citizenship Initiative, a global effort to address the environmental, social, economic, and ethical challenges we face in our business. Among the issues we are addressing are carbon impact, paper specifications and procurement, ethical conduct within our business and among our vendors, and community and charitable support. For more information, please visit our Web-site: **www.wiley.com/go/citizenship**.

The paper in this book was manufactured by a mill whose forest management programs include sustained-yield harvesting of its timberlands. Sustained-yield harvesting principles ensure that the number of trees cut each year does not exceed the amount of new growth.

This book is printed on acid-free paper.

ISBN 13   978-0470-46472-4

Printed in the United States of America.
10 9 8 7 6 5 4

Preface

This textbook has been in continuous classroom use since 1980, and it was again time to polish and refine the material and fill in where needed. It was also an opportunity to carry out suggestions for improvements made by many reviewers and colleagues, as well as those that occurred to the authors while using the preceding edition. Much has been rewritten to be cleaner and clearer, and new features have been introduced. To reduce size and weight, some peripheral topics have been moved to the Web. One chapter previously moved to the Web, *Introduction to Statistics and Probability*, has been returned to the text at the request of reviewers. Also, *Analytic Geometry*, formerly in the calculus version only, is now in both versions.

Features of the Book

Each chapter begins with a listing of **Chapter Objectives** that state specifically what the student should be able to do upon completion of the chapter. Following that, we have tried to present the material as clearly as possible, preferring an intuitive approach rather than an overrigorous one. Realizing that a mathematics book is not easy reading, we have given information in small segments, included many illustrations, and have designed each page with care.

The numerous **Examples** form the backbone of the textbook, and we have added to their number. In many we have added intermediate steps to make them easier to follow. They are fully worked out and are chosen to help the student do the exercises. Examples have markers above and below to separate them clearly from the text discussion.

To give students the essential practice they need to learn mathematics, we include thousands of **Exercises**. Exercises given after each section are graded by difficulty and grouped by type, to allow practice on a particular area. These are indicated by title, as well as by number. The **Chapter Review Problems** are scrambled as to type and difficulty. Answers to all odd-numbered problems are given in the **Answer Key** in the Appendix, every answer is included in the **Annotated Instructor's Edition**, and complete solutions to every problem are contained in the **Instructor's Solutions Manual**. Complete solutions to every other odd problem are given in the **Student's Solutions Manual**.

The book contains hundreds of high-quality, clear **illustrations**, each carefully selected for its inclusion. When the same figure is used twice but in different locations, it is now included in both locations so that students do not have to search for it.

The examples, the text, and the exercises include many **Technical Applications**. We have added applications, especially for the building trades, machine shop, and woodworking shop. Many chapters have a large block of applications, and some of these have been moved forward into the preceding text to provide motivation for the student.

Students are encouraged to try some applications outside their chosen field, and everything they need to work these problems is given right here in this text. However, space does not permit the full discussion of all the background material for each technical field.

We have tried to avoid contrived "school" problems with neat solutions and include many **Problems with Approximate Solutions**. These include expressions and equations with approximate constants, but those that do not yield to many of the exact methods we teach, and must be tackled with an approximate method.

The **Index to Applications** should help in finding specific applications.

We have all seen wild answers on homework and exams, such as "the cost of each pencil is $300." To try to avoid that, we have added **Estimation** steps to many examples, where we have tried to show students how to estimate an answer in order to check their work. We give suggestions for estimation in the chapter on word problems. Thereafter, many applications examples begin with an estimation step or end with a check, either by graphing, by computer, by calculator, by an alternate solution, by making a physical model, or simply to **Examine the Answer for Reasonableness**.

Formulas used in the text are boxed and numbered and listed in the Appendix as the **Summary of Facts and Formulas**. This listing can function as a "handbook" for a mathematics course and for other courses as well and provides a common thread between chapters. We hope it will also help a student see interconnections that might otherwise be overlooked. The formulas are grouped logically in the Summary of Facts and Formulas and are numbered sequentially there. Therefore, the formulas do not necessarily appear in numerical order in the text.

When a listed formula is needed, it is now given right in the text so students do not have to flip through the formula summary to find it.

In addition to mathematical formulas, we include some from technology, motion, electric circuits, and so on. These are grouped together at the end of the Formula Summary and have formula numbers starting with 1000.

We continue the popular feature of **Common Error Boxes** to emphasize some of the pitfalls and traps that "get" students year after year.

The **Graphics Calculator** has been fully integrated throughout. Calculator instruction and examples are given in the text, where appropriate, and calculator problems are given in the exercises. To avoid being too vague and general, we specifically give keystrokes for the Texas Instruments TI-83, TI-84, and TI-89. Our hope is that other calculators are similar enough for these instructions to be useful.

Many problems are given that can be solved practically *only* by a graphics calculator, and the graphics calculator is sometimes used to verify a solution found by another method. However, we have still retained most of the noncalculator methods, such as manual graphing by plotting of point pairs, for those who want to present these methods. **Graphical and calculator methods** are emphasized much more than before. We give **calculator screens**, when a calculator topic is introduced, and perhaps for a few more examples. Screens for those operations are then dropped to avoid cluttering the pages. To our treatment of the arithmetic scientific calculator, we have added the **Symbolic Scientific Calculator** and show screens where appropriate.

We have expanded the use of **Guided Explorations**. Our hope is that they will lead the student to make personal discoveries and gain a more personal appreciation for the concepts.

Every chapter contains optional enrichment activities with the title **Writing**, **Projects**, or **Internet**. These were formerly at the end of each chapter, but we have now moved them to the exercise that is most appropriate. Our hope is that a few students may be attracted to the magic and history of mathematics and welcome a guided introduction into this world. Here we put **Writing Questions** to test and expand a student's knowledge of the material and perhaps explore areas outside of those covered, **Team Projects** to foster "collaborative learning," and **Internet** activities, including references to our companion Web site.

With the margins of the book becoming crowded with calculator screens, in addition to illustrations, we have moved many of the **Marginal Notes** to the text. They are used mostly for encouragement and historical notes.

Teaching and Learning Resources

We provide several supplements to aid both the instructor and the student.

An **Annotated Instructor's Edition** (AIE) of this text contains answers to every exercise and problem. The answers are placed in red right in the exercise or problem. The AIE also has red marginal notes to the instructor, giving teaching tips, applications, and practice problems. ISBN: 978-0470-53495-3

An **Instructor's Solution Manual** contains worked out solutions to every problem in the text and a listing of all computer programs. ISBN: 978-1118-06124-4

The **Student Solutions Manual** gives the solution to every other odd problem. They are usually worked in more detail than in the Instructor's Solution Manual. ISBN: 978-0470-53494-6

WileyPLUS is a powerful online tool that provides instructors with an integrated suite of resources, including an online version of the text, in one easy-to-use Web site. Organized around the essential activities you perform in class, Wiley-PLUS allows you to create class presentations, assign homework and quizzes for automatic grading, and track student progress. Please visit *www.wileyplus.com* or contact your local Wiley representative for a demonstration and further details.

A **Computerized Test Item File** is a bank of test questions with answers. Questions may be mixed, sorted, changed, or deleted. It consists of a test file disk and a test generator disk, ready to run. ISBN: 978-0470-53497-7

Our **Companion Web Site** (www.wiley.com/college/calter) contains all the less frequently used material moved from the preceding edition, as well as complete solutions to every problem in the text.

Acknowledgments

We are extremely grateful to reviewers of this edition and the earlier editions of the book, reviewers of the supplements and the writing questions, and participants in group discussions about the book. They are

A. David Allen, *Ricks College*

Byron Angell, *Vermont Technical College*

David Bashaw, *New Hampshire Technical Institute*

Jim Beam, *Savannah Area Vo-Tech*

Elizabeth Bliss, *Trident Technical College*

Franklin Blou, *Essex County College*

Donna V. Boccio, *Queensboro Community College*

Jacquelyn Briley, *Guilford Technical Community College*

Frank Caldwell, *York Technical College*

James H. Carney, *Lorain County Community College*

Cheryl Cleaves, *State Technical Institute at Memphis*

Ray Collings, *Tri-County Technical College*

Miriam Conlon, *Vermont Technical College*

Robert Connolly, *Algonquin College*

Amy Curry, *College of Lake County*

Kati Dana, *Norwich University*

Linda Davis, *Vermont Technical College*

Dennis Dura, *Cuyahoga Community College*

John Eisley, *Mott Community College*

Walt Granter, *Vermont Technical College*

Crystal Gromer, *Vermont Technical College*

Richard Hanson, *Burnsville, Minnesota*

Tommy Hinson, *Forsythe Community College*

Margie Hobbs, *State Technical Institute at Memphis*

Martin Horowitz, *Queensborough Community College*

Glenn Jacobs, *Greenville Technical College*

Wendell Johnson, *Akron, Ohio*

Joseph Jordan, *John Tyler Community College*

Frank L. Juszli

Rob Kimball, *Wake Technical Community College*

John Knox, *Vermont Technical College*

Bruce Koopika, *Northeast Wisconsin Technical College*

Ellen Kowalczyk, *Madison Area Technical College*

Fran Leach, *Delaware Technical College*

Jon Luke, *Indiana University-Purdue University*

Michelle Maclenar, *Terra Community College*

Paul Maini, *Suffolk County Community College*

Edgar M. Meyer, *St. Cloud State, Minnesota*

David Nelson, *Western Wisconsin Technical College*

Mary Beth Orange, *Erie Community College*

Harold Oxsen, *Walnut Creek, California*

Ursula Rodin, *Nashville State Technical Institute*

Jason Rouvel, *Western Technical College*

Donald Reichman, *Mercer County Community College*

Bob Rosenfeld, *Nassau Community College and University of Vermont*

Nancy J. Sattler, *Terra Technical College*

Frank Scalzo, *Queensborough Community College*

Ned Schillow, *Lehigh Carbondale Community College*

Blin Scatterday, *University of Akron Community and Technical College*

Edward W. Seabloom, *Lane Community College*

Robert Seaver, *Lorain Community College*

Saeed Shaikh, *Miami Dade Community College*

Thomas Stark, *Cincinnati Technical College*

Fereja Tajir, *Illinois Central College*

Dale H. Thielker, *Ranken Technical College*

William N. Thomas, Jr., *Thomas & Associates Group*

Joel Turner, *Blackhawk Technical Institute*

Tingxiu Wang, *Western Missouri State University*

Roy A. Wilson, *Cerritos College*

Jeffrey Willmann, *Maine Maritime Academy*

Douglas Wolansky, *North Alberta Institute of Technology*

Karl Viehe, *University of the District of Columbia*

Henry Zatkis, *New Jersey Institute of Technology*

The solutions to all problems were checked by Susan Porter, who also did developmental editing. Accuracy checking and proofreading were done by John Morin and James Ricci and the copyediting was done by Martha Williams. The authors are grateful to our Project Editor at John Wiley & Sons, Jennifer Brady, Production Editor Sandra Dumas, and Publisher Laurie Rosatone, who have helped to bring this book to completion.

Thank you all.

Michael A. Calter

Middletown, CT

mcalter@wesleyan.edu

www.wesleyan.edu/chem/faculty/calter/

Paul A. Calter

Randolph Center, VT

pcalter@sover.net

http://www.sover.net/~pcalter/

About the Authors

Paul Calter is Professor of Mathematics Emeritus at Vermont Technical College and Visiting Scholar at Dartmouth College. A graduate of the engineering school of The Cooper Union, New York, he received his M.S. in mechanical engineering from Columbia University and a M.F.A. in sculpture from the Vermont College of Fine Arts. Professor Calter has taught Technical Mathematics for over 25 years. In 1987, he was the recipient of the Vermont State College Faculty Fellow Award.

He is member of the American Mathematical Association of Two Year Colleges, the Mathematical Association of America, the National Council of Teachers of Mathematics, the College Art Association, and the Author's Guild.

Calter is involved in the Mathematics Across the Curriculum movement and has developed and taught a course called Geometry in Art and Architecture at Dartmouth College, under an NSF grant.

Professor Calter is the author of several other mathematics textbooks, among which are the *Schaum's Outline of Technical Mathematics, Problem Solving with Computers, Practical Math Handbook for the Building Trades, Practical Math for Electricity and Electronics, Mathematics for Computer Technology, Introductory Algebra* and *Trigonometry, Technical Calculus,* and *Squaring the Circle: Geometry in Art and Architecture.*

Michael Calter is an Associate Professor at Wesleyan University. He received his B.S. from the University of Vermont. After receiving his Ph.D. from Harvard University, he completed a postdoctoral fellowship at the University of California at Irvine. Michael has been working on his father's mathematics texts since 1983, when he completed a set of programs to accompany *Technical Mathematics with Calculus.* Since that time, he has become progressively more involved with his father's writing endeavors, culminating with becoming co-author of the second edition of *Technical Calculus* and the fourth edition of *Technical Mathematics with Calculus.* Michael also enjoys the applications of mathematical techniques to chemical and physical problems as part of his academic research. Michael is a member of the American Mathematical Association of Two Year Colleges, the American Association for the Advancement of Science, and the American Chemical Society.

Michael and Paul enjoy hiking and camping trips together. These have included an expedition up Mt. Washington in January, a hike across Vermont, a walk across England on Hadrian's Wall, and many sketching trips into the mountains.

Contents

Appendices

Indexes

On our web site (www.wiley.com/college/calter)

Binary, Hexadecimal, Octal and BCD Numbers

Boolean Algebra

Graphs on Logarithmic and Semilogarithmic Paper

Inequalities and Linear Programming

Infinite Series

Matrices

Methods of Integration

Simple Equations of Higher Degree

Solving Differential Equations by the Laplace Transform and by Numerical Methods

Review of Numerical Computation

♦♦♦ **OBJECTIVES** ♦♦

When you have completed this chapter, you should be able to

- Perform basic arithmetic operations on signed numbers.
- Perform basic arithmetic operations on approximate numbers.
- Take powers, roots, and reciprocals of signed and approximate numbers.
- Perform combined arithmetic operations to obtain a numerical result.
- Convert numbers between decimal, scientific, and engineering notation.
- Perform basic arithmetic operations on numbers in scientific and engineering notation.
- Convert units of measurement.
- Substitute given values into formulas.
- Solve common percentage problems.

♦♦

We start this first chapter with some definitions to refresh your memory of terms you probably already know. We will point out the difference between exact and *approximate* numbers, a distinction you may not have made in earlier mathematics classes. Then we will perform the ordinary arithmetic operations—addition and subtraction, multiplication and division—but here it may be a bit different from what you are used to. We will use the calculator extensively, which is probably not new to you, but now we will take great care to decide how many digits of the calculator display to keep. Why not keep them all? We will show that when working with approximate numbers keeping too many digits is misleading to anyone who must use the result of your calculation. As a further complication, we will combine both exact and approximate numbers, as well as positive and *negative* numbers, or *signed numbers*. As we proceed, we will point out some rules that will help get us ready for our next chapter on *algebra*, which is a generalization of arithmetic.

Next we will show compact ways to write a very long or very short number, in *scientific notation* or in *engineering notation*. These are important for you to know, not only for your own use but for you to understand them when you come across such numbers in reading technical material.

In technical work, we usually deal with numbers that indicate some measured quantity. Here we show how to convert a number from one unit of measurement to another, say feet to meters, how to use numbers with units of measure in computations, and how to substitute numbers with units into technical formulas. All are vital skills for technical work. Finally we will cover *percentage*. Of all the mathematical topics we cover in this text, probably the one most used in everyday life is percentage.

This is a long chapter. With its many different topics, it may appear choppy and disconnected. The good news is that most of the material should be familiar to you, with perhaps a few new twists. Throughout the chapter, as elsewhere in the book, we will give some help with the use of the calculator. But with so many types of calculators available, we are limited in what we can do, and you will really have to consult the manual that came with your calculator. We urge you to do this now, so by the time you reach Chapter 2 you will be able to calculate with speed and accuracy for the operations shown here. Computations for trigonometry and for logarithms will be covered as we get to them.

1–1   The Real Numbers

In mathematics, as in many other fields, we must learn many new terms. These definitions will make it easier to talk about mathematical ideas later.

Integers

The *integers*

$$\ldots, -4, -3, -2, -1, 0, 1, 2, 3, 4, \ldots$$

are the *whole numbers*, also called the *natural numbers* or *counting numbers*, including zero and negative values. The three dots on the ends indicate that the sequence of numbers continues indefinitely in both directions.

Rational and Irrational Numbers

The *rational numbers* include the integers and all other numbers that can be expressed as the quotient of two integers. Some rational numbers are

$$\frac{1}{2}, \quad -\frac{3}{5}, \quad \frac{57}{23}, \quad -\frac{98}{99}, \quad \text{and} \quad 7 \left(\text{or } \frac{7}{1} \right)$$

Numbers that cannot be expressed as the quotient of two integers are called *irrational*. Some irrational numbers are

$$\sqrt{2}, \quad \sqrt[3]{5}, \quad -\sqrt{7}, \quad \pi, \quad \text{and} \quad e$$

where π is approximately equal to 3.1416 and e is approximately equal to 2.7182. We will have much more to say about the irrational numbers π and e later in the book.

Real and Imaginary Numbers

The rational and irrational numbers together make up the *real numbers*. Numbers such as $\sqrt{-4}$ do not belong to the real number system. They are called *imaginary numbers* and are discussed in a later chapter. Except when otherwise noted, all the numbers we will work with are real numbers.

Decimal Numbers

Most of our computations are with numbers written in the familiar *decimal* system. The names of the places relative to the *decimal point* are shown in Fig. 1–1. We say that the decimal system uses a *base of 10* because it takes 10 units in any place to equal 1 unit in the next-higher place. For example, 10 units in the hundreds position equals 1 unit in the thousands position.

The numbers 10^2, 10^3, etc., are called *powers of 10*. Don't worry if they are unfamiliar to you. We will explain them later.

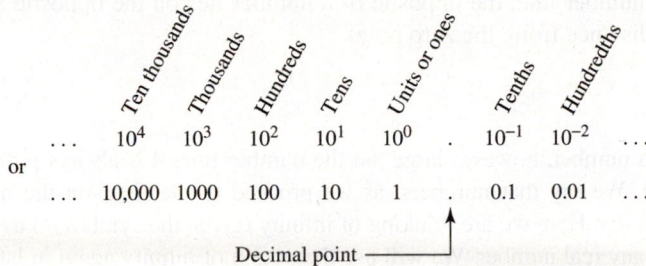

FIGURE 1–1 Values of the positions in a decimal number.

Positional Number Systems and Place Value

A *positional* number system is one in which the *position* of a digit determines its value. Our decimal system is positional.

Each position in a number has a *place value* equal to the base of the number system raised to the position number. The place values in the decimal number system, as well as the place names, are shown in Fig. 1–1.

The Number Line

A mathematical idea is much easier to grasp if shown as a *picture*; thus we will try to picture ideas whenever possible. Such a picture will often be in the form of a *graph*, and the simplest graph is the *number line* (Fig. 1–2). We draw a line on which we mark a zero point, and indicate the direction of increasing values. The line is usually drawn horizontal with increasing values taken to the right, marked with an arrowhead. We next indicate a *scale,* with consecutive numbers equally spaced along the line.

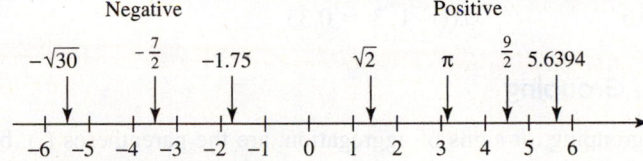

FIGURE 1–2 The number line.

Signed Numbers

A *positive* number is a number that is *greater* than zero, and a *negative* number is *less* than zero. On the number line it is customary to show the positive numbers to the right of zero and the negative numbers to the left of zero. These numbers may be integers, fractions, rational numbers, or irrational numbers.

To distinguish negative numbers from positive numbers, we always place a negative sign $(-)$ in front of a negative number. We usually omit writing the positive sign $(+)$ in front of a positive number. Thus a number without a sign is always assumed to be positive. In this chapter we will often write in a $+$ sign for emphasis.

The Opposite of a Number

The *opposite* of a number n is the number which, when added to n, gives a sum of zero.

♦♦♦ **Example 1:** The opposite of 2 is -2, because $2 + (-2) = 0$. The opposite of -6 is $+6$.   ♦♦♦

On the number line, the opposite of a number lies on the opposite side of, and at an equal distance from, the zero point.

Infinity

If we place a number, however large, on the number line, it is always possible to find a *larger* one. We say that numbers, as we proceed to the right on the number line, *approach infinity*. Here we are thinking of infinity (given the symbol ∞) as some value greater than any real number. We will use the notion of infinity again in later chapters.

Symbols of Equality and Inequality

Several symbols are used to show the relative positions of two quantities on the number line.

$a = b$ means that a *equals* b, and that a and b occupy the same position on the number line.

$a \neq b$ means that a and b are *not equal* and have different locations on the number line.

$a > b$ means that a is *greater than* b, and a lies to the right of b on the number line.

$a < b$ means that a is *less than* b, and a lies to the left of b on the number line.

$a \approx b$ means that a is *approximately equal to* b and that a and b are *near* each other on the number line. Other symbols sometimes used for approximately equal to are $\cong$ and $\simeq$.

♦♦♦ **Example 2:** Here are examples of the use of symbols of equality and inequality.

(a) $4 < 8$ (b) $47 > 24$

(c) $29 > -37$ (d) $-77 < -48$

(e) $^1/_2 \neq 0.6$ (f) $1/3 \approx 0.33$ ♦♦♦

Symbols of Grouping

Symbols of grouping, or signs of aggregation, are the parentheses ( ), brackets [ ], braces { }, and the bar —, also called a *vinculum*. Each symbol means that the terms enclosed are to be treated as a single term.

♦♦♦ **Example 3:** In each of these expressions, the quantity $a + b$ is to be treated collectively.

$$(a + b) \qquad [a + b] \qquad \{a + b\}$$

$$\frac{c}{a + b} \qquad \frac{a + b}{c} \qquad \sqrt{a + b}$$ ♦♦♦

Absolute Value

The *absolute value* of a number n is its *magnitude* regardless of its algebraic sign. It is written $|n|$. It is the distance between n and zero on the number line, without regard to direction.

◆◆◆ **Example 4:** Here is the evaluation of some expressions containing absolute value signs. See if you get the same results.

(a) $|5| = 5$

(b) $|-9| = 9$

(c) $|3 - 7| = |-4| = 4$

(d) $-|-4| = -4$

(e) $-|7 - 21| - |13 - 19| = -|-14| - |-6| = -14 - 6 = -20$ ◆◆◆

Some calculators have a key for evaluating absolute values. Use it to evaluate any of these expressions.

Approximate Numbers

Most of the numbers we deal with in technology are *approximate*.

◆◆◆ **Example 5:**

(a) All numbers that represent *measured* quantities are approximate. A certain shaft, for example, is approximately 1.75 inches in diameter.
(b) Many *fractions* can be expressed only approximately in decimal form. Thus $\frac{2}{3}$ is approximately equal to 0.6667.
(c) *Irrational numbers* can be written only approximately in decimal form. The number $\sqrt{3}$ is approximately equal to 1.732. ◆◆◆

Exact Numbers

An approximate number always has some *uncertainty* in the rightmost digit. That is, we cannot be sure of its exact value. On the other hand, an *exact number* is one that has no uncertainty.

◆◆◆ **Example 6:**

(a) There are exactly 24 hours in a day, no more, no less.
(b) Most automobiles have exactly four wheels.
(c) Exact numbers are usually integers, but not always. For example, there are *exactly* 2.54 cm in an inch, by definition.
(d) On the other hand, not all integers are exact. For example, a certain town has a population of *approximately* 3500 people. ◆◆◆

Significant Digits and Accuracy

In a decimal number, zeros are sometimes used just to locate the decimal point. When zeros are used in that way, we say that they are *not significant*. The remaining digits in the number, including zeros, are called *significant digits*.

◆◆◆ **Example 7:**

(a) The numbers 497.3, 39.05, 8003, 140.3, and 2.008 each have *four* significant digits.
(b) The numbers 1570, 24,900, 0.0583, and 0.000583 each have *three* significant digits. The zeros in these numbers serve only to locate the decimal point.

An overscore is sometimes placed over the last trailing zero that is significant. Thus the numbers 3950̄ and 735,0̄00 each have four significant digits.

(c) The numbers 18.50, 1.490, and 2.000 each have *four* significant digits. The zeros here are not needed to locate the decimal point. They are placed there to show that those digits are in fact zeros, and not some other digit. ◆◆◆

The number of significant digits in a number is often called the *accuracy* of that number. Thus the numbers in Example 7(a) are said to be *accurate to* four significant digits. Knowing the number of significant digits in a number is important for multiplication and division, as we will see in the next section.

◆◆◆ **Example 8:** Verify the number of significant digits in each approximate number.

(a) 39.3   has 3   (b) 274.2   has 4

(c) 3700   has 2   (d) 3.000   has 4

(e) 0.0486   has 3   (f) 3700.0   has 5   ◆◆◆

Decimal Places and Precision

We will see that to add or subtract a number properly, we need to know its *number of decimal places*. To find it, we simply count the number of digits to the right of the decimal point. The number of decimal places is often called the *precision* of the decimal number.

Keep in mind that we are talking about accuracy and precision of *numbers,* not of *measurements.* The *accuracy of a measurement* of some quantity refers to the nearness of the measured value to the "true," correct, or accepted value of that quantity. The *precision of measurements* is a measure of the *repeatability* of a group of measurements, that is, how close together a group of measurements are to *each other.*

◆◆◆ **Example 9:**

(a) The number 395.2 has one decimal place. We say it is precise to one decimal place, or precise to the nearest tenth.
(b) The number 7.284 is precise to three decimal places or precise to the nearest thousandth.
(c) The number 23,800 has no decimal places but is accurate to three significant digits and precise to the nearest hundred.
(d) In the number 18.30, the trailing zero is significant. Therefore it is accurate to four significant digits and precise to the nearest hundredth. ◆◆◆

Thus when using an approximate number, we need to be clear about its number of (a) significant digits and (b) decimal places. These will govern how we treat that number in a calculation. Which of these we call accuracy and which we call precision is not as important, especially as the two words are often confused, even in technical work.

◆◆◆ **Example 10:** Verify the number of decimal places in each approximate number.

(a) 39.3   has 1   (b) 274.2   has 1

(c) 3700   has 0   (d) 3.000   has 3

(e) 0.0486   has 4   (f) 3700.0   has 1   ◆◆◆

Rounding

In the next few sections, we will see that the numbers we get from a computation often contain *worthless digits* that must be *thrown away.* Whenever we do this, we must *round* our answer.

Round down (do not change the last retained digit) when the first discarded digit is 4 or less. *Round up* (increase the last retained digit by 1) when the first discarded digit is 6 or more, or a 5 followed by a nonzero digit in any of the decimal places to the right.

Sometimes we must round to a certain number of decimal places, and other times we must round to a certain number of significant digits. The procedure is no different.

◆◆◆ **Example 11:** Here are some numbers rounded to four significant digits.

Number	Rounded to Four Significant Digits
395.67	395.7
1.09356	1.094
0.0057284	0.005728

◆◆◆

We have seen that the rightmost digit in an approximate number has some uncertainty, but how much? If that last digit is the result of rounding in a previous step, it could be off by as much as *half a unit, either greater or smaller.* This is its *uncertainty.*

◆◆◆ **Example 12:** Here are some examples of rounding to three decimal places.

Number	Rounded to Three Decimal Places
4.3654	4.365
4.3656	4.366
4.365501	4.366
1.764999	1.765
1.927499	1.927

◆◆◆

When the discarded portion is 5 *exactly*, it usually does not matter whether you round up or down. The exception is when you are adding or subtracting a long column of figures, as in statistical computations. If, when discarding a 5, you always rounded up, you could bias the result in that direction. To avoid that, you want to round up about as many times as you round down, and a simple way to do that is to always *round to the nearest even number.* This is just a convention. We could just as well round to the nearest odd number.

◆◆◆ **Example 13:**

Number	Rounded to Two Decimal Places
4.365	4.36
4.355	4.36
7.76500	7.76
7.75500	7.76

◆◆◆

◆◆◆ **Example 14:** The approximate number 35.85, rounded to one decimal place, is 35.8. The number 35.75, rounded to one decimal place, is also 35.8. Thus the number 35.8 could be the rounded value of any number between 35.75 and 35.85. There is simply no way to tell what value may have been in the second decimal place. Now if there is uncertainty in a particular decimal place, it is clear that the values in any places to its right are completely unknown. ◆◆◆

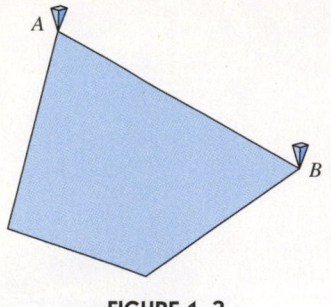

FIGURE 1–3

◆◆◆ **Example 15:** *An Application.* In laying out a ground plan, Fig. 1–3, the distance *AB* is calculated to be 35.8368 ft. Knowing that the surveyors can only measure to a hundredth of a foot, how would you give this dimension on the site plan?

Solution: We would round to two decimal places, getting

$$35.84 \text{ ft}$$

◆◆◆

Exercise 1 ◆ The Real Numbers

Symbols of Equality and Inequality Insert the proper symbol of equality or inequality ($=, \approx, >, <$) between each pair of numbers.

1. 7 and 10
2. 9 and -2
3. -3 and 4
4. -3 and -5
5. $\frac{3}{4}$ and 0.75
6. $\frac{2}{3}$ and 0.667

Absolute Value Evaluate each expression.

7. $|4|$
8. $|-3|$
9. $-|-6|$
10. $-|9 - 23| - |-7 + 3|$
11. $|12 - 5 + 8| - |-6| + |15|$
12. $-|3 - 9| - |5 - 11| + |21 + 4|$

Significant Digits and Decimal Places

Determine the number of significant digits in each approximate number.

13. 78.3
14. 9274
15. 4.008
16. 9400
17. 20,000
18. 5000.0
19. 0.9972
20. 1.0000

Determine the number of decimal places in each approximate number.

21. 39.5
22. 9.55
23. 5.882
24. 193

Rounding

Round each number to two decimal places.

25. 38.468
26. 1.996
27. 96.835001
28. 55.8650
29. 398.372
30. 2.9573

Round each number to one decimal place.

31. 13.98
32. 745.62
33. 5.6501
34. 0.482
35. 398.36
36. 34.927

Round each number to the nearest hundred.

37. 28,583
38. 7550
39. 3,845,240
40. 274,837

Round each number to three significant digits.

41. 9.284
42. 2857
43. 0.04825
44. 483,982
45. 0.08375
46. 29.555

Round each number to five significant digits.

47. 34.9274
48. 827.365
49. 4.03726
50. 0.00365286

51. 5.937254

52. 374.8264

53. Evaluate the expressions in problems 7 through 12 by calculator. On the TI-83/84 and TI-89 it is indicated by **abs(** and is located in the $\boxed{\text{MATH}}$ $\boxed{\text{NUM}}$ menu.

An Application

54. When calculating the required length of a girder, an architect gets a value of 14.8363 ft on her calculator. What dimension should she put on the plans if it is customary to specify griders to the nearest hundredth of a foot?

55. *Team Project:* Make a drawing of a cylindrical steel bar, 1 inch in diameter and 3 in. long. Label the diameter as 1.00 in. Take your drawing to a machine shop and ask for a cost estimate for each of six bars, having lengths of

    3 in.   3.000 in.

    3.0 in.   3.0000 in.

    3.00 in.   3.00000 in.

Before you go, have each member of your team make cost estimates.

56. *Internet:* Systems of numbers having bases other than 10 are used in computer science. They are *binary numbers* (base 2), *octal numbers,* (base 8), and *hexadecimal numbers* (base 16). A complete chapter on these kinds of numbers, which you may download and print, is located on our Web site at www.wiley.com/college/calter

1–2 Addition and Subtraction

Now that we have refreshed our memory about the different kinds of numbers, let's see how they are used in the various arithmetic operations. We will start with addition and subtraction.

Adding and Subtracting Integers by Calculator

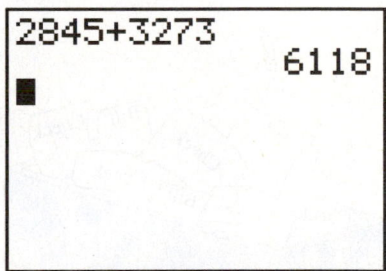

There are many types of calculators in use. In this text we will show screens for the TI-83 Plus calculator, which will usually be the same for the TI-84 Plus, and for TI-89 Titanium, which we shall indicate simply as TI-89.

    To add two numbers by calculator, simply enter the first number; press $\boxed{+}$; enter the second number; and press the *enter* key $\boxed{\text{ENTER}}$ or the *equals* key $\boxed{=}$, or the *execute* key $\boxed{\text{EXE}}$, depending upon your particular calculator. The number we get is called the *sum* of the two numbers.

    For subtraction we use the $\boxed{-}$ key (*not* the $\boxed{(-)}$ key). The result is called the *difference* of the two numbers.

TI-83/84 screen for Example 16. Your calculator display may differ depending upon which numerical format is chosen from the $\boxed{\text{MODE}}$ menu. Here we are in Float mode.

◆◆◆ Example 16: Evaluate 2845 + 3273 by calculator.

Solution: We key in 2845, then press the $\boxed{+}$ key, then 3273, and finally $\boxed{\text{ENTER}}$. The screens for the TI-83 Plus (and TI-84) as well as the TI-89 Titanium are shown.

Changing the Calculator Display

You can select the way a calculator displays numbers from the $\boxed{\text{MODE}}$ menu.

Float (floating) mode on the TI-83 will give the full calculator display, up to ten digits. On the TI-89 you can select the total number of digits to be displayed, including those to the left to the decimal point.

Fix (fixed) mode on either calculator will display a result with the number of decimal places chosen.

TI-89 screen for Example 16.

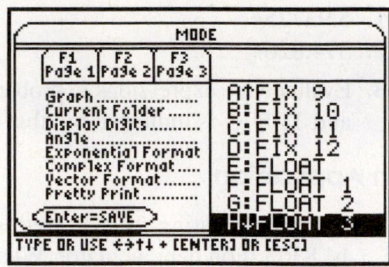

MODE screen for the TI-83/84. You can select either the floating mode (**Float**), or the number of digits to be displayed to the right of the decimal point.

MODE screen for the TI-89. You can select the FIX mode and choose the number of decimal places to be displayed, or a FLOAT mode, and choose the total number of digits to be displayed.

Changing the display *does not* affect the accuracy of a computation, but only the way the result is displayed. However, your mode settings may make your answers look different than those given here.

Adding Signed Numbers

Let us say that we have a shoebox (Fig. 1–4) into which we toss all of our uncashed checks and unpaid bills until we have time to deal with them. Let us further assume that the total checks minus the total bills in the shoebox is $500.

We can think of the amount of a check as a *positive* number because it increases our wealth, and the amount of a bill as a *negative* number because it decreases our wealth. We thus represent a check for $100 as $(+100)$, and a bill for $100 as (-100).

Now, let's *add a check* for $100 to the box. If we had $500 at first, we must now have $600.

$$500 + (+100) = 600$$

or

$$500 + 100 = 600$$

Here we have added a positive number, and our total has increased by that amount. That is easy to understand. But what does it mean to *add a negative number*?

To find out, let us now *add a bill* for $100 to the box. If we had $500 at first, we must now have $100 less, or $400. Representing the bill by (-100), we have

$$500 + (-100) = 400$$

It seems clear that to *add a negative number is no different than subtracting the absolute value of that number*.

$$500 + (-100) = 500 - 100 = 400$$

This gives us our rule of signs for addition of signed numbers.

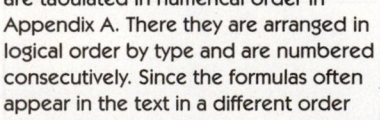

FIGURE 1–4 The shoebox.

All boxed and numbered formulas are tabulated in numerical order in Appendix A. There they are arranged in logical order by type and are numbered consecutively. Since the formulas often appear in the text in a different order than in Appendix A, they may not be in numerical order here in the text.

Rule of Signs for Addition	$a + (-b) = a - b$	1

♦♦♦ **Example 17:** Combine as indicated.

(a) $7 + (-2) = 7 - 2 = 5$

(b) $-8 + (-3) = -8 - 3 = -11$

(c) $-9 + 4 = -5$

♦♦♦

Subtracting Signed Numbers

Let us return to our shoebox. But now instead of adding checks or bills to the box, we will *subtract* (remove) checks or bills from the box.

First we remove (subtract) a check for $100 from the box. If we had $500 at first, we must now have $400.

$$500 - (+100) = 400$$

or

$$500 - 100 = 400$$

Here we have subtracted a positive number, and our total has decreased by the amount subtracted, as expected.

Now let us see what it means to *subtract a negative number*. We will remove (subtract) a bill for $100 from the box. If we had $500 at first, we must now have $100 more, or $600, since we have removed a bill. Representing the bill by (-100), we have

$$500 - (-100) = 600$$

It seems clear that *to subtract a negative number is the same as to add the absolute value of that number*.

$$500 - (-100) = 500 + 100 = 600$$

This gives us our rule of signs for subtraction of signed numbers.

Rule of Signs for Subtraction	$a - (-b) = a + b$	2

◆◆◆ Example 18:

(a) $8 - (-6) = 8 + 6 = 14$ (b) $-7 - (-5) = -7 + 5 = -2$

(c) $-(-16) - 7 - (-9) = 16 - 7 + 9 = 18$ ◆◆◆

Subtracting Negative Numbers by Calculator

Note that two similar-looking calculator keys are used for two different things:

1. To subtract two quantities.
2. To enter a negative quantity.

This difference is clear on the calculator, which has separate keys for these two functions. The $\boxed{-}$ key is used only for subtraction; the $\boxed{(-)}$ key is used to enter a negative quantity.

To enter a negative number on some calculators, simply press the $\boxed{(-)}$ key and then enter in the number.

Try the following examples on your calculator and see if you get the correct answers.

◆◆◆ Example 19: Combine as indicated.

(a) $15 - (-3) = 15 + 3 = 18$ (b) $-5 - (-9) = -5 + 9 = 4$

(c) $-25 - (-5) = -25 + 5 = -20$ ◆◆◆

Common Error	The $\boxed{-}$ key and the $\boxed{(-)}$ key look almost alike. Be careful not to confuse them. Note that the key used to enter a negative quantity has parentheses.

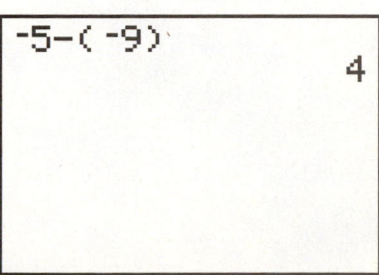

TI-83/84 screen for Example 19(b). Note the different appearances of the negative and the subtraction signs.

Commutative and Associative Laws

These laws are surely familiar to you, even if you do not recognize their names. We will run into them again when studying algebra. The *commutative law* simply says that you can add numbers in *any order*.

Commutative Law for Addition	$a + b = b + a$	3

◆◆◆ **Example 20:** Here is an example of the commutative law with numbers.
$$2 + 3 = 3 + 2$$
$$= 5$$
◆◆◆

The *associative law* says that you can group numbers to be added in several ways.

Associative Law for Addition	$a + (b + c) = (a + b) + c$ $= (a + c) + b$	4

◆◆◆ **Example 21:** The associative law, with numbers, is shown here.
$$2 + 3 + 4 = 2 + (3 + 4) = 2 + 7 = 9$$
$$= (2 + 3) + 4 = 5 + 4 = 9$$
$$= (2 + 4) + 3 = 6 + 3 = 9$$
◆◆◆

Adding and Subtracting Approximate Numbers

Addition and subtraction of integers are simple enough. But now let us tackle the problem mentioned earlier: How many digits do we keep in our answer when adding or subtracting *approximate numbers*?

■ **Explorations:**

(a) A six-foot-tall person stands on a box. How high would you say the person plus the box are if the box is 1.14 ft high?
(b) A certain gasoline tank contains 14.5 gallons. If we siphon 2.585 gallons from a full tank, how many gallons would you say are left in the tank?
(c) A person who weighs 135 pounds picks up a laboratory weight marked 1.750 lb. What would you state as their combined weight?

Keeping in mind that the rightmost digit in an approximate number contains some uncertainty, and that those to its right are unknown, what conclusions can you draw about the addition and subtraction of approximate numbers? Can you say why it is misleading to give the height of the person plus box as 7.14 ft? Can you see the reason for the following rule? ■

Addition and Subtraction	When adding or subtracting approximate numbers, keep as many decimal places in your answer as contained in the number having the fewest decimal places.	5

◆◆◆ **Example 22:** Removing 2.585 gallons from 14.5 gallons gives

$$14.5 - 2.585 = 11.915 \text{ gallons}$$

As 14.5 has just one decimal place, we must round our answer to one decimal place. So we write

$$14.5 - 2.585 = 11.9 \text{ gallons}$$

Here it is common practice to use the equals sign rather than the $\approx$ sign.   ◆◆◆

◆◆◆ **Example 23:** Let us add 32.4 cm and 5.825 cm.

$$32.4 \text{ cm} + 5.825 \text{ cm} = 38.2 \text{ cm} \ (not \ 38.225 \text{ cm})$$

Here we can see the reason for our rule for rounding. In one of the original numbers (32.4 cm), we do not know what digit is to the right of the 4, in the hundredths place. We cannot assume that it is zero, for if we *knew* that it was zero, it would have been written in (as 32.40). Not knowing the digit in the hundredths place in an original number causes uncertainty in the tenths place in the answer. So we discard the hundredths digit and any digits to the right of it. ◆◆◆

◆◆◆ **Example 24:** Here is another example of adding approximate numbers.

$$\begin{array}{r} 25.8 \\ 18.3\ 125 \\ \underline{5.4\ 07} \\ 49.5\ |195 \end{array}$$

discard →

◆◆◆

Common Error	Students *hate* to throw away those last digits. Remember that by keeping worthless digits, you are telling whoever reads that number that it is more precise than it really is.

◆◆◆ **Example 25:** A certain stadium contains about 3,500 people. It starts to rain, and 372 people leave. How many are left in the stadium?

Solution: Subtracting, we obtain

$$3500 - 372 = 3128$$

which we round to 3100, because 3500 here is known to only two significant digits. ◆◆◆

It is safer to round the answer *after* adding (or subtracting), rather than to round the original numbers before adding. If you do round before adding, it is prudent to round each original number to *one more* decimal place than you expect to keep in the rounded answer.

Combining Exact and Approximate Numbers

When you are combining an exact number and an approximate one, the accuracy of the result will be limited by the approximate number. Thus round the result to the number of decimal places found in the approximate number, even though the exact number may *appear* to have fewer decimal places.

◆◆◆ **Example 26:** Express 2 hr and 35.8 min in minutes.

Solution: We must add an exact number (120 minutes) and an approximate number 35.8 minutes).

$$\begin{array}{r} 120 \quad \text{min} \\ +\ \underline{35.8} \quad \text{min} \\ 155.8 \quad \text{min} \end{array}$$

Since 120 is exact, we retain as many decimal places as in the approximate number, 35.8. Our answer is thus 155.8 min. ◆◆◆

Common Error	Be sure to recognize which numbers in a computation are exact; otherwise, you may perform drastic rounding by mistake.

Combining Approximate Numbers by Calculator

The keystrokes for combining approximate numbers are the same as for combining integers. Just be sure not to select a fixed decimal mode with so few decimal places as to cut off significant digits after the decimal point.

To evaluate an expression *containing parentheses,* use the $($ key and the $)$ key.

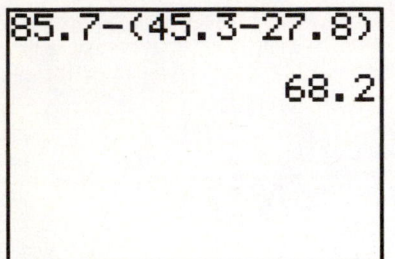

85.7−(45.3−27.8)
                    68.2

TI-83/84 screen for Example 27.

◆◆◆ **Example 27:** The keystrokes used to evaluate

$$85.7 - (45.3 - 27.8)$$

are shown on the screen.   ◆◆◆

Loss of Significant Digits During Subtraction

Subtracting two nearly equal numbers can lead to a drastic loss of significant digits.

◆◆◆ **Example 28:** When we subtract, say,

$$6{,}755 - 6{,}753 = 2$$

we get a result having one significant digit, while our original numbers each had four. While not common, you should be aware that it can happen, and can destroy the accuracy of a computation.   ◆◆◆

Similarly, significant digits can be gained by addition, say,

$$8.0 + 5.0 = 13.0$$

which gives a three significant digit result from numbers having two significant digits each.   ◆◆◆

Exercise 2 ◆ Addition and Subtraction

Adding and Subtracting Signed Numbers Combine as indicated.

1. $926 + 863$
2. $274 + (-412)$
3. $-576 + (-553)$
4. $-207 + (-819)$
5. $-575 - 275$
6. $-771 - (-976)$
7. $1123 - (-704)$
8. $818 - (-207) + 318$

Adding and Subtracting Approximate Numbers Combine each set of approximate numbers as indicated. Round your answer.

9. $4857 + 73.8$
10. $39.75 + 27.4$
11. $296.44 + 296.997$
12. $385.28 - 692.8$
13. $0.000583 + 0.0008372 - 0.00173$

Applications

14. Mt. Blanc is 15,572 ft high, and Pike's Peak is about 14,000 ft high. What is the difference in their heights?
15. California contains 158,933 square miles (mi^2), and Texas 237,321 mi^2. How much larger is Texas than California?

16. A man willed $125,000 to his wife and two children. To one child he gave $44,675, to the other $26,380, and to his wife the remainder. What was his wife's share?

17. A circular pipe has an inside radius r of 10.6 cm and a wall thickness of 2.125 cm. It is surrounded by insulation having a thickness of 4.8 cm (Fig. 1–5). What is the outside diameter D of the insulation?

18. A batch of concrete is made by mixing 267 kg of stone, 125 kg of sand, 75.5 kg of cement, and 25.25 kg of water. Find the total weight of the mixture.

19. Three resistors, having values of 27.3 ohms (Ω), 4.0155 Ω, and 9.75 Ω, are wired in series. What is the total resistance? (See Eq. 1062 which says that the total series resistance is the sum of the individual resistances.)

20. *Writing:* Does your calculator have two separate keys marked with a negative sign? Why? Is there any difference between them, and if so, what? When would you use each? Write a paragraph or two explaining these keys and answering these questions.

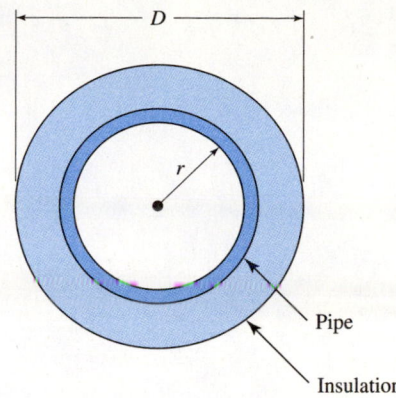

FIGURE 1–5 An insulated pipe.

1–3 Multiplication

Addition and subtraction were easy enough. Let's move on to multiplication.

Symbols and Definitions

Multiplication can be indicated in several ways: by the usual $\times$ symbol; by a dot; or by parentheses, brackets, or braces. Thus the product of b and d could be written

$$b \cdot d \qquad b \times d \qquad b(d) \qquad (b)d \qquad (b)(d)$$

Most common of all is to use no symbol at all. The product of b and d would usually be written bd. When doing algebra avoid using the $\times$ symbol because it could get confused with the letter x.

We get a *product* when we multiply two or more *factors*.

$$(\text{factor})\,(\text{factor})\,(\text{factor}) = \text{product}$$

Multiplying by Calculator

Many calculators use the $\boxed{\times}$ key for multiplication and use an asterisk or star (*) on the screen to represent a multiplication dot.

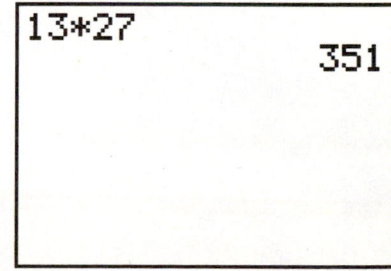

This TI-83/84 screen shows the multiplication of the factors 13 and 27.

Multiplying Signed Numbers

To get our rules of signs for multiplication, we use the idea of *multiplication as repeated addition.* For example, to multiply 3 by 4 means to add four 3's (or three 4's)

$$3 \times 4 = 3 + 3 + 3 + 3$$

or

$$3 \times 4 = 4 + 4 + 4$$

Let us return to our shoebox example. Recall that it contains uncashed checks and unpaid bills. Let's first *add 5 checks* ($+5$), each worth $100 ($+100$), to the box. The value of the contents of the box then increases by $500. Multiplying, we have

$$(\text{number of checks}) \times (\text{value of one check}) = \text{change in value of contents}$$

$$(+5)(+100) = +500$$

Thus a positive number times a positive number gave a positive product. This is nothing new.

Now let's *add 5 bills*, each for $100, to the box, thus decreasing its value by $500. To show this multiplication, we use (+5) for the number of bills, taking (−100) for the value of each bill, and (−500) for the change in value of the box contents.

$$(+5)(-100) = -500$$

Here, a positive number times a negative number gives a negative product.

Next we *remove 5 checks*, each for $100, from the box, thus decreasing its value by $500.

$$(-5)(+100) = -500$$

Here again, the product of a positive number and a negative number is negative. Thus it doesn't matter whether the negative number is the first or the second.

Finally, we *remove 5 bills*, each for $100, from the box, causing its value to increase by $500.

$$(-5)(-100) = +500$$

Here, the product of two negative numbers is positive.

We summarize these findings to get our *rules of signs for multiplication.*

Rules of Signs for Multiplication	$(+a)(+b) = (-a)(-b) = +ab$ *The product of two quantities of like sign is positive.*	6
a and *b* are positive numbers	$(+a)(-b) = (-a)(+b) = -ab$ *The product of two quantities of unlike sign is negative.*	7

◆◆◆ **Example 29:** Multiply.

(a) $2(-3) = -6$ (b) $(-2)3 = -6$ (c) $(-2)(-3) = 6$ ◆◆◆

Multiplying a String of Numbers

When we multiplied two negative numbers, we got a positive product. So when we are multiplying a *string* of numbers, if an even number of them are negative, the answer will be positive, and if an odd number of them are negative, the answer will be negative.

◆◆◆ **Example 30:** Multiply.

(a) $2(-3)(-1)(-2) = -12$ (b) $2(-3)(-1)(2) = 12$ ◆◆◆

Multiplying Negative Numbers by Calculator

As we mentioned earlier, you enter negative numbers into the calculator by using a key marked $(-)$, $+/-$, or CHS . On some calculators you first enter the number and then change its sign using the proper key; on other calculators you press the $(-)$ key first and then enter the number.

◆◆◆ **Example 31:** Use your calculator to multiply −96 by −83.

Solution: You should get

$$(-96)(-83) = 7968 \qquad \text{◆◆◆}$$

A simpler way to do the last problem would be to multiply +96 and +83 and determine the sign by inspection.

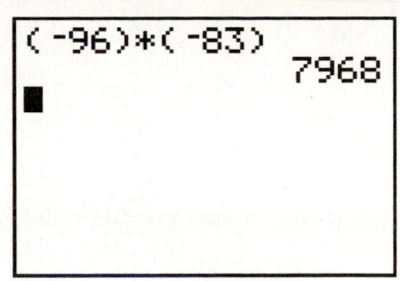

TI-83/84 screen for Example 31. On some calculators the multiplication symbol may be omitted, as the parentheses themselves indicate multiplication.

Common Error	Do not try to use the ⊟ key to enter negative numbers into your calculator. The ⊟ key is only for subtraction.

Multiplication of Approximate Numbers

■ **Exploration:**

Try this. Multiply two approximate numbers, say 5.43 and 4.75, and write down the full calculator display.

$$(5.43)(4.75) = 25.7925$$

But each of the original numbers has some uncertainty: 4.75, for example, could have been any value between 4.746 and 4.754 before it was rounded in some previous step. So repeat the multiplication, replacing 4.75 with 4.746. How does this affect the product? Repeat again, replacing 4.75 with 4.754. Repeat again, now letting 5.43 take on some uncertainty. What can you conclude about whether all those digits in the product should be kept?

Repeat this exploration with other approximate numbers. Do you see the reason for the following rule? ■

Rule	When multiplying two or more approximate numbers, round the result to as many digits as in the factor having the fewest significant digits.	11

◆◆◆ **Example 32:** Here we multiply two numbers, each with 3 significant digits.

$$12.1 \qquad \times \qquad 15.6 \qquad = \qquad 189$$

12.1		15.6		189
↑		↑		↑
three digits		three digits		three digits

◆◆◆

When the factors have different numbers of significant digits, keep the same number of digits in your answer as is contained in the factor that has the *fewest* significant digits.

◆◆◆ **Example 33:** Here we multiply numbers that do not have the same number of significant digits.

$$123.56 \qquad \times \qquad 2.21 \qquad = \qquad 273$$

123.56		2.21		273
↑		↑		↑
five digits		three digits		keep three digits

◆◆◆

<table>
<tr><td>**Common Error**</td><td>Do not confuse *significant digits* with *decimal places*. The number 274.56 has *five* significant digits and *two* decimal places.

Decimal places determine how we round after adding or subtracting. Significant digits determine how we round after multiplying and, as we will soon see, after dividing, raising to a power, or taking roots.</td></tr>
</table>

◆◆◆ **Example 34:** *An Application.* Find the weight of 3.845 cubic feet of stone that has a density of 175 lb/ft^3.

Solution: The weight equals the volume times the density, so,

$$\text{Weight} = 3.845 \times 175 = 673 \text{ lb}$$

after rounding to the three significant digits found in 175.   ◆◆◆

Multiplying Exact and Approximate Numbers

When using *exact numbers* in a computation, treat them as if they had *more* significant figures than any of the approximate numbers in that computation.

◆◆◆ **Example 35:** Multiplying the exact number 3 by the approximate number 6.836 gives

$$3 \times 6.836 = 20.51$$

We kept as many significant digits (4) as found in the approximate number.   ◆◆◆

◆◆◆ **Example 36:** *An Application.* If a certain car tire weighs 32.2 lb, how much will four such tires weigh?

Solution: Multiplying, we obtain

$$32.2(4) = 128.8 \text{ lb}$$

Since the 4 is an exact number, we retain as many significant figures as contained in 32.2, and round our answer to 129 lb.   ◆◆◆

Exercise 3 ◆ Multiplication

Multiplying Signed Numbers

1. $4 \times (-2)$ 2. $(-5) \times (3)$
3. $(-24) \times (-5)$ 4. $(-41) \times (-22)$

Multiplying Approximate Numbers
Multiply each approximate number and retain the proper number of digits in your answer.

5. 3.967×2.84 6. 4.900×59.3
7. 93.9×0.0055908 8. $4.97 \times 9.27 \times 5.78$
9. $69.0 \times (-258)$ 10. $-385 \times (-2.2978)$
11. $2.86 \times (4.88 \times 2.97) \times 0.553$ 12. $(5.93 \times 7.28) \times (8.26 \times 1.38)$

Multiplying Exact and Approximate Numbers Multiply, and keep the proper number of significant digits in your answer. Take each integer as an exact number.

13. 4×2.55

14. -1.46×3

15. $-4.273 \times (-5)$

16. $(-5) \times (-1.022)$

Applications

17. What is the cost of 52.5 tons of cement at $63.25 a ton?

18. If 108 tons of rail is needed for 1 mi of track, how many tons will be required for 476 mi, and what will be its cost at $925 a ton?

19. Three barges carry 26.0 tons of gravel each, and a fourth carries 35.0 tons. What is the value, to the nearest dollar, of the whole shipment, at $12.75 per ton?

20. What will be the cost of installing a telephone line 274 km long, at $5723 per kilometer?

21. The current to a projection lamp is measured at 4.7 A when the line voltage is 115.45 V. Using (power = voltage × current), find the power dissipated in the lamp.

22. A gear in a certain machine rotates at the speed of 1808 rev/min. How many revolutions will it make in 9.500 min?

23. How much will 1000 washers weigh if each weighs 2.375 g?

24. One inch equals exactly 2.54 cm. Convert 385.84 in. to centimeters.

25. If there are 360 degrees per revolution, how many degrees are there in 4.863 revolutions?

1–4 Division

Definitions

The *dividend*, when divided by the *divisor*, gives us the *quotient*.

$$\text{dividend} \div \text{divisor} = \text{quotient}$$

or

$$\frac{\text{dividend}}{\text{divisor}} = \text{quotient}$$

A quantity a/b is also called a *fraction*, where a is called the *numerator* and b is called the *denominator*. It can also be referred to as the *ratio* of a to b. Fractions and ratios are covered in detail in later chapters.

Dividing by Calculator

To divide by calculator, enter the dividend, then press $\boxed{\div}$, then enter the divisor and press $\boxed{\text{ENTER}}$, $\boxed{=}$, or $\boxed{\text{EXE}}$, depending on your particular calculator.

♦♦♦ **Example 37:** Here is the screen for the division of 1305 by 145.

When we multiplied two integers, we always got an integer for an answer. This is not always the case when dividing, as shown in the next example. ♦♦♦

TI-83/84 screen for Example 37.

◆◆◆ **Example 38:** When we divide 2 by 3, we get 0.666666666. . . . We must choose how many digits we wish to retain and round our answer. Rounding to, say, three significant digits, we obtain

$$2 \div 3 \approx 0.667$$

Here it is appropriate to use the $\approx$ symbol.   ◆◆◆

Dividing Signed Numbers

We will now use the rules of signs for multiplication to get the rules of signs for division.

(a) We know that the product of a negative number and a positive number is negative. For example,

$$(-2)(+3) = -6$$

If we divide both sides of this equation by $(+3)$, we get

$$-2 = \frac{-6}{+3}$$

From this we see that *a negative number divided by a positive number gives a negative quotient.*

(b) Again starting with

$$(-2)(+3) = -6$$

we divide both sides by (-2) and get

$$+3 = \frac{-6}{-2}$$

Here we see that *a negative number divided by a negative number gives a positive* quotient.

(c) We also know that the product of two negative numbers is positive. Thus

$$(-2)(-3) = +6$$

Dividing both sides by (-3), we get

$$-2 = \frac{+6}{-3}$$

Thus *a positive number divided by a negative number gives a negative quotient.*

We combine these findings with the fact that the quotient of two positive numbers is positive and get our *rules of signs for division.*

Rules of Signs for Division *a* and *b* are positive numbers	$$\frac{+a}{+b} = \frac{-a}{-b} = \frac{a}{b}$$ *The quotient is positive when dividend and divisor have the same sign.*	**12**
	$$\frac{+a}{-b} = \frac{-a}{+b} = -\frac{a}{b}$$ *The quotient is negative when dividend and divisor have opposite signs.*	**13**

◆◆◆ **Example 39:**

(a) $8 \div (-4) = -2$ (b) $-8 \div 4 = -2$

(c) $-8 \div (-4) = 2$ ◆◆◆

Dividing Approximate Numbers

The rule for rounding after division is almost the same as with multiplication, and is given for the same reason.

Rule	After dividing one approximate number by another, round the quotient to as many digits as there are in the original number having the fewest significant digits.	14

◆◆◆ **Example 40:** Divide 937.5 by 4.75, keeping the proper number of significant digits in the quotient.

Solution: Since 4.75 has fewer significant digits (three) than 937.5, we round our answer to three significant digits, getting 197.   ◆◆◆

◆◆◆ **Example 41:** Divide 846.2 into three equal parts.

Solution: We divide by the integer 3, and since we consider integers to be exact, we retain in our answer the same number of significant digits as in 846.2.

$$846.2 \div 3 = 282.1$$ ◆◆◆

```
937.5/4.75
              197.37
```

TI-83/84 screen for Example 40.

◆◆◆ **Example 42:** Divide 85.4 by -2.5386 by calculator.

Solution: You should get

$$85.4 \div (-2.5386) = -33.6, \text{ rounded}$$ ◆◆◆

As with multiplication, the sign could also have been found by inspection.

◆◆◆ **Example 43:** *An Application.* How fast would an airplane have to travel to go 3895 miles in 5.25 hours? (rate = distance ÷ time)

Solution: Dividing gives

$$\text{rate} = \frac{3895 \text{ mi}}{5.25 \text{ h}} = 742 \text{ mi/h}$$

We have rounded to the three significant digits found in 5.25.   ◆◆◆

```
85.4/(-2.5386)
              -33.64
■
```

TI-83/84 screen for Example 42.

Zero

■ **Exploration:**

Try this. Using your calculator, do the following divisions.

$0 \div 5$ $0 \div 295$ $5 \div 0$ $295 \div 0$ $0 \div 0$

From your results, can you deduce the rules for dividing zero by a number, and for dividing a number by zero?   ■

Division Involving Zero	Zero divided by any quantity (except zero) is zero. Division by zero is not defined. It is an illegal operation in mathematics.	**15**

Reciprocals

The *reciprocal* of any nonzero number n is $1/n$. Thus the product of a quantity and its reciprocal is equal to 1.

◆◆◆ **Example 44:**

(a) The reciprocal of 10 is $1/10$.

(b) The reciprocal of $1/2$ is 2.

(c) The reciprocal of $-\frac{3}{4}$ is $-\frac{4}{3}$.   ◆◆◆

Reciprocals by Calculator

On the TI-83/84, simply enter the number and press the $\boxed{x^{-1}}$ key. On the TI-89 this operation is found in $\boxed{\text{CATALOG}}$ and is indicated by ^ **–1**. Keep as many digits in your result as there are significant digits in the original number.

◆◆◆ **Example 45:**

(a) The reciprocal of 6.38 is 0.157.
(b) The reciprocal of -2.754 is -0.3631.   ◆◆◆

◆◆◆ **Example 46:** *An Application.* The unit of electrical conductance is the *mho* (ohm spelled backward) or *Siemens*. It is the reciprocal of resistance. Find the conductance of a circuit element having a resistance of 598 ohms.

Solution: Taking the reciprocal,

$$\frac{1}{598} = 0.00167 \text{ mhos}$$

◆◆◆

```
-2.754-1
          -.3631
■
```

TI-83/84 screen for Example 45(b).

Exercise 4 ◆ Division

Dividing Signed Numbers Divide, keeping the proper sign on your answer.

1. $14 \div (-2)$ **2.** $(-15) \div (3)$
3. $(-24) \div (-4)$ **4.** $(-49) \div (-7)$

Dividing Approximate Numbers Divide, and then round your answer to the proper number of digits.

5. $947 \div 5.82$ **6.** $0.492 \div 0.00478$
7. $-99.4 \div 286.5$ **8.** $-4.8 \div -2.557$
9. $5836 \div 8264$ **10.** $5.284 \div 3.827$
11. $94{,}840 \div 1.33876$ **12.** $3.449 \div (-6.837)$

Reciprocals Find the reciprocal of each number, retaining the proper number of digits in your answer.

13. 693   **14.** 0.00630
15. -396 **16.** 39.74

17. −0.00573

19. 4.992

18. 938.4

20. −6.93

Applications

21. A stretch of roadway 1858.54 m long is to be divided into 5 equal sections. Find the length of each section.

22. At what rate must a person walk to go 24.5 km in 12.75 h? (rate = distance ÷ time)

23. If three masons can build 245 ft of wall in 4.50 days, how many feet of wall can one mason build in a day? Assume that each mason works at the same rate, and that the same length of wall is built each day.

24. If 867 shares of stock are valued at $84,099, what is the value of each share?

25. The equivalent resistance R of a 475-Ω resistor and a 928-Ω resistor connected in parallel is given by

$$\frac{1}{R} = \frac{1}{475} + \frac{1}{928}$$

Find R.

26. When an object is placed 126 cm in front of a certain thin lens having a focal length f, the image will be formed 245 cm from the lens. The distances are related by

$$\frac{1}{f} = \frac{1}{126} + \frac{1}{245}$$

Find f.

27. The sine of an angle θ (written sin θ) is equal to the reciprocal of the cosecant of θ (csc θ). Find sin θ if csc θ = 3.58.

28. If two straight lines are perpendicular, the slope of one line is the negative reciprocal of the slope of the other. If the slope of a line is −2.55, find the slope of a perpendicular to that line.

1–5 Powers and Roots

Now that we know how to do the four basic arithmetic operations on signed and approximate numbers, let us learn how to find powers and roots.

Powers

■ Exploration:

Try this. Use your calculator to multiply $2 \times 2 \times 2$. Then use it to evaluate 2^3. You can do this using the ⌐∧⌐ key on your calculater, as shown on the screen. Then evaluate $2 \times 2 \times 2 \times 2$, as well as 2^4.

Can you summarize what you have found? ■

In the expression

$$2^4$$

the number 2 is called the *base*, and the number 4 is called the *exponent*. The expression is read "two to the fourth power." Its value is

$$2^4 = 2 \cdot 2 \cdot 2 \cdot 2 = 16$$

To *square* a number means to raise it to the power 2. To *cube* a number means to raise it to the power 3.

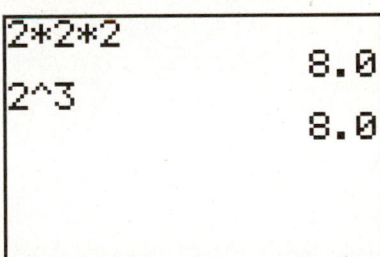

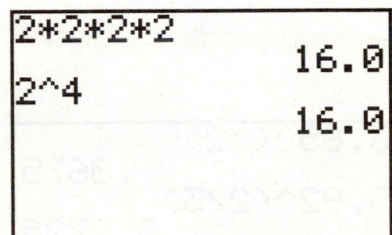

TI-83/84 screens for the Exploration.

Stated as a formula,

Positive Integral Exponent	$x^n = x \cdot x \cdot x \ldots x$ n factors	21

◆◆◆ **Example 47:**

(a) $3^4 = 3 \cdot 3 \cdot 3 \cdot 3 = 81$

(b) $5^3 = 5 \cdot 5 \cdot 5 = 125$

(c) $4^5 = 4 \cdot 4 \cdot 4 \cdot 4 \cdot 4 = 1024$ ◆◆◆

When raising an approximate number to a power, round your answer to the number of significant digits in the base, not the exponent.

◆◆◆ **Example 48:** Use your calculator to verify the following:

(a) $1.65^2 = 2.72$

(b) $1.52^3 = 3.51$

(c) $1.36^5 = 4.65$ ◆◆◆

```
1.65²
             2.72
1.52^3
             3.51
1.36^5
             4.65
```

TI-83/84 screen for Example 48.

Negative Base

A negative base raised to an *even* power gives a *positive* number. A negative base raised to an *odd* power gives a *negative* number.

◆◆◆ **Example 49:** Here are some negative bases raised to various powers.

(a) $(-2)^2 = (-2)(-2) = 4$

(b) $(-2)^3 = (-2)(-2)(-2) = -8$

(c) $(-1)^{24} = 1$

(d) $(-1)^{25} = -1$ ◆◆◆

If you try to do these problems on your calculator, you may get an error indication. Some calculators will not work with a negative base, even though this is a valid operation.

Then how do you do it? Simply enter the base as *positive,* find the power, and determine the sign by inspection.

Negative, Fractional, and Approximate Exponents

From our exploration we know the meaning of, say, 2^4. But what is the meaning of expressions like

$$2^{-4} \qquad \text{or} \qquad 2^{3/5} \qquad \text{or} \qquad 2^{1.55}$$

The explanation will have to wait for later when we have a bit more background. But for now, we can use our calculators to evaluate expressions such as these, in case you need to for, say, your physics course. The keystrokes are no different than for integer exponents.

```
3.85^(-2)
            .0675
5.92^(2/3)
           3.2725
4.46^1.74
          13.4847
```

TI-83/84 screen for Example 50.

◆◆◆ **Example 50:** See if you can verify the following:

(a) $3.85^{-2} = 0.0675$

(b) $5.92^{2/3} = 3.27$

(c) $4.46^{1.74} = 13.5$

Note that we have rounded our answers to the three significant digits found in each base. If the exponent is approximate, round the result to the number of significant digits in the exponent or the base, whichever has the fewest. ◆◆◆

Common Errors	When entering a negative base or negative exponent, be sure to use the key for entering a negative number, not the key for subtraction. Use parentheses when entering a negative base, negative exponent, or a fractional exponent.

◆◆◆ **Example 51:** *An Application.* The volume V of a sphere of radius r is given by

$$V = \frac{4}{3}\pi r^3$$

where $\pi = 3.142$. Find the volume of a sphercial weather balloon having a radius of 2.74 m.

Solution:

$$V = \frac{4}{3}\pi(2.74)^3 = 86.2 \text{ m}^3 \qquad \text{◆◆◆}$$

Roots

■ Exploration:

Try this. Calculate 3×3. Then use your calculator to evaluate $\sqrt{9}$. Then calculate $3 \times 3 \times 3 \times 3 \times 3$. Then evaluate $\sqrt[5]{243}$. On the TI-83/84 you first type 5 and press ENTER , then select the $\sqrt[x]{}$ operation from the MATH menu, type in 243, and press ENTER again. Can you summarize your results? ■

If $a^n = b$, then

$$\sqrt[n]{b} = a$$

which is read "the *n*th root of *b* equals *a*." The symbol $\sqrt{}$ is a *radical sign*, *b* is the *radicand,* and *n* is the *index* of the radical.

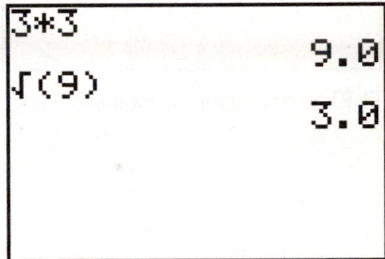

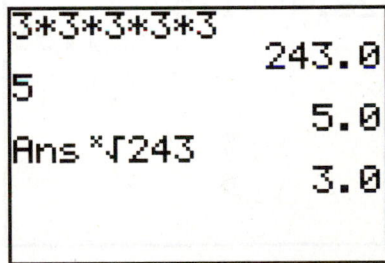

TI-83/84 screens for the Exploration.

◆◆◆ **Example 52:**

(a) $\sqrt{4} = 2$ because $2^2 = 4$

(b) $\sqrt[3]{8} = 2$ because $2^3 = 8$

(c) $\sqrt[4]{81} = 3$ because $3^4 = 81$ ◆◆◆

Principal Root

When we speak about the *root* of a number, we mean the *principal root*, unless otherwise specificed. The *principal root* of a positive number is defined as the *positive root*. Thus $\sqrt{4} = +2$, not ±2.

The principal root is *negative* when we take an *odd* root of a *negative* number.

◆◆◆ **Example 53:**

$$\sqrt[3]{-8} = -2$$

because $(-2)(-2)(-2) = -8$. ◆◆◆

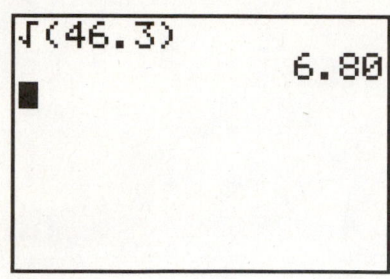

TI-83/84 screen for Example 54.

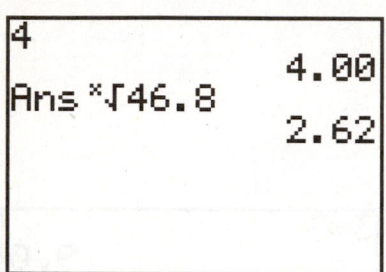

TI-83/84 screen for the Example 55.

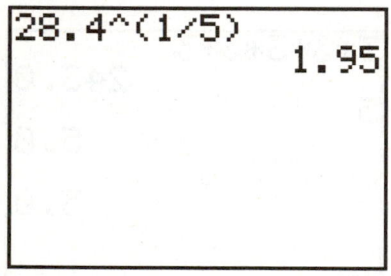

TI-83/84 screen for Example 56.

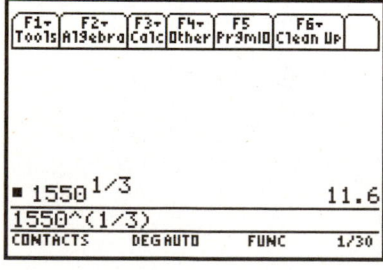

TI-89 screen for Example 58.

Roots by Calculator

To find square roots, we simply use the $\boxed{\sqrt{x}}$ key. Retain as many significant digits as in the radicand.

◆◆◆ **Example 54:** To find $\sqrt{46.3}$ we use the keystrokes shown.   ◆◆◆

For roots other than square roots, your calculator may have a special key, marked something like $\boxed{\sqrt[x]{y}}$. On the TI-83/84, first type the index and press $\boxed{\text{ENTER}}$, next select the $\sqrt[x]{}$ operation from the $\boxed{\text{MATH}}$ menu, then type in the radicand and press $\boxed{\text{ENTER}}$ again.

◆◆◆ **Example 55:** See if you get this result.

$$\sqrt[4]{46.8} = 2.62$$

As with the TI-89, if you have no special root key, you can still find roots using the $\boxed{\wedge}$ key. There is a relationship between powers and roots that we will prove later, but for now we will just use it for computing roots by calculator. It is,

Fractional Exponents	$a^{1/n} = \sqrt[n]{a}$	29

In other words, taking the nth root of a number is the same as raising that number to the $(1/n)$th power. So, for example, to take a fifth root of a number, we simply raise that number to the 1/5 power.

◆◆◆ **Example 56:** The keystrokes for finding $\sqrt[5]{28.4}$, the fifth root of 28.4, are shown.   ◆◆◆

In the preceding example, instead of raising 28.4 to the 1/5 power, we could also have raised it to the 0.2 power, because 0.2 = 1/5. In fact, we will show in the next chapter, that we can raise a base to any decimal or fractional power.

◆◆◆ **Example 57:** See if you can verify the following:

(a) $4.46^{1.74} = 13.5$

(b) $5.92^{2/3} = 3.27$ ◆◆◆

◆◆◆ **Example 58:** *An Application.* A storage bin in the shape of a cube is to have a volume of 1550 cubic feet. Find the dimension of each side.

Solution: The volume of a cube is equal to the cube of the side, so the side s is the cube root of the volume.

$$s = \sqrt[3]{V} = \sqrt[3]{1550} = 11.6 \text{ ft}$$

Odd Roots of Negative Numbers by Calculator

An *even* root of a negative number is *imaginary* (such as $\sqrt{-4}$). We will study these later. But an *odd* root of a negative number is *not* imaginary. It is a real, negative number. As with powers, some calculators will not accept a negative radicand. Fortunately, we can outsmart our calculators and take odd roots of negative numbers anyway.

◆◆◆ **Example 59:** Find $\sqrt[5]{-875}$.

Solution: We know that an odd root of a negative number is real and negative. So we take the fifth root of $+875$, by calculator,

$$\sqrt[5]{+875} = 3.88 \text{ (rounded)}$$

and we only have to place a minus sign before the number.

$$\sqrt[5]{-875} = -3.88$$

◆◆◆

Exercise 5 ◆ Powers and Roots

Evaluate each expression. Retain the proper number of significant digits in your answer.

Powers

1. 2^3	**2.** 5^3	**3.** 9^2
4. 1^3	**5.** 10^3	**6.** 10^1
7. 10^2	**8.** 10^4	

Powers by Calculator

9. $(8.55)^3$	**10.** $(1.07)^5$
11. $(9.55)^3$	**12.** $(84.2)^2$
13. $(3.95)^3$	**14.** $(13.9)^2$
15. $(1.65)^4$	**16.** $(2.98)^2$
17. $(12.5)^2$	**18.** $(1.35)^5$
19. $(2.26)^6$	**20.** $(1.94)^7$

Negative Base

21. $(-3)^3$	**22.** $(-2)^3$	**23.** $(-4)^3$
24. $(-3)^1$	**25.** $(-8.01)^3$	**26.** $(-1.71)^5$
27. $(-5.33)^3$	**28.** $(-12.5)^2$	**29.** $(-1.33)^3$
30. $(-2.34)^5$	**31.** $(-2.84)^3$	**32.** $(-5.84)^2$

Negative Exponent

33. 1^{-3}	**34.** 2^{-3}	**35.** 10^{-2}
36. 10^{-4}	**37.** $(3.85)^{-2}$	**38.** $(1.83)^{-5}$
39. $(3.84)^{-3}$	**40.** $(22.5)^{-2}$	**41.** $(-5.37)^{-3}$
42. $(-2.24)^{-5}$	**43.** $(-1.85)^{-3}$	**44.** $(-4.24)^{-3}$

Fractional and Demical Exponents

45. $(8.55)^{1/3}$	**46.** $(1.07)^{1/5}$	**47.** $(9.55)^{1/3}$
48. $(84.2)^{1/2}$	**49.** $(2.85)^{2/3}$	**50.** $(9.27)^{2/5}$
51. $(1.84)^{2/3}$	**52.** $(4.22)^{1/2}$	**53.** $(8.88)^{2.13}$
54. $(5.27)^{3.25}$	**55.** $(4.38)^{2.63}$	**56.** $(2.48)^{1.42}$

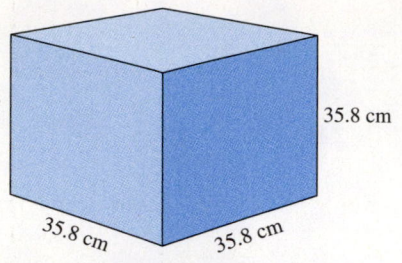

FIGURE 1–6

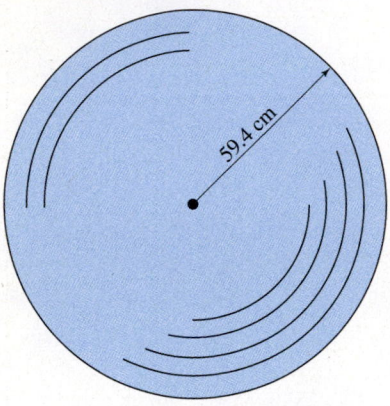

FIGURE 1–7

Applications Involving Powers

57. The distance traveled by a falling body, starting from rest, is equal to $16t^2$, where t is the elapsed time. In 5.448 s, the distance fallen is $16(5.448)^2$ ft. Evaluate this quantity. (Treat 16 here as an approximate number.)

58. The power P dissipated in a resistance R through which is flowing a current I is $P = I^2R$. Therefore the power in a 365-Ω resistor carrying a current of 0.5855 A is $(0.5855)^2(365)$ W. Evaluate this power.

59. The volume of a cube of side 35.8 cm (Fig. 1–6) is $(35.8)^3$. Evaluate this volume.

60. The volume of a 59.4-cm-radius sphere (Fig. 1–7) is $\frac{4}{3}\pi(59.4)^3$ cm^3. Find this volume.

61. An investment of \$2000 at a compound interest rate of $6\frac{1}{4}\%$, left on deposit for $7\frac{1}{2}$ years, will be worth $2000(1.0625)^{7.5}$ dollars. Find this amount, to the nearest cent.

Roots

Find each principal root without using your calculator.

62. $\sqrt{25}$ **63.** $\sqrt[3]{27}$

64. $\sqrt{49}$ **65.** $\sqrt[3]{-27}$

66. $\sqrt[3]{-8}$ **67.** $\sqrt[5]{-32}$

Evaluate each radical by calculator, retaining the proper number of digits in your answer:

68. $\sqrt{49.2}$ **69.** $\sqrt{1.863}$

70. $\sqrt[3]{88.3}$ **71.** $\sqrt{772}$

72. $\sqrt{3875}$ **73.** $\sqrt[3]{7295}$

74. $\sqrt[3]{-386}$ **75.** $\sqrt[5]{-18.4}$

76. $\sqrt[3]{-2.774}$

Applications of Roots

77. The period T (time for one swing) of a simple pendulum (Fig. 1–8) 2.55 ft long is

$$T = 2\pi\sqrt{\frac{2.55}{32.0}} \text{ seconds}$$

Evaluate T.

78. The magnitude Z of the impedance in a circuit having a resistance of 3540 Ω and a reactance of 2750 Ω is

$$Z = \sqrt{(3540)^2 + (2750)^2} \text{ ohms}$$

Find Z.

79. The geometric mean B between 3.75 and 9.83 is

$$B = \sqrt{(3.75)(9.83)}$$

Evaluate B.

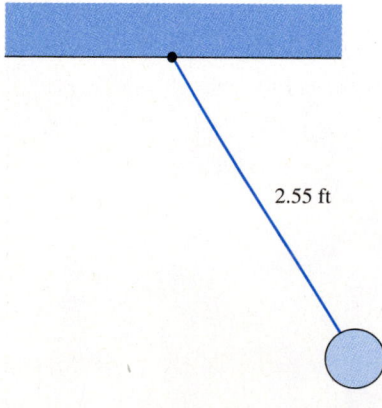

2.55 ft

FIGURE 1–8

1–6 Combined Operations

Order of Operations

If the expression to be evaluated does not contain parentheses, perform the operations in the following order:

1. *Powers and roots, in any order.*
2. *Multiplications and divisions, from left to right.*
3. *Additions and subtractions, from left to right.*

Our first group of calculations will be with integers only, and later we will do some problems that require rounding. We first show a problem containing both addition and multiplication.

◆◆◆ **Example 60:** Evaluate $7 + 3 \times 4$.

Solution: The multiplication is done before the addition.

$$7 + 3 \times 4 = 7 + 12 = 19$$ ◆◆◆

Next we do a calculation having both a power and multiplication.

◆◆◆ **Example 61:** Evaluate 5×3^2.

Solution: We raise to the power before multiplying:

$$5 \times 3^2 = 5 \times 9 = 45$$ ◆◆◆

Parentheses

When an expression contains parentheses, first evaluate the expression within the parentheses and then the entire expression.

◆◆◆ **Example 62:** Evaluate $(7 + 3) \times 4$.

Solution:

$$(7 + 3) \times 4 = 10 \times 4 = 40$$ ◆◆◆

If the sum or difference of two or more numbers is to be raised to a power, those numbers must be enclosed in parentheses.

◆◆◆ **Example 63:** Evaluate $(5 + 2)^2$.

Solution: We combine the numbers inside the parentheses before squaring.

$$(5 + 2)^2 = 7^2 = 49$$ ◆◆◆

◆◆◆ **Example 64:** Evaluate $(2 + 6)(7 + 9)$.

Solution: Evaluate the two quantities in parentheses before multiplying.

$$(2 + 6)(7 + 9) = 8 \times 16 = 128$$ ◆◆◆

◆◆◆ **Example 65:** Evaluate $\dfrac{8 + 4}{9 - 3}$.

Solution: Here the fraction line acts like parentheses, grouping the 8 and 4, as well as the 9 and 3. Written on a single line, this problem would be

$$(8 + 4) \div (9 - 3)$$

or

$$12 \div 6 = 2$$ ◆◆◆

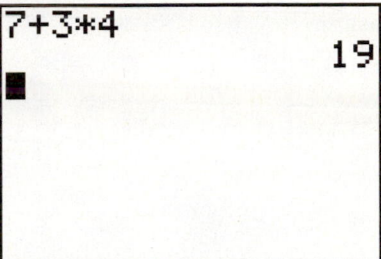

TI-83/84 screen for Example 60.

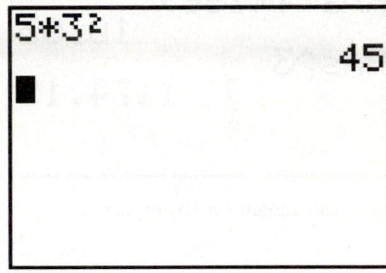

TI-83/84 screen for Example 61.

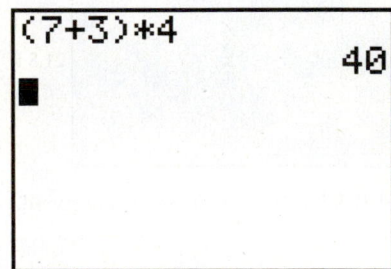

TI-83/84 screen for Example 62.

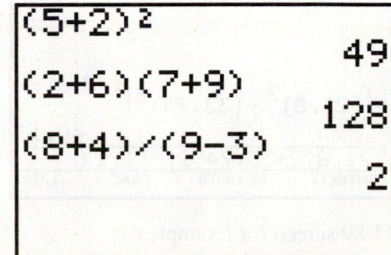

TI-83/84 screen for the Examples 63, 64, and 65.

Combined Operations with Approximate Numbers

Combined operations with approximate numbers are done the same way as those with exact numbers. However, we must round our answer properly using the rules given earlier in this chapter.

If you do a calculation in steps, writing down the intermediate values, it is a good practice to carry *one more* digit in those intermediate values than permitted by our rules, and round the final result.

Instead of writing down intermediate steps, you might keep the calculation entirely in the calculator, and round the final result. However, you will not have a "paper trail" to help check your work. A good compromise is to do the calculation in steps, by calculator, writing down the result of each step to provide a "trail." But instead of clearing the intermediate step from your calculator, use it for the next step. Finally, round the final result. We will show this procedure with an example.

◆◆◆ **Example 66:** Evaluate the expression

$$\left(\frac{118.8 + 4.23}{\sqrt{136}}\right)^3$$

Solution: Let's do the calculation one step at a time, and write down the result of the intermediate steps.

$$\left(\frac{118.8 + 4.23}{\sqrt{136}}\right)^3 = \left(\frac{123.03}{\sqrt{136}}\right)^3$$

$$= (10.55)^3$$

$$= 1174.15$$

Our calculator will give us as many digits as we want, but how many should we keep? In the numerator, we added a number with one decimal place to another with two decimal places, so we are allowed to keep just one. Thus the numerator, after addition, is good to one decimal place, or, in this case, four significant digits. The denominator, however, has just three significant digits, so we round our answer to three significant digits, getting 1170.   ◆◆◆

◆◆◆ **Example 67:** *An Application.* A rectangular courtyard (Fig. 1–9) having sides of 21.8 ft and 33.74 ft has a diagonal measurement *x* given by the expression

$$x = \sqrt{(21.8)^2 + (33.74)^2}$$

Evaluate the expression to find *x*.

Solution: The screen shows the calculation on the TI-89 calculator. We round the answer to 3 significant digits contained in 21.8, and we get $x = 40.2$ ft.   ◆◆◆

```
118.8+4.23
              123.03
Ans/√(136)
               10.55
Ans^3
             1174.15
```

TI-83/84 screen for Example 66.

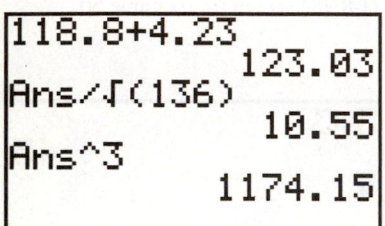

33.74 ft

x

21.8 ft

FIGURE 1–9 A rectangular courtyard.

```
F1▾  F2▾  F3▾  F4▾  F5    F6▾
Tools Algebra Calc Other PrgmIO Clean Up


■[(21.8)² + (33.74)²]^(1/2)
                      40.17
(21.8^2+33.74^2)^(1/2)
CONTACTS   DEG AUTO   FUNC   1/30
```

TI-89 screen for Example 67.

Exercise 6 ◆ Combined Operations

Combined Operations with Exact Numbers

Perform each computation by calculator.

1. $(37)(28) + (36)(64)$

2. $(22)(53) - (586)(4) + (47)(59)$

3. $(63 + 36)(37 - 97)$

4. $(89 - 74 + 95)(87 - 49)$

5. $\dfrac{219}{73} + \dfrac{194}{97}$

6. $\dfrac{228}{38} - \dfrac{78}{26} + \dfrac{364}{91}$

7. $\dfrac{647 + 688}{337 + 108}$

8. $\dfrac{809 - 463 + 1858}{958 - 364 + 508}$

9. $(5 + 6)^2$

10. $(422 + 113 - 533)^4$

11. $(423 - 420)^3$

12. $\left(\dfrac{853 - 229}{874 - 562}\right)^2$

13. $\left(\dfrac{141}{47}\right)^3$

14. $\sqrt{434 + 466}$

15. $\sqrt{(8)(72)}$

16. $\sqrt[3]{657 + 553 - 1085}$

17. $\sqrt[4]{(27)(768)}$

18. $\sqrt{\dfrac{2404}{601}}$

19. $\sqrt[4]{\dfrac{1136}{71}}$

20. $\sqrt{961} + \sqrt{121}$

21. $\sqrt[4]{625} + \sqrt{961} - \sqrt[3]{216}$

22. $\sqrt[4]{256} \times \sqrt{49}$

Combined Operations with Approximate Numbers

Perform each computation, keeping the proper number of digits in your answer.

23. $(7.37)(3.28) + (8.36)(2.64)$

24. $(522)(9.53) - (586)(4.70) + (847)(7.59)$

25. $(63.5 + 83.6)(8.37 - 1.72)$

26. $(8.93 - 3.74 + 9.05)(68.70 - 64.90)$

27. $\dfrac{583}{473} + \dfrac{946}{907}$

28. $\dfrac{6.73}{8.38} - \dfrac{5.97}{8.06} + \dfrac{8.63}{1.91}$

29. $\dfrac{6.47 + 8.604}{3.37 + 90.8}$

30. $\dfrac{809 - 463 + 744}{758 - 964 + 508}$

31. $(5.37 + 2.36)^2$

32. $(4.25 + 4.36 - 5.24)^4$

33. $(6.423 + 1.05)^2$

34. $\left(\dfrac{45.3 - 8.34}{8.74 - 5.62}\right)^{2.5}$

35. $\left(\dfrac{8.90}{4.75}\right)^2$

36. $\sqrt{4.34 + 4.66}$

37. $\sqrt[3]{657 + 553 - 842}$

38. $\sqrt{(28.1)(5.94)}$

39. $\sqrt[5]{(9.06)(4.86)(7.93)}$

40. $\sqrt{\dfrac{653}{601}}$

41. $\sqrt[4]{\dfrac{4.50}{7.81}}$

42. $\sqrt{9.74} + \sqrt{12.5}$

43. $\sqrt[4]{528} + \sqrt{94.2} - \sqrt[3]{284}$

44. $\sqrt[4]{653} + \sqrt{55.3}$

45. *Writing:* Suppose you have submitted a report that contains calculations in which you have rounded the answers according to the rules given in this chapter. Jones, your company rival, has sharply attacked your work, calling it "inaccurate" because you did not keep enough digits, and your boss seems to agree.

Write a memo to your boss defending your rounding practices. Point out why it is misleading to retain too many digits. Do not write more than one page. You may use numerical examples to prove your point.

1–7   Scientific Notation and Engineering Notation

```
500000000*300000
000
              1.5E17
```

Your display might be different, partly depending on your MODE settings.

■ **Exploration:**

Try this. On your calculator, multiply

$$500{,}000{,}000 \times 300{,}000{,}000$$

What did your calculator show for this calculation? Can you explain the meaning of your display? Try multiplying some other very large or very small numbers and try to explain the display.

■

If our answers in the exploration have too many zeros for the display, the calculator will automatically switch into the kind of notation that we will study in this section. The display

$$1.5 \text{ E}17$$

contains two parts: a decimal number, here 1.5, and an integer, 17. We read this as the decimal number times 10 raised to the value of the integer, or

$$1.5 \times 10^{17}$$

Here, 10^{17} is called a *power of ten.*

Powers of 10

We did some work with powers in Sec. 1–5. We saw, for example, that 2^3 meant

$$2^3 = 2 \cdot 2 \cdot 2 = 8$$

Here, the power 3 tells how many 2's are to be multiplied to give the product. For powers of 10, the power tells how many 10's are to be multiplied to give the product.

◆◆◆ **Example 68:** Here are some powers of ten expressed as decimal numbers.

(a) $10^2 = 10 \times 10 = 100$
(b) $10^3 = 10 \times 10 \times 10 = 1000$ ◆◆◆

◆◆◆ **Example 69:** Here are some demical numbers expressed as powers of ten.

(a) $10{,}000 = 10^4$ (b) $1{,}000{,}000 = 10^6$

We can evaluate 10 raised to a *negative* power using a formula that we will derive in the next chapter.

Negative Exponent	$x^{-a} = \dfrac{1}{x^a} \quad (x \neq 0)$	28

◆◆◆ **Example 70:** Here are some examples of 10 raised to a negative power.

(a) $10^{-2} = \dfrac{1}{10^2} = \dfrac{1}{100} = 0.01$

(b) $10^{-5} = \dfrac{1}{10^5} = \dfrac{1}{100{,}000} = 0.00001$ ◆◆◆

Some powers of 10 are summarized in the following table:

Positive Powers	Negative Powers
$1,000,000 = 10^6$	$0.1 = 1/10 = 10^{-1}$
$100,000 = 10^5$	$0.01 = 1/10^2 = 10^{-2}$
$10,000 = 10^4$	$0.001 = 1/10^3 = 10^{-3}$
$1,000 = 10^3$	$0.0001 = 1/10^4 = 10^{-4}$
$100 = 10^2$	$0.00001 = 1/10^5 = 10^{-5}$
$10 = 10^1$	$0.000001 = 1/10^6 = 10^{-6}$
$1 = 10^0$	

Converting Numbers to Scientific Notation

A number is said to be in *scientific notation* when it is written as a number whose absolute value is between 1 and 10, multiplied by a power of 10.

◆◆◆ **Example 71:** The following numbers are written in scientific notation:

(a) 2.74×10^4 (b) 8.84×10^8

(c) 5.4×10^{-7} (d) -1.2×10^{-5} ◆◆◆

To convert a decimal number to scientific notation, first rewrite the given number with a single digit to the left of the demical point, discarding any nonsignificant zeros. Then multiply this number by the power of 10 that will make it equal to the original number.

◆◆◆ **Example 72:** Here we convert to scientific notation.

$$346 = 3.46 \times 100$$
$$= 3.46 \times 10^2 \qquad ◆◆◆$$

◆◆◆ **Example 73:** Another example of conversion to scientific notation.

$$2700 = 2.7 \times 1000$$
$$= 2.7 \times 10^3$$

Note that we have discarded the two nonsignificant zeros.   ◆◆◆

When we are converting a number whose absolute value is less than 1, our power of 10 will be negative, as in Example 74.

◆◆◆ **Example 74:** Here is an example resulting in a negative power.

$$0.00000950 = 9.50 \times 0.000001$$
$$= 9.50 \times 10^{-6}$$

Since the trailing zero is significant in our original number, it is retained in our answer.   ◆◆◆

The sign of the exponent has nothing to do with the sign of the original number. You can convert a negative number to scientific notation just as you would a positive number.

◆◆◆ **Example 75:** Convert $-34,720$ to scientific notation.

Solution: Converting to scientific notation, we obtain

$$-34,720 = -3.472 \times 10,000$$
$$= -3.472 \times 10^4 \qquad ◆◆◆$$

◆◆◆ **Example 76:** *An Application.* A certain tract of land contains 39,700,000 ft^2 of land. Convert this to scientific notation.

Solution: 39,700,000 ft^2 = 3.97 × 10,000,000 ft^2

$\qquad\qquad\qquad\quad$ = 3.97 × 10^7 ft^2 ◆◆◆

Converting Numbers to Scientific Notation

To convert *from* scientific notation, simply reverse the process.

◆◆◆ **Example 77:** Here we convert from scientific notation to decimal form.

$$4.82 \times 10^5 = 4.82 \times 100,000$$
$$= 482,000 \qquad ◆◆◆$$

◆◆◆ **Example 78:** Another example of converting to decimal form.

$$8.25 \times 10^{-3} = 8.25 \times 0.001$$
$$= 0.00825 \qquad ◆◆◆$$

◆◆◆ **Example 79:** *An Application.* The resistance of a certain transmission line is 5.85×10^{-4} Ω (ohms). Write this resistance in decimal notation.

Solution: 5.85×10^{-4} Ω = 0.000585 Ω   ◆◆◆

Converting Numbers to Engineering Notation

Engineering notation is similar to scientific notation. The difference is that
  • the exponent is *a multiple of three;* and
  • there can be one, two, or three digits to the left of the decimal point, rather than just one digit.
Having an exponent that is a multiple of 3 makes it easier to use the *metric prefixes* we will introduce later in this chapter.

◆◆◆ **Example 80:** Some examples of numbers written in engineering notation are as follows:

(a) 66.3×10^3 $\qquad\qquad$ (b) 8.14×10^9
(c) 725×10^{-6} $\qquad\quad$ (d) 28.72×10^{-12} ◆◆◆

Converting to engineering notation is simple if the digits of the decimal number are grouped by commas into sets of three, in the usual way.

◆◆◆ **Example 81:** Here are some conversions to engineering notations.

(a) $21,840 = 21.84 \times 10^3$
(b) $548,000 = 548 \times 10^3$
(c) $72,560,000 = 72.56 \times 10^6$ ◆◆◆

For numbers less than 1, it helps to first separate the digits following the decimal point into groups of three.

◆◆◆ **Example 82:** Try to follow these conversions to engineering notation.

(a) $0.87217 = 0.872\ 17 = 872.17 \times 10^{-3}$
(b) $0.000736492 = 0.000\ 736\ 492 = 736.492 \times 10^{-6}$
(c) $0.0000000472 = 0.000\ 000\ 047\ 2 = 47.2 \times 10^{-9}$ ◆◆◆

◆◆◆ **Example 83:** *An Application.* A certain heating furnace is rated at 2.85×10^5 Btu/h (British thermal units per hour). Express this rating in engineering notation.

Solution: $2.85 \times 10^5 = 285,000 = 285 \times 10^3$ Btu/h   ◆◆◆

Converting Numbers from Engineering Notation

As with converting from scientific notation, we simply reverse the process.

◆◆◆ **Example 84:** Here are some conversions from engineering to decimal notation.

(a) $48.342 \times 10^3 = 48.342 \times 1000 = 48,342$

(b) $8.559 \times 10^6 = 8.599 \times 1,000,000 = 8,559,000$

(c) $8.352 \times 10^{-3} = 8.352 \times 0.001 = 0.008352$

(d) $736 \times 10^{-6} = 736 \times 0.000001 = 0.000736$ ◆◆◆

◆◆◆ **Example 85:** *An Application.* The weight of a certain punch press is 28.56×10^3 lb. Express this weight in decimal notation.

Solution: $28.56 \times 10^3 = 28,560$ lb   ◆◆◆

Addition and Subtraction

Let us turn now to *computations* using scientific and engineering notation. We will first do some simple problems by hand, to show how the powers of ten are combined, and then we will do similar problems by calculator.

If two or more numbers to be added or subtracted have the *same power of 10,* simply combine the numbers and keep the same power of 10.

◆◆◆ **Example 86:** Here we combine numbers that have the same power of 10.

(a) $(2 \times 10^5) + (3 \times 10^5) = 5 \times 10^5$

(b) $(8 \times 10^3) - (5 \times 10^3) + (3 \times 10^3) = 6 \times 10^3$ ◆◆◆

If the sum is greater than 10 or less than 1, it is no longer, strictly speaking, in scientific notation. We change it to scientific notation as we did earlier in this section.

◆◆◆ **Example 87:** These show more examples with numbers having the same power of 10.

(a) $(8.4 \times 10^4) + (7.2 \times 10^4) = 15.6 \times 10^4$

$$= 1.56 \times 10^5$$

(b) $(5.822 \times 10^3) - (5.000 \times 10^3) = 0.822 \times 10^3$

$$= 8.22 \times 10^2$$ ◆◆◆

If the powers of 10 are different, *they must be made equal* before the numbers can be combined. A shift of the decimal point of one place to the *left* will *increase* the exponent by 1. Conversely, a shift of the decimal point one place to the *right* will *decrease* the exponent by 1.

◆◆◆ **Example 88:** Here are examples of combining numbers having different powers of 10.

(a) $(1.5 \times 10^4) + (3 \times 10^3) = (1.5 \times 10^4) + (0.3 \times 10^4)$

$$= 1.8 \times 10^4$$

(b) $(1.25 \times 10^5) - (2.0 \times 10^4) + (4 \times 10^3)$

$$= (1.25 \times 10^5) - (0.20 \times 10^5) + (0.04 \times 10^5)$$

$$= 1.09 \times 10^5$$

(c) $(8.84 \times 10^3) + (9.92 \times 10^4) = (0.884 \times 10^4) + (9.92 \times 10^4)$

$$= 10.80 \times 10^4$$

$$= 1.080 \times 10^5$$

Note that in (c) we started with numbers each having three significant digits, and that our result has four significant digits. This is an example of significant digits being gained during addition. Similarly, significant digits can be lost during subtraction. ◆◆◆

◆◆◆ **Example 89:** *An Application.* Two resistors having resistances of $3.74 \times 10^3 \, \Omega$ and $9.37 \times 10^4 \, \Omega$ are wired in series. Find their combined resistance, which is the sum of the two, in engineering notation.

Solution:

$$(3.74 \times 10^3) + (9.37 \times 10^4) = (3.74 \times 10^3) + (93.7 \times 10^3)$$

$$= 97.4 \times 10^3 \, \Omega \qquad ◆◆◆$$

Multiplication

> We are really using $x^a \cdot x^b = x^{a+b}$. This equation is one of the *laws of exponents* that we will study later.

We multiply powers of 10 by *adding their exponents.*

◆◆◆ **Example 90:**

(a) $10^3 \cdot 10^4 = 10^{3+4} = 10^7$

(b) $10^{-2} \cdot 10^5 = 10^{-2+5} = 10^3$ ◆◆◆

To multiply two numbers in scientific notation, multiply the decimal parts and the powers of 10 *separately.*

◆◆◆ **Example 91:**

(a) $(2 \times 10^5)(3 \times 10^2) = (2 \times 3)(10^5 \times 10^2)$

$$= 6 \times 10^{5+2} = 6 \times 10^7$$

(b) $(2.84 \times 10^3)(7.21 \times 10^4) = (2.84 \times 7.21)(10^3 \times 10^4)$

$$= 20.5 \times 10^{3+4}$$

$$= 20.5 \times 10^7$$

$$= 2.05 \times 10^8 \qquad ◆◆◆$$

◆◆◆ **Example 92:** *An Application.* The dimensions of a rectangular tract are 5.983×10^3 by 1.395×10^4 feet. Find the area of the field.

Solution: Multiplying the two dimensions gives

$$(5.983 \times 10^3)(1.395 \times 10^4) = 8.346 \times 10^{3+4}$$

$$= 8.346 \times 10^7 \text{ ft}^2 \qquad ◆◆◆$$

Division

> We are using a law of exponents for quotients that we will study in Chapter 2.

We divide powers of 10 by subtracting the exponent of the denominator from the exponent of the numerator.

◆◆◆ **Example 93:** Here are some quotients of powers of 10.

(a) $\dfrac{10^5}{10^3} = 10^{5-3} = 10^2$

(b) $\dfrac{10^{-4}}{10^{-2}} = 10^{-4-(-2)} = 10^{-2}$

As with multiplication, we divide the decimal parts and the powers of 10 separately.　　　　　　　　　　　　　　　　　　　　　　　◆◆◆

◆◆◆ **Example 94:** These examples show how to divide numbers in scientific notation.

(a) $\dfrac{8 \times 10^5}{4 \times 10^2} = \dfrac{8}{4} \times \dfrac{10^5}{10^2} = 2 \times 10^{5-2} = 2 \times 10^3$

(b) $\dfrac{12 \times 10^3}{4 \times 10^5} = 3 \times 10^{3-5} = 3 \times 10^{-2}$

(c) $(1.97 \times 10^3) \div (2.52 \times 10^4) = 0.782 \times 10^{-1}$
$$= 7.82 \times 10^{-2}$$
　　　　　　　　　　　　　　　　　　　　　　　　　　　　　　◆◆◆

◆◆◆ **Example 95:** *An Application.* A truck with a capacity of 3.24×10^2 ft^3 contains a load of gravel which weighs 3.77×10^4 lb. Find the density of the sand by dividing the weight by the volume.

Solution:

$$\text{density} = \frac{\text{weight}}{\text{volume}} = \frac{3.77 \times 10^4}{3.24 \times 10^2}$$

$$= 1.16 \times 10^2$$

$$= 116 \text{ lb/ft}^3　　　◆◆◆$$

Scientific and Engineering Notation on the Calculator

Displaying Numbers: We can choose the way a number is *displayed* on a calculator, either in decimal, scientific, or engineering notation. This choice is usually made from a menu.

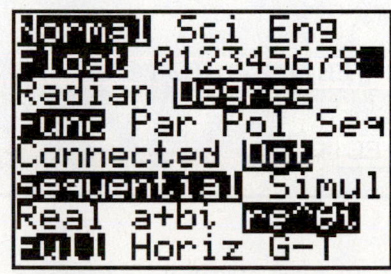

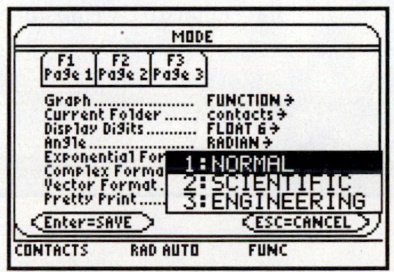

MODE screen for the TI-83/84, showing the Normal, Sci, and Eng modes.

MODE screen for the TI-89 showing the NORMAL, SCIENTIFIC, and ENGINEERING modes.

Entering Numbers: You can *enter* a number in any of these notations, regardless of the mode the calculator is in. However, the entered number will be displayed according to the chosen mode. The power of ten is entered using the *enter exponent* key, usually marked EE , EXP , or EEX .

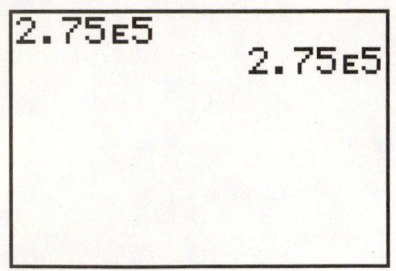

TI-83/84 screen for Example 96. Calculator is in Sci mode.

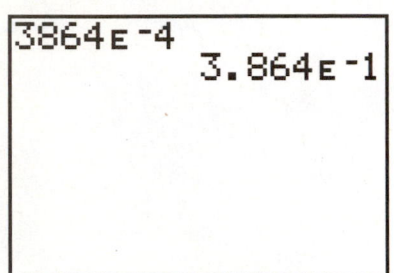

TI-83/84 screen for Example 97. To enter a negative exponent, we must be careful not to use the subtraction key, but the key used for entering a negative number. Calculator is in Sci mode.

♦♦♦ **Example 96:** The keystrokes to enter the number 2.75×10^5 are shown. The number displayed will depend on the mode in which the calculator is set,

275000	if in decimal mode
2.75 E5	if in scientific notation mode
275 E3	if in engineering notation mode

♦♦♦

The calculator will even accept numbers that are not strictly in scientific or engineering notation.

♦♦♦ **Example 97:** The keystrokes to enter 3864×10^{-4} are shown.   ♦♦♦

Calculating in Scientific Notation: The numbers you enter can be in any notation, and can even be mixed in the same calculation.

♦♦♦ **Example 98:** Evaluate

$$\frac{638.2 + (8.117 \times 10^2)}{(27.86 \times 10^{-9})(1255 \times 10^3)}$$

Solution: Here we have some numbers in decimal, scientific, and engineering notation, and another that is in none of these. We key them in just as they appear. Note that the numerator must be enclosed in parentheses. Also be careful not to use the subtraction key to enter the negative exponent. We get, after rounding,

41470	in normal mode
4.147 E4	in scientific notation mode
41.47 E3	in engineering notation mode

♦♦♦

TI-83/84 screen for Example 98, with the calculator in scientific notation mode.

Common Errors	Students often enter powers of 10 (such as 10^4) incorrectly into their calculators, forgetting that 10^4 is really 1×10^4. Thus to enter 10^4, we press
	1 \boxed{EE} 4
	and not
	10 \boxed{EE} 4
	Do not press the times key before the *enter exponent* key. Thus to enter 10^4, we *do not* enter
	1 \boxed{\times} \boxed{EE} 4
	but *do* enter
	1 \boxed{EE} 4

TI-83/84 screen for Example 99.

♦♦♦ **Example 99:** *An Application.* The current in a resistor equals the voltage across the resistor divided by the resistance of that resistor. Find the current in a $4.55 \times 10^4 \, \Omega$ resistor having a voltage across it of 8.25×10^{-2} volts.

Solution: Dividing the voltage by the resistance we get,

$$\text{Current} = \frac{\text{voltage}}{\text{resistance}} = \frac{8.25 \times 10^{-2}}{4.55 \times 10^4}$$

$$= 1.81 \times 10^{-6} \text{ amperes}$$

♦♦♦

Exercise 7 ◆ Scientific and Engineering Notation

Powers of 10

Write each power of 10 as a decimal number.

1. 10^5 **2.** 10^{-2} **3.** 10^{-5}

4. 10^{-1} **5.** 10^4

Write each number as a power of 10.

6. 100 **7.** 1,000,000 **8.** 0.0001

9. 0.001 **10.** 100,000,000

Converting Numbers to Scientific Notation

Write each number in scientific notation.

11. 186,000

12. 0.0035

13. 25,742

14. 8020

15. 98.3×10^3

16. 0.0775×10^{-2}

Converting Numbers from Scientific Notation

Convert each number from scientific notation to decimal notation.

17. 2.85×10^3 **18.** 1.75×10^{-5} **19.** 9×10^4

20. 9.05×10^4 **21.** 3.667×10^{-3}

Converting Numbers to Engineering Notation

Convert each number to engineering notation.

22. 34,382 **23.** 3.58×10^2

24. 26,940 **25.** 0.134

26. 23.48×10^{-2} **27.** 0.00374

Converting Numbers from Engineering Notation

Convert each number from engineering notation to decimal notation.

28. 385×10^3 **29.** $18,640 \times 10^{-3}$ **30.** 8488×10^{-3}

31. 7739×10^{-3} **32.** 6.37×10^3 **33.** 2.66×10^6

Addition and Subtraction

Combine without using a calculator. Give your answer in scientific notation.

34. $(3.0 \times 10^4) + (2.1 \times 10^5)$

35. $(75.0 \times 10^2) + 3210$

36. $(1.557 \times 10^2) + (9.000 \times 10^{-1})$

37. $0.037 - (6.0 \times 10^{-3})$

38. $(7.2 \times 10^4) + (1.1 \times 10^4)$

Multiplication

Multiply the following powers of 10.

39. $10^5 \cdot 10^2$ **40.** $10^4 \cdot 10^{-3}$ **41.** $10^{-5} \cdot 10^{-4}$

42. $10^{-2} \cdot 10^5$ **43.** $10^{-1} \cdot 10^{-4}$

Multiply without using a calculator. Give your answer in scientific notation.

44. $(3.0 \times 10^3)(5.0 \times 10^2)$ **45.** $(5 \times 10^4)(8 \times 10^{-3})$

46. $(2 \times 10^{-2})(4 \times 10^{-5})$ **47.** $(2 \times 10^4)(30{,}000)$

Division

Divide the following powers of 10.

48. $10^8 \div 10^5$ **49.** $10^4 \div 10^6$ **50.** $10^5 \div 10^{-2}$

51. $10^{-3} \div 10^5$ **52.** $10^{-2} \div 10^{-4}$

Divide without using a calculator. Give your answer in scientific notation.

53. $(8 \times 10^4) \div (2 \times 10^2)$ **54.** $(6 \times 10^4) \div 0.03$

55. $(3 \times 10^3) \div (6 \times 10^5)$ **56.** $(8 \times 10^{-4}) \div 400{,}000$

57. $(9 \times 10^4) \div (3 \times 10^{-2})$ **58.** $49{,}000 \div (7.0 \times 10^{-2})$

Scientific and Engineering Notation on the Calculator

Perform the following computations. Display your answer in scientific notation.

59. $(1.58 \times 10^2)(9.82 \times 10^3)$

60. $(9.83 \times 10^5) \div (2.77 \times 10^3)$

61. $(3.87 \times 10^{-2})(5.44 \times 10^5)$

62. $(2.74 \times 10^3) \div (9.13 \times 10^5)$

63. $(5.6 \times 10^2)(3.1 \times 10^{-1})$

64. $(7.72 \times 10^8) \div (3.75 \times 10^{-9})$

Applications

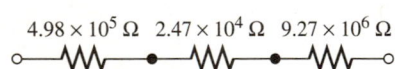

$4.98 \times 10^5\,\Omega$ $2.47 \times 10^4\,\Omega$ $9.27 \times 10^6\,\Omega$

FIGURE 1–10 Resistors in series.

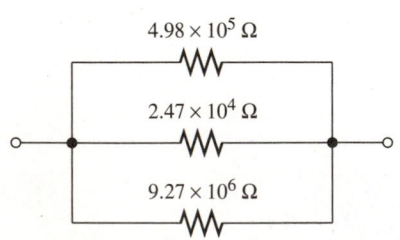

$4.98 \times 10^5\,\Omega$

$2.47 \times 10^4\,\Omega$

$9.27 \times 10^6\,\Omega$

FIGURE 1–11 Resistors in parallel.

65. Three resistors, having resistances of $4.98 \times 10^5\,\Omega$, $2.47 \times 10^4\,\Omega$, and $9.27 \times 10^6\,\Omega$, are wired in series (Fig. 1–10). Find the total resistance, using the formula $R = R_1 + R_2 + R_3$.

66. Find the equivalent resistance if the three resistors of problem 65 are wired in parallel (Fig. 1–11). Use the formula

$$\frac{1}{R} = \frac{1}{R_1} + \frac{1}{R_2} + \frac{1}{R_3}$$

67. Find the power in watts dissipated in a resistor if a current I of 3.75×10^{-3} A produces a voltage drop V of 7.24×10^{-4} V across the resistor. Use the formula $P = VI$.

68. The voltage across an $8.35 \times 10^5\,\Omega$ resistor is 2.95×10^{-3} V. Find the power dissipated in the resistor, using the formula $P = V^2/R$.

69. Three capacitors, 8.26×10^{-6} farad (F), 1.38×10^{-7} F, and 5.93×10^{-5} F, are wired in parallel. Find the equivalent capacitance using the formula $C = C_1 + C_2 + C_3$

70. A wire 4.75×10^3 cm long when loaded is seen to stretch 9.55×10^{-2} cm. Find the strain in the wire, using the formula strain = elongation ÷ length.

71. *Writing:* Study your calculator and its manual specifically on the different display formats (normal, scientific notation, and so forth). List the different formats available to you and explain the diferences between them. Write a few lines explaining how to switch from one format to another.

72. *Project:* In a technical magazine or journal find several uses of scientific or engineering notation. Show why the author has chosen it over decimal notation.

73. *Internet:* Surf the Web for data on the various planets. Find the mean distance of each planet from the sun. Tabulate this distance in decimal, scientific, and engineering notation, in both miles and kilometers. Make a number line, with the sun at zero, and mark the distance of each planet on that line.

1–8 Units of Measurement

Systems of Units

A *unit* is a standard of measurement, such as the meter, inch, hour, or pound. The two main systems in use are the U.S. customary units (feet, pounds, gallons, etc.) and the SI or *metric* system (meters, kilograms, liters, etc.). SI stands for Le Système International d'Unites, or the *International System of Units*. In addition, some special units, such as a *square* of roofing material, and some obsolete units, such as *rods* and *chains,* must occasionally be dealt with.

Most units have an abbreviated form or a symbol, so that we do not have to write the full word. Thus the abbreviation for millimeters is mm, and the symbol for ohms is Ω (capital Greek omega). In this section we will usually give the full word *and* the abbreviation.

Conversion of Units

We convert from one unit of measurement to another by means of a *conversion factor.* Conversion factors for units of measurement, as well as the abbreviations for those units, are given in Appendix B.

◆◆◆ **Example 100:** Convert 1.530 miles (mi) to feet (ft).

Solution: From Appendix B we find the relation between miles and feet.

$$5280 \text{ ft} = 1 \text{ mi}$$

Dividing both sides by 1 mile, we get the conversion factor.

$$\frac{5280 \text{ ft}}{1 \text{ mi}} = 1$$

We know that we can multiply any quantity by 1 without changing the value of that quantity. Thus if we multiply our original quantity (1.530 mi) by the conversion factor (5280 ft/mi), we do not change the value of the original quantity. We will, however, change the units. Multiplying yields

$$1.530 \text{ mi} = 1.530 \text{ m\!i} \times \frac{5280 \text{ ft}}{1 \text{ m\!i}} = 8078 \text{ ft}$$

Note that we have rounded our answer to four significant digits, because all numbers used in the calculation have at least four significant digits (the 5280 is exact).   ◆◆◆

Suppose that in the first step of Example 100, we had divided both sides by 5280 ft instead of by 1 mi. We could have gotten another conversion factor:

$$\frac{1 \text{ mi}}{5280 \text{ ft}} = 1$$

Thus each relation between two units of measurement gives us *two* conversion factors. But suppose, in the preceding example, we had written

$$1.530 \text{ mi} \times \frac{1 \text{ mi}}{5280 \text{ ft}} = ??$$

This is not incorrect but does us no good because *miles* does not cancel.

Common Errors	Choose the conversion factor that will cancel the units you wish to eliminate. Write the units in the equation and make sure they cancel properly.
	Be sure to write the original quantity as a *built-up* fraction, such as $\dfrac{a}{b}$, rather than on a single line, *a/b*. This will greatly reduce your chances of making an error.

The use of conversion factors and making sure that the units in an expression are compatible and cancel properly is part of what is called *dimensional analysis*.

Significant Digits

You should try to use a conversion factor that is exact, or one that contains at least as many significant digits as in your original number. Then you should round your answer to as many significant digits as in the original number.

◆◆◆ **Example 101:** Convert 934 acres to square miles (mi^2).

Solution: From Appendix B we find

$$1 \text{ mi}^2 = 640 \text{ acres}$$

where 640 is an exact number. We must write our conversion factor so that the unwanted unit (acres) is in the denominator, so the acres will cancel. Our conversion factor is thus

$$\frac{1 \text{ mi}^2}{640 \text{ acres}} = 1$$

Multiplying, we obtain

$$934 \text{ acres} = 934 \; \cancel{\text{acres}} \times \frac{1 \text{ mi}^2}{640 \; \cancel{\text{acres}}}$$

$$= 1.46 \text{ mi}^2$$

The conversion factor used here is exact, so we have rounded our answer to the three significant digits of the given number.   ◆◆◆

Converting Areas and Volumes

Length may be given in, say, centimeters (cm), but an *area* may be given in *square* centimeters (cm^2). Similarly, a *volume* may be in *cubic* centimeters (cm^3). So to get a conversion factor for area or volume, if not found in Appendix B, simply square or cube the conversion factor for length. For example, if we take the equation

$$2.54 \text{ cm} = 1 \text{ in.}$$

this conversion is *exact*, by international agreement. Squaring both sides we get

$$(2.54 \text{ cm})^2 = (1 \text{ in.})^2$$

or

$$6.4516 \text{ cm}^2 = 1 \text{ in.}^2$$

This gives us a conversion between square centimeters and square inches. Since 2.54 is an exact number, we may keep all the significant digits in its square.

◆◆◆ **Example 102:** *An Application.* A buliding lot is seen to contain 864 square yards. The deed requires that it be given in acres. Convert this quantity.

Solution: Appendix B has no conversion for square yards. However,

$$1 \text{ yd} = 3 \text{ ft}$$

Squaring yields

$$1 \text{ yd}^2 = (3 \text{ ft})^2 = 9 \text{ ft}^2$$

Also from the table,

$$1 \text{ acre} = 43{,}560 \text{ ft}^2$$

So

$$864 \text{ yd}^2 = 864 \text{ yd}^2 \times \frac{9 \text{ ft}^2}{1 \text{ yd}^2} \times \frac{1 \text{ acre}}{43{,}560 \text{ ft}^2} = 0.179 \text{ acre}$$

◆◆◆

Converting Rates to Other Units

A *rate* is the amount of one quantity expressed *per unit of some other quantity.* Some rates, with typical units, are

rate of travel (mi/h) or (km/h)   *flow rate (gal/min) or (m³/s)*
application rate (lb/acre)   *unit price (dollars/lb)*

Each rate contains *two* units of measure; miles per hour, for example, has *miles* in the numerator and *hours* in the denominator. It may be necessary to convert *either* or *both* of those units to other units. Sometimes a single conversion factor can be found (such as 1 m/h = 1.466 ft/s), but more often you will have to convert each unit with a *separate* conversion factor.

◆◆◆ **Example 103:** A certain chemical is to be added to a pool at the rate of 3.74 oz per gallon of water. Convert this to pounds of chemical per cubic foot of water.

Solution: We write the original quantity as a fraction and multiply by the appropriate factors, themselves written as fractions.

$$3.74 \text{ oz/gal} = \frac{3.74 \text{ oz}}{\text{gal}} \times \frac{1 \text{ lb}}{16 \text{ oz}} \times \frac{7.481 \text{ gal}}{\text{ft}^3} = 1.75 \text{ lb/ft}^3$$

◆◆◆

Using More Than One Conversion Factor

Sometimes you may not be able to find a *single* conversion factor linking the units you want to convert. You may have to use *more than one.*

◆◆◆ **Example 104:** *An Application.* A map shows that the distance to a lighthouse is 7375 yards. We want to lay out this distance on another chart, marked in nautical miles. Convert this quantity.

Solution: In Appendix B we find no conversion factor between nautical miles and yards, but we see that.

$$1 \text{ nau mi} = 6076 \text{ ft} \quad \text{and} \quad 3 \text{ ft} = 1 \text{ yd}$$

So

$$7375 \text{ yd} = 7375 \text{ yd} \times \frac{3 \text{ ft}}{1 \text{ yd}} \times \frac{1 \text{ nau mi}}{6076 \text{ ft}} = 3.641 \text{ nautical mi}$$

◆◆◆

Metric Units

The *metric system* is a system of weights and measures that was developed in France in 1793 and that has since been abopted by most countries of the world. It is widely used in scientific work in the United States.

The basic unit of length in the metric system is the *meter* (m). The unit of area is the *are*, or 100 square meters (m)^2. The unit of volume is the *liter* (L), the volume of a cube one-tenth of a meter on a side. The unit of mass is the *gram* (g), the theoretical weight of a cube of distilled water measuring $\frac{1}{100}$ of a meter on a side.

Metric Prefixes

Converting between metric units is made easy because larger and smaller metric units are related to the basic units by *factors of 10*. These larger or smaller units are indicated by placing a *prefix* before the basic unit. A prefix is a group of letters placed at the beginning of a word to modify the meaning of that word. For example, the prefix *kilo* means 1000, or 10^3. Thus a *kilogram* is 1000 grams. Other metric prefixes are given in Table 1–1.

TABLE 1–1   Metric prefixes.

Amount	Multiples and Submultiples	Prefix	Symbol	Meaning
1 000 000 000 000	10^{12}	tera	T	One trillion times
1 000 000 000	10^9	giga	G	One billion times
1 000 000	10^6	mega	M*	One million times
1 000	10^3	kilo	k*	One thousand times
100	10^2	hecto	h	One hundred times
10	10	deka	da	Ten times
0.1	10^{-1}	deci	d	One tenth of
0.01	10^{-2}	centi	c*	One hundredth of
0.001	10^{-3}	milli	m*	One thousandth of
0.000 001	10^{-6}	micro	μ*	One millionth of
0.000 000 001	10^{-9}	nano	n	One billionth of
0.000 000 000 001	10^{-12}	pico	p	One trillionth of
0.000 000 000 000 001	10^{-15}	femto	f	One quadrillionth of
0.000 000 000 000 000 001	10^{-18}	atto	a	One quintillionth of

*Most commonly used.

◆◆◆ **Example 105:** Here are some uses of metric prefixes.

(a) A *kilo*meter (km) is a thousand meters, because *kilo* means one thousand.
$$1 \text{ km} = 1000 \text{ m}$$

(b) A *centi*meter (cm) is one-hundredth of a meter, because *centi* means one hundredth.
$$1 \text{ cm} = 1/100 \text{ m}$$

(c) A *milli*meter (mm) is one-thousandth of a meter, because *milli* means one thousandth.
$$1 \text{ mm} = 1/1000 \text{ m}$$
◆◆◆

Converting Between Metric Units

Converting from one metric unit to another is usually a matter of multiplying or dividing by a power of 10. Most of the time the names of the units will tell how they are related, so we do not even have to look them up.

◆◆◆ **Example 106:** Convert 72,925 meters (m) to kilometers (km).

Solution: A *kilometer* is a thousand meters, so our conversion factor is

$$\frac{1 \text{ km}}{1000 \text{ m}} = 1$$

So, as before,

$$72{,}925 \text{ m} = 72{,}925 \text{ m} \times \frac{1 \text{ km}}{1000 \text{ m}} = 72.925 \text{ km} \qquad \text{◆◆◆}$$

For more unusual metric units, simply look up the conversion factor in a table.

◆◆◆ **Example 107:** Convert 2.75 newtons (N) to dynes.

Solution: These two metric units of force do not have any basic units in their names, or any prefixes. Thus we cannot tell just from their names how they are related to each other. However, from Appendix B we find that

$$1 \text{ newton} = 10^5 \text{ dynes}$$

Converting in the usual way, we obtain

$$2.75 \text{ newtons} = 2.75 \text{ newtons} \times \frac{10^5 \text{ dynes}}{1 \text{ newton}} = 2.75 \times 10^5 \text{ dynes}$$

or 275,000 dynes. ◆◆◆

Converting Between Customary and Metric Units

We convert between customary and metric units in the same way that we converted within each system.

◆◆◆ **Example 108:** Convert 2.84 U.S. gallons (gal) to liters (L).

Solution: From Appendix B we find

$$1 \text{ gal (U.S.)} = 3.785 \text{ L}$$

Converting gives

$$2.84 \text{ gal} = 2.84 \text{ gal} \times \frac{3.785 \text{ L}}{1 \text{ gal}} = 10.7 \text{ L}$$

rounded to three significant digits. ◆◆◆

Angle Conversions

The degree (°) is a unit of angular measure equal to 1/360 of a revolution; thus 360° = one revolution. A fractional part of a degree may be expressed as a common fraction (such as $36\frac{1}{2}°$), as a decimal (28.74°), or as *minutes* and *seconds*.

A *minute* (′) is equal to 1/60 of a degree; a *second* (″) is equal to 1/60 of a minute, or 1/3600 of a degree.

Another important unit of angular measure is the *radian*. We will learn about radians later.

◆◆◆ **Example 109:** Some examples of angles written in degrees, minutes, and seconds are

$$85°18'42'' \qquad 62°12' \qquad 75°06'03''$$

◆◆◆

Note that minutes or seconds less than 10 are written with an initial zero. Thus 6' is written 06'.

We will sometimes abbreviate "degrees, minutes, and seconds" as DMS.

Conversions involving degrees, minutes, and seconds require several steps. Further, to know how many digits to retain, we note the following:

$$1 \text{ min} = 1/60 \text{ degree} \approx 0.02°$$

and

$$1 \text{ sec} = 1/3600 \text{ degree} \approx 0.0003°$$

Thus an angle known to the nearest minute is about as accurate as a decimal angle known to two decimal places. Also, an angle known to the nearest second is about as accurate as a decimal angle known to the fourth decimal place. Therefore we would treat an angle such as 28°17'37'' as if it had four decimal places and six significant digits.

◆◆◆ **Example 110:** Convert 28°17'37'' to decimal degrees.

Solution: We separately convert the minutes and the seconds to degrees and add them. Since the given angle is known to the nearest second, we will work to four decimal places.

$$37 \text{ sec}\left(\frac{1 \text{ deg}}{3600 \text{ sec}}\right) = 0.0103°$$

$$17 \text{ min}\left(\frac{1 \text{ deg}}{60 \text{ min}}\right) = 0.2833°$$

$$28° = \underline{28.0000°}$$

$$\text{Add:} \qquad 28.2936°$$

◆◆◆

◆◆◆ **Example 111:** Convert 105.2821° to degrees, minutes, and seconds.

Solution: We first convert the decimal part (0.2821°) to minutes,

$$0.2821°\left(\frac{60'}{1°}\right) = 16.93'$$

and convert the decimal part of 16.93' to seconds,

$$0.93'\left(\frac{60''}{1 \text{ min}}\right) = 56''$$

So

$$105.2821° = 105°16'56''$$

◆◆◆

Common Error	When you are reading an angle quickly, it is easy to mistake decimal degrees for degrees and minutes. Don't mistake $$28°50' \quad \text{for} \quad 25.50°$$ $28.50°$ is really $28°30'$, and $28°50'$ is $$28°50' = 28° + \left(\frac{50}{60}\right)° \approx 28.83°$$

Angle Conversions by Calculator

We can convert between decimal degrees and degrees, minutes, seconds by calculator.

◆◆◆ Example 112: Convert 34°44′18″ to decimal degrees.

Solution:

- Put the calculator into degree mode.
- Enter the angle, including the degree, minute, and second symbols. On the TI-83/84, the degree symbol and the minute symbol are found in the ANGLE menu. The $''$ second symbol is an alpha character on the $+$ key.
- Press ENTER.

The angle will be displayed in decimal form, as shown. **◆◆◆**

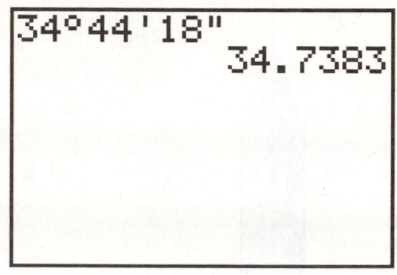

TI-83/84 screen for Example 112. On the TI-89, the degree, minute, and second symbols are alpha characters on the ⃞, =, and 1 keys, respectively.

◆◆◆ Example 113: *An Application.* A surveyor calculates that the angle between two sides of a building site must be 62.8362°. Convert this to degrees, minutes, and seconds so that it can be laid out using a transit or theodolite.

Solution:

- Put the calculator into degree mode.
- Enter the angle.
- Choose ▶ DMS from the ANGLE menu.
- Press ENTER. **◆◆◆**

The angle will be displayed in DMS form, as shown.

TI-83/84 screen for Example 113. On the TI-89, ▶ DMS is the MATH **Angle** menu.

Substituting into Formulas

A *formula* is an equation expressing some general mathematical or physical fact, such as the formula for the area of a circle of radius r.

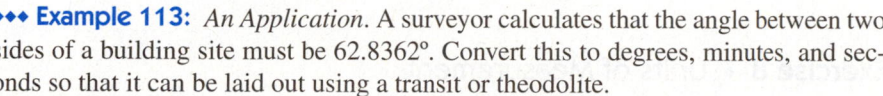

Area of a Circle	$A = \pi r^2$	75

To *substitute into* a formula or other equation means to replace the letter (or literal) quantities in the formula with their numerical values, and evaluate. With formulas, we carry the *unit of measure* along with each numerical value. We must often convert units so that they cancel properly, leaving the answer in the desired units. If the units to be used in a certain formula are specified, convert all quantities to those specified before substituting into the formula.

Further, we usually substitute approximate values, so we must round our answer properly.

◆◆◆ Example 114: *An Application.* A tensile load P of 4510 lb is applied to a bar, Fig. 1–12. It is seen to stretch or elongate by 0.390 mm. Find the modulus of elasticity E, in pounds per square inch, using

$$E = \frac{PL}{ae}$$

where P is the applied load, L is the length of the bar, a is the cross-sectional area of the bar, and e is the elongation.

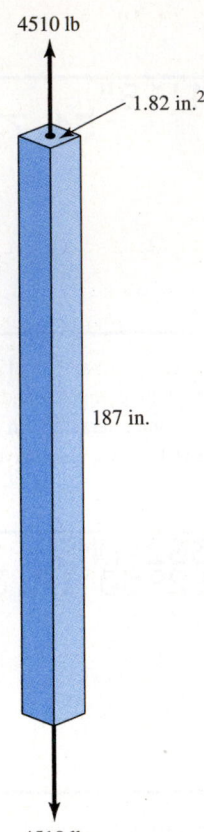

4510 lb

1.82 in.²

187 in.

4510 lb

FIGURE 1-12

Solution: Let us first convert the elongation e so that all units of length will be in inches.

$$e = 0.390 \text{ mm} \left(\frac{1 \text{ in.}}{25.4 \text{ mm}} \right) = 0.01535 \text{ in.}$$

Then substituting, with $P = 4510$ lb, $L = 187$ in. and $a = 1.82$ in.²,

$$E = \frac{4510 \text{ lb} \times 187 \text{ in.}}{1.82 \text{ in.}^2 \times 0.01535 \text{ in.}} = 30,200,000 \text{ lb/in.}^2$$

$$= 30.2 \times 10^6 \text{ lb/in.}^2$$

in engineering notation.   ◆◆◆

Common Error	Students often neglect to include *units* when substituting into a formula, with the result that the units often do not cancel properly.

Exercise 8 ◆ Units of Measurement

Conversion of Units

Convert the following customary units.

1. 152 inches to feet

2. 0.153 mile to yards

3. 762.0 feet to inches

4. 627 feet to yards

5. 29 tons to pounds

6. 88.90 pounds to ounces

7. 89,600 pounds to tons

8. 8552 ounces to pounds

Converting Between Metric Units

Convert the following metric units. Write your answer in scientific notation if the numerical value is greater than 1000 or less than 0.1.

9. 364,000 meters to kilometers

10. 0.000473 volt to millivolts

11. 735,900 grams to kilograms

12. 7.68×10^{-5} kilowatts to watts

13. 6.2×10^9 ohms to megohms

14. 825×10^4 newtons to kilonewtons

15. 9348 picofarads to microfarads

16. 84,398 nanoseconds to milliseconds

Converting Between Customary and Metric Units

Convert between the given customary and metric units.

17. 364.0 meters to feet

18. 6.83 inches to millimeters

19. 7.35 pounds to newtons

20. 2.55 horsepower to kilowatts

21. 4.66 U.S. gallons to liters

22. 825×10^4 dynes to pounds

23. 3.94 yards to meters

24. 834 cubic centimeters to gallons

Converting Areas and Volumes

Convert the following areas and volumes.

25. 2840 square yards to acres

26. 1636 square meters to ares

27. 24.8 square feet to square meters
28. 3.72 square meters to square feet
29. 0.982 square kilometer to acres
30. 5.93 acres to square meters
31. 7.360 cubic feet to cubic inches
32. 4.83 cubic meters to cubic yards

Converting Rates to Other Units

Convert units on the following time rates.

33. 4.86 feet per second to miles per hour
34. 777 gallons per minute to cubic meters per hour
35. 66.2 miles per hour to kilometers per hour
36. 52.0 knots to miles per minute
37. 953 births per year to births per week

Convert units on the following unit prices.

38. $1.25 per gram to dollars per kilogram
39. $800 per acre to cents per square meter
40. $3.54 per pound to cents per ounce
41. $4720 per ton to cents per pound

Angle Conversions

Convert to degrees (decimal).

42. $52°17'$　　　　**43.** $87°25'$　　　　**44.** $118°33'$
45. $72°12'22''$　　　**46.** $29°27'41''$　　**47.** $275°18'35''$

Convert to degrees, minutes, and seconds. Round to the nearest second.

48. $45.257°$　　　**49.** $61.339°$　　　**50.** $27.129°$
51. $177.344°$　　**52.** $185.972°$　　**53.** $128.259°$

Substituting into Formulas

In the following exercises, substitute the given quantities into the indicated formula from technology and finance.

54. Use the formula for simple interest $y = a(1 + nt)$, to find, to the nearest dollar, the amount y to which a principal a of $3000 will accumulate in $t = 5$ years at a simple interest rate n of 6.5%.

55. Using the formula for uniformly accelerated motion,

$$s = v_0 t + \frac{at^2}{2}$$

find the displacement s after $t = 1.30$ s, of a body thrown downward with a speed v_0 of 12.0 ft/s. Take the acceleration a as 32.2 ft/s².

56. Use the formula $C = \dfrac{5}{9}(F - 32)$ to convert 128°F to degrees Celsius

57. A bar (Fig. 1–13) whose length L is 15.2 m has a cross-sectional area a of 12.7 cm². It has an elongation e of 2.75 mm when it is subjected to a tensile load of 22,500 N. Use the equation $E = \dfrac{PL}{ae}$ to find the modulus of elasticity E, in newtons per square centimeter.

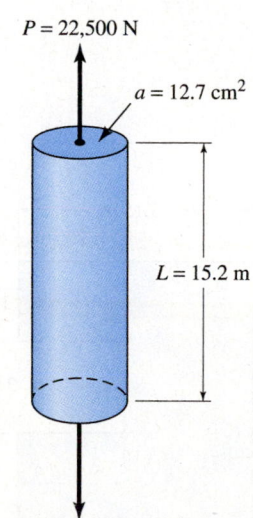

FIGURE 1–13　A bar in tension.

58. Use the formula for compound interest, $y = a(1 + n)^t$, to find, to the nearest dollar, the amount y to which a principal a of $9570 will accumulate in $t = 5$ years at a compound interest rate n of 6.75%.

59. The resistance R_1 of a copper coil is 775 Ω at $t_1 = 20.0°$ C. The temperature coefficient of resistance α is 0.00393 at 20.0° C. Use the formula $R = R_1[1 + \alpha(t - t_1)]$ to find the resistance at 80.0° C.

Applications

60. Convert all of the dimensions for the parts in Fig. 1–14 to inches.

61. A jet fuel tank, in Fig. 1–15, has a volume of 15.7 cubic feet. How many gallons of jet fuel will it hold?

62. A certain circuit board weighs 0.176 pound. Find its weight in ounces.

63. A certain laptop computer weighs 6.35 kilograms. What is its weight in pounds?

64. A generator has an output of 5.34×10^6 millivolts. What is the output in kilovolts?

65. Convert all of the dimension in Fig. 1–16 to centimeters.

66. The surface area of a certain lake, Fig. 1–17, is 7361 square yards. Convert this to square meters.

67. A solar collector, Fig. 1–18, has an area of 8834 square inches. Convert this to square meters.

68. The volume of a balloon, Fig. 1–19, is 8360 cubic feet. Convert this to cubic inches.

69. The volume of a certain gasoline tank, Fig. 1–20, is 9274 cubic centimeters. Convert this to gallons.

70. An airplane is cruising at a speed of 785 miles per hour. Convert this speed to kilometers per hour.

71. *Internet:* On the Web, find today's currency exchange rates. Use them to convert $100 to
(a) Euros   (b) Japanese yen   (c) Canadian dollars

72. *Project:* Scan a newspaper or magazine, noting whether measurements are given in metric units, customary units, or both. Estimate the percent given in metric units. Repeat for a technical journal in your chosen field.

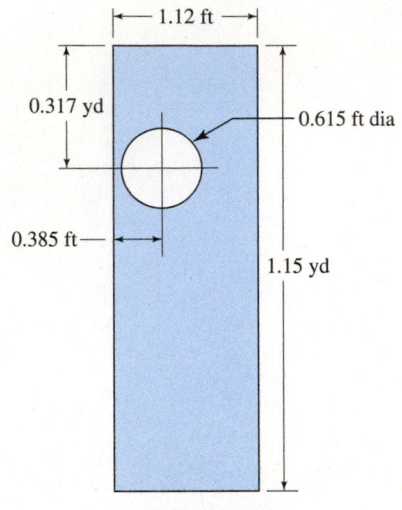

FIGURE 1–14

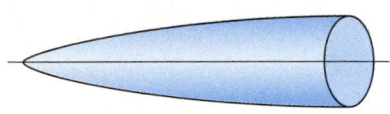

FIGURE 1–15

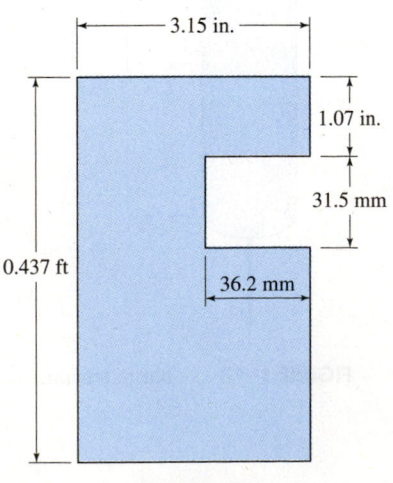

FIGURE 1–16

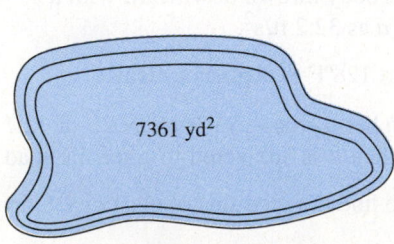

FIGURE 1–17

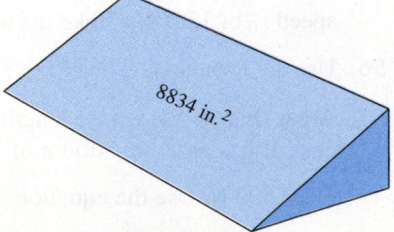

FIGURE 1–18

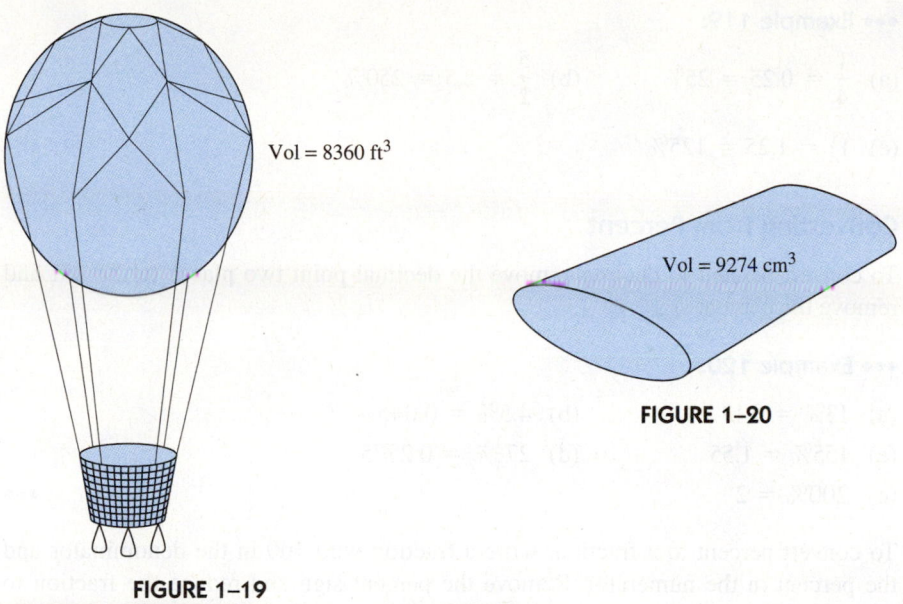

Vol = 8360 ft³

Vol = 9274 cm³

FIGURE 1–20

FIGURE 1–19

1–9 Percentage

Our final topic in this long chapter is the very useful subject of percentage.

Definition of Percent

The word *percent* means *by the hundred,* or *per hundred.* A percent thus gives the number of parts in every hundred.

♦♦♦ **Example 115:** If we say that a certain concrete mix is 12% cement by weight, we mean that 12 lb out of every 100 lb of mix will be cement. ♦♦♦

The word *rate* is often used to indicate a percent, or percentage rate, as in "rejection rate," "rate of inflation," or "growth rate."

♦♦♦ **Example 116:** A failure rate of 2% means that, on average, 2 parts out of every 100 would be expected to fail. ♦♦♦

Percent is another way of expressing a *fraction* having 100 as the denominator.

♦♦♦ **Example 117:** If we say that a builder has finished 75% of a house, we mean that he has finished $\frac{75}{100}$ (or $\frac{3}{4}$) of the house. ♦♦♦

Converting to Percent

Before working some percentage problems, let us first get some practice in converting decimals and fractions to percents, and vice versa. To convert decimals to percent, simply move the decimal point two places to the right and affix the percent symbol (%).

♦♦♦ **Example 118:** Here we convert decimals to percent.

(a) $0.75 = 75\%$ (b) $3.65 = 365\%$

(c) $0.003 = 0.3\%$ (d) $1.05 = 105\%$ ♦♦♦

To convert fractions or mixed numbers to percent, first write the fraction or mixed number as a decimal, and then proceed as above.

◆◆◆ **Example 119:**

(a) $\dfrac{1}{4} = 0.25 = 25\%$ (b) $\dfrac{5}{2} = 2.5 = 250\%$

(c) $1\frac{1}{4} = 1.25 = 125\%$ ◆◆◆

Converting from Percent

To convert percent to decimals, move the decimal point two places to the left and remove the percent sign.

◆◆◆ **Example 120:**

(a) $13\% = 0.13$ (b) $4.5\% = 0.045$

(c) $155\% = 1.55$ (d) $27\frac{3}{4}\% = 0.2775$

(e) $200\% = 2$ ◆◆◆

To convert percent to a fraction, write a fraction with 100 in the denominator and the percent in the numerator. Remove the percent sign and reduce the fraction to lowest terms.

◆◆◆ **Example 121:** Here we show some percents converted to fractions.

(a) $75\% = \dfrac{75}{100} = \dfrac{3}{4}$

(b) $87.5\% = \dfrac{87.5}{100} = \dfrac{875}{1000} = \dfrac{7}{8}$

(c) $125\% = \dfrac{125}{100} = \dfrac{5}{4} = 1\frac{1}{4}$ ◆◆◆

Solving Percentage Problems

Percentage problems always involve three quantities:

1. The *percent rate*, P.
2. The *base*, B, the quantity we are taking the percent of.
3. The *amount*, A, that we get when we take the percent of the base, also called the *percentage*.

In a percentage problem, you will know two of these three quantities (amount, base, or rate), and you will be required to find the third. This is easily done, for the rate, base, and amount are related by the following equation:

Percentage	amount = base × rate $$A = BP$$ where P is expressed as a decimal	**16**

Finding the Amount When the Base and the Rate Are Known

We substitute the given base and rate into Eq. 16 and solve for the amount.

◆◆◆ **Example 122:** What is 35.0 percent of 80.0?

Solution: In this problem the rate is 35.0%, so

$$P = 0.350$$

But is 80.0 the amount or the base?

Tip	In a particular problem, if you have trouble telling which number is the base and which is the amount, look for the key phrase *percent of*. The quantity following this phrase is *always the base*.

Thus we look for the key phrase "percent of."

What is 35.0 percent of 80.0 ?

Since 80.0 immediately follows *percent of,* it is the base. So

$$B = 80.0$$

From Eq. 16,

$$A = PB = (0.350)80.0 = 28.0 \qquad \blacklozenge\blacklozenge\blacklozenge$$

Common Error	Do not forget to convert the percent rate to a *decimal* when using Eq. 16.

◆◆◆ **Example 123:** Find 3.74% of 5710.

Solution: We substitute into Eq. 16 with

$$P = 0.0374 \text{ and } B = 5710$$

So

$$A = PB = (0.0374)(5710) = 214$$

after rounding to three significant digits.   ◆◆◆

◆◆◆ **Example 124:** *An Application.* A proposed beam having a width of 8.50 in. is to have its width increased by 15%. How much width must be added?

Solution:

$$\text{width added} = 0.15 \times 8.50 = 1.28 \text{ in.} \qquad \blacklozenge\blacklozenge\blacklozenge$$

Finding the Base When a Percent of It Is Known

We see from Eq. 16 that the base equals the amount divided by the rate (expressed as a decimal), or $B = A/P$.

◆◆◆ **Example 125:** 12% of what number is 78?

Solution: First find the key phrase.

12 percent of   what number   is 78?
                base

It is clear that we are looking for the base. So

$$A = 78 \quad \text{and} \quad P = 0.12$$

By Eq. 16,

$$B = \frac{A}{P} = \frac{78}{0.12} = 650 \qquad \blacklozenge\blacklozenge\blacklozenge$$

◆◆◆ **Example 126:** 140 is 25% of what number?

Solution: From Eq. 16,

$$B = \frac{A}{P} = \frac{140}{0.25} = 560 \qquad \blacklozenge\blacklozenge\blacklozenge$$

◆◆◆ **Example 127:** *An Application.* How much gravel must we start with if we remove 2.50 cubic yards, which is 35% of the original load?

Solution: The original load is the base B, the quantity removed is the amount A, and the rate P is 0.35. So,

$$\text{Original amount } B = \frac{A}{P} = \frac{2.50}{0.35} = 7.14 \text{ cubic yards} \qquad \text{◆◆◆}$$

Finding the Percent That One Number Is of Another Number

From Eq. 16, the rate equals the amount divided by the base, or $P = A/B$.

◆◆◆ **Example 128:** 42.0 is what percent of 405?

Solution: By Eq. 16, with $A = 42.0$ and $B = 405$,

$$P = \frac{A}{B} = \frac{42.0}{405} = 0.104 = 10.4\% \qquad \text{◆◆◆}$$

◆◆◆ **Example 129:** What percent of 1.45 is 0.357?

Solution: From Eq. 16,

$$P = \frac{A}{B} = \frac{0.357}{1.45} = 0.246 = 24.6\% \qquad \text{◆◆◆}$$

◆◆◆ **Example 130:** *An Application.* A steel beam that used to cost $885 now costs $65 more. By what percent did the cost increase?

Solution: The base B is the old cost of the beam; the amount A is the amount of increase. The percent increase is then

$$P = \frac{A}{B} = \frac{65}{885} = 0.0734$$

or an increase of 7.34%. ◆◆◆

Percent Change

Percentages are often used to compare two quantities. You often hear statements such as the following:

> *The price of steel rose 3% over last year's price.*
>
> *The weights of two cars differed by 20%.*
>
> *Production dropped 5% from last year.*

When the two numbers being compared involve a *change* from one to the other, the *original value* is usually taken as the base.

Percent Change	$\text{percent change} = \dfrac{\text{new value} - \text{original value}}{\text{original value}} \times 100$	**17**

◆◆◆ **Example 131:** A quantity changed from 521 to 835. What was the percent change?

Solution:

$$\text{percent change} = \frac{835 - 521}{521} \times 100 = 60.3\% \text{ increase} \qquad \text{◆◆◆}$$

◆◆◆ **Example 132:** *An Application.* A certain price rose from $1.55 to $1.75. Find the percentage change in price.

Solution: We use the original value, $1.55, as the base. From Eq. 17,

$$\text{percent change} = \frac{1.75 - 1.55}{1.55} \times 100 = 12.9\% \text{ increase} \qquad ◆◆◆$$

Be sure to show the *direction* of change with a plus or a minus sign, or with words such as *increase* or *decrease*.

A common type of problem is to *find the new value* when the original value is changed by a given percent. We see from Eq. 17 that

$$\text{new value} = \text{original value} + (\text{original value}) \times (\text{percent change})$$

◆◆◆ **Example 133:** *An Application.* The voltage across a certain power line dropped from 220 V to 215 V. What was the percent change in voltage?

Solution: Using the original value, 220 V, as the base, we get,

$$\text{percent change} = \frac{220 - 215}{220} \times 100 = 2.27\% \text{ decrease} \qquad ◆◆◆$$

◆◆◆ **Example 134:** *An Application.* Find the cost of a $356.00 suit after the price increases by $2\frac{1}{2}\%$.

Solution: The original value is 356.00, and the percent change, expressed as a decimal, is 0.025. So

$$\text{new value} = 356.00 + 356.00(0.025) = \$364.90 \qquad ◆◆◆$$

Percent Efficiency

The power output of any machine or device is always *less* than the power input, because of inevitable power losses within the device. The *efficiency* of the device is a measure of those losses.

Percent Efficiency	$\text{percent efficiency} = \dfrac{\text{output}}{\text{input}} \times 100$	**20**

◆◆◆ **Example 135:** *An Application.* A certain electric motor consumes 865 W and has an output of 1.12 hp (Fig. 1–21). Find the efficiency of the motor. (1 hp = 746 W.)

Solution: Since output and input must be in the same units, we must convert either to horsepower or to watts. Converting the output to watts, we obtain

$$\text{output} = 1.12 \text{ hp} \left(\frac{746 \text{ W}}{\text{hp}} \right) = 836 \text{ W}$$

By Eq. 20,

$$\text{percent efficiency} = \frac{836}{865} \times 100 = 96.6\% \qquad ◆◆◆$$

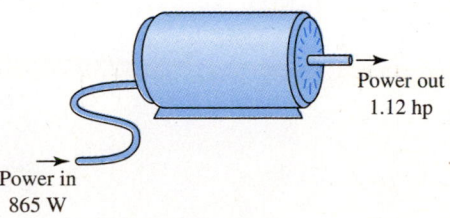

Power out
1.12 hp

Power in
865 W

FIGURE 1–21

Percent Error

The accuracy of measurements is often specified by the *percent error*. The percent error is the difference between the measured value and the known or "true" value, expressed as a percent of the known value.

Percent Error	$$\text{percent error} = \frac{\text{measured value} - \text{known value}}{\text{known value}} \times 100$$	18

◆◆◆ **Example 136:** *An Application.* A laboratory weight that is certified to be 500.0 g is placed on a scale (Fig. 1–22). The scale reading is 507.0 g. What is the percent error in the reading?

Solution: From Eq. 18,

$$\text{percent error} = \frac{507.0 - 500.0}{500.0} \times 100 = 1.4\% \text{ high} \qquad ◆◆◆$$

As with percent change, be sure to specify the *direction* of the error.

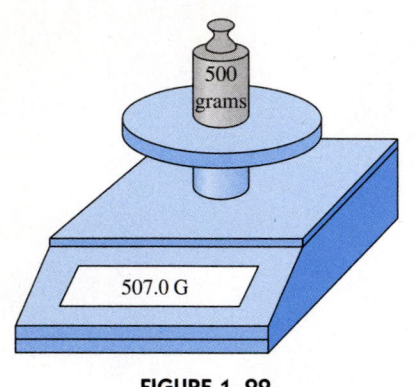

FIGURE 1–22

Percent Concentration

The following equation applies to a mixture of two or more ingredients:

Percent Concentration	$$\text{percent concentration of ingredient } A = \frac{\text{amount of } A}{\text{amount of mixture}} \times 100$$	19

◆◆◆ **Example 137:** *An Application:* A certain fuel mixture contains 18.9 liters of alcohol and 84.7 liters of gasoline. Find the percentage of gasoline in the mixture.

Solution: The total amount of mixture is

$$18.9 + 84.7 = 103.6 \text{ liters}$$

So by Eq. 19,

$$\text{percent gasoline} = \frac{84.7}{103.6} \times 100 = 81.8\% \qquad ◆◆◆$$

Common Error	The denominator in Eq. 19 must be the *total amount* of mixture, or the sum of *all* of the ingredients. Do not use just one of the ingredients.

Exercise 9 ◆ Percentage

Converting to Percent Convert each decimal to a percent.

1. 3.72   **2.** 0.877   **3.** 0.0055   **4.** 0.563

Convert each fraction to a percent. Round to three significant digits.

5. $\dfrac{2}{5}$ **6.** $\dfrac{3}{4}$ **7.** $\dfrac{7}{10}$ **8.** $\dfrac{3}{7}$

Converting from Percent

Convert each percent to a decimal.

9. 23%   **10.** 2.97%   **11.** $287\frac{1}{2}\%$ **12.** $6\frac{1}{4}\%$

Convert each percent to a fraction.

13. 37.5%   **14.** $12\frac{1}{2}\%$ **15.** 150%   **16.** 3%

Finding the Amount Find.

17. 41.1% of 255 tons.   **18.** 15.3% of 326 mi.

19. 33.3% of 662 kg.   **20.** 12.5% of 72.0 gal.

21. 35.0% of 343 liters.   **22.** 50.8% of $245.

23. A resistance, now 7250 Ω, is to be increased by 15.0%. How much resistance should be added?

24. It is estimated that $\frac{1}{2}\%$ of the earth's surface receives more energy than the total projected yearly needs. Assuming the earth's surface area to be 1.97×10^8 mi^2, find the required area in acres.

25. As an incentive to install solar equipment, a tax credit of 42% of the first $1100 and 25% of the next $6400 spent on solar equipment is proposed. How much credit, to the nearest dollar, would a homeowner get when installing $5500 worth of solar equipment?

26. How much metal will be obtained from 375 tons of ore if the metal is 10.5% of the ore?

Finding the Base Find the number of which

27. 86.5 is 16.7%.   **28.** 45.8 is 1.46%.

29. 1.22 is 1.86%.   **30.** 55.7 is 25.2%.

31. 66.6 is 66.6%.   **32.** 58.2 is 75.4%.

33. A Department of Energy report on an experimental electric car gives the range of the car as 161 km and states that this is "49.5% better than on earlier electric vehicles." What was the range of earlier electric vehicles?

34. A man withdrew 25.0% of his bank deposits and spent 45.0% of the amount withdrawn on a car costing $31,100. How much money was originally in the bank?

35. Solar panels provide 65.0% of the heat for a certain building. If $1560 per year is now spent for heating oil, what would have been spent if the solar panels were not used?

36. If the United States imports 9.14 billion barrels (bbl) of oil per day, and if this is 48.2% of its needs, how much oil is needed per day?

Finding the Percentage Rate What percent of

37. 26.8 is 12.3?   **38.** 36.3 is 12.7?

39. 44.8 is 8.27?   **40.** 844 is 428?

41. 455 h is 152 h?   **42.** 483 tons is 287 tons?

43. A 50,500-liter-capacity tank contains 5840 liters of water (Fig. 1–23). Express the amount of water in the tank as a percentage of the total capacity.

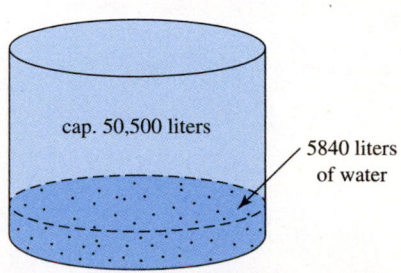

cap. 50,500 liters

5840 liters of water

FIGURE 1–23

44. In a journey of 1560 km, a person traveled 195 km by car and the rest of the distance by rail. What percent of the distance was traveled by rail?

45. A power supply has a dc output of 51 V with a ripple of 0.75 V peak to peak. Express the ripple as a percentage of the dc output voltage.

46. The construction of a building costs $136,000 for materials and $157,000 for labor. What percentage of the total is the labor cost?

Percent Change Find the percent change when a quantity changes

47. from 29.3 to 57.6.   **48.** from 107 to 23.75.
49. from 227 to 298.   **50.** from 0.774 to 0.638.

51. The temperature in a building rose from 19.0°C to 21.0°C during the day. Find the percent change in temperature.

52. A casting initially weighing 115 lb has 22.0% of its material machined off. What is its final weight?

53. A certain common stock rose from a value of $35\frac{1}{2}$ per share to $37\frac{5}{8}$ per share. Find the percent change in value.

54. A house that costs $635 per year to heat has insulation installed in the attic, causing the fuel bill to drop to $518 per year. Find the percent change in fuel cost.

Percent Efficiency

55. A certain device (Fig. 1–24) consumes 18.5 hp and delivers 12.4 hp. Find its efficiency.

56. An electric motor consumes 1250 W. Find the horsepower it can deliver if it is 85.0% efficient. (1 hp = 746 W.)

57. A water pump requires an input of 0.50 hp and delivers 10,100 lb of water per hour to a house 72 ft above the pump. Find its efficiency. (1 hp = 550 ft lb/s.)

58. A certain speed reducer delivers 1.7 hp with a power input of 2.2 hp. Find the percent efficiency of the speed reducer.

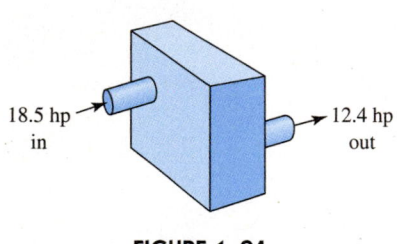

18.5 hp in → → 12.4 hp out

FIGURE 1–24

Percent Error

59. A certain quantity is measured at 125.0 units but is known to be actually 128.0 units. Find the percent error in the measurement.

60. A shaft is known to have a diameter of 35.000 mm. You measure it and get a reading of 34.725 mm. What is the percent error of your reading?

61. A certain capacitor has a working voltage of 125.0 V dc, −10%, +150%. Between what two voltages would the actual working voltage lie?

62. A resistor is labeled as 5500 Ω with a tolerance of ±5%. Between what two values is the actual resistance expected to lie?

Percent Concentration

63. A solution is made by mixing 75.0 liters of alcohol with 125 liters of water. Find the percent concentration of alcohol.

64. 8.0 cubic feet of cement is contained in a concrete mixture that is 12% cement by volume. What is the volume of the total mixture?

65. How many liters of alcohol are contained in 455 liters of a gasohol mixture that is 5.5% alcohol by volume?

66. How many liters of gasoline are there in 155 gal of a methanol–gasoline blend that is 10.0% methanol by volume?

67. *Writing:* We said, "Of all the mathematical topics we cover in this text, probably the one most used in everyday life is percentage." Do you agree? Write a few paragraphs saying if you agree or not, and back your reasons up with specific examples from personal experience.

⬥⬥⬥ CHAPTER 1 REVIEW PROBLEMS ⬥⬥⬥⬥⬥⬥⬥⬥⬥⬥⬥⬥⬥⬥⬥⬥⬥⬥⬥⬥⬥⬥⬥⬥⬥⬥⬥⬥

1. Combine: $1.435 - 7.21 + 93.24 - 4.1116$

2. Give the number of significant digits in:

    (a) 9.886   (b) 1.002   (c) 0.3500   (d) 15,000

3. Multiply: $21.8(3.775 \times 1.07)$

4. Divide: $88.25 \div 9.15$

5. Find the reciprocal of 2.89.

6. Evaluate: $-|-4 + 2| - |-9 - 7| + 5$

7. Evaluate: $(9.73)^2$

8. Evaluate: $(7.75)^{-2}$

9. Evaluate: $\sqrt{29.8}$

10. Evaluate: $(123)(2.75) - (81.2)(3.24)$

11. Evaluate: $(91.2 - 88.6)^2$

12. Evaluate: $\left(\dfrac{77.2 - 51.4}{21.6 - 11.3}\right)^2$

13. Evaluate: $y = 3x^2 - 2x$ when $x = -2.88$

14. Evaluate: $y = 2ab - 3bc + 4ac$ when $a = 5$, $b = 2$, and $c = -6$

15. Evaluate: $y = 2x - 3w + 5z$ when $x = 7.72$, $w = 3.14$, and $z = 2.27$

16. Round to two decimal places.

    (a) 7.977   (b) 4.655   (c) 11.845   (d) 1.004

17. Round to three significant digits.

    (a) 179.2   (b) 1.076   (c) 4.8550   (d) 45,725

18. A news report states that a new hydroelectric generating station will produce 47 million kWh/yr and that this power, for 20 years of operation, is equivalent to 2.0 million barrels of oil. Using these figures, how many kilowatt-hours is each barrel of oil equivalent to?

19. A certain generator has a power input of 2.50 hp and delivers 1310 W. Find its percent efficiency.

20. Use the equation, $\sigma = \dfrac{P}{a}$, find the stress σ, in pounds per square inch, for a force P of 1.17×10^3 N, distributed over an area a of 3.14×10^3 mm^2.

21. Combine: $(8.34 \times 10^5) + (2.85 \times 10^6) - (5.29 \times 10^5)$

22. A train running at 25 mi/h increases its speed by $12\frac{1}{2}\%$. How fast does it then go?

23. The average solar radiation in the continental United States is about 0.206 kW per square meter. How many kilowatts would be collected by 15.0 acres of solar panels?

24. An item rose in price from \$29.35 to \$31.59. Find the percent increase.

25. Find the percent concentration of alcohol if 2.0 liters of alcohol is added to 15 gal of gasoline.

26. A bar, known to be 2.0000 inches in diameter, is measured at 2.0064 in. Find the percent error in the measurement.

27. The Department of Energy estimates that there are 700 billion barrels (bbl) of oil in the oil shale deposits of Colorado, Wyoming, and Utah. Express this amount in scientific notation.

28. Multiply: $(7.23 \times 10^5) \times (1.84 \times 10^{-3})$

29. Divide: -39.2 by -0.003826

30. Convert 6930 Btu/h to foot-pounds per minute.

31. Divide: 8.24×10^{-3} by 1.98×10^7

32. What percent of 40.8 is 11.3?

33. Evaluate: $\sqrt[5]{82.8}$

34. Multiply: $(4.92 \times 10^6) \times (9.13 \times 10^{-3})$

35. Insert the proper sign of equality or inequality between $-\frac{2}{3}$ and -0.660.

36. Convert 0.000426 mA to microamperes.

37. Find 49.2% of 4827.

38. Combine: $-385 - (227 - 499) - (-102) + (-284)$

39. Find the reciprocal of -0.582.

40. Find the percent change in a voltage that increased from 111 V to 118 V.

41. A homeowner added insulation, and her yearly fuel consumption dropped from 628 gal to 405 gal. Her present oil consumption is what percent of the former?

42. Write in decimal notation: 5.28×10^4

43. Convert 49.3 pounds to newtons.

44. Evaluate: $(45.2)^{-0.45}$

45. Using the equation $s = v_0 t + \dfrac{at^2}{2}$, find the distance s (in feet) traveled by a falling object in $t = 5.25$ s, when thrown downward with an initial velocity v_0 of 284 m/min. Here a is the acceleration due to gravity, 32.2 ft/s^2.

46. Write in scientific notation: 0.000374

47. 8460 is what percent of 38,400?

48. The U.S. energy consumption of 37 million barrels of oil equivalent per day is expected to climb to 48 million in 6 years. Find the percent increase in consumption.

49. The population of a certain town is 8118, which is $12\frac{1}{2}\%$ more than it was 3 years ago. What was the population then?

50. The temperature of a room rose from 68.0°F to 73.0°F. Find the percent increase.

51. Combine: $4.928 + 2.847 - 2.836$

52. Give the number of significant digits in 2003.0.

53. Multiply: 2.84(38.4)

54. Divide: $48.3 \div 2.841$

55. Find the reciprocal of 4.82.

56. Evaluate: $|-2| - |3 - 5|$

57. Evaluate: $(3.84)^2$

58. Evaluate: $(7.62)^{-2}$

59. Evaluate: $\sqrt{38.4}$

60. Evaluate: $(49.3 - 82.4)(2.84)$

61. Evaluate: $x^2 - 3x + 2$ when $x = 3$

62. Round 45.836 to one decimal place.

63. Round 83.43 to three significant digits.

64. Multiply: $(7.23 \times 10^5) \times (1.84 \times 10^{-3})$

65. Convert 36.82 in. to centimeters.

66. What percent of 847 is 364?

67. Evaluate: $\sqrt[3]{746}$

68. Find 35.8% of 847.

69. 746 is what percent of 992?

70. Write 0.00274 in scientific notation.

71. Write 73.7×10^{-3} in decimal notation.

72. Evaluate: $(47.3)^{-0.26}$

73. Combine: $6.128 + 8.3470 - 7.23612$

74. Give the number of significant digits in 6013.00.

75. Multiply: $7.184(16.8)$

76. Divide: $78.7 \div 8.251$

77. Find the reciprocal of 0.825.

78. Evaluate: $|-5| - |2 - 7| + |-6|$

2

Introduction to Algebra

◆◆**OBJECTIVES** ◆◆

When you have completed this chapter, you should be able to
- Define common algebraic terms: variable, expression, term, polynomial, and so forth.
- Identify and define an equation.
- Separate a term into variables and constants.
- Simplify an expression by removing symbols of grouping.
- Add and subtract polynomials.
- Use the laws of exponents for multiplication, division, and raising to a power.
- Multiply monomials, binomials, and multinomials.
- Raise a multinomial to a power.
- Divide a polynomial by a monomial or by a polynomial.

◆◆

In Chapter 1 we showed how to raise a number to a power. For example, 3^2, or 3 raised to the power 2, means

$$3^2 = (3)(3)$$

In a similar way, x^2, or x raised to the power 2, means

$$x^2 = (x)(x)$$

where x can stand for any number, not just 3. Going further, we can represent the exponent by a symbol, say n. Thus

$$x^n = \underbrace{(x)(x)(x)(x)...(x)}_{n \text{ factors}}$$

While 3^2 was an *arithmetic* expression, x^n is an *algebraic* expression. We can think of algebra as *a generalization of arithmetic*. Some knowledge of algebra is essential in technical work. Suppose, for example, you see in a handbook that the power P delivered to a resistor, Fig. 2–1, is equal to VI, the voltage times the current. Then in another place you find that P is equal to V^2/R, the square of the voltage divided by the resistance. In a third book you see that the power is equal to I^2R, the square of the current times the resistance! Which is it? Can they all be true? Even in this simple example you must know some algebra to make sense of such information.

We will learn many new words in this chapter but since algebra is generalized arithmetic, some of what was said in Chapter 1 (such as rules of signs) will be repeated here.

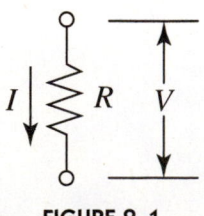

FIGURE 2–1

We will redo the basic operations of addition, subtraction, and so on, but now with symbols rather than numbers. Learn this material well, for it is the foundation on which later chapters rest.

2–1   Algebraic Expressions

Every field has its own special terms, and algebra is no exception. So let's start by learning some new terms and new words that we'll be using throughout our study of mathematics.

Mathematical Expressions

A *mathematical expression* is a grouping of mathematical symbols, such as signs of operation, numbers, and letters.

◆◆◆ **Example 1:** The following are mathematical expressions:

(a) $x^2 - 2x + 3$
(b) $4 \sin 3x$
(c) $5 \log x + e^{2x}$ ◆◆◆

Algebraic Expressions

An *algebraic expression* is one containing only algebraic symbols and operations (addition, subtraction, multiplication, division, roots, and powers), such as in Example 1(a). All other expressions are called *transcendental,* such as Examples 1(b) and (c). We will study those later.

Terms

The plus and the minus signs divide an expression into *terms.*

◆◆◆ **Example 2:** The expression $2x^2 + 5x + 3$ has three terms.

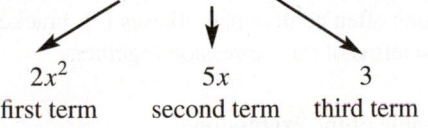

$$2x^2 \qquad 5x \qquad 3$$
first term   second term   third term   ◆◆◆

Equations

None of the expressions in Example 1 contains an equal sign ($=$). When two expressions are set equal to each other, we get an *equation.*

◆◆◆ **Example 3:** The following are equations:

(a) $2x^2 + 3x - 5 = 0$
(b) $6x - 4 = x + 1$
(c) $y = 3x - 5$ ◆◆◆

Constants and Variables

A *constant* is a quantity that does not change in value in a particular problem. It is usually a number, such as 8, 4.67, or π.

A *variable* is a quantity that may change during a particular problem. A variable is usually represented by a letter from the end of the alphabet (x, y, z, etc.).

◆◆◆ **Example 4:** The constants in the expression

$$3x^2 + 4x + 5$$

are 3, 4, and 5, and the variable is x.   ◆◆◆

A constant can also be represented by a letter. Such a letter is usually chosen from the beginning of the alphabet (a, b, c, etc.). The letter k is often used as a constant. An expression in which the constants are represented by letters is called a *literal* expression.

◆◆◆ **Example 5:** The constants in the literal expression

$$ax^2 + bx + c$$

are a, b, and c, and the variable is x.   ◆◆◆

Coefficient

The *coefficient* of a term is the constant part of the term. It is usually written before the variable part of the term.

◆◆◆ **Example 6:**

(a) In the term $5x$, 5 is the coefficient and x is the variable.

(b) In the term $2axy^2$, $2a$ is the coefficient of xy^2.

(c) In the term $-3x$, the coefficient is -3.

(d) In the term x, the coefficient of x is 1.   ◆◆◆

Common Error	Do not forget to include the negative sign with the coefficient.

Symbols of Grouping

Mathematical expressions often contain parentheses ( ), brackets [ ], and braces { }. These are used to group terms of the expression together.

◆◆◆ **Example 7:** The value of the expression

$$2 + 3(4 + 5)$$

is *different* from the value of

$$(2 + 3)4 + 5$$

Recall from our work on combined operations in Chapter 1 that we first evaluate the expression *within* parentheses and then the entire expression. Thus the first expression has a value of 29, and the second expression has a value of 25. As you can see, the placement of symbols of grouping *is* important.   ◆◆◆

We will show how to *remove* parentheses, brackets, and braces later in this chapter and will also learn that any of these symbols can be used to indicate *multiplication*.

The bar (—) is also used to group terms of an expression together.

◆◆◆ **Example 8:** The expressions

$$\frac{x + y}{z} \qquad \text{and} \qquad \frac{z}{x + y}$$

show how the bar is used to group the terms x and y. Here the bar is also used to indicate *division*. Thus

$$\frac{x + y}{z} = (x + y) \div z$$

and

$$\frac{z}{x + y} = z \div (x + y) \qquad \text{◆◆◆}$$

Factors

Any divisor of a term is called a *factor* of that term.

◆◆◆ **Example 9:** The factors of $3axy$ are 3, a, x, and y.   ◆◆◆

◆◆◆ **Example 10:** The expression $2x + 3yz$ has two *terms*, $2x$ and $3yz$. The first term has the factors 2 and x, and the second term has the factors 3, y, and z.   ◆◆◆

A *prime factor of a number* is one which has no divisor other than 1 or the number itself. *A prime factor of an expression* is one which has no divisor other than 1 or the expression itself.

◆◆◆ **Example 11:** Here we show prime factors of a number and of an expression.

(a) The prime factors of 12 are 2, 2, and 3.
(b) The prime factors of x^2 are x and x.   ◆◆◆

Degree

The *degree* of a term refers to the integer power to which the variable is raised. If no power is written, it is understood to be 1.

◆◆◆ **Example 12:** This example shows terms of various degrees.

(a) $2x$ is a first-degree term.
(b) $3x^2$ is a second-degree term.
(c) $5y^9$ is a ninth-degree term.   ◆◆◆

If there is more than one variable, we can give the degree with respect to each variable, and the degree of the entire term, which is the sum of the degrees of each variable.

◆◆◆ **Example 13:** Here we show the degree of a term having more than one variable.

(a) x^2y^3 is of degree two in x and degree three in y. The term as a whole is of degree five.
(b) $3xy^2z^3$ is of degree one in x, degree two in y, and degree three in z. The term as a whole is of degree six.   ◆◆◆

The degree of an *expression* is the same as that of the term having the highest degree.

◆◆◆ **Example 14:** $3x^2 - 2x + 4$ is a second-degree expression.   ◆◆◆

Monomials, Multinomials, and Polynomials

A *monomial* is an algebraic expression having one term.

◆◆◆ **Example 15:** Some monomials are

(a) $3x$
(b) $2xy^2$
(c) $5wz$ ◆◆◆

A *multinomial* is an algebraic expression having *more than one term*.

◆◆◆ **Example 16:** Some multinomials are

(a) $3x + 5$

(b) $2x^3 - 3x^2 + 7$

(c) $\dfrac{1}{x} + 6x$ ◆◆◆

A *polynomial* is a monomial or multinomial in which the powers to which the variables are raised are all *nonnegative integers*. The first two expressions in Example 16 are polynomials, but the third is not.

A *binomial* is a polynomial with *two* terms, and a *trinomial* is a polynomial having *three* terms. In Example 16, the first expression is a binomial, and the second is a trinomial. The third expression is not a polynomial, so the term binomial does not apply.

Exercise 1 ◆ Algebraic Expressions

Mathematical and Algebraic Expressions

Which of the following mathematical expressions are also algebraic expressions?

1. $x + 2y$	**2.** $y - \log x$
3. $3 \sin x$	**4.** $x^2 - z^3$

Literal Expressions

Which of the following algebraic expressions are literal expressions?

5. $5xy - 2x$	**6.** $ax + by$
7. $2az - 3bx$	**8.** $4x^2 + 4y^2$

Terms

How many terms are there in each expression?

9. $x^3 - 2x$	**10.** $5y + y^2 - 5$
11. $ax^2 + bx + c$	**12.** $5 - 2x^3 - 7x + x^2$

Factors

Write the prime factors of each expression.

13. $3ax$	**14.** $9xyz$
15. $7x^2y^3$	**16.** $6a^2bx$

Coefficient

Write the coefficient of each term. Assume that letters from the beginning of the alphabet $(a, b, c, \ldots)$ are constants.

17. $6x^2$	**18.** x
19. $-x$	**20.** $3cx^3$
21. $2ax^5$	

Degree

State the degree of each term.

22. $3x$	**23.** $4y^2$
24. $3xy$	**25.** $5x^2y^3$

State the degree of each expression.

26. $3x + 4$ **27.** $5 - xy$

28. $3x^2 - 2x + 5$

29. $2xy^2 + xy - 4$

2–2   Adding and Subtracting Polynomials

Now that we know some of the language of algebra, let's go on to the basic operations. We will start with addition and subtraction. We will see that much of what we learned in Chapter 1 about adding and subtracting numbers (rules of signs, commutative law, and so forth) applies here as well.

Polynomials

Recall that an algebraic expression in which the power of every variable is a positive integer is called a *polynomial*.

◆◆◆ **Example 17:** Some polynomials are

$$3x^4, \quad 4x^2 - 5x - 4, \quad 2xy + y^3, \quad \text{and} \quad 7$$

Some expressions that are *not* polynomials are

$$1/x, \quad 3x^{-4}, \quad 4x^2 - 5\sqrt{x} - 4, \quad \text{and} \quad 2xy + y^{1/3} \qquad ◆◆◆$$

Combining Like Terms

Terms that differ only in their coefficients are called *like* terms. Their variable parts are the same.

◆◆◆ **Example 18:** Here are some like terms.

(a) $4xz$ and $-5xz$ are like terms.

(b) $3x^2y^3$ and $7x^2y^3$ are like terms.   ◆◆◆

We add and subtract algebraic expressions by *combining like terms*. Like terms are added by adding their coefficients. We also call this *collecting terms*.

◆◆◆ **Example 19:** These examples show the combining of like terms.

(a) $3y + 4y = 7y$

(b) $18z - 9z = 9z$

(c) $7x^2 + 3x^2 = 10x^2$ ◆◆◆

Don't forget that any term with no numerical coefficient has an unwritten coefficient of 1.

◆◆◆ **Example 20:** More examples of combining like terms.

(a) $4x + x = 4x + 1x = 5x$

(b) $-3y + y = -3x - 1y = -2y$

(c) $2z^2 + z^2 = 2z^2 + 1z^2 = 3z^2$ ◆◆◆

Commutative Law of Addition

The commutative law for addition simply states that *you can add quantities in any order*.

Commutative Law for Addition	$a + b = b + a$	3

This law allows us to arrange the terms of an expression to make it easier to simplify.

◆◆◆ **Example 21:** Simplify the expression

$$3x + 2y + 4x + y$$

Solution: We use the commutative law to rearrange the expression to get like terms together. We can then easily combine those like terms.

$$3x + 2y + 4x + y = 3x + 4x + 2y + y$$
$$= 7x + 3y \qquad\qquad ◆◆◆$$

◆◆◆ **Example 22:** Simplify the expression

$$8y + 5x - 5z + 2z - x - 4y$$

Solution: Using the commutative law to rearrange the expression and combining like terms, we get

$$8y + 5x - 5z + 2z - x - 4y = 5x - x + 8y - 4y - 5z + 2z$$
$$= 4x + 4y - 3z \qquad\qquad ◆◆◆$$

The procedure is no different when the terms have *decimal coefficients*. Just be sure to round properly.

◆◆◆ **Example 23:**

$$3.92x - 4.02y - 2.24x + 1.85y = 3.92x - 2.24x - 4.02y + 1.85y$$
$$= 1.68x - 2.17y \qquad\qquad ◆◆◆$$

If there is an expression within parentheses preceded by a plus $(+)$ sign, you may simply remove the parentheses. If the parentheses are preceded by a negative $(-)$ sign, multiply each term within the parentheses by (-1), and then remove the parentheses. Once the parentheses are gone, like terms may be combined.

◆◆◆ **Example 24:** Combine and simplify,

$$-(a - b) - (a + b) = -a + b - a - b = -2a \qquad\qquad ◆◆◆$$

◆◆◆ **Example 25:** Combine and simplify,

$$(4mx + ny) + (mx + 3ny) - (3mx + 2ny)$$

Solution: We first remove parentheses and then rearrange so that like terms are together. Finally we combine like terms.

$$(4mx + ny) + (mx + 3ny) - (3mx + 2ny)$$
$$= 4mx + ny + mx + 3ny - 3mx - 2ny$$
$$= 4mx + mx - 3mx + ny + 3ny - 2ny$$
$$= 2mx + ny \qquad\qquad ◆◆◆$$

◆◆◆ **Example 26:** Combine and simplify,

$$(3a + 5b - 2c) - (2a - b - 4c) - (a + 3b + c)$$
$$= 3a + 5b - 2c - 2a + b + 4c - a - 3b - c$$
$$= 3a - 2a - a + 5b + b - 3b - 2c + 4c - c$$
$$= 3b + c \qquad\qquad ◆◆◆$$

Combining Polynomials by Calculator

Polynomials can be added or subtracted on a calculator or computer that can do symbolic algebra. On the TI-89, for example, no special command is needed. Just enter the expression and it will automatically be displayed in simplified form.

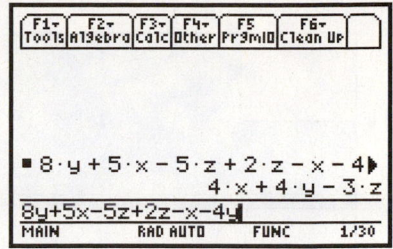

◆◆◆ **Example 27:** Repeat Example 22 by calculator.

Solution: We enter the expression just as shown, and press $\boxed{\text{ENTER}}$. The simplified expression is then displayed.   ◆◆◆

TI-89 screen for Example 27.

Vertical Addition and Subtraction

It is often easier to arrange the expression *vertically,* with like terms in the same column.

◆◆◆ **Example 28:** Combine

$$(3x - 2y + 4z) - (7y + 2w - 5x) - (3w + 2z + 4y - x)$$

Solution: We first remove parentheses.

$$(3x - 2y + 4z) - (7y + 2w - 5x) - (3w + 2z + 4y - x)$$
$$= 3x - 2y + 4z - 7y - 2w + 5x - 3w - 2z - 4y + x$$

We write the expressions one above the other, with like terms in the same colume, and then collect terms.

$$
\begin{array}{l}
3x - 2y + 4z \\
5x - 7y - 2w \\
\underline{x - 4y - 2z - 3w} \\
9x - 13y + 2z - 5w
\end{array}
$$
                                                                                        ◆◆◆

Instructions Given Verbally

In preparation for verbal problems to come later, we give some problems in verbal form.

◆◆◆ **Example 29:** Find the sum of $x - 2y + z$ and $3x + y + z$.

Solution: We have

$$(x - 2y + z) + (3x + y + z)$$
$$= x - 2y + z + 3x + y + z$$
$$= x + 3x - 2y + y + z + z$$
$$= 4x - y + 2z$$
                                                                                        ◆◆◆

◆◆◆ **Example 30:** Subtract $5a - 2b + 6c$ from the sum of $8a - 4b - 3c$ and $a - b + 6c$.

Solution: We add the last two expressions,

$$(8a - 4b - 3c) + (a - b + 6c) = 9a - 5b + 3c$$

Then we subtract the first expression.

$$9a - 5b + 3c - (5a - 2b + 6c)$$
$$= 9a - 5b + 3c - 5a + 2b - 6c$$
$$= 4a - 3b - 3c$$
                                                                                        ◆◆◆

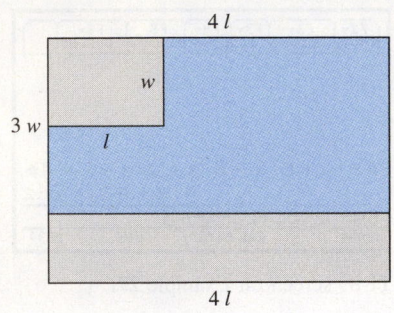

FIGURE 2–2

◆◆◆ **Example 31:** *An Application.* The rectangular building lot, Fig. 2–2, is to be subdivided into smaller plots by fences, as shown. The total lengh of fence needed, including the outer perimeter, is found by adding together the length of each fence.

$$w + l + 3w + 4l + 3w + 4l + 4l$$

Simplify this expression.

Solution: Rearranging and combining like terms gives

$$w + l + 3w + 4l + 3w + 4l + 4l = w + 3w + 3w + l + 4l + 4l + 4l$$
$$= 7w + 13l \qquad ◆◆◆$$

Exercise 2 ◆ Adding and Subtracting Polynomials

Combine as indicated and simplify.

1. $8x + 2x$
2. $6y - 8y$
3. $5a + 2a$
4. $-2m + 5m$
5. $8ab - 2ab$
6. $-9xy + 2xy$
7. $2.8x + 3.2x$
8. $4.56y - 7.38y$
9. $53.5a + 21.6a$
10. $-22.2m + 51.3m$
11. $8.5ab - 21.2ab$
12. $-33.9xy + 58.2xy$
13. $5x - 8x + 2x$
14. $3y + 6y - 4y$
15. $9a - 4a + 2a$
16. $11m - 2m + 6m$
17. $2ab - 8ab - 2ab$
18. $10xy - 9xy + 2xy$
19. $5.2x - 2.8x + 1.2x$
20. $22.5a - 13.5a + 41.6a$
21. $53.6m - 12.2m + 31.3m$
22. $3.73ab - 8.22ab - 1.25ab$
23. $38.3xy - 33.9xy + 58.2xy$
24. $5.88y - 3.56y - 7.18y$
25. $3x + 5x - 4x + 2x$
26. $7y - 6y - 5y + 2y$
27. $5a - 9a - 3a + 2a$
28. $6ab + 2ab - 4ab - 2ab$
29. $9.3x + 5.2x - 1.8x + 1.2x$
30. $2.88y - 5.88y - 2.56y - 5.18y$
31. $11.8m - 43.6m - 32.2m + 31.3m$
32. $22.6xy + 38.3xy - 33.9xy + 58.2xy$
33. $6x - 3x + 7x - 4x + 2x$
34. $7a + 5a - 9a - 3a + 5a$
35. $9ab - 6ab + 5ab - 4ab - 2ab$
36. $5.83y + 2.48y - 5.18y - 2.56y - 3.18y$

37. $32.5m - 11.8m - 23.6m - 32.2m + 11.3m$
38. $47.2xy + 12.6xy + 38.3xy - 39.9xy + 18.2xy$
39. $(x + 2) + (3x - 4)$
40. $(2.84x + 1.32) + (5.88x - 4.44)$
41. $(22.7ab + 21.2) + (83.5 - 48.2ab)$
42. $(4x + 2) + (3x - 4) - (5x + 5)$
43. $(8.33 + 1.05y) - (2.44y + 1.12) + (2.88y - 1.74)$
44. $(28.3 + 1.19xy) - (12.5xy + 44.7) + (2.18xy - 11.6)$

Instructions Given Verbally

45. Add: $a - c + b$ and $b + c - a$.
46. Find the sum: $6bc + n^2 + 3p$ and $-5x + 3n^2$.
47. Subtract: $5a + 7d - 4b + 6c$ from $8b - 10c + 3a - d$.

48. Subtract: $2xy + 4y - 3x$ from $5x - 2xy + 8y$.
49. What is the sum of
    $24by^5 - 14bx^4$, $-72bx^5 + 2by^5 - 3bx^4$ and $9bx^4 + 23by^4 - 21by^5$?

Challenge Problems

50. $(4x + 2y + 4w) + (3w - 4y - 2x) - (5x + 5w + 3y)$
51. $(8.33a + 1.15y - 2.4b) - (2.44y + 5.0b - 1.12a)$
    $+ (3.8b + 2.88y - 1.74a)$
52. $(28.3x + 4.6xy + 37.2y) - (12.5xy + 44.7y - 6.3x)$
    $+ (3.6y + 2.1xy - 11.6x)$
53. $(4x + 2y) + (3w - 4y) - (5x + 5w) + (3x + 2y)$
54. $(1.44a + y) - (2.44y + 1.12a) + (2.88y - 1.74a) - (3.84a - 6.82y)$

55. $(5.22xy + 54.2x) - (28.3x + xy) - (12.5xy + 44.7y) + (2.18xy - 11.6x)$

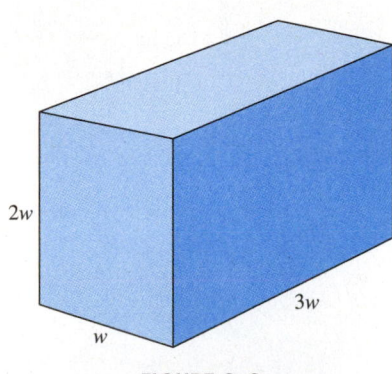

FIGURE 2–3

Applications

56. The surface area of the box in Fig. 2–3 is
$$2[2w^2 + 3w^2 + 6w^2]$$
    Simplify this expression.
57. If a person invests \$5000, x dollars at 12% interest and the rest at 8% interest, the total earnings will be
$$0.12x + 0.08(5000 - x)$$
    Simplify this expression.
58. The distance around the rectangular field, Fig. 2–4, is $w + l + w + l$.
    Simplify this expression.
59. The surface area of the can, Fig. 2–5, obtained by adding the area of the curved side to the area of each end, is $\pi r^2 + 2\pi rh + \pi r^2$. Simplify this expression.
60. The distance s_1 traveled by one falling body in time t is given by $3.74t^2 + 5.83t + 4.22$ ft and the distance s_2 traveled by another falling body in the same time is given by $9.76t^2 + 2.95t + 1.94$ ft. Write an expression for the distance $s_2 - s_1$ between the two bodies, and simplify.

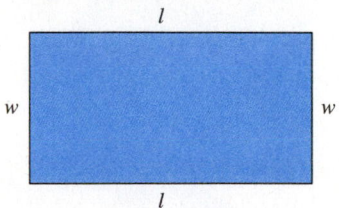

FIGURE 2–4

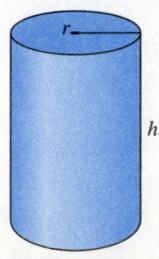

FIGURE 2–5

2–3 Laws of Exponents

Definitions

We have done some work with powers of numbers in Chapter 1. Now we will expand those ideas to include powers of algebraic expressions. Here we will deal only with expressions that have integers (positive or negative, and zero) as exponents. We will need the laws of exponents to multiply and divide algebraic expressions in the following sections.

Recall from Chapter 1 that a positive exponent shows how many times the base is to be multiplied by itself.

◆◆◆ Example 32: In the expression 2^5, the base is 2 and the exponent is 5.

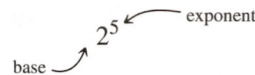

Its meaning is

$$2^5 = (2)(2)(2)(2)(2) = 32$$ ◆◆◆

In general,

Positive Integral Exponent	$x^n = \underbrace{x \cdot x \cdot x \cdot \ldots x}_{n \text{ factors}}$	21

Common Error	An exponent applies only to the symbol directly in front of it. Thus, $$5y^3 = 5(y^3)$$ but $$5y^3 \neq 5^3 y^3$$	

Multiplying Powers

■ Exploration:

Try this. (a) Use Eq. 21 to expand x^3 into its factors, that is, $(x)(x)(x)$. (b) Similarly expand x^4 into its factors. (c) Form the product of x^3 and x^4, with each in its factored form. (d) Simplify your expression and use Eq. 21 again to write your result as x raised to a power.

What did you find? Can you express your findings as a rule?   ■

You should have found that the exponent in the product is the sum of the two original exponents.

$$x^3 \cdot x^4 = x^{3+4} = x^7$$

This gives our first law of exponents.

Products	$x^a \cdot x^b = x^{a+b}$ *When multiplying powers having the same base, add the exponents.*	22

◆◆◆ **Example 33:** Here we show the use of Eq. 21.

$$x^2(x^5) = x^{2+5} = x^7$$ ◆◆◆

If a quantity has no exponent, it is understood to be 1, even though it is not usually written. Thus

$$x = x^1$$

◆◆◆ **Example 34:** Here are more uses of Eq. 21.

(a) $(x)(x^2) = (x^1)(x^2) = (x^{1+2}) = (x^3)$

(b) $(b)(b^3)(b^2) = (b^1)(b^3)(b^2) = (b^{1+3+2}) = (b^6)$

(c) $(y^2)(y^4) = y^{2+4} = y^6$

(d) $(m^3)(m^n) = m^{3+n}$ ◆◆◆

> The "invisible 1" appears again. We saw it before as the unwritten coefficient of every term, and now as the unwritten exponent. It is also in the denominator.
>
> $$x = \frac{1x^1}{1}$$
>
> Do not forget about those invisible 1's. We use them all the time.

Dividing Powers

■ **Exploration:**

Try this. (a) Use Eq. 21 to expand x^4 into its factors. (b) Similarly expand x^2 into its factors. (c) Form the quotient of x^4 divided by x^2 with each in its factored form. (d) Simplify your expression and use Eq. 21 again to write your result as x raised to a power.

What did you find? Can you express your findings as a rule? ■

You should have found that the exponent in the quotient is the difference of the two original exponents.

$$\frac{x^4}{x^2} = x^{4-2} = x^2$$

We state it as another law of exponents:

Quotients	$\dfrac{x^a}{x^b} = x^{a-b} \quad (x \neq 0)$ *When dividing powers having the same base, subtract the exponents.*	23

◆◆◆ **Example 35:** These examples show the use of Eq. 23.

(a) $\dfrac{x^2}{x} = x^{2-1} = x^1 = x$

(b) $\dfrac{y^4}{y^2} = y^{4-2} = y^2$

(c) $\dfrac{a^{5n}}{a^{2n}} = a^{5n-2n} = a^{3n}$

(d) $\dfrac{x^2 y^5}{xy^3} = x^{2-1}y^{5-3} = xy^2$ (e) $\dfrac{x^{a+b}}{x^{a-b}} = x^{a+b-(a-b)} = x^{2b}$ ◆◆◆

Common Error	We *subtract* the exponents, *not divide* them. $\dfrac{4^6}{4^3} \neq 4^2$

Power Raised to a Power

Now let us take a quantity which itself is raised to a power and raise that expression to another power. Let us raise x^3 to the power 2.

$$(x^3)^2$$

■ Exploration:

Try this. (a) Use Eq. 21 to expand $(x^3)^2$ into its factors, that is, $(x^3)(x^3)$. (b) Similarly expand each x^3 into its factors. (c) Simplify your expression and use Eq. 21 again to write your result as x raised to a power. What did you find? Can you express your findings as a rule? ■

You should have found that the exponent in the result is the product of the two original exponents.

In general,

Power Raised to a Power	$(x^a)^b = x^{ab}$	24
	When a power is raised to a power, multiply the exponents.	

◆◆◆ Example 36: Here we show the use of Eq. 24.

(a) $(n^4)^3 = n^{4(3)} = n^{12}$ (b) $(10^2)^4 = 10^{2(4)} = 10^8$ ◆◆◆

	We *multiply* the exponents, *not add* them.
Common Error	$(x^2)^3 \neq x^5$

Product Raised to a Power

Let us now find a rule for raising a product, such as xy, to a power, say 2.

$$(xy)^2$$

■ Exploration:

Try this. As before, use Eq. 21 to expand this expression into its factors, simplify, and use Eq. 21 again to write x and y each raised to a power. Can you express your findings as a rule? ■

You should have found that x and y can each be separately raised to the given power. In general,

	$(xy)^n = x^n \cdot y^n$	25
Product Raised to a Power	*When a product is raised to a power, each factor may be separately raised to the power.*	

◆◆◆ Example 37: These examples use Eq. 25.

(a) $(abc)^4 = a^4b^4c^4$

(b) $(3y)^3 = 3^3y^3 = 27y^3$

(c) $(4.23 \times 10^2)^3 = (4.23)^3 \times (10^2)^3 = 75.7 \times 10^6$

(d) $(2y^4z^n)^3 = 2^3(y^4)^3(z^n)^3 = 8y^{12}z^{3n}$

(e) $(-x^2z)^5 = (-1)^5(x^2)^5(z)^5 = -x^{10}z^5$ ◆◆◆

Common Error	There is no similar rule for the *sum* of two quantities raised to a power. $$(x + y)^n \neq x^n + y^n$$

A good way to test a "rule" that you are not sure of is to *try it with numbers.* In this case, does $(2 + 3)^2$ equal $2^2 + 3^2$? Evaluating each expression, we obtain

$$(5)^2 \overset{?}{=} 4 + 9$$
$$25 \neq 13$$

Quotient Raised to a Power

We can show, in a similar way to that used for products, that

$$\left(\frac{x}{y}\right)^2 = \left(\frac{x}{y}\right)\left(\frac{x}{y}\right) = \frac{x^2}{y^2}$$

Or, in general,

Quotient Raised to a Power	$$\left(\frac{x}{y}\right)^n = \frac{x^n}{y^n} \quad (y \neq 0)$$ *When a quotient is raised to a power, the numerator and denominator may be separately raised to the power.*	26

◆◆◆ **Example 38:** These examples show the use of Eq. 26.

(a) $\left(\dfrac{2}{a}\right)^3 = \dfrac{2^3}{a^3} = \dfrac{8}{a^3}$

(b) $\left(\dfrac{2x}{4y}\right)^2 = \dfrac{2^2x^2}{4^2y^2} = \dfrac{4x^2}{16y^2} = \dfrac{x^2}{4y^2}$

(c) $\left(\dfrac{3c^2}{4d^4}\right)^3 = \dfrac{3^3(c^2)^3}{4^3(d^4)^3} = \dfrac{27c^6}{64d^{12}}$

(d) $\left(-\dfrac{x^2}{y}\right)^4 = \dfrac{(-1)^4(x^2)^4}{y^4} = \dfrac{x^8}{y^4}$ ◆◆◆

Zero Exponent

If we divide x^n by itself we get, by Eq. 23

$$\frac{x^n}{x^n} = x^{n-n} = x^0$$

But any expression divided by itself equals 1, so

Zero Exponent	$$x^0 = 1 \quad (x \neq 0)$$ *Any expression (except 0) raised to the zero power equals 1.*	27

◆◆◆ **Example 39:** Here are some uses of Eq. 27.

(a) $(9626)^0 = 1$

(b) $(abc)^0 = 1$

(c) $(4x^2 + 9x - 35)^0 = 1$

(d) $7a^0 = 7(1) = 7$ ◆◆◆

Negative Exponent

We now divide x^0 by x^a. By Eq. 23

$$\frac{x^0}{x^a} = x^{0-a} = x^{-a}$$

Since $x^0 = 1$, we get

| Negative Exponent | $x^{-a} = \dfrac{1}{x^a} \quad (x \neq 0)$

 When taking the reciprocal of a base raised to a power, change the sign of the exponent. | 28 |

◆◆◆ **Example 40:** Here we show how to write some given expressions with positive exponents only.

(a) $6^{-1} = \dfrac{1}{6^1} = \dfrac{1}{6}$
(b) $x^{-2} = \dfrac{1}{x^2}$
(c) $2x^{-3} = 2\left(\dfrac{1}{x^3}\right) = \dfrac{2}{x^3}$

(d) $\dfrac{3}{x^{-2}} = 3x^2$
(e) $4x^{-2} + 2y^{-3} = \dfrac{4}{x^2} + \dfrac{2}{y^3}$

◆◆◆◆ **Example 41:** Write with positive exponents only,

$$\left(\frac{5}{x}\right)^{-3}$$

Solution: We first apply Law 26 for a quotient raised to a power

$$\left(\frac{5}{x}\right)^{-3} = \frac{5^{-3}}{x^{-3}}$$

Then by Law 28 for negative exponents

$$\frac{5^{-3}}{x^{-3}} = \frac{x^3}{5^3} = \frac{x^3}{125} \qquad\qquad ◆◆◆$$

◆◆◆ **Example 42:** Write with positive exponents only,

$$\left(\frac{3ax}{2z}\right)^{-2}$$

Solution: We first apply Law 26 for a quotient raised to a power.

$$\left(\frac{3ax}{2z}\right)^{-2} = \frac{(3ax)^{-2}}{(2z)^{-2}}$$

Then by Law 28 for negative exponents

$$\frac{(3ax)^{-2}}{(2z)^{-2}} = \frac{(2z)^2}{(3ax)^2}$$

Finally we use Law 25 for a product raised to a power.

$$\frac{(2z)^2}{(3ax)^2} = \frac{2^2 z^2}{3^2 a^2 x^2} = \frac{4z^2}{9a^2 x^2} \qquad\qquad ◆◆◆$$

◆◆◆ **Example 43:** These examples show how to use negative exponents to write an expression without fractions.

(a) $\dfrac{1}{x} = x^{-1}$
(b) $\dfrac{7}{a^3} = 7a^{-3}$
(c) $\dfrac{x^2}{y^2} = x^2 y^{-2}$ ◆◆◆

Summary of the Laws of Exponents

Positive Integral Exponent	$$x^n = \underbrace{x \cdot x \cdot x \cdot \ldots \cdot x}_{n \text{ factors}}$$	21
Products	$$x^a \cdot x^b = x^{a+b}$$ *When multiplying powers having the same base, add the exponents.*	22
Quotients	$$\frac{x^a}{x^b} = x^{a-b} \quad (x \neq 0)$$ *When dividing powers having the same base, subtract the exponents.*	23
Powers	$$(x^a)^b = x^{ab} = (x^b)^a$$ *When a power is raised to a power, multiply the exponents.*	24
Product Raised to a Power	$$(xy)^n = x^n \cdot y^n$$ *When a product is raised to a power, each factor may be separately raised to the power.*	25
Quotient Raised to a Power	$$\left(\frac{x}{y}\right)^n = \frac{x^n}{y^n} \quad (y \neq 0)$$ *When a quotient is raised to a power, the numerator and denominator may be separately raised to the power.*	26
Zero Exponent	$$x^0 = 1 \quad (x \neq 0)$$ *Any expression (except 0) raised to the zero power equals 1.*	27
Negative Exponent	$$x^{-a} = \frac{1}{x^a} \quad (x \neq 0)$$ *When taking the reciprocal of a base raised to a power, change the sign of the exponent.*	28

Exponents on the Calculator

As for addition and subtraction, we need no special instructions for simplifying expressions like those in this section. Simply enter the expression and the calculator will automatically simplify it.

◆◆◆ **Example 44:** Repeat Example 42 by calculator.

Solution: On the TI-89, we enter the expression and press ENTER. ◆◆◆

◆◆◆ **Example 45:** *An Application.* The volume v of a sphere of radius r is given by $v = (4/3)\pi r^3$ where $\pi = 3.1416$. If the radius is tripled, the volume becomes

$$v = \frac{4}{3}\pi(3r)^3$$

Simplify this expression.

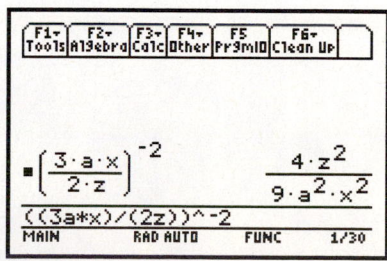

TI-89 screen for Example 44. Note that we need a multiplication symbol between *a* and *x* in the numerator to obtain a correct result.

Solution:

$$v = \frac{4}{3}\pi(3r)^3 = \frac{4}{3}\pi(3)^3 r^3$$

$$= \frac{4}{3}\pi(27)r^3$$

$$= 36\pi r^3 \qquad \text{◆◆◆}$$

Exercise 3 ◆ Laws of Exponents

Definitions: Evaluate each expression.

1. 2^3	**2.** 2^5	**3.** 4^3
4. 3^3	**5.** $(3)^5$	**6.** $(-2)^4$
7. $(-2)^5$	**8.** $(0.001)^3$	**9.** $(-5)^3$

Multiplying Powers: Multiply.

10. $(w^3)(w^2)$	**11.** $(m^5)(m^2)$	**12.** $(p^3)(p^4)$
13. $(x^4)(x^2)$	**14.** $(y^b)(y^3)$	**15.** $(a^3)(a^6)$
16. $(10^5)(10^9)$	**17.** $(10^2)(10^6)$	**18.** $(z^{11})(z^2)$

Dividing Powers: Divide and write your answer without negative exponents.

19. $\dfrac{z^7}{z^5}$	**20.** $\dfrac{a^6}{a^4}$	**21.** $\dfrac{2^4}{2^3}$
22. $\dfrac{10^6}{10^2}$	**23.** $\dfrac{10^4}{10^3}$	**24.** $\dfrac{x^{-6}}{y}$
25. $\dfrac{b^{-4}}{b^{-5}}$	**26.** $\dfrac{y^{a+1}}{y^{a-2}}$	**27.** $\dfrac{10^{b-1}}{10^{b-3}}$

Power Raised to a Power: Simplify.

28. $(x^2)^2$	**29.** $(w^3)^2$	**30.** $(m^5)^2$
31. $(p^3)^4$	**32.** $(a^2)^4$	**33.** $(2^5)^2$
34. $(z^c)^a$	**35.** $(y^{-1})^{-3}$	**36.** $(a^{x-1})^3$

Product Raised to a Power: Raise to the power indicated and remove parentheses.

37. $(2x)^2$	**38.** $(3y)^2$	**39.** $(axy)^3$
40. $(abp^2)^4$	**41.** $(ac)^3$	**42.** $(3a)^2$
43. $(4a^3c^2)^4$	**44.** $(2xyz)^5$	

Quotient Raised to a Power: Raise to the power indicated and remove parentheses.

45. $\left(\dfrac{x}{y}\right)^2$	**46.** $\left(\dfrac{2}{3}\right)^3$	**47.** $\left(-\dfrac{2}{5}\right)^3$
48. $\left(\dfrac{a}{b}\right)^3$	**49.** $\left(\dfrac{4x^2}{3y^2}\right)^2$	**50.** $\left(\dfrac{2ab^3}{3c^2d}\right)^3$

Zero Exponent: Evaluate.

51. $(2x)^0$ **52.** $(2x^2 - 8x + 32)^0$ **53.** $108a^3c^0$

54. $\dfrac{82}{y^0}$ **55.** $\dfrac{c}{y^0}$ **56.** $\dfrac{(z^{-n})(z^2)}{z^{2-n}}$

57. $4\left(\dfrac{a^7}{y^4}\right)^0$

Negative Exponent: Write each expression with positive exponents only.

58. ay^{-1} **59.** x^{-1} **60.** $(-b)^{-3}$

61. $\left(\dfrac{2}{x}\right)^{-4}$ **62.** $ab^{-5}c^{-2}$

63. $4w^{-4} - 3z^{-2}$ **64.** $\left(\dfrac{a}{b}\right)^{-4}$

65. $\left(\dfrac{4x^3}{3y^2}\right)^{-3}$

Express without fractions, using negative exponents where needed.

66. $\dfrac{1}{a}$ **67.** $\dfrac{5}{x^3}$ **68.** $\dfrac{c^2}{d^3}$

69. $\dfrac{w^4z^{-2}}{x^{-3}}$ **70.** $\dfrac{y^2}{x^{-4}}$ **71.** $\dfrac{c^{-1}d^{-4}}{b^{-2}c^{-3}}$

Challenge Problems: Simplify.

72. $\left[2x^2\left(\dfrac{y^3}{w^2}\right)\right]^2$ **73.** $\left[\left(\dfrac{wy}{z}\right)^2(wx)^3\right]^3$

74. $\left[\left(\dfrac{ax}{bz}\right)^2\left(\dfrac{bx}{cz}\right)^3\right]^2$ **75.** $\left[\dfrac{(am)^3(am)^2}{(bn)^2(bn)^4}\right]^3$

Applications

76. We can find the volume of the box in Fig. 2–6 by multiplying length by width by height, getting

$$(3w)(w)(2w)$$

Simplify this expression.

77. A freely falling body, starting from rest, falls a distance of $16.1t^2$ feet in t seconds. In twice that time it will fall

$$16.1(2t)^2 \text{ ft}$$

Simplify this expression.

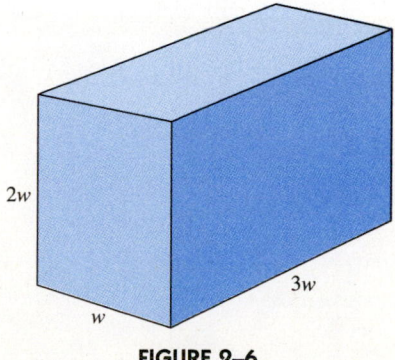

$2w$ $3w$ w

FIGURE 2–6

78. The power in a resistor of resistance R in which a current i flows is i^2R. If the current is reduced to one-third its former value, the power will be

$$\left(\frac{i}{3}\right)^2 R$$

Simplify this expression.

79. The resistance R of two resistors R_1 and R_2 wired in parallel (Fig. 2–7) is found from the equation

$$\frac{1}{R} = \frac{1}{R_1} + \frac{1}{R_2}$$

Write this equation without fractions.

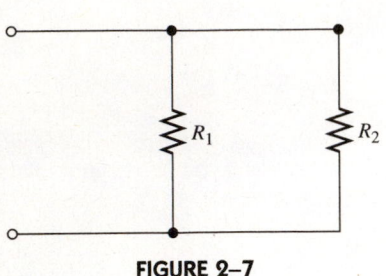

FIGURE 2–7

2–4 Multiplying a Monomial by a Monomial

From addition and subtraction of algebraic expressions we now move on to *multiplication*. We will start with the simplest types and then progress to the more complex.

Symbols and Definitions

Multiplication is indicated in several ways: by the usual $\times$ symbol; by a dot; or by parentheses, brackets, or braces. Thus the product of b and d could be written

$$b \cdot d \qquad b \times d \qquad b(d) \qquad (b)d \qquad (b)(d)$$

We learned earlier that the symbols of grouping (parentheses, brackets, and braces) indicate that the enclosed terms are to be taken as a whole. Here we see that they also indicate *multiplication*.

◆◆◆ Example 46:
(a) $m(n)$ means the product of m and n.
(b) $-3x(2y)$ means the product of $-3x$ and $2y$.   ◆◆◆

Most common of all is to use no symbol at all. The product of b and d would usually be written bd. Avoid using the $\times$ symbol when doing algebra because it could get confused with the letter x.

We get a *product* when we multiply two or more *factors*.

$$(\text{factor})(\text{factor})(\text{factor}) = \text{product}$$

Rules of Signs

When we multiply two factors that have the *same* sign, we get a product that is *positive*. When we multiply two factors that have *opposite* signs, we get a product that is *negative*. Stated as a rule, if a and b are positive quantities, we have

Rules of Signs for Multiplication		
	$(+a)(+b) = (-a)(-b) = +ab$	6
	$(+a)(-b) = (-a)(+b) = -(+a)(+b) = -ab$	7

The product of two factors of like signs is positive, of unlike signs is negative.

◆◆◆ Example 47: The rules of signs are shown in these examples.

(a) $(+x)(+z) = xz$
(b) $(+x)(-z) = -xz$
(c) $(-x)(+z) = -xz$ ◆◆◆

When we multiply *more* than two factors, every pair of negative factors will give a positive product. Thus, if there is an even number of negative factors, the final result will be positive; if there is an odd number of negative factors, the result will be negative.

◆◆◆ Example 48: More examples concerning the rules of signs.

(a) $(a)(-b)(-c) = abc$
(b) $(-p)(-q)(-r) = -pqr$
(c) $(-w)(-x)(-y)(-z) = wxyz$ ◆◆◆

Commutative and Associative Laws for Multiplication

The *commutative law* for multiplication states that the *order* of multiplication is not important.

Commutative Law for Multiplication	$ab = ba$	8

The *associative law* for multiplication allows us to group the numbers to be multiplied in any order.

Associative Law for Multiplication	$a(bc) = (ab)c = (ac)b = abc$	9

◆◆◆ Example 49: It is no surprise that

$$(2)(3) = (3)(2)$$ ◆◆◆

Multiplying Monomials

Recall that a monomial is an algebraic expression having one term, such as the expressions $3y^2$ and $(8x)^3$. To multiply monomials, we use the laws of exponents and the rules of signs.

◆◆◆ Example 50: These examples illustrate how to multiply monomials.

(a) $x^2(x^3) = x^{2+3} = x^5$
(b) $3y(2y^3) = (3)(2)y^{1+3} = 6y^4$
(c) $(-4a^2)(3a^2) = (-4)(3)a^{2+2} = -12a^4$
(d) $(-5xy^2)(2x^3y^2) = (-5)(2)(x)(x^3)(y^2)(y^2)$
$$= -10x^{1+3}y^{2+2}$$
$$= -10x^4y^4$$ ◆◆◆

The procedure is no different when the quantities to be multiplied include approximate numbers. We must, however, retain the proper number of digits in our answer. Recall that *when multiplying approximate numbers we retain as many significant digits in our product as contained in the factor having the fewest significant digits.*

◆◆◆ Example 51: This example shows the multiplication of monomials containing approximate numbers.

$$(3.848x^2)(5.24xy^2) = 20.2x^3y^2$$

Since one of our numerical factors has four significant digits and the other has only three, we round our product to three significant digits. ◆◆◆

It is no harder to multiply three or more monomials than to multiply two monomials.

◆◆◆ **Example 52:** Here are products of three monomials.

(a) $x(x^2)(x^3) = x^{1+2+3} = x^6$

(b) $2a^2(4a)(3a^3) = 2(4)(3)a^{2+1+3} = 24a^6$

(c) $3m(2mn)(5n^2) = 3(2)(5)m^{1+1}n^{1+2} = 30m^2n^3$

(d) $-3a(-2a^2b)(ab^2c^3) = -3(-2)(1)(a)(a^2)(a)(b)(b^2)(c^3)$

$$= 6a^{1+2+1}b^{1+2}c^3$$

$$= 6a^4b^3c^3 \qquad \text{◆◆◆}$$

If there are *literals* (alphebatic letters) in the exponents, simply combine them using the laws of exponents we have already studied.

◆◆◆ **Example 53:** These examples show terms with literals in the exponents.

(a) $x^n(x^2) = x^{n+2}$

(b) $3y^k(2y^{k+1}) = 3(2)y^k y^{k+1}$

$$= 3(2)y^{k+k+1}$$

$$= 6y^{2k+1}$$

(c) $5a^n x^2(-3a^3) = 5(-3)(a^n)(a^3)(x^2)$

$$= -15a^{n+3}x^2 \qquad \text{◆◆◆}$$

◆◆◆ **Example 54:** *An Application.* A cylindrical chemical storage tank, Fig. 2–8, whose height in feet is 4.50 times its base radius contains oil whose density is d lb/ft^3. The weight of the liquid is the volume, $\pi a^2(4.50a)$, times the density d or

$$\text{Weight} = \pi a^2(4.50a)d$$

where $\pi \approx 3.142$. Simplify this expression.

Solution: Multiplying gives

$$\text{Weight} = \pi a^2(4.50a)d$$

$$= (3.142)(4.50)a^{2+1}d$$

$$= 14.1a^3d \qquad \text{◆◆◆}$$

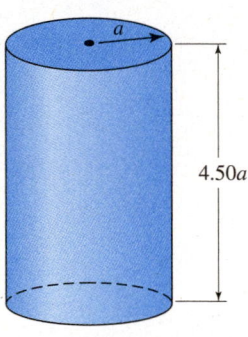

FIGURE 2–8

Exercise 4 ◆ Multiplying a Monomial by a Monomial

Multiply and simplify.

1. $(x^2)(x^4)$

2. $(2a)(3a^2)$

3. $(x^2)(-x^3)$

4. $(5w)(3w^2)$

5. $(2a)(3b^2)$

6. $(3xy)(3xy^2)$

7. $(5m^2n)(3mn^2)$

8. $(2abc^2)(3a^2bc)$

9. $(12.5a)(3.26a^2)$

10. $(15.9x^2)(4.93x^4)$

11. $(3.73xy)(1.77xy^2)$

12. $(3pqr^2)(2p^2qr)$

13. $(2ab)(3b^n)$

14. $(x^{2a})(x^4)$

15. $(1.55a^m)(2.36a^n)$

16. $(2a^xbc^2)(3a^2bc^x)$

17. $(2w^2)(5w)(3w^2)$

18. $(2m^2n)(mn)(3mn^2)$

19. $(4a^2b)(2ab)(3b^n)$

20. $(2.25a)(1.55a^m)(2.36a^n)$

Challenge Problems

21. $(1.84wx^2y)(2.44w^2xy^3)(1.65wx^3y)(2.33w^2xy)$

22. $(3.91a^2b)(1.94ab^3)(2.93ab)(1.43b^2)$

23. $(3.82abc)(a^xbc^2)(1.55a^2bc^x)(ab^3c)$

24. $(2.24m^an)(2.96mn^b)(1.52m^2n)(1.15m^cn)(1.83m^dn^2)$

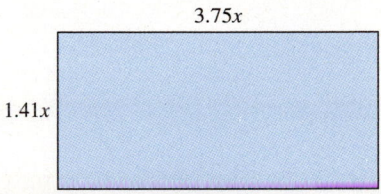

FIGURE 2–9

Applications

25. The area of the field, Fig. 2-9, if doubled, is equal to $2(1.41x)(3.75x)$. Simplify this expression.

26. The volume of the shipping container, Fig. 2-10, is $(h)(1.5h)(3.2h)$. Simplify this expression.

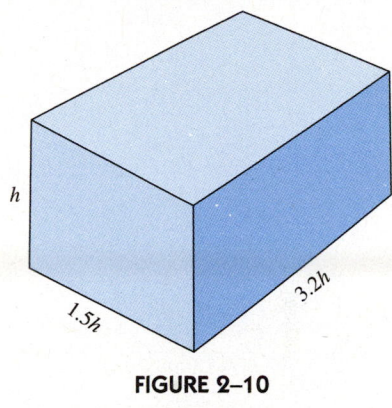

FIGURE 2–10

2–5 Multiplying a Monomial and a Multinomial

We will now use the things we have learned by multiplying monomials to multiply a monomial and a multinomial.

A *multinomial* is an algebraic expression having more than one term.

◆◆◆ **Example 55:** Some multinomials are

(a) $6x - 8$ (b) $7x^2 + 2x - 1$ (c) $x^{-2} + 5$ ◆◆◆

Recall that a *polynomial* is a monomial or a multinomial in which the powers to which the variable is raised are all positive integers. The first two expressions in the preceding example are polynomials, but the third is not. The examples in this chapter will show multiplication of polynomials only, but the methods we show are valid for any multinomials. We have not yet covered the rules needed for multiplying other multinomials, such as those containing radicals, negative exponents, logarithms, and so forth. In later chapters we will show multiplication of such expressions.

A *binomial* is a polynomial with two terms, and a *trinomial* is a polynomial having three terms. In Example 55, the first expression is a binomial and the second is a trinomial.

To multiply a monomial and a multinomial, we use the distributive law (Eq. 10):

Distributive Law for Multiplication	$a(b + c) = ab + ac$	**10**

◆◆◆ **Example 56:** These examples show the use of the distributive law.

(a) $x(x + 1) = x(x) + x(1) = x^2 + x$

(b) $3m(m + m^2) = 3m(m) + 3m(m^2) = 3m^2 + 3m^3$

(c) $2x(x + 1) = 2x(x) + 2x(1) = 2x^2 + 2x$ ◆◆◆

Be especially careful when multiplying negative quantities.

♦♦♦ **Example 57:** Here is an example where we multiply negative quantities.

$$-y(y^2 - 3) = -y(y^2) - y(-3) = -y^3 + 3y$$ ♦♦♦

♦♦♦ **Example 58:** Another example with negative quantities.
$$5x(-2x^2 - 1) = 5x(-2x^2) + 5x(-1)$$
$$= -10x^3 - 5x$$ ♦♦♦

As usual, when the coefficients are approximate numbers we must round our answer to the proper number of significant digits.

♦♦♦ **Example 59:** This example has approximate numbers.

$$2.75w(3.85w + 1.73w^2) = 10.6w^2 + 4.76w^3$$ ♦♦♦

Common Errors	Don't forget to multiply *every* term in the parentheses by the preceding factor. $$-(3x + 4) \neq -3x + 4$$ Instead $$-(3x + 4) = (-1)(3x + 4)$$ $$= (-1)(3x) + (-1)(4)$$ $$= -3x - 4$$
	Multiply the terms within the parentheses only by the factor directly preceding it. $$a - 3(x + y) \neq (a - 3)(x + y)$$

We can extend the distributive law, which was written for a monomial times a binomial, to a monomial times a multinomial having any number of terms. We simply multiply every term in a multinomial by the monomial.

♦♦♦ **Example 60:** Multiply $(b + c + d)$ by a.

Solution:

$$a(b + c + d) = ab + ac + ad$$ ♦♦♦

♦♦♦ **Example 61:** Multiply $2y^2 + 5y - 8$ by $-4y^2$.

Solution:

$$-4y^2(2y^2 + 5y - 8) = -4y^2(2y^2) + (-4y^2)(5y) + (-4y^2)(-8)$$
$$= -8y^4 - 20y^3 + 32y^2$$ ♦♦♦

Multiplication by Calculator

For simple products we usually need no special instructions; just enter the expression and the calculator will automatically simplify it. Otherwise, on the TI-89, we use the **expand** operation from the ALGEBRA menu.

♦♦♦ **Example 62:** Repeat Example 61 by calculator.

Solution: On the TI-89, we select **expand**, enter the expression, and press ENTER.   ♦♦♦

TI-89 screen for Example 62. Here the entered expression does not all fit on the screen.

For more complicated problems having groupings within groupings, start simplifying with the innermost grouping and work outwards.

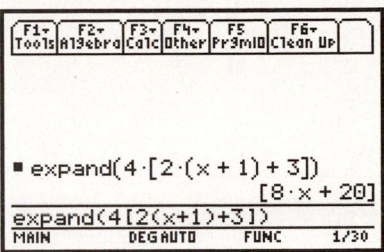

TI 89 screen for Example 63(a).

◆◆◆ **Example 63:**

(a) $4[2(x + 1) + 3] = 4[2x + 2 + 3]$
$$= 4[2x + 5]$$
$$= 8x + 20$$

(b) $a + 4[3 + 2(b - 3)] = a + 4[3 + 2b - 6]$
$$= a + 4[2b - 3]$$
$$= a + 8b - 12$$

(c) $2\{[(1 - m) + (n + 3)] - 5\} - 4 = 2\{[1 - m + n + 3] - 5\} - 4$
$$= 2\{[4 - m + n] - 5\} - 4$$
$$= 2\{-m + n - 1\} - 4$$
$$= -2m + 2n - 2 - 4$$
$$= -2m + 2n - 6 \qquad ◆◆◆$$

Exercise 5 ◆ Multiplying a Monomial and a Multinomial

Remove parentheses and simplify.

1. $-(x + 2)$
2. $-(b - c)$
3. $3 - (-x + 1)$
4. $-(y - 3) + 2$
5. $a + (b + a)$
6. $x + (2 - x)$
7. $x + (x - y)$
8. $3(-2 - x)$
9. $x(b + 2)$
10. $2(a + 3b)$
11. $x(x - 5)$
12. $3.83b(b^2 + 1.27)$
13. $2.03x(1.27x - 2.36)$
14. $3x(-7 - 10x)$
15. $b^4(b^2 + 8)$
16. $a^2b(2a + b - ab)$
17. $-(a + 3.92) - (a - 4.14)$
18. $(4z + 2) + (z - 5)$
19. $(2x + 5) + (x - 2)$
20. $(c - 5) - (6 + 3c)$
21. $(2x + 6a) - (4x - a)$
22. $(0.826c - 5.37) - (2.76 + 0.273c)$
23. $(7262x + 1.26a) - (2844x - 8.23a)$
24. $(5 - 2a^3 + 3a^4) + (4a^4 - 6a^3 - 7)$
25. $(y - z - b) - (b + y + z)$
26. $(5bc + 6c - a) - (8c - 2a + 3bc)$
27. $(2z + 5c - 3a) - (6a + 2c - 4z)$
28. $x - 2.66[y - 3.02(x + 6.22y) + 4.98y]$
29. $\{[3 - (x + 7)] - 3x\} - (x + 7)$

Challenge Problems

30. $-6x^3y^3(3xy + 5x^2y^3 - 2xy^2)$

31. $2ab(9a^2 + 6ab - 3b^2)$

32. $-4.27m^2(2.83m^4 + 6.82m^2 - 3.25m^3 + 2.47)$

33. $-5.16xy(1.23x^2y - 5.83xy^2 + 4.27x^2y^2 - 2.94xy)$

34. $6mn^2(5m^3n + 4mn^2 - 2m^2n + mn)$

35. $6p - \{3p + [2q - ((5p + 4q) + p) - (3p + 2)] - 2p\}$

36. $(2.25c - 9.28b + 3.82a) - (-3.92a - 1.72b - 8.33c)$

37. $(23y^2 + 4y^3 - 12) - (11y^3 - 8y^2 + y)$

38. $24ab - (16ab - 3x^2 + 7z - 2y^2)$

39. $(18y^2 - 12xy) - (6y^2 + xy - a)$

40. $(4y + 2y^2) - (3y - 6b + 4y^2 + 5)$

41. $(-6x - z) - \{3y + [7x - (3z + 8y + x)]\}$

Applications

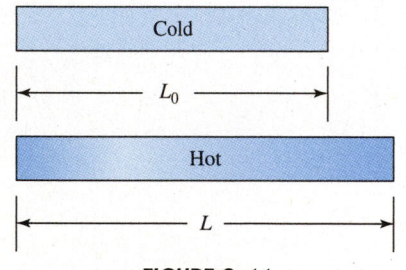

Cold

L_0

Hot

L

FIGURE 2–11

42. When the temperature of a bar of initial length L_0 (Fig. 2–11) rises by an amount Δt, the bar will elongate to a new length L, where L is given by

$$L = L_0(1 + \alpha\Delta t)$$

Here α is the temperature coefficient of expansion for the material from which the bar is made. Simplify this expression by multiplying out.

43. The equivalent resistance between two points in a certain circuit is found to be

$$(R_1 + R_2) - (R_3 - R_4) - (R_5 + R_6 - R_7)$$

Simplify this expression by removing parentheses.

2–6　Multiplying a Binomial by a Binomial

To multiply any two multinomials, we multiply every term in one multinomial by each term in the other multinomial, and combine like terms. Here we apply this rule to binomials and later use it for multinomials having any number of terms.

◆◆◆ **Example 64:** Here we multiply two binomials.

$$(x - 2)(x + 3) = (x)(x) + (x)(3) + (-2)(x) + (-2)(3)$$
$$= x^2 + 3x - 2x - 6$$
$$= x^2 + x - 6$$

◆◆◆

We can also multiply two multinomials by using the distributive rule.

◆◆◆ **Example 65:** Repeating Example 64 using the distributive rule gives

$$(x - 2)(x + 3) = x(x + 3) - 2(x + 3)$$
$$= x^2 + 3x - 2x - 6$$
$$= x^2 + x - 6$$

as before.

◆◆◆

```
F1▾  F2▾  F3▾ F4▾  F5    F6▾
Tools Algebra Calc Other PrgmIO Clean Up



■ expand((x - 2)·(x + 3))
                      x² + x - 6
expand((x-2)*(x+3))
MAIN      RAD AUTO   FUNC   1/30
```

TI-89 screen for Example 64.

FOIL Rule

One way to keep track of the terms when multiplying binomials is by the **FOIL** rule. Multiply the **F**irst terms, then **O**uter, **I**nner, and **L**ast terms. These products can, of course, be done in any order, but we can avoid getting mixed up if we always follow the same FOIL order.

◆◆◆ **Example 66:** Repeating Example 64,

$$
\begin{matrix}
\text{F} & \text{O} & \text{I} & \text{L} \\
\downarrow & \downarrow & \downarrow & \downarrow
\end{matrix}
$$

$$(x - 2)(x + 3) = x^2 + 3x - 2x - 6$$
$$= x^2 + x - 6$$

◆◆◆

Common Error	The FOIL rule is only for multiplying *binomials*.

Exercise 6 ◆ Multiplying a Binomial by a Binomial

Multiply and simplify.

1. $(x + y)(x + z)$
2. $(4a - 3)(a + 2)$
3. $(4m + n)(2m^2 \rightarrow n)$
4. $(y + 2)(y - 2)$
5. $(2x - y)(x + y)$
6. $(a^2 - 3b)(a^2 + 5b)$
7. $(4xy^2 - 3a^3b)(3xy^2 + 4a^3b)$
8. $(2m^2 - 2n^2)(2m^2 + 2n^2)$
9. $(a - 7x)(2a + 3x)$
10. $(3x - z^2)(4x - 3z^2)$
11. $(ax - 5b)(ax + 5b)$
12. $(5y^2 + 3z)(5y^2 - 3z)$

Challenge Problems

13. $(2.93x - 1.11y)(x + y)$
14. $(2.84a^2 - 3.82b)(a^2 + 5.11b)$
15. $(4.03y^2 - 3.92a^3b)(3.26y^2 + 4.73a^3b)$
16. $(2.83m^2 - 2.12n^2)(2.83m^2 + 2.12n^2)$

Applications

17. A rectangle has its length L increased by 2 units and its width W decreased by 3 units. Write an expression for the area of the new rectangle (area = length times width) and multiply out.
18. A car traveling at a rate R for a time T will go a distance equal to RT. If the rate is decreased by 8.5 mi/h and the time is increased by 2.4 h, write an expression for the new distance traveled, and multiply out.

19. *Writing:* Your friend refuses to learn the FOIL rule. "*I want to learn math, not memorize a bunch of tricks!*" he declares. What do you think? Write a paragraph or so giving your opinion on the value or harm in learning devices such as the FOIL rule.

2–7 Multiplying a Multinomial by a Multinomial

We come now to the most general situation, multiplying an expression with any number of terms by another expression having any number of terms.

We multiply multinomials in the same way that we multiplied other expressions. We make use of the distributive law for multiplication and *multiply every term in one multinomial by every term in the other*. Then combine like terms.

◆◆◆ **Example 67:** Here is a binomial times a trinomial.

$$(a + 1)(a^2 + a + 1) = a(a^2) + a(a) + a(1) + 1(a^2) + 1(a) + 1(1)$$
$$= a^3 + a^2 + a + a^2 + a + 1$$
$$= a^3 + 2a^2 + 2a + 1 \qquad ◆◆◆$$

◆◆◆ **Example 68:** Here's another similar to Ex. 67.

$$(x + 2)(x^2 + 4x - 3) = x(x^2) + x(4x) + x(-3) + 2(x^2) + 2(4x) + 2(-3)$$
$$= x^3 + 4x^2 - 3x + 2x^2 + 8x - 6$$
$$= x^3 + 6x^2 + 5x - 6 \qquad ◆◆◆$$

◆◆◆ **Example 69:** This example shows a binomial times a polynomial with four terms.

$$(w - x^2)(w^3 + aw^2 + bx^2 - x^3)$$
$$= w^4 + aw^3 + bwx^2 - wx^3 - w^3x^2 - aw^2x^2 - bx^4 + x^5 \qquad ◆◆◆$$

To multiply *three* multinomials, first multiply two of them. Then multiply that product by the third multinomial. This procedure can, of course, be extended to multiply any number of multinomials.

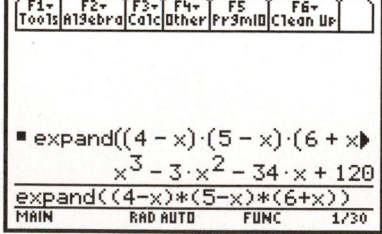

TI-89 screen for Example 70.

◆◆◆ **Example 70:** Multiply $(4 - x)(5 - x)(6 + x)$.

Solution: Let us first multiply one pair of binomials, say, $(5 - x)(6 + x)$.

$$(5 - x)(6 + x) = 30 + 5x - 6x - x^2 = 30 - x - x^2$$

Then let us multiply that product by $(4 - x)$.

$$(4 - x)[(5 - x)(6 + x)] = (4 - x)(30 - x - x^2)$$
$$= 120 - 4x - 4x^2 - 30x + x^2 + x^3$$
$$= 120 - 34x - 3x^2 + x^3 \qquad ◆◆◆$$

To multiply two multinomials that have three or more terms each, we again multiply every term in one multinomial by every term in the other multinomial. To avoid getting confused, try multiplying every term in the second multinomial by the *first* term in the first multinomial; then multiply every term in the second multinomial by the *second* term in the first multinomial, and so forth until all terms have been multiplied.

◆◆◆ **Example 71:**

$$(x^2 + 3x - 1)(x^2 - 3x + 2) = x^4 - 3x^3 + 2x^2$$
$$+3x^3 - 9x^2 + 6x$$
$$-x^2 + 3x - 2$$
$$= x^4 \qquad - 8x^2 + 9x - 2 \qquad ◆◆◆$$

Exercise 7 ◆ Multiplying a Multinomial by a Multinomial

Multiply and simplify.

1. $(x - 3)(x + 4 - y)$

2. $(a - d)(a - 2d + 5)$

3. $(w^2 + w - 5)(4w - 2)$

4. $(a^2 - 5)(3a^2 - 7a - 4)$

5. $(x + 3.88)(x^3 - 2.15x - 6.03)$

6. $(1 + c^2)(4c^2 + 7c - 3)$

7. $(2x^2 - 6xy + 3y^2)(3x + 3y)$

8. $(b^2 - bx + x^2)(b + x)$

9. $(a^2 + 2a - 2)(a + 1)$

10. $(a^2 + 5a - xy)(a + z)$

11. $(c^2 - cm + cn + mn)(c - m)$

12. $(y^2 - x^2)(y^3 + ay^2 - abxy + bx^2 - x^3)$

Challenge Problems

13. $(x - y - z)(x + y + z)$

14. $(5x^3 + 2xy^2 - 2x)(5x^2 - 2x)$

15. $(5x - y + 2x)(4x - y + 6)$

16. $(x + y - z)(x - y - z)$

17. $(a^2 - 5.93a + 31.4)(a^2 - 5.37a + 4.03)$

18. $(m^3 - 4.83 + 32.4m)(m^2 - 3.37m + 2.26)$

19. $(am - ym + yx)(am + ym - yx)$

20. $(1.83b^2 - 2.68bx + 3.82x^2)(1.22b + 2.05x)$

Application

21. A rectangular patio, Fig. 2–12, has a width w and a length l on an architectural drawing. The width is then increased by an amount x and the length by an amount y. In a subsequent change, the width is increased by an amount a and length decreased by an amount b. The resulting area is then $(w + x + a)(l + y - b)$. Simplify this expression by multiplying out.

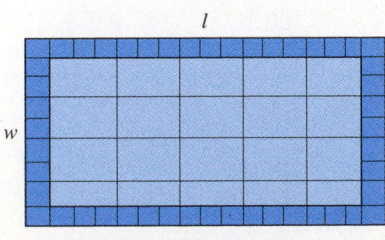

FIGURE 2–12

2–8 Raising a Multinomial to a Power

We see from Eq. 21,

Positive Integral Exponent	$x^n = x \cdot x \cdot x \cdot x \ \ x \ldots x$ n factors	21

that raising an expression to a power is the same as multiplying the expression by itself the proper number of times, provided that the power is a positive integer. We will show how to square a binomial, square a trinomial, and cube a binomial. The same method can, of course, be extended to raise any multinomial to any positive integer power.

◆◆◆ **Example 72:** Square the binomial $(x + 2)$.

Solution:

$$
\begin{aligned}
(x + 2)^2 &= (x + 2)(x + 2) \\
&= x^2 + 2x + 2x + 4 \\
&= x^2 + 4x + 4
\end{aligned}
$$
◆◆◆

◆◆◆ **Example 73:** Square the trinomial $x^2 - 2x + 1$.

Solution:

$$
\begin{aligned}
(x^2 - 2x + 1)^2 &= (x^2 - 2x + 1)(x^2 - 2x + 1) \\
&= x^4 - 2x^3 + x^2 \\
&\quad\ \ - 2x^3 + 4x^2 - 2x \\
&\quad\quad\quad\ + \ x^2 - 2x + 1 \\
&= x^4 - 4x^3 + 6x^2 - 4x + 1
\end{aligned}
$$
◆◆◆

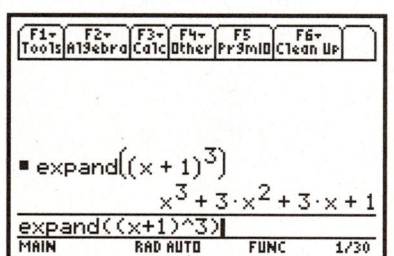

TI-89 screen for Example 74.

◆◆◆ **Example 74:** Cube the binomial $(x + 1)$.

Solution:

$$(x + 1)^3 = (x + 1)(x + 1)(x + 1)$$

Let us first multiply out $(x + 1)(x + 1)$.

$$
\begin{aligned}
(x + 1)^3 &= (x + 1)(x^2 + x + x + 1) \\
&= (x + 1)(x^2 + 2x + 1)
\end{aligned}
$$

Then

$$
\begin{aligned}
(x + 1)^3 &= x(x^2 + 2x + 1) + 1(x^2 + 2x + 1) \\
&= x^3 + 2x^2 + x + x^2 + 2x + 1 \\
&= x^3 + 3x^2 + 3x + 1
\end{aligned}
$$
◆◆◆

Exercise 8 ◆ Raising a Multinomial to a Power

Square each expression.

1. $(x + y)$

2. $(m + n)$

3. $(a - d)$

4. $(z - w)$

5. $(B + D)$

6. $(C - D)$

7. $(4.92y + 3.12z)$

8. $(2.45d - 1.93x)$

9. $(5n + 6x)$

10. $(d^3 + d^2)$

11. $(1 - w)$

12. $(cy^2 - c^3y)$

13. $(b^3 - 13)$

14. $(a^n + b^3)$

15. $(3.88x^2 - 1.33)^2$

16. $(3.84w + 2.14w^2)^2$

Square each trinomial.

17. $(x + y + z)$

18. $(x - y - z)$

19. $(a + b - 1)$

20. $(5a^3 - 3a + 16)$

21. $(c^2 - cd + d^2)$

22. $(w^2 - 5w + 2)$

Challenge Problems

23. $(x - y)^3$

24. $(1.93 - 3.24a)^3$

25. $(3.02m + 2.16n)^3$

26. $(a^2 + 1)^3$

27. $(c + d)^3$

28. $(4p - q)^3$

29. $(3xy^2 + 2x^2y)^3$

30. $(a - b^2)^4$

Applications

31. A square of side x has each side increased by 2 units. Write an expression for the area of the new square, and multiply out.

32. When a current I flows through a resistance R, the power in the resistance is I^2R. If the current is increased by 2.50 amperes, write an expression for the new power, and multiply out.

33. If the radius r of a sphere is decreased by 2 units, write an expression for the new volume of the sphere, and multiply out. (Volume $= \frac{4}{3}\pi r^3$). Use $\pi = 3.142$ and work to three significant digits.

2–9   Dividing a Monomial by a Monomial

Our last basic operation with algebraic expressions is *division*. As always, we will start with the simplest kind, dividing a monomial by a monomial, and then progress to more difficult types.

Symbols for Division

Division may be indicated by any of the following symbols:

$$x \div y \qquad \frac{x}{y} \qquad x/y$$

The names of the parts are

$$\text{quotient} = \frac{\text{dividend}}{\text{divisor}} = \frac{\text{numerator}}{\text{denominator}}$$

The quantity $\dfrac{x}{y}$ is also called a fraction. The horizontal line is the *fraction bar*. It is a symbol of grouping for multiple terms in the numerator or in the denominator.

Reciprocals

As we saw earlier, the *reciprocal* of a number is 1 divided by that number. The reciprocal of n is $1/n$. We can use the idea of a reciprocal to show how division is related to multiplication. We may write the quotient of $x \div y$ as

$$\frac{x}{y} = \frac{x}{1} \cdot \frac{1}{y} = x \cdot \frac{1}{y}$$

We see that to *divide by a number* is the same thing as to *multiply by its reciprocal*. This fact will be especially useful for dividing by a fraction.

Division by Zero

If division by zero were allowed, we could, for example, divide 2 by zero and get a quotient x:

$$\frac{2}{0} = x$$

or

$$2 = 0 \cdot x$$

but there is no number x which, when multiplied by zero, gives 2, so we cannot allow this operation.

Division by zero is not a permissible operation.

◆◆◆ **Example 75:** In the fraction

$$\frac{x + 5}{x - 2}$$

x cannot equal 2, or the illegal operation of division by zero will result.   ◆◆◆

Rules of Signs

The quotient of two terms of *like* sign is *positive*.

$$\frac{+a}{+b} = \frac{-a}{-b} = \frac{a}{b}$$

The quotient of two terms of *unlike* sign is *negative*.

$$\frac{+a}{-b} = \frac{-a}{+b} = -\frac{a}{b}$$

The fraction itself carries a third sign, which, when negative, reverses the sign of the quotient. These three ideas are summarized in the following rules:

Rules of Signs for Division	$$\dfrac{+a}{+b} = \dfrac{-a}{-b} = -\dfrac{-a}{+b} = -\dfrac{+a}{-b} = \dfrac{a}{b}$$	**12**
	$$\dfrac{+a}{-b} = \dfrac{-a}{+b} = -\dfrac{-a}{-b} = -\dfrac{a}{b}$$	**13**

These rules show that any *pair* of negative signs may be removed without changing the value of the fraction.

◆◆◆ **Example 76:** Simplify $-\dfrac{ax^2}{-y}$.

Solution: Removing the pair of negative signs, we obtain

$$-\frac{ax^2}{-y} = \frac{ax^2}{y}$$

◆◆◆

Common Error	Removal of pairs of negative signs pertains only to negative signs that are factors of the *whole* numerator, the *whole* denominator, or the fraction *as a whole*. Do not try to remove pairs of signs that apply only to single terms. Thus $$\frac{x-2}{x-3} \neq \frac{x+2}{x+3}$$

Dividing a Monomial by a Monomial

Any quantity (except 0) divided by itself equals *one*. So if the same factor appears in both the dividend and the divisor, it may be eliminated.

◆◆◆ **Example 77:** Divide $6ax$ by $3a$.

Solution:

$$\frac{6ax}{3a} = \frac{6}{3} \cdot \frac{a}{a} \cdot x = 2x$$

◆◆◆

When there are numerical coefficients that are approxiamte numbers, round the result as usual.

◆◆◆ **Example 78:** This example has approximate coefficients.

$$\frac{8.38bz}{4.22z} = \frac{8.38}{4.22}(b)\left(\frac{z}{z}\right) = 1.99b$$

◆◆◆

To divide quantities having exponents, we use the law of exponents for division.

Quotients	$$\frac{x^a}{x^b} = x^{a-b} \qquad (x \neq 0)$$	**23**

◆◆◆ **Example 79:** Divide y^5 by y^3.

Solution: By Eq. 23,

$$\frac{y^5}{y^3} = y^{5-3} = y^2$$

◆◆◆

◆◆◆ **Example 80:** Here we divide monomials having exponents.

(a) $\dfrac{15x^6}{3x^4} = \dfrac{15}{3} \cdot \dfrac{x^6}{x^4} = 5x^{6-4} = 5x^2$

(b) $\dfrac{8.35y^5}{3.72y} = \dfrac{8.35}{3.72} \cdot \dfrac{y^5}{y} = 2.24y^{5-1} = 2.24y^4$ ◆◆◆

If there is more than one unknown, treat each separately.

◆◆◆ **Example 81:** Divide $18x^5y^2z^4$ by $3x^2yz^3$.

Solution:

$$\frac{18x^5y^2z^4}{3x^2yz^3} = \frac{18}{3} \cdot \frac{x^5}{x^2} \cdot \frac{y^2}{y} \cdot \frac{z^4}{z^3}$$
$$= 6x^3yz$$ ◆◆◆

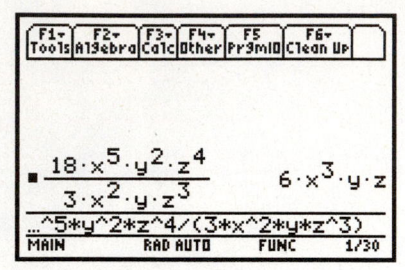

TI-89 screen for Example 81. Here we simply entered the expression and did not need any special instruction.

Sometimes negative exponents will be obtained.

◆◆◆ **Example 82:** This division results in a negative exponent.

$$\frac{6x^2}{x^5} = 6x^{2-5} = 6x^{-3}$$

We can leave the answer in this form, or use Eq. 28, $x^{-a} = 1/x^a$, to eliminate the negative exponent. Thus,

$$6x^{-3} = \frac{6}{x^3}$$ ◆◆◆

The process of dividing a monomial by a monomial is also referred to as *simplifying a fraction*, or *reducing a fraction to lowest terms*.

◆◆◆ **Example 83:** Simplify the fraction

$$\frac{3x^2yz^5}{9xy^4z^2}$$

Solution: The procedure is no different than if we had been asked to divide $3x^2yz^5$ by $9xy^4z^2$.

$$\frac{3x^2yz^5}{9xy^4z^2} = \frac{3}{9}x^{2-1}y^{1-4}z^{5-2}$$
$$= \frac{1}{3}xy^{-3}z^3$$

or

$$= \frac{xz^3}{3y^3}$$ ◆◆◆

Do not be dismayed if the expressions to be divided have negative exponents. Apply Eq. 28, $x^{-a} = 1/x^a$, as before.

◆◆◆ **Example 84:** Divide $21x^2y^{-3}z^{-1}$ by $7x^{-4}y^2z^{-3}$.

Solution: Proceeding as before, we obtain

$$\frac{21x^2y^{-3}z^{-1}}{7x^{-4}y^2z^{-3}} = \frac{21}{7}x^{2-(-4)}y^{-3-2}z^{-1-(-3)}$$
$$= 3x^6y^{-5}z^2$$
$$= \frac{3x^6z^2}{y^5}$$ ◆◆◆

Exercise 9 ◆ Dividing a Monomial by a Monomial

Divide and simplify.

1. x^7 by x^4

2. $21a^4$ by $3a^2$

3. $5xyz$ by xy

4. m^3n by mn

5. $4a^2d$ by $(-2ad)$

6. $-360x^4y^2$ by $(-30x^2y)$

7. $31.4ab^9$ by $2.66ab^8$

8. $54.4x^2z$ by $(-9.11x)$

9. $8.31ad$ by $3.26a$

10. $49.6xyz$ by $(-7.22y)$

11. $42p^5q^4r^2 \div 7p^3qr$

12. $50x^3y^5z^3 \div (-10xy^3z^2)$

13. $-32m^2nx \div 4mx$

14. $48cd^2z^3 \div (-24cd)$

15. $-36a^4b^2c \div 9ab$

16. $45m^2q$ by $(-5mq)$

17. $-24m^3n^3z$ by $4m^3z$

18. $-27a^5p^2$ by $(-9a^3q^2)$

19. $25a^4bcxyz$ by $5a^2bcxz$

20. $18d^3f^2$ by $3d^2f$

21. $32a^2bc \div (-8ab)$

22. $-12m^2n^3 \div 4mn^2$

23. $-36a^2by^2 \div 12a^2y$

24. $44a^2b^3c^4 \div 11a^2bc$

25. $64x^2y^2$ by $8xy$

26. $24pq^2r^3s$ by $8r$

27. $x^2y^6z^2$ by x^2z^2

28. $(a - x)^5$ by $(a - x)^2$

29. a^x by a^y

30. $66c^2dy^3 \div (-22cy)$

31. $-35a^3b^2z \div 7ab^2$

32. $19e^2m^2n^2 \div (-em^2n)$

33. $95abc \div 5a^2b^3c$

34. $45a^2b^2d^2 \div 15abd$

Application

35. The volume enclosed by the proposed spherical radome, Fig. 2–13, is $4/3\pi r^3$ and its surface area is $4\pi r^2$. The radome designer needs the ratio of its volume to area. Find it by dividing its volume by its surface area.

FIGURE 2–13

2–10 Dividing a Polynomial by a Monomial

We continue with division by a monomial, but now we will divide a *polynomial* by a monomial.

To divide a polynomial by a monomial, we simply divide each term of the polynomial by the monomial.

◆◆◆ **Example 85:** Divide $a^2 + a$ by a.

Solution: Dividing gives

$$\frac{a^2 + a}{a} = \frac{a^2}{a} + \frac{a}{a} = a + 1$$

◆◆◆

◆◆◆ **Example 86:** Divide $6x^3 - 3x^2$ by $3x$.

Solution:

$$\frac{6x^3 - 3x^2}{3x} = \frac{6x^3}{3x} - \frac{3x^2}{3x} = 2x^2 - x$$

◆◆◆

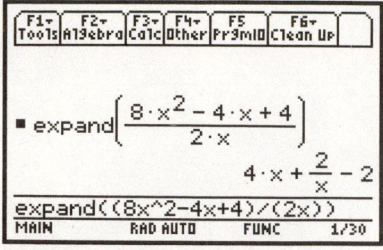

TI-89 screen for Example 87. Here we used **expand** from the ALGEBRA menu.

◆◆◆ **Example 87:** Divide $8x^2 - 4x + 4$ by $2x$.

Solution:

$$\frac{8x^2 - 4x + 4}{2x} = \frac{8x^2}{2x} - \frac{4x}{2x} + \frac{4}{2x}$$

This is really a consequence of the distributive law, Eq. 5.

$$\frac{8x^2 - 4x + 4}{2x} = \frac{1}{2x}(8x^2 - 4x + 4) = \left(\frac{1}{2x}\right)(8x^2) - \left(\frac{1}{2x}\right)(4x) + \left(\frac{1}{2x}\right)(4)$$

Each of these terms is now simplified as in the preceding section.

$$\frac{8x^2}{2x} - \frac{4x}{2x} + \frac{4}{2x} = 4x - 2 + \frac{2}{x}$$

◆◆◆

Keep in mind that the fraction bar is a *symbol of grouping*. It means that the *whole* polynomial is to be divided by the monomial, not just some of its terms.

Common Errors	Do not to forget to divide *every* term of the polynomial by the monomial. $$\frac{4x^2 - 2x + 5}{2x} \neq 2x - 1 + 5$$ Instead, $$\frac{4x^2 - 2x + 5}{2x} = 2x - 1 + \frac{5}{2x}$$
	There is no similar rule for dividing a monomial by a polynomial. $$\frac{a}{b + c} \neq \frac{a}{b} + \frac{a}{c}$$

◆◆◆ **Example 88:** Divide $4xy - 8x^2y + 3xy^2 - 2x^2y^2$ by $16xy$.

Solution:

$$\frac{4xy - 8x^2y + 3xy^2 - 2x^2y^2}{16xy} = \frac{4xy}{16xy} - \frac{8x^2y}{16xy} + \frac{3xy^2}{16xy} - \frac{2x^2y^2}{16xy}$$

$$= \frac{1}{4} - \frac{x}{2} + \frac{3y}{16} - \frac{xy}{8}$$

We may also combine these terms over the common denominator 16, getting

$$\frac{4 - 8x + 3y - 2xy}{16}$$

Alternate Solution: We can get the same result by canceling, noting that every term in both numerator and denominator contains an x and a y.

$$\frac{4xy - 8x^2y + 3xy^2 - 2x^2y^2}{16xy} = \frac{4 - 8x + 3y - 2xy}{16}$$

◆◆◆

When the expression contains approximate coefficients, round your answer properly.

◆◆◆ Example 89: Divide $68.4w^2 - 43.2w^3$ by $2.84w^2$.

Solution:

$$\frac{68.4w^2 - 43.2w^3}{2.84w^2} = 24.1 - 15.2w$$ ◆◆◆

Exercise 10   Dividing a Polynomial by a Monomial

Divide and simplify.

1. $15x^3 + 3x^2$ by x
2. $42m^6 - 2m^3$ by $2m$
3. $36d^5 - 6d^2$ by $3d$
4. $22x^2 + 11y^5$ by 11
5. $48c^4 + 36c^5$ by $12c^2$
6. $27n^3 - 9n^2$ by $(-3n)$
7. $39p^2 + 52p^3$ by $(-13p^2)$
8. $40a^5 - 20a^2$ by $10a$
9. $-25x^3 - 15x^2$ by $5x$
10. $-55.2a^3d + 22.3a^2$ by $(-1.33a^2)$
11. $-8.27bm^3n^2 - 3.22bm^2n$ by $(-1.04m^2n)$
12. $-15.5a^5b + 12.2ab^2$ by $(-4.83ab)$
13. $-21.5a^3b^2 + 31.2a^2b^3$ by $(-14.8ab^2)$
14. $10x^3y - 5xy^4$ by $(-5xy)$
15. $x^2y^3z - xy^4z^2$ by $(-xy^2z)$
16. $16a^3bc^2 + 12a^2b^5c$ by $(-4abc)$
17. $m^2n^2 + m^3n^2 - m^2n^3$ by $(-mn)$
18. $p^5q^2 - p^2q^5 - p^3q^3$ by p^2q^2
19. $x^3y^3 - x^4y + xy^4$ by $(-xy)$
20. $-a^3b^3 - a^2b^2 - ab$ by $(-ab)$
21. $c^3 - 4c^2d^2 + d^3$ by cd^2

Challenge Problems

22. $r^4s^3 - r^2s^2 + r^4s^2$ by $(-r^2s)$
23. $a^4 + 2a^2b^2 - b^4$ by a^2b^2
24. $m^5n^2 + m^2n^2 - m^2n^4$ by $(-m^2n)$
25. $4x^3z + 2xz^2 - 3z^4$ by $(-xz)$
26. $ab^3 + a^3c^2 - b^2c^4$ by $(-abc)$
27. $p^3q^3 + pq^2r^3 - p^2r^4$ by $(-p^2r)$
28. $7m^4n^2 + 7m^3n^3 - 7m^2n^4$ by $(-7m^2n^3)$
29. $4c^4d + 3c^2d^3 - cd^5$ by $(-cd^2)$
30. $3a^2x + 5a^3x^3 - 2ax^2$ by $(-a^2x^2)$
31. $8b^2c^4 + 4b^2c - 12b^3c^3$ by $(-b^3c^2)$

Application

32. The voltage between two points in a certain circuit is

$$6.38R + 8.35R^2 - 3.17R^3$$

Find the current by dividing this voltage by $1.55R$.

2–11   Dividing a Polynomial by a Polynomial

Our final basic operation for this chapter is to divide an expression with two or more terms by another having two or more terms.

To divide one polynomial by another polynomial, follow these steps:

1. Write the divisor and the dividend in the order of descending powers of the variable.
2. Supply any missing terms, using coefficients of zero.
3. Set up the division in long-division form, as in the following example.

Note that this method is used only for polynomials, expressions in which the exponents are all positive integers.

◆◆◆ **Example 90:** Divide $(4x + x^2 + 3)$ by $(x + 1)$.

Solution:

(1) Write the dividend in descending order of the powers.

$$x^2 + 4x + 3$$

(2) There are no missing terms, so we go on to the next step.
(3) Set up in long-division format, making sure that the divisor is written in descending order of the powers as well.

$$(x + 1)\overline{\smash{\big)}\,x^2 + 4x + 3}$$

(4) Divide the first term in the dividend (x^2) by the first term in the divisor (x). The result (x) is written above the dividend, in line with the term having the same power. It is the first term of the quotient.

$$
\begin{array}{r}
x \\
(x + 1)\overline{\smash{\big)}\,x^2 + 4x + 3}
\end{array}
$$

(5) Multiply the divisor by the first term of the quotient. Write the result below the dividend. Subtract it from the dividend.

$$
\begin{array}{r}
x \\
(x + 1)\overline{\smash{\big)}\,x^2 + 4x + 3} \\
\underline{-(x^2 + x)} \\
3x + 3
\end{array}
$$

(6) Repeat steps 4 and 5 until the degree of the remainder is *less* than the degree of the divisor.

$$
\begin{array}{r}
x + 3 \\
(x + 1)\overline{\smash{\big)}\,x^2 + 4x + 3} \\
\underline{-(x^2 + x)} \\
3x + 3 \\
\underline{-(3x + 3)} \\
0
\end{array}
$$

The result is written

$$\frac{x^2 + 4x + 3}{x + 1} = x + 3$$

◆◆◆

Common Error	Remember that the fraction bar is a symbol of grouping. Be sure to divide the dividend *as a whole* by the divisor *as a whole*.

We now try a harder example; one having a remainder.

◆◆◆ **Example 91:** Divide $(2w^2 + 6w^4 - 2)$ by $(w + 1)$.

Solution:

(1) Write the dividend in descending order of the powers.

$$6w^4 + 2w^2 - 2$$

(2) Supply the missing terms with coefficients of zero.

$$6w^4 + 0w^3 + 2w^2 + 0w - 2$$

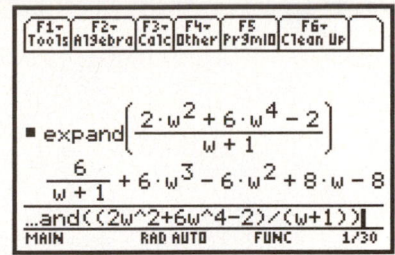

TI-89 screen for Example 91. We use **expand** from the **Algebra** menu.

(3) Set up in long-division format.

$$(w + 1)\overline{\smash{\big)}\,6w^4 + 0w^3 + 2w^2 + 0w - 2}$$

(4) Divide the first term in the dividend ($6w^4$) by the first term in the divisor (w). The result ($6w^3$) is written above the dividend, in line with the term having the same power. It is the first term of the quotient.

$$\begin{array}{r} 6w^3 \\ (w + 1)\overline{\smash{\big)}\,6w^4 + 0w^3 + 2w^2 + 0w - 2} \end{array}$$

(5) Multiply the divisor by the first term of the quotient. Write the result below the dividend. Subtract it from the dividend.

$$\begin{array}{r} 6w^3 \\ (w + 1)\overline{\smash{\big)}\,6w^4 + 0w^3 + 2w^2 + 0w - 2} \\ \underline{-(6w^4 + 6w^3)} \\ -6w^3 + 2w^2 + 0w - 2 \end{array}$$

(6) Repeat steps 4 and 5, each time using the new dividend obtained, until the degree of the remainder is less than the degree of the divisor.

$$\begin{array}{r} 6w^3 - 6w^2 + 8w - 8 \\ (w + 1)\overline{\smash{\big)}\,6w^4 + 0w^3 + 2w^2 + 0w - 2} \\ \underline{-(6w^4 + 6w^3)} \\ -6w^3 + 2w^2 + 0w - 2 \\ \underline{-(-6w^3 - 6w^2)} \\ 8w^2 + 0w - 2 \\ \underline{-(8w^2 + 8w)} \\ -8w - 2 \\ \underline{-(-8w - 8)} \\ 6 \end{array}$$

The result is written

$$\frac{6w^4 + 2w^2 - 2}{w + 1} = 6w^3 - 6w^2 + 8w - 8 + \frac{6}{w + 1}$$

◆◆◆

Common Error

Errors are often made during the subtraction step (step 5 in Example 91).

$$
\begin{array}{r}
6w^3 \\
(w + 1)\overline{)\ 6w^4 + 0w^3\ + 2w^2 + 0w - 2} \\
-(6w^4 + 6w^3) \\
\hline
6w^3 + 2w^2 + 0w - 2 \\
\end{array}
$$

No. Should be $-6w^3$

Exercise 11 ◆ Dividing a Polynomial by a Polynomial

Divide and simplify.

1. $a^2 + 15a + 56$ by $a + 7$

2. $a^4 + 3a^2 + 2$ by $a^2 + 1$

3. $a^2 + a - 56$ by $a - 7$

4. $4x^2 + 23x + 15$ by $4x + 3$

5. $2x^2 + 11x + 5$ by $2x + 1$

6. $6x^2 - 7x - 3$ by $2x - 3$

7. $a^2 - 15a + 56$ by $a - 7$

8. $a^2 - a - 56$ by $a + 7$

9. $3x^2 - 4x - 4$ by $2 - x$

10. $a^8 - 3a^4 + 2$ by $a^4 - 1$

11. $a^3 - 8a - 3$ by $a - 3$

Challenge Problems

12. $27x^3 - 8y^3$ by $3x - 2y$

13. $x^2 - 4x + 3$ by $x + 2$

14. $2 + 4x - x^2$ by $4 - x$

15. $4 + 2x - 5x^2$ by $3 - x$

16. $2x^2 - 5x + 4$ by $x + 1$

17. *Writing:* Suppose your friend was sick and missed the introduction to algebra and is still out of class. Write a note to your friend explaining in your own words what you think algebra is and how it is related to the arithmetic that you both just finished studying.

••• CHAPTER 2 REVIEW PROBLEMS ••••••••••••••••••••••••••••••••••

1. Multiply: $(b^4 + b^2x^3 + x^4)(b^2 - x^2)$
2. Square: $(x + y - 2)$
3. Evaluate: $(7.28 \times 10^4)^3$
4. Square: $(xy + 5)$
5. Multiply: $(3x - m)(x^2 + m^2)(3x - m)$
6. Divide: $-x^6 - 2x^5 - x^4$ by $(-x^4)$
7. Cube: $(2x + 1)$
8. Simplify: $7x - \{-6x - [-5x - (-4x - 3x) - 2]\}$
9. Multiply: $3ax^2$ by $2ax^3$
10. Simplify: $\left(\dfrac{3a^2}{2b^3}\right)^3$
11. Divide: $a^2x - abx - acx$ by ax
12. Divide: $3x^5y^3 - 3x^4y^3 - 3x^2y^4$ by $3x^3y^2$
13. Square: $(4a - 3b)$
14. Multiply: $(x^3 - xy + y^2)(x + y)$
15. Multiply: $(xy - 2)(xy - 4)$
16. Divide: $x^{m+1} + x^{m+2} + x^{m+3} + x^{m+4}$ by x^4
17. Square: $(3x + 2y)$
18. Multiply: $(a^2 - 3a + 8)(a + 3)$
19. Divide: $a^3b^2 - a^2b^5 - a^4b^2$ by a^2b
20. Multiply: $(2x - 5)(x + 2)$
21. Multiply: $(2m - c)(2m + c)(4m^2 + c^2)$
22. Simplify: $\left(\dfrac{8x^5y^{-2}}{4x^3y^{-3}}\right)^3$
23. Divide: $2a^6$ by a^4
24. Multiply: $(a^2 + a^2y + ay^2 + y^3)(a - y)$
25. Simplify: $y - 3[y - 2(4 - y)]$
26. Divide: $-a^7$ by a^5
27. Multiply: $(2x^2 + xy - 2y^2)(3x + 3y)$
28. Divide: $x^4 - \frac{1}{2}x^3 - \frac{1}{3}x^2 - 2x - 1$ by $2x$
29. Simplify: $-2[w - 3(2w - 1)] + 3w$
30. Multiply: $(a^2 + b)(a + b^2)$
31. Multiply: $(a^4 - 2a^3c + 4a^2c^2 - 8ac^3 + 16c^4)(a + 2c)$
32. Divide: $16x^3$ by $4x$
33. Divide: $(a - c)^m$ by $(a - c)^2$
34. Divide: $7 - 8c^2 + 5c^3 + 8c$ by $5c - 3$
35. Divide: $-x^2y - xy^2$ by $(-xy)$
36. Combine: $(-2x^2 - x + 6) - (7x^2 - 2x + 4) + (x^2 - 3)$
37. Cube: $(b - 3)$
38. Simplify: $(2x^2y^3z^{-1})^3$

39. Divide: $2x^{-2}y^3$ by $4x^{-4}y^6$

40. Simplify: $-\{-[-(a - 3) - a]\} + 2a$

41. Multiply: $(x - 2)(x + 4)$

42. Square: $(x^2 + 2)$

43. Divide: $a^2b^2 - 2ab - 3ab^3$ by ab

44. Divide: $3a^3c^3 + 3a^2c - 3ac^2$ by $3ac$

45. Divide: $6a^3x^2 - 15a^4x^2 + 30a^3x^3$ by $(-3a^3x^2)$

46. Divide: $20x^2y^4 - 14xy^3 + 8x^2y^2$ by $2x^2y^2$

47. Evaluate: $(2.83 \times 10^3)^2$

48. Simplify: $(x - 3) - [x - (2x + 3) + 4]$

49. Multiply: $2xy^3$ by $5x^2y$

50. Divide: $27xy^5z^2$ by $3x^2yz^2$

51. Multiply: $(x - 1)(x^2 + 4x)$

52. Square: $(z^2 - 3)$

53. Evaluate: $(1.33 \times 10^4)^2$

54. Simplify: $(y + 1) - [y(y + 1) + (3y - 1) + 5]$

55. Multiply: $2ab^2$ by $3a^2b$

56. Divide: $64ab^4c^3$ by $8a^2bc^3$

57. Divide: $x^8 + x^4 + 1$ by $x^4 - x$

58. Divide: $1 - a^3b^3$ by $1 - ab$

59. To make 750 pounds of a new alloy, we take x pounds of alloy A, which contains 85% copper, and for the remainder use alloy B, which contains 72% copper. The pounds of copper in the final batch will be

$$0.85x + 0.72(750 - x)$$

Simplify this expression.

60. The area of a circle of radius r is πr^2. If the radius is tripled, the area will be

$$\pi(3r)^2$$

Simplify this expression.

61. A freely falling body, starting from rest, falls a distance of $16.1t^2$ feet in t seconds. In half that time it will fall

$$16.1\left(\frac{t}{2}\right)^2 \text{ ft}$$

Simplify this expression.

62. The power in a resistor of resistance R which has a voltage V across it is V^2/R. If the voltage is doubled, the power will be

$$\frac{(2V)^2}{R}$$

Simplify this expression.

63. *Writing:* In the following chapter we will solve simple equations, something you have probably done before. Without peeking ahead, write down in your own words whatever you remember about solving equations. You may give a list of steps or a description in paragraph form.

3

Simple Equations and Word Problems

◆◆◆ **OBJECTIVES** ◆◆

When you have completed this chapter, you should be able to

- Solve simple equations.
- Check an apparent solution to an equation.
- Solve simple fractional equations.
- Solve equations by a graphics calculator and by a calculator that can do symbolic processing.
- Write an algebraic expression to describe a given verbal statement.
- Set up and solve simple word problems.
- Apply the above skills to simple applications in uniform motion, finance, mixtures, and statics.

◆◆

When two mathematical expressions are set equal to each other, we get an *equation*. Much of our work in technical mathematics is devoted to solving equations. We start this chapter with the simplest algebraic types, and in later chapters we cover more difficult ones (quadratic equations, exponential equations, trigonometric equations, etc.).

Why is it so important to solve equations? The main reason is that equations are used to describe or *model* the way certain things happen in the world. Solving an equation often tells us something important that we would not otherwise know. For example, the note produced by a guitar string is not a whim of nature but can be predicted (solved for) when you know the length, the mass, and the tension in the string, and you also know the *equation* relating the pitch, length, mass, and tension.

Thousands of equations exist that link together various quantities in the physical world—in chemistry, in finance, in manufacturing, and so on—and their number is still increasing. To be able to solve and to manipulate such equations is essential for anyone who has to deal with these quantities on the job.

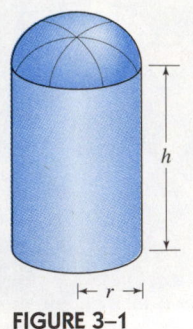

FIGURE 3–1

We *substituted into* equations and formulas in Chapter 1, but here we *solve* equations. For example, the surface area of the silo, Fig. 3–1, is given by

$$A = \frac{1}{2}\pi r^2 + 2\pi rh$$

Given r and h, you already know how to substitute into that equation to find the surface area. But given the surface area and the radius, could you then solve for h? We will show how in this chapter.

3–1   Solving a Simple Equation

Let's start this chapter by learning how to solve a simple equation. Then we will be able to apply those skills to the verbal problems that come a bit later.

Equations

An equation has two *sides* and an *equal sign*.

$$3x^2 - 4x = 2x + 5$$

left side   equal sign   right side

A *conditional equation* is one whose sides are equal only for certain values of the variable.

◆◆◆ **Example 1:** The equation

$$x - 5 = 0$$

is a conditional equation because the sides are equal only when $x = 5$. When we say "equation," we will mean "conditional equation." An equation that is true for *any value* of the variable, such as $x(x + 2) = x^2 + 2x$, is called an *identity*. The symbol $\equiv$ is often used for identities. We would write $x(x + 2) \equiv x^2 + 2x$. ◆◆◆

First-Degree Equations

In this chapter we will limit ourselves to solving *first-degree* equations. Recall that a first-degree term is one in which the variable is raised to the power 1 (which is not written), and that a first-degree equation is one in which no term is higher than first degree. A first-degree equation is also called a *linear* equation. We also limit ourselves here to equations having just one variable.

◆◆◆ **Example 2:** The equation

$$3(x + 4) = 2(x - 5)$$

is a first-degree equation in one variable, the type we will cover in this chapter. ◆◆◆

The Solution of an Equation

The value of the variable that makes the sides of an equation equal to each other is called a *solution* to that equation.

◆◆◆ **Example 3:** The solution of the equation

$$x + 1 = 4$$

is $x = 3$.   ◆◆◆

The solution of an equation is the value of x that makes the left side of the equation equal to the right side. We say that it *satisfies* the equation. The solution of an equation is also called a *root* of the equation.

The variable whose value we seek when we solve an equation is usually called the *unknown quantity*, or simply *the unknown*.

Checking

We should get into the habit of checking our work, for errors creep in everywhere. We can check the proposed solution to an equation by substituting it back into the original equation.

◆◆◆ **Example 4:** Is 12 a solution of the equation $3(x - 4) = 2x$?

Solution: Substituting 12 for x in the equation, we get

$$3(12 - 4) \stackrel{?}{=} 2(12)$$
$$3(8) \stackrel{?}{=} 24$$
$$24 = 24 \quad \text{checks}$$

The value $x = 12$ checks, so 12 is a solution of the given equation. ◆◆◆

Common Error	Check your solution only in the *original* equation. Later versions may already contain errors.

The equal sign with a question mark is not a standard mathematical symbol. However, we will find it useful when we *question* whether one side actually equals the other.

Solving an Equation

To solve an equation, we must get the variable standing alone on one side of the equal sign, with no variable on the other side.

■ Exploration:

Try this. What would you do to get x to stand alone on one side of the equal sign in each of the following:

(a) $x - 4 = 3$ (b) $x + 4 = 3$

(c) $\dfrac{x}{4} = 3$ (d) $4x = 3$

Based on your findings, can you give a general procedure that applies to all four of these examples? ■

You may have concluded that to isolate the variable *we perform the same mathematical operation on both sides of the equation.* That is, we may add the same quantity to both sides, subtract the same quantity from both sides, multiply both sides by the same quantity, and so forth. This will be made clear by examples.

◆◆◆ **Example 5:** Solve the equation $5x = 4x + 7$.

Solution: Subtracting $4x$ from both sides, we obtain

$$5x - 4x = 4x + 7 - 4x$$

Combining like terms yields

$$x = 7$$

Check:
Substituting 7 for x in the original equation gives

$$5(7) \stackrel{?}{=} 4(7) + 7$$
$$35 \stackrel{?}{=} 28 + 7$$
$$35 = 35 \quad \text{checks} \qquad ◆◆◆$$

◆◆◆ **Example 6:** Solve the equation $2x - 3 = 7 - 3x$.

Solution: Adding 3 to both sides,

$$2x - 3 + 3 = 7 - 3x + 3$$
$$2x = 10 - 3x$$

Adding $3x$ to both sides,

$$2x + 3x = 10 - 3x + 3x$$
$$5x = 10$$

Finally we divide both sides by 5.

$$\frac{5x}{5} = \frac{10}{5}$$
$$x = 2$$

With practice you will be able to combine some of the above steps, but don't rush it.

Check:

Substituting 2 for x in the original equation yields

$$2(2) - 3 \stackrel{?}{=} 7 - 3(2)$$
$$4 - 3 \stackrel{?}{=} 7 - 6$$
$$1 = 1 \quad \text{checks} \qquad ◆◆◆$$

Equations Having Symbols of Grouping

We showed earlier that parentheses, brackets, and braces are often used to group quantities within an expression. When the equation contains such symbols of grouping, *remove them early in the solution.*

◆◆◆ **Example 7:** Solve the equation $2 - 4(x + 2) = 5 - 3(2x + 1)$.

Solution: Removing the parentheses, we obtain

$$2 - 4x - 8 = 5 - 6x - 3$$

Combining like terms,

$$-4x - 6 = -6x + 2$$

Adding $6x + 6$ to both sides,

$$\begin{array}{rcr} 6x + 6 = & 6x + 6 \\ \hline 2x \quad = & 8 \end{array}$$

Dividing by 2 gives

$$x = 4$$

Check:

$$2 - 4(4 + 2) \stackrel{?}{=} 5 - 3(8 + 1)$$
$$2 - 24 \stackrel{?}{=} 5 - 27$$
$$-22 = -22 \quad \text{checks} \qquad ◆◆◆$$

Simple Fractional Equations

An equation that contains one or more fractions is called a *fractional equation.*

◆◆◆ **Example 8:** Some fractional equations are

$$\frac{x}{5} = 3x \quad \text{and} \quad \frac{2x}{5} = \frac{2}{7} \qquad ◆◆◆$$

Here we will solve some very simple fractional equations, and in a later chapter we will learn how to solve more complex types.

If an equation contains a single fraction, that fraction can be eliminated by multiplying both sides by the denominator of the fraction.

◆◆◆ **Example 9:** Solve

$$\frac{x}{2} - 4 = 7$$

Solution: Multiplying both sides by 2,

$$2\left(\frac{x}{2} - 4\right) = 2(7)$$

$$2\left(\frac{x}{2}\right) - 2(4) = 2(7)$$

$$x - 8 = 14$$

Adding 8 to both sides,

$$x = 8 + 14 = 22$$

Check:

$$\frac{22}{2} - 4 \stackrel{?}{=} 7$$

$$11 - 4 = 7$$

$$7 = 7 \qquad \text{checks} \qquad \text{◆◆◆}$$

When there are two or more fractions, multiplying both sides by the product of the denominators (called a *common denominator*) will clear the fractions.

◆◆◆ **Example 10:** Solve:

$$\frac{x}{2} + 3 = \frac{x}{3}$$

Solution: Multiplying by 2(3), or 6, we have

$$6\left(\frac{x}{2} + 3\right) = 6\left(\frac{x}{3}\right)$$

$$6\left(\frac{x}{2}\right) + 6(3) = 6\left(\frac{x}{3}\right)$$

$$3x + 18 = 2x$$

Now that we have cleared denominators, we subtract $2x$ from both sides, and also subtract 18 from both sides.

$$3x + 18 - 2x - 18 = 2x - 2x - 18$$

$$3x - 2x = -18$$

$$x = -18$$

Check:

$$\frac{-18}{2} + 3 \stackrel{?}{=} \frac{-18}{3}$$

$$-9 + 3 \stackrel{?}{=} -6$$

$$-6 = -6 \qquad \text{checks} \qquad \text{◆◆◆}$$

Equations with Approximate Numbers

In technical work we usually solve equations that have approximate numbers. The method is no different than before. Just remember to round your answer to the proper number of significant digits. And, of course, the letter representing the variable is not always x, but can be any letter from any alphabet, Greek, Hebrew, and so forth.

◆◆◆ Example 11: Solve for t: $2.93(t + 4.28) = 24.2 - 5.82t$.

Solution: Removing parentheses gives

$$2.93t + 12.54 = 24.2 - 5.82t$$

Combining like terms (collecting terms),

$$2.93t + 5.82t = 24.2 - 12.54$$

$$8.75t = 11.66$$

Dividing,

$$x = \frac{11.66}{8.75} = 1.33$$

rounded to the three significant digits found in the numbers in the given equation. ◆◆◆

Strategy

Remember that our objective when solving an equation is to get the variable by itself on one side of the equation. To do this we can use any valid mathematical operation, as long at it is applied to both sides of the equation. It is not possible to give a procedure that will work for every equation, but the following tips should help. You may have to do these operations in a different order than that shown.

> **Tips**
>
> 1. Eliminate any fractions by multiplying both sides by a common denominator.
> 2. Remove any parentheses by performing the indicated multiplication.
> 3. Get all terms containing x on one side of the equal sign, and all other terms on the other side by adding like quantities to both sides.
> 4. Combine like terms on the same side of the equation at any stage of the solution.
> 5. Remove any coefficient of x by dividing both sides by that coefficient.
> 6. Check the answer by substituting it back into the original equation.

> **Common Error** Students often forget that the mathematical operations you perform must be done to *both sides* of the equation in order to preserve the equality.

◆◆◆ Example 12: Solve

$$\frac{x - 2}{3} = \frac{2x - 4}{2}$$

Solution: We eliminate the fractions by multiplying both sides by 6,

$$6\left(\frac{x - 2}{3}\right) = 6\left(\frac{2x - 4}{2}\right)$$

$$2(x - 2) = 3(2x - 4)$$

Removing parentheses,

$$2x - 4 = 6x - 12$$

We get all x terms on one side by adding 12 and subtracting $2x$ from both sides,

$$2x - 4 + 12 - 2x = 6x - 12 + 12 - 2x$$

Let us now move all the x terms to one side of the equation, and the constants to the other side.

$$2x - 6x = 4 - 12$$

We now combine like terms, getting

$$-4x = -8$$

Dividing both sides by the coefficient of x gives

$$x = 2$$

Check:

$$\frac{2 - 2}{3} \stackrel{?}{=} \frac{2(2) - 4}{2}$$

$$0 = 0 \qquad \text{checks} \qquad \blacklozenge\blacklozenge\blacklozenge$$

Simple Literal Equations

A *literal equation* is one in which some or all of the constants are represented by letters.

♦♦♦ **Example 13:** The following is a literal equation:

$$3x + b = 5 \qquad\qquad \blacklozenge\blacklozenge\blacklozenge$$

where b is a constant. In this chapter we will solve only the simplest kinds of literal equations. Being able to solve a literal equation is important because most of the formulas in technology are literal equations. We often want to solve such a formula for a different quantity. We will do this for a great number of formulas later.

To solve a literal equation for a given quantity means to isolate that quantity on one side of the equal sign. The other side of the equation will, of course, contain the other letter quantities. We do this by following the same procedures we used earlier.

♦♦♦ **Example 14:** Solve the equation $3x + b = 5$ for x.

Solution: Subtracting b from both sides gives

$$3x = 5 - b$$

Dividing both sides by 3, we obtain.

$$x = \frac{5 - b}{3} \qquad\qquad \blacklozenge\blacklozenge\blacklozenge$$

Solving an Equation Using a Calculator's Equation Solver

We can use any calculator's built-in equation solver to solve the equations in this chapter and others.

♦♦♦ **Example 15:** Solve the equation of Example 11,

$$2.93(t + 42.8) = 24.2 - 5.82t$$

using the TI-83/84 equation solver.

Solution:

(a) Make sure the calculator is in **FUNC** (function) mode. We select **Solver** from the $\boxed{\text{MATH}}$ menu and enter the equation, screen (a). It must be in explicit form, with 0 on the left side and all other terms on the right. For the unknown, press the $\boxed{\text{X,T,O,}n}$ key. The equation solver uses X as the unknown, so we use it instead of T. (T is used when in parametric equation mode, which we will cover later.)

(b) Press ENTER . A new screen appears showing a value for X, and a **bound**. Here you can enter a value which you think may be close to the actual value of the root, and a range of values between which the calculator will search for that root. You can accept the default values, or enter new ones. Let us enter a new value of X = 1 and a new bound of 0 to 5, screen (b).

(c) Move the cursor to the line containing "$x =$ " and press SOLVE . (This is ALPHA ENTER on the TI-83/84.)

A root ($x = 1.33$, rounded) falling within the selected bound is shown in screen (c). If there is no root within that bound, or you suspect there is another root, change the initial value of x and the bound.   ◆◆◆

(a) The given equation is entered in the TI-83/84 equation solver.

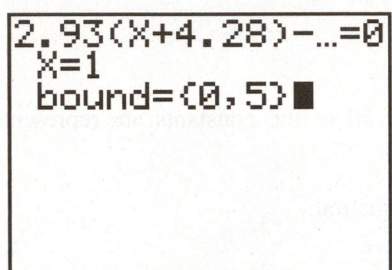

(b) An initial estimate for x and the lower and upper bounds are chosen here.

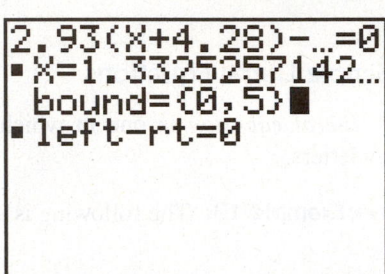

(c) The computed value of x is shown, within the chosen bound. The last line says that the difference between the left and right sides of the equation is zero, showing that the equation is indeed balanced.

Implicit or Literal Equations

To solve an equation by the TI-83/84 equation solver, the equation must be in explicit form. If it is not, simply move all terms to one side of the equal sign, changing signs as appropriate, leaving zero on one side of the equation. However, some calculators that can do symbolic manipulation can solve an equation that is in implicit form. It can also solve literal equations.

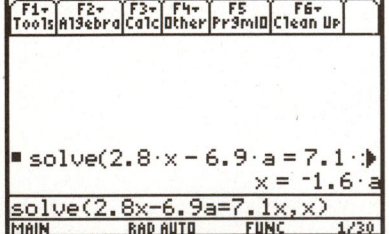

TI-89 screen for Example 16.

◆◆◆ **Example 16:** Solve using the TI-89 equation solver:

$$2.8x - 6.9a = 7.1x$$

Solution:

(a) We select **solve** from the ALGEBRA menu.

(b) Enter the equation, followed by the variable (x) that we want to solve for. Unlike the TI-83/84, the equation *does not* have to be in explicit form.

(b) Pressing ENTER gives the exact solution, while pressing ≈ (located above the ENTER key) gives the approximate decimal solution, as shown. Note that it is not necessary to enter a first guess or bounding values.   ◆◆◆

Exercise 1 ◆ Solving a First-Degree Equation

Solve and check each equation. Treat the constants in these equations as exact numbers. Leave your answers in fractional, rather than decimal, form.

1. $x + 9 = 16$ **2.** $3x - 2 = 10$

3. $30 + 5x = 20x$ **4.** $7x - 29 = 6$

5. $4t + 9 = 11t - 3t$ **6.** $20 - y = 13$

7. $x + 9 = 5$

8. $-6y - 4 = 2y$

9. $5x + 8 = 9x$

10. $4x = 3x - 6$

11. $10 - 5y = 1 - y$

12. $w + 3 = 10$

13. $x - 5 = 6$

14. $8y - 6 = 5y$

15. $7x - 3 = 5x + 1$

16. $25x - 5 = 3x + 6$

17. $y - 4 = 0$

18. $21x = -8 + 5x$

19. $4m - 5 = 10m - 2$

20. $7x + 15 = 8$

21. $6x + 4 = 3x + 19$

22. $16y - 10 = 10y + 14$

23. $17 - 14x = 8 - 11x$

24. $3y - 15 + 4y = 6$

25. $5x - 10 + 13 = 18$

26. $44 - 11p + 4p = -5$

27. $49 - 5y = 3y - 7$

28. $5 + y = 1 + 15y$

29. $4x - 111 = 21 + 14x$

30. $47x - 84 = 2x + 6$

31. $16 + 2z = 11 - 5z$

32. $15x - 9 = 7x - 5$

Equations Having Symbols of Grouping

33. $3(y - 5) + 2(3 + y) = 21$

34. $5(3x - 13) = 10$

35. $4(y - 5) = 2(y - 10)$

36. $2x + 1 = -2(2x + 5)$

37. $3(15 - t) = 3(5 + t)$

38. $4(3x + 20) = 2(x - 5)$

39. $3(2x + 1) = 7$

40. $3(x - 5) = 4(x - 1)$

41. $5x + 6 = 6(x - 3) + 2$

42. $6(2x - 3) + 2(6x + 2) = 2(x + 1)$

43. $19y - 60 = 4(y + 15)$

44. $5(2x - 8) = 2(10 - 5x)$

45. $5 + 3(x - 2) = 58$

46. $3(3m + 13) = 5(m - 1)$

47. $4(x - 5) - 2(6x + 3) = 22$

48. $5(y - 1) + 4(3y + 3) = 3(4y - 6)$

49. $2(4w - 3) = 3(5w + 2)$

50. $3(4x - 2) = 9x$

51. $5a = -2(13 - 9a)$

52. $5 - 3(x - 2) = 4(3x - 1)$

53. $6x + 5 = 7(x - 3)$

54. $3(7 - 2x) + 2 = 3(4x - 1)$

Simple Fractional Equations

55. $\dfrac{1}{3a} = 7$

56. $\dfrac{y}{5} = 4$

57. $\dfrac{x}{4} - 3 = 11$

58. $5 + \dfrac{7x}{3} = \dfrac{x}{2}$

59. $\dfrac{5x}{4} = 2x - 3$

60. $\dfrac{y}{5} - \dfrac{y}{2} + \dfrac{y}{4} = 5$

61. $\dfrac{2x - 4}{7} = \dfrac{2 - 2x}{4}$

Equations with Approximate Numbers

Solve for x. Round your answer to the proper number of significant digits.

62. $8.27x = 4.82$

63. $24.8x - 28.4 = 0$

64. $2.84x + 2.83 = 83.7$

65. $3.82 = 29.3 + 3.28x$

66. $3.82x - 3.28 = 5.29x + 5.82$

67. $382x + 827 = 625 - 846x$

68. $2.94(x + 8.27) = 3.27$

69. $9.38(5.82 + x) = 23.8$

70. $2.34(x - 4.27) = 5.27(x + 3.82)$

71. $92.1(x - 2.34) = 82.7(x - 2.83)$

Simple Literal Equations

Solve for x.

72. $ax + 4 = 7$

73. $6 + bx = b - 3$

74. $c(x - 1) = 5$

75. $ax + 4 = b - 3$

76. $c - bx = a - 3b$

Challenge Problems

77. $6 - 3(2x + 4) - 2x = 7x + 4(5 - 2x) - 8$

78. $3(6 - x) + 2(x - 3) = 5 + 2(3x + 1) - x$

79. $3x - 2 + x(3 - x) = (x - 3)(x + 2) - x(2x + 1)$

80. $(2x - 5)(x + 3) - 3x = 8 - x + (3 - x)(5 - 2x)$

81. $x + (1 + x)(3x + 4) = (2x + 3)(2x - 1) - x(x - 2)$

82. $3 - 6(x - 1) = 9 - 2(1 + 3x) + 2x$

83. $7x + 6(2 - x) + 3 = 4 + 3(6 + x)$

84. $3 - (4 + 3x)(2x + 1) = 6x - (3x - 2)(x + 1) - (6 - 3x)(1 - x)$

85. $7r - 2r(2r - 3) - 2 = 2r^2 - (r - 2)(3 + 6r) - 8$

86. $2w - 3(w - 6)(2w - 2) + 6 = 3w - (2w - 1)(3w + 2) + 6$

Applications

Where did these equations come from, thin air? They are actually from the applications given later in this chapter. There we will learn how to *write* equations such as these to describe technical applications.

87. In order to find the tons x of steel containing 5.25% nickel to be combined with another steel containing 2.84% nickel to make 3.25 tons of steel containing 4.15% nickel, we must solve the equation

$$0.0525x + 0.0284(3.25 - x) = 0.0415(3.25)$$

Solve this equation,

88. To find the time t it takes for a car traveling at 108.0 km/h to overtake a truck traveling at 72.5 km/h with a 1.25 h head start, we must solve the equation

$$108.0t = 72.5(t + 1.25)$$

Solve for t.

89. A hydroelectric generating station, Fig. 3–2, producing 17 MWh (megawatthours) of energy per year adds, after 4.0 months, another generator which, by itself, can produce 11 MWh in 5 months. It takes an additional x months for a total of 25 MWh to be produced, where x can be found from the equation,

$$\frac{17}{12}(4.0 + x) + \frac{11}{5.0}x = 25$$

Solve this equation for x.

90. A consultant had to pay an income tax of $12,386, which was $4867 plus 15% of the amount by which her taxable income x exceeded $32,450. Find x by solving the equation

$$4867 + 0.15(x - 32,450) = 12,386$$

FIGURE 3–2

3–2   Solving Word Problems

In this section we will be asked to

- read a short statement, the so-called *word problem*
- then decide exactly what is being asked for
- then write an equation that describes how the unknown quantity is related to the given quantities
- and finally, solve that equation to obtain the quantity that was asked for.

Keep in mind that we are not solving these problems for their own sake. We don't really care how long it takes runner *A* to reach city *Q*. What we are trying to do is to help you to read a technical statement, to extract the important information, and to use simple mathematics to find a missing quantity. These skills will also help you to deal with the mountains of technical material in written form with which we are all faced: instruction manuals, textbooks, specifications, contracts, insurance policies, building codes, handbooks, and so forth.

It is not possible to give a step-by-step procedure that will enable someone to solve *any* word problem, but we can give some good tips that are almost always helpful. We first give some general tips on how to approach any word problem and later show how to set up specific types.

Math skills are important, but so are reading skills. Most students find word problems difficult, but if you have an *unusual* amount of trouble with them you should seek out a reading teacher and get help.

Study the Problem

Read the problem carefully, more than once if necessary. Be sure you understand the words used. A technical problem will often have unfamiliar terms. Look up their meanings in a dictionary, handbook, or textbook.

◆◆◆ **Example 17:** Word problems may contain words or phrases like

simply supported beam	*angular velocity*	*groundspeed*
concentrated load	*center of gravity*	*kinetic energy*

It goes without saying that you must know the meanings of these words before you can solve a problem in which they appear.   ◆◆◆

Picture the Problem

Try to *visualize* the situation described in the problem. Form a picture in your mind. *Draw a diagram* showing as much of the given information as possible.

◆◆◆ **Example 18:** Suppose that part of a word problem states that "a car goes from city *P* to city *Q*, which is 236 miles away, at an average rate of 55 miles per hour."

At this point you should see not just dead words on paper, but a car moving along a road (perhaps a red sports car cruising down the interstate, stereo playing, you at the wheel). Make a sketch something like Fig. 3–3 showing the road as a line with cities *P* and *Q* at either end, 236 miles apart. Show the car with an arrow giving the direction of travel and the speed.

Of course, we have not solved the problem yet (it isn't even fully stated), but at least we have a good grip on the information given so far.

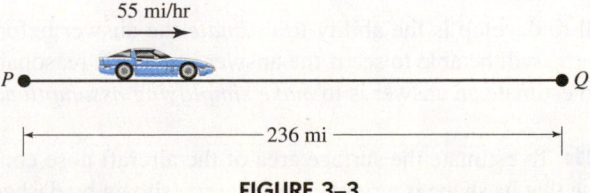

FIGURE 3–3   ◆◆◆

Identify the Unknown(s)

Look for a sentence that asks a question (it may end in a question mark). Find a sentence that starts with *Find . . .* or *Calculate . . .* or *What is . . .* or *How much . . .* or similar phrases. Such a sentence will usually contain the unknown quantity, the thing we want to solve for. Then label this unknown with a statement such as

$$\text{Let } x = \text{cost of each part, dollars}$$

Be sure to include units of measure when defining the unknown.

◆◆◆ **Example 19:** Suppose one sentence in a word problem says, "Find the speed of train *B*." You may be tempted to label the unknown in one of the following ways:

Let x = train B	This is too vague. Are we talking about the speed of train B or the time it takes for train B or the distance traveled by train B or something else?
Let x = speed of the train	Which train?
Let x = speed of train B	Units are missing.
Let x = speed of train B, mi/h	Good

◆◆◆

Define Other Unknowns

If there is more than one unknown, you will often be able to define the additional unknowns *in terms of the original unknown*, rather than introducing another symbol.

◆◆◆ **Example 20:** A word problem contains the following statement: "Find two numbers whose sum is 80 and, ..."

Solution: Here we have two unknowns. We can label them with separate symbols.

$$\text{Let } x = \text{first number}$$
$$\text{Let } y = \text{second number}$$

But a better way is to label the second unknown in terms of the first, thus avoiding the use of a second variable. Since the sum of the two numbers x and y is 80,

$$x + y = 80$$

from which

$$y = 80 - x$$

This enables us to label both unknowns in terms of the single variable x.

$$\text{Let } x = \text{first number}$$
$$\text{Let } 80 - x = \text{second number}$$

◆◆◆

Estimate the Answer

A valuable skill to develop is the ability to *estimate* the answer before starting the problem. Then you will be able to see if the answer you get is reasonable or not.

One way to estimate an answer is to *make simplifying assumptions.*

◆◆◆ **Example 21:** To estimate the surface area of the aircraft nose cone in Fig. 3–4, we could assume that its shape is a right circular cone (shown by dashed lines) rather

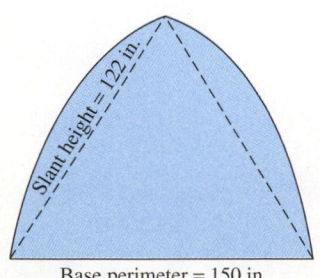

Base perimeter = 150 in.

FIGURE 3–4 Aircraft nose cone.

than the complex shape given. This would enable us to estimate the area using the following simple formula (Eq. 96):

$$\text{Lateral area} = \tfrac{1}{2}(\text{perimeter of base}) \times (\text{slant height})$$

For the nose cone,

$$\text{area} \approx \tfrac{1}{2}(150)(122) = 9150 \text{ in.}^2$$

We would expect our final answer to be a bit greater than 9150 in.^{2} and should be suspicious if our "exact" calculation gives an answer much different from this. ◆◆◆

◆◆◆ **Example 22:** Suppose we want to find the volume of the spindle in Fig. 3–5. To get an estimate, we can find the volume of a cylinder of the same length as the spindle, but that has a diameter somewhere between the largest and smallest diameters of the spindle, say, 17 mm. This will not give us the exact answer, of course, but will put us in the ballpark. If our final answer is very different than our estimate, we would be suspicious.   ◆◆◆

Another technique is to *bracket* the answer. To "bracket" an answer means to find two numbers between which your answer must lie. Don't worry that your two values are far apart. This method will catch more errors than you may suppose.

◆◆◆ **Example 23:** We can *bracket* the volume of the spindle by noting that it must be less than that of a cylinder 19.00 mm in diameter but greater than that of a cylinder 15.00 mm in diameter.   ◆◆◆

◆◆◆ **Example 24:** It is clear that the length of the bridge cable of Fig. 3–6 must be greater than the straight line distances $AB + BC$, but less than the straight line distances $AD + DE + EC$. So the answer must lie between 894 ft and 1200 ft. ◆◆◆

Estimation is not always easy. It takes some time to develop this skill, but it is well worth it. We will show estimation in most of the word problems in this chapter.

Write and Solve an Equation

The next step is to write an equation to relate the given quantities to the unknown quantity. Sometimes the equation will be a formula from mathematics, such as the relationship between the volume of a sphere and its radius. At other times you will need a formula from technology, such as the one relating the strength of a steel rod to its diameter. The relationships you will need for the problems in this book can be found in Appendix A, "Summary of Facts and Formulas." Problems that use some of these formulas are treated later in this chapter. But first, we will solve simple number puzzles, in which the formula is given verbally right in the problem statement.

When you have an answer, label it with the same words you used in defining the unknown.

Check Your Answer

First see if your answer is reasonable. Did you come up with a bridge cable 1 mm in diameter or a person walking at a rate of 54 mi/h? Does your answer agree, within reason, with your estimate? Then you should check your answer in the original problem *statement*. Do *not* check the answer in your equation, which may already contain an error.

Don't Give Up

Be persistent. If you don't "get it" at first, keep trying. Take a break and return to the problem later. Read the problem again. Perhaps try a different approach. You know that the "school problems" that you encounter in a mathematics class all have solutions, so keep at it.

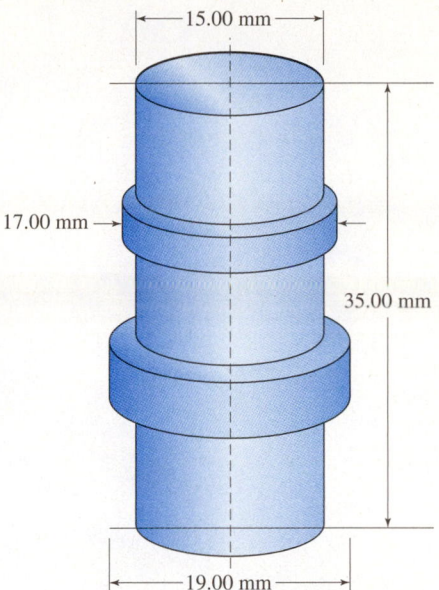

FIGURE 3–5 A spindle.

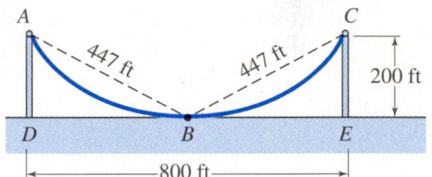

FIGURE 3–6 Suspension bridge.

Here is a summary of the usual steps followed in solving a word problem.

> 1. **Study the Problem.** Look up unfamiliar words. Make a sketch. Try to visualize the situation in your mind.
> 2. **Identify the Unknown(s).** Give it a symbol, such as "Let $x = \ldots$" If there is more than one unknown, try to label the others in terms of the first. Include units.
> 3. **Estimate the Answer.** Make simplifying assumptions or try to bracket the answer.
> 4. **Write and Solve an Equation.** Look for a relationship between the unknown and the known quantities that will lead to an equation. Write and solve that equation for the unknown. Include units in your answer. If there is a second unknown be sure to find it also.
> 5. **Check Your Answer.** See if the answer looks reasonable. See if it agrees with your estimate. Be sure to do a numerical check in the problem statement itself.

Number Puzzles

The number puzzles that follow may not have practical value in themselves, but they have great value as our first word problems. They will give us practice in setting up and solving a word problem and prepare us for the applications that will come later. They can also be worked just for fun.

In word problems, mathematical operations are often indicated by words such as the following:

Addition $(+)$	more, plus, sum, total, add, increased by
Subtraction $(-)$	less, difference of, diminished by, decreased by
Multiplication $(\times)$ or $(\cdot)$	product, times, increased by a factor of
Division $(\div)$	quotient, ratio, divided by, per
Equality $(=)$	is, equals, is equal to, gives, results in

This is not a complete list. The English language is so flexible that the same idea can be given in a great variety of forms. What we must do is to locate such words and replace them with mathematical symbols.

◆◆◆ **Example 25:** If two times a number is increased by seven, the result is equal to five less than four times the number. Find the number.

Solution: Let $x =$ the number. Then,

$$2x = \text{two times the number}$$

From the problem statement,

$$2x + 7 = 4x - 5$$

Solving for x,

$$7 + 5 = 4x - 2x$$
$$2x = 12$$
$$x = 6$$
◆◆◆

◆◆◆ **Example 26:** Let us check the answer from Example 25. Using the words of the problem statement, we get,

Twice six increased by seven is $2(6) + 7 = 12 + 7 = 19$

Five less than four times six is $4(6) - 5 = 24 - 5 = 19$ checks   ◆◆◆

Common Error	Checking an answer by substituting into the equation is *not good enough*. The equation may already contain an error. Check your answer in the problem statement.

Define Other Unknowns

If there is a second unknown, you will often be able to define it in terms of the original unknown, rather than to introduce a new symbol.

◆◆◆ **Example 27:** Find two numbers whose sum is 50 and whose difference is 30.

Solution: This problem has two unknowns. We could choose to label each with a separate symbol,

$$\text{Let } x = \text{larger number}$$

and

$$y = \text{smaller number}$$

but a better way would be to label the second unknown in terms of the first:

$$\text{Let } x = \text{larger number}$$

Then, since the sum of the two numbers is 50, we get

$$50 - x = \text{smaller number}$$

We get the difference of the two numbers by subtracting the smaller number from the larger.

$$x - (50 - x)$$

We are told that this equals 30, so our equation is,

$$x - (50 - x) = 30$$

Removing parentheses gives

$$x - 50 + x = 30$$

Collecting like terms we get,

$$2x - 50 = 30$$

Adding 50 to both sides gives

$$2x = 80$$
$$x = 40 = \text{the larger number}$$
$$50 - x = 10 = \text{the smaller number} \qquad ◆◆◆$$

Exercise 2 ● Solving Word Problems

Identify an unknown and rewrite each expression as an algebraic expression.

1. Ten more than three times a number.
2. Two numbers whose sum is 17.
3. Two numbers whose difference is 42.
4. The amounts of antifreeze and water in 4 gallons of an antifreeze-water solution.
5. A fraction whose denominator is 4 more than 6 times its numerator.
6. The angles in a triangle, if one angle is three times the other.
7. The number of gallons of antifreeze in a radiator containing x gallons of a mixture that is 11% antifreeze.

8. The distance traveled in x hours by a car going 78 km/h.
9. If x equals the length of a rectangle, write an expression for the width of the rectangle if (a) the perimeter is 64; or (b) the area is 320.

Number Puzzles

These "impractical" number puzzles provide practice in reducing a verbal statement to a mathematical equation, without the complications of the technical settings that will come later.

Solve each problem for the required quantity.
10. Four less than 6 times a number is 32. Find the number.
11. Find a number such that the sum of 16 and that number is 3 times that number.
12. Six less than 5 times a number is 19. Find the number.
13. The sum of 45 and some number equals 6 times that number. Find it.
14. Ten less than three times a certain number is 29. Find the number.
15. When three is added to seven times a number, the result is equal to the same as when we subtract seven from nine times that number. Find that number.

3–3   Uniform Motion Applications

Number puzzles were a good warm-up, but now we will do some problems having more application to technology.

Everything you need to solve these problems is given right here. However, with these and others throughout the text, you may search in vain for an example that exactly matches. That is what makes them difficult but also what makes them valuable. Instead of just plugging new numbers into an already worked example, you must dig a bit deeper and apply the basic ideas from the example to the new situation.

Do not be reluctant to try problems outside your chosen field. Some familiarity with other branches of technology will make you more valuable on the job.

We will start with uniform motion problems. The ideas here should be familiar; you already know that if you walk at a rate of 3 miles per hour for 2 hours you will travel 6 miles. The problems in this section are based on that one simple idea.

Motion is called *uniform* when the speed does not change. The distance traveled at constant speed is related to the speed (the *rate* of travel) and the elapsed time by

$$\text{rate} \times \text{time} = \text{distance}$$

or

$$RT = D$$

Be careful not to use this formula for anything but *uniform* motion.

Also, when using this formula or any other, we must be careful with *units of measure*. The units must be consistent. Do not mix feet and miles, for example, or minutes and hours. You may have to convert units, as we did in Chap. 1, so that they cancel properly, and further, leave your answer with the desired units.

◆◆◆ **Example 28:** A plane flies for 2.45 h at 325 km/h. How far does it travel?

Solution: From the equation above,

$$D = 325(2.45) = 796 \text{ km} \qquad \text{◆◆◆}$$

We now do a typical motion problem, in which we will organize our information in table form, to help in the solution, and use the equation (rate × time = distance).

◆◆◆ **Example 29:** A truck leaves a city traveling at a speed of 72.5 km/h, and a car leaves the same city 1.25 h later to overtake the truck. If the car's speed is 108.0 km/h, how long will it take for the car to overtake the truck?

Estimate

The truck has a head start of 1.25 hours and goes at 72.5 km/h, so it is about 91 km ahead when the car starts. But the car goes about 36 km/h faster than the truck, so each hour the truck's lead is cut by 36 km. Thus it would take between 2 and 3 hours to reduce the 91-km lead to zero.

Solution: Let us follow the list of steps suggested for solving any word problem.

Make a Sketch

A simple sketch, Fig. 3–7, shows the city, the road, the meeting point, and the rates of the truck and the car.

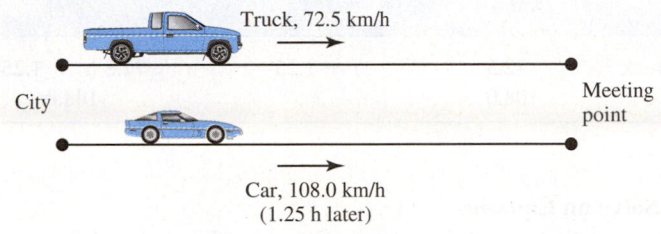

FIGURE 3–7

Identify the Given Information

Let us write down the formula for uniform motion,

$$\text{rate} \times \text{time} = \text{distance}$$

Under each of the three quantities we write the given value of that quantity, both for the truck and for the car. Reading the problem statement, we see that the speeds of the truck and of the car are given.

	Rate (km/h)	×	Time (h)	=	Distance (km)
Truck	72.5				
Car	108.0				

Define the Unknown

What are we looking for? We seek the time for the car to overtake the truck, so we write

Let t = time traveled by car (hours)

and enter this information into our table

	Rate (km/h)	×	Time (h)	=	Distance (km)
Truck	72.5				
Car	108.0		t		

Define the Other Unknowns

We have three empty boxes in our table. First, the time for the truck can be written in terms of t by noting that the truck travels for 1.25 hours longer than the car.

	Rate (km/h)	×	Time (h)	=	Distance (km)
Truck	72.5		$t + 1.25$		
Car	108.0		t		

We complete the table by noting that distance = rate × time.

	Rate (km/h)	×	Time (h)	=	Distance (km)
Truck	72.5		$t + 1.25$		$72.5(t + 1.25)$
Car	108.0		t		$108.0t$

Write and Solve an Equation

An equation says that something is equal to something else. What quantities are equal in this problem? Since, at the instant that the car overtakes the truck, they have both gone the same distance, we set the distance gone by the truck equal to the distance gone by the car.

$$\text{Distance by car} = \text{Distance by truck}$$
$$108.0t = 72.5(t + 1.25)$$

Solve the Equation

$$108.0t = 72.5t + 72.5(1.25)$$
$$108.0t - 72.5t = 90.63$$
$$35.5t = 90.63$$
$$t = 2.55 \text{ hours} = \text{time for car}$$

Check Your Answer

We first note that the answer (2.55 h for the car) agrees with our estimate, which was between 2 and 3 hours. Next, for a more accurate check, let's find the distance traveled by the truck and by the car.

$$\text{Time for truck} = t + 1.25 = 3.80 \text{ h}$$
$$\text{Truck distance} = (72.5 \text{ km/h})(3.80 \text{ h}) = 275.5 \text{ km}$$
$$\text{Car distance} = (108.0 \text{ km/h})(2.55 \text{ h}) = 275.4 \text{ km}$$

These distances agree to three significant digits, the precision to which we are working in this problem. Since most of our original numbers are known only to three significant digits, we cannot expect our answers to check to more significant digits than that.   ◆◆◆

In the preceding example we got our equation from the fact that both vehicles traveled the same distance. This, of course, will not be true for *every* motion problem, and you may have to look for other relationships in order to write an equation.

Exercise 3 ◆ Uniform Motion Applications

1. Two planes start from the same city at the same time. One travels at 252 mi/h and the other at 266 mi/h, in the opposite direction. How long will it take for them to be 1750 miles apart?

2. The pointer of a certain meter can travel to the right at the rate of 10.0 cm/s. What must be the minimum return rate if the total time for the pointer to traverse the full 12.0-cm scale and return to zero must not exceed 2.00 seconds?

3. A train travels from P to Q at a rate of 22.5 km/h. After it has been gone 2.75 hours, an express train leaves P for Q traveling at 85.5 km/h, and reaches Q 1.50 hours ahead of the first train. Find the distance from P to Q, and the time taken by the express train.

4. A freight train leaves A for B, 175 miles away, and travels at the rate of 31.5 mi/h. After 1.50 hours, a train leaves B for A, traveling at 21.5 mi/h. How many miles from B will they meet?

5. Two submarines start from the same spot and travel in opposite directions, one at 115 km/day and the other at 182 km/day. How long will it take for the submarines to be 1470 km apart?

6. A bus travels 87.5 km to another town at a speed of 72.0 km/h. What must be its return rate if the total time for the round trip is to be 2.50 hours?

7. An oil slick from a runaway offshore oil well is advancing toward a beach 354 miles away at the rate of 10.5 mi/day. Two days after the spill, cleanup ships leave the beach and steam toward the slick at a rate of 525 mi/day. At what distance from the beach will they reach the slick?

8. A certain shaper has a forward cutting speed of 115 ft/min and a stroke of 10.5 in. It is observed to make 429 cuts (and returns) in 4.0 minutes. What is the return speed?

9. Spacecraft A is over Houston at noon on a certain day and traveling at a rate of 275 km/h. Spacecraft B, attempting to overtake and dock with A, is over Houston at 1:15 P.M. and is traveling in the same direction as A, at 444 km/h. At what time will B overtake A? At what distance from Houston?

3–4   Money Problems

Our next group of applications is in an area that concerns us all, on the job and especially in our private lives: *money*.

We usually work financial problems to the nearest dollar or nearest penny, regardless of the significant digits in the original numbers.

◆◆◆ **Example 30:** A consultant had to pay income taxes of $4867 plus 15% of the amount by which her taxable income exceeded $32,450. Her tax bill was $12,386. What was her taxable income? Work to the nearest dollar.

Solution: Let x = taxable income (dollars). The amount by which her income exceeded $32,450 is then

$$x - 32,450$$

Her tax is 15% of that amount, plus $4867, so

$$\text{tax} = 4867 + 0.15(x - 32,450) = 12,386$$

Solving for x we get

$$0.15(x - 32{,}450) = 12{,}386 - 4867 = 7519$$

$$x - 32{,}450 = \frac{7519}{0.15} = 50{,}127$$

$$x = 50{,}127 + 32{,}450 = \$82{,}577$$

Check: Her income exceeds \$32,450 by (\$82,577 − \$32,450) or \$50,127. A 15% tax on that amount is 0.15(\$50,127) or \$7,519. Her total tax is then \$7,519 + \$4867 or \$12,386, as required.   ◆◆◆

◆◆◆ **Example 31:** A person invests part of his \$10,000 savings in a bank, at 6%, and part in a certificate of deposit, at 8%, both simple interest. He gets a total of \$750 per year in interest from the two investments. How much is invested at each rate?

Solution: We first define our variables. Let

$$x = \text{amount invested at } 6\%$$

and

$$10{,}000 - x = \text{amount invested at } 8\%$$

If x dollars are invested at 6%, the interest on that investment is $0.06x$ dollars. Similarly, $0.08(10{,}000 - x)$ dollars are earned on the other investment. Since the total earnings are \$750, we write

$$0.06x + 0.08(10{,}000 - x) = 750$$

$$0.06x + 800 - 0.08x = 750$$

$$-0.02x = -50$$

$$x = \$2500 \quad \text{at } 6\%$$

and

$$10{,}000 - x = \$7500 \quad \text{at } 8\%$$

Check: Does this look reasonable? Suppose that half of his savings (\$5000) were invested at each rate. Then the 6% deposit would earn \$300 and the 8% deposit would earn \$400, for a total of \$700. But he got more than that (\$750) in interest, so we would expect more than \$5000 to be invested at 8% and less than \$5000 at 6%, as we have found.   ◆◆◆

Exercise 4 ◆ Money Problems

Financial problems make heavy use of percentage, so you may want to review that material.

1. The labor costs for a certain project were \$3345 per day for 17 technicians and helpers. If each technician earned \$210/day and each helper \$185/day, how many technicians were employed on the project?

2. A company has \$86,500 invested in bonds and earns \$6751 in interest annually. Part of the money is invested at 7.4%, and the remainder at 8.1%, both simple interest. How much is invested at each rate?

3. How much, to the nearest dollar, must a company earn in order to have \$895,000 left after paying 27% in taxes?

4. Three equal batches of fiberglass insulation were bought for \$408: the first for \$17 per ton, the second for \$16 per ton, and the third for \$18 per ton. How many tons of each were bought?

5. A water company changed its rates from \$1.95 per 1000 gal to \$1.16 per 1000 gal plus a service charge of \$45 per month. How much water (to the nearest 1000 gal) can you purchase before your new monthly bill will equal the bill under the former rate structure?

6. What salary should a person receive in order to take home $40,000 after deducting 23% for taxes? Work to the nearest dollar.

7. A student sold used skis and boots for $210, getting 4 times as much for the boots as for the skis. What was the price of each?

8. A used truck and a snowplow attachment are worth $7200, the truck being worth 7 times as much as the plow. Find the value of each.

9. A person spends 1/4 of her annual income for board, 1/12 for clothes, and 1/2 for other expenses, and saves $10,000. What is her income?

10. How much must a person earn to have $45,824 after paying 28% in taxes?

11. A carpenter estimates that a certain deck needs $4285 worth of lumber, if there were no waste. How much should she buy, if she estimates the waste at 7%?

12. The labor costs for a certain brick wall were $1118 per day for 10 masons and helpers. If a mason earned $125/day and a helper $92/day, how many masons were on the job?

13. A company has $173,924 in bonds and from them earns $13,824 in simple interest annually. Part of the money is invested at 6.75% and the remainder at 8.24%. How much is invested at each rate?

14. A company had $528,374 invested, part in stocks that earned 9.45% per year, and the remainder in bonds that earned 6.12% per year, both simple interest. The amount earned from both investments combined was $42,852. How much was in each investment?

15. A student sold a computer and a printer for a total of $995, getting $1\frac{1}{2}$ times as much for the printer as for the computer. What was the price of each?

3–5 Applications Involving Mixtures

We turn now to *mixture* problems, a term which at first may seem very narrow. However, we will not only consider mixtures of liquids, like the oil/gasoline mixture in a snowmobile or chain saw but will also include the mixture of metals in an alloy like brass, the mixture of sand, stone, cement, and water in concrete, the mixture of solids and solvent in paint, and so forth.

Basic Relationships

The total amount of mixture is, obviously, equal to the sum of the amounts of the ingredients.

total amount of mixture = amount of **A** + amount of **B** . . .	**1000**
Final amount of each ingredient = initial amount + amount added − amount removed	**1001**

These two ideas are so obvious that it may seem unnecessary to even write them down. However, it is because they are obvious that they are often overlooked. They state, in other words, that the whole is equal to the sum of its parts.

♦♦♦ **Example 32:** From 100.0 lb of solder, half lead and half zinc, 20.0 lb is removed. Then 30.0 lb of lead is added. How much lead is contained in the final mixture?

Solution:

$$\text{initial weight of lead} = 0.5(100.0) = 50.0 \text{ lb}$$

$$\text{amount of lead removed} = 0.5(20.0) = 10.0 \text{ lb}$$

$$\text{amount of lead added} = 30.0 \text{ lb}$$

By Eq. 1001,

$$\text{final amount of lead} = 50.0 + 30.0 - 10.0 = 70.0 \text{ lb}$$ ♦♦♦

Percent Concentration

The percent concentration of each ingredient is given by the following equation:

$$\text{percent concentration of ingredient } \mathbf{A} = \frac{\text{amount of A}}{\text{amount of mixture}} \times 100 \qquad \boxed{19}$$

♦♦♦ **Example 33:** The total weight of the solder in Example 32 is

$$100.0 - 20.0 + 30.0 = 110.0 \text{ lb}$$

so the percent concentration of lead (of which there is 70.0 lb) is

$$\text{percent lead} = \frac{70.0}{110.0} \times 100 = 63.6\%$$

where 110.0 is the total weight (lb) of the final mixture. ♦♦♦

Two Mixtures

When *two mixtures* are combined to make a *third* mixture, the amount of any ingredient A in the final mixture is given by the following equation:

$$\begin{aligned}\text{final amount of } A = \ &\text{amount of } A \text{ in first mixture} \\ &+ \text{amount of } A \text{ in second mixture}\end{aligned} \qquad \boxed{1002}$$

♦♦♦ **Example 34:** One hundred liters of gasohol containing 12% alcohol is mixed with 200 liters of gasohol containing 8% alcohol. The volume of alcohol in the final mixture is

$$\begin{aligned}\text{final amount of alcohol} &= 0.12(100) + 0.08(200) \\ &= 12 + 16 = 28 \text{ liters}\end{aligned}$$ ♦♦♦

A Typical Mixture Problem

We now use these ideas about mixtures to solve a typical mixture problem.

♦♦♦ **Example 35:** How much steel containing 5.25% nickel must be combined with another steel containing 2.84% nickel to make 3.25 tons of steel containing 4.15% nickel?

Estimate: The final steel needs 4.15% of 3.25 tons of nickel, about 0.135 ton. If we assume that *equal amounts* of each steel were used, the amount of nickel would be 5.25% of 1.625 tons or 0.085 ton from the first alloy, and 2.84% of 1.625 tons or 0.046 ton from the second alloy. This gives a total of 0.131 ton of nickel in the final steel. This is not enough (we need 0.135 ton). Thus *more than half* of the final alloy must come from the higher-nickel alloy, or between 1.625 tons and 3.25 tons.

Solution: Let x = tons of 5.25% steel needed. Fig. 3–8 shows the three alloys and the amount of nickel in each, with the nickel drawn as if it were separated from the rest of the steel. The weight of the 2.84% steel is

$$3.25 - x$$

The weight of nickel that it contains is

$$0.0284(3.25 - x)$$

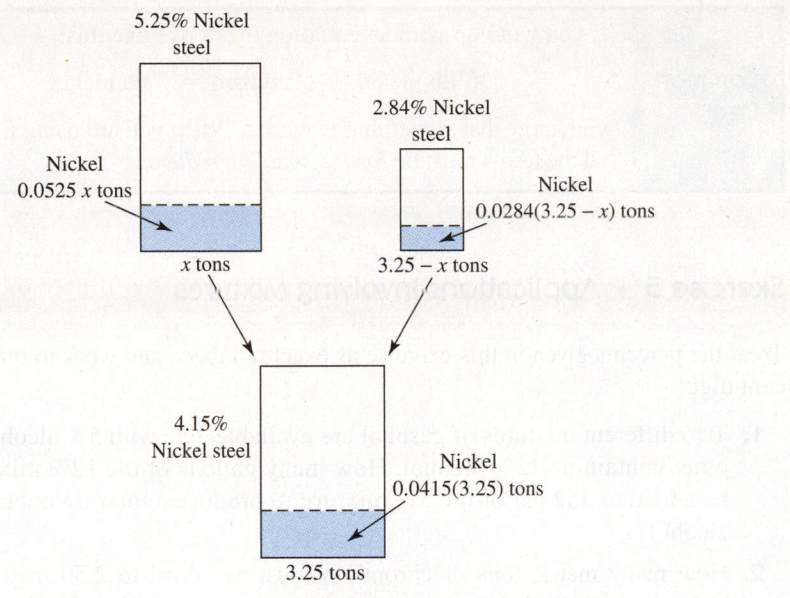

FIGURE 3–8

The weight of nickel in x tons of 5.25% steel is

$$0.0525x$$

The sum of these must give the weight of nickel in the final mixture.

$$0.0525x + 0.0284(3.25 - x) = 0.0415(3.25)$$

Clearing parentheses, we have

$$0.0525x + 0.0923 - 0.0284x = 0.1349$$
$$0.0241x = 0.0426$$
$$x = 1.77 \text{ tons of } 5.25\% \text{ steel}$$
$$3.25 - x = 1.48 \text{ tons of } 2.84\% \text{ steel}$$

Check: First we see that more than half of the final alloy comes from the higher-nickel steel, as predicted in our estimate. Now let us see if the final mixture has the proper percentage of nickel.

$$\text{final tons of nickel} = 0.0525(1.77) + 0.0284(1.48)$$
$$= 0.135 \text{ ton}$$
$$\text{percent nickel} = \frac{0.135}{3.25} \times 100 = 0.0415$$

or 4.15%, as required.

Alternate Solution: This problem can also be set up in table form, as follows:

	Percent Nickel	×	Amount (tons)	=	Amount of Nickel (tons)
Steel with 5.25% nickel	5.25%		x		$0.0525x$
Steel with 2.84% nickel	2.84%		$3.25 - x$		$0.0284(3.25 - x)$
Final steel	4.15%		3.25		$0.0415(3.25)$

We then equate the sum of the amounts of nickel in the original steels with the amount in the final steel, getting

$$0.0525x + 0.0284(3.25 - x) = 0.0415(3.25)$$

as before.   ◆◆◆

Common Error	If you wind up with an equation that looks like this:
	( )lb nickel + ( )lb **iron** = ( )lb nickel
	you know that something is wrong. When you are using Eq. 1002, all the terms must be *for the same ingredient*.

Exercise 5 ◆ Applications Involving Mixtures

Treat the percents given in this exercise as exact numbers, and work to three significant digits.

1. Two different mixtures of gasohol are available, one with 5% alcohol and the other containing 12% alcohol. How many gallons of the 12% mixture must be added to 252 gal of the 5% mixture to produce a mixture containing 9% alcohol?

2. How many metric tons of chromium must be added to 2.50 metric tons of stainless steel to raise the percent of chromium from 11% to 18%?

3. How many kilograms of nickel silver alloy containing 18% zinc and how many kilograms of nickel silver alloy containing 31% zinc must be melted together to produce 706 kg of a new nickel silver alloy containing 22% zinc?

4. A certain bronze alloy containing 4% tin is to be added to 351 lb of bronze containing 18% tin to produce a new bronze containing 15% tin. How many pounds of the 4% bronze are required?

5. How many kilograms of brass containing 63% copper must be melted with 1120 kg of brass containing 72% copper to produce a new brass containing 67% copper?

6. A certain chain saw requires a fuel mixture of 5.5% oil and the remainder gasoline. How many liters of 2.5% mixture and how many of 9.0% mixture must be combined to produce 40.0 liters of 5.5% mixture?

7. A certain automobile cooling system contains 11.0 liters of coolant that is 15% antifreeze. How many liters of mixture must be removed so that, when it is replaced with pure antifreeze, a mixture of 25% antifreeze will result?

8. A vat contains 4110 liters of wine with an alcohol content of 10%. How much of this wine must be removed so that, when it is replaced with wine with a 17% alcohol content, the alcohol content in the final mixture will be 12%?

9. A certain paint mixture weighing 315 lb contains 20% solids suspended in water. How many pounds of water must be allowed to evaporate to raise the concentration of solids to 25%?

10. Fifteen liters of fuel containing 3.2% oil is available for a certain two-cycle engine. This fuel is to be used for another engine requiring a 5.5% oil mixture. How many liters of oil must be added?

11. A concrete mixture is to be made which contains 35% sand by weight, and 642 lb of mixture containing 29% sand is already on hand. How many pounds of sand must be added to this mixture to arrive at the required 35%?

12. How many liters of a solution containing 18% sulfuric acid and how many liters of another solution containing 25% sulfuric acid must be mixed together to make 552 liters of solution containing 23% sulfuric acid? (All percentages are by volume.)

3–6 Statics Applications

The following section should be of great interest to students who expect to be involved in structures of any kind, buildings, bridges, trusses, and so forth.

Moments

The *moment of a force* about some point a (written M_a) is the product of the force F and the perpendicular distance d from the force to the point. The moment of a force is also referred to as *torque*. In Fig. 3–9 it is

Moment of a Force about Point *a*	$M_a = Fd$	1012

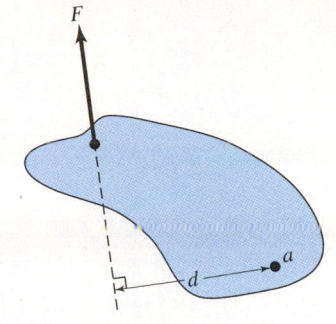

FIGURE 3–9

◆◆◆ **Example 36:** The moment of the force in Fig. 3–10 about point a is

$$M_a = 275 \text{ lb}(1.45 \text{ ft}) = 399 \text{ ft·lb}$$ ◆◆◆

Equations of Equilibrium

If the wagon, Fig. 3–11a, is pushed from the left, it will, of course, move to the right. If it *does not* move, it means there must be an equal force pushing it to the left, Fig. 3–11b. In other words if the wagon does not move, the sum of the horizontal forces acting on the wagon must be zero. When a body is at rest (or moving with a constant velocity) we say that it is in *equilibrium.*

What we said about horizontal forces also applies to vertical forces, and to moments tending to rotate the body. These are formally stated as the *equations of equilibrium.* For a body in equilibrium,

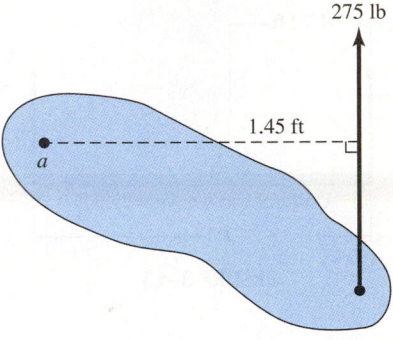

FIGURE 3–10   Moment of a force.

Equations of Equilibrium	The sum of all horizontal forces acting on the body = 0	1013
	The sum of all vertical forces acting on the body = 0	1014
(Newton's First Law of Motion)	The sum of the moments about any point on the body = 0	1015

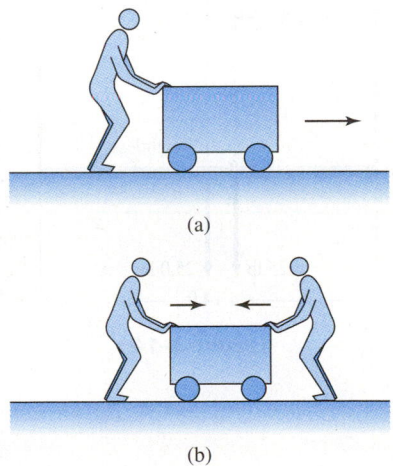

FIGURE 3–11

◆◆◆ **Example 37:** A horizontal uniform beam of negligible weight is 6.35 m long and is supported by columns at either end. A concentrated load of 525 N is applied to the beam. At what distance from one end must this load be located so that the vertical force (called the *reaction*) at that same end is 315 N? What is the reaction at the other end?

Estimate: If the 525-N load were applied at the middle of the beam (the *midspan*), the two reactions would have equal values of $\frac{1}{2}(525)$ or 262.5 N. Since the left reaction (315 N) is greater than that, we deduce that the load is to the left of the midspan and that the reaction at the right will be less than 262.5 N.

Solution: We draw a diagram (Fig. 3–12) and label the required distance as x. By Eq. 1014,

$$R + 315 = 525$$
$$R = 210 \text{ N}$$

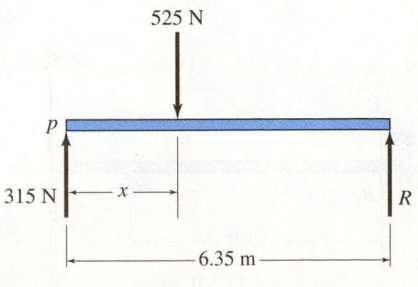

FIGURE 3–12

Taking moments about p, we set the moments that tend to turn the bar in a clockwise (CW) direction equal to the moments that tend to turn the bar in the counterclockwise (CCW) direction. By Eq. 1015,

$$525x = 210(6.35)$$

$$x = \frac{210(6.35)}{525} = 2.54 \text{ m}$$ ♦♦♦

♦♦♦ **Example 38:** Find the reactions R_1 and R_2 in Fig. 3–13.

Estimate: Let's assume that the 125-lb weight is centered on the bar. Then each reaction would equal half the total weight (150 lb ÷ 2 = 75 lb). But since the weight is to the left of center, we expect R_1 to be a bit larger than 75 lb and R_2 to be a bit smaller than 75 lb.

Solution: In a statics problem, we may consider all of the weight of an object to be concentrated at a single point (called the *center of gravity*) on that object. For a uniform bar, the center of gravity is just where you would expect it to be, at the midpoint. Replacing the weights by forces gives the simplified diagram, Fig. 3–14.

The moment of the 125-lb force about p is, by Eq. 1012,

$$125(1.72)\text{ft·lb} \qquad \text{clockwise}$$

Similarly, the other moments about p are

$$25.0(2.07)\text{ft·lb} \qquad \text{clockwise}$$

and

$$4.14\, R_2 \text{ ft·lb} \qquad \text{counterclockwise}$$

But Eq. 1015 says that the sum of the clockwise moments must equal the sum of the counterclockwise moments, so

$$4.14R_2 = 125(1.72) + 25.0(2.07)$$
$$= 266.8$$
$$R_2 = 64.4 \text{ lb}$$

Also, Eq. 1014 says that the sum of the upward forces (R_1 and R_2) must equal the sum of the downward forces (125 lb and 25.0 lb), so

$$R_1 + R_2 = 125 + 25.0$$
$$R_1 = 125 + 25.0 - 64.4$$
$$= 85.6 \text{ lb}$$ ♦♦♦

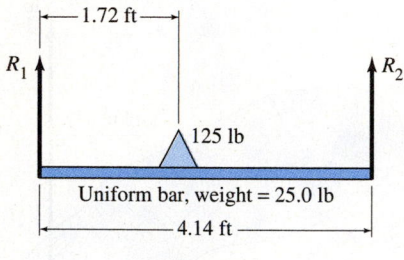

FIGURE 3–13

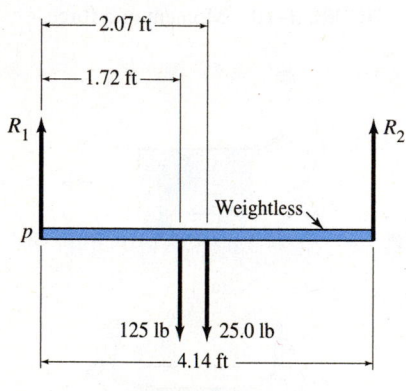

FIGURE 3–14

Exercise 6 ♦ Statics Applications

1. A horizontal beam of negligible weight is 18.0 ft long and is supported by columns at either end. A vertical load of 14,500 lb is applied to the beam at a distance x from the left end. (a) Find x so that the reaction at the left column is 10,500 lb. (b) Find the reaction at the right column.

2. A certain beam of negligible weight is "built-in" at one end and has an additional support 13.1 ft from the left end, as shown in Fig. 3–15. The beam is 17.3 ft long and has a concentrated load of 2350 lb at the free end. Find the vertical reactions R_1 and R_2.

3. A horizontal bar of negligible weight has a 55.1-lb weight hanging from the left end and a 72.0-lb weight hanging from the right end. The bar is seen to balance 97.5 in. from the left end. Find the length of the bar.

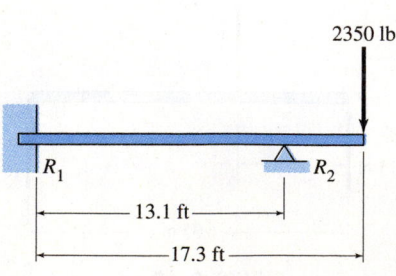

FIGURE 3–15

4. A horizontal bar of negligible weight hangs from two vertical cables, one at each end. When a 624-N force is applied vertically downward from a point 185 cm from the left end of the bar, the right cable is seen to have a tension of 341 N. Find the length of the bar.

5. A uniform horizontal beam is 9.74 ft long and weighs 386 lb. It is supported by columns at either end. A vertical load of 3814 lb is applied to the beam at a distance x from the left end. Find x so that the reaction at the right column is 2000 lb.

6. A uniform horizontal beam is 19.80 ft long and weighs 1360 lb. It is supported at either end. A vertical load of 13,510 lb is applied to the beam 8.450 ft from the left end. Find the reaction at each end of the beam.

7. A bar of uniform cross section is 82.3 in. long and weighs 10.5 lb. A weight of 27.2 lb is suspended from one end. The bar and weight combination is to be suspended from a cable attached at the balance point. How far from the weight should the cable be attached, and what is the tension in the cable?

8. *Project:* Fig. 3–16 shows three beams, each carrying a box of a given weight. The weight of each beam is given, which can be considered to be at the midpoint of each beam. The right end of each of the two upper beams rests on the box on the beam below it. A force of 3150 lb is needed to support the right end of the lowest beam.
Find the reactions R_1, R_2, and R_3, and the distance x.

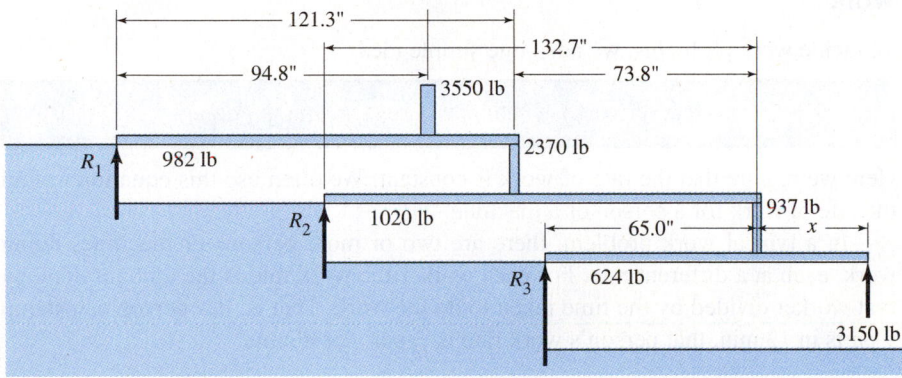

FIGURE 3–16

3–7   Applications to Work, Fluid Flow, and Energy Flow

Solving Rate Problems

We already considered word problems involving constant rates when we learned about uniform motion. This will now make it easier for us to solve other problems in which the quantities are related in the same way. These are *work, fluid flow,* and *energy flow.*

Uniform motion:	*amount traveled = rate of travel × time traveled*
Work:	*amount of work done = rate of work × time worked*
Fluid flow:	*amount of flow = flow rate × duration of flow*
Energy flow:	*amount of energy = rate of energy flow × time*

Notice that these equations are mathematically identical. Many applications involving motion, flow of fluids or solids, flow of heat, electricity, solar energy, mechanical energy, and so on, can be handled in the same way.

To convince yourself of the similarity of all of these types of problems, consider the following work problem:

> If worker M can do a certain job in 5 h, and N can do the same amount in 8 h, how long will it take M and N together to do that same amount?

Are the following problems really any different?

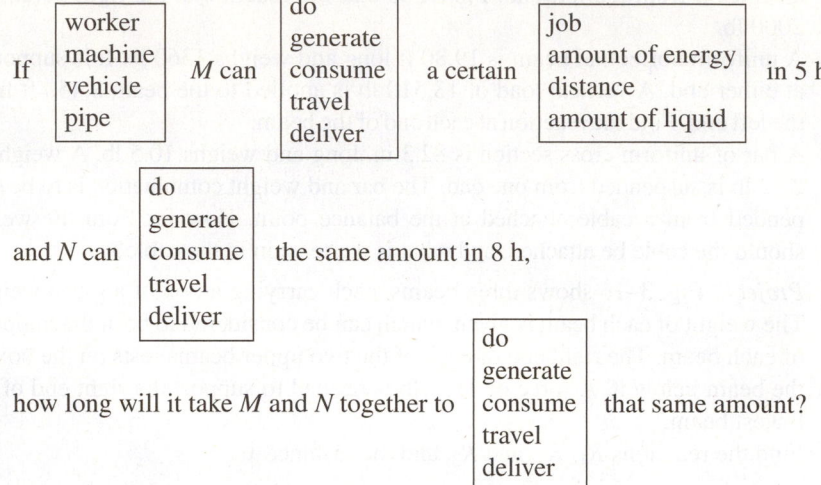

how long will it take M and N together to

| do |
| generate |
| consume |
| travel |
| deliver |

that same amount?

Work

To tackle work problems, we need one simple idea.

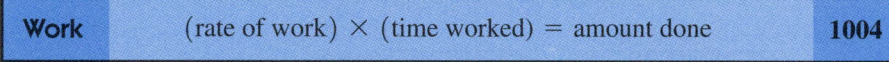

| Work | (rate of work) × (time worked) = amount done | 1004 |

Here we require that the rate of work is constant. We often use this equation to find the rate of work for a person or a machine.

In a typical work problem, there are two or more persons or machines doing work, each at a different rate. For each worker the work rate is the amount done by that worker divided by the time taken to do the work. That is, if a person can stamp 9 parts in 13 min, that person's work rate is $\frac{9}{13}$ part per minute.

◆◆◆ Example 39: Crew A can assemble 2 cars in 5 days, and crew B can assemble 3 cars in 7 days. If both crews together assemble 100 cars, with crew B working 10 days longer than crew A, how many days (rounded to the nearest day) must each crew work?

Estimate: Working alone, crew A does 2 cars in 5 days, or 100 cars in 250 days. Similarly, crew B does 100 cars in about 231 days. Together they do 200 cars in 250 + 231 or 481 days, or about 100 cars in 240 days. Thus we would expect each crew to work about half that, or about 120 days each. But crew B works 10 days longer than A, so we estimate that they work about 5 days longer than 120 days, whereas crew A works about 5 days less than 120 days.

Solution: Let

$$x = \text{days worked by crew } A$$

and

$$x + 10 = \text{days worked by crew } B$$

The work rate of each crew is

$$\text{crew } A: \quad \text{rate} = \frac{2}{5} \text{ car per day}$$

$$\text{crew } B: \quad \text{rate} = \frac{3}{7} \text{ car per day}$$

The amount done by each crew equals their work rate times the number of days worked. The sum of the amounts done by the two crews must equal 100 cars.

$$\frac{2}{5}x + \frac{3}{7}(x + 10) = 100$$

We clear fractions by multiplying by 35

$$14x + 15(x + 10) = 3500$$
$$14x + 15x + 150 = 3500$$
$$29x = 3350$$
$$x = 116 \text{ days for crew } A$$
$$x + 10 = 126 \text{ days for crew } B \quad \text{◆◆◆}$$

Fluid Flow and Energy Flow

For flow problems, we use the simple equation

$$\text{amount of flow} = \text{flow rate} \times \text{time}$$

Here we assume, of course, that the flow rate is constant.

◆◆◆ **Example 40:** A certain small hydroelectric generating station can produce 17 megawatthours (MWh) of energy per year. After 4.0 months of operation, another generator is added which, by itself, can produce 11 MWh in 5.0 months. How many additional months are needed for a total of 25 MWh to be produced?

Solution: Let x = additional months. The original generating station can produce 17/12 MWh per month for $4.0 + x$ months. The new generator can produce 11/5.0 MWh per month for x months. The problem states that the total amount produced (in MWh) is 25, so

$$\frac{17}{12}(4.0 + x) + \frac{11}{5.0}x = 25$$

Multiplying by 60 to clear fractions, we obtain

$$5.0(17)(4.0 + x) + 12(11)x = 25(60)$$
$$340 + 85x + 132x = 1500$$
$$217x = 1160$$
$$x = \frac{1160}{217} = 5.3 \text{ months} \quad \text{◆◆◆}$$

Exercise 7 ◆ Applications to Work, Fluid Flow, and Energy Flow

Work

1. A laborer can do a certain job in 5 days, a second in 6 days, and a third in 8 days. In what time can the three together do the job?

2. Three masons build 318 m of wall. Mason A builds 7.0 m/day, B builds 6.0 m/day, and C builds 5.0 m/day. Mason B works twice as many days as A, and C works half as many days as A and B combined. How many days did each work?

3. If a carpenter can roof a house in 10 days and another can do the same in 14 days, how many days will it take if they work together?

4. A technician can assemble an instrument in 9.5 h. After working for 2.0 h, she is joined by another technician who, alone, could do the job in 7.5 h. How many additional hours are needed to finish the job?

5. A certain screw machine can produce a box of parts in 3.3 h. A new machine is to be ordered having a speed such that both machines working together would produce a box of parts in 1.4 h. How long would it take the new machine alone to produce a box of parts?

Fluid Flow

6. A tank can be filled by a pipe in 3.0 h and emptied by another pipe in 4.0 h. How much time will be required to fill an empty tank if both are running?

7. Two pipes empty into a tank. One pipe can fill the tank in 8.0 h, and the other in 9.0 h. How long will it take both pipes together to fill the tank?

8. A tank has three pipes connected. The first, by itself, could fill the tank in $2\frac{1}{2}$ h, the second in 2.0 h, and the third in 1 h 40 min. In how many minutes will the tank be filled?

9. A tank can be filled by a certain pipe in 18.0 h. Five hours after this pipe is opened, it is supplemented by a smaller pipe which, by itself, could fill the tank in 24.0 h. Find the total time, measured from the opening of the larger pipe, to fill the tank.

10. At what rate must liquid be drained from a tank in order to empty it in 1.50 h if the tank takes 4.70 h to fill at the rate of $3.50 \text{m}^3/\text{min}$?

Energy Flow

11. A certain power plant consumes 1500 tons of coal in 4.0 weeks. There is a stockpile of 10,000 tons of coal available when the plant starts operating. After 3.0 weeks in operation, an additional boiler, capable of using 2300 tons in 3.0 weeks, is put on line with the first boiler. In how many more weeks will the stockpile of coal be consumed?

12. A certain array of solar cells can generate 2.0 megawatthours (MWh) in 5.0 months (under standard conditions). After this array has been operating for 3.0 months, another array of cells is added which alone can generate 5.0 MWh in 7.0 months. How many additional months, after the new array has been added, is needed for the total energy generated from both arrays to be 10 MWh?

13. A landlord owns a house that consumes 2100 gal of heating oil in three winters. He buys another (insulated) house, and the two houses together use 1850 gal of oil in two winters. How many winters would it take the insulated house alone to use 1250 gal of oil?

14. A wind generator can charge 20 storage batteries in 24 h. After the generator has been charging for 6.0 h, another generator, which can charge the batteries in 36 h, is also connected to the batteries. How many additional hours are needed to charge the 20 batteries?

15. A certain solar panel can collect 9000 Btu in 7.0 h. Another panel is added, and together they collect 35,000 Btu in 5.0 h. How long would it take the new panel alone to collect 35,000 Btu?

16. *Writing:* Write down the reasons why you think the word problems in this book are stupid and hardly worth studying, and e-mail them to the authors. We promise to answer you. If you don't want to mail your observations, just file them in your portfolio. If, on the other hand, you think the problems are stupid but still worth studying, state your reasons. Those too can be sent to the author or filed with your notes.

◆◆◆ CHAPTER 3 REVIEW PROBLEMS ◆◆◆◆◆◆◆◆◆◆◆◆◆◆◆◆◆◆◆◆◆◆◆◆◆◆◆◆◆◆◆◆

Solve each equation.

1. $2x - (3 + 4x - 3x + 5) = 4$

2. $5(2 - x) + 7x - 21 = x + 3$

3. $3(x - 2) + 2(x - 3) + (x - 4) = 3x - 1$

4. $x + 1 + x + 2 + x + 4 = 2x + 12$

5. $(2x - 5) - (x - 4) + (x - 3) = x - 4$

6. $4 - 5w - (1 - 8w) = 63 - w$

7. $3z - (z + 10) - (z - 3) = 14 - z$

8. $(2x - 9) - (x - 3) = 0$

9. $3x + 4(3x - 5) = 12 - x$

10. $6(x - 5) = 15 + 5(7 - 2x)$

11. $x^2 - 2x - 3 = x^2 - 3x + 1$

12. $10x - (x - 5) = 2x + 47$

13. $7x - 5 - (6 - 8x) + 2 = 3x - 7 + 106$

14. $3p + 2 = \dfrac{p}{5}$

15. $\dfrac{4}{n} = 3$

16. $\dfrac{2x}{3} - 5 = \dfrac{3x}{2}$

17. $5.90x - 2.80 = 2.40x + 3.40$

18. $4.50(x - 1.20) = 2.80(x + 3.70)$

19. $\dfrac{x - 4.80}{1.50} = 6.20x$

20. $6x + 3 - (3x + 2) = (2x - 1) + 9$

21. $3(x + 10) + 4(x + 20) + 5x - 170 = 15$

22. $20 - x + 4(x - 1) - (x - 2) = 30$

23. $5x + 3 - (2x - 2) + (1 - x) = 6(9 - x)$

24. $3x - (x - 4) + (x + 1) = x - 7$

25. $4(x - 3) = 21 + 7(2x - 1)$

26. $8.20(x - 2.20) = 1.30(x + 3.30)$

27. $5(x - 1) - 3(x + 3) = 12$

28. $3.25x + 4.11(x - 3.75) = 11.2$

29. $5x + (2x - 3) + (x + 9) = x + 1$

30. $1.20(x - 5.10) = 7.30(x - 1.30)$

31. $2(x - 7) = 11 + 2(4x - 5)$

32. $4.40(x - 1.90) = 8.30(x + 1.10)$

33. $8x - (x - 5) + (3x + 2) = x - 14$

34. $7(x + 1) = 13 + 4(3x - 2)$

Solve for x.

35. $bx + 8 = 2$

36. $c + ax = b - 7$

37. $2(x - 3) = 4 + a$

38. $3x + a = b - 5$

39. $8 - ax = c - 5a$

40. Subdivide a meter of tape into two parts so that one part will be 6 cm longer than the other part.

41. A certain mine yields low-grade oil shale containing 18.0 gal of oil per ton of rock, and another mine has shale yielding 30.0 gal/ton. How many tons of each must be sent each day to a processing plant that processes 25,000 tons of rock per day, so that the overall yield will be 23.0 gal/ton?

42. A certain automatic soldering machine requires a solder containing half tin and half lead. How much pure tin must be added to 55 kg of a solder containing 61% lead and 39% tin to raise the tin content to 50%?

43. A person owed to A a certain sum, to B four times as much, to C eight times as much, and to D six times as much. A total of $570 would pay all of the debts. What was the debt to A?

44. A technician spends $\frac{2}{3}$ of his salary for board and $\frac{2}{3}$ of the remainder for clothing, and saves $5000 per year. What is his salary?

45. Find four consecutive odd numbers such that the product of the first and third will be 64 less than the product of the second and fourth.

46. The front and rear wheels of a tractor are 10 ft and 12 ft, respectively, in circumference. How many feet will the tractor have traveled when the front wheel has made 250 revolutions more than the rear wheel?

47. Find the reactions R_1 and R_2 in Fig. 3–17.

48. A carpenter estimates that a certain porch needs $3875 worth of lumber, if there is no waste. How much should he buy, if he estimates the waste on the amount bought is 6%?

49. How many liters of olive oil costing $4.86 per liter must be mixed with 136 liters of corn oil costing $2.75 per liter to make a blend costing $3.00 per liter?

50. The labor costs for a certain wall were $1760 per day for 8 masons and helpers. If a mason earned $280/day and a helper $120/day, how many masons were on the job?

51. According to a tax table, for your filing status, if your taxable income is over $38,000 but not over $91,850, your tax is $5700 plus 28% of the amount over $38,100. If your tax bill was $8126, what was your taxable income? Work to the nearest dollar.

52. A casting weighs 875 kg and is made of a brass that contains 87.5% copper. How many kilograms of copper are in the casting?

53. How much must a person earn to have $60,000 after paying 25% in taxes?

54. 235 gallons of fuel for a two-cycle engine contains 3.15% oil. To this is added 186 gallons of fuel containing 5.05% oil. How many gallons of oil are in the final mixture?

55. How much steel containing 1.15% chromium must be combined with another steel containing 1.50% chromium to make 8.00 tons of steel containing 1.25% chromium?

56. Two gasohol mixtures are available, one with 6.35% alcohol and the other with 11.28% alcohol. How many gallons of the 6.35% mixture must be added to 25.0 gallons of the other mixture to make a final mixture containing 8.00% alcohol?

57. How many tons of tin must be added to 2.75 tons of bronze to raise the percentage of tin from 10.5% to 16.0%?

58. A student sold a computer and a printer for a total of $825, getting half as much for the printer as for the computer. What was the price of each?

59. How many pounds of nickel silver containing 12.5% zinc must be melted with 248 kg of nickel silver containing 16.4% zinc to make a new alloy containing 15.0% zinc?

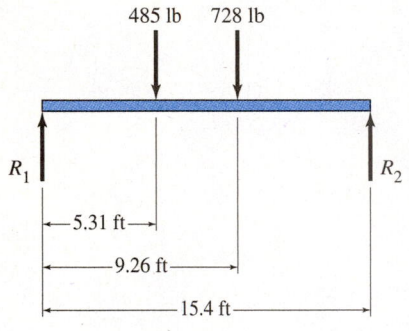

485 lb   728 lb

R_1 R_2

5.31 ft

9.26 ft

15.4 ft

FIGURE 3–17

60. A company has $223,821 invested in two separate accounts and earns $14,817 in simple interest annually from these investments. Part of the money is invested at 5.94%, and the remainder at 8.56%. How much is invested at each rate?

61. From 12.5 liters of coolant containing 13.4% antifreeze, 5.50 liters are removed. If 5.50 liters of pure antifreeze is then added, how much antifreeze will be contained in the final mixture?

62. A person had $125,815 invested, part in a mutual fund that earned 10.25% per year and the remainder in a bank account that earned 4.25% per year, both simple interest. The amount earned from both investments combined was $11,782. How much was in each investment?

63. The labor costs for a certain project were $3971 per day for 15 technicians and helpers. If a technician earned $325/day and a helper $212/day, how many technicians were on the job?

64. 15.6 gallons of cleaner containing 17.8% acid is added to 25.5 gallons of cleaner containing 10.3% acid. How many gallons of acid are in the final mixture?

65. How much brass containing 75.5% copper must be combined with another brass containing 86.3% copper to make 5250 kg of brass containing 80.0% copper?

66. A person walks a certain distance at a rate of 3.8 mi/h. Another person takes 2.1 hours longer to walk the same distance, at a rate of 3.2 mi/h. What distance did they walk?

67. A certain submarine can travel at a rate of 31.5 km/h submerged and 42.3 km/h on the surface. How far can it travel submerged, and return to the starting point on the surface, if the total round trip is to take 8.25 hours?

68. A mixture is made where 50.5 gallons of cleaner containing 18.2% acid is added to 31.6 gallons of cleaner containing 15.2% acid. How many gallons of acid are in the final mixture?

69. A certain space probe traveled at a speed of 1280 km/h, and then, by firing retro-rockets, slowed to 950 km/h. The total distance traveled was 75,300 km in a time of 63.5 hours. Find the distance from the starting point at which the retro-rockets were fired.

70. How many pounds of copper are in a brass ingot that weighs 1550 lb and contains 73.2% copper?

71. A ship leaves port at a rate of 24.5 km/h. After 8.25 hours a launch leaves the same port to overtake the ship and travels at a rate of 35.2 km/h. Find the time traveled by the ship before being overtaken, and the distance from the port at which this occurs.

72. *Writing:* Make up a word problem. It can be very similar to one given in this chapter or, better, very different. Try to solve it yourself. Then swap with a classmate and solve each other's problem. Note down where the problem may be unclear, unrealistic, or ambiguous. Finally, if needed, each of you should rewrite your problem.

73. *Writing:* You probably have some idea what field you plan to enter later: business, engineering, and so on. Look at a book from that field, and find at least one formula commonly used. Write it out; describe what it is used for; and explain the quantities it contains, including their units.

74. *Internet:* In this chapter we have dealt with *equations*, that is, where one expression is *equal to* another expression. For some applications, we must deal with *inequalities*, where two expressions are not equal but where one is greater or less than the other. An example of an inequality is $x + 7 > x - 5$. The solution to an inequality is not a single number, but a *range* of values. In our supplementary material on the Web, we give a chapter on how to graph inequalities and how to solve them. Go to www.wiley.com/college/calter

4

Functions

◆◆◆ **OBJECTIVES** ◆◆◆

When you have completed this chapter, you should be able to

• Distinguish between relations and functions.

• Recognize different forms of a function: equation, table of point pairs, verbal statement, graph.

• Use functional notation.

• Distinguish between implicit and explicit forms of an equation.

• Rewrite a simple implicit equation in explicit form.

• Substitute into a function.

• Manipulate a function.

• Manipulate functions by calculator.

• Write a composite function.

• Find the inverse of a function.

• Find the domain and range of a function.

◆◆

The equations we have been solving in the last few chapters have contained only *one variable*. For example,

$$x(x - 3) = x^2 + 7$$

contains the single variable x.

But many situations involve *two* (or more) variables that are somehow related to each other. Look, for example, at Fig. 4–1, and suppose that you leave your campsite and walk a path up the hill. As you walk, both the horizontal distance x from your camp, and the vertical distance y above your camp, will change. You cannot change x without changing y, and vice versa (unless you jump into the air or dig a hole). The variables x and y are *related*.

In this chapter we study the relation between two variables and introduce the concept of a *function*. The idea of a function provides us with a different way of speaking about mathematical relationships. We could say, for example, that the formula for the area of a circle *as a function of* its radius is $A = \pi r^2$.

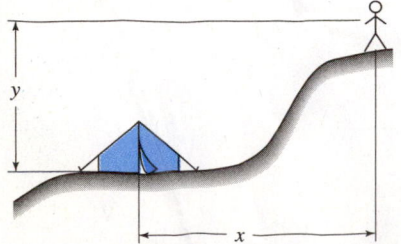

FIGURE 4–1

It may become apparent as you study this chapter that these same problems could be solved without ever introducing the idea of a function. Does the function concept, then, merely give us new jargon for the same old ideas? Not really.

A new way of *speaking* about something can lead to a new way of *thinking* about that thing, and so it is with functions. It also will lead to the powerful and convenient *functional notation*, which will be especially useful if you study calculus and computer programming. In addition, this introduction to functions will prepare us for the later study of *functional variation*.

4–1   Functions and Relations

We noted in the introduction that a function or relation is a way of relating two quantities, such as relating the area of a circle to its radius. We will soon see that a function or relation can take several forms (graphs and tables, for example) but we will usually work in the form of an equation. So for now let's define relations and functions in terms of equations, and expand our definition later.

An equation that enables us to find a value of *y* for a given value of *x* is called a *relation*.

◆◆◆ **Example 1:** The equation

$$y = 3x - 5$$

is a relation. For any value of *x*, say *x* = 2, there is a value of *y* (in this case, *y* = 1).   ◆◆◆

An equation that enables us to find *exactly one* value of *y* for a given value of *x* is called a *function*. Thus while the equation of Example 1 is a relation, it is also a function, because it gives just one value of *y* for any value of *x*.

Thus a function may be in the form of an equation. But *not every equation is a function*.

◆◆◆ **Example 2:** The equation $y = \pm\sqrt{x}$ is not a function. Each value of *x* yields *two* values of *y*. This equation is a relation.   ◆◆◆

We see from Examples 1 and 2 that a relation need not have *one* value of *y* for each value of *x*, as is required for a function. Thus only some equations that are relations can also be called functions.

The fact that a relation is not a function does not imply second-class status. Relations are no less useful when they are not functions.

Domain and Range

A more general definition of a function is usually given in terms of sets. A *set* is a collection of particular things, such as the set of all automobiles. The objects or members of a set are called *elements*.

> Let *x* be an element in set *A*, and *y* an element in set *B*. We say that *y* is a function of *x* if there is a rule that associates exactly one *y* in set *B* with each *x* in set *A*.

Here, set *A* is called the *domain* and set *B* is called the *range*.

We can picture a function as in Fig. 4–2. Each set is represented by a shaded area, and each element by a point within the shaded area. The function *f* associates exactly one *y* in the range *B* with each *x* in the domain *A*.

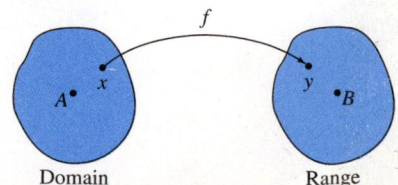

FIGURE 4–2 A similar diagram for a relation would show more than one *y* for some values of *x*.

♦♦♦ **Example 3:** The equation from Example 1,

$$y = 3x - 5$$

can be thought of as a rule that associates exactly one y with any given x. The domain of x is the set of all real numbers, and the range of y is the set of all real numbers. ♦♦♦

Not every real number may be allowed in the domain or the range.

♦♦♦ **Example 4:** For the function

$$y = \sqrt{x}$$

the domain cannot contain any negative values of x, since we are restricting ourselves to the real numbers only. As a result, the range will not contain negative numbers. ♦♦♦

We will show how to find the domain and range of functions later in this chapter.

Different Forms of a Function or Relation

We will deal with a function or relation mostly in the form of an equation. But other ways of relating two variables are as a table of values, a verbal statement, a set of ordered pairs, or a graph, as shown in Fig. 4–3. We will learn about these other forms later in this chapter, except for graphing, which is covered in the next chapter.

Different forms of a relation or function	
Equation: $y = x^2 - 3$	Table of values: $\begin{array}{c\|cccc} x & 0 & 1 & 2 & 3 \\ \hline y & -3 & -2 & 1 & 6 \end{array}$
Verbal statement: "y is equal to the square of x diminished by 3."	
Set of ordered pairs: $(1, -2), (3, 6),$ $(0, -3), (2, 1),$	Graph:

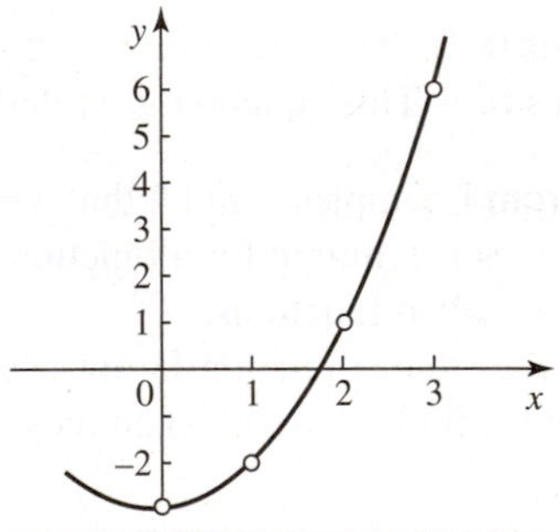

FIGURE 4–3

Implicit and Explicit Forms

When one variable in an equation is isolated on one side of the equal sign, the equation is said to be in *explicit form*.

♦♦♦ **Example 5:** The following equations are all in explicit form:

$$y = 2x^3 + 5$$
$$z = y + 14$$
$$x = 3z^2 + 2z - z$$

♦♦♦

When a variable is *not* isolated, the equation is in *implicit form*.

◆◆◆ **Example 6:** The following equations are all in implicit form:

$$y = x^2 + 4y$$
$$x^2 + y^2 = 25$$
$$w + x = y + z$$
$$x = y + xy - y^2$$

◆◆◆

Dependent and Independent Variables

In the equation

$$y = x + 5$$

y is called the *dependent variable* because its value *depends* on the value of x, and x is called the *independent* variable. Of course, the same equation can be written $x = y - 5$, so that x becomes the dependent variable and y the independent variable.

The terms dependent and independent are used only for an equation in explicit form.

◆◆◆ **Example 7:** In the implicit equation $y - x = 5$, neither x nor y is called dependent or independent.

◆◆◆

Functional Notation

Just as we use the symbol x to represent a *number*, without saying which number we are specifying, we use the notation

$$f(x)$$

to represent a *function* without having to specify which particular function we are talking about.

We can also use functional notation to designate a *particular* function, such as

$$f(x) = x^3 - 2x^2 - 3x + 4$$

Thus a functional relation between two variables x and y, in explicit form, such as

$$y = 5x^2 - 6$$

could be written

$$f(x) = 5x^2 - 6$$

or as

$$y = f(x)$$

It is read "y is a function of x."

Common Error	The expression $y = f(x)$ does not mean "y equals f times x."

The independent variable x is sometimes referred to as the *argument* of the function.

◆◆◆ **Example 8:** *An Application.* We may know that the horsepower P of an engine depends (somehow) on the engine displacement d. We can express this fact by

$$P = f(d)$$

We are saying that P is a function of d, even though we do not know (or perhaps even care, for now) what the relationship is.

◆◆◆

The letter f is usually used to represent a function, but *other letters* can, of course, be used (g and h being common). Subscripts are also used to distinguish one function from another.

◆◆◆ **Example 9:** You may see functions written as

$$y = f_1(x) \qquad\qquad y = g(x)$$
$$y = f_2(x) \qquad\qquad y = h(x)$$
$$y = f_3(x)$$

◆◆◆

The letter y itself is often used to represent a function.

◆◆◆ **Example 10:** The function

$$y = x^2 - 3x$$

will often be written

$$y(x) = x^2 - 3x$$

to emphasize the fact that y is a function of x.   ◆◆◆

Implicit functions can also be represented in functional notation.

◆◆◆ **Example 11:** The equation $x - 3xy + 2y = 0$ can be represented in functional notation by $f(x, y) = 0$.   ◆◆◆

Functions relating *more than two variables* can be represented in functional notation, as in the following example.

◆◆◆ **Example 12:**

$$y = 2x + 3z \qquad\qquad \text{can be written} \qquad y = f(x, z)$$
$$z = x^2 - 2y + w^2 \qquad \text{can be written} \qquad z = f(w, x, y)$$
$$x^2 + y^2 + z^2 = 0 \qquad \text{can be written} \qquad f(x, y, z) = 0$$

◆◆◆

Changing from Implicit to Explicit Form

We often need to rewrite an equation given in the implicit form,

$$f(x, y) = 0$$

in the explicit form,

$$y = f(x)$$

This is necessary, for example, to make a graph of the equation. To rewrite in explicit form, we have to *solve* the given function for y.

◆◆◆ **Example 13:** Rewrite this implicit equation in the explicit form, $y = f(x)$.

$$3x + 2y - 4 = 0$$

Solution: Solving for y we get,

$$2y = 4 - 3x$$
$$y = 2 - \frac{3}{2}x$$

◆◆◆

◆◆◆ **Example 14:** Rewrite this implicit equation in the explicit form, $y = f(x)$.

$$y^2 - x = 0$$

Solution: Solving for y we get,

$$y^2 = x$$

Taking the square root of both side gives

$$y = \pm\sqrt{x}$$

Note that we must keep both the positive and negative values of $\sqrt{x}$, because both values will satisfy the original equation. Since there are two values of y for each x, our resulting equation is a relation, but not a function.   ◆◆◆

Substituting into a Function

If we have a function, say, $f(x)$, then the notation $f(a)$ means to replace x by a in the same function.

◆◆◆ **Example 15:** Given $f(x) = x^3 - 5x$, find $f(2)$.

Solution: The notation $f(2)$ means that 2 is to be *substituted for x* in the given function. Wherever an x appears, we replace it with 2.

$$f(x) = x^3 - 5x$$
$$\updownarrow \qquad \updownarrow \qquad \updownarrow$$
$$f(2) = (2)^3 - 5(2)$$
$$= 8 - 10 = -2 \qquad ◆◆◆$$

◆◆◆ **Example 16:** Given $y(x) = 3x^2 - 2x$, find $y(5)$.

Solution: The notation $y(5)$ means to substitute 5 for x, so

$$y(5) = 3(5)^2 - 2(5)$$
$$= 75 - 10 = 65 \qquad ◆◆◆$$

◆◆◆ **Example 17:** *An Application.* The displacement s, in feet, of an object thrown downward with an initial velocity v_0 is a function of time, given by

$$s = f(t) = v_0 t + \frac{at^2}{2}$$

where $a = 32.2$ ft/s^2. If $v_0 = 12.5$ ft/s, find $f(2.00)$.

Solution: The notation $f(2.00)$ means that we are to find s when $t = 2.00$ s. Substituting gives

$$s = f(t) = f(2.00)$$

$$= \frac{12.5 \text{ ft}}{\text{s}}(2.00 \text{ s}) + \frac{1}{2}\frac{32.2 \text{ ft}}{\text{s}^2}(2.00 \text{ s})^2$$

$$= 12.5(2.00) \text{ ft} + \frac{1}{2}(32.2)(2.00)^2 \text{ ft}$$

$$= 89.4 \text{ ft} \qquad ◆◆◆$$

Functions by Calculator

We can define functions on both the TI-83/84 and TI-89 calculators. We can substitute numbers into a function on the TI-83/84, and substitute both numbers and letters on the TI-89. We will show both.

◆◆◆ **Example 18:** Define the function $y = 2x + 5$ on the TI-83/84. Then find $y(3)$.

Solution: *Screen 1:* Press the ⃞Y= key. Enter $2x + 5$ for Y1, with X obtained by pressing the ⃞X,T,θ,*n* key. *Screen 2:* Press ⃞QUIT to return to the main screen. Press ⃞VARS and select **Y-VARS** and **1: Function**. *Screen 3:* Select **Y1**. *Screen 4:* Then enter (3) following Y1. The value of $y(3)$ will then be given.

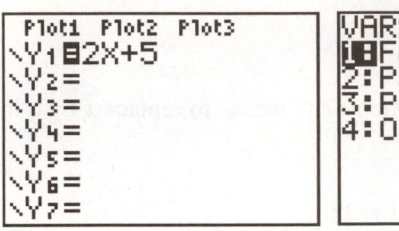   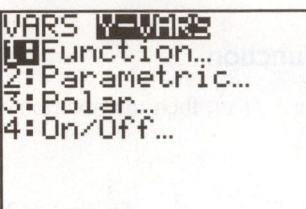

TI-83/84 Screen 1.   TI-83/84 Screen 2.

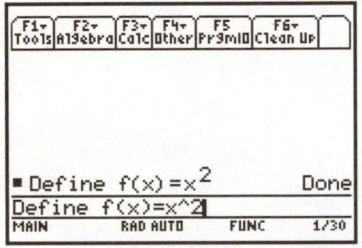

TI-89 screen for Example 19.

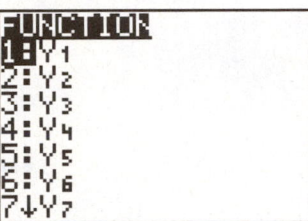   

TI-83/84 Screen 3.   TI-83/84 Screen 4.   ◆◆◆

◆◆◆ **Example 19:** To define the function, say, $f(x) = x^2$, on the TI-89, we select **Define** from the ⃞CATALOG menu. We then type $f(x) = x \wedge 2$, as shown. Here, f is simply an alpha character.   ◆◆◆

◆◆◆ **Example 20:** On the TI-89:

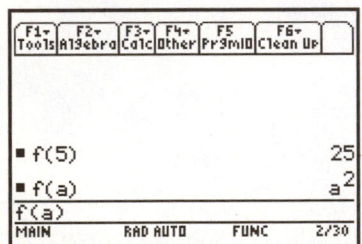

TI-89 screen for Example 20.

(a) To substitute a value into the function $f(x)$, say $x = 5$, we simply enter $f(5)$.

(b) To substitute a literal into the function $f(x)$, say $x = a$, we simply enter $f(a)$. These substitutions are shown.   ◆◆◆

We can combine functions as well.

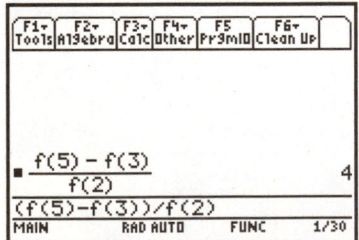

TI-89 screen for Example 21.

◆◆◆ **Example 21:** For the function of the preceding example, evaluate

$$\frac{f(5) - f(3)}{f(2)}$$

Solution: After defining the function as before, we enter

$$\big(f(5) - f(3)\big)/f(2)$$

as shown. We get the result of 4.   ◆◆◆

Functions of more than one variable can be defined.

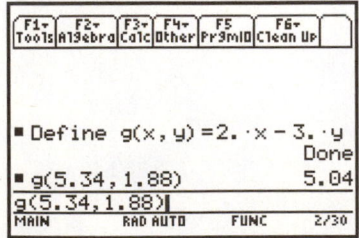

TI-89 screen for Example 22.

◆◆◆ **Example 22:** The screen shows the definition of $g(x, y) = 2x - 3y$. Then we show the substitution of the values, say

$$x = 5.34 \qquad y = 1.88$$

giving a result of 5.04.   ◆◆◆

Exercise 1 ◆ Functions and Relations

Functions vs. Relations

Which of the following relations are also functions? Explain.

1. $y = 3x^2 - 5$
2. $y = \sqrt{2x}$
3. $y = \pm\sqrt{2x}$
4. $y^2 = 3x - 5$
5. $2x^2 = 3y^2 - 4$
6. $x^2 - 2y^2 - 3 = 0$

7. Is the set of ordered pairs $(1, 3)$, $(2, 5)$, $(3, 8)$, $(4, 12)$ a function? Explain.

8. Is the set of ordered pairs $(0, 0)$, $(1, 2)$, $(1, -2)$, $(2, 3)$, $(2, -3)$ a function? Explain.

Implicit and Explicit Forms

Which equations are in explicit form and which in implicit form?

9. $y = 5x - 8$

10. $x = 2xy + y^2$

11. $3x^2 + 2y^2 = 0$

12. $y = wx + wz + xz$

Dependent and Independent Variables

Label the variables in each equation as dependent or independent.

13. $y = 3x^2 + 2x$

14. $x = 3y - 8$

15. $w = 3x + 2y$

16. $xy = x + y$

17. $x^2 + y^2 = z$

Changing from Implicit to Explicit Form

Rewrite the following implicit equations in the explicit form, $y = f(x)$.

18. $x + y = 5$

19. $2x - y + 4 = 0$

20. $x + 2y = 3x - 4y$

21. $2(3x + y) = x + 2$

Substituting into a Function

Substitute the given numerical value into each function.

22. If $f(x) = 2x^2 + 4$, find $f(3)$.

23. If $f(x) = 5x + 1$, find $f(1)$.

24. If $f(x) = 15x + 9$, find $f(3)$.

25. If $f(x) = 5 - 13x$, find $f(2)$.

26. If $g(x) = 9 - 3x^2$, find $g(-2)$.

27. If $h(x) = x^3 - 2x + 1$, find $h(2.55)$.

28. If $f(x) = 7 + 2x$, find $f(3)$.

29. If $f(x) = x^2 - 9$, find $f(-2)$.

30. If $f(x) = 2x + 7$, find $f(-1)$.

Applications

31. The distance traveled by a freely falling body is a function of the elapsed time t:

$$f(t) = v_0t + \tfrac{1}{2}gt^2 \quad \text{ft}$$

where v_0 is the initial velocity and g is the acceleration due to gravity (32.2 ft/s^2). If v_0 is 55.0 ft/s, find $f(10.0)$, $f(15.0)$, and $f(20.0)$.

32. The resistance R of a conductor is a function of temperature:

$$f(t) = R_0(1 + \alpha t)$$

where R_0 is the resistance at 0°C and α is the temperature coefficient of resistance (0.00427 for copper). If the resistance of a copper coil is 9800 Ω at 0°C, find $f(20.0)$, $f(25.0)$, and $f(30.0)$.

33. The maximum deflection in inches of a certain cantilever beam, with a concentrated load applied r feet from the fixed end, is a function of r:

$$f(r) = 0.000030r^2(80 - r) \quad \text{in.}$$

Find the deflections $f(10)$ and $f(15)$.

34. The length of a certain steel bar at temperature t is given by

$$L = f(t) = 96.0\left[1 + 15.0 \times 10^{-6}(t - 75.0)\right] \quad \text{in.}$$

Find $f(155)$.

35. The power P in a resistance R in which a current I flows is given by

$$P = f(I) = I^2R \quad \text{watt} \quad \text{(W)}$$

If $R = 25.6$ Ω, find $f(1.59)$, $f(2.37)$, and $f(3.17)$.

36. *Writing:* The word "function" is used in everyday speech, such as, "The rate of advance in cancer research is a function of the amount of money invested." Find other examples such as this, and relate their meaning to the mathematical meaning of *function* that we have given here.

4–2 More on Functions

In this section we will cover a few more ideas connected with functions that will be useful in later chapters.

A Function as a Set of Point Pairs

We have said that a function or relation can be given in forms other than an equation, such as a set of point pairs, a verbal statement, or a graph. Here we will have another look at sets of point pairs and verbal statements, and do graphing in the next chapter.

■ Exploration:

Try this. Suppose you want to describe how "strong" or how "weak" a certain rubber band is. Hang the band from a shelf, as in Fig. 4–4, tape a ruler alongside, and use a paper clip to attach a plastic bag. Put some pennies into the bag and record the change in length of the band. Then change the number of pennies and record the length. Repeat for different numbers of pennies.

When finished, how did you report your findings? You probably expressed your results not as single numbers, but as *pairs* of numbers, something like the following.

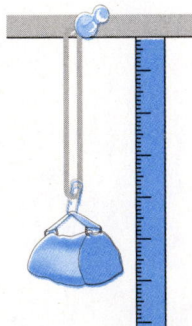

FIGURE 4–4 This may seem trivial, but the same method is used to measure the elasticity of a spring, a steel girder, or other structural member.

Number of Pennies	10	20	30	40	50
Inches Stretched	1.2	2.4	3.5	4.9	6.1

Data such as this, resulting from observation or experiment, is called *empirical data*.

In technical work, we must often deal with *pairs* of numbers, rather than single values. If 10 pennies caused a stretch of 1.2 inches, then 10 and 1.2 in. are called *corresponding values*. Neither number has significance without the other; they only have significance as a pair. Further, if we always write the pair in the same order (say, number of pennies first, followed by inches stretched) it is called an *ordered pair* of numbers. The ordered pair is written (10, 1.2).

We have seen earlier that a function is often given in the form of an equation. However, *a table of point pairs is another way of giving a function.* ◼

◆◆◆ Example 23: Suppose that a similar experiment gave the following results:

Load (kg)	0	1	2	3	4	5	6	7	8
Stretch (cm)	0	0.5	0.9	1.4	2.1	2.4	3.0	3.6	4.0

This table of ordered pairs is a function. The domain is the set of numbers (0, 1, 2, 3, 4, 5, 6, 7, 8), and the range is the set of numbers (0, 0.5, 0.9, 1.4, 2.1, 2.4, 3.0, 3.6, 4.0). The table associates exactly one number in the range with each number in the domain. ◆◆◆

A set of ordered pairs is not always given in table form.

◆◆◆ Example 24: The function of Example 23 can also be written as the set of ordered pairs

(0, 0), (1, 0.5), (2, 0.9), (3, 1.4), (4, 2.1), (5, 2.4), (6, 3.0), (7, 3.6), (8, 4.0) ◆◆◆

A *relation* can also be given as a set of point pairs. But for each value of the independent variable, we may have *two* values of the dependent variable.

◆◆◆ Example 25: *An Application.* A rocket, Fig. 4–5, is fired upward with an initial velocity of 200 ft/s. The times t at which it reaches a height h are given by the table,

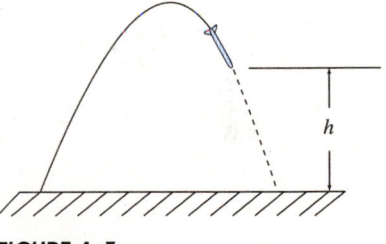

FIGURE 4–5

h (ft)	t_1 (s)	t_2 (s)
0	0.00	7.03
100	0.52	6.97
200	1.10	6.89
300	1.74	6.81
400	2.50	6.72
500	3.45	6.60
600	5.00	6.41

For each height there are *two* times (except when the rocket is at maximum height): one for the rocket on the way up, and the other when it reaches the same height on the way down. ◆◆◆

A Function as a Verbal Statement

A function may also be given in verbal form.

◆◆◆ Example 26: "The shipping charges are 55¢/lb for the first 50 and 45¢ thereafter." This verbal statement is a function relating the shipping costs to the weight of the item. ◆◆◆

We sometimes want to switch from verbal form to an equation, or vice versa.

◆◆◆ **Example 27:** Write y as a function of x if y equals twice the cube of x diminished by half of x.

Solution: We replace the verbal statement by the equation

$$y = 2x^3 - \frac{x}{2}$$

◆◆◆ **Example 28:** The equation $y = 5x^2 + 9$ can be stated verbally as "y equals the sum of 9 and 5 times the square of x."   ◆◆◆

◆◆◆ **Example 29:** Express the volume V of a cone having a base area of 75 units *as a function of* its altitude H.

Solution: This is another way of saying, "Write an equation for *the volume of a cone in terms of its base area and altitude.*"

The formula for the volume of a cone is given in Appendix A, "Summary of Facts and Formulas."

$$V = \frac{1}{3}(\text{base area})(\text{altitude}) = \frac{1}{3}(75)H$$

So

$$V = 25H$$

is the required expression.   ◆◆◆

Substituting into a Function

We previously substituted *numerical* values into a function. We can substitute *literal* values into a function in the same way.

◆◆◆ **Example 30:** Given $f(x) = 3x^2 - 2x + 3$, find $f(5a)$.

Solution: We substitute $5a$ for x.

$$f(5a) = 3(5a)^2 - 2(5a) + 3$$
$$= 3(25a^2) - 10a + 3$$
$$= 75a^2 - 10a + 3 \qquad ◆◆◆$$

◆◆◆ **Example 31:** If $f(x) = x^2 - 2x$, find $f(w^2)$.

Solution:

$$f(w^2) = (w^2)^2 - 2(w^2) = w^4 - 2w^2 \qquad ◆◆◆$$

Sometimes we must substitute into a function containing *more than one variable.*

◆◆◆ **Example 32:** Given $f(x, y, z) = 2y - 3z + x$, find $f(3, 1, 2)$.

Solution: We substitute the given numerical values for the variables. Be sure that the numerical values are taken in the *same order* as the variable names in the functional notation.

$$f(x, y, z)$$
$$\updownarrow \updownarrow \updownarrow$$
$$f(3, 1, 2)$$

Substituting, we obtain

$$f(x, y, z) = 2y - 3z + x$$
$$f(3, 1, 2) = 2(1) - 3(2) + 3 = 2 - 6 + 3$$
$$= -1 \qquad ◆◆◆$$

Manipulating Functions

Functional notation provides us with a convenient way of indicating what is to be done with a function or functions. This includes solving an equation for a different variable, changing from implicit to explicit form, or vice versa, combining two or more functions to make another function, or substituting numerical or literal values into an equation.

◆◆◆ **Example 33:** Write the equation $y = 2x - 3$ in the form $x = f(y)$.

Solution: We are being asked to write the given equation with x, instead of y, as the dependent variable. Solving for x, we obtain

$$2x = y + 3$$

$$x = \frac{y + 3}{2} \qquad \qquad \text{◆◆◆}$$

◆◆◆ **Example 34:** Write the equation $x = 27 - 3y$ in the form $y = f(x)$.

Solution: We are asked to write the equation with y as the dependent variable, so we solve for y. Rearranging gives

$$3y = 27 - x$$

Dividing by 3 we get

$$y = 9 - \frac{x}{3} \qquad \qquad \text{◆◆◆}$$

◆◆◆ **Example 35:** Write the equation $y = 3x^2 - 2x$ in the form $f(x, y) = 0$.

Solution: We are asked here to go from an explicit to an implicit form. Rearranging gives us

$$3x^2 - 2x - y = 0 \qquad \qquad \text{◆◆◆}$$

Composite Functions

Just as we can substitute a constant or a variable into a given function, so we can substitute a *function* into a function.

◆◆◆ **Example 36:** If $g(x) = x + 1$, find

(a) $g(2)$ (b) $g(z^2)$ (c) $g[f(x)]$

Solution:

(a) $g(2) = 2 + 1 = 3$
(b) $g(z^2) = z^2 + 1$
(c) $g[f(x)] = f(x) + 1 \qquad \qquad \text{◆◆◆}$

Being made up of the two functions $g(x)$ and $f(x)$, the function $g[f(x)]$, which we read "g of f of x," is called a *composite function*. If we think of a function as a machine, it is as if we are using the output $f(x)$ of the function machine f as the input of a second function machine g.

$$x \to \boxed{f} \to f(x) \to \boxed{g} \to g[f(x)]$$

We thus obtain $g[f(x)]$ by replacing x in $g(x)$ by the function $f(x)$.

◆◆◆ **Example 37:** Given the functions $g(x) = x + 1$ and $f(x) = x^3$, write the composite function $g[f(x)]$.

Solution: In the function $g(x)$, we replace x by $f(x)$.

$$g(x) = x + 1$$
$$\downarrow \qquad \downarrow$$
$$g[f(x)] = f(x) + 1$$
$$\downarrow$$
$$= x^3 + 1$$

since $f(x) = x^3$.   ◆◆◆

As we have said, the notation $g[f(x)]$ means to substitute $f(x)$ into the function $g(x)$. On the other hand, the notation $f[g(x)]$ means to substitute $g(x)$ into $f(x)$.

$$x \rightarrow \boxed{g} \rightarrow g(x) \rightarrow \boxed{f} \rightarrow f[g(x)]$$

In general, $f[g(x)]$ *will not be the same as* $g[f(x)]$.

◆◆◆ **Example 38:** Given $g(x) = x^2$ and $f(x) = x + 1$, find the following:

(a) $f[g(x)]$ (b) $g[f(x)]$

(c) $f[g(2)]$ (d) $g[f(2)]$

Solution:

(a) $f[g(x)] = g(x) + 1 = x^2 + 1$ (b) $g[f(x)] = [f(x)]^2 = (x + 1)^2$

(c) $f[g(2)] = 2^2 + 1 = 5$ (d) $g[f(2)] = (2 + 1)^2 = 9$

Notice that here $f[g(x)]$ is not equal to $g[f(x)]$.   ◆◆◆

Inverse of a Function

Consider a function f that, given a value of x, returns some value of y.

$$x \rightarrow \boxed{f} \rightarrow y$$

If y is now put into a function g that *reverses the operations* performed in f so that its output is the original x, then g is called the *inverse* of f.

$$x \rightarrow \boxed{f} \rightarrow y \rightarrow \boxed{g} \rightarrow x$$

The inverse of a function $f(x)$ is often designated by $f^{-1}(x)$.

$$x \rightarrow \boxed{f} \rightarrow y \rightarrow \boxed{f^{-1}} \rightarrow x$$

Common Error	Do not confuse f^{-1} with $1/f$.

◆◆◆ **Example 39:** Two such inverse operations are "cube" and "cube root."

$$x \rightarrow \boxed{\text{cube } x} \rightarrow x^3 \rightarrow \boxed{\text{take cube root}} \rightarrow x$$ ◆◆◆

Thus if a function $f(x)$ has an inverse $f^{-1}(x)$ that reverses the operations in $f(x)$, then the *composite* of $f(x)$ and $f^{-1}(x)$ should have no overall effect. If the input is x, then the output must also be x. In symbols, if $f(x)$ and $f^{-1}(x)$ are inverse functions, then

$$f^{-1}[f(x)] = x$$

and

$$f[f^{-1}(x)] = x$$

Conversely, if $g[f(x)] = x$ and $f[g(x)] = x$, then $f(x)$ and $g(x)$ are inverse functions. ♦♦♦

♦♦♦ **Example 40:** Using the example of the cube and cube root, if

$$f(x) = x^3 \quad \text{and} \quad g(x) = \sqrt[3]{x}$$

then

$$g[f(x)] = \sqrt[3]{f(x)} = \sqrt[3]{x^3} = x$$

and

$$f[g(x)] = [g(x)]^3 = (\sqrt[3]{x})^3 = x$$

This shows that here $f(x)$ and $g(x)$ are indeed inverse functions. ♦♦♦

To find the inverse of a function $y = f(x)$:

1. Solve the given equation for x.
2. Interchange x and y.

♦♦♦ **Example 41:** We use the cube and cube root example one more time. Find the inverse $g(x)$ of the function

$$y = f(x) = x^3$$

Solution: We solve for x and get

$$x = \sqrt[3]{y}$$

It is then customary to interchange variables so that the dependent variable is (as usual) y. This gives

$$y = f^{-1}(x) = \sqrt[3]{x}$$

Thus $f^{-1}(x) = \sqrt[3]{x}$ is the inverse of $f(x) = x^3$, as verified earlier. ♦♦♦

♦♦♦ **Example 42:** Find the inverse $f^{-1}(x)$ of the function

$$y = f(x) = 2x + 5$$

Solution: Solving for x gives

$$x = \frac{y - 5}{2}$$

Interchanging x and y, we obtain

$$y = f^{-1}(x) = \frac{x - 5}{2}$$ ♦♦♦

Sometimes the inverse of a function will *not* be a function itself, but it may be a relation.

♦♦♦ **Example 43:** Find the inverse of the function $y = x^2$.

Solution: Take the square root of both sides.

$$x = \pm \sqrt{y}$$

Interchanging x and y, we get

$$y = \pm \sqrt{x}$$

Thus a single value of x (say, 4) gives *two* values of y ($+2$ and -2), so our inverse does not meet the definition of a function. It is, however, a relation.   ◆◆◆

Sometimes the inverse of a function gets a special name. The inverse of the sine function, for example, is called the *arcsin*, and the inverse of an exponential function is a *logarithmic* function.

Finding Domain and Range

The inverse trigonometric functions and the exponential function and its inverse, the logarithmic function, are covered later.

To be strictly correct, the domain should be stated whenever an equation is written. However, this is often not done, so we follow this convention:

> The *domain* of the function $y = f(x)$ is the largest set of x values that will give real values of y. The *range* of the function is the set of all such values of y.

◆◆◆ **Example 44:** The function

$$y = x^2$$

gives a real y value for every real x value. Thus the domain is all of the real numbers, which we can write

$$\text{Domain:} \quad -\infty < x < \infty$$

But notice that there is no x that will make y negative. Thus the range of y includes all the positive numbers and zero. This can be written

$$\text{Range:} \quad y \geq 0$$ ◆◆◆

Our next example is one in which certain values of x result in a *negative number under a radical sign.*

◆◆◆ **Example 45:** Find the domain and the range of the function

$$y = \sqrt{x - 2}$$

Solution: Our method is to see what values of x and y "do not work" (give a non-real result or an illegal operation); then the domain and the range will be those values that "do work."

Any value of x less than 2 will make the quantity under the radical sign negative, resulting in an imaginary y. Thus the domain of x is all the positive numbers equal to or greater than 2.

$$\text{Domain:} \quad x \geq 2$$

An x equal to 2 gives a y of zero. Any x larger than 2 gives a real y greater than zero. Thus the range of y is

$$\text{Range:} \quad y \geq 0$$ ◆◆◆

Our next example shows some values of x that result in *division by zero.*

◆◆◆ **Example 46:** Find the domain and the range of the function

$$y = \frac{1}{x}$$

Solution: Here any value of x but zero will give a real y. Thus the domain is

$$\text{Domain:} \quad x \neq 0$$

Notice here that it is more convenient to state which values of x *do not* work, rather than those that do.

Now what happens to y as x varies over its domain? We see that large x values give small y values, and conversely that small x values give large y values. Also, negative x's give negatives y's. However, there is no x that will make y zero, so

Range: $y \neq 0$ ◆◆◆

Our final example shows both division by zero and negative numbers under a radical sign, for certain x values.

◆◆◆ **Example 47:** Find the domain and the range of the function

$$y = \frac{9}{\sqrt{4 - x}}$$

Solution: Since the denominator cannot be zero, x cannot be 4. Also, any x greater than 4 will result in a negative quantity under the radical sign. The domain of x is then

Domain: $x < 4$

Restricted to these values, the quantity $4 - x$ is positive and ranges from very small (when x is nearly 4) to very large (when x is large and negative). Thus the denominator is positive, since we allow only the principal (positive) root and we can vary from near zero to infinity. The range of y, then, includes all the values greater than zero.

Range: $y > 0$ ◆◆◆

Exercise 2 ◆ More on Functions

A Function as a Verbal Statement

For each of the following, write y as a function of x, where the value of y is equal to the given expression.

1. The cube of x
2. The square root of x, diminished by 5
3. x increased by twice the square of x
4. The reciprocal of the cube of x
5. Two-thirds of the amount by which x exceeds 4

Write the equation called for in each of the following statements. Refer to Appendix A, "Summary of Facts and Formulas," if necessary.

6. Express the area A of a triangle as a function of its base b and altitude h.
7. Express the hypotenuse c of a right triangle as a function of its legs, a and b.
8. Express the volume V of a sphere as a function of the radius r.
9. Express the power P dissipated in a resistor as a function of its resistance R and the current I through the resistor.
10. A car is traveling at a speed of 55 mi/h. Write the distance d traveled by the car as a function of time t.
11. To ship its merchandise, a mail-order company charges 65¢/lb plus $2.25 for handling and insurance. Express the total shipping charge s as a function of the item weight w.
12. A projectile is shot upward with an initial velocity of 125 m/s. Express the height H of the projectile as a function of time.

Substituting into a Function

Substitute the literal values into each function.

13. If $f(x) = 2x^2 + 4$, find $f(a)$.

14. If $f(x) = 2x - \dfrac{1}{x} + 4$, find $f(2a)$.

15. If $f(x) = 5x + 1$, find $f(a + b)$.

16. If $f(x) = 5 - 13x$, find $f(-2c)$.

Substitute the sets of values into each function.

17. If $f(x, y) = 3x + 2y^2 - 4$, find $f(2, 3)$.

18. If $f(x, y, z) = 3z - 2x + y^2$, find $f(3, 1, 5)$.

19. If $g(a, b) = 2b - 3a^2$, find $g(4, -2)$.

20. If $f(x, y) = y - 3x$, find $3f(2, 1) + 2f(3, 2)$.

Manipulating Functions

21. If $y = 5x + 3$, write $x = f(y)$.

22. If $x = \dfrac{2}{y - 3}$, write $y = f(x)$.

23. If $y = \dfrac{1}{x} - \dfrac{1}{5}$, write $x = f(y)$.

24. If $x^2 + y = x - 2y + 3x^2$, write $y = f(x)$.

25. If $5p - q = q - p^2$, write $q = f(p)$.

26. The power P dissipated in a resistor is given by $P = I^2 R$. Write $R = f(P, I)$.

27. Young's modulus E is given by

$$E = \frac{PL}{ae}$$

Write $e = f(P, L, a, E)$.

Composite Functions

28. Given the functions $g(x) = 2x + 3$ and $f(x) = x^2$, write the composite function $g[f(x)]$.

29. Given the functions $g(x) = x^2 - 1$ and $f(x) = 3 + x$, write the composite function $f[g(x)]$.

30. Given the functions $g(x) = 1 - 3x$ and $f(x) = 2x$, write the composite function $g[f(x)]$.

31. Given the functions $g(x) = x - 4$ and $f(x) = x^2$, write the composite function $f[g(x)]$.

Given $g(x) = x^3$ and $f(x) = 4 - 3x$, find

32. $f[g(x)]$ **33.** $f[g(3)]$

34. $g[f(x)]$ **35.** $g[f(3)]$

Inverse of a Function

Find the inverse of

36. $y = 8 - 3x$

37. $y = 5(2x - 3) + 4x$

38. $y = 7x + 2(3 - x)$

39. $y = (1 + 2x) + 2(3x - 1)$

40. $y = 3x - (4x + 3)$

41. $y = 2(4x - 3) - 3x$

42. $y = 4x + 2(5 - x)$

43. $y = 3(x - 2) - 4(x + 3)$

Finding Domain and Range

State the domain and the range of each function.

44. $(0, 2), (1, 4), (2, 8), (3, 16), (4, 32)$

45. $(-10, 20), (5, 7), (-7, 10), (10, 20), (0, 3)$

46.

x	2	4	6	8	10
y	0	-2	-5	-9	-15

Find the domain and the range for each function.

47. $y = \sqrt{x - 7}$

48. $y = \dfrac{3}{\sqrt{x - 2}}$

49. $y = x - \dfrac{1}{x}$

50. $y = \sqrt{x^2 - 25}$

51. $y = \dfrac{4}{\sqrt{1 - x}}$

52. $y = \dfrac{x + 1}{x - 1}$

53. $y = \sqrt{x - 1}$

54. Find the range of the function $y = 3x^2 - 5$ whose domain is $0 \le x \le 5$.

55. *Writing*: State, in your own words, what is meant by "function" and "relation," and describe how a function differs from a relation. You may use examples to help describe these terms.

◆◆◆ CHAPTER 4 REVIEW PROBLEMS ◆◆◆◆◆◆◆◆◆◆◆◆◆◆◆◆◆◆◆◆◆◆◆◆◆◆◆◆◆◆

1. Which of the following relations are also functions?
   (a) $y = 5x^3 - 2x^2$ (b) $x^2 + y^2 = 25$
   (c) $y = \pm\sqrt{2x}$

2. Write y as a function of x if y is equal to half the cube of x, diminished by twice x.

3. Write an equation to express the surface area S of a sphere as a function of its radius r.

4. Find the domain and range for each function.
   (a) $y = \dfrac{5}{\sqrt{3 - x}}$ (b) $y = \dfrac{1 + x}{1 - x}$

5. Label each function as implicit or explicit. If it is explicit, name the dependent and independent variables.
   (a) $w = 3y - 7$
   (b) $x - 2y = 8$

6. Given $y = 3x - 5$, write $x = f(y)$.

7. Given $x^2 + y^2 + 2w = 3$, write $w = f(x, y)$.

8. Write the inverse of the function $y = 9x - 5$.

9. If $y = 3x^2 + 2z$ and $z = 2x^2$, write $y = f(x)$.

10. If $f(x) = 5x^2 - 7x + 2$, find $f(3)$.

11. If $f(x) = 9 - 3x$, find $2f(3) + 3f(1) - 4f(2)$.

12. If $f(x) = 7x + 5$ and $g(x) = x^2$, find $3f(2) - 5g(3)$.

13. If $f(x, y, z) = x^2 + 3xy + z^3$, find $f(3, 2, 1)$.

14. If $f(x) = 8x + 3$ and $g(u) = u^2 - 4$, write $f[g(u)]$.

15. If $f(x) = 5x$, $g(x) = 1/x$, and $h(x) = x^3$, find

$$\frac{5h(1) - 2g(3)}{3f(2)}$$

16. Is the relation $x^2 + 3xy + y^2 = 1$ a function?

17. Replace the function $y = 5x^3 - 7$ by a verbal statement.

18. What are the domain and the range of the function $(10, -8)$, $(20, -5)$, $(30, 0)$, $(40, 3)$, and $(50, 7)$?

19. If $y = 6 - 3x^2$, write $x = f(y)$.

20. If $x = 6w - 5y$ and $y = 3z + 2w$, write $x = f(w, z)$.

21. Write the inverse of the function $y = 3x + 4(2 - x)$.

22. If $7p - 2q = 3q - p^2$, write $q = f(p)$.

23. Given the functions $g(x) = x + x^2$ and $f(x) = 5x$, write the composite function $g[f(x)]$.

24. If $h(p, q, r) = 5r + p + 7q$, find $h(1, 2, 3) + \dfrac{2h(1, 1, 1) - 3h(2, 2, 2)}{h(3, 3, 3)}$

25. If $g_3(w) = 1.74 + 9.25w$, find $g_3(1.44)$.

26. Write the inverse of the function $y = 5(7 - 4x) - x$.

27. If $g(k,l,m) = \dfrac{k + 3m}{2l}$, find $g(4.62, 1.39, 7.26)$.

28. Write the inverse of the function $y = 9x + (x + 5)$.

29. The length of a certain steel bar at temperature t is given by

$$L = f(t) = 112\,(1 + 0.0655t) \quad \text{in.}$$

Find $f(112), f(176)$, and $f(195)$.

30. If $f_1(z) = 3.82z^2 - 2.46$, find $f_1(5.27)$.

31. The power P in a resistance R in which flows a current I is given by

$$P = f(I) = I^2 R \quad \text{W}$$

If $R = 325\ \Omega$, find $f(11.2), f(15.3)$, and $f(21.8)$.

32. Given the functions $g(x) = 3x - 5$ and $f(x) = 7x^2$, write the composite function $f[g(x)]$.

33. Given $g(x) = 5x^2$ and $f(x) = 7 - 2x$ find
    (a) $f[g(x)]$
    (b) $g[f(x)]$
    (c) $f[g(5)]$
    (d) $g[f(5)]$

5

Graphs

◆◆◆ **OBJECTIVES** ◆◆◆

When you have completed this chapter, you should be able to

- Graph a table of point pairs or an empirical function.
- Graph any function given in explicit form.
- Graph some functions given in implicit form.
- Graph a relation.
- Use the graphing calculator to make a complete graph of a function.
- Find the slope of a straight line.
- Graph a straight line given its slope and *y*-intercept.
- Write the equation of a straight line, given its slope and *y*-intercept.
- Write the equation of a line given two points on the line.
- Solve equations graphically.

◆◆◆

It's said that "a picture is worth a thousand words," and in this chapter we will show how to make a picture of an equation, that is, how to *draw a graph*. After introducing the *rectangular coordinate system*, we will plot points, and then data derived from an experiment. We will go on to graph functions and relations, given in both explicit and implicit form, first manually, and then with the graphing calculator. The idea of a *complete graph* will be stressed.

Next we will examine the *straight line* in more detail, giving a quick way to graph it given the equation, and to find its equation, given some information about the line.

Finally covered are methods to approximately solve any equation graphically. This is very important because many equations cannot be solved in any other way.

There will be more on graphing later. For example, we will draw other kinds of graphs in our chapter on statistics: bar charts, pie charts, scatter plots, and so forth. Also, our companion Web site shows how to make graphs on logarithmic and semilogarithmic graph paper.

5–1 Rectangular Coordinates

In Chapter 1 we plotted numbers on the number line. Suppose, now, that we take a second number line and place it at right angles to the first one, so that each intersects the other at the zero mark, as in Fig. 5–1. We call this a *rectangular coordinate system*.

The horizontal number line is called the *x axis*, and the vertical line is called the *y axis*. They intersect at the *origin*. These two axes divide the plane into four *quadrants*, numbered counterclockwise, as shown.

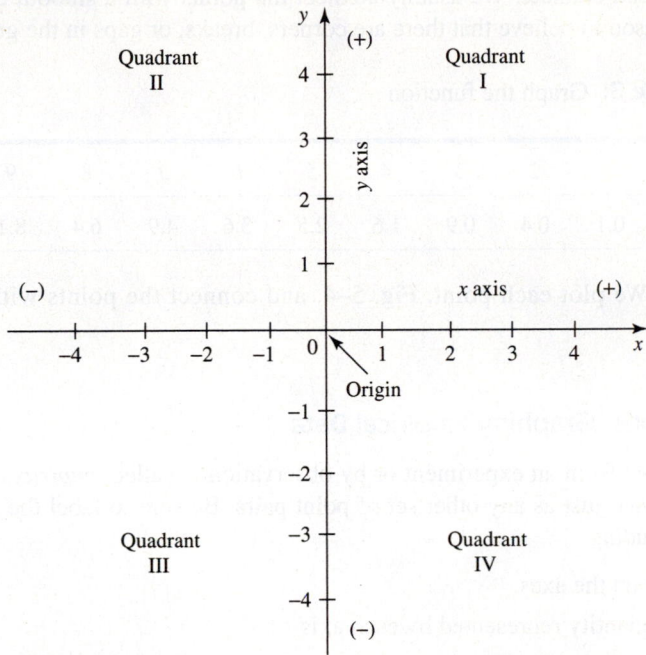

FIGURE 5–1   The rectangular coordinate system. Rectangular coordinates are also called *Cartesian* coordinates, after the French mathematician René Descartes (1596–1650). Another type of coordinate system that we will use later is called the *polar* coordinate system.

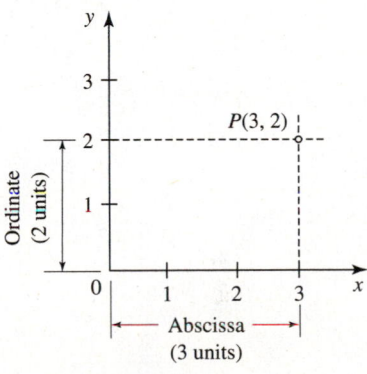

FIGURE 5–2   Is it clear from this figure why we call these *rectangular coordinates?*

Figure 5–2 shows a point *P* in the first quadrant. Its horizontal distance from the origin, called the *x coordinate* or *abscissa* of the point, is 3 units. Its vertical distance from the origin, called the *y coordinate* or *ordinate* of the point, is 2 units. The numbers in the *ordered pair* (3, 2) are called the *rectangular coordinates* (or simply *coordinates*) of the point. They are always written in the same order, with the *x* coordinate first. The letter identifying the point is sometimes written before the coordinates, as in $P(3, 2)$.

Graphing Point Pairs

To plot any ordered pair (h, k), simply place a point at a distance h units from the *y* axis and k units from the *x* axis. Remember that negative values of *x* are located to the left of the origin and that negative *y* values are below the origin.

◆◆◆ **Example 1:** The points

$P(4, 1)$, $Q(-2, 3)$, $R(-1, -2)$, $S(2, -3)$, $T(1.3, 2.7)$, $U(3, 0)$, and $V(0, 2)$

are shown plotted in Fig. 5–3.

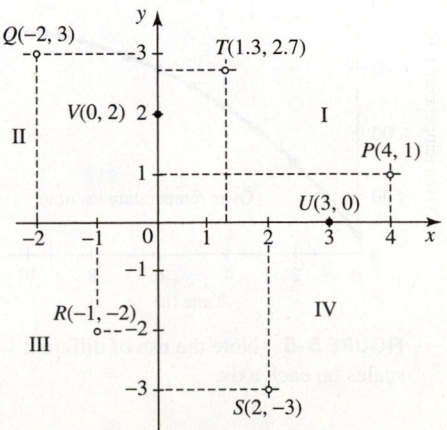

FIGURE 5–3

Notice that the abscissa is negative in the second and third quadrants and that the ordinate is negative in the third and fourth quadrants. Thus the signs of the coordinates of a point tell us the quadrant in which the point lies. ◆◆◆

◆◆◆ **Example 2:** The point $(-3, -5)$ lies in the third quadrant, for that is the only quadrant in which the abscissa and the ordinate are both negative. ◆◆◆

Graphing a Table of Point Pairs

If the function to be graphed is in the form of a table of point pairs, simply plot each point and connect. We usually connect the points with a smooth curve unless we have reason to believe that there are corners, breaks, or gaps in the graph.

◆◆◆ **Example 3:** Graph the function

x	0	1	2	3	4	5	6	7	8	9	10
y	0	0.1	0.4	0.9	1.6	2.5	3.6	4.9	6.4	8.1	10

Solution: We plot each point, Fig. 5–4, and connect the points with a smooth curve. ◆◆◆

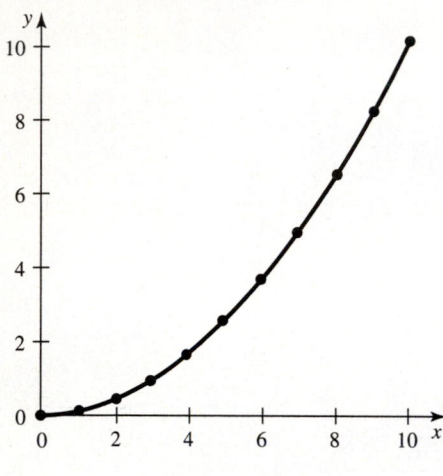

FIGURE 5–4

Applications: Graphing Empirical Data

Data obtained from an experiment or by observation is called *empirical data*. Such data is graphed just as any other set of point pairs. Be sure to label the graph completely, including

- Units on the axes
- The quantity represented by each axis
- The title of the graph

◆◆◆ **Example 4:** (a) Graph the following data for the temperature rise in a certain oven:

Time (h)	0	1	2	3	4	5	6	7	8	9	10
Temperature (°F)	102	463	748	1010	1210	1370	1510	1590	1710	1770	1830

From the graph, estimate
(b) The oven temperature at 7.5 h.
(c) The time at which the oven temperature is 1000°F.
(d) The time it takes for the oven to heat from 500°F to 1000°F.
(e) The time it takes for the oven to heat from 1000°F to 1500°F.
(f) Is the rate of temperature rise increasing or decreasing with time?

Solution: (a) We plot each point, connect the points, and label the graph, as shown in Fig. 5–5.

From the graph, Fig. 5–5, we read, as closely as we can,
(b) Approximately 1600°F at 7.5 h.
(c) Approximately 2.9 h to reach 1000°F.
(d) Approximately $(2.9 - 1.1 = 1.8\ \text{h})$ to heat from 500°F to 1000°F.
(e) Approximately $(5.9 - 2.9 = 3.0\ \text{h})$ to heat from 1000°F to 1500°F.
(f) Decreasing. It took 1.2 h longer to heat from 1000°F to 1500°F than for 500°F to 1000°F. ◆◆◆

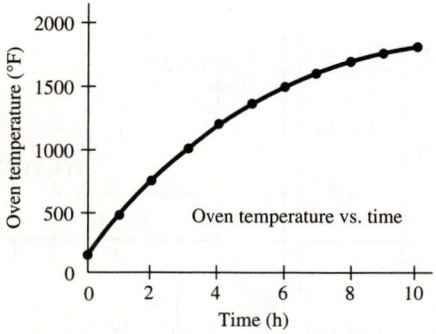

FIGURE 5–5 Note the use of different scales on each axis.

Exercise 1 ◆ Rectangular Coordinates

If h and k are positive quantities, in which quadrants would the following points lie?

1. $(h, -k)$ **2.** (h, k)

3. $(-h, k)$ **4.** $(-h, -k)$

5. Which quadrant contains points having a positive abscissa and a negative ordinate?

6. In which quadrants is the ordinate negative?

7. In which quadrants is the abscissa positive?

8. The ordinate of any point on a certain straight line is -5. Give the coordinates of the point of intersection of that line and the y axis.

9. Find the abscissa of any point on a vertical straight line that passes through the point $(7, 5)$.

Graphing Point Pairs

10. Write the coordinates of points A, B, C, and D in Fig. 5–6.

11. Write the coordinates of points E, F, G, and H in Fig. 5–6.

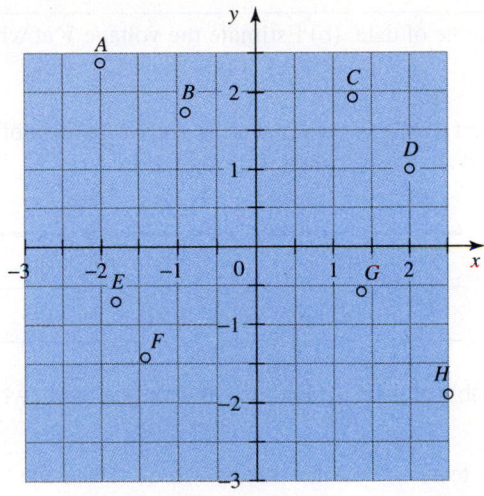

FIGURE 5–6

12. Graph each point.

(a) $(3, 5)$ (b) $(4, -2)$ (c) $(-2.4, -3.8)$

(d) $(-3.5, 1.5)$ (e) $(-4, 3)$ (f) $(-1, -3)$

Graph each set of points, connect them, and identify the geometric figure formed.

13. $(0.7, 2.1)$, $(2.3, 2.1)$, $(2.3, 0.5)$, and $(0.7, 0.5)$

14. $(2, -\frac{1}{2})$, $(3, -1\frac{1}{2})$, $(1\frac{1}{2}, -3)$, and $(\frac{1}{2}, -2)$

15. $(-1\frac{1}{2}, 3)$, $(-2\frac{1}{2}, \frac{1}{2})$, and $(-\frac{1}{2}, \frac{1}{2})$

16. $(-3, -1)$, $(-1, -\frac{1}{2})$, $(-2, -3)$, and $(-4, -3\frac{1}{2})$

17. Three corners of a rectangle have the coordinates $(-4, 9)$, $(8, 3)$, and $(-8, 1)$. Graphically find the coordinate of the fourth corner.

18. The diagonal of a square has the coordinates $(-15, 3)$ and $(5, -3)$. Graphically find the coordinates of the other corners of the square.

Graphing a Table of Point Pairs

Graph each set of ordered pairs. Connect them with a curve that seems to you to best fit the data.

19. $(-3, -2), (9, 6), (3, 2), (-6, -4)$

20. $(-7, 3), (0, 3), (4, 10), (-6, 1), (2, 6), (-4, 0)$

21. $(-10, 9), (-8, 7), (-6, 5), (-4, 3), (-2, 4), (0, 5), (2, 6), (4, 7)$

22. $(0, 4), (3, 3.2), (5, 2), (6, 0), (5, -2), (3, -3.2), (0, -4)$

Applications: Graphing Empirical Data

Graph the following experimental data. Label the graph completely. Take the first quantity in each table as the abscissa, and the second quantity as the ordinate. Connect the points with a smooth curve.

23. The current I (mA) through a tungsten lamp and the voltage V (V) that is applied to the lamp, which are related as shown in the following table:

V	10	20	30	40	50	60	70	80	90	100	110	120
I	158	243	306	367	420	470	517	559	598	639	676	710

(a) Graph this table of data. (b) Estimate the voltage V at which the current I is 500 mA.

24. The melting point T (°C) of a certain alloy and the percent of lead P in the alloy, which are related as shown in the following table:

P	40	50	60	70	80	90
T	186	205	226	250	276	304

(a) Graph this table of data. (b) Estimate the melting point when the percent lead is 65%.

25. A steel wire in tension, with the stress σ (lb/in.^{2}) and the strain ϵ (in./in.), are related as follows:

ϵ	0	0.00019	0.00057	0.00094	0.00134	0.00173	0.00216	0.00256
σ	5000	10,000	20,000	30,000	40,000	50,000	60,000	70,000

(a) Graph this table of data. (b) Estimate the change in stress when the strain increases from 25,000 to 40,000 in./in.

26. The modulus of elasticity E of rubber, in lb/in.^{2}, is related to its durometer hardness number D as follows:

D	27	33	38	43	47	51	53	56	59	62	66	68	69
E	129	150	180	210	240	270	300	330	360	390	450	480	510

(a) Graph this table of data. (b) Estimate the durometer hardness number that will give a modulus of elasticity of 420 lb/in.^{2}.

5–2 Graphing an Equation

In the preceding section we graphed a function that was given in the form of a table of point pairs. Now we will graph a function given in the form of an *equation*. Our first example will be an equation in *explicit form* and our second example will be an equation in *implicit form*. Then we will graph a *relation* given in the form of an equation.

To graph an equation, we first obtain a set of point pairs by substituting values of x into the equation, one at a time, and computing the corresponding values of y. We then plot each point pair. We connect these with a smooth curve, unless we have reason to believe that there are corners, breaks, or gaps in the curve. Our first graph will be of the straight line.

♦♦♦ **Example 5:** Graph the equation

$$y = f(x) = 2x - 1$$

for integer values of x from -2 to $+2$.

Solution: We first obtain a table of point pairs. Substituting into the equation we get,

$$f(-2) = 2(-2) - 1 = -5$$
$$f(-1) = 2(-1) - 1 = -3$$
$$f(0) = 2(0) - 1 = -1$$
$$f(1) = 2(1) - 1 = 1$$
$$f(2) = 2(2) - 1 = 3$$

We plot these point pairs and connect with a smooth curve, Fig. 5–7. Had we known in advance that the graph would be a straight line, we could have saved time by plotting just two points, with perhaps a third as a check. Later in this chapter we will cover the straight line in more detail and show another way to graph it. ♦♦♦

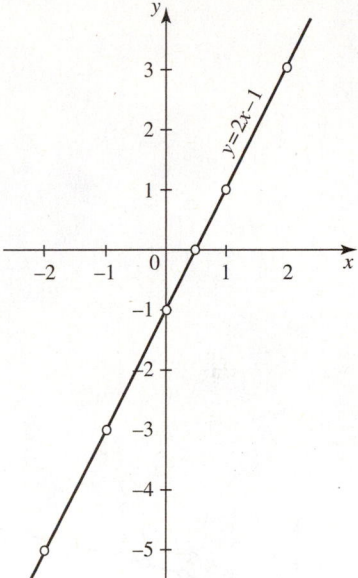

FIGURE 5–7 A first-degree equation will always plot as a straight line—hence the name *linear* equation.

Common Error	Be especially careful when substituting negative values into an equation. It is easy to make an error.

Graphing an Equation Given in Implicit Form

If the equation to be graphed is in implicit form, we must first rewrite it in explicit form, as was shown in the preceding chapter.

♦♦♦ **Example 6:** Graph the equation $x^2 - 4x - y - 3 = 0$ for integer values of x from -5 to $+5$.

Solution: We rewrite this implicit function in explicit form by solving for y, and get

$$y = x^2 - 4x - 3$$

Next we make a table of point pairs. Substituting into the equation we get

$$f(-1) = (-1)^2 - 4(-1) - 3 = 1 + 4 - 3 = 2$$
$$f(0) = 0^2 - 0 - 3 = -3$$
$$f(1) = 1^2 - 4(1) - 3 = 1 - 4 - 3 = -6$$
$$f(2) = 2^2 - 4(2) - 3 = 4 - 8 - 3 = -7$$

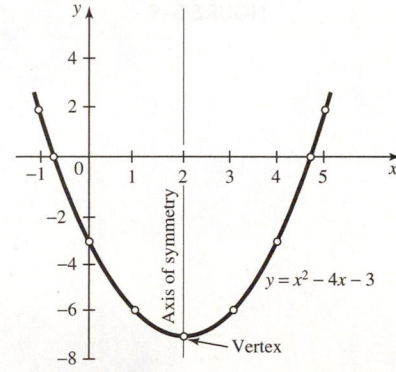

FIGURE 5–8 If we had plotted many more points than these—say, billions of them—they would be crowded so close together that they would seem to form a continuous line. The curve can be thought of as a *collection* of all points that satisfy the equation. Such a curve (or the set of points) is called a *locus* of the equation.

$$f(3) = 3^2 - 4(3) - 3 = 9 - 12 - 3 = -6$$
$$f(4) = 4^2 - 4(4) - 3 = 16 - 16 - 3 = -3$$
$$f(5) = 5^2 - 4(5) - 3 = 25 - 20 - 3 = 2$$

These points are plotted in Fig. 5–8 and connected with a smooth curve. Note that we have used *different scales* for the *x* and *y* axes to make the graph more compact.

This particular curve is called a *parabola*, a very useful and interesting curve that we will study in detail in a later chapter. The parabola is symmetrical about a line called the *axis of symmetry*. That means that any point on one side of that axis has its mirror image on the other side of that axis. The axis of symmetry cuts the parabola at a point called the *vertex*.   ◆◆◆

Graphing a Relation

Graphing a relation is not much different than graphing a function. But here we will get more than one *y* for each value of *x*. Simply plot them both.

◆◆◆ **Example 7:** Graph the relation $y = \pm\sqrt{x}$, for integer values of *x* from -5 to $+5$.

Solution: We make a table of points, and immediately notice that a negative value of *x* will give the square root of a negative number and is not permitted. We start, then, with $x = 0$.

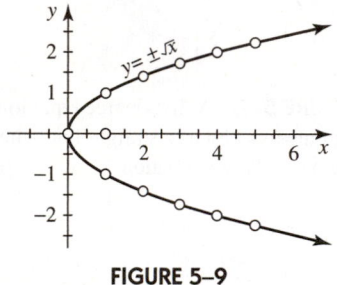

FIGURE 5–9

x	0	1	2	3	4	5
y_1	0	1	1.41	1.73	2	2.24
y_2	0	-1	-1.41	-1.73	-2	-2.24

Plotting each of the two *y* values for each *x*, we get the curve shown in Fig. 5–9. Note that it is also a parabola, as in our preceding example, but it now opens to the right.   ◆◆◆

◆◆◆ **Example 8:** *An Application.* The formula for the power *P* dissipated in a resistor carrying a current of *I* amperes (A) is

$$P = I^2R \text{ watts (W)}$$

where *R* is the resistance in ohms. Graph *P* versus *I* for a resistance of 10,000 Ω. Take *I* from 0 to 10 A.

Solution: A formula is graphed the same way we graph any equation. But here we must be careful to handle the units properly and to label the graph more fully.

Let us choose values of *I* of 0, 1, 2, 3, ..., 10 and make a table of ordered pairs by substituting these into the formula.

I (A)	0	1	2	3	...
P (W)	0	10,000	40,000	90,000	...

At this point we notice that the figures for wattage are so high that it will be more convenient to work in kilowatts (kW), where 1 kW = 1000 W.

I (A)	0	1	2	3	4	5	6	7	8	9	10
P (kW)	0	10	40	90	160	250	360	490	640	810	1000

These points are plotted in Fig. 5–10. Note the labeling of the graph and of the axes.

Exercise 2 ◆ Graphing an Equation

For each equation make a table of point pairs, taking integer values of x from -3 to 3, plot these points, and connect them with a smooth curve.

1. $y = 3x + 1$ 2. $y = 2x - 2$

3. $y = 3 - 2x$ 4. $y = -x + 2$

5. $y = x^2$ 6. $y = 4 - 2x^2$

7. $y = \dfrac{x^2}{x + 3}$ 8. $y = x^2 - 7x + 10$

9. $y = x^2 - 1$ 10. $y = 5x - x^2$

11. $y = x^3$ 12. $y = x^3 - 2$

Graphing an Equation Given in Implicit Form: Rewrite each equation in explicit form and graph for integer values of x from -3 to 3.

13. $x + y - 5 = 0$ 14. $2x + 3y = 10$

15. $x^2 + y - 4 = 0$ 16. $y + x^2 = 12$

Graphing a Relation: Graph each relation for integer values of x from -3 to 3.

17. $y = \pm 2\sqrt{x}$ 18. $y = \pm\sqrt{5x}$

19. $y = 4 \pm \sqrt{2x}$ 20. $y = \pm 3\sqrt{5x} - 3$

Applications

21. A milling machine having a purchase price P of \$15,600 has an annual depreciation A of \$1600. Graph the book value y at the end of each year, for $t = 0$ to 10 years, using the equation $y = P - At$.

22. The force f required to pull a block along a rough surface is given by $f = \mu N$, where N is the normal force and μ is the coefficient of friction. Plot f for values of N from 0 to 100 N, taking μ as 0.45.

23. A 2580-Ω resistor R_2 is wired in parallel with a resistor R_1. Graph the equivalent resistance R for values of R_1 from 0 to 5000 Ω. Use the equation

$$R = \frac{R_1 R_2}{R_1 + R_2}.$$

24. Use the equation $P = I^2 R$ to graph the power P in watts dissipated in a 2500-Ω resistor for values of current I from 0 to 1A.

25. A resistance of 5280 Ω is placed in series with a device that has a reactance of X ohms, Fig. 5–11. Using the equation $Z = \sqrt{R^2 + X^2}$, plot the impedance Z for values of X from 0 to 10,000 Ω.

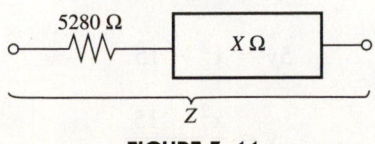

FIGURE 5–11

FIGURE 5–10 Power dissipated in a 10,000-Ω resistor calculated using $P = I^2 R$ with $R = 10,000\ \Omega$.

5–3   Graphing a Function by Calculator

We can quickly graph a function with a *graphing calculator*. The steps are

1.   From the $\boxed{\text{MODE}}$ menu, select the **FUNC** or mode.
2.   Enter the function.
3.   Set the *viewing window*.
4.   Graph.

$\boxed{\text{MODE}}$ screen for the TI-83/84.

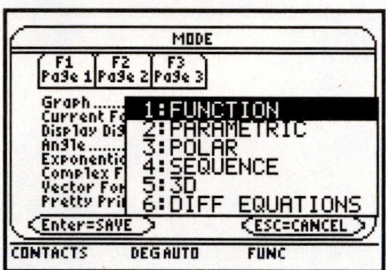

$\boxed{\text{MODE}}$ screen for the TI-89.

We will show these steps using the equation from Example 6. The screens shown will be for the TI-83/84, but are essentially the same for the TI-89.

◆◆◆ **Example 9:** Graph the function $y = x^2 - 4x - 3$.

Solution:

(1)   We *enter the function* by first pressing the $\boxed{\text{Y} =}$ key, screen (1). Enter the function on the first line. Note that there is room for several functions, but for now we will just enter one.

(2)   Next *set the viewing window* by pressing the $\boxed{\text{WINDOW}}$ key. You will get a screen that looks something like screen (2). Here **Xmin** and **Xmax** give the left and right boundaries of the window, and Xscl gives the spacing of the tick marks along the *x* axis. The window is set in a similar way for the vertical direction. The values shown here are the *default values* on some calculators. Yours may show other values or none at all. You can either accept these, to start, or change them to suit your graph.

(3)   To *graph the function* press $\boxed{\text{GRAPH}}$ to get screen (3).

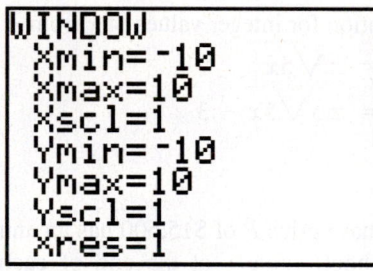

(1) TI-83/84 screen for Example 9.

(2) On TI calculators these values can also be obtained using ZOOM Standard.

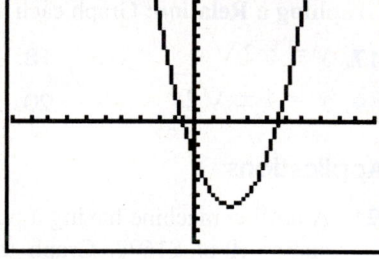

(3) Tick marks are one unit apart on both axes.

◆◆◆

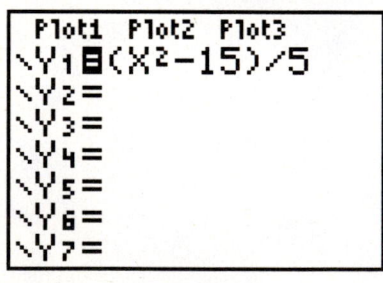

TI-83/84 screen for Example 10.

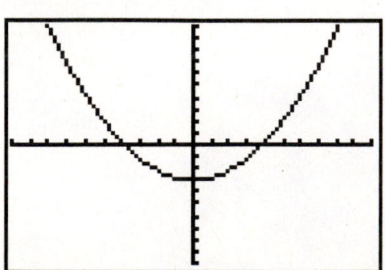

TI-83/84 screen for Example 10. Standard viewing window, with tick marks 1 unit apart.

Graphing an Equation Given in Implicit Form

To graph an equation given in implicit form, we must first put it into explicit form by solving for one variable.

◆◆◆ **Example 10:** Graph the equation

$$x^2 - 5y - 15 = 0$$

Solution: We go to explicit form by solving for *y*.

$$5y = x^2 - 15$$

$$y = \frac{x^2 - 15}{5}$$

We enter this equation, as shown in the screen, and graph.   ◆◆◆

Complete Graph and the Zoom Feature

By *complete graph* we mean that all features of interest are shown. These features include but are not limited to

> Peaks, or *maximum points*, such as *A*, Fig. 5–12.
> Valleys, or *minimum points*, such as point *B*.
> *Intercepts* on the *x* axis, such as *C*, *D*, and *E*.

In the preceding examples the graph appeared to fit nicely into the standard viewing window, with nothing "interesting" occurring outside that window. Further, once you become familiar with the parabola, you can be certain that there are no other points of interest outside of those shown. But if you are not sure that you have a complete graph, you can

- repeatedly increase the size of the *viewing window,* while looking for such points.
- use the ZOOM capability of the calculator to *zoom out* repeatedly.

Then adjust the viewing window to best display all the features of interest.

◆◆◆ **Example 11:** Graph the function $y = x^3 + 15$ and adjust the viewing window to best display the curve.

Solution: The graphing of this function is shown in the following four TI-83/84 screens.

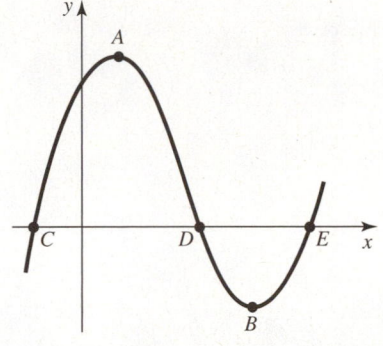

FIGURE 5–12

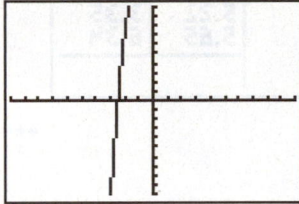 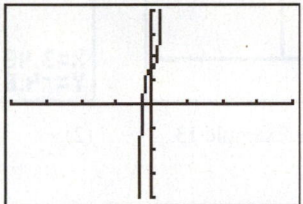 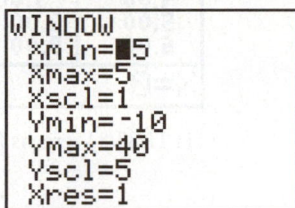 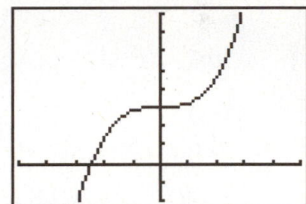

(1) The graph in the standard viewing window. Tick marks are 1 unit apart.

(2) The graph after zooming out. Tick marks are 10 units apart. Some points of interest seem to be above $y = 10$.

(3) New values for the viewing window.

(4) The graph in the new viewing window. Tick marks are 1 unit apart in *x* and 5 units in *y*.

◆◆◆

Exploring Your Graph: Special Operations on the Calculator

Using TRACE and ZOOM to find points of interest is time consuming, and many calculators have built-in operations for quickly finding such points. Some typical operations are

value	gives the value of *y* at a chosen value of *x*
zero	gives a root, that is, the value of *x* at which $y = 0$
maximum	gives the coordinates of a *maximum point*, a point that has a higher y value than any others in its vicinity
minimum	gives the coordinates of a *minimum point*, a point that has a lower y value than any others in its vicinity
intersect	gives the coordinates of the point of intersection of two curves

As a curve may have more than one of any of these points, (two maximums, for example), the calculator will prompt you to give the approximate location of the point you seek.

◆◆◆ **Example 12:** Find the minimum point for the curve of Example 9;

$$y = x^2 - 4x - 3.$$

Solution: We graph the curve as before, and then choose *minimum* (located in the CALC menu on TI calculators). When prompted, we enter the left and right bounds, those values of *x* between which we expect to find our minimum, enter a guess, and press ENTER. The coordinate of the minimum point is then displayed on the screen.

◆◆◆

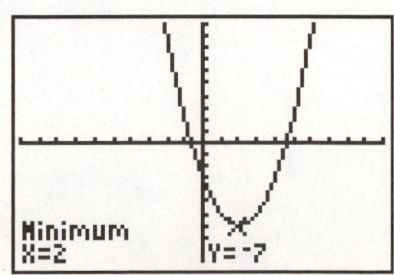

TI-83/84 screen for Example 12. Tick marks are one unit apart.

Table of Point Pairs and Split Screen Display

Some calculators will display a table of values for a function entered in the $\boxed{Y =}$ screen. This feature is very useful if you wish to make a manual plot on ordinary graph paper. You must give the starting value for x and the increment between successive values.

◆◆◆ **Example 13:** For the parabola in Example 9, $y = x^2 - 4x - 3$, we have set the starting value for x at 0 and the increment to 1. We get the table of screen (1). Notice how the y values decrease towards the minimum point and then begin to increase.

Calculators with a *split screen* capability allow the table and the graph to be displayed side-by-side. The wonderful display in screen (2) shows *all three main forms of a function: equation, graph, and table!* It also shows the trace cursor and its coordinates.

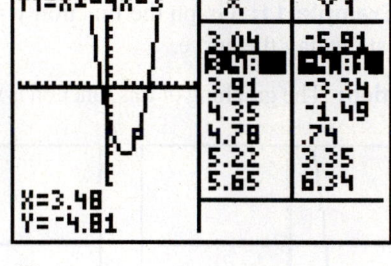

(1) TI-83/84 screens for Example 13.   (2).   ◆◆◆

Exercise 3 ◆ Graphing Functions by Calculator

Graph each function. Set the viewing window for x and y initially from -5 to 5, then resize if needed.

1. $y = x^2$
2. $y = 4 - 2x^2$
3. $y = 3x - 2$
4. $y = 1 - 2x$
5. $y = x^2 - 2x - 1$
6. $y = x^2 + 3x + 1$
7. $y = x^3 - x$
8. $y = 2x - x^3$
9. $y = x^3 - 2$
10. $y = 3.73 - 1.77x^2$
11. $y = 1.74x^2 - 2.35x + 1.84$
12. $x^2 - 2x - y + 2 = 0$
13. $y - 2x^2 - 3x = 3$

Graph each function. Resize the viewing window or use the Zoom feature, if needed, to obtain a complete graph. Then use TRACE and ZOOM or built-in operations to locate any zeros, maximum points, or minimum points.

14. $y = 2x^2 - 14x + 22$

15. $y = 7x^2 + 9x - 14$

16. $y = 4x^3 - 4x^2 + 11x - 24$

17. $y = 5x^4 + 13x^2 - 31$

5–4 The Straight Line

We already learned how to graph any equation, including that for a straight line, by computing and plotting a set of point pairs. But by looking for special features on some curves, such as the straight line and the parabola, our work gets much easier. We will give just an introduction to the straight line here, and we will examine it and the parabola in more detail in our chapter on analytic geometry.

Slope

As a particle moves from point P, Fig. 5–13, to point Q, the horizontal distance moved is called the *run*, and the increase or decrease of the vertical distance between the same points is called the *rise*. For points P and Q in Fig. 5–13, the rise is 6 in a run of 3. The rise divided by the run is called the *slope* of the line and is usually denoted by the letter m.

$$\text{slope } m = \frac{\text{rise}}{\text{run}}$$

♦♦♦ **Example 14:** The slope of the line in Fig. 5–13 is

$$m = \frac{\text{rise}}{\text{run}} = \frac{6}{3} = 2$$

♦♦♦

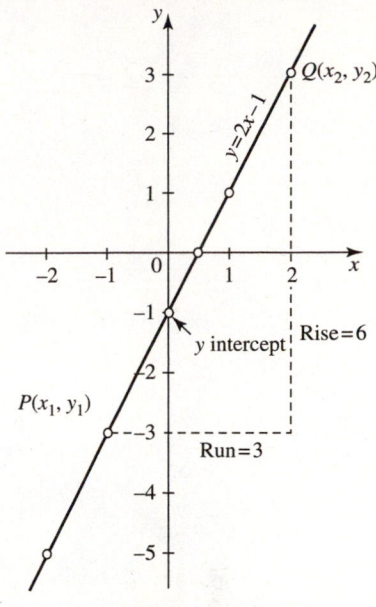

FIGURE 5–13

We can find the slope if given the coefficients of two points on the line. Let the two points be $P(x_1, y_1)$ and $Q(x_2, y_2)$ as shown in Fig. 5–13. Then the rise from P to Q is $y_2 - y_1$ in a run of $x_2 - x_1$. As before, the slope m is the rise divided by the run, so,

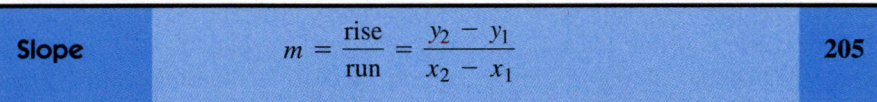

| Slope | $m = \dfrac{\text{rise}}{\text{run}} = \dfrac{y_2 - y_1}{x_2 - x_1}$ | **205** |

The slope of a straight line equals its rise, in a given run, divided by that run.

♦♦♦ **Example 15:** Find the slope of the straight line connecting the points $(3, -2)$ and $(-1, 4)$.

Solution: It doesn't matter which we call point 1 and which we call point 2. Let's choose.

$$x_1 = 3 \qquad y_1 = -2$$
$$x_2 = -1 \qquad y_2 = 4$$

Then from the equation for slope,

$$m = \frac{y_2 - y_1}{x_2 - x_1} = \frac{4 - (-2)}{-1 - 3}$$

$$= \frac{6}{-4} = -\frac{3}{2}$$

Thus the line *drops* by 3 units in a run of 2 units, Fig. 5–14. ♦♦♦

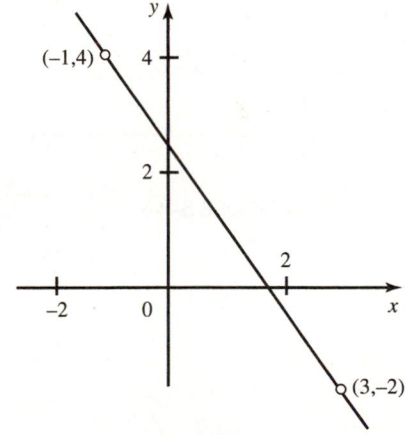

FIGURE 5–14

♦♦♦ **Example 16:** Find the slope of the straight line connecting the points $(-4.23, 3.22)$ and $(1.75, -3.84)$, Fig. 5–15.

Solution: Let's choose

$$x_1 = -4.23 \qquad y_1 = 3.22$$
$$x_2 = 1.75 \qquad y_2 = -3.84$$

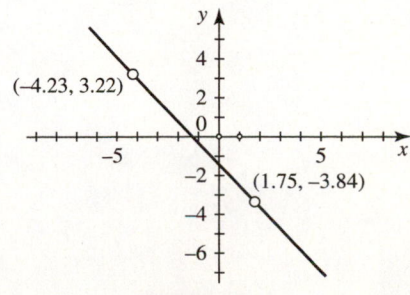

FIGURE 5–15

Then from the equation for slope,

$$m = \frac{y_2 - y_1}{x_2 - x_1} = \frac{-3.84 - 3.22}{1.75 - (-4.23)}$$

$$= \frac{-7.06}{5.98} = -1.18$$

◆◆◆

Common Error	Be careful not to mix up the subscripts. $$m \neq \frac{y_2 - y_1}{x_1 - x_2}$$

Equation of a Straight Line

■ Exploration:

Try this. Graph the function $y = f(x) = 2x - 1$ for values of x from -2 to 2, by computing a table of ordered pairs, plotting each pair, and then connecting them. Then compare your graph with the numbers in the given equation, which are 2 and (-1).

$$y = 2x - 1$$

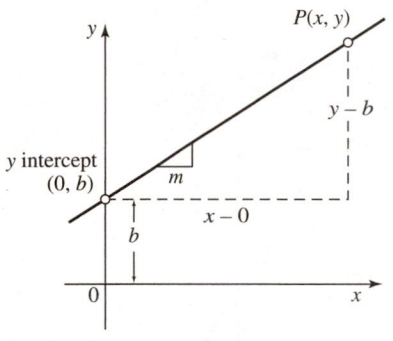

FIGURE 5–16

Are these numbers prominent on your graph? Does this suggest another, faster, way to graph a straight line?   ■

We see from our exploration that the slope and y intercept can be read directly from the given equation when the equation is written in explicit form. But will this always be true, or is it true just for the equation we used in the exploration? To see, let's derive the equation of a straight line.

Suppose that $P(x, y)$ is any point on a straight line, Fig. 5–16. We seek an equation that links x and y in a functional relationship so that, for any x, a value of y can be found. We can get such an equation by applying the definition of slope to our point P and some *known* point on the line. Let us use the y intercept $(0, b)$ as the coordinates for the known point. For P, a general point on the line, we use coordinates (x, y).

For a line that intersects the y axis b units from the origin (Fig. 5–16), the rise is $y - b$ in a run of $x - 0$, so by the equation for slope,

$$m = \frac{y - b}{x - 0}$$

Simplifying, we have $mx = y - b$, which is usually written in the following form:

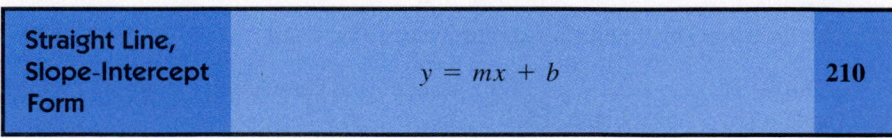

Straight Line, Slope-Intercept Form	$y = mx + b$	210

Not surprisingly, this is called the *slope-intercept* form of the equation of a straight line. Note that it is of *first degree* because neither x nor y has a degree higher than 1. A first degree equation is also called a *linear* equation.

This form of the equation gives us a fast and easy way to graph a straight line.

Graphing the Straight Line

◆◆◆ **Example 17:** Find the slope and the y intercept of the line $y = 3x - 4$. Make a graph.

Solution: By inspection, we see that the slope is the coefficient of x and that the y intercept is the constant term. So

$$m = 3 \text{ and } b = -4$$

We plot the y intercept 4 units below the origin, and through it we draw a line with a rise of 3 units in a run of 1 unit, as shown in Fig. 5–17. ◆◆◆

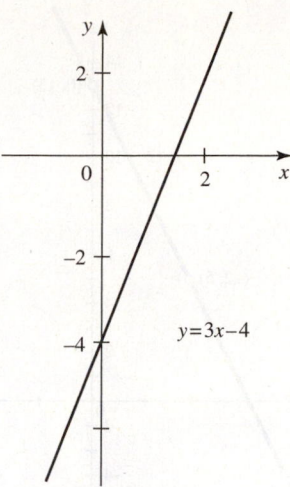

FIGURE 5–17

Writing the Equation of a Line Given the *y* Intercept and Slope

Let's now reverse the process. Given the slope and y intercept, we will write the equation of the line.

◆◆◆ **Example 18:** Write the equation of the straight line, in slope-intercept form, that has a slope of 2 and a y intercept of 3. Make a graph.

Solution: Substituting $m = 2$ and $b = 3$ into Eq. 289 ($y = mx + b$) gives

$$y = 2x + 3$$

To graph the line, we first locate the y intercept, 3 units up from the origin. The slope of the line is 2, so we get another point on the line by moving 1 unit to the right and 2 units up from the y intercept. This brings us to (1, 5). Connecting this point to the y intercept gives us our line as shown in Fig. 5–18. ◆◆◆

◆◆◆ **Example 19:** The equation of a straight line having a slope of -2.75 a y intercept of -3.44 is

$$y = -2.75x - 3.44$$

Its graph is shown in Fig. 5–19. ◆◆◆

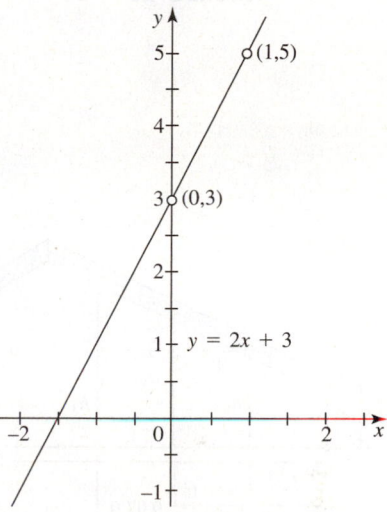

FIGURE 5–18

Writing the Equation Given Two Points on the Line

Now we will show how to write the equation of a straight line if we know the coordinates of two points through which it passes.

Let $P(x_1, y_1)$ and $Q(x_2, y_2)$ be the two known points on the line, and $P(x, y)$ be any other point on the line, Fig. 5–13. The slope of the line, as before, is

$$m = \frac{y_2 - y_1}{x_2 - x_1}$$

But the slope is also equal to

$$m = \frac{y - y_1}{x - x_1}$$

By equating these two expressions we can write the equation of the line, as in the following example.

◆◆◆ **Example 20:** Write the equation, in slope-intercept form, of the straight line passing through the points $(-4, -3)$ and $(-2, 5)$, and make a graph.

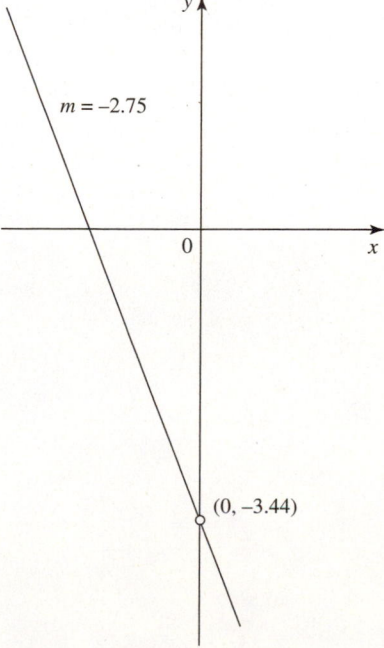

FIGURE 5–19

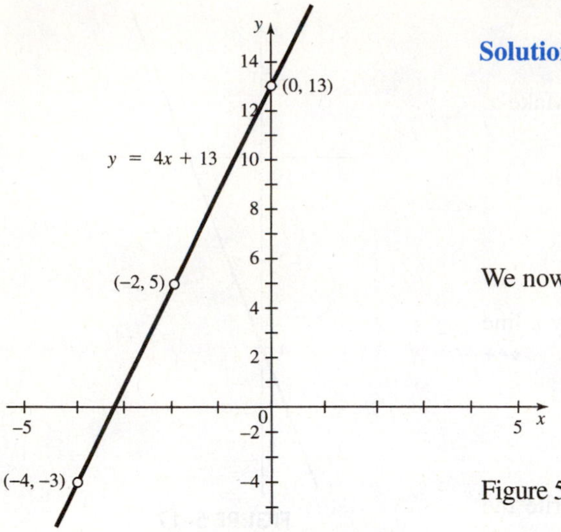

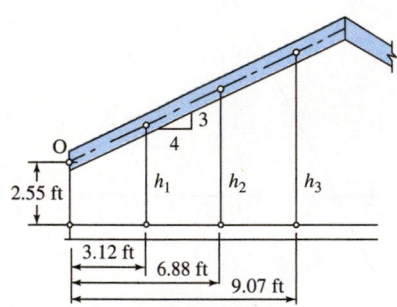

$y = 4x + 13$

$(0, 13)$

$(-2, 5)$

$(-4, -3)$

FIGURE 5–20

Solution: Let's let the first given point be point 1 and the other point 2. Then

$$m = \frac{y - (-3)}{x - (-4)} = \frac{5 - (-3)}{-2 - (-4)}$$

$$\frac{y + 3}{x + 4} = \frac{8}{2} = 4$$

We now solve for y to put the equation into slope-intercept form.

$$y + 3 = 4(x + 4)$$

$$y = 4x + 16 - 3$$

$$= 4x + 13$$

Figure 5–20 shows the graph of this line, having a slope of 4 and a y intercept of 13. ♦♦♦

♦♦♦ **Example 21:** *An Application.* A rafter having a slope of 3/4 supports a horizontal beam by means of three cables, Fig. 5–21. (a) Write an equation for the centerline of the beam, taking the origin at O. (b) Find the lengths of the three cables.

Solution: (a) The slope of the beam is 3/4 and the y intercept is 2.55 ft, so the equation is

$$y = mx + b$$

$$= \frac{3}{4}x + 2.55$$

(b) We get the lengths of the cables by substituting x values into this equation.

$$h_1 = \frac{3}{4}(3.12) + 2.55 = 4.89 \text{ ft}$$

$$h_2 = \frac{3}{4}(6.88) + 2.55 = 7.71 \text{ ft}$$

$$h_3 = \frac{3}{4}(9.07) + 2.55 = 9.35 \text{ ft}$$ ♦♦♦

FIGURE 5–21

Exercise 4 ♦ The Straight Line

Slope: Find the slope of each straight line.

1. Rise = 4; run = 2
2. Rise = 6; run = 4
3. Rise = −4.25, run = 5.33
4. Rise = 7.93, run = −2.66
5. Connecting $(2, 4)$ and $(5, 7)$
6. Connecting $(5, 2)$ and $(3, 6)$
7. Connecting $(−2.84, 5.11)$ and $(5.23, −6.22)$
8. Connecting $(3.88, −3.64)$ and $(−6.93, 2.69)$

Graphing the Straight Line: Find the slope and y intercept of each straight line and make a graph.

9. $y = 3x - 5$
10. $y = 7x + 2$
11. $y = -\dfrac{1}{2}x - \dfrac{1}{4}$
12. $y = -1.75x - 5.44$

Writing the Equation Given y Intercept and Slope: Write the equation of each straight line and make a graph.

13. Slope = 4; y intercept = −3
14. Slope = −1; y intercept = 2

15. Slope $= 3$; y intercept $= -1$

16. Slope $= -2$; y intercept $= 3$

17. Slope $= 2.30$; y intercept $= -1.50$

18. Slope $= -1.50$; y intercept $= 3.70$

Writing the Equation Given Two Points on the Line: Write the equation of each straight line passing through the given points and make a graph.

19. $(2, 3)$ and $(-1, 4)$ **20.** $(-3, 5)$ and $(1, 3)$

21. $(1.22, 2.43)$ and $(-2.11, 3.24)$ **22.** $(3.22, 2.53)$ and $(3.51, -2.54)$

Applications

23. A rafter, Fig. 5–22, has a rise of 9.3 ft and a run of 15.5 ft. What is its slope?

24. *Roadway Grade:* The word *grade* is often used instead of *slope*. Grade, expressed as a percent, is 100 times the slope. Thus a 10% grade has a slope of 0.1, or a rise of 10 ft in every 100 ft of run. For the road in Fig. 5–23 having a 5.0% grade, find the horizontal distance traveled for a 6.0 ft vertical drop.

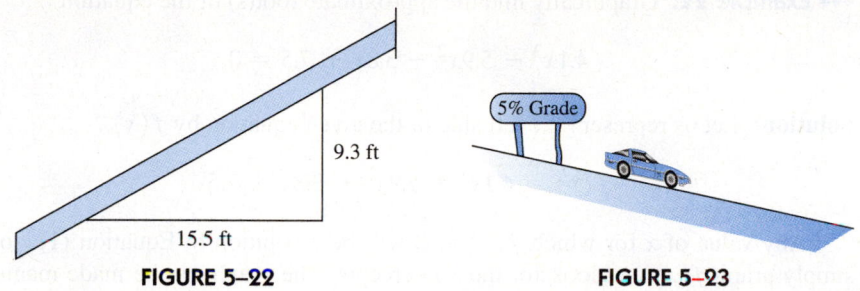

9.3 ft

15.5 ft

5% Grade

FIGURE 5–22   **FIGURE 5–23**

25. *Spring Constant:* The force F needed to stretch the spring, Fig. 5–24, from its unstretched length L_0 to a distance L from the wall is given by $F = kL - kL_0$ where k is called the *spring constant*. Graph this equation. Graphically find the force needed to stretch the spring whose unstretched length is 4.85 in. to a length of 10.5 in. from the wall, using $k = 18.5$ lb/in.

26. *Uniformly Accelerated Motion:* For an object moving with constant acceleration, such as a freely falling body, the velocity v at any time t is given by the equation of a straight line, $v = v_0 + at$, where v_0 is the initial velocity and a is the acceleration. Graph this equation for a ball thrown downward with an initial velocity of 5.83 m/s, Fig. 5–25. Use $a = 9.81$ m/s^2. Graphically find the speed at 3.25 s.

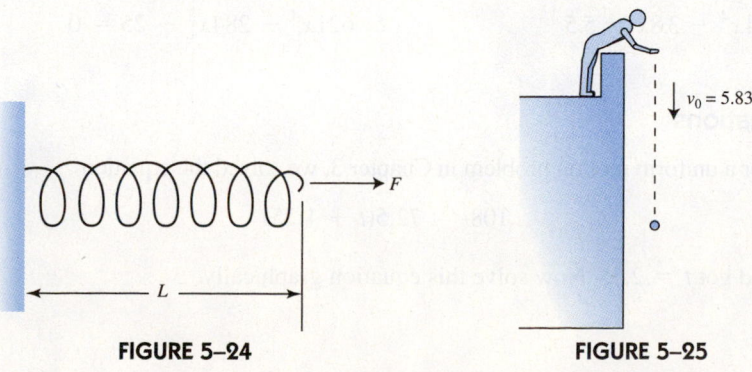

F

L

$v_0 = 5.83$ m/s

FIGURE 5–24   **FIGURE 5–25**

5–5   Solving an Equation Graphically

■ Exploration: *Try this.*

(a) Solve the equation $x^2 - 1 = 0$. What roots do you get?

(b) Graph the equation $y = x^2 - 1$ for values of x from -3 to $+3$. What do you observe? Can you draw a tentative conclusion from your findings from steps (a) and (b)?   ■

We can use our knowledge of graphing functions to solve equations of the form $f(x) = 0$. We mentioned earlier that a point at which a graph of a function $y = f(x)$ crosses or touches the x axis is called an x intercept.

In Fig. 5–26 there are two zeros, since there are two x values for which $y = 0$, and hence $f(x) = 0$. Those x values for which $f(x) = 0$ are called *zeros, roots* or *solutions* to the equation $f(x) = 0$.

Thus if we were to graph the function $y = f(x)$, any value of x at which y is equal to zero would be a solution to $f(x) = 0$. So to solve an equation graphically, we simply put it into the form $f(x) = 0$ and then graph the function $y = f(x)$. Each x intercept is then an approximate solution to the equation.

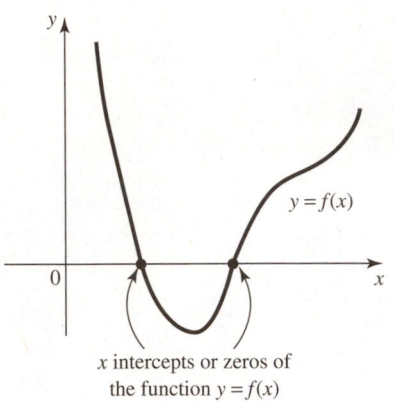

x intercepts or zeros of the function $y = f(x)$

FIGURE 5–26

♦♦♦ **Example 22:** Graphically find the approximate root(s) of the equation

$$4.1x^3 - 5.9x^2 - 3.8x + 7.5 = 0 \qquad (1)$$

Solution: Let us represent the left side of the given equation by $f(x)$.

$$f(x) = 4.1x^3 - 5.9x^2 - 3.8x + 7.5$$

Any value of x for which $f(x) = 0$ will be a solution to Equation (1), so we simply graph $f(x)$ and look for the x intercepts. The graph can be made manually as was shown earlier, or by calculator, as shown here.

We can then find the zero using $\boxed{\text{TRACE}}$ and $\boxed{\text{ZOOM}}$, or by using the built-in **Zero** operation. On TI calculators, choose **Zero** from the $\boxed{\text{CALC}}$ menu. Then enter the left and right bounds between which you expect to find a root, and a guess. Press $\boxed{\text{ENTER}}$ to display the root.   ♦♦♦

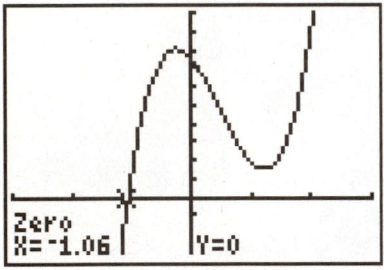

TI-83/84 screen for Example 22. Tick marks are one unit apart.

Exercise 5 ♦ Solving an Equation Graphically

Each of the following equations has at least one root between $x = -10$ and $x = 10$. Graphically find the approximate value of the root(s).

1. $2.4x^3 - 7.2x^2 - 3.3 = 0$ **2.** $9.4x = 4.8x^3 - 7.2$

3. $25x^2 - 19 = 48x + x^3$ **4.** $1.2x + 3.4x^3 = 2.8$

5. $6.4x^4 - 3.8x = 5.5$ **6.** $621x^4 - 284x^3 - 25 = 0$

Applications

7. For a uniform motion problem in Chapter 3, we solved the equation

$$108t = 72.5(t + 1.25)$$

and got $t = 2.55$. Now solve this equation graphically.

8. For a financial problem in Chapter 3, we solved the equation

$$276x + 120(12 - x) = 2220$$

and got $x = 5.00$. Verify this solution graphically.

9. For a mixture problem in Chapter 3, we solved the equation

$$0.0525x + 0.0284(3.25 - x) = 0.0415(3.25)$$

and got $x = 1.77$. Solve this equation graphically.

••• CHAPTER 5 REVIEW PROBLEMS •••••••••••••••••••••••••••••••••••

1. Graph the following points and connect them with a smooth curve:

x	0	1	2	3	4	5	6
y	2	$2\frac{1}{4}$	3	4	6	9	13

2. Plot the function $y = x^3 - 2x$ for $x = -3$ to $+3$. Label any roots or intercepts.

3. Graph the function $y = 3x - 2x^2$ from $x = -3$ to $x = 3$. Label any roots or intercepts.

4. Graphically find the approximate value of the roots of the equation

$$(x + 3)^2 = x - 2x^2 + 4$$

5. Find the slope of the straight line
(a) having a rise of 8 in a run of 2
(b) connecting $(-3, 5)$ and $(2, 7)$

6. Make a graph of the pressure p (lb/in.^{2}) vs. v in an engine cylinder, where v is the volume (in.^{3}) above the piston.

p	39.6	44.7	53.8	73.5	85.8	113.2	135.8	178.2
v	10.61	9.73	8.55	7.00	6.23	5.18	4.59	3.87

7. Write the equation of the line connecting $(-2, 5)$ and $(1, -3)$.

8. Write the equation of the line connecting $(3.22, -1.43)$ and $(2.51, 4.24)$.

For problems 9 through 12, make a complete graph of each function, choosing domain and a range that include all of the features of interest.

9. $y = 5x^2 + 24x - 12$

10. $y = 9x^2 + 22x - 17$

11. $y = 6x^3 + 3x^2 - 14x - 21$

12. $y = 3x^4 + 21x^2 + 21$

For problems 13 and 14, write the equation of each straight line, and make a graph.

13. Slope $= -2$; y intercept $= 4$
14. Slope $= 4$; y intercept $= -6$

For problems 15 and 16, find the slope and the y intercept of each straight line, and make a graph.

15. $y = 12x - 4$

16. $y = -2x + 11$

17. *Resistance Change with Temperature:* The resistance of a coil of wire is 10.4 Ω at 10.5°C and increases to 12.7 Ω at 91.1°C. Assuming the graph of resistance vs. temperature is a straight line, make this graph. Graphically find the resistance at 42.7°C.

Geometry

♦♦♦ **OBJECTIVES** ♦♦♦

When you have completed this chapter, you should be able to

- Find the angles formed by intersecting straight lines.
- Solve practical problems that require finding the sides and angles of right triangles.
- Solve practical problems in which the area of a triangle or a quadrilateral must be found.
- Solve applications problems involving the circumference, diameter, and area of a circle, or the tangent to a circle.
- Compute surface areas and volumes of spheres, cylinders, cones, and other solid figures.

♦♦♦

Geometry is a very old branch of mathematics. The word *geometry*, or geo-metry, means *earth measure* and probably refers to the ancient Egyptians' use of knotted ropes to measure the land so that boundary markers could be replaced after the annual flooding of the Nile. Geometry was developed by Pythagoras, Euclid, Archimedes, and many others and was included in the *quadrivium,* the four subjects needed for a bachelor's degree in the Middle Ages. We can only hint at that rich history here, confining ourselves mostly to the practical and very important computation of dimensions, areas, and volumes of the plane and solid geometric figures that we encounter in technical work. For example, how would you compute the volume, and hence the weight, of the bored hexagonal stock of Fig. 6–1? Here we will show how.

Further, our introduction to angles and triangles will prepare us for our later study of trigonometry. We will touch only on *Euclidean geometry* in this chapter and cover some *analytic geometry* later. Other geometries, those of more than three dimensions, non-Euclidean geometries, projective geometry, fractal geometry, and more, are subjects for other texts.

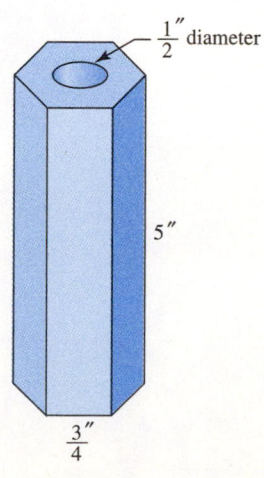

FIGURE 6–1

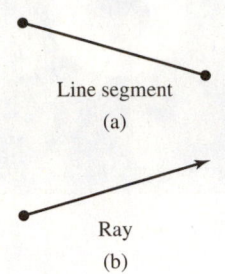

Line segment

(a)

Ray

(b)

FIGURE 6–2

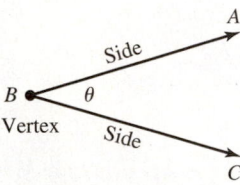

FIGURE 6–3   An angle.

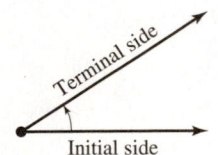

Terminal side

Initial side

FIGURE 6–4   An angle formed by rotation.

6–1   Straight Lines and Angles

The straight line is not new to us, for we studied it in some detail in the last chapter. Here we will add a second straight line which intersects the first at an *angle*.

Angles

A *line segment* is that portion of a straight line lying between two endpoints [Fig. 6–2(a)]. A *ray*, or *half-line*, is the portion of a line lying to one side of a point (end-point) on the line [Fig. 6–2(b)].

An *angle* is formed when two rays intersect at their endpoints, Fig. 6–3. The point of intersection is called the *vertex* of the angle, and the two rays are called the *sides* of the angle.

The angle shown in Fig. 6–3 can be designated in any of the following ways:

$$\text{angle } ABC \quad \text{angle } CBA \quad \text{angle } B \quad \text{angle } \theta$$

The symbol $\angle$ means *angle*, so $\angle B$ means angle B.

An angle can also be thought of as having been *generated* by a ray turning from some *initial position* to a *terminal position* (Fig. 6–4). We adopt the usual convention that an angle is positive when formed by a counterclockwise rotation, as in the figure, and negative when formed by a clockwise rotation.

One *revolution* is the amount a ray would turn to return to its original position.

The *units of angular measure* in common use are the *degree* and the *radian*.

The *measure* of an angle is the number of units of measure it contains. Two angles are equal if they have the same measure. For brevity we will usually say, for example, that "angle A equals angle B," rather than the more correct "the measure of angle A equals the measure of angle B."

The degree (°) is a unit of angular measure equal to 1/360 of a revolution. Thus there are 360° in one complete revolution. Recall that we learned how to convert between degrees, minutes, and seconds and decimal degrees in Chap. 1.

Figure 6–5 shows a *right* angle ($\frac{1}{4}$ revolution, or 90°), usually marked with a small square at the vertex; an *acute* angle (less than $\frac{1}{4}$ revolution); an *obtuse* angle

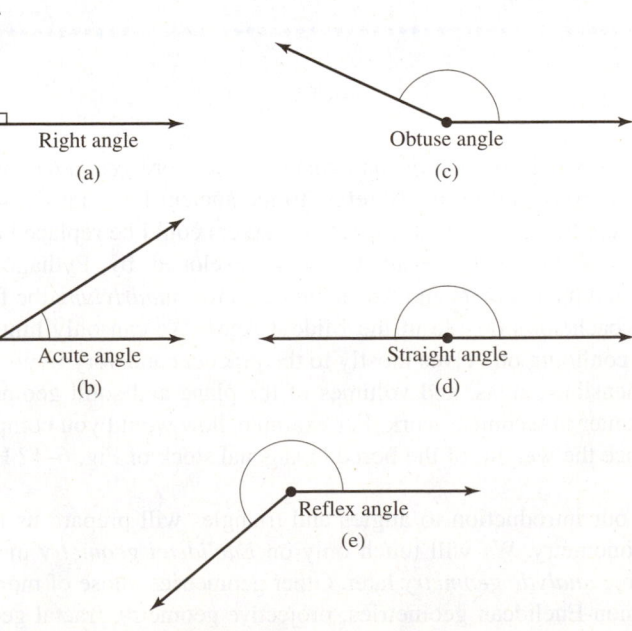

Right angle

(a)

Obtuse angle

(c)

Acute angle

(b)

Straight angle

(d)

Reflex angle

(e)

FIGURE 6–5   Types of angles.

(greater than $\frac{1}{4}$ revolution but less than $\frac{1}{2}$ revolution); and a *straight* angle ($\frac{1}{2}$ revolution, or 180°). Also shown is a *reflex* angle (greater than $\frac{1}{2}$ revolution but less than one revolution).

Two lines at right angles to each other are said to be *perpendicular*.

Two lines that never intersect are said to be *parallel*.

Two angles are called *complementary* if the sum of their measures is a right angle.

Two angles are called *supplementary* if the sum of their measures is a straight angle.

Two more words we use in reference to angles are *intercept* and *subtend*. In Fig. 6–6 we say that the angle θ *intercepts* the section PQ of the curve. Conversely, we say that angle θ is *subtended* by PQ.

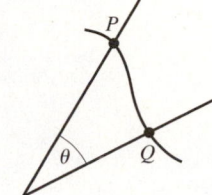

FIGURE 6–6

Angles Between Intersecting Lines

■ Exploration:

Try this. Draw two intersecting lines as in Fig. 6–7, either by hand or with a CAD (computer-assisted drawing) program. Measure the angles A, B, C, and D (or have them displayed if using CAD). How do they appear to be related? Change the angle between the lines (or drag, if using CAD) and again observe the angles. What do you conclude? ■

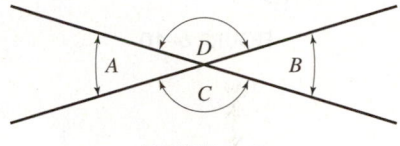

FIGURE 6–7

Angles A and B in Fig. 6–7 are called *opposite* angles or *vertical* angles. You may have concluded from your exploration that A and B are equal. Angles C and D are also opposite angles.

> Opposite (vertical) angles of two intersecting straight lines are equal. **64**

Two angles are called *adjacent* when they have a common side and a common vertex, such as angles A and C of Fig. 6–7. When two lines intersect, the adjacent angles are *supplementary*; that is, their sum is 180°.

◆◆◆ **Example 1:** *An Application.* Find (in degrees) angles A and B in the structure shown in Fig. 6–8.

Solution: We see from Fig. 6–8 that RQ and PS are straight intersecting lines. Since angle A is opposite the 34° angle, angle $A = 34°$. Angle B and the 34° angle are supplementary, so

$$B = 180° - 34° = 146°$$

◆◆◆

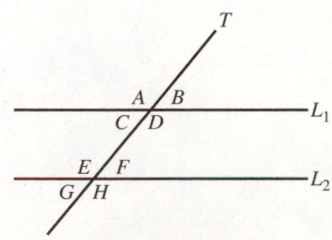

FIGURE 6–8

Families of Lines and Transversals

A *family of lines* is a set of lines that are somehow related to each other. Examples include a family of lines that are parallel, or a family of lines that meet at a point.

A *transversal* is a line that intersects a family of lines. In Fig. 6–9, two parallel lines L_1 and L_2 are cut by transversal T. Angles A, B, G, and H are called *exterior* angles, and C, D, E, and F are called *interior* angles. Angles A and E are called *corresponding* angles. Other corresponding angles in Fig. 6–9 are C and G, B and F, and D and H. Angles C and F are called *alternate interior angles*, as are D and E. We have the theorem:

FIGURE 6–9

> If two parallel straight line are cut by a transversal, corresponding angles are equal, and alternate interior angles are equal.   **65**

Thus in Fig. 6–9, $\angle A = \angle E$, $\angle C = \angle G$, $\angle B = \angle F$, and $\angle D = \angle H$. Also, $\angle D = \angle E$ and $\angle C = \angle F$.

◆◆◆ **Example 2:** *An Application.* The top girder PQ in the structure of Fig. 6–8 is parallel to the ground, and angle C is 73°. Find angle D.

Solution: We have two parallel lines, PQ and RS, cut by transversal PS. By statement 65,

$$\angle D = \angle C = 73°$$ ◆◆◆

Another useful theorem applies when a number of parallel lines are cut by *two* transversals, such as in Fig. 6–10. The portions of the transversals lying between the same parallels are called *corresponding segments*. (In Fig. 6–10, a and b are corresponding segments, c and d are corresponding segments, and e and f are corresponding segments.)

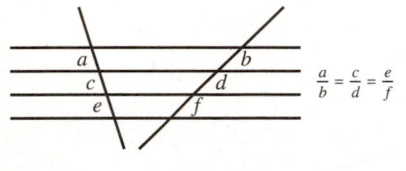

$$\frac{a}{b} = \frac{c}{d} = \frac{e}{f}$$

FIGURE 6–10

> If two lines are cut by a number of parallels, the corresponding segments are proportional.   **66**

In Fig. 6–10,

$$\frac{a}{b} = \frac{c}{d} = \frac{e}{f}$$

◆◆◆ **Example 3:** *An Application.* A portion of a street map is shown in Fig. 6–11. Find the distances PQ and QR.

Solution: From statement 66,

$$\frac{PQ}{172} = \frac{402}{355}$$

$$PQ = \frac{402}{355}(172) = 195 \text{ ft}$$

Similarly,

$$\frac{QR}{448} = \frac{402}{355}$$

$$QR = \frac{402}{355}(448) = 507 \text{ ft}$$ ◆◆◆

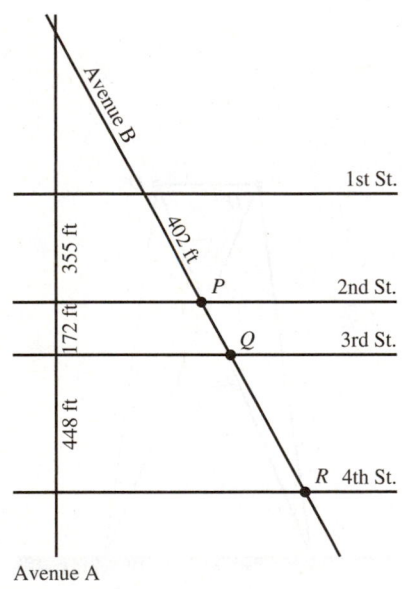

FIGURE 6–11

Exercise 1 ◆ Straight Lines and Angles

1. Find angle θ in Figs. 6–12 (a), (b), and (c).
2. Find angles A, B, C, D, E, F, and G in Fig. 6–13.
3. Find distance x in Fig. 6–14.

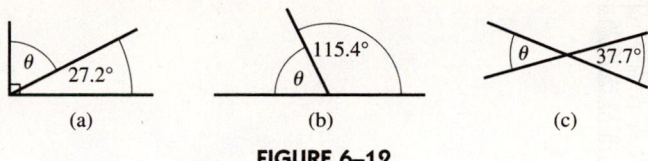

FIGURE 6–12

Applications

4. Find angles *A* and *B* in Fig. 6–15. Assume that each beam is of uniform width.
5. On a certain day, the angle of elevation of the sun (the angle that the sun's rays make with the horizontal) is 46.3°, as shown in Fig. 6–16. Find the angles *A*, *B*, *C*, and *D* that a ray of sunlight makes with a horizontal sheet of glass. Assume the glass to be so thin as to not bend the ray.
6. Three parallel steel cables hang from a girder to the deck of a bridge (Fig. 6–17). Find distance *x*.

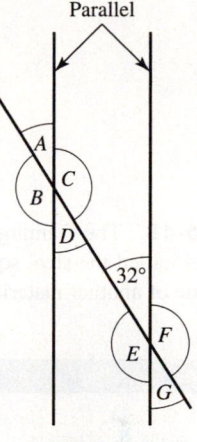

FIGURE 6–13

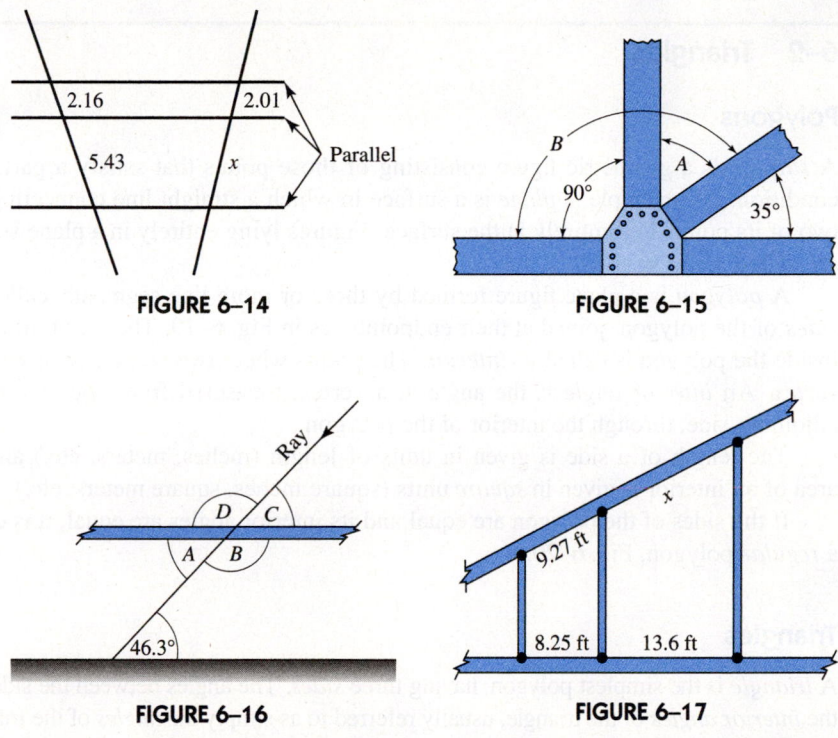

FIGURE 6–14   **FIGURE 6–15**

FIGURE 6–16   **FIGURE 6–17**

7. *Project: The Framing Square.* A framing square, Fig. 6–18, consists of a 24-in.-long *blade* set at a right angle to an 18-in.-long *tongue*. Both arms are graduated in inches both on the inside edge and outside edge and are engraved with various scales for board feet, rafter lengths, and so forth. Obtain a framing square and figure out how to use it to

   (a) bisect an angle
   (b) subdivide a line into a given number of equal parts

   We will have more uses for the framing square later in this chapter.

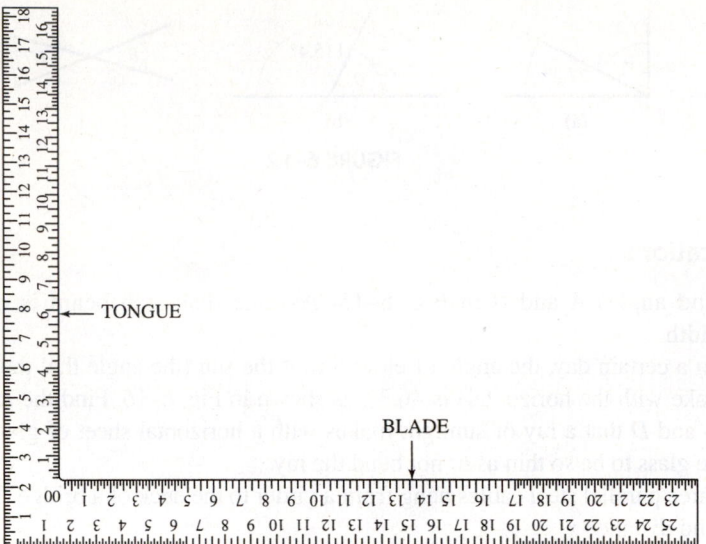

FIGURE 6–18 The framing square is sometimes called the steel square, even when made of another material.

6–2   Triangles

Polygons

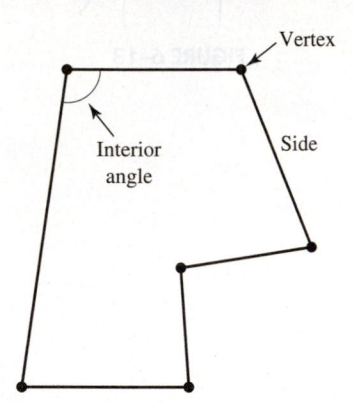

FIGURE 6–19   A polygon.

A *surface* is a geometric figure consisting of those points that satisfy a particular condition. For example, a *plane* is a surface in which a straight line connecting any two of its points lies entirely in the surface. Figures lying entirely in a plane surface are called *plane figures*.

A *polygon* is a plane figure formed by three or more line segments, called the *sides* of the polygon, joined at their endpoints, as in Fig. 6–19. The set of all points inside the polygon is called its *interior*. The points where two sides meet is called a *vertex*. An *interior angle* is the angle at a vertex, measured from one side to the adjoining side, through the interior of the polygon.

The length of a side is given in units of length (inches, meters, etc.) and the area of an interior is given in *square* units (square inches, square meters, etc.).

If the sides of the polygon are equal and its interior angles are equal, it is called a *regular* polygon, Fig. 6–20.

Triangles

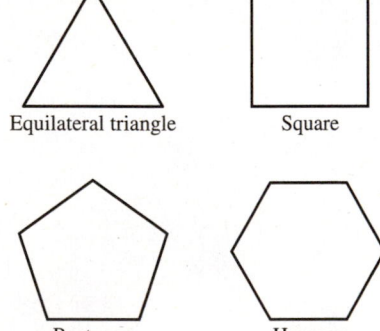

FIGURE 6–20   Some regular polygons.

A *triangle* is the simplest polygon, having three *sides*. The angles between the sides are the *interior angles* of the triangle, usually referred to as simply the *angles* of the triangle.

As shown in Fig. 6–21, a *scalene* triangle has no equal sides; an *isoceles* triangle has two equal sides; and an *equilateral* triangle has three equal sides.

An *acute* triangle has three acute angles; an *obtuse* triangle has one obtuse angle and two acute angles; and a *right* triangle has one right angle and two acute angles. A triangle that has no right angle is called an *oblique* triangle. All the triangles shown are oblique, except for the right triangle.

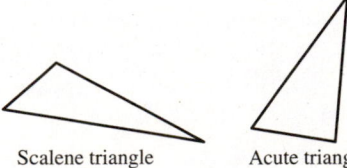
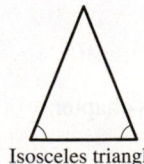
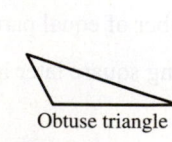
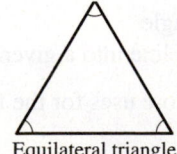
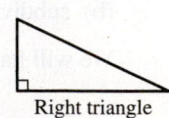

Scalene triangle   Acute triangle   Isosceles triangle   Obtuse triangle   Equilateral triangle   Right triangle

FIGURE 6–21   Types of triangles.

Don't confuse the words *oblique* and *obtuse*. They look similar but have different meanings.

Altitude and Base

The *altitude* of a triangle is the perpendicular distance from a vertex to the opposite side, called the *base*, or an extension of that side (Fig. 6–22).

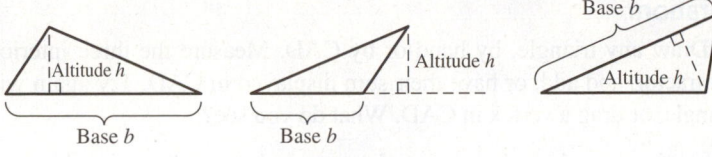

FIGURE 6–22

Area of a Triangle

■ **Exploration:** *Try this.*

(a) Draw a rectangle of width *b* and height *h*, Fig. 6–23(a). The area of this rectangle is, of course, its width times its height, or *bh*. Bisect the rectangle with a diagonal. What is the area of triangle *ABC?*

(b) In a new sketch, draw oblique triangle *DEF*, Fig. 6–23(b), with base *b* and altitude *h*. Using your conclusion from (a), subtract the area of triangle *EFG* from that of triangle *DEG*. What do you conclude about the area of triangle *DEF?* ■

From your exploration, you probably arrived at the familiar formula for the area of a triangle:

| **Area of a Triangle** | *Area equals one-half the product of the base and the altitude to that base.* $$A = \frac{bh}{2}$$ | **102** |

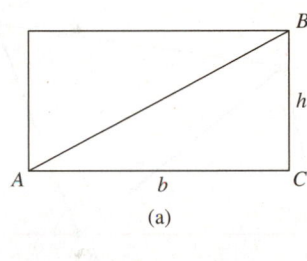

(a)

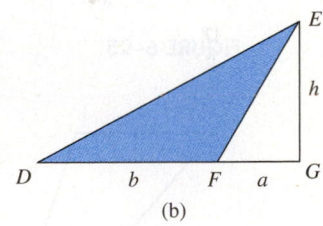

(b)

FIGURE 6–23

◆◆◆ **Example 4:** Find the area of the shaded triangle in Fig. 6–23(b) if its base *b* is 52.0 and the altitude *h* is 48.0.

Solution: By Eq. 102,

$$\text{area} = \frac{52.0(48.0)}{2} = 1250 \text{ sq. units (rounded)} \qquad ◆◆◆$$

If the altitude is not known but we have instead the lengths of the three sides, we may use *Hero's formula*. If *a*, *b*, and *c* are the lengths of the sides, we can find the area using the following formula:

| **Hero's Formula** | $$\text{Area of triangle} = \sqrt{s(s - a)(s - b)(s - c)}$$ where *s* is half the perimeter, or $$s = \frac{a + b + c}{2}$$ | **103** |

This formula is named for Hero (or Heron) of Alexandria, a Greek mathematician and physicist of the 1st century C.E.

Example 5: Find the area of a triangle having sides of lengths 3.25, 2.16, and 5.09.

Solution: We first find s, which is half the perimeter.

$$s = \frac{3.25 + 2.16 + 5.09}{2} = 5.25$$

Thus the area is

$$\text{area} = \sqrt{5.25(5.25 - 3.25)(5.25 - 2.16)(5.25 - 5.09)} = 2.28 \text{ sq. units}$$

Sum of the Angles

■ Exploration:

Try this. Draw any triangle, by hand or by CAD. Measure the three interior angles with a protractor, and add, or have their sum displayed in CAD. Try again with a different triangle, or drag a vertex in CAD. What do you see? ■

Your exploration may have led you to the extremely useful relationship among the interior angles of any triangle.

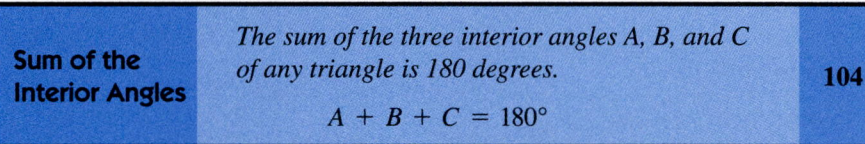

| Sum of the Interior Angles | The sum of the three interior angles A, B, and C of any triangle is 180 degrees. $A + B + C = 180°$ | 104 |

Example 6: Find angle A in a triangle if the other two interior angles are 38° and 121°.

Solution: By Eq. 104,

$$A = 180 - 121 - 38 = 21°$$

Exterior Angles

■ Exploration:

An *exterior angle* is the angle between the side of a triangle and an extension of an adjacent side, such as angle theta in Fig. 6–24. *Try this.* Using the facts that the sum of the interior angles is 180° and that the sum of theta and angle C is 180°, see if you can verify this formula:

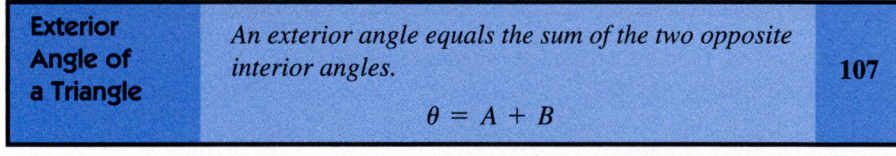

| Exterior Angle of a Triangle | An exterior angle equals the sum of the two opposite interior angles. $\theta = A + B$ | 107 |

■

Example 7: *An Application.* Find angle ϕ in the section of truss, Fig. 6–25.

Solution: Angle ϕ is an exterior angle to triangle PQR, so

$$\phi = 78° + 63° = 141°$$

Congruent and Similar Triangles

Two triangles (or any other polygons, for that matter) are said to be *congruent* if the angles and sides of one are equal to the angles and sides of the other, as in Fig. 6–26. Two triangles are said to be *similar* if they have the *same shape,* even if one triangle is larger than the other. This means that the angles of one of the triangles must equal the angles of the other triangle, as in Fig. 6–27. Sides that lie between the same pair of equal angles are called *corresponding sides*, such as sides a and d.

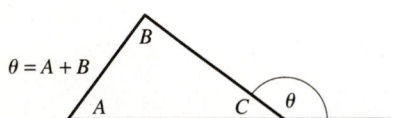

$\theta = A + B$

FIGURE 6–24 An exterior angle.

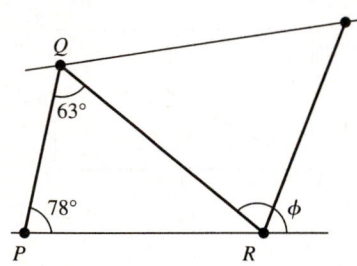

FIGURE 6–25

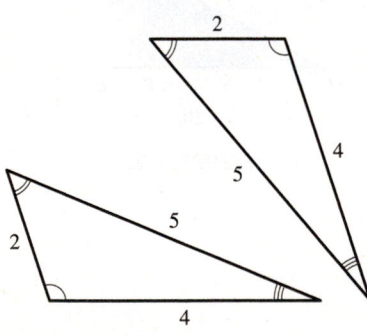

FIGURE 6–26 Congruent triangles. Angles marked with the same number of small arcs are equal.

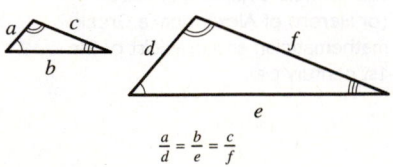

$$\frac{a}{d} = \frac{b}{e} = \frac{c}{f}$$

FIGURE 6–27 Similar triangles.

Sides b and e, as well as sides c and f, are also corresponding sides. We have the following two theorems:

Similar Triangles	If two angles of a triangle equal two angles of another triangle, their third angles must be equal, and the triangles are similar.	**108**
	If two triangles are similar, the ratios of corresponding sides are equal.	**109**

We will see in a later chapter that relationship 109 holds for similar figures *other than* triangles, and for similar *solids* as well.

◆◆◆ **Example 8:** *An Application.* Two beams, AB and CD, in the framework of Fig. 6–28 are parallel. Find distance AE.

Solution: By statement 64, we know that angle AEB equals angle DEC. Also, by statement 65, angle ABE equals angle ECD. Thus triangle ABE is similar to triangle CDE. Since AE and ED are corresponding sides, the ratio of one pair of corresponding sides must equal the ratio of another pair of corresponding sides.

$$\frac{AE}{5.87} = \frac{5.14}{7.25}$$

$$AE = 5.87\left(\frac{5.14}{7.25}\right) = 4.16 \text{ m} \qquad ◆◆◆$$

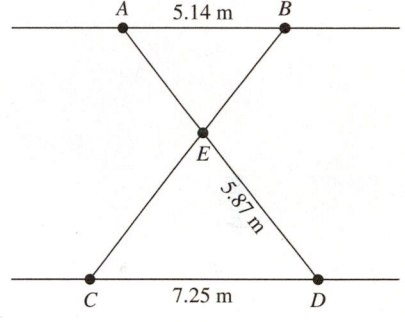

FIGURE 6–28

Right Triangles

In a right triangle ABC, Fig. 6–29, the side opposite the right angle is called the *hypotenuse,* and the other sides are called the *legs.* Since the sum of the interior angles must be 180°, angles A and B must add up to 90°; that is, they are *complementary.*

■ **Exploration:**

Try this. Using a computer drawing program, draw any right triangle. Then construct a square on each side. Have the computer display the area of the square on the hypotenuse, and the sum of the areas of the squares on each leg. How are they related? Then drag a corner of the triangle (making sure it stays a right triangle) to different positions. What can you say about the areas? ■

Your exploration may have led you to the well-known *Pythagorean theorem:*

Pythagorean Theorem	*The square of the hypotenuse of a right triangle is equal to the sum of the squares of the two legs.* $$a^2 + b^2 = c^2$$	**110**

◆◆◆ **Example 9:** A right triangle has legs of length 6.00 units and 11.0 units. Find the length of the hypotenuse.

Solution: Letting c = the length of the hypotenuse, we have

$$c^2 = 6.00^2 + 11.0^2 = 36.0 + 121 = 157$$

$$c = \sqrt{157} = 12.5 \text{ units (rounded)} \qquad ◆◆◆$$

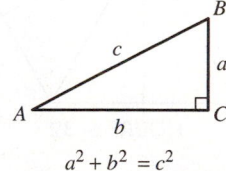

FIGURE 6–29 The Pythagorean theorem is named for the Greek mathematician Pythagoras (ca. 580–500 B.C.E.).

Common Error	Remember that the Pythagorean theorem applies *only to right triangles.* Later we will use trigonometry to find the sides and angles of oblique triangles (triangles with no right angles).

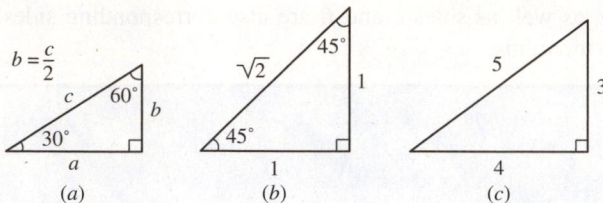

FIGURE 6–30   Some special right triangles.

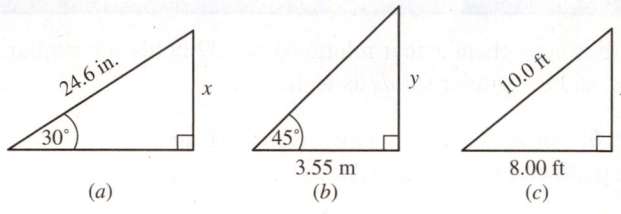

FIGURE 6–31

Special Right Triangles

In a *30–60–90* right triangle [Fig. 6–30(a)], the side opposite the 30° angle is half the length of the hypotenuse.

A *45° right triangle* [Fig. 6–30(b)] is also *isosceles*, and the hypotenuse is $\sqrt{2}$ times the length of either side.

A *3–4–5 triangle* [Fig. 6–30(c)] is a right triangle in which the sides are in the ratio of 3 to 4 to 5.

◆◆◆ **Example 10:** By inspection, we can say that

(a) Side x in Fig. 6–31(a) is 12.3 in.
(b) Side y in Fig. 6–31(b) is 3.55 m.
(c) Side z in Fig. 6–31(c) is 6.00 ft.   ◆◆◆

The 30–60–90 triangle is useful in solving problems involving the regular hexagon, which can be subdivided into 30–60–90 right triangles, Fig. 6–32.

◆◆◆ **Example 11:** *An Application.* A hex-head bolt measures 0.750 in. across the flats, Fig. 6–33. Find the shortest distance x from the center of the bolt to an obstruction that will just allow the bolt to turn.

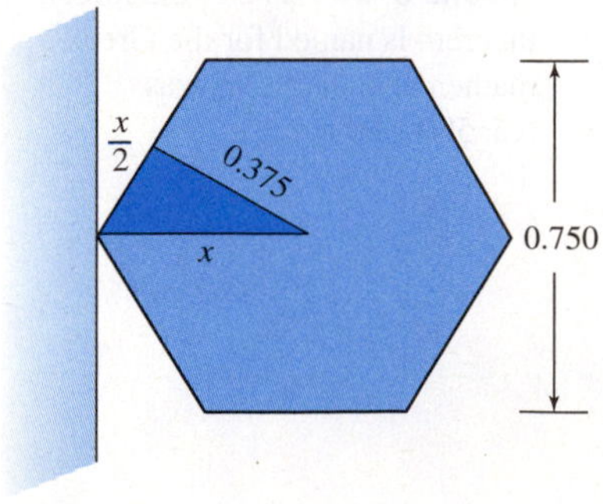

Solution: We note that x is the hypotenuse of a 30–60–90 triangle whose long leg is $0.750 \div 2$ or 0.375 in. The short leg of a 30–60–90 triangle is half the hypotenuse, or $x/2$. Then by the Pythagorean theorom,

$$x^2 = \left(\frac{x}{2}\right)^2 + (0.375)^2$$

$$x^2 - \frac{x^2}{4} = 0.1406$$

$$\frac{3x^2}{4} = 0.1406$$

$$x^2 = \frac{4(0.1406)}{3} = 0.1875$$

$$x = 0.433 \text{ in.}$$ ◆◆◆

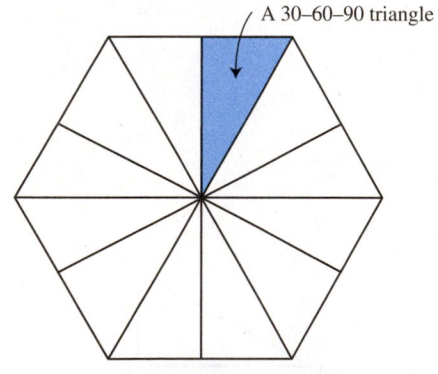

A 30–60–90 triangle

FIGURE 6–32

FIGURE 6–33

Exercise 2 ◆ Triangles

1. Find angles θ and ϕ in Fig. 6–34.
2. Find the area of the triangle in Fig. 6–35.
3. Find the area of the triangle in Fig. 6–36.
4. Find the area of the triangle in Fig. 6–37.
5. The two triangles in Fig. 6–38 are similar. Find sides a and b.
6. Find the hypotenuse in the right triangle of Fig. 6–39.
7. Find side a in the right triangle of Fig. 6–40.
8. Find side b in the right triangle of Fig. 6–41.

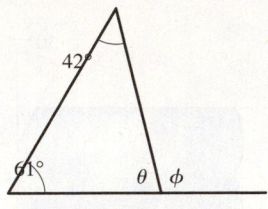

FIGURE 6–34

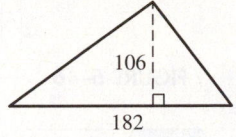

FIGURE 6–35

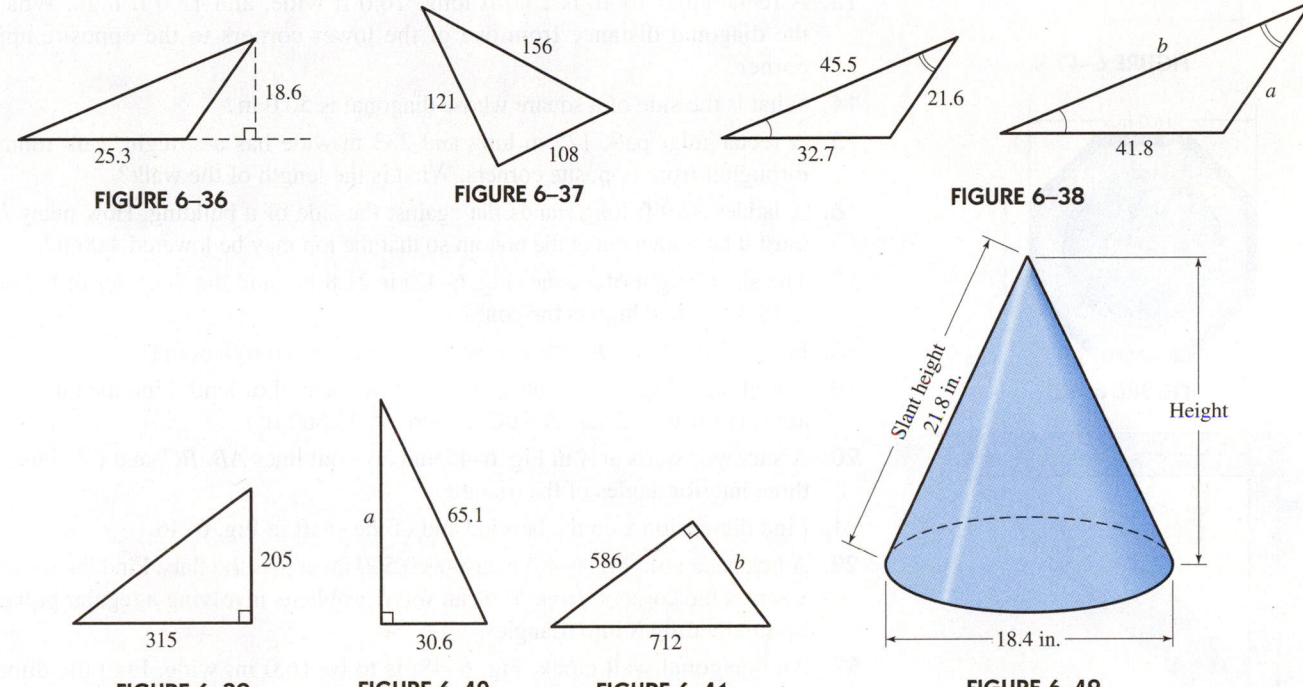

FIGURE 6–36 **FIGURE 6–37** **FIGURE 6–38**

FIGURE 6–39 **FIGURE 6–40** **FIGURE 6–41** **FIGURE 6–42**

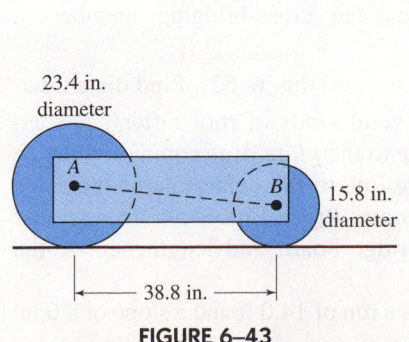

FIGURE 6–43

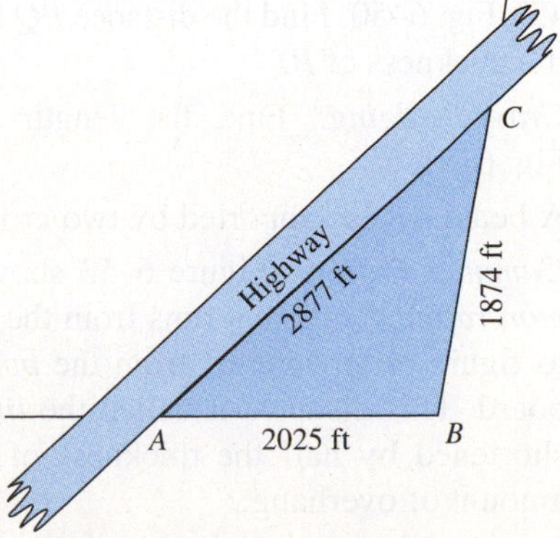

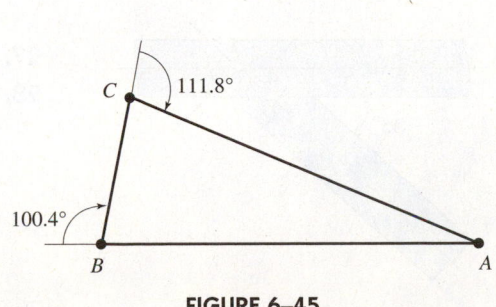

FIGURE 6–44   Note: Angle B is not a right angle.

FIGURE 6–45

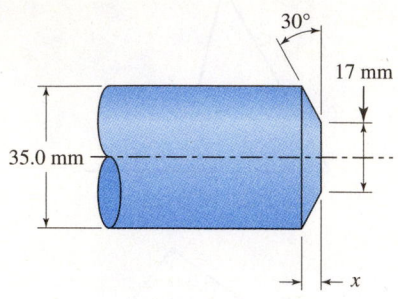

FIGURE 6–46

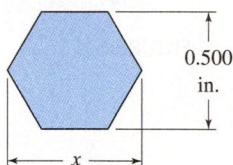

FIGURE 6–47

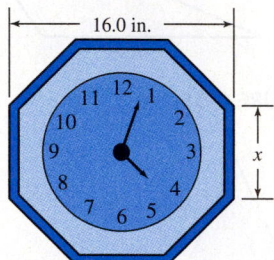

FIGURE 6–48

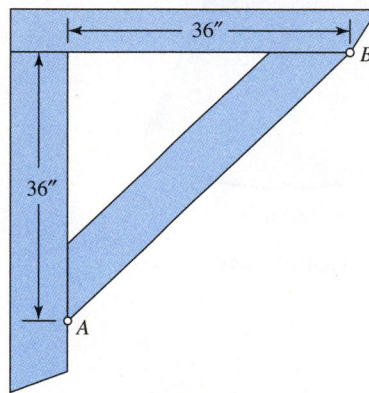

FIGURE 6–49

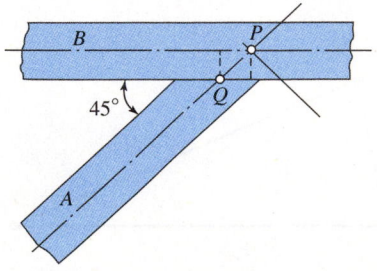

FIGURE 6–50

Applications

To help you solve each problem, draw a diagram and label it completely. Look for special triangles or right triangles contained in the diagram. Be sure to look up any word that is unfamiliar.

9. What is the cost, to the nearest dollar, of a triangular piece of land whose base is 828 ft and altitude is 412 ft at $1125 an acre? (1 acre = 43,560 ft^2)

10. A vertical pole 45.0 ft high stands on level ground and is supported by three guy wires attached to the top and reaching the ground at distances of 60.0 ft, 108.0 ft, and 200.0 ft from the foot of the pole. What are the lengths of the wires?

11. A ladder 39.0 ft long reaches to the top of a building when its foot stands 15.0 ft from the building. How high is the building?

12. Two streets, one 16.2 m and the other 31.5 m wide, cross at right angles. What is the diagonal distance between the opposite corners?

13. A rectangular room is 20.0 ft long, 16.0 ft wide, and 12.0 ft high. What is the diagonal distance from one of the lower corners to the opposite upper corner?

14. What is the side of a square whose diagonal is 50.0 m?

15. A rectangular park 125 m long and 233 m wide has a straight walk running through it from opposite corners. What is the length of the walk?

16. A ladder 32.0 ft long stands flat against the side of a building. How many feet must it be drawn out at the bottom so that the top may be lowered 4.00 ft?

17. The slant height of a cone (Fig. 6–42) is 21.8 in., and the diameter of the base is 18.4 in. How high is the cone?

18. Find the distance *AB* between the centers of the two rollers in Fig. 6–43.

19. A highway (Fig. 6–44) cuts a corner from a parcel of land. Find the number of acres in the triangular lot *ABC*. (1 acre = 43,560 ft^2)

20. A surveyor starts at *A* in Fig. 6–45 and lays out lines *AB, BC,* and *CA*. Find the three interior angles of the triangle.

21. Find dimension *x* on the beveled end of the shaft in Fig. 6–46.

22. A hex head bolt, Fig. 6–47, measures 0.500 in. across the flats. Find the distance *x* across the corners. *Hint*: You can solve problems involving a regular polygon by subdividing it into triangles.

23. An octagonal wall clock, Fig. 6–48, is to be 16.0 in. wide. Find the dimension *x*.

24. *Diagonal Brace:* Find the length *AB* of the diagonal brace in Fig. 6–49.

25. *Diagonal Brace:* A brace *A* is to join rafter *B* (width = 3.25 in.) at an angle of 45°, Fig. 6–50. Find the distance *PQ* by which *A* must be shortened to allow for the thickness of *B*.

26. *Cross-Bridging:* Find the length *AB* of the cross-bridging member in Fig. 6–51.

27. A beam *AB* is supported by two crossed beams (Fig. 6–52). Find distance *x*.

28. *Common Rafters:* Figure 6–53 shows several kinds of roof rafters. A *common rafter* is one that runs from the *ridge* to the *plate*. It is common practice to figure rafter lengths from the *building line* to the centerline of the ridge board. This distance is called the *line length, PQ* in Fig. 6–54. This is later shortened by half the thickness of the ridge board and lengthened by the amount of overhang.

Find the line length of a rafter that has a run of 14.0 ft and a slope of 8.0 in. per foot.

29. *Project: Framing Square:* Obtain a framing square and figure out how to use it to
    (a) lay out a miter cut (a cut of 45°)
    (b) draw an equilateral triangle
    (c) locate the center of a triangle
    (d) lay out an angle of 30° or 60°.

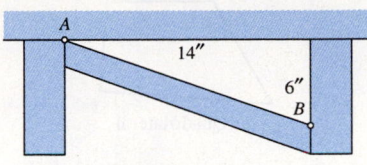

FIGURE 6–51

30. *Project:* A *dissection proof* is one in which a geometric figure is dissected or cut up, and the pieces rearranged to prove something. There are many dissection proofs of the Pythagorean theorem. Find a few of these, and choose one on which to make a classroom presentation.

31. *Project:* There are various kinds of "centers" of a triangle, including the centroid, the incenter, the circumcenter, and the orthocenter. Research and describe how each is found. Can you find a use for any of them?

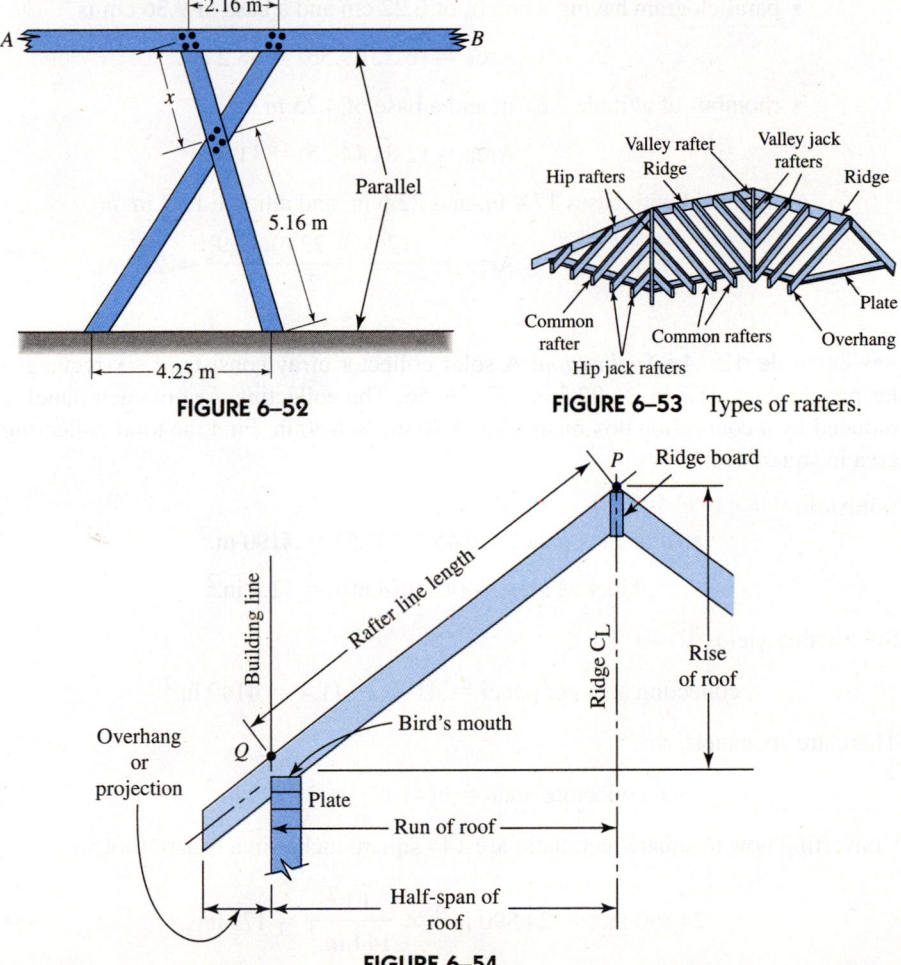

FIGURE 6–52

FIGURE 6–53   Types of rafters.

FIGURE 6–54

6–3 Quadrilaterals

A *quadrilateral* is a polygon having four sides. They are the familiar figures shown in Fig. 6–55. The formula for the area of each interior is given right on the figures (b)–(f).

For the *parallelogram*, opposite sides are parallel and equal. Opposite angles are equal, and each diagonal cuts the other diagonal into two equal parts (they *bisect* each other).

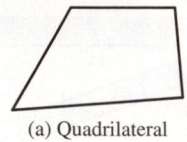

(a) Quadrilateral

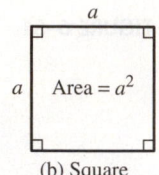

(b) Square

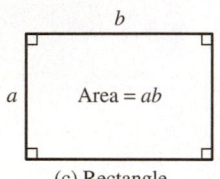

(c) Rectangle

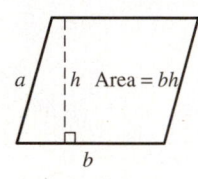

(d) Parallelogram

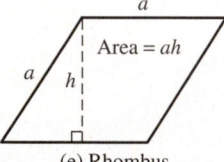

(e) Rhombus

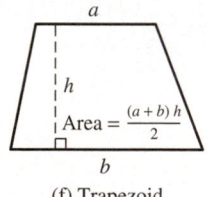

(f) Trapezoid

FIGURE 6–55 Quadrilaterals.

The *rhombus* is also a parallelogram, so the previous facts apply to it as well. In addition, its diagonals bisect each other at *right angles* and bisect the angles of the rhombus.

The *trapezoid* has two parallel sides, which are called the *bases*, and the altitude is the distance between the bases.

◆◆◆ **Example 12:** The area of a

- square of side 5.84 m is

$$\text{Area} = (5.84)^2 = 34.1 \text{ m}^2$$

- rectangle measuring 3.85 ft by 7.88 ft is

$$\text{Area} = (3.85)(7.88) = 30.3 \text{ ft}^2$$

- parallelogram having a height of 6.22 cm and a base of 9.36 cm is

$$\text{Area} = (6.22)(9.36) = 58.2 \text{ cm}^2$$

- rhombus of altitude 2.81 m and a base of 4.25 m is

$$\text{Area} = (2.81)(4.25) = 11.9 \text{ m}^2$$

- trapezoid with bases 17.4 in. and 22.6 in. and altitude 12.9 in. is

$$\text{Area} = \frac{(17.4 + 22.6)(12.9)}{2} = 258 \text{ in.}^2 \quad \text{◆◆◆}$$

◆◆◆ **Example 13:** *An Application.* A solar collector array consists of six rectangular panels, each 45.3 in. × 92.5 in., Fig. 6–56. The collecting area of each panel is reduced by a connection box measuring 4.70 in. × 8.80 in. Find the total collecting area in square feet.

Solution:

$$\text{area of each panel} = (45.3)(92.5) = 4190 \text{ in.}^2$$

$$\text{blocked area} = (4.70)(8.80) = 41.4 \text{ in.}^2$$

Subtracting yields

$$\text{collecting area per panel} = 4190 - 41.4 = 4149 \text{ in.}^2$$

There are six panels, so

$$\text{total collecting area} = 6(4149) = 24,890 \text{ in.}^2$$

Converting now to square feet, there are 144 square inches in a square foot so

$$24,890 \text{ in.}^2 = 24,890 \text{ in.}^2 \times \frac{1 \text{ ft}^2}{144 \text{ in.}^2} = 173 \text{ ft}^2 \quad \text{◆◆◆}$$

Sum of the Interior Angles

■ **Exploration:** *Try this.*

(a) Draw any quadrilateral. Then select a point P inside the quadrilateral and connect it to each vertex, Fig. 6–57. The sum of the angles in each triangle is 180°, so the four triangles contain a total of $4 \times 180°$, or 720°. From this subtract the four angles around P to get the sum of the interior angles of the quadrilateral. What do you get?

(b) Try this again with a pentagon, a polygon of 5 sides.

(c) Try this again with a polygon of n sides. Can you generalize your result? ■

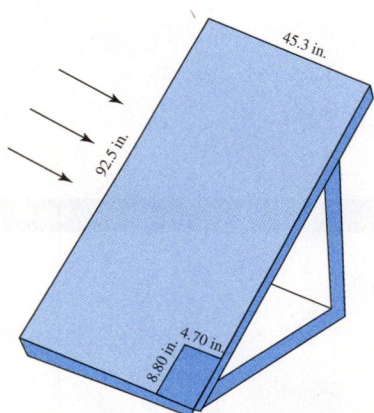

FIGURE 6–56 One solar panel.

You may have found that for a quadrilateral the sum of the interior angles is 360°, for a pentagon it is 540°, and for a polygon of n sides it is

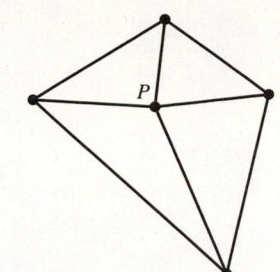

FIGURE 6–57

| Interior Angles of Any Polygon | Sum of angles $= (n - 2)180°$ | 72 |

◆◆◆ Example 14: Find angle θ in Fig. 6–58.

Solution: The polygon shown has seven sides, so $n = 7$. By Eq. 72,

$$\text{sum of angles} = (7 - 2)(180°) = 900°$$

Adding the six given angles gives us

$$278° + 62° + 123° + 99° + 226° + 43° = 831°$$

So

$$\theta = 900° - 831° = 69° \qquad \text{◆◆◆}$$

◆◆ Example 15: *An Application.* The *miter angle* is the angle at which a saw or miter box must be set to cut a piece of stock to form a joint. It is equal to half the angle between the pieces to be joined. Find the miter angle θ for the pentagonal window in Fig. 6–59.

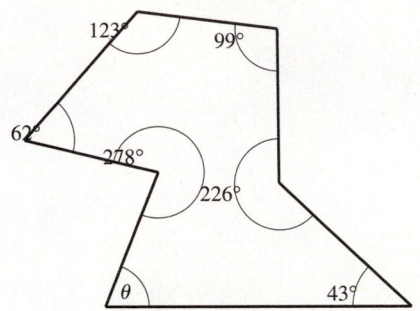

FIGURE 6–58

Solution: The sum of the angles of the pentagon are, with $n = 5$,

$$\text{sum of angles} = (n - 2)180°$$
$$= (5 - 2)180° = 540°$$

Each interior angle is then $540/5 = 108°$. The miter angle is half of that, or

$$\theta = 54° \qquad \text{◆◆◆}$$

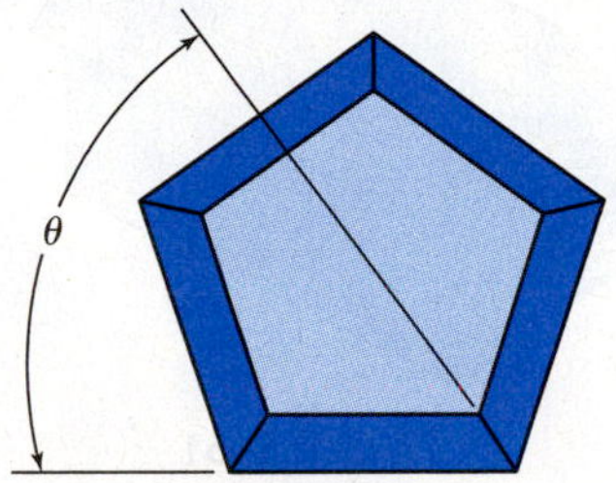

FIGURE 6–59

Exercise 3 ◆ Quadrilaterals

1. Find the area and perimeter of
   (a) a square of side 5.83 in.
   (b) a square of side 4.82 m.
   (c) a rectangle measuring 384 cm × 734 cm.
   (d) a rectangle measuring 55.4 in. by 73.5 in.

2. Find the area of
   (a) a parallelogram whose base is 4.52 ft and whose altitude is 2.95 ft.
   (b) a parallelogram whose base is 16.3 m and whose altitude is 22.6 m.
   (c) a rhombus whose base is 14.2 cm and whose altitude is 11.6 cm
   (d) a rhombus whose base is 382 in. and whose altitude is 268 in.
   (e) a trapezoid whose bases are 3.83 m and 2.44 m and whose altitude is 1.86 m.
   (f) a trapezoid whose bases are 33.6 ft and 24.7 ft and whose altitude is 15.3 ft.

Applications

3. What will be the cost, to the nearest dollar, of flagging a sidewalk 312 ft long and 6.5 ft wide, at $13.50 per square yard?
4. How many 9-in.-square tiles will cover a floor 48 ft by 12 ft?
5. What will it cost to carpet a floor, 6.25 m by 7.18 m, at $7.75 per square meter?
6. How many rolls of paper, each 8.00 yd long and 18.0 in. wide, will paper the sides of a room 16.0 ft by 14.0 ft and 10.0 ft high, deducting 124 ft² for doors and windows?

7. What is the cost of plastering the walls and ceiling of a room 40 ft long, 36 ft wide, and 22 ft high, at $8.50 per square yard, allowing 1375 ft² for doors, windows, and baseboard?

8. What will it cost to cement the floor of a cellar 25.3 ft long and 18.4 ft wide, at $3.50 per square foot?

9. Find the cost of lining a topless rectangular tank 68 in. long, 54 in. wide, and 48 in. deep with zinc, weighing 5.2 lb per square foot, at $1.55 per pound installed.

10. A parcel of land lies between two parallel streets, Fig. 6–60. How many acres will remain in the parcel after the strip shown is taken for a new street? (1 acre = 43,560 ft²)

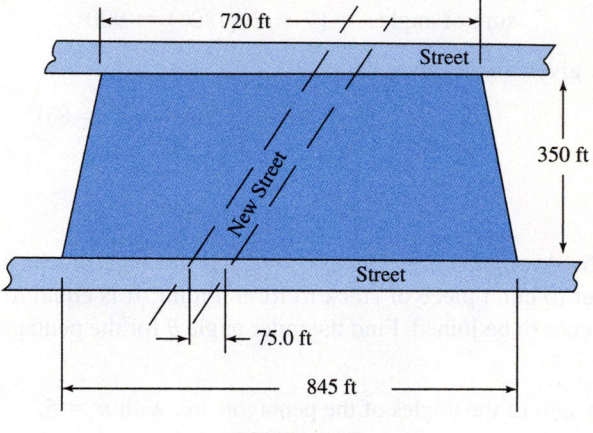

FIGURE 6–60

11. *Brick Requirement.* Estimate the number of bricks needed for a wall 13 × 18 ft, having one door 4 ft × 9 ft and two windows each 3 ft × 6 ft. Each brick has faces measuring 8 × 2 ¹/₄ inches, with half-inch joints between bricks.

12. *Squares of Roofing Material.* Roofing material is often specified in *squares*, where one square equals 100 ft². How many squares of shingles are needed to cover the roof of Fig. 6–61, neglecting waste?

13. *Miter Angles:* Calculate the miter angle θ for a
    (a) triangular window (equilateral)
    (b) square or rectangular frame
    (c) hexagonal window
    (d) octagonal wall clock

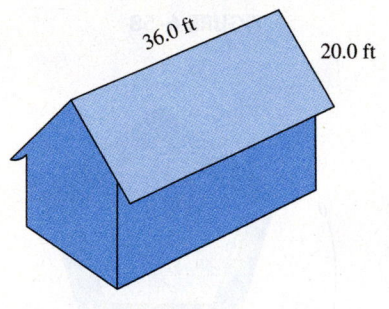

FIGURE 6–61

6–4   The Circle

A *circle*, Fig. 6–62, is a plane curve in which all points are at a given distance (called the *radius*) from a fixed point (called the *center*). The *diameter* is twice the radius. The diameter cuts the circle into two *semicircles*.

Circumference and Pi (π)

The *circumference* of a circle is its total length, or the *distance around*.

■ Exploration:

Try this. Wrap a strip of paper around a circular object, such as a jar lid, and mark the point where it starts to overlap. The length from the mark to the end of the strip is the *circumference* of the circular object. Next measure the diameter of the lid, divide that number into the circumference, and record. Repeat for several circular objects. What did you find? ■

You should have gotten a quotient a bit larger than three, regardless of the size of the lid. The ratio of the circumference C of a circle to its diameter d is the same *for all circles*. It is denoted by the Greek letter π (pi).

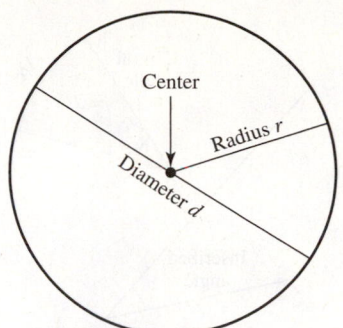

FIGURE 6–62 A circle.

Definition of π	$\pi = \dfrac{\text{Circumference}}{\text{Diameter}} = \dfrac{C}{d} \approx 3.1416$ *Pi is the ratio of the circumference of any circle to its diameter.*	**73**

Pi (π) is an *irrational number* with the approximate value 3.1416. It is stored in your calculator to more decimal places than you will probably ever need. Look for a key marked $\boxed{\pi}$.

We can use the definition of π to find the circumference of a circle. If $\pi = C/d$, then

Circumference of a Circle	$C = 2\pi r = \pi d$ *The circumference of a circle equals π times its diameter.*	**74**

◆◆◆ **Example 16:** The circumference C of a circle having a diameter of 5.54 in. is

$$C = 5.54\pi \approx 17.4 \text{ in.}$$ ◆◆◆

◆◆◆ **Example 17:** The radius r of a circle having a circumference of 854 cm is

$$r = \frac{854}{2\pi} \approx 136 \text{ cm}$$ ◆◆◆

Area of a Circle

For a circle of radius r,

Area of a Circle	$A = \pi r^2 = \dfrac{\pi d^2}{4}$ *The area of a circle (that is, its interior) equals π times the square of the radius.*	**75**

◆◆◆ **Example 18:** The area of a circle having a radius of 3.75 in. is

$$A = \pi(3.75 \text{ in.})^2 = 44.2 \text{ in.}^2$$ ◆◆◆

◆◆◆ **Example 19:** The area of a circle is 583 cm². Find its radius.

Solution: By Eq. 75,

$$r^2 = \frac{583 \text{ cm}^2}{\pi} = 185.6 \text{ cm}^2$$

$$r = \sqrt{185.6} = 13.6 \text{ cm}$$ ◆◆◆

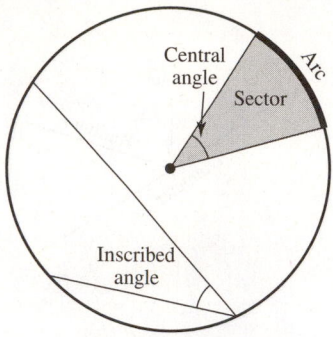

FIGURE 6–63   A circle.

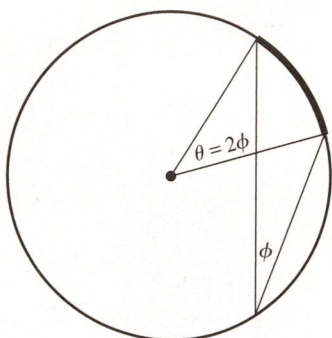

FIGURE 6–64

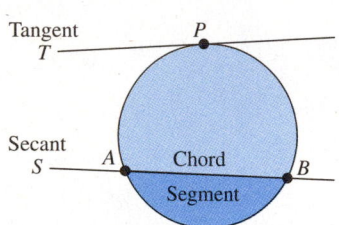

FIGURE 6–65

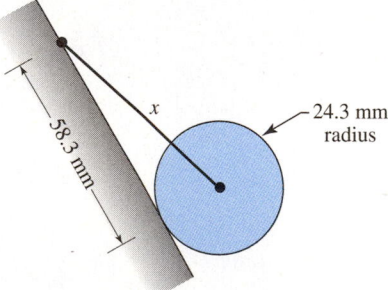

FIGURE 6–66

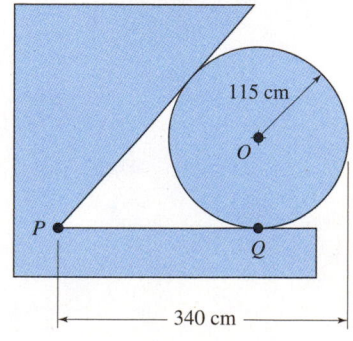

FIGURE 6–67

Arc and Sector

At this point we need a few more definitions, Fig. 6–63.

(a) A *central angle* is one whose vertex is at the center of the circle.
(b) An *inscribed angle* is one whose vertex is on the circle.
(c) An *arc* is a portion of the circle between two points on the circle.
(d) A *sector* is a plane region bounded by two radii and one of the arcs intercepted by those radii.

We have the following relationship between a central angle and an inscribed angle, Fig. 6–64.

An Inscribed and a Central Angle Subtending the Same Arc	*If an inscribed angle ϕ and a central angle θ subtend the same arc, the central angle is twice the inscribed angle.* $$\theta = 2\phi$$	**81**

Tangent, Secant, and Chord

Let us now add some straight lines touching our circle.

* A *tangent* to a circle is a line that touches the circle in just one point.
* A *secant* to a circle is a line that intersects the circle in two points.
* A *chord* is that portion of a secant joining two points on the circle.

These terms are the same when applied to curves other than the circle.

In Fig. 6–65, the tangent T touches the circle at P; the secant line S cuts the circle at two points A and B. Line segment AB is a chord. A *segment of a circle* is that portion cut off by a chord. We will show how to compute areas of segments in our chapter on radian measure.

We have two theorems about chords and tangents.

Perpendicular Bisector of a Chord	The perpendicular bisector of a chord passes through the center of the circle.	**86**
Tangent and Radius	A tangent to a circle is perpendicular to the radius drawn through the point of contact.	**83**

◆◆◆ **Example 20:** *An Application.* A roller in a printing press hangs from a link of length x and touches an inclined plate, Fig. 6–66. Find x.

Solution: By statement 83 we know that the angle between the plane and the radius drawn to the point of contact is a right angle. This enables us to use the Pythagorean theorem.

$$x^2 = (24.3)^2 + (58.3)^2 = 3989$$
$$x = 63.2 \text{ mm}$$

◆◆◆

◆◆◆ **Example 21:** *An Application.* Find the distance OP in Fig. 6–67.

Solution: By statement 83, we know that angle PQO is a right angle. Also,

$$PQ = 340 - 115 = 225 \text{ cm}$$

So, by the Pythagorean theorem,

$$OP = \sqrt{(115)^2 + (225)^2} = 253 \text{ cm}$$

◆◆◆

Our second theorem concerns two tangents to a circle drawn from an external point, Fig. 6–68.

Two Tangents Drawn to a Circle	Two tangents drawn to a circle from a point outside the circle make equal angles with a line drawn from the circle's center to the external point. The distances from the external point to each point of tangency are equal. Here, *PA* equals *PB*. Also, angle *APC* equals angle *BPC*.	**84**

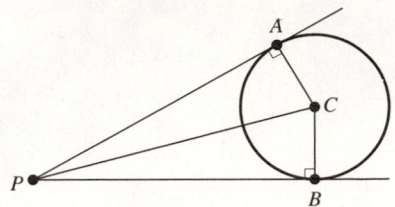

FIGURE 6–68

Turning now from tangents to chords, Fig. 6–69, we have the theorem

Intersecting Chords	If two chords in a circle intersect, the product of the parts of one chord equals the product of the parts of the other chord Here, $ab = cd$	**85**

◆◆◆ Example 22: Find *x* in Fig. 6–70.

Solution: By theorem 85 we can write

$$16.3x = 10.1(12.5)$$

$$x = \frac{10.1(12.5)}{16.3} = 7.75 \qquad ◆◆◆$$

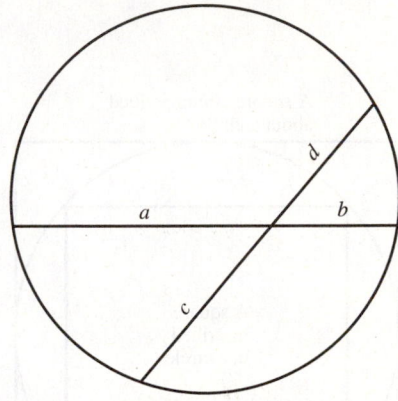

FIGURE 6–69

Semicircle

Any angle inscribed in a semicircle is a right angle.	**82**

◆◆◆ Example 23: Find the distance *x* in Fig. 6–71.

Solution: By statement 82, we know that $\theta = 90°$. Then, by the Pythagorean theorem,

$$x = \sqrt{(250)^2 - (148)^2} = 201 \qquad ◆◆◆$$

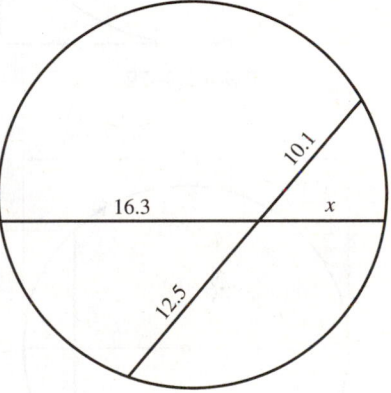

FIGURE 6–70

Exercise 4 ◆ The Circle

1. Find the circumference and area of a circle of radius 4.82 cm.

2. Find the circumference and area of a circle of radius 2.385 in.

3. Find the radius and area for a circle whose circumference is 74.8 in.

4. Find the radius and area for a circle whose circumference is 2.73 m.

5. Find the radius and circumference of a circle whose area is 39.5 ft².

6. Find the radius and circumference of a circle whose area is 2.74 m².

7. *Inscribed and Circumscribed.* A polygon is said to be *inscribed in* a circle if every vertex of the polygon lies on the circle. The circle is then said to be *circumscribed about* the polygon. A polygon is said to be circumscribed about a circle if the circle is tangent to each side of the polygon. The circle is then said to be inscribed in the polygon.

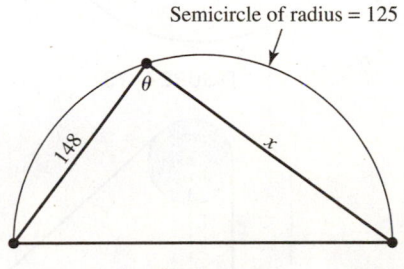

FIGURE 6–71

How much larger is the side of a square circumscribing a circle 155 cm in diameter than a square inscribed in the same circle, Fig. 6–72?

8. The radius of a circle is 5.00 m. Find the diameter of another circle containing 4 times the area of the first.

9. Find the distance x in Fig. 6–73.

10. In Fig. 6–74, distance OP is 8.65 units. Find the distance PQ.

11. Figure 6–75 shows a semicircle with a diameter of 105. Find distance PQ.

Applications

12. In a park is a circular fountain whose basin is 22.5 m in circumference. What is the diameter of the basin?

13. The area of the bottom of a circular pan is 196 in.^{2}. What is its diameter?

14. Find the diameter of a circular solar pond that has an area of 125 m^2.

15. What is the circumference of a circular lake 33.0 m in diameter?

16. The distance around a circular park is 2.50 mi. How many acres does it contain? (1 sq. mi = 640 acres)

17. A woodcutter uses a tape measure and finds the circumference of a tree to be 95.0 in. Assuming the tree cross-section to be circular, what length of chain-saw bar is needed to fell the tree? (By cutting from both sides, one can fell a tree whose diameter is twice the length of the chain-saw bar.)

18. What must be the diameter d of a cylindrical piston so that a pressure of 125 lb/in.^{2} on its circular end will result in a total force of 3610 lb? *Hint:* The force on a surface is equal to the area of that surface times the pressure on that surface.

19. Find the length of belt needed to connect the three 15.0-cm-diameter pulleys in Fig. 6–76. *Hint:* The total *curved* portion of the belt is equal to the circumference of one pulley.

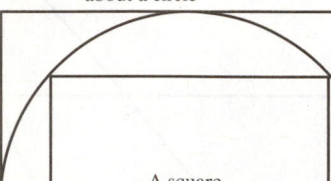

A square circumscribed about a circle

A square inscribed in a circle

FIGURE 6–72

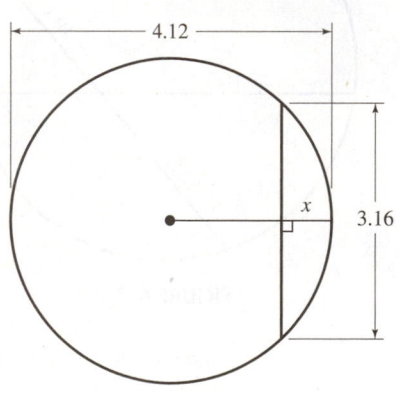

FIGURE 6–73

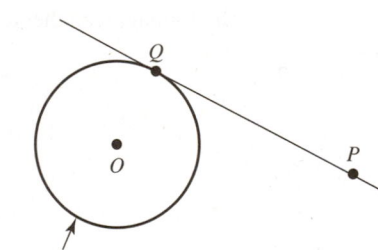

6.35 diameter

FIGURE 6–74

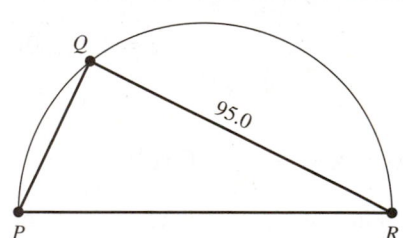

95.0

FIGURE 6–75

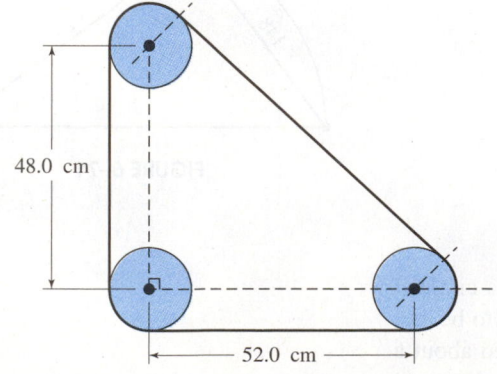

48.0 cm

52.0 cm

FIGURE 6–76

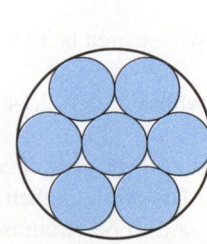

FIGURE 6–77

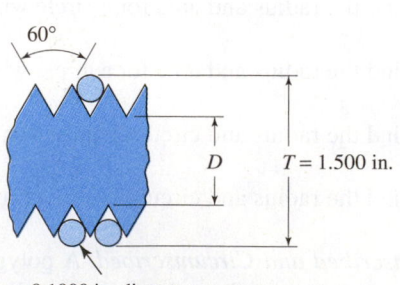

60°

D $T = 1.500$ in.

0.1000 in. diameter

FIGURE 6–78 Measuring a screw thread "over wires."

20. Seven cables of equal diameter are contained within a circular conduit, as in Fig. 6–77. If the inside diameter of the conduit is 25.6 cm, find the cross-sectional area *not* occupied by the cables.

21. A 60° screw thread is measured by placing three wires on the thread and measuring the distance *T*, as in Fig. 6–78. Find the distance *D* if the wire diameters are 0.1000 in. and the distance *T* is 1.500 in., assuming that the root of the thread is a sharp V shape. *Hint:* In a right triangle whose angles are 30° and 60°, the hypotenuse is twice the length of the shortest side.

22. Figure 6–79 shows two of the supports for a hemispherical dome. Find the length of girder *AB*.

23. A certain car tire is 78.5 cm in diameter. How far will the car move forward with one revolution of the wheel?

24. Figure 6–80 shows a circular window under the eaves of a roof. Find the distance *x*.

25. Figure 6–81 shows a round window in a dormer the shape of an equilateral triangle. Find the radius *r* of the window.

26. Figure 6–82 shows a circular design, 2.04 m in diameter, located in the angle of a roof. Find distances *AB*, *AD*, and *AC*.

27. Figure 6–83 shows some of the internal framing in a rose window. Find *AB*, given that *BC* = 7.38 ft, *DB* = 4.77 ft, and *BE* = 5.45 ft.

28. Figure 6–84 shows a square window surmounted by a circular arch. Find the following:
(a) The radius *r* of the arch
(b) The length *s* of the curved underside of the arch
(c) The total open area of the window, including the area of the square and the area under the arch
Hint: Note that four of the arcs *s* would form a complete circle.

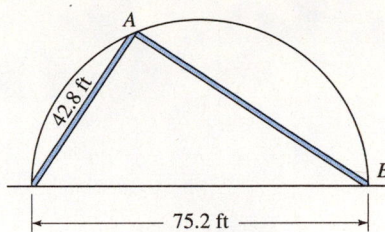

FIGURE 6–79

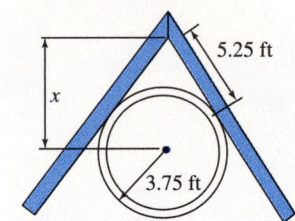

FIGURE 6–80   Design for a Circular Window.

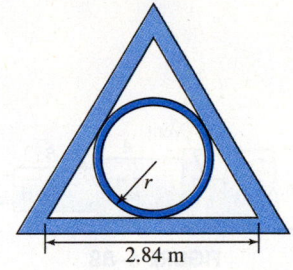

FIGURE 6–81   A Window in a Dormer.

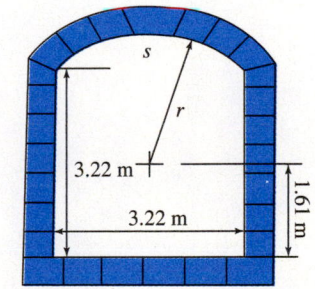

FIGURE 6–84   A Square Window Topped by a Circular Arch.

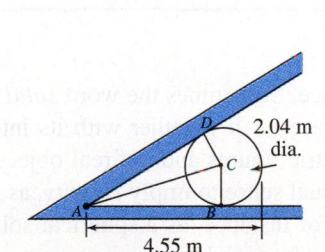

FIGURE 6–82   A Circular Design.

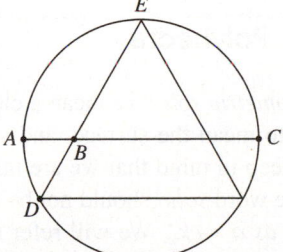

FIGURE 6–83   A Rose Window.

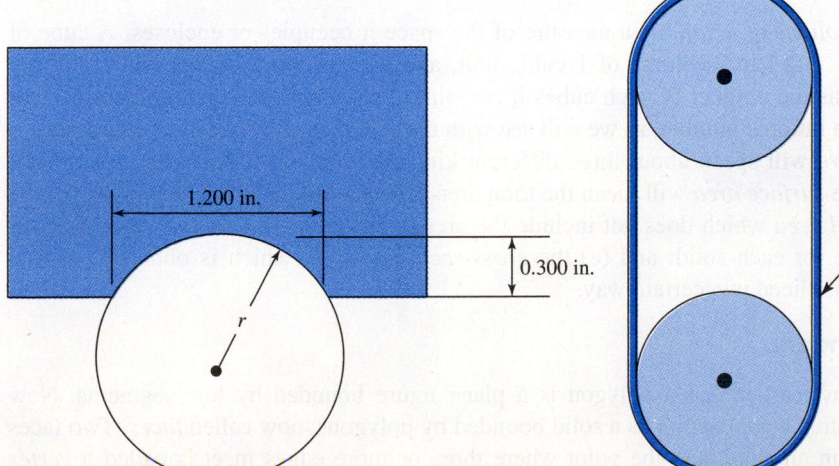

FIGURE 6–85

FIGURE 6–86

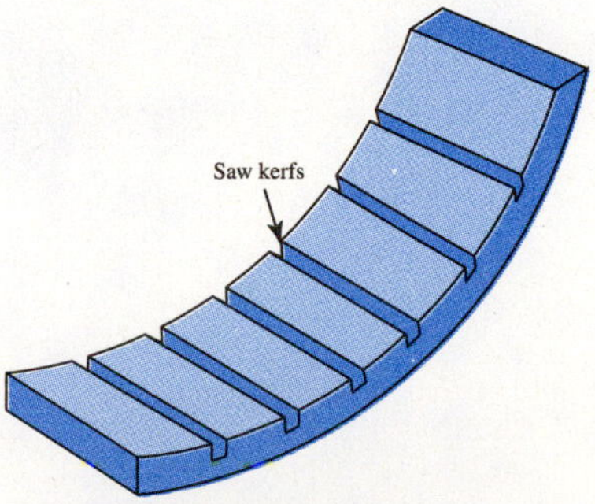

FIGURE 6–87   Kerfing.

29. Find the radius *r* of the cutter needed to machine a circular arc having a depth of 0.300 in. and a width of 1.200 in., as shown in Fig. 6–85.

30. You buy a used band saw, Fig. 6–86, but the blade and owner's manual are missing. You measure the wheels at 12.8 in. in diameter and the distance between their centers at 20.0 in. Calculate the length of band saw blade needed.

31. *Framing Square:* It is sometimes difficult to locate the center of a circle, say, a circular hoop. Using the fact that "any angle inscribed in a semicircle is a right angle," how would you use a framing square to quickly find the ends of a diameter of a circle (provided that the circle is 26 inches or less in diameter)? How would you then locate the center of the circle?

32. *Kerfing:* A *kerf* is the notch or groove left by a saw. Kerfing is a carpentry method for bending a board into a curve by cutting equally spaced, parallel kerfs along one face, as in Fig. 6–87, and then bending the board until the kerfs just close. A board 48.0 in. long and 0.750 in. thick is to be bent into a semicircle. How many cuts, each 0.100 in. wide, are needed?

33. *Project:* Calculate the kerfing needed to bend a board into a circular arc, then actually do it in a woodworking shop. Demonstrate your results to your class.

34. *Writing:* Write a section in an instruction manual for machinists on how to find the root diameter of a screw thread by measuring "over wires" (Fig. 6–78). Your entry should have two parts: first, a "how-to" section giving step-by-step instructions, and then a "theory" section explaining why this method works. Keep the entry to one page or less.

35. *Project:* A ventilation system, Fig. 6–88, consists of circular ducts with rectangular vents. Each vent measures 8 in. by 15 in. The diameter of each circular duct is chosen so that its cross-sectional area is equal to that of the preceding duct, minus the area of the preceding vent. Find the diameters of ducts *A* and *B*.

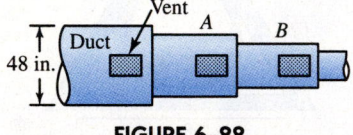

FIGURE 6–88

6–5   Polyhedra

By *geometric solid* we mean a closed surface in space. Sometimes the word *solid* is taken to mean the surface, and sometimes the surface itself together with its interior. Keep in mind that we are talking about geometric figures and not real objects, and the word *solid* should not be taken here in its usual sense to imply *rigidity*, as in *"solid as a rock."* We will refer to a soap bubble, for instance, as a spherical solid even though the soap film and the enclosed air are far from rigid.

Volume and Area of a Solid

The *volume of a solid* is a measure of the space it occupies or encloses. A cube of side 1 unit has a volume of 1 cubic unit, and we can think of the volume of any solid as the number of such cubes it contains. This may not be a whole number or even a rational number, as we will see with the volumes of cylinders and spheres.

We will speak about three different kinds of *areas* in connection with solids: (a) the *surface area* will mean the total area of the solid, including any ends; (b) the *lateral area* which does not include the area of the ends or *base*(s), which we will define for each solid; and (c) the *cross-sectional area*, which is obtained when a solid is sliced in a certain way.

Polyhedra

We saw earlier that a polygon is a plane figure bounded by line segments. Now we define a *polyhedron* as a solid bounded by polygons, now called *faces*. Two faces meet in an *edge*, and the point where three or more edges meet is called a *vertex* (Fig. 6–89). We will discuss the polyhedra one at a time.

Prism

A *prism* is a polyhedron with two parallel, identical faces (called *bases*), and whose remaining faces (called *lateral faces*) are parallelograms, Fig. 6–90, The lateral faces are formed by joining corresponding vertices of the bases. The *altitude* of a prism is the perpendicular distance between the bases.

A prism is named according to the shape of its bases (i.e., triangular prism, quadrangular prism, and so forth). A prism is also called *right* if its bases are perpendicular to its lateral edges and the lateral faces are all rectangles, otherwise it is called *oblique*.

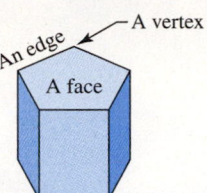

FIGURE 6–89   A polyhedron. The name *polyhedron* is from *poly* = many and *hedron* = faces.

Any Prism	Volume = (area of base)(altitude)	91
Right Prism	Lateral area = (perimeter of base)(altitude) Does not include the bases.	92

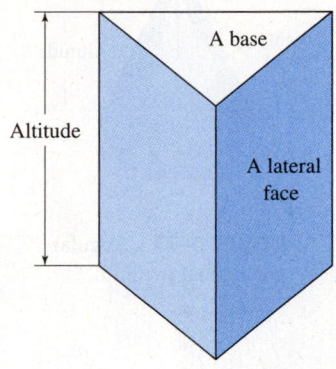

FIGURE 6–90   A right triangular prism.

◆◆◆ **Example 24:** Find the volume and lateral area of a prism whose altitude is 22.8 mm and whose bases are triangles, each with a perimeter of 16.9 mm and area of 147 mm^2.

Solution: Using the prism formulas gives

$$\text{Volume} = \text{base area} \times \text{altitude}$$
$$= 147(22.8) = 3350 \text{ mm}^3$$

$$\text{Lateral area} = \text{base perimeter} \times \text{altitude}$$
$$= 16.9(22.8) = 385 \text{ mm}^2 \qquad ◆◆◆$$

Rectangular Parallelepiped and Cube

A *rectangular parallelepiped* is a right rectangular prism (Fig. 6–91). All six faces are rectangles. It is often called a *rectangular solid*. It is simply the familiar square-cornered *box*.

Rectangular Parallelepiped	Volume = lwh	89
	Surface area = $2(lw + hw + lh)$	90

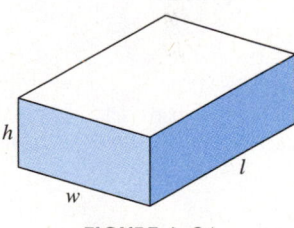

FIGURE 6–91

◆◆◆ **Example 25:** Find the volume and surface area of a crate measuring 2.70 ft by 3.40 ft by 5.90 ft.

Solution: We get

$$\text{Volume} = \text{length} \times \text{width} \times \text{height}$$
$$= 2.70(3.40)(5.90) = 54.2 \text{ ft}^3$$

$$\text{Surface area} = 2[2.70(3.40) + 2.70(5.90) + 3.40(5.90)]$$
$$= 90.3 \text{ ft}^2 \qquad ◆◆◆$$

The *cube* is a rectangular parallelepiped having all sides equal (Fig. 6–92). It has 6 square faces, 12 equal edges, and 8 vertices. Unlike the other solids we have covered so far, it is a *regular* polyhedron, so-called because all its faces are identical regular polygons.

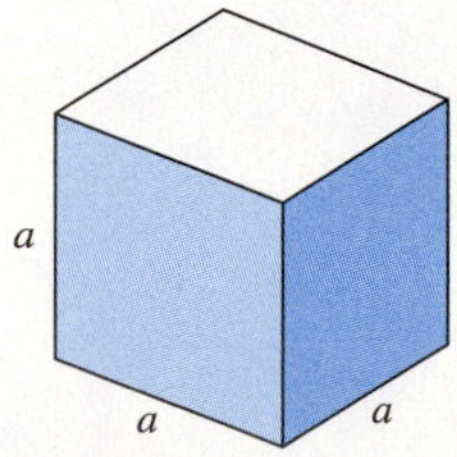

FIGURE 6–92   Cube.

Cube	Volume $= a^3$	87
	Surface area $= 6a^2$	88

◆◆◆ **Example 26:** *An Application.* Find the volume and surface area of a cubical room that is 10.5 ft on a side.

Solution: By Eq. 87,

$$\text{Volume} = (10.5)^3 = 1160 \text{ ft}^3$$

and by Eq. 88,

$$\text{Surface area} = 6(10.5)^2 = 662 \text{ ft}^2 \qquad \text{◆◆◆}$$

The Pyramid

A *pyramid* is a polyhedron whose base is a polygon, and whose other faces are triangles formed by connecting vertices of the base to a common point (the *vertex* of the pyramid) (Fig. 6–93).

A pyramid is named for the shape of its base: *triangular pyramid, quadrangular pyramid,* and so forth. A *regular* pyramid is one whose base is a regular polygon and whose altitude passes through the center of that polygon. A *frustrum* of a pyramid is the portion of the pyramid between its base and a plane section parallel to its base (Fig. 6–94).

The *altitude* of a regular pyramid is the perpendicular distance from base to vertex. The altitude of the frustum of a regular pyramid is the perpendicular distance between its base and that plane section.

The *slant height* of a regular pyramid is the altitude of each triangular face. The slant height of the frustum of a regular pyramid is the perpendicular distance between one edge of the base and one edge of the plane section.

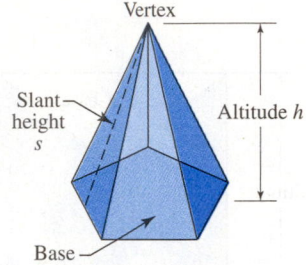

FIGURE 6–93 Regular pentagonal pyramid.

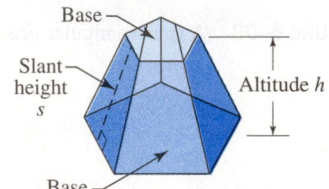

FIGURE 6–94 Frustum of a regular pentagonal pyramid.

Any Pyramid	$\text{Volume} = \left(\dfrac{h}{3}\right)(\text{area of base})$	95
Regular Pyramid	$\text{Lateral area} = \left(\dfrac{s}{2}\right)(\text{perimeter of base})$ Does not include the base.	96
Frustum of Any Pyramid	$\text{Volume} = \dfrac{h}{3}\left(A_1 + A_2 + \sqrt{A_1 A_2}\right)$ Where A_1 and A_2 are the areas of the bases	97
Frustum of Regular Pyramid	$\text{Lateral area} = \left(\dfrac{s}{2}\right)(\text{sum of base perimeters})$	98

◆◆◆ **Example 27:** Find the volume of a pyramid having a square base 4.51 cm on a side and an altitude of 3.72 cm.

Solution: The area of the base is $(4.51)^2 = 20.3 \text{ cm}^2$. The volume of the pyramid is then

$$\text{Volume} \frac{(3.72)(20.3)}{3} = 25.2 \text{ cm}^3 \qquad \text{◆◆◆}$$

Exercise 5 ◆ Polyhedra

The Prism

1. An oblique prism has a base area of 5.63 in.² and an altitude of 4.72 in. Find its volume.
2. The base of a right prism is an equilateral triangle 3.74 mm on a side. Its altitude is 8.35 mm. Find its volume and lateral area.
3. Find the volume of the triangular prism in Fig. 6–95.
4. Find the lateral area, total area, and volume of each prism in Fig. 6–96. Assume that each base is a regular polygon.

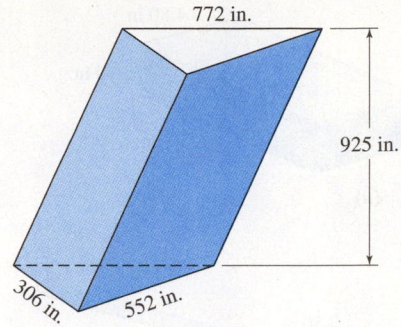

FIGURE 6–95

Prism Applications

5. The support strut shown in Figure 6–97 has a rectangular base that is 12 in. wide and 18 in. deep. Find the lateral area and volume of the strut.
6. A roof, shown simplified in Figure 6–98, is in the shape of a triangular prism. Find the area of the shingled surface and the volume enclosed by the roof.
7. Two intersecting roofs are simplified in Figure 6–99. Find the area of the shingled surface and the volume enclosed by the roofs. Make sure you count the portion contained under both roofs only once and not twice.

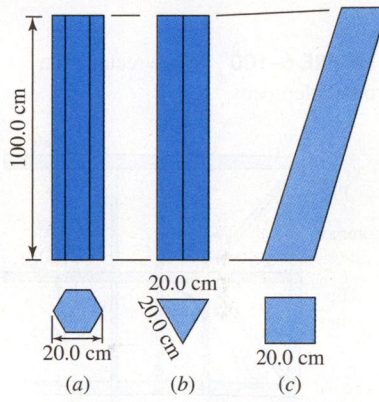

FIGURE 6–96 Some prisms.

Rectangular Parallelepiped and Cube

8. Find the lateral area, total area, and volume of each rectangular parallelepiped in Figure 6–100.

9. Find the surface area and volume of a cube with the following sides:
   a. 3.75 in.   b. 26.3 cm   c. 2.24 ft

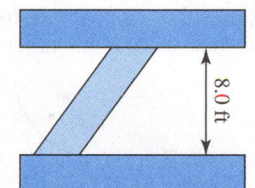

FIGURE 6–97 A support strut.

10. Find the volume and surface area of a rectangular parallelepiped that measures 26.8 cm × 83.4 cm × 55.3 cm.

Rectangular Parallelepiped and Cube Applications

11. A 1.00-in. cube of steel is placed in a surface grinding machine, and the vertical feed is set so that 0.0050 in. of metal is removed from the top of the cube at each cut. How many cuts are needed to reduce the volume of the cube by 0.50 in.³?
12. A rectangular tank is being filled with liquid, with each cubic meter of liquid increasing the depth by 2.0 cm. The length of the tank is 12.0 m.
    (a) What is the width of the tank?
    (b) How many cubic meters will be required to fill the tank to a depth of 3.0 m?
13. How many loads of gravel will be needed to cover 2.0 mi of roadbed, 35 ft wide, to a depth of 3.0 in. if one truckload contains 8.0 yd³ of gravel?
14. *Board Feet*. A *board foot* of lumber is the volume of wood contained in a board 1 in. thick, 1 ft long, and 1 ft wide. How many cubic inches and cubic feet are in a board foot?

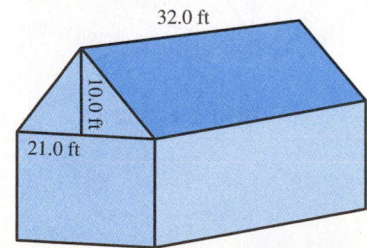

FIGURE 6–98 A prismatic roof.

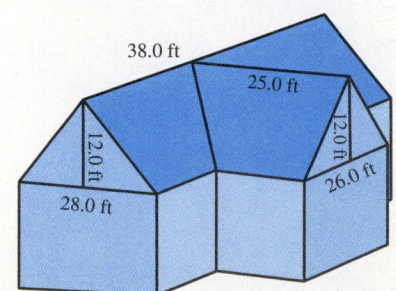

FIGURE 6–99 Intersecting roofs.

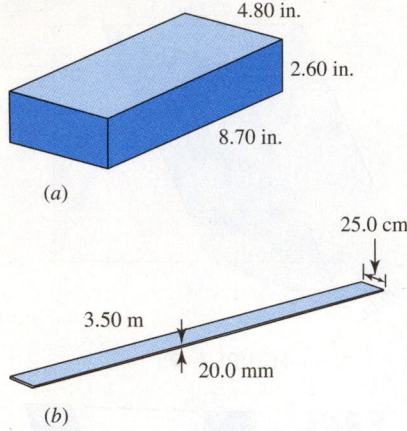

(a)

(b)

FIGURE 6–100   Some rectangular parallelepipeds.

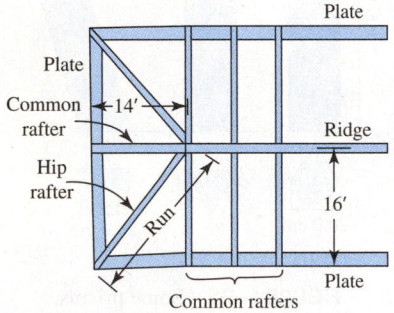

FIGURE 6–101   Top view of roof showing placement of hip rafters.

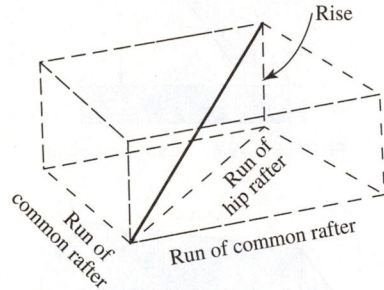

FIGURE 6–102

15. *Board Feet.* How many board feet are in a plank that is $2^1/_2$ in. thick, 15 in. wide, and 18 ft long?

16. *Board Feet.* How many board feet are contained by 12 two-by-fours, each 8 ft long? (Board feet calculations normally use the rough or unplaned dimensions, so here use the nominal 2×4 dimensions rather than the actual ones).

17. *Volume of Excavations.* How many cubic yards of earth must be removed for a rectangular cellar hole dug on level ground, 48 ft long, 36 ft wide, and 12 ft deep?

18. *Mortar Requirement.* Estimate the cubic feet of mortar needed for a 1000 square foot wall made of brick, each $3^3/_4$ deep and with faces measuring $8^1/_2 \times 2^1/_4$ inches, with half-inch-joints.

19. *Cords of Firewood.* A standard *cord* of wood is a " well-stacked" pile measuring 4 ft by 4 ft by 8 ft. How many cubic feet of wood are in a cord?

20. *Truck Capacity.* The bed of a certain pickup truck measures 58 in. by 74 in. and is 16 in. deep. How many cubic feet of sand can it hold, filled level with the sides, allowing 1.2 ft^3 for each wheel well?

21. *Truck Capacity.* Using the pickup truck of Problem 20, determine how many truckloads of firewood are needed to make a cord, if the wood is "well stacked" level with the sides of the truck.

22. Find the volume of packed gravel (weighing 110 lb/ft^3) needed to cover a driveway that is 15 ft wide and 30 ft long to a depth of 3.0 inches.

23. How much *loose* gravel (weighing 100 lb/ft^3) is needed (before packing) for the driveway in the preceding problem, to be spread and then packed to a depth of 3.0 in?

24. *Hip Rafters.* A hip rafter runs from the ridge to a corner of the roof. Find the line length of the hip rafter in Fig. 6–101. The common rafters on one side of the hip have a run of 16.0 ft and a slope of 8.00 in. per foot, and on the other side have a run of 14.0 ft. *Hint*: The length of the hip rafter is the diagonal of a box whose base is 16.0 ft by 14.0 ft, and whose height is equal to the rise of the roof, as in Fig. 6–102.

25. *Project:* At a country fair in Vermont a prize was given to whoever most closely guessed the weight of a rough block of marble. It was a prism, more or less, whose ends were roughly parallelograms with bases of about 5 ft and with a height of about 4 ft. Its lateral edges averaged about 7 ft in length. Estimate its weight, assuming the density of marble to be 170 lb/ft^3. Try doing the computation in your head, as you would have to at a fair, before using your calculator.

The Pyramid

26. Find the volume and lateral area of each pyramid or frustum in Figure 6–103.

27. Find the volume and lateral area of a regular pyramid having a square base 6.83 in. on a side and an altitude of 7.93 in.

28. The frustum of a regular pyramid has square bases, one 4.83 mm on a side and the other 2.84 mm on a side. Its altitude is 3.88 mm. Find its volume and lateral area.

29. The slant height of a right pyramid is 11.0 in., and the base is a 4.00-in. square. Find the area of the entire surface.

Pyramid Applications

30. How many cubic feet are in a piece of timber 30.0 ft long, one end being a 15.0-in. square and the other a 12.0-in. square?

31. a. Find the volume enclosed by the pyramidal roof on a square tower. Take the base as 22.0 ft on a side and the height as 24.5 ft, and ignore the overhang.
    b. Find the lateral area of the roof.

32. The pyramidal roof shown in Figure 6–104 has an octagonal base of 4.50 ft on a side and a slant of 14.0 ft. How many square feet of shingles are needed to cover the roof, not counting any waste?

33. A house that is 50.0 ft long and 40.0 ft wide has a pyramidal roof whose height is 15.0 ft. Find the length of a hip rafter that reaches from a corner of the building to the vertex of the roof.

34. A simplified hip roof is shown in Figure 6–105. Find the area of the shingled surface and the volume enclosed by the roof.

35. *Project:* A certain rain gauge is in the shape of an inverted frustum of a square pyramid. The lower end is $\frac{1}{2}$ inch on a side (inside dimension), the upper end is 2 inches on a side, and its height is 5 inches. Calculate where to place marks on the gauge so that it gives the same readings as would a gauge with parallel vertical sides.

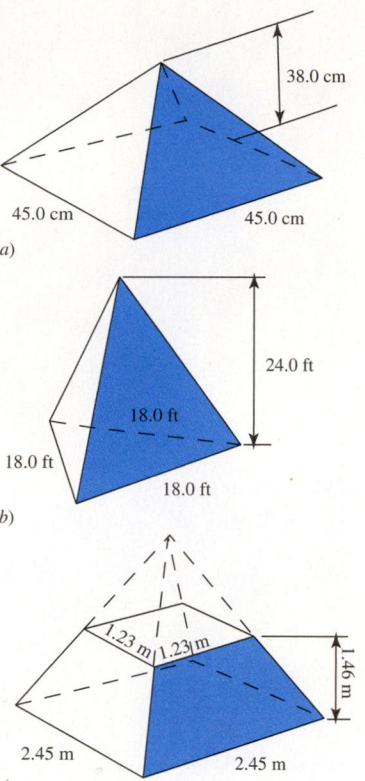

(a)

(b)

(c)
FIGURE 6–103   Some pyramids.

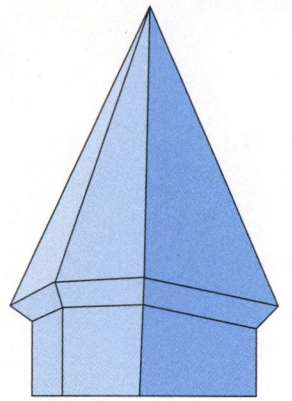

FIGURE 6–104   A pyramid roof.

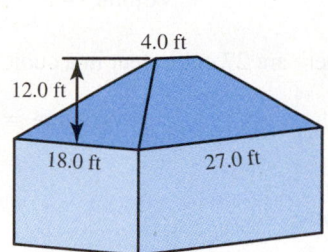

FIGURE 6–105   A hip roof.

6–6   Cylinder, Cone, and Sphere

Let us turn now from the polyhedra, solids bounded by planes, to solids bounded by curved surfaces. They are the cylinder, cone, and sphere.

The Cylinder

A cylinder is a solid with two parallel, identical faces (called *bases*), whose lateral surface is formed by joining corresponding points on the bases (Fig. 6–106). A cylinder is named according to the shape of its bases (i.e., circular cylinder, elliptical cylinder, and so forth). The *axis* of a cylinder is the line connecting the centers of its bases. A cylinder is called *right* if its bases are perpendicular to its axis; otherwise it is called *oblique*. The *altitude* of a cylinder is the perpendicular distance between the bases.

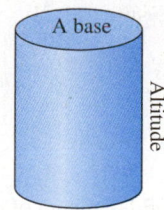

FIGURE 6–106   Right circular cylinder.

Any Cylinder	Volume = (area of base)(altitude)	91
Right Cylinder	Lateral area = (perimeter of base)(altitude) Does not include the bases.	92

◆◆◆ **Example 28:** Find (a) the volume and (b) the lateral area of a right circular cylinder having a base radius of 5.73 units and an altitude of 8.24 units.

Solution: (a) The area of the circular base is

$$A = \pi(5.73)^2 = 103 \text{ square units}$$

So the volume of the cylinder is

$$V = (\text{area of base})(\text{altitude})$$
$$= 103(8.24) = 849 \text{ cubic units}$$

(b) The perimeter of the base is
$$P = 2\pi(5.73) = 36.0 \text{ units}$$
so the lateral area is then
$$\text{Lateral area} = (\text{perimeter of base})(\text{altitude})$$
$$= 36.0(8.24) = 297 \text{ square units.} \qquad \blacklozenge\blacklozenge\blacklozenge$$

◆◆◆ **Example 29:** *An Application.* A cylindrical form for a right circular concrete column is 22.5 in. in diameter and 18.5 ft high. How many cubic yards of concrete are needed to fill this form?

Solution: The column diameter in feet is
$$\frac{22.5}{12} = 1.875 \text{ ft}$$
Then by Eq. 91,
$$\text{Volume} = \frac{\pi(1.875)^2(18.5)}{4} = 51.1 \text{ ft}^3$$

Since there are 27 cubic feet in a cubic yard,
$$\text{Volume} = \frac{51.1}{27} = 1.89 \text{ yd}^3 \qquad \blacklozenge\blacklozenge\blacklozenge$$

The Cone

A *cone* is a solid bounded by a plane region (the *base*) and the surface formed by line segments joining a point (the *vertex*) to points on the boundary of the base. If the base is circular and a perpendicular through its center passes through the vertex, we have a *right circular cone* (Fig. 6–107). Otherwise it is called *oblique*. A *frustum* of a cone is defined just like the frustum of a pyramid (Fig. 6–108).

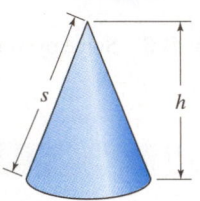

FIGURE 6–107   Right circular cone.

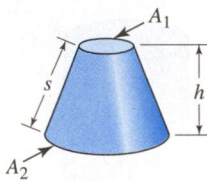

FIGURE 6–108   Frustum of right circular cone.

Any Cone	$\text{Volume} = \left(\dfrac{h}{3}\right)(\text{area of base})$	**95**
Right Circular Cone	$\text{Lateral area} = \left(\dfrac{s}{2}\right)(\text{circumference of base})$ Does not include the base.	**96**
Frustum of Any Cone	$\text{Volume} = \dfrac{h}{3}\left(A_1 + A_2 + \sqrt{A_1 A_2}\right)$ where A_1 and A_2 are the areas of the bases	**97**
Frustum of Right Circular Cone	$\text{Lateral area} = \left(\dfrac{s}{2}\right)(\text{sum of base circumferences})$	**98**

◆◆◆ **Example 30:** *An Application:* (a) Find the volume of the conical cupola atop a round tower, Fig. 6–109. (b) How many square feet of shingles are needed to cover the cupola?

Solution:

(a) The area of the cone's base is $\pi(7.30)^2 = 167 \text{ ft}^2$.
    Then by Eq. 95,
$$\text{Volume} = \frac{(22.8)(167)}{3} = 1270 \text{ ft}^3$$

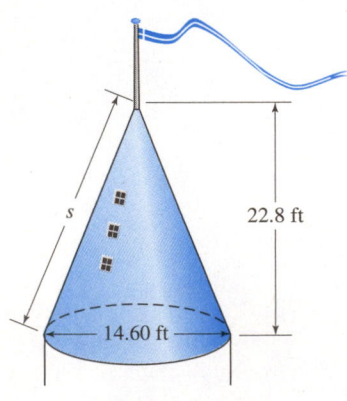

FIGURE 6–109

(b) We find the slant height *s* by the Pythagorean theorem

$$s^2 = (7.30)^2 + (22.8)^2 = 573$$

so

$$s = 23.9 \text{ ft}$$

The circumference of the base is $14.6\pi = 45.9$ ft, so by Eq. 96,

$$\text{Lateral area} = \frac{23.9(45.9)}{2} = 549 \text{ ft}^2$$

which is the square footage of shingles needed (not accounting for waste). ◆◆◆

The Sphere

A sphere is the set of points in space at a given distance from a fixed point (Fig. 6–110). The given point is, of course, the *center* of the sphere and the fixed distance is the *radius r*. The *diameter d* is twice the radius.

A section of a sphere cut by a plane is a circle. If the plane passes through the center of the sphere, we get a *great circle,* whose radius equals that of the sphere. A great circle divides the sphere into two *hemispheres.*

The surface area and volume of a sphere of radius *r* are given by

FIGURE 6–110 Sphere.

Surface Area of a Sphere	$\text{Area} = 4\pi r^2$	**94**
Volume of a Sphere	$\text{Volume} = \left(\dfrac{4}{3}\right)\pi r^3$	**93**

◆◆◆ **Example 31:** For a sphere of radius 2.55 cm,

$$\text{Surface area} = 4\pi(2.55)^2 = 81.7 \text{ cm}^2$$

$$\text{Volume} = \left(\frac{4}{3}\right)\pi(2.55)^3 = 69.5 \text{ cm}^3 \qquad \text{◆◆◆}$$

Exercise 6 ◆ Cylinder, Cone, and Sphere

Cylinder

1. Find the lateral area, total area, and volume of each of the following right circular cylinders:
   (a) Base diameter = 34.5 in., height = 26.4 in.
   (b) Base diameter = 134 cm, height = 226 cm
   (c) Base diameter = 3.65 cm, height = 2.76 m
2. Find the volume and the lateral area of a right circular cylinder having a base radius of 128 and a height of 285.
3. Find the volume of the cylinder in Fig. 6–111.

Cylinder Applications

4. A 7.00-ft-long piece of iron pipe has an outside diameter of 3.50 in. and the volume of iron in the pipe is equal to 449 in.^{3}. Find the wall thickness.
5. A certain steel bushing is in the shape of a hollow cylinder 18.0 mm in diameter and 25.0 mm long, with an axial hole 12.0 mm in diameter. Find the volume of steel in one bushing.

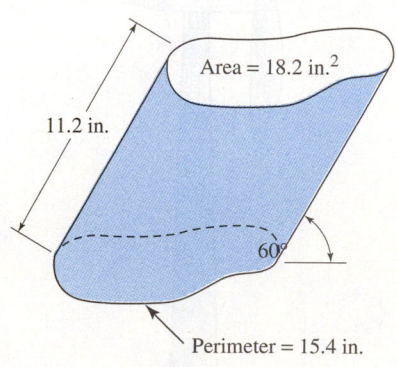

FIGURE 6–111

6. A steel gear is to be lightened by drilling holes through the gear. The gear is 3.50 in. thick. Find the diameter d of the holes if each is to remove 12.0 oz. Use a density of 485 lb/ft³ for iron.

7. A certain gasoline engine has four cylinders, each with a bore of 82.0 mm and a piston stroke of 95.0 mm. Find the engine displacement in liters. (The engine *displacement* is the total volume swept out by all of the pistons.)

8. Find the lateral area and the volume enclosed by a cylindrical tower having a round base 18.0 ft in diameter and a height of 31.5 ft.

9. A cylindrical chimney has a round base with a 2.55 m outside diameter, a 2.00 m inside diameter, and a height of 7.54 m. Find the volume of masonry in the chimney.

10. A circular hole for a wading pool, 3.00 ft deep and 32.0 ft in diameter, is to be dug in level ground. How many cubic yards of earth must be removed?

11. A cylindrical oil tank is 4.50 ft in diameter and 7.20 ft long. Find its capacity, in gallons. (1 ft³ = 7.48 gal)

12. *Archimedes on the Cylinder*: Archimedes stated that the lateral area of a cylinder is equal to the area of a circle with a radius that is the mean proportional between the cylinder's height and its base diameter. Verify this using modern notation.

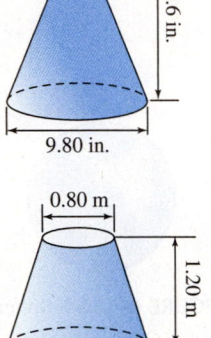

FIGURE 6–112   Some cones.

Cone

13. Find the lateral area, total area, and volume of each right circular cone or frustum in Fig. 6–112.

14. The circumference of the base of a right circular cone is 40.0 in., and the slant height is 38.0 in. What is the area of the lateral surface?

15. Find the volume of a circular cone whose altitude is 24.0 cm and whose base diameter is 30.0 cm.

Cone Applications

16. How many cubic feet are in a tapered timber column 30.0 ft long if one end has an area of 225 sq. in. and the other end has an area of 144 sq. in.?

17. Estimate the volume of a conical pile of sand that is 12.5 ft high and has a base diameter of 14.2 ft.

18. Tapered wooden columns are used to support a tent-like structure at an exposition. Each column is 50 ft high and has a circumference of 5.0 ft at one end and a circumference of 3.0 ft at the other. Find the volume of each column.

19. Find the volume of a tapered steel roller 12.0 ft long and having end diameters of 12.0 in. and 15.0 in.

20. A certain support column for a wind generator is 65.5 ft tall and has diameters of 8.55 ft at its base and 4.17 ft at its upper end, Fig. 6–113. It is made of steel, 1.25 in. thick. Find the volume of steel in the column.

21. *Writing:* A cylindrical surface and a conical surface can each be *generated* by means of a straight line that moves in a certain way. Find out how this is done and describe it in a short paper.

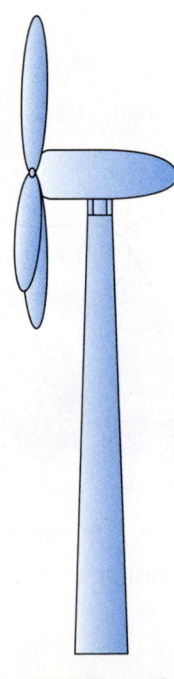

FIGURE 6–113

Sphere

22. Find the volume and the radius of a sphere having a surface area of 462.

23. Find the surface area and the radius of a sphere that has a volume of 5.88.

24. Find the volume and surface area of a sphere having each radius:
   a. 744 in.
   b. 1.55 m
25. Find the volume and radius of a sphere having a surface area of 46.0 cm^2.
26. Find the surface area and radius of a sphere that has a volume of 462 ft^3.

Sphere Applications

27. How many great circles of a sphere would have the same area as that of the surface of the sphere?
28. Find the weight in pounds of 100 steel balls each 2.50 inches in diameter (density = 485 lb/ft^3).
29. A spherical radome encloses a volume of 9000 m^3. Assume that the sphere is complete,
   (a) Find the radome radius, r.
   (b) If the radome is constructed of a material weighing 2.00 kg/m^2, find its weight.

••• CHAPTER 6 REVIEW PROBLEMS ••••••••••••••••••••••••••••••••••

1. A rocket ascends in a straight path at constant velocity at an angle of 60.0° with the horizontal. After 1.00 min, it is directly over a point that is a horizontal distance of 12.0 mi from the launch point. Find the speed of the rocket.

2. A rectangular beam 16 in. thick is cut from a log 20 inches in diameter. Find the greatest depth beam that can be obtained.

3. A cylindrical tank 4.00 m in diameter is placed with its axis vertical and is partially filled with water. A spherical diving bell is then completely immersed in the tank, causing the water level to rise 1.00 m. Find the diameter of the diving bell.

4. Two vertical piers are 240 ft apart and support a circular bridge arch. The highest point of the arch is 30.0 ft higher than the piers. Find the radius of the arch.

5. Two antenna masts are 10 m and 15 m high and are 12 m apart. How long a wire is needed to connect the tops of the two masts?

6. Find the area and the side of a rhombus whose diagonals are 100 and 140.

7. Find the area of a triangle that has sides of length 573, 638, and 972.

8. Two concentric circles have radii of 5.00 and 12.0. Find the length of a chord of the larger circle which is tangent to the smaller circle.

9. A belt that does not cross goes around two pulleys, each with a radius of 4.00 in. and whose centers are 9.00 in. apart. Find the length of the belt.

10. A regular triangular pyramid has an altitude of 12.0 m, and a base 4.00 m on a side. Find the area of a section made by a plane parallel to the base and 4.00 m from the vertex.

11. A fence parallel to one side of a triangular field cuts a second side into segments of 15.0 m and 21.0 m long. The length of the third side is 42.0 m. Find the length of the shorter segment of the third side.

12. When the design in Fig. 6–114 is rotated about axis AB, we generate a cone inscribed in a hemisphere which is itself inscribed in a cylinder. Show that the volumes of these three solids are in the ratio 1:2:3. (Archimedes was so pleased with this discovery, it is said, that he ordered this figure to engraved on his tomb.)

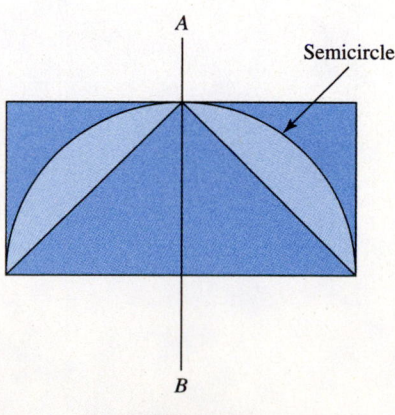

FIGURE 6–114

13. Four interior angles of a certain irregular pentagon are 38°, 96°, 112°, and 133°. Find the fifth interior angle.

14. Find the area of a trapezoid whose bases have lengths of 837 m and 583 m and are separated by a distance of 746 m.

15. A plastic drinking cup has a base diameter of 49.0 mm, is 63.0 mm wide at the top, and is 86.0 mm tall. Find the volume of the cup.

16. A spherical balloon is 17.4 ft in diameter and is made of a material that weighs 2.85 lb per 100 ft^2 of surface area. Find the volume and the weight of the balloon.

17. Find the area of a triangle whose base is 38.4 in. and whose altitude is 53.8 in.

18. Find the volume of a sphere of radius 33.8 cm.

19. Find the area of a parallelogram of base 39.2 m and height 29.3 m.

20. Find the volume of a cylinder with base radius 22.3 cm and height 56.2 cm.

21. Find the surface area of a sphere having a diameter of 39.2 in.

22. Find the volume of a right circular cone with base diameter 2.84 ft and height 5.22 ft.

23. Find the volume of a box measuring 35.8 in. × 37.4 in. × 73.4 in.

24. Find the volume of a triangular prism having a length of 4.65 cm if the area of one end is 24.6 cm^2.

25. To find a circle diameter when the center is inaccessible, you can place a scale as in Fig. 6–115 and measure the chord c and the perpendicular distance h. Find the radius of the curve if the chord length is 8.25 cm and h is 1.16 cm.

26. A circular pool, Fig. 6–116, is surrounded by a concrete apron 4.50 in. thick. Find the number of cubic yards of concrete needed.

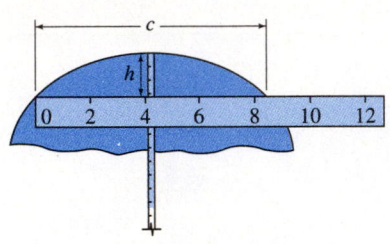

FIGURE 6–115

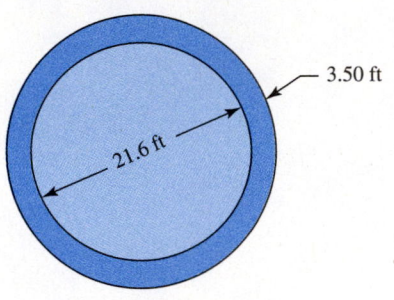

FIGURE 6–116

7

Right Triangles and Vectors

◆◆◆ **OBJECTIVES** ◆◆

When you have completed this chapter, you should be able to
- Find the trigonometric functions of an angle.
- Find the acute angle that has a given trigonometric function.
- Find the missing sides and angles of a right triangle.
- Solve practical problems involving the right triangle.
- Resolve a vector into components and, conversely, combine components into a resultant vector.
- Solve practical problems using vectors.

◆◆

With this chapter we begin our study of *trigonometry*, the branch of mathematics that enables us to solve triangles. The *trigonometric functions* are introduced here and are used to solve right triangles. Other kinds of triangles (*oblique* triangles) are discussed in Chapter 8 and more applications of trigonometry are given later.

We also build on what we learned about angles and triangles in Chapter 6, mainly the Pythagorean theorem and the fact that the sum of the angles of a triangle is 180°. Briefly introduced in this chapter are *vectors*, the study of which will be continued in Chapter 8.

Triangles are everywhere: those we can see, like the triangles in a steel truss or formed by the support cables for an antenna, and those we *cannot* see, like the triangles a surveyor uses for triangulation, or the impedance triangle used to solve problems in electronics. For dealing with a wide range of applications you must become proficient in the material of this chapter. For example, you might be called upon to compute the lengths of the diagonal struts in the framework of Fig. 7–1. In this chapter we will show how to do such calculations.

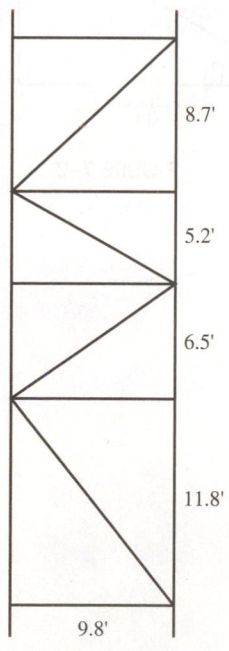

FIGURE 7–1

7–1   The Trigonometric Functions

We introduced the right triangle in our chapter on geometry and continue with it here. The difference is that now we will work with the *angles* of a triangle, and not just the sides, as we did before.

Sine, Cosine, and Tangent

■ Exploration:

Try this. Draw a right triangle with sides of any length. Do this with drafting instruments or with a CAD program. Then

- Find the ratio of any two sides by measuring them and dividing one by the other. Recall that a ratio of two quantities is their quotient, that is, one divided by the other.
- Next, *enlarge* the triangle on a photocopier and find the ratio of the same two sides.
- Then, *reduce* the triangle on a photocopier and find the ratio of the same two sides.
- Rotate, move, or flip your triangle, without changing the magnitude of the angles, and again find the ratio of the same two sides.   ■

What do you conclude? Repeat by taking the ratio of two other sides. Are your conclusions the same?

You may have found that if you do not change the angles of a right triangle, *the ratios of the sides are always the same*, regardless of how long the sides are.

For the right triangle of Fig. 7–2, we first note that one side is opposite to acute angle θ and the other side is adjacent to θ, and we recall that the hypotenuse is always the side opposite the right angle. There are six ways to form ratios of the three sides. The three most important are defined as follows:

- The *sine* of angle θ (sin θ) is the ratio of the opposite side to the hypotenuse.
- The *cosine* of angle θ (cos θ) is the ratio of the adjacent side to the hypotenuse.
- The *tangent* of angle θ (tan θ) is the ratio of the opposite side to the adjacent side.

(We will cover the remaining three ratios, the *cotangent*, *secant*, and *cosecant* in the next chapter.)

We have just defined the sine, cosine, and tangent as the ratios of sides of a right triangle, and these are often called the *trigonometric* ratios. But we have seen in our exploration that the value of sin θ, for example, depends only on θ and not on the size or orientation of the sides. In fact, we will see that the value of sin θ does not even require that we draw a right triangle. Further, for any θ there is one and only one value for sin θ. That is exactly our definition of a *function*. Hence we refer to the sine, cosine, and tangent as *trigonometric functions*.

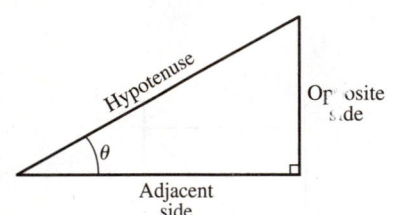

FIGURE 7–2

Trigonometric Functions		
	$\sin \theta = \dfrac{\text{opposite side}}{\text{hypotenuse}}$	111
	$\cos \theta = \dfrac{\text{adjacent side}}{\text{hypotenuse}}$	112
	$\tan \theta = \dfrac{\text{opposite side}}{\text{adjacent side}}$	113

••• Example 1: Find the sine, cosine, and tangent of angle θ in Fig. 7–3.

Solution: By their definitions, we get

$$\sin \theta = \frac{\text{opposite side}}{\text{hypotenuse}} = \frac{123}{205} = 0.600$$

$$\cos \theta = \frac{\text{adjacent side}}{\text{hypotenuse}} = \frac{164}{205} = 0.800$$

$$\tan \theta = \frac{\text{opposite side}}{\text{adjacent side}} = \frac{123}{164} = 0.750 \qquad \text{•••}$$

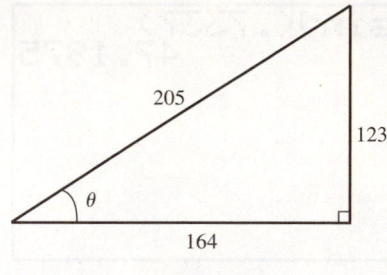

FIGURE 7–3

Common Error	Do not omit the angle when writing a trigonometric function. To write "sin = 0.543," for example, has no meaning.

Trigonometric Functions by Calculator

Since the trigonometric functions are always the same for a given angle, they can be built into calculators. On most calculators, we simply press the key for the desired ratio, $\boxed{\sin}$, $\boxed{\cos}$, or $\boxed{\tan}$, enter the angle, and press $\boxed{\text{ENTER}}$.

We can enter an angle in either degrees or radians, so be sure that the calculator $\boxed{\text{MODE}}$ is set for the angular units you wish to use. In this chapter we will work only in degrees and do radians later.

Common Error	It is easy to forget to set your calculator in the proper Degree/Radian mode. Be sure to check it each time.

TI-83/84 screen for Example 2.

••• Example 2: To find sin 53.4°, we first check that we are in **DEGREE** mode. Then press $\boxed{\sin}$ 53.4 $\boxed{\text{ENTER}}$. The screen is shown.

Finding a trigonometric function is usually an intermediate step in a calculation and not the final answer. If so, we follow the usual practice of rounding intermediate values to one more significant digit than in the original data. Here we would usually write

$$\sin 53.4° = 0.8028 \qquad \text{•••}$$

If the angle is given in degrees, minutes, and seconds, it can either be converted to decimal degrees as was shown in Chapter 1, or entered directly. On the TI-83/84, the symbols (°) and (′) are found in the $\boxed{\text{ANGLE}}$ menu. The (″) symbol is entered as an $\boxed{\text{ALPHA}}$ character found on the $\boxed{+}$ key.

••• Example 3: The screen for finding cos 48°27′ is shown. $\qquad \text{•••}$

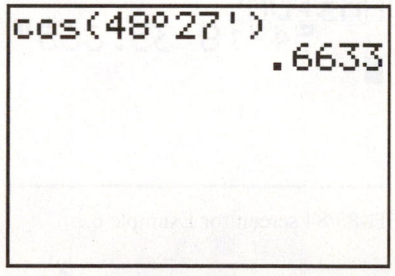

TI-83/84 screen for Example 3. On the TI-89, the degree, minute, and second symbols are alpha characters on the $\boxed{\ }$, $\boxed{=}$, and $\boxed{1}$ keys, respectively.

Finding the Angle

The operation of finding the angle when a trigonometric function is given is the *inverse* of finding the function when the angle is given. There is special notation to indicate the inverse trigonometric function. If

$$\sin \theta = x$$

we write

$$\theta = \arcsin x$$

or

$$\theta = \sin^{-1} x$$

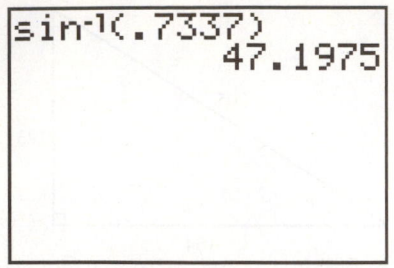

TI-83/84 screen for Example 4.

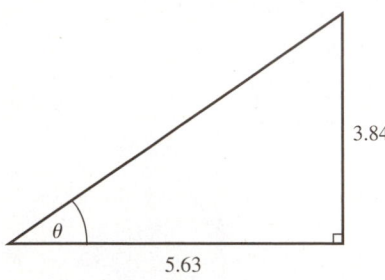

FIGURE 7–4

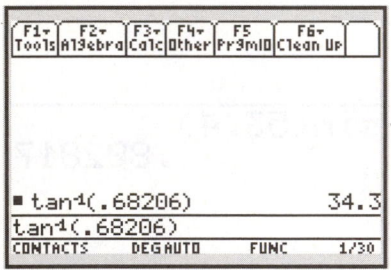

TI-89 screen for Example 5.

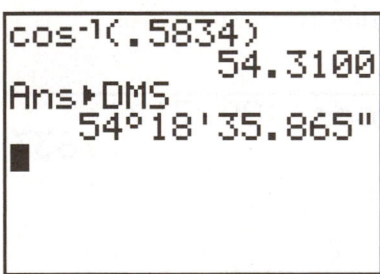

TI-83/84 screen for Example 6.

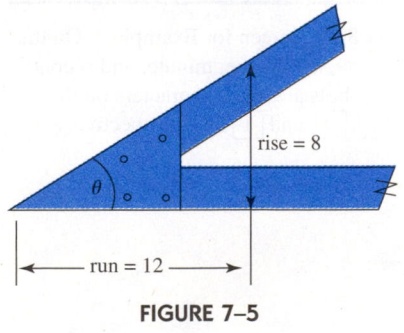

FIGURE 7–5

which is read "θ is the angle whose sine is x." Similarly, we use the symbols arccos x, $\cos^{-1} x$, arctan x, and so on.

On the calculator, we use the inverse trigonometric function keys, $\boxed{\sin^{-1}}$, $\boxed{\cos^{-1}}$, or $\boxed{\tan^{-1}}$.

◆◆◆ **Example 4:** If $\sin \theta = 0.7337$, find θ in degrees to four significant digits.

Solution: Switch the calculator into Degree mode. Then the keystrokes are: $\boxed{\sin^{-1}}$ .7337 $\boxed{\text{ENTER}}$ and we get the screen shown. So, to four significant digits,

$$\theta = \sin^{-1} 0.7337 = 47.20°$$ ◆◆◆

◆◆◆ **Example 5:** Find angle θ in Fig. 7–4 to three significant digits.

Solution: We note that we are given the sides opposite and adjacent to θ. The trigonometric function linking the opposite and adjacent sides is the tangent, so

$$\tan \theta = \frac{3.84}{5.63} = 0.68206$$

We using the $\boxed{\text{TAN}^{-1}}$ on the calculator and get

$$\theta = \tan^{-1} 0.68206 = 34.3°   \text{rounded.}$$

The TI-89 screen for this operation is shown.   ◆◆◆

If your calculator is in Degree mode, the angle will be displayed in decimal degrees. If you want the angle in degrees, minutes, and seconds, first find the angle in decimal degrees and then convert it using ▶DMS from the $\boxed{\text{ANGLE}}$ menu.

◆◆◆ **Example 6:** The screen shows the steps for finding the angle, in DMS, whose cosine is 0.5834. Here, we would round our answer to the nearest minute.   ◆◆◆

Common Error	Do not confuse the inverse with the reciprocal. The *inverse* of sin θ is: $\sin^{-1} \theta$. The *reciprocal* of sin θ is: $\dfrac{1}{\sin \theta} = (\sin \theta)^{-1}$ They *are not equal*! $\sin^{-1} \theta \neq (\sin \theta)^{-1}$

◆◆◆ **Example 7:** *An Application.* Fig. 7–5 shows a gusset plate for a prefabricated roof truss whose rafter has a rise of 8 in a run of 12. Find the angle θ.

Solution: We need to find the angle that has a tangent equal to 8/12. By calculator,

$$\theta = \tan^{-1} \frac{8}{12} = 33.7°$$ ◆◆◆

Exercise 1 ◆ The Trigonometric Functions

Sine, Cosine, and Tangent

1. Write the sine, cosine, and tangent of angle θ for each triangle in Fig. 7–6. Keep four decimal places.

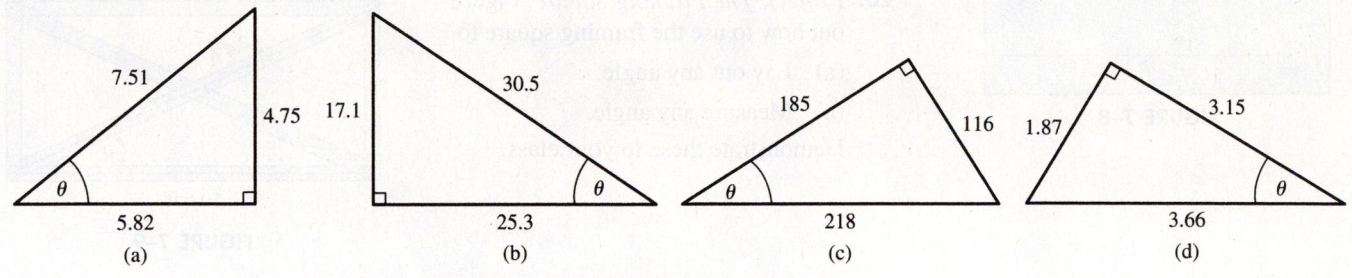

FIGURE 7–6

2. Find the missing side for each triangle in Fig. 7–7. Then write the sine, cosine, and tangent of angle θ, keeping four decimal places.

3. Find the sine, cosine, and tangent. Keep four decimal places.

   (a) 49.3°   (b) 38.9°   (c) 18.3°   (d) 2.07°

   (e) 85.3°   (f) 28.7°   (g) 73.7°   (h) 43.9°

   (i) 3.345°   (j) 58.49°   (k) 78.37°   (l) 22.05°

   (m) 83°43′   (n) 78°27′   (o) 33°47′   (p) 63°29′

Finding the Angle

Find the angle in decimal degrees whose trigonometric function is given. Keep three significant digits.

   4. $\sin A = 0.500$ **5.** $\tan D = 1.53$ **6.** $\sin G = 0.528$

   7. $\cos K = 0.770$ **8.** $\sin B = 0.483$ **9.** $\cos E = 0.847$

Evaluate the following, giving your answer in decimal degrees to three significant digits.

   10. arcsin 0.635   **11.** arccos 0.862   **12.** $\tan^{-1} 2.85$

   13. $\sin^{-1} 0.175$ **14.** $\cos^{-1} 0.229$ **15.** arctan 4.26

16. Find angle θ for each triangle in Fig. 7–6. Keep three significant digits.

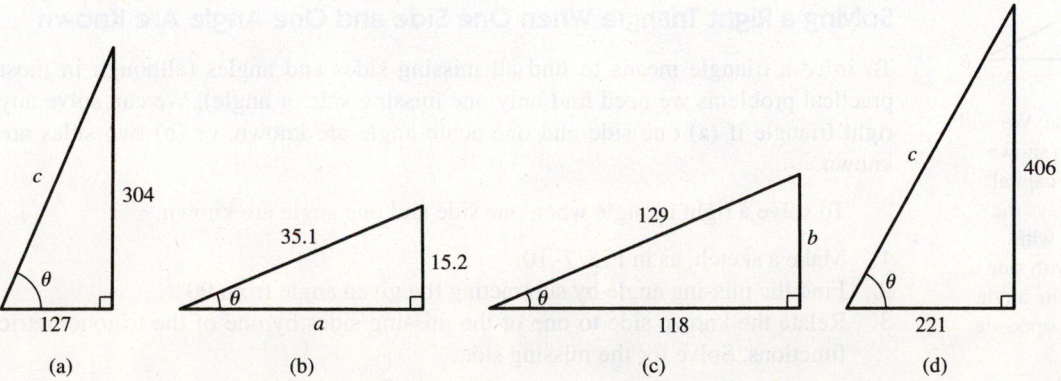

FIGURE 7–7

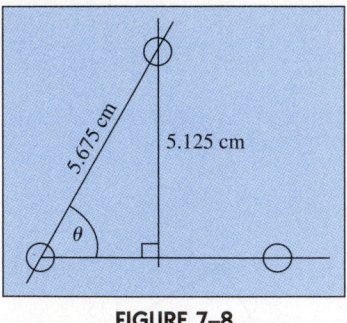

FIGURE 7–8

Applications

17. A certain roof has a rise of 9 in a run of 12. What angle does it make with the horizontal?

18. Find the angle θ in the machined plate shown in Fig. 7–8.

19. Find the angle θ in the truss shown in Fig. 7–9.

20. *Project: The Framing Square:* Figure out how to use the framing square to

   (a) Lay out any angle.

   (b) Measure any angle.

Demonstrate these to your class.

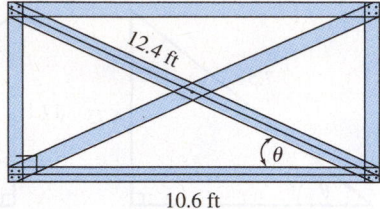

FIGURE 7–9

7–2   Solving a Right Triangle

We will soon see that a great number of applications require us to solve a right triangle. It is an essential skill for technical work. Our tools for solving right triangles consist of the trigonometric functions just introduced and, from Chapter 6, the Pythagorean theorem and the fact that, for a right triangle, the sum of the two acute angles must be 90°.

Pythagorean Theorem	$a^2 + b^2 = c^2$	**110**
Sum of the Acute Angles	$A + B = 90°$	**104**
Trigonometric Functions	$\sin \theta = \dfrac{\text{side opposite to } \theta}{\text{hypotenuse}}$	**111**
	$\cos \theta = \dfrac{\text{side adjacent to } \theta}{\text{hypotenuse}}$	**112**
	$\tan \theta = \dfrac{\text{side opposite to } \theta}{\text{side adjacent to } \theta}$	**113**

Solving a Right Triangle When One Side and One Angle Are Known

To *solve* a triangle means to find all missing sides and angles (although in most practical problems we need find only one missing side or angle). We can solve any right triangle if (a) one side and one acute angle are known, or (b) two sides are known.

To solve a right triangle when one side and one angle are known,

1. Make a sketch, as in Fig. 7–10.
2. Find the missing angle by subtracting the given angle from 90°.
3. Relate the known side to one of the missing sides by one of the trigonometric functions. Solve for the missing side.

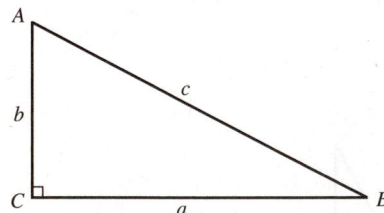

FIGURE 7–10   A right triangle. We will usually *label* a right triangle as shown here. We label the angles with capital letters A, B, and C, with C always the right angle. We label the sides with lowercase letters a, b, and c, with side a opposite angle A, side b opposite angle B, and side c (the hypotenuse) opposite angle C (the right angle).

4. Repeat step 3 to find the second missing side.
5. Check your work with the Pythagorean theorem.

◆◆◆ Example 8: Solve right triangle ABC if $A = 18.6°$ and $c = 135$.

Solution:

(1) We make a sketch as shown in Fig. 7–11.
(2) Then, since the sum of the acute angles must be 90°,

$$B = 90° - 18.6° = 71.4°$$

(3) Let us now find side a. We must use one of the trigonometric functions. *But how do we know which one to use?* And further, *which of the two angles, A or B, should we write the trig function for?*

It is simple. First, always work with the *given angle*, because if you made a mistake in finding angle B and then used it to find the sides, they would be wrong also. Then, to decide which trigonometric function to use, we note that side a is *opposite* angle A and that the given side is the *hypotenuse*. Thus, our trig function must be one that relates the *opposite side* to the *hypotenuse*. Our choice is the sine.

$$\sin A = \frac{\text{opposite side}}{\text{hypotenuse}}$$

Substituting the given values, we obtain

$$\sin 18.6° = \frac{a}{135}$$

Solving for a yields

$$a = 135 \sin 18.6°$$
$$= 135(0.3190) = 43.1$$

(4) We now find side b. Note that side b is *adjacent* to angle A. We therefore use the cosine.

$$\cos 18.6° = \frac{b}{135}$$

so

$$b = 135 \cos 18.6°$$
$$= 135(0.9478) = 128$$

We have thus found all missing parts of the triangle.

(5) For a check, we see if the three sides will satisfy the Pythagorean theorem.

Check:

$$(43.1)^2 + (128)^2 \stackrel{?}{=} (135)^2$$
$$18,242 \stackrel{?}{=} 18,225$$

Since we are working to three significant digits, this is close enough for a check. **◆◆◆**

As a rough check of any triangle, see if the longest side is opposite the largest angle and if the shortest side is opposite the smallest angle. Also check that the hypotenuse is greater than either leg but less than their sum.

◆◆◆ Example 9: In right triangle ABC, angle $B = 55.2°$ and $a = 207$. Solve the triangle.

Solution:

(1) We make a sketch as shown in Fig. 7–12.
(2) Then

$$A = 90° - 55.2° = 34.8°$$

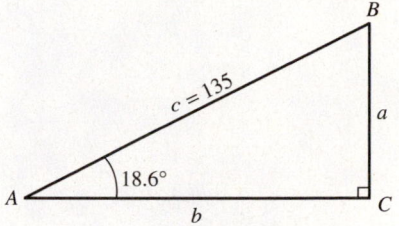

FIGURE 7–11 Realize that either of the two legs can be called *opposite* or *adjacent*, depending on which angle we are referring them to. Here b is adjacent to angle A but opposite to angle B. But there is no doubt about the hypotenuse. It is *always* the longest side.

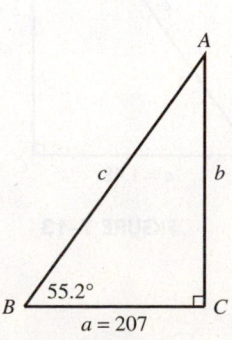

FIGURE 7–12

(3) Using the cosine gives

$$\cos 55.2° = \frac{207}{c}$$

$$c = \frac{207}{\cos 55.2°} = \frac{207}{0.5707} = 363$$

(4) Then using the tangent,

$$\tan 55.2° = \frac{b}{207}$$

$$b = 207 \tan 55.2° = 207(1.439) = 298$$

(5) Checking with the Pythagorean theorem, we have

$$(363)^2 \stackrel{?}{=} (207)^2 + (298)^2$$

$$131{,}769 \stackrel{?}{=} 131{,}653 \text{ (checks to within three significant digits)} \quad \blacklozenge\blacklozenge\blacklozenge$$

> **Tip**   Whenever possible, use the *given information* for each computation, rather than some quantity previously calculated. This way, any errors in the early computation will not be carried along.
>
> This, of course, is good advice for performing *any* computation, not just for solving right triangles.

Solving a Right Triangle When Two Sides Are Known

1. Draw a diagram of the triangle.
2. Write the trigonometric function that relates one of the angles to the two given sides. Solve for the angle.
3. Subtract the angle just found from 90° to get the second angle.
4. Find the missing side by the Pythagorean theorem.
5. Check the computed side and angles with trigonometric functions, as shown in the following example.

◆◆◆ **Example 10:** Solve right triangle *ABC* if $a = 1.48$ and $b = 2.25$.

Solution:

(1) We sketch the triangle as shown in Fig. 7–13.
(2) To find angle *A*, we note that the 1.48 side is *opposite* angle *A* and that the 2.25 side is *adjacent* to angle *A*. The trig function *relating opposite* and *adjacent* is the *tangent*

$$\tan A = \frac{1.48}{2.25} = 0.6578$$

from which

$$A = 33.3°$$

(3) Solving for angle *B*, we have

$$B = 90° - A = 90 - 33.3 = 56.7°$$

(4) We find side *c* by the Pythagorean theorem.

$$c^2 = (1.48)^2 + (2.25)^2 = 7.253$$

$$c = 2.69$$

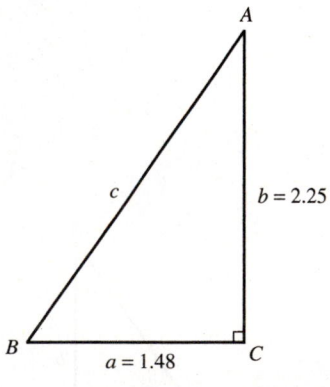

FIGURE 7–13

Check:

$$\sin 33.3° \overset{?}{=} \frac{1.48}{2.69}$$

$$0.549 \approx 0.550 \qquad \text{(checks)}$$

and

$$\sin 56.7° \overset{?}{=} \frac{2.25}{2.69}$$

$$0.836 = 0.836 \qquad \text{(checks)}$$

(Note that we could just as well have used another trigonometric function, such as the cosine, for our check.) ◆◆◆

◆◆◆ **Example 11:** *An Application.* Fig. 7–14 shows a swampy area in a park. To order materials for an elevated footpath across the swamp the distance x must be found by calculation, because it cannot be measured directly. A surveyor at S sights a stake at A, turns the transit 90°, and locates a stake at B. The distance SB is measured. Then after moving the transit to B, Angle ABS is measured. Find x.

Solution: In right triangle SAB we know the side adjacent to the known angle and seek the side opposite the known angle, so we use the tangent function.

$$\tan 68.3° = \frac{x}{55.2}$$

$$x = 55.2 \tan 68.3° = 139 \text{ ft}$$

◆◆◆

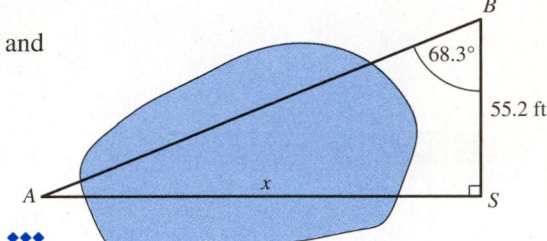

FIGURE 7–14

We have given just one application here. We will have many more in the next section.

Exercise 2 ◆ Solving a Right Triangle

Right Triangles With One Side and One Angle Known

Sketch each right triangle and find all of the missing parts. Assume the triangles to be labeled as in Fig. 7–10. Work to three significant digits.

1. $a = 155$ $A = 42.9°$ **2.** $b = 82.6$ $B = 61.4°$

3. $a = 1.74$ $B = 31.9°$ **4.** $b = 7.74$ $A = 22.5°$

5. $a = 284$ $A = 64.7°$ **6.** $b = 73.2$ $B = 37.5°$

7. $b = 9.26$ $B = 55.2°$ **8.** $a = 1.73$ $A = 39.3°$

9. $b = 82.4$ $A = 31.4°$ **10.** $a = 18.3$ $B = 44.1°$

Right Triangles With Two Sides Known

Sketch each right triangle and find all missing parts. Work to three significant digits and express the angles in decimal degrees.

11. $a = 382$ $b = 274$ **12.** $a = 3.88$ $c = 5.37$

13. $b = 3.97$ $c = 4.86$ **14.** $a = 63.9$ $b = 84.3$

15. $a = 27.4$ $c = 37.5$ **16.** $b = 746$ $c = 957$

17. $a = 41.3$ $c = 63.7$ **18.** $a = 4.82$ $b = 3.28$

19. $b = 228$ $c = 473$ **20.** $a = 274$ $c = 429$

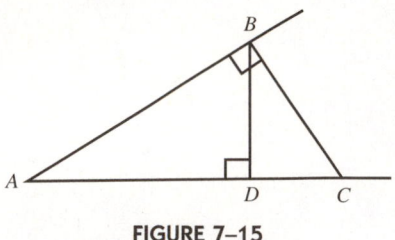

FIGURE 7–15

21. *Challenge Problem*: Referring to Fig. 7–15, prove the following:

Altitude Drawn to the Hypotenuse	In a right triangle, the altitude drawn to the hypotenuse forms two right triangles that are similar to each other and to the original triangle.

22. *Team Project*: We mentioned that the ancient Egyptians had probably used stretched ropes to do simple surveying. Demonstrate how a long loop of rope marked to form a 3–4–5 right triangle (called the *Egyptian triangle* or *rope-stretcher's triangle*) can be used to lay out right angles. Check it against a corner of a basketball court or some other right angle on campus. Take pictures or videos of your work and show them in class.

23. *Computer*: Using CAD, (1) draw a right triangle, and label one acute angle as θ; (2) have the program measure each side, and compute and display the ratios of the sides, as related to θ; (3) using the built-in trigonometric functions, compute and display the sine, cosine, and tangent of θ; (4) compare these to the ratios of the sides computed in step 2; (5) drag a vertex of the triangle, making sure it stays a right triangle, and observe the new values. What do you conclude?

7–3   Applications of the Right Triangle

There are, of course, a huge number of applications for the right triangle, a few of which are given in the following examples and exercises. A typical application is that of finding a distance that cannot be measured directly, as shown in the following example.

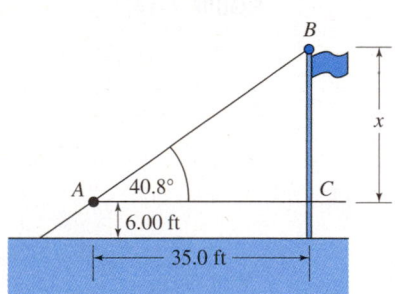

FIGURE 7–16

◆◆◆ **Example 12:** To find the height of a flagpole (Fig. 7–16), a person measures 35.0 ft from the base of the pole and then measures an angle of 40.8° from a point 6.00 ft above the ground to the top of the pole. Find the height of the flagpole.

Estimate: If angle *A* were 45°, then *BC* would be the same length as *AC*, or 35 ft. But our angle is a bit less than 45°, so we expect *BC* to be less than 35 ft, say, 30 ft. Thus our guess for the entire height is about 36 ft.

Solution: In right triangle *ABC*, *BC* is opposite the known angle, and *AC* is adjacent. Using the tangent, we get

$$\tan 40.8° = \frac{x}{35.0}$$

where *x* is the height of the pole above the observer. Then

$$x = 35.0 \tan 40.8° = 30.2 \text{ ft}$$

Adding 6.00 ft, we find that the total pole height is 36.2 ft, measured from the ground.   ◆◆◆

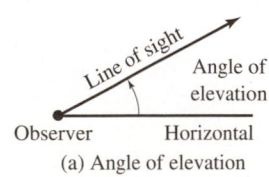

(a) Angle of elevation

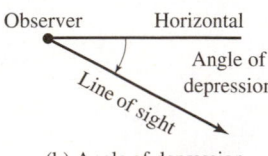

(b) Angle of depression

FIGURE 7–17   Angles of elevation and depression.

◆◆◆ **Example 13:** From a plane at an altitude of 2750 ft, the pilot observes the angle of depression of a lake to be 18.6°. How far is the lake from a point on the ground directly beneath the plane?

Solution: We first note that an *angle of depression*, as well as an *angle of elevation*, is defined as the angle between a line of sight and the *horizontal*, as shown in Fig. 7–17. We then make a sketch for our problem (Fig. 7–18). Since the ground and the horizontal line drawn through the plane are parallel, angle A and the angle of depression (18.6°) are alternate interior angles. Therefore $A = 18.6°$. Then

$$\tan 18.6° = \frac{2750}{x}$$

$$x = \frac{2750}{\tan 18.6°} = 8170 \text{ ft} \qquad \blacklozenge\blacklozenge\blacklozenge$$

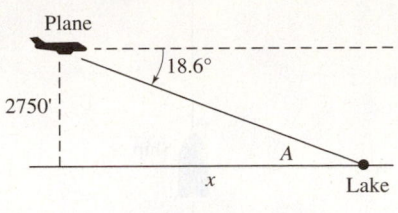

FIGURE 7–18

◆◆◆ **Example 14:** A frame is braced by wires AC and BD, as shown in Fig. 7–19. Find the length of each wire.

Estimate: We can make a quick estimate if we remember that the sine of 30° is 0.5. Thus we expect wire AC to be a little shorter than twice side CD, or around 30 ft.

Solution: Noting that in right triangle ACD, CD is opposite the given angle and AC is the hypotenuse, we choose the sine.

$$\sin 31.6° = \frac{15.4}{AC}$$

$$AC = \frac{15.4}{\sin 31.6°} = 29.4 \text{ ft} \qquad \blacklozenge\blacklozenge\blacklozenge$$

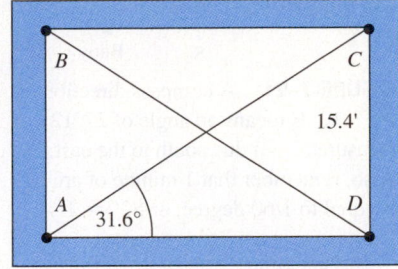

FIGURE 7–19

Exercise 3 ◆ Applications of the Right Triangle

In this group of exercises you may assume that lines that appear vertical or horizontal in the figures are so, unless otherwise noted.

As with other applications and verbal problems, don't expect to find an example in the text that exactly matches. Instead, you must apply the *ideas* from the text to each new situation.

Measuring an Inaccessible Distance

1. From a point on the ground 255 m from the base of a tower, the angle of elevation to the top of the tower is 57.6°. Find the height of the tower.
2. A pilot 4220 m directly above the front of a straight train observes that the angle of depression of the end of the train is 68.2°. Find the length of the train.
3. From the top of a lighthouse 156 ft above the surface of the water, the angle of depression of a boat is observed to be 28.7°. Find the horizontal distance from the boat to the lighthouse.
4. An observer in an airplane 1520 ft above the surface of the ocean observes that the angle of depression of a ship is 28.8°. Find the straight-line distance from the plane to the ship.
5. The distance PQ across a swamp is desired. A line PR of length 59.3 m is laid off at right angles to PQ, Fig. 7–20. Angle PRQ is measured at 47.6°. Find the distance PQ.
6. The angle of elevation of the top of a building from a point on the ground 275 ft from its base is 51.3°. Find the height of the building.
7. The angle of elevation of the top of a building from a point on the ground 75.0 yd from its base is 28.0°. How high is the building?
8. From the top of a hill 125 ft above a stream, the angles of depression of a point on the near shore and of a point on the opposite shore are 42.3° and 40.6°. Find the width of the stream between these two points.
9. From the top of a tree 15.0 m high on the shore of a pond, the angle of depression of a point on the other shore is 6.70°. What is the width of the pond?

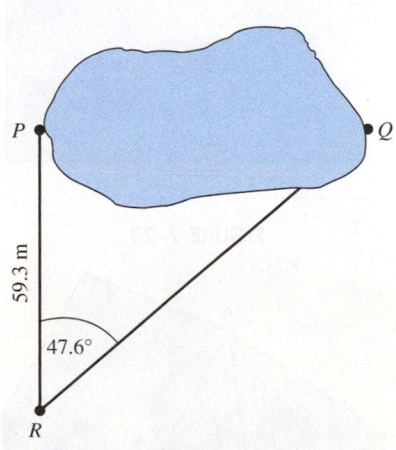

FIGURE 7–20

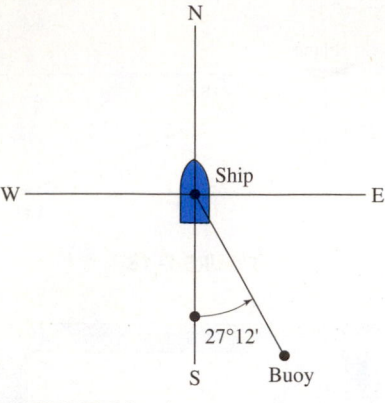

FIGURE 7–21   A compass direction of S 27°12′ E means an angle of 27°12′ measured from due south to the east. Also, remember that 1 minute of arc (1′) is equal to 1/60 degree, or 60′ = 1°.

Navigation

10. A passenger on a ship sailing due north at 8.20 km/h noticed that at 1 P.M., a buoy was due east of the ship. At 1:45 P.M., the bearing of the buoy from the ship was S 27°12′ E (Fig. 7–21). How far was the ship from the buoy at 1:00 P.M.?

11. An observer at a point P on a coast sights a ship in a direction N 43°15′ E. The ship is at the same time directly east of a point Q, 15.6 km due north of P. Find the distance of the ship from point P and from Q.

12. A ship sailing parallel to a straight coast is directly opposite one of two lights on the shore. The angle between the lines of sight from the ship to these lights is 27°50′, and it is known that the lights are 355 m apart. Find the perpendicular distance of the ship from the shore.

13. An airplane left an airport and flew 315 mi in the direction N 15°18′ E, then turned and flew 296 mi in the direction S 74°42′ E. How far, and in what direction, should it fly to return to the airport?

14. A ship is sailing due south at a speed of 7.50 km/h when a light is sighted with a bearing S 35°28′ E. One hour later the ship is due west of the light. Find the distance from the ship to the light at that instant.

15. After leaving port, a ship holds a course N 46°12′ E for 225 mi. Find how far north and how far east of the port the ship is now located.

Structures

16. A guy wire from the top of an antenna is anchored 53.5 ft from the base of the antenna and makes an angle of 85.2° with the ground. Find (a) the height h of the antenna and (b) the length L of the wire.

17. Find the angle θ and length AB in the truss shown in Fig. 7–22.

18. A house is 8.75 m wide, and its roof has an angle of inclination of 34.0° (Fig. 7–23). Find the length L of the rafters.

19. A guy wire 82.0 ft long is stretched from the ground to the top of a telephone pole 65.0 ft high. Find the angle between the wire and pole.

20. Common rafter AB in Fig. 7–24 makes an angle of 35.0° with the level. Find the angle that hip rafter AC makes with the level. *Hint:* First find the rise of the common rafter, which is also the rise of the hip rafter. Then use the Pythagorean theorem *twice*, first in the horizontal plane and then in the vertical.

21. A quantity of gusset plates, Fig. 7–25, are to be prefabricated for a construction job. Find (a) angles A and B, (b) length AB, and (c) the area of the gusset.

22. A 4.00-ft diameter cylindrical water tank is to be located in a crawl space under a roof, as in Fig. 7–26. Find the distrance x.

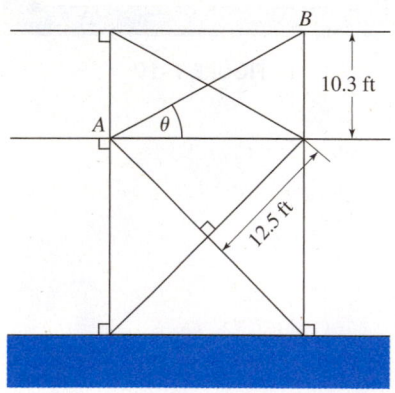

FIGURE 7–22

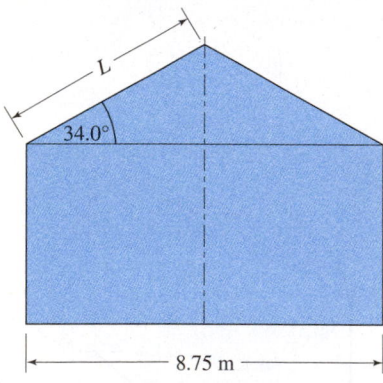

FIGURE 7–23

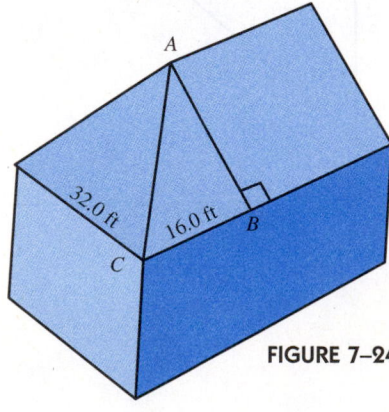

FIGURE 7–24

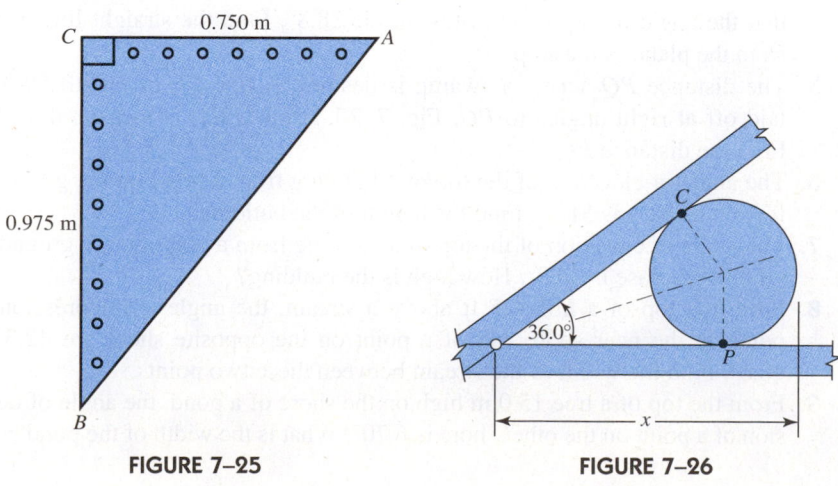

FIGURE 7–25   **FIGURE 7–26**

Geometry

23. What is the angle θ, to the nearest tenth of a degree, between a diagonal AB of a cube and a diagonal AC of a face of that cube (Fig. 7–27)?
24. Two of the sides of an isosceles triangle have a length of 150 units, and each of the base angles is 68.0°. Find the altitude and the base of the triangle.
25. The diagonal of a rectangle is 3 times the length of the shorter side. Find the angle, to the nearest tenth of a degree, between the diagonal and the longer side.
26. Find the angles between the diagonals of a rectangle whose dimensions are 580 units × 940 units.
27. Find the area of a parallelogram if the lengths of the sides are 255 units and 482 units and if one angle is 83.2°.
28. Find the length of a side of regular hexagon inscribed in a 125-cm-radius circle.
29. Find the length of the side of a regular pentagon circumscribed about a circle of radius 244 in. *Hint*: Draw lines from the center of the pentagon to each vertex and to the midpoints of each side, to form right triangles.

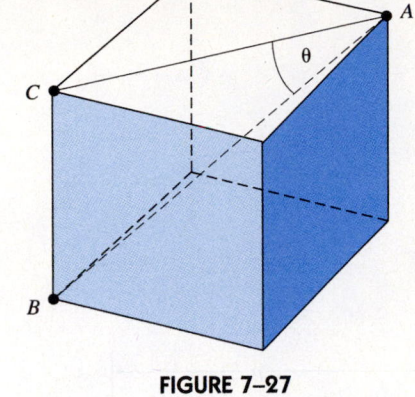

FIGURE 7–27

Shop Trigonometry

Trigonometry finds many uses in the machine shop. Given here are some standard shop calculations.

30. *Bolt spacing*: Find the dimension x in Fig. 7–28.
31. *Bolt circle*: A bolt circle (Fig. 7–29) is to be made on a jig borer. Find the dimensions x_1, y_1, x_2, and y_2.
32. *Bolt circle*: A bolt circle with a radius of 36.000 cm contains 24 equally spaced holes. Find the straight-line distance between the holes.
33. *Tapered shaft*: A 9.000-in.-long shaft is to taper 3.500° and be 1.000 inch in diameter at the narrow end (Fig. 7–30). Find the diameter d of the larger end. *Hint*: Draw a line (shown dashed in the figure) from the small end to the large, to form a right triangle. Solve that triangle.
34. *Tapered groove*: A groove machined in a block, Fig. 7–31, is inspected by measuring the protrusion of a cylindrical pin. Find the dimension x.
35. *Pulley fragment*: A measuring square having an included angle of 60.0° is placed over the pulley fragment of Fig. 7–32. The distance d from the corner of the square to the pulley rim is measured at 5.53 in. Find the pulley radius r.

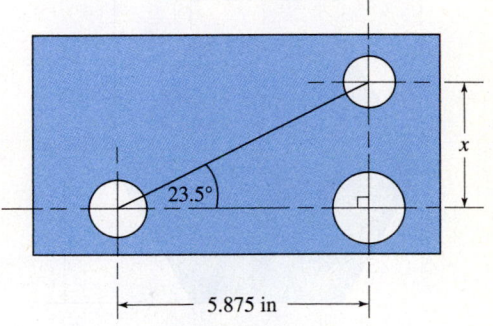

FIGURE 7–28

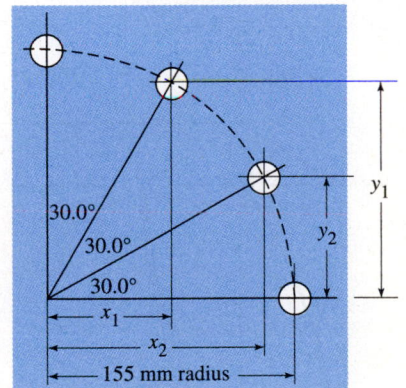

FIGURE 7–29 A bolt circle.

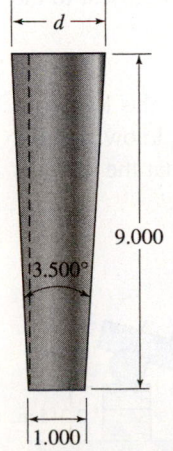

FIGURE 7–30
Tapered shaft.

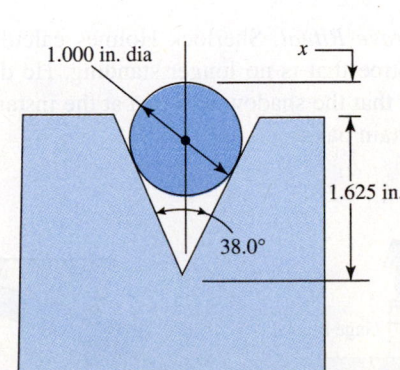

FIGURE 7–31

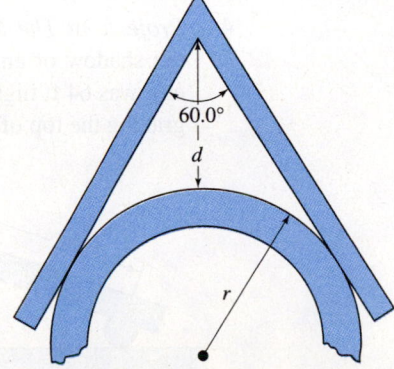

FIGURE 7–32 Measuring the radius of a broken pulley.

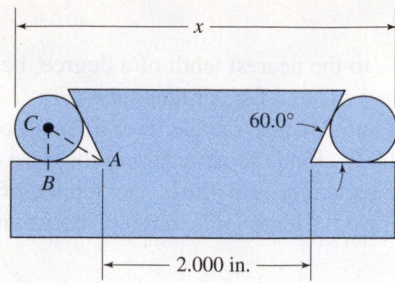

FIGURE 7–33   Measuring a dovetail.

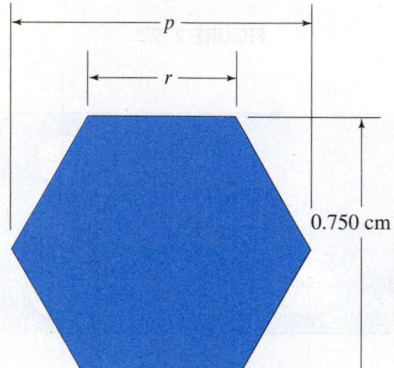

FIGURE 7–34

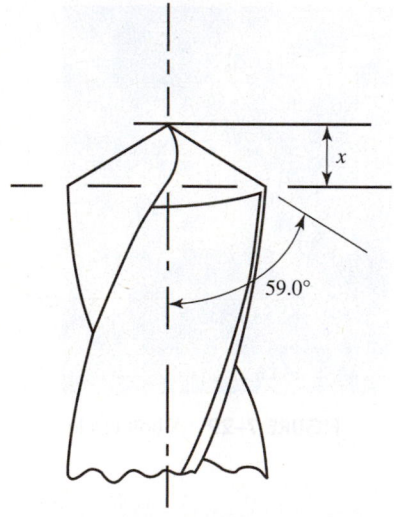

FIGURE 7–35

36. *Dovetail*: A common way of measuring dovetails is with the aid of round plugs, as in Fig. 7–33. Find the distance *x* over the 0.5000-in.-diameter plugs. *Hint*: Find the length of side *AB* in the 30–60–90 triangle *ABC*, and then use *AB* to find *x*.

37. *Hexagonal stock*: A bolt head (Fig. 7–34) measures 0.750 cm across the flats. Find the distance *p* across the corners and the width *r* of each flat.

38. *Twist drill*: For general work, twist drills are sharpened to a point angle of 59.0°, Fig. 7–35. Find the distance *x* for a drill having a diameter of 0.875 in.

39. *Sine bar*: A *sine bar*, Fig. 7–36, is used to position a workpiece at a precise angle for cutting or for inspection. Its length is commonly 5 or 10 inches, and is usually used with *gage blocks*. For a 5-inch sine bar (distance between plugs = 5.000 in.), (a) what height of gage block stack is needed to produce an angle of 25.60° and (b) what angle is produced by a gage block stack of 1.554 in.? (c) A tapered wedge is placed on a sine bar and the gage blocks inserted so that its upper surface is parallel to the surface plate, Fig. 7–37. Find the wedge angle θ.

40. *Team Project*: Make a device for measuring angles in the horizontal plane, using a sighting tube of some sort and a protractor. Use this "transit" and a rope of known length to find some distance on your campus, such as the distance across a ball field, using the methods of this chapter. Compare your results with the actual taped distance. Give a prize to the team that gets closest to the taped value.

41. *Project*: In *The Musgrave Ritual*, Sherlock Holmes calculates the length of the shadow of an elm tree that is no longer standing. He does know that the elm was 64 ft high and that the shadow was cast at the instant that the sun was grazing the top of a certain oak tree.

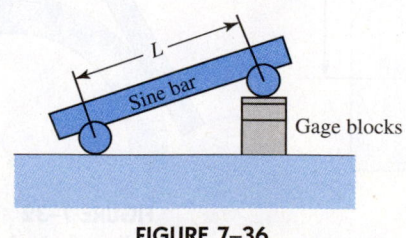

FIGURE 7–36

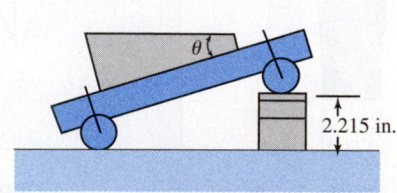

FIGURE 7–37

Holmes held a 6-ft-long fishing rod vertical and measured the length of its shadow at the proper instant. It was 9 ft long. He then said, "Of course the calculation now was a simple one. If a rod of six feet threw a shadow of nine, a tree of sixty-four feet would throw one of _____." How long was the shadow of the elm?

42. *Team Project*: Actually use the method that Sherlock Holmes used in *The Musgrave Ritual* to find the height of a flagpole or building on your campus. Find some way of checking your result.

7–4   An Angle in Standard Position

Early in this chapter we defined the trigonometric funcions in terms of the sides of a right triangle. For many purposes, such as the vectors we will work with in the next section, it is more useful to define the trig functions in terms of an angle θ drawn on coordinate axes, as shown in Fig. 7–38. When drawn in this way, with the vertex at the origin O and with one side along the x axis, the angle is said to be in *standard position*. We may think of an angle as being *generated* by a line rotating counterclockwise from an initial position on the x axis to some terminal position.

Next we select any point P on the terminal side of the angle, with rectangular coordinates x and y. We form a right triangle OPQ by dropping a perpendicular from P to the x axis. This side OQ is *adjacent* to angle θ and has a length x. Similarly side PQ is *opposite* to angle θ and has a length y. The side OP is the *hypotenuse* of the right triangle, and we label its length r. Note that by the Pythagoran theorem, $r^2 = x^2 + y^2$.

As before, we define the sine of θ as the ratio of opposite to hypotenuse. But here it is also the ratio of the distance y to the distance r. Similarly the cosine and tangent can be defined in terms of x, y, and r. Our three trignometric functions, defined both as ratios of the sides of a right triangle and as the coordinates of a point on the terminal side of an angle in standard position, are then

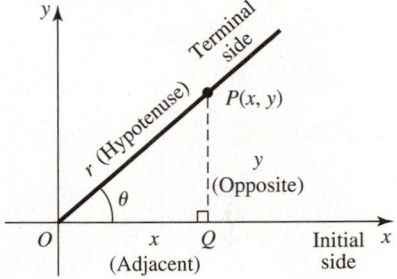

FIGURE 7–38   An angle in standard position.

	$\text{sine } \theta = \sin \theta = \dfrac{y}{r} = \dfrac{\text{opposite side}}{\text{hypotenuse}}$	**111**
Trigonometric Functions	$\text{cosine } \theta = \cos \theta = \dfrac{x}{r} = \dfrac{\text{adjacent side}}{\text{hypotenuse}}$	**112**
	$\text{tangent } \theta = \tan \theta = \dfrac{y}{x} = \dfrac{\text{opposite side}}{\text{adjacent side}}$	**113**

◆◆◆ **Example 15:** A point on the terminal side of angle θ has the coordinates (2.75, 3.14). (a) Find r, (b) write the sine, cosine, and tangent of θ, and (c) find θ, all to three significant digits.

Solution: (a) We have $x = 2.75$ and $y = 3.14$. We find r using the Pythagorean theorem.

$$r^2 = (2.75)^2 + (3.14)^2 = 17.4$$

$$r = 4.17$$

(b) Then by the definitions of the trigonometric functions,

$$\sin \theta = \frac{y}{r} = \frac{3.14}{4.17} = 0.753$$

$$\cos \theta = \frac{x}{r} = \frac{2.75}{4.17} = 0.659$$

$$\tan \theta = \frac{y}{x} = \frac{3.14}{2.75} = 1.14$$

(c) Since $\tan \theta = 1.14$,

$$\theta = \tan^{-1} 1.14 = 48.8°$$

◆◆◆

Exercise 4 ◆ Angles in Standard Position

The terminal side of an angle in standard position passes through the given point. Sketch the angle, compute the distance r from the orgin to the point, write the six trigonometric functions of the angle, and find the angle. Work to three significant digits.

1. (2.25, 4.82)
2. (1.74, 2.88)
3. (3.72, 5.49)
4. (7.93, 8.27)
5. (1.93, 4.83)
6. (7.27, 3.77)

7–5   Introduction to Vectors

Vector and Scalar Quantities

Many quantities in technology cannot be described fully without giving their *direction* as well as their magnitude. It is not always useful to know how fast something is moving, for example, without knowing the *direction* in which it is moving. Velocity is called a *vector quantity*, having both magnitude and direction, as opposed to, say, weight, which is called a *scalar quantity*. The weight of a football is a scalar quantity, but its velocity at any instant is a vector quantity. Other vector quantities include force and acceleration; other scalar quantities include time and volume.

Representation of a Vector

A vector is represented by an arrow whose length is proportional to the magnitude of the vector and whose direction is the same as the direction of the vector quantity (Fig. 7–39).

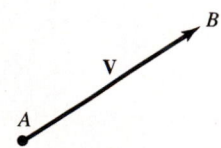

FIGURE 7–39
Representation of a vector.

Vectors are represented differently in different textbooks, but they are usually written in boldface type. Here, we'll use the most common notation: **boldface** Roman capitals to represent vectors, and nonboldface *italic* capitals to represent scalar quantities. So in this textbook, **B** is understood to be a vector quantity, having both magnitude and direction, while *B* is understood to be a scalar quantity, having magnitude but no direction. Boldface letters, however, are not practical in handwritten work. Instead, it is customary to place an *arrow* over vector quantities.

In practical problems we usually have to

(a) replace a single vector by two other vectors, called *components*, that produce the same effect as the original vector, or

(b) replace two or more vectors by a single vector, called the *resultant*, that produces the same effect as the original vectors.

We will first show how to find components, and then the resultant.

Components of a Vector

■ Exploration:

Try this. Hang an object, say a stapler, from two long rubber bands attached to a single point. Note the length of each rubber band. Now separate the ends of the two rubber bands, as shown in Fig. 7–40. What happens to the lengths of the rubber bands as their ends are moved farther apart? Can you explain what has happened? ■

In general, any vector can be replaced by two or more vectors which, acting together, exactly duplicate the effect of the original vector. They are called the *components* of the vector. Thus the single vector (the force exerted by the two rubber bands when together) was replaced by two components (the forces exerted by the rubber bands when apart) that duplicated the effect (the supporting of the stapler) of the single vector.

Replacing a vector by its components is called *resolving* a vector. If the two components are perpendicular to each other (not the case with our rubber bands) they are called *rectangular* components. In this chapter we will resolve a vector only into rectangular components, saving other cases for later.

We can represent a vector and its components by means of a *vector diagram*. There are two ways to do this, the *tip-to-tail* (or *head-to-tail*) method and the *parallelogram* method. With the tip-to-tail method, we simply draw the vectors **A** and **B** with the tail of one starting at the tip of the other, Fig. 7–41(a). The resultant **R** is the vector that completes the triangle.

In the parallelogram method, Fig. 7–41(b), we draw the vectors **A** and **B** *tail-to-tail* and complete the parallelogram by drawing lines from the tip of each vector parallel to the other vector. The resultant **R** is then the diagonal of the parallelogram. For rectangular components, the parallelogram is a rectangle.

Either way, vectors **A, B,** and **R** *form a triangle*. If **A** and **B** are perpendicular, we have a *right* triangle. Thus we can resolve a vector into its rectangular components using the right triangle trigonometry of this chapter, as shown in the following example.

◆◆◆ Example 16: A vector has a magnitude of 248 units and makes an angle of 38.2° with the horizontal. Find the magnitudes of its horizontal and vertical components.

Solution: We complete the parallelogram, Fig. 7–42, by drawing a horizontal and a vertical line from the tip of the given vector, drawing the horizontal and vertical components, and choosing labels for the three vectors.

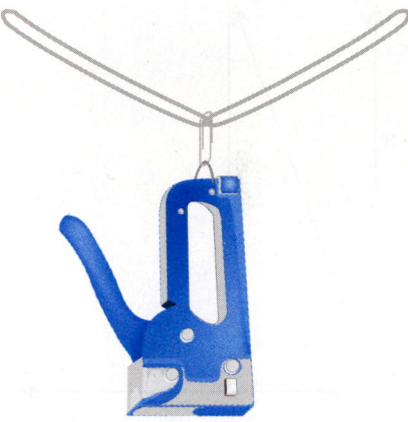

FIGURE 7–40

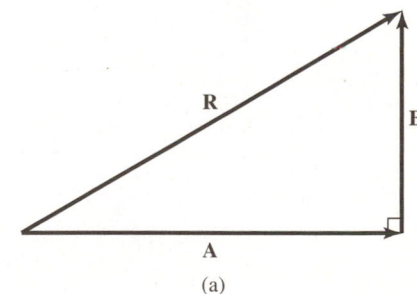

(a)

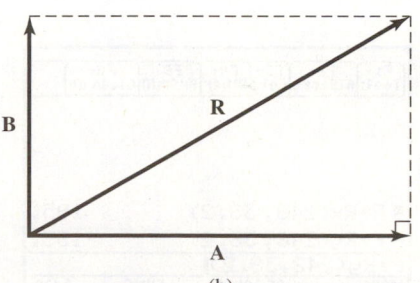

(b)

FIGURE 7–41 (a) and (b)

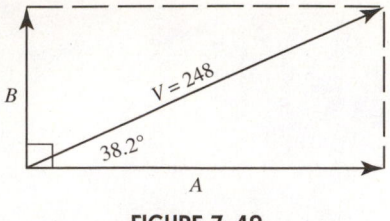

FIGURE 7–42

We then note that V and A are related by the cosine of the given angle.

$$\cos 38.2° = \frac{A}{248}$$

$$A = 248 \cos 38.2° = 195$$

Similarly,

$$\sin 38.2° = \frac{B}{248}$$

$$B = 248 \sin 38.2° = 153 \qquad \text{♦♦♦}$$

The ∠ Symbol

The symbol $\angle$ means "*at an angle of*." It provides a convenient way to specify the angle a vector makes with some reference direction. The reference direction is usually taken as the horizontal or the x axis. Some calculators have a $\boxed{\angle}$ key, which we will show how to use in a later example.

♦♦♦ **Example 17:** Find the x and y components of the vector $55.4 \angle 63.5°$.

Solution: The expression $55.4 \angle 63.5$ means "a vector of magnitude 55.4 at an angle of 63.5°." We sketch our vector, Fig. 7–43, with a magnitude of 55.4, at an angle of 63.5° with the x axis. We complete the parallelogram and label the components, choosing A and B. Then

$$\cos 63.5° = \frac{A}{55.4}$$

$$A = 55.4 \cos 63.5° = 24.7$$

$$\sin 63.5° = \frac{B}{55.4}$$

$$B = 55.4 \sin 63.5° = 49.6 \qquad \text{♦♦♦}$$

FIGURE 7–43

Rectangular Components by Calculator

Some calculators can easily find components of a vector.

♦♦♦ **Example 18:** Find the components for the vector in Example 16 by calculator. Recall that the given vector had a magnitude of 248 at an angle of 38.2°.

Solution: One method on the TI-89, for example, uses the $P \blacktriangleright R_x$ instruction to find the component from which the given angle is referenced, and $P \blacktriangleright R_y$ to find the component perpendicular to that one. Both instructions are found in the $\boxed{\text{MATH}}$ **Angle** menu.

Enter $P \blacktriangleright R_x$ or $P \blacktriangleright R_y$. Then in parentheses, enter the magnitude of the vector, and then the angle, separated by a comma. The angle is taken by the calculator as degrees or radians, depending on the current mode setting. It can be overridden temporarily by affixing (°) or (ʳ) from the $\boxed{\text{MATH}}$ **Angle** menu. Finally, use $\boxed{\approx}$ on the $\boxed{\text{ENTER}}$ key to get a decimal answer. Here are the keystrokes for Example 16, with the calculator in **DEGREE** mode. ♦♦♦

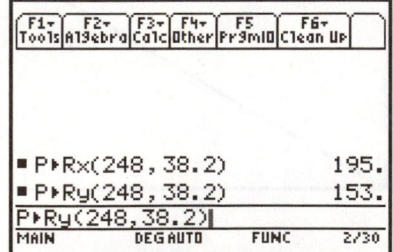

TI-89 screen for Example 18.

◆◆◆ Example 19: Another method on the TI-89 uses the **Rect** instruction from the [MATH] **Matrix/Vector ops** menu. Enter the magnitude and angle, in brackets, separated by a comma. The angle is entered using the [∠] symbol, one of the keyboard characters. Then enter ▸**Rect**. The keystokes using the same vector as in the preceding example are shown on the screen. The calculator is in Degree mode.   **◆◆◆**

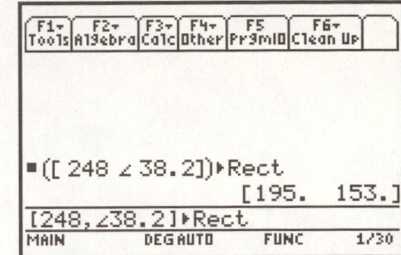

TI-89 screen for Example 19.

Resultant of Two Perpendicular Vectors

Just as any vector can be *resolved* into components, so can several vectors be *combined* into a single vector called the *resultant*, or *vector sum*. The process of combining vectors into a resultant is called *vector addition*.

When combining or adding two perpendicular vectors, we use right-triangle trigonometry, just as we did for resolving a vector into rectangular components.

◆◆◆ Example 20: Find the resultant of two perpendicular vectors whose magnitudes are 485 and 627. Also find the angle that it makes with the 627-magnitude vector.

Solution: We draw a vector diagram as shown in Fig. 7–44. Then by the Pythagorean theorem,

$$R = \sqrt{(485)^2 + (627)^2} = 793$$

Then,

$$\tan \theta = \frac{485}{627} = 0.774$$

$$\theta = 37.7°$$ **◆◆◆**

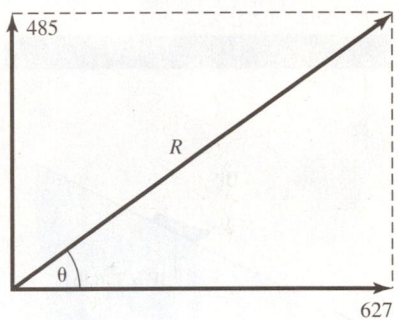

FIGURE 7–44   Resultant of two vectors.

Resultants by Calculator

We will now show a calculator method for finding resultants. We will use it to find the resultant of two perpendicular vectors, but later we will use the same method to find the resultant of any number of vectors at various angles.

◆◆◆ Example 21: Find the resultant of the vectors in Example 20 by calculator. Recall that the given perpendicular vectors had magnitudes of 485 and 627.

Solution: We simply add the two given vectors. We enter each vector in parentheses, using the [∠] symbol for the angle, and combine them using a (+) sign. Here are the keystrokes, with the calculator in Degree mode.   **◆◆◆**

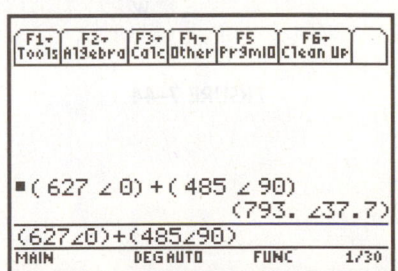

TI-89 screen for Example 21. We could also use the R▸Pr, R▸Pθ, or R▸Polar features on this calculator. However, they would work only for two perpendicular vectors. This method is good for any number of vectors at any angle.

Exercise 5 ◆ Introduction to Vectors

Components of a Vector

Given the magnitude of each vector and the angle θ that it makes with the x axis, find the x and y components.

1. Magnitude = 4.93 $\theta = 48.3°$
2. Magnitude = 835 $\theta = 25.8°$
3. Magnitude = 1.884 $\theta = 58.24°$
4. Magnitude = 362 $\theta = 13.8°$
5. Magnitude = 836 $\theta = 45.2°$

Find the rectangular components of each vector.

6. 3.85 ∠22.2°   **7.** 22.7 ∠64.9°

8. 943 ∠18.4°   **9.** 18.4 ∠77.3°

10. 283 ∠38.5°

In the following problems, the magnitudes A and B of two perpendicular vectors are given. Find the resultant and the angle that it makes with B.

11. $A = 483$ $B = 382$ **12.** $A = 2.85$ $B = 4.82$

13. $A = 7364$ $B = 4837$ **14.** $A = 46.8$ $B = 38.6$

15. $A = 1.25$ $B = 2.07$ **16.** $A = 274$ $B = 529$

17. $A = 6.82$ $B = 4.83$ **18.** $A = 58.3$ $B = 37.2$

19. $A = 2.27$ $B = 3.97$

7–6   Applications of Vectors

Any vector quantity such as force, velocity, or impedance can be resolved or combined by the methods of the preceding section. We will start with force vectors, and then cover velocity vectors and impedance vectors. As before, everything you need to solve these problems is given in this text, and you are encouraged to try problems outside your chosen field.

Force Vectors

Our first problem is simply to resolve a vector into its components.

◆◆◆ **Example 22:** A sled whose weight W is 386 lb is on an icy incline making an angle of 27.4° with the horizontal, Fig. 7–44(a). Find (a) the normal component N and (b) the tangential component T of the sled's weight. (c) What force F parallel to the incline is needed to keep the sled from sliding down the hill?

Solution: We draw a vector diagram, Fig. 7–44(b), showing W acting vertically downwards, N at a right angle to the plane, and, T and F parallel to the plane. Then we resolve W into its components N and T.

$$\text{(a)} \quad N = W \cos \theta = 386 \cos 27.4° = 343 \text{ lb}$$

$$\text{(b)} \quad T = W \sin \theta = 386 \sin 27.4° = 178 \text{ lb}$$

(c) The force F needed to hold the sled in place is just equal to the tangential component T, or

$$F = T = 178 \text{ lb} \qquad ◆◆◆$$

Our next problem requires the statics equations which we first gave in Chap. 3. They are

> The moment of a force about some point is the product of the force F and the perpendicular distance from the force to the point,
>
> and the equations of equilibrium,
>
> The sum of all horizontal forces acting on a body = 0
> The sum of all vertical forces acting on a body = 0
> The sum of all moments acting on a body = 0

You may want to flip through that material before starting these problems.

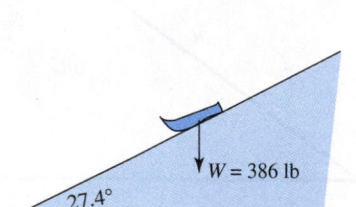

(a)

(b)

FIGURE 7–44

◆◆◆ **Example 23:** A cable running from the top of a telephone pole creates a horizontal pull of 875 Newtons (N), as shown in Fig. 7–45. A support cable running to the ground is inclined 71.5° from the horizontal. Find the tension in the support cable.

Solution: We draw the forces acting at the top of the pole as shown in Fig. 7–46. We see that the horizontal component of the tension T in the support cable must equal the horizontal pull of 875 N. So

$$\cos 71.5° = \frac{875}{T}$$

$$T = \frac{875}{\cos 71.5°} = 2760 \text{ N} \qquad ◆◆◆$$

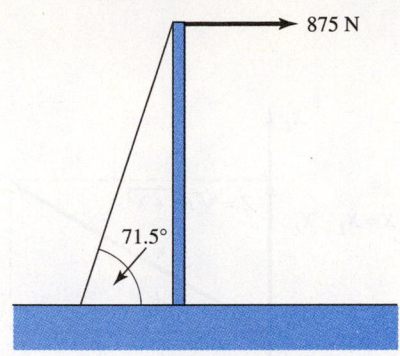

FIGURE 7–45

Velocity Vectors

◆◆◆ **Example 24:** A river flows at the rate of 4.70 km/h. A rower, who can travel 7.51 km/h in still water, heads directly across the current. That is, the boat remains pointed perpendicular to the current while being carried downstream by it. Find the actual rate and direction of travel of the boat.

Solution: The boat is crossing the current and at the same time is being carried downstream (Fig. 7–47). Thus the velocity V has one component of 7.51 km/h across the current and another component of 4.70 km/h downstream. We find the magnitude of the resultant of these two components by the Pythagorean theorem.

$$V^2 = (7.51)^2 + (4.70)^2$$

from which

$$V = 8.86 \text{ km/h}$$

Now finding the angle θ yields

$$\tan \theta = \frac{4.70}{7.51}$$

from which

$$\theta = 32.0° \qquad ◆◆◆$$

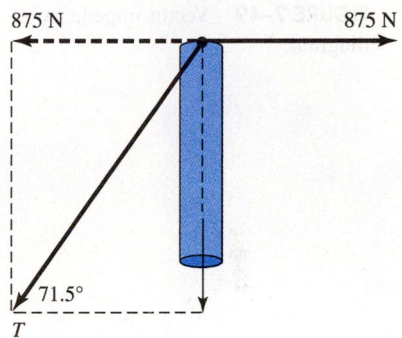

FIGURE 7–46

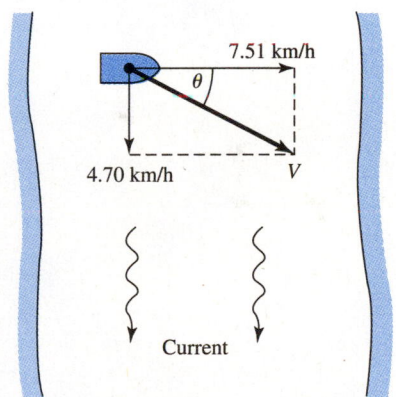

FIGURE 7–47

Impedance Vectors

Vectors find extensive use in electrical technology, and one of the most common applications is in the calculation of impedances. We will not attempt to teach ac circuits in a few paragraphs but only hope to reinforce concepts learned in your other courses. For non-electrical students, everything you need to work these problems is given right here.

Fig. 7–48 shows a resistor, an inductor, and a capacitor, connected in series with an ac source. The *reactance X* is a measure of how much the capacitance and inductance retard the flow of current in such a circuit. It is the difference between the *capacitive reactance* X_C and the *inductive reactance* X_L.

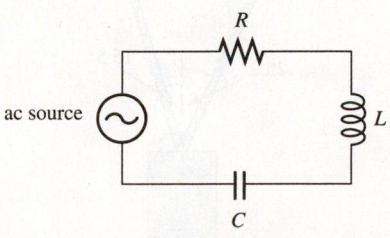

FIGURE 7–48

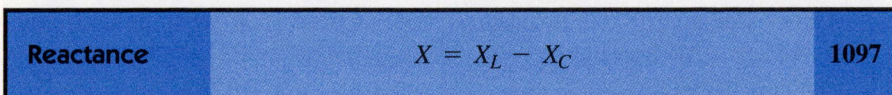

| Reactance | $X = X_L - X_C$ | 1097 |

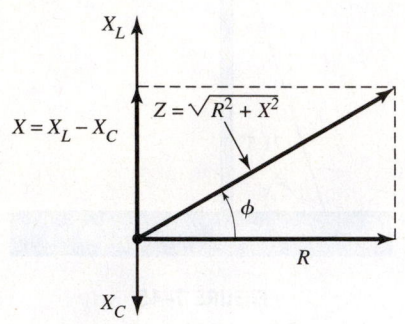

FIGURE 7–49 Vector impedance diagram.

The *impedance Z* is a measure of how much the flow of current in an ac circuit is retarded by all circuit elements, including the resistance. The magnitude of the impedance is related to the total resistance R and reactance X by the following formula:

Impedance	$Z = \sqrt{R^2 + X^2}$	1098

The impedance, resistance, and reactance form the three sides of a right triangle, the *vector impedance diagram* (Fig. 7–49). The angle ϕ between Z and R is called the *phase angle*.

Phase Angle	$\phi = \arctan \dfrac{X}{R}$	1099

♦♦♦ **Example 25:** The capacitive reactance of a certain circuit is 2720 Ω, the inductive reactance is 3260 Ω, and the resistance is 1150 Ω. Find the reactance, the magnitude of the impedance of the circuit, and the phase angle.

Solution: By Eq. 1097,

$$X = 3260 - 2720 = 540 \ \Omega$$

By Eq. 1098,

$$Z = \sqrt{(1150)^2 + (540)^2} = 1270 \ \Omega$$

and by Eq. 1099,

$$\phi = \arctan \frac{540}{1150} = 25.2° \qquad ♦♦♦$$

Exercise 6 ♦ Applications of Vectors

Force Vectors

1. What force, neglecting friction, must be exerted to drag a 56.5-N weight up a slope inclined 12.6° from the horizontal?
2. What is the largest weight that a tractor can drag up a slope that is inclined 21.2° from the horizontal if it is able to pull along the incline with the force of 2750 lb? Neglect the force due to friction.
3. Two ropes hold a crate as shown in Fig. 7–50. The tension in one is 994 N in a direction 15.5° with the vertical, and the other has a tension of 624 N in a direction 25.2° with the vertical. How much does the crate weigh?
4. A truck weighing 7280 lb is on a bridge inclined 4.80° from the horizontal. Find the force of the truck normal (perpendicular) to the bridge.
5. A person has just enough strength to pull a 1270-N weight up a certain slope. Neglecting friction, find the angle at which the slope is inclined to the horizontal if the person is able to exert a pull of 551 N.
6. A person wishes to pull a 255-lb weight up an incline to the top of a wall 14.5 ft high. Neglecting friction, what is the length of the shortest incline (measured along the incline) that can be used if the person's pulling strength is 145 lb?
7. A truck weighing 18.6 tons stands on a hill inclined 15.4° from the horizontal. How large a force must be counteracted by brakes to prevent the truck from rolling downhill?

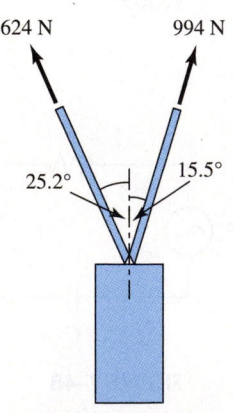

FIGURE 7–50

Velocity Vectors

8. A plane is headed due west with an air speed of 212 km/h (Fig. 7–51). It is driven from its course by a wind from due north blowing at 23.6 km/h. Find the ground speed of the plane and the actual direction of travel. Refer to Fig. 7–52 for definitions of flight terminology.

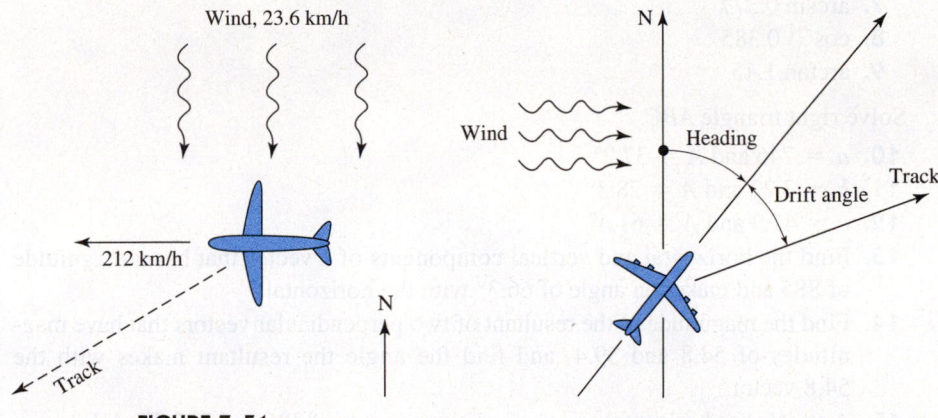

FIGURE 7–51

FIGURE 7–52 Flight terminology. The *heading* of an aircraft is the direction in which the craft is *pointed.* Due to air current, it usually will not travel in that direction but in an actual path called the *track.* The angle between the heading and the track is the *drift angle.* The *air speed* is the speed relative to the surrounding air, and the *ground speed* is the craft's speed relative to the earth.

9. At what air speed should an airplane head due west in order to follow a course S 80° 15′ W if a 20.5-km/h wind is blowing from due north?

10. A pilot heading his plane due north finds the actual direction of travel to be N 5° 12′ E. The plane's air speed is 315 mi/h, and the wind is from due west. Find the plane's ground speed and the velocity of the wind.

11. A certain escalator travels at a rate of 10.6 m/min, and its angle of inclination is 32.5°. What is the vertical component of the velocity? How long will it take a passenger to travel 10.0 m vertically?

12. A projectile is launched at an angle of 55.6° to the horizontal and follows a straight path with a speed of 7550 m/min. Find the vertical and horizontal components of this velocity.

13. At what speed with respect to the water should a ship head due north in order to follow a course N 5° 15′ E if a current is flowing due east at the rate of 10.6 mi/h?

Impedance Vectors

14. A circuit has a reactance of 2650 Ω and a phase angle of 44.6°. Find the resistance and the magnitude of the impedance.

15. A circuit has a resistance of 115 Ω and a phase angle of 72.0°. Find the reactance and the magnitude of the impedance.

16. A circuit has an impedance of 975 Ω and a phase angle of 28.0°. Find the resistance and the reactance.

17. A circuit has a reactance of 5.75 Ω and a resistance of 4.22 Ω. Find the magnitude of the impedance and the phase angle.

18. A circuit has a capacitive reactance of 1776 Ω, an inductive reactance of 5140 Ω, and a total impedance of 5560 Ω. Find the resistance and the phase angle.

••• CHAPTER 7 REVIEW PROBLEMS •••••••••••••••••••••••••••••••

Write the sine, cosine, and tangent of each angle. Keep four decimal places.

1. 72.9°
2. 35.226°
3. 60.16°

Find angle θ in decimal degrees, if

4. $\sin \theta = 0.574$

5. $\cos \theta = 0.824$

6. $\tan \theta = 1.345$

Evaluate each expression. Give your answer in degrees.

7. arcsin 0.377

8. $\cos^{-1} 0.385$

9. arctan 1.43

Solve right triangle *ABC*.

10. $a = 746$ and $A = 37.2°$

11. $b = 3.72$ and $A = 28.5°$

12. $c = 45.9$ and $A = 61.4°$

13. Find the horizontal and vertical components of a vector that has a magnitude of 885 and makes an angle of 66.3° with the horizontal.

14. Find the magnitude of the resultant of two perpendicular vectors that have magnitudes of 54.8 and 39.4, and find the angle the resultant makes with the 54.8 vector.

15. A vector has horizontal and vertical components of 385 and 275. Find the magnitude and the direction of that vector.

16. A vertical pole on horizontal ground casts a shadow 13.5 m long when the angle of elevation of the sun is 15.4°. Find the height of the pole.

17. From a point 125 ft in front of a church, the angles of elevation of the top and base of its steeple are 22.5° and 19.6°, respectively. Find the height of the steeple.

18. A circuit has a resistance of 125 Ω, an impedance of 256 Ω, an inductive reactance of 312 Ω, and a positive phase angle. Find the capacitive reactance and the phase angle.

Evaluate to four decimal places.

19. cos 59.2°   **20.** sin 19.3°

21. tan 52.8°   **22.** tan 37.2°

23. cos 24.7°   **24.** sin 42.9°

Find, to the nearest tenth of a degree, the angle whose trigonometric function is given.

25. $\tan \theta = 1.7362$ **26.** $\tan \theta = 2.9914$

27. $\sin \theta = 0.7253$ **28.** $\tan \theta = 3.2746$

29. $\cos \theta = 0.9475$ **30.** $\cos \theta = 0.3645$

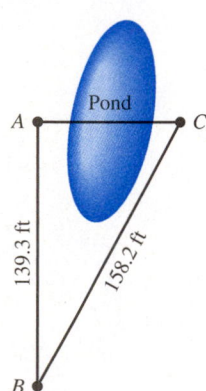

A ●——Pond——● C

139.3 ft

158.2 ft

B ●

FIGURE 7–53

31. Point *C* is due east of point *A* across a pond, as shown in Fig. 7–53. To measure the distance *AC*, a person walks 139.3 ft due south to point *B*, then walks 158.2 ft to *C*. Find the distance *AC* and the direction the person was walking when going from *B* to *C*.

32. Figure 7–54 shows the design for the front of a tower. Find the outside radius of the largest circular clock that will just fit in that triangular space.

33. *Project:* Finding the height of a flagpole may not be a very useful activity, but can you think of situations where *the same mathematics* can be used in a more practical way, such as monitoring the height of a weather balloon? List as many applications as you can.

34. *Writing:* Write a short paragraph, with examples, on how you might possibly use an idea from this chapter in a real-life situation: on the job, in a hobby, or around the house.

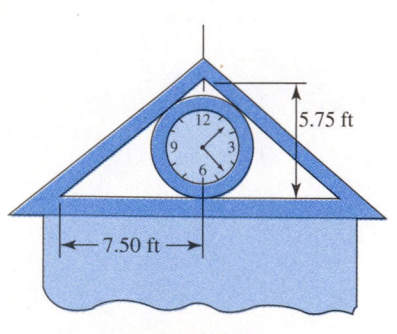

5.75 ft

←—7.50 ft—→

FIGURE 7–54

8

Oblique Triangles and Vectors

••• OBJECTIVES •••

When you have completed this chapter, you should be able to

• Write the trig function of any angle given a point on the terminal side.

• Find a trig function for any angle using a calculator.

• Identify the algebraic sign of a given trig function for any angle in any quadrant.

• Determine the reference angle for a given angle.

• Determine the angle(s), given the value of a trig function.

• Solve oblique triangles using the law of sines.

• Solve oblique triangles using the law of cosines.

• Solve applied problems involving oblique triangles.

• Find the resultant of vectors at any angle.

• Determine the resultant of any number of vectors.

• Resolve any vector into its components.

• Solve applied problems involving vectors.

In the preceding chapter we learned just enough trigonometry to solve right triangles. Even with those few tools you saw that we were able to handle a great variety of applications, including vectors at right angles.

However, many applications require us to solve an *oblique triangle*, such as for finding distance *PQ* in Fig. 8–1. For these, we need to be able to write the trigonometric functions of an *obtuse* angle. Further, two vectors are not always perpendicular but can be at any angle, such as the force vectors in Fig. 8–2. For such applications we need to be able to write the trigonometric functions of *any* angle, including those greater than 180°.

To start, we will expand our definition of the trigonometric functions to include angles in any quadrant. We also introduce three more trigonometric functions, the *cotangent, secant*, and *cosecant*, and will see that they are simply the reciprocals of our three original functions. We then learn how to find the angle when the function is given, a process made more complicated because there is more than one angle that has a given function.

Next we solve oblique triangles. We cannot use our right triangle trigonometry directly because we have no right triangle, but we can use it to derive the *law of sines* and the *law of cosines*, our main tools for solving oblique triangles.

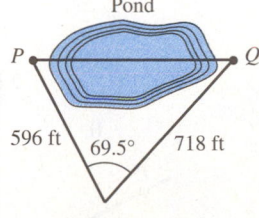

FIGURE 8–1

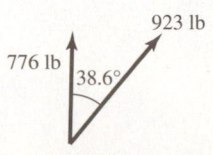

FIGURE 8–2

Finally we give a wide range of problems requiring the solution of oblique triangles, including those involving vectors.

8–1   Trigonometric Functions of Any Angle

In Chap. 7 we defined the sine, cosine, and tangent of acute angles. We did this in terms of the sides of a right triangle, and also for an angle in standard position. Let us now expand this definition to angles of any size.

We also introduce three new trigonometric functions, the *cotangent*, *secant*, and *cosecant*. Their definitions, along with the sine, cosine, and tangent, are

Trigonometric Functions	$\text{sine } \theta = \sin \theta = \dfrac{y}{r} = \dfrac{\text{opposite side}}{\text{hypotenuse}}$	**111**
	$\text{cosine } \theta = \cos \theta = \dfrac{x}{r} = \dfrac{\text{adjacent side}}{\text{hypotenuse}}$	**112**
	$\text{tangent } \theta = \tan \theta = \dfrac{y}{x} = \dfrac{\text{opposite side}}{\text{adjacent side}}$	**113**
	$\text{cotangent } \theta = \cot \theta = \dfrac{x}{y} = \dfrac{\text{adjacent side}}{\text{opposite side}}$	**114**
	$\text{secant } \theta = \sec \theta = \dfrac{r}{x} = \dfrac{\text{hypotenuse}}{\text{adjacent side}}$	**115**
	$\text{cosecant } \theta = \csc \theta = \dfrac{r}{y} = \dfrac{\text{hypotenuse}}{\text{opposite side}}$	**116**

Figure 8–3 shows angles in the second, third, and fourth quadrants. The trigonometric functions of any of these angles are defined exactly as for an acute angle in quadrant I. From any point P on the terminal side of the angle we drop

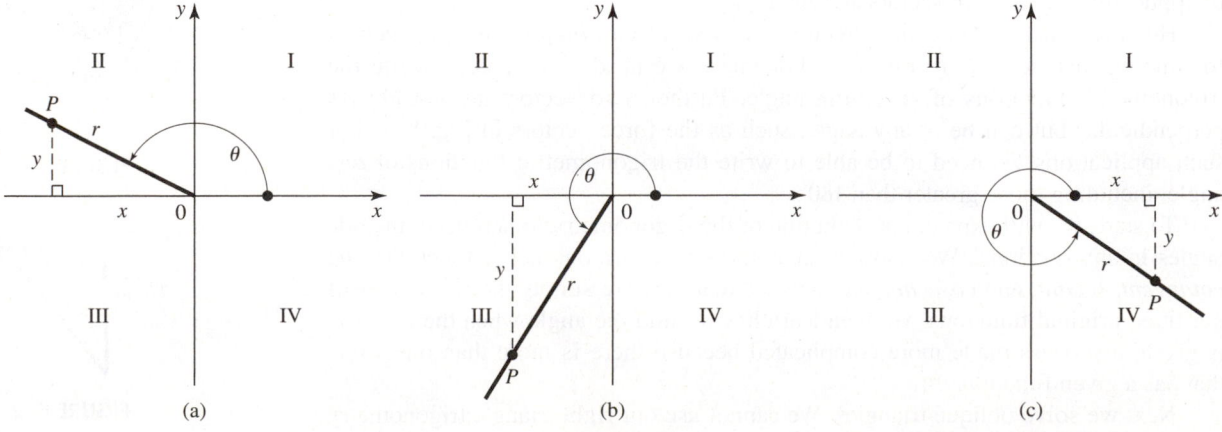

(a)   (b)   (c)

FIGURE 8–3

a perpendicular to the x axis, forming a right triangle with legs x and y and with a hypotenuse r. The six trigonometric functions are then given by Eqs. 111 through 113, just as before, except that *some of the functions are now negative.*

♦♦♦ **Example 1:** A point on the terminal side of angle θ has the coordinates $(-3, -5)$. Write the six trigonometric functions of θ to three significant digits.

Solution: We sketch the angle as shown in Fig. 8–4 and see that it lies in the third quadrant. We find distance r by the Pythagorean theorem.

$$r^2 = (-3)^2 + (-5)^2 = 9 + 25 = 34$$
$$r = 5.83$$

Then,

$$\sin \theta = \frac{y}{r} = \frac{-5}{5.83} = -0.858$$

$$\cos \theta = \frac{x}{r} = \frac{-3}{5.83} = -0.515$$

$$\tan \theta = \frac{y}{x} = \frac{-5}{-3} = 1.67$$

$$\cot \theta = \frac{x}{y} = \frac{-3}{-5} = 0.600$$

$$\sec \theta = \frac{r}{x} = \frac{5.83}{-3} = -1.94$$

$$\csc \theta = \frac{r}{y} = \frac{5.83}{-5} = -1.17 \qquad ♦♦♦$$

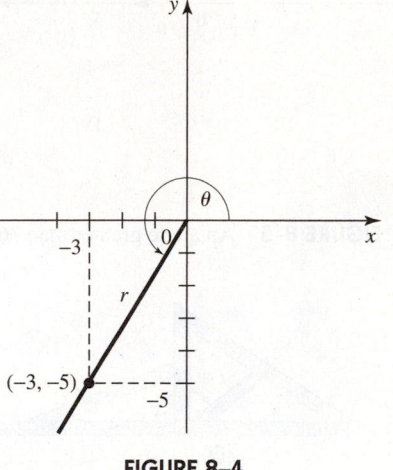

FIGURE 8–4

Trigonometric Functions of Any Angle by Calculator

In Sec. 7–1 we learned how to find the trigonometric functions of an acute angle by calculator. Now we will find the trigonometric functions of any angle, acute or obtuse, positive or negative, or greater than 360°.

The keystrokes are exactly the same as those we used for acute angles, and the calculator will automatically give the correct algebraic sign. But you must put it into the proper mode, DEGREE or RADIAN, before doing the calculation.

♦♦♦ **Example 2:** Evaluate sin 212° to four significant digits.

Solution: We place the calculator into DEGREE mode, and simply key in

$$\boxed{\text{SIN}} \quad 212 \quad \boxed{\text{ENTER}}$$

The screen is shown. ♦♦♦

Negative angles are entered using the $\boxed{(-)}$ key.

♦♦♦ **Example 3:** The keystrokes to find the tangent of $-35°$ are

$$\boxed{\text{TAN}} \quad \boxed{(-)} \quad 35 \quad \boxed{\text{ENTER}}$$

Your calculator screen should look something like the screen shown. Remember that negative angles are measured *clockwise* from the positive x direction. ♦♦♦

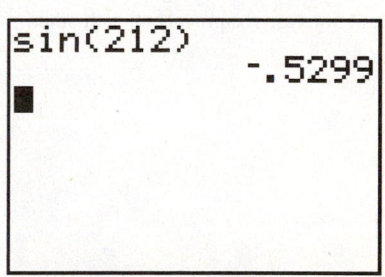

TI-83/84 screen for Example 2.

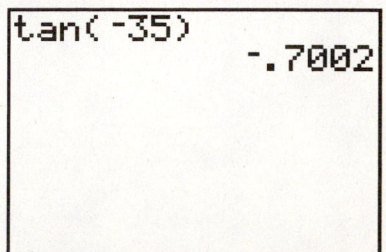

TI-83/84 screen for Example 3.

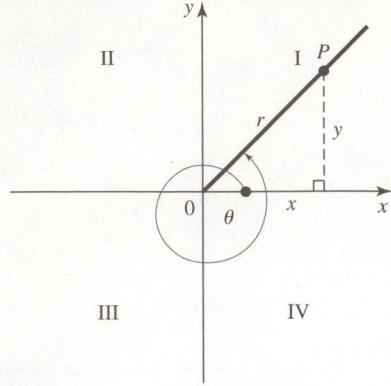

FIGURE 8–5 An angle greater than 360°.

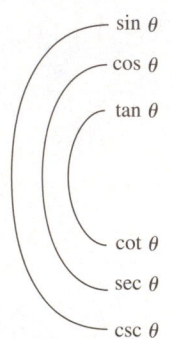

FIGURE 8–6

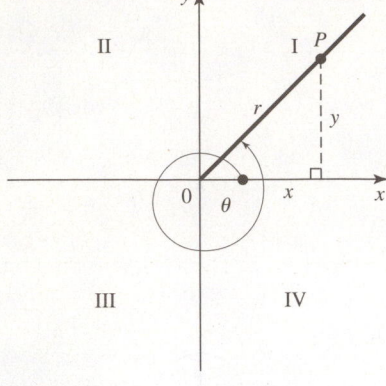

FIGURE 8–7 This diagram shows which functions are reciprocals of each other.

Angles greater than 360°, Fig. 8–5, are handled in the same manner as any other angle.

◆◆◆ **Example 4:** Verify that

(a) $\cos 412° = 0.6157$ (b) $\tan 555° = 0.2679$ ◆◆◆

◆◆◆ **Example 5:** *An Application.* In order to analyze forces in the truss in Fig. 8–6, we need the trigonometric functions of angle θ. Find the sine and cosine of θ, to four significant digits.

Solution: By calculator,

$$\sin 138° = 0.6691$$

$$\cos 138° = -0.7431 \qquad ◆◆◆$$

Reciprocal Relationships

From the trigonometric functions we see that

$$\sin \theta = \frac{y}{r}$$

and that

$$\csc \theta = \frac{r}{y}$$

Obviously, $\sin \theta$ and $\csc \theta$ are reciprocals.

$$\csc \theta = \frac{1}{\sin \theta}$$

Inspection of the trigonometric functions shows two more sets of reciprocals (see also Fig. 8–7).

Reciprocal Relationships	$\csc \theta = \dfrac{1}{\sin \theta}$	**117a**
	$\sec \theta = \dfrac{1}{\cos \theta}$	**117b**
	$\cot \theta = \dfrac{1}{\tan \theta}$	**117c**

◆◆◆ **Example 6:** Find the cosecant of θ if its sine is 0.6349.

Solution: By Eq. 117a,

$$\csc \theta = \frac{1}{\sin \theta} = \frac{1}{0.6349} = 1.575 \qquad ◆◆◆$$

Cotangent, Secant, and Cosecant by Calculator

Most calculators do not have keys for the cot, sec, and csc. We find them instead by using the reciprocal relationships.

```
cos(72.6)
              .2990
1/Ans
             3.3440
```

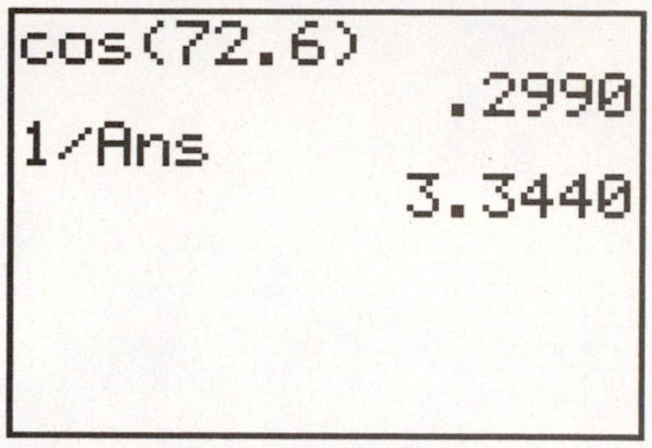

TI-83/84 screen for Example 7.

◆◆◆ **Example 7:** Find sec 72.6° by calculator.

Solution: We first find cos 72.6° and then take its reciprocal, as shown in the screen.   ◆◆◆

Cofunctions

The sine of angle A in Fig. 8–8 is

$$\sin A = \frac{a}{c}$$

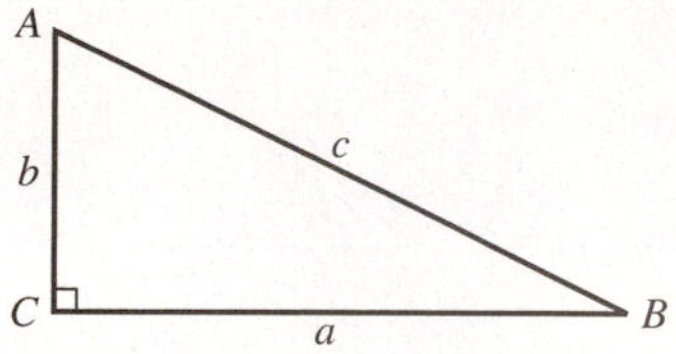

But a/c is *also* the cosine of the complementary angle B. Thus we can write

$$\sin A = \cos B$$

FIGURE 8–8

Here $\sin A$ and $\cos B$ are called *cofunctions*. Similarly, $\cos A = \sin B$. These cofunctions and others in the same right triangle are given in the following boxes.

Cofunctions Where $A + B = 90°$		
	$\sin A = \cos B$	118a
	$\cos A = \sin B$	118b
	$\tan A = \cot B$	118c
	$\cot A = \tan B$	118d
	$\sec A = \csc B$	118e
	$\csc A = \sec B$	118f

In general, *a trigonometric function of an acute angle is equal to the corresponding cofunction of the complementary angle.*

◆◆◆ **Example 8:** If the cosine of the acute angle A in a right triangle ABC is equal to 0.725, find the sine of the other acute angle, B.

Solution: By Eq. 118b,

$$\sin B = \cos A = 0.725 \qquad ◆◆◆$$

The keystrokes for cot, sec, and csc are no different for obtuse angles than for acute angles.

◆◆◆ **Example 9:** Verify the following calculations to four significant digits:

(a) $\cot 101° = -0.1944$ (b) $\sec 213° = -1.192$

(c) $\csc 294° = -1.095$ ◆◆◆

Exercise 1 ◆ Trigonometric Functions of Any Angle

The terminal side of an angle in standard position passes through the given point. Sketch the angle, compute the distance r from the origin to the point, and write the six trigonometric functions of the angle. Work to three significant digits.

1. $(3.00, -5.00)$
2. $(-4.00, 12.0)$
3. $(24.0, -7.00)$
4. $(-15.0, -8.00)$
5. $(1.59, -3.11)$
6. $(-5.13, -1.17)$

Trigonometric Functions of Any Angle by Calculator

Write, to four significant digits, the sine, cosine, and tangent of each angle.

7. $101°$	8. $216°$	9. $331°$
10. $125.8°$	11. $-62.85°$	12. $-227.4°$
13. $486°$	14. $-527°$	15. $114°23'$
16. $-11°18'$	17. $412°$	18. $238°$

Reciprocal Relationships

Evaluate to four decimal places.

19. Find $\csc \theta$ if $\sin \theta = 0.7352$
20. Find $\cot \theta$ if $\tan \theta = 1.4638$
21. Find $\sec \theta$ if $\cos \theta = 0.7354$

Cotangent, Secant, and Cosecant by Calculator

Evaluate to four decimal places.

22. $\sec 158.3°$	23. $\cot 153.6°$	24. $\csc 122.7°$
25. $\csc 207.4°$	26. $\sec 215.4°$	27. $\cot 228.7°$

Cofunctions

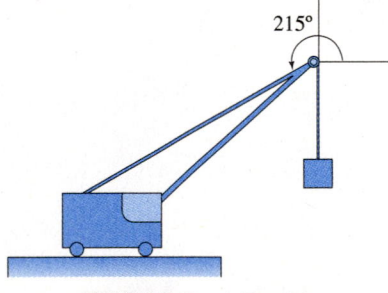

215°

FIGURE 8–9 A Derrick

Express as a function of the complementary angle.

28. $\sin 38°$	29. $\cos 73°$	30. $\tan 19°$
31. $\sec 85.6°$	32. $\cot 63.2°$	33. $\csc 82.7°$
34. $\tan 35°14'$		

35. *An Application*: To analyze the force vectors at the end of the derrick, Fig. 8–9, we need to find the sine and cosine of the given angle. Find them to three decimal places.

8–2 Finding the Angle When the Trigonometric Function Is Known

Given an angle, we can find its trigonometric functions. Now we reverse the operation and find the angle when given the trigonometric function. We did the same in Chap. 7 with acute angles, but here we will see that it is slightly more complicated with angles greater than 90°.

■ **Exploration:** *Try this.*

(a) Use your calculator to find the following:

$$\sin 35° \qquad \sin 145° \qquad \sin 395°$$

What did you find? Can you explain your findings?

(b) Now find the angle whose sine is 0.573576. What can you say about your result? ■

When you took the sine of an angle, you got a *single* answer. But your exploration should have shown you that there is *more than one* angle that has the same sine (or cosine and tangent). In fact, there are infinitely many, if you count angles greater than one revolution. So how do we find the one that we need? We use something called the *reference angle*.

Reference Angle

For an angle in standard position on coordinate axes, the *acute* angle that its terminal side makes with the *x* axis is called the *reference angle, θ′*. It is always taken as *positive*.

◆◆◆ **Example 10:** The reference angle θ' for an angle of 125° is

$$\theta' = 180° - 125° = 55°$$

as shown in Fig. 8–10. ◆◆◆

◆◆◆ **Example 11:** The reference angle θ' for an angle of 236° is

$$\theta' = 236° - 180° = 56°$$

as in Fig. 8–11. ◆◆◆

◆◆◆ **Example 12:** The reference angle θ' for an angle of 331.6° (Fig. 8–12) is

$$\theta' = 360° - 331.6° = 28.4°$$ ◆◆◆

Note that we treat the quadrantal angles, 180° and 360° as *exact* numbers. Thus the rounding of our answer is determined by the decimal places in the given angle.

Common Error	The reference angle is measured always from the *x* axis, never from the *y* axis. It is always positive.

Algebraic Signs of the Trigonometric Functions

We saw in Chap. 7 that the trigonometric functions of first-quadrant angles were always positive. From Fig. 8–3, it is clear that some of the trigonometric functions of angles in the second, third, and fourth quadrants are negative, because *x* or *y* can be negative (*r* is always positive). Figure 8–13 shows the quadrant in which a trigonometric functions is positive. Otherwise it is negative.

Instead of trying to remember which trigonometric functions are negative in which quadrants, just sketch the angle and note whether *x* or *y* is negative. From this information you can figure out whether the function you want is positive or negative.

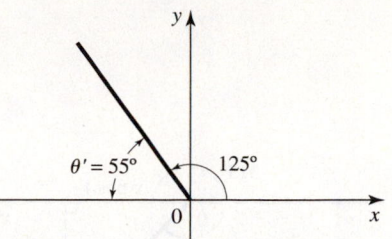

FIGURE 8–10 Reference angle θ'. It is also called the *working* angle.

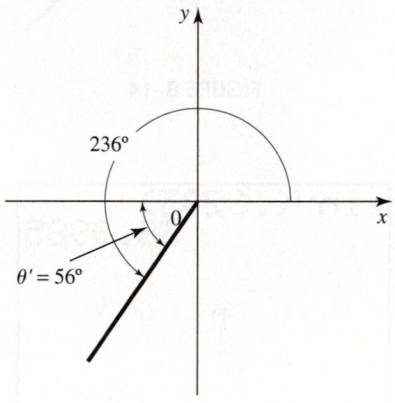

FIGURE 8–11

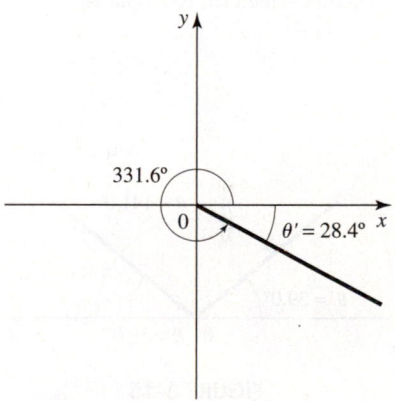

FIGURE 8–12

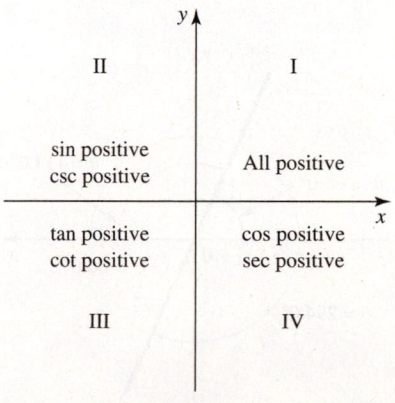

FIGURE 8–13

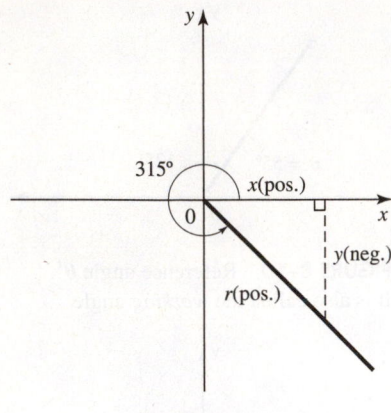

FIGURE 8–14

TI-83/84 screen for Example 14.

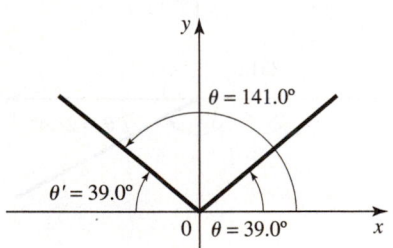

FIGURE 8–15

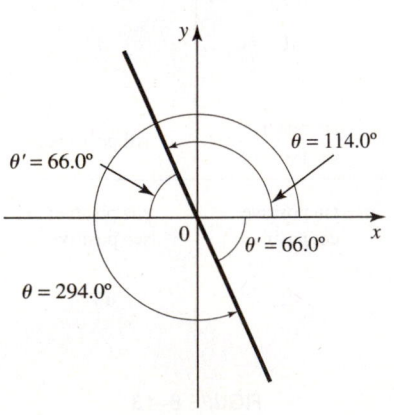

FIGURE 8–16

◆◆◆ **Example 13:** What is the algebraic sign of csc 315°?

Solution: We make a sketch, such as in Fig. 8–14. It is not necessary to draw the angle accurately, but it must be shown in the proper quadrant, quadrant IV in this case. We note that y is negative and that r is (always) positive. So

$$\csc 315° = \frac{r}{y} = \frac{(+)}{(-)} = \text{negative}$$ ◆◆◆

Finding the Angle by Calculator

- As with acute angles, we use the inverse trig keys on our calculators.
- The angle then given by the calculator *is taken as the reference angle*. Note its algebraic sign.
- Sketch the angle to determine the quadrants in which the given trig function is positive or negative. We usually want only the two positive angles that have the given trig function.

◆◆◆ **Example 14:** Find two positive values for θ less than 360°, if $\sin \theta = 0.6293$. Work to the nearest tenth of a degree.

Solution: We key in $\boxed{\text{SIN}^{-1}}$ 0.6293 $\boxed{\text{ENTER}}$ and get 39.0°, rounded as shown. We take this as our reference angle.

$$\theta' = 39.0°$$

A sketch of the angle shows that the sine is positive in quadrants I and II. In each of those quadrants we draw an angle of 39.0° with the x axis, as in Fig. 8–15. The quadrant I angle is simply 39.0°. To find the quadrant II angle, it is clear we must subtract the reference angle from 180°.

$$\theta = 180° - 39.0° = 141.0°$$

So our two positive angles less than 360° are

$$\theta = 39.0° \text{ and } 141.0°$$

Check: We can check our results by taking the sine of our angles.

$$\sin 39.0° = 0.6293 \quad \text{Checks.}$$

$$\sin 141.0° = 0.6293 \quad \text{Checks.}$$ ◆◆◆

◆◆◆ **Example 15:** Find, to the nearest tenth of a degree, the two positive angles less than 360° that have a tangent of −2.25.

Solution: From the calculator, $\tan^{-1}(-2.25) = -66.0°$, so our reference angle is +66.0°. The tangent is negative in the second and fourth quadrants. As shown in Fig. 8–16, our second-quadrant angle is

$$180° - 66.0° = 114.0°$$

and our fourth-quadrant angle is

$$360° - 66.0° = 294.0°$$ ◆◆◆

◆◆◆ **Example 16:** Find $\cos^{-1} 0.575$ to the nearest tenth of a degree.

Solution: From the calculator,

$$\cos^{-1} 0.575 = 54.9°$$

The cosine is positive in the first and fourth quadrants. Our fourth-quadrant angle is

$$360° - 54.9° = 305.1°$$ ◆◆◆

Inverse of the Cotangent, Secant, and Cosecant

When the cotangent, secant, or cosecant is given, we use the reciprocal relationships (Eqs. 117) as in the following example.

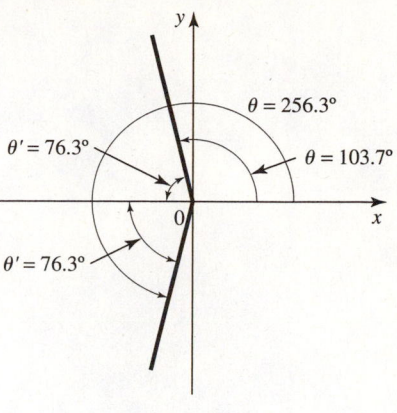

FIGURE 8–17

◆◆◆ **Example 17:** Find two positive angles less than 360° that have a secant of −4.22. Work to the nearest tenth of a degree.

Solution: If we let the angle be θ, then, by Eq. 117b,

$$\cos \theta = \frac{1}{\sec \theta} = \frac{1}{-4.22} = -0.237$$

$$\theta = \cos^{-1}(-0.237)$$

$$= 103.7° \quad \text{by calculator}$$

The secant is also negative in the third quadrant. Our reference angle (Fig. 8–17) is

$$\theta' = 180° - 103.7° = 76.3°$$

so the third-quadrant angle is

$$\theta = 180° + 76.3° = 256.3° \qquad \text{◆◆◆}$$

◆◆◆ **Example 18:** Evaluate arcsin (−0.528) to the nearest tenth of a degree.

Solution: As before, we seek only two positive angles less than 360°. By calculator,

$$\arcsin(-0.528) = -31.9°$$

which is a fourth-quadrant angle. As a positive angle, it is

$$\theta = 360° - 31.9° = 328.1°$$

The sine is also negative in the third quadrant. Using 31.9° as our reference angle, we have

$$\theta = 180° + 31.9° = 211.9° \qquad \text{◆◆◆}$$

Exercise 2 ◆ Finding the Angle When the Trigonometric Function Is Known

Reference Angle

Find the reference angle for each given angle.

1. 163°
2. 274°
3. 305°
4. 138.6°
5. 249.3°

Algebraic Signs of the Trigonometric Functions

If θ is an angle in standard position, state in what quadrants its terminal side can lie if

6. $\theta = 123°$.
7. $\theta = 272°$.
8. $\theta = -47°$.
9. $\theta = -216°$.
10. $\theta = 415°$.
11. $\theta = -415°$.
12. $\theta = 845°$.
13. $\sin \theta$ is positive.
14. $\cos \theta$ is negative.
15. $\sec \theta$ is positive.

State whether the following expressions are positive or negative. Do not use your calculator, and try not to refer to your book.

16. $\sin 174°$ **17.** $\cos 110°$ **18.** $\tan 315°$

19. $\sec 332°$ **20.** $\cot 206°$ **21.** $\csc 196°$

Give the algebraic signs of the sine, cosine, and tangent of the following. Do not use your calculator.

22. $110°$ **23.** $206°$ **24.** $335°$

25. $-48°$ **26.** $500°$

Finding the Angle by Calculator

Find two positive angles less than 360° whose trigonometric function is given. Round your angles to a tenth of a degree.

27. $\sin \theta = 0.7761$

28. $\tan \theta = -0.1587$

29. $\cos \theta = 0.8372$

30. $\cos \theta = 0.3215$

31. $\tan \theta = 6.372$

32. $\cos \theta = 0.4476$

33. $\sin \theta = -0.6358$

Inverse of the Cotangent, Secant, and Cosecant

Find two positive angles less than 360° whose trigonometric function is given. Round your angles to a tenth of a degree.

34. $\cot \theta = 2.8458$

35. $\sec \theta = 1.7361$

36. $\cot \theta = -0.2315$

37. $\csc \theta = -3.852$

38. $\cot \theta = 0.3315$

8–3   Law of Sines

We cannot use the trigonometric functions directly to solve an oblique triangle, but we will use them to derive the *law of sines*, which can be used for any triangle.

Derivation

We will derive the *law of sines* for an oblique triangle in which all three angles are acute such as in Fig. 8–18. We start by breaking the given triangle into two right triangles by drawing altitude h to side AB.

Then right triangle ACD,

$$\sin A = \frac{h}{b} \quad \text{or} \quad h = b \sin A$$

And right triangle BCD,

$$\sin B = \frac{h}{a} \quad \text{or} \quad h = a \sin B$$

So

$$b \sin A = a \sin B$$

Dividing by $\sin A \sin B$, we have

$$\frac{a}{\sin A} = \frac{b}{\sin B}$$

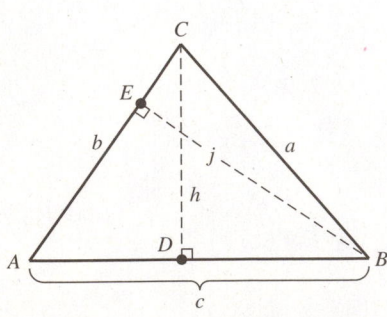

FIGURE 8–18 Derivation of the law of sines.

Similarly, drawing altitude j to side AC, and using triangles BEC and AEB, we get

$$j = a \sin C = c \sin A$$

or

$$\frac{a}{\sin A} = \frac{c}{\sin C}$$

Combining this with the previous result, we obtain the following equation:

| **Law of Sines** | $$\dfrac{a}{\sin A} = \dfrac{b}{\sin B} = \dfrac{c}{\sin C}$$ *The sides of a triangle are proportional to the sines of the opposite angles.* | **105** |

We have just derived the law of sines for a triangle having all acute angles. The law of sines also holds when one of the angles is obtuse, and the derivation is nearly the same. See the projects at the end of this section.

Solving a Triangle When Two Angles and One Side Are Known (AAS or ASA)

Recall that "solving a triangle" means to find all missing sides and angles. Here we use the law of sines to solve an oblique triangle. To do this, we must have *a known side opposite to a known angle*, as well as another angle. We abbreviate these given conditions as AAS (angle–angle–side) or ASA (angle–side–angle).

◆◆◆ **Example 19:** Solve triangle ABC where $A = 32.5°$, $B = 49.7°$, and $a = 226$.

Solution: We first sketch the triangle as shown in Fig. 8–19. We want to find the unknown angle C and sides b and c. The missing angle is found by substracting the two known angles from 180°.

$$C = 180° - 32.5° - 49.7° = 97.8°$$

Then, by the law of sines,

$$\frac{a}{\sin A} = \frac{b}{\sin B}$$

$$\frac{226}{\sin 32.5°} = \frac{b}{\sin 49.7°}$$

Solving for b, we get

$$b = \frac{226 \sin 49.7°}{\sin 32.5°} = 321$$

Again using the law of sines, we have

$$\frac{a}{\sin A} = \frac{c}{\sin C}$$

$$\frac{226}{\sin 32.5°} = \frac{c}{\sin 97.8°}$$

So

$$c = \frac{226 \sin 97.8°}{\sin 32.5°} = 417$$

◆◆◆

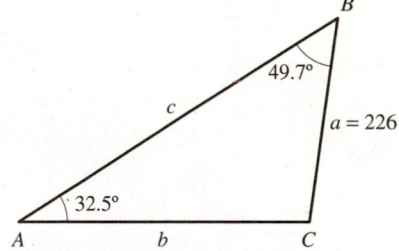

FIGURE 8–19 A diagram drawn more or less to scale can serve as a good check of your work and reveal inconsistencies in the given data. As another rough check, see that the longest side is opposite the largest angle and that the shortest side is opposite the smallest angle.

Notice that if two angles are given, the third can easily be found, since all angles of a triangle must have a sum of 180°. This means that if two angles and an included side are given (ASA), the problem can be solved as shown above for AAS.

◆◆◆ Example 20: *An Application.* A factory is prefabricating a quantity of roof trusses, as in Fig. 8–20. Find the length of member *AB*.

Solution: We have a known side opposite a known angle, so we can use the law of sines.

$$\frac{AB}{\sin 21.5°} = \frac{11.4}{\sin 37.2°}$$

$$AB = \frac{11.4 \sin 21.5°}{\sin 37.2°} = 6.91 \text{ ft}$$

◆◆◆

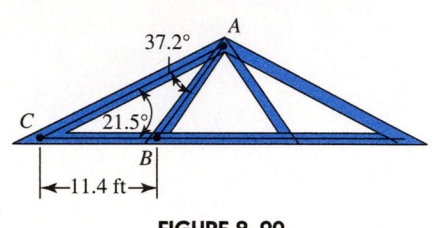

FIGURE 8–20

Solving a Triangle When Two Sides and One Angle Are Given (SSA): The Ambiguous Case

■ **Exploration:**

Try this. By hand or by computer drafting program, draw the following triangle *ABC*.

$$C = 28.4° \qquad a = 171 \qquad c = 107$$

Is it possible to draw a triangle, given this data? Is more than one triangle possible?   ■

In Example 19, *two angles and one side* were given. We can also use the law of sines when *one angle and two sides* are given, provided that the given angle is opposite one of the given sides. But we must be careful here. Sometimes we will get no solutions, one solution, or two solutions, depending on the given information. The possibilities are given in Table 8–1.

We see that we will sometimes get *two* correct solutions (as in the exploration), both of which may be reasonable in a given application.

Common Error	If it appears that you will get no solution when doing an application with an oblique triangle, it probably means that the data are incorrect or that the problem is not properly set up. Don't give up at that point, but go back and check your work.

A simple way to check for the number of solutions is to make a sketch. But the sketch must be fairly accurate, as in the exploration, and as in the following example.

◆◆◆ Example 21: Solve triangle *ABC* where $A = 27.6°$, $a = 112$, and $c = 165$.

Solution: Let's first calculate the altitude *h* to find out how many solutions we may have.

$$h = c \sin A$$
$$= 165 \sin 27.6° = 76.4$$

We see that side *a* (112) is greater than *h* (76.4), but less than *c* (165), so we have the ambiguous case with two solutions, as verified by a sketch Fig. 8–21. We will solve for both possible triangles using the law of sines. However sometimes, like here, it is more convenient to use the *reciprocals* of the expressions in the law of sines.

$$\frac{\sin C}{165} = \frac{\sin 27.6°}{112}$$

$$\sin C = \frac{165 \sin 27.6°}{112} = 0.6825$$

$$C = 43.0°$$

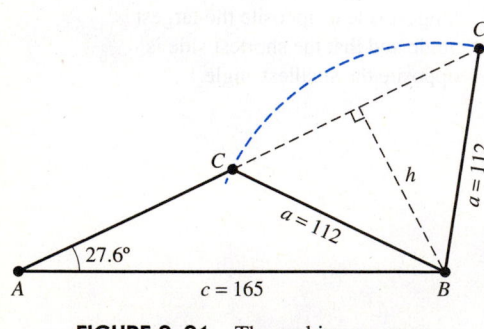

FIGURE 8–21   The ambiguous case.

TABLE 8–1 Possible Solutions When Using the Law of Sines

Given angle A opposite to given side a;

When angle A is obtuse, and:	Then:	Example:
1. $a > c$	Side a can intersect side b in only one place, so we get *one solution*.	
2. $a \leq c$	Side a is too short to intersect side b, so there is *no solution*.	

When angle A is acute, and :	Then:	Example:
1. $a \geq c$	Side a is too long to intersect side b in more than one place, so there is only *one solution*.	
2. $a = h < c$ where $h = c \sin A$	Side a just reaches side b, so we get a right triangle having *one solution*.	
3. $a < h < c$ where the altitude h is $h = c \sin A$	Side a is too short to touch side b, so we get *no solution*.	
4. $h < a < c$ where $h = c \sin A$	Side a can intersect side b in two places, giving *two solutions*. This is the *ambiguous case*.	

This is *one* of the possible values for C. But recall from Sec. 8–2 that there are *two* angles less than 180° for which the sine is positive. One of them, θ, is in the first quadrant, and the other, $180° - \theta$, is in the second quadrant. So the other possible value for C is

$$C = 180° - 43.0° = 137.0°$$

We now find the two corresponding values for side b and angle B.
When $C = 43.0°$,

$$B = 180° - 27.6° - 43.0° = 109.4°$$

So

$$\frac{b}{\sin 109.4°} = \frac{112}{\sin 27.6°}$$

from which $b = 228$.

When $C = 137.0°$,

$$B = 180° - 27.6° - 137.0° = 15.4°$$

So

$$\frac{b}{\sin 15.4°} = \frac{112}{\sin 27.6°}$$

from which $b = 64.2$. So our two solutions are given in the following table:

	A	B	C	a	b	c
1.	27.6°	109.4°	43.0°	112	228	165
2.	27.6°	15.4°	137.0°	112	64.2	165

◆◆◆

Common Error	In a problem such as the preceding one, it is easy to forget the second possible solution ($C = 137.0°$), especially since a calculator will give only the acute angle when computing the arc sine.

Not every SSA problem has two solutions, as we'll see in the following example.

◆◆◆ **Example 22:** *An Application.* Find angles B and C, and side b, on the gusset plate of Fig. 8–22.

Solution: When we sketch the triangle, we see that the given information allows us to draw the triangle in only one way. Thus will get a unique solution. By the law of sines,

$$\frac{\sin C}{412} = \frac{\sin 35.2°}{525}$$

$$\sin C = \frac{412 \sin 35.2°}{525} = 0.4524$$

Thus the two possible values for C are

$$C = 26.9° \quad \text{and} \quad C = 180° - 26.9° = 153.1°$$

Our sketch, even if crudely drawn, shows that C cannot be obtuse, so we discard the 153.1° value. Then since the sum of the three interior angles must be 180°.

$$B = 180° - 35.2° - 26.9° = 117.9°$$

We use the law of sines once again.

$$\frac{b}{\sin 117.9°} = \frac{525}{\sin 35.2°}$$

$$b = \frac{525 \sin 117.9°}{\sin 35.2°} = 805 \text{ cm}$$

◆◆◆

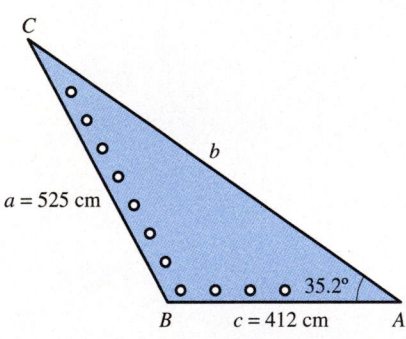

$a = 525$ cm

b

$c = 412$ cm

$35.2°$

C B A

FIGURE 8–22

Exercise 3 ◆ Law of Sines

Two Angles and One Side Known

1. Solve each triangle in Fig. 8–23.

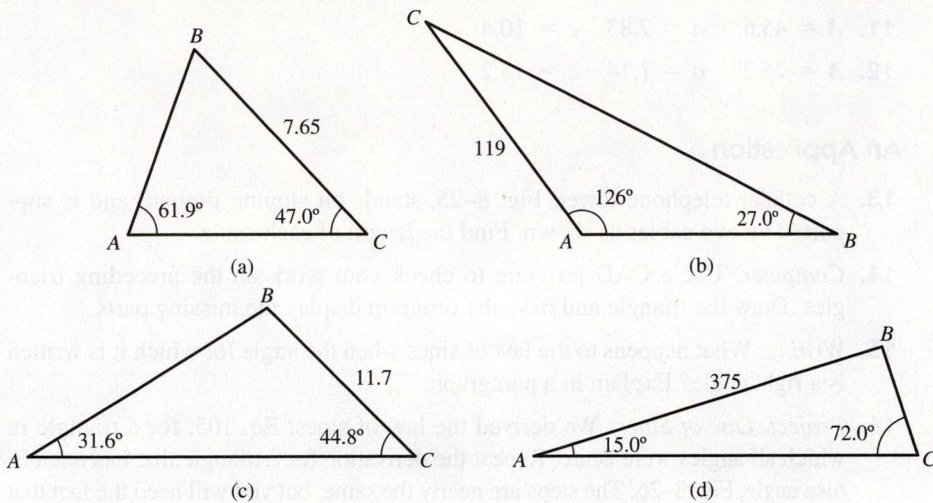

FIGURE 8–23

Solve triangle *ABC*.

2. $B = 125°$, $C = 32.0°$, $c = 58.0$
3. $A = 24.14°$, $B = 38.27°$, $a = 5562$
4. $B = 55.38°$, $C = 18.20°$, $b = 77.85$
5. $A = 44.47°$, $C = 63.88°$, $c = 1.065$
6. $A = 18.0°$, $B = 12.0°$, $a = 50.7$

Two Sides and One Angle Known

7. Solve each triangle in Fig. 8–24.

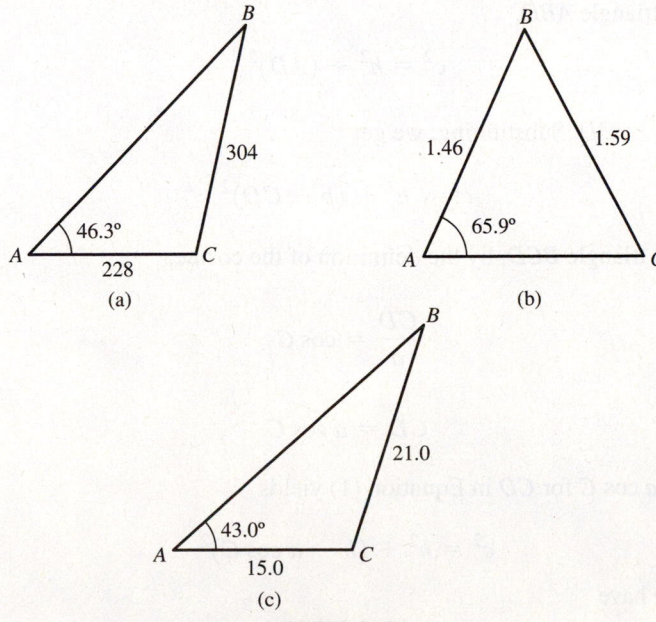

FIGURE 8–24

Solve triangle *ABC*.

8. $A = 47.9°$ $a = 3.28$ $c = 2.35$

9. $C = 61.7°$ $b = 284$ $c = 382$

10. $C = 51.8°$ $b = 25.6$ $c = 24.9$

11. $A = 45.6°$ $a = 7.83$ $c = 10.4$

12. $A = 25.2°$ $a = 7.14$ $c = 13.2$

An Application

13. A cellular telephone tower, Fig. 8–25, stands on sloping ground, and is supported by two cables as shown. Find the length of each cable.

14. *Computer:* Use a CAD program to check your work on the preceding triangles. Draw the triangle and have the program display the missing parts.

15. *Writing:* What happens to the law of sines when the angle for which it is written is a right angle? Explain in a paragraph.

16. *Project, Law of Sines:* We derived the law of sines, Eq. 105, for a triangle in which all angles were acute. Repeat the derivation for a triangle that has one obtuse angle, Fig. 8–26. The steps are nearly the same, but you will need the fact that $\sin(180° - B) = \sin B$.

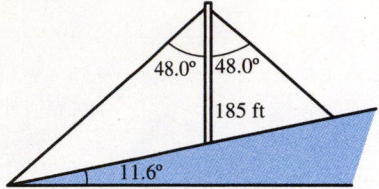

FIGURE 8–25

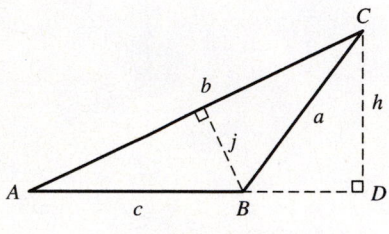

FIGURE 8–26

8–4   Law of Cosines

To use the law of sines we need a known side opposite a known angle. Sometimes we do not have that information, as when, for example, we know three sides and no angle. We can still solve such a triangle using the *law of cosines.*

Derivation

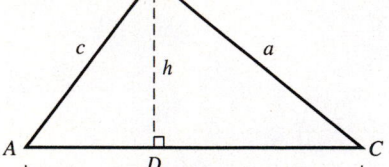

FIGURE 8–27   Derivation of the law of cosines.

Consider an oblique triangle *ABC* as shown in Fig. 8–27. As we did for the law of sines, we start by dividing the triangle into two right triangles by drawing an altitude *h* to side *AC*.

In right triangle *ABD*,

$$c^2 = h^2 + (AD)^2$$

But $AD = b - CD$. Substituting, we get

$$c^2 = h^2 + (b - CD)^2 \qquad (1)$$

Now, in right triangle *BCD*, by the definition of the cosine,

$$\frac{CD}{a} = \cos C$$

or

$$CD = a \cos C$$

Substituting $a \cos C$ for *CD* in Equation (1) yields

$$c^2 = h^2 + (b - a \cos C)^2$$

Squaring, we have

$$c^2 = h^2 + b^2 - 2ab \cos C + a^2 \cos^2 C \qquad (2)$$

Let us leave this expression for the moment and write the Pythagorean theorem for the same triangle *BCD*.

$$h^2 = a^2 - (CD)^2$$

Again substituting $a \cos C$ for *CD*, we obtain

$$h^2 = a^2 - (a \cos C)^2$$

$$= a^2 - a^2 \cos^2 C$$

Substituting this expression for h^2 back into (2), we get

$$c^2 = a^2 - a^2 \cos^2 C + b^2 - 2ab \cos C + a^2 \cos^2 C$$

Cancelling $a^2 \cos^2 C$ and collecting terms, we get the law of cosines.

$$c^2 = a^2 + b^2 - 2ab \cos C$$

If we repeated the derivation two more times, with perpendiculars drawn to side *AB* and then to side *BC*, we would get two more forms of the law of cosines.

Law of Cosines	$$a^2 = b^2 + c^2 - 2bc \cos A$$ $$b^2 = a^2 + c^2 - 2ac \cos B$$ $$c^2 = a^2 + b^2 - 2ab \cos C$$ **106** *The square of any side equals the sum of the squares of the other two sides minus twice the product of the other sides and the cosine of the opposite angle.*

◆◆◆ **Example 23:** Find side *a* in Fig. 8–28.

Solution: By the law of cosines,

$$a^2 = (1.24)^2 + (1.87)^2 - 2(1.24)(1.87) \cos 42.8°$$

$$= 5.03 - 4.64(0.7337) = 1.63$$

$$a = 1.28 \text{ m}$$

◆◆◆

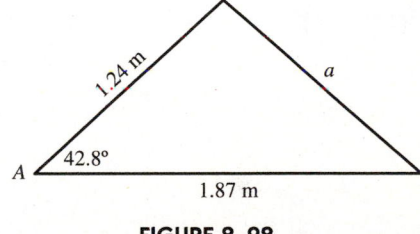

FIGURE 8–28

When to Use the Law of Sines or the Law of Cosines

It is sometimes not clear whether to use the law of sines or the law of cosines to solve a triangle. We use the law of sines when we have a *known side opposite a known angle*. We use the law of cosines only when the law of sines does not work, that is, for all other cases. In Fig. 8–29, the heavy lines indicate the known information and may help in choosing the proper law.

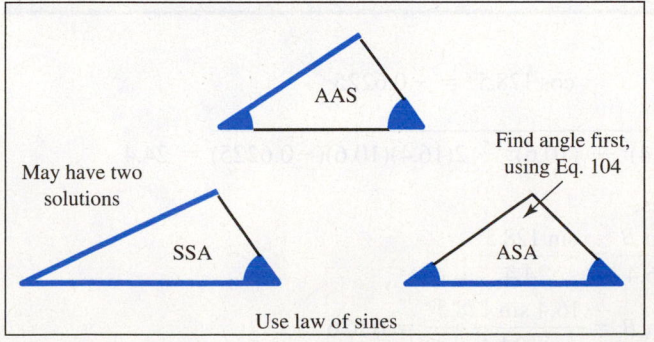

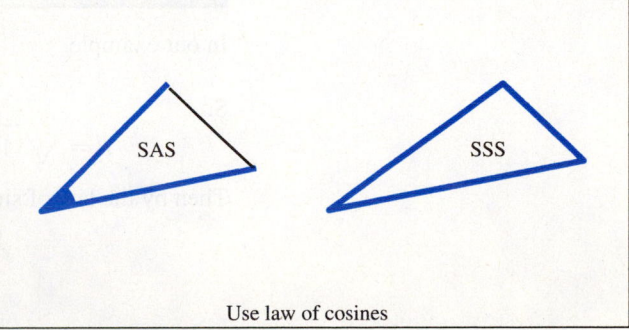

FIGURE 8–29 When to use the law of sines or the law of cosines.

Using the Law of Cosines When Two Sides and the Included Angle Are Known

We can solve triangles by the law of cosines if we know *two sides and the angle between them,* or if we know *three sides.* We consider the first of these cases in the following example.

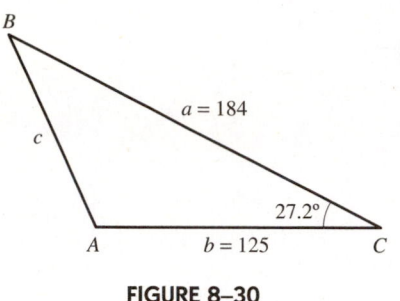

FIGURE 8–30

◆◆◆ **Example 24:** Solve triangle ABC where $a = 184$, $b = 125$, and $C = 27.2°$.

Solution: We make a sketch as shown in Fig. 8–30. Notice that we cannot initially use the law of sines because we do not have a known side opposite a known angle. Instead, we use the law of cosines to find side c.

$$c^2 = a^2 + b^2 - 2ab \cos C$$
$$= (184)^2 + (125)^2 - 2(184)(125) \cos 27.2° = 8568$$
$$c = 92.6$$

Now that we have a known side opposite a known angle, we can use the law of sines to find angle A or angle B. Which shall we find first?

Use the law of sines to *find the acute angle first* (angle B in this example). If, instead, you solve for the obtuse angle first, you may forget to subtract the angle obtained by calculator from 180°. Further, if one of the angles is so close to 90° that you cannot tell from your sketch if it is acute or obtuse, find the other angle first, and then subtract the two known angles from 180° to obtain the third angle.

So using the law sines to find angle B,

$$\frac{\sin B}{125} = \frac{\sin 27.2°}{92.6}$$
$$\sin B = \frac{125 \sin 27.2°}{92.6} = 0.617$$
$$B = 38.1° \text{ and } B = 180 - 38.1 = 141.9°$$

We drop the larger value because our sketch shows us that B must be acute. Then subtracting the known angles from 180°,

$$A = 180° - 27.2° - 38.1° = 114.7°$$ ◆◆◆

In our next example, the given angle is *obtuse.*

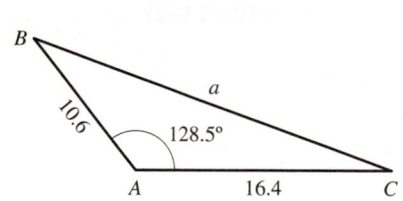

FIGURE 8–31

◆◆◆ **Example 25:** Solve triangle ABC where $b = 16.4$, $c = 10.6$, and $A = 128.5°$.

Solution: We make a sketch as shown in Fig. 8–31. Then, by the law of cosines,

$$a^2 = (16.4)^2 + (10.6)^2 - 2(16.4)(10.6) \cos 128.5°$$

Common Error	The cosine of an obtuse angle is *negative.* Be sure to use the proper algebraic sign when applying the law of cosines to an obtuse angle.

In our example,

$$\cos 128.5° = -0.6225$$

So

$$a = \sqrt{(16.4)^2 + (10.6)^2 - 2(16.4)(10.6)(-0.6225)} = 24.4$$

Then by the law of sines,

$$\frac{\sin B}{16.4} = \frac{\sin 128.5°}{24.4}$$
$$\sin B = \frac{16.4 \sin 128.5°}{24.4} = 0.526$$
$$B = 31.7° \quad \text{and} \quad B = 180 - 31.7 = 148.3°$$

We drop the larger value because our sketch shows us that B must be acute. Then subtracting the known angles from 180°,

$$C = 180° - 31.7° - 128.5° = 19.8°$$ ◆◆◆

◆◆◆ **Example 26:** *An Applicaiton.* Find the length of girder AB in Fig. 8–32.

Solution: We know two sides and their included angle, and we want the side opposite to the known angle. This is ideal for use of the law of cosines.

$$(AB)^2 = (5.25)^2 + (18.6)^2 - 2(5.25)(18.6)\cos 98.2°$$

$$= 401.4$$

$$AB = 20.0 \text{ ft}$$ ◆◆◆

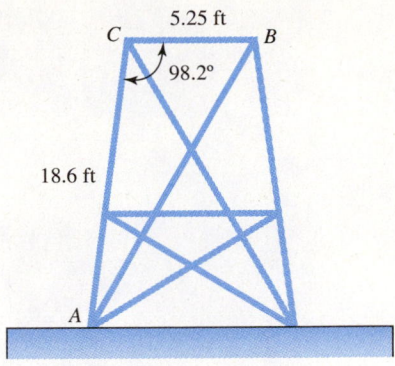

FIGURE 8–32

Using the Law of Cosines When Three Sides Are Known

When three sides of an oblique triangle are known, we can use the law of cosines to solve for one of the angles. A second angle is found using the law of sines, and the third angle is found by subtracting the other two from 180°.

◆◆◆ **Example 27:** Solve triangle ABC in Fig. 8–33, where $a = 128$, $b = 146$, and $c = 222$.

Solution: We start by writing the law of cosines for any of the three angles. A good way to avoid ambiguity is to *find the largest angle first* (the law of cosines will tell us if it is acute or obtuse). Then we are sure that the other two angles are acute. Writing the law of cosines for angle C gives

$$(222)^2 = (128)^2 + (146)^2 - 2(128)(146)\cos C$$

Solving for $\cos C$ gives

$$\cos C = -0.3099$$

Since the cosine is negative, C must be obtuse, so by calculator

$$C = 108.1°$$

Then by the law of sines

$$\frac{\sin A}{128} = \frac{\sin 108.1°}{222}$$

from which $\sin A = 0.548$. Since we know that A is acute, we get

$$A = 33.2°$$

Finally,

$$B = 180° - 108.1° - 33.2° = 38.7°$$ ◆◆◆

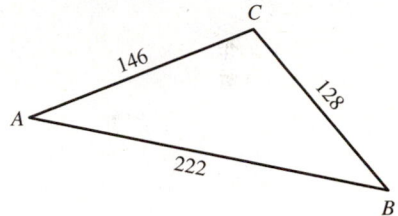

FIGURE 8–33

◆◆◆ **Example 28:** *An Application.* To analyze the performance of an internal combustion engine, the crank angle θ is needed for various positions of the piston, Fig. 8–34. Find the crank angle when the piston is at the height shown.

Solution: We have an oblique triangle in which three sides are known. Let's use the law of cosines for angle θ.

$$(18.7)^2 = (6.72)^2 + (15.4)^2 - 2(6.72)(15.4)\cos\theta$$

$$2(6.72)(15.4)\cos\theta = (6.72)^2 + (15.4)^2 - (18.7)^2$$

$$\cos\theta = -0.3255$$

$$\theta = 109°$$ ◆◆◆

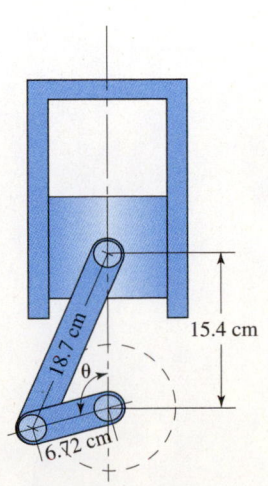

FIGURE 8–34

Exercise 4 ◆ Law of Cosines

Two Sides and One Angle Known

1. Solve each triangle in Fig. 8–35.

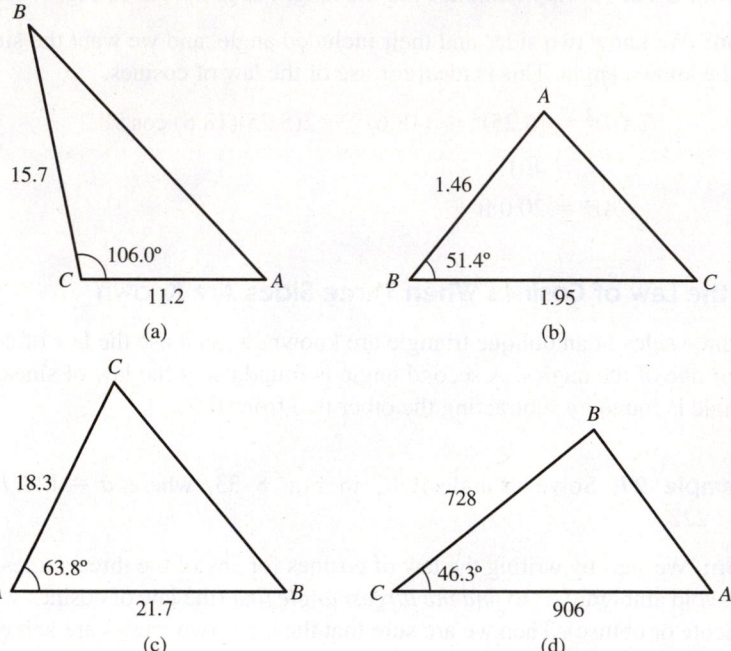

(a)

(b)

(c)

(d)

FIGURE 8–35

Solve triangle *ABC*.

2. $B = 41.7°$, $a = 199$ $c = 202$
3. $A = 115°$, $b = 46.8$ $c = 51.3$
4. $C = 67.0°$, $a = 9.08$ $b = 6.75$
5. $B = 129°$, $a = 186$ $c = 179$
6. $A = 158°$, $b = 1.77$ $c = 1.99$

Three Sides Known

7. Solve each triangle in Fig. 8–36.

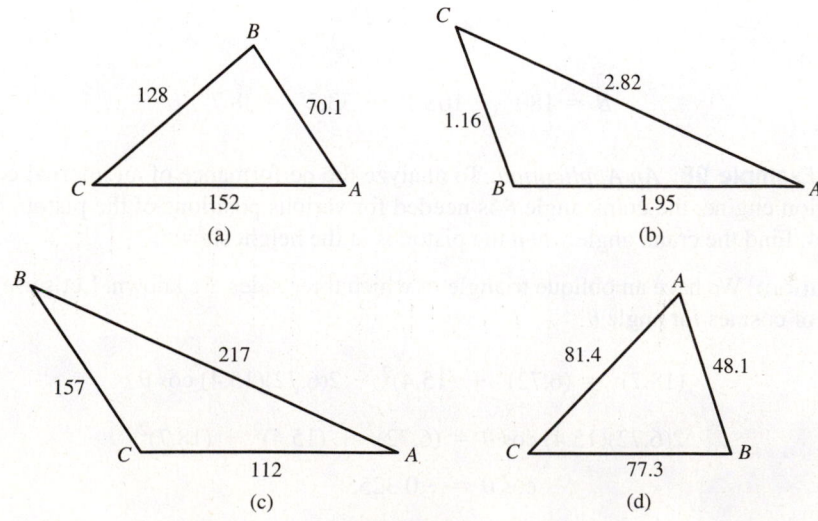

(a)

(b)

(c)

(d)

FIGURE 8–36

Solve triangle *ABC*.

8. $a = 11.3$ $b = 15.6$ $c = 12.8$

9. $a = 1.475$ $b = 1.836$ $c = 2.017$

10. $a = 369$ $b = 177$ $c = 199$

11. $a = 18.6$ $b = 32.9$ $c = 17.9$

12. $a = 5311$ $b = 6215$ $c = 7112$

An Application

13. Find the distance *PQ* across the pond in Fig. 8–37, given in the introduction to this chapter.

Computer

14. Use a CAD program to check your work on the preceding applications. Draw the figure and have the program display the dimensions of the missing parts.

15. *Project, Hero's Formula:* Find a derivation of Hero's formula for the area of a triangle, Eq. 103, and try to reproduce it. The derivation uses the law of cosines, Eq. 106 but is not easy. See, for example, the entry for "Hero's formula" in Weisstein's *CRC Concise Encyclopedia of Mathematics*.

16. *Team Project:* Use a handheld GPS to find the coordinates of a field, at least a few acres in area. Use triangulation to compute the area of that field. Then taking into account the possible errors in the GPS readings, estimate the accuracy of your computed area.

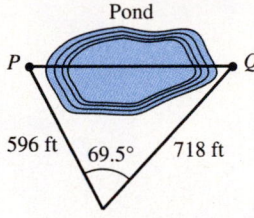

FIGURE 8–37

8–5 Applications

As with right triangles, oblique triangles have many applications in technology, as you will see in the exercises for this section. Follow the same procedures for setting up these problems as we used for other word problems, and solve the resulting triangle by the law of sines or the law of cosines, or both.

If an *area* of an oblique triangle is needed, either compute all the sides and use Hero's formula (Eq. 103), or find an altitude with right-triangle trigonometry and use Area $= \frac{1}{2}$(base)(altitude).

♦♦♦ **Example 29:** Find the area of the gusset in Fig. 8–38(a).

Solution: We first find θ.

$$\theta = 90° + 35.0° = 125.0°$$

We now know two sides and the angle between them, so can find side x by the law of cosines.

$$x^2 = (18.0)^2 + (20.0)^2 - 2(18.0)(20.0) \cos 125° = 1137$$

$$x = 33.7 \text{ in.}$$

Finally we find the area of the gusset by Hero's formula (Eq. 103).

$$s = \frac{1}{2}(18.0 + 20.0 + 33.7) = 35.9$$

$$\text{area} = \sqrt{35.9(35.9 - 18.0)(35.9 - 20.0)(35.9 - 33.7)}$$

$$= 150 \text{ in.}^2$$

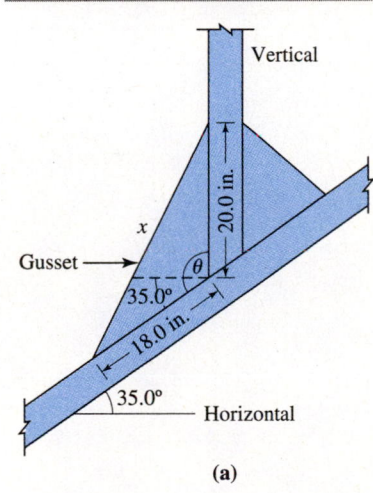

(a)

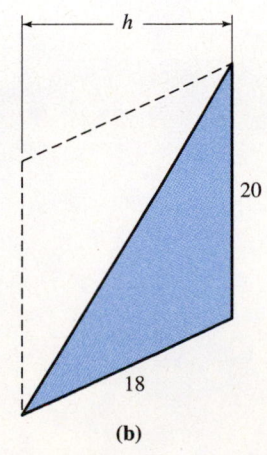

(b)

FIGURE 8–38

Check: Does the answer look reasonable? Let's place the gusset inside the parallelogram, as shown in Fig. 8–38(b), and estimate the height h by eye at about 15 in. Thus the area of the parallelogram would be 15×20, or 300 in.^{2}, just double that found for the triangle.   ◆◆◆

◆◆◆ **Example 30:** A ship takes a sighting on two buoys. At a certain instant, the bearing of buoy A is N 44.23° W, and that of buoy B is N 62.17° E. The distance between the buoys is 3.60 km, and the bearing of B from A is N 87.87° E. Find the distance of the ship from each buoy.

Estimate: Let us draw the figure with a ruler and protractor as shown in Fig. 8–39. Notice how the compass directions are laid out, starting from the north and turning in the indicated direction. Measuring, we get $SA = 1.7$ units and $SB = 2.8$ units. If you try it, you will probably get slightly different values.

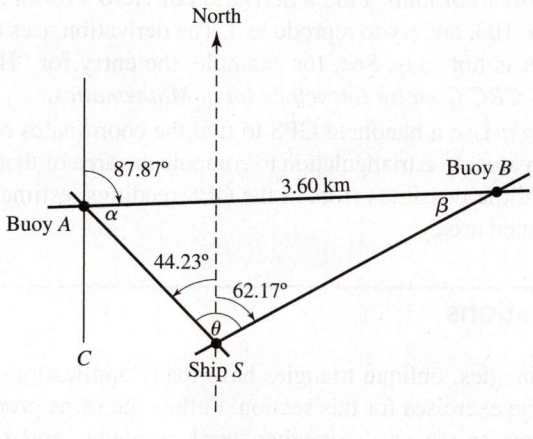

FIGURE 8–39

Solution: Calculating the angles of triangle ABS gives

$$\theta = 44.23° + 62.17° = 106.40°$$

Since angle SAC is 44.23°,

$$\alpha = 180° - 87.87° - 44.23° = 47.90°$$

and

$$\beta = 180° - 106.40° - 47.90° = 25.70°$$

From the law of sines,

$$\frac{SA}{\sin 25.70°} = \frac{SB}{\sin 47.90°} = \frac{3.60}{\sin 106.40°}$$

from which

$$SA = \frac{3.60 \sin 25.70°}{\sin 106.40°} = 1.63 \text{ km}$$

and

$$SB = \frac{3.60 \sin 47.90°}{\sin 106.40°} = 2.78 \text{ km}$$

both of which agree with our estimated values.   ◆◆◆

Exercise 5 ◆ Applications

Determining Inaccessible Distances

1. Two stakes, *A* and *B*, are 88.6 m apart. From a third stake *C*, the angle *ACB* is 85.4°, and from *A*, the angle *BAC* is 74.3°. Find the distance from *C* to each of the other stakes.

2. From a point on level ground, the angles of elevation of the top and the bottom of an antenna standing on top of a building are 32.6° and 27.8°, respectively. If the building is 125 ft high, how tall is the antenna? Remember that angles of elevation or depression are always measured from the *horizontal*.

3. The sides of a triangular lot measure 115 m, 187 m, and 215 m. Find the angles between the sides.

4. Two boats are 45.5 km apart. Both are traveling toward the same point, which is 87.6 km from one of them and 77.8 km from the other. Find the angle at which their paths intersect.

Navigation

5. A ship is moving at 15.0 km/h in the direction N 15.0° W. A helicopter with a speed of 22.0 km/h is due east of the ship. In what direction should the helicopter travel if it is to meet the ship? *Hint:* Draw your diagram after *t* hours have elapsed.

6. City *A* is 215 miles N 12.0° E from city *B*. The bearing of city *C* from *B* is S 55.0° E. The bearing of *C* from *A* is S 15.0° E. How far is *C* from *A*? From *B*?

7. A ship is moving in a direction S 24.25° W at a rate of 8.60 mi/h. If a launch that travels at 15.4 mi/h is due west of the ship, in what direction should it travel in order to meet the ship?

8. A ship is 9.50 km directly east of a port. If the ship sails exactly southeast for 2.50 km, how far will it be from the port?

9. From a plane flying due east the bearing of a radio station is S 31.0° E at 1:00 P.M. and S 11.0°E at 1:20 P.M. The ground speed of the plane is 625 km/h. Find the distance of the plane from the station at 1:00 P.M.

Structures

10. A tower for a wind generator stands vertically on sloping ground whose inclination with the horizontal is 11.6°. From a point 42.0 m downhill from the tower (measured along the slope), the angle of elevation of the top of the tower is 18.8°. How tall is the tower?

11. A vertical cellular phone antenna stands on a slope that makes an angle of 8.70° with the horizontal. From a point directly uphill from the antenna, the angle elevation of its top is 61.0°. From a point 16.0 m farther up the slope (measured along the slope), the angle of elevation of its top is 38.0°. How tall is the antenna?

12. A power pole on level ground is supported by two wires that run from the top of the pole to the ground (Fig. 8–40). One wire is 18.5 m long and makes an angle of 55.6° with the ground, and the other wire is 17.8 m long. Find the angle that the second wire makes with the ground.

13. A 71.6-m-high antenna mast is to be placed on sloping ground, with the cables making an angle of 42.5° with the top of the mast (Fig. 8–41). Find the length of each cable.

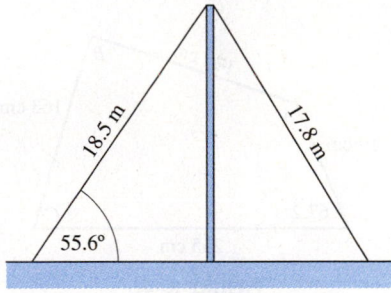

FIGURE 8–40

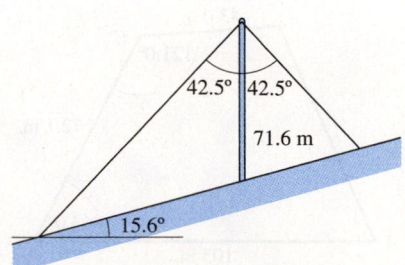

FIGURE 8–41

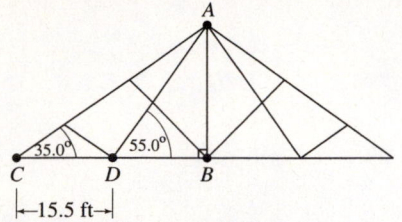

FIGURE 8–42 Roof truss.

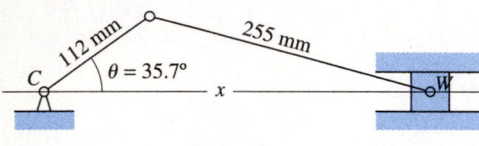

FIGURE 8–43

14. In the roof truss in Fig. 8–42, find the lengths of members *AB*, *BD*, *AC*, and *AD*.

15. From a point on level ground between two power poles of the same height, cables are stretched to the top of each pole. One cable is 52.6 ft long, the other is 67.5 ft long, and the angle of intersection between the two cables is 125°. Find the distance between the poles.

16. A pole standing on level ground makes an angle of 85.8° with the horizontal. The pole is supported by a 22.0-ft prop whose base is 12.5 ft from the base of the pole. Find the angle made by the prop with the horizontal.

Mechanisms

17. In the slider crank mechanism of Fig. 8–43, find the distance *x* between the wrist pin *W* and the crank center *C* when $\theta = 35.7°$.

18. In the four-bar linkage of Fig. 8–44, find angle θ when angle *BAD* is 41.5°.

19. Two links, *AC* and *BC*, are pivoted at *C*, as shown in Fig. 8–45. How far apart are *A* and *B* when angle *ACB* is 66.3°?

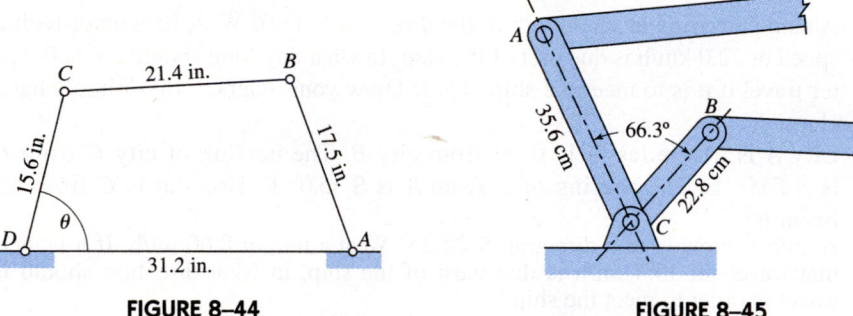

FIGURE 8–44

FIGURE 8–45

Geometry

20. Find angles *A*, *B*, and *C* in the quadrilateral in Fig. 8–46.

21. Find side *AB* in the quadrilateral in Fig. 8–47.

22. Two sides of a parallelogram are 22.8 and 37.8 m, and one of the diagonals is 42.7 m. Find the angles of the parallelogram.

23. Find the lengths of the sides of a parallelogram if its diagonal, which is 125 mm long, makes angles with the sides of 22.7° and 15.4°.

24. Find the lengths of the diagonals of a parallelogram, two of whose sides are 3.75 m and 1.26 m; their included angle is 68.4°.

25. A *median* of a triangle is a line joining a vertex to the midpoint of the opposite side. In triangle *ABC*, $A = 62.3°$, $b = 112$, and the median from *C* to the midpoint of *c* is 186. Find *c*.

26. The sides of a triangle are 124, 175, and 208. Find the length of the median drawn to the longest side.

27. The angles of a triangle are in the ratio 3:4:5, and the shortest side is 994. Solve the triangle.

28. The sides of a triangle are in the ratio 2:3:4. Find the cosine of the largest angle.

29. Two solar panels are to be placed as shown in Fig. 8–48. Find the minimum distance *x* so that the first panel will not cast a shadow on the second when the angle of elevation of the sun is 18.5°.

30. Find the overhang *x* so that the window in Fig. 8–49 will be in complete shade when the sun is 60° above the horizontal.

31. *Computer:* For the slider crank mechanism of Fig. 8–44, compute and print the position *x* of the wrist pin for values of θ from 0 to 180° in 15° steps.

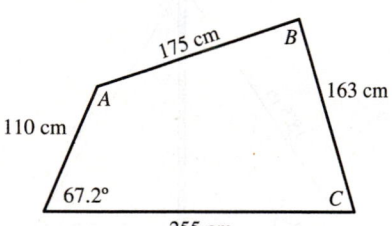

FIGURE 8–46

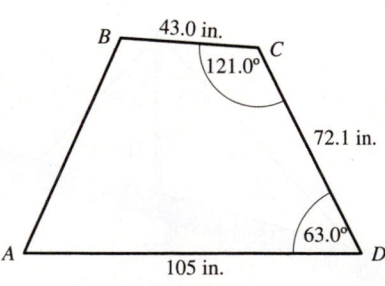

FIGURE 8–47

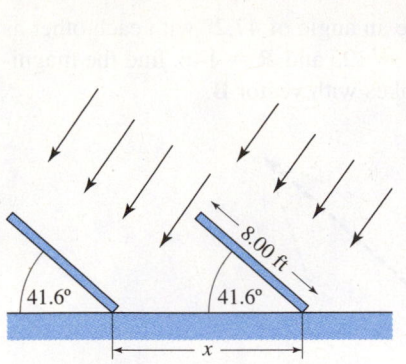

FIGURE 8–48 Solar panels.

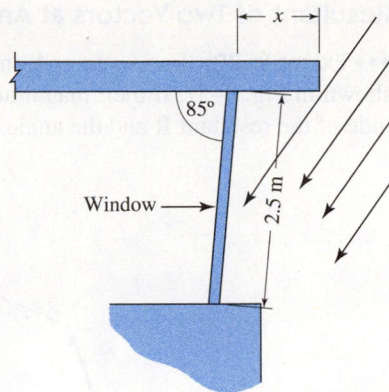

FIGURE 8–49

8–6 Non-Perpendicular Vectors

In Chap. 7 we found the components of a vector in the first quadrant. Also in Chap. 7 we found the resultant of *perpendicular* vectors. Here we will find the components of a vector in *any* quadrant, and the resultant of vectors at *any* angle. We will start by finding components.

Finding x and y Components of a Vector

The procedure is no different than in Chap. 7, except now the vector can be in any quadrant.

◆◆◆ **Example 31:** Find the x and y components of the vector $\mathbf{V} = 374 \underline{/137°}$.

Solution: We draw the magnitude of the vector and its components on coordinate axes, Fig. 8–50.

Then,

$$V_x = V \cos 137°$$

$$= 374 \cos 137° = -274$$

Similarly,

$$V_y = 374 \sin 137° = 255$$

By calculator: We showed two ways to find components on the TI-89 Titanium in Chap. 7. *First method*: Enter $P \blacktriangleright R_x$ or $P \blacktriangleright R_y$, from the **MATH** Angle menu. Then in parentheses, enter the magnitude of the vector and then the angle, separated by a comma. Finally, use $\approx$ on the **ENTER** keys to get a decimal answer.

Second method: In brackets, enter the magnitude and then the angle, using the $\angle$ symbol, one of the keyboard characters. Next enter ▸**Rect** from the **MATH** **Matrix/Vector ops** menu. Then use $\approx$ on the **ENTER** key to get a decimal answer.

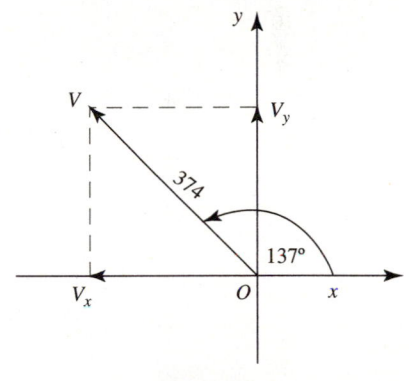

FIGURE 8–50

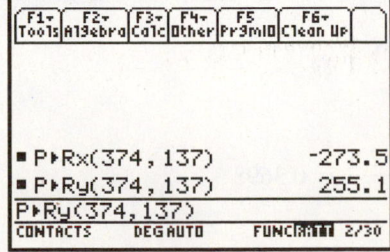

TI-89 screen for the first method.

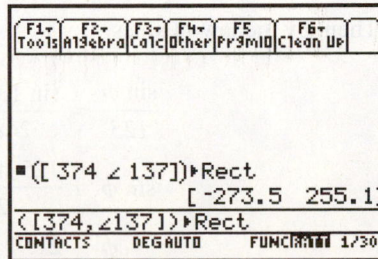

TI-89 screen for the second method. ◆◆◆

Resultant of Two Vectors at Any Angle

◆◆◆ **Example 32:** Two vectors, **A** and **B**, make an angle of 47.2° with each other as shown in Fig. 8–51. If their magnitudes are $A = 125$ and $B = 146$, find the magnitude of the resultant **R** and the angle that **R** makes with vector **B**.

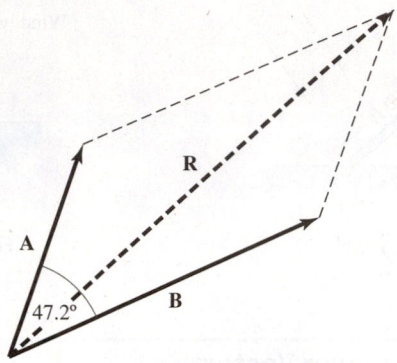

FIGURE 8–51

Solution: We make a vector diagram, either tip to tail Fig. 8–52(a) or by the parallelogram method. Fig. 8–52(b). Either way, we must solve the oblique triangle in Fig. 8–52(c) for R and ϕ. First finding θ we get

$$\theta = 180° - 47.2° = 132.8°$$

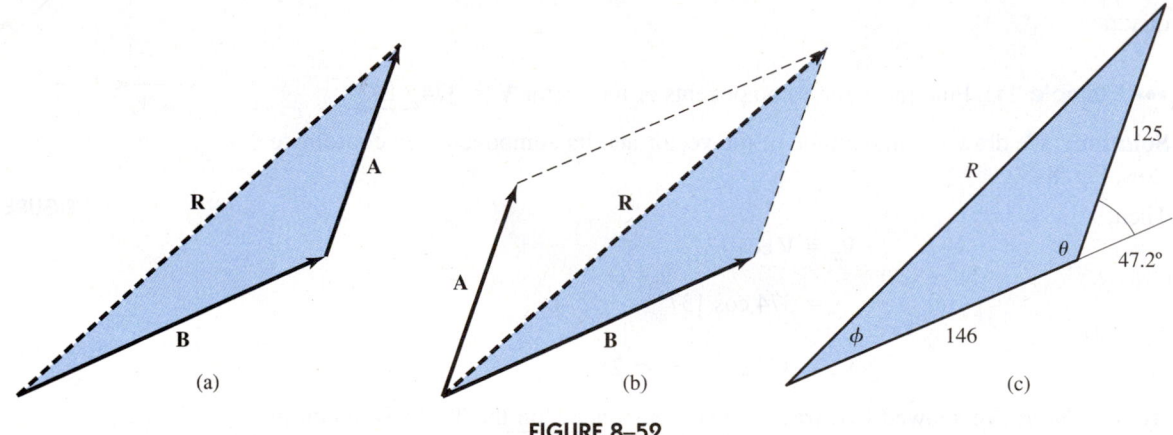

FIGURE 8–52

By the law of cosines,

$$R^2 = (125)^2 + (146)^2 - 2(125)(146) \cos 132.8° = 61{,}740$$

$$R = 248$$

Then, by the law of sines,

$$\frac{\sin \phi}{125} = \frac{\sin 132.8}{248}$$

$$\sin \phi = \frac{125 \sin 132.8}{248} = 0.3698$$

$$\phi = 21.7°$$

◆◆◆

Common Error	Remember from Chap. 7 that we name a vector using a bold Roman letter, like **A** and the magnitude of a vector by a lightface italic letter, such as *A*. Thus *A* would be the magnitude of vector **A**. It's easy to confuse the two.

Resultant of Several Vectors at Any Angles

The law of sines and the law of cosines are good for adding *two* nonperpendicular vectors. However, when *several* vectors are to be added, we usually break each into its *x* and *y* components and combine them, as in the following example.

◆◆◆ **Example 33:** Find the resultant of the vectors shown in Fig. 8–53(a).

Solution: The *x* component of a vector of magnitude *V* at any angle θ is

$$V \cos \theta$$

and the *y* component is

$$V \sin \theta$$

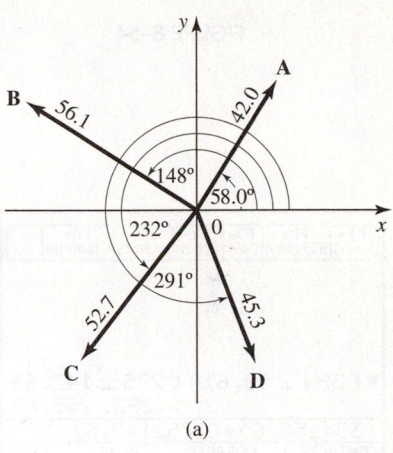

These equations apply for an angle in any quadrant. We compute and tabulate the *x* and *y* components of each original vector and find the sums of each as shown in the following table.

Vector	x Component	y Component
A	$42.0 \cos 58.0° = 22.3$	$42.0 \sin 58.0° = 35.6$
B	$56.1 \cos 148° = -47.6$	$56.1 \sin 148° = 29.7$
C	$52.7 \cos 232° = -32.4$	$52.7 \sin 232° = -41.5$
D	$45.3 \cos 291° = 16.2$	$45.3 \sin 291° = -42.3$
R	$R_x = -41.5$	$R_y = -18.5$

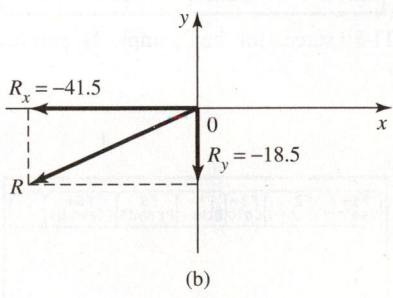

(b)

FIGURE 8–53

The magnitudes of the rectangular components of the resultant, R_x and R_y are shown in Fig. 8–53(b). We find the magnitude *R* of the resultant by the Pythagorean theorem.

$$R^2 = (-41.5)^2 + (-18.5)^2 = 2065$$
$$R = 45.4$$

We find the angle θ by

$$\theta = \arctan \frac{R_y}{R_x}$$
$$= \arctan \frac{-18.5}{-41.5}$$
$$= 24.0° \text{ or } 204°$$

Since our resultant is in the third quadrant, we drop the 24.0° value. Thus the resultant has a magnitude of 45.4 and a direction of 204°. This is often written in the form

$$R = 45.4 \underline{/204°}$$

◆◆◆

Resultants by Calculator

We will now use the calculator method for finding resultants that we introduced in the preceding chapter. But here we use the method to find the resultant of any number of vectors at various angles.

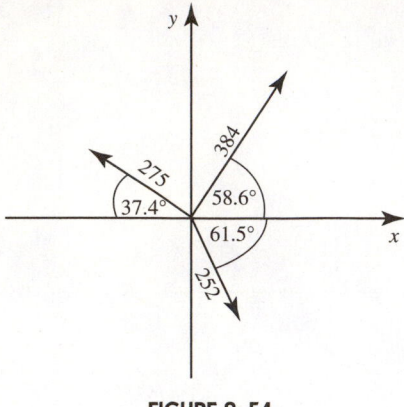

FIGURE 8–54

◆◆◆ **Example 34:** (a) Find the resultant of the vectors in Fig. 8–54, by TI-89 calculator, and (b) resolve that resultant into rectangular components.

Solution: (a) We first compute the angle that each vector makes with the positive x direcion.

$$384\underline{/58.6°}$$
$$275\underline{/(180 - 37.4)} = 275\underline{/142.6°}$$
$$252\underline{/(360 - 61.5)} = 252\underline{/298.5°}$$

We enter each vector in parentheses, using the $\boxed{\angle}$ symbol for the angle (a keyboard character on the T1-89), and combine them using a $\boxed{(+)}$ sign. The keystrokes, with the calculator in Polar and Degree modes are shown. We get a resultant of $292\underline{/69.6°}$.

(b) We now convert this vector to rectangular form as we showed in the preceding chapter. Enter the magnitude and angle, in brackets, separated by a comma. Then enter ▸**Rect** from the $\boxed{\text{MATH}}$ **Matrix/Vector ops** menu. The components of our resultant are then 102 in the x direction and 274 in the y direction.   ◆◆◆

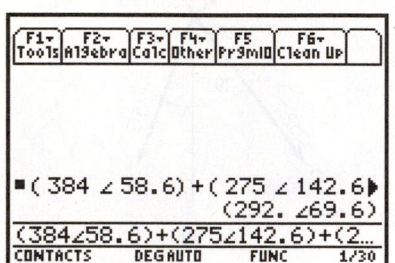

TI-89 screen for the Example 34, part (a).

Common Error	Before adding vectors, make sure their angles are all measured in the same direction from the same axis. We usually measure these angles countereclockwise from the positive x direction, as in the preceding examle.

◆◆◆ **Example 35:** *An Application.* A horizontal cantilever beam braced by two cables supports a load, as shown in Fig. 8–55(a). Find the resultant of the three forces on the end of the beam. (b) Resolve that resultant into horizontal and vertical components.

Solution: (a) By calculator, we add the three force vectors

$$1570\underline{/45.6°}$$
$$1820\underline{/(180 - 29.7)} = 1820\underline{/150.3°}$$
$$2550\underline{/270°}$$

and get a resultant of $714\underline{/-132.5°}$

(b) We resolve this vector into components by calculator and get

Horizontal force $= -482$ lb   Vertical force $= -526$ lb.

Thus the forces on the end of the beam are 482 lb to the left and 526 lb downward, as in Fig. 8.55(b).   ◆◆◆

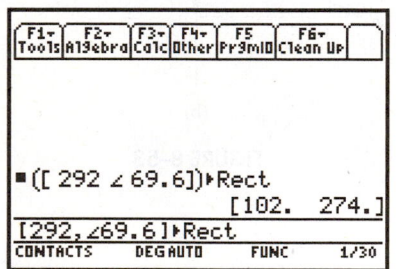

TI-89 screen for Example 34, part (b).

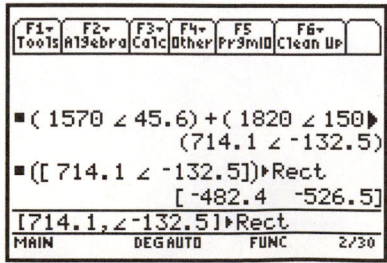

TI-89 screen for Example 35, in Polar and degree modes.

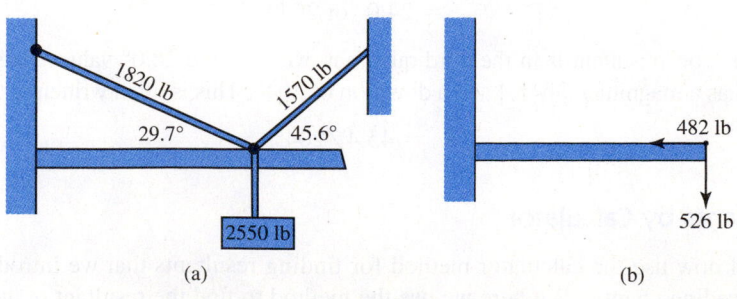

(a)   (b)

FIGURE 8–55

Exercise 6 ◆ Non-Perpendicular Vectors

Resultant of Two Vectors

The magnitudes of vectors **A** and **B** are given in the following table, as well as the angle between the vectors. For each, find the magnitude R of the resultant and the angle that resultant makes with vector **B**.

	Magnitudes		
	A	*B*	Angle
1.	244	287	21.8°
2.	1.85	2.06	136°
3.	55.9	42.3	55.5°
4.	1.006	1.745	148.4°
5.	4483	5829	100.0°
6.	35.2	23.8	146°

Find the resultant of each pair of vectors.

7. $4.83 / 18.3°$ and $5.99 / 83.5°$

8. $13.5 / 29.3°$ and $27.8 / 77.2°$

9. $635 / 22.7°$ and $485 / 48.8°$

10. $83.2 / 49.7°$ and $52.5 / 66.3°$

Resultant of Several Vectors

Find the resultant of each of the following sets of vectors.

11. $273 / 34.0°$, $179 / 143°$, $203 / 225°$, $138 / 314°$

12. $72.5 / 284°$, $28.5 / 331°$, $88.2 / 104°$, $38.9 / 146°$

Force Vectors

13. Two forces of 18.6 N and 21.7 N are applied to a point on a body. The angle between the forces is 44.6°. Find the magnitude of the resultant and the angle that it makes with the larger force.

14. Two forces whose magnitudes are 187 lb and 206 lb act on an object. The angle between the forces is 88.4°. Find the magnitude of the resultant force.

15. A force of 125 N pulls due west on a body, and a second force pulls N 28.7° W. The resultant force is 212 N. Find the second force and the direction of the resultant.

16. Forces of 675 lb and 828 lb act on a body. The smaller force acts due north; the larger force acts N 52.3° E. Find the direction and the magnitude of the resultant.

17. Two forces of 925 N and 1130 N act on an object. Their lines of action make an angle of 67.2° with each other. Find the magnitude and the direction of their resultant.

18. Two forces of 136 lb and 251 lb act on an object with an angle of 53.9° between their lines of action. Find the magnitude of their resultant and its direction.

19. The resultant of two forces of 1120 N and 2210 N is 2870 N. What angle does the resultant make with each of the two forces?

20. Three forces are in equilibrium: 212 N, 325 N, and 408 N. Find the angles between their lines of action.

Recall from Chap.7 that the *heading* of an aircraft is the direction in which the craft is *pointed*. Due to air current, it will usually not travel in that direction, but in an actual path called the *track*. The angle between the heading and the track is the *drift angle*. The *air speed* is the speed relative to the surrounding air, and the *ground speed* is the craft's speed relative to the ground.

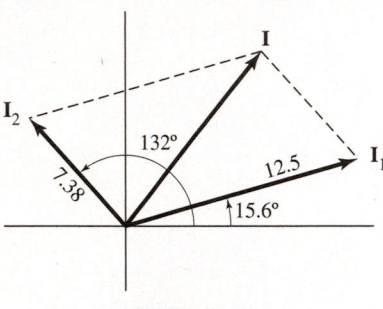

FIGURE 8–56

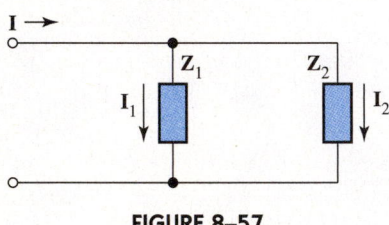

FIGURE 8–57

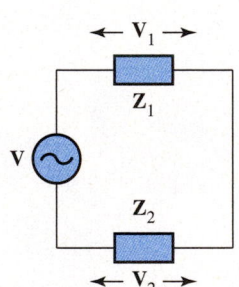

FIGURE 8–58

Velocity Vectors

21. As an airplane heads west with an air speed of 325 mi/h, a wind with a speed of 35.0 mi/h causes the plane to travel slightly south of west with a ground speed of 305 mi/h. In what direction is the wind blowing? In what direction does the plane travel?

22. A boat heads S 15.0° E on a river that flows due west. The boat travels S 11.0° W with a speed of 25.0 km/h. Find the speed of the current and the speed of the boat in still water.

23. A pilot wishes to fly in the direction N 45.0° E. The wind is from the west at 36.0 km/h, and the plane's speed in still air is 388 km/h. Find the heading and the ground speed.

24. The heading of a plane is N 27.7° E, and its air speed is 255 mi/h. If the wind is blowing from the south with a velocity of 42.0 mi/h, find the actual direction of travel of the plane, and its ground speed.

25. A plane flies with a heading of N 48.0° W and an air speed of 584 km/h. It is driven from its course by a wind of 58.0 km/h from S 12.0° E. Find the ground speed and the drift angle of the plane.

Current and Voltage Vectors

26. We will see later that it is possible to represent an alternating current or voltage by a vector whose length is equal to the maximum amplitude of the current or voltage, placed at an angle that we later define as the *phase angle*. Then to add two alternating currents or voltages, we *add the vectors* representing those voltages or currents in the same way that we add force or velocity vectors.

A current I_1 is represented by a vector of magnitude 12.5 A at an angle of 15.6°, and a second current I_2 is represented by a vector of magnitude 7.38 A at an angle of 132°, as shown in Fig. 8–56. Find the magnitude and the direction of the sum of these currents, represented by the vector I.

27. Figure 8–57 shows two impedances in parallel, with the currents in each represented by

$$I_1 = 18.4 \text{ A} \quad \text{at } 51.5°$$

and

$$I_2 = 11.3 \text{ A} \quad \text{at } 0°$$

The current I will be the vector sum of I_1 and I_2. Find the magnitude and the direction of the vector representing I.

28. Figure 8–58 shows two impedances in series, with the voltage drop V_1 equal to 92.4 V at 71.5° and V_2 equal to 44.2 V at −53.8°. Find the magnitude and the direction of the vector representing the total drop V.

••• CHAPTER 8 REVIEW PROBLEMS •••••••••••••••••••••••••••••

Solve oblique triangle *ABC* if

1. $C = 135°$ $a = 44.9$ $b = 39.1$
2. $A = 92.4°$ $a = 129$ $c = 83.6$
3. $B = 38.4°$ $a = 1.84$ $c = 2.06$
4. $B = 22.6°$ $a = 2840$ $b = 1170$
5. $A = 132°$ $b = 38.2$ $c = 51.8$

In what quadrant(s) will the terminal side of θ lie if

6. $\theta = 227°$
7. $\theta = -45°$
8. $\theta = 126°$
9. $\theta = 170°$
10. $\tan \theta$ is negative

Without using book or calculator, state the algebraic sign of

11. tan 275°   **12.** sec (−58°)

13. cos 183°   **14.** cos 45°

15. sin 300°

Write the sin, cos, and tan, to three significant digits, for the angle whose terminal side passes through the given point.

16. (−2, 5)   **17.** (−3, −4)   **18.** (5, −1)

Two vectors of magnitudes A and B are separated by an angle θ. Find the resultant and the angle that the resultant makes with vector **B**.

19. $A = 837$ $B = 527$ $\theta = 58.2°$

20. $A = 2.58$ $B = 4.82$ $\theta = 82.7°$

21. $A = 44.9$ $B = 29.4$ $\theta = 155°$

22. $A = 8374$ $B = 6926$ $\theta = 115.4°$

23. From a ship sailing due north at the rate of 18.0 km/h, the bearing of a lighthouse is N 18°15′ E. Ten minutes later the bearing is N 75°46′ E. How far is the ship from the lighthouse at the time of the second observation?

Write the sin, cos, and tan, to four decimal places, of

24. 273°   **25.** 175°   **26.** 334°36′

27. 127°22′   **28.** 114°

Evaluate each expression to four significant digits.

29. sin 35°cos 35°

30. $\tan^2 68°$

31. $(\cos 14° + \sin 14°)^2$

32. What angle does the slope of a hill make with the horizontal if a vertical tower 18.5 m tall, located on the slope of the hill, is found to subtend an angle of 25.5° from a point 35.0 m directly downhill from the foot of the tower, measured along the slope?

33. Three forces are in equilibrium. One force of 457 lb acts in the direction N 28.0° W. The second force acts N 37.0° E. Find the direction of the third force of magnitude 638 lb. *Hint*: Draw the first two vectors tip-to-tail, and the third to complete the triangle.

Find to the nearest tenth of a degree all nonnegative values of θ less than 360°.

34. $\cos \theta = 0.736$

35. $\tan \theta = -1.16$

36. $\sin \theta = 0.774$

Evaluate to the nearest tenth of a degree.

37. arcsin 0.737

38. $\tan^{-1} 4.37$

39. $\cos^{-1} 0.174$

40. A ship wishes to travel in the direction N 38.0° W. The current is from due east at 4.20 mi/h, and the speed of the ship in still water is 18.5 mi/h. Find the direction in which the ship should head and the speed of the ship in the actual direction of travel.

41. Two forces of 483 lb and 273 lb act on a body. The angle between the lines of action of these forces is 48.2°. Find the magnitude of the resultant and the angle that it makes with the 483-lb force.

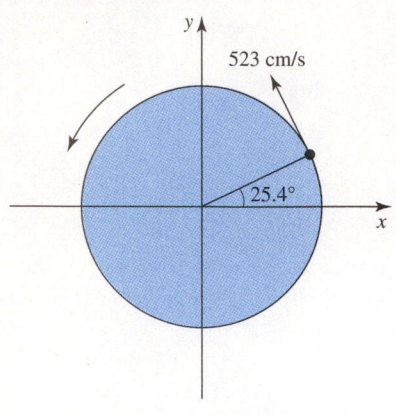

FIGURE 8–59

42. Find the vector sum of two voltages, $\mathbf{V}_1 = 134\underline{/24.5°}$ and $\mathbf{V}_2 = 204\underline{/85.7°}$.

43. A point on a rotating wheel has a tangential velocity of 523 cm/s. Find the x and y components of the velocity when the point is in the position shown in Fig. 8–59.

44. To find the distance from stake P to stake Q on the other side of a river, Fig. 8–60, a surveyor lays out a line PR at an angle of 110° to the line of sight PQ. Angle PRQ is measured at 41.6°. Find distance PQ.

45. Find distance x in the bracket shown in Fig. 8–61.

46. *Writing:* Suppose you had analyzed a complex bridge truss made up of many triangles, sometimes using the law of sines and sometimes the more time-consuming law of cosines. Your client's accountant, angry over the size of your bill, has attacked your report for using the longer law of cosines when it is clear to him that the shorter law of sines is also "good for solving triangles." Write a letter to your client explaining why you sometimes had to use one law and sometimes the other.

47. *Project:* Four mutually tangent circles are shown in Fig. 8–62. Find the radius of the shaded circle.

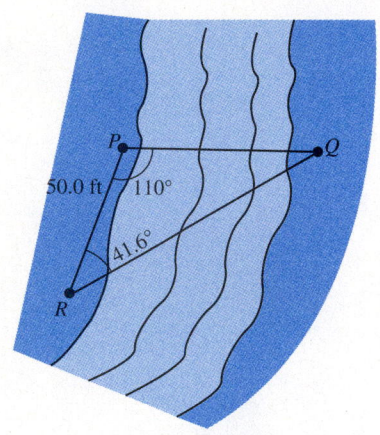

FIGURE 8–60

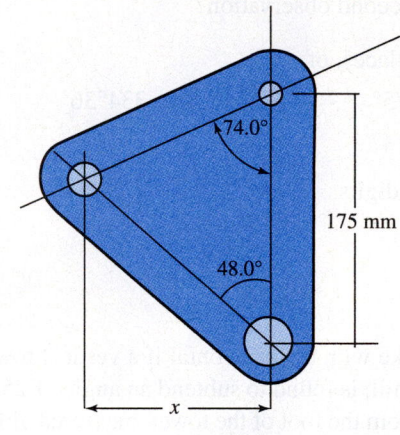

FIGURE 8–61

FIGURE 8–62

9

Systems of Linear Equations

◆◆◆ OBJECTIVES ◆◆

When you have completed this chapter, you should be able to

- Find an approximate graphical solution to a system of two equations.
- Solve a system of two equations in two unknowns by the addition-subtraction method or by substitution.
- Solve a system of two equations by calculator.
- Solve a system of two equations having fractional coefficients or having the unknowns in the denominators.
- Solve a system of three equations by addition-subtraction, by substitution, or by calculator.
- Write a system of equations to describe an application problem and solve those equations.

◆◆◆

Let us leave trigonometry for a while and return to our study of algebra. In an earlier chapter we learned how to solve a linear (first-degree) equation which has one unknown. Here we will show how to solve a set of *two* linear equations in which there are *two* unknowns.

Why? Because some problems in technology can only be described by two equations. For example, to find the two currents I_1 and I_2 in the circuit of Fig. 9–1, we must solve the two equations

$$98.0\, I_1 - 43.0\, I_2 = 5.00$$

$$-43.0\, I_1 + 115.0\, I_2 = 10.0$$

We must find values for I_1 and I_2 that satisfy *both* equations at the same time.

Just as some applications need two equations for their description, others need *three* equations. We also study those in this chapter.

Here we will solve sets of equations using graphical or algebraic techniques, or both, and by calculator. In the next chapter we will learn how to solve systems of any number of equations using determinants.

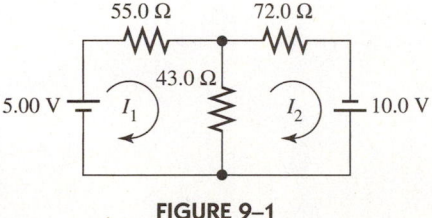

FIGURE 9–1

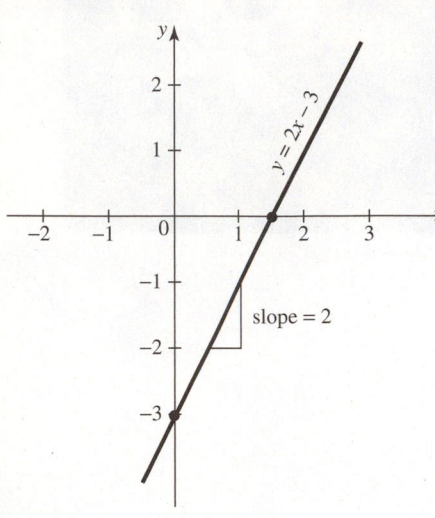

FIGURE 9–2

9–1   Systems of Two Linear Equations

Linear Equations

We have previously defined a *linear equation* as one of *first degree*.

◆◆◆ **Example 1:** The equation

$$3x + 5 = 20$$

is a linear equation in *one unknown*. No term has a degree higher than 1. We learned how to solve these in Chap. 3.   ◆◆◆

◆◆◆ **Example 2:** The equation

$$y = 2x - 3$$

is a linear equation in *two unknowns*. If we graph this equation, we get a *straight line*, as shown in Fig. 9–2. Glance back at Chapter 5 if you've forgotten how to make such a graph.

Systems of Linear Equations

A set of two or more equations is called a *system of equations*. They are also called *simultaneous* equations.   ◆◆◆

◆◆◆ **Example 3:**

(a) $3x - 2y = 5$
     $x + 4y = 1$
     is a system of two linear equations in two unknowns.

(b) $x - 2y + 3z = 4$
     $3x + y - 2z = 1$
     $2x + 3y - z = 3$
     is a system of three linear equations in three unknowns.

(c) $2x - y = 5$
     $x - 2z = 3$
     $3y - z = 1$
     is also a system of three linear equations in three unknowns. Note that some variables may have coefficients of zero and not appear in every equation. This system can also be written

$$2x - 1y + 0z = 5$$
$$1x + 0y - 2z = 3$$
$$0x + 3y - 1z = 1$$

The systems of equations in Example 3 are said to be in *standard form:* all variables in alphabetical order on one side and the constant term on the other side.   ◆◆◆

Solution to a System of Equations

The *solution to a system of equations* is a set of values of the unknowns that will *satisfy every equation* in the system.

◆◆◆ **Example 4:** The system of equations

$$x + y = 5$$
$$x - y = 3$$

is satisfied *only* by the values $x = 4$, $y = 1$, and by *no other* set of values. Thus the ordered pair (4, 1) is the solution to the system. These equations are also said to be *independent* from each other. That means that we can find values that satisfy the first equation, say $x = 3$ and $y = 2$, that do *not* satisfy the second equation. On the other hand, the equations

$$x + y = 5$$
$$2x + 2y = 10$$

are not independent, as the second is obtained from the first simply by multiplying by 2. *Any* values of x and y that satisfy one will satisfy the other. These equations are called *dependent*.

To get a numerical solution for all of the unknowns in a system of linear equations, if one exists, *there must be as many independent equations as there are unknowns*. We will first solve two equations in two unknowns, then later, three equations in three unknowns. But for a solution to be possible, the number of equations must always equal the number of unknowns.

Approximate Graphical Solution to a System of Two Equations

■ Exploration:

Try this. Either by hand or with a graphing calculator,

(a) Graph the straight line: $y = x - 1$.
(b) On the same axes, graph the straight line: $y = 3 - x$.
(c) Find the coordinates of the point where the two lines intersect.

What do you suppose is the significance of the intersection point? Do those coordinates satisfy the first equation? The second? Both? What does this mean? Are there other intersection points? ■

In your exploration, you may have found that the coordinates of the point of intersection, $x = 2$ and $y = 1$, are the only values that satisfy *both* equations. This is exactly what is meant by a *solution* to a system of equations.

The exploration also shows how to find an approximate graphical solution to a system of two equations. Simply graph the two equations and find their point of intersection.

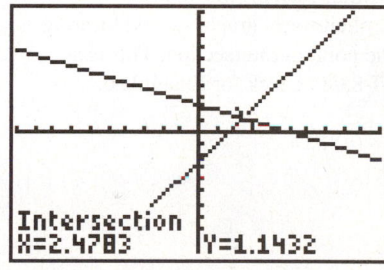

TI-83/84 screen for Example 5.

◆◆◆ **Example 5:** Graphically solve the pair of equations

$$1.53x + 3.35y = 7.62$$
$$2.84x - 1.94y = 4.82$$

Solution: The graphing calculator requires that equations be entered in explicit form. We thus solve each for y, temporarily keeping an extra digit so as to maintain accuracy.

$$y = -0.4567x + 2.275$$
$$y = 1.464x - 2.485$$

We enter these equations into the calculator and get two straight lines that appear to intersect somewhere around (2.5, 1.0).

We can now use TRACE and ZOOM to locate the point of intersection as closely as we wish. In addition, some calculators can automatically locate the point of intersection. On the TI-83/84 this function is called **intersect** and is found on the CALC menu. On the TI-89, press GRAPH, **F5 Math**, and choose **Intersection**. You must indicate the two curves whose intersection point is wanted, enter a guess, and the calculator will find the point of intersection. The screen shows a point of intersection at

$$x = 2.48 \quad \text{and} \quad y = 1.14$$

rounded to as many digits as in the original equations. These values are then the solution to the given system of equations, as can be verified by substituting back. ◆◆◆

Solving a Pair of Linear Equations by the Addition-Subtraction Method

The method of *addition-subtraction*, and the method of *substitution* that follows, both have the object of *eliminating* one of the unknowns.

In the addition-subtraction method, we eliminate one of the unknowns by first (if necessary) multiplying each equation by such numbers that will make the coefficients of one unknown in both equations equal in absolute value. The two equations are then added or subtracted so as to eliminate that variable.

◆◆◆ **Example 6:** Solve by the addition-subtraction method:

$$3x - y = 1$$
$$x + y = 3$$

Solution: Simply adding the two equations causes y to drop out.

$$3x - y = 1$$
$$\underline{x + y = 3}$$
Adding: $4x \quad\;\; = 4$
$$x = 1$$

We now get y by substituting $x = 1$ back into one of the original equations, usually choosing the simplest one for this. Here we choose the second equation, getting

$$1 + y = 3$$
$$y = 2$$

Our solution is then $x = 1$, $y = 2$

Check: We substitute $(1, 2)$ into the first equation and get

$$3(1) - 2 = 1 \qquad \text{Checks}$$

and then into the second equation

$$1 + 2 = 3 \qquad \text{Checks} \qquad\qquad ◆◆◆$$

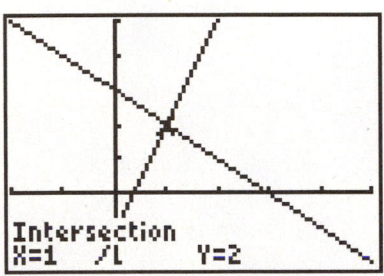

We can check our solutions by writing each given equation in explicit form, graphing, and locating the point of intersection. This is a TI-83/84 check for Example 6.

In the next example we must multiply one equation by a constant before adding.

◆◆◆ **Example 7:** Solve by the addition-subtraction method:

$$2x - 3y = -4$$
$$x + \;\; y = \;\; 3$$

Solution: We multiply the second equation by 3.

$$2x - 3y = -4$$
$$\underline{3x + 3y = \;\;\; 9}$$
Adding: $5x \qquad\;\; = 5$

We have thus reduced our two original equations to a single equation in one unknown. Solving for x gives

$$x = 1$$

Substituting into the second original equation, we have

$$1 + y = 3$$
$$y = 2$$

So the solution is $x = 1$, $y = 2$.

Check: Substituting into the first original equation.

$$2(1) - 3(2) \stackrel{?}{=} -4$$
$$2 - 6 = -4 \text{ (checks)}$$

Also substituting into the second original equation.

$$1 + 2 = 3 \text{ (checks)}$$

Another, graphical, check is shown. ◆◆◆

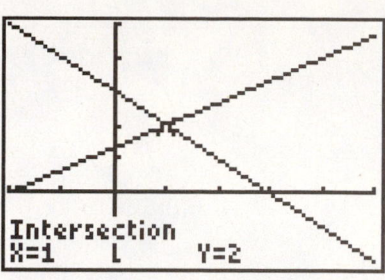

TI-83/84 screen for Example 7.

Often it is necessary to multiply *both* given equations by suitable factors, as shown in the following example.

◆◆◆ Example 8: Solve by addition or subtraction:

$$5x - 3y = 19$$
$$7x + 4y = 2$$

Solution: Multiplying the first equation by 4 and the second by 3,

$$20x - 12y = 76$$
$$\underline{21x + 12y = 6}$$
$$\text{Adding:} \quad 41x \qquad = 82$$
$$x = 2$$

Substituting $x = 2$ into the first given equation gives

$$5(2) - 3y = 19$$
$$-3y = 19 - 10$$
$$3y = -9$$
$$y = -3$$

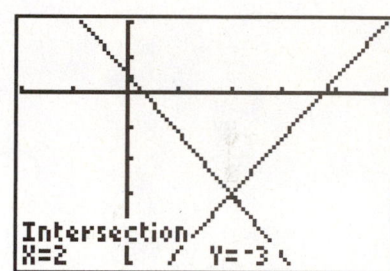

TI-83/84 screen for Example 8.

So the solution is $x = 2$, $y = -3$.

These values check when substituted into each of the original equations (work not shown). Notice that we could have eliminated the x terms by multiplying the first equation by 7 and the second by -5, and adding. The results, of course, would have been the same. ◆◆◆

The coefficients in the preceding examples were integers, but in applications they will usually be approximate numbers. If so, we must retain the proper number of significant digits, as in the following example.

◆◆◆ Example 9: Solve for x and y:

$$2.64x + 8.47y = 3.72$$
$$1.93x + 2.61y = 8.25$$

Solution: Let's choose eliminate y. Since we are less likely to make a mistake by adding rather than subtracting, let us multiply each equation by numbers that will make the y terms have opposite signs and add the resulting equations. So let's multiply the first equation by 2.61 and the second equation by -8.47. We will also carry some extra digits and round our answer at the end.

$$6.89x + 22.11y = 9.71$$
$$\underline{-16.35x - 22.11y = -69.88}$$
$$-9.46x \qquad = -60.17$$
$$x = 6.36$$

Substituting into the first given equation yields

$$2.64(6.36) + 8.47y = 3.72$$
$$8.47y = 3.72 - 16.79 = -13.07$$
$$y = -1.54 \qquad \text{◆◆◆}$$

Another approach is to divide each equation by the coeffcients of its x term, thus making each x coefficient equal to 1. We then subtract one equation from the other.

◆◆◆ Example 10: Solve the equations given in Example 9 by first eliminating the x terms.

Solution: We divide the first equation by 2.64 and the second equation by 1.93.

$$x + 3.208y = 1.409$$
$$x + 1.352y = 4.275$$

Next we subtract the second equation from the first.

$$1.856y = -2.866$$
$$y = -1.54$$

as in the preceding example. The value of x is then found by substituting back into either given equation. ◆◆◆

Substitution Method

The addition-subtraction method works well when both given equations are in the same form. If they are in different forms it's often easier to use the substitution method.

To use the *substitution method* to solve a pair of linear equations, first solve either original equation for one unknown in terms of the other unknown. Then substitute this expression into the other equation, thereby eliminating one unknown.

◆◆◆ Example 11: Solve by substitution:

$$7x - 9y = 1 \qquad (1)$$
$$y = 5x - 17 \qquad (2)$$

Solution: We substitute $(5x - 17)$ for y in the first equation and get

$$7x - 9(5x - 17) = 1$$
$$7x - 45x + 153 = 1$$
$$-38x = -152$$
$$x = 4$$

Substituting $x = 4$ into Eq. 2 gives

$$y = 5(4) - 17$$
$$y = 3$$

So our solution is $x = 4$, $y = 3$. ◆◆◆

Intersection
X=4   L   Y=3

TI-83/84 screen for Example 11.

Systems Having No Solution

Certain systems of equations have no unique solution. If you try to solve either of these types, *both variables will vanish.*

◆◆◆ Example 12: Solve the system

$$2x + 3y = 5 \qquad (1)$$
$$6x + 9y = 2 \qquad (2)$$

Solution: Multiplying the first equation by 3,

$$6x + 9y = 15$$

$$\frac{6x + 9y = 2}{0x + 0y = 13}$$

Subtracting

We have no solution, as there are no values of x and y that when multiplied by zero can give 13.

◆◆◆

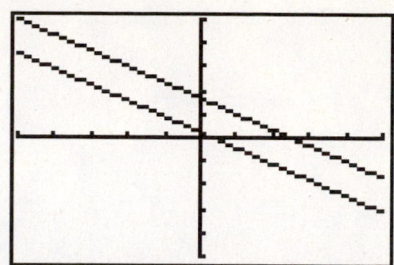

TI-83/84 screen for Example 12. The given equations graph as parallel lines, with no point of intersection.

If both variables vanish and an *inequality* results, as in this example, the system is called *inconsistent*. The equations would plot as two *parallel lines* as shown in the screen. There is no point of intersection and hence no solution.

If both variables vanish and an *equality* results (such as $4 = 4$), the system is called *dependent*. The two equations would plot as a *single line*, indicating that there are infinitely many solutions.

It does not matter much whether a system is inconsistent or dependent; in either case we get no useful solution. But the practical problems that we solve here will always have numerical solutions, so if your variables vanish, go back and check your work.

Solving Sets of Equations Symbolically by Calculator

Some calculators can symbolically solve a system of equations. On the TI-89, for example, we use the **solve** operation from the ALGEBRA menu, as we did for solving a single equation. Now, however, we enter more than one equation, separated by **and**, followed by the variables we wish to solve for.

◆◆◆ **Example 13:** Solve the pair of equations from Example 8 by calculator.

$$5x - 3y = 19$$
$$7x + 4y = 2$$

Solution: We enter

$$\text{solve } (5x - 3y = 19 \text{ and } 7x + 4y = 2, \{x, y\})$$

Pressing ENTER gives us the solution.

◆◆◆

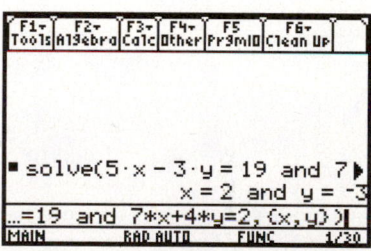

TI-89 screen for Example 13. The **and** instruction is located in the MATH Test menu.

Exercise 1 ◆ Systems of Two Linear Equations

Graphical Solution

Graphically find the approximate solution to each system of equations. If you have a graphics calculator, use the ZOOM and TRACE, or INTERSECT features to find the solution.

1. $2x - y = 5$
$x - 3y = 5$

2. $x + 2y = -7$
$5x - y = 9$

3. $x - 2y = -3$
$3x + y = 5$

4. $4x + y = 8$
$2x - y = 7$

5. $2x + 5y = 4$
$5x - 2y = -3$

6. $x - 2y + 2 = 0$
$3x - 6y + 2 = 0$

Algebraic Solution

Solve each system of equations by addition-subtraction, or by substitution. Check some by graphing.

7. $2x + y = 11$
$3x - y = 4$

8. $x - y = 7$
$3x + y = 5$

9. $4x + 2y = 3$
$-4x + y = 6$

10. $2x - 3y = 5$
$3x + 3y = 10$

11. $3x - 2y = -15$
$5x + 6y = 3$

12. $7x + 6y = 20$
$2x + 5y = 9$

13. $x + 5y = 11$
$3x + 2y = 7$

14. $4x - 5y = -34$
$2x - 3y = -22$

15. $x = 11 - 4y$
$5x - 2y = 11$

16. $2x - 3y = 3$
$4x + 5y = 39$

17. $7x - 4y = 81$
$5x - 3y = 57$

18. $3x + 4y = 85$
$5x + 4y = 107$

19. $3x - 2y = 1$
$2x + y = 10$

20. $5x - 2y = 3$
$2x + 3y = 5$

21. $y = 9 - 3x$
$x = 8 - 2y$

22. $y = 2x - 3$
$x = 19 - 3y$

23. $2x + 3y = 9$
$5x + 4y = 5$

24. $x - 2y = 11$
$y = 5x - 10$

25. $29.1x - 47.6y = 42.8$
$11.5x + 72.7y = 25.8$

26. $4.92x - 8.27y = 2.58$
$6.93x + 2.84y = 8.36$

27. $4n = 18 - 3m$
$m = 8 - 2n$

28. $5p + 4q - 14 = 0$
$17p = 31 + 3q$

29. $3w = 13 + 5z$
$4w - 7z - 17 = 0$

30. $3u = 5 + 2v$
$5v + 2u = 16$

31. $3.62x = 11.7 + 4.73y$
$4.95x - 7.15y - 12.8 = 0$

32. $3.03a = 5.16 + 2.11b$
$5.63b + 2.26a = 18.8$

33. $4.17w = 14.7 - 3.72v$
$v = 8.11 - 2.73w$

34. $5.66p + 4.17q - 16.9 = 0$
$13.7p = 32.2 + 3.61q$

9–2   Applications

Many applications contain two or more unknowns that must be found. To solve such problems, we must write *as many independent equations as there are unknowns*. Otherwise, it is not possible to obtain numerical answers.

Set up these problems as we did in Chap. 3, and solve the resulting system of equations by any of the methods of this chapter.

We give applications from several branches of technology, so you can find problems that apply to your field. However, you may want to try some applications outside your own field. Everything you need to tackle such problems is given in these pages. Having some familiarity with branches of technology other than your own will make you more valuable on the job.

Uniform Motion Applications

Recall from Chap. 3 that motion is called uniform when the speed does not change. These problems can be set up using the simple formula,

$$\boxed{\text{rate} \times \text{time} = \text{distance}}$$

◆◆◆ **Example 14:** A delivery truck is traveling at 40 mi/h. After the truck has a 35-mile head start, a car leaves from the same place traveling at 65 mi/h, to overtake the truck. (a) How long will it take the car to overtake the truck? (b) How far from the starting point will the car overtake the truck?

Solution: Let t = time for the car to overtake the truck, in hours, and d = distance from starting point to where car overtakes truck, in miles.

(a) For each vehicle, rate $\times$ time = distance, so noting that in t hours the truck travels 35 fewer miles than the car, we write;

$$\begin{array}{lll} \text{Truck:} & 40t = d - 35 & (1) \\ \text{Car:} & 65t = d & (2) \end{array}$$

Subtracting Eq. 1 from Eq. 2 we get

$$25t = 35$$
$$t = 1.4 \text{ h}$$

(b) To find the distance we substitute into Eq. 2, getting

$$d = 65(1.4) = 91 \text{ miles}$$

So the car overtakes the truck in 1.4 hours, at a distance of 91 miles from the starting point. ◆◆◆

Money Applications

Money applications usually involve percentage,

$$\boxed{\text{amount} = \text{base} \times \text{rate}}$$

and sometimes the formula for simple interest, given in the following example.

◆◆◆ **Example 15:** A certain investment in bonds had a value of \$248,000 after 4 years, and of \$260,000 after 5 years, at simple interest (Fig. 9–3). Find the amount invested and the interest rate. [The formula $y = P(1 + nt)$ gives the amout y obtained by investing an amount P for t years at an interest rate n.]

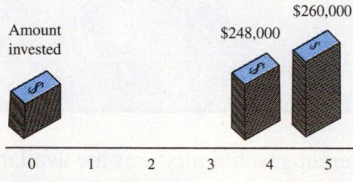

FIGURE 9–3

Solution: For the 4-year investment, t = 4 years and y = \$248,000. Substituting into the given formula yields.

$$\$248,000 = P(1 + 4n)$$

We substitute again, with $t = 5$ years and $y = \$260,000$.

$$\$260,000 = P(1 + 5n)$$

Thus we get two equations in two unknowns: P (the amount invested) and n (the interest rate). Next we remove parentheses.

$$\$248,000 = P + 4\,Pn \tag{1}$$
$$\$260,000 = P + 5\,Pn \tag{2}$$

At this point we may be tempted to subtract one of these equations from the other to eliminate P. But this will not work because P remains in the term containing Pn. Instead, we multiply Eq. 1 by 5 and Eq. 2 by -4 to eliminate the Pn term.

$$
\begin{aligned}
1{,}240{,}000 &= 5P + 20Pn \\
-1{,}040{,}000 &= -4P - 20Pn \\
\hline
\$200{,}000 &= P
\end{aligned}
$$

Adding,

Substituting back into Eq.1 gives us

$$248{,}000 = 200{,}000 + 4(200{,}000)n$$

from which

$$n = 0.06$$

Thus a sum of $\$200,000$ was invested at 6%.

Check: For the 4-year period,

$$y = 200{,}000[1 + 4(0.06)] = \$248{,}000 \quad \text{(checks)}$$

and for the 5-year period,

$$y = 200{,}000[1 + 5(0.06)] = \$260{,}000 \quad \text{(checks)} \qquad ◆◆◆$$

Applications Involving Mixtures

> The basic ideas used for mixture applications were given in Chap. 3. They are,
>
> For a mixture of several ingredients A, B, . . .
>
> total amount of mixture = amount of A + amount of B + . . .
>
> For each ingredient, say ingredient A,
>
> Final amount of A = initial amount of A + amount of A added − amount of A removed
>
> And the percent concentration of any ingredient, say ingredient A,
>
> $$\text{percent concentration of A} = \frac{\text{amount of A}}{\text{amount of mixture}} \times 100$$

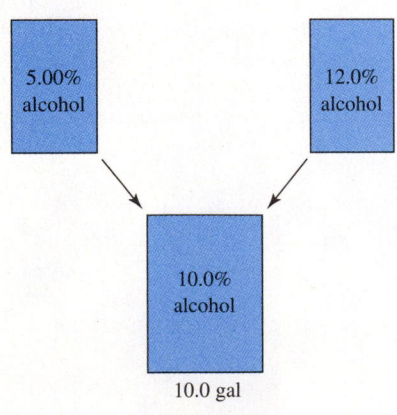

5.00% alcohol

12.0% alcohol

10.0% alcohol

10.0 gal

FIGURE 9–4

◆◆◆ **Example 16:** Two different gasohol mixtures are available, one containing 5.00% alcohol, and the other 12.0% alcohol. How much of each, to the nearest gallon, should be mixed to obtain 10.0 gal of gasohol containing 10.0% alcohol (Fig. 9–4)?

Estimate: Note that using 5 gallons of each original mixture would give the 10 gallons needed, but with a percent alcohol midway between 5% and 12%, or around 8.5%. This is lower than needed, so we reason that we need more than 5 gallons of the 12% mixture and less than 5 gallons of the 5% mixture.

Solution: We let

$$x = \text{gal of 5.00\% gasohol needed}$$
$$y = \text{gal of 12.0\% gasohol needed}$$

So

$$x + y = 10.0$$

and

$$0.0500x + 0.120y = 0.100(10.0) = 1.00$$

are our two equations in two unknowns. Multiplying the first equation by 5 and the second equation by 100, we have

$$5x + 5y = 50.0$$
$$5.00x + 12.0y = 100$$

Subtracting the first equation from the second yields

$$7.0y = 50.0$$
$$y = 7.14 \text{ gal of 12.0\% mixture}$$
$$x = 10.0 - 7.14 = 2.86 \text{ gal of 5.00\% mixture}$$

This agrees with our estimate. ♦♦♦

Statics Applications

The ability to solve systems of equations is of great use in statics when there are more than one unknown forces acting on a body. For these applications we will use the formulas first given in Chap. 3.

> The movement of a force about some point a is the product of the force F and the perpendicular distance d from the force to the point,
>
> $$M_a = Fd$$
>
> and the equations of equilibirium,
>
> The sum of all horizontal forces acting on a body $= 0$
>
> The sum of all vertical forces acting on a body $= 0$
>
> The sum of all moments acting on a body $= 0$

♦♦♦ **Example 17:** Find the forces F_1 and F_2 in Fig. 9–5.

Solution: We resolve each vector into its x and y components and arrange the values in a table, for convenience.

Force	x component	y component
F_1	$-F_1 \cos 41.5° = -0.749F_1$	$F_1 \sin 41.5° = 0.663F_1$
F_2	$F_2 \cos 27.6° = 0.886F_2$	$-F_2 \sin 27.6° = -0.463F_2$
1520 lb	$1520 \cos 51.3° = 950$	$1520 \sin 51.3° = 1186$
2130 lb	$-2130 \cos 25.8° = -1918$	$-2130 \sin 25.8° = -927$

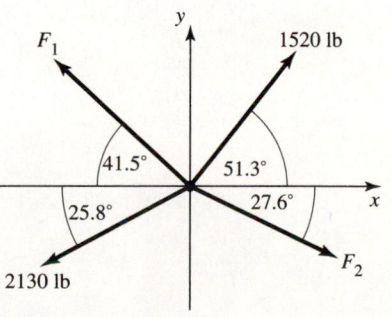

FIGURE 9–5

We then set the sum of the x components to zero,

$$-0.749F_1 + 0.886F_2 + 950 - 1918 = 0$$

or

$$-0.749F_1 + 0.886F_2 = 968 \qquad (1)$$

and set the sum of the y components to zero,

$$0.663F_1 - 0.463F_2 + 1186 - 927 = 0$$

or

$$0.663F_1 - 0.463F_2 = -259 \qquad (2)$$

Now we divide Eq. 1 by 0.749 and divide Eq. 2 by 0.663 and get

$$-F_1 + 1.183F_2 = 1292 \qquad (3)$$
$$\underline{F_1 - 0.698F_2 = -391}$$

Adding:
$$0.485F_2 = 901$$
$$F_2 = 1860 \text{ lb}$$

Substituting back into Eq. 3 gives

$$F_1 = 1.183F_2 - 1292$$
$$= 1.183(1860) - 1292$$
$$= 908 \text{ lb} \qquad ◆◆◆$$

Applications to Work, Fluid Flow, and Energy Flow

Here we repeat the simple ideas from Chap. 3 that we used to set up applications of this sort.

Work:	amount of work done = rate of work × time worked
Fluid flow:	amount of flow = flow rate × duration of flow
Energy flow:	amount of energy transmitted = rate of energy flow × time

◆◆◆ **Example 18:** During a certain day, two computer printers are observed to process 1705 from letters, with the slower printer in use for 5.5 h and the faster for 4.0 h (Fig. 9–6). On another day the slower printer works for 6.0 h and the faster for 6.5 h, and together they print 2330 from letters. How many letters can each print in an hour, working alone?

Estimate: Assume for now that both printers work at the same rate. On the first day they work a total of 9.5 h and print 1705 letters, or $1705 \div 9.5 \approx 180$ letters per hour, and on the second day, $2330 \div 12.5 \approx 186$ letters per hour. But we expect

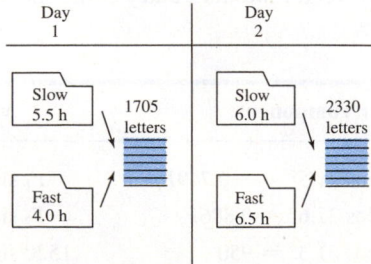

FIGURE 9–6

more from the fast printer, say, around 200 letters/h, and less from the slow printer, say, around 150 letters/h.

Solution: We let

$$x = \text{rate of slow printer, letters/h}$$

$$y = \text{rate of fast printer, letters/h}$$

We write two equations to express, for each day, the total amount of work produced, remembering from Chap. 3 that

$$\text{amount of work} = \text{work rate} \times \text{time}$$

On the first day, the slow printer produces $5.5x$ letters while the fast printer prodces $4.0y$ letters. Together they produce

$$5.5x + 4.0y = 1705$$

Similarly, for the second day,

$$6.0x + 6.5y = 2330$$

Using the addition-subtraction method, we multiply the first equation by 6.5 and the second by -4.0

$$
\begin{aligned}
35.75x + 26y &= 11{,}083 \qquad (3)\\
-24x - 26y &= -9{,}320
\end{aligned}
$$

Add: $\quad 11.75x \qquad\quad = \quad 1{,}763$

$$x = 150 \text{ letters/h}$$

Substituting back gives

$$5.5(150) + 4.0y = 1705$$
$$y = 220 \text{ letters/h}$$

◆◆◆

Electrical Applications

For our final example let us finish the problem posed in the introduction to this chapter. The ideas and formulas needed will be familiar to electrical students, and would be handy for nonelectricals to know as well.

> When a voltage V is applied to a conductor, Fig. 9–7, a current I will flow through that conductor. The current and voltage are related by *Ohms Law*,
>
> $$\text{Current } I = \frac{\text{voltage}}{\text{resistance}} = \frac{V}{R}$$
>
> where R is called the *resistance* of the conductor. The current is given in amperes (A), the voltage in volts (V), and the resistance in ohms (Ω). The other fact that we will need is called *Kirchhoff's voltage law*, Eq.1067, which says,
>
> The sum of the voltage rises and drops around any closed loop is zero.

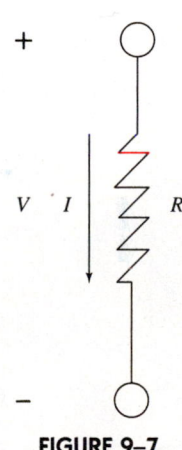

$V \quad I \qquad R$

FIGURE 9–7

◆◆◆ **Example 19:** Find the currents I_1 and I_2 in the two-loop network of Fig. 9–8.

Solution: Before solving the equations given in the introduction, we will briefly show how they were obtained.

According to Kirchhoff's laws the sum of the voltages around each loop must equal zero. That is, the voltage rise due to the battery must equal the sum of the voltage drops through each resistor. For loop 1

$$55.0 I_1 + 43.0 I_1 - 43.0 I_2 = 5.00 \qquad (1)$$

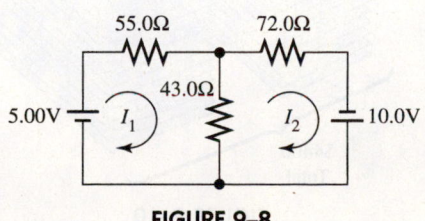

FIGURE 9–8

Note that the 43.0-ohm resistor lies in *both* loops. The current I_2 in that resistor causes a voltage rise in loop 1 that must be included. Similarly, when writing the equation for loop 2 we must take into account the voltage rise in the 43.0-ohm resistor caused by I_1. So for loop 2,

$$72.0\,I_2 + 43.0\,I_2 - 43.0\,I_1 = 10.0 \tag{2}$$

Eqs. 1 and 2 simplify to

$$98.0\,I_1 - 43.0\,I_2 = 5.00 \tag{3}$$
$$-43.0\,I_1 + 115.0\,I_2 = 10.0 \tag{4}$$

We now solve Eqs. 3 and 4 simultaneously. Let us multiply Eq. 3 by 115 and Eq. 4 by 43, and add.

$$
\begin{aligned}
11{,}270\,I_1 - 4945\,I_2 &= 575 \\
-1849\,I_1 + 4945\,I_2 &= 430 \\
\hline
9421\,I_1 \qquad\quad &= 1005 \\
I_1 &= 0.107 \text{ A}
\end{aligned}
$$

Adding:

Substituting back into Eq. 3 gives

$$43.0\,I_2 = 98.0(0.107) - 5.00 = 5.48$$
$$I_2 = 0.128 \text{ A} \qquad \blacklozenge\blacklozenge\blacklozenge$$

Exercise 2 ◆ Applications

Uniform Motion Applications

1. To determine the speed of a boat, it is clocked, with the current, to go a distance of 18.5 miles in 1.31 hours. Returning the same distance against the current took 3.32 h. Find (a) the speed of the boat in still water and (b) the speed of the current.

2. A certain river has a speed of 2.50 mi/h. A rower travels downstream for 1.50 h and returns in 4.50 h. Find his rate in still water, and find the one-way distance traveled.

3. A canoeist can paddle 20.0 mi down a certain river and back in 8.0 h (40.0 mi round trip). She can also paddle 5.0 mi down river in the same time as she paddles 3.0 mi up river. Find her rate in still water, and find the rate of the current.

4. A space shuttle and a damaged satellite are 384 mi apart (Fig. 9–9). The shuttle travels at 836 mi/h, and the satellite at 682 mi/h, in the same direction. How long will it take the shuttle to overtake the satellite, and how far will the shuttle be from its original position when it does catch up to the satellite?

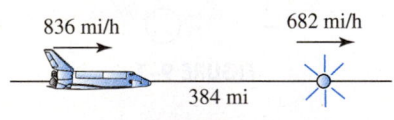

836 mi/h   682 mi/h

384 mi

FIGURE 9–9

Money Applications

5. A certain investment, at simple interest, amounted in 5 years to $3000 and in 6 years to $3100. Find the amount invested, to the nearest dollar, and the rate of interest. Use the simple interest formula, $y = a(1 + nt)$.

6. A shipment of 21 computer keyboards and 33 monitors cost $35,564.25. Another shipment of 41 keyboards 36 monitors cost $49,172.50. Find the cost of each keyboard and each monitor.

7. A person invested $4400, part of it in railroad bonds bearing 6.2% interest and the remainder in state bonds bearing 9.7% interest (Fig. 9–10), and she received the same income from each. How much, to the nearest dollar, was invested in each?

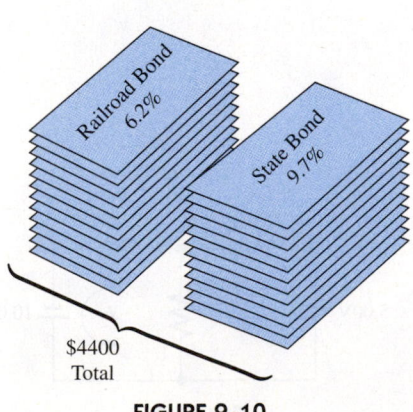

Railroad Bond 6.2%   State Bond 9.7%

$4400 Total

FIGURE 9–10

8. A farmer bought 100 acres of land, part at $370 an acre and part at $450 (Fig. 9–11), paying $42,200 for the 100 acres. How much land was there in each part?

9. If I lend my money at 6% simple interest for given time, I shall receive $720 interest; but if I lend it for 3 years longer, I shall receive $1800. Find the amount of money and the time.

Applications Involving Mixtures

10. A certain brass alloy contains 35% zinc and 3.0% lead (Fig. 9–12). Then x kg of zinc and y kg of lead are added to 200 kg of the original alloy to make a new alloy that is 40% zinc and 4.0% lead.

    (a) Verify that the amount of zinc is given by

    $$0.35(200) + x = 0.40(200 + x + y)$$

    and that the amount of lead is given by

    $$0.03(200) + y = 0.04(200 + x + y)$$

    (b) Solve for x and y.

11. A certain concrete mixture contains 5.00% cement and 8.00% sand (Fig. 9–13). How many pounds of this mixture and how many pounds of sand should be combined with 255 lb of cement to make a batch that is 12.0% cement and 15.0% sand?

12. A distributor has two gasohol blends: one that contains 5.00% alcohol and another with 11.0% alcohol. How many gallons of each must be mixed to make 500 gal of gasohol containing 9.50% alcohol?

13. A potting mixture contains 12.0% peat moss and 6.00% vermiculite. How much peat and how much vermiculite must be added to 100 lb of this mixture to produce a new mixture having 15.0% peat and 15.0% vermiculite?

Statics Applications

14. Find the forces F_1 and F_2 in Fig. 9–14.
15. Find the tensions in the ropes in Fig. 9–15.
16. When the 45.3-kg mass in Fig. 9–16 is increased to 100 kg, the balance point shifts 15.4 cm. Find the length of the bar and the original distance from the balance point to the 45.3-kg mass.

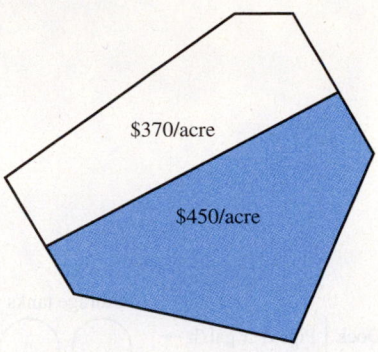

FIGURE 9–11 One hundred acres for $42,200.

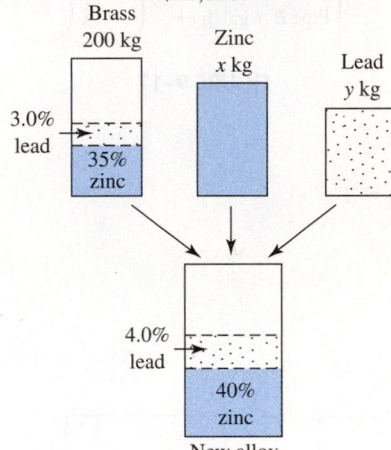

FIGURE 9–12

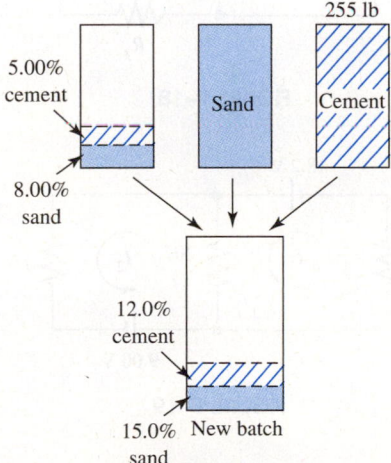

FIGURE 9–13

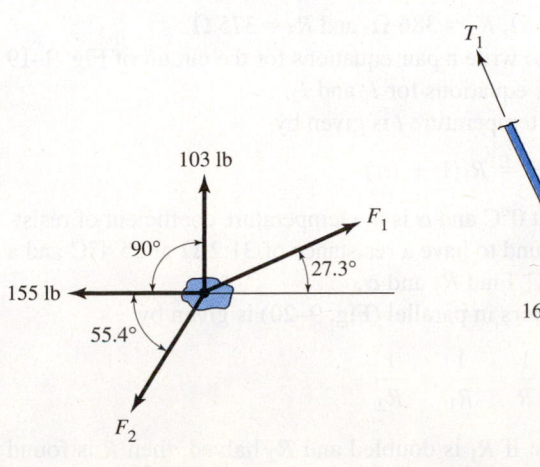

FIGURE 9–14

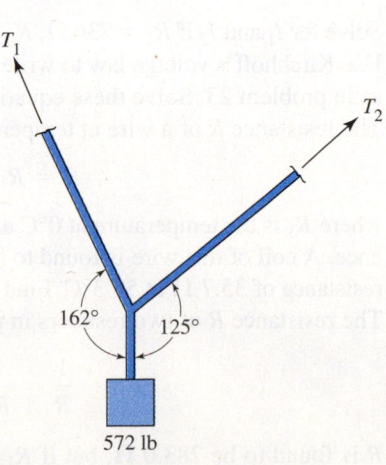

FIGURE 9–15

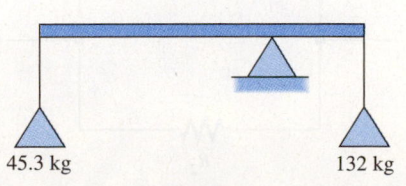

FIGURE 9–16

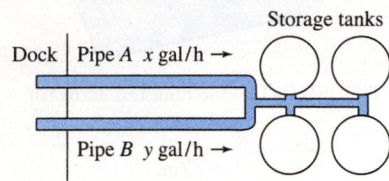

Dock | Pipe A x gal/h →

Pipe B y gal/h →

Storage tanks

FIGURE 9–17

Applications to Work, Fluid Flow, and Energy Flow

17. A carpenter and a helper can do a certain job in 15.0 days. If the carpenter works 1.50 times as fast as the helper, how long would it take each, working alone, to do the job?

18. During one week, two machines produce a total of 27,210 parts, with the faster machine working for 37.5 h and the slower for 28.2 h. During another week, they produce 59,830 parts, with the faster machine working 66.5 h and the slower machine working 88.6 h. How many parts can each, working alone, produce in and hour?

19. Two different-sized pipes lead from a dockside to a group of oil storage tanks (Fig. 9–17). On one day the two pipes are seen to deliver 117,000 gal, with pipe A in use for 3.5 h and pipe B for 4.5 h. On another day the two pipes deliver 151,200 gal, with pipe A opreating for 5.2 h and pipe B for 4.8 h. Assume that pipe A can deliver x gal/h and pipe B can deliver y gal/h.
(a) Verify that the total gallons of oil delivered for the two days are given by the equations

$$3.5x + 4.5y = 117{,}000$$
$$5.2x + 4.8y = 151{,}200$$

(b) Solve for x and y.

20. Working together, two conveyors can fill a certain bin in 6.00 h. If one conveyor works 1.80 times as fast as the other, how long would it take each to fill the bin working alone?

21. A hydroelectric generating plant and a coal-fired generating plant togerther supply a city of 255,000 people, with the hydro plant producing 1.75 times the power of the coal plant. How many people could each service alone?

22. During a certain week, a small wind generator and a small hydro unit together produce 5880 kWh, with the wind generator operating only 85.0% of the time. During another week, the two units produce 6240 kWh, with the wind generator wording 95.0% of the time and hydro unit down 7.50 h for repairs. Assuming that each unit has a constant output when operating, find the number of kilowatts produced by each in 1.00 h.

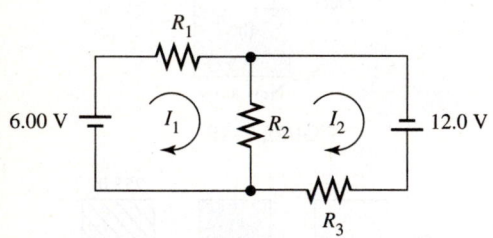

R_1

6.00 V I_1 R_2 I_2 12.0 V

R_3

FIGURE 9–18

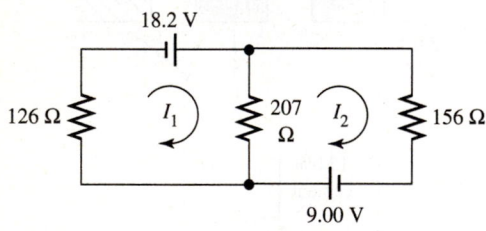

18.2 V

126 Ω I_1 207 Ω I_2 156 Ω

9.00 V

FIGURE 9–19

Electrical Applications

23. To find the currents I_1 and I_2 in Fig. 9–18, we use Kirchhoff's voltage law in each loop and get the following pair of equations:

$$6.00 - R_1 I_1 - R_2 I_1 + R_2 I_2 = 0$$
$$12.0 - R_3 I_2 - R_2 I_2 + R_2 I_1 = 0$$

Solve for I_1 and I_2 if $R_1 = 736$ Ω, $R_2 = 386$ Ω, and $R_3 = 375$ Ω.

24. Use Kirchhoff's voltage law to write a pair equations for the circuit of Fig. 9–19 as in problem 23. Solve these equations for I_1 and I_2.

25. The resistance R of a wire at temperature t is given by

$$R = R_1(1 + \alpha t)$$

where R_1 is the temperature at 0°C and α is the temperature coefficient of resistance. A coil of this wire is found to have a resistance of 31.2 Ω at 25.4°C and a resistance of 35.7 Ω at 57.3°C. Find R_1 and α.

26. The resistance R of two resistors in parallel (Fig. 9–20) is given by

$$\frac{1}{R} = \frac{1}{R_1} + \frac{1}{R_2}$$

R is found to be 283.0 Ω, but if R_1 is doubled and R_2 halved, then R is found to be 291.0 Ω. Find R_1 and R_2.

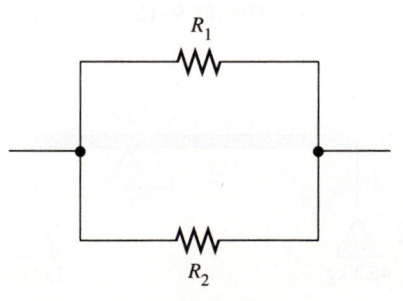

R_1

R_2

FIGURE 9–20

Miscellaneous Applications

27. A surveyor measures the angle of elevation of a hill at 15.8° (Fig. 9–21). She then moves 346 ft closer, on level ground, and measures the angle of elevation at 21.4°. Find the height h of the hill and the distance d. (*Hint:* Write the expression for the tangent of the angle of elevation, at each location, and solve the two resulting equations simultaneously.)

28. A ship traveling a straight course sights a lighthouse at an angle of 32.8° (Fig. 9–22). After the ship sails another 2.78 mi, the lighthouse is at an angle of 77.2°. Find the distances x and y.

29. The arm of an industrial robot starts at a speed v_0 and drops 34.8 cm in 4.28 s, at constant acceleration a. Its motion is described by

$$s = v_0 t + at^2/2$$

or

$$34.8 = 4.28 v_0 + a(4.28)^2/2$$

In another trial, the arm is found to drop 58.3 cm in 5.57 s, with the same initial speed and acceleration. Find v_0 and a.

30. Crates start at the top of an inclined roller ramp with a speed of v_0 (Fig. 9–23). They roll down with constant acceleration, reaching a speed of 16.3 ft/s after 5.58 s. The motion is described by

$$v = v_0 + at$$

or

$$16.3 = v_0 + 5.58a$$

The crates are also seen to reach a speed of 18.5 ft/s after 7.03 s. Find v_0 and a.

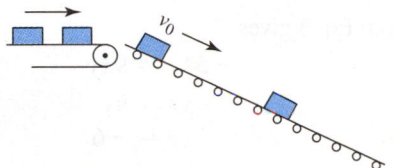

FIGURE 9–23

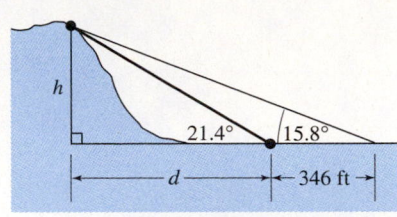

FIGURE 9–21

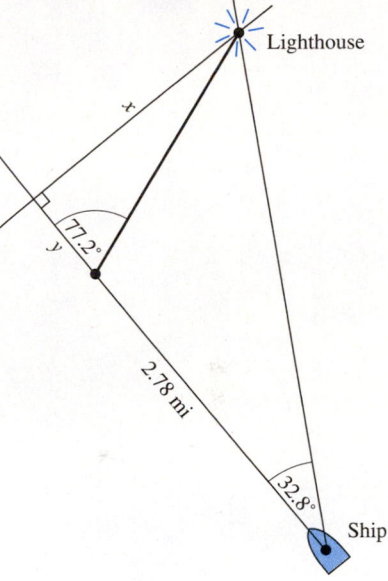

FIGURE 9–22

9–3 Other Systems of Equations

A System with Fractional Coefficients

When an equation in our system has fractional coefficients, simply multiply that entire equation by some number that is divisible by each of the denominators in the equation. Such a number is called a *common denominator*. The smallest such number is called a *least common denominator* (LCD).

◆◆◆ **Example 20:** Solve for x and y:

$$\frac{x}{2} + \frac{y}{3} = \frac{5}{6} \quad (1)$$

$$\frac{x}{4} - \frac{y}{2} = \frac{7}{4} \quad (2)$$

Solution: We "clear" denominators by multiplying Eq. 1 by 6 and Eq. 2 by 4.

$$3x + 2y = 5 \quad (3)$$
$$x - 2y = 7 \quad (4)$$

Adding Eqs. 3 and 4: $4x \quad = 12$
$$x = 3$$

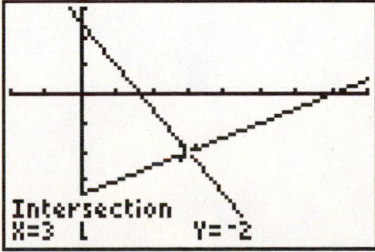

TI-83/84 screen for Example 20.

Having x, we can now substitute back to get y. It is not necessary to substitute back into one of the original equations. We choose instead the easiest place to substitute, such as Eq. 4. (However, when *checking*, be sure to substitute your answers into both of the *original* equations.) Substituting $x = 3$ into Eq. 4 gives

$$3 - 2y = 7$$
$$2y = 3 - 7 = -4$$
$$y = -2 \qquad \text{◆◆◆}$$

Fractional Equations with Unknowns in the Denominator

The same method (multiplying by a common denominator) can be used to clear fractions when the unknowns appear in the denominators. Note that such equations are not linear (that is, not of first degree) as were the equations that we have solved so far, but we are able to solve them by the same methods. Of course, neither x nor y can equal zero in these equations, or we will have division by zero.

◆◆◆ Example 21: Solve the system

$$\frac{10}{x} - \frac{9}{2y} = 6 \qquad (1)$$

$$\frac{20}{3x} + \frac{15}{y} = 1 \qquad (2)$$

Solution: We multiply Eq. 1 by $2xy$ and Eq. 2 by $3xy$.

$$20y - 9x = 12xy \qquad (3)$$
$$20y + 45x = 3xy \qquad (4)$$

Subtracting Eq. 4 from Eq. 3 gives

$$-54x = 9xy$$
$$-54 = 9y$$
$$y = -6$$

Substituting $y = -6$ back into Eq. 3, we have

$$20(-6) - 9x = 12(-6)x$$
$$-120 - 9x = -72x$$
$$63x = 120$$
$$x = \frac{120}{63} = \frac{40}{21}$$

So our solution is $x = \dfrac{40}{21}$, $y = -6$. $\qquad \text{◆◆◆}$

A convenient way to solve nonlinear systems such as these is to *substitute new variables* that will make the equations linear. Solve in the usual way and then substitute back.

◆◆◆ Example 22: Solve the nonlinear system of equations

$$\frac{2}{3x} + \frac{3}{5y} = 17 \qquad (1)$$

$$\frac{3}{4x} + \frac{2}{3y} = 19 \qquad (2)$$

Solution: We substitute $m = 1/x$ and $n = 1/y$ and get the linear system

$$\frac{2m}{3} + \frac{3n}{5} = 17 \qquad (3)$$

$$\frac{3m}{4} + \frac{2n}{3} = 19 \qquad (4)$$

Again, we clear fractions by multiplying each equation by a common denominator. Multiplying Eq. 3 by 15 gives

$$10m + 9n = 255 \tag{5}$$

and multiplying Eq. 4 by 12 gives

$$9m + 8n = 228 \tag{6}$$

Using the addition-subtraction method, we multiply Eq. 5 by 8 and multiply Eq. 6 by -9.

$$
\begin{aligned}
80m + 72n &= 2040 \\
-81m - 72n &= -2052 \\
\hline
\text{Adding:} \quad -m &= -12 \\
m &= 12
\end{aligned}
$$

Substituting $m = 12$ into Eq. 5 yields

$$
\begin{aligned}
120 + 9n &= 255 \\
9n &= 135 \\
n &= 15
\end{aligned}
$$

Finally, we substitute back to get x and y.

$$x = \frac{1}{m} = \frac{1}{12} \quad \text{and} \quad y = \frac{1}{n} = \frac{1}{15} \qquad \text{♦♦♦}$$

Common Error	Students often forget that last step. We are solving not for m and n, but for x and y.

Literal Equations

We use the method of addition-subtraction or the method of substitution for solving systems of equations with literal coefficients, treating the literals as if they were numbers.

♦♦♦ Example 23: Solve for x and y in terms of m and n:

$$2mx + ny = 3 \tag{1}$$
$$mx + 3ny = 2 \tag{2}$$

Solution: We will use the addition-subtraction method. Multiply the second equation by -2.

$$
\begin{aligned}
2mx + ny &= 3 \\
-2mx - 6ny &= -4 \\
\hline
\text{Adding:} \quad -5ny &= -1 \\
y &= \frac{1}{5n}
\end{aligned}
$$

Substituting back into the second original equation, we obtain

$$mx + 3n\left(\frac{1}{5n}\right) = 2$$

$$mx + \frac{3}{5} = 2$$

$$mx = 2 - \frac{3}{5} = \frac{7}{5}$$

$$x = \frac{7}{5m} \qquad \text{♦♦♦}$$

We can use the solution of a system of equations with literal coefficients to derive a *general* formula for the solution of a pair of linear equations.

◆◆◆ **Example 24:** Solve for x and y by the addition-subtraction method:

$$a_1x + b_1y = c_1 \tag{1}$$
$$a_2x + b_2y = c_2 \tag{2}$$

Solution: Multiplying the first equation by b_2 and the second equation by b_1, we obtain

$$a_1b_2x + b_1b_2y = b_2c_1$$
$$a_2b_1x + b_1b_2y = b_1c_2$$

Subtracting: $(a_1b_2 - a_2b_1)x = b_2c_1 - b_1c_2$

Dividing by $a_1b_2 - a_2b_1$ gives

$$x = \frac{b_2c_1 - b_1c_2}{a_1b_2 - a_2b_1}$$

Now solving for y, we multiply the first equation by a_2 and the second equation by a_1. Writing the second equation above the first, we get

$$a_1a_2x + a_1b_2y = a_1c_2$$
$$a_1a_2x + a_2b_1y = a_2c_1$$

Subtracting: $(a_1b_2 - a_2b_1)y = a_1c_2 - a_2c_1$

Dividing by $a_1b_2 - a_2b_1$ yields

$$y = \frac{a_1c_2 - a_2c_1}{a_1b_2 - a_2b_1}$$ ◆◆◆

This result may be summarized as follows:

Two Linear Equations in Two Unknowns **(Cramer's Rule)**	The solution to the set of equations $a_1x + b_1y = c_1$ $a_2x + b_2y = c_2$ is $x = \dfrac{b_2c_1 - b_1c_2}{a_1b_2 - a_2b_1} \qquad y = \dfrac{a_1c_2 - a_2c_1}{a_1b_2 - a_2b_1}$ $(a_1b_2 - a_2b_1 \neq 0)$ **52**

Thus Eq. 52 is a formula for solving a pair of linear equations. We simply have to identify the six numbers $a_1, b_1, c_1, a_2, b_2,$ and c_2 and substitute them into the equations. But be sure to put the equations into standard form first.

◆◆◆ **Example 25:** Solve using Eq. 52:

$$9y = 7x - 15$$
$$-17 + 5x = 8y$$

Solution: Rewriting the equations in the form given in Eq. 52,

$$7x - 9y = 15$$
$$5x - 8y = 17$$

We then substitute, with

$$a_1 = 7 \quad b_1 = -9 \quad c_1 = 15$$
$$a_2 = 5 \quad b_2 = -8 \quad c_2 = 17$$

$$x = \frac{b_2 c_1 - b_1 c_2}{a_1 b_2 - a_2 b_1} = \frac{-8(15) - (-9)(17)}{7(-8) - 5(-9)}$$

$$\frac{33}{-11} = -3$$

$$y = \frac{a_1 c_2 - a_2 c_1}{a_1 b_2 - a_2 b_1} = \frac{7(17) - (5)(15)}{-11}$$

$$= -4$$

◆◆◆

Exercise 3 ◆ Other Systems of Equations

Solve simultaneously. Check some by calculator.

Fractional Coefficients

1. $\dfrac{x}{5} + \dfrac{y}{6} = 18$

$\dfrac{x}{2} - \dfrac{y}{4} = 21$

2. $\dfrac{x}{2} + \dfrac{y}{3} = 7$

$\dfrac{x}{3} + \dfrac{y}{4} = 5$

3. $\dfrac{x}{3} + \dfrac{y}{4} = 8$

$x - y = -3$

4. $\dfrac{x}{2} + \dfrac{y}{3} = 5$

$\dfrac{x}{3} + \dfrac{y}{2} = 5$

5. $\dfrac{3x}{5} + \dfrac{2y}{3} = 17$

$\dfrac{2x}{3} + \dfrac{3y}{4} = 19$

6. $\dfrac{x}{7} + 7y = 251$

$\dfrac{y}{7} + 7x = 299$

7. $\dfrac{m}{2} + \dfrac{n}{3} - 3 = 0$

$\dfrac{n}{2} + \dfrac{m}{5} = \dfrac{23}{10}$

8. $\dfrac{p}{6} - \dfrac{q}{3} + \dfrac{1}{3} = 0$

$\dfrac{2p}{3} - \dfrac{3q}{4} - 1 = 0$

9. $\dfrac{r}{6.20} - \dfrac{s}{4.30} = \dfrac{1}{3.10}$

$\dfrac{r}{4.60} - \dfrac{s}{2.30} = \dfrac{1}{3.50}$

Unknowns in the Denominator

10. $\dfrac{8}{x} + \dfrac{6}{y} = 3$

$\dfrac{6}{x} + \dfrac{15}{y} = 4$

11. $\dfrac{1}{x} + \dfrac{3}{y} = 11$

$\dfrac{5}{x} + \dfrac{4}{y} = 22$

12. $\dfrac{5}{x} + \dfrac{6}{y} = 7$

$\dfrac{7}{x} + \dfrac{9}{y} = 10$

13. $\dfrac{2}{x} + \dfrac{4}{y} = 14$

$\dfrac{6}{x} - \dfrac{2}{y} = 14$

14. $\dfrac{6}{x} + \dfrac{8}{y} = 1$

$\dfrac{7}{x} - \dfrac{11}{y} = -9$

15. $\dfrac{2}{5x} + \dfrac{5}{6y} = 14$

$\dfrac{2}{5x} - \dfrac{3}{4y} = -5$

16. $\dfrac{5}{3a} + \dfrac{2}{5b} = 7$

$\dfrac{7}{6a} - 3 = \dfrac{1}{10b}$

17. $\dfrac{1}{5z} + \dfrac{1}{6w} = 18$

$\dfrac{1}{4w} - \dfrac{1}{2z} + 21 = 0$

18. $\dfrac{8.10}{5.10t} + \dfrac{1.40}{3.60s} = 1.80$

$\dfrac{2.10}{1.40s} - \dfrac{1.30}{5.20t} + 3.10 = 0$

Literal Equations

Solve for x and y in terms of the other literal quantities.

19. $ax + 2by = 1$

$3ax + by = 2$

20. $mx + 3ny = 2$

$2mx + ny = 1$

21. $2px + 3qy = 3$

$3px + 2qy = 4$

22. $7cx + 3dy = 5$

$2cx + 8dy = 6$

23. $3x - 2y = a$

$2x + y = b$

24. $ax + by = r$

$ax + cy = s$

25. $ax - dy = c$

$mx - ny = c$

26. $px - qy + pq = 0$

$2px - 3qy = 0$

9–4   Systems of Three Equations

Our strategy here is to reduce a given system of three equations in three unknowns to a system of two equations in two unknowns, which we already know how to solve.

Addition-Subtraction Method

We take any two of the given equations and, by addition-subtraction or substitution, eliminate one variable, obtaining a single equation in two unknowns. We then take another pair of equations (which must include the one not yet used, as well as one of those already used) and similarly obtain a second equation in the *same two* unknowns. This pair of equations can then be solved simultaneously, and the values obtained substituted back to obtain the third variable.

◆◆◆ **Example 26:** Solve the following:

$$6x - 4y - 7z = 17 \tag{1}$$
$$9x - 7y - 16z = 29 \tag{2}$$
$$10x - 5y - 3z = 23 \tag{3}$$

Solution: It is a good idea to number your equations, as in this example, to help keep track of your work. Let us start by eliminating x from Eqs. 1 and 2.

Multiply Eq. 1 by 3:	$18x - 12y - 21z = 51$	(4)
Multiply Eq. 2 by -2:	$-18x + 14y + 32z = -58$	(5)
Add Eqs. 4 and 5:	$2y + 11z = -7$	(6)

We now eliminate the same variable, x, from Eqs. 1 and 3.

Multiply Eq. 1 by -5:	$-30x + 20y + 35z = -85$	(7)
Multiply Eq. 3 by 3:	$30x - 15y - 9z = 69$	(8)
Add Eqs. 7 and 8:	$5y + 26z = -16$	(9)

Now we solve Eqs. 6 and 9 simultaneously.

Multiply Eq. 6 by 5:	$10y + 55z = -35$	
Multiply Eq. 9 by -2:	$-10y - 52z = 32$	
Add:	$3z = -3$	
	$z = -1$	

Substituting $z = -1$ into Eq. 6 gives us

$$2y + 11(-1) = -7$$
$$y = 2$$

Substituting $y = 2$ and $z = -1$ into Eq. 1 yields

$$6x - 4(2) - 7(-1) = 17$$
$$x = 3$$

Our solution is then $x = 3$, $y = 2$, and $z = -1$.

Check: We check a system of three (or more) equations in the same way that we checked a system of two equations: by substituting back into the original equations.

Substituting into Eq. 1 gives

$$6(3) - 4(2) - 7(-1) = 17$$
$$18 - 8 + 7 = 17 \quad \text{(checks)}$$

Substituting into Eq. 2, we get

$$9(3) - 7(2) - 16(-1) = 29$$
$$27 - 14 + 16 = 29 \quad \text{(checks)}$$

Finally we substitute into Eq. 3.

$$10(3) - 5(2) - 3(-1) = 23$$
$$30 - 10 + 3 = 23 \quad \text{(checks)} \qquad ◆◆◆$$

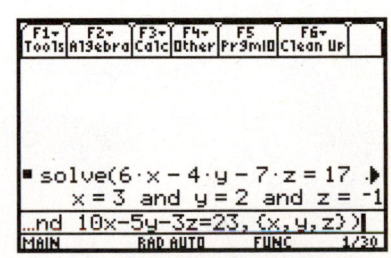

TI-89 screen for Example 26. Here we entered the three equations separated by the **and** instruction from the MATH **Test** menu. This was followed by $\{x, y, z\}$, the variables for which we are solving.

Substitution Method

A *sparse* system (one in which many terms are missing) is often best solved by substitution, as in the following example.

◆◆◆ **Example 27:** Solve by substituting:

$$x - y = 4$$
$$x + z = 8$$
$$x - y + z = 10$$

Solution: From the first two equations, we can write both y and z in terms of x.

$$y = x - 4$$

and

$$z = 8 - x$$

Substituting these back into the third equation yields

$$x - (x - 4) + (8 - x) = 10$$

from which

$$x = 2$$

Substituting back gives

$$y = x - 4 = 2 - 4 = -2$$

and

$$z = 8 - x = 8 - 2 = 6$$

Our solution is then $x = 2$, $y = -2$, and $z = 6$.   ◆◆◆

Fractional Equations

We use the same techniques for solving a set of three fractional equations that we did for two equations:

1. Multiply each equation by a common denominator to eliminate fractions.
2. If the unknowns are in the denominators, substitute new variables that are the reciprocals of the originals

◆◆◆ **Example 28:** Solve for x, y, and z:

$$\frac{4}{x} + \frac{9}{y} - \frac{8}{z} = 3 \tag{1}$$

$$\frac{8}{x} - \frac{6}{y} + \frac{4}{z} = 3 \tag{2}$$

$$\frac{5}{3x} + \frac{7}{2y} - \frac{2}{z} = \frac{3}{2} \tag{3}$$

Solution: We make the substitution

$$p = \frac{1}{x}, \quad q = \frac{1}{y} \quad \text{and} \quad r = \frac{1}{z}$$

and also multiply Eq. 3 by the common denominator, 6, to clear fractions.

$$4p + 9q - 8r = 3 \tag{4}$$
$$8p - 6q + 4r = 3 \tag{5}$$
$$10p + 21q - 12r = 9 \tag{6}$$

We multiply Eq. 5 by 2 and add it to Eq. 4, eliminating the r terms.

$$20p - 3q = 9 \tag{7}$$

Then we multiply Eq. 5 by 3 and add it to Eq. 6.

$$34p + 3q = 18 \tag{8}$$

Adding Eqs. 7 and 8 gives

$$54p = 27$$
$$p = \frac{1}{2}$$

Then substituting back into Eq. 8,

$$3q = 18 - 17 = 1$$
$$q = \frac{1}{3}$$

and from Eq. 5,

$$4r = 3 - 4 + 2 = 1$$
$$r = \frac{1}{4}$$

Returning to our original variables, we have, $x = 2$, $y = 3$, $z = 4$.   ◆◆◆

Literal Equations

The same techniques apply to literal equations, as shown in the following example.

♦♦♦ Example 29: Solve for x, y, and z:

$$x + 3y = 2c \tag{1}$$
$$y + 2z = a \tag{2}$$
$$2z - x = 3 \tag{3}$$

Solution: We first notice that adding Eqs. 1 and 3 will eliminate x.

$$x + 3y = 2c \tag{1}$$
$$-x + 2z = + 3 \tag{3}$$

Adding Eqs. 1 + 3: $\qquad 3y + 2z = 2c + 3 \tag{4}$

From this equation we subtract Eq. 2, eliminating z:

$$3y + 2z = 2c + 3 \tag{4}$$
$$y + 2z = + a \tag{2}$$

Subtracting Eq. 2: $\qquad 2y = 2c + 3 - a$

from which we obtain

$$y = \frac{2c + 3 - a}{2}$$

Substituting back into Eq. 1, we have

$$x = 2c - 3\left(\frac{2c + 3 - a}{2}\right)$$

which simplifies to

$$x = \frac{3a - 2c - 9}{2}$$

Substituting this expression for x back into Eq. 3 gives

$$2z = 3 + \left(\frac{3a - 2c - 9}{2}\right)$$

which simplifies to

$$z = \frac{3a - 2c - 3}{4} \qquad \text{♦♦♦}$$

♦♦♦ Example 30: *An Application.* Find the forces F_1, F_2, and F_3 in Fig. 9–24.

Solution: We show the three unknown forces resolved into x and y components in the figure. Then we apply the three equations of equilibrium.

The sum of the vertical forces must be zero

$$F_{1y} + F_{2y} - F_{3y} - 995 = 0 \tag{1}$$

The sum of the horizontal forces must be zero

$$F_{1x} - F_{2x} - F_{3x} = 0 \tag{2}$$

The sum of the moments about any point, p in this example, must be zero:

$$12.0(995) + 26.0F_{3y} - 46.0F_{2y} = 0 \tag{3}$$

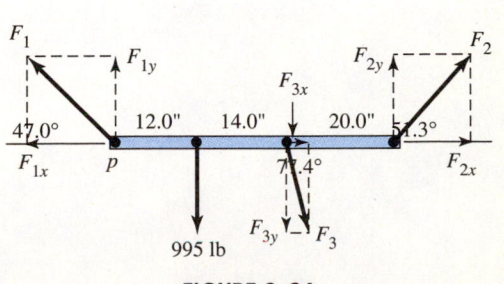

FIGURE 9–24

Using the trigonometric functions, we write these equations in terms of F_1, F_2, and F_3.

$$F_1 \sin 47.0° + F_2 \sin 51.3° - F_3 \sin 77.4° - 995 = 0 \tag{4}$$

$$F_1 \cos 47.0° - F_2 \cos 51.3° - F_3 \cos 77.4° = 0 \tag{5}$$

$$12.0(995) + 26.0\,F_3 \sin 77.4° - 46.0\,F_2 \sin 51.3° = 0 \tag{6}$$

After evaluating the trigonometric functions and simplifying, we have

$$0.7314F_1 + 0.7804F_2 - 0.9759F_3 = 995 \tag{7}$$

$$0.6820F_1 - 0.6252F_2 - 0.2181F_3 = 0 \tag{8}$$

$$-35.90F_2 + 25.37F_3 = -11{,}940 \tag{9}$$

Solving Eq. 9 for F_3 gives

$$F_3 = 1.415F_2 - 470.6 \tag{10}$$

Substituting this into Eqs. 7 and 8 and simplifying, we get

$$F_1 - 0.821F_2 = 732.4 \tag{11}$$

$$F_1 - 1.37F_2 = -150.5$$

Subtracting, we get $0.549F_2 = 883.0$, from which

$$F_2 = 1611 \text{ lb}$$

Substituting back into Eq. 11 gives

$$F_1 = 732.5 + 0.8210(1611) = 2055 \text{ lb}$$

and substituting into Eq. 10 we get

$$F_3 = 1.415(1611) - 470.6 = 1809 \text{ lb} \qquad \text{◆◆◆}$$

Exercise 4 ◆ Systems of Three Equations

Solve each systems of equations by any method.

1. $x + y = 35$
$x + z = 40$
$y + z = 45$

2. $x + y + z = 12$
$x - y = 2$
$x - z = 4$

3. $3x + y = 5$
$2y - 3z = -5$
$x + 2z = 7$

4. $x - y = 5$
$y - z = -6$
$2x - z = 2$

5. $x + y + z = 18$
$x - y + z = 6$
$x + y - z = 4$

6. $x + y + z = 90$
$2x - 3y = -20$
$2x + 3z = 145$

7. $x + 2y + 3z = 14$
$2x + y + 2z = 10$
$3x + 4y - 3z = 2$

8. $x + y + z = 35$
$x - 2y + 3z = 15$
$y - x + z = -5$

9. $x - 2y + 2z = 5$
$5x + 3y + 6z = 57$
$x + 2y + 2z = 21$

10. $1.21x + 1.48y + 1.63z = 6.83$
$4.94x + 4.27y + 3.63z = 21.7$
$2.88x + 4.15y - 2.79z = 2.76$

11. $5a + b - 4c = -5$
$3a - 5b - 6c = -20$
$a - 3b + 8c = -27$

12. $p + 3q - r = 10$
$5p - 2q + 2r = 6$
$3p + 2q + r = 13$

Fractional Equations

13. $x + \dfrac{y}{3} = 5$

$x + \dfrac{z}{3} = 6$

$y + \dfrac{z}{3} = 9$

14. $\dfrac{1}{x} + \dfrac{1}{y} = 5$

$\dfrac{1}{y} + \dfrac{1}{z} = 7$

$\dfrac{1}{x} + \dfrac{1}{z} = 6$

15. $\dfrac{x}{10} + \dfrac{y}{5} + \dfrac{z}{20} = \dfrac{1}{4}$

$x + y + z = 6$

$\dfrac{x}{3} + \dfrac{y}{2} + \dfrac{z}{6} = 1$

16. $\dfrac{1}{x} + \dfrac{2}{y} - \dfrac{1}{z} = -3$

$\dfrac{3}{x} + \dfrac{1}{y} + \dfrac{1}{z} = 4$

$\dfrac{1}{x} - \dfrac{1}{y} + \dfrac{2}{z} = 6$

Literal Equations

Solve for x, y, and z.

17. $x - y = a$

$y + z = 3a$

$5z - x = 2a$

18. $x + a = y + z$

$y + a = 2x + 2z$

$z + a = 3x + 3y$

19. $ax + by = (a + b)c$

$by + cz = (c + a)b$

$ax + cz = (b + c)a$

20. $x + y + 2z = 2(b + c)$

$x + 2y + z = 2(a + c)$

$2x + y + z = 2(a + b)$

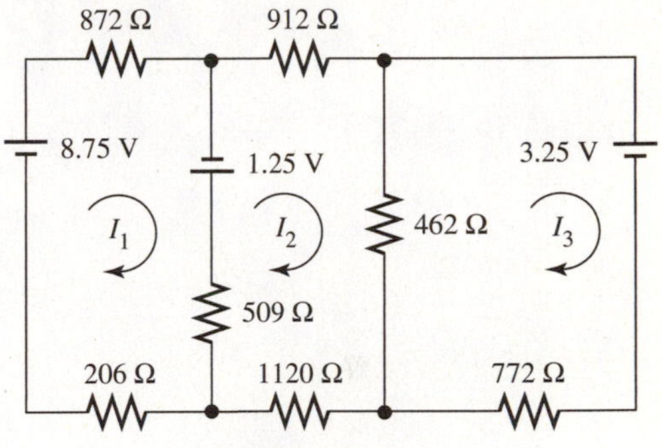

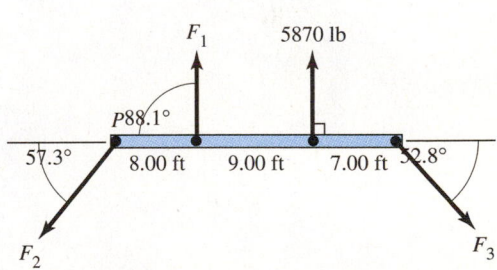

FIGURE 9–25

Electrical Applications

21. When writing Kirchhoff's voltage law for a certain three-loop network, we get the set of equations

$$\begin{aligned} 3I_1 + 2I_2 - 4I_3 &= 4 \\ I_1 - 3I_2 + 2I_3 &= -5 \\ 2I_1 + I_2 - I_3 &= 3 \end{aligned}$$

where I_1, I_2, and I_3 are the loop currents in amperes. Solve for these currents.

22. For the three-loop network of Fig. 9–25:

(a) Use Kirchhoff's law to show that the currents may be found from

$$\begin{aligned} 159I_1 - 50.9I_2 \quad &= 1 \\ 407I_1 - 2400I_2 + 370I_3 &= 1 \\ 142I_2 - 380I_3 &= 1 \end{aligned}$$

(b) Solve this set of equations for the three currents.

FIGURE 9–26

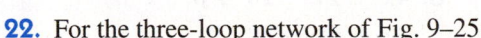

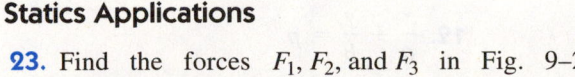

Statics Applications

23. Find the forces F_1, F_2, and F_3 in Fig. 9–26.

24. Find the forces F_1, F_2, and F_3 in Fig. 9–27.

FIGURE 9–27

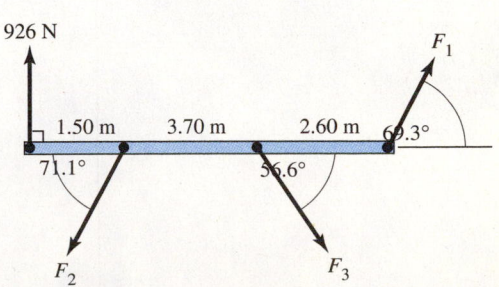

25. *Project*: A certain nickel silver alloy contains the following:

Silver	55.9%	Lead	0.10%
Zinc	31.25%	Nickel	12.00%
Tin	0.50%	Manganese	0.25%

How many pounds of zinc, tin, lead, nickel, and manganese must be added to 400 lb of this alloy to make a new "leaded nickel silver" with the following composition:

Silver	44.50%	Lead	1.00%
Zinc	42.00%	Nickel	10.00%
Tin	0.50%	Manganese	2.00%

26. *Project:* Solve the following set of equations. This will give us formulas for solving a set of three equations, which we will find useful in the next chapter.

$$a_1x + b_1y + c_1z = k_1$$
$$a_2x + b_2y + c_2z = k_2$$
$$a_3x + b_3y + c_3z = k_3$$

••• CHAPTER 9 REVIEW PROBLEMS •••••••••••••••••••••••••••••••••

Solve each system of equations by any method.

1. $4x + 3y = 27$
$2x - 5y = -19$

2. $\dfrac{x}{3} + \dfrac{y}{2} = \dfrac{4}{3}$
$\dfrac{x}{2} + \dfrac{y}{3} = \dfrac{7}{6}$

3. $\dfrac{15}{x} + \dfrac{4}{y} = 1$
$\dfrac{5}{x} - \dfrac{12}{y} = 7$

4. $2x + 4y - 3z = 22$
$4x - 2y + 5z = 18$
$6x + 7y - z = 63$

5. $5x + 3y - 2z = 5$
$3x - 4y + 3z = 13$
$x + 6y - 4z = -8$

6. $3x - 5y - 2z = 14$
$5x - 8y - z = 12$
$x - 3y - 3z = 1$

7. $\dfrac{3}{x+y} + \dfrac{4}{x-z} = 2$
$\dfrac{6}{x+y} + \dfrac{5}{y-z} = 1$
$\dfrac{4}{x-z} + \dfrac{5}{y-z} = 2$

8. $4x + 2y - 26 = 0$
$3x + 4y = 39$

9. $2x - 3y + 14 = 0$
$3x + 2y = 44$

10. $\dfrac{2}{x} + \dfrac{1}{y} = \dfrac{4}{3}$
$\dfrac{3}{x} + \dfrac{5}{y} = \dfrac{19}{6}$

11. $\dfrac{9}{x} + \dfrac{8}{y} = \dfrac{43}{6}$
$\dfrac{3}{x} + \dfrac{10}{y} = \dfrac{29}{6}$

12. $\dfrac{x}{a} + \dfrac{y}{b} = p$
$\dfrac{x}{b} + \dfrac{y}{a} = q$

13. $x + y = a$
$x + z = b$
$y + z = c$

14. $\dfrac{5x}{6} + \dfrac{2y}{5} = 14$

$\dfrac{3x}{4} - \dfrac{2y}{5} = 5$

15. $\dfrac{2x}{7} + \dfrac{2y}{3} = \dfrac{16}{3}$
$x + y = 12$

16. $x + y + z = 35$
$5x + 4y + 3z = 22$
$3x + 4y - 3z = 2$

17. $2x - 4y + 3z = 10$
$3x + y - 2z = 6$
$x - 3y - z = 20$

18. $5x + y - 4z = -5$
$3x - 5y - 6z = -20$
$x - 3y + 8z = -27$

19. $x + 3y - z = 10$
$5x - 2y + 2z = 6$
$3x + 2y + z = 13$

20. $x + 21y = 2$
$2x + 27y = 19$

21. $2x - y = 9$
$5x - 3y = 14$

22. $6x - 2y + 5z = 53$
$5x + 3y + 7z = 33$
$x + y + z = 5$

23. A sum of money was divided between A and B so that A's share was to B's share as 5 is to 3. Also, A's share exceeded $\frac{5}{9}$ of the whole sum by \$50. What was each share?

24. A and B together can do a job in 12 days, and B and C together can do the same job in 16 days. How long would it take them all working together to do the job if A does $1\frac{1}{2}$ times as much as C?

25. If the numerator of a certain fraction is increased by 2 and its denominator diminished by 2, its value will be 1. If the numerator is increased by the denominator and the denominator is diminished by 5, its value will be 5. Find the fraction.

26. If the width of a certain rectangle is increased by 3 and the length decreased by 3, the area is seen to increase by 6 square units [Fig. 9–28(a)]. But if the width is reduced by 5 units and the length increased by 3 units, the area decreases by 90 square units [Fig. 9–28(b)]. Find the original dimensions.

27. If the width of a certain rectangle is increased by 3 units and the length reduced by 4 units, we get a square with the same area as the original rectangle. Find the length and the width of the original rectangle.

28. *Writing:* Suppose that you have a number of pairs of equations to solve in order to find the loop currents in a circuit that you are designing. You must leave for a week but want your assistant to solve the equations in your absence. Write step-by-step instructions to be followed, using the addition-subtraction method, including instructions as to the number of digits to be retained.

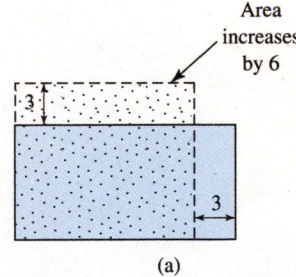

Area increases by 6

(a)

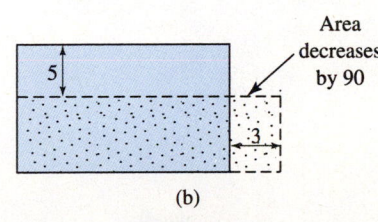

Area decreases by 90

(b)

FIGURE 9–28

10

Matrices and Determinants

◆◆◆ **OBJECTIVES** ◆◆◆

When you have completed this chapter you should be able to

• Identify various types of arrays and matrices.

• Solve a system of equations of any size by calculator using the unit matrix method.

• Evaluate determinants of any size by calculator or by minors.

• Solve a system of equations of any size by determinants, using a calculator or by hand computation.

◆◆

In the preceding chapter we learned how to solve sets of equations graphically and by a few algebraic methods. Here we give other powerful tools—matrices and determinants. We will see that they are more useful than our other methods for systems of more than three equations.

For example, we saw in the preceding chapter that analysis of a two-loop network gave a system of two equations. Similarly, a four-loop network (Fig. 10–1) would result in a system of four equations, perhaps something like these.

$$57.2I_1 + 92.5I_2 - 23.0I_3 - 11.4I_4 = 38.2$$
$$95.3I_1 - 14.9I_2 + 39.0I_3 + 59.9I_4 = 29.3$$
$$66.3I_1 + 81.4I_2 - 91.5I_3 + 33.4I_4 = -73.6$$
$$38.2I_1 - 46.6I_2 + 30.1I_3 + 93.2I_4 = 55.7$$

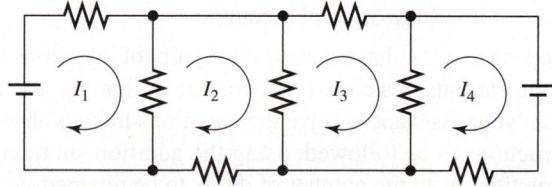

FIGURE 10–1   A four-loop network.

How would you solve this system? We will show how in this chapter.

We will start with some definitions pertaining to matrices, followed by a powerful matrix method for solving systems of equations. We will show how to evaluate the smallest determinant and how it can be used to solve a set of two linear equations. We then expand the method to larger determinants and the solution of systems with any number of equations. We will use the calculator in addition to manual calculation and, as usual, we will include applications.

10–1 Introduction to Matrices

Arrays

A set of numbers, called *elements*, arranged in a pattern, is called an *array*. Arrays are named for the shape of the pattern made by the elements.

◆◆◆ Example 1:

$$\begin{pmatrix} 3 & 6 & 4 \\ 1 & 3 & 6 \\ 9 & 2 & 4 \end{pmatrix} \qquad \begin{pmatrix} 6 & 2 & 8 \\ 3 & 2 & 7 \end{pmatrix} \qquad \begin{pmatrix} 7 & 3 & 9 & 1 \\ & 8 & 3 & 2 \\ & & 9 & 1 \\ & & & 5 \end{pmatrix}$$

(a) *square array* (b) *rectangular array* (c) *triangular array*   ◆◆◆

In everyday language, an array is called a *table*.

Matrices

A *matrix* is a *rectangular* array.

◆◆◆ Example 2: In Example 1, (a) and (b) are matrices. Further, (a) is a *square* matrix. For a square matrix, the diagonal running from the upper-left element to the lower-right element is called the *main diagonal*. In (a), the elements on the main diagonal are 3, 3, and 4. The diagonal from upper right to lower left is called the *secondary* diagonal. In our example, the elements 4, 3, and 9 are on the secondary diagonal.   ◆◆◆

Subscripts

Each element in an array is located in a horizontal *row* and a vertical *column*. We indicate the row and column by means of *subscripts*.

◆◆◆ Example 3: The element a_{25} is located in row 2 and column 5.   ◆◆◆

Thus an element in an array needs *double subscripts* to give its location.

Dimensions

A matrix will have, in general, m rows and n columns. The numbers m and n are the *dimensions* of the matrix, as, for example, a 4×5 matrix. The dimensions are *ordered*, with the number of rows written first, and the number of columns second.

◆◆◆ Example 4: The matrix

$$\begin{pmatrix} 2 & 3 & 5 & 9 \\ 1 & 0 & 8 & 3 \end{pmatrix}$$

has the dimensions 2×4, and the matrix

$$\begin{pmatrix} 5 & 3 & 8 \\ 1 & 0 & 3 \\ 2 & 5 & 2 \\ 1 & 0 & 7 \end{pmatrix}$$

has the dimensions 4×3.   ◆◆◆

Scalars and Vectors

A single number, as opposed to an array of numbers, is called a *scalar*.

◆◆◆ **Example 5:** Some scalars are 5, 693.6, and −24.3.   ◆◆◆

A scalar can also be thought of as an array having just one row and one column.

A *vector* is an array consisting of a single row or a single column. In everyday language a vector is called a *list*.

◆◆◆ **Example 6:** The array $(2, 6, -2, 8)$ is called a *row vector*, and the array

$$\begin{pmatrix} 7 \\ 3 \\ 9 \\ 2 \end{pmatrix}$$

is called a *column vector*.   ◆◆◆

Naming a Matrix

We will often denote or *name* a matrix with a single letter.

◆◆◆ **Example 7:** We can let

$$\mathbf{A} = \begin{pmatrix} 2 & 5 & 1 \\ 0 & 4 & 3 \end{pmatrix}$$

Thus we can represent an entire array by a single symbol.   ◆◆◆

The Unit Matrix and the Null Matrix

A *unit matrix* (or *identity matrix*) is a square matrix having ones along its main diagonal and zeros elsewhere.

◆◆◆ **Example 8:** A 3×3 unit matrix is

$$\begin{pmatrix} 1 & 0 & 0 \\ 0 & 1 & 0 \\ 0 & 0 & 1 \end{pmatrix}$$ ◆◆◆

In the next section we will show how to use the unit matrix to solve a system of equations.

A *null matrix* is one in which every element is zero.

Entering a Matrix into a Calculator

We will later use our calculators to solve equations using square matrices, so now let us see how to enter one. The steps are

- Enter the dimensions of the matrix: the number of rows and of columns.
- Enter each element of the matrix, row by row.

◆◆◆ **Example 10:** Enter the following matrix: into a TI-83/84 calculator

$$\begin{pmatrix} 4 & -2 \\ -3 & 5 \end{pmatrix}$$

Solution: We will show the procedure by a series of TI-83/84 calculator screens.

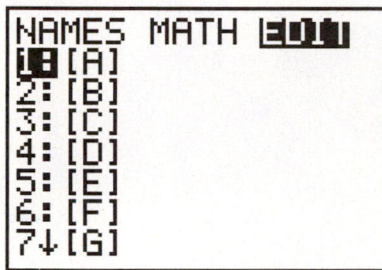

(1) Enter this matrix menu by pressing MATRIX. Select **EDIT**.

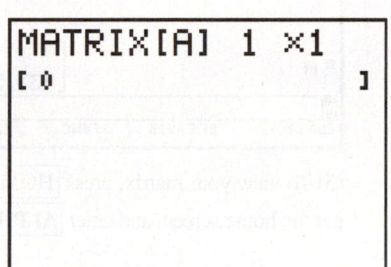

(2) Select a matrix name from the list. We choose [A] and press ENTER. This matrix editing screen appears.

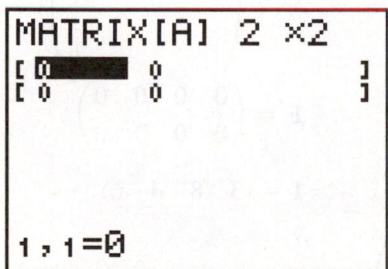

(3) Enter the matrix dimensions by pressing 2 ENTER 2.

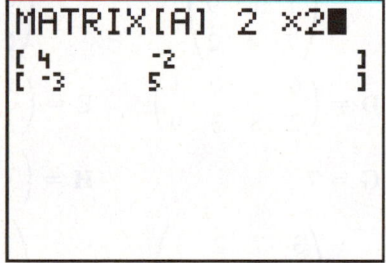

(4) Enter the elements row by row, left to right, pressing ENTER after each entry.

Our given matrix is now stored in the calculator as matrix [A].　　◆◆◆

We will now show how to enter and name a matrix into a TI-89 calculator.

◆◆◆ **Example 11:** Enter the matrix from Example 10 into a TI-89 calculator.

Solution: Again the best way to show the procedure is with a series of screens.

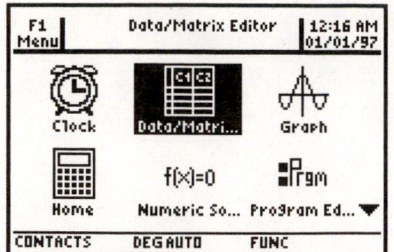

(1) Select **Data/Matrices** on the **APPS** screen.

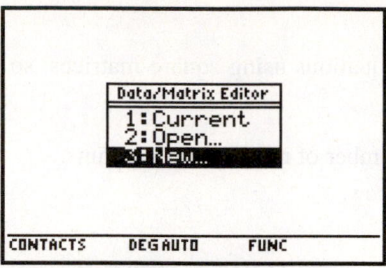

(2) Choose **3: New.**

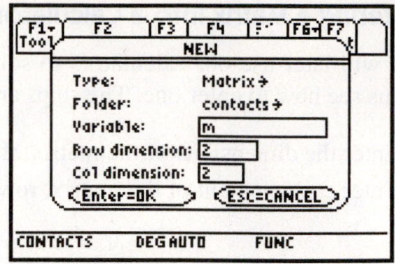

(3) Enter the following information:
Type: **Matrix**
Variable: Name the matrix, say
ALPHA m
Row and column dimensions: **2 and 2**

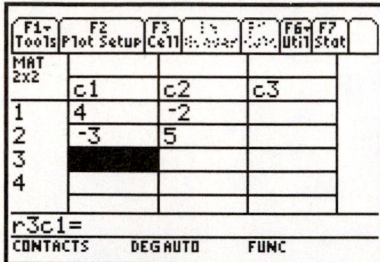

(4) Press ENTER twice to get this screen. Then enter the numerical values in their proper cells.

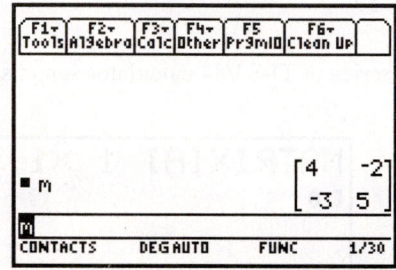

(5) To view your matrix, press HOME to get the home screen, and enter ALPHA m.

◆◆◆

Exercise 1 ◆ Definitions

Given the following arrays:

$$A = \begin{pmatrix} 2 & 5 & 1 \\ 6 & 3 & 7 \\ 1 & 6 & 9 \\ 7 & 4 & 2 \end{pmatrix} \qquad B = \begin{pmatrix} 7 \\ 3 \\ 9 \\ 2 \end{pmatrix} \qquad C = \begin{pmatrix} f & i & q & w \\ & g & w & k \\ & & c & z \\ & & & b \end{pmatrix}$$

$$D = \begin{pmatrix} 6 & 2 & 0 & 1 \\ 2 & 8 & 3 & 9 \end{pmatrix} \qquad E = \begin{pmatrix} x & y \\ z & w \end{pmatrix} \qquad F = \begin{pmatrix} 0 & 0 & 0 & 0 \\ 0 & 0 & 0 & 0 \end{pmatrix}$$

$$G = 7 \qquad H = \begin{pmatrix} 3 & 8 \\ & 5 \end{pmatrix} \qquad I = (3 \quad 8 \quad 4 \quad 6)$$

$$J = \begin{pmatrix} 3 & 7 & 2 & 1 \\ 5 & 2 & 9 & 3 \\ 5 & 1 & 7 & 2 \\ 7 & 3 & 9 & 1 \end{pmatrix} \qquad K = \begin{pmatrix} 1 & 0 & 0 & 0 \\ 0 & 1 & 0 & 0 \\ 0 & 0 & 1 & 0 \\ 0 & 0 & 0 & 1 \end{pmatrix}$$

Which of the 11 arrays shown is

1. a rectangular array?
2. a square array?
3. a triangular array?
4. a column vector?
5. a row vector?
6. a table?
7. a list?
8. a scalar?
9. a null matrix?
10. a unit matrix?

For matrix **A**, find each of the following elements:

11. a_{32} **12.** a_{41}

Give the dimensions of

13. matrix **A**.   **14.** matrix **B**.
15. matrix **D**.   **16.** matrix **E**.

Enter the following matrices into your calculator.

17. $\begin{pmatrix} 8 & 3 \\ 3 & 1 \end{pmatrix}$ **18.** $\begin{pmatrix} 7 & 8 & 3 \\ 5 & 8 & 6 \\ 3 & 0 & 1 \end{pmatrix}$

19. $\begin{pmatrix} 5 & 8 & 2 & 6 \\ 7 & 1 & 8 & 3 \\ 4 & 3 & 0 & 1 \end{pmatrix}$ **20.** $\begin{pmatrix} 4 & 3 & 0 & 1 \\ 5 & 8 & 2 & 6 \\ 3 & 1 & 2 & 7 \\ 7 & 1 & 8 & 3 \end{pmatrix}$

10–2 Solving Systems of Equations by the Unit Matrix Method

This method is perhaps the fastest and easiest way to solve a set of equations by calculator. We will show it for a set of three equations, but it can be used for any number of equations.

Let us solve a system of equations by the addition-subtraction method given in Chap. 9. At the same time we will show a matrix of the coefficients and the column of constants to the right of the equations.

$$\begin{array}{rl} x + 4y + 3z = 1 & \quad (1) \\ 2x + 5y + 4z = 4 & \quad (2) \\ x - 3y - 2z = 5 & \quad (3) \end{array} \qquad \left(\begin{array}{ccc|c} 1 & 4 & 3 & 1 \\ 2 & 5 & 4 & 4 \\ 1 & -3 & -2 & 5 \end{array}\right)$$

First we multiply Eq. 1 by 2 and subtract it from Eq. 2.
We also subtract Eq. 1 from Eq. 3. We get

$$\begin{array}{rl} x + 4y + 3z = 1 & \quad (4) \\ - 3y - 2z = 2 & \quad (5) \\ - 7y - 5z = 4 & \quad (6) \end{array} \qquad \left(\begin{array}{ccc|c} 1 & 4 & 3 & 1 \\ 0 & -3 & -2 & 2 \\ 0 & -7 & -5 & 4 \end{array}\right)$$

Next we multiply Eq. 5 by 4/3 and add the result to Eq. 4.
We also multiply Eq. 5 by $-7/3$ and add the result to Eq. 6.

$$\begin{array}{rl} x + 0 + \frac{1}{3}z = \frac{11}{3} & \quad (7) \\ y + \frac{2}{3}z = -\frac{2}{3} & \quad (8) \\ -\frac{1}{3}z = -\frac{2}{3} & \quad (9) \end{array} \qquad \left(\begin{array}{ccc|c} 1 & 0 & \frac{1}{3} & \frac{11}{3} \\ 0 & 1 & \frac{2}{3} & -\frac{2}{3} \\ 0 & 0 & -\frac{1}{3} & -\frac{2}{3} \end{array}\right)$$

Finally, we multiply Eq. 9 by -3, getting $z = 2$ (Eq. 12). Then we use Eq. 12 to eliminate z from Eqs. 7 and 8.

$$\begin{array}{rl} x + 0 + 0 = 3 & \quad (10) \\ y + 0 = -2 & \quad (11) \\ z = 2 & \quad (12) \end{array} \qquad \left(\begin{array}{ccc|c} 1 & 0 & 0 & 3 \\ 0 & 1 & 0 & -2 \\ 0 & 0 & 1 & 2 \end{array}\right)$$

Our solution is then

$$x = 3; \quad y = -2; \quad z = 2$$

Now notice the final matrix. The square matrix made up of the coefficients of the variables has 1's along the main diagonal and zeros elsewhere. This we had earlier defined as a *unit matrix*. Thus *if we transform our original matrix of the coefficients into a unit matrix, then the column of constants is transformed into the solutions to the equations.*

We can do the transformations by hand, exactly as we did previously, or let the calculator do it, as in the following example.

◆◆◆ **Example 12:** Solve the above set of equations using the TI-83/84 calculator.

Solution: We choose a matrix name, say [A], enter the dimensions, 3 rows and 4 columns, and enter the coefficients and the constants, row by row, from left to right as was shown Example 10. We then select **rref** from the $\boxed{\text{MATRIX}}$ MATH menu. This stands for *reduced row-echelon form*, a name sometimes given to the form we seek. Then enter matrix [A], and press $\boxed{\text{ENTER}}$. We then read the solutions in the rightmost column,

$$x = 3; \; y = -2; \; z = 2$$

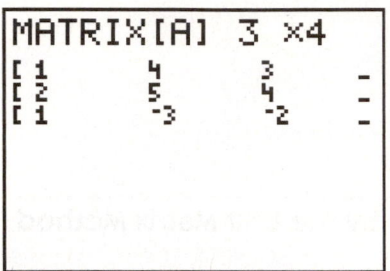

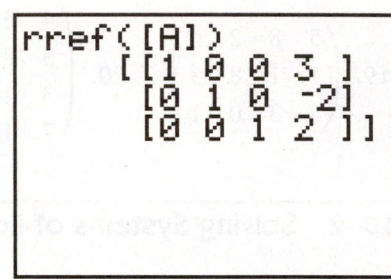

TI-83/84 screen for the given matrix in Example 12. The rightmost column of constants is off the screen.

Screen for Example 12, showing the solutions.   ◆◆◆

◆◆◆ **Example 13:**

Let us now solve the set of four equations given for the four-loop network in the introduction. We select matrix [A], enter the dimensions 4×5, and enter the elements. The entire matrix does not fit on one screen so we show them in the first three TI-83/84 screens.

We choose **rref** from the $\boxed{\text{MATRIX}}$ MATH menu, enter matrix [A], and press $\boxed{\text{ENTER}}$. We then read the solutions in the rightmost column,

$$I_1 = -1.01$$
$$I_2 = 1.69$$
$$I_3 = 2.01$$
$$I_4 = 1.20$$

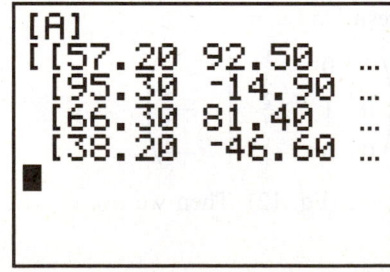

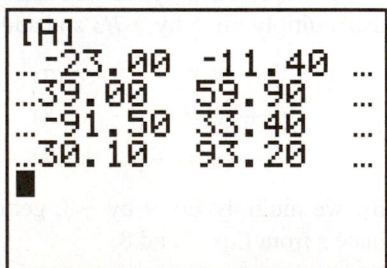

Columns 1 and 2.

Columns 3 and 4.

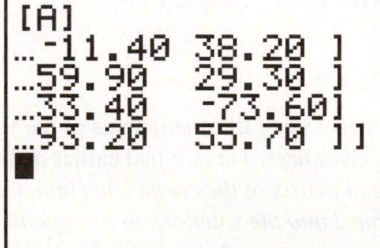

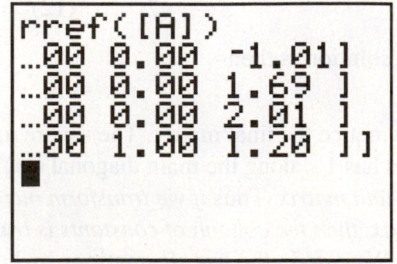

Columns 4 and 5.

Screen for Example 13. Columns 4 and 5, after transforming to a unit matrix.   ◆◆◆

◆◆◆ **Example 14:** Repeat Example 12 with the TI-89 calculator.

Solution: We enter the matrix as was shown in Example 11, and name it **a**. To display the matrix, type ALPHA a on the entry line of the HOME screen. To obtain the unit matrix, enter **rref**, from the MATH Matrix menu, followed by the name of the matrix as shown.

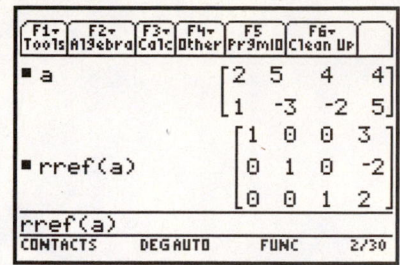

TI-89 screen for Example 14.

Exercise 2 ◆ Solving Systems of Equations by the Unit Matrix Method

Solve each system of equations by calculator using the unit matrix method.

Two Equations in Two Unknowns

1. $2x + y = 11$
$3x - y = 4$

2. $5x + 7y = 101$
$7x - y = 55$

3. $3x - 2y = -15$
$5x + 6y = 3$

4. $7x + 6y = 20$
$2x + 5y = 9$

5. $x + 5y = 11$
$3x + 2y = 7$

6. $4x - 5y = -34$
$2x - 3y = -22$

7. $x + 4y = 11$
$5x - 2y = 11$

8. $2x - 3y = 3$
$4x + 5y = 39$

9. $7x - 4y = 81$
$5x - 3y = 57$

10. $3x + 4y = 85$
$5x + 4y = 107$

11. $3x - 2y = 1$
$2x + y = 10$

12. $5x - 2y = 3$
$2x + 3y = 5$

13. $y = 9 - 3x$
$x = 8 - 2y$

14. $y = 2x - 3$
$x = 19 - 3y$

15. $29.1x - 47.6y = 42.8$
$11.5x + 72.7y = 25.8$

16. $4.92x - 8.27y = 2.58$
$6.93x + 2.84y = 8.36$

17. $4n = 18 - 3m$
$m = 8 - 2n$

18. $5p + 4q - 14 = 0$
$17p = 31 + 3q$

19. $3w = 13 + 5z$
$4w - 7z - 17 = 0$

20. $3u = 5 + 2v$
$5v + 2u = 16$

21. $\dfrac{x}{3} + \dfrac{y}{2} = 1$
$\dfrac{x}{2} - \dfrac{y}{3} = -1$

22. $\dfrac{x}{2} - \dfrac{3y}{4} = 0$
$\dfrac{x}{3} + \dfrac{y}{4} = 0$

Three Equations in Three Unknowns

23. $x + y + z = 18$
$x - y + z = 6$
$x + y - z = 4$

24. $x + y + z = 12$
$x - y = 2$
$x - z = 4$

25. $x + y = 35$
$x + z = 40$
$y + z = 45$

26. $x + y + z = 35$
$x - 2y + 3z = 15$
$y - x + z = -5$

27. $x + 2y + 3z = 14$
$2x + y + 2z = 10$
$3x + 4y - 3z = 2$

28. $x + y + z = 90$
$2x - 3y = -20$
$2x + 3z = 145$

29. $x - 2y + 2z = 5$
$5x + 3y + 6z = 57$
$x + 2y + 2z = 21$

30. $3x + y = 5$
$2y - 3z = -5$
$x + 2z = 7$

31. $2x - 4y + 3z = 10$
$3x + y - 2z = 6$
$x + 3y - z = 20$

32. $x - y = 5$
$y - z = -6$
$2x - z = 2$

33. $1.15x + 1.95y + 2.78z = 15.3$
$2.41x + 1.16y + 3.12z = 9.66$
$3.11x + 3.83y - 2.93z = 2.15$

34. $1.52x - 2.26y + 1.83z = 4.75$
$4.72x + 3.52y + 5.83z = 45.2$
$1.33x + 2.61y + 3.02z = 18.5$

35. $x + \dfrac{y}{3} = 5$

$x + \dfrac{z}{3} = 6$

$y + \dfrac{z}{3} = 9$

36. $\dfrac{x}{10} + \dfrac{y}{5} + \dfrac{z}{20} = \dfrac{1}{4}$

$x + y + z = 6$

$\dfrac{x}{3} + \dfrac{y}{2} + \dfrac{z}{6} = 1$

Four Equations in Four Unknowns

37. $x + y + 2z + w = 18$
$x + 2y + z + w = 17$
$x + y + z + 2w = 19$
$2x + y + z + w = 16$

38. $2x - y - z - w = 0$
$x - 3y + z + w = 0$
$x + y - 4z + w = 0$
$x + y \qquad + w = 36$

39. $3x - 2y - z + w = -3$
$-x - y + 3z + 2w = 23$
$x + 3y - 2z + w = -12$
$2x - y - z - 3w = -22$

40. $x + 2y = 5$
$y + 2z = 8$
$z + 2u = 11$
$u + 2x = 6$

41. $x + y = a + b$
$y + z = b + c$
$z + w = a - b$
$w - x = c - b$

42. $2x - 3y + z - w = -6$
$x + 2y - z \qquad = 8$
$3y + z + 3w = 0$
$3x - y \qquad + w = 0$

Five Equations in Five Unknowns

43. $x + y = 9$
$y + z = 11$
$z + w = 13$
$w + u = 15$
$u + x = 12$

44. $3x + 4y + z \qquad = 35$
$3z + 2y - 3w = 4$
$2x - y + 2w = 17$
$3z - 2w + v = 9$
$w + y = 13$

45. $w + v + x + y = 14$
$w + v + x + z = 15$
$w + v + y + z = 16$
$w + x + y + z = 17$
$v + x + y + z = 18$

Applications

46. A shipment of 4 cars and 2 trucks cost $172,172. Another shipment of 3 cars and 5 trucks cost $209,580. Find the cost of each car and each truck.

47. An airplane and a helicopter are 125 mi apart. The airplane is traveling at 226 mi/h, and the helicopter at 85.0 mi/h, both in the same direction. How long will it take the airplane to overtake the helicopter, and in what distance from the initial position of the airplane?

48. A certain alloy contains 31.0% zinc and 3.50% lead. Then x kg of zinc and y kg of lead are added to 325 kg of the original alloy to make a new alloy that is 45.0% zinc and 4.80% lead. The amount of zinc is given by

$$0.310(325) + x = 0.450(325 + x + y)$$

and the amount of lead is given by

$$0.0350(325) + y = 0.0480(325 + x + y)$$

Solve for x and y.

49. Applying Kirchhoff's law to a certain three-loop network gives

$$283I_1 - 274I_2 + 163I_3 = 352$$
$$428I_1 + 163I_2 + 373I_3 = 169$$
$$338I_1 - 112I_2 - 227I_3 = 825$$

Solve this set of equations for the three currents

50. Find the forces F_1, F_2, and F_3 in Fig. 10–2. Neglect the weight of the beam.

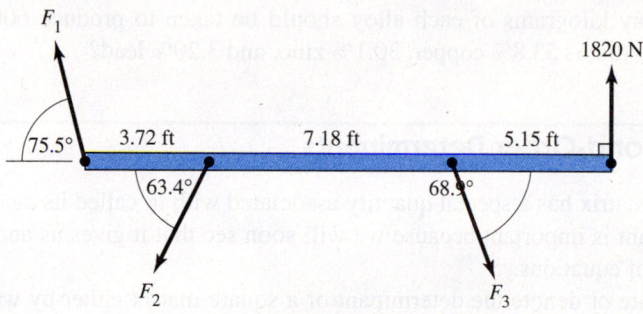

FIGURE 10–2

51. Applying Kirchhoff's law to a certain four-loop network gives the following equations:

$$57.2I_1 + 92.5I_2 - 23.0I_3 - 11.4I_4 = 38.2$$
$$95.3I_1 - 14.9I_2 + 39.0I_3 + 59.9I_4 = 29.3$$
$$66.3I_1 + 81.4I_2 - 91.5I_3 + 33.4I_4 = -73.6$$
$$38.2I_1 - 46.6I_2 + 30.1I_3 + 93.2I_4 = 55.7$$

Solve for the four loop currents by calculator.

52. The following equations result when Kirchhoff's law is applied to a certain four-loop network:

$$4.27I_1 - 5.27I_2 + 4.27I_3 + 9.63I_4 = 6.82$$
$$7.92I_1 + 9.36I_2 - 9.72I_3 + 4.14I_4 = -8.83$$
$$8.36I_1 - 2.27I_2 + 4.77I_3 + 7.33I_4 = 3.93$$
$$7.37I_1 + 9.36I_2 - 3.82I_3 - 2.73I_4 = 5.04$$

Find the four currents by calculator.

53. Four types of computers are made by a company, each requiring the following numbers of hours for four manufacturing steps:

	Assembly	Burn-In	Inspection	Testing
Model *A*	4.50 h	12 h	1.25 h	2.75 h
Model *B*	5.25 h	12 h	1.75 h	3.00 h
Model *C*	6.55 h	24 h	2.25 h	3.75 h
Model *D*	7.75 h	36 h	3.75 h	4.25 h
Available	15,157 h	43,680 h	4824 h	8928 h

Also shown in the table are the monthly production hours available for each step. How many of each type of computer, rounded to the nearest unit, can be made each month?

54. We have available four bronze alloys containing the following percentages of copper, zinc, and lead:

	Alloy 1	Alloy 2	Alloy 3	Alloy 4
Copper	52.0	53.0	54.0	55.0
Zinc	30.0	38.0	20.0	38.0
Lead	3.00	2.00	4.00	3.00

How many kilograms of each alloy should be taken to produce 600 kg of a new alloy that is 53.8% copper, 30.1% zinc, and 3.20% lead?

10–3   Second-Order Determinants

Every square matrix has a special quantity associated with it, called its *determinant*. The determinant is important because we will soon see that it gives us another way to solve a set of equations.

We indicate or denote the determinant of a square matrix either by writing **det** before the matrix, or by a symbol consisting of the elements of the matrix enclosed between vertical bars. Thus

$$\det\begin{pmatrix} 1 & 2 \\ 3 & 4 \end{pmatrix} = \begin{vmatrix} 1 & 2 \\ 3 & 4 \end{vmatrix}$$

The words *element, row, column, principal* and *secondary diagonal* for a determinant have the same meaning as for a square matrix.

◆◆◆ **Example 15:** In the determinant

$$\begin{vmatrix} 2 & 5 \\ 6 & 1 \end{vmatrix}$$

each of the numbers 2, 5, 6, and 1 is called an *element*. There are two *rows*, the first row containing the elements 2 and 5, and the second row having the elements 6 and 1. There are also two *columns*, the first with elements 2 and 6, and the second with elements 5 and 1. The elements along the principal diagonal are 2 and 1, and along the secondary diagonal they are 5 and 6.   ◆◆◆

Value of a Determinant

We find the *value* of a second order determinant by applying the following rule:

> *The value of a second-order determinant is equal to the product of the elements on the principal diagonal, minus the product of the elements on the secondary diagonal.*

or

Second-Order Determinant	$\begin{vmatrix} a_1 & b_1 \\ a_2 & b_2 \end{vmatrix} = a_1b_2 - a_2b_1$	**54**

◆◆◆ **Example 16:** The value of the determinant of Example 15 is

$$\begin{vmatrix} 2 & 5 \\ 6 & 1 \end{vmatrix} = 2(1) - 6(5) = -28 \qquad \text{◆◆◆}$$

It is common practice to use the word *determinant* to refer to the *symbol*, $\begin{vmatrix} a_1 & b_1 \\ a_2 & b_2 \end{vmatrix}$ but also to the *rule* for its evaluation, as well as to the *value* $(a_1b_2 - a_2b_1)$.

Here we will usually refer to the symbol as the determinant, and to $(a_1b_2 - a_2b_1)$ as the value of the determinant.

◆◆◆ **Example 17:** Here are a few more second-order determinants. See if you get the same resultss.

(a) $\begin{vmatrix} 2 & -1 \\ 3 & 5 \end{vmatrix} = 2(5) - 3(-1) = 13$

(b) $\begin{vmatrix} 4 & a \\ x & 3b \end{vmatrix} = 4(3b) - x(a) = 12b - ax$

(c) $\begin{vmatrix} 0 & 3 \\ 5 & 2 \end{vmatrix} = 0(2) - 5(3) = -15$

◆◆◆

Common Error	Students sometimes *add* the numbers on a diagonal instead of *multiplying*. Be careful not to do that.
	Don't just ignore a zero element. It causes the product along its diagonal to be zero.

Determinants by Calculator

In the preceding section we showed how to enter a matrix into a calculator. Now we will find the determinant of a square matrix.

◆◆◆ **Example 18:** Find the determinant of the following matrix by TI-83/84 calculator.

$$\begin{vmatrix} 4 & -2 \\ -3 & 5 \end{vmatrix}$$

Solution: We first enter the matrix as was described in Example 10. Then we perform the steps shown in the following three screens.

```
NAMES MATH EDIT
1▪det(
2: ᵀ
3: dim(
4: Fill(
5: identity(
6: randM(
7↓augment(
```

```
det(■
```

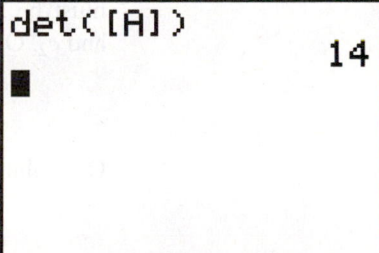

```
det([A])

        14
■
```

TI-83/84 screens for Example 18.
(1) Enter the MATRIX menu and choose MATH.

(2) Select **det**, which stands for determinant.

(3) Return to the MATRIX menu and select the matrix for which we want the determinant. We select [A] and press ENTER.

◆◆◆

The TI-89 can evaluate determinants, even when the elements contain letter values.

◆◆◆ **Example 19:** Evaluate the following determinant using the TI-89 calculator.

$$\begin{vmatrix} x & 3 \\ 2y & w \end{vmatrix}$$

Solution: We enter the matrix as was shown in Example 11, and name it **q**. To display the matrix, type ALPHA q on the entry line of the HOME screen. To obtain the determinant we enter **det** from the MATH Matrix menu, followed by the name of the matrix, as shown.

◆◆◆

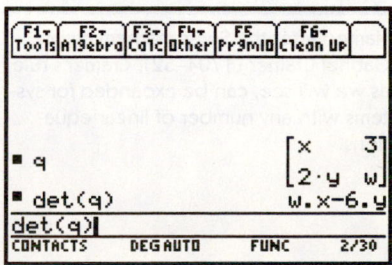

TI-89 screen for Example 19.

Solving a System of Two Linear Equations by Determinants

■ **Exploration:**

Try this. We saw in the preceding chapter that the solution to the set of equations

$$a_1x + b_1y + c_1 = 0$$
$$a_2x + b_2y + c_2 = 0 \tag{1}$$

is

$$x = \frac{b_2c_1 - b_1c_2}{a_1b_2 - a_2b_1} \tag{2}$$

$$y = \frac{a_1c_2 - a_2c_1}{a_1b_2 - a_2b_1} \tag{3}$$

and in this section that the value of a second-order determinant is

$$\begin{vmatrix} a_1 & b_1 \\ a_2 & b_2 \end{vmatrix} = a_1b_2 - a_2b_1 \tag{4}$$

Now compare Eqs. 2, 3, and 4. What similarities do you see? Could you replace any parts of Eqs. 2 and 3 by Eq. 4? ■

You have probably noticed that the denominators in both Eqs. 2 and 3 are identical to the determinant in Eq. 4. This denominator is called the *determinant of the coefficients,* or sometimes the *determinant of the system.* It is usually given the Greek capital letter delta (Δ). If this determinant equals zero, the set of equations has no unique solution.

Thus we can rewrite the solution for x as

$$x = \frac{b_2c_1 - b_1c_2}{\begin{vmatrix} a_1 & b_1 \\ a_2 & b_2 \end{vmatrix}}$$

Now look at the numerator of Eq. 2, which is $b_2c_1 - b_1c_2$. If this expression were the value of some determinant, it is clear that the elements on the principal diagonal must be b_2 and c_1 and that the elements on the secondary diagonal must be b_1 and c_2. One such determinant is

$$\begin{vmatrix} c_1 & b_1 \\ c_2 & b_2 \end{vmatrix} = b_2c_1 - b_1c_2$$

Our solution for x can then be expressed as the quotient of two determinants

$$x = \frac{\begin{vmatrix} c_1 & b_1 \\ c_2 & b_2 \end{vmatrix}}{\begin{vmatrix} a_1 & b_1 \\ a_2 & b_2 \end{vmatrix}}$$

An expression for y can be developed in a similar way. Thus the solution to the system of Eqs. 1, in terms of determinants, is given by the following rule:

Named after the Swiss mathematician Gabriel Cramer (1704–52), Cramer's rule, as we will see, can be expanded for systems with any number of linear equations.

| Cramer's Rule | $x = \dfrac{\begin{vmatrix} c_1 & b_1 \\ c_2 & b_2 \end{vmatrix}}{\begin{vmatrix} a_1 & b_1 \\ a_2 & b_2 \end{vmatrix}}$ and $y = \dfrac{\begin{vmatrix} a_1 & c_1 \\ a_2 & c_2 \end{vmatrix}}{\begin{vmatrix} a_1 & b_1 \\ a_2 & b_2 \end{vmatrix}}$ | **59** |

In the array made from the coeffiecients in the set of Eqs. 1

$$\begin{pmatrix} a_1 & b_1 & c_1 \\ a_2 & b_2 & c_2 \end{pmatrix}$$

We will refer to the values

$$\begin{vmatrix} a_1 \\ a_2 \end{vmatrix}$$ as the column of x coefficients,

$$\begin{vmatrix} b_1 \\ b_2 \end{vmatrix}$$ as the column of y coefficients, and

$$\begin{vmatrix} c_1 \\ c_2 \end{vmatrix}$$ as the column of constants.

Cramer's rule, in words, is then

Cramer's Rule	The solution for each variable is a fraction; its denominator is the determinant of the coefficients Δ; its numerator is also Δ but with its column of coefficients for the variable we seek replaced by the column of constants.

Of course, the denominator Δ cannot be zero, or we get division by zero, indicating that the set of equations has no unique solution.

◆◆◆ Example 20: Solve by determinants:

$$2x - 3y = 1$$
$$x + 4y = -5$$

Solution: The array of the coefficient is

$$\begin{pmatrix} 2 & -3 & 1 \\ 1 & 4 & -5 \end{pmatrix}$$

We first evaluate the determinant Δ, because if it is zero there is no unique solution, and we need not proceed further.

$$\Delta = \begin{vmatrix} 2 & -3 \\ 1 & 4 \end{vmatrix} = 2(4) - 1(-3) = 8 + 3 = 11$$

Solving for x, we have

$$x = \frac{\begin{vmatrix} 1 & -3 \\ -5 & 4 \end{vmatrix}}{\Delta} = \frac{1(4) - (-5)(-3)}{11} = \frac{4 - 15}{11} = -1$$

column of constants

It is easiest to find y by substituting back into one of the preceding equations, but we will use determinants instead to show how it is done.

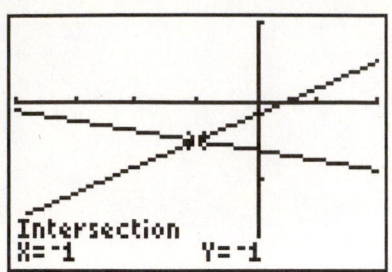

Graphical check for Example 20.

Solving for y yields

$$y = \frac{\begin{vmatrix} 2 & 1 \\ 1 & -5 \end{vmatrix}}{\Delta} = \frac{2(-5) - 1(1)}{11} = \frac{-10 - 1}{11} = -1$$

column of constants

Check: The screen shows a graphical check. The equations are graphed, after being rewritten in explicit form, and show an intersection at $(-1, -1)$. We could, of course, check our answers by substituting back into the original equations.

◆◆◆

Common Error	Don't forget to first arrange the given equations into the form $$a_1 x + b_1 y = c_1$$ $$a_2 x + b_2 y = c_2$$ before writing the determinants.

Calculator Solution Using Determinants

We can have our calculators evaluate the determinants and perform the divisions in one operation, as in the following example.

◆◆◆ **Example 21:** Solve the equations in the preceding example by calculator.

Solution: We first name and enter each of the three arrays as was shown earlier. Let us choose

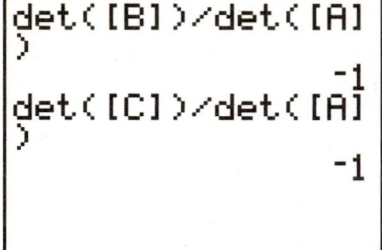

TI-83/84 screen for Example 21.

[A] as the determinant of the coefficients

[B] as the determinant used to find x

[C] as the determinant used to find y

We then find x by dividing the determinant of [B] by the determinant of [A], and find y by dividing the determinant of [C] by the determinant of [A], as shown in the calculator screen.

◆◆◆

Literal Equations

The procedure is no different when our equation contains literal quantities. We will not, of course, get a numerical answer, but one that contains the letter quantities.

◆◆◆ **Example 22:** Solve for x and y by determinants:

$$2ax - by + 3a = 0$$
$$4y = 5x - a$$

Solution: We first rearrange our equations.

$$2ax - by = -3a$$
$$5x - 4y = a$$

The determinant of the coefficients is

$$\Delta = \begin{vmatrix} 2a & -b \\ 5 & -4 \end{vmatrix} = 2a(-4) - 5(-b) = 5b - 8a$$

So

$$x = \frac{\begin{vmatrix} -3a & -b \\ a & -4 \end{vmatrix}}{\Delta} = \frac{(-3a)(-4) - a(-b)}{5b - 8a}$$

$$= \frac{12a + ab}{5b - 8a}$$

and

$$y = \frac{\begin{vmatrix} 2a & -3a \\ 5 & a \end{vmatrix}}{\Delta} = \frac{2a(a) - 5(-3a)}{5b - 8a}$$

$$= \frac{2a^2 + 15a}{5b - 8a}$$

◆◆◆

Exercise 3 ◆ Second-Order Determinants

Value of a Determinant

Find the value of each determinant. Do and/or check some by calculator.

1. $\begin{vmatrix} 3 & 2 \\ 1 & -4 \end{vmatrix}$　　**2.** $\begin{vmatrix} -2 & 5 \\ 3 & -3 \end{vmatrix}$　　**3.** $\begin{vmatrix} 0 & 5 \\ -3 & -4 \end{vmatrix}$

4. $\begin{vmatrix} 8 & 3 \\ -1 & 2 \end{vmatrix}$　　**5.** $\begin{vmatrix} -4 & 5 \\ 7 & -2 \end{vmatrix}$　　**6.** $\begin{vmatrix} 7 & -6 \\ -5 & 4 \end{vmatrix}$

7. $\begin{vmatrix} 4.82 & 2.73 \\ 2.97 & 5.28 \end{vmatrix}$　　**8.** $\begin{vmatrix} 48.7 & -53.6 \\ 4.93 & 9.27 \end{vmatrix}$　　**9.** $\begin{vmatrix} -2/3 & 2/5 \\ -1/3 & 4/5 \end{vmatrix}$

10. $\begin{vmatrix} 1/2 & 1/4 \\ -1/2 & -1/4 \end{vmatrix}$　　**11.** $\begin{vmatrix} a & b \\ c & d \end{vmatrix}$　　**12.** $\begin{vmatrix} 2m & 3n \\ -m & 4n \end{vmatrix}$

Solving a System of Two Linear Equations by Determinants

Solve by determinants.

13. $4x + 2y = 3$
$-4x + y = 6$

14. $2x - 3y = 5$
$3x + 3y = 10$

15. $3x - 2y = -15$
$5x + 6y = 3$

16. $7x + 6y = 20$
$2x + 5y = 9$

17. $x + 5y = 11$
$3x + 2y = 7$

18. $4x - 5y = -34$
$2x - 3y = -22$

19. $x = 11 - 4y$
$5x - 2y = 11$

20. $2x - 3y = 3$
$4x + 5y = 39$

21. $\dfrac{x}{3} + \dfrac{y}{4} = 8$

$x - y = -3$

22. $\dfrac{x}{2} + \dfrac{y}{3} = 5$

$\dfrac{x}{3} + \dfrac{y}{2} = 5$

23. $\dfrac{3x}{5} + \dfrac{2y}{3} = 17$

$\dfrac{2x}{3} + \dfrac{3y}{4} = 19$

24. $\dfrac{x}{7} + 7y = 251$

$\dfrac{y}{7} + 7x = 299$

25. $7x - 4y = 81$
$5x - 3y = 57$

26. $3x + 4y = 85$
$5x + 4y = 107$

27. $3x - 2y = 1$
$2x + y = 10$

28. $5x - 2y = 3$
$2x + 3y = 5$

29. $y = 9 - 3x$
$x = 8 - 2y$

30. $y = 2x - 3$
$x = 19 - 3y$

31. $2x + 3y = 9$
$5x + 4y = 5$

32. $x - 2y = 11$
$y = 5x - 10$

33. $\dfrac{m}{2} + \dfrac{n}{3} - 3 = 0$
$\dfrac{n}{2} + \dfrac{m}{5} = \dfrac{23}{10}$

34. $\dfrac{p}{6} - \dfrac{q}{3} + \dfrac{1}{3} = 0$
$\dfrac{2p}{3} - \dfrac{3q}{4} - 1 = 0$

35. $4n = 18 - 3m$
$m = 8 - 2n$

36. $5p + 4q - 14 = 0$
$17p = 31 + 3q$

37. $3w = 13 + 5z$
$4w - 7z - 17 = 0$

38. $3u = 5 + 2v$
$5v + 2u = 16$

39. $4.17w = 14.7 - 3.72v$
$v = 8.11 - 2.73w$

40. $5.66p + 4.17q - 16.9 = 0$
$13.7p = 32.2 + 3.61q$

Literal Equations

Solve for x and y by determinants.

41. $ax + by = p$
$cx + dy = q$

42. $px - qy + pq = 0$
$2px - 3qy = 0$

43. $2x + by = 3$
$cx + 4y = d$

44. $4mx - 3ny = 6$
$3mx + 2ny = -7$

45. *Writing:* On the surface, a determinant looks completely different from a set of equations. What has one to do with the other? Explain this in your own words in a paragraph or two.

10–4   Higher-Order Determinants

To solve a system of more than two equations by determinants, we must be able to write and evaluate determinants of orders greater than 2. We show how to do that here, first by calculator and then by the use of *minors*. The calculator method is, of course, the fastest and easiest but can only be used with equations that have numerical coefficients. The method of minors will work for literal equations as well as numerical.

Evaluating a Higher-Order Determinant by Calculator

The procedure for evaluating a higher-order determinant by calculator is no different from that for a second-order determinant.

◆◆◆ **Example 23:** Evaluate the determinant

$$\begin{vmatrix} 2 & 1 & 0 \\ 3 & -2 & 1 \\ 5 & 1 & 0 \end{vmatrix}$$

Solution: The calculator screens are as shown.   ◆◆◆

Development by Minors

The following method may be used to evaluate determinants of any size. It is more difficult than using a calculator but can be used for determinants that contain letters.

The *minor of an element in a determinant* is a determinant of next lower order, obtained by deleting the row and the column in which that element lies.

◆◆◆ **Example 24:** We find the minor of element c in the determinant

$$\begin{vmatrix} a & b & c \\ d & e & f \\ g & h & i \end{vmatrix}$$

by striking out the first row and the third column.

$$\begin{vmatrix} a & b & \cancel{c} \\ d & e & f \\ g & h & i \end{vmatrix} \quad \text{or} \quad \begin{vmatrix} d & e \\ g & h \end{vmatrix} \qquad ◆◆◆$$

The *sign factor* for an element depends upon the position of the element in the determinant. To find the sign factor, add the row number and the column number of the element. If the sum is even, the sign factor is $+1$; if the sum is odd, the sign factor is -1.

◆◆◆ **Example 25:**

(a) For the determinant in Example 24, element c is in the first row and the third column. The sum of the row and column numbers

$$1 + 3 = 4$$

is even, so the sign factor for the element c is $+1$.

(b) The sign factor for the element b in Example 24 is -1.   ◆◆◆

The sign factor for the element in the upper left-hand corner of the determinant is $+1$, and the signs alternate according to the pattern

$$\begin{vmatrix} + & - & + & - & + & - & \cdot \\ - & + & - & + & - & \cdot & \cdot \\ + & - & + & - & \cdot & \cdot & \cdot \\ - & + & - & \cdot & \cdot & \cdot & \cdot \\ + & - & \cdot & \cdot & \cdot & \cdot & \cdot \\ - & \cdot & \cdot & \cdot & \cdot & \cdot & \cdot \\ \cdot & \cdot & \cdot & \cdot & \cdot & \cdot & \cdot \end{vmatrix}$$

so that the sign factor may be found simply by *counting off* from the upper-left corner.

A minor with the sign factor attached is called a *signed minor*. Thus

TI-83/84 screens for Example 23.

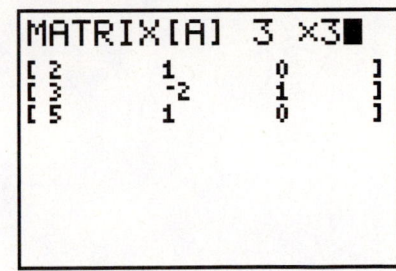

(1) In the MATRIX EDIT screen we enter the matrix dimensions and the nine elements.

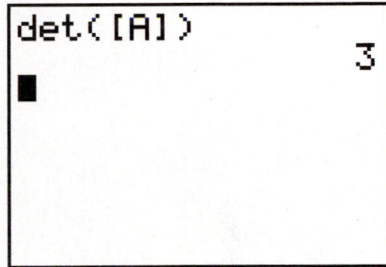

(2) We then choose **det** from the MATRIX MATH menu and select matrix [A]. Press ENTER to get the value, 3, of the determinant.

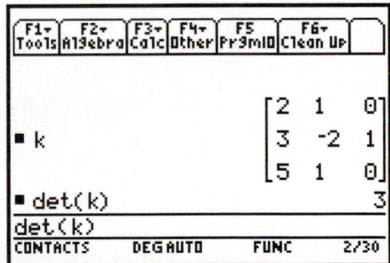

TI-89 screen for Example 23. Here we have named the given determinant k.

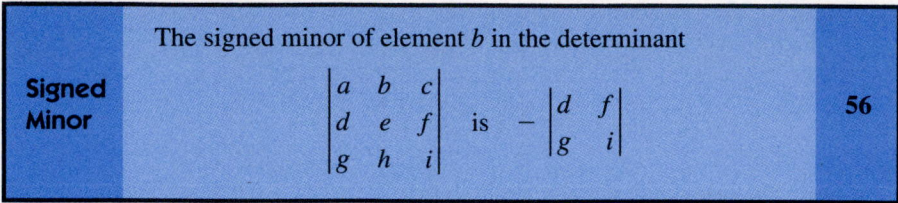

<table>
<tr><td rowspan="1">Signed Minor</td><td>The signed minor of element b in the determinant

$\begin{vmatrix} a & b & c \\ d & e & f \\ g & h & i \end{vmatrix}$ is $- \begin{vmatrix} d & f \\ g & i \end{vmatrix}$</td><td>56</td></tr>
</table>

◆◆◆ **Example 26:** The element in the second row and the first column in the following determinant is 3.

$$\begin{vmatrix} -2 & 5 & 1 \\ 3 & -2 & 4 \\ 1 & -4 & 2 \end{vmatrix}$$

The sign factor of that element is -1 so the signed minor of that element is

$$- \begin{vmatrix} 5 & 1 \\ -4 & 2 \end{vmatrix}$$

◆◆◆

We now define the *value of a determinant* as follows:

Value of a Determinant	To find the value of a determinant, 1. Choose any row or any column to develop by minors. 2. Write the product of every element in that row or column and its signed minor. 3. Add these products to get the value of the determinant.	57

◆◆◆ **Example 27:** Evaluate by minors:

$$\begin{vmatrix} 3 & 2 & 4 \\ 1 & 6 & -2 \\ -1 & 5 & -3 \end{vmatrix}$$

Solution: We first choose a row or a column for development. The work of expansion is greatly reduced if that row or column contains zeros. Our given determinant has no zeros, so let us choose column 1, which at least contains some 1's.

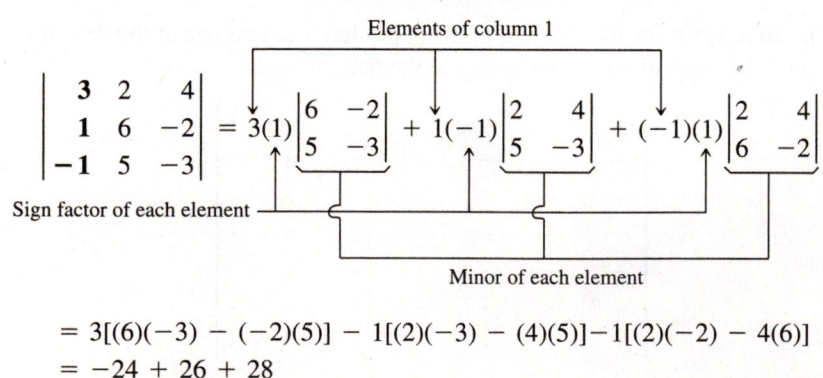

$$= 3[(6)(-3) - (-2)(5)] - 1[(2)(-3) - (4)(5)] - 1[(2)(-2) - 4(6)]$$
$$= -24 + 26 + 28$$
$$= 30$$

Thus, 30 is the value of the given determinant. We would have gotten the same result if we had chosen any other row or column for development.   ◆◆◆

◆◆◆ **Example 28:** Evaluate by minors:

$$\begin{vmatrix} a_1 & b_1 & c_1 \\ a_2 & b_2 & c_2 \\ a_3 & b_3 & c_3 \end{vmatrix}$$

Solution: We can choose any row or column for development and would get the same result for each. Let us choose the first row for development. We get

$$a_1 \begin{vmatrix} b_2 & c_2 \\ b_3 & c_3 \end{vmatrix} - b_1 \begin{vmatrix} a_2 & c_2 \\ a_3 & c_3 \end{vmatrix} + c_1 \begin{vmatrix} a_2 & b_2 \\ a_3 & b_3 \end{vmatrix}$$

$$= a_1(b_2c_3 - b_3c_2) - b_1(a_2c_3 - a_3c_2) + c_1(a_2b_3 - a_3b_2)$$
$$= a_1b_2c_3 - a_1b_3c_2 - a_2b_1c_3 + a_3b_1c_2 + a_2b_3c_1 - a_3b_2c_1$$
$$= a_1b_2c_3 + a_3b_1c_2 + a_2b_3c_1 - a_3b_2c_1 - a_1b_3c_2 - a_2b_1c_3 \qquad ◆◆◆$$

The value of the determinant is thus given by the following formula:

Third-Order Determinant	$\begin{vmatrix} a_1 & b_1 & c_1 \\ a_2 & b_2 & c_2 \\ a_3 & b_3 & c_3 \end{vmatrix} = \begin{aligned} & a_1b_2c_3 + a_3b_1c_2 + a_2b_3c_1 \\ & - a_3b_2c_1 - a_1b_3c_2 - a_2b_1c_3 \end{aligned}$	55

Solving a System of Equations by Determinants

If we were to *algebraically* solve the system of equations

$$a_1x + b_1y + c_1z = k_1$$
$$a_2x + b_2y + c_2z = k_2$$
$$a_3x + b_3y + c_3z = k_3$$

we would get the following solution:

Three Equations in Three Unknowns	$x = \dfrac{b_2c_3k_1 + b_1c_2k_3 + b_3c_1k_2 - b_2c_1k_3 - b_3c_2k_1 - b_1c_3k_2}{a_1b_2c_3 + a_3b_1c_2 + a_2b_3c_1 - a_3b_2c_1 - a_1b_3c_2 - a_2b_1c_3}$ $y = \dfrac{a_1c_3k_2 + a_3c_2k_1 + a_2c_1k_3 - a_3c_1k_2 - a_1c_2k_3 - a_2c_3k_1}{a_1b_2c_3 + a_3b_1c_2 + a_2b_3c_1 - a_3b_2c_1 - a_1b_3c_2 - a_2b_1c_3}$ $z = \dfrac{a_1b_2k_3 + a_3b_1k_2 + a_2b_3k_1 - a_3b_2k_1 - a_1b_3k_2 - a_2b_1k_3}{a_1b_2c_3 + a_3b_1c_2 + a_2b_3c_1 - a_3b_2c_1 - a_1b_3c_2 - a_2b_1c_3}$	53

Notice that the three denominators are identical and are equal to the value of the determinant formed from the coefficients of the unknowns (Eq. 55). As before, we call it the *determinant of the coefficients,* Δ.

Furthermore, we can obtain the numerator of each from the determinant of the coefficients by replacing the coefficients of the variable in question with the constants k_1, k_2, and k_3, as we did with a set of two equations. The solution to our set of equations

$$a_1x + b_1y + c_1z = k_1$$
$$a_2x + b_2y + c_2z = k_2$$
$$a_3x + b_3y + c_3z = k_3$$

is then given by

| Cramer's Rule | $$x = \dfrac{\begin{vmatrix} k_1 & b_1 & c_1 \\ k_2 & b_2 & c_2 \\ k_3 & b_3 & c_3 \end{vmatrix}}{\Delta} \qquad y = \dfrac{\begin{vmatrix} a_1 & k_1 & c_1 \\ a_2 & k_2 & c_2 \\ a_3 & k_3 & c_3 \end{vmatrix}}{\Delta} \qquad z = \dfrac{\begin{vmatrix} a_1 & b_1 & k_1 \\ a_2 & b_2 & k_2 \\ a_3 & b_3 & k_3 \end{vmatrix}}{\Delta}$$ where $$\Delta = \begin{vmatrix} a_1 & b_1 & c_1 \\ a_2 & b_2 & c_2 \\ a_3 & b_3 & c_3 \end{vmatrix} \neq 0$$ | **60** |

Although we do not prove it, Cramer's rule works for higher-order systems as well. We restate it now in words.

| Cramer's Rule | The solution for any variable is a fraction whose denominator is the determinant of the coefficients and whose numerator is the same determinant, except that the column of coefficients for the variable for which we are solving is replaced by the column of constants. | **58** |

♦♦♦ **Example 29:** Solve by determinants:

$$x + 2y + 3z = 14$$
$$2x + y + 2z = 10$$
$$3x + 4y - 3z = 2$$

Solution: We first write the determinant of the coefficients

$$\Delta = \begin{vmatrix} 1 & 2 & 3 \\ 2 & 1 & 2 \\ 3 & 4 & -3 \end{vmatrix}$$

We now make a new determinant by replacing the coefficients of x,

$\begin{vmatrix} 1 \\ 2 \\ 3 \end{vmatrix}$, by the column of constants $\begin{vmatrix} 14 \\ 10 \\ 2 \end{vmatrix}$, getting $\begin{vmatrix} 14 & 2 & 3 \\ 10 & 1 & 2 \\ 2 & 4 & -3 \end{vmatrix}$

Next we replace the coefficients of y with the column of constants and get

$$\begin{vmatrix} 1 & 14 & 3 \\ 2 & 10 & 2 \\ 3 & 2 & -3 \end{vmatrix}$$

We can evaluate these three determinants by minors or by calculator, as shown here. We enter the three matrices as shown earlier. Here [A] is the determinant of the coefficients, [B] is the determinant used for finding x, and [C] is the determinant used for finding y, screens (1) – (3).

Next we use **det** from the MATRIX MATH menu to evaluate each determinant, screen (4), and then calculate x and y.

$$x = \frac{[B]}{[A]} = \frac{28}{28} = 1$$

$$y = \frac{[C]}{[A]} = \frac{56}{28} = 2$$

We can get z also by determinants or, more easily, by substituting back. Substituting $x = 1$ and $y = 2$ into the first equation, we obtain

$$1 + 2(2) + 3z = 14$$
$$3z = 9$$
$$z = 3$$

The solution is thus $(1, 2, 3)$.   ◆◆◆

Often, some terms will have zero coefficients or decimal coefficients. Further, the terms may be out of order, as in the following example. We will show a solution by minors, but a calculator could be used as well.

◆◆◆ **Example 30:** Solve for x, y, and z:

$$23.7y + 72.4x = 82.4 - 11.3x$$
$$25.5x - 28.4z + 19.3 = 48.2y$$
$$13.4 + 66.3z = 39.2x - 10.5$$

Solution: We rewrite each equation in the form $ax + by + cz = k$, combining like terms as we go and putting in the missing terms with zero coefficients. We do this *before* writing the determinant.

$$83.7x + 23.7y + \quad 0z = 82.4 \qquad (1)$$
$$25.5x - 48.2y - 28.4z = -19.3$$
$$-39.2x + \quad 0y + 66.3z = -23.9 \qquad (2)$$

The determinant of the system is then

$$\Delta = \begin{vmatrix} 83.7 & 23.7 & 0 \\ 25.5 & -48.2 & -28.4 \\ -39.2 & 0 & 66.3 \end{vmatrix}$$

Let us develop the first row by minors.

$$\Delta = 83.7 \begin{vmatrix} -48.2 & -28.4 \\ 0 & 66.3 \end{vmatrix} - 23.7 \begin{vmatrix} 25.5 & -28.4 \\ -39.2 & 66.3 \end{vmatrix} + 0$$

$$= 83.7(-48.2)(66.3) - 23.7[25.5(66.3) - (-28.4)(-39.2)]$$

$$= -281,000 \text{ (to three significant digits)}$$

Now solving for x gives us

$$x = \frac{\begin{vmatrix} 82.4 & 23.7 & 0 \\ -19.3 & -48.2 & -28.4 \\ -23.9 & 0 & 66.3 \end{vmatrix}}{\Delta}$$

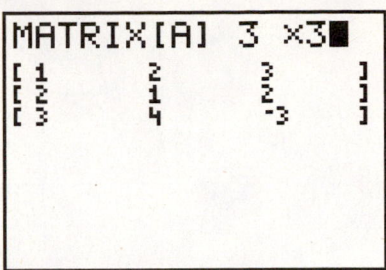

(1)

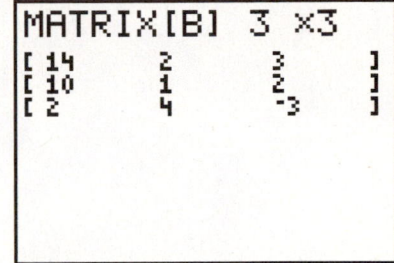

(2)

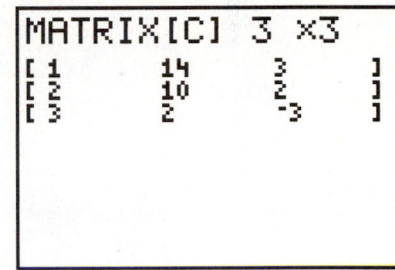

(3)

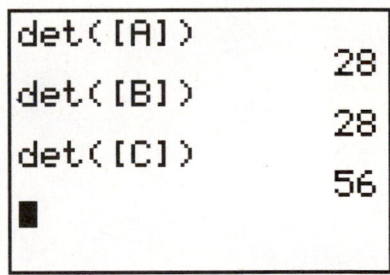

(4)

TI-83/84 screens for Example 29.

Let us develop the first row of the determinant in the numerator by minors. We get

$$82.4 \begin{vmatrix} -48.2 & -28.4 \\ 0 & 66.3 \end{vmatrix} - 23.7 \begin{vmatrix} -19.3 & -28.4 \\ -23.9 & 66.3 \end{vmatrix} + 0$$

$$= 82.4(-48.2)(66.3) - 23.7[(-19.3)(66.3) - (-28.4)(-23.9)]$$
$$= -217,000$$

Dividing by Δ yields

$$x = \frac{-217,000}{-281,000} = 0.772$$

For this system, it is easiest to solve for y and z by substituting back. From Eq.1, we get

$$23.7y = 82.4 - 83.7(0.772) = 17.8$$
$$y = 0.751$$

and from Eq. 2

$$66.3z = -23.9 + 39.2(0.772) = 6.36$$
$$z = 0.0959$$

The solution is then $x = 0.772$, $y = 0.751$, and $z = 0.0959$.   ◆◆◆

Systems of More than Three Equations

We can evaluate a determinant of any order by repeated use of the method of minors. Thus any row or column of a fifth-order determinant can be developed, thereby reducing the determinant to five fourth-order determinants. Each of these can then be developed into four third-order determinants, and so on, until only second-order determinants remain.

Obviously this is a lot of work, which can be made easier by first simplifying the determinant by various means. Instead, we will use the calculator to evaluate large determinants, or use the unit matrix method shown earlier.

Exercise 4 ◆ Higher-Order Determinants

Evaluate each determinant by calculator or by minors.

1. $\begin{vmatrix} 1 & 0 & 2 \\ 3 & 1 & 0 \\ 1 & 2 & 1 \end{vmatrix}$ **2.** $\begin{vmatrix} 2 & -1 & 3 \\ 0 & 2 & 1 \\ 3 & -2 & 4 \end{vmatrix}$

3. $\begin{vmatrix} -3 & 1 & 2 \\ 0 & -1 & 5 \\ 6 & 0 & 1 \end{vmatrix}$ **4.** $\begin{vmatrix} -1 & 0 & 3 \\ 2 & 0 & -2 \\ 1 & -3 & 4 \end{vmatrix}$

5. $\begin{vmatrix} 5 & 1 & 2 \\ -3 & 2 & -1 \\ 4 & -3 & 5 \end{vmatrix}$ **6.** $\begin{vmatrix} 1.0 & 2.4 & -1.5 \\ -2.6 & 0 & 3.2 \\ -2.9 & 1.0 & 4.1 \end{vmatrix}$

7. $\begin{vmatrix} 2 & 1 & 3 \\ 0 & -2 & 4 \\ 0 & 1 & 5 \end{vmatrix}$ **8.** $\begin{vmatrix} 1 & 5 & 4 \\ -3 & 6 & -2 \\ -1 & 5 & 3 \end{vmatrix}$

9.
$$\begin{vmatrix} 4 & 3 & 1 & 0 \\ -1 & 2 & -3 & 5 \\ 0 & 1 & -1 & 2 \\ 0 & 2 & -3 & 5 \end{vmatrix}$$

10.
$$\begin{vmatrix} -1 & 3 & 0 & 2 \\ 2 & -1 & 1 & 0 \\ 5 & 2 & -2 & 0 \\ 1 & -1 & 3 & 1 \end{vmatrix}$$

11.
$$\begin{vmatrix} 2 & 0 & -1 & 0 \\ 0 & 0 & 2 & -1 \\ 1 & 3 & 2 & 1 \\ 3 & 1 & 1 & -2 \end{vmatrix}$$

12.
$$\begin{vmatrix} 1 & 2 & -1 & 1 \\ -1 & 1 & 2 & 3 \\ 3 & -1 & 1 & 2 \\ 1 & 2 & -1 & 1 \end{vmatrix}$$

13.
$$\begin{vmatrix} 3 & 1 & 0 & 2 & 4 \\ 1 & 2 & 4 & 0 & 1 \\ 2 & 3 & 1 & 4 & 2 \\ 1 & 2 & 0 & 2 & 1 \\ 3 & 4 & 1 & 3 & 1 \end{vmatrix}$$

14.
$$\begin{vmatrix} 2 & 1 & 5 & 3 & 6 \\ 1 & 4 & 2 & 4 & 3 \\ 3 & 1 & 2 & 4 & 1 \\ 5 & 2 & 3 & 1 & 4 \\ 4 & 5 & 2 & 3 & 1 \end{vmatrix}$$

Solving a System of Equations by Determinants

Solve by determinants. Evaluate the determinants by calculator or by minors.

15. $x + y + z = 18$
$x - y + z = 6$
$x + y - z = 4$

16. $x + y + z = 12$
$x - y = 2$
$x - z = 4$

17. $x + y = 35$
$x + z = 40$
$y + z = 45$

18. $x + y + z = 35$
$x - 2y + 3z = 15$
$y - x + z = -5$

19. $x + 2y + 3z = 14$
$2x + y + 2z = 10$
$3x + 4y - 3z = 2$

20. $x + y + z = 90$
$2x - 3y = -20$
$2x + 3z = 145$

21. $x - 2y + 2z = 5$
$5x + 3y + 6z = 57$
$x + 2y + 2z = 21$

22. $3x + y = 5$
$2y - 3z = -5$
$x + 2z = 7$

23. $2x - 4y + 3z = 10$
$3x + y - 2z = 6$
$x + 3y - z = 20$

24. $x - y = 5$
$y - z = -6$
$2x - z = 2$

25. $1.15x + 1.95y + 2.78z = 15.3$
$2.41x + 1.16y + 3.12z = 9.66$
$3.11x + 3.83y - 2.93z = 2.15$

26. $1.52x - 2.26y + 1.83z = 4.75$
$4.72x + 3.52y + 5.83z = 45.2$
$1.33x + 2.61y + 3.02z = 18.5$

27. $x + \dfrac{y}{3} = 5$
$x + \dfrac{z}{3} = 6$
$y + \dfrac{z}{3} = 9$

28. $\dfrac{x}{10} + \dfrac{y}{5} + \dfrac{z}{20} = \dfrac{1}{4}$
$x + y + z = 6$
$\dfrac{x}{3} + \dfrac{y}{2} + \dfrac{z}{6} = 1$

29. $x + y + 2z + w = 18$
$x + 2y + z + w = 17$
$x + y + z + 2w = 19$
$2x + y + z + w = 16$

30. $2x - y - z - w = 0$
$x - 3y + z + w = 0$
$x + y - 4z + w = 0$
$x + y + w = 36$

31. $3x - 2y - z + w = -3$
$-x - y + 3z + 2w = 23$
$x + 3y - 2z + w = -12$
$2x - y - z - 3w = -22$

32. $x + 2y = 5$
$y + 2z = 8$
$z + 2u = 11$
$u + 2x = 6$

33. $x + y = a + b$
$y + z = b + c$
$z + w = a - b$
$w - x = c - b$

34. $2x - 3y + z - w = -6$
$x + 2y - z = 8$
$3y + z + 3w = 0$
$3x - y + w = 0$

35. $x + y = 9$
$y + z = 11$
$z + w = 13$
$w + u = 15$
$u + x = 12$

36. $3x + 4y + z = 35$
$3z + 2y - 3w = 4$
$2x - y + 2w = 17$
$3z - 2w + v = 9$
$w + y = 13$

37. $w + v + x + y = 14$
$w + v + x + z = 15$
$w + v + y + z = 16$
$w + x + y + z = 17$
$v + x + y + z = 18$

38. *Team Project:* The following equations result when Kirchhoff's law is applied to a certain four-loop network:

$$14.7I_1 - 25.7I_2 + 14.7I_3 + 19.3I_4 = 26.2$$
$$17.2I_1 + 19.6I_2 - 19.2I_3 + 24.4I_4 = -28.3$$
$$18.6I_1 - 22.7I_2 + 24.7I_3 + 17.3I_4 = 23.3$$
$$27.7I_1 + 19.6I_2 - 33.2I_3 - 42.3I_4 = 25.4$$

Find the four currents by a calculator.

39. *Team Project*: Solve the following by any method:

$$28.3x + 29.2y - 33.1z + 72.4u + 29.4v = 39.5$$
$$73.2x - 28.4y + 59.3z - 27.4u + 49.2v = 82.3$$
$$33.7x + 10.3y + 72.3z + 29.3u - 21.2v = 28.4$$
$$92.3x - 39.5y + 29.5z - 10.3u + 82.2v = 73.4$$
$$88.3x + 29.3y + 10.3z + 84.2u + 29.3v = 39.4$$

◆◆◆ CHAPTER 10 REVIEW PROBLEMS ◆◆◆◆◆◆◆◆◆◆◆◆◆◆◆◆◆◆◆◆◆◆◆◆◆◆◆◆◆◆◆◆

Evaluate.

1. $\begin{vmatrix} 6 & \frac{1}{2} & -2 \\ 3 & \frac{1}{4} & 4 \\ 2 & -\frac{1}{2} & 3 \end{vmatrix}$

2. $\begin{vmatrix} 2 & 7 & -2 & 8 \\ 4 & 1 & 1 & -3 \\ 0 & 3 & -1 & 4 \\ 6 & 4 & 2 & -8 \end{vmatrix}$

3. $\begin{vmatrix} 0 & n & m \\ -n & 0 & l \\ -m & -l & 0 \end{vmatrix}$

4. $\begin{vmatrix} 8 & 2 & 0 & 1 & 4 \\ 0 & 1 & 4 & 2 & 7 \\ 2 & 6 & 3 & 8 & 0 \\ 1 & 4 & 2 & 6 & 5 \\ 4 & 6 & 8 & 3 & 5 \end{vmatrix}$

5. $\begin{vmatrix} 25 & 23 & 19 \\ 14 & 11 & 9 \\ 21 & 17 & 14 \end{vmatrix}$

6. $\begin{vmatrix} x+y & x-y \\ x-y & x+y \end{vmatrix}$

7. $\begin{vmatrix} 9 & 13 & 17 \\ 11 & 15 & 19 \\ 17 & 21 & 25 \end{vmatrix}$

8. $\begin{vmatrix} 5 & -3 & -2 & 0 \\ 4 & 1 & -6 & 2 \\ -1 & 4 & 3 & -5 \\ 0 & 6 & -4 & 2 \end{vmatrix}$

9. $\begin{vmatrix} 1 & 2 & 3 \\ 3 & 1 & 2 \\ 2 & 3 & 1 \end{vmatrix}$

10. $\begin{vmatrix} 2 & -3 & 1 \\ -2 & 4 & 5 \\ 3 & -1 & -4 \end{vmatrix}$

11. $\begin{vmatrix} 1 & 2 & 2 & 4 \\ 1 & 4 & 4 & 1 \\ 1 & 1 & 2 & 2 \\ 4 & 8 & 11 & 13 \end{vmatrix}$

12. $\begin{vmatrix} a-b & -2a \\ 2b & a-b \end{vmatrix}$

13. $\begin{vmatrix} 3 & 1 & 5 & 2 \\ 4 & 10 & 14 & 6 \\ 8 & 9 & 1 & 4 \\ 6 & 15 & 21 & 9 \end{vmatrix}$

14. $\begin{vmatrix} 7 & 8 & 9 \\ 28 & 35 & 40 \\ 21 & 26 & 30 \end{vmatrix}$

15. $\begin{vmatrix} 1 & 5 & 2 \\ 4 & 7 & 3 \\ 9 & 8 & 6 \end{vmatrix}$

16. $\begin{vmatrix} 3 & -6 \\ -5 & 4 \end{vmatrix}$

17. $\begin{vmatrix} 6 & 4 & 7 \\ 9 & 0 & 8 \\ 5 & 3 & 2 \end{vmatrix}$

18. $\begin{vmatrix} 3 & 2 & 1 & 3 \\ 4 & 2 & 2 & 4 \\ 2 & 3 & 1 & 6 \\ 10 & 4 & 5 & 8 \end{vmatrix}$

Solve by any method.

19. $\begin{aligned} x + y + z + w &= -4 \\ x + 2y + 3z + 4w &= 0 \\ x + 3y + 6z + 10w &= 9 \\ x + 4y + 10z + 20w &= 24 \end{aligned}$

20. $\begin{aligned} x + 2y + 3z &= 4 \\ 3x + 5y + z &= 18 \\ 4x + y + 2z &= 12 \end{aligned}$

21. $\begin{aligned} 4x + 3y &= 27 \\ 2x - 5y &= -19 \end{aligned}$

22. $\begin{aligned} 8x - 3y - 7z &= 85 \\ x + 6y - 4z &= -12 \\ 2x - 5y + z &= 33 \end{aligned}$

23. $\begin{aligned} x + 2y &= 10 \\ 2x - 3y &= -1 \end{aligned}$

24. $\begin{aligned} x + y + z + u &= 1 \\ 2x + 3y - 4z + 5u &= -31 \\ 3x - 4y + 5z + 6u &= -22 \\ 4x + 5y - 6z - u &= -13 \end{aligned}$

25. $2x - 3y = 7$
$5x + 2y = 27$

26. $6x + y = 60$
$3x + 2y = 39$

27. $2x + 5y = 29$
$2x - 5y = -21$

28. $4x + 3y = 7$
$2x - 3y = -1$

29. $4x - 5y = 3$
$3x + 5y = 11$

30. $x + 5y = 41$
$3x - 2y = 21$

31. $x + 2y = 7$
$x + y = 5$

32. $8x - 3y = 22$
$4x + 5y = 18$

33. $3x + 4y = 25$
$4x + 3y = 21$

34. $37.7x = 59.2 + 24.6y$
$28.3x - 39.8y - 62.5 = 0$

35. $5.03a = 8.16 + 5.11b$
$3.63b + 7.26a = 28.8$

36. $541x + 216y + 412z = 866$
$211x + 483y - 793z = 315$
$215x + 495y + 378z = 253$

37. $72.3x + 54.2y + 83.3z = 52.5$
$52.2x - 26.6y + 83.7z = 75.2$
$33.4x + 61.6y + 30.2z = 58.5$

38. A link in a certain mechanism starts at a speed v_0 and travels 27.5 inches in 8.25 s, at constant acceleration a. Its motion is described by

$$s = v_0 t + \tfrac{1}{2}at^2$$

or

$$27.5 = 8.25v_0 + \tfrac{1}{2}a(8.25)^2$$

In another trial, the link is found to travel 34.6 inches in 11.2 s, with the same initial speed and acceleration. Find v_0 and a.

39. *Writing:* We have had a number of methods for solving a system of equations, both in this chapter and the preceding one. List them, give a one-line description of each, and briefly state the advantages and disadvantages of each.

40. *On our Web site:* Our treatment of matrices here has just scratched the surface. A much fuller treatment is available at our companion Web site at www.wiley.com/college/calter

There you will learn how to

Transpose a matrix
Add and subtract matrices
Multiply a scalar and a matrix
Multiply a vector and a matrix
Multiply two matrices
Multiply a row vector and a matrix
Find the inverse of a matrix
Solve a system of equations by matrix inversion

Factoring and Fractions

◆◆◆ **OBJECTIVES** ◆◆

When you have completed this chapter, you should be able to

- Factor expressions by removing common factors.
- Factor binomials that are the difference of two squares, or the sum or difference of two cubes.
- Factor expressions by grouping.
- Use factoring techniques in applications.
- Determine values of variables that result in undefined fractions.
- Simplify algebraic fractions.
- Add, subtract, and divide algebraic fractions.
- Simplify complex fractions.
- Solve fractional equations more complex than those solved in earlier chapters.
- Solve literal equations and formulas containing fractions.
- Factor expressions, simplify fractions, and solve fractional equations by calculator.

◆◆

Continuing with our study of algebra, we will first learn how to simplify algebraic expressions by means of factoring. This skill will help us to manipulate algebraic expressions as well as the fractions and fractional equations that follow.

Although the calculator, computer, and metric system (all of which use decimal notation) have somewhat reduced the use of *common* fractions, *algebraic* fractions are as important as ever. We must be able to handle them in order to solve fractional equations and to manipulate the formulas that are so important in technology. For example, Fig. 11–1 shows a bar attached to a sphere. The distance x to the center of gravity is given by

$$x = \frac{10m_1 + 25m_2}{m_1 + m_2}$$

Could you solve this equation for, say, m_1? In this chapter we show how to solve this type of equation.

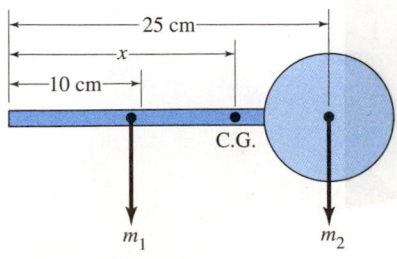

FIGURE 11–1

Here we will show traditional pencil-and-paper methods of solution alongside solutions by a calculator such as the TI-89 that can manipulate algebraic expressions.

11–1   Common Factors

As usual, we must start with definitions of the terms we will be using in the following sections.

Factors of an Expression

The *factors of an expression* are those quantities whose product is the original expression.

◆◆◆ **Example 1:** The factors of $x^2 - 9$ are $x + 3$ and $x - 3$, because

$$(x + 3)(x - 3) = x^2 + 3x - 3x - 9$$
$$= x^2 - 9$$ ◆◆◆

Many expressions have no factors other than 1 and themselves. Such expressions are called *prime*.

Factoring is the process of finding the factors of an expression. It is *the reverse of finding the product* of two or more quantities.

Multiplication (Finding the Product)	$x(x + 4) = x^2 + 4x$
Factoring (Finding the Factors)	$x^2 + 4x = x(x + 4)$

We usually factor an expression by *recognizing the form* of that expression. In the first type of factoring we will cover, we look for *common factors*.

Common Factors

■ **Exploration:**

Try this. Expand this expression by multiplying out.

$$a(b + c + d)$$

Now how would you return the expression you just got back to its original form? Can you state your findings as a general rule?   ■

If each term of an expression contains the same quantity (called the *common factor*), the quantity may be *factored out*.

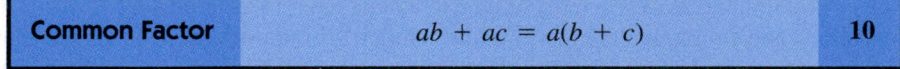

Common Factor	$ab + ac = a(b + c)$	10

This is nothing but the *distributive law* that we studied earlier.

◆◆◆ **Example 2:** In the expression

$$x^3 - 3x$$

each term contains an x as a common factor. So we write

$$x^3 - 3x = x(x^2 - 3)$$

Most of the factoring we will do will be of this type.   ◆◆◆

◆◆◆ **Example 3:** Here are some examples of the factoring out of common factors. See if you get the same results.

(a) $2x^2 + x = x(2x + 1)$

(b) $3xy^2 - 9x^3y + 6x^2y^2 = 3xy(y - 3x^2 + 2xy)$

(c) $3x^3 - 6x^2y + 9x^4y^2 = 3x^2(x - 2y + 3x^2y^2)$

(d) $3x^3 - 2x + 5x^4 = x(3x^2 - 2 + 5x^3)$ ◆◆◆

Common Error	Students are sometimes puzzled over the "1" in Example 3(a). Why should it be there? After all, when you remove a chair from a room, it is *gone*; there is nothing (zero) remaining where the chair used to be. If you remove an x by factoring, you might assume that nothing (zero) remains where the x used to be. $$2x^2 + x = x(2x + 0)?$$ Prove to yourself that this is not correct by multiplying the factors to see if you get back the original expression.

Factors in the Denominator

Common factors may appear in the denominators of the terms as well as in the numerators.

◆◆◆ **Example 4:** Here we show some examples of the factoring out of common factors from the denominators.

(a) $\dfrac{1}{x} + \dfrac{2}{x^2} = \dfrac{1}{x}\left(1 + \dfrac{2}{x}\right)$

(b) $\dfrac{x}{y^2} + \dfrac{x^2}{y} + \dfrac{2x}{3y} = \dfrac{x}{y}\left(\dfrac{1}{y} + x + \dfrac{2}{3}\right)$ ◆◆◆

Checking

To check if factoring has been done correctly, simply multiply your factors together and see if you get back the original expression. This check will tell whether you have factored correctly, but not whether you have factored *completely*.

◆◆◆ **Example 5:** Let us check Example 4(b) by multiplying out.

$$\frac{x}{y}\left(\frac{1}{y} + x + \frac{2}{3}\right) = \left(\frac{x}{y}\right)\left(\frac{1}{y}\right) + \left(\frac{x}{y}\right)x + \left(\frac{x}{y}\right)\left(\frac{2}{3}\right)$$

$$= \frac{x}{y^2} + \frac{x^2}{y} + \frac{2x}{3y}$$

Checks.   ◆◆◆

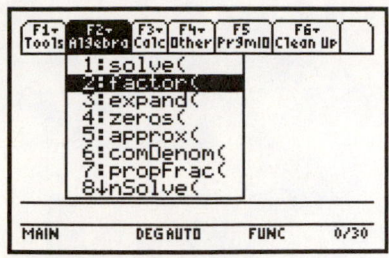

(1) TI-89 screen for Example 6.

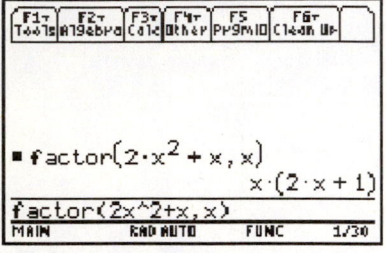

(2) TI-89 screen for Example 6.

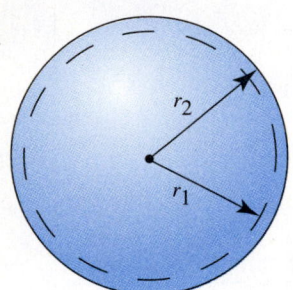

FIGURE 11–2

Factoring by Calculator

A calculator that can manipulate algebraic symbols can be used to factor an expression. We will do a simple factoring here, with more complex types to come later.

◆◆◆ Example 6: Factor the expression from Example 3(a): $2x^2 + x$ on the TI-89 or similar calculator.

Solution: From the **F2 Algebra** menu in the home screen, select **factor**, as shown in screen (1).

Enter the expression to be factored, the variable in the expression that we are factoring with respect to, close parentheses, and press $\boxed{\text{ENTER}}$.   ◆◆◆

◆◆◆ Example 7: *An Application.* The mass of a spherical shell, Fig. 11–2, having an outside radius of r_2 and an inside radius of r_1 is

$$\text{mass} = \frac{4}{3}\pi r_2{}^3 D - \frac{4}{3}\pi r_1{}^3 D$$

where D is the mass density of the material. Factor the right side of this equation.

Solution: We remove the common factors $\frac{4}{3}\pi D$ and get

$$\text{mass} = \frac{4}{3}\pi D(r_2{}^3 - r_1{}^3)$$ ◆◆◆

Exercise 1 ◆ Common Factors

Factor completely, by hand or by calculator. Check your results.

1. $3y^2 + y^3$
2. $6x - 3y$
3. $x^5 - 2x^4 + 3x^3$
4. $9y - 27xy$
5. $3a + a^2 - 3a^3$
6. $8xy^3 - 6x^2y^2 + 2x^3y$
7. $4pq + 6q^2 + 2q^3$
8. $\dfrac{a}{3} - \dfrac{a^2}{4} + \dfrac{a^3}{5}$
9. $\dfrac{3}{x} + \dfrac{2}{x^2} - \dfrac{5}{x^3}$

Challenge Problems

10. $\dfrac{3ab^2}{y^3} - \dfrac{6a^2b}{y^2} + \dfrac{12ab}{y}$
11. $\dfrac{5m}{2n} + \dfrac{15m^2}{4n^2} - \dfrac{25m^3}{8n}$
12. $\dfrac{16y^2}{9x^2} - \dfrac{8y^3}{3x^3} + \dfrac{24y^4}{9x}$
13. $5a^2b + 6a^2c$
14. $a^2c + b^2c + c^2d$
15. $4x^2y + cxy^2 + 3xy^3$
16. $4abx + 6a^2x^2 + 8ax$
17. $3a^3y - 6a^2y^2 + 9ay^3$

18. $2a^2c - 2a^2c^2 + 3ac$

19. $5acd - 2c^2d^2 + bcd$

20. $4b^2c^2 - 12abc - 9c^2$

21. $8x^2y^2 + 12x^2z^2$

22. $6xyz + 12x^2y^2z$

23. $3a^2b + abc - abd$

24. $5a^3x^2 - 5a^2x^3 + 10a^2x^2z$

Applications

25. When a bar of length L_0, Fig. 11–3, is changed in temperature by an amount t, its new length L will be $L = L_0 + L_0\alpha t$, where α is the coefficient of thermal expansion. Factor the right side of this equation.

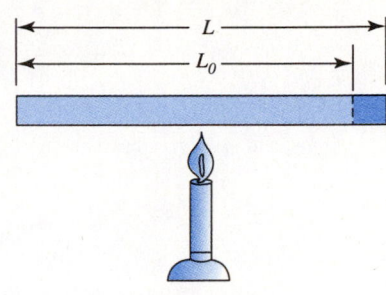

FIGURE 11–3

26. A sum of money a when invested for t years at an interest rate n will accumulate to an amount y, where $y = a + ant$. Factor the right side of this equation.

27. When a resistance R_1 is heated from a temperature t_1 to a new temperature t, it will increase in resistance by an amount $\alpha(t - t_1)R_1$, where α is the temperature coefficient of resistance. The final resistance will then be $R = R_1 + \alpha(t - t_1)R_1$. Factor the right side of this equation.

28. An item costing P dollars is reduced in price by 15%. The resulting price C is then $C = P - 0.15P$. Factor the right side of this equation.

29. The displacement of a uniformly accelerated body is given by Eq. 1018

$$s = v_0t + \frac{a}{2}t^2$$

Factor the right side of this equation.

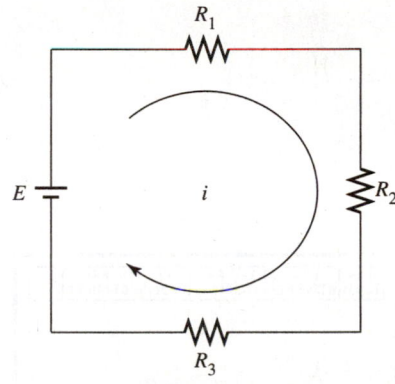

30. The sum of the voltage drops across the resistors in Fig. 11–4 must equal the battery voltage E.

$$E = iR_1 + iR_2 + iR_3$$

Factor the right side of this equation.

FIGURE 11–4

11–2 Difference of Two Squares

The next type of expression that we will factor is a binomial in which one square is subtracted from another.

■ Exploration:

Try this. Multiply $(2x + 3)$ by $(2x - 3)$.

Examine the product you just obtained. How many terms does it have? When you multiply two binomials, do you usually get that number of terms? How are the terms related to those in the two original binomials? Can you express your results in general terms? ■

An expression of the form

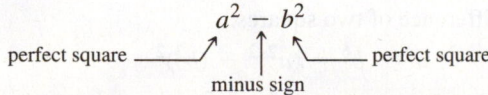

where one perfect square is subtracted from another, is called a *difference of two squares*. It arises when $(a - b)$ and $(a + b)$ are multiplied together.

| Difference of Two Squares | $a^2 - b^2 = (a + b)(a - b)$ | 33 |

This is one example of what is called a *special product*.

Factoring the Difference of Two Squares

Once we recognize its form, the difference of two squares is easily factored.

♦♦♦ **Example 8:** This example shows how to factor a difference of two squares.

$$4x^2 - 9 = (2x)^2 - (3)^2$$

$$(2x)^2 - (3)^2 = (2x \qquad)(2x \qquad)$$

square root of the first term

$$(2x)^2 - (3)^2 = (2x \quad 3)(2x \quad 3)$$

square root of the last term

$$4x^2 - 9 = (2x + 3)(2x - 3)$$

opposite signs ♦♦♦

♦♦♦ **Example 9:** Here are more examples of the factoring of a difference of two squares.

(a) $y^2 - 1 = (y + 1)(y - 1)$

(b) $9a^2 - 16b^2 = (3a + 4b)(3a - 4b)$

(c) $49x^2 - 9a^2y^2 = (7x - 3ay)(7x + 3ay)$

(d) $1 - a^2b^2c^2 = (1 - abc)(1 + abc)$ ♦♦♦

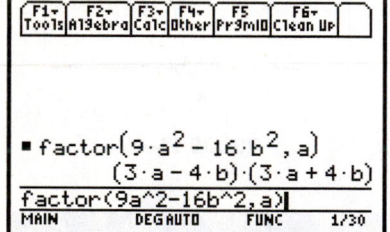

TI-89 calculator solution for Example 9(b).

| Common Error | There is no similar rule for factoring the *sum* of two squares, such as $a^2 + b^2$ |

The difference of any two quantities that have *even powers* can be factored as the difference of two squares. *Express each quantity as a square* by means of the following law of exponents (Eq. 24) for a power raised to a power, that we learned in Chap. 2,

$$x^{ab} = (x^a)^b$$

♦♦♦ **Example 10:** Factor the expression $a^4 + b^6$.

Solution: We write each term as a square, as follows.

$$a^4 = (a^2)^2 \quad \text{and} \quad b^6 = (b^3)^2$$

We now factor the difference of two squares.

$$a^4 - b^6 = (a^2)^2 - (b^3)^2$$
$$= (a^2 + b^3)(a^2 - b^3)$$ ♦♦♦

Sometimes one or both terms in the given expression may be a fraction.

◆◆◆ Example 11: Factor the expression

$$\frac{1}{x^2} - \frac{1}{y^2}$$

Solution: We use the law of exponents, $\frac{x^n}{y^n} = \left(\frac{x}{y}\right)^n$, for a quotient raised to a power to write each term as a square.

$$\frac{1}{x^2} = \left(\frac{1}{x}\right)^2 \quad \text{and} \quad \frac{1}{y^2} = \left(\frac{1}{y}\right)^2$$

We can then factor the expression as a difference of two squares.

$$\frac{1}{x^2} - \frac{1}{y^2} = \left(\frac{1}{x}\right)^2 - \left(\frac{1}{y}\right)^2$$

$$= \left(\frac{1}{x} + \frac{1}{y}\right)\left(\frac{1}{x} - \frac{1}{y}\right)$$ ◆◆◆

Factoring Completely

After factoring an expression, see if any of the factors themselves can be factored *again*.

◆◆◆ Example 12: Factor $a - ab^2$.

Solution: Always remove any common factors first. Factoring, we obtain

$$a - ab^2 = a(1 - b^2)$$

Now factoring the difference of two squares, we get

$$a - ab^2 = a(1 + b)(1 - b)$$ ◆◆◆

◆◆◆ Example 13: *An Application.* The volume of the steel frame, Fig. 11–5, is given by $t\delta S^2 - t\delta s^2$, where t is the thickness of the frame and δ is the density of steel. Factor this expression.

Solution: We first remove the common factor $t\delta$ and then factor the difference of two squares.

$$t\delta S^2 - t\delta s^2 = t\delta(S^2 - s^2)$$
$$= t\delta(S + s)(S - s)$$ ◆◆◆

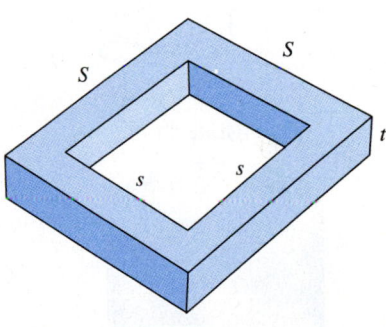

FIGURE 11–5

Exercise 2 ◆ Difference of Two Squares

Factor completely.

1. $4 - x^2$
2. $x^2 - 9$
3. $9a^2 - x^2$
4. $25 - x^2$
5. $4x^2 - 4y^2$
6. $9x^2 - y^2$
7. $x^2 - 9y^2$
8. $16x^2 - 16y^2$
9. $9c^2 - 16d^2$
10. $25a^2 - 9b^2$
11. $9y^2 - 1$
12. $4x^2 - 9y^2$

Challenge Problems

13. $m^4 - n^4$
14. $a^8 - b^8$
15. $4m^2 - 9n^4$
16. $9^2 - 4b^4$
17. $a^{16} - b^8$
18. $9a^2b^2 - 4c^4$
19. $25x^4 - 16y^6$
20. $36y^2 - 49z^6$

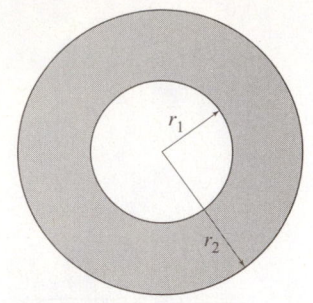

FIGURE 11–6   Circular washer.

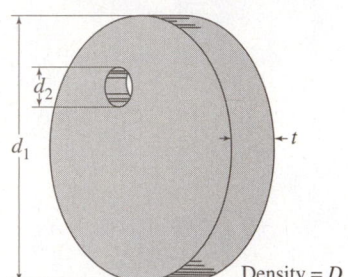

FIGURE 11–7

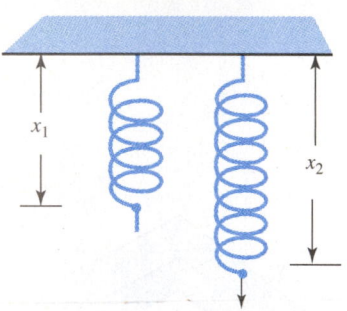

FIGURE 11–8

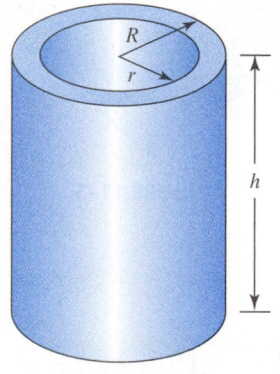

FIGURE 11–9

21. $16a^4 - 121$

22. $121a^4 - 16$

23. $25a^4b^4 - 9$

24. $121a^2 - 36b^4$

25. $\dfrac{1}{a^2} - \dfrac{1}{b^2}$

26. $\dfrac{4}{m^2} - \dfrac{9}{n^2}$

27. $\dfrac{a^2}{x^2} - \dfrac{b^2}{y^2}$

28. $\dfrac{25}{x^4} - \dfrac{9}{y^6}$

Applications

29. The thrust washer in Fig. 11–6 has a surface area of

$$\text{area} = \pi r_2{}^2 - \pi r_1{}^2$$

Factor this expression.

30. A flywheel of diameter d_1 (Fig. 11–7) has a balancing hole of diameter d_2 drilled through it. The mass M of the flywheel is then

$$\text{mass} = \frac{\pi d_1{}^2}{4}Dt - \frac{\pi d_2{}^2}{4}Dt$$

Factor this expression.

31. A spherical balloon shrinks from radius r_1 to radius r_2. The change in surface area is $4\pi r_1{}^2 - 4\pi r_2{}^2$. Factor this expression.

32. When an object is released from rest, the distance fallen between time t_1 and time t_2 is $\frac{1}{2}gt_2^2 - \frac{1}{2}gt_1^2$, where g is the acceleration due to gravity. Factor this expression.

33. When a body of mass m slows down from velocity v_1 to velocity v_2, the decrease in kinetic energy is $\frac{1}{2}mv_1^2 - \frac{1}{2}mv_2^2$. Factor this expression.

34. The work required to stretch a spring, Fig. 11–8, from a length x_1 (measured from the unstretched position) to a new length x_2 is $\frac{1}{2}kx_2^2 - \frac{1}{2}kx_1^2$. Factor this expression.

35. The formula for the volume of a cylinder of radius r and height h is $v = \pi r^2 h$. Write an expression for the volume of a hollow cylinder, Fig. 11–9, that has an inside radius r, and outside radius R, and a height h. Then factor the expression completely.

36. A body at temperature T will radiate an amount of heat kT^4 to its surroundings and will absorb from the surroundings an amount of heat kT_s^4, where T_s is the temperature of the surroundings. Write an expression for the net heat transfer by radiation (amount radiated minus amount absorbed), and factor this expression completely.

11–3   Factoring Trinomials

A *trinomial*, you may recall, is a polynomial having *three terms*, and a *quadratic trinomial* in x has an x^2 term, an x term, and a constant term. The coefficient of x^2 is called the *leading coefficient* and the coefficient of x is called the *middle coefficient*.

◆◆◆ **Example 14:** $4x^2 + 3x - 5$ is a quadratic trinomial.

The leading coefficient is 4, the middle coefficient is 3, and the constant term is -5.   ◆◆◆

■ Exploration:

Try this. Multiply the two binomials $(2x + 1)$ and $(3x + 4)$.

What kind of expression did you get? What are the values of the two coefficients and the constant term? How are they related to the numbers in the given binomials? Can you state it as a general rule? ■

When we multiply the two binomials $(ax + b)$ and $(cx + d)$, we get a trinomial with a leading coefficient of ac, a middle coefficient of $(ad + bc)$, and a constant term of bd.

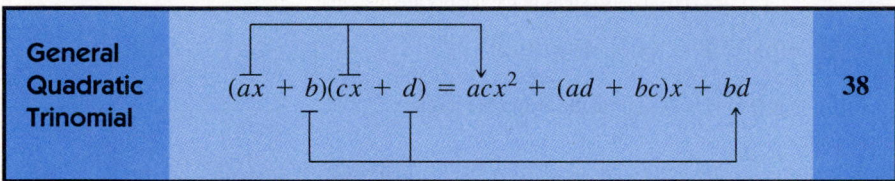

| General Quadratic Trinomial | $(ax + b)(cx + d) = acx^2 + (ad + bc)x + bd$ | 38 |

Test for Factorability

Not all quadratic trinomials can be factored. We test for factorability as follows:

| Test for Factorability | The trinomial $ax^2 + bx + c$ (where a, b, and c are constants) is factorable if $b^2 - 4ac$ is a perfect square. | 36 |

◆◆◆ **Example 15:** Can the trinomial $6x^2 + x - 12$ be factored?

Solution: Using the test for factorability, Eq. 44, with $a = 6$, $b = 1$, and $c = -12$, we have

$$b^2 - 4ac = 1^2 - 4(6)(-12) = 289$$

By taking the square root of 289 on the calculator, we see that it is a perfect square ($289 = 17^2$), so the given trinomial is factorable. We will see later that its factors are $(2x + 3)$ and $(3x - 4)$. ◆◆◆

Factoring by Trial and Error

To factor a quadratic trinomial, we must find two binomials $(ax + b)$ and $(cx + d)$ for which

$$ac = \text{the leading coefficient of the given trinomial}$$
$$ad + bc = \text{the middle coefficient}$$
$$bd = \text{the constant term}$$

We will start with the simplest case, a trinomial with *a leading coefficient of* 1.

◆◆◆ **Example 16:** Factor the trinomial $x^2 + 8x + 15$.

Solution: From Eq. 38, this trinomial, if factorable, will factor as $(x + b)(x + d)$, where b and d have a sum of 8 and a product of 15. The integers 5 and 3 have a sum of 8 and a product of 15. Thus

$$x^2 + 8x + 15 = (x + 5)(x + 3)$$ ◆◆◆

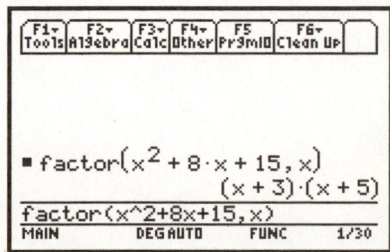

TI-89 calculator solution for Example 16.

Using the Signs to Aid Factoring

The signs of the terms of the trinomial can tell you the signs of the factors.

1. If the sign of the last term is *positive*, both signs in the factors will be the same (both positive or both negative). The sign of the middle term of the trinomial will tell whether both signs in the factors are positive or negative.

♦♦♦ **Example 17:** These examples show how a positive last term affects the signs in the factors.

positive, so signs in the factors are the same

(a) $x^2 + 6x + 8 = (x + 4)(x + 2)$

positive, so both signs in the factors are positive

positive, so signs in the factors are the same

(b) $x^2 - 6x + 8 = (x - 4)(x - 2)$

negative, so both signs in the factors are negative   ♦♦♦

2. If the sign of the last term of the trinomial is *negative*, the signs of b and d in the factors will differ (one positive, one negative). The sign of the middle term of the trinomial will tell which quantity, b or d, is larger (in absolute value), the positive one or the negative one.

♦♦♦ **Example 18:** This example show how a negative last term affects the signs in the factors.

negative, so the signs in the factors differ, one positive and one negative

$x^2 - 2x - 8 = (x - 4)(x + 2)$

negative, so the larger number (4) has a negative sign   ♦♦♦

♦♦♦ **Example 19:** Factor $x^2 + 2x - 15$.

Solution: We first look at the sign of the last term (-15). The negative sign tells us that the signs of the factor will differ. So we may write

$$x^2 + 2x - 15 = (x + \quad)(x - \quad)$$

We then find two numbers whose product is -15 and whose sum is 2. These two numbers must have opposite signs in order to have a negative product. Also, since the middle coefficient is $+2$, the positive number must be 2 greater than the negative number. Two numbers that meet these conditions are $+5$ and -3. So

$$x^2 + 2x - 15 = (x + \quad)(x - \quad)$$ ♦♦♦

Next let us factor a trinomial whose leading coefficient is *not* 1, using the signs of the terms to help us.

♦♦♦ **Example 20:** Factor $6x^2 - 13x - 5$.

Solution: We assume that this expression will factor into the form $(ax + b)(cx + d)$. The leading coefficient is 6, so a and c can be (1 and 6), (6 and 1), (2 and 3), or (3 and 2). Let us try (3 and 2).

$$(3x + b)(2x + d)$$

The sign of the constant term (−5) is negative, which says that b and d have opposite signs. The sign of the middle coefficient is also negative, telling us that the greater of the two, in absolute value, must be negative. So b and d can be (1 and −5) or (−5 and 1). Let us try (−5 and 1).

$$(3x - 5)(2x + 1)$$

We check by multiplying out, getting $6x^2 - 7x - 5$. No good, but everything matches but the middle term. Note that a switch of a and c will affect only the middle term, so we next try

$$(2x - 5)(3x + 1)$$

which does check. ◆◆◆

Remember to *remove any common factors* before factoring a trinomial.

◆◆◆ **Example 21:** Factor $18x^2 - 15 - 39x$.

Solution: Factoring out a 3 from the expression gives

$$3(6x^2 - 5 - 13x)$$

$$= 3(6x^2 - 13x - 5)$$

after rearranging terms. We see that the expression in parentheses is the same as in the preceding example, so the factors are

$$3(2x - 5)(3x + 1)$$ ◆◆◆

◆◆◆ **Example 22:** *An Application.* We can find the thickness t of the angle iron in Fig. 11–10 by solving the equation, $t^2 - 14t + 24 = 0$. Factor the left side of this equation.

Solution: The last term is positive, so the signs in both factors will be the same. The middle term is negative, so the signs in both factors will be negative. So we can write

$$t^2 - 14t + 24 = (t - \quad)(t - \quad)$$

We now seek two numbers whose product is 24 and whose sum is −14. The numbers −2 and −12 will work, so we get

$$t^2 - 14t + 24 = (t - 2)(t - 12)$$ ◆◆◆

Grouping Method

Some students have a knack for factoring and can quickly factor a trinomial by trial and error. Others rely on the longer but surer grouping method. The grouping method eliminates the need for trial and error.

◆◆◆ **Example 23:** Factor $3x^2 - 16x - 12$.

Solution:

1. Multiply the leading coefficient and the constant term.

$$3(-12) = -36$$

2. Find two numbers whose product equals −36 and whose sum equals the middle coefficient, −16. Two such numbers are 2 and −18.
3. Rewrite the trinomial, splitting the middle term according to the selected factors $(-16x = 2x - 18x)$.

$$3x^2 + 2x - 18x - 12$$

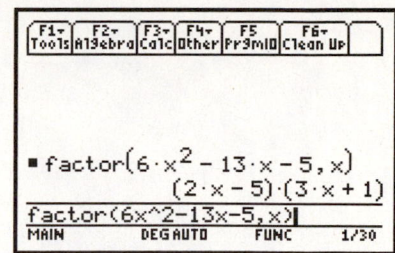

TI-89 calculator solution for Example 20.

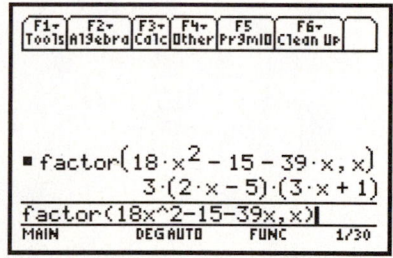

TI-89 calculator solution for Example 21.

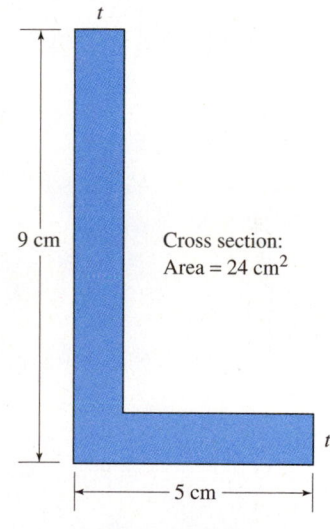

FIGURE 11–10

Group the first two terms together and the last two terms together.

$$(3x^2 + 2x) + (-18x - 12)$$

4. Remove common factors from each grouping.

$$x(3x + 2) - 6(3x + 2)$$

5. Remove the common factor $(3x + 2)$ from the entire expression and you get:

$$(3x + 2)(x - 6)$$

which are the required factors.   ◆◆◆

Common Error	It is easy to make a mistake when factoring out a negative quantity. Thus when we factored out a -6 in going from step 3 to step 4 in Example 23, we got $$(-18x - 12) = -6(3x + 2)$$ but not $$-6(3x - 2)$$ ⟍ incorrect!

The Perfect Square Trinomial

■ **Exploration:**

Try this. Square the binomial $(2x + 3)$.

What kind of expression did you get? What are the values of the two coefficients and the constant term? How are they related to the numbers in the given binomial? Can you state your result as a general rule?   ■

The expression obtained when a binomial is squared is called a *perfect square trinomial*, another special product. In your exploration, you may have observed that the first and last terms of the trinomial are the squares of the first and last terms of the binomial,

$$(2x + 3)^2$$
square ↙   ↘ square
$$4x^2 + 12x + 9$$

and that the middle term is twice the product of the terms of the binomial.

$$(2x + 3)^2$$
product
$$6x$$
twice the product
$$4x^2 + 12x + 9$$

Also, the constant term is *always positive*.

Perfect Square Trinomial	$(a + b)^2 = a^2 + 2ab + b^2$	**39**
	$(a - b)^2 = a^2 - 2ab + b^2$	**40**

We can factor a perfect square trinomial in the same way we factored the general quadratic trinomial. However, the work is faster if we recognize that a trinomial is a perfect square. If it is, its factors will be the square of a binomial. The terms of that binomial are the square roots of the first and last terms of the trinomial. The sign in the binomial will be the same as the sign of the middle term of the trinomial.

◆◆◆ **Example 24:** Factor $a^2 - 4a + 4$.

Solution: The first and last terms are both perfect squares, and the middle term is twice the product of the square roots of the first and last terms. Thus the trinomial is a perfect square. Factoring, we obtain

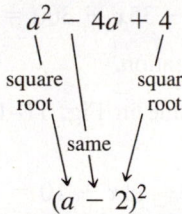

$$(a - 2)^2$$

◆◆◆

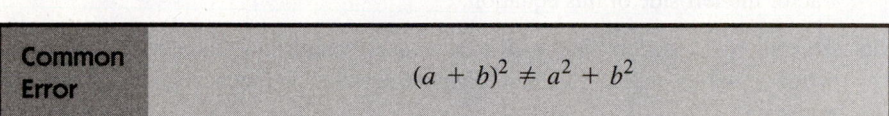

TI-89 calculator solution for Example 24.

Common Error	$(a + b)^2 \neq a^2 + b^2$

Exercise 3 ◆ Factoring Trinomials

Factor completely, by hand or by calculator. Check your results.

Trinomials with a Leading Coefficient of 1

1. $x^2 - 10x + 21$
2. $x^2 - 15x + 56$
3. $x^2 - 10x + 9$
4. $x^2 + 13x + 30$
5. $x^2 + 7x - 30$
6. $x^2 + 3x + 2$
7. $x^2 + 7x + 12$
8. $c^2 + 9c + 18$
9. $x^2 - 4x - 21$
10. $x^2 - x - 56$
11. $x^2 + 6x + 8$
12. $x^2 + 12x + 32$
13. $b^2 - 8b + 15$
14. $b^2 + b - 12$
15. $b^2 - b - 12$
16. $3b^2 - 30b + 63$
17. $2y^2 - 26y + 60$
18. $4z^2 - 16z - 84$

The General Quadratic Trinomial

19. $4x^2 - 13x + 3$
20. $5a^2 - 8a + 3$
21. $5x^2 + 11x + 2$
22. $7x^2 + 23x + 6$
23. $12b^2 - b - 6$
24. $6x^2 - 7x + 2$
25. $2a^2 + a - 6$
26. $2x^2 + 3x - 2$
27. $5x^2 - 38x + 21$
28. $4x^2 + 7x - 15$
29. $3x^2 + 6x + 3$
30. $2x^2 + 11x + 12$
31. $3x^2 - x - 2$
32. $7x^2 + 123x - 54$
33. $4x^2 - 10x + 6$
34. $3x^2 + 11x - 20$
35. $4a^2 + 4a - 3$
36. $9x^2 - 27x + 18$
37. $9a^2 - 15a - 14$
38. $16c^2 - 48c + 35$

The Perfect Square Trinomial

39. $x^2 + 4x + 4$
40. $x^2 - 30x + 225$
41. $y^2 - 2y + 1$
42. $x^2 + 2x + 1$
43. $2y^2 - 12y + 18$
44. $9 - 12a + 4a^2$
45. $9 + 6x + x^2$
46. $4y^2 - 4y + 1$

47. $9x^2 + 6x + 1$ **48.** $16x^2 + 16x + 4$

49. $9y^2 - 18y + 9$ **50.** $16n^2 - 8n + 1$

51. $16 + 16a + 4a^2$ **52.** $1 + 20a + 100a^2$

Applications

53. To find the dimensions of a rectangular field having a perimeter of 70 ft and an area of 300 ft^2, we must solve the quadratic equation

$$x^2 - 35x + 300 = 0$$

Factor the left side of this equation.

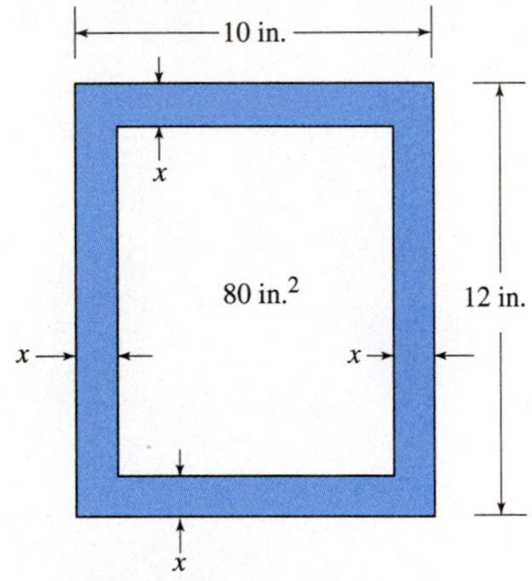

54. To find the width of the frame in Fig. 11–11, we must solve the quadratic equation

$$x^2 - 11x + 10 = 0$$

Factor the left side of this equation.

55. To find two resistors that will give an equivalent resistance of 400 Ω when wired in series and 75 Ω when wired in parallel, we must solve the quadratic equation

$$R^2 - 400R + 30{,}000 = 0$$

Factor the left side of this equation.

56. An object is thrown upward with an initial velocity of 32 ft/s from a building 128 ft above the ground. The height s of the object above the ground at any time t is given by

$$s = 128 + 32t - 16t^2$$

Factor the right side of this quadratic equation.

57. An object is thrown into the air with an initial velocity of 82 ft/s. To find the time it takes for the object to reach a height of 45 ft, we must solve the quadratic equation

$$16t^2 - 82t + 45 = 0$$

Factor the left side of this equation.

58. To find the depth of cut h needed to produce a flat of a certain width on a 1-in.-radius bar (Fig. 11–12), we must solve the equation

$$4h^2 - 8h + 3 = 0$$

Factor the left side of this equation.

59. To find the width $2m$ of a road that will give a sight distance of 1000 ft on a curve of radius 500 ft, we must solve the equation

$$m^2 - 1000\,m + 250{,}000 = 0$$

Factor the left side of this equation.

60. *Project:* Some trinomials that have two variables (such as $x^2 + 5xy + 6y^2$) can be factored by temporarily dropping one variable (y in this example), factoring the remaining trinomial ($x^2 + 5x + 6$) into $(x + 3)(x + 2)$, and then putting back the second variable, getting $(x + 3y)(x + 2y)$. Try this technique on the following trinomials:

$$x^2 - 13xy + 36y^2 \qquad x^2 + 19xy + 84y^2 \qquad x^2 - 9xy + 20y^2$$

FIGURE 11–11

10 in.

12 in.

80 in.^{2}

x

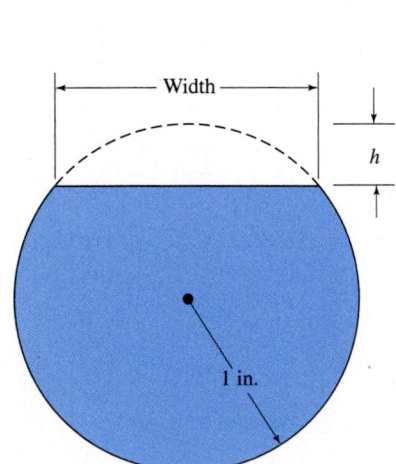

FIGURE 11–12

Width

h

1 in.

11–4   Other Factorable Expressions

More Factoring by Grouping

We have used the grouping method to factor trinomials. Now we extend it to expressions containing *four* terms. Here we try to arrange terms into smaller groups of two, each of which has a common factor. We will show the method by examples.

◆◆◆ **Example 25:** Factor $ab + 4a + 3b + 12$.

Solution: Group the two terms containing the factor a, and group the two containing the factor 3. Remove the common factor from each pair of terms.

$$(ab + 4a) + (3b + 12) = a(b + 4) + 3(b + 4)$$

Both terms now have the common factor $(b + 4)$, which we factor out, getting

$$(b + 4)(a + 3) \qquad ◆◆◆$$

◆◆◆ **Example 26:** Factor $x^2 - y^2 + 2x + 1$.

Solution: Taking our cue from Example 25, we try grouping the x terms together.

$$(x^2 + 2x) + (-y^2 + 1)$$

But we are no better off than before. We then might notice that if the $+1$ term were grouped with the x terms, we would get a trinomial that could be factored. Thus

$$(x^2 + 2x + 1) - y^2$$
$$(x + 1)(x + 1) - y^2$$

or

$$(x + 1)^2 - y^2$$

We now have the difference of two squares. Factoring again gives

$$(x + 1 + y)(x + 1 - y) \qquad ◆◆◆$$

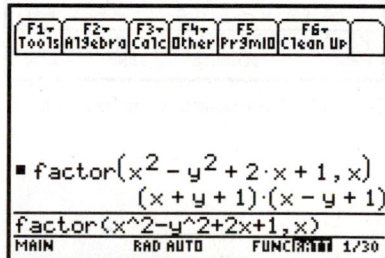

TI-89 calculator solution for Example 26.

Sum or Difference of Two Cubes

An expression such as

$$x^3 + 27$$

is called the *sum of two cubes* (x^3 and 3^3). In general, when we multiply the binomial $(a + b)$ and the trinomial $(a^2 - ab + b^2)$, we obtain

$$(a + b)(a^2 - ab + b^2) = a^3 - a^2b + ab^2 + a^2b - ab^2 + b^3$$
$$= a^3 + b^3$$

All but the cubed terms drop out, leaving the sum of two cubes.

Sum of Two Cubes	$a^3 + b^3 = (a + b)(a^2 - ab + b^2)$	34
Difference of Two Cubes	$a^3 - b^3 = (a - b)(a^2 + ab + b^2)$	35

When we recognize that an expression is the sum (or difference) of two cubes, we can write the factors immediately.

◆◆◆ **Example 27:** Factor $x^3 + 27$.

Solution: This expression is the sum of two cubes, $x^3 + 3^3$. Substituting into Eq. 34, with $a = x$ and $b = 3$, yields

$$x^3 + 27 = x^3 + 3^3 = (x + 3)(x^2 - 3x + 9)$$

same sign

opposite sign

always +

Common Error	The middle term of the trinomials in Eqs. 34 and 35 is often mistaken as $2ab$. $$a^3 + b^3 \neq (a + b)(a^2 - 2ab + b^2)$$ no!

◆◆◆ **Example 28:** Factor $108 - 32x^3$.

Solution: As usual, we first try to remove a common factor.

$$108 - 32x^3 = 4(27 - 8x^3)$$

We then recognize the expression in parentheses as the difference of two cubes, so

$$108 - 32x^3 = 4(3 - 2x)(9 + 6x + 4x^2)$$

◆◆◆

■ factor$(108 - 32 \cdot x^3, x)$
  $-4 \cdot (2 \cdot x - 3) \cdot (4 \cdot x^2 + 6 \cdot x + 9)$
factor(108-32x^3,x)
MAIN   DEG AUTO   FUNC   1/30

TI-89 calculator solution for Example 28.

Exercise 4 ◆ Other Factorable Expressions

Factor completely, by hand or by calculator. Check your results.

Factoring by Grouping

1. $a^3 + 3a^2 + 4a + 12$
2. $x^3 + x^2 + x + 1$
3. $x^3 - x^2 + x - 1$
4. $2x^3 - x^2 + 4x - 2$
5. $x^2 - bx + 3x - 3b$
6. $ab + a - b - 1$
7. $3x - 2y - 6 + xy$
8. $x^2y^2 - 3x^2 - 4y^2 + 12$
9. $x^2 + y^2 + 2xy - 4$
10. $x^2 - 6xy + 9y^2 - a^2$
11. $m^2 - n^2 - 4 + 4n$
12. $p^2 - r^2 - 6pq + 9q^2$

Sum or Difference of Two Cubes

13. $64 + x^3$
14. $1 - 64y^3$
15. $2a^3 - 16$
16. $a^3 - 27$
17. $x^3 - 1$
18. $x^3 - 64$
19. $x^3 + 1$
20. $a^3 - 343$
21. $a^3 + 64$
22. $x^3 + 343$
23. $x^3 + 125$
24. $64a^3 + 27$
25. $216 - 8a^3$
26. $343 - 27y^3$

Applications

27. The volume of a hollow spherical shell having an inside radius of r_1 and an outside radius r_2 is $\frac{4}{3}\pi r_2{}^3 - \frac{4}{3}\pi r_1{}^3$. Factor this expression completely.

28. A cistern is in the shape of a hollow cube whose inside dimension is s and whose outside dimension is S. If it is made of concrete of density d, its weight is $dS^3 - ds^3$. Factor completely.

29. *Writing*: We have studied the factoring of seven different types of expressions in this chapter. List them and give an example of each. State in words how to recognize each and how to tell one from the other. Also list at least four other expressions that are *not* one of the given seven, and state why each is different from those we have studied.

11–5 Simplifying Fractions

From factoring, we move on to our study of fractions and fractional equations. As usual we start with a reminder of some definitions.

A fraction has a *numerator*, a *denominator*, and a *fraction line*, or bar.

$$\text{fraction line} \rightarrow \frac{a}{b} \begin{array}{l} \leftarrow \text{ numerator} \\ \leftarrow \text{ denominator} \end{array}$$

This fraction can also be written on a single line as *a/b*. A fraction is a way of indicating a *quotient* of two quantities. Thus the fraction *a/b* can be read as "*a divided by b*." Another way of indicating this same division is $a \div b$. The quotient of two quantities is also spoken of as the *ratio* of those quantities. Thus *a/b* is the ratio of *a* to *b*.

Recall that the bar or fraction line is a *symbol of grouping*. The quantities in the numerator must be treated as a whole, and the quantities in the denominator must be treated as a whole.

◆◆◆ **Example 29:** In the fraction

$$\frac{x + 4}{2}$$

the numerator $x + 4$ must be treated as a whole. The 4 in the numerator, for example, *cannot* be divided by the 2 in the denominator, without also dividing the x by 2.

Division by Zero

Since division by zero is not permitted, it should be understood in our work with fractions that *the denominator cannot be zero*.

◆◆◆ **Example 30:** What values of x are not permitted in the following fraction

$$\frac{3x}{x^2 + x - 6}$$

Solution: Factoring the denominator, we get

$$\frac{3x}{x^2 + x - 6} = \frac{3x}{(x - 2)(x + 3)}$$

We see that an x equal to 2 or to -3 will make $(x - 2)$ or $(x + 3)$ equal to zero. This will result in division by zero, so these values are not permitted. ◆◆◆

Common Fractions and Algebraic Fractions

A *common* fraction is one whose numerator and denominator are *both integers*. An *algebraic* fraction is one whose numerator and/or denominator contain *literal* quantities.

◆◆◆ **Example 31:**

(a) The following are common fractions.

$$\frac{2}{3}, \quad \frac{-9}{5}, \quad \frac{-124}{125}, \quad \text{and} \quad \frac{18}{-11}$$

(b) The following are algebraic fractions:

$$\frac{x}{y}, \quad \frac{\sqrt{x+2}}{x}, \quad \frac{3}{y}, \quad \text{and} \quad \frac{x^2}{x-3} \qquad \text{◆◆◆}$$

Rational Algebraic Fractions

An algebraic fraction is called *rational* if the numerator and the denominator are both *polynomials*. Recall that a polynomial is an expression in which the exponents are nonnegative integers.

◆◆◆ **Example 32:** The following are rational fractions:

$$\frac{3}{y}, \quad \frac{3}{w^3}, \quad \text{and} \quad \frac{x^2}{x-3}$$

But $\dfrac{\sqrt{x+2}}{x}$ is not a rational fraction. ◆◆◆

Proper and Improper Fractions

A *proper* common fraction is one whose numerator is smaller than its denominator.

◆◆◆ **Example 33:** $\frac{3}{5}, \frac{1}{3}$, and $\frac{9}{11}$ are proper fractions, whereas $\frac{8}{5}, \frac{3}{2}$, and $\frac{7}{4}$ are *improper* fractions. ◆◆◆

A proper *algebraic* fraction is a rational fraction whose numerator is of *lower degree* than the denominator.

◆◆◆ **Example 34:** The following are proper fractions:

$$\frac{x}{x^2+2} \quad \text{and} \quad \frac{x^2+2x-3}{x^3+9}$$

The following are improper fractions:

$$\frac{x^3-2}{x^2+x-3} \quad \text{and} \quad \frac{3+2y^2}{y} \qquad \text{◆◆◆}$$

Mixed Form

A *mixed number* is the sum of an integer and a fraction.

◆◆◆ **Example 35:**

(a) The following are mixed numbers:

$$2\frac{1}{2}, \quad 5\frac{3}{4}, \quad \text{and} \quad 3\frac{1}{3}$$

(b) To change an improper fraction to a mixed number, we divide numerator by denominator.

$$\frac{45}{7} = 6 + \frac{3}{7} = 6\frac{3}{7}$$

(c) To change a mixed number to an improper fraction, write the integer part as a fraction, and add it to the fractional part.

$$5\frac{2}{3} = 5 + \frac{2}{3} = \frac{15}{3} + \frac{2}{3} = \frac{17}{3}$$ ◆◆◆

A *mixed expression* is the sum or difference of a polynomial and a rational algebraic fraction.

◆◆◆ **Example 36:** The following are mixed expressions:

$$3x - 2 + \frac{1}{x} \quad \text{and} \quad y - \frac{y}{y^2 + 1}$$ ◆◆◆

Simplifying a Fraction by Reducing to Lowest Terms

We reduce a fraction to lowest terms by dividing both numerator and denominator by any factor that is contained in both.

$$\frac{ad}{bd} = \frac{a}{b} \qquad\qquad \mathbf{41}$$

◆◆◆ **Example 37:** Reduce the following to lowest terms. Write the answer without negative exponents.

(a) $\dfrac{9}{12} = \dfrac{3(3)}{4(3)} = \dfrac{3}{4}$

(b) $\dfrac{3x^2yz}{9xy^2z^3} = \dfrac{3}{9} \cdot \dfrac{x^2}{x} \cdot \dfrac{y}{y^2} \cdot \dfrac{z}{z^3} = \dfrac{x}{3yz^2}$ ◆◆◆

When possible, factor the numerator and the denominator. Then divide both numerator and denominator by any factors common to both.

◆◆◆ **Example 38:** Here are some examples where factoring the numerator and/or the denominator will aid in simplifying a fraction.

(a) $\dfrac{2x^2 + x}{3x} = \dfrac{(2x + 1)x}{3(x)} = \dfrac{2x + 1}{3}$

(b) $\dfrac{ab + bc}{bc + bd} = \dfrac{b(a + c)}{b(c + d)} = \dfrac{a + c}{c + d}$

(c) $\dfrac{2x^2 - 5x - 3}{4x^2 - 1} = \dfrac{(2x + 1)(x - 3)}{(2x + 1)(2x - 1)} = \dfrac{x - 3}{2x - 1}$

(d) $\dfrac{x^2 - ax + 2bx - 2ab}{2x^2 + ax - 3a^2} = \dfrac{x(x - a) + 2b(x - a)}{(x - a)(2x + 3a)}$

$$= \dfrac{(x - a)(x + 2b)}{(x - a)(2x + 3a)} = \dfrac{x + 2b}{2x + 3a}$$ ◆◆◆

The process of striking out the same factors from numerator and denominator is called *canceling*.

> **Common Errors**
>
> Most students love canceling because they think they can cross out any term that stands in their way. If you use canceling, use it carefully! If a factor is missing from *even one term* in the numerator or denominator, that factor *cannot* be canceled.
>
> $$\frac{xy - z}{wx} \neq \frac{y - z}{w}$$
>
> We may divide (or multiply) the numerator and denominator by the same quantity (Eq. 41), but we *may not add or subtract* the same quantity in the numerator and denominator, as this will change the value of the fraction. For example,
>
> $$\frac{3}{5} \neq \frac{3 + 1}{5 + 1} = \frac{4}{6} = \frac{2}{3}$$

Simplifying Fractions by Calculator

Many fractions can be simplified by a computer algebra system. On the TI-89 calculator, we use the **comDenom** (common denominator) operation found in the **Algebra** menu.

◆◆◆ **Example 39:** Simplify the fraction $\frac{2x^2 + x}{3x}$.

Solution: The keystrokes are shown. We select **comDenom**, enter the expression in parentheses, and press ENTER.   ◆◆◆

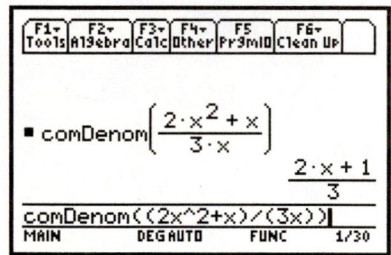

TI-89 screen for Example 39.

Simplifying a Fraction by Manipulating Signs

Sometimes a fraction can be reduced simply by working with its algebraic signs.

◆◆◆ **Example 40:** Simplify $\frac{x - a}{a - x}$.

Solution: Let us factor a (-1) from the numerator.

$$\frac{x - a}{a - x} = \frac{(-1)(-x + a)}{a - x}$$

$$= \frac{-(a - x)}{a - x} = -1 \qquad ◆◆◆$$

When there is a minus sign on the entire fraction, it is the same as multiplying the *numerator* of the fraction by (-1), as in the following example.

◆◆◆ **Example 41:** Simplify the fraction $-\frac{3x - 2}{2 - 3x}$.

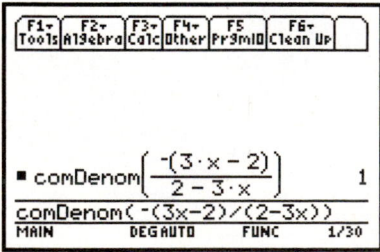

TI-89 calculator solution for Example 41.

Solution: Let us rewrite the minus sign on the entire fraction as $\frac{-1}{1}$.

$$-\frac{3x - 2}{2 - 3x} = \frac{-1}{1} \cdot \frac{3x - 2}{2 - 3x}$$

$$= \frac{-1(3x - 2)}{2 - 3x}$$

$$= \frac{-3x + 2}{2 - 3x} = 1 \qquad ◆◆◆$$

Exercise 5 ◆ Simplifying Fractions

In each fraction, what values of x, if any, are not permitted?

1. $\dfrac{12}{x}$

2. $\dfrac{x}{12}$

3. $\dfrac{18}{x - 5}$

4. $\dfrac{5x}{x^2 - 49}$

5. $\dfrac{7}{x^2 - 3x + 2}$

6. $\dfrac{3x}{8x^2 - 14x + 3}$

Hint: Factor the denominators in problems 4, 5, and 6.

Simplify each fraction by manipulating the algebraic signs.

7. $\dfrac{a - b}{b - a}$

8. $-\dfrac{2x - y}{y - 2x}$

9. $\dfrac{(a - b)(c - d)}{b - a}$

10. $\dfrac{w(x - y - z)}{y - x + z}$

Reduce to lowest terms. Write your answers without negative exponents. Do some algebraic fractions by calculator.

11. $\dfrac{14}{21}$

12. $\dfrac{81}{18}$

13. $\dfrac{75}{35}$

14. $\dfrac{36}{44}$

15. $\dfrac{2ab}{6b}$

16. $\dfrac{12m^2n}{15mn^2}$

17. $\dfrac{21m^2p^2}{28mp^4}$

18. $\dfrac{abx - bx^2}{acx - cx^2}$

19. $\dfrac{x^2 - 4}{x^3 - 8}$

◆◆◆

Challenge Problems

20. $\dfrac{2a^3 + 6a^2 - 8a}{2a^3 + 2a^2 - 4a}$

21. $\dfrac{2m^3n - 2m^2n - 24mn}{6m^3 + 6m^2 - 36m}$

22. $\dfrac{9x^3 - 30x^2 + 25x}{3x^4 - 11x^3 + 10x^2}$

23. $\dfrac{2a^2 - 2}{a^2 - 2a + 1}$

24. $\dfrac{3a^2 - 4ab + b^2}{a^2 - ab}$

25. $\dfrac{x^2 - z^2}{x^3 - z^3}$

26. $\dfrac{2x^2}{6x - 4x^2}$

27. $\dfrac{2a^2 - 8}{2a^2 - 2a - 12}$

28. $\dfrac{2a^2 + ab - 3b^2}{a^2 - ab}$

29. $\dfrac{x^2 - 1}{2xy + 2y}$

30. $\dfrac{x^3 - a^2x}{x^2 - 2ax + a^2}$

31. $\dfrac{2x^4y^4 + 2}{3x^8y^8 - 3}$

32. $\dfrac{18a^2c - 6bc}{42a^2d - 14bd}$

33. *Writing:* In simplifying a fraction, we are careful not to change its value. If that's the case, and the new fraction is "the same" as the old, then why bother? Write a few paragraphs explaining why you think it's valuable to simplify fractions (and other expressions as well) or whether there is some advantage to just leaving them alone.

11–6   Multiplying and Dividing Fractions

Now we will do the four basic operations with fractions: adding, subtracting, multiplying and dividing. We will start with multiplication and division.

Multiplying Fractions

We multiply a fraction $\dfrac{a}{b}$ by another fraction $\dfrac{c}{d}$ as follows:

Multiplying Fractions	$$\frac{a}{b} \cdot \frac{c}{d} = \frac{ac}{bd}$$ *The product of two or more fractions is a fraction whose numerator is the product of the numerators of the original fractions and whose denominator is the product of the denominators of the original fractions.*	**42**

◆◆◆ **Example 42:** These examples show the multiplication of fractions. Note that in (b) we have canceled out the common factors 2 and 3 from the numerator and denominator after multiplying, and in (c) we have converted the mixed numbers to improper fractions before multiplying.

(a) $\dfrac{2}{3} \cdot \dfrac{5}{7} = \dfrac{2(5)}{3(7)} = \dfrac{10}{21}$ (b) $\dfrac{1}{2} \cdot \dfrac{2}{3} \cdot \dfrac{3}{5} = \dfrac{1(2)(3)}{2(3)(5)} = \dfrac{1}{5}$

(c) $5\dfrac{2}{3} \cdot 3\dfrac{1}{2} = \dfrac{17}{3} \cdot \dfrac{7}{2} = \dfrac{119}{6}$ ◆◆◆

Common Error	When multiplying mixed numbers, students sometimes try to multiply the whole parts and the fractional parts separately. $$2\frac{2}{3} \times 4\frac{3}{5} \neq 8\frac{6}{15}$$ The correct way is to write each mixed number as an improper fraction and to multiply as shown. $$2\frac{2}{3} \times 4\frac{3}{5} = \frac{8}{3} \times \frac{23}{5} = \frac{184}{15} = 12\frac{4}{15}$$

We multiply algebraic fractions in the same way. It's a good idea to leave any products in *factored form* until after you simplify.

◆◆◆ **Example 43:** Try multiplying and simplifying these algebraic fractions and see if you get the same result.

(a) $\dfrac{2a}{3b} \cdot \dfrac{5c}{4a} = \dfrac{10ac}{12ab} = \dfrac{5c}{6b}$

(b) $\dfrac{x}{x+2} \cdot \dfrac{x^2-4}{x^3} = \dfrac{x(x^2-4)}{(x+2)x^3}$

$\qquad\qquad\qquad = \dfrac{x(x+2)(x-2)}{(x+2)x^3} = \dfrac{x-2}{x^2}$

(c) $\dfrac{x^2 + x - 2}{x^2 - 4x + 3} \cdot \dfrac{2x^2 - 3x - 9}{2x^2 + 7x + 6}$

$= \dfrac{(x - 1)(x + 2)(x - 3)(2x + 3)}{(x - 1)(x - 3)(x + 2)(2x + 3)} = 1$ ♦♦♦

Dividing Fractions

To divide one fraction, *a/b*, by another fraction, *c/d*,

$$\dfrac{\dfrac{a}{b}}{\dfrac{c}{d}}$$

we multiply numerator and denominator by *d/c*, as follows:

$$\dfrac{\dfrac{a}{b} \cdot \dfrac{d}{c}}{\dfrac{c}{d} \cdot \dfrac{d}{c}} = \dfrac{\dfrac{a}{b} \cdot \dfrac{d}{c}}{\dfrac{cd}{cd}} = \dfrac{\dfrac{a}{b} \cdot \dfrac{d}{c}}{1} = \dfrac{a}{b} \cdot \dfrac{d}{c} = \dfrac{ad}{bc}$$

We see that dividing by a fraction is the same as *multiplying by the reciprocal* of that fraction.

Division of Fractions	$\dfrac{a}{b} \div \dfrac{c}{d} = \dfrac{a}{b} \cdot \dfrac{d}{c} = \dfrac{ad}{bc}$ When dividing fractions, invert the divisor and multiply.	**43**

♦♦♦ **Example 44:** These following examples show the division and simplification of common fractions and mixed numbers.

(a) $\dfrac{2}{3} \div \dfrac{5}{7} = \dfrac{2}{3} \cdot \dfrac{7}{5} = \dfrac{14}{15}$

(b) $3\dfrac{2}{5} \div 2\dfrac{4}{15} = \dfrac{17}{5} \div \dfrac{34}{15}$

$= \dfrac{17}{5} \times \dfrac{15}{34} = \dfrac{17(5)(3)}{5(2)(17)} = \dfrac{3}{2}$ ♦♦♦

♦♦♦ **Example 45:** *An Application.* A $2^3/_4$ inch long bolt is to be threaded into a plate to a depth *x* equal to $^3/_8$ of its length. Find *x*.

Solution: We change the mixed number to an improper fraction, multiply, and simplify.

$$2\dfrac{3}{4} \times \dfrac{3}{8} = \dfrac{11}{4} \times \dfrac{3}{8}$$

$$= \dfrac{33}{32} = 1\dfrac{1}{32} \text{ in.}$$ ♦♦♦

♦♦♦ **Example 46:** *An application.* The *pitch* of a screw thread is the distance between adjacent threads. The bolt, Fig. 11–13 has a pitch of $\frac{1}{16}$. Find the number of threads in $1\frac{3}{8}$ in.

Solution: We must divide the length by the pitch to get the number of threads. We change the mixed number to an improper fraction, invert the divisor, multiply, and simplify.

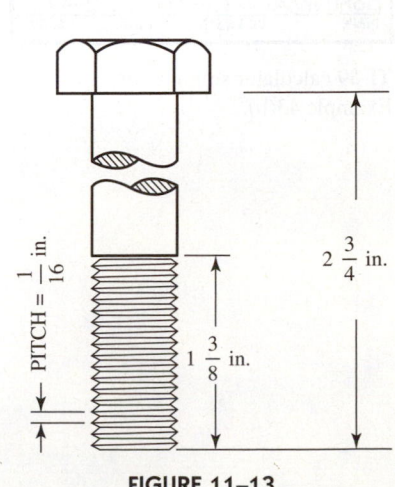

PITCH = $\dfrac{1}{16}$ in.

$1\dfrac{3}{8}$ in.

$2\dfrac{3}{4}$ in.

FIGURE 11–13

$$1\frac{3}{8} \div \frac{1}{16} = \frac{11}{8} \div \frac{1}{16}$$

$$= \frac{11}{8} \times \frac{16}{1}$$

$$= 22 \text{ threads} \qquad \blacklozenge\blacklozenge\blacklozenge$$

◆◆◆ **Example 47:** These examples show the division of algebraic fractions.

(a) $\dfrac{x}{y} \div \dfrac{x+2}{y-1} = \dfrac{x}{y} \cdot \dfrac{y-1}{x+2}$

$$= \frac{x(y-1)}{y(x+2)} = \frac{xy - x}{xy + 2y}$$

(b) $\dfrac{x^2 + x - 2}{x} \div \dfrac{x+2}{x^2} = \dfrac{x^2 + x - 2}{x} \cdot \dfrac{x^2}{x+2}$

$$= \frac{(x+2)(x-1)x^2}{x(x+2)} = x(x-1)$$

(c) $x \div \dfrac{\pi r^2 x}{4} = \dfrac{x}{1} \div \dfrac{\pi r^2 x}{4} = \dfrac{x}{1} \cdot \dfrac{4}{\pi r^2 x} = \dfrac{4}{\pi r^2}$ ◆◆◆

Common Error	We invert the *divisor* and multiply. Be sure not to invert the dividend.
	Do not do any canceling until *after* inverting the divisor.

Multiplying and Dividing Fractions by Calculator

To multiply or divide fractions, we again use the **comDenom** operation from the **Algebra** menu.

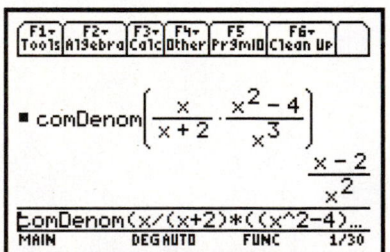

TI-89 calculator solution for Example 43(a).

TI-89 calculator solution for Example 43(b).

◆◆◆ **Example 48:** At left are the TI-89 calculator screens for Examples 43(a) and 43(b). ◆◆◆

Exercise 6 ◆ Multiplying and Dividing Fractions

Multiply and reduce. Do some by calculator.

1. $\dfrac{1}{3} \times \dfrac{2}{5}$

2. $\dfrac{3}{7} \times \dfrac{21}{24}$

3. $\dfrac{2}{3} \times \dfrac{9}{7}$

4. $\dfrac{11}{3} \times 7$

5. $\dfrac{2}{3} \times 3\dfrac{1}{5}$

6. $3 \times 7\dfrac{2}{5}$

7. $3\dfrac{3}{4} \times 2\dfrac{1}{2}$

8. $\dfrac{3}{5} \times \dfrac{2}{7} \times \dfrac{5}{9}$

9. $3 \times \dfrac{5}{8} \times \dfrac{4}{5}$

10. $\dfrac{15a^2}{7b^2} \cdot \dfrac{28ab}{9a^3c}$

11. $\dfrac{a^4 b^4}{2a^2 y^n} \cdot \dfrac{a^2 x}{xy^n}$

12. $\dfrac{x+y}{x-y} \cdot \dfrac{x^2 - y^2}{(x+y)^2}$

13. $\dfrac{x^2 - a^2}{xy} \cdot \dfrac{xy}{x+a}$

14. $\dfrac{a}{x-y} \cdot \dfrac{b}{x+y}$

15. $\dfrac{x+y}{10} \cdot \dfrac{ax}{3(x+y)}$

16. $\dfrac{c}{x^2 - y^2} \cdot \dfrac{d}{x^2 - y^2}$

Divide and reduce. Try some by calculator.

17. $\dfrac{7}{9} \div \dfrac{5}{3}$

18. $3\dfrac{7}{8} \div 2$

19. $\dfrac{7}{8} \div 4$

20. $\dfrac{9}{16} \div 8$

21. $24 \div \dfrac{5}{8}$

22. $\dfrac{5}{8} \div 2\dfrac{1}{4}$

23. $2\dfrac{7}{8} \div 1\dfrac{1}{2}$

24. $2\dfrac{5}{8} \div \dfrac{1}{2}$

25. $50 \div 2\dfrac{3}{5}$

26. $\dfrac{5abc^3}{3x^2} \div \dfrac{10ac^3}{6bx^2}$

27. $\dfrac{7x^2y}{3ad} \div \dfrac{2xy^2}{3a^2d}$

28. $\dfrac{4a^3x}{6dy^2} \div \dfrac{2a^2x^2}{8a^2y}$

29. $\dfrac{5x^2y^3z}{6a^2b^2c} \div \dfrac{10xy^3z^2}{8ab^2c^2}$

30. $\dfrac{3an + cm}{x^2 - y^2} \div (x^2 + y^2)$

31. $\dfrac{a^2 + 4a + 4}{d + c} \div (a + 2)$

32. $\dfrac{ac + ad + bc + bd}{c^2 - d^2} \div (a + b)$

33. $\dfrac{5(x + y)^2}{x - y} \div (x + y)$

34. $\dfrac{1}{x^2 + 17x + 30} \div \dfrac{1}{x + 15}$

35. $\dfrac{5xy}{a - x} \div \dfrac{10xy}{a^2 - x^2}$

36. $\dfrac{3x - 6}{2x + 3} \div \dfrac{x^2 - 4}{4x^2 + 2x - 6}$

37. $\dfrac{a^2 - a - 2}{5a - 1} \div \dfrac{a - 2}{10a^2 + 13a - 3}$

Applications

38. An $8\dfrac{3}{4}$ in. length of a certain steel bar weighs $1\dfrac{1}{8}$ lb. Find the weight of a similar bar of length $22\dfrac{2}{3}$ in. in length.

39. A machine part, Fig. 11–14, is to have 5 holes equally spaced by $1\dfrac{7}{8}$ in. Find the distance x.

40. The framing in Fig. 11–15 has 7 studs equally spaced. Find the distance x, between their centers.

41. For a thin lens Fig. 11–16 the relationship between the focal length f, the object distance p, and the image distance q is

$$f = \frac{pq}{p + q}$$

A second lens of focal length f_1 has the same object distance p but a different image distance q_1.

$$f_1 = \frac{pq_1}{p + q_1}$$

Find the ratio f/f_1 and simplify.

42. The mass density of an object is its mass divided by its volume. Write and simplify an expression for the density of a sphere having a mass m and a volume equal to $4\pi r^3/3$.

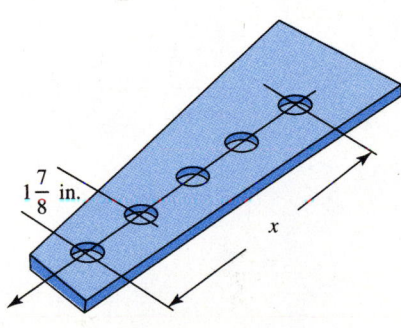

$1\dfrac{7}{8}$ in.

x

FIGURE 11–14

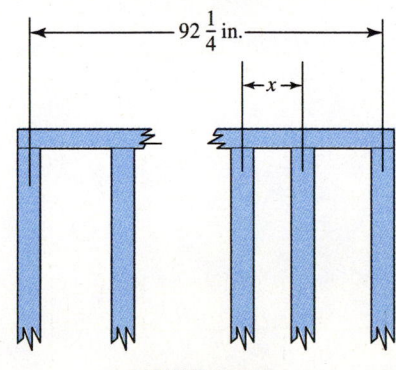

$92\dfrac{1}{4}$ in.

x

FIGURE 11–15

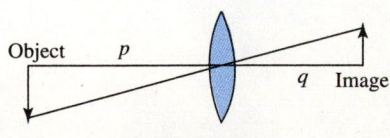

Object p

q Image

FIGURE 11–16

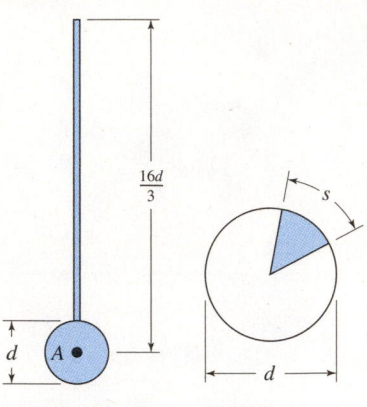

FIGURE 11–17   FIGURE 11–18

43. The pressure on a surface is equal to the total force divided by the area. Write and simplify an expression for the pressure on a circular surface of area $\pi d^2/4$ subjected to a distributed load F.

44. The stress on a bar in tension is equal to the load divided by the cross-sectional area. Write and simplify an expression for the stress in a bar having a trapezoidal cross section of area $(a + b)h/2$, subject to a load P.

45. The acceleration on a body is equal to the force on the body divided by its mass, and the mass equals the volume of the object times the density. Write and simplify an expression for the acceleration of a sphere having a volume $4\pi r^3/3$ and a density D, subjected to a force F.

46. To find the moment of the area A in Fig. 11–17, we must multiply the area $\pi d^2/4$ by the distance to the pivot, $16d/3$. Multiply and simplify.

47. The circle in Fig. 11–18 has an area $\pi d^2/4$, and the sector has an area equal to $ds/4$. Find the ratio of the area of the circle to the area of the sector.

11–7   Adding and Subtracting Fractions

From multiplying and dividing fractions, we now move on the operations of addition and subtraction.

Similar Fractions

Similar fractions (also called *like* fractions) are those having the same (common) denominator.

Addition and Subtraction of Fractions	$$\frac{a}{b} \pm \frac{c}{b} = \frac{a \pm c}{b}$$ *To add or subtract similar fractions, combine the numerators and place them over their common denominator.*	44

◆◆◆ **Example 49:** These examples show the addition and subtraction of common and algebraic fractions having the same denominators.

(a) $\dfrac{2}{3} + \dfrac{5}{3} = \dfrac{2 + 5}{3} = \dfrac{7}{3}$

(b) $\dfrac{1}{x} + \dfrac{3}{x} = \dfrac{1 + 3}{x} = \dfrac{4}{x}$

(c) $\dfrac{3x}{x + 1} - \dfrac{5}{x + 1} + \dfrac{x^2}{x + 1} = \dfrac{3x - 5 + x^2}{x + 1}$ ◆◆◆

Least Common Denominator

The *least common denominator*, or LCD (also called the *lowest* common denominator), is the smallest expression that is exactly divisible by each of the denominators. Thus the LCD must contain all the prime factors of each of the denominators. The common denominator of two or more fractions is simply the product of the denominators of those fractions. To find the *least* common denominator, drop any prime factor from one denominator that also appears in another denominator.

◆◆◆ **Example 50:** Find the LCD for the two fractions $\frac{3}{8}$ and $\frac{1}{18}$.

Solution: Factoring each denominator, we obtain

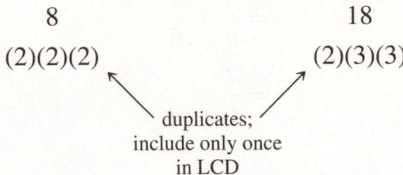

$$8 \qquad\qquad\qquad 18$$
$$(2)(2)(2) \qquad\qquad (2)(3)(3)$$

duplicates;
include only once
in LCD

Dropping one of the 2's that appear in both sets of factors, we find that our LCD is then the product of these factors.

$$\text{LCD} = (2)(2)(2)(3)(3) = 72 \qquad\qquad ◆◆◆$$

For this simple problem, you probably found the LCD by inspection in less time than it took to read this example, and you are probably wondering what all the fuss is about. What we are really doing is developing a *method* that we can use on algebraic fractions when the LCD is *not* obvious.

◆◆◆ **Example 51:** Find the LCD for the fractions

$$\frac{5}{x^2 + x}, \quad \frac{x}{x^2 - 1}, \quad \text{and} \quad \frac{9}{x^3 - x^2}$$

Solution: The denominator $x^2 + x$ has the prime factors x and $x + 1$.
The denominator $x^2 - 1$ has the prime factors $x + 1$ and $x - 1$.
The denominator $x^3 - x^2$ has the prime factors x, x, and $x - 1$.
The factor $(x + 1)$ appears in two denominators, as do the factors $(x - 1)$ and x. Dropping the duplicates, our LCD is then

$$(x)(x)(x + 1)(x - 1) \quad \text{or} \quad x^2(x^2 - 1) \qquad\qquad ◆◆◆$$

Combining Fractions with Different Denominators

■ Exploration:

Try this. You join a party in the next dorm room, bringing with you $\frac{3}{4}$ of a pizza. They already have $\frac{2}{3}$ of the same size pizza. How would you slice the two pizzas so that every slice from both was the same size? If you did this, how large would each slice be? What does this have to do with adding fractions? ■

To combine fractions with different denominators, first find the LCD. Then multiply numerator and denominator of each fraction by that quantity that will make the denominator equal to the LCD. Finally, combine as shown previously and simplify.

The method for adding and subtracting unlike fractions can be summarized as follows.

Combining Unlike Fractions	$\dfrac{a}{b} \pm \dfrac{c}{d} = \dfrac{ad}{bd} \pm \dfrac{bc}{bd} = \dfrac{ad \pm bc}{bd}$	45

◆◆◆ **Example 52:** Add $\frac{1}{2}$ and $\frac{2}{3}$.

Solution: The LCD is 6, so

$$
\begin{aligned}
\frac{1}{2} + \frac{2}{3} &= \frac{1}{2}\left(\frac{3}{3}\right) + \frac{2}{3}\left(\frac{2}{2}\right) \\
&= \frac{3}{6} + \frac{4}{6} \\
&= \frac{7}{6}
\end{aligned}
$$

◆◆◆

To combine integers and fractions, treat the integer as a fraction having 1 as a denominator, and combine as shown above. The same procedure may be used to change a *mixed number* to an *improper fraction*.

◆◆◆ **Example 53:** This example shows the addition of an integer and a common fraction.

$$
\begin{aligned}
3 + \frac{2}{9} &= \frac{3}{1} + \frac{2}{9} = \frac{3}{1} \cdot \frac{9}{9} + \frac{2}{9} \\
&= \frac{27}{9} + \frac{2}{9} = \frac{29}{9}
\end{aligned}
$$

◆◆◆

◆◆◆ **Example 54:** Combine: $\dfrac{x}{2y} - \dfrac{5}{x}$.

Solution: The LCD will be the product of the two denominators, or $2xy$. So

$$
\begin{aligned}
\frac{x}{2y} - \frac{5}{x} &= \frac{x}{2y}\left(\frac{x}{x}\right) - \frac{5}{x}\left(\frac{2y}{2y}\right) \\
&= \frac{x^2}{2xy} - \frac{10y}{2xy} \\
&= \frac{x^2 - 10y}{2xy}
\end{aligned}
$$

◆◆◆

It is not necessary to write the fractions with the *least* common denominator; any common denominator will work as well. But your final result will then have to be reduced to lowest terms.

The procedure is the same, of course, even when the denominators are more complicated.

◆◆◆ **Example 55:** Combine the fractions

$$
\frac{x + 2}{x - 3} + \frac{2x + 1}{3x - 2}
$$

Solution: Our LCD is $(x - 3)(3x - 2)$, so

$$
\begin{aligned}
\frac{x + 2}{x - 3} + \frac{2x + 1}{3x - 2} &= \frac{(x + 2)(3x - 2)}{(x - 3)(3x - 2)} + \frac{(2x + 1)(x - 3)}{(3x - 2)(x - 3)} \\
&= \frac{(3x^2 + 4x - 4) + (2x^2 - 5x - 3)}{(x - 3)(3x - 2)} \\
&= \frac{5x^2 - x - 7}{3x^2 - 11x + 6}
\end{aligned}
$$

◆◆◆

Addition and Subtraction by Calculator

We can use **comDenom** again to add and subtract fractions.

◆◆◆ **Example 56:** The screen for adding the fractions in Example 55 is shown. We select **comDenom**, from the **Algebra** menu enter the expression in parentheses, and press ENTER. ◆◆◆

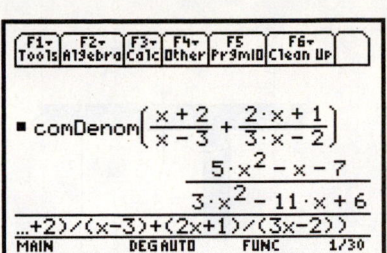

TI-89 screen for Example 56.

An Application

◆◆◆ **Example 57:** The day shift in a certain factory can assemble 26 computers in 5 days, and the night shift can assemble 13 computers in 4 days. How many computers can both shifts assemble in one day?

Solution: The day shift can assemble $\frac{26}{5}$ of a computer per day. The night shift can assemble $\frac{13}{4}$ of a computer per day. Together they can assemble

$$\frac{26}{5} + \frac{13}{4}$$

computers per day. Combining fractions over the LCD, 20, we get

$$\frac{26}{5} + \frac{13}{4} = \frac{26(4)}{20} + \frac{13(5)}{20}$$

$$= \frac{104 + 65}{20} = \frac{169}{20}$$

$$= 8\frac{9}{20} \text{ computers per day} \qquad \text{◆◆◆}$$

Exercise 7 ◆ Adding and Subtracting of Fractions

Common Fractions and Mixed Numbers Combine and simplify.

Don't use your calculator for these numerical problems. The practice you get working with common fractions will help you when doing algebraic fractions.

1. $\frac{3}{5} + \frac{2}{5}$

2. $\frac{1}{8} - \frac{3}{8}$

3. $\frac{2}{7} + \frac{5}{7} - \frac{6}{7}$

4. $\frac{5}{9} + \frac{7}{9} - \frac{1}{9}$

5. $\frac{1}{3} - \frac{7}{3} + \frac{11}{3}$

6. $\frac{1}{5} - \frac{9}{5} + \frac{12}{5} - \frac{2}{5}$

7. $\frac{1}{2} + \frac{2}{3}$

8. $\frac{3}{5} - \frac{1}{3}$

9. $\frac{3}{4} + \frac{7}{16}$

10. $\frac{2}{3} + \frac{3}{7}$

11. $\frac{5}{9} - \frac{1}{3} + \frac{3}{18}$

12. $\frac{1}{2} + \frac{1}{3} + \frac{1}{5}$

13. $2 + \frac{3}{5}$

14. $3 - \frac{2}{3} + \frac{1}{6}$

Algebraic Fractions Combine and simplify. Try some by calculator.

15. $\frac{1}{a} + \frac{5}{a}$

16. $\frac{3}{x} + \frac{2}{x} - \frac{1}{x}$

17. $\frac{2a}{y} + \frac{3}{y} - \frac{a}{y}$

18. $\frac{x}{3a} - \frac{y}{3a} + \frac{z}{3a}$

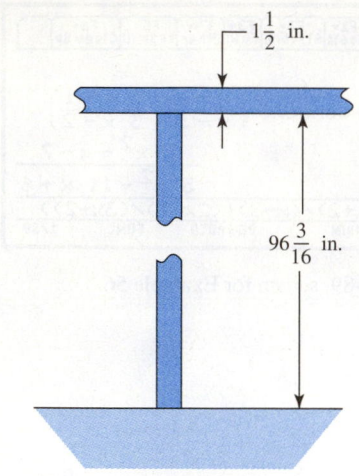

FIGURE 11–19

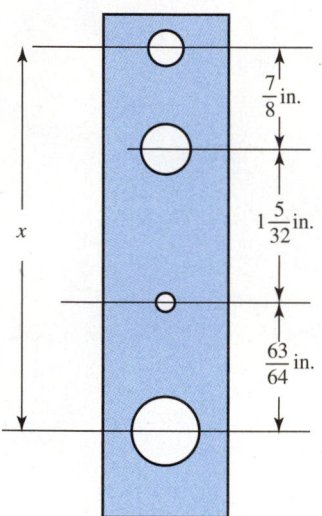

FIGURE 11–20

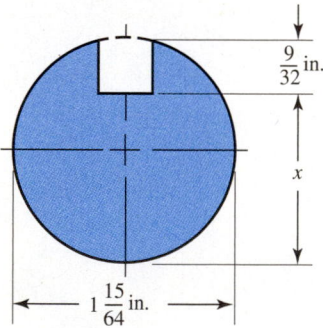

FIGURE 11–21

19. $\dfrac{5x}{2} - \dfrac{3x}{2}$

20. $\dfrac{7}{x + 2} - \dfrac{5}{x + 2}$

21. $\dfrac{3x}{a - b} + \dfrac{2x}{b - a}$

22. $\dfrac{a}{x^2} - \dfrac{b}{x^2} + \dfrac{c}{x^2}$

23. $\dfrac{3a}{2x} + \dfrac{2a}{5x}$

24. $\dfrac{1}{x} + \dfrac{1}{y} + \dfrac{1}{z}$

25. $\dfrac{a + b}{3} - \dfrac{a - b}{2}$

26. $\dfrac{3}{a + b} + \dfrac{2}{a - b}$

27. $\dfrac{4}{x - 1} - \dfrac{5}{x + 1}$

28. $\dfrac{x + 1}{x - 1} - \dfrac{x - 1}{x + 1}$

Applications

Treat the given numbers in these problems as exact, and leave your answers in fractional form. Do not use your calculator.

29. A stud $96\,\dfrac{3}{16}$ in. long is capped by a plate $1\dfrac{1}{2}$ in. thick, Fig. 11–19. Find their combined height.

30. Find the distance x in the drilled plate, Fig. 11–20.

31. There boards having lengths of $42\,\dfrac{3}{8}$ in. and $38\,\dfrac{5}{16}$ in. are cut from board $97\,\dfrac{3}{4}$ in. long. The saw kerf is $\dfrac{5}{32}$ in. What is the length of the remaining piece?

32. Find the distance x in the shaft, Fig. 11–21.

33. A certain work crew can grade 7 mi of roadbed in 3 days, and another crew can do 9 mi in 4 days. How much can both crews together grade in 1 day? *Hint:* First find the amount that each crew can do in one day (e.g., the first crew can grade $\dfrac{7}{3}$ mi per day). Then add the separate amounts to get the daily total. You can use a similar approach to the other work problems in this group.

34. Liquid is running into a tank from a pipe, Fig. 11–22, that can fill four tanks in 3 days. Meanwhile, liquid is running out from a drain that can empty two tanks in 4 days. What will be the net change in volume in 1 day?

35. A planer makes a 1-m cutting stroke at a rate of 15 m/min and returns at 75 m/min. How long does it take for the cutting stroke and return? (distance = rate × time)

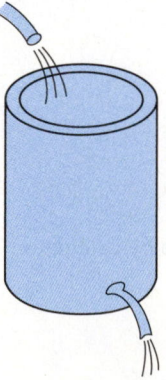

FIGURE 11–22

36. The equivalent resistance R of two resistors R_1 and R_2 is given by $\dfrac{1}{R} = \dfrac{1}{R_1} + \dfrac{1}{R_2}$. What is the equivalent resistance of a 5-Ω and a 15-Ω resistor wired in parallel, Fig. 11–23?

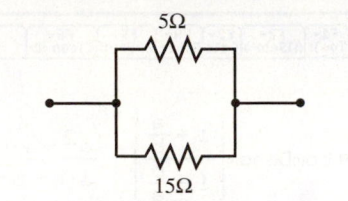

FIGURE 11–23

37. One crew can put together five machines in 8 days. Another crew can assemble three of these machines in 4 days. How many machines can both crews together assemble in 1 day?

38. One crew can assemble M machines in p days. Another crew can assemble N of these machines in q days. Write an expression for the number of machines that both crews together can assemble in 1 day, combine into a single term, and simplify.

39. A steel plate in the shape of a trapezoid (Fig. 11–24) has a hole of diameter d. The area of the plate, less the hole, is

$$\frac{(a + b)h}{2} - \frac{\pi d^2}{4}$$

Combine these two terms and simplify.

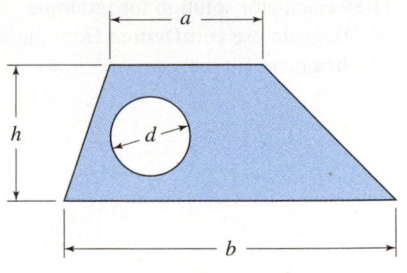

FIGURE 11–24

40. The total resistance R in Fig. 11–25 is

$$\frac{R_1 R_2}{R_1 + R_2} + R_3$$

Combine into a single term and simplify.

41. If a car travels a distance d at a constant rate V, the time required will be d/V. The car then continues for a distance d_1 at a rate V_1, and a third distance d_2 at rate V_2. Write an expression for the total travel time; then combine the three terms into a single term and simplify.

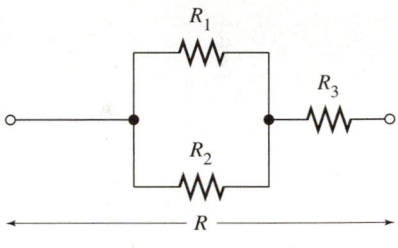

FIGURE 11–25

42. *Project*: Fractions are sometimes taught using *manipulatives*: rods, circles, blocks, and so fourth. Explore how this is done, make some manipulatives out of cardboard, and give a presentation to your class.

11–8   Complex Fractions

Fractions that have only *one* fraction line are called *simple* fractions. Fractions with *more than one* fraction line are called *complex* fractions. We show how to simplify complex fractions in the following examples.

◆◆◆ **Example 58:** Simplify the complex fraction

$$\frac{\dfrac{1}{2} + \dfrac{2}{3}}{3 + \dfrac{1}{4}}$$

Solution: We can simplify this fraction by multiplying numerator and denominator by the least common denominator for all of the individual fractions. The denominators are 2, 3, and 4, so the LCD is 12. Multiplying, we obtain

$$\frac{\left(\dfrac{1}{2} + \dfrac{2}{3}\right)12}{\left(3 + \dfrac{1}{4}\right)12} = \frac{6 + 8}{36 + 3} = \frac{14}{39} \qquad ◆◆◆$$

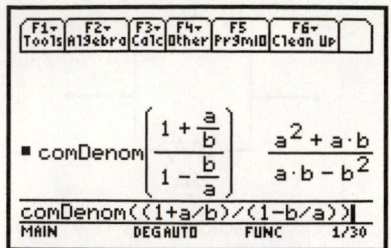

TI-89 calculator solution for Example 59. We again use **comDenom** from the **Algebra** menu.

◆◆◆ **Example 59:** Simplify the complex fraction

$$\frac{1 + \dfrac{a}{b}}{1 - \dfrac{b}{a}}$$

Solution: The LCD for the two small fractions a/b and b/a is ab. Multiplying, we obtain

$$\frac{\left(1 + \dfrac{a}{b}\right)ab}{\left(1 - \dfrac{b}{a}\right)ab} = \frac{ab + a^2}{ab - b^2}$$

or, in factored form,

$$\frac{a(b + a)}{b(a - b)}$$ ◆◆◆

Exercise 8 ◆ Complex Fractions

Simplify. Leave your answers as improper fractions.

1. $\dfrac{\dfrac{2}{3} + \dfrac{3}{4}}{\dfrac{1}{5}}$

2. $\dfrac{\dfrac{3}{4} - \dfrac{1}{3}}{\dfrac{1}{2} + \dfrac{1}{6}}$

3. $\dfrac{\dfrac{1}{2} + \dfrac{1}{3} + \dfrac{1}{4}}{3 - \dfrac{4}{5}}$

4. $\dfrac{\dfrac{4}{5}}{\dfrac{1}{5} + \dfrac{2}{3}}$

5. $\dfrac{5 - \dfrac{2}{5}}{6 + \dfrac{1}{3}}$

6. $\dfrac{1}{2} + \dfrac{3}{\dfrac{2}{5} + \dfrac{1}{3}}$

7. $\dfrac{x + \dfrac{y}{4}}{x - \dfrac{y}{3}}$

8. $\dfrac{\dfrac{a}{b} + \dfrac{x}{y}}{\dfrac{a}{z} - \dfrac{x}{c}}$

9. $\dfrac{1 + \dfrac{x}{y}}{1 - \dfrac{x^2}{y^2}}$

10. $\dfrac{x + \dfrac{a}{c}}{x + \dfrac{b}{d}}$

11. $\dfrac{a^2 + \dfrac{x}{3}}{4 + \dfrac{x}{5}}$

12. $\dfrac{3a^2 - 3y^2}{\dfrac{a + y}{3}}$

13. $\dfrac{x + \dfrac{2d}{3ac}}{x + \dfrac{3d}{2ac}}$

14. $\dfrac{4a^2 - 4x^2}{\dfrac{a + x}{a - x}}$

15. $\dfrac{x^2 - \dfrac{y^2}{2}}{\dfrac{x - 3y}{2}}$

16. $\dfrac{\dfrac{ab}{7} - 3d}{3c - \dfrac{ab}{d}}$

17. $\dfrac{1 + \dfrac{1}{x + 1}}{1 - \dfrac{1}{x - 1}}$

18. $\dfrac{xy - \dfrac{3x}{ac}}{\dfrac{ac}{x} + 2c}$

Applications

19. A car travels a distance d_1 at a rate V_1, then another distance d_2 at a rate V_2. The average speed for the entire trip is

$$\text{average speed} = \dfrac{d_1 + d_2}{\dfrac{d_1}{V_1} + \dfrac{d_2}{V_2}}$$

Simplify this complex fraction.

20. The equivalent resistance of two resistors in parallel is

$$\dfrac{R_1 R_2}{R_1 + R_2}$$

If each resistor is made of wire of resistivity ρ, with R_1 using a wire of length L_1 and cross-sectional area A_1, and R_2 having a length L_2 and area A_2, our expression becomes

$$\dfrac{\dfrac{\rho L_1}{A_1} \cdot \dfrac{\rho L_2}{A_2}}{\dfrac{\rho L_1}{A_1} + \dfrac{\rho L_2}{A_2}}$$

Simplify this complex fraction.

21. The complex fraction

$$\dfrac{\dfrac{1}{x + h} - \dfrac{1}{x}}{h}$$

occurs when you are determining the derivative of $1/x$ in calculus. Simplify this fraction.

22. *Writing:* Suppose that a vocal member of your local school board says that the study of fractions is no longer important now that we have calculators and computers and insists that it be cut from the curriculum to save money.

Write a short letter to the editor of your local paper in which you agree or disagree. Give your reasons for retaining or eliminating the study of fractions.

23. *Project, Golden Ratio:* The golden ratio Φ has the value

$$\Phi = \dfrac{1 + \sqrt{5}}{2} \approx 1.61803 \ldots$$

which is also given by the following repeated fraction. Demonstrate by calculation, by hand, or with a spreadsheet that this is true.

$$\Phi = 1 + \cfrac{1}{1 + \cfrac{1}{1 + \cfrac{1}{1 + \cdots}}}$$

11–9   Fractional Equations

Solving a Fractional Equation

An equation in which one or more terms is a fraction is called a *fractional equation*. To solve a fractional equation, first eliminate the fractions by multiplying both sides of the equation by the least common denominator (LCD) of *every* term. We can do this because multiplying both sides of an equation by the same quantity (the LCD in this case) does not unbalance the equation. With the fractions thus eliminated, the equation is then solved as any nonfractional equation.

◆◆◆ **Example 60:** Solve for *x:*

$$\frac{3x}{5} - \frac{x}{3} = \frac{2}{15}$$

Solution: Multiplying both sides of the equation by the LCD (15), we obtain

$$15\left(\frac{3x}{5} - \frac{x}{3}\right) = 15\left(\frac{2}{15}\right)$$

$$9x - 5x = 2$$

$$x = \frac{1}{2}$$

Check: We substitute our answer $x = 1/2$, into our original equation.

$$\frac{3\left(\frac{1}{2}\right)}{5} - \frac{\frac{1}{2}}{3} \stackrel{?}{=} \frac{2}{15}$$

$$\frac{3\left(\frac{1}{2}\right)}{5}\left(\frac{3}{3}\right) - \frac{\frac{1}{2}}{3}\left(\frac{5}{5}\right) \stackrel{?}{=} \frac{2}{15}$$

$$\frac{\frac{9}{2} - \frac{5}{2}}{15} \stackrel{?}{=} \frac{2}{15}$$

$$\frac{\frac{4}{2}}{15} = \frac{2}{15} \quad \text{Checks} \qquad ◆◆◆$$

■ solve$\left(\dfrac{3 \cdot x}{5} - \dfrac{x}{3} = 2/15, x\right)$

$x = 1/2$

solve(3x/5-x/3=2/15,x)

TI-89 calculator solution for Example 60.

Solving a Fractional Equation by Calculator

We select **solve** from the **Algebra** menu, enter the equation, enter the variable to be solved for, and press ENTER .

◆◆◆ **Example 61:** The calculator screen for Example 60 is shown.   ◆◆◆

Equations with the Unknown in the Denominator

The procedure is the same when the unknown appears in the denominator of one or more terms. However, the LCD will now contain the unknown. Here it is understood that x cannot have a value that will make any of the denominators in the problem equal to zero. Such forbidden values are sometimes stated with the problem, but often they are not.

◆◆◆ **Example 62:** Solve for x

$$\frac{2}{3x} = \frac{5}{x} + \frac{1}{2} \quad (x \neq 0)$$

Assume the integers in this equation to be exact numbers and leave the answer in fractional form.

Solution: The LCD is $6x$. Multiplying both sides of the equation yields

$$6x\left(\frac{2}{3x}\right) = 6x\left(\frac{5}{x} + \frac{1}{2}\right)$$

$$6x\left(\frac{2}{3x}\right) = 6x\left(\frac{5}{x}\right) + 6x\left(\frac{1}{2}\right)$$

$$4 = 30 + 3x$$

$$3x = -26$$

$$x = -\frac{26}{3}$$

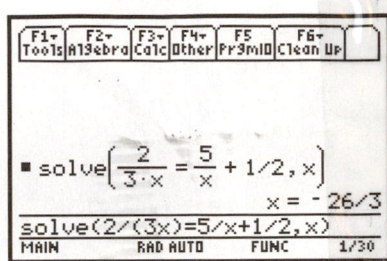

TI-89 calculator solution for Example 62.

Check:

$$\frac{2}{3\left(-\frac{26}{3}\right)} \stackrel{?}{=} \frac{5}{-\frac{26}{3}} + \frac{1}{2}$$

$$-\frac{2}{26} \stackrel{?}{=} -\frac{15}{26} + \frac{13}{26}$$

$$-\frac{2}{26} = -\frac{2}{26} \quad \text{(checks)}$$ ◆◆◆

◆◆◆ **Example 63:** Solve for x:

$$\frac{8x + 7}{5x + 4} = 2 - \frac{2x}{5x + 1}$$

Solution: Multiplying by the LCD $(5x + 4)(5x + 1)$ yields

$$(8x + 7)(5x + 1) = 2(5x + 4)(5x + 1) - 2x(5x + 4)$$

$$40x^2 + 35x + 8x + 7 = 2(25x^2 + 5x + 20x + 4) - 10x^2 - 8x$$

$$40x^2 + 43x + 7 = 50x^2 + 50x + 8 - 10x^2 - 8x$$

$$43x + 7 = 42x + 8$$

$$x = 1$$ ◆◆◆

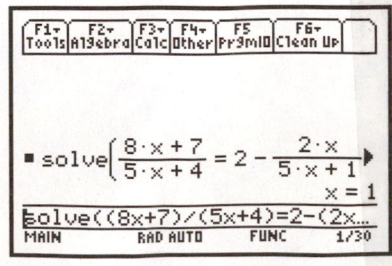

TI-89 calculator solution for Example 63.

Common Error	The technique of multiplying by the LCD in order to eliminate the denominators is *valid only when we have an equation.* Do not multiply through by the LCD when there is no equation!

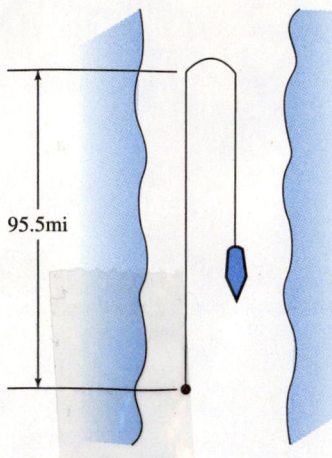

95.5mi

FIGURE 11–26

◆◆◆ **Example 64:** *An Application.* An excursion boat, Fig. 11–26, travels up river a distance of 95.5 mi at a speed of 10.3 mi/h and then returns, with the current, to its starting point. What must the return rate x be so that the total round trip time is 15.0 h?

Solution: The total time is

$$\text{time upstream} + \text{return time} = 15.0 \text{ h}$$

For uniform motion, time = distance ÷ rate, so

$$\frac{\text{distance}}{\text{upstream rate}} + \frac{\text{distance}}{\text{return rate}} = 15.0$$

$$\frac{95.5}{10.3} + \frac{95.5}{x} = 15.0$$

where x is the downstream rate. We multiply by $10.3x$ to clear fractions and solve for x.

$$95.5x + 95.5(10.3) = 15.0(10.3x)$$

$$59.0x = 984$$

$$x = 16.7 \text{ mi/h}$$ ◆◆◆

Exercise 9 ◆ Fractional Equations

Solve for x. Assume the integers in these equations to be exact numbers, and leave your answers in fractional form.

1. $2x + \dfrac{x}{3} = 28$

2. $4x + \dfrac{x}{5} = 42$

3. $x + \dfrac{x}{5} = 24$

4. $\dfrac{x}{6} + x = 21$

5. $3x - \dfrac{x}{7} = 40$

6. $x - \dfrac{x}{6} = 25$

7. $\dfrac{2x}{3} - x = -24$

8. $\dfrac{3x}{5} + 7x = 38$

9. $\dfrac{x}{2} + \dfrac{x}{3} + \dfrac{x}{4} = 26$

10. $\dfrac{x-1}{2} = \dfrac{x+1}{3}$

11. $\dfrac{3x-1}{4} = \dfrac{2x+1}{3}$

12. $x + \dfrac{2x}{3} + \dfrac{3x}{4} = 29$

13. $2x + \dfrac{x}{3} - \dfrac{x}{4} = 50$

14. $3x - \dfrac{2x}{3} - \dfrac{5x}{6} = 18$

15. $3x - \dfrac{x}{6} + \dfrac{x}{12} = 70$

16. $\dfrac{x}{4} + \dfrac{x}{6} + \dfrac{x}{8} = 26$

17. $\dfrac{6x-19}{2} = \dfrac{2x-11}{3}$

18. $\dfrac{7x-40}{8} = \dfrac{9x-80}{10}$

19. $\dfrac{3x-116}{4} + \dfrac{180-5x}{6} = 0$

20. $\dfrac{3x-4}{2} - \dfrac{3x-1}{16} = \dfrac{6x-5}{8}$

21. $\dfrac{x-1}{8} - \dfrac{x+1}{18} = 1$

22. $\dfrac{3x}{4} + \dfrac{180-5x}{6} = 29$

23. $\dfrac{15x}{4} = \dfrac{9}{4} - \dfrac{3-x}{2}$

24. $\dfrac{x}{4} + \dfrac{x}{10} + \dfrac{x}{8} = 19$

Equations with Unknown in Denominator

25. $\dfrac{2}{3x} + 6 = 5$

26. $9 - \dfrac{4}{5x} = 7$

27. $4 + \dfrac{1}{x + 3} = 8$

28. $\dfrac{x + 5}{x - 2} = 5$

29. $\dfrac{x - 3}{x + 2} = \dfrac{x + 4}{x - 5}$

30. $\dfrac{3}{x + 2} = \dfrac{5}{x} + \dfrac{4}{x^2 + 2x}$

31. $\dfrac{9}{x^2 + x - 2} = \dfrac{7}{x - 1} - \dfrac{3}{x + 2}$

32. $\dfrac{3x + 5}{2x - 3} = \dfrac{3x - 3}{2x - 1}$

33. $\dfrac{4}{x^2 - 1} + \dfrac{1}{x - 1} + \dfrac{1}{x + 1} = 0$

34. $\dfrac{x}{3} - \dfrac{x^2 - 5x}{3x - 7} = \dfrac{2}{3}$

35. The pointer of a certain meter can travel to the right at the rate of 10.0 cm/s. What must be the minimum return rate if the total time for the pointer to traverse the full 12.0-cm scale and return to zero must not exceed 2.00 seconds?

36. A bus travels 87.5 km to another town at a speed of 72.0 km/h. What must be its return rate if the total time for the round trip is to be 2.50 hours?

37. Three masons build 318 m of wall. Mason *A* builds 7.0 m/day, *B* builds 6.0 m/day, and *C* builds 5.0 m/day. Mason *B* works twice as many days as *A*, and *C* works half as many days as *A* and *B* combined. How many days did each work?

38. If a carpenter can roof a house in 10 days and another can do the same in 14 days, how many days will it take if they work together?

39. A certain shaper has a forward cutting speed of 115 ft/min and a stroke of 10.5 in. It is observed to make 429 cuts (and returns) in 4.0 minutes. What is the return speed?

40. A certain screw machine can produce a box of parts in 3.3 h. A new machine is to be ordered having a speed such that both machines working together would produce a box of parts in 1.4 h. How long would it take the new machine alone to produce a box of parts?

41. A landlord owns a house that consumes 2100 gal of heating oil in three winters. He buys another (insulated) house, and the two houses together use 1850 gal of oil in two winters. How many winters would it take the insulated house alone to use 1250 gal of oil?

11–10   Literal Equations and Formulas

A *literal equation* is one in which some or all of the constants are represented by letters.

◆◆◆ **Example 65:** The following is a literal equation:

$$a(x + b) = b(x + c) \qquad\qquad ◆◆◆$$

A *formula* is a literal equation that relates two or more mathematical or physical quantities. These are the equations that describe the workings of the physical world. In Chap. 1 we *substituted into* formulas. Here we *solve* formulas or other literal equations for one of its quantities.

Solving Literal Equations and Formulas

When we solve a literal equation or formula, we cannot, of course, get a *numerical* answer, as we could with a numerical equation. Our object here is to *isolate* one of the letters on one side of the equal sign. We "solve for" one of the literal quantities.

◆◆◆ Example 66: Solve for x:

$$a(x + b) = b(x + c)$$

Solution: Our goal is to isolate x on one side of the equation. Removing parentheses, we obtain

$$ax + ab = bx + bc$$

Subtracting bx and then ab will place all of the x terms on one side of the equation.

$$ax - bx = bc - ab$$

Factoring to isolate x yields

$$x(a - b) = b(c - a)$$

Dividing by $(a - b)$, where $a \neq b$, gives us

$$x = \frac{b(c - a)}{a - b}$$

◆◆◆

Check: We substitute our answer into our original equation.

$$a\left[\frac{b(c - a)}{a - b} + b\right] \stackrel{?}{=} b\left[\frac{b(c - a)}{a - b} + c\right]$$

$$a\left[\frac{bc - ab}{a - b} + \frac{ab - b^2}{a - b}\right] \stackrel{?}{=} b\left[\frac{bc - ab}{a - b} + \frac{ac - bc}{a - b}\right]$$

$$\frac{a}{a - b}[b(c - b)] \stackrel{?}{=} \frac{b}{a - b}[a(c - b)]$$

$$\frac{ab(c - b)}{a - b} = \frac{ab(c - b)}{a - b} \quad \text{(checks)}$$

◆◆◆

We solve literal equations by calculator in the same way we solved numerical equations. Select **solve** from the **Algebra** menu, enter the equation, enter the variable to be solved for, and press ENTER .

◆◆◆ Example 67: Here we show the TI-89 screen for Example 66.   ◆◆◆

TI-89 calculator solution for Example 66.

◆◆◆ Example 68: Solve for x.

$$b\left(b + \frac{x}{a}\right) = d$$

Solution: Dividing both sides by b, we have

$$b + \frac{x}{a} = \frac{d}{b}$$

Subtracting b yields

$$\frac{x}{a} = \frac{d}{b} - b$$

Multiplying both sides by a gives us

$$x = a\left(\frac{d}{b} - b\right)$$

TI-89 calculator solution for Example 68.

Check: Substituting $a(d/b - b)$ for x in the original equation,

$$b\left(b + \frac{x}{a}\right) = d$$

$$b\left[b + \frac{a(d/b - b)}{a}\right] \stackrel{?}{=} d$$

$$b\left(b + \frac{d}{b} - b\right) \stackrel{?}{=} d$$

$$b\left(\frac{d}{b}\right) = d \quad \text{(checks)} \qquad \blacklozenge\blacklozenge\blacklozenge$$

To solve many literal *fractional* equations, the procedure is the same as for other fractional equations: Multiply by the LCD to eliminate the fractions.

◆◆◆ **Example 69:** Solve the following for x in terms of a, b, and c:

$$\frac{x}{b} - \frac{a}{c} = \frac{x}{a}$$

Solution: Multiplying by the LCD (abc) yields

$$acx - a^2b = bcx$$

Rearranging so that all x terms are together on one side of the equation, we obtain

$$acx - bcx = a^2b$$

Factoring gives us

$$x(ac - bc) = a^2b$$

Dividing by $(ac - bc)$ gives

$$x = \frac{a^2b}{ac - bc} \qquad \blacklozenge\blacklozenge\blacklozenge$$

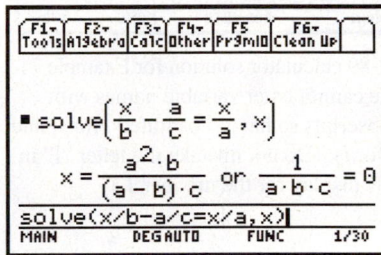

TI-89 calculator solution for Example 69. The calculator will often return a second solution, as here. Since we are looking for x in terms of the other literals, this second solution is of no use to us.

◆◆◆ **Example 70:** Solve for x.

$$\frac{x}{x - a} - \frac{x + 2b}{x + a} = \frac{a^2 + b^2}{x^2 - a^2}$$

Solution: The LCD is $x^2 - a^2$, that is, $(x - a)(x + a)$. Multiplying each term by the LCD gives

$$x(x + a) - (x + 2b)(x - a) = a^2 + b^2$$

Removing parentheses and rearranging so that all of the x terms are together, we obtain

$$x^2 + ax - (x^2 + 2bx - ax - 2ab) = a^2 + b^2$$

$$x^2 + ax - x^2 - 2bx + ax + 2ab = a^2 + b^2$$

$$2ax - 2bx = a^2 - 2ab + b^2$$

Then we factor:

$$2x(a - b) = (a - b)^2$$

and divide by $2(a - b)$:

$$x = \frac{(a - b)^2}{2(a - b)} = \frac{a - b}{2} \qquad \blacklozenge\blacklozenge\blacklozenge$$

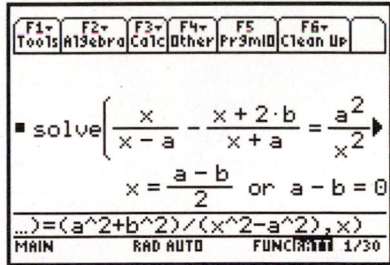

TI-89 calculator solution for Example 70. The given equation does not fit on the screen. Here again, we disregard the second solution.

Formulas

An important application of literal equations is the vast number of *formulas* used in technology. We have already shown, in Chap. 1, how to substitute numbers into a formula. Here we will show how to solve a formula for a different quantity.

When dealing with formulas, be careful to distinguish between capital letters and lowercase letters. In the same formula, T and t, for example, will represent different quantities.

◆◆◆ **Example 71:** *An Application*: The formula for the amount of heat q flowing by conduction through a wall of thickness L, conductivity k, and cross-sectional A is

$$q = \frac{kA(t_1 - t_2)}{L}$$

where t_1 and t_2 are the temperatures of the warmer and cooler sides, respectively. Solve this equation for t_1.

Solution: Multiplying both sides by L/kA gives

$$\frac{qL}{kA} = t_1 - t_2$$

then adding t_2 to both sides, we get

$$t_1 = \frac{qL}{kA} + t_2$$

◆◆◆

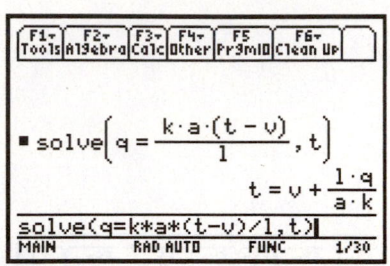

TI-89 calculator solution for Example 71. We cannot enter variable names with subscripts so have substituted t for t_1 and v for t_2. (Do not mistake the letter "l" in this display for the number 1.)

Exercise 10 ◆ Literal Equations and Formulas

Solve for x. Try some by calculator.

1. $2ax = bc$
2. $ax + dx = a - c$
3. $a(x + y) = b(x + z)$
4. $4x = 2x + ab$
5. $4acx - 3d^2 = a^2d - d^2x$
6. $a(2x - c) = a + c$
7. $a^2x - cd = b - ax + dx$
8. $3(x - r) = 2(x + p)$
9. $\dfrac{a}{2}(x - 3w) = z$
10. $cx - x = bc - b$
11. $3x + m = b$
12. $ax + m = cx + n$
13. $ax - bx = c + dx - m$
14. $3m + 2x - c = x + d$
15. $ax - ab = cx - bc$
16. $p(x - b) = qx + d$

Solve for either w, y, or z.

17. $5dw = ab$
18. $7ay - 3 = 4pq$
19. $5mn = mz - 2$
20. $2w + 3abw = 16$
21. $m(mn + y) = 3 - y$
22. $3(pz - 2) = 4(z + 5)$

Literal Fractional Equations

23. $\dfrac{w + x}{x} = w(w + y)$
24. $\dfrac{ax + b}{c} = bx - a$
25. $\dfrac{p - q}{x} = 3p$
26. $\dfrac{bx - c}{ax - c} = 5$

27. $\dfrac{a - x}{5} = \dfrac{b - x}{2}$

28. $\dfrac{x}{a} - b = \dfrac{c}{d} - x$

29. $\dfrac{x}{a} - a = \dfrac{a}{c} - \dfrac{x}{c - a}$

30. $\dfrac{x}{a - 1} - \dfrac{x}{a + 1} = b$

31. $\dfrac{x - a}{x - b} = \left(\dfrac{2x - a}{2x - b}\right)^2$

32. $\dfrac{a - b}{bx + c} + \dfrac{a + b}{ax - c} = 0$

Formulas

33. The correction C for the sag in a surveyor's tape weighing w lb/ft and pulled with a force of P lb is

$$C = \frac{w^2 L^3}{24P^2} \text{ feet}$$

Solve this equation for the distance measured, L.

34. When a bar of length L_0 having a coefficient of linear thermal expansion α is increased in temperature by an amount Δt, it will expand to a new length L, where

$$L = L_0(1 + \alpha\Delta t) \qquad \textbf{1056}$$

Solve this equation for Δt.

35. Solve the equation given in problem 34 for the initial length L_0.

36. A rod of cross-sectional area a and length L will stretch by an amount e when subject to a tensile load of P, Fig. 11–27. The modulus of elasticity is given by

$$E = \frac{PL}{ae} \qquad \textbf{1053}$$

Solve this equation for a.

37. The formula for the displacement s of a freely falling body having an initial velocity v_0 and acceleration a is

$$s = v_0 t + \tfrac{1}{2}at^2 \qquad \textbf{1018}$$

Solve this equation for a.

38. The formula for the amount of heat flowing through a wall by conduction, Fig. 11–28, is

$$q = \frac{kA(t_1 - t_2)}{L}$$

where k is the conductivity of the wall material and A is the cross-sectional area. Solve this equation for t_2.

39. An amount a invested at a simple interest rate n for t years will accumulate to an amount y, where $y = a + ant$. Solve for a.

40. The formula for the equivalent resistance R for the parallel combination of two resistors, R_1 and R_2, is

$$\frac{1}{R} = \frac{1}{R_1} + \frac{1}{R_2} \qquad \textbf{1063}$$

Solve this formula for R_2.

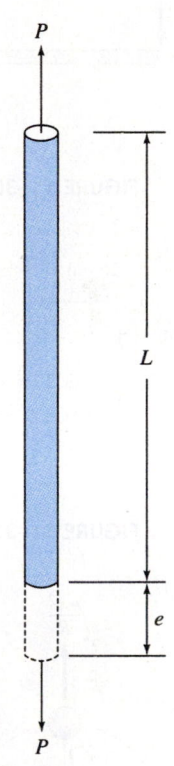

FIGURE 11–27

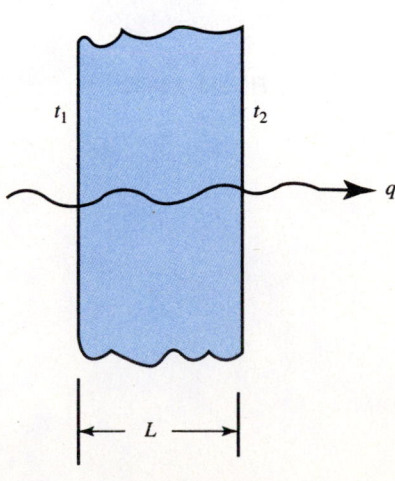

FIGURE 11–28

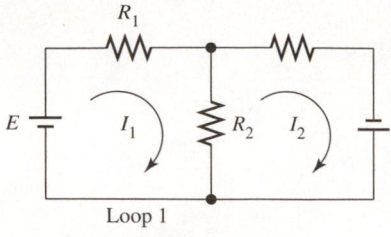

FIGURE 11–29

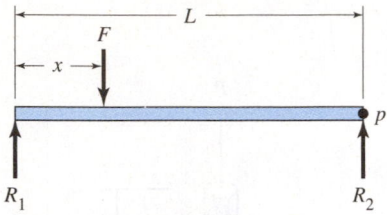

FIGURE 11–30

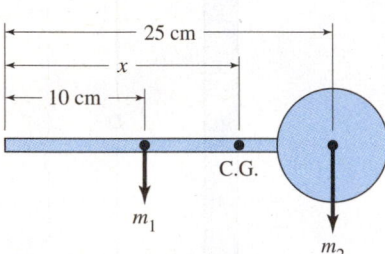

FIGURE 11–31

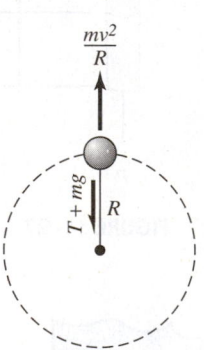

FIGURE 11–32

41. Kirchhoff's voltage law, *the sum of the voltage rises and drops around any closed loop is zero*, applied to loop 1 in Fig. 11–29, gives

$$E = I_1 R_1 + I_1 R_2 - I_2 R_2$$

Solve for I_1.

42. If the resistance of a conductor is R_1 at temperature t_1, the resistance will change to a value R when the temperature changes to t, where

$$R = R_1[1 + \alpha(t - t_1)]$$	**1069**

and α is the temperature coefficient of resistance at temperature t_1. Solve this equation for t_1.

43. Solve the equation given in problem 42 for the initial resistance R_1.

44. Taking the moment M about point p in Fig. 11–30, we get

$$M = R_1 L - F(L - x)$$

Solve for L.

45. Three masses, m_1, m_2, and m_3, are attached together and accelerated by means of a force F, where

$$F = m_1 a + m_2 a + m_3 a$$

Solve for the acceleration a.

46. Solve the problem given in the introduction to this chapter: A bar of mass m_1 is attached to a sphere of mass m_2 (Fig. 11–31). The distance x to the center of gravity (C. G.) is

$$x = \frac{10m_1 + 25m_2}{m_1 + m_2}$$

Solve for m_1.

47. A ball of mass m is swung in a vertical circle (Fig. 11–32). At the top of its swing, the tension T in the cord plus the ball's weight mg is just balanced by the centrifugal force mv^2/R. Thus,

$$T + mg = \frac{mv^2}{R}$$

Solve for m.

48. The total energy of a body of mass m, moving with velocity v and located at a height y above some datum, is the sum of the potential energy mgy and the kinetic energy $\frac{1}{2}mv^2$. So,

$$E = mgy + \tfrac{1}{2}mv^2$$

Solve for m.

••• CHAPTER 11 REVIEW PROBLEMS ••••••••••••••••••••••••••••••••••

Factor completely.

1. $x^2 - 2x - 15$ **2.** $2a^2 + 3a - 2$

3. $x^6 - y^4$ **4.** $2ax^2 + 8ax + 8a$

5. $2x^2 + 3x - 2$ **6.** $a^2 + ab - 6b^2$

7. $8x^3 - \dfrac{y^3}{27}$ **8.** $\dfrac{x^2}{y} - \dfrac{x}{y}$

9. $2ax^2y^2 - 18a$

10. $xy - 2y + 5x - 10$

11. $3a^2 - 2a - 8$

12. $x - bx - y + by$

13. $\dfrac{2a^2}{12} - \dfrac{8b^2}{27}$

14. $(y + 2)^2 - z^2$

15. $2x^2 - 20ax + 50a^2$

16. $x^2 - 7x + 12$

17. $4a^2 - (3a - 1)^2$

18. $1 - 16x^2$

19. $a^2 - 2a - 8$

20. $9x^4 - x^2$

21. $x^2 - 21x + 110$

22. $27a^3 - 8w^3$

23. $3x^2 - 6x - 45$

24. $16x^2 - 16xy + 4y^2$

25. $64m^3 - 27n^3$

26. $6ab + 2ay + 3bx + xy$

27. $15a^2 - 11a - 12$

28. $2y^4 - 18$

29. $ax - bx + ay - by$

Solve for x.

30. $cx - 5 = ax + b$

31. $a(x - 3) - b(x + 2) = c$

32. $\dfrac{5 - 3x}{4} + \dfrac{3 - 5x}{3} = \dfrac{3}{2} - \dfrac{5x}{3}$

33. $\dfrac{3x - 1}{11} - \dfrac{2 - x}{10} = \dfrac{6}{5}$

34. $\dfrac{x + 3}{2} + \dfrac{x + 4}{3} + \dfrac{x + 5}{4} = 16$

35. $\dfrac{2x + 1}{4} - \dfrac{4x - 1}{10} + \dfrac{5}{4} = 0$

36. $x^2 - (x - p)(x + q) = r$

37. $mx - n = \dfrac{nx - m}{p}$

38. $p = \dfrac{q - rx}{px - q}$

39. $m(x - a) + n(x - b) + p(x - c) = 0$

40. $\dfrac{x - 3}{4} - \dfrac{x - 1}{9} = \dfrac{x - 5}{6}$

41. $\dfrac{1}{x - 5} - \dfrac{1}{4} = \dfrac{1}{3}$

42. $\dfrac{2}{x - 4} + \dfrac{5}{2(x - 4)} + \dfrac{9}{2(x - 4)} = \dfrac{1}{2}$

43. $\dfrac{2}{x - 2} = \dfrac{5}{2(x - 1)}$

44. $\dfrac{6}{(2x + 1)} = \dfrac{4}{(x - 1)}$

45. $\dfrac{x + 2}{x - 2} - \dfrac{x - 2}{x + 2} = \dfrac{x + 7}{x^2 - 4}$

Perform the indicated operations and simplify.

46. $\dfrac{a^2 + b^2}{a - b} - a + b$

47. $\dfrac{3}{5x^2} - \dfrac{2}{15xy} + \dfrac{1}{6y^2}$

48. $(a^2 + 1 + a)\left(1 - \dfrac{1}{a} + \dfrac{1}{a^2}\right)$

49. $\dfrac{\dfrac{a - 1}{6} - \dfrac{2a - 7}{2}}{\dfrac{3a}{4} - 3}$

50. $\dfrac{1 + \dfrac{a - c}{a + c}}{1 - \dfrac{a - c}{a + c}}$

51. $\left(1 + \dfrac{x + y}{x - y}\right)\left(1 - \dfrac{x - y}{x + y}\right)$

52. $\dfrac{a^3 - b^3}{a^3 + b^3} \times \dfrac{a^2 - ab + b^2}{a - b}$

53. $\dfrac{3wx^2y^3}{7axyz} \times \dfrac{4a^3xz}{6aw^2y}$

54. $\dfrac{4a^2 - 9c^2}{4a^2 + 6ac}$

55. $\dfrac{b^2 - 5b}{b^2 - 4b - 5}$

56. $\dfrac{3a^2 + 6a}{a^2 + 4a + 4}$

57. $\dfrac{20(a^3 - c^3)}{4(a^2 + ac + c^2)}$

58. $\dfrac{x^2 - y^2 - 2yz - z^2}{x^2 + 2xy + y^2 - z^2}$

59. $\dfrac{2a^2 + 17a + 21}{3a^2 + 26a + 35}$

60. The reduction in power in a resistance R caused by lowering the voltage across the resistor from V_2 to V_1 is

$$\frac{V_2{}^2}{R} - \frac{V_1{}^2}{R}$$

Factor this expression.

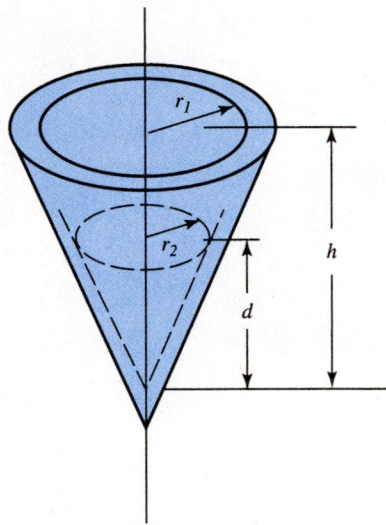

FIGURE 11–33

61. A conical tank of height h is filled to a depth d, Fig. 11–33. The volume of liquid that can still be put into the tank is

$$V = \frac{\pi}{3}r_1{}^2h - \frac{\pi}{3}r_2{}^2d$$

where r_1 is the radius of the top of the tank and r_2 is the radius of the surface of the liquid. Factor the right side of this equation.

12

Quadratic Equations

♦♦♦ **OBJECTIVES** ♦♦
When you have completed this chapter you should be able to
- Solve quadratics using a calculator's equation solver.
- Solve quadratics using a calculator that can do symbolic algebra.
- Solve quadratics by graphing, either manually or by calculator.
- Solve quadratics by the quadratic formula.
- Apply quadratics to a variety of applications.

♦♦♦

So far we have only solved first-degree (linear) equations, as well as sets of linear equations. Now we move on to equations of second degree, or *quadratic equations*.

We have already solved equations using calculators that have a built-in equation solver, and using calculators that can do symbolic processing. Also, in our chapter on graphing, we learned how to graphically find the approximate solution to any equation. We will start by applying those methods to quadratic equations. The methods are no different here, except we must look for *two* solutions instead of one. We will also show the more traditional manual method, the use of the quadratic formula.

As usual, we will follow the mathematics with numerous applications. Take, for example, a simple falling-body problem, Fig.12–1: "If an object is thrown downward with a speed of 15.5 ft/s, how long will it take to fall 125 ft?" If we substitute $s = 125$ ft, $v_0 = 15.5$ ft/s, and $a = 32.2$ ft/s^2 into the equation for a freely falling body

$$s = v_0 t + \frac{1}{2} at^2$$

we get

$$125 = 15.5t + 16.1t^2$$

This is called a *quadratic equation*. How shall we solve it for t? We will show how in this chapter.

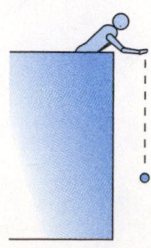

FIGURE 12–1

12–1 Solving a Quadratic Equation Graphically and by Calculator

Recall that a polynomial equation is one in which all powers of x are positive integers. A *quadratic equation* is a polynomial equation of second degree. That is, the highest power of x in the equation is 2. It is common practice to refer to a quadratic equation simply as a *quadratic*.

◆◆◆ **Example 1:** The following equations are quadratic equations:

(a) $4x^2 - 5x + 2 = 0$ (b) $x^2 = 58$

(c) $9x^2 - 5x = 0$ (d) $2x^2 - 7 = 0$

(e) $3.27 + 1.85x = 2.29x^2$ ◆◆◆

A *quadratic function* is one whose highest-degree term is of second degree.

◆◆◆ **Example 2:** The following functions are quadratic functions:

(a) $f(x) = 5x^2 - 3x + 2$ (b) $f(x) = 9 - 3x^2$

(c) $f(x) = x(x + 7)$ (d) $f(x) = x - 4 - 3x^2$ ◆◆◆

Some quadratic equations have a term missing. A quadratic that has no x term is called a *pure* quadratic; one that has no constant term is called an *incomplete* quadratic.

◆◆◆ **Example 3:**

(a) $x^2 - 9 = 0$ is a *pure* quadratic.

(b) $x^2 - 4x = 0$ is an *incomplete* quadratic.   ◆◆◆

Solving a Quadratic Graphically

We will show several ways to solve a quadratic; first graphically and by calculator, and, in the next section, by formula.

■ **Exploration:** In our chapter on graphing we plotted the quadratic function

$$f(x) = x^2 - 4x - 3$$

getting a curve that we called a *parabola*.

Try this. Either graph this function again or look back at our earlier graph. Does the curve intercept the x axis, and if so, how many times? Can you imagine a parabola that has more x-intercepts? Zoom out far enough to convince yourself that the curve will not turn and re-cross the x axis at some other place. Can you imagine a parabola that has *fewer* x-intercepts?

Using ┃TRACE┃, ┃ZOOM┃ or **zero**, find the value of x at an intercept and substitute it into the given function. Then state the significance of an intercept. ■

Your exploration may have shown you that a quadratic function in x can have 0, 1, or 2 x-intercepts, but no more than two. Also, the value of x at an intercept, when substituted back into the function, makes that function equal to zero.

Earlier we saw that a graphical solution to the equation

$$f(x) = 0$$

could be obtained by plotting the function

$$y = f(x)$$

and locating the points where the curve crosses the *x* axis. Those points are called the *zeros* of the function. Those points are then the *solution, or roots*, of the equation.

Many graphing calculators can find roots on a graph, as we saw in earlier chapters. Recall that on the TI-83/84, press the CALC key and select **2: zero**. On the TI-89, press GRAPH, then **F5 Math,** and finally **2: zero**. You must give the interval on the *x* axis bracketing the zero and, on the TI-83/84, a guess for the value of the zero.

♦♦♦ **Example 4:** Graphically find the approximate roots of the equation

$$x^2 + x - 3 = 0$$

Solution: The function

$$y = x^2 + x - 3$$

is plotted in Fig. 12–2. Reading the *x*-intercepts as accurately as possible, we get for the roots

$$x \approx -2.3 \quad \text{and} \quad x \approx 1.3$$

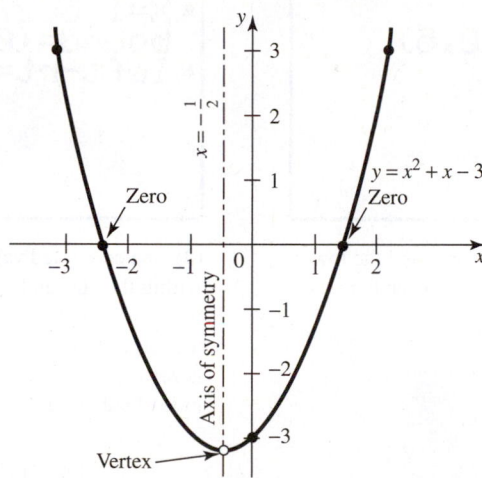

FIGURE 12–2
The graph of the quadratic function is a parabola, one of the four *conic sections* (the curves obtained when a cone is intersected by a plane at various angles). It has many interesting properties and applications. We take only a very brief look at the parabola here, with the main treatment saved for our study of the conic sections.

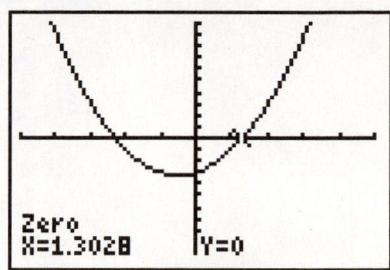

TI-83/84 Calculator solution for Example 4. We used the **zero** operation to find this root. To find the other root we would change the search interval to include that root (say, -5 to 0) and make our guess closer to that root, say -3.

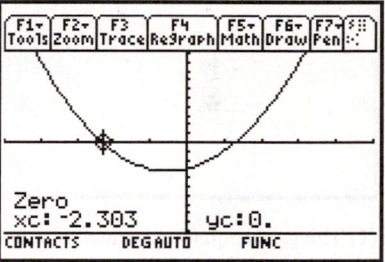

TI-89 solution for Example 4. Here we have found the left-most root using the zero operation and by changing the search interval to -5, 0.

Note that the curve in this example is symmetrical about the line $x = -\frac{1}{2}$. This line is called the *axis of symmetry*. That means that for any point P on the curve there is another point Q on the curve such that the axis of symmetry is the perpendicular bisector of line PQ. The point where the parabola crosses the axis of symmetry is called the *vertex*.

Recall that earlier we said that a low point in a curve is called a *minimum point,* and that a high point is called a *maximum point*. Thus the vertex of this parabola is also a minimum point. ♦♦♦

Solving a Quadratic Using a Calculator's Equation Solver

We can use any built-in equation solver to solve a quadratic, just as we used it earlier to solve other equations.

◆◆◆ **Example 5:** Solve the equation of Example 4 using the TI-83/84 calculator's equation solver.

Solution:

(a) We select **Solver** from the $\boxed{\text{MATH}}$ menu and enter the equation, Screen (1). The equation must be in explicit form, with all terms on one side and 0 on the other side.

(b) Press $\boxed{\text{ENTER}}$. A new screen is displayed, Screen (2), showing a guess value of x and a **bound**. You may accept the default values or enter new ones. Let us enter a new guess of $x = 3$ and a new bound of 0 to 5.

(c) Move the cursor to the line containing "**X =**" and press $\boxed{\text{SOLVE}}$. (This is $\boxed{\text{ALPHA}}$ $\boxed{\text{ENTER}}$ on the TI-83/84.)

A root falling within the selected bound is displayed, Screen (3). To find the other root we would change the initial guess and the bound.

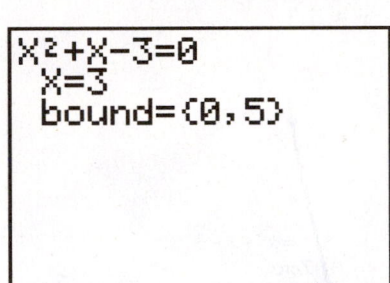

(1) The given equation is entered in the TI-83/84 equation solver.

(2) An initial guess for x, and the lower and upper bounds, are chosen here.

(3) The computed value of x is shown, within the chosen bound. The last line says that the difference between the left and right sides of the equation is zero, showing that the equation is indeed balanced. ◆◆◆

Implicit Functions

To solve a quadratic or other equation by graphing or by a calculator's equation solver, the equation must be in explicit form. If it is not, simply move all terms to one side of the equal sign, changing signs as appropriate, leaving zero on the other side of the equation.

However, some calculators that can do symbolic manipulation can solve an equation that is in implicit form, as we have seen earlier in this text.

◆◆◆ **Example 6:** Solve using the TI-89 equation solver:

$$1.8x^2 - 4.9 = 2.1x$$

Solution:

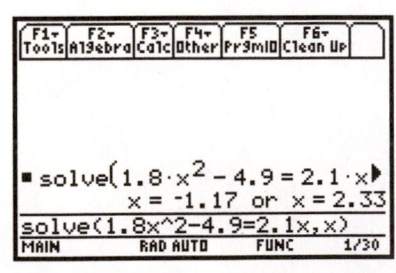

TI–89 screen for Example 6.

(a) We select **solve** from the **Algebra** menu.

(b) Enter the equation, followed by the variable (x) that we want to solve for. Unlike on the TI-83/84, the equation *does not* have to be in explicit form.

(c) Pressing $\boxed{\text{ENTER}}$ would give the exact solution, while pressing $\boxed{\approx}$ (located above the $\boxed{\text{ENTER}}$ key) gives the approximate decimal solution, as shown. Note that both solutions are given and it was not necessary to enter a first guess or bounding values. We get

$$x = -1.17 \quad \text{and} \quad x = 2.33$$ ◆◆◆

An Application

We will give just one application here, to be followed by many later in the chapter.

◆◆◆ Example 7: Given the equation from the introduction to this chapter

$$125 = 15.5t + 16.1t^2$$

use a calculator to solve for a positive value of t (a) by graphing, and (b) using the calculator's equation solver.

Solution: We could enter this equation directly into the TI-89, or put it into explicit form for the TI-83/84, as,

$$16.1t^2 + 15.5t - 125 = 0$$

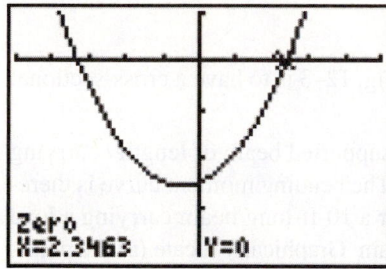

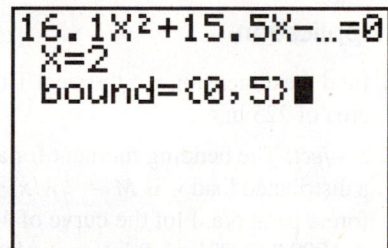

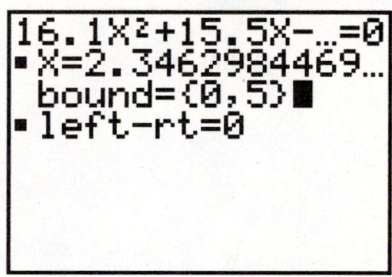

Graphical check for Example 7.
We have used the variable x because
we cannot enter t on this calculator.

TI-83/84 solution for Example 7.
We have set the bounds at $x = 0$ and
$x = 5$, and made a guess of $x = 2$.

TI-83/84 Screen for Example 7.
Again we have used the variable x.

(a) We graph this equation on the TI-83/84 and use the **Zero** operation to locate a positive root at $t = 2.3463$.

(b) We enter the equation into the **solver**, enter a guess for t and the bounds of t, move the cursor to the **X =** line, and press $\boxed{\text{SOLVE}}$. We get a value of $t = 2.3463$, as we did by graphing.   **◆◆◆**

Exercise 1 ◆ Solving a Quadratic Equation Graphically and by Calculator

Find the roots of each quadratic by any of the methods shown in this section. Keep three significant digits. For some, use more than one method and compare results.

Explicit Functions

1. $x^2 - 12x + 28 = 0$ **2.** $x^2 - 6x + 7 = 0$

3. $x^2 + x - 19 = 0$ **4.** $x^2 - x - 13 = 0$

5. $3x^2 + 12x - 35 = 0$ **6.** $29.4x^2 - 48.2x - 17.4 = 0$

7. $36x^2 + 3x - 7 = 0$ **8.** $28x^2 + 29x + 7 = 0$

9. $49x^2 + 21x - 5 = 0$ **10.** $16x^2 - 16x + 1 = 0$

11. $3x^2 - 10x + 4 = 0$ **12.** $x^2 - 34x + 22 = 0$

Implicit Functions

13. $3x^2 + 5x = 7$ **14.** $4x + 5 = x^2 + 2x$

15. $x^2 - 4 = 4x + 7$ **16.** $x^2 - 6x - 14 = 3$

17. $6x - 300 = 205 - 3x^2$ **18.** $3x^2 - 25x = 5x - 73$

19. $2x^2 + 100 = 32x - 11$ **20.** $33 - 3x^2 = 10x$

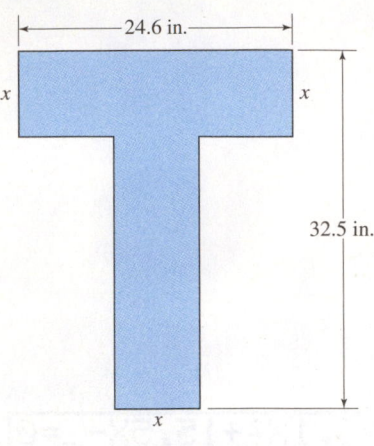

24.6 in.

x x

32.5 in.

x

FIGURE 12–3

Challenge Problems

21. $4.26x + 5.74 = 1.27x^2 + 2.73x$

22. $1.83x^2 - 4.26 = 4.82x + 7.28$

23. $x^2 - 6.27x - 14.4 = 3.17$

24. $6.47x - 338 = 205 - 3.73x^2$

25. $x(2x - 3) = 3x(x + 4) - 2$

26. $2.95(x^2 + 8.27x) = 7.24x(4.82x - 2.47) + 8.73$

27. $(4.20x - 5.80)(7.20x - 9.20) = 8.20x + 9.90$

28. $(2x - 1)^2 + 6 = 6(2x - 1)$

An Application

29. Find the dimension x if the steel T-beam in Fig. 12–3 is to have a cross-sectional area of 225 in.^{2}

30. *Project:* The bending moment for a simply supported beam of length l carrying a distributed load w is $M = \frac{1}{2}wlx - \frac{1}{2}wx^2$. The bending moment curve is therefore a parabola. Plot the curve of M vs. x for a 10-ft-long beam carrying a load of 1000 lb/ft. Take 1-ft intervals along the beam. Graphically locate (a) the points of zero bending moment and (b) the point of maximum bending moment.

12–2   Solving a Quadratic by Formula

We have learned several calculator methods for solving a quadratic, and here we present a manual method, the quadratic formula. It will work for any quadratic, regardless of the type of roots, can be used for literal quadratic equations, and can easily be programmed for the computer. To use the formula, we must first put a quadratic equation into *general form.*

General Form of a Quadratic

A quadratic is in *general form* when it is written in the following form, where a, b, and c are constants:

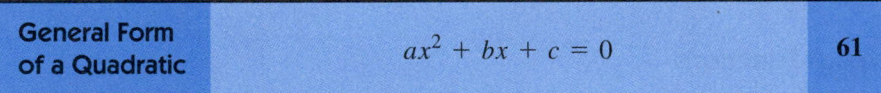

| General Form of a Quadratic | $ax^2 + bx + c = 0$ | 61 |

◆◆◆ **Example 8:** Write the quadratic equation

$$7 - 4x = \frac{5x^2}{3}$$

in general form, and identify a, b, and c.

Solution: Subtracting $5x^2/3$ from both sides and writing the terms in descending order of the exponents, we obtain

$$-\frac{5x^2}{3} - 4x + 7 = 0$$

Quadratics in general form are usually written without fractions and with the first term positive. Multiplying by -3, we get

$$5x^2 + 12x - 21 = 0$$

The equation is now in general form, with $a = 5$, $b = 12$, and $c = -21$. ◆◆◆

The Quadratic Formula

We can find the roots of any quadratic equation $ax^2 + bx + c = 0$ by using the well-known *quadratic formula*

Quadratic Formula	$x = \dfrac{-b \pm \sqrt{b^2 - 4ac}}{2a}$	62

We will derive this formula towards the end of this section, but now we will show its use. Simply put the given equation into general form (Eq. 61); list a, b, and c; and substitute them into the formula.

◆◆◆ **Example 9:** Solve $2x^2 - 5x - 3 = 0$ by the quadratic formula.

Solution: The equation is already in general form, with

$$a = 2 \qquad b = -5 \qquad c = -3$$

Substituting into Eq. 62, we obtain

$$x = \frac{-(-5) \pm \sqrt{(-5)^2 - 4(2)(-3)}}{2(2)}$$

$$= \frac{5 \pm \sqrt{25 + 24}}{4}$$

$$= \frac{5 \pm \sqrt{49}}{4} = \frac{5 \pm 7}{4}$$

Thus we get two answers,

$$x = \frac{5 + 7}{4} = 3 \quad \text{and} \quad x = \frac{5 - 7}{4} = -\frac{1}{2}$$

Check: If you are skeptical about a yet-unproven formula, as you should be, you can always check your answer. We will check by back-substitution.

Substituting 3 into the original equation gives

$$2(3)^2 - 5(3) - 3 = 0$$
$$18 - 15 - 3 = 0 \quad \text{Checks.}$$

Substituting $\left(-\frac{1}{2}\right)$ into the original equation we get

$$2\left(-\frac{1}{2}\right)^2 - 5\left(-\frac{1}{2}\right) - 3 = 0$$

$$2\left(\frac{1}{4}\right) + \frac{5}{2} - 3 = 0$$

$$\frac{1}{2} + \frac{5}{2} - \frac{6}{2} = 0 \quad \text{Checks.}$$

◆◆◆

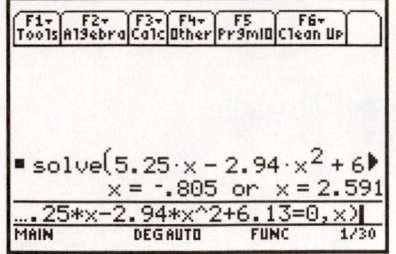

TI-89 Calculator solution for Example 10.

Common Error	Always rewrite a quadratic in general form before trying to use the quadratic formula.

◆◆◆ **Example 10:** Solve the equation $5.25x - 2.94x^2 + 6.13 = 0$ by the quadratic formula.

Solution: The constants a, b, and c are *not* 5.25, −2.94, and 6.13, as you might think at first glance. We must rearrange the terms into general form.

$$-2.94x^2 + 5.25x + 6.13 = 0$$

Although not a necessary step, dividing through by the coefficient of x^2 simplifies the work a bit.

$$x^2 - 1.79x - 2.09 = 0$$

Substituting into the quadratic formula with $a = 1$, $b = -1.79$, and $c = -2.09$, we have

$$x = \frac{1.79 \pm \sqrt{(-1.79)^2 - 4(1)(-2.09)}}{2(1)}$$

$$= 2.59 \quad \text{and} \quad -0.805 \qquad \qquad ◆◆◆$$

Completing the Square

To derive the quadratic formula, we will manipulate the general quadratic equation into the form of a *perfect square trinomial*. This method is called *completing the square*. Recall from our chapter on factoring that a perfect square trinomial is one in which

1. The first and last terms are perfect squares.
2. The middle term is twice the product of the square roots of the outer terms.

◆◆◆ **Example 11:** Make the left side of this equation a perfect square trinomial by completing the square and solve.

$$x^2 - 8x = 0$$

Solution: We need a third term on the left to make the expression a trinomial.

$$x^2 - 8x + (\ ) = 0$$

We complete the square by *adding the square of half the middle coefficient to both sides*.

The middle coefficient is (-8); half the middle coefficient is (-4); the square of half the middle coefficient is then 16. We thus add 16 to both sides.

$$x^2 - 8x + 16 = 16$$

Common Error	When you are adding the quantity needed to complete the square to the left-hand side, it is easy to forget to add the same quantity to the right-hand side. $x^2 - 8x \boxed{+ 16} = \boxed{+16}$ └──── don't forget

Now the first and last terms of the trinomial are perfect squares, and the middle term is twice the product of the square roots of the outer terms. The left side of this equation is now a perfect square trinomial. Factoring and solving for x gives

$$(x - 4)^2 = 16$$

$$x - 4 = \pm 4$$

$$x = 4 \pm 4$$

$$x = 4 - 4 = 0 \quad and \quad x = 4 + 4 = 8 \qquad \text{◆◆◆}$$

This is not the easiest way to solve a quadratic, which could have been solved faster by factoring the given equation. However, it gives us the tools to derive a formula for solving any quadratic.

Derivation of the Quadratic Formula

Given the quadratic equation

$$ax^2 + bx + c = 0$$

we start by subtracting c from both sides and dividing by a.

$$x^2 + \frac{b}{a}x = -\frac{c}{a}$$

We complete the square by *adding the square of half the middle coefficient to both sides* of the equation. The middle coefficient is (b/a); half the middle coefficient is $(b/2a)$; the square of half the middle coefficient is then $(b/2a)^2$. We add this quantity to both sides.

$$x^2 + \frac{b}{a}x + \left(\frac{b}{2a}\right)^2 = \frac{b^2}{4a^2} - \frac{c}{a}$$

$$= \frac{b^2 - 4ac}{4a^2}$$

after combining the terms on the right over a common denominator. The left side of this equation is now a perfect square trinomial. Factoring gives

$$\left(x + \frac{b}{2a}\right)^2 = \frac{b^2 - 4ac}{4a^2}$$

Taking the square root of both sides yields

$$x + \frac{b}{2a} = \pm \frac{\sqrt{b^2 - 4ac}}{2a}$$

Rearranging, we get the formula for finding the roots of the quadratic equation, $ax^2 + bx + c = 0$:

Quadratic Formula	$x = \dfrac{-b \pm \sqrt{b^2 - 4ac}}{2a}$	62

Exercise 2 ◆ Solving a Quadratic by Formula

Solve by quadratic formula. Give your answers in decimal form to three significant digits. Check some by calculator.

1. $x^2 + 5x - 6 = 0$

2. $x^2 - 22x + 8 = 0$

3. $x^2 - 12x + 3 = 0$

4. $x^2 + 2x - 7 = 0$

5. $2x^2 - 15x + 9 = 0$

6. $3x^2 - 10x + 6 = 0$

7. $5x^2 - 25x + 4 = 0$

8. $5x^2 + 22x + 3 = 0$

Challenge Problems

9. $1.22x^2 - 11.5x + 9.89 = 0$

10. $5.11x^2 + 18.6x + 3.88 = 0$

11. $2.96x^2 - 33.2x + 4.05 = 0$

12. $3.22x^2 + 9.66x + 2.85 = 0$

13. $9 + 2x^2 = 25x$

14. $3x^2 = 17x - 6$

15. $3.25 - 31.0x^2 = 4.99x - 63.5$

16. $5.82x + 4.99 = 2.04x^2 + 3.11$

17. $3.88(x^2 + 7.72) = 6.34x(3.99x - 3.81) + 7.33$

18. $(3.99x - 4.22)(6.34x - 8.34) = 7.24x + 8.55$

19. *Project, The Discriminant:* In the quadratic formula, the quantity under the radical sign, $b^2 - 4ac$, is called the *discriminant*. It can be used to predict whether the roots are real and equal, real and unequal, or not real. Try different values of *a, b,* and *c* to give different values for the discriminant. See if you can arrive at some rules for predicting roots, based on the value of the discriminant.

12–3   Applications

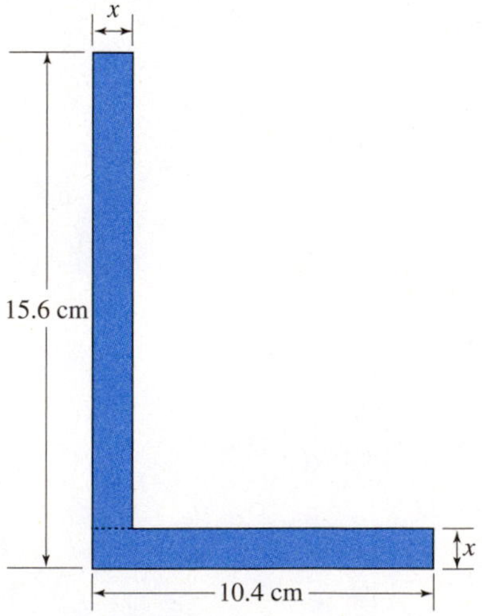

Now that we have the tools to solve any quadratic, let us go on to problems from technology that require us to solve these equations. At this point you might want to take a quick look at Chap. 3 and review some of the suggestions for setting up and solving applications problems. You should set up these problems just as you did then.

When you solve the resulting quadratic, you will usually get two roots. If one of the roots does not make sense in the physical problem (such as a beam having a length of -2000 ft), throw it away. But do not be too hasty. Often a second root will give an unexpected but valid answer.

◆◆◆ **Example 12:** The angle iron in Fig. 12–4 has a cross-sectional area of 53.4 cm^2. Find the thickness *x*.

Estimate: Let's assume that we have two rectangles of width *x*, with lengths of 15.6 cm and 10.4 cm. Setting their combined area equal to 53.4 cm^2 gives

$$x(15.6 + 10.4) = 53.4$$

FIGURE 12–4

Thus we get the approximation $x \approx 2.05$ cm. But since our assumed lengths were too great, because we have counted the small square in the corner twice, our estimated value of x must be too small. We thus conclude that $x > 2.05$ cm.

Solution: We divide the area into two rectangles, as shown by the dashed line at the bottom of Fig. 12–4. One rectangle has an area of $10.4x$, and the other has an area of $(15.6 - x)x$. Since the sum of these areas must be 53.4,

$$10.4x + (15.6 - x)x = 53.4$$

Putting this equation into standard form, we get

$$10.4x + 15.6x - x^2 = 53.4$$
$$x^2 - 26.0x + 53.4 = 0$$

We solve this equation both by graphing and by the equation solver on the TI-83/84. We get one root at

$$x = 2.25 \text{ cm}$$

We get a second root at about 24 cm, which we discard because it is an impossible solution to the given problem.

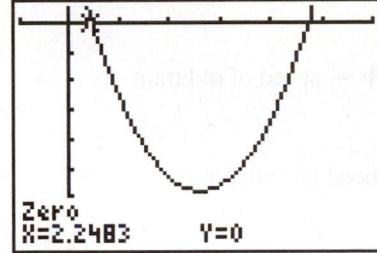
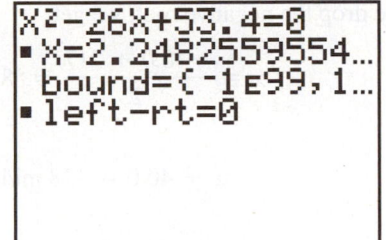

Graphical check for Example 12. Graph of $y = x^2 - 26.0x + 53.4 = 0$. Ticks are spaced 5 units in x and 10 units in y. Note one root at $(2.2483, 0)$ and another at about $(24, 0)$.

Calculator solution: The equation solver on the TI-83/84 shows a solution of $x = 2.2483$ for Example 12.

Check: Does our answer meet the requirements of the original statement? Let us compute the area of the angle iron using our value of 2.25 cm for the thickness. We get

$$\text{area} = 2.25(15.6 - 2.25) + 2.25(10.4)$$
$$= 30.0 + 23.4 = 53.4 \text{ cm}^2$$

which is the required area. Our answer is also a little bigger than our 2.05 cm estimate, as expected. ◆◆◆

◆◆◆ **Example 13:** A certain train is to be replaced with a "bullet" train that goes 40.0 mi/h faster than the old train and that will make its regular 850-mi run in 3.00 h less time. Find the speed of each train.

Solution: Let

$$x = \text{rate of old train (mi/h)}$$

Then

$$x + 40.0 = \text{rate of bullet train (mi/h)}$$

The time it takes the old train to travel 850 mi at x mi/h is, by Eq. 1017,

$$\text{time} = \frac{\text{distance}}{\text{rate}} = \frac{850}{x} \text{ (h)}$$

The time for the bullet train is then $(850/x - 3.00)$ h. Applying Eq. 1017 for the bullet train gives us

$$\text{rate} \times \text{time} = \text{distance}$$

$$(x + 40.0)\left(\frac{850}{x} - 3.00\right) = 850 \tag{1}$$

Removing parentheses, we have

$$850 - 3.00x + \frac{34,000}{x} - 120 = 850$$

Collecting terms and multiplying through by x gives

$$-3.00x^2 + 34,000 - 120x = 0$$

or

$$x^2 + 40.0x - 11,330 = 0$$

Solving for x by the quadratic formula yields

$$x = \frac{-40.0 \pm \sqrt{1600 - 4(-11,330)}}{2}$$

If we drop the negative root, we get

$$x = \frac{-40.0 + 217}{2} = 88.3 \text{ mi/h} = \text{speed of old train}$$

and

$$x + 40.0 = 128 \text{ mi/h} = \text{speed of bullet train} \qquad \blacklozenge\blacklozenge\blacklozenge$$

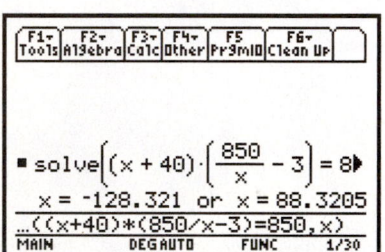

TI-89 Calculator solution for Example 13. We have entered Eq. 1 directly, before any simplifying. We drop the negative root.

Exercise 3 ◆ Applications

Number Puzzles   The "numbers" in problems 2 through 6 are all positive integers.

1. What fraction added to its reciprocal gives $2\frac{1}{6}$?
2. Find three consecutive numbers such that the sum of their squares will be 434.
3. Find two numbers whose difference is 7 and the difference of whose cubes is 1267.
4. Find two numbers whose sum is 11 and whose product is 30.
5. Find two numbers whose difference is 10 and the sum of whose squares is 250.
6. A number increased by its square is equal to 9 times the next higher number. Find the number.

Geometry Problems

7. A rectangle is to be 2 m longer than it is wide and have an area of 24 m². Find its dimensions.
8. One leg of a right triangle is 3 cm greater than the other leg, and the hypotenuse is 15 cm. Find the legs of the triangle.
9. A rectangular sheet of brass is twice as long as it is wide. Squares, 3 cm × 3 cm, are cut from each corner (Fig. 12–5), and the ends are turned up to form an open box having a volume of 648 cm³. What are the dimensions of the original sheet of brass?
10. The length, width, and height of a cubical shipping container are all decreased by 1.0 ft, thereby decreasing the volume of the cube by 37 ft³. What was the volume of the original container?
11. Find the dimensions of a rectangular field that has a perimeter of 724 m and an area of 32,400 m².

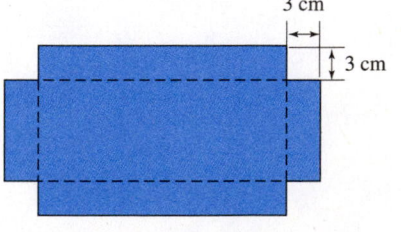

3 cm

3 cm

FIGURE 12–5

12. A flat of width w is to be cut on a bar of radius r (Fig. 12–6). Show that the required depth of cut x is given by the formula

$$x = r \pm \sqrt{r^2 - \frac{w^2}{4}}$$

13. A casting in the shape of a cube is seen to shrink 0.175 in. on a side, with a reduction in volume of 2.48 in.³. Find the original dimensions of the cube.

14. The cylinder in Fig. 12–7 has a surface area of 846 cm², including the ends. Find its radius.

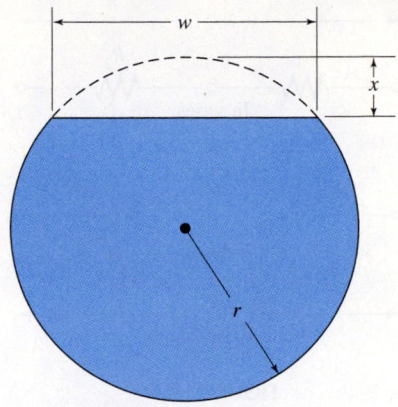

FIGURE 12–6

Uniform Motion

15. A truck travels 350 mi to a delivery point, unloads, and, now empty, returns to the starting point at a speed 8.00 mi/h greater than on the outward trip. What was the speed of the outward trip if the total round-trip driving time was 14.4 h?

16. An airplane flies 355 mi to city A. Then, with better winds, it continues on to city B, 448 mi from A, at a speed 15.8 mi/h greater than on the first leg of the trip. The total flying time was 5.20 h. Find the speed at which the plane traveled to city A.

17. An express bus travels a certain 250-mi route in 1.0 h less time than it takes a local bus to travel a 240-mi route. Find the speed of each bus if the speed of the local is 10 mi/h less than that of the express.

18. A trucker calculates that if he increased his average speed by 20 km/h, he could travel his 800-km route in 2.0 h less time than usual. Find his usual speed.

19. A boat sails 30 km at a uniform rate. If the rate had been 1 km/h more, the time of the sailing would have been 1 h less. Find the rate of travel.

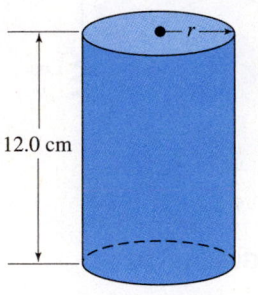

12.0 cm

FIGURE 12–7

Work Problems

20. A certain punch press requires 3 h longer to stamp a box of parts than does a newer-model punch press. After the older press has been punching a box of parts for 5 h, it is joined by the newer machine. Together, they finish the box of parts in 3 additional hours. How long does it take each machine, working alone, to punch a box of parts?

21. Two water pipes together can fill a certain tank in 8.40 h. The smaller pipe alone takes 2.50 h longer than the larger pipe to fill that same tank. How long would it take the larger pipe alone to fill the tank?

22. A laborer built 35 m of stone wall. If she had built 2 m less each day, it would have taken her 2 days longer. How many meters did she build each day, working at her usual rate?

23. A woman worked part-time a certain number of days, receiving for her pay $1800. If she had received $10 per day less than she did, she would have had to work 3 days longer to earn the same sum. How many days did she work?

Simply Supported Beam

24. For a simply supported beam of length l having a distributed load of w lb/ft (Fig. 12–8), the bending moment M at any distance x from one end is given by

$$M = \frac{1}{2}wlx - \frac{1}{2}wx^2$$

Find the locations on the beam where the bending moment is zero.

25. A simply supported beam, 25.0 ft long, carries a distributed load of 1550 lb/ft. At what distances from an end of the beam will the bending moment be 112,000 ft · lb? (Use the equation from problem 24.)

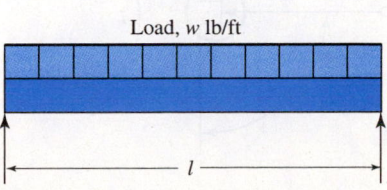

Load, w lb/ft

FIGURE 12–8 Simply supported beam with a uniformly distributed load.

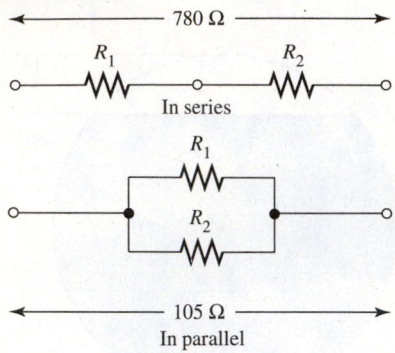

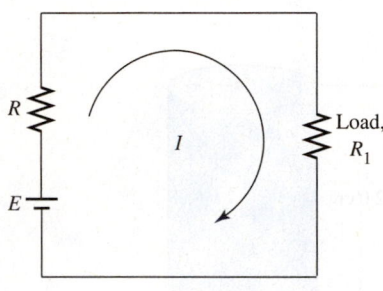

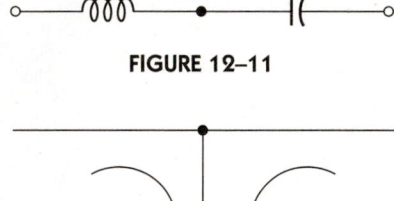

FIGURE 12–9

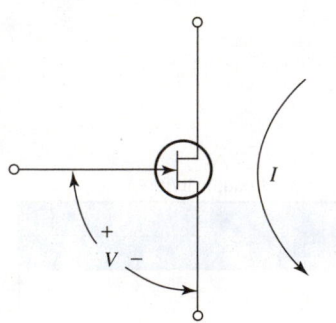

FIGURE 12–10

$$ \begin{array}{c} L \qquad\qquad C \\ \text{⦿━━ooo━━━●━━━┤├━━o} \end{array} $$

FIGURE 12–11

FIGURE 12–12

FIGURE 12–13

Freely Falling Body   Use $s = v_0 t + \frac{1}{2} g t^2$ for these falling-body problems, but be careful of the signs. If you take the upward direction as positive, g will be *negative*.

26. An object is thrown upward with a velocity of 145 ft/s. When will it be 85.0 ft above its initial position? ($g = 32.2$ ft/s^2)

27. An object is thrown upward with an initial speed of 120 m/s. Find the time for it to return to its starting point. (In the metric system, $g = 980$ cm/s^2.)

Electrical Problems

28. Referring to Fig. 12–9, (a) determine what two resistances will give a total resistance of 780 Ω when wired in series and 105 Ω when wired in parallel.
(b) Find two resistances that will give an equivalent resistance of 9070 Ω in series and 1070 Ω in parallel. (Recall that the equivalent resistance for two resistors R_1 and R_2 is $R_1 + R_2$ when wired in series, and $R_1 R_2 /(R_1 + R_2)$ when wired in parallel.)

29. In the circuit of Fig. 12–10, the power P dissipated in the load resistor R_1 is

$$ P = EI - I^2 R $$

If the voltage E is 115 V, and if $R = 100$ Ω, find the current I needed to produce a power of 29.3 W in the load.

30. The reactance X of a capacitance C and an inductance L in series (Fig. 12–11) is

$$ X = \omega L - \frac{1}{\omega C} $$

where ω is the angular frequency in rad/s, L is in henries, and C is in farads. Find the angular frequency needed to make the reactance equal to 1500 Ω, if $L = 0.5$ henry and $C = 0.2 \times 10^{-6}$ farad (0.2 microfarad).

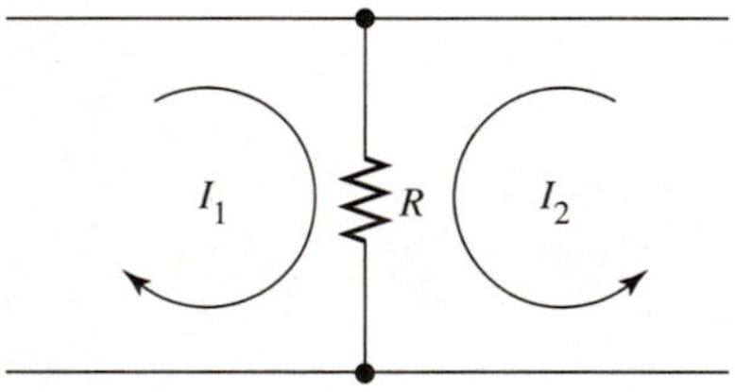

31. Figure 12–12 shows two currents flowing in a single resistor R. The total current in the resistor will be $I_1 + I_2$, so the power dissipated is

$$ P = (I_1 + I_2)^2 R $$

If $R = 100$ Ω and $I_2 = 0.2$ A, find the current I_1 needed to produce a power of 9.0 W.

32. A *square-law device* is one whose output is proportional to the square of the input. A junction field-effect transistor (JFET) (Fig. 12–13) is such a device. The current I that will flow through an *n*-channel JFET when a voltage V is applied is

$$ I = A\left(1 - \frac{V}{B}\right)^2 $$

where A is the drain saturation current and B is the gate source pinch-off voltage.
(a) Solve this equation for V.
(b) A certain JFET has a drain saturation current of 4.8 mA and a gate source pinch-off voltage of -2.5 V. What input voltage is needed to produce a current of 1.5 mA?

33. *Project:* A cylindrical tank having a diameter of 75.5 in. is placed so that it touches a wall, as in Fig. 12–14. Find the radius x of the largest pipe that can fit into the space between the tank, the wall, and the floor. *Hint:* First use the Pythagorean theorem to show that $OC = 53.4$ in. and that $OP = 15.6$ in. Then in triangle OQR, $(OP - x)^2 = x^2 + x^2$.

Check your work by making a CAD drawing, and have the program give the missing dimension.

••• CHAPTER 12 REVIEW PROBLEMS •••••••••••••••••••••••••••••••

Solve each equation. Give any approximate answers to three significant digits.

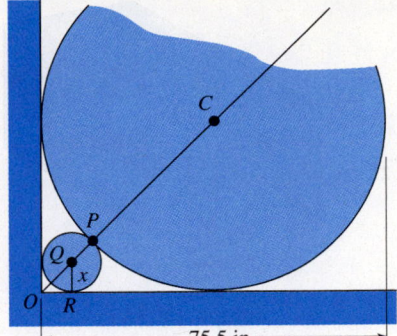

FIGURE 12–14

1. $y^2 - 5y - 6 = 0$

2. $2x^2 - 5x = 2$

3. $w^2 - 5w = 0$

4. $x(x - 2) = 2(-2 - x + x^2)$

5. $\dfrac{r}{3} = \dfrac{r}{r + 5}$

6. $6y^2 + y - 2 = 0$

7. $3t^2 - 10 = 13t$

8. $2.73x^2 + 1.47x - 5.72 = 0$

9. $\dfrac{1}{t^2} + 2 = \dfrac{3}{t}$

10. $3w^2 + 2w - 11 = 0$

11. $9 - x^2 = 0$

12. $\dfrac{2y}{3} = \dfrac{3}{5} + \dfrac{2}{y}$

13. $2x(x + 2) = x(x + 3) + 5$

14. $18 + w^2 + 11w = 0$

15. $9y^2 + y = 5$

16. $\dfrac{5}{w} - \dfrac{4}{w^2} = \dfrac{3}{5}$

17. $\dfrac{z}{2} = \dfrac{5}{z}$

18. $3x - 6 = \dfrac{5x + 2}{4x}$

19. $2x^2 + 3x = 2$

20. $x^2 + Rx - R^2 = 0$

21. $27x = 3x^2 - 5$

22. $6w^2 + 13w + 6 = 0$

23. $5y^2 = 125$

24. $\dfrac{t + 2}{t} = \dfrac{4t}{3}$

25. $1.26x^2 - 11.8 = 1.13x$

26. $2.62z^2 + 4.73z - 5.82 = 0$

27. $5.12y^2 + 8.76y - 9.89 = 0$

28. $3w^2 + 2w - 2 = 0$

29. $27.2w^2 + 43.6w = 45.2$

30. Plot the parabola $y = 3x^2 + 2x - 6$. Label the vertex, the axis of symmetry, and any zeros.

31. A person purchased some bags of insulation for $1000. If she had purchased 5 more bags for the same sum, they would have cost 12 cents less per bag. How many did she buy?

32. The perimeter of a rectangular field is 184 ft, and its area 1920 ft². Find its dimensions.

33. The rectangular yard of Fig. 12–15 is to be enclosed by fence on three sides, and an existing wall is to form the fourth side. The area of the yard is to be 450 m², and its length is to be twice its width. Find the dimensions of the yard.

34. A fast train runs 8.0 mi/h faster than a slow train and takes 3.0 h less to travel 288 mi. Find the rates of the trains.

35. A man started to walk 3 mi, intending to arrive at his destination at a certain time. After walking 1 mi, he was detained 10 min and had to walk the rest of the way 1 mi/h faster in order to arrive at the intended time. What was his original speed?

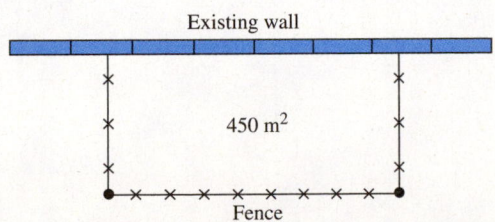

FIGURE 12–15

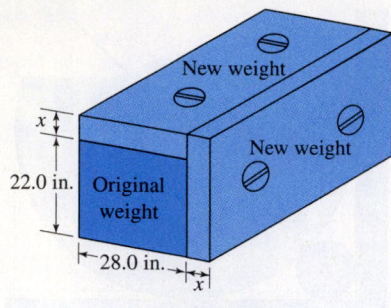

FIGURE 12–16

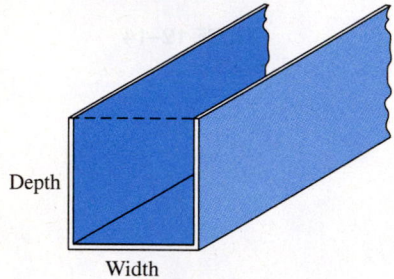

FIGURE 12–17

36. A rectangular field is 12 m longer than it is wide and has an area of 448 m². What are the lengths of its sides?

37. A tractor wheel, 15 ft in circumference, makes one revolution in a certain number of seconds. If it slowed so as to take 1.0 s longer per revolution, the tractor would travel 14,400 ft less in 1.0 h. In how many seconds does it make one revolution?

38. The iron counterweight in Fig. 12–16 is to have its weight increased by 50% by plates of iron bolted along the top and side (but not at the ends). The top plate and the side plate have the same thickness. Find their thickness.

39. A 26-in.-wide strip of steel is to have its edges bent up at right angles to form an open trough, as in Fig. 12–17. The cross-sectional area is to be 80 in². Disregarding the thickness of the steel sheet, find the width and the depth of the trough.

40. A boat sails 30 mi at a uniform rate. If the rate had been 1 mi/h less, the time of the sailing would have been 1 h more. Find the rate of travel.

41. In a certain number of hours a woman traveled 36.0 km. If she had traveled 1.50 km more per hour, it would have taken her 3.00 h less to make the journey. How many kilometers did she travel per hour?

42. The length of a rectangular court exceeds its width by 2 m. If the length and the width were each increased by 3 m, the area of the court would be 80 m². Find the dimensions of the court.

43. The area of a certain square will double if its length and width are increased by 6.0 ft and 4.0 ft, respectively. Find its original dimensions.

44. A mirror 18 in. by 12 in. is to be set in a frame of uniform width, and the area of the frame is to be equal to that of the glass. Find the width of the frame.

45. *Writing:* We have given several ways to solve a quadratic. List them and explain the advantages and disadvantages of each.

46. *On our Web site:* For methods of solution of equations of quadratic type, simple equations of higher degree, and systems of quadratics, see **Equations of Higher Degree** at our Web site: www.wiley.com/college/calter

13

Exponents and Radicals

♦♦♦ **OBJECTIVES** ♦♦♦

When you have completed this chapter, you should be able to

- Use the laws of exponents to simplify and combine expressions having integral exponents, by hand or by calculator.
- Simplify radicals by removing perfect powers, by rationalizing the denominator, and by reducing the index.
- Add, subtract, multiply, and divide radicals.
- Solve radical equations, manually or by calculator.

♦♦

We introduced exponents and gave the laws of exponents in our "Introduction to Algebra" chapter. We review those laws here, give more advanced examples of their use, with applications, and show how to manipulate expressions having exponents by calculator.

Next we make a strong connection between exponents and radicals, and show how to simplify, add, subtract, multiply, and divide radical expressions. We did some calculation of roots in Chapter 1, but with only *numbers* under the radical sign. Here we show how to handle expressions with *literals* under the radical sign. The ability to manipulate both exponents and radicals is needed to work with many formulas found in technology.

Finally we add another kind of equation to our growing list, the *radical equation*. As with quadratics, we start with methods of solution that we already know, solution by graphing and by calculator. This is followed by methods for an algebraic solution, and of course, applications. For example, the natural frequency f_n of the weight bouncing at the end of a spring, Fig. 13–1, is given by

$$f_n = \frac{1}{2\pi}\sqrt{\frac{kg}{W}}$$

where g is the gravitational constant, k is the spring constant, and W is the weight. You would have no problem finding f_n, given the other quantities, but how would you solve for, say, W? You will learn how in this chapter.

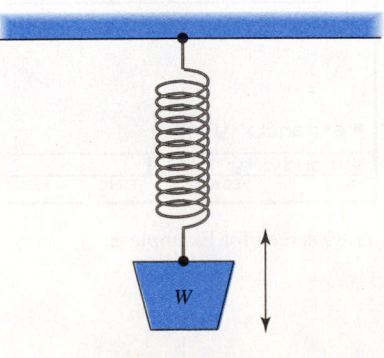

FIGURE 13–1

13–1   Integral Exponents

In this section, we continue the study of exponents that we started in Chap. 2. We repeat the laws of exponents derived there and use them to simplify more difficult expressions than before. You should glance back at that section before starting here.

Negative Exponents

The law that we derived in Chap. 2 for a negative exponent is repeated here.

Negative Exponent	$$x^{-a} = \frac{1}{x^a} \quad (x \neq 0)$$ *When taking the reciprocal of a base raised to a power, change the sign of the exponent.*	**28**

We'll use the laws of exponents mainly to simplify expressions, to make them easier to work with in later computations, such as solving equations containing exponents. For example, we use the law for negative exponents to rewrite an expression so that it does not contain a negative exponent.

◆◆◆ **Example 1:** In these examples we use Eq. 28 to eliminate negative exponents.

(a) $7^{-1} = \dfrac{1}{7}$ (b) $x^{-1} = \dfrac{1}{x}$

(c) $z^{-3} = \dfrac{1}{z^3}$ (d) $xy^{-1} = \dfrac{x}{y}$

(e) $\dfrac{ab^{-2}}{c^{-3}d} = \dfrac{ac^3}{b^2 d}$ ◆◆◆

Common Error	$$x^{-a} \neq x^{1/a}$$

Integral Exponents by Calculator

A calculator that can do symbolic algebra may be used to simplify expressions containing exponents. On the TI-89, we use **expand** from the **Algebra** menu.

◆◆◆ **Example 2:** Simplify the expression from Example 1(d),

$$xy^{-1}$$

Solution: We choose **expand** and enter the expression as shown. Pressing ENTER gives the simplified expression. ◆◆◆

We will show calculator screens in many of the following examples.

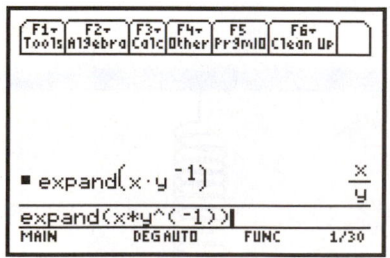

TI-89 screen for Example 2.

Zero Exponents

The law that we derived in Chap. 2 for a zero exponent is

Zero Exponent	$x^0 = 1 \quad (x \neq 0)$ *Any quantity (except 0) raised to the zero power equals 1.*	27

We use this law to rewrite an expression so that it contains no zero exponents.

♦♦♦ **Example 3:** Here we have eliminated zero exponents using Eq. 27.

(a) $367^0 = 1$

(b) $(\sin 37.2°)^0 = 1$

(c) $x^2 y^0 z = x^2 z$

(d) $(abc)^0 = 1$ ♦♦♦

We'll use these laws frequently in the examples to come.

Power Raised to a Power

Another law that we derived in Chap. 2 is for raising a power to a power.

Power Raised to a Power	$(x^a)^b = x^{ab} = (x^b)^a$ *When raising a power to a power, keep the same base and **multiply the exponents.***	24

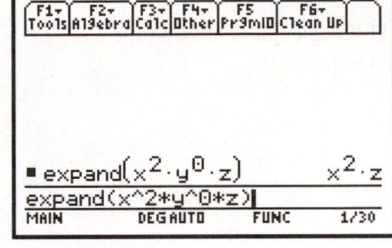

TI-89 calculator solution for Example 3(c).

We use this law to simplify expressions, as follows.

♦♦♦ **Example 4:** These examples show expressions simplified using Eq. 24.

(a) $(2^2)^2 = 2^4 = 16$

(b) $(3^{-1})^4 = 3^{-4} = \dfrac{1}{3^4} = \dfrac{1}{81}$

(c) $(a^2)^3 = a^6$

(d) $(z^{-3})^{-2} = z^6$

(e) $(x^5)^{-2} = x^{-10} = \dfrac{1}{x^{10}}$ ♦♦♦

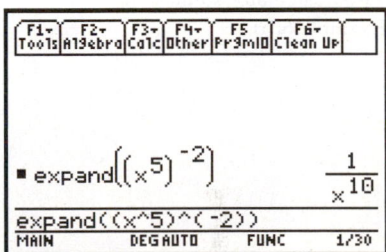

TI-89 calculator solution for Example 4(e).

Products

The two laws of exponents that apply to products are given here again.

Products	$x^a \cdot x^b = x^{a+b}$ *When multiplying powers of the same base, keep the same base and **add the exponents.***	22
Product Raised to a Power	$(xy)^n = x^n \cdot y^n$ *When a product is raised to a power, each factor may be **separately** raised to that power.*	25

The use of these laws in simplifying expressions is shown in the following examples.

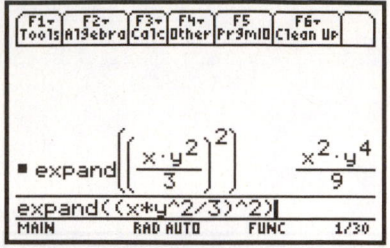

TI-89 calculator solution for Example 5(c).

◆◆◆ **Example 5:** Here we simplify using Eqs. 22 and 25.

(a) $2axy^3z(3a^2xyz^3) = 6a^3x^2y^4z^4$ (b) $(3bx^3y^2)^3 = 3^3b^3(x^3)^3(y^2)^3 = 27b^3x^9y^6$

(c) $\left(\dfrac{xy^2}{3}\right)^2 = \dfrac{(xy^2)^2}{3^2} = \dfrac{x^2(y^2)^2}{9} = \dfrac{x^2y^4}{9}$ ◆◆◆

Common Error	Parentheses are important when we are raising products to a power. $\quad (2x)^0 = 1 \qquad$ but $\qquad 2x^0 = 2(1) = 2$ $\quad (4x)^{-1} = \dfrac{1}{4x} \qquad$ but $\qquad 4x^{-1} = \dfrac{4}{x}$

We try to leave our expressions without zero or negative exponents, as in the following examples.

◆◆◆ **Example 6:** Here we show how to eliminate zero or negative exponents.

(a) $3p^3q^2r^{-4}(4p^{-5}qr^4) = 12p^{-2}q^3r^0 = \dfrac{12q^3}{p^2}$

(b) $(2m^2n^3x^{-2})^{-3} = 2^{-3}m^{-6}n^{-9}x^6 = \dfrac{x^6}{8m^6n^9}$ ◆◆◆

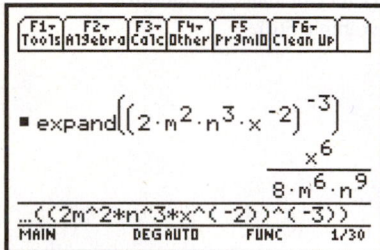

TI-89 calculator solution for Example 6(b).

Letters in the exponents are handled just as if they were numbers.

◆◆◆ **Example 7:** This example has letters in the exponents.

$$x^{n+1}y^{1-n}(x^{2n-3}y^{n-1}) = x^{n+1+2n-3}y^{1-n+n-1}$$
$$= x^{3n-2}y^0 = x^{3n-2} \qquad ◆◆◆$$

Quotients

Our two laws of exponents that apply to quotients are repeated here.

Quotients	$\dfrac{x^a}{x^b} = x^{a-b} \quad (x \neq 0)$ *When dividing powers of the same base, keep the same base and **subtract the exponents**.*	23
Quotient Raised to a Power	$\left(\dfrac{x}{y}\right)^n = \dfrac{x^n}{y^n} \quad (y \neq 0)$ *When a quotient is raised to a power, the numerator and the denominator may be **separately** raised to the power.*	26

We also use these laws to simplify expressions, as follows.

◆◆◆ **Example 8:** Here we simplify expressions using Eqs. 23 and 26.

(a) $\dfrac{x^5}{x^3} = x^2$ (b) $\dfrac{16p^5q^{-3}}{8p^2q^{-5}} = 2p^3q^2$

(c)
$$\frac{x^{2p-1}y^2}{x^{p+4}y^{1-p}} = x^{2p-1-p-4}y^{2-1+p}$$
$$= x^{p-5}y^{1+p}$$

(d)
$$\left(\frac{x^3y^2}{p^2q}\right)^3 = \frac{x^9y^6}{p^6q^3}$$

◆◆◆

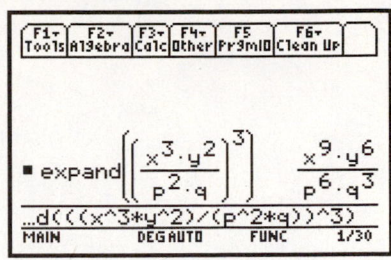

TI-89 calculator solution for Example 8(d).

As before, we try to leave our expressions without zero or negative exponents.

◆◆◆ **Example 9:** These examples show how to eliminate zero and negative exponents follow.

(a)
$$\frac{8a^3b^{-5}c^3}{4a^5b^{-1}c^3} = 2a^{-2}b^{-4}c^0 = \frac{2}{a^2b^4}$$

(b)
$$\left(\frac{a}{b}\right)^{-2} = \frac{a^{-2}}{b^{-2}} = \frac{b^2}{a^2} = \left(\frac{b}{a}\right)^2$$

◆◆◆

Note in Example 9(b) that the negative exponent had the effect of inverting the fraction.

Don't forget our hard-won skills of factoring, of combining fractions over a common denominator, and of simplifying complex fractions. Use them where needed, as shown here.

◆◆◆ **Example 10:** This example involves adding fractions and simplifying a compound fraction.

$$\frac{1}{(5a)^2} + \frac{1}{(2a^2b^{-4})^2} = \frac{1}{25a^2} + \frac{1}{\dfrac{4a^4}{b^8}}$$

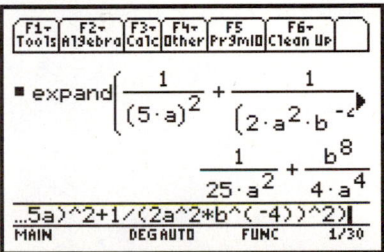

TI-89 calculator solution for Example 10. We do this problem in two parts. First we use **expand** on the given expression.

We simplify the compound fraction and combine the two fractions over a common denominator, getting

$$\frac{1}{25a^2} + \frac{b^8}{4a^4} = \frac{4a^2}{100a^4} + \frac{25b^8}{100a^4}$$
$$= \frac{4a^2 + 25b^8}{100a^4}$$

◆◆◆

◆◆◆ **Example 11:** Here is another example with a compound fraction.

$$(x^{-2} - 3y)^{-2} = \frac{1}{(x^{-2} - 3y)^2} = \frac{1}{\left(\dfrac{1}{x^2} - 3y\right)^2}$$

$$= \frac{1}{\left(\dfrac{1 - 3x^2y}{x^2}\right)^2}$$

$$= \frac{x^4}{(1 - 3x^2y)^2}$$

◆◆◆

Next we use **comDenom** to combine the two fractions over a common denominator.

Common Error	Do *not* distribute the –2 power across the two terms in parentheses. $$(x^{-2} - 3y)^{-2} \neq (x^{-2})^{-2} - (3y)^{-2}$$

◆◆◆ **Example 12:** This example shows how factoring can be used to simplify an expression.

$$\frac{p^{4m} - q^{2n}}{p^{2m} + q^n} = \frac{(p^{2m})^2 - (q^n)^2}{p^{2m} + q^n}$$

Factoring the difference of two squares in the numerator and canceling gives us

$$\frac{(p^{2m} + q^n)(p^{2m} - q^n)}{p^{2m} + q^n} = p^{2m} - q^n$$

◆◆◆

Exercise 1 ◆ Integral Exponents

Simplify, and write without negative exponents. Do some by calculator.

1. $3x^{-1}$

2. $4a^{-2}$

3. $2p^0$

4. $(3m)^{-1}$

5. $a(2b)^{-2}$

6. x^2y^{-1}

7. a^3b^{-2}

8. m^2n^{-3}

9. p^3q^{-1}

10. $(2x^2y^3z^4)^{-1}$

11. $(3m^3n^2p)^0$

12. $(5p^2q^2r^2)^3$

13. $(4a^3b^2c^6)^{-2}$

14. $(a^p - b^q)^2$

15. $(x + y)^{-1}$

16. $(2a + 3b)^{-1}$

17. $(m^{-2} - 6n)^{-2}$

18. $(4p^2 + 5q^{-4})^{-2}$

19. $2x^{-1} + y^{-2}$

20. $p^{-1} - 3q^{-2}$

21. $\left(\dfrac{3}{b}\right)^{-1}$

22. $\left(\dfrac{x}{4}\right)^{-1}$

23. $\left(\dfrac{2x}{3y}\right)^{-2}$

24. $\left(\dfrac{3p}{4y}\right)^{-1}$

25. $\left(\dfrac{2px}{3qy}\right)^{-3}$

26. $\left(\dfrac{p^2}{q^0}\right)^{-1}$

27. $\left(\dfrac{3a^4b^3}{5x^2y}\right)^2$

28. $\left(\dfrac{x^2}{y}\right)^{-3}$

Challenge Problems

29. $(3m)^{-3} - 2n^{-2}$

30. $(5a)^{-2} + (2a^2b^{-4})^{-2}$

31. $(x^n + y^m)^2$

32. $(x^{a-1} + y^{a-2})(x^a + y^{a-1})$

33. $(16x^6y^0 \div 8x^4y) \div 4xy^6$

34. $(a^{n-1} + b^{n-2})(a^n + b^{n-1})$

35. $(72p^6q^7 \div 9p^4q) \div 8pq^6$

36. $\dfrac{(p^2 - pq)^8}{(p - q)^4}$

37. $\left(\dfrac{-2a^3x^3}{3b^2y}\right)^{2n}$

38. $\left(\dfrac{3p^2y^3}{4qx^4}\right)^{-2}$

39. $\left(\dfrac{2p^{-2}z^3}{3q^{-2}x^{-4}}\right)^{-1}$

40. $\left(\dfrac{9m^4n^3}{3p^3q}\right)^3$

41. $\left(\dfrac{5w^2}{2z}\right)^p$

42. $\left(\dfrac{5a^5b^3}{2x^2y}\right)^{3n}$

43. $\left(\dfrac{3n^{-2}y^3}{5m^{-3}x^4}\right)^{-2}$

44. $\left(\dfrac{z^0z^{2n-2}}{z^n}\right)^3$

Applications

45. The resistance R of two resistors R_1 and R_2 wired in parallel, as shown in Fig. 13–2, is given by Eq. 1063.

$$\frac{1}{R} = \frac{1}{R_1} + \frac{1}{R_2}$$

Write this equation using negative exponents.

46. When a current I causes a power P in a resistance, R, that resistance is given by

$$R = \frac{P}{I^2}$$

Write this equation without fractions.

47. The equation for power to a resistor is given in Problem 46. If the current is then doubled and the resistance is halved, we have

$$(2I)^2\left(\frac{R}{2}\right)$$

Simplify this expression.

48. The volume of a cube of side a is a^3. If we double the length of the side, the volume becomes

$$(2a)^3$$

Simplify this expression.

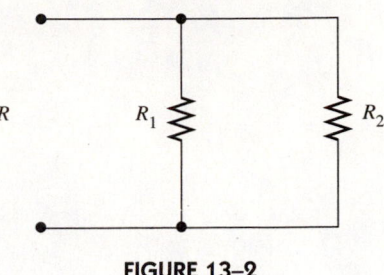

FIGURE 13–2

13–2   Simplification of Radicals

Here we will review and expand upon material first given in earlier chapters. Recall that a radical consists of a *radical sign*, a quantity under the radical sign called the *radicand*, and the *index* of the radical.

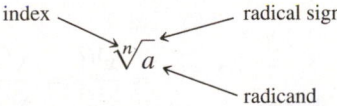

where n is an integer. For now we will restrict the radicand to a positive number when the index is even. We will consider even roots of negative numbers when we cover imaginary numbers.

Relation between Fractional Exponents and Radicals

We have seen that when a number, say 5, is multiplied by itself

$$5 \times 5 = 25$$

then 5 is called the *square root* of 25.

$$5 = \sqrt{25}$$

In other words, *when two equal factors are multiplied together, one of these factors is the square root of the product.* Let us try this with the number $a^{1/2}$.

$$a^{1/2} \times a^{1/2} = a^{1/2 + 1/2} = a^1 = a$$

Here we have two equal factors, $a^{1/2}$, and $a^{1/2}$, multiplied together, so one of these factors is the square root of the product a.

Fractional Exponent	$a^{1/2} = \sqrt{a}$	29

Raising to an exponent $^1/_2$ means the same as taking the square root. Recall that we used this relation in Chap. 1 to take the root of a number using a calculator that did not have keys for taking roots.

◆◆◆ **Example 13:** Some expressions with fractional exponents are rewritten here as radicals.

(a) $4^{1/2} = \sqrt{4} = 2$ (b) $8^{1/3} = \sqrt[3]{8} = 2$

(c) $x^{1/2} = \sqrt{x}$ (d) $y^{1/4} = \sqrt[4]{y}$

(e) $w^{-1/2} = \dfrac{1}{\sqrt{w}}$ ◆◆◆

Now let us raise both sides of Eq. 29 to the power m.

$$(a^{1/n})^m = (\sqrt[n]{a})^m \tag{1}$$

Applying the law of exponents for a power raised to a power gives

$$(a^{1/n})^m = a^{m/n} = (a^m)^{1/n} = \sqrt[n]{a^m} \tag{2}$$

Combining (1) and (2) gives the following formula,

Fractional Exponent	$a^{m/n} = \sqrt[n]{a^m} = (\sqrt[n]{a})^m$	**30**

◆◆◆ **Example 14:** These examples show the application of Eq. 30.

(a) $8^{2/3} = (\sqrt[3]{8})^2 = (2)^2 = 4$

(b) $\sqrt[3]{x^4} = x^{4/3}$

(c) $(\sqrt[5]{y^2})^3 = (y^{2/5})^3 = y^{6/5}$ ◆◆◆

We use these definitions to switch between *exponential form* and *radical form*.

◆◆◆ **Example 15:** Express $x^{1/2}y^{-1/3}$ in radical form.

Solution:

$$x^{1/2}y^{-1/3} = \frac{x^{1/2}}{y^{1/3}} = \frac{\sqrt{x}}{\sqrt[3]{y}}$$

◆◆◆

◆◆◆ **Example 16:** Express $\sqrt[3]{y^2}$ in exponential form.

Solution:

$$\sqrt[3]{y^2} = (y^2)^{1/3} = y^{2/3}$$

◆◆◆

Common Error	Don't confuse the **coefficient** of a radical with the **index** of a radical. $$3\sqrt{x} \neq \sqrt[3]{x}$$

Root of a Product

We have several *rules of radicals*, which are similar to the laws of exponents and, in fact, are derived from them. The first rule is for products. By our definition of a radical,

$$\sqrt[n]{ab} = (ab)^{1/n}$$

Using the law of exponents for a product raised to a power, and then returning to radical form,

$$(ab)^{1/n} = a^{1/n}b^{1/n} = \sqrt[n]{a}\,\sqrt[n]{b}$$

So our first rule of radicals is:

| Root of a Product | $\sqrt[n]{ab} = \sqrt[n]{a}\sqrt[n]{b}$ *The root of a product equals the product of the roots of the factors.* | 31 |

◆◆◆ Example 17: We may split the radical $\sqrt{9x}$ into two radicals, as follows:
$$\sqrt{9x} = \sqrt{9}\ \sqrt{x} = 3\sqrt{x}$$
since $\sqrt{9} = 3$. ◆◆◆

◆◆◆ Example 18: Write as a single radical $\sqrt{7}\ \sqrt{2}\ \sqrt{x}$.

Solution: By Eq. 31,
$$\sqrt{7}\ \sqrt{2}\ \sqrt{x} = \sqrt{7(2)x} = \sqrt{14x}$$ ◆◆◆

Common Errors

There is no rule similar to Eq. 31 for the square root of a *sum*. Recall that the bar is symbol of grouping. In the expression
$$\sqrt{a + b}$$
a and b must be treated *as a whole*, and *not individually*. In other words,
$$\sqrt{a + b} \neq \sqrt{a} + \sqrt{b}$$

Equation 31 does not hold when a and b are both *negative* and the index is even.
$$\left(\sqrt{-4}\right)^2 = \sqrt{-4}\sqrt{-4} \neq \sqrt{(-4)(-4)}$$
$$\neq \sqrt{16} = +4$$

Instead, we convert to *imaginary numbers*, as we will show in a later chapter.

Root of a Quotient

We saw that the root of a product can be split up into the roots of the individual factors. Similarly, the root of a quotient can be expressed as the root of the numerator divided by the root of the denominator. We first write the quotient in exponential form.
$$\sqrt[n]{\frac{a}{b}} = \left(\frac{a}{b}\right)^{1/n} = \frac{a^{1/n}}{b^{1/n}}$$

by the law of exponents for a quotient raised to a power. Returning to radical form, we have the following:

| Root of a Quotient | $\sqrt[n]{\frac{a}{b}} = \frac{\sqrt[n]{a}}{\sqrt[n]{b}}$ *The root of a quotient equals the quotient of the roots of numerator and denominator.* | 32 |

◆◆◆ **Example 19:** The radical $\sqrt{\dfrac{w}{25}}$ can be written $\dfrac{\sqrt{w}}{\sqrt{25}}$ or $\dfrac{\sqrt{w}}{5}$. ◆◆◆

Simplest Form for a Radical

Suppose you solve a problem and get an answer of $\dfrac{5}{\sqrt{x}}$, but the expression in the answer key is given as $\dfrac{5\sqrt{x}}{x}$. Are these equivalent? (They are.) Suppose you look up the formula for the natural frequency of a weight bouncing at the end of a spring and find it in two different books as

$$\frac{1}{2\pi}\sqrt{\frac{kg}{W}} \quad \text{or} \quad \frac{\sqrt{kgW}}{2\pi W}$$

Are these the same? (They are.)

We usually give a mathematical expression in some agreed upon simplified format.

You would not, for example, give a result as $\dfrac{x^2}{x}$, but would simplify it to x.

Here we will learn how to put radicals into a form so that they can easily be compared or combined. This is called *simplest form*. A radical is said to be in *simplest form* when

1. The radicand has been reduced as much as possible.
2. There are no radicals in the denominator and no fractional radicands.
3. The index has been made as small as possible.

Reducing the Radicand

If our radical is a *square root*, we see if there is a perfect square under the radical sign. It can then be moved outside the radical sign, as in the following examples.

We are doing a few numerical problems here as a way of learning the rules. If you simply want the decimal value of a radical expression containing only numbers, use your calculator.

◆◆◆ **Example 20:** These examples show how to remove a perfect square from under the radical sign.

(a) $\sqrt{49} = \sqrt{7^2} = 7$

(b) $\sqrt{50} = \sqrt{(25)(2)} = \sqrt{25}\sqrt{2} = 5\sqrt{2}$

(c) $\sqrt{x^3} = \sqrt{x^2 x} = \sqrt{x^2}\sqrt{x} = x\sqrt{x}$ ◆◆◆

◆◆◆ **Example 21:** Simplify $\sqrt{50x^3}$.

Solution: We factor the radicand so that some factors are perfect squares.

$$\sqrt{50x^3} = \sqrt{(25)(2)x^2 x}$$

Then, by Eq. 31,

$$= \sqrt{25}\sqrt{x^2}\sqrt{2x} = 5x\sqrt{2x}$$ ◆◆◆

◆◆◆ **Example 22:** Simplify $\sqrt{24y^5}$.

Solution: We look for factors of the radicand that are perfect squares.

$$\sqrt{24y^5} = \sqrt{4(6)y^4 y} = 2y^2\sqrt{6y}$$ ◆◆◆

When the radicand contains more than one term, try to *factor out* a perfect *n*th power (where *n* is the index).

◆◆◆ Example 23: Simplify $\sqrt{4x^2y + 12x^4z}$.

Solution: We factor $4x^2$ from the radicand and then remove it from under the radical sign.

$$\sqrt{4x^2y + 12x^4z} = \sqrt{4x^2(y + 3x^2z)}$$
$$= 2x\sqrt{y + 3x^2z}$$

◆◆◆

If our radical is a *cube root*, we try to remove perfect cubes from under the radical sign. If the radical is a *fourth root*, we look for perfect fourth powers, and so on.

◆◆◆ Example 24: Simplify $\sqrt[3]{24y^5}$.

Solution: We look for factors of the radicand that are perfect cubes.

$$\sqrt[3]{24y^5} = \sqrt[3]{8(3)y^3y^2}$$
$$= \sqrt[3]{8y^3}\sqrt[3]{3y^2} = 2y\sqrt[3]{3y^2}$$

◆◆◆

◆◆◆ Example 25: Here we remove perfect fourth powers.

$$\sqrt[4]{24y^5} = \sqrt[4]{24y^4y} = y\sqrt[4]{24y}$$

◆◆◆

Removing Radicals from the Denominator

A fractional expression is considered in simpler form when its denominators contain no radicals. To put it into this form is called *rationalizing* the denominator. We will show how to rationalize the denominator when it is a square root, a cube root, or a root with any index, and when it has more than one term.

If the denominator is a square root, multiply numerator and denominator of the fraction by a quantity that will make the radicand in the denominator a perfect square. Note that we are eliminating radicals from the *denominator* and that the numerator may still contain radicals. Further, even though we call this process *simplifying*, the resulting radical may look more complicated than the original.

◆◆◆ Example 26: Here we rationalize the denominator.

$$\frac{5}{\sqrt{2}} = \frac{5}{\sqrt{2}} \cdot \frac{\sqrt{2}}{\sqrt{2}} = \frac{5\sqrt{2}}{\sqrt{4}} = \frac{5\sqrt{2}}{2}$$

◆◆◆

When the *entire* fraction is under the radical sign, we make the denominator of that fraction a perfect square and remove it from under the radical sign.

◆◆◆ Example 27: Again we rationalize the denominator.

$$\sqrt{\frac{3x}{2y}} = \sqrt{\frac{3x(2y)}{2y(2y)}} = \sqrt{\frac{6xy}{4y^2}} = \frac{\sqrt{6xy}}{2y}$$

◆◆◆

If the denominator is a *cube root*, we must multiply numerator and denominator by a quantity that will make the denominator under the radical sign a perfect cube.

◆◆◆ **Example 28:** Simplify

$$\frac{7}{\sqrt[3]{4}}$$

Solution: Now 4 is not a perfect cube, so it cannot be removed from the radical. However, 8 is a perfect cube (2^3). We can get an 8 under the radical sign by multiplying numerator and denominator by $\sqrt[3]{2}$.

$$\frac{7}{\sqrt[3]{4}} = \frac{7}{\sqrt[3]{4}} \cdot \frac{\sqrt[3]{2}}{\sqrt[3]{2}} = \frac{7\sqrt[3]{2}}{\sqrt[3]{8}} = \frac{7\sqrt[3]{2}}{2}$$

◆◆◆

The same principle applies regardless of the index. In general, if the index is n, we must make the quantity under the radical sign (in the denominator) a perfect nth power.

◆◆◆ **Example 29:** In this example the index is 5.

$$\frac{2y}{3\sqrt[5]{x}} = \frac{2y}{3\sqrt[5]{x}} \cdot \frac{\sqrt[5]{x^4}}{\sqrt[5]{x^4}} = \frac{2y\sqrt[5]{x^4}}{3\sqrt[5]{x^5}} = \frac{2y\sqrt[5]{x^4}}{3x}$$

◆◆◆

◆◆◆ **Example 30:** Sometimes the denominator will have more than one term, as in this example.

$$\sqrt{\frac{a}{a^2 + b^2}} = \sqrt{\frac{a}{a^2 + b^2} \cdot \frac{a^2 + b^2}{a^2 + b^2}} = \sqrt{\frac{a(a^2 + b^2)}{(a^2 + b^2)^2}} = \frac{\sqrt{a(a^2 + b^2)}}{a^2 + b^2}$$

◆◆◆

Reducing the Index

We can sometimes reduce the index by writing the radical in exponential form and then reducing the fractional exponent, as in the next example.

◆◆◆ **Example 31:** These examples show how to reduce the index.

(a) $\sqrt[6]{x^3} = x^{3/6} = x^{1/2} = \sqrt{x}$

(b) $\sqrt[4]{4x^2y^2} = \sqrt[4]{(2xy)^2}$

$$= (2xy)^{2/4} = (2xy)^{1/2}$$

$$= \sqrt{2xy}$$

◆◆◆

Exercise 2 ◆ Simplification of Radicals

Exponential and Radical Forms

Express in radical form.

1. $a^{1/4}$

2. $x^{1/2}$

3. $z^{3/4}$

4. $a^{1/2}b^{1/4}$

5. $(m - n)^{1/2}$

6. $(x^2y)^{-1/2}$

7. $\left(\dfrac{x}{y}\right)^{-1/3}$

8. $a^0 b^{-3/4}$

Express in exponential form.

9. $\sqrt{b}$

10. $\sqrt[3]{x}$

11. $\sqrt{y^2}$

12. $4\sqrt[3]{xy}$

13. $\sqrt[n]{a+b}$

14. $\sqrt[n]{x^m}$

15. $\sqrt{x^2y^2}$

16. $\sqrt[n]{a^n b^{3n}}$

Simplifying Radicals

Write in simplest form. Do not use your calculator for any numerical problems.
Leave your answers in radical form.

17. $\sqrt{18}$

18. $\sqrt{75}$

19. $\sqrt{63}$

20. $\sqrt[3]{16}$

21. $\sqrt[3]{56}$

22. $\sqrt[4]{48}$

23. $\sqrt{a^3}$

24. $3\sqrt{50x^5}$

25. $\sqrt{36x^2y}$

26. $\sqrt[3]{x^2y^5}$

27. $\sqrt{\dfrac{3}{7}}$

28. $\sqrt{\dfrac{2}{3}}$

29. $\sqrt[3]{\dfrac{1}{4}}$

30. $\sqrt{\dfrac{5}{8}}$

31. $\sqrt[3]{\dfrac{2}{9}}$

32. $\sqrt[4]{\dfrac{7}{8}}$

33. $\sqrt{\dfrac{1}{2x}}$

34. $\sqrt{\dfrac{5m}{7n}}$

Challenge Problems

Simplify.

35. $x\sqrt[3]{16x^3y}$

36. $\sqrt[4]{64m^2n^4}$

37. $3\sqrt[5]{32xy^{11}}$

38. $6\sqrt[3]{16x^4}$

39. $\sqrt{a^3 - a^2b}$

40. $x\sqrt{x^4 - x^3y^2}$

41. $\sqrt{9m^3 + 18n}$

42. $\sqrt{2x^3 + x^4y}$

43. $\sqrt{\dfrac{3a^3}{5b}}$

44. $\sqrt{\dfrac{5ab}{6xy}}$

45. $\sqrt[3]{\dfrac{1}{x^2}}$

46. $\sqrt[3]{\dfrac{81x^4}{16yz^2}}$

47. $\sqrt[6]{\dfrac{4x^6}{9}}$

48. $\sqrt{x^2 - \left(\dfrac{x}{2}\right)^2}$

Applications

49. The period ω_n for simple harmonic motion is given by

$$\omega_n = \sqrt{\dfrac{kg}{W}}$$

where k is the spring constant and W is the weight. Write this equation in exponential form.

50. Rationalize the denominator of the equation in problem 49 to obtain a different form of that equation.

51. The magnitude of the impedance Z of a series RLC circuit is given by

$$Z = \sqrt{R^2 + X^2}$$

Write this equation in exponential form.

52. Given the equation from problem 50, write the expression for Z when $X = 2R$, and simplify.

53. The hypotenuse in right triangle ABC, shown in Fig. 13–3, is given by the Pythagorean theorem.

$$c = \sqrt{a^2 + b^2}$$

Write an expression for c when $b = 3a$, and simplify.

54. A stone is thrown upward with a horizontal velocity of 40 ft/s and an upward velocity of 60 ft/s. At t seconds it will have a horizontal displacement H equal to $40t$ and a vertical displacement V equal to $60t - 16t^2$. The straight-line distance S from the stone to the launch point is found by the Pythagorean theorem. Write an equation for S in terms of t, and simplify.

55. *Writing:* Explain how exponents and radicals are really two different ways of writing the same expression. Also explain why, if they are the same, we need both.

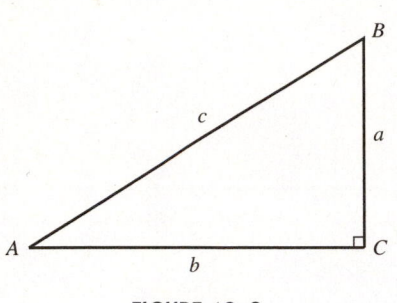

FIGURE 13–3

13–3　Operations with Radicals

Adding and Subtracting Radicals

One reason we learned to simplify radicals is to be able to combine them. Radicals are called *similar* if they have the same index and the same radicand, such as $5\sqrt[3]{2x}$ and $3\sqrt[3]{2x}$. We add and subtract radicals by *combining similar radicals*.

◆◆◆ **Example 32:** Here's how to combine similar radicals.

$$5\sqrt{y} + 2\sqrt{y} - 4\sqrt{y} = 3\sqrt{y}$$　　　　　◆◆◆

Radicals that may look similar at first glance may not actually be similar.

◆◆◆ **Example 33:** The radicals

$$\sqrt{2x} \quad \text{and} \quad \sqrt{3x}$$

are *not* similar.　　　　　◆◆◆

Common Error	Do not try to combine radicals that are not similar. $$\sqrt{2x} + \sqrt{3x} \neq \sqrt{5x}$$

Radicals that do not appear to be similar at first may turn out to be so after simplification.

◆◆◆ **Example 34:** In these examples we simplify and then combine radicals.

(a) $\sqrt{18x} - \sqrt{8x} = 3\sqrt{2x} - 2\sqrt{2x} = \sqrt{2x}$

(b) $\sqrt[3]{24y^4} + \sqrt[3]{81x^3y} = 2y\sqrt[3]{3y} + 3x\sqrt[3]{3y}$

$$= (2y + 3x)\sqrt[3]{3y} \qquad \text{after factoring.}$$

(c) $5y\sqrt{\dfrac{x}{y}} - 2\sqrt{xy} + x\sqrt{\dfrac{y}{x}}$

$$= 5y\sqrt{\dfrac{xy}{y^2}} - 2\sqrt{xy} + x\sqrt{\dfrac{yx}{x^2}}$$

$$= 5\sqrt{xy} - 2\sqrt{xy} + \sqrt{xy} = 4\sqrt{xy}$$ ◆◆◆

Multiplying Radicals

Radicals having the *same index* can be multiplied by using Eq. 31 $\sqrt[n]{a}\sqrt[n]{b} = \sqrt[n]{ab}$.

◆◆◆ **Example 35:** Here we show how to multiply radicals having the same index.

(a) $\sqrt{x}\sqrt{2y} = \sqrt{x(2y)} = \sqrt{2xy}$

(b) $(5\sqrt[3]{3a})(2\sqrt[3]{4b}) = (5)(2)\sqrt[3]{(3a)(4b)} = 10\sqrt[3]{12ab}$

(c) $(2\sqrt{m})(5\sqrt{n})(3\sqrt{mn}) = (2)(5)(3)\sqrt{m(n)(mn)} = 30\sqrt{m^2n^2} = 30mn$

(d) $\sqrt{2x}\sqrt{3x} = \sqrt{(2x)(3x)} = \sqrt{6x^2} = x\sqrt{6}$

(e) $\sqrt{3x}\sqrt{3x} = 3x$

(f) $\sqrt[3]{2x}\sqrt[3]{4x^2} = \sqrt[3]{8x^3} = 2x$ ◆◆◆

We multiply radicals having *different indices* by first going to exponential form, then multiplying using Eq. 22 $(x^a x^b = x^{a+b})$, and finally returning to radical form.

◆◆◆ **Example 36:** This example shows how to multiply radicals having different indices.

$$\sqrt{a}\sqrt[4]{b} = a^{1/2}b^{1/4}$$

$$= a^{2/4}b^{1/4} = (a^2b)^{1/4}$$

Or, in radical form,

$$= \sqrt[4]{a^2b}$$ ◆◆◆

When the radicands are the same, the work is even easier.

◆◆◆ **Example 37:** Multiply $\sqrt[3]{x}$ by $\sqrt[6]{x}$.

Solution:

$$\sqrt[3]{x}\sqrt[6]{x} = x^{1/3}x^{1/6} = x^{2/6}x^{1/6}$$

By Eq. 22, $x^a x^b = x^{a+b}$.

$$= x^{3/6} = x^{1/2}$$

$$= \sqrt{x}$$ ◆◆◆

Multinomials containing radicals are multiplied in the way we learned in Chap. 2.

◆◆◆ **Example 38:** Multiply and simplify $\sqrt{50x}\,(\sqrt{2x} - \sqrt{xy})$.

Solution: Here we are multiplying a monomial by a binomial. We multiply each term in the binomial by the monomial.

$$\sqrt{50x}(\sqrt{2x} - \sqrt{xy}) = \sqrt{50x}\sqrt{2x} - \sqrt{50x}\sqrt{xy} = \sqrt{100x^2} - \sqrt{50x^2y}$$

$$= 10x - 5x\sqrt{2y}$$ ◆◆◆

◆◆◆ Example 39: Multiply and simplify $(3 + \sqrt{x})(2\sqrt{x} - 4\sqrt{y})$.

Solution: Here we have the product of two binomials, which we covered in Chapter 2. We multiply each term in the first binomial by each term in the second binomial, or use the FOIL rule.

$$(3 + \sqrt{x})(2\sqrt{x} - 4\sqrt{y}) = 3(2\sqrt{x}) + 3(-4\sqrt{y}) + \sqrt{x}(2\sqrt{x}) - \sqrt{x}(4\sqrt{y})$$
$$= 6\sqrt{x} - 12\sqrt{y} + 2x - 4\sqrt{xy} \qquad\qquad ◆◆◆$$

◆◆◆ Example 40: Multiply and simplify $(\sqrt{x} + \sqrt{y})(\sqrt{x} - \sqrt{y})$.

Solution: Again we have the product of two binomials, but now you may recognize them as the difference of two squares.

$$(\sqrt{x} - \sqrt{y})(\sqrt{x} + \sqrt{y}) = \sqrt{x^2} + \sqrt{xy} - \sqrt{xy} - \sqrt{y^2}$$
$$= \sqrt{x^2} - \sqrt{y^2}$$
$$= x - y \qquad\qquad ◆◆◆$$

To raise a radical to a *power*, we simply multiply the radical by itself the proper number of times.

◆◆◆ Example 41: Cube and simplify the expression

$$2x\sqrt{y}$$

Solution:

$$(2x\sqrt{y})^3 = (2x\sqrt{y})(2x\sqrt{y})(2x\sqrt{y})$$
$$= 2^3 x^3 (\sqrt{y})^3$$
$$= 8x^3 \sqrt{y^3} = 8x^3 y\sqrt{y} \qquad\qquad ◆◆◆$$

Dividing Radicals

Radicals having the same indices can be divided using Eq. 32.

$$\frac{\sqrt[n]{a}}{\sqrt[n]{b}} = \sqrt[n]{\frac{a}{b}}$$

◆◆◆ Example 42: Divide and simplify

$$\frac{\sqrt{4x^5}}{\sqrt{2x}}$$

Solution: Using Eq. 32, we place the quantities to be divided under a single radical sign, and simplify.

$$\frac{\sqrt{4x^5}}{\sqrt{2x}} = \sqrt{\frac{4x^5}{2x}} = \sqrt{2x^4} = x^2\sqrt{2} \qquad\qquad ◆◆◆$$

◆◆◆ Example 43: Divide and simplify

$$\frac{\sqrt{3a^7} + \sqrt{12a^5} - \sqrt{6a^3}}{\sqrt{3a}}$$

Solution: We divide each term in the numerator by the denominator and then simplify each term as in the preceding example.

$$\frac{\sqrt{3a^7} + \sqrt{12a^5} - \sqrt{6a^3}}{\sqrt{3a}} = \sqrt{\frac{3a^7}{3a}} + \sqrt{\frac{12a^5}{3a}} - \sqrt{\frac{6a^3}{3a}}$$

$$= \sqrt{a^6} + \sqrt{4a^4} - \sqrt{2a^2}$$

$$= a^3 + 2a^2 - a\sqrt{2} \qquad \text{◆◆◆}$$

It is a common practice to rationalize the denominator after division.

◆◆◆ **Example 44:** Divide and simplify

$$\frac{\sqrt{10x}}{\sqrt{5y}} = \sqrt{\frac{10x}{5y}} = \sqrt{\frac{2x}{y}}$$

Rationalizing the denominator, we obtain

$$\sqrt{\frac{2x}{y}} = \sqrt{\frac{2x}{y} \cdot \frac{y}{y}} = \frac{\sqrt{2xy}}{y} \qquad \text{◆◆◆}$$

If the indices are different, we go to exponential form, as we did for multiplication. Divide using Eq. 23 $\left(\dfrac{x^a}{x^b} = x^{a-b}\right)$ and then return to radical form.

◆◆◆ **Example 45:** Here we divide two radicals having different indices.

$$\frac{\sqrt[3]{a}}{\sqrt[4]{b}} = \frac{a^{1/3}}{b^{1/4}} = \frac{a^{4/12}}{b^{3/12}} = \left(\frac{a^4}{b^3}\right)^{1/12}$$

Returning to radical form, we obtain

$$\left(\frac{a^4}{b^3}\right)^{1/12} = \sqrt[12]{\frac{a^4}{b^3}}$$

Next we rationalize the denominator of the expression under the radical sign. Since the index of the radical is 12, we want to make the denominator a 12th power. We do this by multiplying numerator and denominator by b^9.

$$\sqrt[12]{\frac{a^4}{b^3}} = \sqrt[12]{\frac{a^4}{b^3} \cdot \frac{b^9}{b^9}}$$

$$= \sqrt[12]{\frac{a^4 b^9}{b^{12}}}$$

$$= \frac{\sqrt[12]{a^4 b^9}}{b} \qquad \text{◆◆◆}$$

When dividing by a binomial containing square roots, multiply the divisor and the dividend by the *conjugate* of that binomial. The *conjugate* of a binomial is a binomial having the same two terms, but differ only in the sign of one term. Thus the conjugate of $(a + b)$ is $(a - b)$. Recall that when we multiply a binomial, say, $(a + b)$ by its conjugate $(a - b)$, we get

$$(a + b)(a - b) = a^2 - ab + ab - b^2$$

$$= a^2 - b^2$$

we get an expression where the cross-product terms ab and $-ab$ drop out, and the remaining terms are both squares. We have the difference of two squares. Thus if a and b were square roots, our final expression would have no square roots.

We use this operation to remove square roots from the denominator, as shown in the following example.

◆◆◆ **Example 46:** Divide $(3 + \sqrt{x})$ by $(2 - \sqrt{x})$ and simplify.

Solution: The conjugate of the divisor $2 - \sqrt{x}$ is $2 + \sqrt{x}$. Multiplying divisor and dividend by $2 + \sqrt{x}$, we get

$$\frac{3 + \sqrt{x}}{2 - \sqrt{x}} = \frac{3 + \sqrt{x}}{2 - \sqrt{x}} \cdot \frac{2 + \sqrt{x}}{2 + \sqrt{x}}$$

$$= \frac{6 + 3\sqrt{x} + 2\sqrt{x} + \sqrt{x}\sqrt{x}}{4 + 2\sqrt{x} - 2\sqrt{x} - \sqrt{x}\sqrt{x}}$$

$$= \frac{6 + 5\sqrt{x} + x}{4 - x}$$

after combining like terms.   ◆◆◆

Exercise 3 ◆ Operations with Radicals

As in Exercise 2, do not use your calculator for any numerical problems. Leave your answers in radical form.

Addition and Subtraction of Radicals. Combine as indicated and simplify.

1. $2\sqrt{24} - \sqrt{54}$
2. $\sqrt{300} + \sqrt{108} - \sqrt{243}$
3. $\sqrt{24} - \sqrt{96} + \sqrt{54}$
4. $\sqrt{128} - \sqrt{18} + \sqrt{32}$
5. $2\sqrt{50} + \sqrt{72} + 3\sqrt{18}$
6. $\sqrt[3]{384} - \sqrt[3]{162} + \sqrt[3]{750}$
7. $2\sqrt[3]{2} - 3\sqrt[3]{16} + \sqrt[3]{54}$
8. $3\sqrt[3]{108} + 2\sqrt[3]{32} - \sqrt[3]{256}$
9. $\sqrt[3]{625} - 2\sqrt[3]{135} - \sqrt[3]{320}$
10. $5\sqrt[3]{320} + 2\sqrt[3]{40} - 4\sqrt[3]{135}$
11. $\sqrt[4]{768} - \sqrt[4]{48} - \sqrt[4]{243}$
12. $3\sqrt{\dfrac{5}{4}} + 2\sqrt{45}$
13. $\sqrt{128x^2y} - \sqrt{98x^2y} + \sqrt{162x^2y}$
14. $3\sqrt{\dfrac{1}{3}} - 2\sqrt{\dfrac{3}{4}} + 4\sqrt{3}$
15. $7\sqrt{\dfrac{27}{50}} - 3\sqrt{\dfrac{2}{3}}$
16. $4\sqrt{50} - 2\sqrt{72} + \dfrac{3}{\sqrt{2}}$
17. $\sqrt{a^2x} + \sqrt{b^2x}$

Multiplication of Radicals. Multiply and simplify.

18. $3\sqrt{3}$ by $5\sqrt{3}$
19. $2\sqrt{3}$ by $3\sqrt{8}$
20. $\sqrt{8}$ by $\sqrt{160}$
21. $\sqrt{\dfrac{5}{8}}$ by $\sqrt{\dfrac{3}{4}}$
22. $4\sqrt[3]{45}$ by $2\sqrt[3]{3}$
23. $3\sqrt{3}$ by $2\sqrt[3]{2}$
24. $3\sqrt{2}$ by $2\sqrt[3]{3}$
25. $2\sqrt{3}$ by $\sqrt[4]{5}$

26. $2\sqrt[3]{3}$ by $5\sqrt[4]{4}$

27. $2\sqrt[3]{24}$ by $\sqrt[9]{\frac{8}{27}}$

28. $2x\sqrt{3a}$ by $3\sqrt{y}$

29. $3\sqrt[3]{9a^2}$ by $\sqrt[3]{3abc}$

30. $\sqrt[5]{4xy^2}$ by $\sqrt[5]{8x^2y}$

31. $\sqrt{\frac{a}{b}}$ by $\sqrt{\frac{c}{d}}$

32. $\sqrt{a}$ by $\sqrt[4]{b}$

33. $\sqrt[3]{x}$ by $\sqrt{y}$

34. $\sqrt{xy}$ by $2\sqrt{xz}$ and $\sqrt[3]{x^2y^2}$

35. $\sqrt[3]{a^2b}$ by $\sqrt[3]{2a^2b^2}$ and $\sqrt{3a^3b^2}$

Powers. Square the following expressions and simplify.

36. $3\sqrt{y}$

37. $4\sqrt[3]{4x^2}$

38. $3x\sqrt[3]{2x^2}$

39. $5 + 4\sqrt{x}$

40. $3 - 5\sqrt{a}$

41. $\sqrt{a} + 5a\sqrt{b}$

Division of Radicals. Divide and simplify.

42. $8 \div 3\sqrt{2}$

43. $6\sqrt{72} \div 12\sqrt{32}$

44. $2 \div \sqrt[3]{6}$

45. $\sqrt{72} \div 2\sqrt[4]{64}$

46. $8 \div 2\sqrt[3]{4}$

47. $8\sqrt[3]{ab} \div 4\sqrt{ac}$

48. $\sqrt[3]{4ab} \div \sqrt[4]{2ab}$

49. $(3 + \sqrt{2}) \div (2 - \sqrt{2})$

50. $5 \div \sqrt{3x}$

51. $4\sqrt{x} \div \sqrt{a}$

52. $10 \div \sqrt[3]{9x^2}$

53. $12 \div \sqrt[3]{4x^2}$

Challenge Problems. Perform the indicated operation and simplify.

54. $\sqrt{2b^2xy} - \sqrt{2a^2xy}$

55. $\sqrt{x^2y} - \sqrt{4a^2y}$

56. $\sqrt{80a^3} - 3\sqrt{20a^3} - 2\sqrt{45a^3}$

57. $4\sqrt{3a^2x} - 2a\sqrt{48x}$

58. $\sqrt[3]{125x^2} - 2\sqrt[3]{8x^2}$

59. $\sqrt[3]{1250a^3b} + \sqrt[3]{270c^3b}$

60. $2\sqrt[3]{ab^7} + 3\sqrt[3]{a^7b} + 2\sqrt[3]{8a^4b^4}$

61. $\sqrt[5]{a^{13}b^{11}c^{12}} - 2\sqrt[5]{a^8bc^2} + \sqrt[5]{a^3b^6c^7}$

62. $\sqrt{\frac{9x}{16}} - \frac{3\sqrt{x}}{4}$

63. $(\sqrt{5} - \sqrt{3}) \times 2\sqrt{3}$

64. $(x + \sqrt{y}) \times \sqrt{y}$

65. $\sqrt{x^3 - x^4y} \times \sqrt{x}$

66. $(\sqrt{x} - \sqrt{2}) \times (2\sqrt{x} + \sqrt{2})$

67. $(a + \sqrt{b}) \times (a - \sqrt{b})$

68. $(\sqrt{x} + \sqrt{y}) \times (\sqrt{x} + \sqrt{y})$

69. $(4\sqrt{x} + 2\sqrt{y}) \times (4\sqrt{x} - 5\sqrt{y})$

70. $(x + \sqrt{xy}) \times (\sqrt{x} - \sqrt{y})$

71. $\sqrt{x} \div (\sqrt{x} + \sqrt{y})$

72. $a \div (a + \sqrt{b})$

73. $(a + \sqrt{b}) \div (a - \sqrt{b})$

74. $(\sqrt{x} - \sqrt{y}) \div (\sqrt{x} + \sqrt{y})$

75. $(3\sqrt{m} - \sqrt{2n}) \div (\sqrt{3n} + \sqrt{m})$

76. $(5\sqrt{2x})^3$

77. $(2x\sqrt{3x})^3$

78. $(5\sqrt[3]{2ax})^3$

13–4   Radical Equations

A *radical equation* is one in which the unknown is under the radical sign.

◆◆◆ **Example 47:** The equation

$$\sqrt{x + 8} = 2$$

is a radical equation.   ◆◆◆

■ Exploration:

Try this. Use your graphing calculator to graph the function

$$y = \sqrt{x + 8} - 2$$

Does the curve cross the x axis? What is the meaning of any such x-intercept? If you substitute it back into the given equation, what do you find?   ■

Approximate Solution of Radical Equations by Graphing

We learned how to find the approximate solution of equations by graphing in Chap. 5. Recall that we moved all terms of the given equation to one side and graphed that expression. Then by zooming in and using $\boxed{\text{TRACE}}$ or **zero** we were able to find the value of x at which the curve crossed the x axis to any precision we wanted. That method works for any equation, and we use it here to solve a radical equation.

◆◆◆ **Example 48:** Find an approximate solution to the equation given in Example 47.

Solution: Rearranging the equation gives

$$\sqrt{x + 8} - 2 = 0$$

So we graph the function

$$y = \sqrt{x + 8} - 2$$

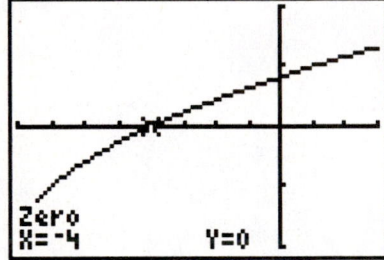

TI-83/84 screen for Example 48. Graph of $y = \sqrt{x + 8} - 2$, showing a root at $x = -4$. The **zero** operation is found in the $\boxed{\text{CALC}}$ menu.

as shown. Using $\boxed{\text{TRACE}}$, $\boxed{\text{ZOOM}}$, or **zero**, we verify that the root is at $x = -4$.   ◆◆◆

Solving a Radical Equation Using a Calculator's Equation Solver

We can use any built-in equation solver to solve a radical equation, just as we used it earlier to solve other equations.

◆◆◆ **Example 49:** Solve the equation of Example 47 using the TI-83/84 equation solver.

Solution:

(a) We select **Solver** from the $\boxed{\text{MATH}}$ menu and enter the equation, Screen (1). The equation must be in explicit form, with all terms on the right side and 0 on the left side.

(b) Press $\boxed{\text{ENTER}}$. A new screen is displayed, (2), showing a guess value of x and a bound. You may accept the default values or enter new ones. Let us enter a new guess of $x = -2$ and a new bound of -5 to 0.

(c) Move the cursor to the line containing "X = " and press $\boxed{\text{SOLVE}}$. (This is $\boxed{\text{ALPHA}}$ $\boxed{\text{ENTER}}$ on the TI-83/84.)

A root falling within the selected bound is displayed, screen (3).   ◆◆◆

Screens for Example 49.

```
EQUATION SOLVER
eqn:0=√(X+8)-2
```

```
√(X+8)-2=0
X=-2
bound={-5,0}
```

```
√(X+8)-2=0
■X=-4.000000000...
 bound={-5,0}■
■left-rt=0
```

(1) The given equation is entered in the TI-83/84 equation solver.

(2) An initial guess for x, and the lower and upper bounds, are chosen here.

(3) The computed value of x is shown, within the chosen bound. The last line says that the difference between the left and right sides of the equation is zero, showing that the equation is indeed balanced.

Implicit Functions

To solve a quadratic or other equation by graphing or by the TI-83/84 equation solver, the quadratic must be in explicit form. If it is not, simply move all terms to one side of the equal sign, changing signs as appropriate, leaving zero on one side of the equation. However, some calculators that can do symbolic manipulation can solve an equation that is in implicit form, as we have seen earlier in this text.

◆◆◆ **Example 50:** Solve using the TI-89 equation solver:

$$\sqrt{3x + 2} = \sqrt{x + 4}$$

Solution:

(a) We select **solve** from the **Algebra** menu.

(b) Enter the equation, followed by the variable (x) that we want to solve for. Enter the radicals in exponential form. The equation *does not* have to be in explicit form.

(c) Pressing $\boxed{\text{ENTER}}$ would give the exact solution, while pressing $\boxed{\approx}$ (located above the $\boxed{\text{ENTER}}$ key) gives an approximate decimal solution.

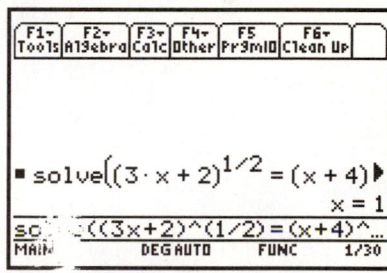

TI-89 screen for Example 50.

◆◆◆

Solving a Radical Equation Algebraically

To solve a radical equation algebraically, it is necessary to isolate the radical term on one side of the equal sign and then raise both sides to whatever power will eliminate the radical.

◆◆◆ **Example 51:** Solve for x:

$$\sqrt{x - 5} - 4 = 0$$

Solution: Rearranging yields

$$\sqrt{x - 5} = 4$$

Squaring both sides: $\qquad x - 5 = 16$

$$x = 21$$

Check:

$$\sqrt{21 - 5} - 4 \overset{?}{=} 0$$

$$\sqrt{16} - 4 \overset{?}{=} 0$$

$$4 - 4 = 0 \quad \text{(checks)}$$

◆◆◆

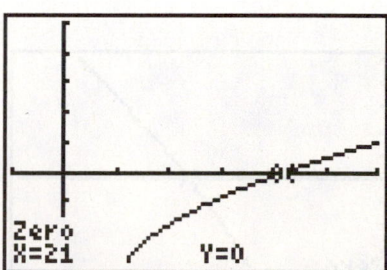

Graphical check for Example 51. Graph of $f(x) = \sqrt{x - 5} - 4$ showing a root at $x = 21$.

◆◆◆ **Example 52:** Solve for x:

$$\sqrt{x^2 - 6x} = x - 9$$

Solution: Squaring, we obtain

$$x^2 - 6x = x^2 - 18x + 81$$
$$12x = 81$$
$$x = \frac{81}{12} = \frac{27}{4}$$

Check:

$$\sqrt{\left(\frac{27}{4}\right)^2 - 6\left(\frac{27}{4}\right)} \stackrel{?}{=} \frac{27}{4} - 9$$

$$\sqrt{\frac{729}{16} - \frac{162}{4}} \stackrel{?}{=} \frac{27}{4} - \frac{36}{4}$$

$$\sqrt{\frac{81}{16}} = \frac{9}{4} \neq -\frac{9}{4} \quad \text{(does } not \text{ check)}$$

Thus the given equation has no solution.   ◆◆◆

Common Error	The squaring process often introduces *extraneous roots*. These are discarded because they do not satisfy the original equation. Check your answer in the original equation.

If the equation has more than one radical, isolate one at a time and square both sides. It is usually better to isolate and square the *most complicated* radical first, as in the following example.

◆◆◆ **Example 53:** Solve for x.

$$\sqrt{x - 32} + \sqrt{x} = 16$$

Solution: Rearranging gives

$$\sqrt{x - 32} = 16 - \sqrt{x}$$

Squaring yields

$$x - 32 = (16)^2 - 32\sqrt{x} + x$$

We rearrange again to isolate the radical.

$$32\sqrt{x} = 256 + 32 = 288$$
$$\sqrt{x} = 9$$

Squaring again, we obtain

$$x = 81$$

Check:

$$\sqrt{81 - 32} + \sqrt{81} \stackrel{?}{=} 16$$
$$\sqrt{49} + 9 \stackrel{?}{=} 16$$
$$7 + 9 = 16 \quad \text{(checks)}$$ ◆◆◆

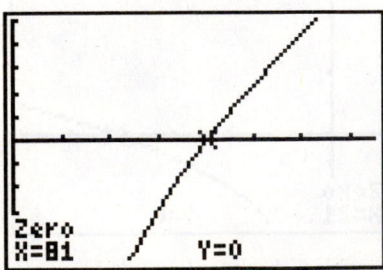

Graphical check for Example 53.
Graph of $f(x) = \sqrt{x - 32} + \sqrt{x} - 16$
showing a root at $x = 81$.

If a radical equation contains a fraction, we proceed as we did with other fractional equations and multiply both sides by the least common denominator.

◆◆◆ **Example 54:** Solve the following:

$$\sqrt{x - 3} + \frac{1}{\sqrt{x - 3}} = \sqrt{x}$$

Solution: Multiplying both sides by $\sqrt{x - 3}$ gives

$$x - 3 + 1 = \sqrt{x}\sqrt{x - 3}$$
$$x - 2 = \sqrt{x^2 - 3x}$$

Squaring both sides yields

$$x^2 - 4x + 4 = x^2 - 3x$$
$$x = 4$$

Check:

$$\sqrt{4 - 3} + \frac{1}{\sqrt{4 - 3}} \stackrel{?}{=} \sqrt{4}$$

$$1 + \frac{1}{1} = 2 \quad \text{(checks)} \qquad \text{◆◆◆}$$

Radical equations having indices other than 2 are solved in a similar way.

◆◆◆ **Example 55:** Solve for x.

$$\sqrt[3]{x - 5} = 2$$

Solution: Cubing both sides, we obtain

$$x - 5 = 8$$
$$x = 13$$

Check:

$$\sqrt[3]{13 - 5} \stackrel{?}{=} 2$$
$$\sqrt[3]{8} = 2 \quad \text{(checks)} \qquad \text{◆◆◆}$$

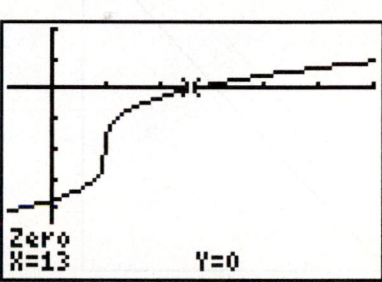

Graphical check for Example 55. Graph of $f(x) = \sqrt[3]{x - 5} - 2$ showing a root at $x = 13$.

◆◆◆ **Example 56:** *An Application.* Let us return to the problem posed in the introduction to this chapter. That is, to solve for the weight W in the equation for the natural frequency of a weight bobbing at the end of a spring, Fig. 13–4, given

$$f_n = \frac{1}{2\pi}\sqrt{\frac{kg}{W}}$$

Solution: Multiplying by 2π,

$$2\pi f_n = \sqrt{\frac{kg}{W}}$$

Squaring both sides, we get

$$4\pi^2 f_n^2 = \frac{kg}{W}$$

Finally we multiply by W and divide by $4\pi^2 f_n^2$,

$$W = \frac{kg}{4\pi^2 f_n^2}$$

◆◆◆

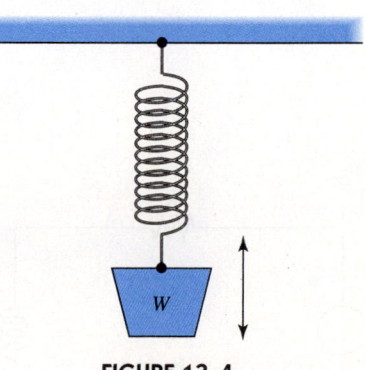

FIGURE 13–4

Exercise 4 ◆ Radical Equations

Radical Equations

Solve for x and check.

1. $\sqrt{x} = 6$

2. $\sqrt{x} + 5 = 9$

3. $\sqrt{7x + 8} = 6$

4. $\sqrt{3x - 2} = 5$

5. $\sqrt{2.95x - 1.84} = 6.23$

6. $\sqrt{5.88x + 4.92} = 7.72$

7. $\sqrt{x + 1} = \sqrt{2x - 7}$

8. $2 = \sqrt[4]{1 + 3x}$

9. $\sqrt{3x + 1} = 5$

10. $\sqrt[3]{2x} = 4$

11. $\sqrt{x - 3} = \dfrac{4}{\sqrt{x - 3}}$

12. $x + 2 = \sqrt{x^2 + 6}$

13. $\sqrt{x^2 - 7.25} = 8.75 - x$

14. $\sqrt{x - 15.5} = 5.85 - \sqrt{x}$

15. $\dfrac{6}{\sqrt{3 + x}} = \sqrt{x + 3}$

16. $\sqrt{12 + x} = 2 + \sqrt{x}$

Applications

17. The right triangle in Fig. 13–5 has one side equal to 25.3 cm and a perimeter of 68.4 cm, so

$$x + 25.3 + \sqrt{x^2 + (25.3)^2} = 68.4$$

Solve this equation for x. Then find the hypotenuse.

18. Find the missing side and the hypotenuse of a right triangle that has one side equal to 293 in. and a perimeter of 994 in.

19. Find the side and the hypotenuse of a right triangle that has a side of 2.73 m and a perimeter of 11.4 m.

20. The natural frequency f_n of a body under simple harmonic motion is found from the equation

$$f_n = \frac{1}{2\pi}\sqrt{\frac{kg}{W}}$$

where g is the gravitational constant, k is the spring constant, and W is the weight. Solve for k.

21. The magnitude Z of the impedance in an RLC circuit (Fig. 13–6) having a resistance R, an inductance L, and a capacitance C, to which is applied a voltage of frequency ω, is given by Eq. 1098.

$$Z = \sqrt{R^2 + \left(\omega L - \frac{1}{\omega C}\right)^2}$$

Solve for C.

22. The resonant frequency ω_n for the circuit shown in Fig. 13–6 is given by Eq. 1090.

$$\omega_n = \frac{1}{\sqrt{LC}}$$

Solve for L.

23. *Writing:* Explain in your own words how to solve a radical equation graphically. Be sure to give the reasons behind this method. Also say if the method can or cannot be used for other types of equations, and why.

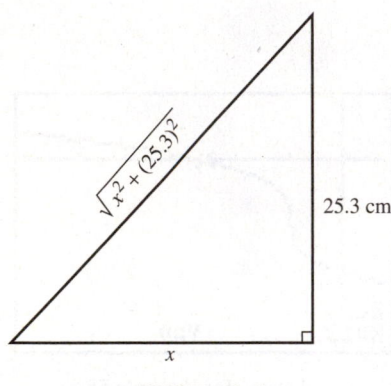

25.3 cm

FIGURE 13–5

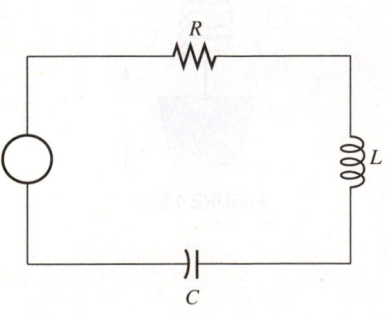

FIGURE 13–6

24. *Project:* Sometimes when solving a radical equation, the squaring operation will result in a quadratic equation, which can then be solved by the methods of the preceding chapter. Try solving these equations, and be sure to check all solutions.

(a) $3\sqrt{x-1} - \dfrac{4}{\sqrt{x-1}} = 4$

(b) $\sqrt{5x^2 - 3x - 41} = 3x - 7$

(c) $\sqrt{x+5} = \dfrac{12}{\sqrt{x+12}}$

(d) $\sqrt{7x+8} - \sqrt{5x-4} = 2$

25. *Project, Golden Ratio:* The value of the golden ratio Φ is given by $\Phi = \dfrac{1+\sqrt{5}}{2}$.

It is also given by the following radical equation. Demonstrate by calculation, by hand, or with a spreadsheet that this is true.

$$\Phi = \sqrt{1 + \sqrt{1 + \sqrt{1 + \sqrt{1 + \cdots}}}}$$

••• CHAPTER 13 REVIEW PROBLEMS •••••••••••••••••••••••••••••••

Simplify. Don't use a calculator.

1. $\sqrt{52}$

2. $\sqrt{108}$

3. $\sqrt[3]{162}$

4. $\sqrt[4]{9}$

5. $\sqrt[6]{4}$

6. $\sqrt{81a^2x^3y}$

7. $3\sqrt[4]{81x^5}$

8. $\sqrt{ab^2 - b^3}$

9. $\sqrt[3]{(a-b)^5 x^4}$

Perform the indicated operations and simplify. Don't use a calculator.

10. $4b\sqrt{3y} \cdot 5\sqrt{x}$

11. $\sqrt{x} \cdot \sqrt{x^3 - x^4 y}$

12. $(\sqrt{x} + \sqrt{y})(\sqrt{x} - \sqrt{y})$

13. $\dfrac{3 - 2\sqrt{3}}{2 - 5\sqrt{2}}$

14. $\dfrac{\sqrt{x} - \sqrt{y}}{\sqrt{x} + \sqrt{y}}$

15. $\sqrt{98x^2y^2} - \sqrt{128x^2y^2}$

16. $4\sqrt[3]{125x^2} + 3\sqrt[3]{8x^2}$

17. $\sqrt{a}\sqrt[3]{b}$

18. $\sqrt[3]{2abc^2} \cdot \sqrt{abc}$

19. $(3 + 2\sqrt{x})^2$

20. $(4x\sqrt{2x})^3$

21. $3\sqrt{50} - 2\sqrt{32}$

22. $2\sqrt{2} \div \sqrt[3]{2}$

23. $\sqrt{2ab} \div \sqrt{4ab^2}$

24. $9 \div \sqrt[3]{7x^2}$

25. $3\sqrt{9} \cdot 4\sqrt{8}$

Solve for x and check:

26. $\dfrac{\sqrt{x} - 8}{\sqrt{x} - 6} = \dfrac{\sqrt{x} - 4}{\sqrt{x} + 2}$

27. $\sqrt{x+6} = 4$

28. $\sqrt{2x-7} = \sqrt{x-3}$

29. $\sqrt[5]{2x+4} = 2$

30. $\sqrt{4x^2 - 3} = 2x - 1$

31. $\sqrt{x} + \sqrt{x - 9.75} = 6.23$

32. $\sqrt[3]{21.5x} = 2.33$

Simplify, and write without negative exponents.

33. $(x^{n-1} + y^{n-2})(x^n + y^{n-1})$ **34.** $(72a^6b^7 \div 9a^4b) \div 8ab^6$

35. $\left(\dfrac{9x^4y^3}{6x^3y}\right)^3$ **36.** $\left(\dfrac{5w^2}{2x}\right)^n$

37. $\left(\dfrac{2x^5y^3}{x^2y}\right)^3$ **38.** $5a^{-1}$

39. $3w^{-2}$ **40.** $2r^0$

41. $(3x)^{-1}$ **42.** $2a^{-1} + b^{-2}$

43. $x^{-1} - 2y^{-2}$ **44.** $(5a)^{-3} - 3b^{-2}$

45. $(3x)^{-2} + (2x^2y^{-4})^{-2}$ **46.** $(a^n + b^m)^2$

47. $(p^{a-1} + q^{a-2})(p^a + q^{a-1})$ **48.** $(16a^6b^0 \div 8a^4b) \div 4ab^6$

49. p^2q^{-1} **50.** x^3y^{-2}

51. r^2s^{-3} **52.** a^3b^{-1}

53. $(3x^3y^2z)^0$ **54.** $(5a^2b^2c^2)^3$

55. The volume V of a sphere of radius r is given

$$V = \frac{4}{3}\pi r^3$$

If the radius is tripled, the volume is then

$$V = \frac{4}{3}\pi(3r)^3$$

Simplify this expression.

56. Find the missing side and the hypotenuse of a right triangle that has one side equal to 154 in. and a perimeter of 558 in.

57. The geometric mean B between two numbers A and C is

$$B = \sqrt{AC}$$

Use the rules of radicals to write this equation in a different form.

58. Equation 1040 gives the damped angular velocity ω_d of a suspended weight as

$$\omega_d = \sqrt{\omega_n^2 - \frac{c^2g^2}{\omega^2}}$$

Rationalize the denominator and simplify.

Radian Measure, Arc Length, and Rotation

14

♦♦♦ **OBJECTIVES** ♦♦

When you have completed this chapter, you should be able to

- Convert angles between radians, degrees, and revolutions.
- Write the trigonometric functions of an angle given in radians.
- Compute arc length, radius, or central angle.
- Compute the angular velocity of a rotating body.
- Compute the linear speed of a point on a rotating body.
- Solve applied problems involving arc length or rotation.

♦♦

Many objects in technology spin at constant speed: wheels, CDs, hard drives, gears, pulleys, motors, shafts, and the earth itself, rotating about its axis. In this short chapter we will develop the math to deal with these applications. We also need to be able to find the distance between two points on a circle, such as two locations on a band of a computer disk, or two points on a great or small circle on the globe. We should be able to find out, for example, how far the rack of Fig. 14–1 will move when the pinion (the small gear) rotates, say, 300°.

Our main tool for both arc length and rotation will be *radian measure*. We introduced radian measure earlier, but here we give it full treatment. Up to now we have usually used degrees as our unit of angular measure, but we will see that the radian is more useful in many cases. The reason is that the radian is not an arbitrary unit like the degree (why 360 degrees in a revolution rather than, say, 300?). The radian uses a part of the circle itself, (the radius) as a unit.

This chapter also marks our return to trigonometry, with which we last studied right and oblique triangles. It will be followed by a few more chapters on trigonometry.

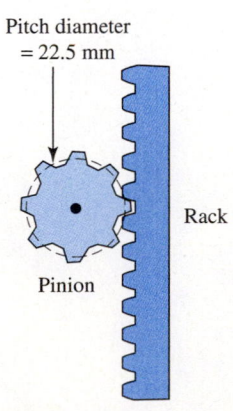

Pitch diameter
= 22.5 mm

Rack

Pinion

FIGURE 14–1 Rack and pinion.

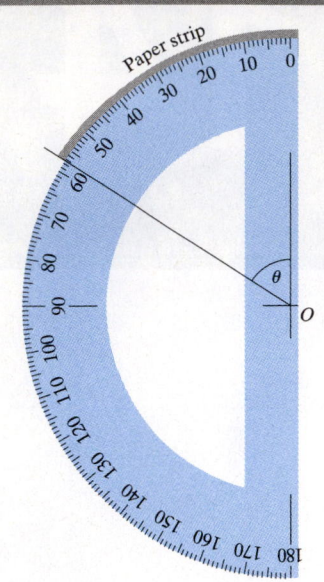

FIGURE 14–2

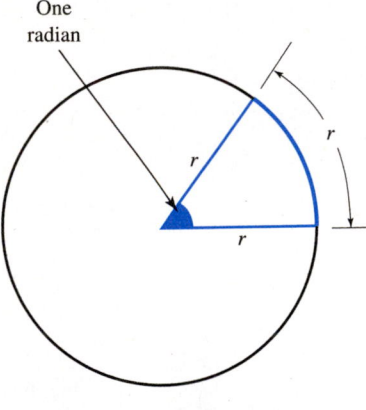

One radian

FIGURE 14–3

14–1   Radian Measure

■ Exploration:

Try this. (a) Make a photocopy of a protractor, Fig. 14–2. (b) Cut a strip of paper whose length equals the radius of the protractor circle. (c) Starting at 0°, bend the strip around the edge of the protractor and mark its end point (this will be easier if you paste the photocopy to cardboard and cut it out). (d) Draw a line from the end of the strip to the center O, forming angle θ.

What is the significance of θ? About how large is θ in degrees? Estimate roughly how many angles of size θ will fit completely around the circle.   ■

The angle θ you constructed in the exploration is defined as one *radian*. A radian is the central angle subtended by an arc whose length is one radius, Fig. 14–3.

We will see that the radian is a *dimensionless ratio* and not a unit of measure like the degree or the inch. Thus it is not strictly correct to write "radians" after the angle, but we will do so anyway because it makes it easier to keep track of our angles during a computation.

Angle Conversion

By definition, an arc having a length equal to the radius of the circle subtends a central angle of one radian. It follows that an arc having a length of twice the radius subtends a central angle of two radians, and so on. Thus an arc with a length of 2π times the radius (the entire circumference) subtends a central angle of 2π radians. Therefore 2π radians is equal to 1 revolution, or 360°. This gives us the following conversions for angular measure:

Angle Conversions	1 rev = 360° = 2π rad	78

If we divide 360° by 2π, we get
$$1 \text{ rad} \approx 57.3°$$
Some prefer to use Eq. 78 in this approximate form instead of the one given. You may also prefer to use π rad = 180°.

Using Eq. 78, we convert angular units in the same way that we converted other units in Chapter 1.

◆◆◆ Example 1: Convert 47.6° to radians and revolutions.

Solution: By Eq. 78,
$$47.6° \left(\frac{2\pi \text{ rad}}{360°} \right) = 0.831 \text{ rad}$$

Note that 2π and 360° in Eq. 78 are exact numbers, so we keep the same number of digits in our answer as in the given angle. Now converting to revolutions, we obtain
$$47.6° \left(\frac{1 \text{ rev}}{360°} \right) = 0.132 \text{ rev} \qquad ◆◆◆$$

◆◆◆ Example 2: Convert 1.8473 rad to degrees and revolutions.

Solution: By Eq. 78,
$$1.8473 \text{ rad} \left(\frac{360°}{2\pi \text{ rad}} \right) = 105.84°$$

and

$$1.8473 \text{ rad}\left(\frac{1 \text{ rev}}{2\pi \text{ rad}}\right) = 0.29401 \text{ rev}$$ ◆◆◆

Angle Conversion by Calculator

We showed how use a calculator to convert between decimal degrees and degrees, minutes, and seconds in Chapter 1. Here we will show conversions between decimal degrees and radians.

◆◆◆ **Example 3:** Convert 2.865 radians to degrees, on the TI-83/84 calculator.

Solution:

- Put the calculator into degree mode.
- Enter the angle, in radians. On the TI-83/84, the radian symbol is found in the Angle menu.
- Press ENTER.

The angle will be displayed in degrees. So,

$$2.865 \text{ rad} = 164.2° \quad \text{(rounded)}$$ ◆◆◆

TI-83/84 screen for Example 3.

◆◆◆ **Example 4:** Convert 38.52° to radians, on the TI-89 calculator.

Solution:

- Put the calculator into radian mode.
- Enter the angle, in degrees, getting the degree symbol from the MATH Angle menu.
- Press ENTER.

The angle will be displayed in radians. So,

$$38.52° = 0.6723 \text{ rad}$$ ◆◆◆

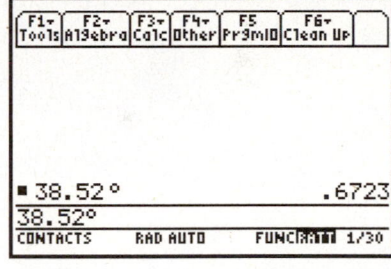

TI-89 screen for Example 4. Notice the RAD indication at the bottom of the screen.

Radian Measure in Terms of π

■ **Exploration:**

Try this. (a) On your cardboard protractor, lay off two more arcs of one radius each, getting angles of 2 and 3 radians. (b) Mark these on your protractor, as in Fig. 14–4. Bisect these angles to get angles of 0.5, 1.5 and 2.5 radians, and mark. (c) Since 180° = π radians, write "π" below 180° on your protractor. (d) Then subdivide 180° into four equal parts and label each with the appropriate fraction of π, as shown. You have just constructed a protractor with three scales:

(a) degrees (b) radians in decimal form (c) radians in terms of π ■

Not only can we express radians in decimal form, but it is also very common to express radian measure in terms of π. We know that 180° equals π radians, so

$$90° = \frac{\pi}{2} \text{ rad}$$

$$45° = \frac{\pi}{4} \text{ rad}$$

$$15° = \frac{\pi}{12} \text{ rad}$$

and so on. Thus to convert an angle from degrees to radians in terms of π, multiply the angle by (π rad/180°), and reduce to lowest terms.

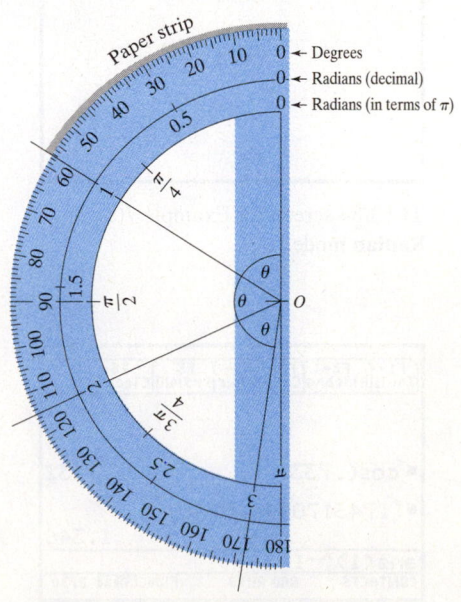

FIGURE 14–4

◆◆◆ **Example 5:** Express 135° in radian measure in terms of π.

Solution:

$$135°\left(\frac{\pi \text{ rad}}{180°}\right) = \frac{135\pi}{180} \text{ rad} = \frac{3\pi}{4} \text{ rad}$$

◆◆◆

Common Error

Students sometimes confuse the decimal value of π with the degree equivalent of π radians.

What is the value of π? 180 or 3.1416 . . .?

Remember that the approximate decimal value of π is always

$$\pi \approx 3.1416$$

but that π radians converted to degrees equals

$$\pi \text{ radians} = 180°$$

Note that we write "radians" or "rad" after π when referring to the angle, but not when referring to the decimal value.

To convert an angle from radians to degrees, multiply the angle by $(180°/\pi \text{ rad})$. Cancel the π in numerator and denominator, and reduce.

◆◆◆ **Example 6:** Convert $7\pi/9$ rad to degrees.

Solution:

$$\frac{7\pi}{9} \text{ rad}\left(\frac{180°}{\pi \text{ rad}}\right) = \frac{7(180)}{9} \text{ deg} = 140°$$

◆◆◆

Trigonometric Functions of Angles in Radians

We use a calculator to find the trigonometric functions of angles in radians just as we did for angles in degrees. However, we first *switch the calculator into Radian mode*. Consult your calculator manual for how to switch to this mode. Some calculators will display an R or the word RAD when in Radian mode.

On some calculators you enter the angle and then press the required trigonometric function key. On other calculators you must press the trigonometric function key *first* and then enter the angle. Be sure you know which way your own calculator works.

◆◆◆ **Example 7:** Use your calculator to verify the following to four decimal places:

(a) sin 2.83 rad = 0.3066   (b) cos 1.52 rad = 0.0508

(c) tan 0.463 rad = 0.4992

◆◆◆

To find the cotangent, secant, or cosecant, we use the reciprocal relations, just as when working in degrees.

◆◆◆ **Example 8:** Find sec 0.733 rad to three decimal places.

Solution: We put the calculator into Radian mode. The reciprocal of the secant is the cosine, so we take the cosine of 0.733 rad and get

$$\cos 0.733 \text{ rad} = 0.7432$$

Then taking the reciprocal of 0.7432 gives

$$\sec 0.733 = \frac{1}{\cos 0.733} = \frac{1}{0.7432} = 1.346$$

◆◆◆

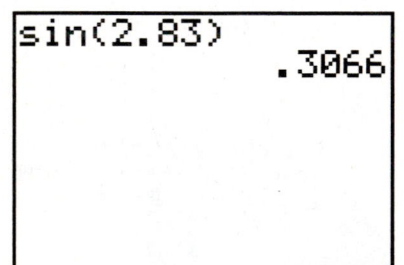

TI-83/84 screen for Example 7(a), in **Radian** mode.

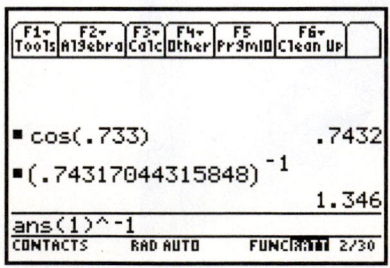

TI-89 screen for Example 8, in **Radian** mode.

◆◆◆ **Example 9:** Use your calculator to verify the following to four decimal places:

(a) csc 1.33 rad = 1.0297   (b) cot 1.22 rad = 0.3659

(c) sec 0.726 rad = 1.3372   ◆◆◆

The Inverse Trigonometric Functions

The inverse trigonometric functions are found the same way as when working in degrees. Just be sure that your calculator is in radian mode.

As with the trigonometric functions, some calculators require that you press the function key before entering the number, and some require that you press it after entering the number.

Remember that the inverse trigonometric function can be written in two different ways. Thus the inverse sine can be written

$$\arcsin \theta \quad \text{or} \quad \sin^{-1} \theta$$

Also recall that there are infinitely many angles that have a particular value of a trigonometric function. Of these, we are finding just the smallest positive angle.

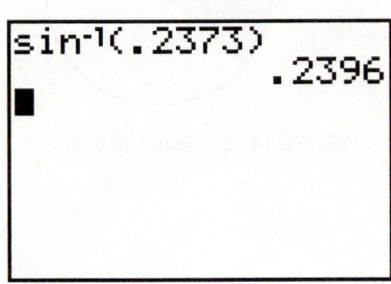

TI-83/84 screen for Example 10(a), in **RADIAN** mode.

◆◆◆ **Example 10:** Use your calculator to verify the following, in radians, to four decimal places:

(a) arcsin 0.2373 = 0.2396 rad   (b) $\cos^{-1} 0.5152 = 1.0296$ rad

(c) arctan 3.246 = 1.2720 rad   ◆◆◆

To find the arccot, arcsec, and arccsc, we first take the reciprocal of the given function and then find the inverse function, as shown in the following example.

◆◆◆ **Example 11:** If $\theta = \cot^{-1} 2.745$, find θ in radians to four decimal places.

Solution: If the cotangent of θ is 2.745, then

$$\cot \theta = 2.745$$

so

$$\frac{1}{\tan \theta} = 2.745$$

Taking reciprocals of both sides gives

$$\tan \theta = \frac{1}{2.745} = 0.3643$$

So

$$\theta = \tan^{-1} 0.3643 = 0.3494 \qquad \text{◆◆◆}$$

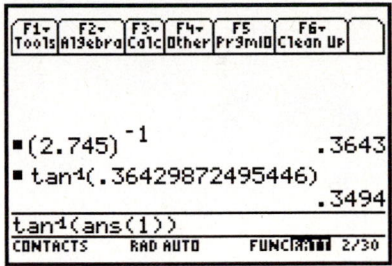

TI-89 screen for Example 11, in **RADIAN** mode.

It is no harder to find the trigonometric function of an angle in radians expressed in terms of π.

◆◆◆ **Example 12:** Find $\cos(5\pi/12)$ to four significant digits.

Solution: With our calculator in radian mode,

$$\cos \frac{5\pi}{12} = 0.2588 \qquad \text{◆◆◆}$$

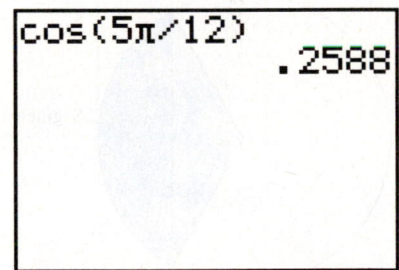

TI-83/84 screen for Example 12, in **RADIAN** mode.

◆◆◆ **Example 13:** Evaluate to four significant digits:

$$5 \cos\left(\frac{2\pi}{5}\right) + 4 \sin^2\left(\frac{3\pi}{7}\right)$$

Solution: We can evaluate this expression in steps, or key it directly into the calculator as shown.

$$5 \cos \frac{2\pi}{5} + 4 \sin^2 \frac{3\pi}{7} = 5.347 \qquad \text{◆◆◆}$$

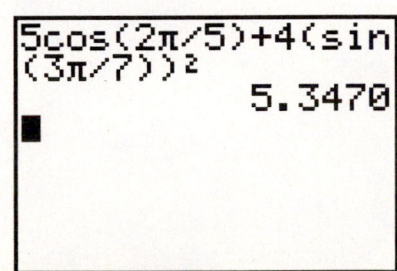

TI-83/84 screen for Example 13, in **RADIAN** mode.

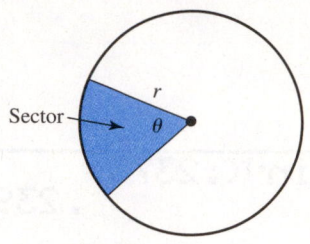

FIGURE 14–5   Sector of a circle.

Areas of Sectors and Segments

Sectors and segments of a circle (Fig. 14–5) were defined in Chap. 6. Now we will compute their areas.

The area of a circle of radius r is given by πr^2, so the area of a semicircle, of course, is $\pi r^2/2$; the area of a quarter circle is $\pi r^2/4$, and so on. The segment area is the same fractional part of the whole area as the central angle is of a whole revolution.

Thus if the central angle is 1/4 revolution, the sector area is also 1/4 of the total circle area.

If the central angle (in radians) is $\theta/2\pi$ revolution, the sector area is also $\theta/2\pi$ of the total circle area. So

$$\text{area of sector} = \pi r^2 \left(\frac{\theta}{2\pi} \right)$$

or

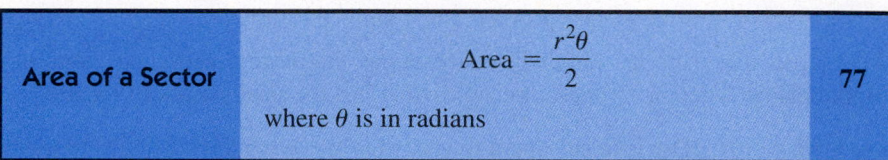

Area of a Sector	$\text{Area} = \dfrac{r^2\theta}{2}$ where θ is in radians	77

♦♦♦ Example 14: Find the area of a sector having a radius of 8.25 m and a central angle of 46.8°.

Solution: We first convert the central angle to radians.

$$46.8° \left(\frac{\pi \text{ rad}}{180°} \right) = 0.8168 \text{ rad}$$

Then, by Eq. 77,

$$\text{area} = \frac{(8.25)^2(0.8168)}{2} = 27.8 \text{ m}^2 \qquad ♦♦♦$$

The area of a *segment* of a circle (Fig. 14–6) is

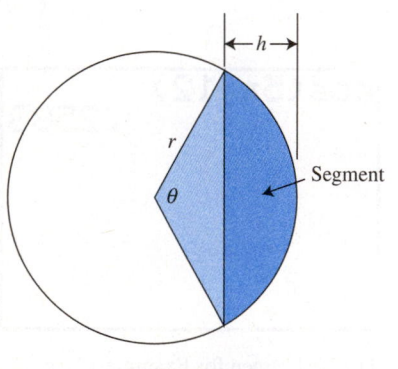

FIGURE 14–6

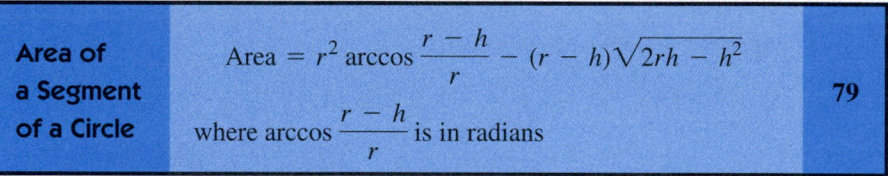

Area of a Segment of a Circle	$\text{Area} = r^2 \arccos \dfrac{r-h}{r} - (r-h)\sqrt{2rh - h^2}$ where $\arccos \dfrac{r-h}{r}$ is in radians	79

These two formulas for the area of a segment of a circle are given here without proof. You will be asked to derive Eq. 79 at the end of this section, and to derive Eq. 80 later.

♦♦♦ Example 15: Compute the area of a segment having a height of 10.0 cm in a circle of radius 25.0 cm.

Solution: Substituting into Eq. 79 gives

$$\text{Area} = (25.0)^2 \arccos \frac{25.0 - 10.0}{25.0} - (25.0 - 10.0)\sqrt{2(25.0)(10.0) - (10.0)^2}$$

$$= 625 \arccos 0.600 - 15.0\sqrt{500 - 100}$$

$$= 625(0.9273) - 15.0(20.0) = 280 \text{ cm}^2 \qquad ♦♦♦$$

A *second* formula for the area of a segment is:

Area of a Segment of a Circle	$\text{Area} = \dfrac{1}{2}r^2(\theta - \sin\theta)$ where θ is the central angle subtended by the segment, in radians	80

◆◆◆ **Example 16:** A chord in a 24.8-cm-diameter circle subtends a central angle of 1.22 rad. Find the area of the segment cut off by the chord.

Solution: By Eq. 80,

$$\text{Area} = \frac{1}{2}(12.4)^2\,(1.22 - \sin 1.22)$$

$$= 76.88(1.22 - 0.9391) = 21.6 \text{ cm}^2 \qquad ◆◆◆$$

Exercise 1 ◆ Radian Measure

Convert to radians.

1. 47.8°
2. 18.7°
3. 35.25°
4. 0.370 rev
5. 1.55 rev
6. 1.27 rev

Convert to revolutions.

7. 1.75 rad
8. 2.30 rad
9. 3.12 rad
10. 0.0633 rad
11. 1.12 rad
12. 0.766 rad

Convert to degrees (decimal).

13. 2.83 rad
14. 4.275 rad
15. 0.372 rad
16. 0.236 rad
17. 1.14 rad
18. 0.116 rad

Convert each angle given in degrees to radian measure in terms of π.

19. 60°
20. 130°
21. 66°
22. 240°
23. 126°
24. 105°
25. 78°
26. 305°
27. 400°
28. 150°
29. 81°
30. 189°

Convert each angle given in radian measure to degrees. Give approximate values to one decimal place.

31. $\dfrac{\pi}{8}$
32. $\dfrac{2\pi}{3}$
33. $\dfrac{9\pi}{11}$
34. $\dfrac{3\pi}{5}$
35. $\dfrac{\pi}{9}$
36. $\dfrac{4\pi}{5}$
37. $\dfrac{7\pi}{8}$
38. $\dfrac{5\pi}{9}$
39. $\dfrac{2\pi}{15}$
40. $\dfrac{6\pi}{7}$
41. $\dfrac{\pi}{12}$
42. $\dfrac{8\pi}{9}$

Evaluate to four significant digits.

43. $\sin \dfrac{\pi}{3}$ **44.** $\tan 0.442$ **45.** $\cos 1.063$

46. $\tan\left(-\dfrac{2\pi}{3}\right)$ **47.** $\cos \dfrac{3\pi}{5}$ **48.** $\sin\left(-\dfrac{7\pi}{8}\right)$

49. $\sec 0.355$ **50.** $\csc \dfrac{4\pi}{3}$ **51.** $\cot \dfrac{8\pi}{9}$

52. $\tan \dfrac{9\pi}{11}$ **53.** $\cos\left(-\dfrac{6\pi}{5}\right)$ **54.** $\sin 1.075$

55. $\cos 1.832$ **56.** $\cot 2.846$ **57.** $\sin 0.6254$

58. $\csc 0.8163$ **59.** $\arcsin 0.7263$ **60.** $\arccos 0.6243$

61. $\cos^{-1} 0.2320$ **62.** $\text{arccot } 1.546$ **63.** $\sin^{-1} 0.2649$

64. $\csc^{-1} 2.6263$ **65.** $\arctan 3.7253$ **66.** $\text{arcsec } 2.8463$

67. $\sin^2 \dfrac{\pi}{6} + \cos \dfrac{\pi}{6}$ **68.** $7 \tan^2 \dfrac{\pi}{9}$

69. $\cos^2 \dfrac{3\pi}{4}$ **70.** $\dfrac{\pi}{6} \sin \dfrac{\pi}{6}$

71. $\sin \dfrac{\pi}{8} \tan \dfrac{\pi}{8}$ **72.** $3 \sin \dfrac{\pi}{9} \cos^2 \dfrac{\pi}{9}$

73. Find the area of a sector having a radius of 5.92 in. and a central angle of 62.5°.

74. Find the area of a sector having a radius of 3.15 m and a central angle of 28.3°.

75. Find the area of a segment of height 12.4 cm in a circle of radius 38.4 cm.

76. Find the area of a segment of height 55.4 inches in a circle of radius 122.6 in.

77. A chord in a 128-cm-diameter circle subtends a central angle of 1.55 rad. Find the area of the segment cut off by the chord.

78. A chord in a 36.9-in.-diameter circle subtends a central angle of 2.23 rad. Find the area of the segment cut off by the chord.

Applications

79. A grinding machine for granite uses a grinding disk that has four abrasive pads with the dimensions shown in Fig. 14–7. Find the area of each pad.

80. A partial pulley is in the form of a sector with a cylindrical hub, as shown in Fig. 14–8. Using the given dimensions, find the volume of the pulley, including the hub.

81. A weight bouncing on the end of a spring moves with *simple harmonic motion* according to the equation $y = 4 \cos 25t$, where y is in inches. Find the displacement y when $t = 2.00$ s. (In this equation, the angle $25t$ must be in radians.)

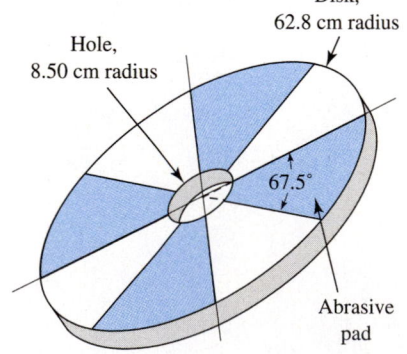

Disk, 62.8 cm radius

Hole, 8.50 cm radius

67.5°

Abrasive pad

FIGURE 14–7

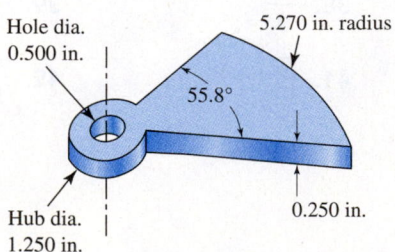

Hole dia. 0.500 in.

Hub dia. 1.250 in.

5.270 in. radius

55.8°

0.250 in.

FIGURE 14–8

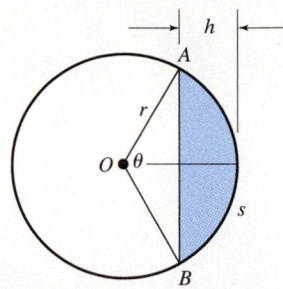

FIGURE 14–9 We will derive the other formula for the area of a segment in a later chapter.

82. *Small Angle Approximations:* For the angles from 0° to 10°, with steps every $\frac{1}{2}°$, use a spreadsheet to compute and print each angle in radians, and the sine and tangent of that angle, to four decimal places. What do you notice about these three columns of figures? What is the largest angle for which the sine and tangent do not differ from the angle itself by more than three significant digits? How could you use this information?

83. *Project: Area of a Segment of a Circle:* Derive Eq. 79 for the area of a segment of a circle

$$A = r^2 \arccos \frac{r-h}{r} - (r-h)\sqrt{2rh-h^2}$$

(where $\arccos (r-h)/r$ is in radians) in terms of the radius r and h, Fig. 14-9.

Hint: Start with the formula, area of a sector $= \dfrac{rs}{2} = \dfrac{r^2\theta}{2}$, and subtract the area

of the triangle *OAB*.

14–2 Arc Length

■ Exploration:

Try this. Locate the center of a circular object, such as a paper plate, and measure its radius. Then with a strip of paper, lay off an arc along the edge of the plate, measure it, and divide it by the radius. Next measure the central angle subtended by the arc, and convert to radians. What do you notice?

Repeat with different lengths of arc. Repeat with different sized plates. Can you make a general statement about your findings? ■

If we measure arc length and radius *in the same units*, those units will cancel when we divide one by the other. We are left with a *dimensionless ratio*. In your exploration, you may have seen that when dividing arc length by radius, you got a dimensionless ratio that is *equal to the angle in radians*. Stated as a formula,

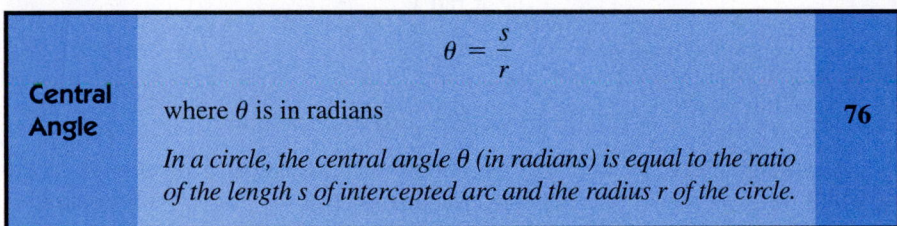

Central Angle	$$\theta = \frac{s}{r}$$ where θ is in radians	76
	In a circle, the central angle θ (in radians) is equal to the ratio of the length s of intercepted arc and the radius r of the circle.	

Thus the radian is not a unit of measure like the degree or inch, although we usually carry the word "radian" or "rad" along as if it were a unit of measure.

We can use Eq. 76 to find any of the quantities θ, r, or s, Fig. 14–10, when the other two are known.

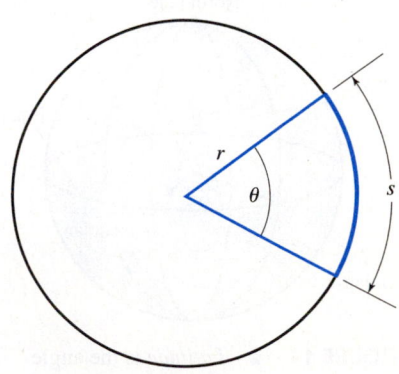

FIGURE 14–10 Relationship between arc length, radius, and central angle.

◆◆◆ Example 17: Find the angle that would intercept an arc of 27.0 ft in a circle of radius 21.0 ft.

Solution: From Eq. 76,

$$\theta = \frac{s}{r} = \frac{27.0 \text{ ft}}{21.0 \text{ ft}} = 1.29 \text{ rad} \qquad \text{◆◆◆}$$

◆◆◆ Example 18: Find the arc length intercepted by a central angle of 62.5° in a 10.4-cm-radius circle.

Solution: Converting the angle to radians, we get

$$62.5° \left(\frac{\pi \text{ rad}}{180°} \right) = 1.09 \text{ rad}$$

By Eq. 76,

$$s = r\theta = 10.4 \text{ cm}(1.09) = 11.3 \text{ cm}$$ ◆◆◆

◆◆◆ **Example 19:** Find the radius of a circle in which an angle of 2.06 rad intercepts an arc of 115 ft.

Solution: By Eq. 76,

$$r = \frac{s}{\theta} = \frac{115 \text{ ft}}{2.06} = 55.8 \text{ ft}$$ ◆◆◆

Common Error	Be sure that r and s have the same units. Convert if necessary.

◆◆◆ **Example 20:** Find the angle that intercepts a 35.8-in. arc in a circle of radius 49.2 cm.

Solution: We divide s by r and convert units at the same time.

$$\theta = \frac{s}{r} = \frac{35.8 \text{ in.}}{49.2 \text{ cm}} \cdot \frac{2.54 \text{ cm}}{1 \text{ in.}} = 1.85 \text{ rad}$$ ◆◆◆

Applications

The ability to find arc lengths has many applications in technology. Let us start with the problem from the introduction to this chapter.

◆◆◆ **Example 21:** How far will the rack in Fig. 14–11 move when the pinion rotates 300.0°?

Solution: Converting to radians gives us

$$\theta = 300.0°\left(\frac{\pi \text{ rad}}{180°}\right) = 5.236 \text{ rad}$$

Then, by Eq. 76,

$$s = r\theta = 11.25 \text{ mm}(5.236) = 58.91 \text{ mm}$$

As a rough check, we note that the pinion rotates less than 1 revolution, so the rack will travel a distance less than the circumference of the pinion. This circumference is 22.5π or about 71 mm. So our answer of 58.91 mm seems reasonable. ◆◆◆

◆◆◆ **Example 22:** How many miles north of the equator is a town of latitude 43.6° N? Assume that the earth is a sphere of radius 3960 mi, and refer to the definitions of latitude and longitude shown in Fig. 14–12.

Estimate: Our given latitude angle is about 1/8 of a circle, so the required distance must be about 1/8 of the earth's circumference (8000π or 25,000 mi), or about 3120 mi.

Solution: The latitude angle, in radians, is

$$\theta = 43.6°\left(\frac{\pi \text{ rad}}{180°}\right) = 0.761 \text{ rad}$$

Then, by Eq. 76,

$$s = r\theta = 3960 \text{ mi}(0.761) = 3010 \text{ mi}$$ ◆◆◆

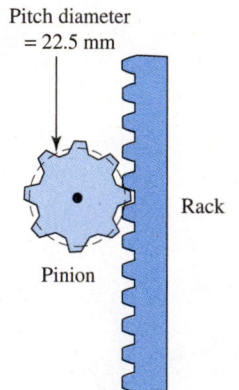

Pitch diameter = 22.5 mm

Rack

Pinion

FIGURE 14–11   Rack and pinion.

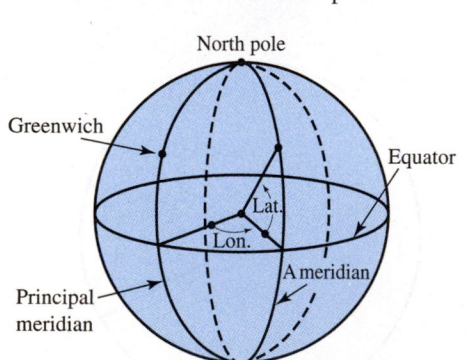

North pole

Greenwich

Equator

Lat.

Lon.

A meridian

Principal meridian

FIGURE 14–12 *Latitude* is the angle (measured at the earth's center) between a point on the earth and the equator. *Longitude* is the angle between the meridian passing through a point on the earth and the principal (or prime) meridian passing through Greenwich, England.

Exercise 2 ◆ Arc Length

In the following exercises, *s* is the length of arc subtended by a central angle θ in a circle of radius *r*.

1. $r = 4.83$ in., $\theta = 2\pi/5$.   Find *s*.
2. $r = 11.5$ cm, $\theta = 1.36$ rad.   Find *s*.
3. $r = 284$ ft, $\theta = 46.4°$.   Find *s*.
4. $r = 2.87$ m, $\theta = 1.55$ rad.   Find *s*.
5. $r = 64.8$ in., $\theta = 38.5°$.   Find *s*.
6. $r = 28.3$ ft, $s = 32.5$ ft.   Find θ.
7. $r = 263$ mm, $s = 582$ mm.   Find θ.
8. $r = 21.5$ ft, $s = 18.2$ ft.   Find θ.
9. $r = 3.87$ m, $s = 15.8$ ft.   Find θ.
10. $\theta = \pi/12$, $s = 88.1$ in.   Find *r*.
11. $\theta = 77.2°$, $s = 1.11$ cm.   Find *r*.
12. $\theta = 2.08$ rad, $s = 3.84$ m.   Find *r*.
13. $\theta = 12°55'$, $s = 28.2$ ft.   Find *r*.
14. $\theta = 5\pi/6$, $s = 125$ mm.   Find *r*.

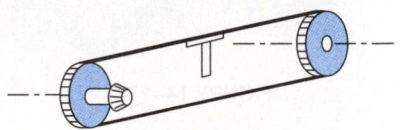

FIGURE 14–13

Applications

Where needed, assume the earth to be a sphere with a radius of 3960 mi. Actually, the distance from pole to pole is about 27 mi less than the diameter at the equator.

15. A certain town is at a latitude of 35.2°N. Find the distance in miles from the town to the north pole.
16. Find the latitude of a city that is 1265 mi from the equator.
17. A satellite is in a circular orbit 225 mi above the equator of the earth. How many miles must it travel for its longitude to change by 85.0°?
18. City *B* is due north of city *A*. City *A* has a latitude of 14°37′ N, and city *B* has a latitude of 47°12′ N. Find the distance in kilometers between the cities.
19. The hour hand of a clock is 85.5 mm long. How far does the tip of the hand travel between 1:00 A.M. and 11:00 A.M.?
20. Find the radius of a circular railroad track that will cause a train to change direction by 17.5° in a distance of 180 m.
21. The pulley attached to the tuning knob of a radio (Fig. 14–13) has a radius of 35 mm. How far will the needle move if the knob is turned a quarter of a revolution?
22. Find the length of contact *ABC* between the belt and pulley in Fig. 14–14.
23. One circular "track" on a magnetic disk used for computer data storage is located at a radius of 155 mm from the center of the disk. If 1000 "bits" of data can be stored in 1 mm of this track, how many bits can be stored in the length of this track subtending an angle of $\pi/12$ rad ?
24. If we assume the earth's orbit around the sun to be circular, with a radius of 93 million mi, how many miles does the earth travel (around the sun) in 125 days?
25. A 1.25-m-long pendulum swings 5.75° on each side of the vertical. Find the length of arc traveled by the end of the pendulum.
26. A brake band is wrapped around a drum (Fig. 14–15). If the band has a width of 92.0 mm, find the area of contact between the band and the drum.
27. Sheet metal is to be cut from the pattern of Fig. 14–16(a) and bent to form the frustum of a cone [Fig. 14–16(b)], with top and bottom open. Find the dimensions *r* and *R* and the angle θ in degrees.

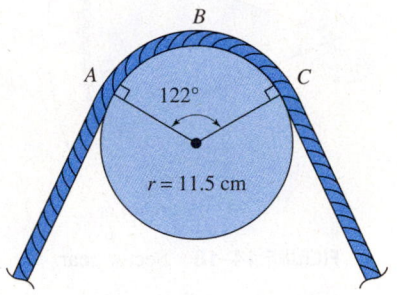

FIGURE 14–14   Belt and pulley.

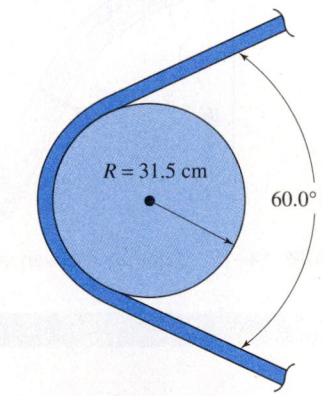

FIGURE 14–15   Brake drum.

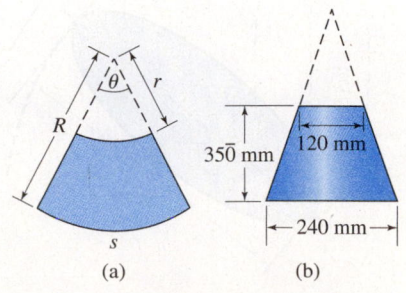

FIGURE 14–16

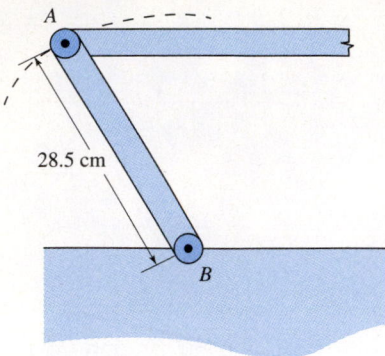

FIGURE 14–17

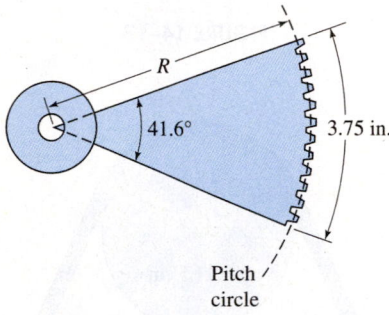

FIGURE 14–18   Sector gear.

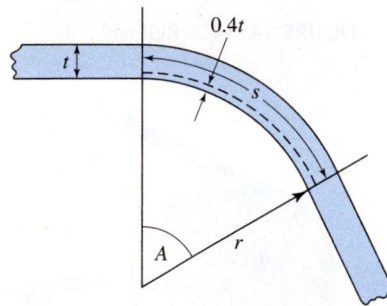

FIGURE 14–19   Bending allowance.

28. An isosceles triangle is to be inscribed in a circle of radius 1.000. Find the angles of the triangle if its base subtends an arc of length 1.437.

29. The link AB in the mechanism of Fig. 14–17 rotates through an angle of 28.3°. Find the distance traveled by point A.

30. Find the radius R of the sector gear of Fig. 14–18.

31. A circular highway curve has a radius of 325.500 ft and a central angle of 15°25′05″ measured to the centerline of the road. Find the length of the curve.

32. When metal is in the process of being bent, the amount s that must be allowed for the bend is called the *bending allowance*, as shown in Fig. 14–19. Assume that the neutral axis (the line at which there is no stretching or compression of the metal) is at a distance from the inside of the bend equal to 0.4 of the metal thickness t.

(a) Show that the bending allowance is

$$s = \frac{A(r + 0.4t)\pi}{180}$$

where A is the angle of bend in degrees.

(b) Find the bending allowance for a 60° bend in $\frac{1}{4}$-in.-thick steel with a radius of 1.50 in.

33. Earlier we gave the formula for the area of a circular sector of radius r and central angle θ: area $= r^2\theta/2$ (Eq. 77). Using Eq. 76, $\theta = s/r$, show that the area of a sector is also equal to $rs/2$, where s is the length of the arc intercepted by the central angle.

34. Find the area of a sector having a radius of 34.8 cm and an arc length of 14.7 cm.

35. *Project:* The angle D (measured at the earth's center) between two points on the earth's surface is found by

$$\cos D = \sin L_1 \sin L_2 + \cos L_1 \cos L_2 \cos(M_1 - M_2)$$

where L_1 and M_1 are the latitude and longitude, respectively, of one point, and L_2 and M_2 are the latitude and longitude of the second point. Find the angle between Pittsburgh (latitude 41.0° N, longitude 80.0° W) and Houston (latitude 29.8° N, longitude 95.3° W) by substituting into this equation.

36. *Project:* Using your answer from the preceding project, compute the distance in miles between Pittsburgh and Houston. Check your answer by measuring a map or a globe. Can you explain any differences?

14–3   Uniform Circular Motion

Angular Velocity

Let us consider a rigid body that is rotating about a point O, as shown in Fig. 14–20. The *angular velocity* ω is a measure of the *rate* at which the object rotates. The motion is called *uniform* when the angular velocity is constant. The units of angular velocity are degrees, radians, or revolutions, per unit time.

Angular Displacement

The angle θ through which a body rotates in time t is called the *angular displacement*. It is related to ω and t by the following equation:

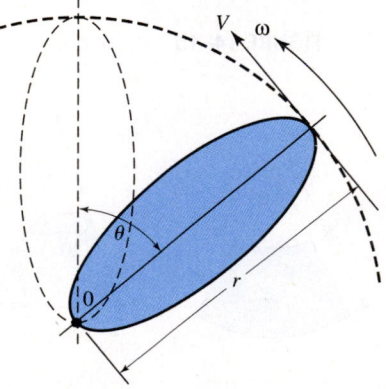

FIGURE 14–20   Rotating body. The symbol ω is lowercase Greek omega, not the letter w.

Angular Displacement	$\theta = \omega t$ *The angular displacement is the product of the angular velocity and the elapsed time.*	1026

This equation is similar to our old formula (distance = rate × time) for linear motion.

◆◆◆ **Example 23:** A wheel is rotating with an angular velocity of 1800 rev/min. How many revolutions does the wheel make in 1.5 s?

Solution: We first make the units of time consistent. Converting yields

$$\omega = \frac{1800 \text{ rev}}{\text{min}} \cdot \frac{1 \text{ min}}{60 \text{ s}} = 30 \text{ rev/s}$$

Then, by Eq. 1026,

$$\theta = \omega t = \frac{30 \text{ rev}}{\text{s}} (1.5 \text{ s}) = 45 \text{ rev} \qquad \text{◆◆◆}$$

◆◆◆ **Example 24:** Find the angular velocity in revolutions per minute of a pulley that rotates 275° in 0.750 s.

Solution: By Eq. 1026,

$$\omega = \frac{\theta}{t} = \frac{275°}{0.750 \text{ s}} = 367 \text{ deg/s}$$

Converting to rev/min, we obtain

$$\omega = \frac{367°}{\text{s}} \cdot \frac{60 \text{ s}}{\text{min}} \cdot \frac{1 \text{ rev}}{360°} = 61.2 \text{ rev/min} \qquad \text{◆◆◆}$$

◆◆◆ **Example 25:** How long will it take a spindle rotating at 3.55 rad/s to make 1000 revolutions?

Solution: Converting revolutions to radians, we get

$$\theta = 1000 \text{ rev} \cdot \frac{2\pi \text{ rad}}{1 \text{ rev}} = 6280 \text{ rad}$$

Then, by Eq. 1026,

$$t = \frac{\theta}{\omega} = \frac{6280 \text{ rad}}{3.55 \text{ rad/s}} = 1770 \text{ s} = 29.5 \text{ min} \qquad \text{◆◆◆}$$

Linear Speed

For any point on a rotating body, the linear displacement per unit time along the circular path is called the *linear speed*. The linear speed is zero for a point at the center of rotation and is directly proportional to the distance r from the point to the center of rotation. If ω is expressed in radians per unit time, the linear speed v is given by the following equation:

	$v = \omega r$	
Linear Speed	*The linear speed of a point on a rotating body is equal to the angular velocity of the body times the distance of the point from the center of rotation.*	**1027**

Applications

◆◆◆ **Example 26:** A wheel is rotating at 2450 rev/min. Find the linear speed of a point 35.0 cm from the center.

Solution: We first express the angular velocity in terms of radians.

$$\omega = \frac{2450 \text{ rev}}{\text{min}} \cdot \frac{2\pi \text{ rad}}{\text{rev}} = 15{,}400 \text{ rad/min}$$

Then, by Eq. 1027,

$$v = \omega r = \frac{15{,}400 \text{ rad}}{\text{min}}(35.0 \text{ cm}) = 539{,}000 \text{ cm/min}$$

$$= 89.8 \text{ m/s}$$

What became of "radians" in our answer? Shouldn't the final units be rad · m/s? No. Remember that radians is a dimensionless ratio; it is the ratio of two lengths (arc length and radius) whose units cancel.

Alternate Solution: For each revolution, a point at radius r travels a distance equal to the circumference of a circle of radius r, or $2\pi r$ cm/rev. Its linear speed is then,

$$v = \omega(2\pi r)$$

$$= \frac{2450 \text{ rev}}{\text{min}} \cdot \frac{2\pi(35.0) \text{ cm}}{\text{rev}}$$

$$= 539{,}000 \text{ cm/min}$$

◆◆◆

Common Error	Remember when using Eq. 1027 that the angular velocity must be expressed in *radians* per unit time.

◆◆◆ **Example 27:** A belt having a speed of 885 in./min turns a 12.5-in.-radius pulley. Find the angular velocity of the pulley in rev/min.

Solution: By Eq. 1027,

$$\omega = \frac{v}{r} = \frac{885 \text{ in./min}}{12.5 \text{ in.}} = 70.8 \text{ rad/min}$$

Converting to revolutions we get

$$\omega = \frac{70.8 \text{ rad}}{\text{min}} \cdot \frac{1 \text{ rev}}{2\pi \text{ rad}} = 11.3 \text{ rev/min}$$

◆◆◆

Cutting Speed: For machine shop operations, the *cutting speed* is the linear speed at which the outermost edge of a cutting tool meets a workpiece. Sometimes the tool is rotating, as in a milling machine or drill, and sometimes the workpiece is rotating, as in a lathe.

Cutting speed is usually given in feet per minute (ft/min) or meters per minute (m/min). It is often called SFPM, or surface feet per minute. Ranges of typical cutting speeds are given in Fig. 14–21. Machinists will often refer to tables to compute cutting speeds or rotational speeds, but we will show how to find them using basic principles.

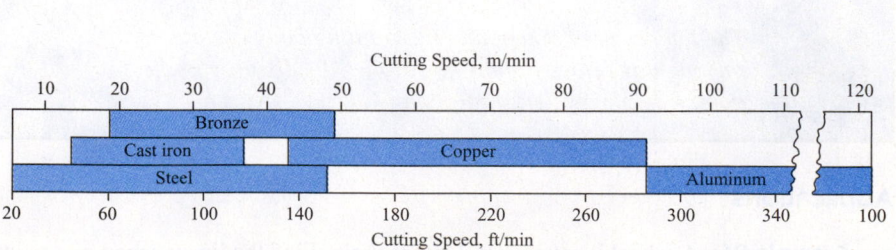

FIGURE 14–21

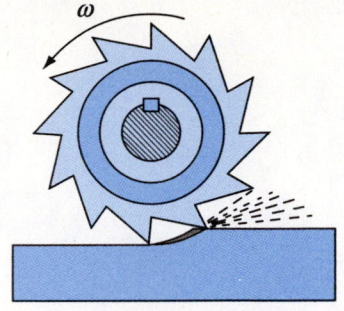

FIGURE 14–22

♦♦♦ **Example 28:** *An Application.* A 4.00-in.-dia. milling machine cutter is used to mill a bronze casting, Fig. 14–22. The recommended cutting speed for that particular alloy, cutter type, and lubrication is 120 ft/min. What angular velocity of the cutter will give that cutting speed?

Solution: Converting 2.00 in. to feet gives $r = 0.1667$ ft. Then by our equation for linear speed, we get

$$\omega = \frac{v}{r} = \left(\frac{120 \text{ ft}}{\text{min}} \right) \div 0.1667 \text{ ft}$$

$$= 720 \text{ rad/min}$$

Converting to revolutions per minute,

$$\omega = \frac{720 \text{ rad}}{\text{min}} \times \frac{\text{rev}}{2\pi \text{ rad}} = 115 \text{ rev/min}$$

Not all rotational speeds will usually be available on a machine, so a value close to this would be chosen. ♦♦♦

Exercise 3 ♦ Uniform Circular Motion

Angular Velocity

1. Convert 1850 rev/min to radians per second and degrees per second.
2. Convert 5.85 rad/s to revolutions per minute and degrees per second.
3. Convert 77.2 deg/s to revolutions per minute and radians per second.
4. Convert $3\pi/5$ rad/s to revolutions per minute and degrees per second.
5. Convert 48.1 deg/s to revolutions per minute and radians per second.
6. Convert 22,600 rev/min to radians per second and degrees per second.

Linear Speed

7. A disk is rotating at 334 rev/min. Find the linear speed, in ft/min, of a point 3.55 in. from the center.
8. A wheel rotates at 46.8 rad/s. Find the linear speed, in cm/s, of a point 36.8 cm from the center.
9. A point 1.14 ft from the center of a rotating wheel has a linear speed of 56.3 ft/min. Find the angular velocity of the wheel in rev/min.
10. The rim of a rotating wheel 83.4 cm in diameter has a linear speed of 58.3 m/min. Find the angular velocity of the wheel in rev/min.

Applications

11. A flywheel makes 725 revolutions in a minute. How many degrees does it rotate in 1.00 s?
12. A propeller on a wind generator rotates 60.0° in 1.00 s. Find the angular velocity of the propeller in revolutions per minute.
13. A gear is rotating at 2550 rev/min. How many seconds will it take to rotate through an angle of 2.00 rad?
14. A sprocket 3.00 inches in diameter is driven by a chain that moves at a speed of 55.5 in./s. Find the angular velocity of the sprocket in rev/min.
15. A capstan on a magnetic tape drive rotates at 3600 rad/min and drives the tape at a speed of 45.0 m/min. Find the diameter of the capstan in millimeters.

16. A blade on a water turbine turns 155° in 1.25 s. Find the linear speed of a point on the tip of the blade 0.750 m from the axis of rotation.

17. A steel bar 6.50 inches in diameter is being turned in a lathe. The surface speed of the bar is 55.0 ft/min. How many revolutions will the bar make in 10.0 s?

18. Assuming the earth to be a sphere 7920 mi in diameter, calculate the linear speed in miles per hour of a point on the equator due to the rotation of the earth about its axis.

19. Assuming the earth's orbit about the sun to be a circle with a radius of 93.0×10^6 mi, calculate the linear speed of the earth around the sun.

20. A car is traveling at a rate of 65.5 km/h and has tires that have a radius of 31.6 cm. Find the angular velocity of the wheels of the car.

21. A wind generator has a propeller 21.7 ft in diameter, and the gearbox between the propeller and the generator has a gear ratio of 1:44 (with the generator shaft rotating faster than the propeller). Find the tip speed of the propeller when the generator is rotating at 1800 rev/min.

22. A milling machine cutter has a diameter of 75.0 mm and is rotating at 56.5 rev/min. What is the linear speed at the edge of the cutter?

23. For a lathe tool cutting a particular aluminum alloy, Fig. 14–23, the radius of the workpiece is 2.45 in. and the recommended cutting speed is 110 ft/min. What spindle speed will give that value?

24. You are programming a numerically controlled drill press to drill holes in a cast iron block, whose recommended cutting speed is 30.0 m/min. What rotational speed of a 25.4 mm diameter drill will give that cutting speed?

25. *Writing:* When we multiply an angular velocity in radians per second by length in feet, we get a linear speed in feet per second. Explain why radians do not appear in the units for linear speed.

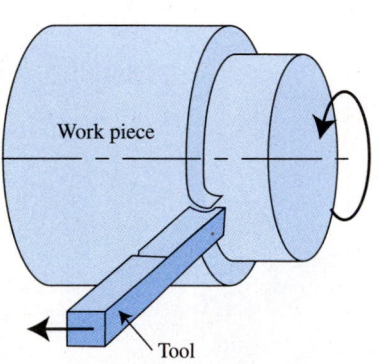

FIGURE 14–23

Labels: Work piece, Tool

••• CHAPTER 14 REVIEW PROBLEMS •••••••••••••••••••••••••••••••••

Convert from radians to degrees.

1. $\dfrac{3\pi}{7}$ 2. $\dfrac{9\pi}{4}$ 3. $\dfrac{\pi}{9}$

4. $\dfrac{2\pi}{5}$ 5. $\dfrac{11\pi}{12}$

6. Find the central angle in radians that would intercept an arc of 5.83 m in a circle of radius 7.29 m.

7. Find the angular velocity in rad/s of a wheel that rotates 33.5 revolutions in 1.45 min.

8. A wind generator has blades 3.50 m long. Find the tip speed when the blades are rotating at 35 rev/min.

Convert to radians in terms of π.

9. 300°   10. 150°
11. 230°   12. 145°

Evaluate to four significant digits.

13. $\sin \dfrac{\pi}{9}$ 14. $\cos^2 \dfrac{\pi}{8}$

15. $\sin \left(\dfrac{2\pi}{5}\right)^2$ 16. $3 \tan \dfrac{3\pi}{7} - 2 \sin \dfrac{3\pi}{7}$

17. $4 \sin^2\dfrac{\pi}{9} - 4 \sin\left(\dfrac{\pi}{9}\right)^2$

18. What arc length is subtended by a central angle of 63.4° in a circle of radius 4.85 in.?

19. A winch has a drum 30.0 cm in diameter. The steel cable wrapped around the drum is to be pulled in at a rate of 8.00 ft/s. Ignoring the thickness of the cable, find the angular speed of the drum in revolutions per minute.

20. Find the linear velocity of the tip of a 5.50-in.-long minute hand of a clock.

21. A satellite has a circular orbit around the earth with a radius of 4250 mi. The satellite makes a complete orbit of the earth in 3 days, 7 h, and 35 min. Find the linear speed of the satellite.

Evaluate to four decimal places. (Angles are in radians.)

22. cos 1.83

23. tan 0.837

24. csc 2.94

25. sin 4.22

26. cot 0.371

27. sec 3.38

28. $\cos^2 1.74$

29. $\sin(2.84)^2$

30. $(\tan 0.475)^2$

31. $\sin^2(2.24)^2$

32. Find the area of a sector with a central angle of 29.3° and a radius of 37.2 in.

33. Find the radius of a circle in which an arc of length 384 mm intercepts a central angle of 1.73 rad.

34. Find the arc length intercepted in a circle of radius 3.85 m by a central angle of 1.84 rad.

35. A parts-storage tray, shown in Fig. 14–24, is in the form of a partial sector of a circle. Using the dimensions shown, find the inside volume of the tray in cubic centimeters.

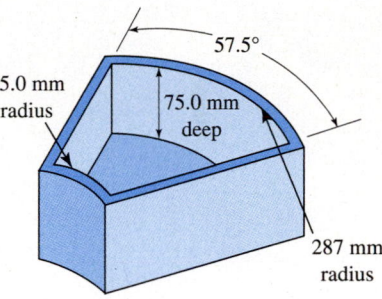

FIGURE 14–24

15

Trigonometric, Parametric, and Polar Graphs

••• **OBJECTIVES** •••
When you have completed this chapter you should be able to
- Graph the sine wave, by calculator or manually.
- Find the amplitude, period, frequency, and phase shift for a sine wave.
- Find roots or instantaneous values on a sine wave.
- Write the equation of a given sine wave.
- Graph and analyze a sine wave as a function of time.
- Graph the cosine, tangent, cotangent, secant, and cosecant functions.
- Graph the inverse trigonometric functions.
- Graph parametric equations.
- Graph points and equations in polar coordinates.
- Convert between polar and rectangular form.

••

So far we have dealt with curves that rise, or fall, or perhaps rise *and* fall a few times. Now we will introduce curves that *oscillate*, repeating the same shape indefinitely, the *periodic functions*. These are the sort of curves we find in alternating current, or the mechanical vibrations that could cause a bridge to collapse. We find periodic motion in mechanical devices, such as the pistons in an automobile engine, the motions of the celestial bodies, sound waves, and in radio, radar, and television signals. Periodic signals are crucial to the operation of many of the exciting technological devices of the twenty-first century, from computers to satellite telephones.

In this chapter we give a small introduction to the world of periodic functions. Our main focus will be on the sine function, which has wide applications to alternating current, mechanical vibrations, and so forth. Our task in this chapter will be to *graph* such functions, building upon our earlier methods for graphing and to extract useful information from the function. For example, given the alternating current

$$I = 37.5(\sin 284t - 22°)$$

you should be able to make a graph, find the amplitude and frequency of the current, and find the instantaneous current at any given instant.

We will make heavy use of the graphing calculator and also show some manual methods. In addition to making a graph, we will *find roots*, just as we did with other functions, and find the ordinate of the function at any abscissa, the so-called *instantaneous value*. Next we will describe functions in which both the abscissa and ordinate are described in terms of a third variable, called a *parameter*. This gives us what are called *parametric equations*, and we show how to graph them, both by calculator and manually.

All the above graphing, including parametric equations, is done in our familiar rectangular coordinate system. Finally we introduce *polar coordinates*, which are more useful for graphing certain kinds of functions. We will do graphs in polar coordinates both manually and by calculator.

15–1   Graphing the Sine Wave by Calculator

■ Exploration:

Try this. Using your graphing calculator, graph $y = \sin x$. A trigonometric function is graphed the same way as we graphed other functions in Chapter 5. Enter the function in the $\boxed{Y=}$ editor, put the calculator into **DEGREE** mode, set the viewing window from -100 to $+800$ on the *x* axis, and -2 to $+2$ on the *y* axis, and press $\boxed{GRAPH}$.

What is different about this curve than others we have studied? What are the maximum and minimum values of *y*? Are there maximum and minimum values of *x*? Increase the width of the window. Does the curve seem to repeat? How often? ■

Periodic Functions

A curve that *repeats* its shape over and over, like the sine wave in our exploration, is called a *periodic curve* or *periodic waveform*. A function whose graph is periodic is called a *periodic function*. The horizontal axis represents either an *angle* (in degrees or radians) or *time* (usually in seconds or milliseconds). In this section we will show waveforms for which the horizontal axis is an angle and cover those as a function of time later in this chapter.

Each repeated portion of the curve is called a *cycle*. The *period* of a periodic waveform is the horizontal distance occupied by one cycle. Depending on the units on the horizontal axis, the period is expressed in degrees per cycle, radians per cycle, seconds per cycle, and so forth. Using $\boxed{TRACE}$ on the graph obtained in your exploration, you should find that the period is 360° per cycle. Recall that a *complete graph* of a function is one that contains all features of interest. Thus the complete graph of a periodic waveform must contain at least one cycle. Let us now do an example in radians.

◆◆◆ **Example 1:** Graph at least one cycle of the sine function $y = \sin x$, where *x* is in radians.

Solution: We put the calculator into **RADIAN** mode. Let us keep the height of the viewing window as before, from -2 to $+2$ on the *y* axis. The width, however, has to be the radian equivalent of 360°, or 2π radians. Let us thus set the width of the viewing window from $x = -1$ to $+7$. Set **X scale** and **Y scale** to 1. We get the graph shown.   ◆◆◆

The graphs we obtained in the preceding examples show the typical shape of the sine wave. However, another sine wave may have a different height or *amplitude*, a different *period*, or may have a horizontal offset, called a *phase shift*. In the next few sections we will show how slight changes in the sine function will result in changes in the curve.

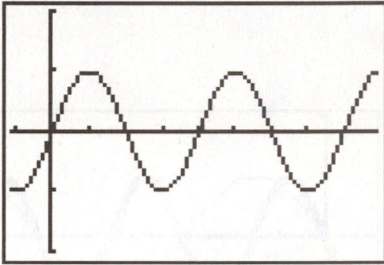

Screen for $y = \sin x$ from the exploration. Ticks on the vertical axis are one unit apart, and on the horizontal axis are 90° apart, so the period is 360° per cycle. Some calculators can automatically set the viewing window for trigonometric graphs. On TI calculators, it is called **ZTrig** and is in the $\boxed{ZOOM}$ menu. However, that window may not always be the best for a particular graph.

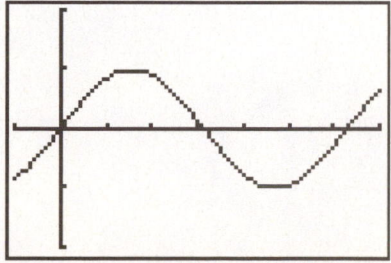

Screen for Example 1. Graph of $y = \sin x$, in radians. Tick marks on the horizontal axis are 1 radian apart.

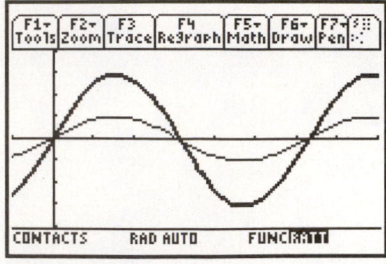

TI-89 graphs of $y = \sin x$ (thin) and $y = 3 \sin x$ (thick). Tick marks are spaced 1 radian on the x axis and 1 on the y axis. To change the line thickness on the TI-89, select **F6, Style,** when in the $\boxed{Y=}$ editor, and select from a variety of styles.

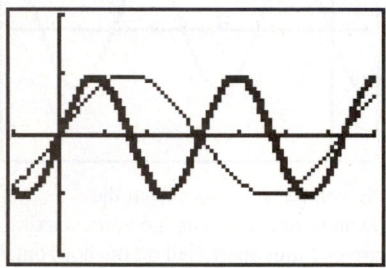

TI-83/84 graphs of $y = \sin x$ and $y = \sin 2x$ (shown heavy) in radians. Tick marks on the horizontal axis are 1 radian apart. On the TI-83/84, you change line style in the $\boxed{Y=}$ editor. Move the cursor to the desired function and then to the left of the screen, to the graph style icon. Press $\boxed{\text{ENTER}}$ repeatedly until you find the line style you want.

The Sine Wave: Amplitude

■ Exploration:

Try this. In the same viewing window, graph one cycle of

$$y = \sin x \quad \text{and} \quad y = 3 \sin x$$

either in degrees or radians. How do the two sine waves differ? What is the effect of the coefficient 3 in the second equation? Then in the same viewing window, graph $y = \sin x$, $y = 4 \sin x$, and $y = -2 \sin x$. What conclusions do you draw? ■

The *amplitude* of a periodic function is half the difference between its greatest and least values. Amplitude is what distinguishes between an alternating voltage that will either tickle you or kill you, or between a seismic disturbance that will rock your house or bring it down.

In the preceding exploration, the first curve had an amplitude of 1 and the second had an amplitude of 3.

In general, the sine function has a coefficient a, and the absolute value of a is the amplitude of the sine wave.

$$y = a \sin x$$
$$\text{amplitude} = |a|$$

The Sine Wave: Period and Frequency

■ Exploration:

Try this. In the same viewing window, graph

$$y = \sin x \quad \text{and} \quad y = \sin 2x$$

either in degrees or in radians.

How do the two waves differ? What is the effect of the x-coefficient 2 in the second equation? Then in the same viewing window, graph $y = \sin x$, $y = \sin 3x$, and $y = \sin x/2$. What conclusions do you draw? ■

We earlier defined the *period* as the interval it takes for a periodic waveform to complete one cycle, after which it repeats. We saw that the period for the sine function $y = a \sin x$ was 360° or 2π radians.

In our preceding exploration, the curve $y = \sin 2x$ completed *two* cycles in the same interval that it took $y = \sin x$ to complete a *single* cycle. This is not surprising. Since $2x$ is twice as large as x, then $2x$ will reach a full cycle twice as soon as x.

The *frequency f* is the reciprocal of the period. It is frequency that lets you tune to different stations on an FM (frequency modulation) radio dial, that makes the difference between different colors of light, and that distinguishes sounds that people can hear from others that only dogs can hear.

In general,

$$y = \sin bx$$

will have a period of $360°/b$, if x is in degrees, or $2\pi/b$ radians, if x is in radians.

Period		
	$P = \dfrac{360}{b}$ degrees/cycle	163
	$P = \dfrac{2\pi}{b}$ radians/cycle	

The *frequency f* is the reciprocal of the period.

Frequency		
	$f = \dfrac{b}{360}$ cycles/degree	164
	$f = \dfrac{b}{2\pi}$ cycles/radian	

◆◆◆ **Example 2:** Find the period and frequency of the function $y = \sin 6x$, both in degrees and radians, and graph one cycle.

Solution: Here, $b = 6$, so the period is

$$P = \frac{360°}{6} = 60°/\text{cycle}$$

or

$$P = \frac{2\pi}{6} = \frac{\pi}{3} \text{ radians/cycle} \approx 1.047 \text{ rad/cycle}$$

The frequency is the reciprocal of the period, so

$$f = \frac{1}{P} = \frac{1}{60°} \approx 0.01667 \text{ cycles/degree}$$

or

$$f = \frac{1}{P} = \frac{3}{\pi} \approx 0.9549 \text{ cycles/radian}.$$

The graph is shown.   ◆◆◆

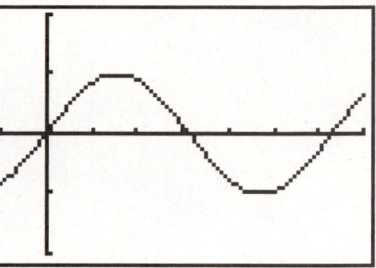

Screen for Example 2. Graph of $y = \sin 6x$. Each tick mark on the x axis equals 10°.

The Sine Wave: Phase Shift

■ **Exploration:**

Try this. In the same viewing window, graph, in degrees,

$$y = \sin x \qquad \text{and} \qquad y = \sin(x + 45°)$$

How do the two graphs differ? What is the effect of adding 45° to x in the second equation? What do you think would be the effect of *subtracting* 45° from x? ■

We see that the graph of $y = \sin x$ passes through the origin but that the graph of $y = \sin(x + 45°)$ is shifted in the x direction. The amount of such shift is called the *phase shift*, *phase angle* or *phase displacement*. In the graph from our exploration, this phase shift was 45° in the negative direction.

Phase shifts occur even in the simplest AC circuits. The alternating current through an inductor, for example, will be shifted with respect to the alternating voltage across that inductor. A capacitor will cause a similar phase shift, in a different direction. These differences in phase are important in analyzing such a circuit.

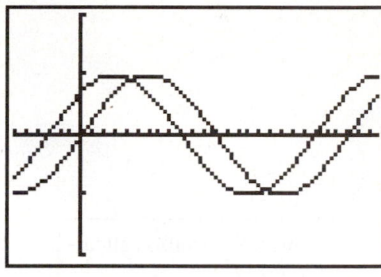

Graphs of $y = \sin x$ and $y = \sin(x + 45°)$. Ticks on the x axis are 15° apart.

◆◆◆ **Example 3:** In the same viewing window, graph

$$y = \sin x \qquad \text{and} \qquad y = \sin(3x - 45°)$$

from $x = -30°$ to 180°. Graphically find the phase shift.

Solution: Our calculator screen is shown with tick marks on the x axis spaced 15° apart. We see that the graph of $y = \sin(3x - 45°)$ is shifted 15° in the positive direction.   ◆◆◆

In general, for the sine function

$$y = \sin(bx + c)$$

we define the phase shift as

$$\text{Phase shift} = -\frac{c}{b}$$

In Example 3, we had $c = -45°$ and $b = 3$, so the phase shift was $-(-45°)/3$ or 15°.

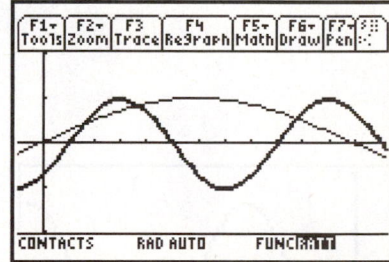

TI-89 graphs for Example 3, with $y = \sin(3x - 45°)$ shown thick. Ticks are spaced 15° in x and 1 unit in y.

Amplitude and Phase Shift Related

■ **Exploration:**

Try this. In the same viewing window, graph

$$y = \sin x$$
$$y = -\sin(x + 180)$$
$$y = -\sin(x - 180)$$

with your calculator in **DEGREE** mode. What do you see?

You should have observed that changing the sign of a has the same effect as shifting the sine wave by half the period, in either direction. ■

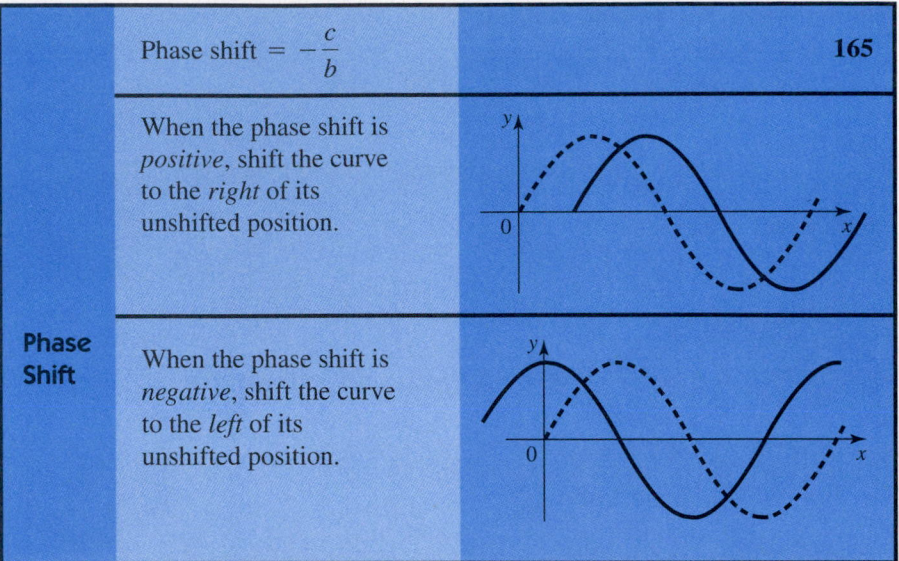

The phase shift will have the same units, radians or degrees, as the constant c.

The General Sine Wave

In the preceding sections we introduced the sine function constants a, b, and c separately. Now we will consider a sine function that has all three constants in place.

A sine wave is also called a *sinusoid*.

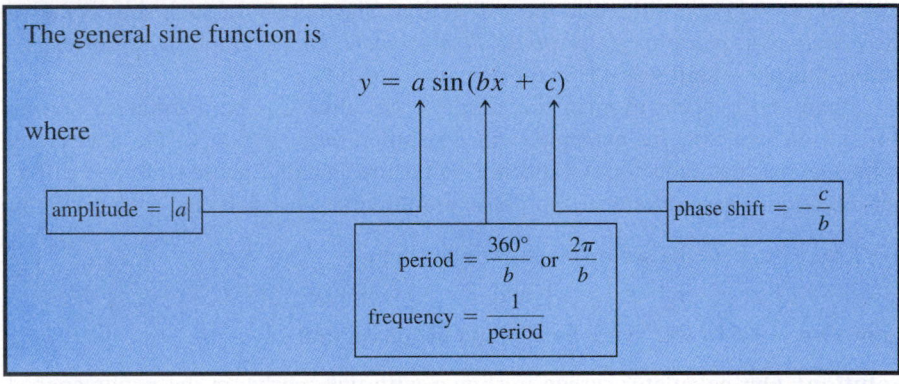

◆◆◆ **Example 4:** (a) Find the amplitude, period, frequency, and phase shift. (b) Graph one cycle of the sine wave

$$y = 3 \sin(2x - 90°)$$

Solution: (a) We have $a = 3$, $b = 2$, and $c = -90°$. Therefore,

amplitude $= |a| = 3$

period $P = 360°/b = 360°/2 = 180°$/cycle

frequency $f = 1/P = 1/180°$ or 0.00556 cycles/degree

phase shift $= -c/b = -(-90°)/2 = 45°$

(b) The graph is shown.   ◆◆◆

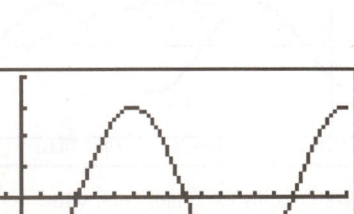

Screen for Example 4. Graph of $y = 3 \sin(2x - 90°)$. Tick marks on the x axis are 15° apart. As expected, the amplitude is 3, the period is 180°, and the curve is shifted to the right by 45°.

Zeros and Instantaneous Value

The *zero* or *root* of a periodic function is the value of x when y is zero, just as for our other functions. The *instantaneous value* of the function is the value of y at a particular value of x. On a graphing calculator we find zeros and instantaneous values as before, using TRACE and ZOOM. On some calculators we can use **zero** to find a zero, or **value** to find an instantaneous value.

◆◆◆ **Example 5:** Graph the function

$$y = 2 \sin(3x + 15°)$$

and find the value of y when $x = 45°$.

Solution: We graph the function as shown, with the calculator in **Degree** mode. On the TI-83/84, we select **value** from the $\boxed{\text{CALC}}$ menu, and enter the value of $x = 45°$. The value of y at that point is displayed. ◆◆◆

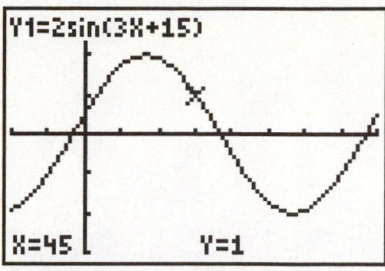

TI-83/84 screen for Example 5. We get a value of $y = 1$ at $x = 45°$.

An Application

◆◆◆ **Example 6:** In the mechanism of Fig. 15–1, rotating wheel W contains a pin P, located at a radius of 8.75 in. from the wheel's center. Follower F moves up and down in bearing B as the wheel rotates, the upper face of F being kept in contact with the pin by a spring.

(a) Write an equation for the vertical displacement y of the follower, as a function of θ. Neglect the diameter of the pin.
(b) Graph the equation, in degrees, for at least one cycle.
(c) Graphically find the zeros of the equation within one cycle.
(d) Graphically find the instantaneous value of y when $\theta = 55.0°$.

Solution:

(a) Distance y is related to the angle θ by

$$\sin \theta = \frac{y}{8.75}$$

where 8.75 in. is the distance from the pin to the wheel center. So our function is

$$y = 8.75 \sin \theta \qquad \text{inches}$$

(b) The graph is shown.
(c) We see zeros at $0°$, $180°$, and $360°$.
(d) When $\theta = 55.0°$, we get $y \approx 7.17$ in. That is, when the wheel turns by $55.0°$, the follower drops by 7.17 in. ◆◆◆

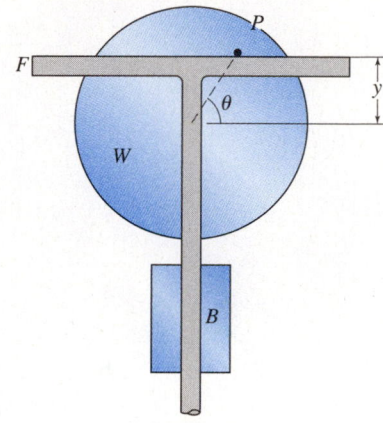

FIGURE 15–1

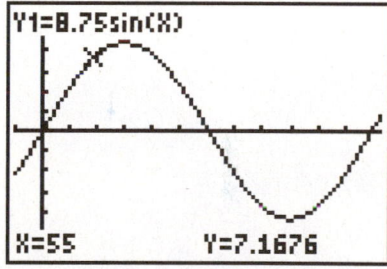

TI-83/84 screen for Example 6. Graph of $y = 8.75 \sin \theta$. Tick marks are spaced $30°$ on the horizontal axis and 2 inches on the vertical axis. Note that the calculator will always show the horizontal axis as the x axis, even though it is θ in our example.

Exercise 1 ◆ Graphing the Sine Wave by Calculator

Graph each sine wave. Find the amplitude, period, and phase shift.

1. $y = 2 \sin x$

2. $y = -3 \sin x$

3. $y = \sin 2x$

4. $y = \sin 3x$

5. $y = 3 \sin 2x$

6. $y = 2 \sin 3x$

7. $y = \sin(x + 15°)$

8. $y = \sin(x - 45°)$

9. $y = \sin(x - \pi/2)$
10. $y = \sin(x + \pi/8)$
11. $y = 3 \sin(x + 45°)$
12. $y = 2 \sin(x - 35°)$
13. $y = -4 \sin(x - \pi/4)$
14. $y = 5 \sin(x + \pi/2)$
15. $y = \sin(2x + 55°)$
16. $y = \sin(3x - 25°)$
17. $y = \sin(3x - \pi/3)$
18. $y = \sin(4x + \pi/6)$
19. $y = 3 \sin(2x + 55°)$
20. $y = 2 \sin(3x - 25°)$
21. $y = -2 \sin(3x - \pi/2)$
22. $y = 4 \sin(4x + \pi/6)$
23. $y = 3.73 \sin(4.32x + 55.4°)$
24. $y = 1.82 \sin(4.43x - 22.5°)$

Zeros and Instantaneous Value

Find the zeros in one cycle of the given functions, and the value of y at the given value of x.

25. $y = \sin(2x + 15°)$ $x = 15°$
26. $y = 3 \sin(x - 35°)$ $x = 45°$
27. $y = 2 \sin(3x - 3)$ $x = 1$ radian
28. $y = 4 \sin(x + 2)$ $x = 1$ radian

Applications

29. As the ladder in Fig. 15–2 is raised, the height h increases as θ increases.
    (a) Write the function $h = f(\theta)$.
    (b) Graph $h = f(\theta)$ for $\theta = 0$ to $90°$.

30. Figure 15–3 shows a Scotch yoke mechanism, or Scotch crosshead.
    (a) Write an expression for the displacement y of the rod PQ as a function of θ and of the length R of the rotating arm.
    (b) Graph $y = f(\theta)$ for $\theta = 0$ to $360°$.

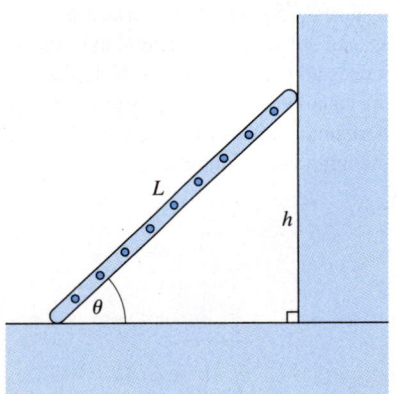

FIGURE 15–2   A ladder.

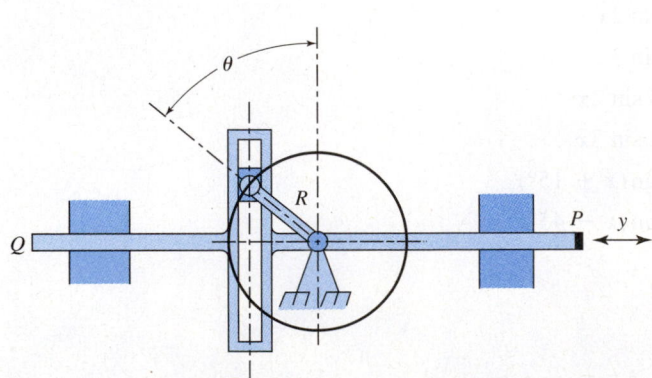

FIGURE 15–3   Scotch crosshead.

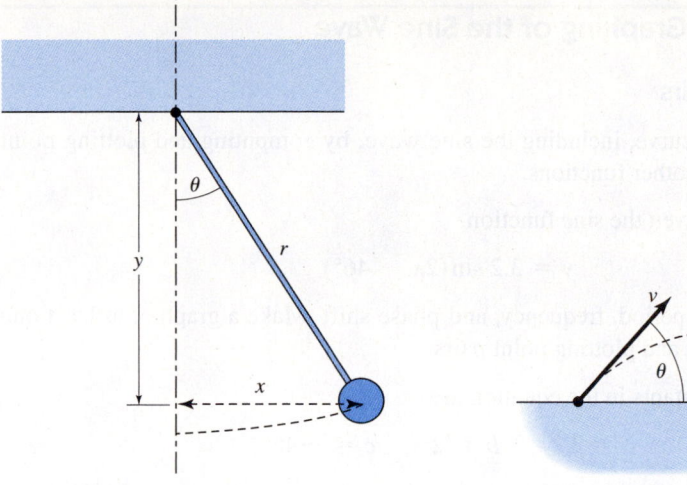

FIGURE 15–4 A pendulum.

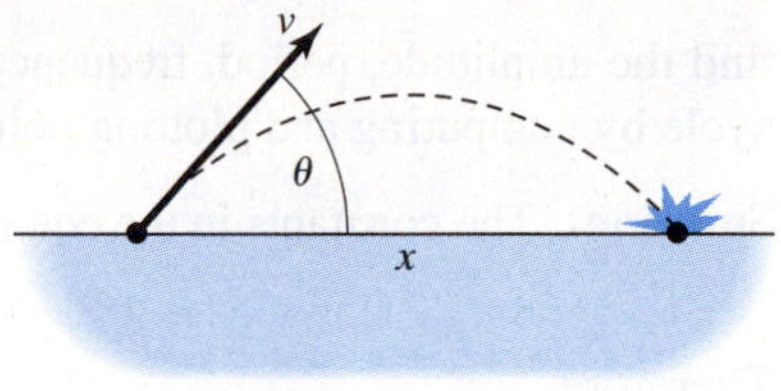

FIGURE 15–5 A projectile.

31. For the pendulum of Fig. 15–4:
   (a) Write an equation for the horizontal distance x as a function of θ.
   (b) Graph $x = f(\theta)$ for $\theta = 0$ to $30°$.

32. The projectile shown in Fig. 15–5 is fired with an initial velocity v at an angle θ with the horizontal. Its range (the horizontal distance traveled before hitting the ground) is given by

$$x = \frac{v^2}{g} \sin 2\theta$$

where g is the acceleration due to gravity.
   (a) Graph the range of the projectile for $\theta = 0$ to $90°$. Take $v = 100$ ft/s and $g = 32.2$ ft/s^2.
   (b) What angle gives a maximum range?

33. The vertical distance x for cog A in Fig. 15–6 is given by $x = r \sin \theta$. Write an expression for the vertical distance to cog B.

34. In the mechanism shown in Fig. 15–7, the rotating arm A lifts the follower F vertically.
   (a) Write an expression for the vertical displacement y as a function of θ. Take $\theta = 0$ in the extreme clockwise position shown.
   (b) Graph $y = f(\theta)$ for $\theta = 0$ to $90°$.

35. *Writing:* Observe and compile a list of periodic phenomena, with a one-line description of each. Examples do not all have to be from technology and, for this exercise, do not have to repeat exactly from cycle to cycle.

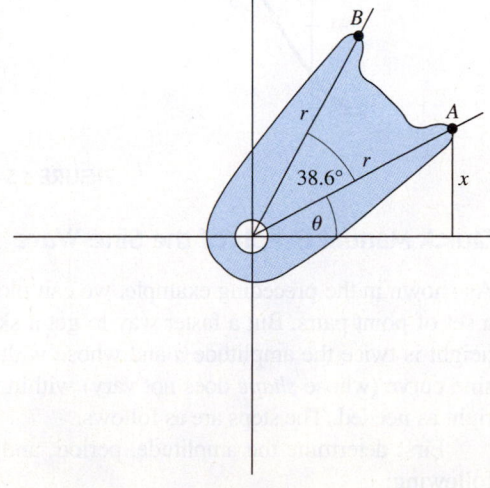

FIGURE 15–6

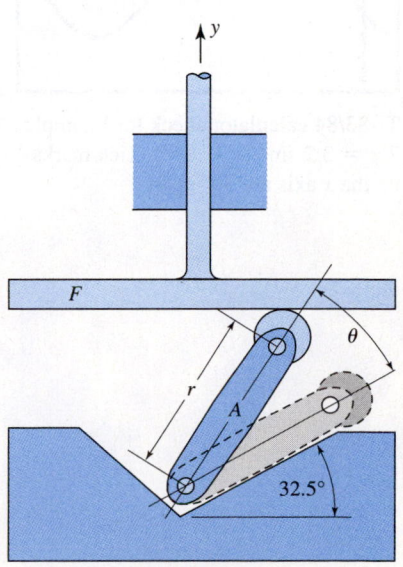

FIGURE 15–7

15–2   Manual Graphing of the Sine Wave

Plotting Point Pairs

We can graph any curve, including the sine wave, by computing and plotting point pairs, as we did for other functions.

◆◆◆ Example 7: Given the sine function

$$y = 3.2 \sin(2x - 46°)$$

find the amplitude, period, frequency, and phase shift. Make a graph of at least one cycle by computing and plotting point pairs.

Solution: The constants in the equation are

$$a = 3.2 \qquad b = 2 \qquad c = -46°$$

Therefore,

$$\text{amplitude} = |a| = 3.2$$

$$\text{period } P = \frac{360°}{b} = \frac{360°}{2} = 180°/\text{cycle}$$

$$\text{frequency } f = \frac{1}{P} = \frac{1}{180°/\text{cycle}} = 0.00556 \text{ cycles/degree}$$

$$\text{phase shift} = -\frac{c}{b} = -\frac{-46}{2} = 23°$$

Let us compute points from $x = -15°$ to $195°$, in steps of $15°$.

x	$-15°$	$0°$	$15°$	$30°$	$45°$	$60°$	$75°$	$90°$	$105°$	$120°$	$135°$	$150°$	$165°$	$180°$	$195°$
y	-3.10	-2.30	-0.882	0.774	2.22	3.08	3.10	2.30	0.88	-0.77	-2.22	-3.08	-3.10	-2.30	-0.88

These points are plotted in Fig. 15–8 and connected with a smooth curve. For comparison, we show a calculator screen.

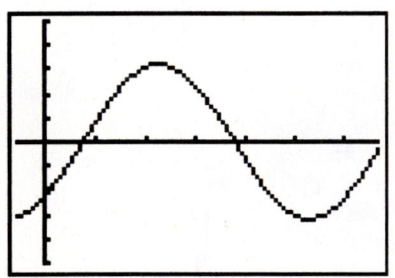

TI-83/84 calculator check for Example 7 $y = 3.2 \sin(2x - 46°)$. Tick marks on the x axis are $30°$ apart.

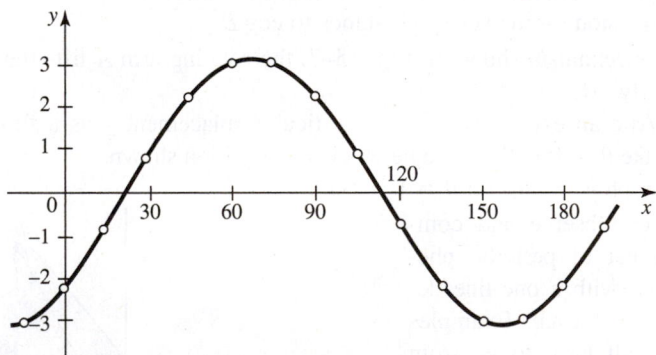

FIGURE 15–8   ◆◆◆

Quick Manual Graph of the Sine Wave

As shown in the preceding example, we can plot any curve by computing and plotting a set of point pairs. But a faster way to get a sketch is to first draw a rectangle whose height is twice the amplitude a and whose width equals the period P. Then sketch the sine curve (whose *shape* does not vary) within that box. Finally shift the curve left or right as needed. The steps are as follows.

First determine the amplitude, period, and phase shift of the curve. Then do the following:

(a) Draw two horizontal lines, each at a distance equal to the amplitude *a* from the *x* axis.

(b) Draw a vertical line at a distance from the origin equal to the period *P*. We now have a rectangle of width *P* and height 2*a*.

(c) Subdivide the period *P* into four equal parts. Label the *x* axis at these points, and draw vertical lines through them.

(d) Lightly sketch in the sine curve.

(e) Shift the curve by the amount of the phase shift.

◆◆◆ **Example 8:** Make a quick sketch of $y = 2\sin(3x + 60°)$.

Solution: We have $a = 2$, $b = 3$, and $c = 60°$. So the amplitude $= |a| = 2$, the period is $P = 360°/3 = 120°$ and phase shift $= -60°/3 = -20°$. The steps for sketching the curve are shown in Fig. 15–9.

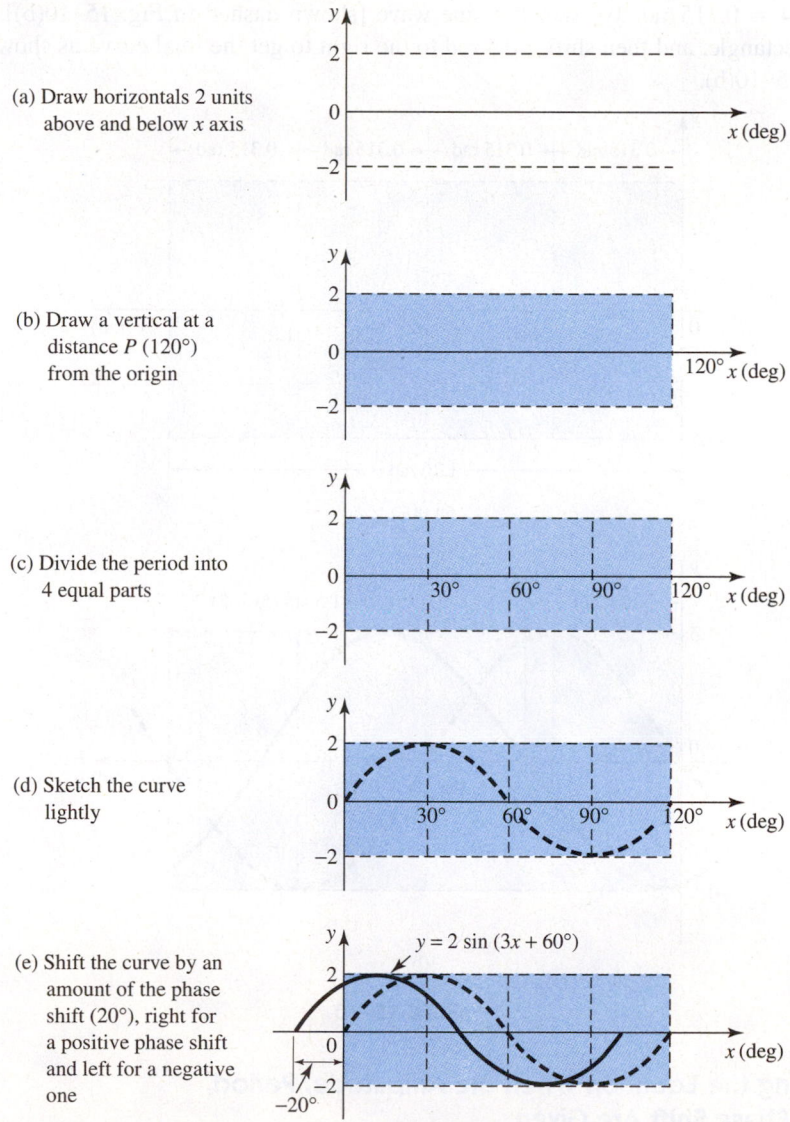

(a) Draw horizontals 2 units above and below *x* axis

(b) Draw a vertical at a distance *P* (120°) from the origin

(c) Divide the period into 4 equal parts

(d) Sketch the curve lightly

(e) Shift the curve by an amount of the phase shift (20°), right for a positive phase shift and left for a negative one

FIGURE 15–9 Quick sketching of the sine curve. We have just taken most of a page to describe the so-called *quick* method! You are right to be skeptical of its speed, but try a few and you will see that is really is faster.

We now do an example where the units are radians.   ◆◆◆

◆◆◆ **Example 9:** Make a quick sketch of

$$y = 1.5 \sin(5x - 2)$$

Solution: From the given equation,

$$a = 1.5, \qquad b = 5, \qquad \text{and} \qquad c = -2 \text{ rad}$$

so

$$\text{Amplitude} = 1.5$$

$$P = \frac{2\pi}{5} = 1.26 \text{ rad}$$

and

$$\text{phase shift} = -\left(\frac{-2}{5}\right) = 0.4 \text{ rad}$$

As shown in Fig. 15–10(a), we draw a rectangle whose height is $2 \times 1.5 = 3$ units and whose width is 1.26 rad; then we draw three verticals spaced apart by $1.26/4 = 0.315$ rad. We sketch a sine wave [shown dashed in Fig. 15–10(b)] into this rectangle, and then shift it 0.4 rad to the right to get the final curve as shown in Fig. 15–10(b).

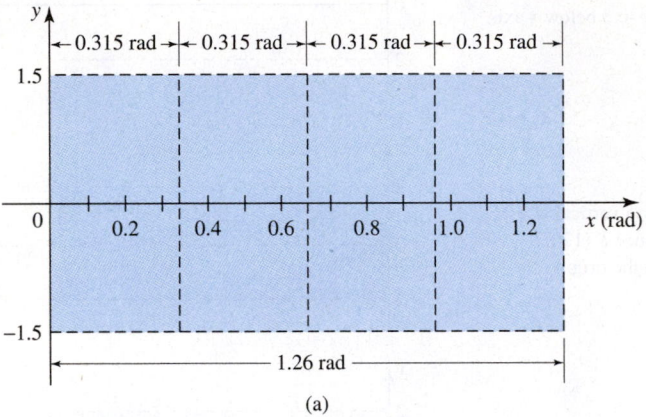

(a)

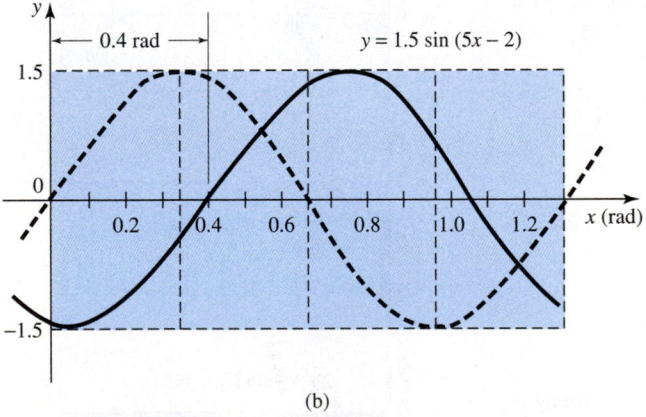

(b)

FIGURE 15–10   ◆◆◆

Writing the Equation When the Amplitude, Period, and Phase Shift Are Given

Let us now reverse the process, and write the equation when the three constants are given.

◆◆◆ **Example 10:** Write the equation of a sine wave for which $a = 5$, having a period of π, and a phase shift of -0.5 rad.

Solution: The period is given as π, so

$$P = \pi = \frac{2\pi}{b}$$

so

$$b = \frac{2\pi}{\pi} = 2$$

Then

$$\text{phase shift} = -0.5 = -\frac{c}{b}$$

Since $b = 2$, we get

$$c = -2(-0.5) = 1$$

Substituting into the general equation of the sine function

$$y = a \sin(bx + c)$$

with $a = 5$, $b = 2$, and $c = 1$ gives

$$y = 5 \sin(2x + 1) \qquad \blacklozenge\blacklozenge\blacklozenge$$

Writing the Equation When the Curve Is Given

Finally, let us write the equation from a given graph of the sine wave. This is useful, say, when interpreting an oscilloscope display. We will show this with an example.

◆◆◆ **Example 11:** Write the equation of the sine wave, Fig. 15–11.

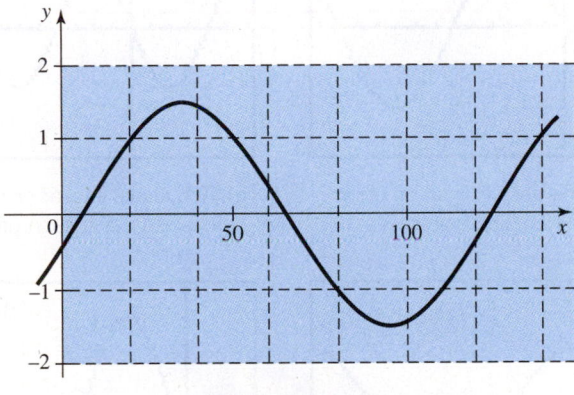

FIGURE 15–11

Solution: *Amplitude:* We read the amplitude from the graph

$$\text{amplitude} = |a| = 1.5$$

We could chose $a = +1.5$ or $a = -1.5$. Either is fine but would change the value of the phase shift we will get later. Let us choose $+1.5$ for simplicity. Our equation, so far, is then

$$y = 1.5 \sin (bx + c)$$

Period: We see from the graph that the curve repeats every $120°$, so the period P is $120°$. Then

$$b = \frac{360°}{P} = \frac{360°}{120°} = 3$$

Our equation, so far, is then

$$y = 1.5 \sin (3x + c)$$

Phase Shift: From the graph we see that the sine wave is shifted 5° to the right, so the phase shift = 5°. Then

$$c = -b \times \text{(phase shift)} = -3(5°) = -15°$$

Our final equation is then

$$y = 1.5 \sin(3x - 15°)$$ ◆◆◆

Exercise 2 ◆ Manual Graphing of the Sine Wave

1. Do a manual graph, either by plotting point-pairs or by the "quick" method, of any of the functions in Exercise 1, as directed by your instructor.

Writing the Equation, Given a, the Period, and the Phase Shift

2. Write the equation of a sine curve with $a = 3$, a period of 4π, and a phase shift of zero.
3. Write the equation of a sine curve with $a = -2$, a period of 6π, and a phase shift of $-\pi/4$.

Writing the Equation When the Curve Is Given

4. Write the equation of each sine wave.

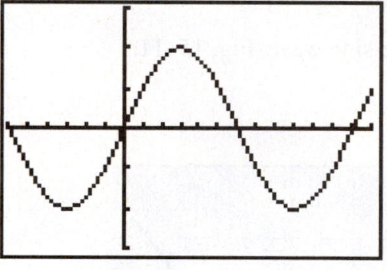

(a) Tick marks are 15° apart on the *x* axis and 1 unit apart on the *y* axis.

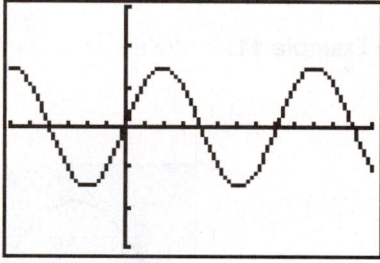

(b) Tick marks are 15° apart on the *x* axis and 1 unit apart on the *y* axis.

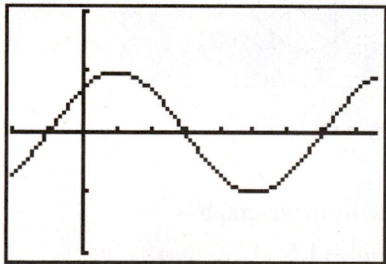

(c) Tick marks are 45° apart on the *x* axis and 1 unit apart on the *y* axis.

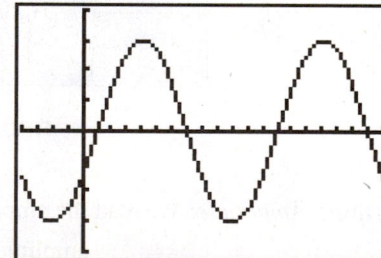

(d) Tick marks are 15° apart on the *x* axis and 1 unit apart on the *y* axis.

5. *Project, Composite Curves, Addition of Ordinates:* One way to graph a function containing several terms is to graph each term separately and then add (or subtract) them on the graph paper. This method, called the method of *addition of ordinates,* is especially useful when one or more terms of the expression to be graphed are trigonometric functions.

    Graph the curve $y = 8/x + \cos x$ by separately graphing $y = 8/x$ and $y = \cos x$, and graphically adding the ordinates of each to obtain the final curve. Here *x* is in radians.

15–3 The Sine Wave as a Function of Time

In our graphs of the sine wave so far, we have plotted y as the function of an *angle*. The units on the horizontal axis have been degrees or radians. But for many applications we have a periodic function that varies with *time*, rather than an angle. For example, an alternating voltage varies with time, so many cycles per *second*. The same is true of mechanical vibrations and other periodic phenomena. Fortunately we can rewrite all our previous definitions and formulas in terms of time.

The sine curve can be generated in a simple geometric way. Figure 15–12 shows a vector *OP* rotating counterclockwise with a constant angular velocity ω. A rotating vector is called a *phasor*. Its angular velocity ω is almost always given in radians per second (rad/s).

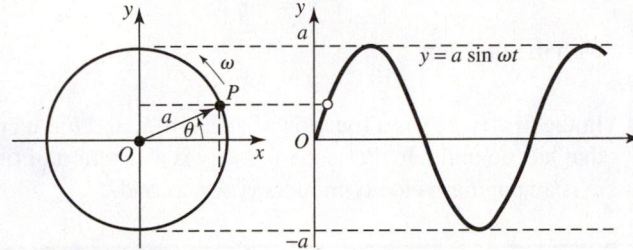

FIGURE 15–12 The sine curve generated by a rotating vector.

If the length of the phasor is a, then its projection on the y axis is

$$y = a \sin \theta$$

But since the angle θ at any instant t is equal to ωt,

$$y = a \sin \omega t$$

Further, if the phasor does not start from the x axis but has some *phase angle* ϕ, we get

$$y = a \sin(\omega t + \phi)$$

Notice in this equation that y *is a function of time*, rather than of an angle.

◆◆◆ **Example 12:** Write the equation for a sine wave generated by a phasor of length 8 rotating with an angular velocity of 300 rad/s and with a phase angle of 0.

Solution: Substituting, with $a = 8$, $\omega = 300$ rad/s, and $\phi = 0$, we have

$$y = 8 \sin 300t \qquad\qquad\qquad ◆◆◆$$

Period

Recall that the *period* was defined as the distance along the x axis taken by one cycle of the waveform. When the units on the x axis represented an angle, the period was in radians or degrees. Now that the x axis is time, the period will be in seconds. From Eq. 163,

$$P = \frac{2\pi}{b}$$

But in the equation $y = a \sin \omega t$, $b = \omega$, so we have the following formula:

Period of a Sine Wave	$P = \dfrac{2\pi}{\omega}$	1076

where P is in seconds and ω is in rad/s.

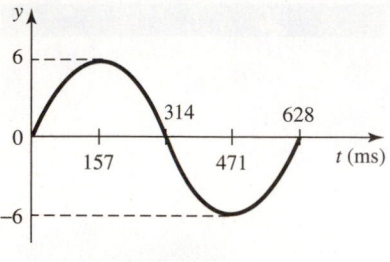

FIGURE 15–13

♦♦♦ Example 13: Find the period and the amplitude of the sine wave $y = 6 \sin 10t$, and make a graph with time t as the horizontal axis.

Solution: The period, from Eq. 1076, is

$$P = \frac{2\pi}{\omega} = \frac{2\pi}{10} = 0.628 \text{ s}$$

Thus it takes 628 ms for a full cycle, 314 ms for a half-cycle, 157 ms for a quarter-cycle, and so forth. This sine wave, with amplitude 6, is plotted in Fig. 15–13.   **♦♦♦**

Common Error	Don't confuse the function $$y = \sin bx$$ with the function $$y = \sin \omega t$$ In the first case, y is a function of an angle x, and b is a coefficient that has no units. In the second case, y is a function of time t, and ω is an angular velocity in radians per second.

Frequency

We earlier saw that the frequency f of a periodic waveform is equal to the reciprocal of the period P. So the frequency of a sine wave is given by the following equation:

Frequency of a Sine Wave	$$f = \frac{1}{P} = \frac{\omega}{2\pi} \text{ (hertz)}$$ where P is in seconds and ω is in rad/s. The unit of frequency is therefore cycles/s, or *hertz* (Hz). $$1 \text{ Hz} = 1 \text{ cycle/s}$$	**1077**

Higher frequencies are often expressed in kilohertz (kHz), where

$$1 \text{ kHz} = 10^3 \text{ Hz}$$

or in megahertz (MHz), where

$$1 \text{ MHz} = 10^6 \text{ Hz}$$

♦♦♦ Example 14: The frequency of the sine wave of Example 13 is

$$f = \frac{1}{P} = \frac{1}{0.628} = 1.59 \text{ Hz}$$ **♦♦♦**

When the period P is not wanted, the angular velocity can be obtained directly from Eq. 1077 by noting that $\omega = 2\pi f$.

♦♦♦ Example 15: Find the angular velocity of a 1000-Hz source.

Solution: From Eq. 1077,

$$\omega = 2\pi f = 2\pi(1000) = 6280 \text{ rad/s}$$ **♦♦♦**

A sine wave as a function of time can also have a phase shift, expressed in either degrees or radians, as in the following example.

◆◆◆ **Example 16:** Given the sine wave $y = 5.83 \sin(114t + 15°)$, find (a) the amplitude, (b) the angular velocity, (c) the period, (d) the frequency, and (e) the phase shift; and (f) make a graph.

Solution:

(a) The amplitude is 5.83 units.

(b) The angular velocity is $\omega = 114$ rad/s.

(c) From Eq. 1076, the period is

$$P = \frac{2\pi}{\omega} = \frac{2\pi}{114} = 0.0551 \text{ s} = 55.1 \text{ ms}$$

(d) From Eq. 1077, the frequency is

$$f = \frac{1}{P} = \frac{1}{0.0551} = 18.1 \text{ Hz}$$

(e) The phase shift o phase angle is 15°. It is not unusual to see a sine wave given with ωt in radians and the phase angle in degrees. Even though it is often done, some object to mixing degrees and radians in the same equation. If you do it, work with extra care. We find the phase shift, in units of time, the same way that we earlier found the phase shift for the general sine function. To find the value of t at which the positive half-cycle starts, we set y equal to zero and solve for t.

$$y = \sin(\omega t + \phi) = 0$$
$$\omega t = -\phi$$
$$t = -\frac{\phi}{\omega} = \text{phase shift}$$

Substituting our values for ω and ϕ (which we first convert to 0.262 radian) gives

$$\text{phase shift} = -\frac{0.262 \text{ rad}}{114 \text{ rad/s}} = -0.00230 \text{ s}$$

So our curve is shifted 2.30 milliseconds to the left.

(f) This sine wave is graphed in Fig. 15–14. We draw a rectangle of height 2(5.83) and width 55.1 ms. We subdivide the rectangle into four rectangles of equal width and sketch in the unshifted sine wave, shown dashed. We then shift the sine wave 2.30 ms to the left to get the final (solid) curve.

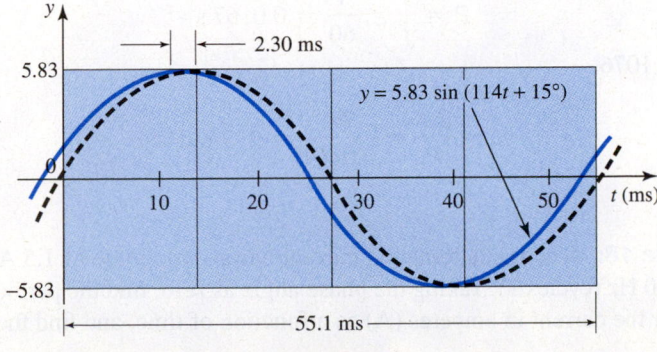

FIGURE 15–14

◆◆◆

Common Error	It's easy to get confused as to which direction to shift the curve. Check your work by substituting the value of an *x*-intercept (such as -0.00230 s in Example 16) back into the original equation. You should get a value of zero for *y*.

Alternating Current

When a loop of wire rotates in a magnetic field, any portion of the wire cuts the field while traveling first in one direction and then in the other direction. Since the polarity of the voltage induced in the wire depends on the direction in which the field is cut, the induced current will travel first in one direction and then in the other: it *alternates*.

The same is true with an armature more complex than a simple loop. We get an *alternating current*. Just as the rotating point *P* in Fig. 15–12 generated a sine wave, the current induced in the rotating armature will have the shape of a sine wave. This voltage or current can be described by the equation

$$y = a \sin(\omega t + \phi)$$

or, in electrical terms,

$$i = I_m \sin(\omega t + \phi_2) \qquad \text{(Eq. 1075)}$$

where *i* is the current at time *t*, I_m the maximum current (the amplitude), and ϕ_2 the phase angle. We have given the phase angle ϕ subscripts because, in a given circuit, the current and voltage waves will usually have different phase angles. Also, if we let V_m stand for the amplitude of the voltage wave, the instantaneous voltage *v* becomes

$$v = V_m \sin(\omega t + \phi_1) \qquad \text{(Eq. 1074)}$$

Equations 1076 and 1077 for the period and frequency still apply here.

$$\text{period} = \frac{2\pi}{\omega} \quad \text{(s)} \qquad \text{(Eq. 1076)}$$

$$\text{frequency} = \frac{1}{P} = \frac{\omega}{2\pi} \quad \text{(Hz)} \qquad \text{(Eq. 1077)}$$

◆◆◆ **Example 17:** Utilities in the United States supply alternating current at a frequency of 60 Hz. Find the angular velocity ω and the period *P*.

Solution: By Eq. 1077,

$$P = \frac{1}{f} = \frac{1}{60} = 0.0167 \text{ s}$$

and, by Eq. 1076,

$$\omega = \frac{2\pi}{P} = \frac{2\pi}{0.0167} = 377 \text{ rad/s} \qquad \qquad \text{◆◆◆}$$

◆◆◆ **Example 18:** A certain alternating current has an amplitude of 1.5 A and a frequency of 60 Hz (cycles/s). Taking the phase angle as zero, find the period, write the equation for the current in amperes (A) as a function of time, and find the current at $t = 0.01$ s.

Solution: From Example 17,

$$P = 0.0167 \text{ s} \quad \text{and} \quad \omega = 377 \text{ rad/s}$$

so the equation is

$$i = 1.5 \sin(377t)$$

When $t = 0.01$ s,

$$i = 1.5 \sin(377)(0.01) = -0.882 \text{ A}$$

as shown in Fig. 15–15. A calculator check is also shown, with the calculator in **RADIAN** mode. ◆◆◆

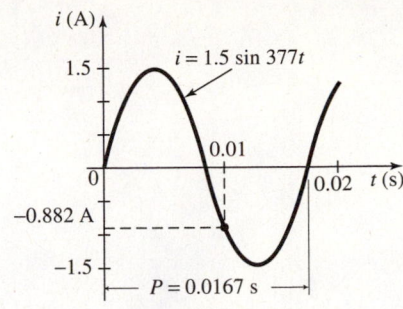

FIGURE 15–15

Phase Shift

When writing the equation of a *single* alternating voltage or current, we are usually free to choose the origin anywhere along the time axis, so we can place it to make the phase angle equal to zero. However, when there are *two* curves on the same graph that are *out of phase*, we usually locate the origin so that the phase angle of one curve is zero.

If the difference in phase between two sine waves is t seconds, then we often say that one curve *leads* the other by t seconds. Conversely, we could say that one curve *lags* the other by t seconds.

◆◆◆ **Example 19:** Figure 15–16 shows a voltage wave and a current wave, with the voltage wave leading the current wave by 5 ms. Write the equations for the two waves.

Solution: For the voltage wave,

$$V_m = 190, \quad \phi = 0, \quad \text{and} \quad P = 0.02 \text{ s}$$

By Eq. 1076,

$$\omega = \frac{2\pi}{0.02} = 100\pi = 314 \text{ rad/s}$$

so

$$v = 190 \sin 314t$$

where v is expressed in volts. For the current wave, $I_m = 2.5$ A. Since the curve is shifted to the right by 5 ms,

$$\text{phase shift} = 0.005 \text{ s} = -\frac{\phi}{\omega}$$

so the phase angle ϕ is

$$\phi = -(0.005 \text{ s})(100\pi \text{ rad/s}) = -\frac{\pi}{2} \text{ rad} = -90°$$

The angular frequency is the same as before, so with ϕ in degrees,

$$i = 2.5 \sin(314t - 90°)$$

where i is expressed in amperes. With ϕ in radians, our eequation becomes

$$i = 2.5 \sin\left(314t - \frac{\pi}{2}\right).$$ ◆◆◆

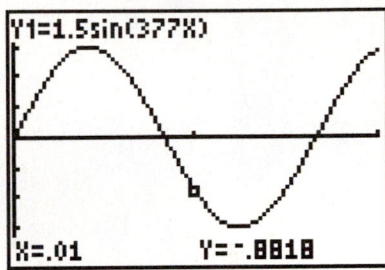

TI-83/84 check for Example 18, using TRACE.

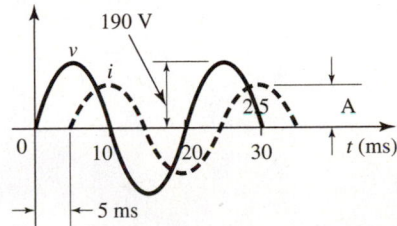

FIGURE 15–16

Exercise 3 ◆ The Sine Wave as a Function of Time

Find the period and the angular velocity of a repeating waveform that has a frequency of

1. 68.0 Hz.   **2.** 10.0 Hz.   **3.** 5000 Hz.

Find the frequency (in hertz) and the angular velocity of a repeating waveform whose period is

4. 1.00 s.   **5.** $\frac{1}{8}$ s.   **6.** 95.0 ms.

7. If a periodic waveform has a frequency of 60.0 Hz, how many seconds will it take to complete 200 cycles?

8. Find the frequency in Hz for a wave that completes 150 cycles in 10 s.

Find the period and the frequency of a sine wave that has an angular velocity of

9. 455 rad/s.   **10.** 2.58 rad/s.   **11.** 500 rad/s.

Find the period, amplitude, and phase angle for

12. The sine wave shown in Fig. 15–17(a).

13. The sine wave shown in Fig. 15–17(b).

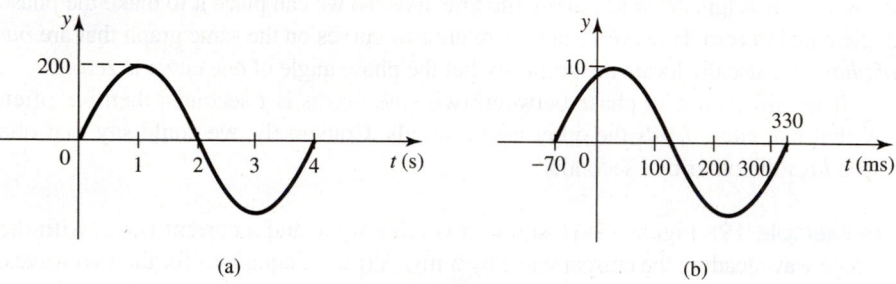

(a)   (b)

FIGURE 15–17

14. Write an equation for a sine wave generated by a phasor of length 5 rotating with an angular velocity of 750 rad/s and with a phase angle of 0°.

15. Repeat problem 14 with a phase angle of 15°.

Plot each sine wave.

16. $y = \sin t$ **17.** $y = 3 \sin 377t$

18. $y = 54 \sin(83t - 20°)$ **19.** $y = 375 \sin\left(55t + \dfrac{\pi}{4}\right)$

Mechanical Applications

20. The weight Fig. 15–18 is pulled down 2.50 in. and released. The distance x is given by

$$x = 2.50 \sin\left(\omega t + \frac{\pi}{2}\right)$$

Graph x as a function of t for two complete cycles. Take $\omega = 42.5$ rad/s.

21. The arm in the Scotch crosshead of Fig. 15–3 is rotating at a rate of 2.55 rev/s.
   (a) Write an equation for the displacement y as a function of time.
   (b) Graph that equation for two complete cycles.

Alternating Current

22. An alternating current has the equation

$$i = 25 \sin(635t - 18°)$$

where i is given in amperes, A. Find the maximum current, period, frequency, phase angle, and the instantaneous current at $t = 0.01$ s.

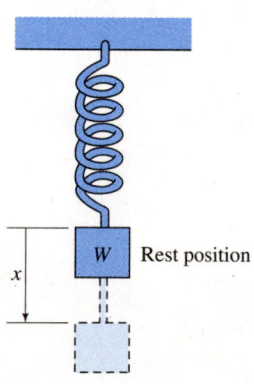

FIGURE 15–18 Weight hanging from a spring.

23. Given an alternating voltage $v = 4.27 \sin(463t + 27°)$, find the maximum voltage, the period, frequency, and phase angle, and the instantaneous voltage at $t = 0.12$ s.

24. Write an equation of an alternating voltage that has a peak value of 155 V, a frequency of 60 Hz, and a phase angle of 22°.

25. Write an equation for an alternating current that has a peak amplitude of 49.2 mA, a frequency of 35 Hz, and a phase angle of 63.2°.

26. *Project: Adding Sine Waves of the Same Frequency:* We have seen that $A \sin \omega t$ is the *y* component of a phasor of magnitude *A* rotating at angular velocity ω. Similarly, $B \sin(\omega t + \phi)$ is the *y* component of a phasor of magnitude *B* rotating at the same angular velocity ω, but with a phase angle ϕ between *A* and *B*. Since each is the *y* component of a phasor, their sum is equal to the sum of the *y* components of the two phasors, in other words, simply the *y* component of the resultant of those phasors.

Thus to add two sine waves of the same frequency, we simply *find the resultant of the phasors representing those sine waves.*

Verify that the sum of the two sine waves

$$y = 2.00 \sin \omega t \qquad \text{and} \qquad y = 3.00 \sin (\omega t + 60°)$$

is equal to

$$y = 4.43 \sin (\omega t + 36.6°)$$

27. *Writing: Doppler Effect:* As a train approaches, the pitch of its whistle seems higher than when the train recedes. This is called the *Doppler effect.* Read about and write a short paper on the Doppler effect, including an explanation of why it occurs.

28. *Project: Beats:* Beats occur when two sine waves of slightly different frequencies are combined.

    (a) Read about beats, and write a short paper on the subject.
    (b) Graph two sine waves, having frequencies of 24 Hz and 30 Hz, for about 20 cycles. Then graph the sum of those two sine waves.
    (c) From the preceding graphs, what can you deduce about the frequency of the beats?
    (d) Try to produce beats, using a musical instrument, tuning forks, or some other device.
    (e) Name one practical use for beats.

29. *Project:* Graph the function that is the sum of the following four functions:

$$y = \sin x$$
$$y = (1/3) \sin 3x$$
$$y = (1/5) \sin 5x$$
$$y = (1/7) \sin 7x$$

Use a graphics calculator or computer or manual addition of ordinates on graph paper.

What can you say about the shape of the composite curve? What do you think would happen if you added in more sine waves [$y = (1/9) \sin 9x$) and so on]?

15–4 Graphs of the Other Trigonometric Functions

The Cosine Wave

■ Exploration:

Try this. Graph at least one cycle of $y = \sin x$. Then on the same axes, graph $y = \cos x$.

How are the two waves similar? How do they differ? How would you obtain one from the other? ■

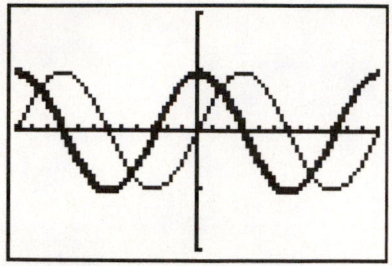

Calculator graphs of $y = \sin x$, shown light, and $y = \cos x$, shown heavy. Tick marks are 30° apart on the x axis and one unit apart on the y axis.

Cosine and Sine Curves Related

Note in Fig. 15–19 and in the calculator screen that the cosine curve and the sine curve have the *same shape*. In fact, the cosine curve appears to be identical to a sine curve shifted 90° to the left, or

$$\cos \theta = \sin(\theta + 90°) \tag{1}$$

We can show that Eq. 1 is true. We lay out the two angles θ and $\theta + 90°$ (Fig. 15–20), choose points P and Q so that $OP = OQ$, and drop perpendiculars PR and QS to the x axis. Since triangles OPR and OQS are congruent, we have $OR = QS$. The cosine of θ is then

$$\cos \theta = \frac{OR}{OP} = \frac{QS}{OQ} = \sin(\theta + 90°)$$

which verifies Eq. 1.

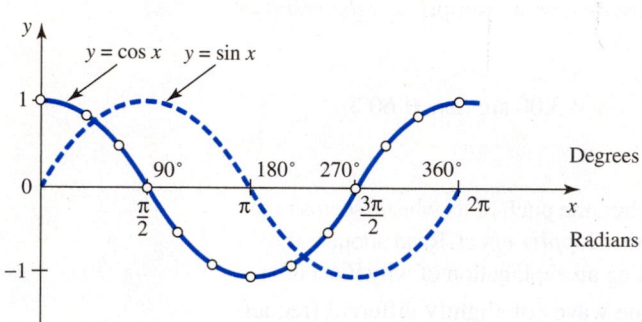

FIGURE 15–19   Manual graphs of $y = \sin x$ and $y = \cos x$.

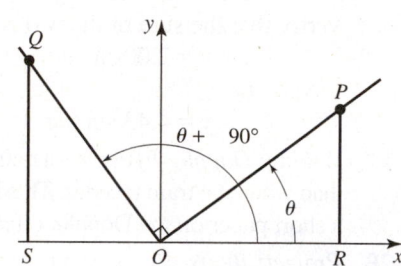

FIGURE 15–20

Graph of the General Cosine Function $y = a\cos(bx + c)$

The equation for the general cosine function is

$$y = a\cos(bx + c)$$

The amplitude, period, frequency, and phase shift are found the same way as for the sine curve, and, of course, the quick plotting method works here, too.

◆◆◆ **Example 20:** Find the amplitude, period, frequency, and phase shift for the curve $y = 3\cos(4x - 120°)$, and make a graph.

Solution: From the equation, $|a|$ = amplitude = 3. Since $b = 4$, the period is

$$\text{period} = \frac{360°}{4} = 90°/\text{cycle}$$

and

$$\text{frequency} = \frac{1}{P} = 0.0111 \text{ cycle/degree}$$

Since $c = -120°$,

$$\text{phase shift} = -\frac{c}{b} = -\frac{-120°}{4} = 30°$$

So we expect the curve to be shifted 30° to the right. The graph can be drawn manually or by calculator, as shown. ◆◆◆

Since the sine and cosine curves are identical except for phase shift, we can use either one to describe periodically varying quantities.

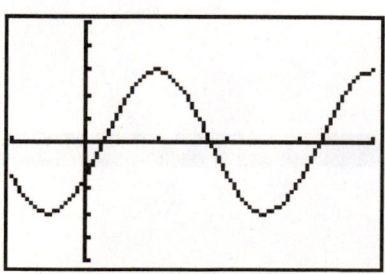

Screen for Example 20 shows the graph of $y = 3\cos(4x - 120°)$. Tick marks are 30° apart on the x axis and one unit apart on the y axis.

Graphs of the Tangent, Cotangent, Secant, and Cosecant Functions

For completeness, the graphs of all six trigonometric functions are shown in Fig. 15–21. To make them easier to compare, the same horizontal scale is used for each curve. Only the sine and cosine curves find much use in technology. Notice that

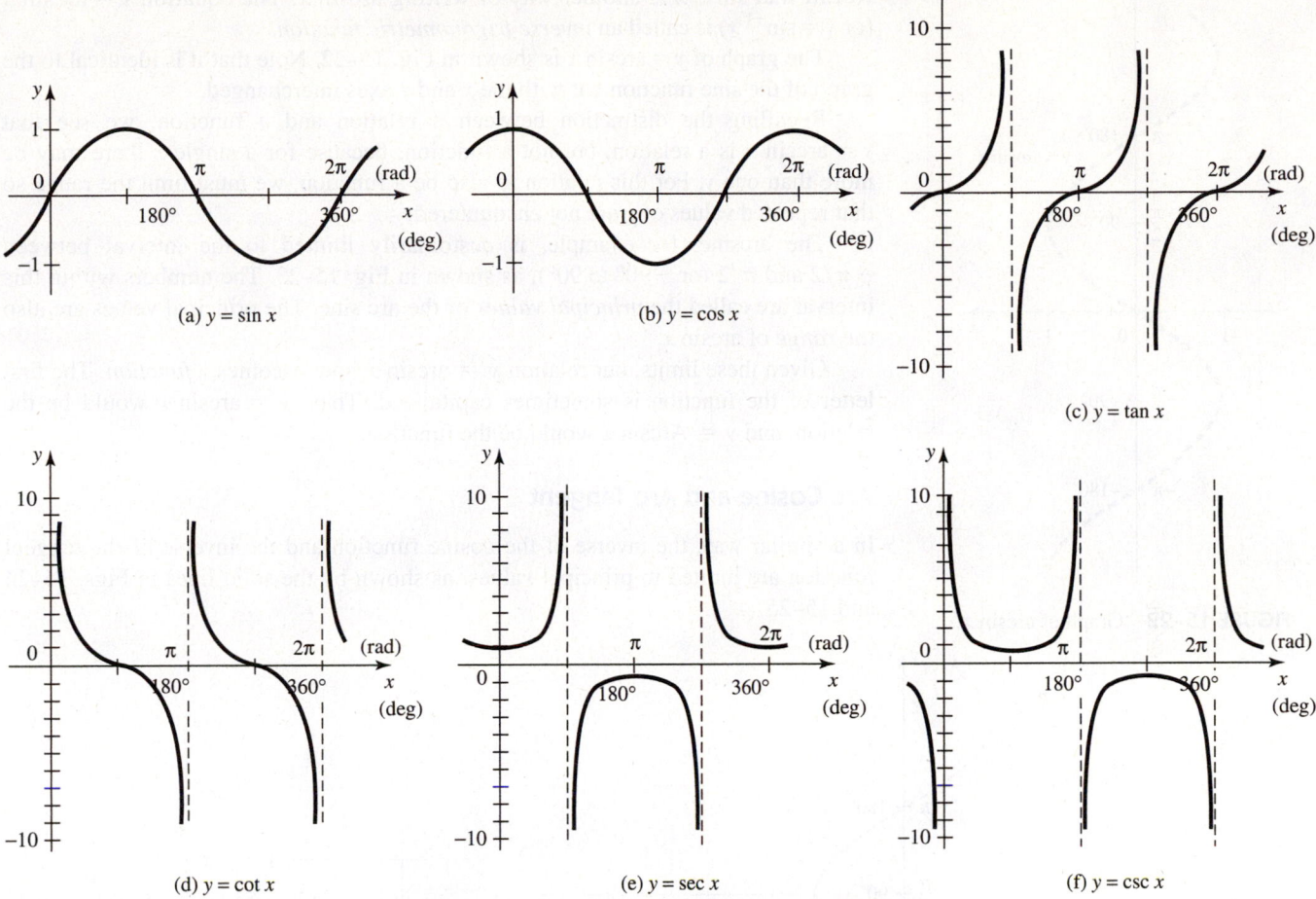

FIGURE 15–21 Graphs of the six trigonometric functions.

of the six, they are the only curves that have no "gaps" or "breaks." They are called *continuous* curves, while the others are called *discontinuous*.

If the need should arise to graph a tangent, cotangent, secant, or cosecant function, simply make a table of point pairs and plot them, or use a graphics calculator or graphing utility on a computer. With the calculator or the computer, you will need to use the reciprocal relations.

The Inverse Trigonometric Functions

If we interchange the variables of an equation, the new equation we get is called the *inverse* of the original.

◆◆◆ **Example 21:** The equation

$$x = y^2 + 5$$

is the inverse of the equation

$$y = x^2 + 5 \qquad \qquad ◆◆◆$$

◆◆◆ **Example 22:** The equation

$$y = \sin x$$

has the inverse

$$x = \sin y$$

Or, solving for *y*,

$$y = \arcsin x = \sin^{-1}x \qquad \qquad ◆◆◆$$

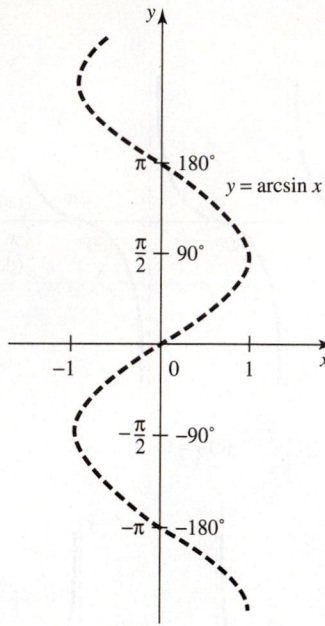

FIGURE 15–22   Graph of arcsin *x*.

Recall that $\sin^{-1} x$ is another way of writing arcsin *x*. The equation $y = $ arcsin *x* (or $y = \sin^{-1} x$) is called an *inverse trigonometric function*.

The graph of $y = $ arcsin *x* is shown in Fig. 15–22. Note that it is identical to the graph of the sine function but with the *x* and *y* axes interchanged.

Recalling the distinction between a relation and a function, we see that $y = $ arcsin *x* is a relation, but not a function, because for a single *x* there may be more than one *y*. For this relation to also be a function, we must limit the range so that repeated values of *y* are not encountered.

The arcsine, for example, is customarily limited to the interval between $-\pi/2$ and $\pi/2$ (or $-90°$ to $90°$), as shown in Fig. 15–23. The numbers within this interval are called the *principal values* of the arc sine. The principal values are also the *range* of arcsin *x*.

Given these limits, our relation $y = $ arcsin *x* now becomes a *function*. The first letter of the function is sometimes capitalized. Thus, $y = $ arcsin *x* would be the relation, and $y = $ Arcsin *x* would be the function.

Arc Cosine and Arc Tangent

In a similar way, the inverse of the cosine function and the inverse of the tangent function are limited to principal values, as shown by the solid lines in Figs. 15–24 and 15–25.

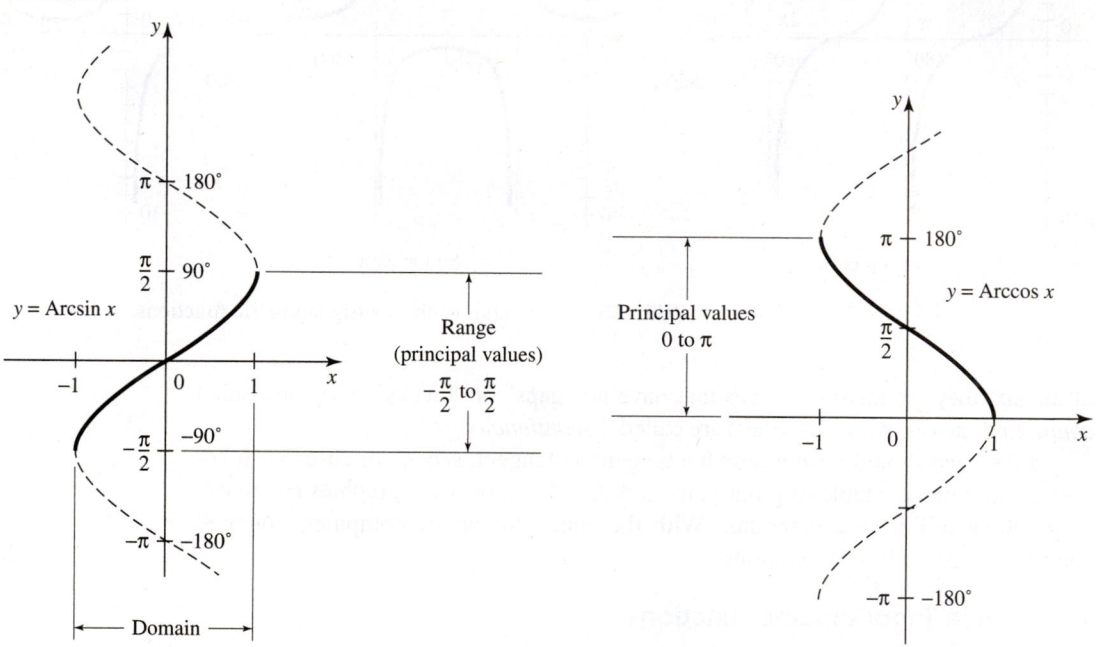

FIGURE 15–23   Principal values.   **FIGURE 15–24**

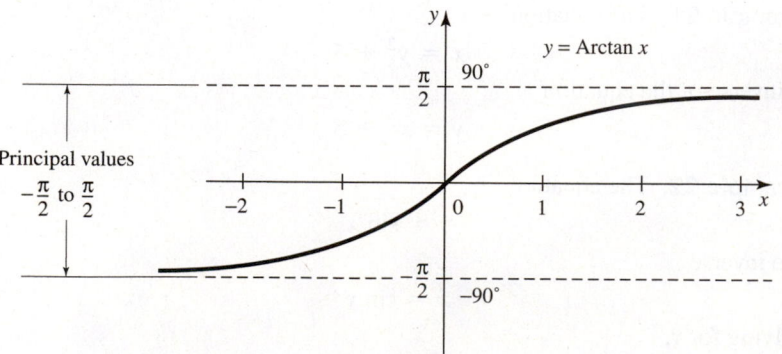

FIGURE 15–25

Graphing an Inverse Trigonometric Function by Calculator

We can graph an inverse trigonometric function manually by computing and plotting points, or by graphics calculator. Simply enter the function, as we did many times before, choose a suitable range, and graph. Be sure that the calculator is in **RADIAN** mode. Most calculators will automatically give the principal values.

◆◆◆ **Example 23:** Graph the inverse trigonometric function $y = \text{Arcsin } x$ by calculator.

Solution: We put the calculator into **RADIAN** mode and enter the function. We find a suitable viewing window by trial and error, and get the graph shown. ◆◆◆

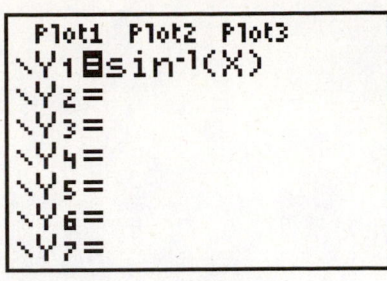

TI-83/84 screen for Example 23.

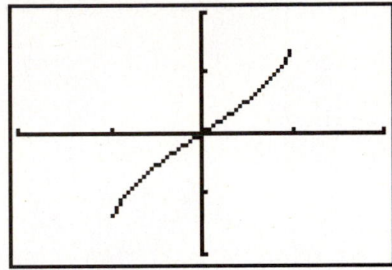

Calculator graph of $y = \text{Arcsin } x$. Tick marks on the x axis are 1 unit apart and on the y axis are one radian apart.

Exercise 4 ◆ Graphs of the Other Trigonometric Functions

The Cosine Curve

Make a complete graph of each function. Find the amplitude, period, and phase shift.

1. $y = 3 \cos x$　　　**2.** $y = -2 \cos x$　　　**3.** $y = \cos 3x$

4. $y = \cos 2x$　　　**5.** $y = 2 \cos 3x$　　　**6.** $y = 3 \cos 2x$

7. $y = \cos(x - 1)$　　**8.** $y = \cos(x + 2)$　　**9.** $y = 3 \cos\left(x - \dfrac{\pi}{4}\right)$

10. $y = 2 \cos(3x + 1)$

The Tangent Curve

Make a complete graph of each function.

11. $y = 2 \tan x$ **12.** $y = \tan 4x$ **13.** $y = 3 \tan 2x$

14. $y = \tan\left(x - \dfrac{\pi}{2}\right)$ **15.** $y = 2 \tan(3x - 2)$ **16.** $y = 4 \tan\left(x + \dfrac{\pi}{6}\right)$

Cotangent, Secant, and Cosecant Curves

Make a complete graph of each function.

17. $y = 2 \cot 2x$ **18.** $y = 3 \sec 4x$ **19.** $y = 3 \csc 3x$

20. $y = \cot(x - 1)$ **21.** $y = 2 \sec(3x + 1)$ **22.** $y = \csc(2x + 0.5)$

Applications

23. For the pendulum of Fig. 15–4:
    (a) Write an equation for the vertical distance y as a function of θ.
    (b) Graph $y = f(\theta)$ for $\theta = 0$ to $30°$.
24. Repeat problem 23 taking $\theta = 0$ when the pendulum is $20°$ from the vertical and increasing counterclockwise.
25. A rocket is rising vertically, as shown in Fig. 15–26, and is tracked by a radar dish at R.
    (a) Write an expression for the altitude y of the rocket as a function of θ.
    (b) Graph $y = f(\theta)$ for $\theta = 0$ to $60°$.
26. The ship in Fig. 15–27 travels in a straight line while keeping its searchlight trained on a reef.
    (a) Write an expression for the distance d as a function of the angle θ.
    (b) Graph $d = f(\theta)$ for $\theta = 0$ to $90°$.

27. If a weight W is dragged along a surface with a coefficient of friction f, as shown in Fig. 15–28, the force needed is

$$F = \frac{fW}{f \sin \theta + \cos \theta}$$

Graph F as a function of θ for $\theta = 0$ to $90°$. Take $f = 0.55$ and $W = 5.35$ kg.

28. *Project, Adding Sine and Cosine Waves of the Same Frequency:* In Sec. 15–4 we verified that

$$\cos \omega t = \sin(\omega t + 90°)$$

and saw that a cosine wave $A \cos \omega t$ is identical to the sine wave $A \sin \omega t$, except for a phase difference of $90°$ between the two curves (see Fig. 15–19). Thus we can find the sum $(A \sin \omega t + B \cos \omega t)$ of a sine and a cosine wave the same way that we added two sine waves in Exercise 3, problem 26: by finding the resultant of the phasors representing each sine wave.

Show that

Addition of a Sine Wave and a Cosine Wave	$A \sin \omega t + B \cos \omega t = R \sin(\omega t + \phi)$ where $R = \sqrt{A^2 + B^2}$ and $\phi = \arctan \dfrac{B}{A}$	166

29. *Project:* Verify that the sum of the two waves

$$y = 274 \sin \omega t \quad \text{and} \quad y = 371 \cos \omega t$$

is equal to

$$y = 461 \sin(\omega t + 53.6°)$$

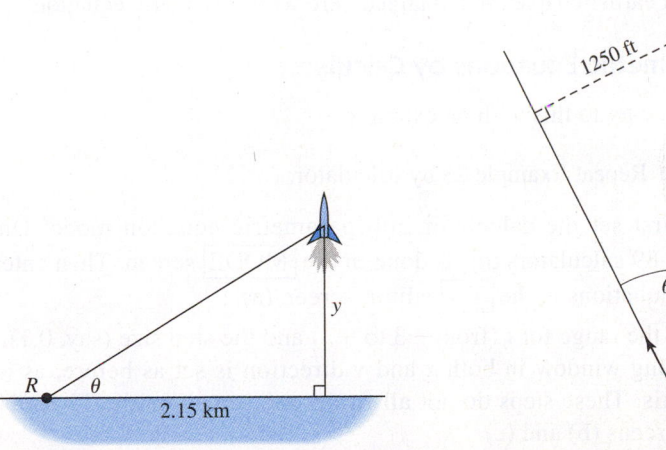

FIGURE 15–26 Tracking a rocket.

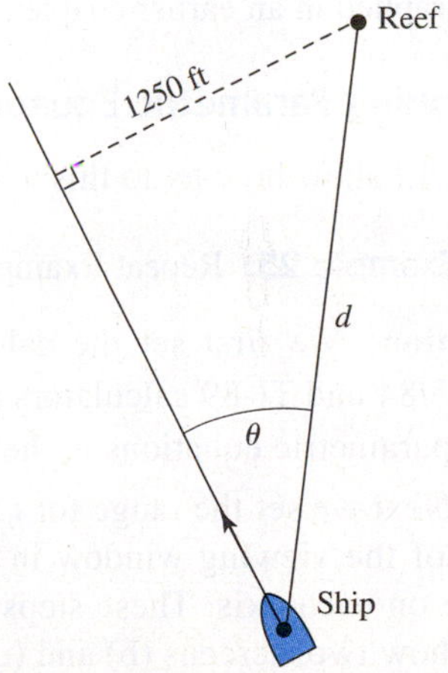

FIGURE 15–27

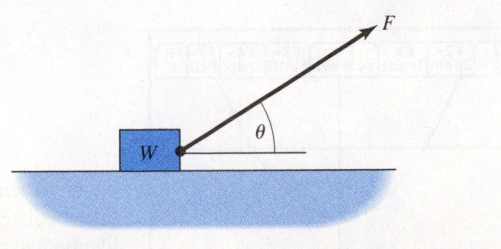

FIGURE 15–28 A weight dragged along a surface.

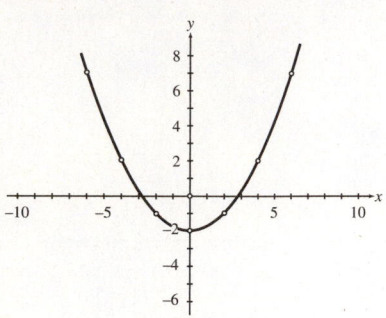

FIGURE 15–29

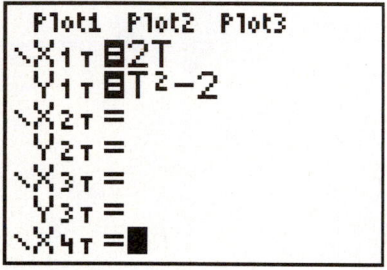

(a)

```
WINDOW
 Tmin=-3█
 Tmax=3
 Tstep=.1
 Xmin=-6
 Xmax=6
 Xscl=1
↓Ymin=-3
```

(b)

```
WINDOW
↑Tstep=.1
 Xmin=-6
 Xmax=6
 Xscl=1
 Ymin=-3
 Ymax=2
 Yscl=1
```

(c)

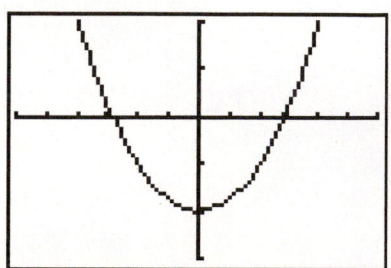

(d) Graph of $x = 2t$ and $y = t^2 - 2$. Tick marks are 1 unit apart on both axes.

15–5   Graphing a Parametric Equation

For the equations we have graphed so far, in rectangular coordinates, y has been expressed as a function of x.

$$y = f(x)$$

But x and y can also be related to each other by means of a third variable, say, t, if both x and y are given as functions of t.

$$x = g(t)$$

and

$$y = h(t)$$

Such equations are called *parametric equations*. The third variable t is called the *parameter*.

To graph parametric equations, we assign values to the parameter t and compute x and y for each t. We then plot the table of (x, y) pairs.

◆◆◆ **Example 24:** Graph the parametric equations

$$x = 2t \quad \text{and} \quad y = t^2 - 2$$

for $t = -3$ to 3.

Solution: We make a table with rows for t, x, and y. We take values of t from -3 to 3, and for each we compute x and y.

t	-3	-2	-1	0	1	2	3
x	-6	-4	-2	0	2	4	6
y	7	2	-1	-2	-1	2	7

We now plot the (x, y) pairs, $(-6, 7)$, $(-4, 2)$, . . . , $(6, 7)$ and connect them with a smooth curve (Fig. 15–29). The curve obtained is a parabola, the same curve we graphed in an earlier chapter, but obtained here with parametric equations.

◆◆◆

Graphing Parametric Equations by Calculator

We will show how to do this with an example.

◆◆◆ **Example 25:** Repeat Example 23 by calculator.

Solution: We first set the calculator into parametric equation mode. On the TI-83/84 and TI-89 calculators this is done in the MODE screen. Then enter the two parametric equations in the Y= editor, screen (a).

Next we set the range for t (from -3 to $+3$) and the step size (say, 0.1). The size of the viewing window in both x and y direction is set as before, as is the scale on each axis. These steps do not all fit on one screen on the TI-83/84. So we show two, screens (b) and (c).

Pressing GRAPH gives the same curve as we obtained by hand in the preceding example, screen (d). Note that we are in *rectangular* coordinates, not polar.

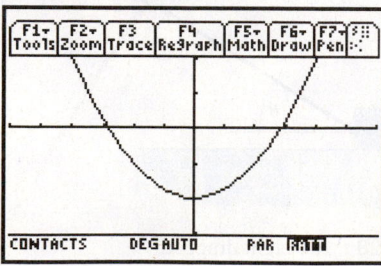

TI-89 screen for Example 25. The intermediate screens are almost identical to those for the TI-83/84 so we will not repeat them here.

◆◆◆

Graphing a Trigonometric Equation in Parametric Form

The procedure is the same when our parametric equations contain trigonometric expressions. We will do an example of graphing by hand, with a calculator graph for comparison.

◆◆◆ **Example 26:** Plot these parametric equations for θ from 0 to 2π radians.

$$x = 3 \cos \theta$$
$$y = 2 \sin 2\theta$$

Solution: We select values for θ and compute x and y.

Point	1	2	3	4	5	6	7	8	9
θ	0	$\dfrac{\pi}{4}$	$\dfrac{\pi}{2}$	$\dfrac{3\pi}{4}$	π	$\dfrac{5\pi}{4}$	$\dfrac{3\pi}{2}$	$\dfrac{7\pi}{4}$	2π
x	3	2.12	0	−2.12	−3	−2.12	0	2.12	3
y	0	2	0	−2	0	2	0	−2	0

Each (x, y) pair is plotted in Fig. 15–30.

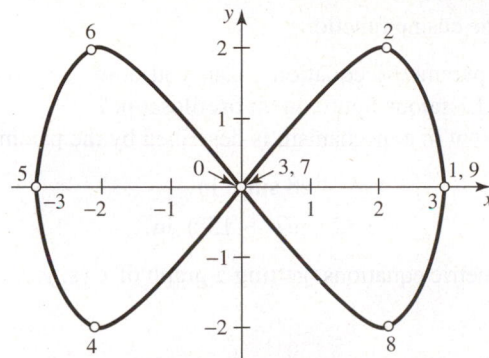

FIGURE 15–30 Patterns of this sort can be obtained by applying ac voltages to both the horizontal and the vertical deflection plates of an oscilloscope. Called *Lissajous figures,* they can indicate the relative amplitudes, frequencies, and phase angles of the two applied voltages.

◆◆◆

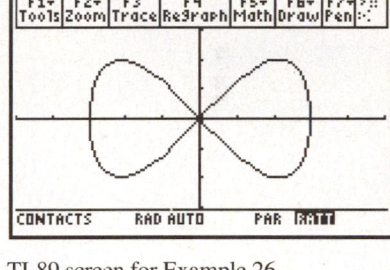

TI-83/84 screen for Example 26. Note that the calculator had to be set into **RADIAN** mode for this graph, as θ was given in radians.

TI-89 screen for Example 26.

Exercise 5 ◆ Graphing a Parametric Equation

Graphing a Parametric Equation by Calculator

Graph the following parametric equations for values of t from −3 to 3.

1. $x = t, y = t$
2. $x = 3t, y = t^2$
3. $x = -t, y = 2t^2$
4. $x = -2t, y = t^2 + 1$

Graphing a Trigonometric Equation in Parametric Form

Graph each pair of parametric equations.

5. $x = \sin \theta$
   $y = \sin \theta$

6. $x = \sin \theta$
   $y = 2 \sin \theta$

7. $x = \sin \theta$
   $y = \sin 2\theta$

8. $x = \sin \theta$
   $y = \sin 3\theta$

9. $x = \sin \theta$
$y = \sin(\theta + \pi/4)$

10. $x = \sin \theta$
$y = \sin(2\theta - \pi/6)$

11. $x = \sin 2\theta$
$y = \sin 3\theta$

12. $x = \sin \theta$
$y = 1.32 \sin 2\theta$

13. $x = 5.83 \sin 2\theta$
$y = 4.24 \sin \theta$

14. $x = 6.27 \sin \theta$
$y = 4.83 \sin(\theta + 0.527)$

Applications

15. Create a series of Lissajous figures as follows: For each, let our reference wave be $y_1 = \sin t$. Then for y_2 choose

  (a) $y_2 = 2 \sin t$ (double the amplitude of the reference wave)
  (b) $y_2 = \sin 2t$ (double the frequency of the reference wave)
  (c) $y_2 = \sin 3t$ (triple the frequency of the reference wave)
  (d) $y_2 = \sin 4t$ (quadruple the frequency of the reference wave)
  (e) $y_2 = \sin(t + 45°)$ (a phase shift of 45° from the reference wave)
  (f) $y_2 = \sin(t + 135°)$ (a phase shift of 135° from the reference wave)
  (g) $y_2 = \sin(t + 90°)$ (a phase shift of 90° from the reference wave)
  (h) $y_2 = \cos t$ (the cosine function)

Plot each pair of parametric equations. Can you draw any conclusions about how to interpret a Lissajous figure on an oscilloscope?

16. The motion of a point in a mechanism is described by the parametric equations:

$$x = 4.88 \sin t \text{ in.}$$
$$y = 3.82 \sin(t + 15°) \text{ in.}$$

Graph these parametric equations, getting a graph of x vs. y.

17. *Trajectories:* If air resistance is neglected, a projectile will move horizontally with constant velocity and fall with constant acceleration like any falling body. Thus if the projectile is launched with an initial horizontal velocity of 453 ft/s and an initial vertical velocity of 593 ft/s, the parametric equations of motion will be:

$$x = 453t \text{ ft} \quad \text{and} \quad y = 593t - 16.1t^2 \text{ ft}$$

(a) Graph these equations to get the trajectory of the projectile.
From the graph, determine

(b) the projectile's maximum height.

(c) the x distance for which the height is a maximum.

(d) the projectile's maximum distance, assuming that the ground is level.

(e) the height when $x = 5000$ ft.

18. Many parametric curves, Fig 15–31, find use in gearing and in machine design. Using the given equations, make a graph of the

(a) cycloid   (b) trochoid   (c) prolate cycloid

(d) hypocycloid   (e) involute of a circle

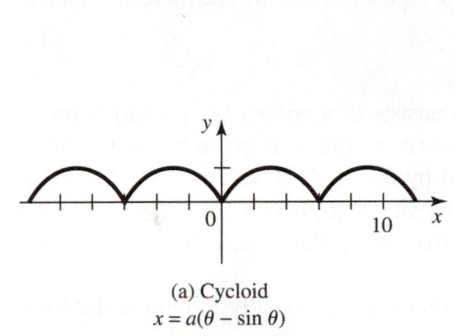

(a) Cycloid
$x = a(\theta - \sin \theta)$
$y = a(1 - \cos \theta)$

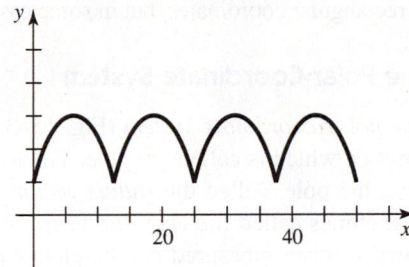

(b) Trochoid
$x = a\theta - b \sin \theta$
$y = a - b \cos \theta$
where $b < a$. (Graphed with $a = 2$, $b = 1$)

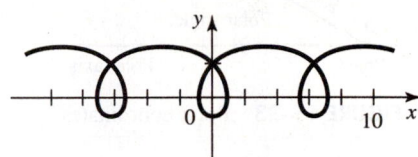

(c) Prolate cycloid
$x = a\theta - b \sin \theta$
$y = a - b \cos \theta$
(graphed with $a = 1$, $b = 2$)

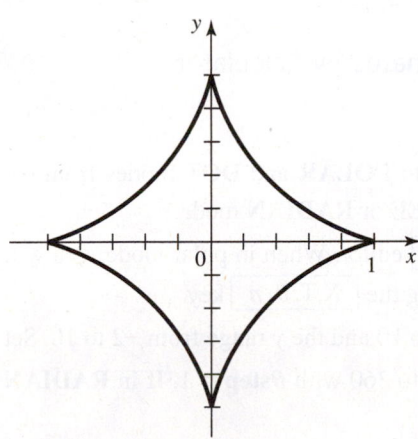

(d) Hypocycloid or astroid
$x = a \cos^3 \theta$
$y = a \sin^3 \theta$

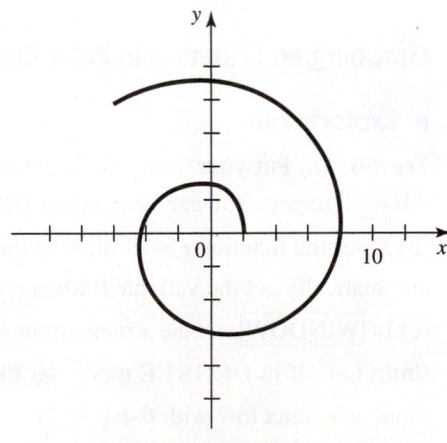

(e) Involute of a circle
$x = a(\cos \theta + \theta \sin \theta)$
$y = a(\sin \theta - \theta \cos \theta)$

FIGURE 15–31 Graphs of some parametric equations, in rectangular coordinates. Unless otherwise noted, the curves were graphed with $a = 1$. When a circle of radius r rolls without slipping along a straight line,

(a) a point on the circumference generates a *cycloid*

(b) a point on the radius, located between 0 and r, generates a *trochoid*

(c) a point on the extension of the radius generates a *prolate cycloid*

(d) The *hypocycloid* is the path of a point on the circumference of a circle as it rolls without slipping on the inside of a larger circle.

(e) The *involute of a circle* is the path of the end of a taut string as it is unwound from a spool.

19. *Project:* Many of the curves, such as the cycloid, can be demonstrated using simple models made of cardboard or thin wood. Make one or more and show them in class.

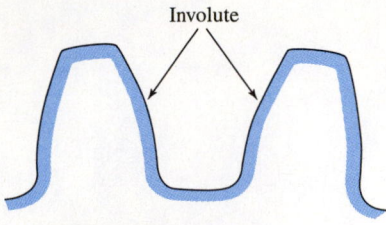

FIGURE 15–32 Spur gear teeth.

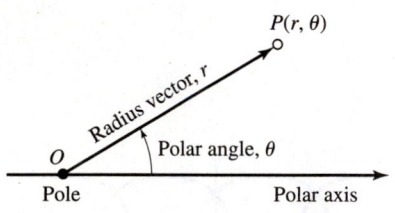

FIGURE 15–33 Polar coordinates.

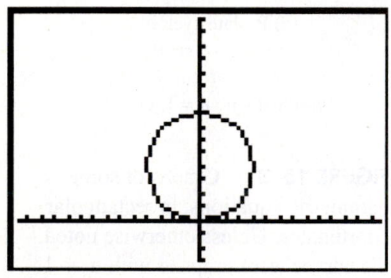

TI-83/84 graph of $r = 8 \sin \theta$, a circle. On TI calculators, polar mode is **Pol** on the MODE menu. Selecting **ZSquare** will adjust the scales so that circles appear circular.

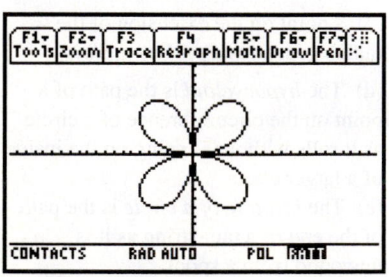

TI-89 screen for Example 27. When entering the equation choose the variable θ, a second function on the ^ key. Choose ZOOM **Square** to get the display shown

20. *CAD:* Rather than use a model, generate a curve like the cycloid using a computer drawing program, like the Geometer's Sketchpad®. Such a program allows you to put figures into motion, giving a live demonstration of how the curve is generated.

21. *Project:* Learn how to use Lissajous figures on the oscilloscope to analyze sine waves. Then borrow an oscilloscope and a signal generator and give a demonstration to your class.

22. *Writing:* The *involute* is the curve used for the shape of the teeth of a spur gear, Fig. 15–32. Write a short paper explaining what an involute is, and why it is chosen for this purpose.

15–6 Graphing in Polar Coordinates

Up to this point we have done all our graphing in the familiar rectangular coordinate system. We now introduce a new coordinate system, which is more useful than rectangular coordinates for some kinds of graphing. Most of our graphing will continue to be in rectangular coordinates, but in some cases polar coordinates will be more convenient.

The Polar Coordinate System

The *polar coordinate system* (Fig. 15–33) consists of a *polar axis*, passing through point O, which is called the *pole*. The location of a point P is given by its distance r from the pole, called the *radius vector*, and by the angle θ, called the *polar angle* (sometimes called the *vectorial* angle or *reference* angle). The polar angle is called *positive* when measured counterclockwise from the polar axis, and *negative* when measured clockwise.

The *polar coordinates* of a point P are thus r and θ, usually written in the form $P(r, \theta)$, or as $r\underline{/\theta}$ (read "r at an angle of θ").

◆◆◆ **Example 27:** A point at a distance 5 from the origin with a polar angle of 28° can be written

$$P(5, 28°) \quad \text{or} \quad 5\underline{/28°} \qquad \qquad ◆◆◆$$

Graphing an Equation in Polar Coordinates by Calculator

■ **Exploration:**

Try this. (a) Put your TI-83/84 calculator into **POLAR** and **DOT** modes from the MODE screen. You can be in either **DEGREE** or **RADIAN** mode.

(b) Enter the function $r = 8 \sin \theta$ in the Y= editor. When in polar mode, you will automatically get the variable θ when pressing the X,T,θ, n key.

(c) In WINDOW, set the x range from −10 to 10 and the y range from −2 to 10. Set θ**min** to 0. If in **DEGREE** mode, set θ**max** to 360 with θ**step** = 1. If in **RADIAN** mode set θ**max** to 7 with θ**step** = .01.

(d) Choose ZOOM **ZSquare** to make the scale spacings in x equal to the spacing in y.

(e) Press GRAPH. ■

Describe the graph you get. Does it look anything like the sine waves we graphed earlier?

◆◆◆ **Example 28:** Make a polar graph of the four-petaled rose, $r = 8 \sin 2\theta$ on the TI-89.

Solution:

(a) Put your calculator into **POLAR** mode from the MODE screen.

(b) Enter the function $r = 8 \sin 2\theta$ in the Y= editor. Use the variable θ, a second function on the ∧ key. While still in the Y= editor, press F6 and choose **Dot**.

(c) In WINDOW, set the x range from −10 to 10 and the y range from −10 to 10. Set **θmin** to 0. If in **DEGREE** mode, set **θmax** to 360 with **θstep** = 1. If in **RA-DIAN** mode set **θmax** to 7 with **θstep** = .01.

(d) Choose ZOOM **ZoomSqr** to make the scale spacings in x equal to the spacing in y.

(e) Press GRAPH.   ◆◆◆

Graphing an Equation in Polar Coordinates Manually

For manual graphing in polar coordinates, it is convenient, although not essential, to have *polar coordinate graph paper*, as in Fig. 15–35. This paper has equally

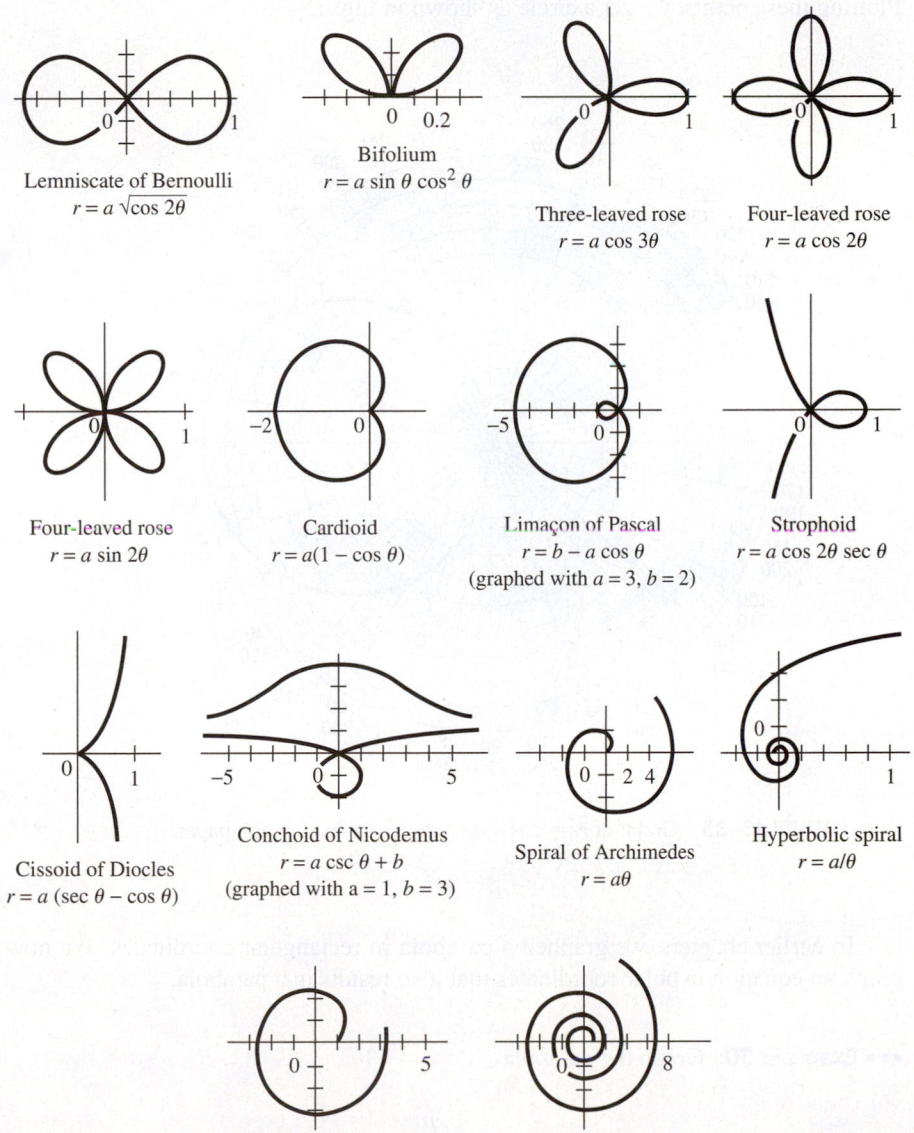

Lemniscate of Bernoulli
$r = a \sqrt{\cos 2\theta}$

Bifolium
$r = a \sin \theta \cos^2 \theta$

Three-leaved rose
$r = a \cos 3\theta$

Four-leaved rose
$r = a \cos 2\theta$

Four-leaved rose
$r = a \sin 2\theta$

Cardioid
$r = a(1 - \cos \theta)$

Limaçon of Pascal
$r = b - a \cos \theta$
(graphed with $a = 3$, $b = 2$)

Strophoid
$r = a \cos 2\theta \sec \theta$

Cissoid of Diocles
$r = a (\sec \theta - \cos \theta)$

Conchoid of Nicodemus
$r = a \csc \theta + b$
(graphed with $a = 1$, $b = 3$)

Spiral of Archimedes
$r = a\theta$

Hyperbolic spiral
$r = a/\theta$

Parabolic spiral
$r = a \sqrt{\theta} + b$
(graphed with $a = b = 1$)

Logarithmic spiral
$r = e^{a\theta}$
(graphed with $a = 0.1$)

FIGURE 15–34 Some curves in polar coordinates. Unless otherwise noted, the curves were graphed with $a = 1$.

spaced concentric circles for the radii, and an angular scale for the angle, in degrees or radians. To plot a point $P(r, \theta)$, first assign a suitable scale to the radii. Then place a point on the graph at a radius r and angle θ. A point with a radius of $(-r)$ is plotted in the *opposite direction* to $(+r)$.

To graph a function $r = f(\theta)$, simply assign convenient values to θ and compute the corresponding value for r. Then plot the resulting table of point pairs.

◆◆◆ Example 29: Graph the function $r = \cos \theta$.

Solution: Let us take values for θ every 30° and make a table.

θ	0	30°	60°	90°	120°	150°	180°	210°	240°	270°	300°	330°	360°
r	1	0.87	0.5	0	−0.5	−0.87	−1	−0.87	−0.5	0	0.5	0.87	1

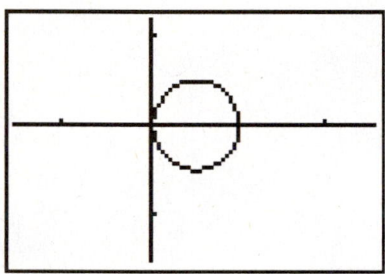

Calculator screen for Example 29. For comparison, here is a calculator graph of $r = \cos \theta$. Tick marks are one unit apart.

Plotting these points, we get a circle as shown in Fig. 15–35.

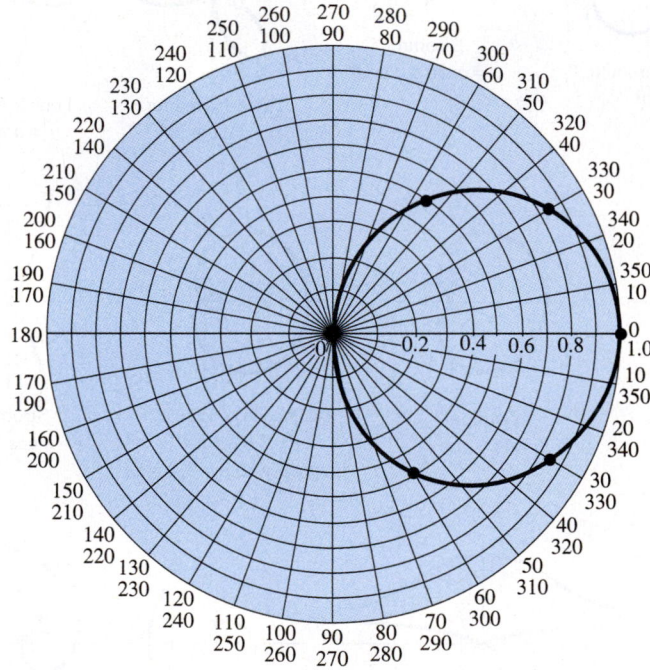

FIGURE 15–35   Graph of $r = \cos \theta$ on polar coordinate graph paper.   ◆◆◆

In earlier chapters, we graphed a parabola in rectangular coordinates. We now graph an equation in polar coordinates that also results in a parabola.

◆◆◆ Example 30: Graph the parabola

$$r = \frac{p}{1 - \cos \theta}$$

with $p = 1$.

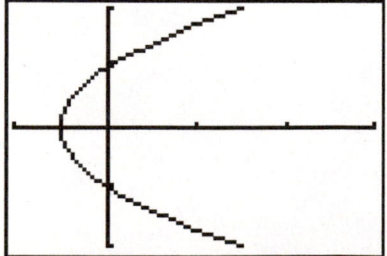

Calculator screen for Example 30: A graph of the parabola, for comparison. Tick marks are 1 unit apart.

Solution: As before, we compute r for selected values of θ.

θ	0°	30°	60°	90°	120°	150°	180°	210°	240°	270°	300°	330°	360°
r		7.46	2.00	1.00	0.667	0.536	0.500	0.536	0.667	1.00	2.00	7.46	

Note that we get division by zero at $\theta = 0°$ and 360°, so that the curve does not exist there. Plotting these points (except for 7.46, which is off the graph), we get the parabola shown in Fig. 15–36. A calculator plot is also shown. ◆◆◆

Transforming Between Polar and Rectangular Coordinates

We can easily see the relationships between rectangular coordinates and polar coordinates when we put both systems on a single diagram (Fig. 15–37). Using the trigonometric functions and the Pythagorean theorem, we get the following equations:

Rectangular	$x = r \cos \theta$	119
	$y = r \sin \theta$	120
Polar	$r = \sqrt{x^2 + y^2}$	121
	$\theta = \arctan \dfrac{y}{x}$	122

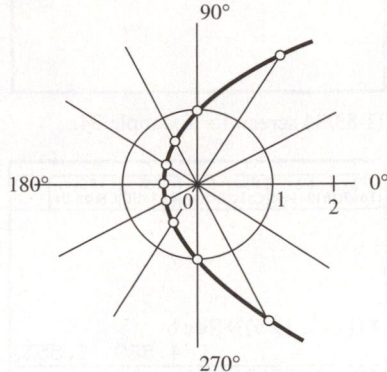

FIGURE 15–36

◆◆◆ **Example 31:** What are the polar coordinates of $P(3, 4)$?

Solution:

$$r = \sqrt{9 + 16} = 5$$
$$\theta = \arctan \frac{4}{3} = 53.1°$$

So the polar coordinates of P are $(5, 53.1°)$.

Recall that we used the same computation to find the resultant of two perpendicular vectors. Thus the resultant of the vector having an x component of 3 and a y component of 4 is $5 \underline{/53.1°}$. ◆◆◆

◆◆◆ **Example 32:** The polar coordinates of a point are $(8, 125°)$. What are the rectangular coordinates?

Solution:

$$x = 8 \cos 125° = -4.59$$
$$y = 8 \sin 125° = 6.55$$

So the rectangular coordinates are $(-4.59, 6.55)$.

Recall that we used the same method to find the x and y components of a vector. Thus the vector $8 \underline{/125°}$ has an x component of -4.59 and a y component of 6.55. ◆◆◆

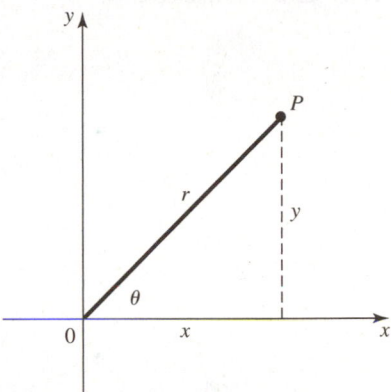

FIGURE 15–37 Rectangular and polar coordinates of a point.

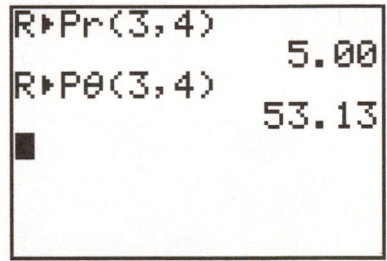

TI-83/84 screen for Example 33.

Transforming Between Rectangular and Polar Coordinates by Calculator

Some calculators can convert between rectangular and polar coordinates.

◆◆◆ **Example 33:** Repeat Example 31 by calculator.

Solution: *On the TI-83/84:* (a) Select **R ▶ Pr** from the ANGLE menu. Enter the rectangular coordinates to be converted, in parentheses. Press ENTER to get the value of r.

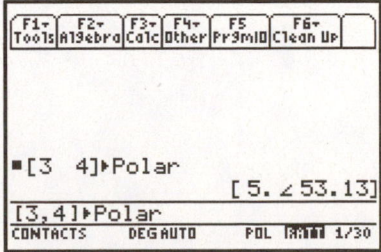

TI-89 screen for Example 33.

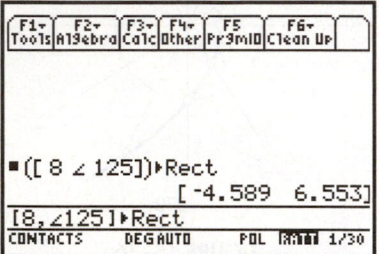

```
P▸Rx(8,125)
                -4.59
P▸Ry(8,125)
                 6.55
```

TI-83/84 screen for Example 34.

```
F1▾  F2▾ F3▾ F4▾ F5   F6▾
Tools Algebra Calc Other PrgmIO Clean Up



■([ 8 ∠ 125])▸Rect
                [-4.589   6.553]
[8,∠125]▸Rect
CONTACTS    DEG AUTO    POL [RAD] 1/30
```

TI-89 screen for Example 34.

(b) Select **R ▶ Pθ** from the ANGLE menu. Enter the rectangular coordinates to be converted, in parentheses. Press ENTER to get the value of θ. The angle will be in degrees or radians depending upon your MODE setting.

On the TI-89: (a) Enter the rectangular coordinates to be converted, in brackets. (b) Select **R ▶ Polar** from the MATH **Matrix/Vector ops** menu. (c) Press ≈ to get decimal values. The angle will be in degrees or radians, depending upon your MODE setting.   ◆◆◆

◆◆◆ **Example 34:** Repeat Example 32 by calculator.

Solution: *On the TI-83/84:* (a) Select **R ▶ Px** from the ANGLE menu. Enter the polar coordinates to be converted, in parentheses. The angle will enter in degrees or radians, depending upon your MODE setting. Press ENTER to get the value of *x*.
(b) Select **R ▶ Py** from the ANGLE menu. Enter the polar coordinates to be converted, in parentheses. Press ENTER to get the value of *y*.
On the TI-89: (a) Enter the polar coordinates to be converted, in brackets. The angle will enter in degrees or radians, depending upon your MODE setting
(b) Select **P ▶ Rect** from the MATH **Matrix/Vector ops** menu. (c) Press ≈ to get decimal values.   ◆◆◆

Transforming an Equation

We may also use Eqs. 119 through 122 to transform *equations* from one system of coordinates to the other.

◆◆◆ **Example 35:** Transform the polar equation $r = \cos \theta$ to rectangular coordinates.

Solution: Multiplying both sides by *r* yields

$$r^2 = r \cos \theta$$

But, by Eq. 121, $r^2 = x^2 + y^2$, and by Eq. 119, $r \cos \theta = x$, so

$$x^2 + y^2 = x \qquad \text{◆◆◆}$$

◆◆◆ **Example 36:** Transform the rectangular equation $2x + 3y = 5$ into polar form.

Solution: By Eqs. 119 and 120,

$$2r \cos \theta + 3r \sin \theta = 5$$

or

$$r(2 \cos \theta + 3 \sin \theta) = 5$$

$$r = \frac{5}{2 \cos \theta + 3 \sin \theta} \qquad \text{◆◆◆}$$

Exercise 6 ♦ Graphing in Polar Coordinates

Plot each point in polar coordinates.

1. $(4, 35°)$ **2.** $(3, 120°)$ **3.** $(2.5, 215°)$

4. $(3.8, 345°)$ **5.** $\left(2.7, \dfrac{\pi}{6}\right)$ **6.** $\left(3.9, \dfrac{7\pi}{8}\right)$

7. $\left(-3, \dfrac{\pi}{2}\right)$ **8.** $\left(4.2, \dfrac{2\pi}{5}\right)$ **9.** $(3.6, -20°)$

10. $(-2.5, -35°)$ **11.** $\left(-1.8, -\dfrac{\pi}{6}\right)$ **12.** $\left(-3.7, -\dfrac{3\pi}{5}\right)$

Graphing an Equation in Polar Coordinates

Graph each function in polar coordinates.

13. $r = 2 \cos \theta$ **14.** $r = 3 \sin \theta$ **15.** $r = 3 \sin \theta + 3$

16. $r = 2 \cos \theta - 1$ **17.** $r = 3 \cos 2\theta$ **18.** $r = 2 \sin 3\theta$

19. $r = \sin 2\theta - 1$ **20.** $r = 2 + \cos 3\theta$

21. Graph any of the curves in Fig. 15–34.

Transforming Between Rectangular and Polar Coordinates

Write the polar coordinates of each point.

22. (2.00, 5.00)

23. (3.00, 6.00)

24. (1.00, 4.00)

25. (4.00, 3.00)

26. (2.70, −1.80)

27. (−4.80, −5.90)

28. (207, 186)

29. (−312, −509)

30. (1.08, −2.15)

Write the rectangular coordinates of each point.

31. (5.00, 47.0°)

32. (6.30, 227°)

33. (445, 312°)

34. $\left(3.60, \dfrac{\pi}{5} \right)$

35. $\left(-4.00, \dfrac{3\pi}{4} \right)$

36. $\left(18.3, \dfrac{2\pi}{3} \right)$

37. (15.0, −35.0°)

38. (−12.0, −48.0°)

39. $\left(-9.80, -\dfrac{\pi}{5} \right)$

Transforming an Equation

Write each polar equation in rectangular form.

40. $r = 6$

41. $r = 2 \sin \theta$

42. $r = \sec \theta$

43. $r^2 = 1 - \tan \theta$

44. $r(1 - \cos \theta) = 1$

45. $r^2 = 4 - r \cos \theta$

Write each rectangular equation in polar form.

46. $x = 2$ **47.** $y = -3$
48. $x = 3 - 4y$ **49.** $x^2 + y^2 = 1$
50. $3x - 2y = 1$ **51.** $y = x^2$

Applications

52. The hole locations for the steel plate, Fig. 15–38, are to be programmed into a numerically controlled jig borer. The turntable on the borer requires that the holes be located by giving the angle θ of each hole and the distance r to each hole, as measured from hole H. Convert the dimensions of each hole.

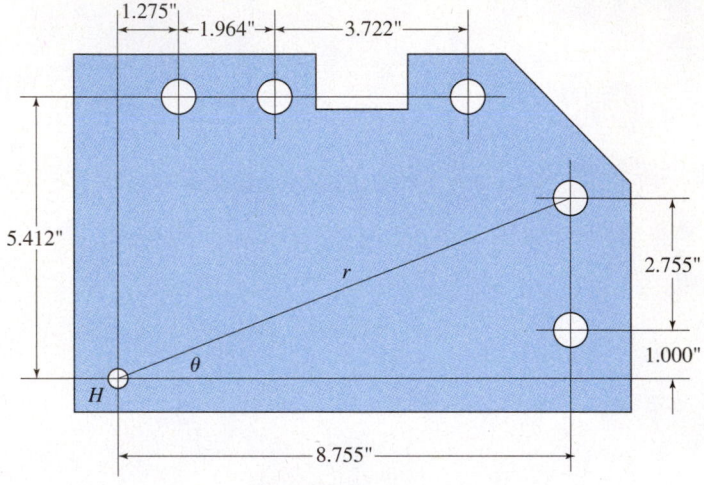

FIGURE 15–38

53. The plate, Fig. 15–39, is to have seven tack welds at the given locations. The programmable welder uses an X-Y positioning system, with coordinates as shown. Convert the location of each weld to rectangular coordinates.

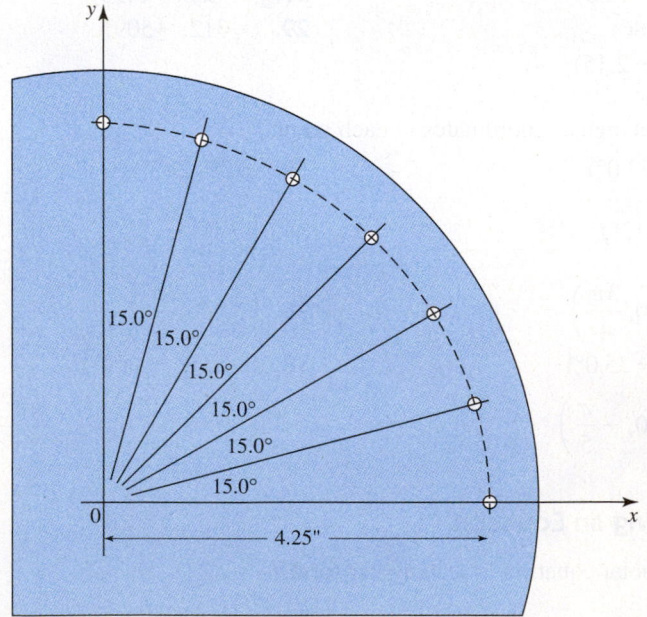

FIGURE 15–39

54. Each vane in an impeller, Fig. 15–40, has the polar equation

$$r = 4.25 \cos \theta$$

Transform this polar equation to rectangular form so that it may be used in a numerically controlled milling machine that uses X-Y positioning.

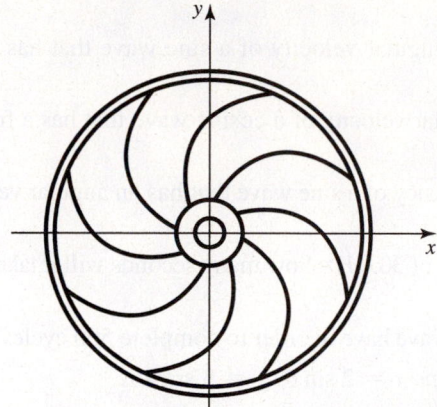

FIGURE 15–40

••• **CHAPTER 15 REVIEW PROBLEMS** ••••••••••••••••••••••••••••••••••••

Graph one cycle of each curve and find the period, amplitude, and phase shift.

1. $y = 3 \sin 2x$

2. $y = 5 \cos 3x$

3. $y = 1.5 \sin \left(3x + \dfrac{\pi}{2} \right)$

4. $y = -5 \cos \left(x - \dfrac{\pi}{6} \right)$

5. $y = 2.5 \sin \left(4x + \dfrac{2\pi}{9} \right)$

6. $y = 4 \tan x$

7. Write the equation of a sine curve with $a = 5$, a period of 3π, and a phase shift of $-\dfrac{\pi}{6}$.

Plot each point in polar coordinates.

8. $(3.4, 125°)$

9. $(-2.5, 228°)$

10. $\left(1.7, -\dfrac{\pi}{6} \right)$

Graph each equation in polar coordinates.

11. $r^2 = \cos 2\theta$

12. $r + 2 \cos \theta = 1$

Write the polar coordinates of each point, to three significant digits.

13. $(7, 3)$

14. $(-5.30, 3.80)$

15. $(-24, -52)$

Write the rectangular coordinates of each point.

16. $(3.80, 48.0°)$

17. $\left(-65, \dfrac{\pi}{9} \right)$

18. $(3.80, -44.0°)$

Transform into rectangular form.

19. $r(1 - \cos \theta) = 2$ **20.** $r = 2 \cos \theta$

Transform into polar form.

21. $x - 3y = 2$

22. $5x + 2y = 1$

23. Find the frequency and the angular velocity of a sine wave that has a period of 2.50 s.

24. Find the period and the angular velocity of a cosine wave that has a frequency of 120 Hz.

25. Find the period and the frequency of a sine wave that has an angular velocity of 44.8 rad/s.

26. If a sine wave has a frequency of 30.0 Hz, how many seconds will it take to complete 100 cycles?

27. What frequency must a sine wave have in order to complete 500 cycles in 2.0 s?

28. Graph the parametric equations $x = 2 \sin \theta$, $y = 3 \sin 4\theta$.

29. Write an equation for an alternating current that has a peak amplitude of 92.6 mA, a frequency of 82.0 Hz, and a phase angle of 28.3°.

30. Write the equation for a mechanical vibration that has an amplitude of 1.5 mm, a frequency of 155 Hz, and a phase angle of 0.

31. Given an alternating voltage $v = 27.4 \sin (736t + 37.0°)$, find the maximum voltage, period, frequency, phase angle, and the instantaneous voltage at $t = 0.250$ s.

32. *Project:* A force F (Fig. 15–41) pulls the weight along a horizontal surface. If f is the coefficient of friction, then

$$F = \frac{fW}{f \sin \theta + \cos \theta}$$

Graphically find the value of θ, to two significant digits, so that the least force is required. Do this for values of f of 0.50, 0.60, and 0.70.

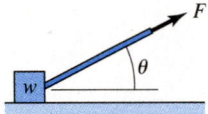

FIGURE 15–41

Trigonometric Identities and Equations

◆◆◆ OBJECTIVES ◆◆

When you have completed this chapter, you should be able to

- Write a trigonometric expression in terms of the sine and cosine.
- Simplify a trigonometric expression using the fundamental identities.
- Prove trigonometric identities using the fundamental identities.
- Simplify expressions or prove identities using the sum or difference formulas, the double-angle formulas, or the half-angle formulas.
- Evaluate trigonometric expressions.
- Solve trigonometric equations.

◆◆◆

In mathematics we usually try to simplify expressions as much as possible. In earlier chapters, we simplified algebraic expressions of all sorts. In this, our final chapter on trigonometry, we will simplify *trigonometric* expressions. For example, an expression such as tan x cos x simplifies to sin x.

For this we need to know how various trigonometric functions are related. We will start with the simplest (and most useful) *fundamental identities*. These identities are equations relating one trigonometric expression to another. Using them, we can replace one expression with another that will lead to a simpler result. We then proceed to trigonometric expressions containing sums and differences of two angles, double angles, and half angles.

This is followed by a short section on evaluating trigonometric expressions and another on solving trigonometric equations. We approximately found roots of a trigonometric equation by calculator in the preceding chapter, and here we learn how to do an exact solution. For example, we know how to find the vertical and horizontal displacements of a projectile, Fig. 16–1, given the initial velocity and the launch angle θ. But how would we solve for θ, given the other quantities? We will learn how in this chapter.

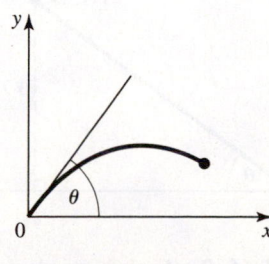

FIGURE 16–1

16–1 Fundamental Identities

In this section we will give the identities that we will use throughout this chapter. We will start with some we already know and go on to derive some new ones.

Reciprocal Relations

We have already encountered the reciprocal relations earlier, and we repeat them here.

Reciprocal Relations	$\sin \theta = \dfrac{1}{\csc \theta}$ or $\csc \theta = \dfrac{1}{\sin \theta}$ or $\sin \theta \csc \theta = 1$	**117a**
	$\cos \theta = \dfrac{1}{\sec \theta}$ or $\sec \theta = \dfrac{1}{\cos \theta}$ or $\cos \theta \sec \theta = 1$	**117b**
	$\tan \theta = \dfrac{1}{\cot \theta}$ or $\cot \theta = \dfrac{1}{\tan \theta}$ or $\tan \theta \cot \theta = 1$	**117c**

In our next set of examples, we will show how to *simplify* a trigonometric expression, that is, to rewrite it as an equivalent expression with fewer terms, and eliminating fractions where possible. A good way to simplify many expressions is to change all their functions to only sines and cosines.

♦♦♦ Example 1: Rewrite the expression

$$\frac{\cos \theta}{\sec^2 \theta}$$

to one containing only sines and cosines, and simplify. Recall that $\sec^2 \theta$ is another way of writing $(\sec \theta)^2$.

Solution: Using Eq. 117b gives us

$$\frac{\cos \theta}{\sec^2 \theta} = \cos \theta \, (\cos^2 \theta) = \cos^3 \theta \qquad \text{♦♦♦}$$

Quotient Relations

■ Exploration:

Try this. In the same viewing window, graph

$$y_1 = \frac{\sin x}{\cos x}$$

and

$$y_2 = \tan x$$

using a heavier line for y_2. What do you see? Can you propose a trigonometric identity based on your observations? ■

Figure 16–2 shows an angle θ in standard position, as when we first defined the trigonometric functions. We see that

$$\sin \theta = \frac{y}{r} \quad \text{and} \quad \cos \theta = \frac{x}{r}$$

Dividing yields

$$\frac{\sin \theta}{\cos \theta} = \frac{\dfrac{y}{r}}{\dfrac{x}{r}} = \frac{y}{x}$$

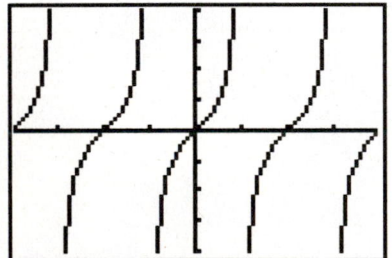

Screen for the exploration.

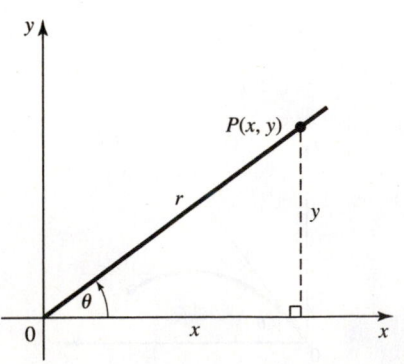

FIGURE 16–2 Angle in standard position.

But $y/x = \tan\theta$, so we have the following identities:

Quotient Relations	$\tan\theta = \dfrac{\sin\theta}{\cos\theta}$	123
	$\cot\theta = \dfrac{\cos\theta}{\sin\theta}$	124

where $\cot\theta$ is found by taking the reciprocal of $\tan\theta$.

◆◆◆ Example 2: Rewrite the expression

$$\frac{\cot\theta}{\csc\theta} - \frac{\tan\theta}{\sec\theta}$$

so that it contains only the sine and cosine, and simplify.

Solution:

$$\frac{\cot\theta}{\csc\theta} - \frac{\tan\theta}{\sec\theta} = \frac{\cos\theta}{\sin\theta}\cdot\frac{\sin\theta}{1} - \frac{\sin\theta}{\cos\theta}\cdot\frac{\cos\theta}{1}$$

$$= \cos\theta - \sin\theta \qquad\qquad \text{◆◆◆}$$

Pythagorean Relations

■ **Exploration:**

Try this. In the same viewing window, graph

$$y_1 = \sin^2 x$$
$$y_2 = \cos^2 x$$
$$y_3 = y_1 + y_2$$

Describe what you see. Do you find your result surprising? Can you suggest a trigonometric identity based on your observations? ■

We can get three more relations by applying the Pythagorean theorem to the triangle in Fig. 16–2.

$$x^2 + y^2 = r^2$$

Dividing through by r^2, we get

$$\frac{x^2}{r^2} + \frac{y^2}{r^2} = 1$$

or

$$\left(\frac{x}{r}\right)^2 + \left(\frac{y}{r}\right)^2 = 1$$

But $x/r = \cos\theta$ and $y/r = \sin\theta$, so we get:

Pythagorean Relation	$\sin^2\theta + \cos^2\theta = 1$	125

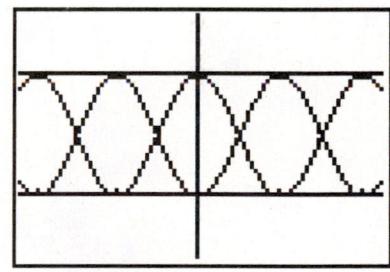

Screen for the exploration. On the TI-83/84, you will find Y1 and Y2 on the VARS Y-VARS, **Function** menu. On the TI-89, simply type Y1(x) and Y2(x) from the keyboard.

We can get a second Pythagorean relation by dividing Eq. 125 through by $\cos^2 \theta$.

$$\sin^2 \theta + \cos^2 \theta = 1$$

$$\frac{\sin^2 \theta}{\cos^2 \theta} + \frac{\cos^2 \theta}{\cos^2 \theta} = \frac{1}{\cos^2 \theta}$$

or

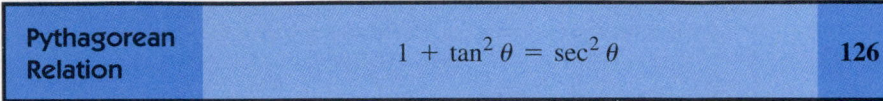

Pythagorean Relation	$1 + \tan^2 \theta = \sec^2 \theta$	126

Finally, we get a third Pythagorean relation by dividing Eq. 125 through by $\sin^2 \theta$.

$$\sin^2 \theta + \cos^2 \theta = 1$$

$$\frac{\sin^2 \theta}{\sin^2 \theta} + \frac{\cos^2 \theta}{\sin^2 \theta} = \frac{1}{\sin^2 \theta}$$

or

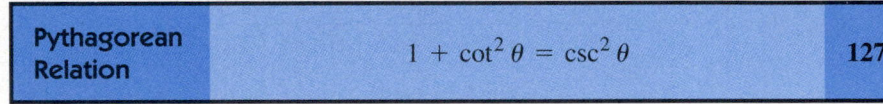

Pythagorean Relation	$1 + \cot^2 \theta = \csc^2 \theta$	127

◆◆◆ **Example 3:** Simplify

$$\sin^2 \theta - \csc^2 \theta - \tan^2 \theta + \cot^2 \theta + \cos^2 \theta + \sec^2 \theta$$

Solution: By the Pythagorean relations,

$$\sin^2 \theta - \csc^2 \theta - \tan^2 \theta + \cot^2 \theta + \cos^2 \theta + \sec^2 \theta$$
$$= (\sin^2 \theta + \cos^2 \theta) - (\tan^2 \theta - \sec^2 \theta) + (\cot^2 \theta - \csc^2 \theta)$$
$$= 1 - (-1) + (-1) = 1$$ ◆◆◆

Simplifying a Trigonometric Expression

One use of the trigonometric identities is the simplification of expressions, as in the preceding examples. We now give a few more examples.

◆◆◆ **Example 4:** Simplify $(\cot^2 \theta + 1)(\sec^2 \theta - 1)$.

Solution: We start by replacing $(\cot^2 \theta + 1)$ by $\csc^2 \theta$ and $(\sec^2 \theta - 1)$ by $\tan^2 \theta$

$$(\cot^2 \theta + 1)(\sec^2 \theta - 1) = \csc^2 \theta \tan^2 \theta$$

But $\csc^2 \theta = \dfrac{1}{\sin^2 \theta}$ and $\tan^2 \theta = \dfrac{\sin^2 \theta}{\cos^2 \theta}$, so

$$\csc^2 \theta \tan^2 \theta = \frac{1}{\sin^2 \theta} \cdot \frac{\sin^2 \theta}{\cos^2 \theta}$$

$$= \frac{1}{\cos^2 \theta}$$

Finally, since $\dfrac{1}{\cos \theta} = \sec \theta$,

$$(\cot^2 \theta + 1)(\sec^2 \theta - 1) = \sec^2 \theta$$ ◆◆◆

Common Error	It easy to forget *algebraic* operations when working with trigonometric expressions. We still need to factor, to combine fractions over a common denominator, and so on.

◆◆◆ **Example 5:** Simplify $\dfrac{1 - \sin^2 \theta}{\sin \theta + 1}$.

Solution: Factoring the difference of two squares in the numerator gives

$$\frac{1 - \sin^2 \theta}{\sin \theta + 1} = \frac{(1 - \sin \theta)(1 + \sin \theta)}{\sin \theta + 1} = 1 - \sin \theta \qquad ◆◆◆$$

◆◆◆ **Example 6:** Simplify $\dfrac{\cos \theta}{1 - \sin \theta} + \dfrac{\sin \theta - 1}{\cos \theta}$.

Solution: Combining the two fractions over a common denominator, we have

$$\frac{\cos \theta}{1 - \sin \theta} + \frac{\sin \theta - 1}{\cos \theta} = \frac{\cos^2 \theta + (\sin \theta - 1)(1 - \sin \theta)}{(1 - \sin \theta) \cos \theta}$$

$$= \frac{\cos^2 \theta + \sin \theta - \sin^2 \theta - 1 + \sin \theta}{(1 - \sin \theta) \cos \theta}$$

Replacing $\cos^2\theta$ with $1 - \sin^2\theta$ and collecting terms gives

$$\frac{-2 \sin^2 \theta + 2 \sin \theta}{(1 - \sin \theta) \cos \theta}$$

Factoring the numerator,

$$\frac{2 \sin \theta(-\sin \theta + 1)}{(1 - \sin \theta) \cos \theta}$$

$$= \frac{2 \sin \theta}{\cos \theta}$$

Finally, since $\dfrac{\sin \theta}{\cos \theta} = \tan \theta$, we get

$$\frac{\cos \theta}{1 - \sin \theta} + \frac{\sin \theta - 1}{\cos \theta} = 2 \tan \theta \qquad ◆◆◆$$

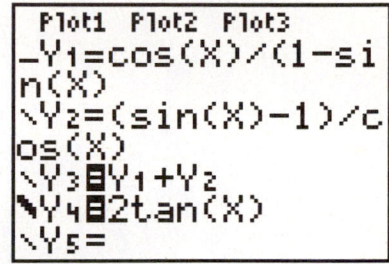

TI-83/84 check for Example 6. We defined the two parts of the given expression as Y1 and Y2, and combined them as Y3, finding them in the $\boxed{\text{VARS}}$ menu. We deselected Y1 and Y2 so that they will not graph. We do this by moving the cursor to the equals sign in the function to be deselected and press $\boxed{\text{ENTER}}$. Then we have printed Y3 thin and Y4 heavy.

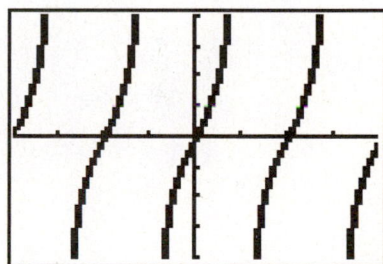

Plot of the given identity, thin, overlaid by its presumed equivalent, heavy.

Simplifying a Trigonometric Expression by Calculator

Some calculators such as the TI-89 can simplify a trigonometric expression. They may even simplify them automatically, as in the following example.

◆◆◆ **Example 7:** Simplify by calculator,

$$\frac{\sin x}{\cos x}$$

Solution: We simply enter the expression and press $\boxed{\text{ENTER}}$. The simplified expression tan x appears in the display. ◆◆◆

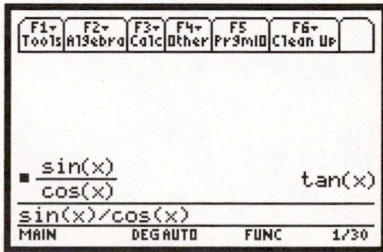

TI-89 screen for Example 7.

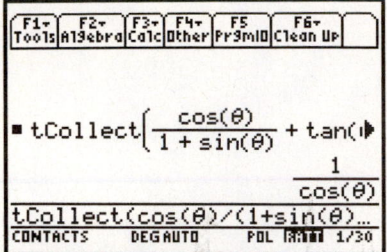

TI-89 screen for Example 8.

If the calculator will not simplify an expression by default, try using **tCollect** or **tExpand** from the **Algebra/Trig** menu. These operation works best if the calculator is in **RADIAN** mode.

♦♦♦ **Example 8:** Simplify

$$\frac{\cos \theta}{1 + \sin \theta} + \tan \theta$$

Solution: We enter the expression and notice that it does not simplify by default, so we use **tCollect**. We get

$$\frac{\cos \theta}{1 + \sin \theta} + \tan \theta = \frac{1}{\cos \theta}$$

♦♦♦

Since a trigonometric expression can take so many different forms, the calculator may not simplify it in a useful way.

Proving a Trigonometric Identity

■ Exploration:

Try this. In the same window, graph

$$y_1 = \tan x \cos x \qquad \text{and} \qquad y_2 = \sin x$$

using a heavier line for y_2. What do you see?   ■

In some problems we will be asked to manipulate a trigonometric expression so that it matches another expression. For example, you may be asked to verify or prove that the expressions in our exploration are equal, that is,

$$\tan x \cos x = \sin x$$

This equation is called an *identity* because it is true for all values of the variable x for which the functions are defined (for example, the tangent is not defined at $x = 90°$).

To prove an identity, we manipulate one or both sides until both sides match. For this we use the fundamental identities and basic operations, such as factoring, reducing fractions, and so forth. We work each side *separately* and do not treat the identity as if it was an equation, for which we could transpose terms and multiply both sides by the same quantity, for example.

If one side of the identity is more complicated than the other, it is a good idea to start by simplifying that side.

♦♦♦ **Example 9:** Prove the following identity.

$$\frac{1}{\sin x \sec x \cot x} = 1$$

Solution: Let us try expressing each trigonometric function in terms of the sine and cosine. Thus, $\sec x = 1/\cos x$ and $\cot x = \cos x / \sin x$.

$$\frac{1}{\sin x \sec x \cot x} = 1$$

$$\frac{1}{\sin x \left(\dfrac{1}{\cos x}\right)\left(\dfrac{\cos x}{\sin x}\right)} = 1$$

The denominator of the fraction is thus equal to 1, and the identity is proved.   ♦♦♦

◆◆◆ **Example 10:** Prove the identity

$$\frac{\cos x \csc x}{\tan x} = \cot^2 x$$

Solution: We write each expression on the left in terms of sines and cosines, and simplify.

$$\frac{\cos x \csc x}{\tan x} \;\Big|\; \cot^2 x$$

$$\frac{\cos x \left(\dfrac{1}{\sin x}\right)}{\dfrac{\sin x}{\cos x}} \;\Big|\; \cot^2 x$$

$$\frac{\cos^2 x}{\sin^2 x} \;\Big|\; \cot^2 x$$

$$\cot^2 x = \cot^2 x$$

◆◆◆

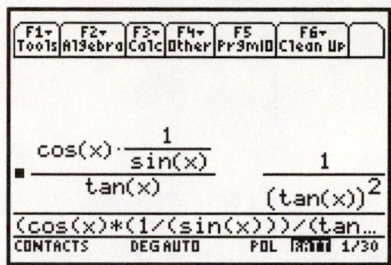

TI-89 screen for Example 10. Note the use of the reciprocal relations.

Exercise 1 ◆ Fundamental Identities

Change to an expression containing only sin and cos.

1. $\tan x - \sec x$ **2.** $\cot x + \csc x$ **3.** $\tan \theta \csc \theta$

4. $\sec \theta - \tan \theta \sin \theta$ **5.** $\dfrac{\tan \theta}{\csc \theta} + \dfrac{\sin \theta}{\tan \theta}$ **6.** $\cot x + \tan x$

Simplify.

7. $1 - \sec^2 x$ **8.** $\dfrac{\csc \theta}{\sin \theta}$ **9.** $\dfrac{\cos \theta}{\cot \theta}$

10. $\sin \theta \csc \theta$ **11.** $\tan \theta \csc \theta$ **12.** $\dfrac{\sin \theta}{\csc \theta}$

13. $\sec x \sin x$ **14.** $\sec x \sin x \cot x$

15. $\csc \theta \tan \theta - \tan \theta \sin \theta$

16. $\dfrac{\cos x}{\cot x \sin x}$ **17.** $\cot \theta \tan^2 \theta \cos \theta$ **18.** $\dfrac{\tan x(\csc^2 x - 1)}{\sin x + \cot x \cos x}$

19. $\dfrac{\sin \theta}{\cos \theta \tan \theta}$ **20.** $\dfrac{\sin^2 x + \cos^2 x}{1 - \cos^2 x}$

21. $\dfrac{1}{\sec^2 x} + \dfrac{1}{\csc^2 x}$ **22.** $\sin \theta(\csc \theta + \cot \theta)$

23. $\csc x - \cot x \cos x$ **24.** $1 + \dfrac{\tan^2 \theta}{1 + \sec \theta}$

25. $\dfrac{\sec x - \csc x}{1 - \cot x}$ **26.** $\dfrac{1}{1 + \sin x} + \dfrac{1}{1 - \sin x}$

27. $\sec^2 x(1 - \cos^2 x)$ **28.** $\tan x + \dfrac{\cos x}{\sin x + 1}$

29. $\cos \theta \sec \theta - \dfrac{\sec \theta}{\cos \theta}$ **30.** $\cot^2 x \sin^2 x + \tan^2 x \cos^2 x$

Prove each identity. (All identities in this chapter *can* be proven.)

31. $\tan x \cos x = \sin x$ **32.** $\tan x = \dfrac{\sec x}{\csc x}$

33. $\dfrac{\sin x}{\csc x} + \dfrac{\cos x}{\sec x} = 1$ **34.** $\sin \theta = \dfrac{1}{\cot \theta \sec \theta}$

35. $(\cos^2\theta + \sin^2\theta)^2 = 1$

36. $\tan x = \dfrac{\tan x - 1}{1 - \cot x}$

37. $\dfrac{\csc\theta}{\sec\theta} = \cot\theta$

38. $\cot^2 x = \dfrac{\cos x}{\tan x \sin x}$

39. $\cos x + 1 = \dfrac{\sin^2 x}{1 - \cos x}$

40. $\csc x - \sin x = \cot x \cos x$

41. $\cot^2 x - \cos^2 x = \cos^2 x \cot^2 x$

42. $\csc x = \cos x \cot x + \sin x$

43. $1 = (\csc x - \cot x)(\csc x + \cot x)$

44. $\tan x = \dfrac{\tan x + \sin x}{1 + \cos x}$

45. $\dfrac{\tan x + 1}{1 - \tan x} = \dfrac{\sin x + \cos x}{\cos x - \sin x}$

46. $\cot x = \cot x \sec^2 x - \tan x$

47. $\dfrac{\sin\theta + 1}{1 - \sin\theta} = (\tan\theta + \sec\theta)^2$

48. $\dfrac{1 + \sin\theta}{1 - \sin\theta} = \dfrac{1 + \csc\theta}{\csc\theta - 1}$

49. $(\sec\theta - \tan\theta)(\tan\theta + \sec\theta) = 1$

50. $\dfrac{1 + \cot\theta}{\csc\theta} = \dfrac{\tan\theta + 1}{\sec\theta}$

Applications

51. Figure 16–3 shows a crate of weight W hanging from a rope, and being held in place by a horizontal force F and the tension T in the cable. (a) Show that the equations of equilibrium are $F = T \sin\theta$ and $W = T \cos\theta$, and (b) that $F = W \tan\theta$.

52. *Inclined Plane:* A block of weight W rests on an inclined plane, Fig. 16–4. The tangential and normal forces it exerts on the plane are

$$T = W \sin\theta \qquad \text{and} \qquad N = W \cos\theta$$

When the block just begins to slide, the frictional force f is equal and opposite to T. Show that the coefficient of friction μ, the frictional force divided by the normal force, is equal to $\tan\theta$.

53. The parametric equations of an ellipse, Fig. 16–5, in terms of a parameter α, are

$$x = a \cos\alpha$$
$$y = b \sin\alpha$$

The ellipse may also be described by the equation

$$\frac{x^2}{a^2} + \frac{y^2}{b^2} = 1 \tag{1}$$

Substitute the parametric equations into Eq. 1 and show that the left side does in fact equal 1.

54. In calculus, we take the derivative of $\tan x$ by taking the derivative of $\sin x / \cos x$, and get the expression

$$\frac{(\cos x)(\cos x) - (\sin x)(-\sin x)}{\cos^2 x}$$

Show that this expression is equal to $\sec^2 x$.

55. *Alternating Current:* Given $R = Z \cos\theta$ and $X_L - X_C = Z \sin\theta$, where R is the resistance, X_L is the inductive reactance and $X_C =$ capacitive reactance, evaluate and simplify

(a) $R^2 + (X_L - X_C)^2$

(b) $\dfrac{X_L - X_C}{R}$

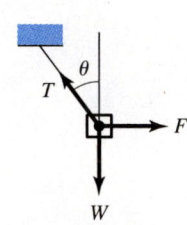

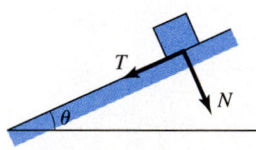

θ

T

F

W

FIGURE 16–3

T

N

θ

FIGURE 16–4

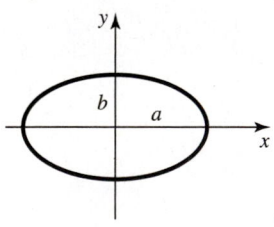

y

b

a

x

FIGURE 16–5

16–2 Sum or Difference of Two Angles

Having established the fundamental trignometric identities, let us now move on to others that are very useful in technical work, starting with the trig functions of the sum or difference of two angles.

We now wish to derive a formula for the sine of the sum of two angles, for example, $\sin(\alpha + \beta)$.

■ **Exploration:**

Try this. Use your calculator to evaluate

$$\sin 20° \qquad \sin 30° \qquad \sin 50°$$

Does the sum of sin 20° and sin 30° equal the sine of 50°? ■

	The sine of the sum of two angles is *not* the sum of the sine of each angle.
Common Error	$$\sin(\alpha + \beta) \neq \sin \alpha + \sin \beta$$ We will show that $$\sin(\alpha + \beta) = \sin \alpha \cos \beta + \cos \alpha \sin \beta$$

We start by drawing two positive acute angles, α and β (Fig. 16–6), small enough so that their sum $(\alpha + \beta)$ is also acute. From any point P on the terminal side of β we draw perpendicular AP to the x axis and draw perpendicular BP to line OB. Since the angle between two lines equals the angle between the perpendiculars to those two lines (can you demonstrate that this is true?), we note that angle APB is equal to α.

Then

$$\sin(\alpha + \beta) = \frac{AP}{OP} = \frac{AD + PD}{OP} = \frac{BC + PD}{OP}$$
$$= \frac{BC}{OP} + \frac{PD}{OP}$$

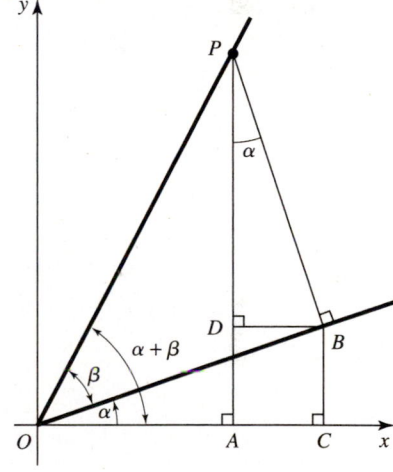

FIGURE 16–6

But in triangle OBC,

$$BC = OB \sin \alpha$$

and in triangle PBD,

$$PD = PB \cos \alpha$$

Substituting, we obtain

$$\sin(\alpha + \beta) = \frac{OB \sin \alpha}{OP} + \frac{PB \cos \alpha}{OP}$$

But in triangle OPB,

$$\frac{OB}{OP} = \cos \beta$$

and

$$\frac{PB}{OP} = \sin \beta$$

Thus,

$$\sin(\alpha + \beta) = \sin \alpha \cos \beta + \cos \alpha \sin \beta$$

Convince yourself that this is true by using your calculator to compute $\sin(45° + 30°)$ using this identity, and comparing it with $\sin(75°)$.

Cosine of the Sum of Two Angles

Again using Fig. 16–6, we can derive an expression for $\cos(\alpha + \beta)$.
In triangle OAP

$$\cos(\alpha + \beta) = \frac{OA}{OP} = \frac{OC - AC}{OP} = \frac{OC - BD}{OP}$$

$$= \frac{OC}{OP} - \frac{BD}{OP}$$

Now, in triangle OBC,

$$OC = OB \cos \alpha$$

and in triangle PDB,

$$BD = PB \sin \alpha$$

Substituting, we obtain

$$\cos(\alpha + \beta) = \frac{OB \cos \alpha}{OP} - \frac{PB \sin \alpha}{OP}$$

As before,

$$\frac{OB}{OP} = \cos \beta \qquad \text{and} \qquad \frac{PB}{OP} = \sin \beta$$

Therefore,

$$\cos(\alpha + \beta) = \cos \alpha \cos \beta - \sin \alpha \sin \beta$$

Difference of Two Angles

We can obtain a formula for the sine of the difference of two angles merely by substituting $-\beta$ for β in the equation previously derived for $\sin(\alpha + \beta)$.

$$\sin[\alpha + (-\beta)] = \sin \alpha \cos(-\beta) + \cos \alpha \sin(-\beta)$$

But for β, which is still in the first quadrant, $(-\beta)$ is in the fourth, so

$$\cos(-\beta) = \cos \beta \qquad \text{and} \qquad \sin(-\beta) = -\sin \beta$$

Therefore,

$$\sin(\alpha - \beta) = \sin \alpha \cos \beta + \cos \alpha(-\sin \beta)$$

which is rewritten as follows:

$$\sin(\alpha - \beta) = \sin \alpha \cos \beta - \cos \alpha \sin \beta$$

We see that the result is identical to the formula for $\sin(\alpha + \beta)$ except for a change in sign. This enables us to write the two identities in a single expression using the $\pm$ sign. When the double signs are used, it is understood that the upper signs correspond and the lower signs correspond.

Sine of Sum or Difference of Two Angles	$\sin(\alpha \pm \beta) = \sin \alpha \cos \beta \pm \cos \alpha \sin \beta$	128

Similarly, finding $\cos(\alpha - \beta)$, we have

$$\cos[\alpha + (-\beta)] = \cos\alpha\cos(-\beta) - \sin\alpha\sin(-\beta)$$

Thus,

$$\cos(\alpha - \beta) = \cos\alpha\cos\beta + \sin\alpha\sin\beta$$

or

Cosine of Sum or Difference of Two Angles	$\cos(\alpha \pm \beta) = \cos\alpha\cos\beta \mp \sin\alpha\sin\beta$	129

◆◆◆ **Example 11:** Expand the expression $\sin(x + 3y)$.

Solution: Since $\sin(\alpha + \beta) = \sin\alpha\cos\beta + \cos\alpha\sin\beta$, we write

$$\sin(x + 3y) = \sin x\cos 3y + \cos x\sin 3y \qquad \text{◆◆◆}$$

◆◆◆ **Example 12:** Simplify

$$\cos 5x\cos 3x - \sin 5x\sin 3x$$

Solution: This is similar in form to $\cos(\alpha + \beta) = \cos\alpha\cos\beta - \sin\alpha\sin\beta$, where $\alpha = 5x$ and $\beta = 3x$, so

$$\cos 5x\cos 3x - \sin 5x\sin 3x = \cos(5x + 3x)$$
$$= \cos 8x \qquad \text{◆◆◆}$$

◆◆◆ **Example 13:** Prove that

$$\cos(180° - \theta) = -\cos\theta$$

Solution: We expand the left side using the identity for $\cos(\alpha - \beta)$, getting

$$\cos(180° - \theta) = \cos 180°\cos\theta + \sin 180°\sin\theta$$

But $\cos 180° = -1$ and $\sin 180° = 0$, so

$$\cos(180° - \theta) = (-1)\cos\theta + (0)\sin\theta = -\cos\theta \qquad \text{◆◆◆}$$

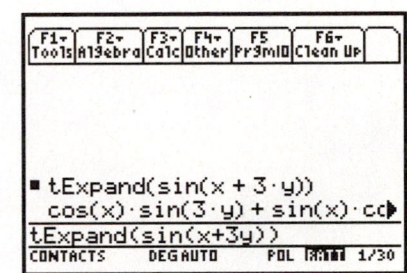

TI-89 check for Example 11, using **tExpand** from the Math Angle menu. Note that part of the result is off the screen, requiring that we scroll to the right.

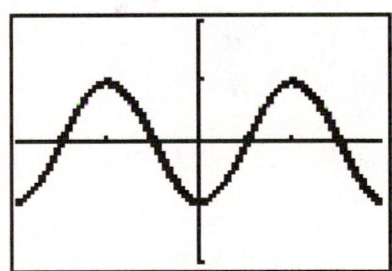

Graphical check of Example 13. We plotted each side of the given identity and see that one graph overlays the other.

Tangent of the Sum or Difference of Two Angles

We have derived formulas for the sine and cosine of the sum or difference of two angles. We now do the same for the *tangent*.
Since

$$\tan\theta = \frac{\sin\theta}{\cos\theta}$$

we can get an expression for $\tan(\alpha + \beta)$ simply by dividing $\sin(\alpha + \beta)$ by $\cos(\alpha + \beta)$.

$$\tan(\alpha + \beta) = \frac{\sin(\alpha + \beta)}{\cos(\alpha + \beta)} = \frac{\sin\alpha\cos\beta + \cos\alpha\sin\beta}{\cos\alpha\cos\beta - \sin\alpha\sin\beta}$$

Dividing numerator and denominator by $\cos \alpha \cos \beta$ yields

$$\tan(\alpha + \beta) = \frac{\dfrac{\sin \alpha}{\cos \alpha} + \dfrac{\sin \beta}{\cos \beta}}{1 - \dfrac{\sin \alpha}{\cos \alpha} \cdot \dfrac{\sin \beta}{\cos \beta}}$$

$$= \frac{\tan \alpha + \tan \beta}{1 - \tan \alpha \tan \beta}$$

A similar derivation (which we will not do) will show that $\tan(\alpha - \beta)$ is identical to the expression just derived, except, as we might expect, for a reversal of signs. We combine the two expressions using double signs as follows:

Tangent of Sum or Difference of Two Angles	$\tan(\alpha \pm \beta) = \dfrac{\tan \alpha \pm \tan \beta}{1 \mp \tan \alpha \tan \beta}$	130

♦♦♦ **Example 14:** Simplify

$$\frac{\tan 3x + \tan 2x}{\tan 2x \tan 3x - 1}$$

Solution: This will match the form of $\tan(\alpha + \beta)$ if we factor (-1) from the denominator.

$$\frac{\tan 3x + \tan 2x}{\tan 2x \tan 3x - 1} = \frac{\tan 3x + \tan 2x}{-(-\tan 2x \tan 3x + 1)}$$

$$= -\frac{\tan 3x + \tan 2x}{1 - \tan 3x \tan 2x}$$

$$= -\tan(3x + 2x)$$

$$= -\tan 5x \qquad\qquad ♦♦♦$$

♦♦♦ **Example 15:** Prove that

$$\tan(45° + x) = \frac{1 + \tan x}{1 - \tan x}$$

Solution: Expanding the left side by using the identity for $\tan(\alpha + \beta)$ gives

$$\tan(45° + x) = \frac{\tan 45° + \tan x}{1 - \tan 45° \tan x}$$

But $\tan 45° = 1$, so

$$\tan(45° + x) = \frac{1 + \tan x}{1 - \tan x} \qquad\qquad ♦♦♦$$

♦♦♦ **Example 16:** Prove that

$$\frac{\cot y - \cot x}{\cot x \cot y + 1} = \tan(x - y)$$

Solution: Since the cotangent is the reciprocal of the tangent, we write

$$\frac{\dfrac{1}{\tan y} - \dfrac{1}{\tan x}}{\dfrac{1}{\tan x} \cdot \dfrac{1}{\tan y} + 1} = \tan(x - y)$$

Multiplying numerator and denominator by tan x tan y,

$$\frac{\tan x - \tan y}{1 + \tan x \tan y} = \tan(x - y)$$

This now matches the form of $\tan(\alpha - \beta)$, so

$$\tan(x - y) = \tan(x - y) \qquad \blacklozenge\blacklozenge\blacklozenge$$

Exercise 2 ◆ Sum or Difference of Two Angles

Expand by means of the addition and subtraction formulas, and simplify.

1. $\sin(\theta + 30°)$ **2.** $\cos(45° - x)$

3. $\sin(x + 60°)$ **4.** $\tan(\pi + \theta)$

5. $\cos\left(x + \dfrac{\pi}{2}\right)$ **6.** $\tan(2x + y)$

7. $\sin(\theta + 2\phi)$ **8.** $\tan(2\theta - 3\alpha)$

Simplify.

9. $\cos 2x \cos 9x + \sin 2x \sin 9x$ **10.** $\cos(\pi + \theta) + \sin(\pi + \theta)$

11. $\sin 3\theta \cos 2\theta - \cos 3\theta \sin 2\theta$

Prove each identity.

12. $\cos x = \sin(x + 90°)$

13. $\sin(\alpha + \beta) + \sin(\alpha - \beta) = 2 \sin \alpha \cos \beta$

14. $\cos(2\pi - x) = \cos x$ **15.** $\sin\left(x + \dfrac{\pi}{6}\right) - \sin\left(x - \dfrac{\pi}{6}\right) = \cos x$

16. $\tan(360° - \beta) = -\tan \beta$ **17.** $\cos(x + 60°) + \cos(60° - x) = \cos x$

18. $\dfrac{\cos(x - y)}{\sin x \cos y} = \tan y + \cot x$ **19.** $\dfrac{1 + \tan x}{1 - \tan x} = \tan\left(\dfrac{\pi}{4} + x\right)$

Applications

20. *Simple Harmonic Motion:* The weight, Fig. 16–7, moves with what is called *simple harmonic motion*. Its vertical displacement is given by $y = 2.51 \cos(2t + 35.4°)$. Show that this motion is equivalent to

$$y = 2.05 \cos 2t - 1.45 \sin 2t$$

21. *Inclined Plane:* A horizontal force of *P* pounds is applied to a body on an inclined plane, Fig. 16–8. When the body just begins to move up the plane, (a) show that the equations of equilibrium are:

$$P \cos \theta - W \sin \theta = f \qquad N - W \cos \theta = P \sin \theta \qquad f = N \tan \phi$$

where *N* is the component of the weight normal to the plane, *f* is the frictional force, and $\tan \phi = \mu$, the coefficient of friction. (b) Solve these equations for *P*, showing that $P = W \tan(\theta + \phi)$.

22. *Inclined Plane:* The car, Fig 16–9, decelerates while going down a hill. (a) Show that its deceleration *a* is given by $a = g \tan \alpha \cos \beta - g \sin \beta$, where $\tan \alpha = f/N$ and β is the angle of inclination of the roadway. (b) Show that this expression is equivalent to

$$a = \frac{g \sin(\alpha - \beta)}{\cos \alpha}$$

23. *Project, Adding a Sine Wave and a Cosine Wave of the Same Frequency:* Prove that the sum of the sine wave $A \sin \omega t$ and the cosine wave $B \cos \omega t$ is

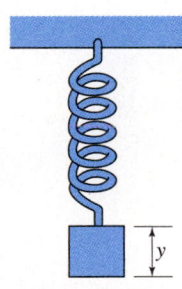

FIGURE 16–7

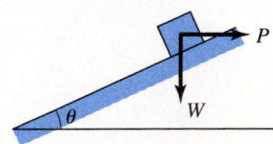

FIGURE 16–8

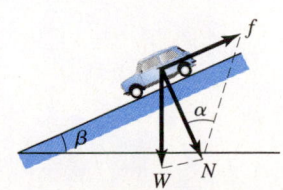

FIGURE 16–9

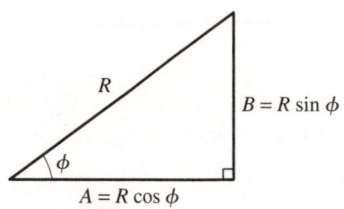

FIGURE 16–10

Addition of a Sine Wave and a Cosine Wave	$A \sin \omega t + B \cos \omega t = R \sin(\omega t + \phi)$ where $R = \sqrt{A^2 + B^2}$ and $\phi = \arctan\dfrac{B}{A}$	166

Hint: Let $A \sin \omega t$ be a sine wave of amplitude A, and $B \cos \omega t$ be a cosine wave of amplitude B, each of frequency $\omega/2\pi$. If we draw a right triangle (Fig. 16–10) with sides A and B and hypotenuse R, then

$$A = R \cos \phi \qquad \text{and} \qquad B = R \sin \phi$$

The sum of the sine wave and the cosine wave is then

$$A \sin \omega t + B \cos \omega t = R \sin \omega t \cos \phi + R \cos \omega t \sin \phi$$
$$= R(\sin \omega t \cos \phi + \cos \omega t \sin \phi)$$

Do you recognize the form of the expression on the right? Try to use an identity from this chapter to complete the derivation.

24. *Project:* Use the formula for the addition of a sine wave and a cosine wave to express each following expression as a single sine function.
(a) $y = 47.2 \sin \omega t + 64.9 \cos \omega t$
(b) $y = 8470 \sin \omega t + 7360 \cos \omega t$
(c) $y = 1.83 \sin \omega t + 2.74 \cos \omega t$
(d) $y = 84.2 \sin \omega t + 74.2 \cos \omega t$

16–3   Functions of Double Angles and Half-Angles

We come now to our last batch of identities, those involving two times an angle and those involving half an angle.

Functions of Double Angles

■ Exploration:

Is the sine of twice an angle equal to twice the sine of that angle? Is $\sin 2\alpha = 2 \sin \alpha$? *Try this:* By calculator, evaluate

$$\sin 2(40°) \qquad 2 \sin(40°)$$

Are these equal? Try it again with other angles. Try it for the cosine and tangent. What do you conclude?   ■

Common Error	The sine of twice an angle is *not* twice the sine of the angle. Nor is the cosine (or tangent) of twice an angle equal to twice the cosine (or tangent) of that angle. Remember to use the formulas from this section for all of the trig functions of double angles.

Sine of Twice an Angle

An equation for the $\sin 2\alpha$ may be derived by setting $\beta = \alpha$ in the identity for $\sin(\alpha + \beta)$.

$$\sin(\alpha + \alpha) = \sin \alpha \cos \alpha + \cos \alpha \sin \alpha$$

which we can rewrite as:

Sine of Twice an Angle	$\sin 2\alpha = 2 \sin \alpha \cos \alpha$	131

Cosine of Twice an Angle

Similarly, setting $\beta = \alpha$ in the identity for $\cos(\alpha + \beta)$ gives

$$\cos(\alpha + \alpha) = \cos\alpha\cos\alpha - \sin\alpha\sin\alpha$$

which can also be written,

Cosine of Twice an Angle	$\cos 2\alpha = \cos^2\alpha - \sin^2\alpha$	132a

There are two alternative forms to this identity. Since $\sin^2\alpha + \cos^2\alpha = 1$,

$$\cos^2\alpha = 1 - \sin^2\alpha$$

Substituting yields

$$\cos 2\alpha = 1 - \sin^2\alpha - \sin^2\alpha$$

Thus,

Cosine of Twice an Angle	$\cos 2\alpha = 1 - 2\sin^2\alpha$	132b

Similarly, we can use the same identity, $\sin^2\alpha + \cos^2\alpha = 1$, to eliminate the $\sin^2\alpha$ term, getting

$$\cos 2\alpha = \cos^2\alpha - (1 - \cos^2\alpha)$$

Thus,

Cosine of Twice an Angle	$\cos 2\alpha = 2\cos^2\alpha - 1$	132c

◆◆◆ **Example 17:** Prove that

$$\cos 2A + \sin(A - B) = 0$$

where A and B are the two acute angles of a right triangle.

Solution: Using the identities for $\cos 2\alpha$ and for $\sin(\alpha - \beta)$ we get

$$\cos 2A + \sin(A - B) = \cos^2 A - \sin^2 A + \sin A \cos B - \cos A \sin B$$

But angles A and B are complementary, so we may use the cofunctions

$$\cos B = \sin A \quad \text{and} \quad \sin B = \cos A$$

So

$$\cos 2A + \sin(A - B) = \cos^2 A - \sin^2 A + \sin A \sin A - \cos A \cos A$$
$$= \cos^2 A - \sin^2 A + \sin^2 A - \cos^2 A = 0 \qquad \text{◆◆◆}$$

◆◆◆ **Example 18:** Simplify the expression

$$\frac{\sin 2x}{1 + \cos 2x}$$

Solution: Using the identities for $\sin 2\alpha$ and $\cos 2\alpha$ gives

$$\frac{\sin 2x}{1 + \cos 2x} = \frac{2\sin x\cos x}{1 + 2\cos^2 x - 1}$$
$$= \frac{2\sin x\cos x}{2\cos^2 x}$$
$$= \frac{\sin x}{\cos x} = \tan x \qquad \text{◆◆◆}$$

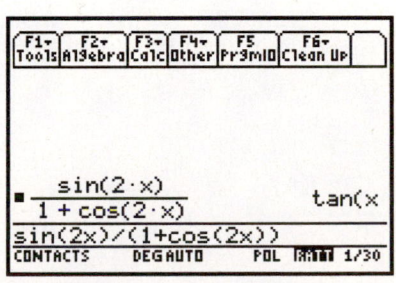

TI-89 check of Example 18. The calculator simplified this expression by default, so no instruction was needed.

Tangent of Twice an Angle

Setting $\beta = \alpha$ in the identity for $\tan(\alpha + \beta)$ gives

$$\tan(\alpha + \alpha) = \frac{\tan \alpha + \tan \alpha}{1 - \tan \alpha \tan \alpha}$$

Therefore,

Tangent of Twice an Angle	$\tan 2\alpha = \dfrac{2 \tan \alpha}{1 - \tan^2\alpha}$	133

♦♦♦ **Example 19:** Prove that

$$\frac{2 \cot x}{\csc^2 x - 2} = \tan 2x$$

Solution: We use the identity for $\tan 2\alpha$ to expand the right side.

$$\frac{2 \cot x}{\csc^2 x - 2} = \frac{2 \tan x}{1 - \tan^2 x}$$

Replacing $\tan x$ by $1/\cot x$,

$$= \frac{\dfrac{2}{\cot x}}{1 - \dfrac{1}{\cot^2 x}}$$

Simplifying this compound fraction by multiplying numerator and denominator by $\cot^2 x$,

$$= \frac{2 \cot x}{\cot^2 x - 1}$$

Finally, since $\cot^2 x = \csc^2 x - 1$, we get

$$\frac{2 \cot x}{\csc^2 x - 2} = \frac{2 \cot x}{\csc^2 x - 2}$$

♦♦♦

Functions of Half-Angles

■ **Exploration:**

Is the sine of half an angle equal to half the sine of that angle? *Try this:* By calculator, evaluate

$$\sin \frac{60°}{2} \qquad \tfrac{1}{2} \sin 60°$$

Are these equal? Try it again with other angles. Try it for the cosine and tangent. What do you conclude? ■

Sine of Half an Angle

The double-angle identities we just derived can also be thought of as half-angle identities. If one angle is double another, the second must be half the first. Let us start with the double-angle identity

$$\cos 2\theta = 1 - 2 \sin^2 \theta$$

We solve for $\sin \theta$.

$$2 \sin^2 \theta = 1 - \cos 2\theta$$

$$\sin \theta = \pm \sqrt{\frac{1 - \cos 2\theta}{2}}$$

For emphasis, we replace θ by $\alpha/2$.

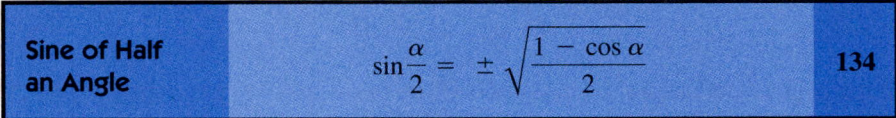

Sine of Half an Angle	$\sin \dfrac{\alpha}{2} = \pm \sqrt{\dfrac{1 - \cos \alpha}{2}}$	134

The $\pm$ sign in this identity is to be read as plus *or* minus, but *not both*. This sign is different from the $\pm$ sign in the quadratic formula, for example, where we took *both* the positive and the negative values.

The reason for this difference is clear from Fig. 16–11, which shows a graph of $\sin \alpha/2$ and a graph of $+ \sqrt{(1 - \cos \alpha)/2}$. Note that the two curves are the same only when $\sin \alpha/2$ is positive. When $\sin \alpha/2$ is negative, it is necessary to use the negative of $\sqrt{(1 - \cos \alpha)/2}$. This occurs when $\alpha/2$ is in the third or fourth quadrant. Thus we choose the plus or the minus according to the quadrant in which $\alpha/2$ is located.

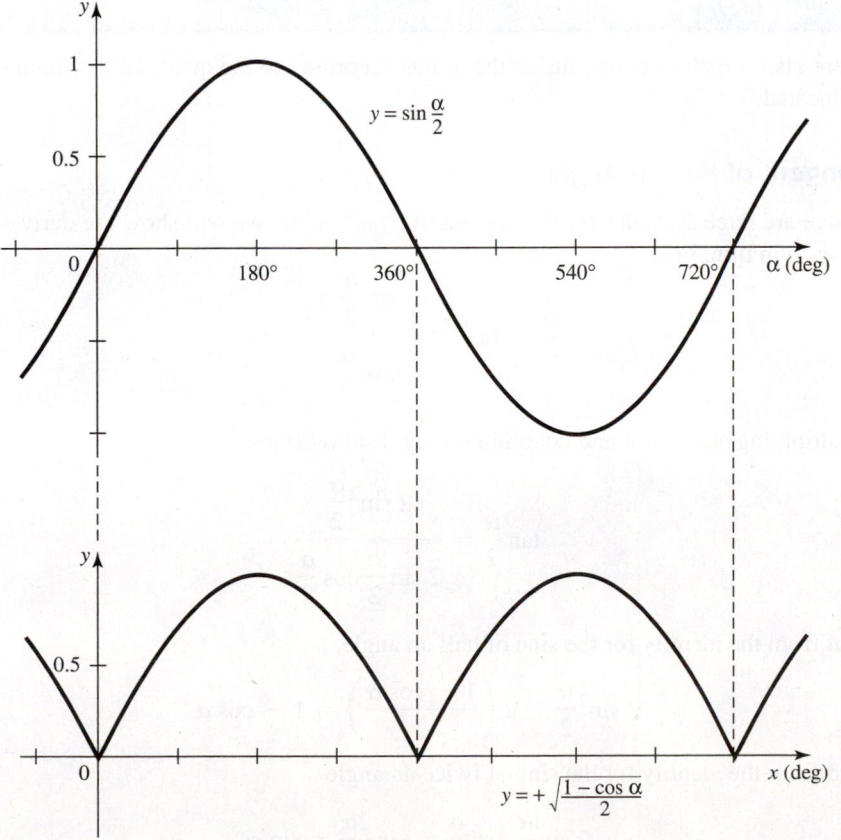

FIGURE 16–11

◆◆◆ **Example 20:** Given that $\cos 200° = -0.9397$, find $\sin 100°$.

Solution: From Eq. 134,

$$\sin \frac{200°}{2} = \pm\sqrt{\frac{1 - \cos 200°}{2}} = \pm\sqrt{\frac{1 - (-0.9397)}{2}} = \pm 0.9848$$

Since $100°$ is in the second quadrant, the sine is positive, so we drop the minus sign and get

$$\sin 100° = 0.9848 \qquad\qquad ◆◆◆$$

We would not dream of finding $\sin 100°$ in this way. We are simply trying to illustrate the use of the $\pm$ sign.

Cosine of $\alpha/2$

We start with the double-angle identity

$$\cos 2\theta = 2\cos^2 \theta - 1$$

Then we solve for $\cos \theta$.

$$2\cos^2 \theta = 1 + \cos 2\theta$$

$$\cos \theta = \pm\sqrt{\frac{1 + \cos 2\theta}{2}}$$

We next replace θ by $\alpha/2$ and obtain the following:

Cosine of Half an Angle	$\cos\dfrac{\alpha}{2} = \pm\sqrt{\dfrac{1 + \cos \alpha}{2}}$	135

Here also, we choose the plus *or* the minus according to the quadrant in which $\alpha/2$ is located.

Tangent of Half an Angle

There are three formulas for the tangent of a half-angle; we will show the derivation of each in turn. First,

$$\tan \frac{\alpha}{2} = \frac{\sin \dfrac{\alpha}{2}}{\cos \dfrac{\alpha}{2}}$$

Multiplying numerator and denominator by $2\sin(\alpha/2)$ gives

$$\tan \frac{\alpha}{2} = \frac{2\sin^2 \dfrac{\alpha}{2}}{2\sin \dfrac{\alpha}{2} \cos \dfrac{\alpha}{2}} \qquad\qquad (1)$$

But from the identity for the sine of half an angle,

$$2\sin^2 \frac{\alpha}{2} = 2\left(\frac{1 - \cos \alpha}{2}\right) = 1 - \cos \alpha$$

and from the identity for the sine of twice an angle

$$2\sin \frac{\alpha}{2} \cos \frac{\alpha}{2} = \sin \frac{2\alpha}{2} = \sin \alpha$$

Substituting into (1) gives

Tangent of Half an Angle	$\tan\dfrac{\alpha}{2} = \dfrac{1 - \cos\alpha}{\sin\alpha}$	136a

Another form of this identity is obtained by multiplying numerator and denominator by $1 + \cos\alpha$

$$\tan\frac{\alpha}{2} = \frac{1 - \cos\alpha}{\sin\alpha} \cdot \frac{1 + \cos\alpha}{1 + \cos\alpha}$$

$$= \frac{1 - \cos^2\alpha}{\sin\alpha(1 + \cos\alpha)} = \frac{\sin^2\alpha}{\sin\alpha(1 + \cos\alpha)}$$

which simplifies to

Tangent of Half an Angle	$\tan\dfrac{\alpha}{2} = \dfrac{\sin\alpha}{1 + \cos\alpha}$	136b

We can obtain a third formula for the tangent by dividing the sine by the cosine.

$$\tan\frac{\alpha}{2} = \frac{\sin\dfrac{\alpha}{2}}{\cos\dfrac{\alpha}{2}} = \frac{\pm\sqrt{\dfrac{1 - \cos\alpha}{2}}}{\pm\sqrt{\dfrac{1 + \cos\alpha}{2}}}$$

Tangent of Half an Angle	$\tan\dfrac{\alpha}{2} = \pm\sqrt{\dfrac{1 - \cos\alpha}{1 + \cos\alpha}}$	136c

◆◆◆ **Example 21:** Prove that

$$\frac{3 - \cos\theta}{3 + \cos\theta} = \frac{1 + \sin^2\dfrac{\theta}{2}}{1 + \cos^2\dfrac{\theta}{2}}$$

Solution: Using the identities for the sine and cosine of half an angle we get

$$\frac{3 - \cos\theta}{3 + \cos\theta} = \frac{1 + \dfrac{1 - \cos\theta}{2}}{1 + \dfrac{1 + \cos\theta}{2}}$$

Multiplying numerator and denominator by 2

$$= \frac{2 + 1 - \cos\theta}{2 + 1 + \cos\theta}$$

$$= \frac{3 - \cos\theta}{3 + \cos\theta} \qquad\qquad ◆◆◆$$

Exercise 3 ◆ Functions of Double Angles and Half-Angles

Double Angles

Simplify.

1. $2 \sin^2 x + \cos 2x$

2. $2 \sin 2\theta \cos 2\theta$

3. $\dfrac{2 \tan x}{1 + \tan^2 x}$

4. $\dfrac{2 - \sec^2 x}{\sec^2 x}$

Prove each identity.

5. $\dfrac{2 \tan \theta}{1 - \tan^2 \theta} = \tan 2\theta$

6. $\dfrac{1 - \tan^2 x}{1 + \tan^2 x} = \cos 2x$

7. $\tan \theta + \cot \theta = 2 \csc 2\theta$

8. $\cot x - \tan x = 2 \cot 2x$

9. $\dfrac{1 + \cot^2 x}{\cot^2 x - 1} = \sec 2x$

10. $\dfrac{\sin 2\theta + \sin \theta}{1 + \cos \theta + \cos 2\theta} = \tan \theta$

11. $\dfrac{2 \cos 2x}{\sin 2x - 2 \sin^2 x} = 1 + \cot x$

12. $\dfrac{\sin 2\alpha + 1}{\cos \alpha + \sin \alpha} = \sin \alpha + \cos \alpha$

13. $\dfrac{\cot^2 x - 1}{2 \cot x} = \cot 2x$

Half-Angles

Prove each identity.

14. $2 \sin^2 \dfrac{\theta}{2} + \cos \theta = 1$

15. $4 \cos^2 \dfrac{x}{2} \sin^2 \dfrac{x}{2} = 1 - \cos^2 x$

16. $\csc \theta + \cot \theta = \cot \dfrac{\theta}{2}$

17. $\dfrac{\cos^2 \dfrac{\theta}{2} - \cos \theta}{\sin^2 \dfrac{\theta}{2}} = 1$

18. $\tan \dfrac{\theta}{2} \tan \theta + 1 = \sec \theta$

19. $\left(\cos \dfrac{x}{2} + \sin \dfrac{x}{2} \right)^2 = \sin x + 1$

20. $\dfrac{1 - \cos x}{\sin x} = \tan \dfrac{x}{2}$

21. $\dfrac{1 - \tan^2 \dfrac{\theta}{2}}{1 + \tan^2 \dfrac{\theta}{2}} = \cos \theta$

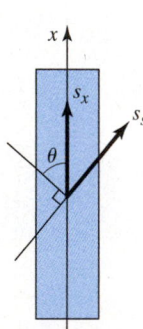

FIGURE 16–12

Applications

22. *Shear Stress:* The shear stress s_s on a cross section of a bar in tension, Fig. 16–12, is related to the axial stress s_x by the formula

$$s_s = s_x \sin \theta \cos \theta$$

where θ is the angle between the axis of the bar and the normal to the cross section. Use the double-angle identities to write this expression with just a single trigonometric ratio.

23. *Trajectories:* A projectile is launched on level ground at an angle θ and with an initial velocity of v_0, Fig. 16–13. In time t the x and y displacements are

$$x = (v_0 \cos \theta)t \qquad \text{and} \qquad y = (v_0 \sin \theta)t - \tfrac{1}{2}gt^2$$

(a) Write an expression for the time elapsed before the projectile hits the ground.

(b) Show that the expression for x at that time (this is the *range R*) is

$$R = (v_0^2/g) \sin 2\theta$$

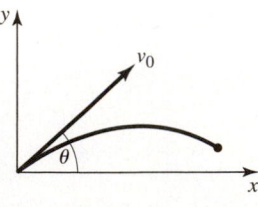

FIGURE 16–13

24. One way to determine the index of refraction of a material is to measure the total refraction of a light ray through a prism made of that material, Fig. 16–14. The index of refraction n, relative to that of air, is given by

$$n = \frac{\sin \frac{1}{2}(\alpha + \delta)}{\sin \frac{\alpha}{2}}$$

Rewrite this expression so that it does not contain functions of half-angles, showing that

$$n = \sqrt{\frac{1 - \cos \alpha \cos \delta + \sin \alpha \sin \delta}{1 - \cos \alpha}}$$

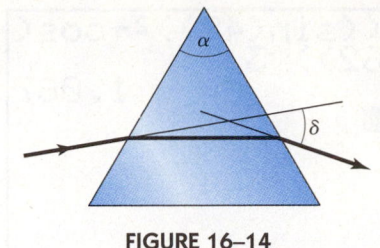

FIGURE 16–14

16–4 Evaluating a Trigonometric Expression

Before we solve trigonometric equations in the following section, let us first evaluate some trigonometric expressions. By that we mean to find the numerical value of the expression. We will do that by calculator.

Simply perform the indicated operations. Just remember that angles are assumed to be in *radians* unless marked otherwise. The following examples show the calculations with intermediate steps written down; for comparison, the screens show the work done in a single step. Use the method with which you feel most comfortable.

♦♦♦ **Example 22:** Evaluate the following to three significant digits:

$$7.82 - 3.15 \cos 67.8°$$

Solution: By calculator, we find that $\cos 67.8° = 0.3778$, so our expression becomes

$$7.82 - 3.15(0.3778) = 6.63$$

As usual when doing such calculations, we keep all the figures in the calculator whenever possible. If you must write down an intermediate result, as we did here just to show the steps, retain 1 or 2 digits more than required in the final answer.

♦♦♦

```
7.82-3.15cos(67.
8)
           6.6298
```

TI-83/84 screen for Example 22. The calculator was in **DEGREE** mode for this example.

♦♦♦ **Example 23:** Evaluate the following to four significant digits:

$$1.836 \sin 2 + 2.624 \tan 3$$

Solution: Here no angular units are indicated, so we assume them to be radians. Switching to radian mode, we get

$$1.836 \sin 2 + 2.624 \tan 3 = 1.836(0.90930) + 2.624(-0.14255)$$

$$= 1.295 \qquad ♦♦♦$$

```
1.836sin(2)+2.62
4tan(3)
           1.2954
■
```

TI-83/84 screen for Example 23, done in a single step. Here the calculator was in **RADIAN** mode.

♦♦♦ **Example 24:** Evaluate the expression

$$(\sin^2 48° + \cos 62°)^3$$

to four significant digits.

TI-83/84 screen for Example 24, with the calculator in **DEGREE** mode. The computation is done here in one step, but with the extensive use of parentheses. You may prefer to do it in two or more steps.

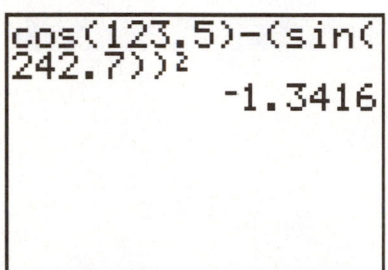

TI-83/84 screen for Example 25, done in a single step. The calculator was in **DEGREE** mode.

Solution: The notation $\sin^2 48°$ is the same as $(\sin 48°)^2$. Let us carry five digits and round to four in the last step.

$$(\sin^2 48° + \cos 62°)^3 = [(0.74314)^2 + 0.46947]^3$$
$$= (1.0217)^3 = 1.067 \qquad \text{◆◆◆}$$

Common Error	Do not confuse an exponent that is on the angle with one that is on the entire function. $$(\sin \theta)^2 = \sin^2 \theta \neq \sin \theta^2$$

◆◆◆ Example 25: Evaluate the expression

$$\cos 123.5° - \sin^2 242.7°$$

to four significant digits.

Solution: Since

$$\cos 123.5° = -0.55194$$

and

$$\sin 242.7° = -0.88862$$

we get

$$\cos 123.5° - \sin^2 242.7° = -0.55194 - (-0.88862)^2$$
$$= -1.342$$

rounded to four digits. ◆◆◆

An Application: Compound cuts:

Figure 16–15 shows a board lying flat on a workbench, with a *miter angle* μ, the acute angle between an *edge* and the end of the board. Figure 16–16 shows a *bevel angle* β, the acute angle between the *face* and end of the board. When the end has both a miter angle and a bevel angle, we say the end is cut at a *compound angle*, and that the cut that produces it is a *compound cut*. Compound cuts are needed for frames, tapered columns, crown moldings, hoppers, and so forth.

Figure 16–17 shows part of a frame, lying flat, having a *corner angle* of γ. The corner angle is 90° for a rectangular frame, 60° for an equilateral triangle, 108° for a regular pentagon, and so forth. For any regular polygon of n sides, the sum of the interior angles is given by Eq. 72, sum of the interior angles $= (n-2)180°$, so each corner angle γ is

$$\gamma = \frac{(n-2)180°}{n}$$

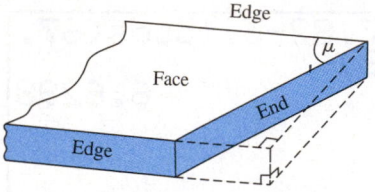

FIGURE 16–15   Miter angle μ.

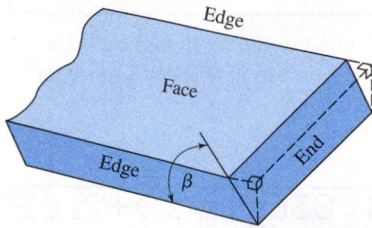

FIGURE 16–16   Bevel angle β.

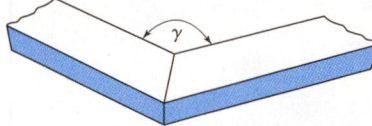

FIGURE 16–17   Corner angle γ.

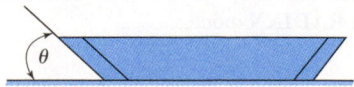

FIGURE 16–18   Tilt angle θ.

For a frame lying flat on the bench, the miter angle is equal to half the corner angle and the bevel angle is 90°.

Now let us *tilt* the frame up from the bench by a *tilt angle* θ, Fig. 16–18, so that its faces no longer lie in the same plane. Here the end of each board will now have a miter angle different from half the corner angle, and a bevel angle other than 90°. We can find the miter and bevel angles for a compound cut with a tilt angle θ and corner angle γ from the following formulas, which we give without proof:

$$\text{Miter angle } \mu = \tan^{-1}\left(\cos \theta \tan \frac{\gamma}{2}\right)$$

$$\text{Bevel angle } \beta = \sin^{-1}\left(\sin \theta \sin \frac{\gamma}{2}\right)$$

◆◆◆ **Example 26:** Find the miter and bevel angles for a rectangular frame whose faces are titled at an angle of 15.0°.

Solution: We substitute into the formulas with $\gamma = 90.0°$ and $\theta = 15.0°$.

$$\text{Miter angle } \mu = \tan^{-1}\left(\cos 15° \tan\frac{90°}{2}\right)$$

$$= \tan^{-1}(0.966) = 44.0°$$

$$\text{Bevel angle } \beta = \sin^{-1}\left(\sin 15° \sin\frac{90°}{2}\right)$$

$$= \sin^{-1}(0.183) = 10.5° \qquad ◆◆◆$$

Exercise 4 ◆ Evaluating a Trigonometric Expression

Evaluate each trigonometric expression to three significant digits.

1. $5.27 \sin 45.8° - 1.73$

2. $2.84 \cos 73.4° - 3.83 \tan 36.2°$

3. $3.72 (\sin 28.3° + \cos 72.3°)$

4. $11.2 \tan 5 + 15.3 \cos 3$

5. $2.84(5.28 \cos 2 - 2.82) + 3.35$

6. $2.63 \sin 2.4 + 1.36 \cos 3.5 + 3.13 \tan 2.5$

7. $\sin 35° + \cos 35°$

8. $\sin 125° \tan 225°$

9. $\cos 270° \cos 150° + \sin 270° \sin 150°$

10. $\dfrac{\sin^2 155°}{1 + \cos 155°}$

11. $\sin^2 75°$

12. $\tan^2 125° - \cos^2 125°$

13. $(\cos^2 206° + \sin 206°)^2$

14. $\sqrt{\sin^2 112° - \cos 112°}$

Applications

15. *Trajectories:* An object thrown at an angle θ and with an initial velocity of v_0 follows the path given by

$$y = x \tan \theta - \frac{16.1x^2}{v_0^2} \sec^2 \theta \qquad \text{feet}$$

If $v_0 = 376$ ft/s and $\theta = 35.5°$, find y when $x = 125$ ft.

16. *Simple Harmonic Motion:* The weight, Fig. 16–19, moves with simple harmonic motion, its displacement given by $x = 4.52 \cos(2.55t + 30.8°)$ inch. Find x when $t = 110$ s.

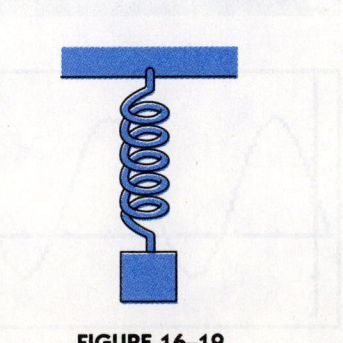

FIGURE 16–19

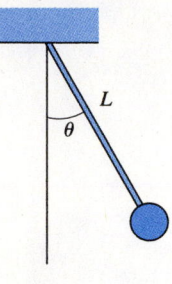

FIGURE 16–20

17. *Pendulum:* The period T for a pendulum of length L, Fig. 16–20, is approximately given by

$$T \cong 2\pi \sqrt{\frac{L}{g}\left(1 + \frac{1}{4}\sin^2\frac{\theta}{2} - \frac{9}{64}\sin^4\frac{\theta}{2}\right)}$$

where $g = 32.2$ ft/s^2 and θ is the angle between the pendulum and the vertical. For a pendulum of length 1.25 ft, find T when $\theta = 7.83°$.

18. For the engine crank and connecting rod, Fig. 16–21, (a) show that

$$x = r\cos\theta + \sqrt{L^2 - r^2\sin^2\theta} \quad \text{inch}$$

(b) If $L = 8.75$ in. and $r = 3.28$ in., find x when $\theta = 30°$.

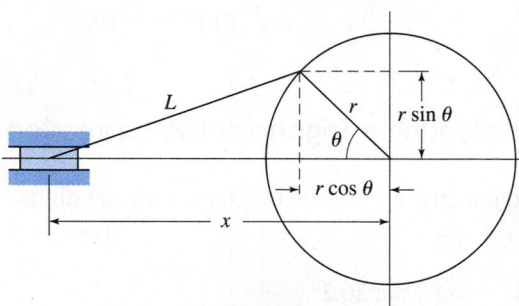

FIGURE 16–21

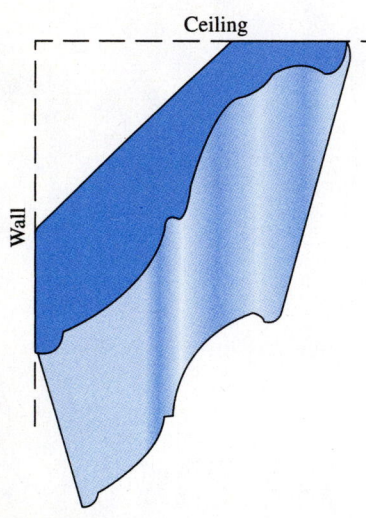

FIGURE 16–22   A typical crown molding.

19. *Crown Molding:* A *crown molding*, Fig. 16–22, is the trim between a wall and ceiling. Find the compound cut angles if the molding is tilted at 45.0° at a corner of the room where the walls meet at (a) 90.0° and (b) 125°.

20. *Skylight Framing:* The rectangular opening in a roof, Fig. 16–23, is flared to admit more light. The tilt angle is 75.0°. Find the compound cut angles at the corners.

21. *Hexagonal Window:* The faces of a regular hexagonal window are tilted by an angle of 18.0°. Find the compound cut angles.

22. *Equilateral Triangular Church Window:* Find the compound cut angles for a church window whose faces are tilted by an angle of 25.0°.

23. *Project:* Make a picture frame in a shop, using angles you have calculated. Show it in class and explain what you did.

24. *Computer:* Make a spreadsheet showing the compound cut angles for a range of tilts and corner angles. Hang it in your shop for reference.

25. *Project:* The compound cut formulas are not easy to derive. Find a derivation on the Web and see if you can follow it.

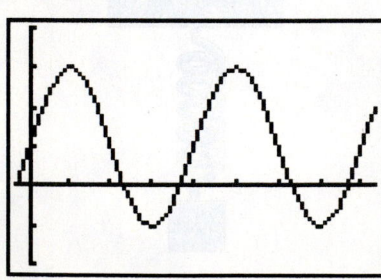

FIGURE 16–23   Skylight framing.

16–5 Solving a Trigonometric Equation

We will now use our ability to manipulate trigonometric functions to solve trigonometric equations. But first we will solve trigonometric equations graphically and then by calculator.

■ Exploration:

Try this. Use your calculator to graph the trigonometric function

$$y = 2\sin x + 1$$

in degrees, for $x = -50°$ to $+700°$.

Earlier when you graphically found roots of equations, what did you look for? Do you see any roots here? Where, approximately? Can you list *all* the roots? Why or why not?

Graph of $y = 2\sin x + 1$. Tick marks on the x axis are 90° apart.

In your exploration, you probably found that the given equation has an infinite number of roots, both positive and negative. However, it is customary to list only *nonnegative values* of the roots, and only those between 0° and 360°.

Graphical Solution of Trigonometric Equations

In an earlier chapter, we used a graphics calculator or a graphing utility on the computer to get an approximate solution to an algebraic equation. Here we use exactly the same procedure to solve a trigonometric equation.

We put the given equation into the form $f(x) = 0$ and then graph $y = f(x)$. We then use the $\boxed{\text{TRACE}}$ and $\boxed{\text{ZOOM}}$ or **zero** features to locate the zeros as accurately as we wish.

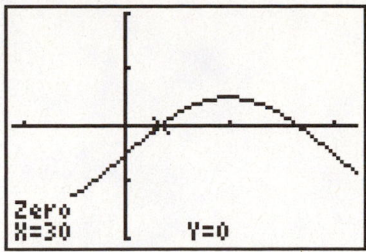

TI-83/84 screen for Example 27: $y = \sin x - \frac{1}{2}$, showing roots at 30° and 150°. We have used the **zero** feature to locate the root at $x = 30°$. Tick marks on the x axis are 90° apart.

◆◆◆ **Example 27:** Find, to the nearest tenth of a degree, any nonnegative values less than 360° of the zeros of the equation

$$\sin x = \frac{1}{2}$$

Solution: We rewrite the given equation in the form

$$\sin x - \frac{1}{2} = 0$$

and graph the function $y = \sin x - \frac{1}{2}$ as shown in the screen at the right. We see zeros at about 30° and 150°. Using $\boxed{\text{TRACE}}$ and $\boxed{\text{ZOOM}}$ or **zero** at each zero, we get

$$x = 30.0°, \ 150.0°$$

to the nearest tenth of a degree.   ◆◆◆

Solving a Trigonometric Equation by Calculator

We solve a trigonometric equation on the TI-89 just as we solved algebraic equations. Enter **solve** from $\boxed{\text{CATALOG}}$, followed by the equation, a comma, and the variable to be solved for.

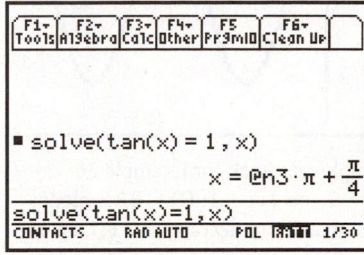

TI-89 screen for Example 28.

◆◆◆ **Example 28:** Solve the trigonometric equation, $\tan x = 1$, on the TI-89, in radians.

Solution: We enter **solve (tan (x) = 1, x)**. To restrict the answer to values less than one revolution, we can enter a range for x on the command line, $\mathbf{x \geq 0}$ **and** $\mathbf{x < 2\pi}$ (the **and** instruction is found in $\boxed{\text{CATALOG}}$). Then press $\boxed{\text{ENTER}}$.   ◆◆◆

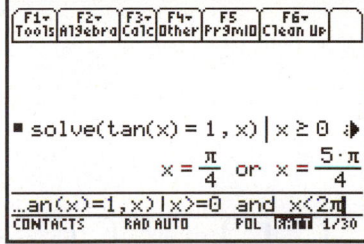

Second TI-89 screen for Example 28, where we have restricted the range of x from 0 to one revolution.

Equations Containing a Single Trigonometric Function and a Single Angle

Next we will solve some trigonometric equations using our knowledge of trigonometry and the trigonometric identities, rather than by graphing. However, we will graph each as a check. It is difficult to give general rules for solving the great variety of possible trigonometric equations, but we will show the approach by means of examples.

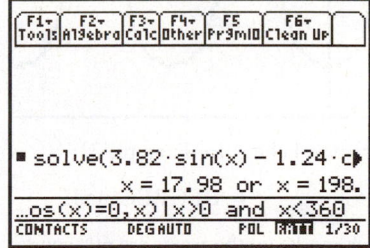

TI-89 check for Example 29.

◆◆◆ **Example 29:** Solve the equation

$$3.82 \sin x - 1.24 \cos x = 0$$

Solution: We may notice that if we divide through by $\cos x$ we will get an equation with just one trigonometric function. Dividing gives

$$3.82 \tan x - 1.24 = 0$$

Isolating $\tan x$ we get

$$\tan x = \frac{1.24}{3.82} = 0.3246$$

Taking the inverse function by calculator

$$x = \tan^{-1} 0.3246 = 17.98°$$

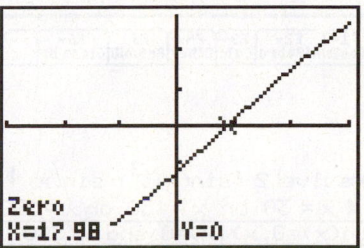

TI-83/84 graph of $y = 3.82 \sin x - 1.24 \cos x$ for Example 29, showing a root at $x = 17.98°$. The calculator is in **DEGREE** mode with ticks spaced 20° in x.

But keep in mind that the tangent is positive not only in the first quadrant, but also in the third quadrant. Using 17.9° as our reference angle, we get a second value for x,

$$x = 17.98° + 180° = 197.98°$$

Check: We graph the function $y = 3.83 \tan x - 1.24$ and find zeros at 17.98° and 197.98°.   ◆◆◆

Our next example contains a double angle.

◆◆◆ **Example 30:** Solve the equation $2 \cos 2x - 1 = 0$.

Solution: Rearranging and dividing, we have

$$\cos 2x = \tfrac{1}{2}$$
$$2x = 60°, 300°, 420°, 660°, \ldots$$
$$x = 30°, 150°, 210°, \text{ and } 330°$$

if we limit our solution to angles less than 360°. The screen shows a graph of the function $y = 2 \cos 2x - 1$ and its zeros. Note that although Example 29 contained a double angle, *we did not need the double-angle formula.* It would have been needed, however, if the same problem contained both a double angle *and* a single angle.   ◆◆◆

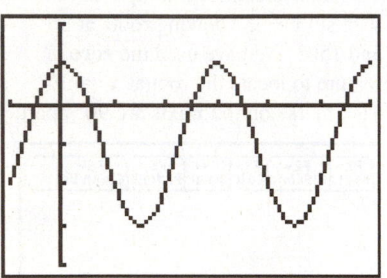

Graphical check for Example 30: $y = 2 \cos 2x - 1$. **DEGREE** mode; tick marks on the x axis are 60° apart

Common Error	It is easy to forget to find *all the angles* less than 360°, especially when the given equation contains a double angle, such as in Example 30.

If one of the functions is *squared*, we may have an equation in *quadratic form*, which can be solved by the methods for solving any quadratic equation.

◆◆◆ **Example 31:** Solve the equation $\sec^2 x = 4$.

Solution: Taking the square root of both sides gives us

$$\sec x = \pm \sqrt{4} = \pm 2$$

We thus have two solutions. By the reciprocal relations, we get

$$\frac{1}{\cos x} = 2 \qquad \text{and} \qquad \frac{1}{\cos x} = -2$$
$$\cos x = \tfrac{1}{2} \qquad \text{and} \qquad \cos x = -\tfrac{1}{2}$$
$$x = 60°, 300° \qquad \text{and} \qquad x = 120°, 240°$$

Our solution is then

$$x = 60°, 120°, 240°, 300°$$

The screen shows a graph of the function $y = \dfrac{1}{\cos^2 x} - 4$ and its zeros.   ◆◆◆

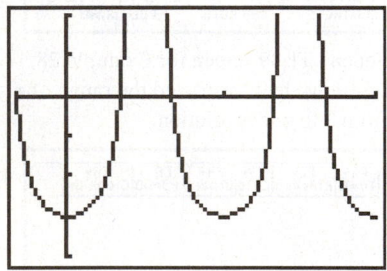

Graphical check for Example 31: $y = \dfrac{1}{\cos^2 x} - 4$. **DEGREE** mode; Tick marks on the x axis are 60° apart.

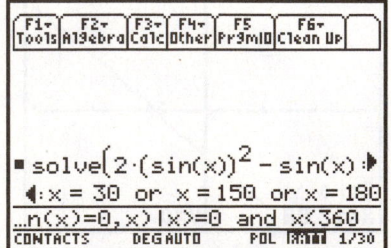

TI-89 check for Example 32. The value $x = 0$ is off the screen.

◆◆◆ **Example 32:** Solve the equation $2 \sin^2 x - \sin x = 0$.

Solution: This is an *incomplete quadratic* in $\sin x$. Factoring, we obtain

$$\sin x (2 \sin x - 1) = 0$$

Setting each factor equal to zero we get

$$\sin x = 0 \qquad\qquad 2\sin x - 1 = 0$$
$$x = 0°, 180° \qquad\qquad \sin x = \tfrac{1}{2}$$

Since sine is positive in the first and second quadrants, $x = 30°, 150°$

The screen shows a graph of the function $y = 2\sin^2 x - \sin x$ and its zeros. ◆◆◆

If an equation is in the form of a quadratic that cannot be factored, use the quadratic formula.

◆◆◆ **Example 33:** Solve the equation $\cos^2 x = 3 + 5\cos x$.

Solution: Rearranging into standard quadratic form, we have

$$\cos^2 x - 5\cos x - 3 = 0$$

This cannot be factored, so we use the quadratic formula.

$$\cos x = \frac{5 \pm \sqrt{25 - 4(-3)}}{2}$$

There are two values for $\cos x$.

$$\cos x = 5.54 \qquad\qquad \cos x = -0.541$$

(not possible)

Since cosine is negative in the second and third quadrants,

$$x = 123°, 237°$$

The screen shows a graph of the function $y = \cos^2 x - 5\cos x - 3$ and its zeros. ◆◆◆

Equations with One Angle But More Than One Function

If an equation contains two or more trigonometric functions of the same angle, first transpose all the terms to one side and try to factor that side into factors, *each containing only a single function*, and proceed as above.

◆◆◆ **Example 34:** Solve $\sin x \sec x - 2\sin x = 0$.

Solution: Factoring, we have

$$\sin x(\sec x - 2) = 0$$

Setting each factor equal to zero gives

$$\sin x = 0 \qquad\qquad \sec x = 2$$
$$\qquad\qquad\qquad \cos x = \tfrac{1}{2}$$

Since cosine is positive in the first and fourth quadrants,

$$x = 0°, 180° \qquad\qquad x = 60°, 300°$$

The screen shows a graph of the function $y = \sin x \sec x - 2\sin x$ and its zeros. ◆◆◆

If an expression is *not factorable* at first, use the fundamental identities to *express everything in terms of a single trigonometric function*, and proceed as above.

◆◆◆ **Example 35:** Solve $\sin^2 x + \cos x = 1$.

Solution: By Eq. 125, $\sin^2 x = 1 - \cos^2 x$. Substituting gives

$$1 - \cos^2 x + \cos x = 1$$
$$\cos x - \cos^2 x = 0$$

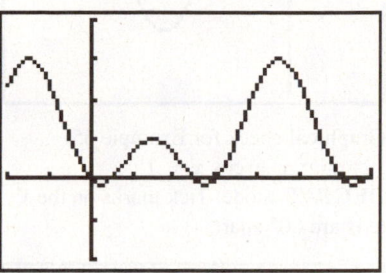

Graphical check for Example 32:
$2\sin^2 x - \sin x$. **DEGREE** mode:
Tick marks on the x axis are 60° apart.

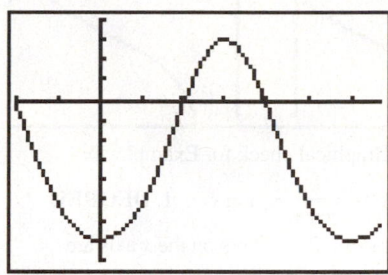

Graphical check for Example 33:
$y = \cos^2 x - 5\cos x - 3$.
Degree mode: Tick marks on the x axis are 60° apart.

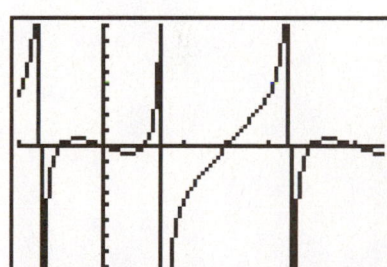

Graphical check for Example 34:
$y = \sin x \sec x - 2\sin x$. Degree mode:
Tick marks on the x axis are 60° apart.

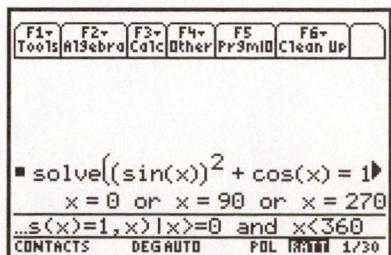

TI-89 check for Example 35.

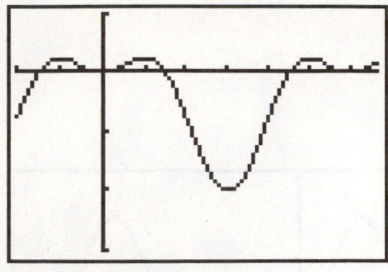

Graphical check for Example 35:
$y = \sin^2 x + \cos x - 1$.
DEGREE mode: Tick marks on the x axis are 60° apart.

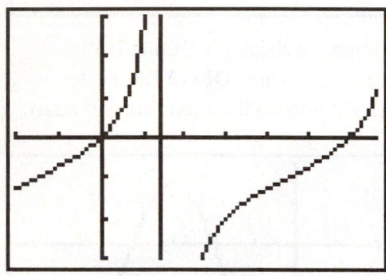

Graphical check for Example 36:
$y = \dfrac{1}{\cos x} + \tan x - 1$. **DEGREE**
mode: Tick marks on the x axis are 60° apart.

Factoring, we obtain

$$\cos x(1 - \cos x) = 0$$

$\cos x = 0$	$\cos x = 1$
$x = 90°, 270°$	$x = 0°$

The screen shows a graph of the function $y = \sin^2 x + \cos x - 1$ and its zeros.

◆◆◆

In order to simplify an expression using the Pythagorean relations, it is often necessary to *square both sides*.

◆◆◆ Example 36: Solve $\sec x + \tan x = 1$.

Solution: We have no identity that enables us to write $\sec x$ in terms of $\tan x$, but we do have an identity for $\sec^2 x$. We rearrange and square both sides, getting

$$\sec x = 1 - \tan x$$
$$\sec^2 x = 1 - 2\tan x + \tan^2 x$$

Replacing $\sec^2 x$ by $1 + \tan^2 x$ gives

$$1 + \tan^2 x = 1 - 2\tan x + \tan^2 x$$
$$\tan x = 0$$
$$x = 0, 180°$$

Since squaring can introduce extraneous roots that do not satisfy the original equation, we substitute back to check our answers. We find that the only angle that satisfies the given equation is $x = 0°$. The screen shows a graph of $y = \frac{1}{\cos x} + \tan x - 1$ and the zeros.

◆◆◆

Exercise 5 ◆ Solving a Trigonometric Equation

Solve each equation for all nonnegative values of x less than 360°. Do some by calculator.

1. $\sin x = \frac{1}{2}$
2. $2\cos x - \sqrt{3} = 0$
3. $1 - \tan x = 0$
4. $\sin x = \sqrt{3}\cos x$
5. $4\sin^2 x = 3$
6. $2\sin 3x = \frac{1}{2}$
7. $3\sin x - 1 = 2\sin x$
8. $\csc^2 x = 4$
9. $2\cos^2 x = 1 + 2\sin^2 x$
10. $2\sec x = -3 - \cos x$
11. $4\sin^4 x = 1$
12. $2\csc x - \cot x = \tan x$
13. $1 + \tan x = \sec^2 x$
14. $1 + \cot^2 x = \sec^2 x$
15. $3\cot x = \tan x$
16. $\sin^2 x = 1 - 6\sin x$
17. $3\sin(x/2) - 1 = 2\sin^2(x/2)$
18. $\sin x = \cos x$
19. $4\cos^2 x + 4\cos x = -1$
20. $1 + \sin x = \sin x\cos x + \cos x$
21. $3\tan x = 4\sin^2 x\tan x$
22. $3\sin x\tan x + 2\tan x = 0$
23. $\sec x = -\csc x$
24. $\sin x = 2\cos(x/2)$
25. $\sin x = 2\sin x\cos x$
26. $\cos x\sin 2x = 0$

Applications

27. *Alternating Current:* For the current

$$i = 274\sin(144t + 35.0°) \qquad \text{mA}$$

find the smallest time t, greater than 0, at which $i = 100$ mA.

28. *Snell's Law:* For a ray of light passing between glass and air, Fig. 16–24, the angles θ and ϕ are related by Snell's law,

$$\frac{\sin \theta}{\sin \phi} = \text{constant}$$

The constant is called the *index of refraction* of one material relative to the other, and is approximately 1.50 for glass relative to air. Find ϕ for a ray of light passing from air to glass and striking the glass surface at $\theta = 45.0°$.

29. *Trajectories:* An object thrown at an angle θ from the horizontal and with an initial velocity of v_0 follows the path given by

$$y = x \tan \theta - \frac{16.1x^2}{v_0^2} \sec^2 \theta \qquad \text{ft}$$

If the initial velocity is 125 ft/s and the object is to land 224 ft from the launch point, assuming level ground, at what angle(s) should the object be thrown?

30. *Moment of Inertia:* The moment of inertia, about the x axis, of the segment of Fig. 16–25 is

$$I = \frac{r^4}{4}(\alpha - \sin \alpha \cos \alpha) \qquad \text{inch}^4$$

where α is in radians. For a sector of radius 4.85 in., use a graphical method to find the angle α that will give a moment of inertia of 57.0 in.^{4}.

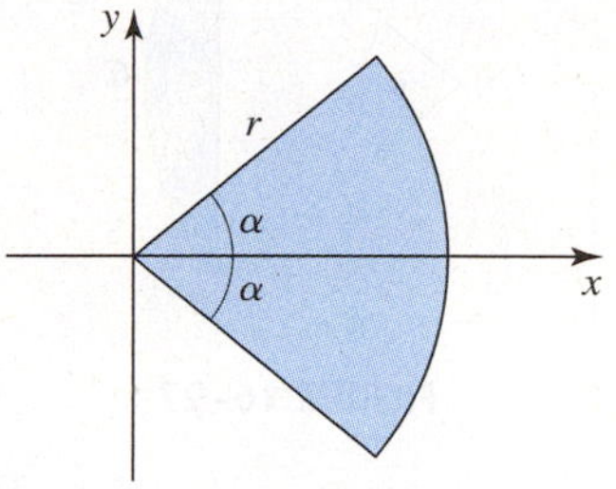

31. *Writing:* Explain in your own words the difference between a trigonometric identity and a trigonometric equation.

32. *Writing:* Describe how to approximately solve a trigonometric equation by graphics calculator. How is this different or the same as the method used earlier to solve other equations?

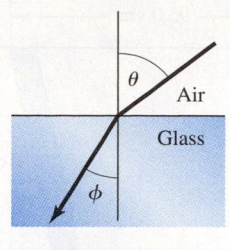

FIGURE 16–24

FIGURE 16–25

◆◆◆ CHAPTER 16 REVIEW PROBLEMS ◆◆◆◆◆◆◆◆◆◆◆◆◆◆◆◆◆◆◆◆◆◆◆◆◆◆◆◆◆◆◆◆

Prove.

1. $\dfrac{\cot \alpha \cot \beta + 1}{\cot \beta - \cot \alpha} = \cot(\alpha - \beta)$ **2.** $\dfrac{1 + \tan \theta}{1 - \tan \theta} = \sec 2\theta + \tan 2\theta$

3. $\sin \theta \cot \dfrac{\theta}{2} = \cos \theta + 1$ **4.** $\sin^4 \theta + 2 \sin^2 \theta \cos^2 \theta + \cos^4 \theta = 1$

5. $\dfrac{\sin \theta \sec \theta}{\tan \theta} = 1$ **6.** $\sec^2 \theta - \sin^2 \theta \sec^2 \theta = 1$

7. $\dfrac{\sec \theta}{\sec \theta - 1} + \dfrac{\sec \theta}{\sec \theta + 1} = 2 \csc^2 \theta$

8. $\cos \theta \tan \theta \csc \theta = 1$ **9.** $(\sec \theta - 1)(\sec \theta + 1) = \tan^2 \theta$

10. $\dfrac{\cot \theta}{1 + \cot^2 \theta} \cdot \dfrac{\tan^2 \theta + 1}{\tan \theta} = 1$

11. $\dfrac{\cos \theta}{1 - \tan \theta} + \dfrac{\sin \theta}{1 - \cot \theta} = \sin \theta + \cos \theta$

12. $(1 - \sin^2 \theta) \sec^2 \theta = 1$ **13.** $\tan^2 \theta(1 + \cot^2 \theta) = \sec^2 \theta$

14. $\dfrac{\sin \theta - 2 \sin \theta \cos \theta}{2 - 2 \sin^2 \theta - \cos \theta} = -\tan \theta$

Solve for all positive values of θ less than 360°.

15. $1 + 2 \sin^2 \theta = 3 \sin \theta$ **16.** $3 + 5 \cos \theta = 2 \cos^2 \theta$

17. $\cos \theta - 2 \cos^3 \theta = 0$ **18.** $\sin \theta + \cos 2\theta = 1$

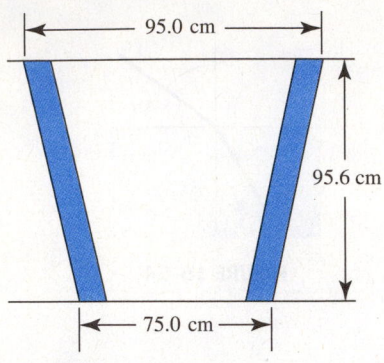

FIGURE 16–26   A hopper. Compound cuts are sometimes called "hopper cuts."

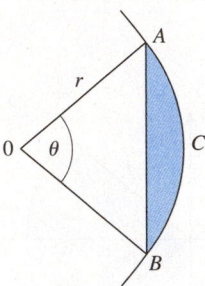

FIGURE 16–27

19. $\sin \theta = 1 - 3 \cos \theta$ **20.** $\sin^2 \theta = 1 + 6 \sin \theta$

21. $16 \cos^4 \dfrac{\theta}{2} = 9$

Evaluate each trigonometric expression to three significant digits.

22. $63.4 \cos 4.11 + 72.4 \tan 5.73$

23. $3.85(\cos 52.5° + \sin 22.6°)$

24. $\cos^2 46.2°$

25. $\sin^2 3.53 + \cos^2 1.77$

26. Find the compound cut angles for the rectangular hopper, Fig. 16–26.

27. *Project:* We have proven that the identities in this chapter were true for *acute* angles. They are, in fact, true for any angles, obtuse or acute. Choose any identity and prove this assertion.

28. *Project, Area of a Segment of a Circle:* Prove that the area of the segment that subtends an angle θ in a circle of radius r, Fig. 16–27, is

Area of a Segment	$\text{Area} = \dfrac{r^2}{2}(\theta - \sin \theta)$	**80**

Hint: From the area of sector *OACB* subtract the area of triangle *OAB*, and simplify. You will need to use the identity for the sine of twice an angle.

29. *Project:* Several sections of rail were welded together to form a continuous straight rail 255 ft long. It was installed when the temperature was 0°F, and the crew neglected to allow a gap for thermal expansion. The ends of the rail are fixed so that they cannot move outward, so the rail took the shape shown in Fig. 16–28 when the temperature rose to 120°F. Assuming the curve to be circular, find the height x at the midpoint of the track.

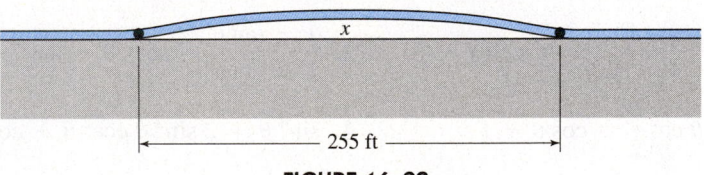

FIGURE 16–28

17

Ratio, Proportion, and Variation

◆◆◆ OBJECTIVES ◆◆

When you have completed this chapter, you should be able to

- Set up and solve a proportion for a missing quantity.
- Solve applied problems using proportions.
- Find dimensions, areas, and volumes of similar geometric figures.
- Set up and solve problems involving direct variation, inverse variation, joint variation, and combined variation.
- Solve applied problems involving variation.
- Set up and solve power function problems.

We earlier described a ratio as the quotient of two quantities, say a/b. Here we will set one ratio equal to another to get a *proportion*. We will learn how to solve proportions for a missing quantity.

We next apply the idea of proportions to the very important subject of *similar figures*. In technology we often have to relate an actual object to a drawing or model of that object—from a machine part to an engineering drawing of that part, from a geographical area to a map of that area, from a building to a scale model of that building. How do dimensions, areas, and volumes on one relate to those on the other? For example, if a one-fourth scale model of a space probe, Fig. 17–1, has a surface area of 13.5 m², what is the surface area of the actual probe? Here we give you the tools to handle such problems.

When two quantities, say, x and y, are connected by some functional relation, $y = f(x)$, some variation in the independent variable x will cause a variation in the dependent variable y. In our study of *functional variation*, we study the changes in y brought about by changes in x, for several simple functions. We often want to know the amount by which y will change when we make a certain change in x. The quantities x and y will not be abstract quantities but quantities we care about, such as; By how much will the deflection y of a beam increase when we decrease the beam thickness x by 50%?

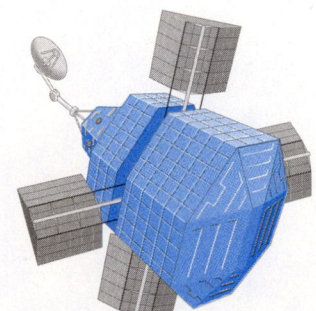

FIGURE 17–1

We will see that the methods in this chapter also give us another powerful tool for making *estimates*. In particular, we will be able to estimate a new value for one variable when we change the other by a certain amount. Again, we will be interested in such real quantities as the velocity of an object or the resistance of a circuit.

We have, in fact, already studied some aspects of functional variation—first when we substituted numerical values for x into a function and computed corresponding values of y, and later when we graphed functions. We do some graphing here, too, and also learn some better techniques for solving numerical problems.

17–1   Ratio and Proportion

Ratio

In our work so far, we have often dealt with expressions of the form

$$\frac{a}{b}$$

There are several different ways of looking at such an expression. We can say that a/b is

- A *fraction,* with a numerator a and a denominator b
- A *quotient,* where a is *divided* by b
- The *ratio* of a to b

Thus a ratio can be thought of as a fraction or as the quotient of two quantities. Another way to write a ratio is to use a colon (:) instead of a fraction line. Thus the ratio a/b can also be written

$$a : b$$

Dimensionless Ratios

For a ratio of two physical quantities, it is usual to express the numerator and denominator *in the same units,* so that they cancel and leave the ratio *dimensionless.*

◆◆◆ **Example 1:** A corridor is 8 ft wide and 12 yd long. Find the ratio of length to width.

Solution: We first express the length and width in the same units, say, feet.

$$12 \text{ yd} = 36 \text{ ft}$$

So the ratio of length to width is

$$\frac{36}{8} = \frac{9}{2}$$

Note that this ratio carries no units.   ◆◆◆

Dimensionless ratios are handy because you do not have to worry about units. For this reason they are often used in technology.

◆◆◆ **Example 2:** The *fuel-air ratio* is the ratio of the mass of fuel to air in a combustion chamber. The *turns ratio* is the ratio of turns of wire in the secondary winding of a transformer to the number of turns in the primary winding. Some other dimensionless ratios are:

Poisson's ratio	load ratio	gear ratio
endurance ratio	pi	trigonometric ratio
radian measure of angles		

◆◆◆

The word *specific* is often used to denote a ratio when there is a standard unit to which a given quantity is being compared under standard conditions.

◆◆◆ **Example 3:** The *specific gravity* of a substance is the ratio of the weight of a given volume of that substance to the weight of the same volume of water. *The specific heat* is the ratio of the amount of heat necessary to raise the temperature of a given mass of a substance by 1°C to the amount necessary to raise the temperature of an equal mass of water by the same amount. Other examples are:

specific weight	specific conductivity
specific volume	specific speed

◆◆◆

◆◆◆ **Example 4:** Aluminum has a density of 165 lb/ft³, and water has a density of 62.4 lb/ft³. Find the specific gravity (SG) of aluminum.

Solution: Dividing gives

$$SG = \frac{165 \text{ lb/ft}^3}{62.4 \text{ lb/ft}^3} = 2.64$$

Note that specific gravity is a *dimensionless ratio*.   ◆◆◆

Proportion

A *proportion* is an equation obtained when one ratio is set equal to another. If the ratio $a:b$ is equal to the ratio $c:d$, we have the proportion

$$a:b = c:d$$

which reads "the ratio of a to b equals the ratio of c to d" or "a is to b as c is to d." The quantities in the proportion (here a, b, c, and d) are called the *terms* of the proportion. The two inside terms of a proportion are called the *means,* and the two outside terms are the *extremes*.

$$a:b \quad = \quad c:d$$

We will often write such a proportion in the form

$$\frac{a}{b} = \frac{c}{d}$$

Finding a Missing Term

Solve a proportion just as you would any other fractional equation.

◆◆◆ **Example 5:** Find x if $\dfrac{3}{x} = \dfrac{7}{9}$.

Solution: Multiplying both sides by the LCD, $9x$, we obtain

$$27 = 7x$$

$$x = \frac{27}{7}$$ ◆◆◆

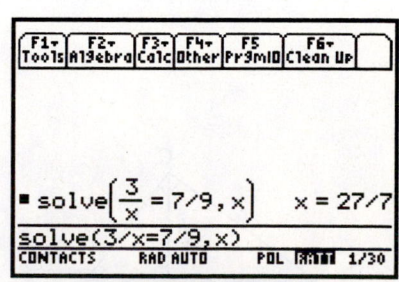

Of course, you can use certain calculators to solve these problems. Here is the T1-89 solution for Example 3.

◆◆◆ **Example 6:** Find x if $\dfrac{x+2}{3} = \dfrac{x-1}{5}$.

Solution: Multiplying through by 15 gives

$$5(x+2) = 3(x-1)$$
$$5x + 10 = 3x - 3$$
$$2x = -13$$
$$x = -\frac{13}{2} \qquad \qquad \text{◆◆◆}$$

◆◆◆ **Example 7:** Insert the missing quantity z in the following equation:

$$\frac{x-3}{2x} = \frac{z}{4x^2}$$

Solution: Solving for z gives us

$$z = \frac{4x^2(x-3)}{2x}$$
$$= 2x(x-3) \qquad \qquad \text{◆◆◆}$$

Mean Proportional

We sometimes encounter the term *mean proportional* or *geometric mean*. For example, from geometry we have the theorem:

The altitude drawn to the hypotenuse of a right triangle is the mean proportional between the segments into which it divides the hypotenuse. What does that mean?

When the means (the two inside terms) of a proportion are equal, as in

$$a:b = b:c$$

the term b is called the *mean proportional* between a and c. Solving for b, we get

$$b^2 = ac$$

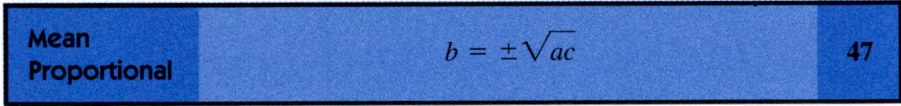

Mean Proportional	$b = \pm\sqrt{ac}$	47

The mean proportional b is also called the *geometric mean* between a and c, because a, b, and c form a *geometric progression* (a series of numbers in which each term is obtained by multiplying the previous term by the same quantity).

◆◆◆ **Example 8:** Find the mean proportional between 3 and 12.

Solution: From Eq. 47,

$$b = \pm\sqrt{3(12)} = \pm 6 \qquad \qquad \text{◆◆◆}$$

Returning to our triangle, Fig. 17–2, and with our definition of mean proportional, we can interpret the theorem to read,

$$\frac{AD}{h} = \frac{h}{DB}$$

Try to prove this theorem by using similar triangles. We will use the mean proportional in a later chapter, to insert a geometric mean between two given terms.

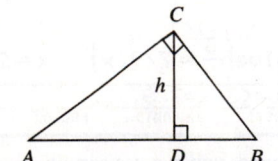

FIGURE 17–2 An altitude drawn to the hypotenuse of a right triangle.

Applications

All the remaining sections in this chapter will use ratio and proportion to solve problems from technology. We will give a few simple applications here.

◆◆◆ **Example 9:** *Gear Ratio:* The ratio of the speeds of two gears, Fig. 17–3, is found from the ratio of the number of teeth in each, with the smaller gear always turning faster than the larger. Find the speed N of gear A, having 44 teeth, if gear B has 12 teeth and rotates at 486 rev/min.

Solution: We set up the proportion

$$\frac{N}{486} = \frac{12}{44}$$

from which

$$N = \frac{(12)(486)}{44} = 133 \text{ rev/min} \qquad ◆◆◆$$

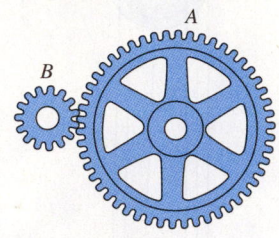

FIGURE 17–3

◆◆◆ **Example 10:** *Wire Resistance:* The resistance of a wire to the flow of current is proportional to its length. If a 15.0 ft length of wire has a resistance of 0.500 Ω, find the resistance of 12.0 ft of the same wire.

Solution: If we let R be the resistance of the 12.0 ft length, we can write the proportion

$$\frac{R}{12.0} = \frac{0.500}{15.0}$$

Solving for R we get

$$R = \frac{12(0.500)}{15.0} = 0.400 \text{ Ω} \qquad ◆◆◆$$

◆◆◆ **Example 11:** *Actual Mechanical Advantage:* A simple machine, such as the lever in Fig. 17–4, can be thought of as a *force-multiplying* device. Here a force F applied to the lever can lift a weight W. The ratio of W to F is called the *actual mechanical advantage* R_a of the machine.

$$\text{Actual mechanical advantage } R_a = \frac{\text{output force}}{\text{input force}} = \frac{W}{F}$$

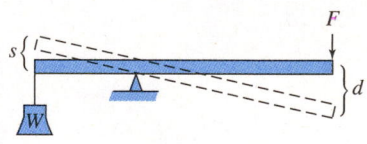

FIGURE 17–4

Find the actual mechanical advantage of the lever shown if a force of 125 lb can lift a weight of 258 lb.

Solution: We take the ratio of the output force (the weight) to the input force,

$$\text{Actual mechanical advantage } R_a = \frac{\text{output force}}{\text{input force}} = \frac{W}{F} = \frac{258}{125}$$

$$= 2.06$$

Note that mechanical advantage is a *dimensionless ratio*. ◆◆◆

◆◆◆ **Example 12:** *Ideal Mechanical Advantage:* The *work* done by a constant force acting on an object is the product of the force and the distance that the object moves. Thus the work done by the force F in Fig. 17–4 is equal to Fd, where d is the distance traveled by the force. The work done lifting the weight is Ws, where s is the

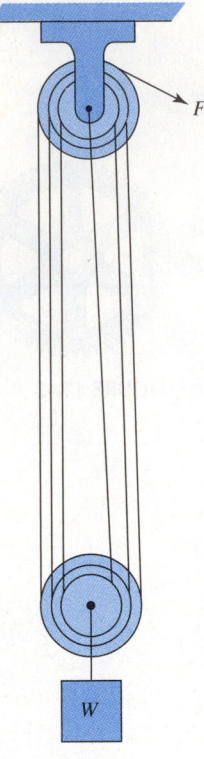

FIGURE 17–5

distance traveled by the weight. The *ideal mechanical advantage* R_i is the dimensionless ratio of the distance d traveled by the force to the distance s traveled by the weight,

$$\text{Ideal mechanical advantage } R_i = \frac{\text{distance traveled by the force}}{\text{distance traveled by the weight}} = \frac{d}{s}$$

The weight lifted by the pulley system, Fig. 17–5, moves 1.00 meter for every 6.00 meters of rope reeled in. The ideal mechanical advantage R_i is thus 6.00.   ♦♦♦

♦♦♦ **Example 13:** *Efficiency:* In an ideal machine the ideal and actual mechanical advantages would be equal, but in a real machine having friction the actual is less than the ideal.

We had earlier defined *efficiency*, in general, as output ÷ input (Eq. 19). The *efficiency E* of a *machine* is defined as the work output divided by the work input,

$$E = \frac{\text{work output}}{\text{work input}}$$

But the work done by a constant force equals the force times the distance traveled by that force (Eq. 1005). So the work output is Ws and the work input is Fd.

$$E = \frac{Ws}{Fd} = \frac{W/d}{F/s} = \frac{R_a}{R_i}$$

Thus the efficiency is the dimensionless ratio of the actual mechanical advantage to the ideal mechanical advantage.

For the pulley system in Fig. 17-5, it takes a force of 115 lb to lift a weight of 600 lb. Find the efficiency of the pulley system.

Solution: The actual mechanical advantage is

$$R_a = \frac{600}{115}$$
$$= 5.22$$

We had already determined, in Example 12, that the ideal mechanical advantage R_i was 6.00, so the efficiency is

$$E = \frac{R_a}{R_i} = \frac{5.22}{6.00}$$
$$= 0.870 \quad \text{or} \quad 87.0\%$$ ♦♦♦

Exercise 1 ♦ Ratio and Proportion

Find the value of x.

1. $3:x = 4:6$ **2.** $x:5 = 3:10$ **3.** $4:6 = x:4$

4. $3:x = x:12$ **5.** $x:(14 - x) = 4:3$ **6.** $x:12 = (x - 12):3$

7. $x:6 = (x + 6):10\frac{1}{2}$ **8.** $(x - 7):(x + 7) = 2:9$

Insert the missing quantity.

9. $\dfrac{x}{3} = \dfrac{?}{9}$ **10.** $\dfrac{?}{4x} = \dfrac{7}{16x}$ **11.** $\dfrac{5a}{7b} = \dfrac{?}{-7b}$

12. $\dfrac{a - b}{c - d} = \dfrac{b - a}{?}$ **13.** $\dfrac{x + 2}{5x} = \dfrac{?}{5}$

Find the mean proportional between the following.

14. 2 and 50   **15.** 3 and 48   **16.** 6 and 150

17. 5 and 45   **18.** 4 and 36

Applications

19. *Transformer Turns Ratio:* For the transformer shown in Fig. 17–6, the ratio of the number of turns in the secondary winding to the number of turns in the primary winding is 15. The secondary winding has 4500 turns. Find the number of turns in the primary.

20. *Gear Ratio:* Find the speed of gear G, Fig. 17–7.

21. *Actual Mechanical Advantage:* Find the actual mechanical advantage for the crank and axle machine, Fig. 17–8.

22. *Ideal Mechanical Advantage:* Find the ideal mechanical advantage for the screw jack, Fig. 17–9, if the end of the jack handle moves 57.3 cm to raise the weight 1.57 cm.

23. *Efficiency:* It takes a force of 34.8 lb to raise a weight of 74.8 lb along the inclined plane, Fig. 17–10, and the crate moves 28.7 in. along the plane while rising vertically by 12.8 in. Find (a) the actual mechanical advantage, (b) the ideal mechanical advantage, and (c) the efficiency of the plane.

24. *Golden Ratio:* A line, as shown in Fig. 17–11, is subdivided into two segments a and b such that the ratio of the smaller segment a to the larger segment b equals the ratio of the larger b to the whole $(a + b)$. The ratio a/b is called the *golden ratio* or *golden section*. Set up a proportion, based on the above definition, and compute the numerical value of this ratio.

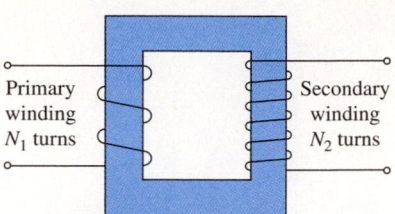

FIGURE 17–6 A transformer.

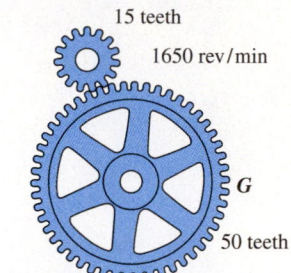

FIGURE 17–7 Spur gears.

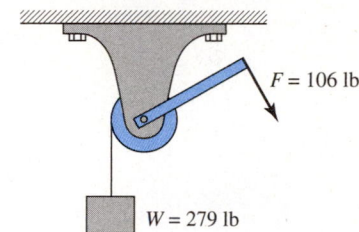

FIGURE 17–8 Crank and axle.

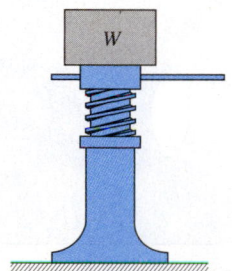

FIGURE 17–9 Screw jack.

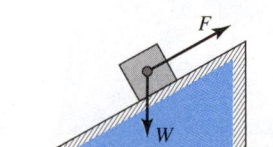

FIGURE 17–10 Inclined plane.

FIGURE 17–11 A line subdivided by the golden ratio.

17–2 Similar Figures

Our main use for ratio and proportion will be for *similar figures,* with applications to scale drawings, maps, and scale models.

■ **Exploration:** *Try this.*

(a) Arrange four equal cardboard squares to form a square whose side is twice the side of a single square, Fig. 17–12(a). Measure the diagonal of a single square and also that of the four-square array. How do they compare? How does the area of the four-square array compare with the area of a single square?

(b) Arrange eight children's blocks to form a cube, Fig. 17–12(b), each side of which is twice that of a single block. How does the volume of the eight-block cube compare with the volume of a single block?

Now suppose you had a scale drawing of a computer on which all dimensions were half those on the actual object, and the area of the screen measures 42 in.² on the drawing. What would you suppose the actual screen area would be?

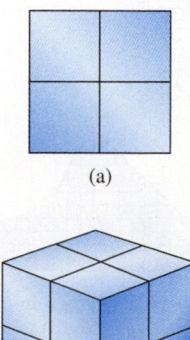

(a)

(b)

FIGURE 17–12

Again given the half-scale drawing just mentioned, and the volume of the computer's case that measures 288 in.³ on the drawing, what would you suppose its actual volume would be?   ■

We considered similar triangles in our chapter on geometry and said that corresponding sides were in proportion. We now expand the idea to cover similar plane figures of *any* shape, and also similar *solids*.

Similar figures (plane or solid) are those in which the distance between any two points on one of the figures is a *constant multiple* of the distance between two corresponding points on the other figure. In other words, if two corresponding dimensions are in the ratio of, say, 2 : 1, all other corresponding dimensions must be in the ratio of 2 : 1. In other words,

Dimensions of Similar Figures	Corresponding dimensions of plane or solid similar figures are in proportion.	**99**

We are using the term *dimension* to refer only to linear dimensions, such as lengths of sides or perimeters. It does not refer to angles, areas, or volumes.

♦♦♦ **Example 14:** Two similar solids are shown in Fig. 17–13. Find the hole diameter *D*.

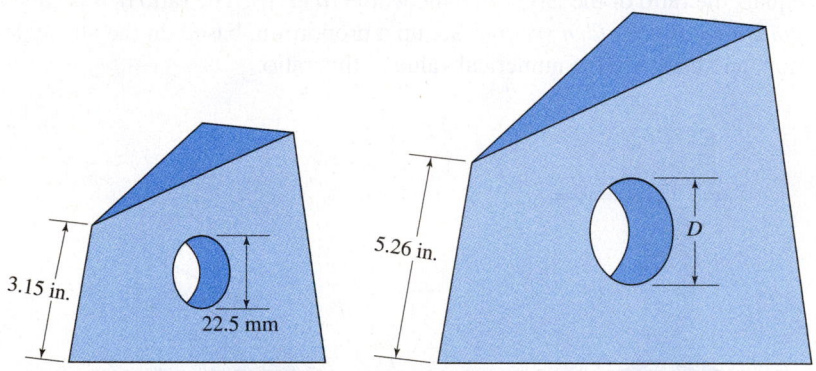

FIGURE 17–13   Similar solids.

Solution: By Statement 99, we write the proportion

$$\frac{D}{22.5} = \frac{5.26}{3.15}$$

$$D = \frac{22.5(5.26)}{3.15} = 37.6 \text{ mm}$$

Note that since we are dealing with *ratios* of corresponding dimensions, it was *not necessary* to convert all dimensions to the same units.   ♦♦♦

Areas of Similar Figures

The area of a square of side *s* is $s \times s$ or s^2. Thus if a side is multiplied by a factor *k*, the area of the larger square is $(ks) \times (ks)$ or k^2s^2. We see that the area has increased by a factor of k^2.

An area more complicated than a square can be thought of as being made up of many small squares, as shown in Fig. 17–14. Then, if a dimension of that area is multiplied by *k*, the area of each small square increases by a factor of k^2, and hence the entire area of the figure increases by a factor of k^2. Thus,

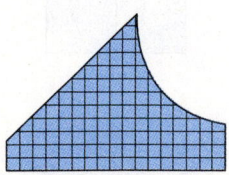

FIGURE 17–14   We can, in our minds, make the squares so small that they completely fill any irregular area. We use a similar idea in calculus when finding areas by integration.

Areas of Similar Figures	Areas of similar plane or solid figures are proportional to the *squares* of any two corresponding dimensions.	**100**

Thus if a figure has its sides doubled, the new area would be *four* times the original area. This relationship is valid not only for plane areas, but also for surface areas and cross-sectional areas of solids.

◆◆◆ **Example 15:** The triangular top surface of the smaller solid shown in Fig. 17–13 has an area of 4.84 in.^{2}. Find the area A of the corresponding surface on the larger solid.

Estimate: If each dimension on the larger solid were twice those on the smaller, then corresponding areas would be four times as large. The area of the top surface would then be about 19 in.^{2}. But the larger dimensions are *less* than double that of the smaller, so we expect an area less than 19 in.^{2} (and, of course, greater than 4.84 in.^{2}).

Solution: By Statement 100, the area A is to 4.84 as the *square* of the ratio of 5.26 to 3.15. We write the proportion

$$\frac{A}{4.84} = \left(\frac{5.26}{3.15}\right)^2$$

$$A = 4.84\left(\frac{5.26}{3.15}\right)^2 = 13.5 \text{ in.}^2 \qquad \text{◆◆◆}$$

Scale Drawings

An important application of similar figures is in the use of *scale drawings,* such as maps, engineering drawings, surveying layouts, and so on. The ratio of distances on the drawing to corresponding distances on the actual object is called the *scale* of the drawing. Use Statement 100 to convert between the areas on the drawing and areas on the actual object.

◆◆◆ **Example 16:** A certain map has a scale of 1 : 5000. How many acres on the land are represented by 168 in.^{2} on the map?

Solution: If $A =$ the area on the land, then, by Statement 100,

$$\frac{A}{168} = \left(\frac{5000}{1}\right)^2 = 25{,}000{,}000$$

$$A = 168(25{,}000{,}000) = 4.20 \times 10^9 \text{ in.}^2$$

Converting to acres, we have

$$A = 4.20 \times 10^9 \text{ in.}^2 \left(\frac{1 \text{ ft}^2}{144 \text{ in.}^2}\right)\left(\frac{1 \text{ acre}}{43{,}560 \text{ ft}^2}\right) = 670 \text{ acres} \qquad \text{◆◆◆}$$

Volumes of Similar Solids

Just as we thought of an irregular area as made up of many small squares, we can think of any solid as being made up of many tiny cubes, each of which has a volume equal to the cube of its side. Thus if the dimensions of the solid are multiplied by a factor of k, the volume of each cube (and hence the entire solid) will increase by a factor of k^3.

Volumes of Similar Figures	Volumes of similar solid figures are proportional to the *cubes* of any two corresponding dimensions.	**101**

◆◆◆ **Example 17:** If the volume of the smaller solid in Fig. 17–13 is 15.6 in.^{3}, find the volume V of the larger solid.

Solution: By Statement 101, the volume V is to 15.6 as the *cube* of the ratio of 5.26 to 3.15. So we have the proportion

$$\frac{V}{15.6} = \left(\frac{5.26}{3.15}\right)^3$$

$$V = 15.6\left(\frac{5.26}{3.15}\right)^3 = 72.6 \text{ in.}^3$$

Note that we know very little about the size and shape of these solids, yet we are able to compute the volume of the larger solid. The methods of this chapter give us another powerful tool for making *estimates*.   ◆◆◆

Common Error	Students often forget to *square* corresponding dimensions when finding areas and to *cube* corresponding dimensions when finding volumes.

A *scale model* of something, such as a building, and the object itself, are similar solids. Thus we can use proportions to find dimensions, areas, and volumes on one given the scale and the corresponding quantity on the other.

◆◆◆ **Example 18:** The model of a rocket fuel tank has a scale of $1:5$. The model has a surface area of 536 in.^{2} and a volume of 875 in.^{3}. Find (a) the surface area S and (b) the volume V of the full-sized tank.

Solution: (a) Areas are proportional to the squares of corresponding dimensions, so we write the proportion

$$\frac{S}{536} = \left(\frac{5}{1}\right)^2$$

$$S = 25(536) = 13,400 \quad \text{square inches}$$

or $13,400 \div 144 = 93.1 \text{ ft}^2$.

(b) Volumes are proportional to the cubes of corresponding dimensions, so we write the proportion

$$\frac{V}{875} = \left(\frac{5}{1}\right)^3$$

$$V = 125(875) = 109,375 \text{ cubic inches}$$

or $109,375 \div 1728 = 63.3 \text{ ft}^3$.   ◆◆◆

Exercise 2 ◆ Similar Figures

1. A container for storing propane gas is 0.755 m high and contains 20.0 liters. How high must a container of similar shape be to have a volume of 40.0 liters?
2. A certain wood stove has a firebox volume of 4.25 ft^3. What firebox volume would be expected if all dimensions of the stove were increased by a factor of 1.25?
3. If the stove in problem 2 weighed 327 lb, how much would the larger stove be expected to weigh? Remember that the weight of an object is proportional to its volume.
4. A certain solar house stores heat in 155 metric tons of stone which are in a chamber beneath the house. Another solar house is to have a chamber of

similar shape but with all dimensions increased by 15%. How many metric tons of stone will it hold?

5. Each side of a square is increased by 15.0 mm, and the area is seen to increase by 2450 mm^2. What were the dimensions of the original square?

6. The floor plan of a certain building has a scale of $\frac{1}{4}$ in. = 1 ft and shows a room having an area of 40 in.^{2}. What is the actual room area in square feet?

7. The area of a window of a car is 18.2 in.^{2} on a drawing having a scale of 1:4. Find the actual window area in square feet.

8. A pipe 3.00 inches in diameter discharges 500 gal of water in a certain time. What must be the diameter of a pipe that will discharge 750 gal in the same time? Assume that the amount of flow through a pipe is proportional to its cross-sectional area.

9. If it cost $756 to put a fence around a circular pond, what will it cost to enclose another pond having $\frac{1}{5}$ the area?

10. A triangular field whose base is 215 m contains 12,400 m^2. Find the area of a field of similar shape whose base is 328 m.

11. The site for an industrial park has an area of 2.75 in.^{2} on a map drawn to a scale of 1:10,000. Find the area of the park, in acres.

12. A U.S. Geological Survey map has a scale of 1:24,000. If two buildings are 3.86 miles apart, how many inches apart are they on the map?

13. How many acres are there in a woodland that measures 1.75 in.^{2} on the USGS map of the preceding problem?

14. A scale model of a building has a scale of 1:25. The roof on the model measures 225 in.^{2}. Find the area, in square feet, of the actual roof.

15. A room in the model of the preceding problem has a volume of 837 in.^{3} What is the volume, in cubic feet, of the corresponding room in the actual building?

16. A scale model of a space capsule has a scale of 1:8. Its volume is found, by immersion in water, to be 556,000 cm^3. Find the volume of the actual capsule, in cubic meters.

17. *Writing:* Suppose that your company makes plastic trays and is planning new ones with dimensions, including thickness of the material, double those now being made. Your company president is convinced that they will need only twice as much plastic as the older version. "Twice the size, twice the plastic," he proclaims, and no one is willing to challenge him. Your job is to make a presentation to the president where you tactfully point out that he is wrong and where you explain that the new trays will require eight times as much plastic. Write your presentation.

18. *Team Project:* We saw that the volumes (and hence the weights) of solids are proportional to the *cube* of corresponding dimensions. Applying that idea, suppose that a sporting goods company has designed a new line of equipment (ski packages, wind surfers, diving gear, jet skis, clothing, etc.) based on the following statement:

The weights of people of similar build are proportional to the cube of their heights.

The goal of this project is to prove or disprove the given statement. Your team will use data gathered from students on your campus in reaching a conclusion.

17–3   Direct Variation

■ **Exploration:**

Try this. In the same viewing window, graph

$$y = x \qquad y = 2x \qquad y = 3x$$

(a) What happens to y, for each function, as x increases? (b) What is the effect of the coefficient of x?   ■

If two variables are related, as in our exploration, by an equation of the form

$$y = kx$$

where k is a constant, we say that

$$y \text{ varies directly as } x$$

or that

$$y \text{ is directly proportional to } x$$

The constant k is called the *constant of proportionality*. Direct variation can also be indicated by using the special symbol $\propto$, which means *is proportional to*. With either symbol, we have what is called *direct variation*.

Direct Variation	$y = kx \qquad$ or $\qquad y \propto x$	49

Solving Variation Problems

Variation problems can be solved with or without evaluating the constant of proportionality. We first show a solution in which the constant *is* found by substituting the given values into Eq. 49.

◆◆◆ **Example 19:** If y is directly proportional to x, and y is 27 when x is 3, find y when x is 6.

Solution: Since y varies directly as x, we use Eq. 49.

$$y = kx$$

To find the constant of proportionality, we substitute the given values for x and y, 3 and 27.

$$27 = k(3)$$

So $k = 9$. Our equation is then

$$y = 9x$$

When $x = 6$,

$$y = 9(6) = 54 \qquad\qquad ◆◆◆$$

We now show how to solve such a problem *without* finding the constant of proportionality.

◆◆◆ **Example 20:** Solve Example 17 *without* finding the constant of proportionality.

Solution: When quantities vary directly with each other, we can set up a proportion and solve it. Let us represent the initial values of x and y by x_1 and y_1, and the second set of values by x_2 and y_2. Substituting each set of values into Eq. 49 gives us

$$y_1 = kx_1$$

and

$$y_2 = kx_2$$

Note that k has the same value in both equations. We divide the second equation by the first, and k cancels.

$$\frac{y_2}{y_1} = \frac{x_2}{x_1}$$

The proportion says: "The new y is to the old y as the new x is to the old x." We now substitute the old x and y (3 and 27), as well as the new x (6).

$$\frac{y_2}{27} = \frac{6}{3}$$

Solving yields

$$y_2 = 27\left(\frac{6}{3}\right) = 54 \quad \text{as before.} \qquad\qquad ◆◆◆$$

Of course, the same proportion can also be written in the form $y_1/x_1 = y_2/x_2$. Two other forms are also possible:

$$\frac{x_1}{y_1} = \frac{x_2}{y_2}$$

and

$$\frac{x_1}{x_2} = \frac{y_1}{y_2}$$

◆◆◆ **Example 21:** If y varies directly as x, fill in the missing numbers in the table of values.

x	1	2		5	
y			16	20	28

Solution: We find the constant of proportionality from the given pair of values (5, 20). Starting with Eq. 49, we have

$$y = kx$$

and substituting gives

$$20 = k(5)$$
$$k = 4$$

so

$$y = 4x$$

With this equation we find the missing values.

When $x = 1$: $y = 4$

When $x = 2$: $y = 8$

When $y = 16$: $x = \dfrac{16}{4} = 4$

When $y = 28$: $x = \dfrac{28}{4} = 7$

So the completed table is

x	1	2	4	5	7
y	4	8	16	20	28

◆◆◆

Applications

Many practical problems can be solved using direct variation. Once you know that two quantities are directly proportional, you may assume an equation of the form of Eq. 49. Substitute the two given values to obtain the constant of proportionality, which you then put back into Eq. 49 to obtain the complete equation. From it you may find any other corresponding values.

Alternatively, you may decide not to find k, but to form a proportion in which three values will be known, enabling you to find the fourth.

◆◆◆ **Example 22:** The force F needed to stretch a spring (Fig. 17–15) is directly proportional to the distance x stretched. If it takes 15 N to stretch a certain spring 28 cm, how much force is needed to stretch it 34 cm?

Estimate: We see that 15 N will stretch the spring 28 cm, or about 2 cm per newton. Thus a stretch of 34 cm should take about $34 \div 2$, or 17 N.

Solution: We assume an equation of the form

$$F = kx$$

where k is a constant (called the *spring constant*) for a particular spring, and depends upon its material, wire thickness, heat treatment, and so forth. Substituting the first set of values, we have

$$15 = k(28)$$
$$k = \frac{15}{28}$$

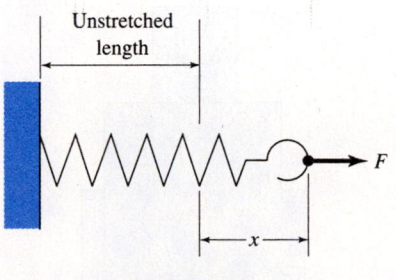

FIGURE 17–15

So the equation is $F = \dfrac{15}{28}x$. When $x = 34$,

$$F = \left(\frac{15}{28}\right)34 = 18 \text{ N (rounded)}$$

which is close to our estimated value of 17 N.   ◆◆◆

Exercise 3 ◆ Direct Variation

1. If y varies directly as x, and y is 56 when x is 21, find y when x is 74.
2. If w is directly proportional to z, and w has a value of 136 when z is 10.8, find w when z is 37.3.
3. If p varies directly as q, and p is 846 when q is 135, find q when p is 448.
4. If y is directly proportional to to x, and y has a value of 88.4 when x is 23.8,
   (a) Find the constant of proportionality.
   (b) Write the equation $y = f(x)$.
   (c) Find y when $x = 68.3$.
   (d) Find x when $y = 164$.

Assuming that y varies directly as x, fill in the missing values in each table of ordered pairs.

5.
x	9	11	
y	45		75

6.
x	3.40	7.20		12.3
y		50.4	68.6	

7.
x	115	125		154
y			167	187

Applications

8. The distance between two cities is 828 km, and they are 29.5 cm apart on a map. Find the distance between two points 15.6 cm apart on the same map.
9. If the weight of 2500 steel balls is 3.65 kg, find the number of balls in 10.0 kg.
10. If 80 transformer laminations make a stack 1.75 cm thick, how many laminations are contained in a stack 3.00 cm thick?
11. If your car now gets 21.0 mi/gal of gas, and if you can go 251 mi on a tank of gasoline, how far could you drive with the same amount of gasoline with a car that gets 35.0 mi/gal?
12. A certain automobile engine delivers 53 hp and has a displacement (the total volume swept out by the pistons) of 3.0 liters. If the power is directly proportional to the displacement, what horsepower would you expect from a similar engine that has a displacement of 3.8 liters?
13. The resistance of a conductor is directly proportional to its length. If the resistance of 2.60 mi of a certain transmission line is 155 Ω, find the resistance of 75.0 mi of that line.
14. The resistance of a certain spool of wire is 1120 Ω. A piece 10.0 m long is found to have a resistance of 12.3 Ω. Find the length of wire on the spool.
15. If a certain machine can make 1850 parts in 55 min, how many parts can it make in 7.5 h? Work to the nearest part.
16. In Fig. 17–16, the constant force on the plunger keeps the pressure of the gas in the cylinder constant. The piston rises when the gas is heated and falls when

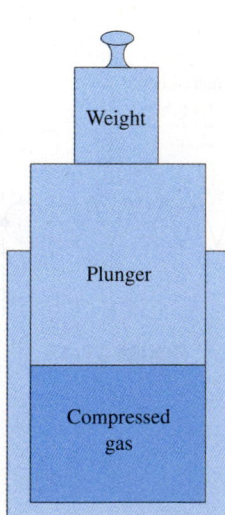

FIGURE 17–16

the gas is cooled. If the volume of the gas is 1520 cm³ when the temperature is 302 K, find the volume when the temperature is 358 K.

17. The power generated by a hydroelectric plant is directly proportional to the flow rate through the turbines, and a flow rate of 5625 gallons of water per minute produces 41.2 MW. How much power would you expect when a drought reduces the flow to 5000 gal/min?

For problem 16, use Charles's law: *The volume of a gas at constant pressure is directly proportional to its absolute temperature.* K is the abbreviation for kelvin, the SI absolute temperature scale. Add 273.15 to Celsius temperatures to obtain temperatures on the kelvin scale.

17–4   The Power Function

Definition

Earlier in this chapter we saw that we could represent the statement "*y* varies directly as *x*" by Eq. 49,

$$y = kx$$

Similarly, if *y* varies directly as the *square* of *x*, we have

$$y = kx^2$$

or if *y* varies directly as the *square root* of *x*,

$$y = k\sqrt{x} = kx^{1/2}$$

These are all examples of the *power function*.

Power Function	$y = kx^n$	148

The constants can, of course, be represented by any letter. Appendix A shows *a* instead of *k*.

The constants *k* and *n* can be any positive or negative number. This simple function gives us a great variety of relations that are useful in technology and whose forms depend on the value of the exponent, *n*. A few of these are shown in the following table:

When:	We Get This Function:		Whose Graph Is a:
$n = 1$	Linear function	$y = kx$ (direct variation)	Straight line
$n = 2$	Quadratic function	$y = kx^2$	Parabola
$n = 3$	Cubic function	$y = kx^3$	Cubical parabola
$n = -1$		$y = \dfrac{k}{x}$ (inverse variation)	Hyperbola

For exponents of 4 and 5, we have the *quartic* and *quintic* functions, respectively.

Graph of the Power Function

The graph of a power function varies greatly, depending on the exponent. We first show some power functions whose exponents are positive integers.

♦♦♦ Example 23: Graph the power functions $y = x^2$ and $y = x^3$ for a range of *x* from −3 to +3, by calculator.

Solution: The curves $y = x^2$ and $y = x^3$ are plotted as shown. Notice that the plot of $y = x^2$ has no negative values of *y*. This is typical of power functions that have

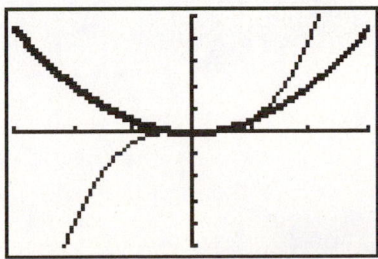

Screen for Example 21: $y = x^2$, shown heavy, and $y = x^3$, shown light.

even positive integers for exponents. The plot of $y = x^3$ does have negative y's for negative values of x. The shape of this curve is typical of a power function that has an *odd* positive integer for an exponent.   ♦♦♦

When we are graphing a power function with a *fractional* exponent, the curve may not exist for negative values of x.

♦♦♦ **Example 24:** Graph the function $y = x^{1/2}$ for $x = -2$ to 9 by calculator.

Solution: We see that for negative values of x, there are no real number values of y. For the nonnegative values, we get the graph as shown in the screen at left.   ♦♦♦

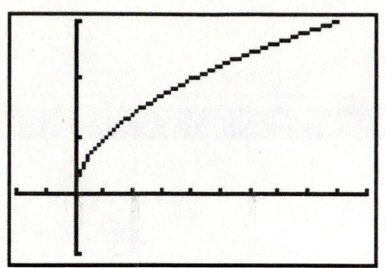

Screen for Example 22: $y = x^{1/2}$.
Tick marks are 1 unit apart.

Solving Power Function Problems

As with direct variation (and with inverse variation and combined variation, treated later), we can solve problems involving the power function with or without finding the constant of proportionality. The following example will illustrate both methods.

♦♦♦ **Example 25:** If y varies directly as the $\frac{3}{2}$ power of x, and y is 54 when x is 9, find y when x is 25.

Solution by Solving for the Constant of Proportionality: We let the exponent n in Eq. 148 be equal to $\frac{3}{2}$.

$$y = kx^{3/2}$$

As before, we evaluate the constant of proportionality by substituting the known values into the equation. Using $x = 9$ and $y = 54$, we obtain

$$54 = k(9)^{3/2} = k(27)$$

so $k = 2$. Our power function is then

$$y = 2x^{3/2}$$

This equation can then be used to find other pairs of corresponding values. For example, when $x = 25$,

$$y = 2(25)^{3/2} = 250$$

Solution by Setting Up a Proportion: Here we set up a proportion in which three values are known, and we then solve for the fourth. If

$$y_1 = kx_1^{3/2}$$

and

$$y_2 = kx_2^{3/2}$$

then y_2 is to y_1 as $kx_2^{3/2}$ is to $kx_1^{3/2}$.

$$\frac{y_2}{y_1} = \left(\frac{x_2}{x_1}\right)^{3/2}$$

Note that the constant k has canceled out. Substituting $y_1 = 54$, $x_1 = 9$, and $x_2 = 25$ we get

$$\frac{y_2}{54} = \left(\frac{25}{9}\right)^{3/2}$$

from which

$$y_2 = 54\left(\frac{25}{9}\right)^{3/2} = 250$$

as before   ♦♦♦

An Application

◆◆◆ **Example 26:** The horizontal distance S traveled by a projectile is directly proportional to the square of its initial velocity V. If the distance traveled is 1250 m when the initial velocity is 190 m/s, find the distance traveled when the initial velocity is 250 m/s.

Solution: The problem statement implies a power function of the form $S = kV^2$. Substituting the initial set of values gives us

$$1250 = k(190)^2$$

$$k = \frac{1250}{(190)^2} = 0.0346$$

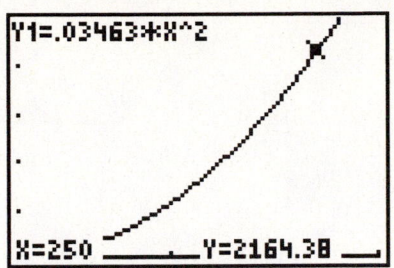

T1-83/84 check for Example 24.

So our relationship is

$$S = 0.0346V^2$$

When $V = 250$ m/s,

$$S = 0.0346(250)^2 = 2160 \text{ m}$$

Check: We use our graphics calculator to plot $S = 0.03463V^2$. Then using the $\boxed{\text{TRACE}}$ feature, we check that $S = 2160$ when $V = 250$, that $S = 1250$ when $V = 190$, and that $S = 0$ when $V = 0$. ◆◆◆

Some problems may contain no numerical values at all, as in the following example.

◆◆◆ **Example 27:** If y varies directly as the cube of x, by what factor will y change if (a) x is doubled, and (b) x is increased by 25%?

Solution: The relationship between x and y is

$$y = kx^3$$

If we give subscripts 1 to the initial values and subscripts 2 to the final values, we may write the proportion

$$\frac{y_2}{y_1} = \frac{kx_2^3}{kx_1^3} = \left(\frac{x_2}{x_1}\right)^3$$

(a) If the new x is twice the old x, we have

$$x_2 = 2x_1$$

Substituting,

$$\frac{y_2}{y_1} = \left(\frac{x_2}{x_1}\right)^3 = \left(\frac{2x_1}{x_1}\right)^3 = 2^3 = 8$$

So

$$y_2 = 8y_1$$

Thus y has increased by a factor of 8.

(b) If the new x is 25% greater than the old x, then $x_2 = x_1 + 0.25x_1$ or $1.25x_1$. We thus substitute

$$x_2 = 1.25x_1$$

so

$$\frac{y_2}{y_1} = \left(\frac{1.25x_1}{x_1}\right)^3 = \left(\frac{1.25}{1}\right)^3 = 1.95$$

or

$$y_2 = 1.95y_1$$

So y has increased by a factor of 1.95. ◆◆◆

Exercise 4 ◆ The Power Function

1. If y varies directly as the square of x, and y is 726 when x is 163, find y when x is 274.

2. If y is directly proportional to the square of x, and y is 5570 when x is 172, find y when x is 382.

3. If y varies directly as the cube of x, and y is 4.83 when x is 1.33, find y when x is 3.38.

4. If y is directly proportional to the cube of x, and y is 27.2 when x is 11.4, find y when x is 24.9

5. If y varies directly as the square of x, and y is 285.0 when x is 112.0, find y when x is 351.0.

6. If y varies directly as the square root of x, and y is 11.8 when x is 342, find y when x is 288.

7. If y is directly proportional to the cube of x, and y is 638 when x is 145, find y when x is 68.3

8. If y is directly proportional to the five-halves power of x, and y has the value 55.3 when x is 17.3,
   (a) Find the constant of proportionality.
   (b) Write the equation $y = f(x)$.
   (c) Find y when $x = 27.4$.
   (d) Find x when $y = 83.6$.

9. If y varies directly as the fourth power of x, fill in the missing values in the following table of ordered pairs:

x	18.2	75.6	
y	29.7		154

10. If y is directly proportional to the $\frac{3}{2}$ power of x, fill in the missing values in the following table of ordered pairs:

x	1.054	1.135	
y		4.872	6.774

11. If y varies directly as the cube root of x, fill in the missing values in the following table of ordered pairs:

x	315		782
y		148	275

12. Graph the power function $y = 1.04x^2$ for $x = -5$ to 5.

13. Graph the power function $y = 0.553x^5$ for $x = -3$ to 3.

14. Graph the power function $y = 1.25x^{3/2}$ for $x = 0$ to 5.

Freely Falling Body For these problems, we use Eq. 1018, $s = v_0t + at^2/2$, which says that, if assume the initial velocity is zero, the distance s fallen by a body (from rest) varies directly as the square of the elapsed time t.

15. If a body falls 176 m in 6.00 s, how far will it fall in 9.00 s?

16. If a body falls 4.90 m during the first second, how far will it fall during the third second?

17. If a body falls 129 ft in 2.00 s, how many seconds will it take to fall 525 ft?

18. If a body falls 738 units in 3.00 s, how far will it fall in 6.00 s?

Power in a Resistor From Eq. 1066, $P = I^2 R$, we see that power P dissipated in a resistor varies directly as the square of the current I in the resistor.

19. If the power dissipated in a resistor is 486 W when the current is 2.75 A, find the power when the current is 3.45 A.

20. If the current through a resistor is increased by 28%, by what percent will the power increase?

21. By what factor must the current in an electric heating coil be increased to triple the power consumed by the heater?

Photographic Exposures

22. The exposure time for a photograph is directly proportional to the square of the f stop. (The f stop of a lens is its focal length divided by its diameter.) A certain photograph will be correctly exposed at a shutter speed of $1/100$ s with a lens opening of f5.6. What shutter speed is required if the lens opening is changed to f8?

23. For the photograph in problem 22, what lens opening is needed for a shutter speed of $1/50$ s?

24. Most cameras have the following f stops: f2.8, f4, f5.6, f8, f11, and f16. To keep the same correct exposure, by what factor must the shutter speed be increased when the lens is opened one stop? (Photographer's rule of thumb: *Double the exposure time for each increase in f stop.*)

25. A certain enlargement requires an exposure time of 24 s when the enlarger's lens is set at f22. What lens opening is needed to reduce the exposure time to 8 s?

26. *Computer:* The power function $y = 11.9x^{2.32}$ has been proposed as an equation to fit the following data:

x	1	2	3	4	5	6	7	8	9	10
y	11.9	59.4	152	292	496	754	1097	1503	1901	2433

Write a program or use a spreadsheet that will compute and print the following for each given x:

(a) The value of y from the formula.

(b) The difference between the value of y from the table and that obtained from the formula. This is called a *residual*.

17–5   Inverse Variation

■ **Exploration:** The two functions

$$y = x \qquad \text{and} \qquad y = 1/x$$

look somewhat alike. (a) Can you predict, for each, what happens to y as x gets larger? (b) What happens to y for each as x gets very large? (c) What happens to y for each as x gets very small?

Try this. Graph the two functions in the same viewing window, for $x = 0$ to 3. Does your graph bear out your predictions?

For the bar in tension, Fig.17–17, the stress σ is equal to the applied force P divided by the cross-sectional area a of the bar, or $\sigma = \dfrac{P}{a}$. What happens mathematically to the stress as a increases? As a gets very small? Does the mathematics agree with what you know about the stress in a bar? ■

When we say that "y varies inversely as x" or that "y is inversely proportional to x," we mean that x and y are related by the following equation, where, as before, k is a constant of proportionality:

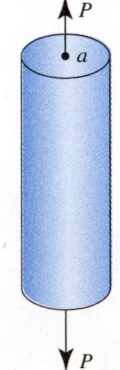

FIGURE 17–17   A bar in tension.

Inverse Variation	$y = \dfrac{k}{x}$ or $y \propto \dfrac{1}{x}$	50

The equation $y = k/x$ can also be written as

$$y = kx^{-1}$$

that is, a *power function with a negative exponent*. Another form is obtained by multiplying both sides of $y = k/x$ by x, getting

$$xy = k$$

Each of these three forms indicate inverse variation. Inverse variation problems are solved by the same methods as for any other power function. As before, we can work these problems with or without finding the constant of proportionality.

◆◆◆ **Example 28:** If y is inversely proportional to x, and y is 8 when x is 3, find y when x is 12.

Solution: The variables x and y are related by $y = k/x$. We find the constant of proportionality by substituting the given values of x and y (3 and 8).

$$8 = \frac{k}{3}$$
$$k = 3(8) = 24$$

So $y = 24/x$. When $x = 12$,

$$y = \frac{24}{12} = 2$$ ◆◆◆

Common Error	Do not confuse inverse variation with the inverse of a function.

◆◆◆ **Example 29:** If y is inversely proportional to the square of x, by what factor will y change if x is tripled?

Solution: From the problem statement we know that $y = k/x^2$ or $k = x^2 y$. If we use the subscript 1 for the original values and 2 for the new values, we can write

$$k = x_1^2 y_1 = x_2^2 y_2$$

Since the new x is triple the old, we substitute $3x_1$ for x_2.

$$x_1^2 y_1 = (3x_1)^2 y_2$$
$$= 9x_1^2 y_2$$
$$y_1 = 9y_2$$

or

$$y_2 = \frac{y_1}{9}$$

So the new y is one-ninth of its initial value.   ◆◆◆

◆◆◆ **Example 30:** If y varies inversely as x, fill in the missing values in the table:

x	1			9	12
y			12	4	

Solution: We substitute the ordered pair (9, 4) into Eq. 50.

$$4 = \frac{k}{9}$$
$$k = 36$$

So

$$y = \frac{36}{x}$$

Filling in the table, we have

x	1	3	9	12
y	36	12	4	3

◆◆◆

An Application

◆◆◆ **Example 31:** The number of oscillations N that a pendulum (Fig. 17–18) makes per unit time is inversely proportional to the square root of the length L of the pendulum, and $N = 2.00$ oscillations per second when $L = 85.0$ cm.

(a) Write the equation $N = f(L)$.

(b) Find N when $L = 115$ cm.

Solution:

(a) From the problem statement, the equation will be of the form

$$N = \frac{k}{\sqrt{L}}$$

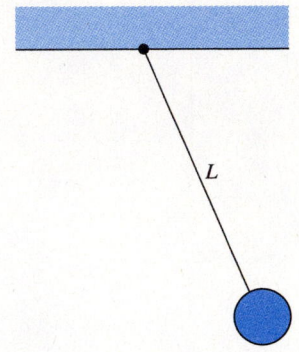

FIGURE 17–18 A simple pendulum.

Substituting $N = 2.00$ and $L = 85.0$ gives us

$$2.00 = \frac{k}{\sqrt{85.0}}$$

$$k = 2.00\sqrt{85.0} = 18.4$$

So the equation is

$$N = \frac{18.4}{\sqrt{L}}$$

(b) When $L = 115$,

$$N = \frac{18.4}{\sqrt{115}} = 1.72 \text{ oscillations per second}$$

Check: Does the answer seem reasonable? We see that N has decreased, which we expect with a longer pendulum, but it has not decreased by much. As a check, we note that L has increased by a factor of $115 \div 85$, or 1.35, so we expect N to decrease by a factor of $\sqrt{1.35}$, or 1.16. Dividing, we find that $2.00 \div 1.72 = 1.16$, so the answer checks. ◆◆◆

Exercise 5 ◆ Inverse Variation

1. If y varies inversely as x, and y is 385 when x is 832, find y when x is 226.
2. If y is inversely proportional to the square of x, and y has the value 1.55 when x is 7.38, find y when x is 44.2.
3. If y is inversely proportional to x, how does y change when x is doubled?
4. If y varies inversely as x, and y has the value 104 when x is 532,
   (a) Find the constant of proportionality.
   (b) Write the equation $y = f(x)$.
   (c) Find y when x is 668.
   (d) Find x when y is 226.

5. If y is inversely proportional to x, fill in the missing values.

x	306	622	
y	125		418

6. Fill in the missing values, assuming that y is inversely proportional to the square of x.

x	2.69			7.83
y		1.16	1.52	

7. If y varies inversely as the square root of x, fill in the missing values.

x			3567	5725
y	1136	1828		

8. If y is inversely proportional to the cube root of x, by what factor will y change when x is tripled?

9. If y is inversely proportional to the square root of x, by what percentage will y change when x is decreased by 50.0%?

10. Plot the curve $y = 5/x^2$ for $x = 0$ to 5.

11. Plot the function $y = 1/x$ for $x = 0$ to 10.

Boyle's Law

Boyle's law states that for a confined gas at a constant temperature, the product of the pressure and the volume is a constant. Another way of stating this law is that the pressure is inversely proportional to the volume, or that the volume is inversely proportional to the pressure. Assume a constant temperature in the following problems.

12. A certain quantity of gas, when compressed to a volume of 2.50 m^3, has a pressure of 184 Pa. The *pascal* (Pa) is the SI unit of pressure. It equals 1 newton per square meter. Find the pressure resulting when that gas is further compressed to 1.60 m^3.

13. The air in a cylinder is at a pressure of 14.7 lb/in.^{2} and occupies a volume of 175 in.^{3} Find the pressure when it is compressed to 25.0 in.^{3}.

14. A balloon contains 320 m^3 of gas at a pressure of 140,000 Pa. What would the volume be if the same quantity of gas were at a pressure of 250,000 Pa?

Gravitational Attraction

Newton's law of gravitation states that any two bodies attract each other with a force that is inversely proportional to the square of the distance between them.

15. The force of attraction between two certain steel spheres is 3.75×10^{-5} dyne when the spheres are placed 18.0 cm apart. Find the force of attraction when they are 52.0 cm apart.

16. How far apart must the spheres in problem 15 be placed to cause the force of attraction between them to be 3.00×10^{-5} dyne?

17. The force of attraction between the earth and some object is called the *weight* of that object. The law of gravitation states, then, that the weight of an object is inversely proportional to the square of its distance from the center of the earth. If a person weighs 150 lb on the surface of the earth (assume this to be 3960 mi from the center), how much will he weigh 1500 mi above the surface of the earth?

18. How much will a satellite, whose weight on earth is 675 N, weigh at an altitude of 250 km above the surface of the earth?

Illumination

The *inverse square law* states that for a surface illuminated by a light source, the intensity of illumination on the surface is inversely proportional to the *square* of the distance between the source and the surface.

19. A certain light source produces an illumination of 800 lux (a lux is 1 lumen per square meter) on a surface. Find the illumination on that surface if the distance to the light source is doubled.

20. A light source located 2.75 m from a surface produces an illumination of 528 lux on that surface. Find the illumination if the distance is changed to 1.55 m.

21. A light source located 7.50 m from a surface produces an illumination of 426 lux on that surface. At what distance must that light source be placed to give an illumination of 850 lux?

22. When a document is photographed on a certain copy stand, an exposure time of $\frac{1}{25}$ s is needed, with the light source 0.750 m from the document. At what distance must the light be located to reduce the exposure time to $\frac{1}{100}$ s?

Electrical

23. The current in a resistor is inversely proportional to the resistance. By what factor will the current change if a resistance increases 10.0% due to heating?

24. The resistance of a wire is inversely proportional to the square of its diameter. If an AWG (American wire gauge) size 12 conductor (0.0808-in. diameter) has a resistance of 14.8 Ω, what will be the resistance of an AWG size 10 conductor (0.1019-in. diameter) of the same length and material?

25. The capacitive reactance X_c of a circuit varies inversely as the capacitance C of the circuit. If the capacitance of a certain circuit is decreased by 25.0%, by what percentage will X_c change?

Graphics Calculator

26. Use a graphics calculator or computer to graph

$$y = 1/x \quad \text{and} \quad y = 1/x^2$$

in the same viewing window from $x = -4$ to 4. How do you explain the different behavior of the two functions for negative values of x?

17–6 Functions of More Than One Variable

So far in this chapter, we have considered only cases where y was a function of a *single* variable x. In functional notation, this is represented by $y = f(x)$. In this section we cover functions of two or more variables, such as

$$y = f(x, w)$$

and

$$y = f(x, w, z)$$

and so forth.

Joint Variation

When y varies directly as both x and w, we say that y varies *jointly* as x and w. The three variables are related by the following equation, where, as before, k is a constant of proportionality:

Joint Variation	$y = kxw \quad$ or $\quad y \propto xw$	51

◆◆◆ **Example 32:** If y varies jointly as both x and w, how will y change when x is doubled and w is one-fourth of its original value?

Solution: Let y' be the new value of y obtained when x is replaced by $2x$ and w is replaced by $w/4$, while the constant of proportionality k, of course, does not change. Substituting in Eq. 51, we obtain

$$y' = k(2x)\left(\frac{w}{4}\right)$$

$$= \frac{kxw}{2}$$

But, since $kxw = y$, then

$$y' = \frac{y}{2}$$

So the new y is half as large its former value.   ◆◆◆

Combined Variation

Relationships more complicated than that in Eq. 51 are usually referred to as *combined* variation. The way they are related can be found by careful reading of the problem statement.

◆◆◆ **Example 33:** If y varies directly as the cube of x and inversely as the square of z, we write

$$y = \frac{kx^3}{z^2}$$ ◆◆◆

◆◆◆ **Example 34:** If y is directly proportional to x and the square root of w, and inversely proportional to the square of z, we write

$$y = \frac{kx\sqrt{w}}{z^2}$$ ◆◆◆

Once you have an equation, the problem is solved in the same way as we did simpler types.

◆◆◆ **Example 35:** If y varies directly as the square root of x and inversely as the square of w, by what factor will y change when x is made four times larger and w is tripled?

Solution: First write the equation linking y to x and w, including a constant of proportionality k. From the problem statement

$$y = \frac{k\sqrt{x}}{w^2}$$

We get a new value for y (let's call it y') when x is replaced by $4x$ and w is replaced by $3w$. Thus

$$y' = \frac{k\sqrt{4x}}{(3w)^2} = k\frac{2\sqrt{x}}{9w^2}$$

$$= \frac{2}{9}\left(\frac{k\sqrt{x}}{w^2}\right) = \frac{2}{9}y$$

We see that y' is $\frac{2}{9}$ as large as the original y.   ◆◆◆

◆◆◆ **Example 36:** If y varies directly as the square of w and inversely as the cube root of x, and $y = 225$ when $w = 103$ and $x = 157$, find y when $w = 126$ and $x = 212$.

Solution: From the problem statement,

$$y = k\frac{w^2}{\sqrt[3]{x}}$$

Solving for k, we have

$$225 = k\frac{(103)^2}{\sqrt[3]{157}} = 1967k$$

or $k = 0.1144$. So

$$y = \frac{0.1144w^2}{\sqrt[3]{x}}$$

When $w = 126$ and $x = 212$,

$$y = \frac{0.1144(126)^2}{\sqrt[3]{212}} = 305 \quad \blacklozenge\blacklozenge\blacklozenge$$

◆◆◆ **Example 37:** Assume y is directly proportional to the square of x and inversely proportional to the cube of w. By what factor will y change if x is increased by 15% and w is decreased by 20%?

Solution: The relationship between x, y, and w was given verbally. As an equation,

$$y = k\frac{x^2}{w^3}$$

so we can write the proportion

$$\frac{y_2}{y_1} = \frac{k\dfrac{x_2^2}{w_2^3}}{k\dfrac{x_1^2}{w_1^3}} = \frac{x_2^2}{w_2^3}\cdot\frac{w_1^3}{x_1^2} = \left(\frac{x_2}{x_1}\right)^2\left(\frac{w_1}{w_2}\right)^3$$

We now replace x_2 with $1.15x_1$ and replace w_2 with $0.8w_1$.

$$\frac{y_2}{y_1} = \left(\frac{1.15x_1}{x_1}\right)^2\left(\frac{w_1}{0.8w_1}\right)^3 = \frac{(1.15)^2}{(0.8)^3} = 2.58$$

So y will increase by a factor of 2.58. ◆◆◆

An Application

◆◆◆ **Example 38:** If the strength S of a rectangular beam, Fig. 17–19, varies jointly as its width w, the square of its depth d, and inversely as its length L, how will the strength change if all three dimensions are increased by a factor of 1.50?

Solution: The strength of the original beam may be written

$$S_1 = \frac{kwd^2}{L}$$

and the strength of the new beam is

$$S_2 = \frac{k(1.50w)(1.50d)^2}{1.50L}$$

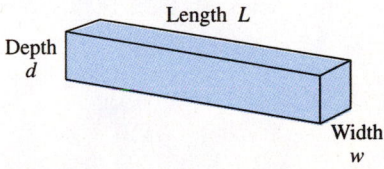

FIGURE 17–19 A rectangular beam.

Dividing one by the other we get

$$\frac{S_2}{S_1} = \frac{\dfrac{k(1.50w)(1.50d)^2}{1.50L}}{\dfrac{kwd^2}{L}} = (1.50)^2 = 2.25$$

Thus the strength of the new beam is 2.25 times that of the old. ◆◆◆

Exercise 6 ◆ Functions of More Than One Variable

Joint Variation

1. If y varies jointly as w and x, and y is 483 when x is 742 and w is 383, find y when x is 274 and w is 756.
2. If y varies jointly as x and w, by what factor will y change if x is tripled and w is halved?
3. If y varies jointly as w and x, by what percent will y change if w is increased by 12% and x is decreased by 7.0%?
4. If y varies jointly as w and x, and y is 3.85 when w is 8.36 and x is 11.6, evaluate the constant of proportionality, and write the complete expression for y in terms of w and x.
5. If y varies jointly as w and x, fill in the missing values.

w	x	y
46.2	18.3	127
19.5	41.2	
	8.86	155
12.2		79.8

Combined Variation

6. If y is directly proportional to the square of x and inversely proportional to the cube of w, and y is 11.6 when x is 84.2 and w is 28.4, find y when x is 5.38 and w is 2.28.
7. If y varies directly as the square root of w and inversely as the cube of x, by what factor will y change if w is tripled and x is halved?
8. If y is directly proportional to the cube root of x and to the square root of w, by what percent will y change if x and w are both increased by 7.0%?
9. If y is directly proportional to the $\frac{3}{2}$ power of x and inversely proportional to w, and y is 284 when x is 858 and w is 361, evaluate the constant of proportionality, and write the complete equation for y in terms of x and w.
10. If y varies directly as the cube of x and inversely as the square root of w, fill in the missing values in the table.

w	x	y
1.27		3.05
	5.66	1.93
4.66	2.75	3.87
7.07	1.56	

Geometry

11. The area of a triangle varies jointly as its base and altitude. By what percent will the area change if the base is increased by 15% and the altitude decreased by 25%?
12. If the base and the altitude of a triangle are both halved, by what factor will the area change?

Electrical

13. When an electric current flows through a wire, the resistance to the flow varies directly as the length and inversely as the cross-sectional area of the wire. If the length and the diameter are both tripled, by what factor will the resistance change?
14. If 750 m of 3.00-mm-diameter wire has a resistance of 27.6 Ω, what length of similar wire 5.00 mm in diameter will have the same resistance?

Gravitation

15. Newton's law of gravitation states that every body in the universe attracts every other body with a force that varies directly as the product of their masses and inversely as the square of the distance between them. By what factor will the force change when the distance is doubled and each mass is tripled?

16. If both masses are increased by 60% and the distance between them is halved, by what percent will the force of attraction increase?

Illumination

17. The intensity of illumination at a given point is directly proportional to the intensity of the light source and inversely proportional to the square of the distance from the light source. If a desk is properly illuminated by a 75.0-W lamp 8.00 ft from the desk, what size lamp will be needed to provide the same lighting at a distance of 12.0 ft?

18. How far from a 150-candela light source would a picture have to be placed so as to receive the same illumination as when it is placed 12 m from an 85-candela source?

Gas Laws

19. The volume of a given weight of gas varies directly as its absolute temperature t and inversely as its pressure p. If the volume is 4.45 m^3 when $p = 225$ kilopascals (kPa) and $t = 305$ K, find the volume when $p = 325$ kPa and $t = 354$ K.

20. If the volume of a gas is 125 ft^3, find its volume when the absolute temperature is increased 10% and the pressure is doubled.

Work

21. The amount paid to a work crew varies jointly as the number of persons working and the length of time worked. If 5 workers earn \$5123.73 in 3.0 weeks, in how many weeks will 6 workers earn a total of \$6148.48?

22. If 5 bricklayers take 6.0 days to finish a certain job, how long would it take 7 bricklayers to finish a similar job requiring 4 times the number of bricks?

Strength of Materials

23. The maximum safe load of a rectangular beam (Fig. 17–20) varies jointly as its width and the square of its depth and inversely as the length of the beam. If a beam 8.00 in. wide, 11.5 in. deep, and 16.0 ft long can safely support 15,000 lb, find the safe load for a beam 6.50 in. wide, 13.4 in. deep, and 21.0 ft long made of the same material.

24. If the width of a rectangular beam is increased by 11%, the depth decreased by 8%, and the length increased 6%, by what percent will the safe load change?

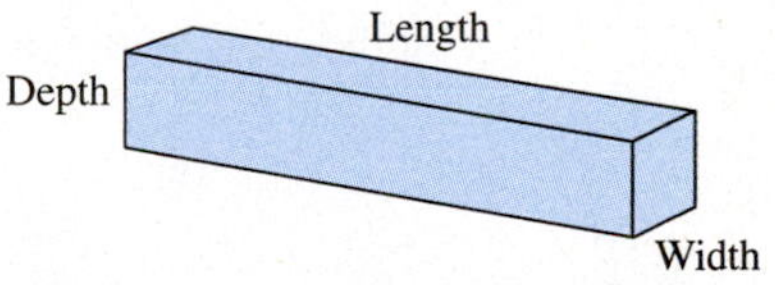

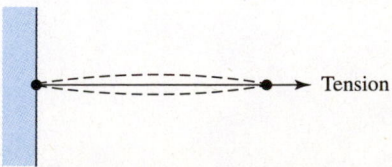

FIGURE 17–20   A rectangular beam.

Mechanics

25. The number of vibrations per second made when a stretched wire (Fig. 17–21) is plucked varies directly as the square root of the tension in the wire and inversely as its length. If a 1.00-m-long wire will vibrate 325 times a second when the tension is 115 N, find the frequency of vibration if the wire is shortened to 0.750 m and the tension is decreased to 95.0 N.

26. The kinetic energy of a moving body is directly proportional to its mass and the square of its speed. If the mass of a bullet is halved, by what factor must its speed be increased to have the same kinetic energy as before?

FIGURE 17–21   A stretched wire.

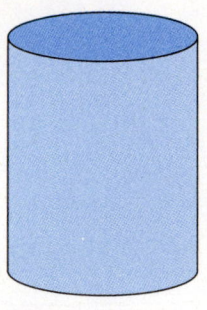

FIGURE 17–22
A vertical cylindrical tank.

Fluid Flow

27. The time needed to empty a vertical cylindrical tank (Fig. 17–22) varies directly as the square root of the height of the tank and the square of its radius. By what factor will the emptying time change if the height is doubled and the radius increased by 25%?

28. The power available in a jet of liquid is directly proportional to the cross-sectional area of the jet and to the cube of the velocity. By what factor will the power increase if the area and the velocity are both increased 50%?

••• CHAPTER 17 REVIEW PROBLEMS •••••••••••••••••••••••••••••••

1. If y varies inversely as x, and y is 736 when x is 822, find y when x is 583.

2. If y is directly proportional to the $\frac{5}{2}$ power of x, by what factor will y change when x is tripled?

3. If y varies jointly as x and z, by what percent will y change when x is increased by 15.0% and z is decreased by 4.00%?

4. The braking distance of an automobile varies directly as the square of the speed. If the braking distance of a certain automobile is 34.0 ft at 25.0 mi/h, find the braking distance at 55.0 mi/h.

5. The rate of flow of liquid from a hole in the bottom of a tank is directly proportional to the square root of the liquid depth. If the flow rate is 225 L/min when the depth is 3.46 m, find the flow rate when the depth is 1.00 m.

6. The power needed to drive a ship varies directly as the cube of the speed of the ship, and a 77.4-hp engine will drive a certain ship at 11.2 knots. Find the horsepower needed to propel that ship at 18.0 knots.

7. If the tensile strength of a cylindrical steel bar varies directly as the square of its diameter, by what factor must the diameter be increased to triple the strength of the bar?

8. The life of an incandescent lamp varies inversely as the 12th power of the applied voltage, and the light output varies directly as the 3.5th power of the applied voltage. By what factor will the life increase if the voltage is lowered by an amount that will decrease the light output by 10%?

9. One of Kepler's laws states that the time for a planet to orbit the sun varies directly as the $\frac{3}{2}$ power of its distance from the sun. How many years will it take for Saturn, which is about $9\frac{1}{2}$ times as far from the sun as is the earth, to orbit the sun?

10. The volume of a cone varies directly as the square of its base radius and as its altitude. By what factor will the volume change if its altitude is doubled and its base radius is halved?

11. The number of oscillations made by a pendulum in a given time is inversely proportional to the length of the pendulum. A certain clock with a 30.00-in.-long pendulum is losing 15.00 min/day. Should the pendulum be lengthened or shortened, and by how much?

12. A trucker usually makes a trip in 18.0 h at an average speed of 60.0 mi/h. Find the traveling time if the speed were reduced to 50.0 mi/h.

13. The force on the vane of a wind generator varies directly as the area of the vane and the square of the wind velocity. By what factor must the area of a vane be increased so that the wind force on it will be the same in a 12-mi/h wind as it was in a 35-mi/h wind?

14. The maximum deflection of a rectangular beam varies inversely as the product of the width and the cube of the depth. If the deflection of a beam having a width of 15 cm and a depth of 35 cm is 7.5 mm, find the deflection if the width is made 20 cm and the depth 45 cm.

15. When an object moves at a constant speed, the travel time is inversely proportional to the speed. If a satellite now circles the earth in 26 h 18 min, how long will it take if booster rockets increase the speed of the satellite by 15%?

16. The life of an incandescent lamp is inversely proportional to the 12th power of the applied voltage. If a lamp has a life of 2500 h when run at 115 V, what will its life expectancy be when it is run at 110 V?

17. How far from the earth will a spacecraft be equally attracted by the earth and the moon? The distance from the earth to the moon is approximately 239,000 mi, and the mass of the earth is about 82.0 times that of the moon.

18. Ohm's law states that the electric current flowing in a circuit varies directly as the applied voltage and inversely as the resistance. By what percent will the current change when the voltage is increased by 25% and the resistance increased by 15%?

19. The allowable strength F of a column varies directly as its length L and inversely as its radius of gyration r. If $F = 35$ kg when L is 12 m and r is 3.6 cm, find F when L is 18 m and r is 4.5 cm.

20. The time it takes a pendulum to go through one complete oscillation (the *period*) is directly proportional to the square root of its length. If the period of a 1-m pendulum is 1.25 s, how long must the pendulum be to have a period of 2.5 s?

21. If the maximum safe current in a wire is directly proportional to the $\frac{3}{2}$ power of the wire diameter, by what factor will the safe current increase when the wire diameter is doubled?

22. Four workers take ten 6-h days to finish a job. How many workers are needed to finish a similar job which is 3 times as large, in five 8-h days?

23. When a jet of water strikes the vane of a water turbine and is deflected through an angle θ, the force on the vane varies directly as the square of the jet velocity and the sine of $\theta/2$. If θ is decreased from 55° to 40°, and if the jet velocity increases by 40%, by what percentage will the force change?

24. By what factor will the kinetic energy change if the speed of a projectile is doubled and its mass is halved? Use the fact that the kinetic energy of a moving body is directly proportional to the mass and the square of the velocity of the body.

25. The fuel consumption for a ship is directly proportional to the cube of the ship's speed. If a certain tanker uses 1584 gal of diesel fuel on a certain run at 15.0 knots, how much fuel would it use for the same run when the speed is reduced to 10.0 knots? A *knot* is equal to 1 nautical mile per hour. (1 nautical mile $=$ 1.151 statute miles.)

26. The diagonal of a certain square is doubled. By what factor does the area change?

27. A certain parachute is made from 52.0 m² of fabric. If all of its dimensions are increased by a factor of 1.40, how many square meters of fabric will be needed?

28. The rudder of a certain airplane has an area of 7.50 ft². By what factor must the dimensions be scaled up to triple the area?

29. A 1.5-in.-diameter pipe fills a cistern in 5.0 h. Find the diameter of a pipe that will fill the same cistern in 9.0 h.

30. If 315 m of fence will enclose a circular field containing 2.0 acres, what length will enclose 8.0 acres?

31. The surface area of a one-quarter model of an automobile measures 1.10 m². Find the surface area of the full-sized car.

32. A certain water tank is 25.0 ft high and holds 5000 gal. How much would a 35.0-ft-high tank of similar shape hold?

33. A certain tractor weighs 5.80 tons. If all of its dimensions were scaled up by a factor of 1.30, what would the larger tractor be expected to weigh?

34. If a trough 5.0 ft long holds 12 pailfuls of water, how many pailfuls will a similar trough hold that is 8.0 ft long?

35. A ball 4.50 inches in diameter weighs 18.0 oz. What is the weight of another ball of the same density that is 9.00 inches in diameter?

36. A ball 4.00 inches in diameter weighs 9.00 lb. What is the weight of a ball 25.0 cm in diameter made of the same material?

37. A worker's contract has a cost-of-living clause that requires that his salary be proportional to the cost-of-living index. If he earned $1600 per week when the index was 94.0, how much should he earn when the index is 127?

38. A woodsman pays a landowner $8.00 for each cord of wood he cuts on the landowner's property, and he sells the wood for $75.00 per cord. When the price of firewood rose to $95.00, the landowner asked for a proportional share of the price. How much should he get per cord?

39. The series of numbers 3.5, 5.25, 7.875, . . . form a *geometric progression*, where the ratio of any term to the one preceding it is a constant. This is called the *common ratio*. Find this ratio.

40. The *specific gravity* (SG) of a solid or liquid is the ratio of the density of the substance to the density of water at a standard temperature (Eq. 1044). Taking the density of water as 62.4 lb/ft^3, find the density of copper having a specific gravity of 8.89.

41. An iron casting having a surface area of 746 cm^2 is heated so that all its dimensions increase by 1.00%. Find the new area.

42. A certain boiler pipe will permit a flow of 35.5 gal/min. Buildup of scale inside the pipe eventually reduces its diameter to three-fourths of its previous value. Assuming that the flow rate is proportional to the cross-sectional area, find the new flow rate for the pipe.

18

Exponential and Logarithmic Functions

OBJECTIVES

When you have completed this chapter, you should be able to

- Graph the exponential function.
- Solve exponential growth and decay problems by formula or by the universal growth and decay curves.
- Convert expressions between exponential and logarithmic form.
- Evaluate common and natural logarithms and antilogarithms.
- Evaluate, manipulate, and simplify logarithmic expressions.
- Solve exponential and logarithmic equations.
- Solve applied problems involving exponential or logarithmic equations.

According to legend, the inventor of chess asked that the reward for his invention be a single grain of wheat on the first square of a chessboard, then two grains on the second square, four on the next, and so forth, each square having double the number on the preceding square. If that were done, the total amount of wheat given would be greater than that grown by mankind since the beginning of agriculture. This is an illustration of the astounding properties of exponential growth.

We will study the exponential function here and give other, more practical examples of its use. We also introduce logarithms and study their properties. Logarithms were once widely used for computation but now have been replaced by calculators and computers. We will use logarithms mainly to solve exponential equations. Of course, we will also use our calculators to solve these equations, as well as the logarithmic equations to follow.

This chapter is loaded with interesting applications from many fields. These include computation of compound interest and other financial quantities, temperature changes with time, the catenary, a curve used in suspension bridges, population

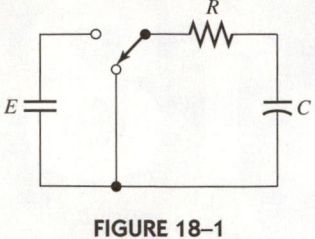

FIGURE 18–1

growth and decline, speed changes in a rotating wheel, decrease in light intensity as it passes through glass or water, change in atmospheric pressure with altitude, radioactive decay, star magnitudes, decibels gained or lost in terms of power, sound level, voltage, or current, heat losses from pipes, swings of a pendulum, density change of seawater with depth, the Richter scale for earthquakes, and the pH value of liquids.

For example, the current in the capacitor, Fig. 18–1, at any time t is given by the exponential equation

$$ i = \frac{E}{R} e^{-t/RC} $$

Could you solve for the time at which a given current is reached, given the other quantities? We will show how in this chapter.

18–1   The Exponential Function

Definition

An *exponential function* is one in which the independent variable appears in the exponent. The quantity that is raised to the power is called the *base*.

◆◆◆ **Example 1:** The following are exponential functions if a, b, and e represent positive constants:

$$ y = 5^x \qquad y = b^x \qquad y = e^x \qquad y = 10^x $$

$$ y = 3a^x \qquad y = 7^{x-3} \qquad y = 5e^{-2x} $$
◆◆◆

Realize that these exponential functions are different from the *power function,* $y = ax^n$. In the power function the exponent is a *constant*. Here the *unknown is in the exponent*. We shall see that e, like π, has a specific value which we will discuss later in this chapter.

Graph of the Exponential Function

■ **Exploration:**

Try this. Use your graphing calculator to graph the exponential function

$$ y = 2^x $$

For comparison, graph the power function $y = x^2$ in the same viewing window. Both functions contain the same three quantities, y, x, and 2. Are the two curves the same? What are some important differences? ■

Note in the screen shown at left that the graph of the exponential function is completely different from the graph of the power function. The power function graphs as a parabola, which is symmetrical about the y axis, while the exponential function has no symmetry. The power function has no asymptote (a line that a curve approaches ever closer as the distance from the origin increases) while the exponential function has the x axis as an asymptote.

Now let us see what happens if an exponential function has a *negative* exponent.

■ **Exploration:**

Try this. Graph $y = 2^x$ and $y = 2^{-x}$ in the same viewing window.
What is the effect of making the exponent negative? ■

When the exponent is *positive*, we say that the graph *increases exponentially*, and that when the exponent is *negative*, the graph *decreases exponentially*. Also

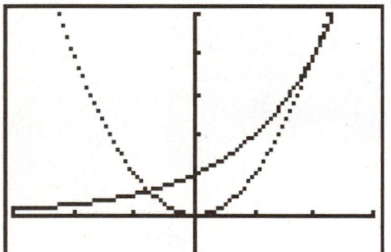

Graphs of $y = 2^x$ (solid) and $y = x^2$ (dotted). Tick marks on the horizontal axis are 1 unit apart.

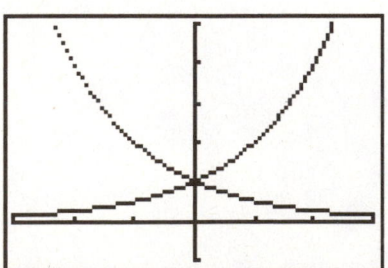

Graphs of $y = 2^x$ (solid) and $y = 2^{-x}$ (dotted). Tick marks on the horizontal axis are 1 unit apart.

notice that neither curve reaches the x axis. Thus for the exponential functions $y = 2^x$ and $y = 2^{-x}$, y cannot equal zero.

Compound Interest

We refer to an exponential increase as *exponential growth*, and to exponential decrease as *exponential decay*. We want to be able to write an equation describing exponential growth or decay so we can use that equation to predict the behavior of that quantity over time. For example, if a radioactive substance decays exponentially by 2% per year, how much will be left after 50 years?

To get such an equation, we will start with a familiar example of *money deposited at compound interest*. Then in the next section we will extend our findings to the exponential growth or decay of other quantities.

When an amount of money a (called the *principal*) is invested at an interest rate n (expressed as a decimal), it will earn an amount of interest (an) at the end of the first interest period. If this interest is then added to the principal, and if both continue to earn interest, we have what is called *compound interest*, the type of interest commonly given to bank accounts.

■ Exploration:

Try this. Say you deposit $1000 in a bank account giving an interest of 5% at the end of each year. How much will you earn in interest in one year? How much will you have at the end of the first year?

Now say that you keep the interest earned, plus the initial $1000, in the account for another year. How much will you have at the end of the second year?

Make a table showing your balance at the end of each year for several years. It should look something like this.

Starting amount: $1000

End of year one: $1000 + 5% of $1000 = $1000(1.05) = $1050

End of year two: $1050 + 5% of $1050 = $1050(1.05) = $1102.50

What can you say about the growth of your money? ■

You should have found that the amount y_t you have at the end of any year t is equal to the amount at the start of that year multiplied by $(1 + n)$. Stated as a formula,

Recursion Relation for Compound Interest	$y_t = y_{t-1}(1 + n)$	1010

This equation is called a *recursion relation*, from which each term in a series of numbers may be found from the term immediately preceding it. Let us use Eq. 1010 to make a table showing the balance at the end of a number of interest periods.

Period	Balance y at End of Period
1	$a(1 + n)$
2	$a(1 + n)^2$
3	$a(1 + n)^3$
4	$a(1 + n)^4$
.	.
.	.
.	.
t	$a(1 + n)^t$

We thus get the following exponential function, with a equal to the principal, n equal to the rate of interest, and t the number of interest periods:

Compound Interest Computed Once per Period	$y = a(1 + n)^t$	**1009**

◆◆◆ **Example 2:** *An Application:* To what amount (to the nearest cent) will an initial deposit of \$500 accumulate if invested at a compound interest rate of $6\frac{1}{2}\%$ per year for 8 years?

Solution: We are given

> Principal $a = \$500$
> Interest rate $n = 0.065$, expressed in decimal form
> Number of interest periods $t = 8$

Substituting into Eq. 1009 gives

$$y = 500(1 + 0.065)^8$$
$$= \$827.50 \qquad \qquad ◆◆◆$$

Compound Interest Computed *m* Times per Period

We can use Eq. 1009 to compute interest on a savings account for which the interest is computed once a year. But what if the interest is computed more often, as most banks do, for example, monthly, weekly, or even daily?

   Let us modify the compound interest formula. Instead of computing interest *once* per period, let us compound it *m times* per period. The exponent in Eq. 1009 thus becomes *mt*. The interest rate *n*, however, has to be reduced to $(1/m)$th of its old value, or to n/m. Our equation then becomes:

Compound Interest Computed *m* Times per Period	$y = a\left(1 + \dfrac{n}{m}\right)^{mt}$	**1011**

◆◆◆ **Example 3:** Repeat the computation of Example 2 with the interest computed *monthly* rather than annually.

Solution: As in Example 2, $a = \$500$, annual interest rate $n = 0.065$, and $t = 8$. But n is an annual rate, so we must use Eq. 1011 with $m = 12$.

$$y = 500\left(1 + \frac{0.065}{12}\right)^{12(8)} = 500(1.00542)^{96}$$
$$= \$839.83$$

or \$12.33 more than in Example 2. ◆◆◆

Exponential Growth

Equation 1011, $y = a(1 + n/m)^{mt}$, allows us to compute the amount obtained, with interest compounded any number of times per period that we choose. We could use this formula to compute the interest if it were compounded every week, or every day, or every second. But what about *continuous* compounding, or continuous growth? What would happen to Eq. 1011 if m got very, very large, in fact, infinite? Let us make m increase to large values; but first, to simplify the work, we make a substitution. Let

$$k = \frac{m}{n}$$

Equation 1011 then becomes

$$y = a\left(1 + \frac{1}{k}\right)^{knt}$$

which can be written

$$y = a\left[\left(1 + \frac{1}{k}\right)^{k}\right]^{nt}$$

Then as m grows large, so will k.

■ Exploration:

Try this. Given the expression

$$\left(1 + \frac{1}{k}\right)^{k}$$

(a) Try to predict what will happen to the value of this expression as k grows without bound. Will the entire expression also grow without bound? Reach a limiting value? Vanish? Then, (b) set this expression equal to y and graph for positive values of k (except on the graphics calculator you would use x rather than k) or (c) using a calculator or spreadsheet, compute values of this expression for ever increasing values of k.

What did you find from your graph or computation? Did you predict correctly? ■

In your exploration you may have gotten the surprising result that as m (and k) continue to grow infinitely large, the value of $(1 + 1/k)^k$ does *not* grow without limit, but approaches a specific value: 2.7183. . . . This important number is given the special symbol e. We can express the same idea in *limit notation* by writing the following equation:

Definition of e	$$\lim_{k \to \infty}\left(1 + \frac{1}{k}\right)^{k} = e$$ *As k grows without bound, the value of $(1 + 1/k)^k$ approaches $e\,(\approx 2.7183\ldots)$.*	137

The number e is named after a Swiss mathematician, Leonhard Euler (1707–83). Its value has been computed to thousands of decimal places. In the exercise we show how to find e using series.

Thus when m gets infinitely large, the quantity $(1 + 1/k)^k$ approaches e, and the formula for continuous growth becomes the following:

Exponential Growth	$$y = ae^{nt}$$	151

◆◆◆ **Example 4:** Repeat the computation of Example 2 with the interest compounded *continuously* rather than monthly. Use the $\boxed{e^x}$ key on your calculator.

Solution: From Eq. 151,

$$y = 500e^{0.065(8)}$$
$$= 500e^{0.52}$$
$$= \$841.01$$

or $1.18 more than when compounded monthly.   ◆◆◆

Common Error	When you are using Eq. 151, the time units of n and t *must agree*.

◆◆◆ **Example 5:** *An Application:* A colony of bacteria used to clean up an oil spill grows exponentially at the rate of 2% per minute. By how many times will it have increased after 4 h?

Solution: The units of n and t do *not* agree (yet).

$$n = 2\% \text{ per } \textit{minute} \quad \text{and} \quad t = 4 \textit{ hours}$$

So we convert one of them to agree with the other.

$$t = 4 \text{ h} = 240 \text{ min}$$

Using Eq. 151,

$$y = ae^{nt} = ae^{0.02(240)} = ae^{4.8} = 122a$$

So the final amount is 122 times as great as the initial amount.   ◆◆◆

Exponential Decay

When a quantity *decreases* so that the amount of decrease is proportional to the present amount, we have what is called *exponential decay*. Take, for example, a piece of steel just removed from a furnace. The amount of heat leaving the steel depends upon its temperature: The hotter it is, the faster heat will leave, and the faster it will cool. But as the steel cools, its lower temperature causes slower heat loss. Slower heat loss causes slower rate of temperature drop, and so on. The result is the typical *exponential decay curve* shown in Fig. 18–2. Finally, the temperature is said to *asymptotically approach* room temperature.

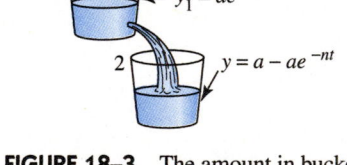

FIGURE 18–2 Exponential decrease in temperature. This is called *Newton's law of cooling*.

The equation for exponential decay is the same as for exponential growth, except that the exponent is negative.

Exponential Decay	$y = ae^{-nt}$	153

◆◆◆ **Example 6:** *An Application:* A room initially 80°F above the outside temperature cools exponentially at the rate of 25% per hour. Find the temperature of the room (above the outside temperature) at the end of 135 min.

Solution: We first make the units consistent: 135 min = 2.25 h. Then, by Eq.153,

$$y = 80e^{-0.25(2.25)} = 45.6°\text{F}$$ ◆◆◆

Exponential Growth to an Upper Limit

Suppose that the bucket in Fig. 18–3 initially contains a gallons of water above the hole. Then assume that y_1, the amount remaining in the bucket, decreases exponentially. By Eq. 153,

$$y_1 = ae^{-nt}$$

FIGURE 18–3   The amount in bucket 2 grows exponentially to an upper limit a.

The amount of water in bucket 2 thus increases from zero to a final amount a. The amount y in bucket 2 at any instant is equal to the amount that has left bucket 1, or

$$y = a - y_1$$
$$= a - ae^{-nt}$$

Factoring, we get

Exponential Growth to an Upper Limit	$y = a(1 - e^{-nt})$	154

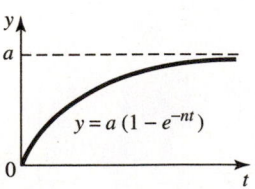

FIGURE 18–4   Exponential growth to an upper limit.

So the equation for this type of growth to an upper limit is exponential, where a equals the upper limit quantity, n is equal to the rate of growth, and t represents time (see Fig. 18–4).

◆◆◆ Example 7: *An Application:* The voltage across a certain capacitor grows exponentially from 0 V to a maximum of 75 V, at a rate of 7% per second. Write the equation for the voltage at any instant, and find the voltage at 2.0 s.

Solution: From Eq. 154, with $a = 75$ and $n = 0.07$,

$$v = 75(1 - e^{-0.07t})$$

where v represents the instantaneous voltage. When $t = 2.0$ s,

$$v = 75(1 - e^{-0.07(2.0)})$$
$$= 9.80 \text{ V} \qquad \text{◆◆◆}$$

◆◆◆ Example 8: *An Application:* A colony of "oil-eating" bacteria colonizing a new area of a spill grows at a rate of 7% per day, with an upper limit of 212,000 bacteria per liter. (a) Write the equation for the population y at any instant. (b) Find the population at five days.

Solution: (a) From Eq. 154, with $a = 212{,}000$ and $n = 0.07$,

$$y = 212{,}000(1 - e^{-0.07t})$$

(b) When $t = 5$ days,

$$y = 212{,}000(1 - e^{-0.07(5)})$$
$$= 62{,}600 \text{ bacteria/liter} \qquad \text{◆◆◆}$$

◆◆◆ Example 9: *An Application; The Catenary:* A chain or flexible rope or cable will hang in a shape called a *catenary*, Fig. 18–5. Its equation is

$$y = \frac{a(e^{x/a} + e^{-x/a})}{2}$$

Graph the catenary equation for values of x from -5 to 5, for $a = 1$, for $a = 2$, and $a = 3$.

Solution: We use a graphing utility or make and plot a table of point pairs, getting the three curves shows in Fig. 18–6.

Note that a is the y-intercept and it also determines the shape of the curve.   ◆◆◆

Time Constant

Our equations for exponential growth and decay all contained a rate of growth n. Often it is more convenient to work with the *reciprocal* of n, rather than n itself. This reciprocal is called the *time constant, T.*

Time Constant	$T = \dfrac{1}{\text{growth rate } n}$	155

◆◆◆ Example 10: If a quantity decays exponentially at a rate of 25% per second, what is the time constant?

Solution:

$$T = \frac{1}{0.25} = 4 \text{ s}$$

Notice that the time constant has the units of *time.*   ◆◆◆

FIGURE 18–5 A rope or chain hangs in the shape of a catenary. Mount your finished graph for Example 9 vertically, and hang a fine chain from the endpoints. Does its shape (after you have adjusted its length) match that of your graph?

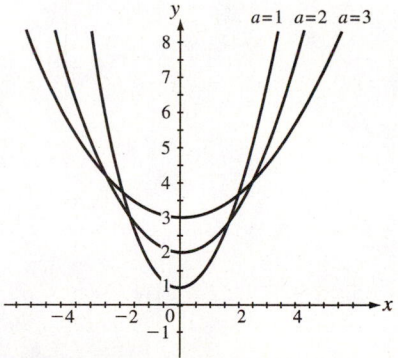

FIGURE 18–6 Graph of
$$y = \frac{a(e^{x/a} + e^{-x/a})}{2}.$$

Universal Growth and Decay Curves

If we replace n by $1/T$ in Eq. 153, we get

$$y = ae^{-t/T}$$

Then we divide both sides by a.

$$\frac{y}{a} = e^{-t/T}$$

We plot this curve with the dimensionless ratio t/T for our horizontal axis and the dimensionless ratio y/a for the vertical axis (Fig. 18–7). Thus the horizontal axis is *the number of time constants* that have elapsed, and the vertical axis is *the ratio of the final amount to the initial amount*. The resulting graph is called the *universal decay curve*.

We can use the universal decay curve to obtain *graphical solutions* to exponential decay problems.

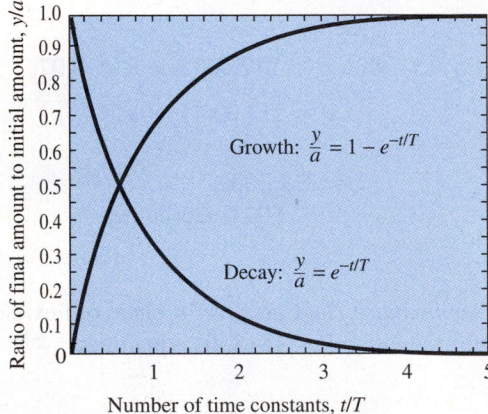

FIGURE 18–7   Universal growth and decay curves.

◆◆◆ **Example 11:** *An Application:* A current decays from an initial value of 300 mA at an exponential rate of 20% per second. Using the universal decay curve, graphically find the current after 7 s.

Solution: We have $n = 0.20$, so the time constant is

$$T = \frac{1}{n} = \frac{1}{0.20} = 5 \text{ s}$$

At our given time $t = 7$ s the number of time constants elapsed is then equal to

$$\frac{t}{T} = \frac{7}{5} = 1.4 \text{ time constants}$$

From the universal decay curve, at $t/T = 1.4$, we read

$$\frac{y}{a} = 0.25$$

Since $a = 300$ mA,

$$y = 300(0.25) = 75 \text{ mA}$$

Figure 18–7 also shows the *universal growth curve*

$$\frac{y}{a} = 1 - e^{-t/T}$$

It is obtained and used in the same way as the universal decay curve.

Exercise 1 ◆ The Exponential Function

Exponential Function

Graph each exponential function, manually or by calculator, for the given values of x. Take $e = 2.718$.

1. $y = 0.2(3.2)^x$ $(x = -4 \text{ to } +4)$

2. $y = 3(1.5)^{-2x}$ $(x = -1 \text{ to } 5)$

3. $y = 5(1 - e^{-x})$ $(x = 0 \text{ to } 10)$

4. $y = 4e^{x/2}$ $(x = 0 \text{ to } 4)$

The Catenary

5. A cable hangs from the tops of two poles spaced 10.00 m apart, Fig. 18–8. (a) Use the equation $y = (a/2)(e^{x/a} + e^{-x/a})$ with $a = 4$ to determine the value of y at the ends of the cable. (b) Find the sag s from the cable ends to its low points. (c) Determine the pole height h so that the cable clears the ground by 3.00 m.

Compound Interest

6. Find the amount to which $500 will accumulate in 6 years at a compound interest rate of 6% per year compounded annually.

7. If our compound interest formula is solved for a, we get

$$a = \frac{y}{(1 + n)^t}$$

where a is the amount that must be deposited now at interest rate n to produce an amount y in t years. The amount a is called the *present worth* or *present value* of the future amount y. This formula, and the three that follow, are standard formulas used in business and finance. How much money (to the nearest dollar) would have to be invested at $7\frac{1}{2}$% per year to accumulate to $10,000 in 10 years?

8. What annual compound interest rate (compounded annually) is needed to enable an investment of $5000 to accumulate to $10,000 in 12 years? Use Eq. 1009, $y = a(1+n)^t$.

9. Using Eq. 1011, $y = a\left(1 + \dfrac{n}{m}\right)^{mt}$, find the amount (to the nearest cent) to which $1 will accumulate in 20 years at a compound interest rate of 10% per year (a) compounded annually, (b) compounded monthly, and (c) compounded daily.

10. If an amount of money R is deposited *every year* for t years at a compound interest rate n, it will accumulate to an amount y, where

$$y = R\left[\frac{(1 + n)^t - 1}{n}\right]$$

A series of annual payments such as these is called an *annuity*. If a worker pays $2000 per year into a retirement plan yielding 6% annual interest, what will the value of this annuity be after 25 years? (Work to the nearest dollar.)

11. If the equation for an annuity in problem 10 is solved for R, we get

$$R = \frac{ny}{(1 + n)^t - 1}$$

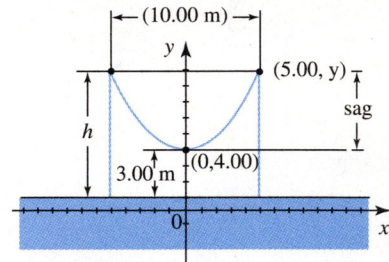

FIGURE 18–8 A cable hanging between two poles.

where R is the annual deposit required to produce a total amount y at the end of t years. This is called a *sinking fund*. How much, to the nearest dollar, would the worker in problem 10 have to deposit each year (at 6% per year) to have a total of $100,000 after 25 years?

12. The yearly payment R that can be obtained for t years from a present investment a is

$$R = a\left[\frac{n}{(1 + n)^t - 1} + n\right]$$

This is called *capital recovery*. If a person has $85,000 in a retirement fund, how much (to the nearest dollar) can be withdrawn each year so that the fund will be exhausted in 20 years, if the amount remaining in the fund is earning $6\frac{1}{4}\%$ per year?

Exponential Growth

Use either the formulas or the universal growth and decay curves, as directed by your instructor.

13. A quantity grows exponentially at the rate of 5.00% per year for 7 years. Find the final amount if the initial amount is 201 units.

14. A yeast manufacturer finds that the yeast will grow exponentially at a rate of 15.0% per hour. How many pounds of yeast must the manufacturer start with to obtain 499 lb at the end of 8 h?

15. If the population of a country was 11.4 million in 2010 and grows at an annual rate of 1.63%, find the population by the year 2015.

16. If the rate of inflation is 12.0% per year, how much could a camera that now costs $225 be expected to cost 5 years from now?

17. The oil consumption of a certain country was 12.0 million barrels in 2010, and it grows at an annual rate of 8.30%. Find the oil consumption by the year 2015.

18. A pharmaceutical company makes a vaccine-producing bacterium that grows at a rate of 2.5% per hour. How many units of this organism must the company have initially to have 1000 units after 5 days?

Exponential Growth to an Upper Limit

19. A steel casting initially at 0°C is placed in a furnace at 801°C. If it increases in temperature at the rate of 5.00% per minute, find its temperature after 20.0 min.

20. Equation 154, $y = a(1 - e^{-nt})$, approximately describes the hardening of concrete, where y is the compressive strength (lb/in.^{2}) t days after pouring. Using values of $a = 4000$ and $n = 0.0696$ in the equation, find the compressive strength after 14 days.

21. When the switch in Fig. 18–9 is closed, the current i will grow exponentially according to the following equation:

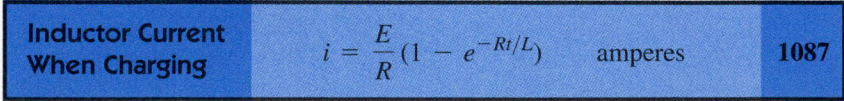

Inductor Current When Charging	$i = \dfrac{E}{R}(1 - e^{-Rt/L})$ amperes	1087

where L is the inductance in henries and R is the resistance in ohms. Find the current at 0.0750 s if $R = 6.25\ \Omega$, $L = 186$ H, and $E = 249$ V.

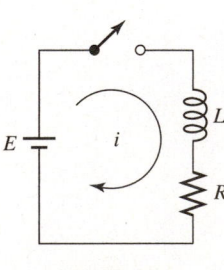

FIGURE 18–9

Exponential Decay

22. A flywheel is rotating at a speed of 1805 rev/min. When the power is disconnected, the speed decreases exponentially at the rate of 32.0% per minute. Find the speed after 10.0 min.

23. A steel forging is 1495°F above room temperature. If it cools exponentially at the rate of 2.00% per minute, how much will its temperature drop in 1 h?

24. As light passes through glass or water, its intensity decreases exponentially according to the equation

$$I = I_0 e^{-kx}$$

where I is the intensity at a depth x and I_0 is the intensity before entering the glass or water. If, for a certain filter glass, $k = 0.500/\text{cm}$ (which means that each centimeter of filter thickness removes half the light reaching it), find the fraction of the original intensity that will pass through a filter glass 2.00 cm thick.

25. The atmospheric pressure p decreases exponentially with the height h (in miles) above the earth according to the function $p = 29.92e^{-h/5}$ in. of mercury. Find the pressure at a height of 30,100 ft.

26. A certain radioactive material loses its radioactivity at the rate of $2\frac{1}{2}\%$ per year. What fraction of its initial radioactivity will remain after 10.0 years?

27. When a capacitor C, charged to a voltage E, is discharged through a resistor R (Fig. 18–10), the current i will decay exponentially according to the following equation:

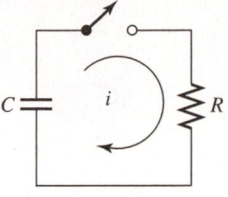

FIGURE 18–10

Capacitor Current When Charging or Discharging	$i = \dfrac{E}{R} e^{-t/RC}$ amperes	1082

Find the current after 45 ms (0.045 s) in a circuit where $E = 220$ V, $C = 130\ \mu\text{F}$ (130×10^{-6} F), and $R = 2700\ \Omega$.

28. When a fully discharged capacitor C (Fig. 18–11) is connected across a battery, the current i flowing into the capacitor will decay exponentially according to Eq. 1082. If $E = 115$ V, $R = 351\ \Omega$, $C = 0.000750$ F, find the current after 75.0 ms.

29. In a certain fabric mill, cloth is removed from a dye bath and is then observed to dry exponentially at the rate of 24% per hour. What percent of the original moisture will still be present after 5 h?

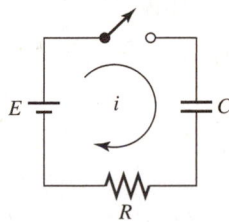

FIGURE 18–11

30. *Team Project, Computer:* Use a computer to make a large-scale duplicate of the universal growth and decay curves, Fig. 18–7. It should have a grid fine enough to make it useful in solving the problems in this section.

31. *Computer:* Assuming that the present annual world oil consumption is 17×10^9 barrels/yr, that this rate of consumption is increasing at a rate of 5% per year, and that the total world oil reserves are 1700×10^9 barrels, compute and print the following table:

Year	Annual Consumption	Oil Remaining
0	17	1700
1	17.85	1682.15
.	.	.
.	.	.
.	.	.

Have the computation stop when the oil reserves are almost gone.

32. *Computer:* We can get an approximate value for e from the following infinite series:

Series Approximation for e	$e = 2 + \dfrac{1}{2!} + \dfrac{1}{3!} + \dfrac{1}{4!} + \cdots$	158

where 4! (read "4 factorial") is $4(3)(2)(1) = 24$. Write a program to compute e using the first five terms of the series.

33. *Computer:* The infinite series often used to calculate e^x is as follows:

| Series Approximation for e^x | $e^x = 1 + x + \dfrac{x^2}{2!} + \dfrac{x^3}{3!} + \dfrac{x^4}{4!} + \cdots$ | 159 |

Write a program to compute e^x by using the first 15 terms of this series, and use it to find e^5.

34. *Computer:* The series expansion for b^x is as follows:

| Series Approximation for b^x | $b^x = 1 + x \ln b + \dfrac{(x \ln b)^2}{2!} + \dfrac{(x \ln b)^3}{3!} + \cdots$
 $(b > 0)$ | 150 |

Write a program to compute b^x using the first x terms of the series and use it to find 2.6^3.

18–2 Logarithms

Let us look at the exponential function

$$y = b^x$$

This equation contains three quantities. Given any two, we should be able to find the third.

■ Exploration:

Try this. Find each missing quantity by calculator.

(a) $y = 2.48^{3.50}$ (b) $24.0 = b^{3.50}$ (c) $24.0 = 2.48^x$

Were you able to find y, b, and x? ■

We already have the tools to solve the first two of these. By calculator,

(a) $y = 2.48^{3.50} = 24.0$

(b) $b = 24.0^{1/3.50} = 2.48$

However, none of the math we have learned so far will enable us to find x. For this and similar problems, we need *logarithms*.

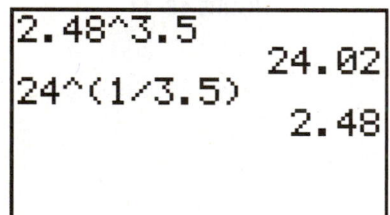

TI-83/84 screen for the exploration.

Definition of a Logarithm

The logarithm of some positive number y *is the exponent* to which a base b must be raised to obtain y.

◆◆◆ **Example 12:** Since $10^2 = 100$, we say that "2 is the logarithm of 100, to the base 10" because 2 is the exponent to which the base 10 must be raised to obtain 100. This is written

$$\log_{10} 100 = 2$$

Notice that the base is written as a subscript after the word *log*. The statement is read "The log of 100 (to the base 10) is 2." This means that 2 is an exponent because *a logarithm is an exponent;* 100 is the result of raising the base to that power. Therefore these two expressions are equivalent:

$$10^2 = 100$$
$$\log_{10} 100 = 2$$
◆◆◆

Given the exponential function $y = b^x$, we say that "x is the logarithm of y, to the base b" because x is the exponent to which the base b must be raised to obtain y.

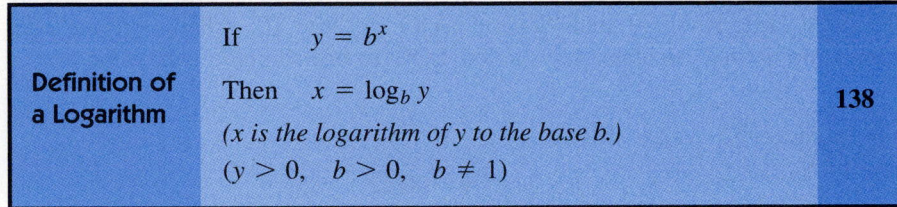

Definition of a Logarithm	If $\quad y = b^x$ Then $\quad x = \log_b y$ *(x is the logarithm of y to the base b.)* $(y > 0, \quad b > 0, \quad b \neq 1)$	**138**

In practice, the base b is taken as either 10 for common logarithms or e for natural logarithms. If "log" is written without a base, the base is assumed to be 10.

◆◆◆ Example 13:

$$\log 5 = \log_{10} 5 \qquad \text{◆◆◆}$$

Converting between Logarithmic and Exponential Forms

In working with logarithmic and exponential expressions, we often have to switch between exponential and logarithmic forms. When converting from one form to the other, remember that (1) the base in exponential form is also the base in logarithmic form and (2) "the *log* is the *exponent*."

◆◆◆ Example 14: Change the exponential expression $3^2 = 9$ to logarithmic form.

Solution: The base (3) in the exponential expression remains the base in the logarithmic expression.

$$3^2 = 9$$
$$\downarrow$$
$$\log_3 (\ ) = (\ )$$

The log is the exponent.

$$3^2 \searrow = 9$$
$$\log_3 (\ ) = 2$$

So our expression is

$$\log_3 9 = 2$$

This is read "the log *of* 9 (to the base 3) *is* 2." (We could also write the same expression in *radical form* as $\sqrt{9} = 3$.) $\qquad$ ◆◆◆

◆◆◆ Example 15: Change $\log_e x = 3$ to exponential form.

Solution:

$$\log_e x = 3$$

$$e^3 = x \qquad \text{◆◆◆}$$

◆◆◆ Example 16: Change $10^3 = 1000$ to logarithmic form.

Solution:

$$\log_{10} 1000 = 3 \qquad \text{◆◆◆}$$

Common logarithms are also called *Briggs' logarithms,* after Henry Briggs (1561–1630).

Common Logarithms

Although we may write logarithms to any base, only two bases are in regular use: the base 10 and the base *e*. Logarithms to the base *e*, called *natural* logarithms, are discussed later. Now we study logarithms to the base 10, called *common* logarithms. We use the calculator key marked LOG to find common logarithms.

◆◆◆ **Example 17:** Verify that

$$\log 74.85 = 1.8742$$

◆◆◆

A logarithm such as 1.8742 can be expressed as the sum of an integer, called the *characteristic,* and positive decimal number 0.8742, called the *mantissa.*

$$1 + 0.8742$$

Characteristic ──↑   ↑── Mantissa

◆◆◆ **Example 18:** Verify that

(a) $\log 7.485 = 0.8742$

(b) $\log 74.85 = 1.8742$

(c) $\log 748.5 = 2.8742$

(d) $\log 7485 = 3.8742$

◆◆◆

TI-83/84 screen for Example 18. The first calculation has scrolled out of view.

It's clear from this example that the characteristic, which locates only the decimal point, is *not* a significant digit. Thus, we round the *mantissa* of a logarithm to as many significant digits as in the original number. (When a logarithm or antilogarithm is used as an intermediary step in a longer calculation, it is a good idea to carry one more significant digit, and round the final answer.)

$$\text{Log } \boxed{74.85} = 1.\boxed{8742}$$

4 significant digits ──↑   ↑── 4 significant digits

◆◆◆ **Example 19:** Verify the following:

(a) $\log 8372 = 3.9228$

(b) $\log 52.5 = 1.720$

(c) $\log 4.80 = 0.681$

(d) $\log 136 = 2.134$

Here we have rounded the mantissa of each logarithm to the same number of significant digits as in the original number.

◆◆◆

On the TI-89, **log** is found in CATALOG.

◆◆◆ **Example 20:** *An Application.* The number of years *t* it will take for an investment of $1000 to reach a value of $3000 at an interest rate of 2.5% is given by

$$t = \frac{\log 3000 - \log 1000}{\log 1.025}$$

Find *t*.

Solution: By calculator,

$$\frac{\log 3000 - \log 1000}{\log 1.025} = \frac{3.4771 - 3.000}{0.0107}$$

$$= 44.6 \text{ years}$$

◆◆◆

◆◆◆ **Example 21:** Try to take the common logarithm of -2. You should get an error indication. We cannot take the logarithm of a negative number.   ◆◆◆

Common Error	Many calculators will supply the *open parentheses* when a logarithm key is pressed, and it is usually not necessary to close parentheses to evaluate a logarithm. Thus log(55 will correctly return 1.740. However, if you want to evaluate, say, log 55 + 2, you must *remember to close parentheses* after entering 55, getting $$\log(55) + 2 = 3.74$$ Otherwise you will erroneously get $$\log(55 + 2 = 1.76$$ which is actually the log of 57.

Antilogarithms

Suppose that we have the logarithm of some number N and want to find the number itself. This process is called *finding the antilogarithm (antilog)*. If, for example,

$$\log N = 1.8742$$

we can find N by switching to exponential form.

$$N = 10^{1.8742}$$

This is evaluated using the $\boxed{10^x}$ key on the calculator. Round an antilogarithm to as many significant digits as in *the mantissa* (the digits to the right of the decimal point).

Antilog of 1.8742 = 74.85

4 significant digits ⎯⎯⎯⎯⎯↑   ↑⎯⎯ 4 significant digits

◆◆◆ **Example 22:** Find x if $\log x = 2.7415$.

Solution: By calculator,

$$10^{2.7415} = 551.4$$

rounded to the four significant digits found in the mantissa of the logarithm 2.7415.
◆◆◆

◆◆◆ **Example 23:** Verify the following antilogarithms.

(a) The antilogarithm of 1.815 is 65.3
(b) The antilogarithm of 0.8462 is 7.018
(c) The antilogarithm of 2.322 is 210
(d) The antilogarithm of 1.375 is 23.7

We have rounded each antilog to the same number of significant digits as in the mantissa of the logarithm.
◆◆◆

◆◆◆ **Example 24:** *An Application.* The atmospheric pressure B (in inches of mercury, in. Hg) at an altitude of 35,000 ft can be found from the equation

$$60{,}470 \log\!\left(\frac{29.92}{B}\right) = 35{,}000$$

Evaluate B.

Solution: Dividing by 60,470 and taking the antilogarithm we get

$$\frac{29.92}{B} = 10^{0.5788} = 3.793$$

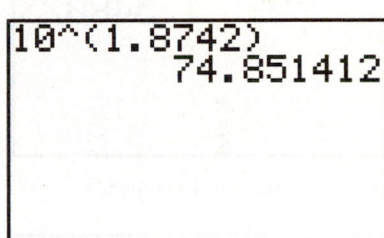

TI-83/84 screen.

TI-83/84 screen for Example 22.

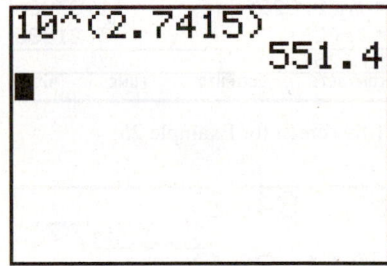

TI-89 screen for Example 23. On the TI-89, 10^x is found in $\boxed{\text{CATALOG}}$.

Solving for B gives

$$B = \frac{29.92}{3.793} = 7.888 \text{ in. Hg} \qquad \blacklozenge\blacklozenge\blacklozenge$$

Natural Logarithms

Natural logarithms are also called *Napierian logarithms*, after John Napier (1550–1617), who first wrote about logarithms.

You may want to refer to Sec. 18–1 for our earlier discussion of the number e (2.718 approximately). Since we have already defined e, we can use it as a base; logarithms using e as a base are called *natural* logarithms. They are written $\ln x$. Thus $\ln x$ is understood to mean $\log_e x$.

Use the same rules to round *natural* logarithms and antilogarithms as for common logs and antilogs.

♦♦♦ **Example 25:** Verify the following natural logarithms.

(a) $\ln 2.446 = 0.8945$

(b) $\ln 3.85 = 1.348$

(c) $\ln 2.826 = 1.0389$

(d) $\ln 25.0 = 3.219$

We have rounded the mantissa of each logarithm to the same number of significant digits as in the original number. ♦♦♦

TI-83/84 screen for Example 25.

TI-89 screen for Example 25.

♦♦♦ **Example 26:** *An Application.* A certain pendulum, after being released from a height of 4.00 cm, reaches a height y after 3.00 seconds, where y can be found from

$$\ln\left(\frac{y}{4.00}\right) = -0.50$$

Evaluate y.

Solution: Taking the natural antilogarithm of both sides gives

$$\frac{y}{4.00} = e^{-0.50} = 0.607$$

From which

$$y = 4.00(0.607) = 2.43 \text{ cm} \qquad \blacklozenge\blacklozenge\blacklozenge$$

Natural Antilogarithms

If we know that the natural logarithm of some number N is, say, 3.825

$$\ln N = 3.825$$

we can find N by switching to exponential form, where the base is e.

$$N = e^{3.825}$$

TI-83/84 screen for Example 27.

We evaluate this using the $\boxed{e^x}$ key on the calculator.

♦♦♦ **Example 27:** Verify the following natural antilogarithms.

(a) The natural antilogarithm of 0.8462 is 2.331

(b) The natural antilogarithm of 1.884 is 6.58

(c) The natural antilogarithm of 2.174 is 8.79

(d) The natural antilogarithm of 1.4945 is 4.457

TI-89 screen for Example 27.

We have rounded each antilog to the same number of significant digits as in the mantissa of the logarithm. ♦♦♦

◆◆◆ **Example 28:** *An Application.* The temperature T of a bronze casting initially at $1250°\,\text{F}$ decreases exponentially with time. Its temperature after 5.00 s can be found from the equation

$$\ln\left(\frac{T}{1250}\right) = -1.55$$

Evaluate T.

Solution: Taking the natural antilogarithm gives

$$\frac{T}{1250} = e^{-1.55} = 0.212$$

From which

$$T = 1250\,(0.212) = 265°\text{F}$$ ◆◆◆

Exercise 2 ◆ Logarithms

Converting Between Logarithmic and Exponential Forms

Convert to logarithmic form.

1. $3^4 = 81$ **2.** $5^3 = 125$ **3.** $4^6 = 4096$

4. $7^3 = 343$ **5.** $x^5 = 995$ **6.** $a^3 = 6.83$

Convert to exponential form.

7. $\log_{10} 100 = 2$ **8.** $\log_2 16 = 4$ **9.** $\log_5 125 = 3$

10. $\log_4 1024 = 5$ **11.** $\log_3 x = 57$ **12.** $\log_x 54 = 285$

Common Logarithms

Find the common logarithm of each number.

13. 27.6   **14.** 4.83   **15.** 5.93

16. 9.26   **17.** 48.3   **18.** 385

19. 836   **20.** 1.84   **21.** 27.4

Find the number whose common logarithm is given.

22. 1.584   **23.** 2.957   **24.** 5.273

25. 0.886   **26.** 2.857   **27.** 0.366

28. 2.227   **29.** 4.974   **30.** 3.979

Natural Logarithms

Find the natural logarithm of each number.

31. 48.3   **32.** 846   **33.** 2365

34. 1.285   **35.** 1.845   **36.** 4.77

37. 1.374   **38.** 45,900   **39.** 1.364

Find the number whose natural logarithm is given.

40. 2.846   **41.** 4.264   **42.** 0.879

43. 1.845   **44.** 0.365   **45.** 5.937

46. 0.936   **47.** 4.9715   **48.** 3.8422

Applications

49. An amount of money a invested at a compound interest rate of n per year will take t years to accumulate to an amount y, where t is

$$t = \frac{\log y - \log a}{\log(1 + n)} \quad \text{(years)}$$

How many years will it take an investment of $10,000 to triple in value when deposited at 8.00% per year?

(The equation in this problem is derived from the compound interest formula. The equations in problems 50 and 51 are obtained from the equations for an annuity and for capital recovery from Exercise 1.)

50. If an amount R is deposited once every year at a compound interest rate n, the number of years it will take to accumulate to an amount y is

$$t = \frac{\log\left(\dfrac{ny}{R} + 1\right)}{\log(1 + n)} \quad \text{(years)}$$

How many years will it take an annual payment of $1500 to accumulate to $13,800 at 9.0% per year?

51. If an amount a is invested at a compound interest rate n, it will be possible to withdraw a sum R at the end of every year for t years until the deposit is exhausted. The number of years is given by

$$t = \frac{\log\left(\dfrac{an}{R - an} + 1\right)}{\log(1 + n)} \quad \text{(years)}$$

If $200,000 is invested at 12% interest, for how many years can an annual withdrawal of $30,000 be made before the money is used up?

52. The *magnitude M* of a star of intensity I is

$$M = 2.5 \log \frac{I_1}{I} + 1$$

where I_1 is the intensity of a first-magnitude star. What is the magnitude of a star whose intensity is one-tenth that of a first-magnitude star?

53. The difference in elevation h (ft) between two locations having barometer readings of B_1 and B_2 inches of mercury (in. Hg) is given by the logarithmic equation

$$h = 60,470 \log(B_2/B_1)$$

where B_1 is the pressure at the *upper* station. Find the difference in elevation between two stations having barometer readings of 29.14 in. Hg at the lower station and 26.22 in. Hg at the upper.

54. The heat loss q per foot of cylindrical pipe insulation (Fig. 18–12) having an inside radius r_1 and outside radius r_2 is given by the logarithmic equation

$$q = \frac{2\pi k(t_1 - t_2)}{\ln(r_2/r_1)} \quad \text{(Btu/h)}$$

where t_1 and t_2 are the inside and outside temperatures (°F) and k is the conductivity of the insulation. Find q for a 4.0-in.-thick insulation having a conductivity of 0.036 and wrapped around a 9.0-in.-diameter pipe at 550°F if the surroundings are at 90.0°F.

55. If a resource is being used up at a rate that increases exponentially, the time it takes to exhaust the resource (called the *exponential expiration time,* EET) is

$$\text{EET} = \frac{1}{n} \ln\left(\frac{nR}{r} + 1\right)$$

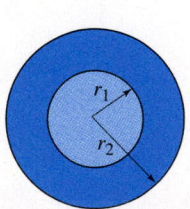

FIGURE 18–12 An insulated pipe.

where n is the rate of increase in consumption, R the total amount of the resource, and r the initial rate of consumption. If we assume that the United States has oil reserves of 207×10^9 barrels and that our present rate of consumption is 6.00×10^9 barrels/yr, how long will it take to exhaust these reserves if our consumption increases by 7.00% per year?

56. *Computer:* The series approximation for the natural logarithm of a positive number x is as follows:

Series Approximation for ln x	$\ln x = 2a + \dfrac{2a^3}{3} + \dfrac{2a^5}{5} + \dfrac{2a^7}{7} + \cdots$ where $a = \dfrac{x-1}{x+1}$	**161**

Write a program to compute $\ln x$ using the first 10 terms of the series and use it to find $\ln 4$.

57. *Writing, History of Logarithms:* Logarithms have an interesting history. For example, the French mathematician Pierre Laplace (1749–1827) wrote that *"the method of logarithms, by reducing to a few days the labor of many months, doubles, as it were, the life of the astronomer, besides freeing him from the errors and disgust inseparable from long calculation."* Write a short paper on the history of logarithms.

18–3   Properties of Logarithms

Products

We will soon be solving exponential and logarithmic equations. For them we need to know some properties of logarithms. For example, how do we find the logarithm of the product of two numbers?

■ **Exploration:**

Try this. Find by calculator, using common logs:

$$\log 4 \qquad \log 2 \qquad \log 8$$

(a) Can you find any relationship between the three numbers you have found?
(b) Can you express this relationship, if it turns out to be true, as a general rule?
(c) Repeat, using natural logs. Does the same relationship seem to hold? ■

We wish to write an expression for the log of the product of two positive numbers, say, M and N.

$$\log_b MN = ?$$

Let us substitute: $M = b^c$ and $N = b^d$ where $b > 0$ and $b \neq 1$. Then

$$MN = b^c b^d = b^{c+d}$$

by the laws of exponents. Writing this expression in logarithmic form, we have

$$\log_b MN = c + d$$

But, since $b^c = M$, $c = \log_b M$. Similarly, $d = \log_b N$. Substituting, we have the following equation:

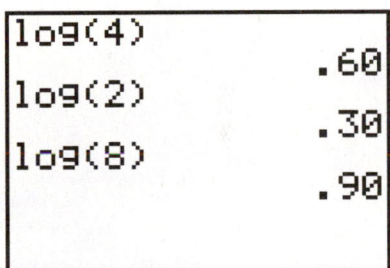

T1-83/84 screen for the exploration. You may have observed in the exploration that log 8 is the sum of log 4 and log 2. The calculator is set to show two decimal places.

Log of a Product	$\log_b MN = \log_b M + \log_b N$ *The log of a product equals the sum of the logs of the factors.*	139

◆◆◆ Example 29: Here are examples showing the log of a product.

(a) $\log 7x = \log 7 + \log x$

(b) $\log 3 + \log x + \log y = \log 3xy$ ◆◆◆

Common Errors	The log of a sum is *not* equal to the sum of the logs. $\log_b(M + N) \neq \log_b M + \log_b N$
	The product of two logs is *not* equal to the sum of the logs. $(\log_b M)(\log_b N) \neq \log_b M + \log_b N$

Quotients

■ Exploration:

Try this. Using either natural or common logs, find by calculator

$$\log 12 \qquad \log 3 \qquad \log 4$$

How do the three numbers you have found appear to be related? Can you express this relationship, if it turns out to be true, as a general rule?   ■

Let us now consider the quotient M divided by N. Using the same substitution as above, we obtain

$$\frac{M}{N} = \frac{b^c}{b^d} = b^{c-d}$$

Going to logarithmic form yields

$$\log_b \frac{M}{N} = c - d$$

Finally, substituting $c = \log_b M$ and $d = \log_b N$ gives us the following equation:

Log of a Quotient	$\log_b \frac{M}{N} = \log_b M - \log_b N$ *The log of a quotient equals the log of the dividend minus the log of the divisor.*	140

◆◆◆ Example 30: Here we show the log of a quotient.

$$\log \frac{3}{4} = \log 3 - \log 4$$ ◆◆◆

◆◆◆ Example 31: Express the following as the sum or difference of two or more logarithms:

$$\log \frac{ab}{xy}$$

Solution: By Eq. 140,

$$\log \frac{ab}{xy} = \log ab - \log xy$$

and by Eq. 139,

$$\log ab - \log xy = \log a + \log b - (\log x + \log y)$$
$$= \log a + \log b - \log x - \log y \qquad \blacklozenge\blacklozenge\blacklozenge$$

◆◆◆ **Example 32:** The expression $\log 10x^2 - \log 5x$ can be expressed as a single logarithm as follows:

$$\log \frac{10x^2}{5x} \qquad \text{or} \qquad \log 2x \qquad \blacklozenge\blacklozenge\blacklozenge$$

Common Errors	Similar errors are made with quotients as with products.
	$\log_b (M - N) \neq \log_b M - \log_b N$
	$\dfrac{\log_b M}{\log_b N} \neq \log_b M - \log_b N$

Powers

■ Exploration:

Try this. Using either natural or common logs, find by calculator

$$\log 7 \qquad \log 7^2$$

Can you find any relationship between the two numbers you have found? Can you express it, if it turns out to be true, as a general rule? ■

Consider now the quantity M raised to a power p. Substituting b^c for M, as before, we have

$$M^p = (b^c)^p = b^{cp}$$

In logarithmic form,

$$\log_b M^p = cp$$

Substituting $c = \log_b M$ gives us the following equation:

Log of a Power	$\log_b M^p = p \log_b M$	141
	The log of a number raised to a power equals the power times the log of the number.	

◆◆◆ **Example 33:** Here we show how to rewrite the log of a power as a product.

$$\log 3.85^{1.4} = 1.4 \log 3.85 \qquad \blacklozenge\blacklozenge\blacklozenge$$

◆◆◆ **Example 34:** The expression $7 \log x$ can be expressed as a single logarithm with a coefficient of 1 as follows:

$$\log x^7 \qquad \blacklozenge\blacklozenge\blacklozenge$$

◆◆◆ **Example 35:** Express $2 \log 5 + 3 \log 2 - 4 \log 1$ as a single logarithm.

Solution:

$$2 \log 5 + 3 \log 2 - 4 \log 1 = \log 5^2 + \log 2^3 - \log 1^4$$

$$= \log \frac{5^2 + 2^3}{1^4}$$

$$= \log \frac{25(8)}{1} = \log 200$$

◆◆◆

◆◆◆ **Example 36:** Express as a single logarithm with a coefficient of 1:

$$2 \log \frac{2}{3} + 4 \log \frac{a}{b} - 3 \log \frac{x}{y}$$

Solution:

$$2 \log \frac{2}{3} + 4 \log \frac{a}{b} - 3 \log \frac{x}{y} = \log \frac{2^2}{3^2} + \log \frac{a^4}{b^4} - \log \frac{x^3}{y^3}$$

$$= \log \frac{4a^4 y^3}{9b^4 x^3}$$

◆◆◆

◆◆◆ **Example 37:** Rewrite the following equation without logarithms.

$$\log (a^2 - b^2) + 2 \log a = \log (a - b) + 3 \log b$$

Solution: We first try to combine all of the logarithms into a single logarithm. Rearranging gives

$$\log (a^2 - b^2) - \log (a - b) + 2 \log a - 3 \log b = 0$$

By the properties of logarithms,

$$\log \frac{a^2 - b^2}{a - b} + \log \frac{a^2}{b^3} = 0$$

or

$$\log \frac{a^2 (a - b)(a + b)}{b^3 (a - b)} = 0$$

Simplifying yields

$$\log \frac{a^2 (a + b)}{b^3} = 0$$

We then eliminate logarithms by going from logarithmic to exponential form.

$$\frac{a^2 (a + b)}{b^3} = 10^0 = 1$$

or

$$a^2 (a + b) = b^3$$

◆◆◆

◆◆◆ **Example 38:** *An Application.* We have seen that the formula of compound interest, Eq. 1009, will give the value y of an invesment of an initial amount a for a period of t years at interest rate n.

$$y = a(1 + n)^t$$

But suppose we want to know how many years it will take for a given investment to reach a certain amount. For this we have to solve this equation for t.

Solution: We first divide by a, and then take the logarithms of both sides.

$$(1 + n)^t = \frac{y}{a}$$

$$\log(1 + n)^t = \log \frac{y}{a}$$

We simplify by using the property of the log of a power (Eq. 141) and for the log of a quotient (Eq. 140).

$$t \log(1 + n) = \log y - \log a$$

Finally we solve for t.

$$t = \frac{\log y - \log a}{\log(1 + n)}$$

◆◆◆

Roots

■ Exploration:

Try this. Using either natural or common logs, find by calculator

$$\log \sqrt[3]{64} \qquad \log 64$$

Is there any relationship that you can see between the numbers you have found? If true, can you express it as a general rule? ■

We can write a given radical expression in exponential form and then use the rule for powers. Thus, by Eq. 29,

$$\sqrt[q]{M} = M^{1/q}$$

We may take the logarithm of both sides of an equation, just as we took the square root of both sides of an equation or took the sine of both sides of an equation. Taking the logarithm of both sides, we obtain

$$\log_b \sqrt[q]{M} = \log_b M^{1/q}$$

Then, by Eq. 141,

Log of a Root	$$\log_b \sqrt[q]{M} = \frac{1}{q} \log_b M$$ *The log of the root of a number equals the log of the number divided by the index of the root.*	142

◆◆◆ **Example 39:** These examples show the use of Eq. 142.

(a) $\log \sqrt[5]{8} = \frac{1}{5} \log 8 \approx \frac{1}{5}(0.9031) \approx 0.1806$

(b) $\dfrac{\log x}{2} + \dfrac{\log z}{4} = \log \sqrt{x} + \log \sqrt[4]{z}$

◆◆◆

Log of 1

Let us take the log to the base b of 1 and call it x. Thus

$$x = \log_b 1$$

Switching to exponential form, we have

$$b^x = 1$$

This equation is satisfied, for any positive b, by $x = 0$. Therefore:

Log of 1	$\log_b 1 = 0$ *The log of 1 is zero.*	143

◆◆◆ **Example 40:**

(a) $\log 1 = 0$ (b) $\ln 1 = 0$ ◆◆◆

Log of the Base

We now take the $\log_b$ of its own base b. Let us call this quantity x.

$$\log_b b = x$$

Going to exponential form gives us

$$b^x = b$$

So x must equal 1.

Log of the Base	$\log_b b = 1$ *The log (to the base b) of b is equal to 1.*	144

◆◆◆ **Example 41:** These examples show the log of the base.

(a) $\log_5 5 = 1$ (b) $\log 10 = 1$ (c) $\ln e = 1$ ◆◆◆

Log of the Base Raised to a Power

Consider the expression $\log_b b^n$. We set this expression equal to x and, as before, go to exponential form.

$$\log_b b^n = x$$
$$b^x = b^n$$
$$x = n$$

Therefore:

Log of the Base Raised to a Power	$\log_b b^n = n$ *The log (to the base b) of b raised to a power is equal to the power.*	145

◆◆◆ **Example 42:** Here we show the use of Eq. 145.

(a) $\log_2 2^{4.83} = 4.83$ (b) $\log_e e^{2x} = 2x$

(c) $\log_{10} 0.0001 = \log_{10} 10^{-4} = -4$ ◆◆◆

Base Raised to a Logarithm of the Same Base

We want to evaluate an expression of the form

$$b^{\log_b x}$$

Setting this expression to y and taking the logarithms of both sides

$$y = b^{\log_b x}$$

Taking logs of both sides,

$$\log y = \log_b b^{\log_b x}$$

But, by Eq. 141, the log of a number raised to a power equals the power times the log of the number, so

$$\log y = \log_b x \, \log_b b$$

Since $\log_b b = 1$, we get

$$\log_b y = \log_b x$$

Taking the antilog of both sides we get

$$y = x$$

But $y = b^{\log_b x}$ so

Base Raised to a Logarithm of the Same Base	$b^{\log_b x} = x$	146

◆◆◆ **Example 43:**

(a) $10^{\log w} = w$ (b) $e^{\log_e 3x} = 3x$ ◆◆◆

Change of Base

We can convert between natural logarithms and common logarithms (or between logarithms to any base) by the following procedure. Suppose that $\log N$ is the common logarithm of some number N. We set it equal to x.

$$\log N = x$$

Going to exponential form, we obtain

$$10^x = N$$

Now we take the natural log of both sides.

$$\ln 10^x = \ln N$$

By Eq. 141,

$$x \ln 10 = \ln N$$

$$x = \frac{\ln N}{\ln 10}$$

But $x = \log N$, so we have the following equation:

Change of Base	$\log N = \dfrac{\ln N}{\ln 10}$ where $\ln 10 \approx 2.3026$ *The common logarithm of a number is equal to the natural log of that number divided by the natural log of 10.*	147

Some computer languages enable us to find the natural log but not the common log. This equation lets us convert from one to the other.

◆◆◆ **Example 44:** Find $\log N$ if $\ln N = 5.824$.

Solution: By Eq. 147,

$$\log N = \frac{5.824}{2.3026} = 2.529$$

◆◆◆

◆◆◆ **Example 45:** What is $\ln N$ if $\log N = 3.825$?

Solution: By Eq. 147,

$$\ln N = \ln 10\,(\log N)$$
$$= 2.3026\,(3.825) = 8.807 \qquad\text{◆◆◆}$$

Exercise 3 ◆ Properties of Logarithms

Write as the sum or difference of two or more logarithms.

1. $\log \frac{2}{3}$

2. $\log 4x$

3. $\log ab$

4. $\log \frac{x}{2}$

5. $\log xyz$

6. $\log 2ax$

7. $\log \frac{3x}{4}$

8. $\log \frac{5}{xy}$

9. $\log \frac{1}{2x}$

10. $\log \frac{2x}{3y}$

11. $\log \frac{abc}{d}$

12. $\log \frac{x}{2ab}$

Express as a single logarithm with a coefficient of 1. Assume that the logarithms in each problem have the same base.

13. $\log 3 + \log 4$

14. $\log 7 - \log 5$

15. $\log 2 + \log 3 - \log 4$

16. $3 \log 2$

17. $4 \log 2 + 3 \log 3 - 2 \log 4$

18. $\log x + \log y + \log z$

19. $3 \log a - 2 \log b + 4 \log c$

20. $\frac{\log x}{3} + \frac{\log y}{2}$

21. $\log \frac{x}{a} + 2 \log \frac{y}{b} + 3 \log \frac{z}{c}$

Rewrite each equation so that it contains no logarithms.

22. $\log x + 3 \log y = 0$

23. $\log_2 x + 2 \log_2 y = x$

24. $2 \log (x - 1) = 5 \log (y + 2)$

25. $\log (p^2 - q^2) - \log (p + q) = 2$

Simplify each expression.

26. $\log_2 2$

27. $\log_e e$

28. $\log_{10} 10$

29. $\log_3 3^2$

30. $\log_e e^x$

31. $\log_{10} 10^x$

32. $e^{\log_e x}$

33. $2^{\log_2 3y}$

34. $10^{\log x^2}$

Change of Base

Find the common logarithm of the number whose natural logarithm is the given value.

35. 8.36

36. -3.846

37. 3.775

38. 15.36

39. 5.26

40. -0.638

Find the natural logarithm of the number whose common logarithm is the given value.

41. 84.9

42. 2.476

43. -3.82

44. 73.9

45. 2.37

46. -2.63

47. *Project:* Use a graphing utility to separately graph each side of the equation

$$\log 2x \overset{?}{=} \log 2 + \log x$$

from $x = 0$ to 10. Then graph both sides of

$$\log(2 + x) \overset{?}{=} \log 2 + \log x$$

and

$$(\log 2)(\log x) \overset{?}{=} \log 2 + \log x$$

What do your graphs tell you about these three expressions?

48. *Project, Logarithmic and Semilogarithmic Graph Paper:* Graphing data on paper ruled with logarithmic spacing along one or both axes, one can deduce certain things about those data. To find out more, check our textbook Web site at www.wiley.com/college/calter.

18–4   Solving an Exponential Equation

We now return to the problem we started in the exploration of Sec. 18–2 but could not finish. We tried to solve the following equation for x.

$$24.0 = 2.48^x$$

We will show several methods for solving this equation:

(a) Graphical solution
(b) Using a calculator's equation solver
(c) Using a symbolic algebra calculator
(d) Algebraically

Graphical Solution

We can, of course, find an approximate graphical solution to this equation in the usual way.

◆◆◆ **Example 46:** Graphically solve $24.0 = 2.48^x$ for x.

Solution: We move both terms to one side of the equals sign and set the resulting expression equal to y.

$$y = 2.48^x - 24.0$$

We graph this function, as shown, and find a root at $x \approx 3.50$.   ◆◆◆

Solution by Calculator

We can use any built-in equation solver to solve an exponential equation just as we used it earlier to solve other equations.

◆◆◆ **Example 47:** Solve the equation of Example 46 using the TI-83/84 equation solver.

Solution:

(a) We select **Solver** from the **Math** menu and enter the equation (Screen 1). As for graphing, the equation must be in explicit form, with all terms on one side and 0 on the other side.
(b) Press **ENTER**. A new screen is displayed (Screen 2), showing a guess value of x and a **bound**. You may accept the default values or enter new ones. Let us enter a new guess of $x = 3$ and a new bound of 0 to 5.
(c) Move the cursor to the line containing "$x=$" and press **SOLVE**. (This is **ALPHA** **ENTER** on the TI-83/84.) A root falling within the selected bound is displayed (Screen 3).

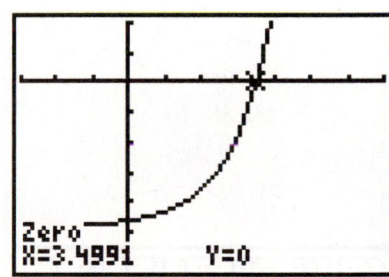

T1-83/84 screen for Example 46: $y = 2.48^x - 24.0$. Tick marks on the horizontal axis are 1 unit apart.

 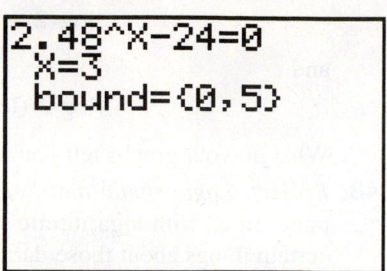

(1) The given equation is entered in the TI-83/84 equation solver.

(2) An initial guess for x, and the lower and upper bounds, are chosen here.

(3) The computed value of x is shown, within the chosen bound. The last line says that the difference between the left and right sides of the equation is zero, showing that the equation is indeed balanced.   ◆◆◆

To solve an exponential equation or any other equation by graphing or by the TI-83/84 equation solver, the equation must be in explicit form. However, some calculators that can do symbolic manipulation such as the T1-89 can solve an equation that is in implicit form, as we have seen earlier in this text.

◆◆◆ **Example 48:** Solve the equation from the preceding example using the TI-89.

Solution:

(a) We select **solve** from the **Algebra** menu.
(b) Enter the equation, followed by the variable (x) for which we want to solve. Unlike the TI-83/84, the equation *does not* have to be in explicit form.
(c) Pressing $\boxed{\text{ENTER}}$ would give the exact solution, while pressing $\boxed{\approx}$ (located above the $\boxed{\text{ENTER}}$ key) gives the approximate decimal solution, as shown.

Note that it was not necessary to enter a first guess or bounding values.   ◆◆◆

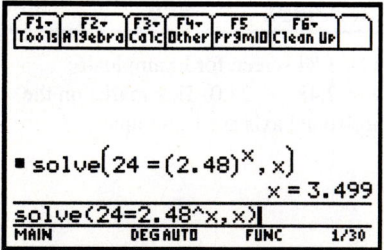

TI-89 screen for Example 48.

Algebraic Solution

For an *algebraic* solution to this type of equation we need logarithms. The key to solving exponential equations is to *take the logarithm of both sides*. This enables us to use Eq. 141, $\log_b M^p = p \log_b M$, to extract the unknown from the exponent. We used common logs in this example, but natural logs would work as well. Taking the logarithm of both sides gives

$$24.0 = 2.48^x$$
$$\log 24.0 = \log 2.48^x$$

By Eq. 141,

$$\log 24.0 = x \log 2.48$$

$$x = \frac{\log 24.0}{\log 2.48} = \frac{1.380}{0.3945} = 3.50$$

This agrees with our graphical and computer solutions.

Solving an Exponential Equation with Base *e*

◆◆◆ Example 49: Solve for *x*:

$$157 = 112e^{3x+2}$$

Solution: Dividing by 112, we have

$$1.402 = e^{3x+2}$$

If an exponential equation contains the base *e*, we take *natural* logs rather than common logarithms. However, simply switching from exponential to logarithmic form will sometimes be the best approach, as in the following example. Going from exponential to logarithmic form,

$$\ln 1.402 = 3x + 2$$

$$x = \frac{\ln 1.402 - 2}{3} = -0.554 \qquad \text{◆◆◆}$$

◆◆◆ Example 50: Solve for *x* to three significant digits:

$$3e^x + 2e^{-x} = 4(e^x - e^{-x})$$

Solution: Removing parentheses gives

$$3e^x + 2e^{-x} = 4e^x - 4e^{-x}$$

Combining like terms yields

$$e^x - 6e^{-x} = 0$$

Factoring gives

$$e^{-x}(e^{2x} - 6) = 0$$

This equation will be satisfied if either of the two factors is equal to zero. So we set each factor equal to zero (just as when solving a quadratic by factoring) and get

$$e^{-x} = 0 \qquad \text{and} \qquad e^{2x} = 6$$

There is no value of *x* that will make e^{-x} equal to zero, so we get no root from that equation. We solve the other equation by taking the natural log of both sides.

$$2x \ln e = \ln 6 = 1.792$$

$$x = 0.896 \qquad \text{◆◆◆}$$

◆◆◆ Example 51: Solve for *x* to three significant digits.

$$2e^{3x+2} = 4e^{x-1}$$

Solution: Dividing, we get,

$$\frac{e^{3x+2}}{e^{x-1}} = 2$$

Then by the laws of exponents,

$$e^{(3x+2)-(x-1)} = 2$$

$$e^{(2x+3)} = 2$$

Taking the natural logarithm of both sides gives

$$2x + 3 = \ln 2$$

$$x = \frac{\ln 2 - 3}{2} = -1.15 \qquad \text{◆◆◆}$$

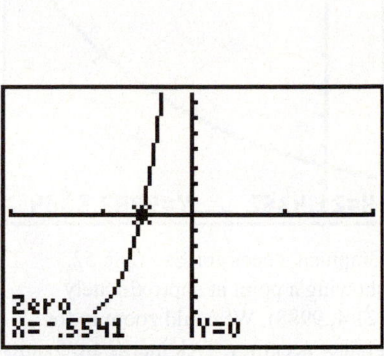

Graphical check for Example 49: $y = 112\,e^{3x+2} - 157$. Tick marks on the horizontal axis are 1 unit apart.

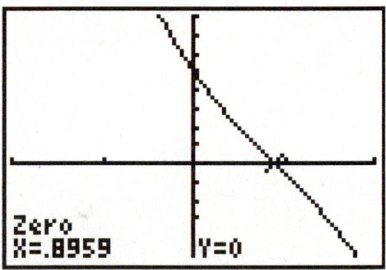

Graphical check for Example 50: $y = 3e^x + 2e^{-x} - 4(e^x - e^{-x})$. Tick marks on the horizontal axis are 1 unit apart.

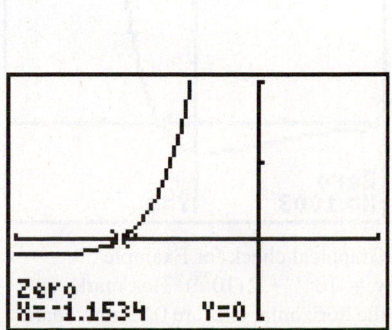

Graphical check for Example 51: Graph of $y = 2e^{3x+2} - 4e^{x-1}$. Tick marks on the *x* axis are 1 unit apart.

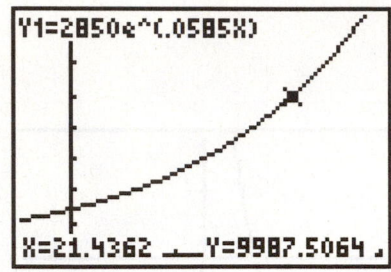

Graphical check for Example 52, showing a point at approximately (21.4, 9988). We could zoom in for greater accuracy. Tick marks are 5 units apart on the horizontal axis and 2000 units apart on the vertical axis.

♦♦♦ **Example 52:** *An Application. Population Growth.* The equations for exponential growth can naturally be applied to living organisms. We would expect this, as the number of children produced in a generation depends on the number of parents. The populations of organisms for which we can control the conditions of growth (temperature, food supply, available space) grow at nearly ideal exponential rates. In particular, a population of a colony of bacteria will grow exponentially if we provide them with enough food.

A bacterial sample contains 2850 bacteria and grows at a rate of 5.85% per hour. Assuming that the bacteria grow exponentially,

(a) Write the equation for the number N of bacteria as a function of time.
(b) Find the time for the sample to grow to 10,000 bacteria.

Solution:

(a) We substitute into the equation for exponential growth, $y = ae^{nt}$, with $y = N$, $a = 2850$, and $n = 0.0585$, getting

$$N = 2850e^{0.0585t}$$

(b) Setting $N = 10,000$ and solving for t, we have

$$10,000 = 2850e^{0.0585t}$$

Dividing through by 2850 and switching sides gives

$$e^{0.0585t} = 3.51$$

Going to logarithmic form,

$$0.0585t \ln e = \ln 3.51$$

$$t = \frac{\ln 3.51}{0.0585} = 21.5 \text{ h}$$

♦♦♦

Solving an Exponential Equations with Base 10

If an equation contains 10 as a base, the work will be simplified by taking *common* logarithms.

♦♦♦ **Example 53:** Solve $10^{5x} = 2(10^{2x})$ to three significant digits.

Solution: Dividing both sides by 10^{2x} and simplifying using the law of exponents for quotients,

$$\frac{10^{5x}}{10^{2x}} = 2$$

$$10^{5x-2x} = 10^{3x} = 2$$

We then take the common log of both sides, or switch to logarithmic form.

$$3x \log 10 = \log 2$$

$$x = \frac{\log 2}{3} = 0.100$$

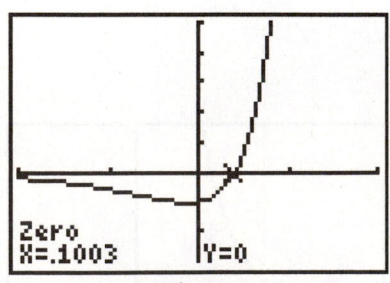

Graphical check for Example 53: $y = 10^{5x} - 2(10^{2x})$. Tick marks on the horizontal axis are 0.25 unit apart.

since $\log 10 = 1$.   ♦♦♦

Half-Life

Being able to solve an exponential equation for the exponent allows us to return to the formulas for exponential growth and decay (Sec. 18–1) and to derive two interesting quantities: half-life and doubling time.

When a material decays exponentially according to Eq. 153,

$$y = ae^{-nt}$$

The time it takes for the material to be half gone is called the *half-life*. If we let $y = a/2$,

$$\frac{1}{2} = e^{-nt} = \frac{1}{e^{nt}}$$

$$2 = e^{nt}$$

Taking natural logarithms gives us

$$\ln 2 = \ln e^{nt} = nt$$

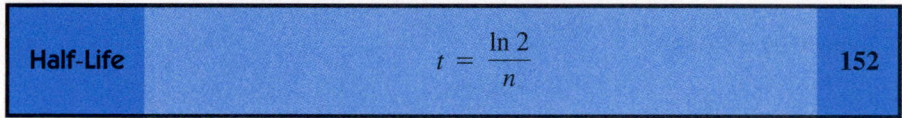

Half-Life	$t = \dfrac{\ln 2}{n}$	152

◆◆◆ **Example 54:** *An Application.* Find the half-life of a radioactive material that decays exponentially at the rate of 2.0% per year.

Solution:

$$\text{half-life} = \frac{\ln 2}{0.020} = 35 \text{ years} \qquad \text{◆◆◆}$$

Doubling Time

If a quantity grows exponentially according to the following function Eq. 151,

$$y = ae^{nt}$$

it will eventually double (y will be twice a). Setting $y = 2a$, we get

$$2a = ae^{nt}$$

or $2 = e^{nt}$. Taking natural logs, we have

$$\ln 2 = \ln e^{nt} = nt \ln e = nt$$

Solving for t we get

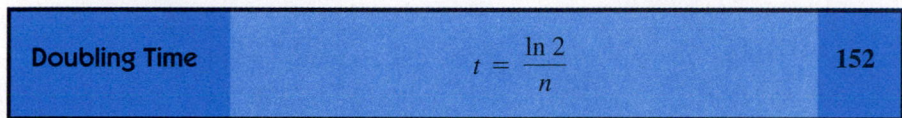

Doubling Time	$t = \dfrac{\ln 2}{n}$	152

Notice that the equations for half-life and doubling time are identical. Further, if we let P be the rate of growth expressed as a percent (not as a decimal), and since $\ln 2 \approx 0.7$, we get the following *rule of thumb*; a valuable tool for *estimation*.

$$\text{Half-Life or Doubling Time} \approx \frac{70}{P}$$

◆◆◆ **Example 55:** *An Application. Radioactive Decay.* Radioactive materials decay exponentially because the rate of decay depends only on how many atoms are present to decay. Certain isotopes of certain elements have half-lives so short that we can watch them decay on a human time scale, while others take thousands of years. We use radioactive decay to generate electricity and also to date archeological artifacts. This takes advantage of the presence of a small amount of carbon-14 in the atmosphere. Plants take in this carbon-14 and after the plants dies, it decays with a half-life of 5730 years. We measure the amount of carbon-14 left in a plant sample found with an artifact to tell how long the plant has been dead.

Remember, there is still half of the original radioactivity left after one half-life has passed. It takes over three half-lives for the original radioactivity to decay by 90%.

How long does it take for 90% of a sample of strontium-90 to decay?

Solution: After 90% of the sample has decayed, 10% or 1/10, remains, so $a/y = 10$. From the equation for exponential decay, Eq. 153.

$$\frac{y}{a} = e^{-nt}$$

Taking the natural log of both sides,

$$\ln \frac{y}{a} = -nt \ln e$$

Solving for t we get,

$$t = \frac{-\ln(y/a)}{n} = \frac{\ln(a/y)}{n}$$

where $-\ln(y/a) = \ln(a/y)$. The time t_{10} to reach the 10% level is then Eq. 152,

$$t_{1/10} = \frac{\ln 10}{n}$$

Now we need to use the half-life of strontium-90 (28.8 years) to find n. From Eq. 152,

$$n = \frac{\ln 2}{t_{1/2}} = \frac{0.693}{28.8} = 0.0241 \text{ per year}$$

and

$$t_{1/10} = \frac{\ln 10}{0.0241} = 95.5 \text{ years}$$

◆◆◆

Exercise 4 ◆ Solving an Exponential Equation

Solve for x to three significant digits.

1. $2^x = 7$
2. $(7.26)^x = 86.8$
3. $(1.15)^{x+2} = 12.5$
4. $(2.75)^x = (0.725)^{x^2}$
5. $(15.4)^{\sqrt{x}} = 72.8$
6. $e^{5x} = 125$
7. $5.62e^{3x} = 188$
8. $1.05e^{4x+1} = 5.96$
9. $e^{2x-1} = 3e^{x+3}$
10. $14.8e^{3x^2} = 144$
11. $5^{2x} = 7^{3x-2}$
12. $3^{x^2} = 175^{x-1}$
13. $10^{3x} = 3(10^x)$
14. $e^x + e^{-x} = 2(e^x - e^{-x})$
15. $2^{3x+1} = 3^{2x+1}$
16. $5^{2x} = 3^{3x+1}$
17. $7e^{1.5x} = 2e^{2.4x}$

Applications

18. The current i in a certain circuit is given by

$$i = 6.25e^{-125t} \text{ (amperes)}$$

where t is the time in seconds. At what time will the current be 1.00 A?

19. The current through a charging capacitor is given by

$$i = \frac{E}{R} e^{-t/RC} \qquad \text{(Eq. 1082)}$$

If $E = 325$ V, $R = 1.35\ \Omega$, and $C = 3210\ \mu F$, find the time at which the current through the capacitor is 0.0165 A.

20. The voltage across a charging capacitor is given by

$$v = E(1 - e^{-t/RC}) \qquad \text{(Eq. 1083)}$$

If $E = 20.3$ V and $R = 4510$ Ω, find the time when the voltage across a 545-μF capacitor is equal to 10.1 V.

21. The temperature above its surroundings of an iron casting initially at 2005°F will be

$$T = 2005e^{-0.0620t}$$

after t seconds. Find the time for the casting to be at a temperature of 500°F above its surroundings.

22. A certain long pendulum, released from a height y_0 above its rest position, will be at a height

$$y = y_0 e^{-0.75t}$$

at t seconds. If the pendulum is released at a height of 15 cm, at what time will the height be 5.0 cm?

23. The population of a certain city growing at a rate of 2.0% per year from an initial population of 9000 will grow in t years to an amount

$$P = 9000e^{0.02t}$$

How many years (to the nearest year) will it take the population to triple?

24. The barometric pressure in inches of mercury at a height of h feet above sea level is

$$p = 30.0e^{-kh}$$

where $k = 3.83 \times 10^{-5}$. At what height will the pressure be 10.0 in. of mercury?

25. The approximate density of seawater at a depth of h mi is

$$d = 64.0e^{0.00676h} \text{ (lb/ft}^3\text{)}$$

Find the depth h at which the density will be 64.5 lb/ft^3.

26. A rope passing over a rough cylindrical beam (Fig. 18–13) supports a weight W. The force F needed to hold the weight is

$$F = We^{-\mu\theta}$$

where μ is the coefficient of friction and θ is the angle of wrap in radians. If $\mu = 0.150$, what angle of wrap is needed for a force of 100 lb to hold a weight of 200 lb?

27. Using the formula for compound interest, Eq. 1009, $y = a(1 + n)^t$, calculate the number of years it will take a sum of money to triple when invested at a rate of 12% per year.

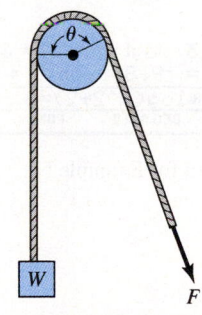

FIGURE 18–13

28. Using the formula for present worth, $a = \dfrac{y}{(1 + n)^t}$, in how many years will it take for $50,000 accumulate to $70,000 at 15% interest?

29. Find the half-life of a material that decays exponentially at the rate of 3.50% per year.

30. How long will it take the U.S. annual oil consumption to double if it is increasing exponentially at a rate of 7.0% per year?

31. How long will it take the world population to double at an exponential growth rate of 1.64% per year?

32. What is the maximum annual growth in energy consumption permissible if the consumption is not to double in the next 20 years?

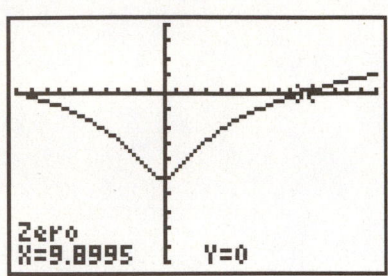

TI-83/84 graph of $y = 3 \log(x^2 + 2) - 6$. Tick marks on the horizontal axis are 1 unit apart.

TI-83/84 screen for Example 56. The computed value of x is shown, within the chosen bound.

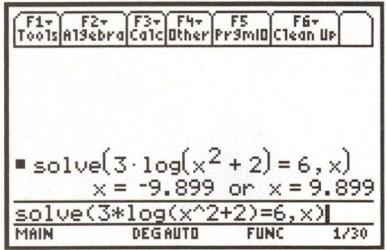

TI-89 screen for Example 57.

18–5   Solving a Logarithmic Equation

Graphical Solution

■ Exploration:

Try this. We have used our graphics calculators to find approximate solutions to many kinds of equations. Now use your calculator to graphically find the roots of the equation

$$3 \log(x^2 + 2) = 6$$ ■

In your exploration you probably graphed the function

$$y = 3 \log(x^2 + 2) - 6$$

as shown, and found roots at $x \approx 9.90$ and -9.90.

Using a Calculator's Equation Solver

We will now use a built-in equation solver to solve a logarithmic equation, just as we solved exponential equations earlier in this chapter.

◆◆◆ **Example 56:** Solve the equation from the exploration using the TI-83/84 equation solver.

Solution: We select **Solver** from the $\boxed{\text{MATH}}$ menu and enter the equation. Enter a new guess of $x = 5$ and a new bound of 0 to 20. Move the cursor to the line containing "$x =$" and press $\boxed{\text{SOLVE}}$.

The root falling within the selected bound is displayed. To find the other root we would change the initial guess and the bound to include negative values.   ◆◆◆

Solving Equations That Are in Implicit Form

◆◆◆ **Example 57:** Solve the equation from the preceding example using the TI-89 equation solver:

Solution: We select **solve** from the **Algebra** menu, enter the equation, followed by the variable (x) that we want to solve for, and press $\boxed{\approx}$ to get the approximate decimal solution. Note that both solutions are given.   ◆◆◆

Algebraic Solution

Often a logarithmic equation can be solved simply by rewriting it in exponential form. If just one term in our equation contains a logarithm, we isolate that term on one side of the equation before going to exponential form.

◆◆◆ **Example 58:** Algebraically solve the logarithmic equation of the preceding examples, to three significant digits.

Solution: Rearranging and dividing by 3 gives

$$\log(x^2 + 2) = 2$$

Going to exponential form, we obtain

$$x^2 + 2 = 10^2 = 100$$
$$x^2 = 98$$
$$x = \pm 9.90$$ ◆◆◆

If every term contains "log," we use the properties of logarithms to combine those terms into a single logarithm on each side of the equation and then take the antilog of both sides.

◆◆◆ **Example 59:** Solve for x to three significant digits:

$$3 \log x - 2 \log 2x = 2 \log 5$$

Solution: Using the properties of logarithms for quotients (Eq. 140) and for powers (Eq. 141) gives

$$\log \frac{x^3}{(2x)^2} = \log 5^2$$

$$\log \frac{x}{4} = \log 25$$

Taking the antilog of both sides, we have

$$\frac{x}{4} = 25$$

$$x = 100 \qquad\qquad ◆◆◆$$

If one or more terms do not contain a log, combine all of the terms that do contain a log on one side of the equation. Then go to exponential form.

◆◆◆ **Example 60:** Solve for x:

$$\log(5x + 2) - 1 = \log(2x - 1)$$

Solution: Rearranging yields

$$\log(5x + 2) - \log(2x - 1) = 1$$

Using the property of logarithms for quotients (Eq. 140),

$$\log \frac{5x + 2}{2x - 1} = 1$$

Expressing in exponential form gives

$$\frac{5x + 2}{2x - 1} = 10^1 = 10$$

Solving for x, we have

$$5x + 2 = 20x - 10$$

$$12 = 15x$$

$$x = \frac{4}{5} \qquad\qquad ◆◆◆$$

An Application

◆◆◆ **Example 61:** *The Richter Scale.* The Richter magnitude R, used to rate the intensity of an earthquake, is given by

$$R = \log \frac{a}{a_0}$$

where a and a_0 are the vertical amplitudes of the ground movement of the measured earthquake and of another earthquake taken as reference. Figure 18–14 shows the Richer magnitude for earthquakes of different severities.

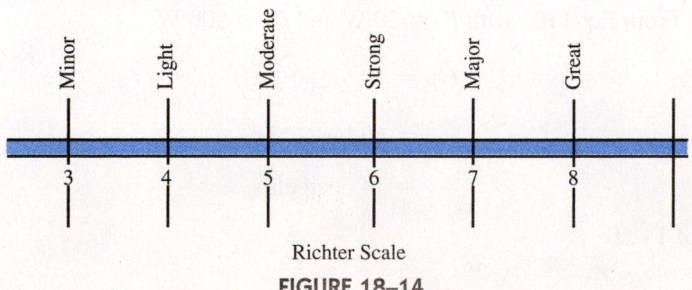

Richter Scale
FIGURE 18–14

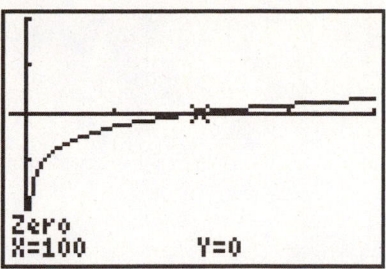

Graphical check for Example 59:
$y = 3 \log x - 2 \log 2x - 2 \log 5$
showing a root at $x = 100$. Tick marks on the horizontal axis are 50 units apart.

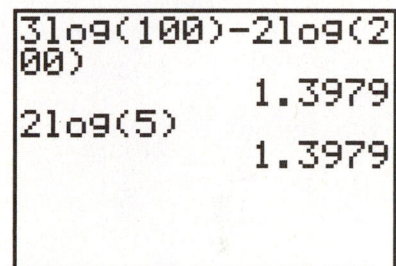

TI-83/84 check for Example 59. As usual, we must beware of extraneous roots. Here we have checked our answer of $x = 100$ by substituting it separately into each side of the given equation.

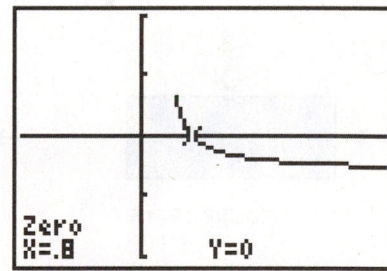

Graphical check for Example 60:
$y = \log(5x + 2) - 1 - \log(2x - 1)$
showing a root at $x = 0.8$. Tick marks on the horizontal axis are 1 unit apart.

If two earthquakes measure 4 and 5 on the Richer scale, by what factor is the amplitude of the stronger quake greater than that of the weaker?

Solution: Let us rewrite the equation for the Richter magnitude using the laws of logarithms.

$$R = \log a - \log a_0$$

Then, for the first earthquake,

$$4 = \log a_1 - \log a_0$$

and for the second,

$$5 = \log a_2 - \log a_0$$

Subtracting the first equation from the second gives

$$1 = \log a_2 - \log a_1$$
$$= (\log a_2/a_1)$$

Going to exponential form, we have

$$\frac{a_2}{a_1} = 10^1 = 10$$

$$a_2 = 10\, a_1$$

So the second earthquake has 10 times the amplitude of the first.

Check: Does this seem reasonable? From our study of logarithms, we know that as a number increases tenfold, its common logarithm increases only by 1. For example, $\log 45 = 1.65$ and $\log 450 = 2.65$. Since the Richter magnitude is proportional to the common log of the amplitude, it seems reasonable that a tenfold increase in amplitude increases the Richter magnitude by just 1 unit.   ◆◆◆

◆◆◆ **Example 62:** *An Application: Decibels and Power Gain or Loss*
An important use of logarithms is to compare power levels, voltage levels, or sound levels. Let us first do electrical power and voltage, and then sound levels.

If the power input to a certain network or device, Fig. 18–15, is P_1 and the power output is P_2, the power gain or loss G_P in the device is defined by the following logarithmic equation:

$$G_P = \log_{10} \frac{P_2}{P_1} \quad \text{bels}$$

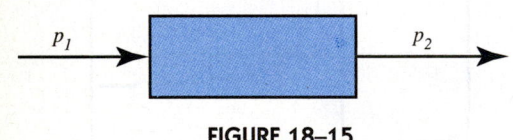

FIGURE 18–15

The bel is too large a unit for practical use so it is customary to use the *decibel* (dB) instead, where a decibel is 1/10 of a bel. Our equation then becomes:

Power Gain or Loss	$G_P = 10 \log_{10} \dfrac{P_2}{P_1}$ decibels	**1103**

A certain amplifier has a power output of 500 W for an input of 20 W. Find the decibels gained or lost.

Solution: From Eq. 1103 with $P_1 = 20$, W and $P_2 = 500$ W,

$$G_P = 10 \log_{10} \frac{500}{20}$$

$$= 10 \log_{10} 25$$

$$= 14 \text{ decibels}$$

or a gain of 14 dB.   ◆◆◆

♦♦♦ **Example 63:** *An Application:* To find P_1 or P_2 given the gain, we must solve the logarithmic Eq. 1103 by methods given earlier in this chapter.

A certain device has a gain of 2.50 dB with an input of 75.0 W. What is the power output?

Solution: From Eq.1103,

$$2.50 = 10 \log_{10} \frac{P_2}{75.0}$$

$$\frac{2.50}{10} = \log_{10} \frac{P_2}{75.0}$$

Going from logarithmic to exponential form,

$$10^{0.250} = \frac{P_2}{75.0}$$

$$P_2 = 75.0 \times 10^{0.250}$$

$$= 133 \text{ W}$$

♦♦♦

Using the fact (Eq. 1065) that power P is equal to $P = \dfrac{V^2}{R}$, Eq. 1103 becomes:

$$G_V = 10 \log_{10} \frac{V_2{}^2/R}{V_1{}^2/R}$$

$$= 10 \log_{10} \left(\frac{V_2}{V_1} \right)^2$$

Where G_V is the voltage gain or loss.

Voltage Gain or Loss	$G_V = 20 \log_{10} \dfrac{V_2}{V_1}$ decibels	**1104**

Here we have used Eq. 141, *the log of a number raised to a power equals the power times the log of the number*. We also use the subscript V to distinguish voltage gain from power gain.

♦♦♦ **Example 64:** *An Application: Voltage Gain or Loss*

The voltage of a section of transmission line is 2550 V at the near end and 2440 V at the far end. Find the decibels lost.

Solution: From Eq. 1104, we get

$$G_V = 20 \log_{10} \frac{2440}{2550}$$

$$= -0.38 \text{ decibel}$$

or a loss of 0.38 dB.

♦♦♦

♦♦♦ **Example 65:** *An Application: Sound Decibel Level*

A common use for the decibel scale is for indicating sound levels. That is because the human ear does not respond to sound in a linear fashion, but in a logarithmic fashion. For example, a *doubling* in sound level is not perceived by the ear as a doubling in loudness, but a *tenfold* increase *is* perceived by the ear as doubling in loudness. If the output of an audio amplifier is doubled from 10 W to 20 W, it will *not* sound twice as loud, but if the output of an audio amplifier is increased tenfold from 10 W to 100 W, it *will* sound twice as loud.

The loudness of a sound is perceived as approximately proportional to the *logarithm* of the intensity level. If we have two sources of sound having intensities of I_1 and I_2,

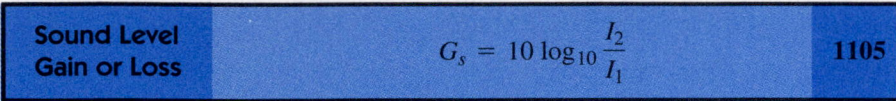

Sound Level Gain or Loss	$G_s = 10 \log_{10} \dfrac{I_2}{I_1}$	1105

Find the decibel level of the loudest tolerable sound (threshold of pain) that has an intensity level of 10^{-4} watts/cm^2 relative to the faintest audible sound (threshold of hearing) that has an intensity level of 10^{-16} watts/cm^2.

Solution: Substituting the given valves,

$$G_s = 10 \log_{10} \frac{10^{-4}}{10^{-16}} = 120 \text{ dB}$$

◆◆◆

The span between these two values is called the *dynamic range*. Some approximate decibel levels are given in Fig. 18–16.

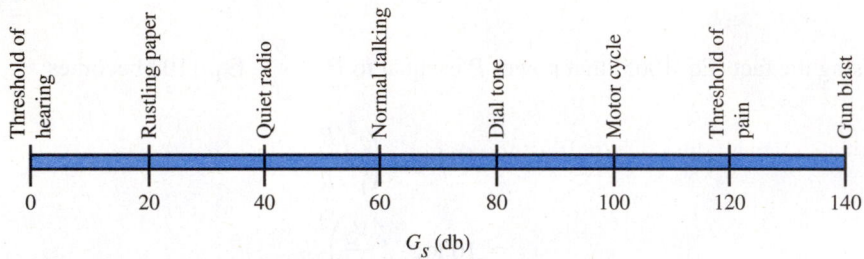

FIGURE 18–16   Approximate sound levels.

Exercise 5 ◆ Solving a Logarithmic Equation

Find the value of x in each expression.

1. $x = \log_3 9$
2. $x = \log_2 8$
3. $x = \log_8 2$
4. $x = \log_9 27$
5. $x = \log_{27} 9$
6. $x = \log_4 8$
7. $x = \log_8 4$
8. $x = \log_{27} 81$
9. $\log_x 8 = 3$
10. $\log_3 x = 4$
11. $\log_x 27 = 3$
12. $\log_x 16 = 4$
13. $\log_5 x = 2$
14. $\log_x 2 = \frac{1}{2}$
15. $\log_{36} x = \frac{1}{2}$
16. $\log_2 x = 3$
17. $x = \log_{25} 125$
18. $x = \log_5 125$

Solve for x. Give any approximate results to three significant digits. Check your answers.

19. $\log(2x + 5) = 2$
20. $2 \log(x + 1) = 3$
21. $\log(2x + x^2) = 2$
22. $\ln x - 2 \ln x = \ln 64$
23. $\ln 6 + \ln(x - 2) = \ln(3x - 2)$
24. $\ln(x + 2) - \ln 36 = \ln x$
25. $\ln(5x + 2) - \ln(x + 6) = \ln 4$
26. $\log x + \log 4x = 2$
27. $\ln x + \ln(x + 2) = 1$
28. $\log 8x^2 - \log 4x = 2.54$
29. $2 \log x - \log(1 - x) = 1$
30. $3 \log x - 1 = 3 \log(x - 1)$
31. $\log(x^2 - 4) - 1 = \log(x + 2)$
32. $2 \log x - 1 = \log(20 - 2x)$
33. $\log(x^2 - 1) - 2 = \log(x + 1)$
34. $\ln 2x - \ln 4 + \ln(x - 2) = 1$

Applications

35. The barometer reading is 29.66 in. Hg at a lower station, 3909 ft below the upper station. Find the barometer reading at the upper station. An equation giving difference in elevation h between two stations having different barometer readings B_1 and B_2 is given by $h = 60{,}470 \log(B_2/B_1)$.
36. What will be the barometer reading 815.0 ft above a station having a reading of 28.22 in. Hg?

37. Use Eq. 1103, $G_p = 10 \log_{10} \dfrac{P_2}{P_1}$, to find the power transmitted by a transmission line with an input of 2750 kW and a loss of 3.25 dB.

38. What power input is needed to produce a 250-W output with an amplifier having a 50 dB gain?

39. For a pipe diameter of 11.4 in., a temperature difference of 528°F, and an insulation conductivity of 0.044, find the outside radius r_2 of the insulation that will result in a heat loss of 215 Btu/h. Use the expression for the heat loss from a pipe $q = \dfrac{2\pi k(t_1 - t_2)}{\ln(r_1/r_2)}$ Btu/h.

40. *pH:* The pH value of a solution having a concentration C of hydrogen ions is given by the following equation:

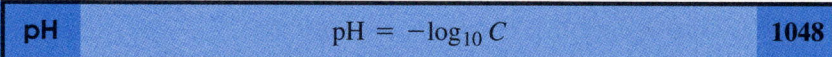

pH	$pH = -\log_{10} C$	1048

The units of C are moles of hydrogen ions per liter of solution. A mole is a unit used in chemistry for expressing large quantities of very small entities. A mole of hydrogen ions, for example, has a mass of one gram. Find the concentration of hydrogen ions in a solution having a pH of 4.65.

41. A pH of 7 is considered *neutral*, while a lower pH is *acidic* and a higher pH is *alkaline*. What is the hydrogen ion concentration at a pH of 7.0?

42. The acid rain during a certain storm had a pH of 3.5. Find the hydrogen ion concentration. How does it compare with that for a pH of 7.0? Show that
   (a) when the pH doubles, the hydrogen ion concentration is *squared*.
   (b) when the pH increases by a factor of n, the hydrogen ion concentration is raised *to the nth power*.

43. Two earthquakes have Richter magnitudes of 4.6 and 6.2. By what factor is the amplitude of the stronger quake greater than that of the weaker? Use the equation $R = \log \dfrac{a}{a_0}$.

44. A certain amplifier is to give a power output of 1000 W for an input of 50 W. Find the dB gain.

45. The power output of a certain device is half the input. What is the change in decibel level?

46. For a certain device, $P_1 = 4.0$ W and $P_2 = 280$ W. Find the change in decibel level.

47. A 20-W speaker is replaced with a 400-W speaker. Find the increase in decibel level.

48. The power delivered by a transmission line is 80% of the initial power. Find the change in decibel level.

49. Find the voltage gain of an amplifier that raises the voltage from 1.0 V to 84 V.

50. The voltage in a section of transmission line drops from 1555 V to 1524 V. Find the change in decibel level.

51. The voltage at the end of a certain transmission line is 90% of the initial voltage. Find the change in decibel level.

52. The voltage at the end of a certain transmission line is three-fourths the initial voltage. Find the change in decibel level.

53. *Decibel Level Decrease with Distance:* The decibel decrease G when moving from a sound source at a distance d_n to a distance d_f is given by

$$G = 20 \log_{10} \dfrac{d_f}{d_n} \text{ dB}$$

Find the decibel drop when moving twice as far from a sound source.

54 Find the decibel drop when moving three times as far from a sound source.

••• CHAPTER 18 REVIEW PROBLEMS •••••••••••••••••••••••••

Convert to logarithmic form.

1. $x^{5.2} = 352$ **2.** $4.8^x = 58$

3. $24^{1.4} = x$

Convert to exponential form.

4. $\log_3 56 = x$ **5.** $\log_x 5.2 = 124$

6. $\log_2 x = 48.2$

Solve for x.

7. $\log_{81} 27 = x$ **8.** $\log_3 x = 4$

9. $\log_x 32 = -\frac{5}{7}$

Write as the sum or difference of two or more logarithms.

10. $\log xy$ **11.** $\log \dfrac{3x}{z}$

12. $\log \dfrac{ab}{cd}$

Express as a single logarithm.

13. $\log 5 + \log 2$ **14.** $2 \log x - 3 \log y$

15. $\frac{1}{2} \log p - \frac{1}{4} \log q$

Find the common logarithm of each number to four decimal places.

16. 6.83   **17.** 364

18. 0.00638   **19.** 18.6

20. 87.4

Find x if $\log x$ is equal to the given value.

21. 2.846   **22.** 1.884

23. −0.473   **24.** 3.361

Find the natural logarithm of each number to four decimal places.

25. 84.72   **26.** 2846

27. 0.00873

Find the number whose natural logarithm is the given value.

28. 5.273   **29.** 1.473

30. −4.837

Find $\log x$ if

31. $\ln x = 4.837$ **32.** $\ln x = 8.352$

Find $\ln x$ if

33. $\log x = 5.837$ **34.** $\log x = 7.264$

Solve for x to three significant digits.

35. $(4.88)^x = 152$ **36.** $e^{2x+1} = 72.4$

37. $3 \log x - 3 = 2 \log x^2$

38. $\log(3x - 6) = 3$

39. A sum of \$1500 is deposited at a compound interest rate of $6\frac{1}{2}\%$ compounded quarterly. How much to the nearest dollar will it accumulate to in 5 years?

40. A quantity grows exponentially at a rate of 2.00% per day for 10.0 weeks. Find the final amount if the initial amount is 500 units.

41. A flywheel decreases in speed exponentially at the rate of 5.00% per minute. Find the speed after 20.0 min if the initial speed is 2250 rev/min.

42. Find the half-life of a radioactive substance that decays exponentially at the rate of 3.50% per year.

43. Find the doubling time for a population growing at the rate of 3.0% per year.

44. *Computer:* An iron ball with a mass of 1 kg is cut into two equal pieces. It is cut again, the second cut producing four equal pieces, the third cut making eight pieces, and so on. Write a program to compute the number of cuts needed to make each piece smaller than one atom (mass = 9.4×10^{-26} kg). Assume that no material is removed during cutting.

45. *Computer: Recursion Relation for Exponential Growth:* Our equation for exponential growth, $y = ae^{nt}$, gives the final amount y of a quantity growing at a rate n for t time periods, starting with amount a. Another equation for exponential growth is the recursion relation

$$y_t = By_{t-1} \qquad \text{(Eq. 156)}$$

where the amount y_t is obtained by multiplying the preceding amount y_{t-1} by the constant B, whose value is e^n. Use a spreadsheet to show that the two formulas give the same values over 20 time periods, choosing suitable values for the variables.

46. A transmission line has a loss of 1.50 dB with an input of 1175 V. What is the voltage output?

47. A transmission line has a loss of 0.50 dB with an output of 2500 V. What is the voltage input?

48. *Computer: Nonlinear Growth Equation:* The equations we have given for exponential growth represent a population that can grow without being limited by shortage of food or space, by predators, and so forth. This is rarely the case, so biologists have modified the equation to take those limiting factors into account. One model of the growth equation that takes these limits into account is obtained by multiplying the right side of Eq. 156 by the factor $(1 - y_{t-1})$.

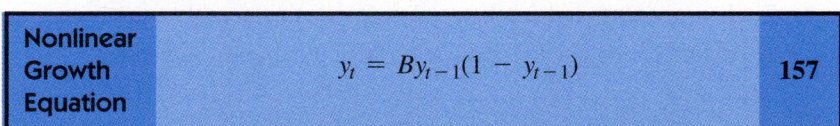

Nonlinear Growth Equation	$y_t = By_{t-1}(1 - y_{t-1})$	157

This equation is one form of what is called the *Logistic Function*, or the *Verhulst Population Model.* As y_{t-1} gets larger, the factor $(1 - y_{t-1})$ gets smaller. This equation will cause the predicted population y to stabilize at some value, for a range of values of the birthrate B from 1 to 3, regardless of the starting value for y.

Use a spreadsheet to compute the population for 20 time periods using both the nonlinear growth equation and the recursion relation for exponential growth, and compare. Use a starting value of 0.8 and a birthrate B of 2.

49. *Computer Chaos:* Repeat the preceding computation three more times with birthrates of 0.9, 3, and 4. What do you observe for each? *At which birthrate does the population*
(a) die out?
(b) reach a stable value?
(c) oscillate between values?
(d) give unpredictable results, or *chaos*?

50. *Writing:* In this chapter we have introduced two more kinds of equations: the *exponential* and the *logarithmic* equations. Earlier we had five other types. List the seven types, give an example of each, and write one sentence for each telling how it differs from the others.

19

Complex Numbers

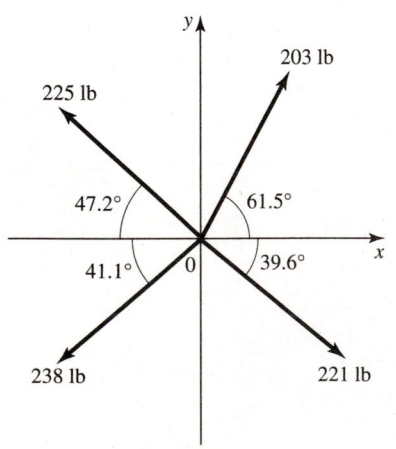

FIGURE 19–1

◆◆◆ **OBJECTIVES** ◆◆◆◼◆◆◆◆◆◆◆◆◆◆◆◆◆◆◆◆◆◆◆◆◆◆◆◆◆◆◆◆◆◆◆◆◆◆◆◆◆

When you have completed this chapter, you should be able to

- Write complex numbers in rectangular, polar, trigonometric, and exponential forms.
- Find the sums, differences, products, quotients, and powers of complex numbers.
- Find components and resultants of vectors using complex numbers.
- Solve alternating current applications using complex numbers.

◆◆◆

Up to now we have avoided square roots of negative numbers, expressions like $\sqrt{-4}$. We deal with them here by introducing imaginary numbers and complex numbers. We will show that an imaginary number is *not* imaginary and a complex number is not complicated, as its unfortunate name implies, but actually simplifies many computations.

A complex number can be written in many different forms, each with its own advantages. They are easily manipulated by calculator, which will greatly simplify our work. We will do the usual operations, addition, subtraction, multiplication, and so forth.

A graph of a complex number will show how the different forms of a complex number are related, and why they can be used to represent vectors. We will repeat the operations with vectors we did in our trigonometry chapters, resolving a vector into components, and finding the resultant of several vectors. For example, finding the resultant of the vectors in Fig. 19–1 can be a long process, but we will see later how complex numbers simplify the work.

Complex numbers find extensive use in alternating current computations. We will briefly show this application in the last section of this chapter to reinforce what electrical students learn in circuits classes.

19–1   Complex Numbers in Rectangular Form

As usual with any new topic, we start with some definitions.

Imaginary Numbers

Recall that in the real number system, the equation

$$x^2 = -1$$

had no solution because there was no real number such that its square was -1. Now we extend our number system to allow the quantity $\sqrt{-1}$ to have a meaning. We define the *imaginary unit* as the square root of -1 and represent it by the symbol i.

$$\text{Imaginary unit: } i = \sqrt{-1}$$

As complex numbers find wide use in electric circuits, the letter j is sometimes used for the imaginary unit, reserving i for current.

An *imaginary number* is the imaginary unit multiplied by a real number. An example of an imaginary number is $9i$. In general, we usually denote the real number by b, so the imaginary number would be written as bi.

Complex Numbers

A *complex number* is any number, real or imaginary, that can be written in the form

$$a + bi$$

where a and b are real numbers, and $i = \sqrt{-1}$ is the *imaginary unit*. A complex number written this way is said to be in *rectangular form*.

◆◆◆ **Example 1:** The following numbers are complex numbers in rectangular form.

$$4 + 2i \qquad -7 + 8i \qquad 5.92 - 2.93i \qquad 83 \qquad 27i \qquad \text{◆◆◆}$$

When $b = 0$ in a complex number $a + bi$, we have a *real number*. When $a = 0$, the number is called a *pure imaginary number*.

◆◆◆ **Example 2:**

(a) The complex number 48 is also a real number.
(b) The complex number $9i$ is also a pure imaginary number.   ◆◆◆

The two parts of a complex number are called the *real part* and the *imaginary part*.

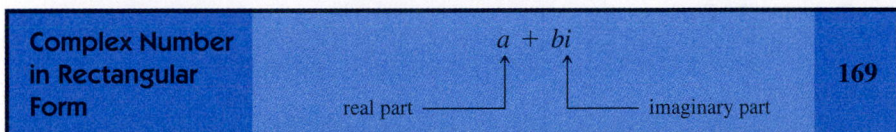

| Complex Number in Rectangular Form | $a + bi$ real part ⎯⎯⎯ imaginary part | 169 |

Addition and Subtraction of Complex Numbers

To combine complex numbers, separately combine the real parts, then combine the imaginary parts, and express the result in the form $a + bi$.

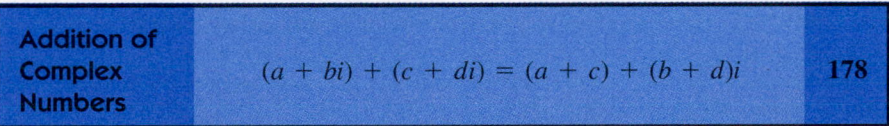

| Addition of Complex Numbers | $(a + bi) + (c + di) = (a + c) + (b + d)i$ | 178 |

Subtraction of Complex Numbers	$(a + bi) - (c + di) = (a - c) + (b - d)i$	179

◆◆◆ Example 3: Here are some examples of the addition and subtraction of complex numbers in rectangular form.

(a) $3i + 5i = 8i$

(b) $2i + (6 - 5i) = 6 - 3i$

(c) $(2 - 5i) + (-4 + 3i) = (2 - 4) + (-5 + 3)i = -2 - 2i$

(d) $(-6 + 2i) - (4 - i) = (-6 - 4) + [2 - (-1)]i = -10 + 3i$ ◆◆◆

Powers of *i*

We often have to evaluate powers of the imaginary unit, especially the square of i. Since

$$i = \sqrt{-1}$$

then

$$i^2 = \sqrt{-1}\,\sqrt{-1} = -1$$

Higher powers are easily found.

$$i^3 = i^2 i \quad = (-1)i = -i$$
$$i^4 = (i^2)^2 = (-1)^2 = 1$$
$$i^5 = i^4 i \quad = (1)i \quad = i$$
$$i^6 = i^4 i^2 = (1)i^2 \quad = -1$$

We see that the values are starting to repeat: $i^5 = i$, $i^6 = i^2$, and so on. The first four values keep repeating. Note that when the exponent n is a multiple of 4, then $i^n = 1$.

Powers of *i*	$i = \sqrt{-1}$ $i^2 = -1$ $i^3 = -i$	177
	$i^4 = 1$ $i^5 = i$ . . .	

◆◆◆ Example 4: Evaluate i^{17}.

Solution: Using the laws of exponents, we express i^{17} in terms of one of the first four powers of i.

$$i^{17} = i^{16}i = (i^4)^4 i = (1)^4 i = i$$ ◆◆◆

Multiplying by Imaginary Numbers

Multiply as with ordinary numbers, but use Eq. 177 to simplify any powers of i.

◆◆◆ Example 5: Here are examples of multiplying by imaginary numbers.

(a) $5 \times 3i = 15i$

(b) $2i \times 4i = 8i^2 = 8(-1) = -8$

(c) $3 \times 4i \times 5i \times i = 60i^3 = 60(-i) = -60i$ ◆◆◆

Multiplying Complex Numbers

We multiply complex numbers in the same way we multiplied binomials in Chapter 2.

◆◆◆ **Example 6:** Multiply the complex numbers

$$(a + bi)(c + di)$$

Solution: We multiply just like any other binomials

$$(a + bi)(c + di) = ac + adi + bci + bdi^2$$

Next we replace i^2 by -1 and collect terms

$$(a + bi)(c + di) = ac + (ad + bc)i + bd(-1)$$

$$= (ac - bd) + (ad + bc)i \qquad ◆◆◆$$

So, in general,

Multiplication of Complex Numbers	$(a + bi)(c + di) = (ac - bd) + (ad + bc)i$	180

◆◆◆ **Example 7:** Here we give some examples of multiplication of complex numbers in rectangular form.

(a) $3(5 + 2i) = 15 + 6i$

(b) $(3i)(2 - 4i) = 6i - 12i^2 = 6i - 12(-1) = 12 + 6i$

(c) $(3 - 2i)(-4 + 5i) = 3(-4) + 3(5i) + (-2i)(-4) + (-2i)(5i)$

$$= -12 + 15i + 8i - 10i^2$$

$$= -12 + 15i + 8i - 10(-1)$$

$$= -2 + 23i \qquad ◆◆◆$$

We will see in the next section that multiplication and division are easier in *polar* form.

To multiply radicals that contain negative quantities under the radical sign, first express all quantities in terms of i; then proceed to multiply. Always be sure to convert radicals to imaginary numbers *before* performing other operations, or contradictions may result.

◆◆◆ **Example 8:** Multiply $\sqrt{-4}$ by $\sqrt{-4}$.

Solution: Converting to imaginary numbers, we obtain

$$\sqrt{-4}\,\sqrt{-4} = (2i)(2i) = 4i^2$$

Since $i^2 = -1$,

$$4i^2 = -4 \qquad ◆◆◆$$

Common Error	It is *incorrect* to write $$\sqrt{-4}\,\sqrt{-4} = \sqrt{(-4)(-4)} = \sqrt{16} = +4$$ Our previous rule of $\sqrt{a} \cdot \sqrt{b} = \sqrt{ab}$ applied only to *positive a and b.*

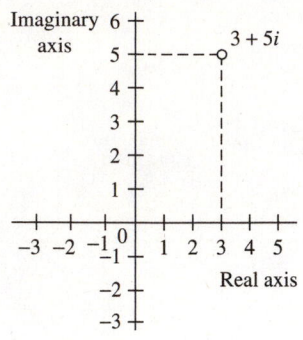

FIGURE 19–2 The complex plane.

Graph of a Complex Number

To graph a complex number we use axes similar to our familiar rectangular coordinate system (Fig. 19–2). Now the horizontal axis is the *real* axis, and the vertical axis is the *imaginary* axis. Such a coordinate system defines what is called the *complex plane*. Real numbers are graphed as points on the horizontal axis; pure imaginary numbers are graphed as points on the vertical axis. Complex numbers, such as $3 + 5i$ (Fig. 19–2), are graphed elsewhere within the plane.

To plot a complex number $a + bi$ in the complex plane, simply locate a point with a horizontal coordinate of a and a vertical coordinate of b.

◆◆◆ **Example 9:** Plot the complex numbers $(2 + 3i)$, $(-1 + 2i)$, $(-3 - 2i)$, and $(1 - 3i)$.

Solution: The points are plotted in Fig. 19–3.   ◆◆◆

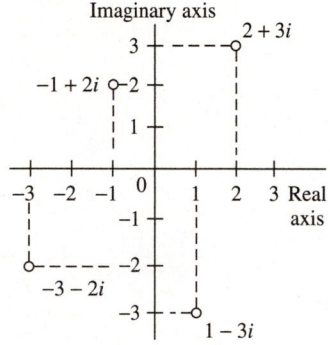

FIGURE 19–3 Such a plot is called an *Argand diagram*, named for Jean Robert Argand (1768–1822).

The Conjugate of a Complex Number

Two numbers that are located symmetrically across the real axis, as in Fig. 19–4, are called *complex conjugates*. Thus the conjugate of a complex number is obtained by changing the sign of the imaginary part.

◆◆◆ **Example 10:** Here are some conjugates.

(a) The conjugate of $2 + 3i$ is $2 - 3i$.
(b) The conjugate of $-5 - 4i$ is $-5 + 4i$.
(c) The conjugate of $a + bi$ is $a - bi$.   ◆◆◆

◆◆◆ **Example 11:** Multiply the complex number $2 + 3i$ by its conjugate.

Solution: The conjugate of $2 + 3i$ is $2 - 3i$. Multiplying,

$$(2 + 3i)(2 - 3i) = 4 - 6i + 6i - 9i^2 = 4 - (9)(-1) = 13$$ ◆◆◆

Note that the form of a complex number times its conjugate is the same as for the difference of two squares. Here, as there, two of the products are equal but of opposite sign and hence are eliminated.

Conjugates are needed when we divide complex numbers, as we will now see.

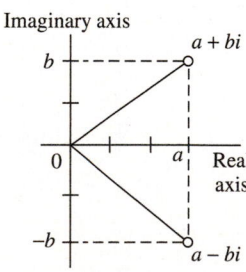

FIGURE 19–4

Dividing Complex Numbers

Division involving *single terms*, real or imaginary, is shown by examples.

◆◆◆ **Example 12:**

(a) $8i \div 2 = 4i$ (b) $6i \div 3i = 2$
(c) $(4 - 6i) \div 2 = 2 - 3i$ ◆◆◆

◆◆◆ **Example 13:** Divide 6 by $3i$.

Solution:

$$\frac{6}{3i} = \frac{6}{3i} \times \frac{3i}{3i} = \frac{18i}{9i^2} = \frac{18i}{-9} = -2i$$ ◆◆◆

To divide by a complete complex number, *multiply dividend and divisor by the conjugate of the divisor.* This will make the divisor a real number, as in the following example.

◆◆◆ **Example 14:** Divide $3 - 4i$ by $2 + i$.

Solution: We multiply numerator and denominator by the conjugate $(2 - i)$ of the denominator.

$$\frac{3 - 4i}{2 + i} = \frac{3 - 4i}{2 + i} \cdot \frac{2 - i}{2 - i}$$

$$= \frac{6 - 3i - 8i + 4i^2}{4 + 2i - 2i - i^2}$$

$$= \frac{2 - 11i}{5}$$

$$= \frac{2}{5} - \frac{11}{5}i$$

◆◆◆

This is very similar to *rationalizing the denominator* of a fraction containing radicals. In general,

Division of Complex Numbers	$\dfrac{a + bi}{c + di} = \dfrac{ac + bd}{c^2 + d^2} + \dfrac{bc - ad}{c^2 + d^2}i$	181

Exercise 1 ◆ Complex Numbers in Rectangular Form

Combine and simplify.

1. $(3 - 2i) + (-4 + 3i)$

2. $(-1 - 2i) - (i + 6)$

3. $(a - 3i) + (a + 5i)$

4. $(p + qi) + (q + pi)$

5. $\left(\dfrac{1}{2} + \dfrac{i}{3}\right) + \left(\dfrac{1}{4} - \dfrac{i}{6}\right)$

6. $(-84 + 91i) - (28 + 72i)$

7. $(2.28 - 1.46i) + (1.75 + 2.66i)$

Evaluate each power of i.

8. i^{11}

9. i^5

10. i^{10}

11. i^{21}

12. i^{14}

Multiply and simplify.

13. $7 \times 2i$

14. $9 \times 3i$

15. $3i \times 5i$

16. $i \times 4i$

17. $4 \times 2i \times 3i \times 4i$

18. $i \times 5 \times 4i \times 3i \times 5$

19. $(5i)^2$

20. $(3i)^2$

21. $2(3 - 4i)$

22. $-3(7 + 5i)$

23. $4i(5 - 2i)$

24. $5i(2 + 3i)$

25. $(3 - 5i)(2 + 6i)$

26. $(5 + 4i)(4 + 2i)$

27. $(6 + 3i)(3 - 8i)$

28. $(5 - 3i)(8 + 2i)$

29. $(5 - 2i)^2$

30. $(3 + 6i)^2$

Graph each complex number.

31. $2 + 5i$

32. $-1 - 3i$

33. $3 - 2i$

34. $2 - i$

35. $5i$

36. $2.25 - 3.62i$

Write the conjugate of each complex number.

37. $2 - 3i$ **38.** $-5 - 7i$ **39.** $p + qi$

40. $-5i + 6$ **41.** $-mi + n$ **42.** $5 - 8i$

Divide and simplify.

43. $8i \div 4$ **44.** $9 \div 3i$

45. $12i \div 6i$ **46.** $44i \div 2i$

47. $(4 + 2i) \div 2$ **48.** $8 \div (4 - i)$

49. $(-2 + 3i) \div (1 - i)$ **50.** $(5 - 6i) \div (-3 + 2i)$

51. $(7i + 2) \div (3i - 5)$ **52.** $(-9 + 3i) \div (2 - 4i)$

19–2   Complex Numbers in Polar Form

In an earlier chapter we saw that a point could be located by *polar coordinates*, as well as by rectangular coordinates. Similarly, a complex number can be given in *polar form* as well as in rectangular form.

Why another form? We will see that while addition and subtraction of complex numbers are best done in rectangular form, multiplication and division are easier in polar form.

Figure 19–5 shows how the rectangular and polar forms are related. There we have plotted the complex number $a + bi$. If we now connect that point to the origin by a line segment of length r, it makes an angle θ with the horizontal axis. We can express the same complex number in terms of r and θ, if we note that

$$\cos \theta = \frac{a}{r}$$

So

$$a = r \cos \theta$$

Similarly,

$$b = r \sin \theta$$

Substituting gives

$$a + bi = r \cos \theta + ri \sin \theta$$
$$= r(\cos \theta + i \sin \theta) \tag{1}$$

We can also write this expression using the simpler notation,

$$a + bi = r\underline{/\theta} \tag{2}$$

A third polar form can be obtained from Euler's formula, Eq. 185, which we give without proof.

$$\cos \theta + i \sin \theta = e^{i\theta}$$

Here e is the base of natural logarithms and θ is in radians. Multiplying both sides by r, we get

$$r(\cos \theta + i \sin \theta) = re^{i\theta}$$

So,

$$a + bi = re^{i\theta} \tag{3}$$

While Eqs. 1, 2, and 3 are polar forms of a complex number, we will refer to Eq. 1 as the *trigonometric* form, Eq. 2 as the *polar* form, and Eq. 3 as the *exponential* form. In all three polar forms, r is called the *magnitude* or *absolute value* of the complex number, and the angle θ is called the *argument* of the complex number.

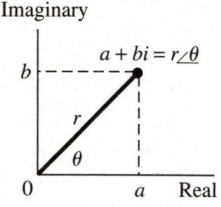

FIGURE 19–5   Polar form of a complex number shown on a complex plane.

Named for the Swiss mathematician Leonhard Euler (1707–1783).

Our various forms of a complex number are then

Forms of a Complex Number	Rectangular	$a + bi$	169
	Polar	$r\underline{/\theta}$	170
	Trignometric	$r(\cos\theta + i\sin\theta)$	171
	Exponential	$r\,e^{i\theta}$	172
	where $r = \sqrt{a^2 + b^2}$		173
	$\theta = \arctan\dfrac{b}{a}$		174
	$a = r\cos\theta$		175
	$b = r\sin\theta$		176

In practice, we will just use the rectangular form for addition and subtraction, and the polar form for multiplication and division.

Converting between Rectangular and Polar Form

To convert from rectangular to polar we use Eqs. 173 and 174 to calculate r and θ.

◆◆◆ **Example 15:** Write the complex number $2 + 3i$ in polar form. Keep three significant digits.

Solution: With $a = 2$ and $b = 3$ we have

$$r = \sqrt{2^2 + 3^2} = 3.61$$

and

$$\tan\theta = \frac{3}{2}$$
$$\theta = 56.3°$$

So

$$2 + 3i = 3.61\underline{/56.3°}$$

◆◆◆

To convert from polar to rectangular, we use Eqs. 175 and 176 to calculate a and b.

◆◆◆ **Example 16:** Write $6\underline{/30°}$ in rectangular form. Keep three significant digits.

Solution: By inspection,

$$r = 6 \quad \text{and} \quad \theta = 30°$$

So

$$a = r\cos\theta = 6\cos 30° = 5.20$$

and

$$b = r\sin\theta = 6\sin 30° = 3.00$$

Thus our complex number in rectangular form is $5.20 + 3.00i$.   ◆◆◆

Arithmetic Operations in Polar Form

We saw that addition and subtraction are fast and easy in rectangular form, and we now show that multiplication, division, and raising to a power are best done in polar form.

Products

Let us multiply $r(\cos \theta + i \sin \theta)$ by $r'(\cos \theta' + i \sin \theta')$.

$$[r(\cos \theta + i \sin \theta)][r'(\cos \theta' + i \sin \theta')]$$
$$= rr'(\cos \theta \cos \theta' + i \cos \theta \sin \theta' + i \sin \theta \cos \theta' + i^2 \sin \theta \sin \theta')$$
$$= rr'[(\cos \theta \cos \theta' - \sin \theta \sin \theta') + i(\sin \theta \cos \theta' + \cos \theta \sin \theta')]$$

But by the trigonometric identities for the sine and cosine of the sum of two angles

$$\sin(\theta + \theta') = \sin \theta \cos \theta' + \cos \theta \sin \theta' \qquad (128)$$

$$\cos(\theta + \theta') = \cos \theta \cos \theta' - \sin \theta \sin \theta' \qquad (129)$$

So

$$r(\cos \theta + i \sin \theta) \cdot r'(\cos \theta' + i \sin \theta') = rr'[\cos(\theta + \theta') + i \sin(\theta + \theta')]$$

Rewriting this result using our simpler notation gives

	$$r\underline{/\theta} \cdot r'\underline{/\theta'} = rr'\underline{/\theta + \theta'}$$	
Products	*The absolute value of the product of two complex numbers is the product of their absolute values, and the argument is the sum of the individual arguments.*	**182**

◆◆◆ **Example 17:** Multiply $5\underline{/30°}$ by $3\underline{/20°}$.

Solution: The absolute value of the product will be $5(3) = 15$, and the argument of the product will be $30° + 20° = 50°$. So

$$5\underline{/30°} \cdot 3\underline{/20°} = 15\underline{/50°}$$ ◆◆◆

The angle θ is not usually written greater than 360°. Subtract multiples of 360° if necessary.

◆◆◆ **Example 18:** Multiply $6.27\underline{/300°}$ by $2.75\underline{/125°}$.

Solution:

$$6.27\underline{/300°} \times 2.75\underline{/125°} = (6.27)(2.75)\underline{/300° + 125°}$$
$$= 17.2\underline{/425°}$$
$$= 17.2\underline{/65°}$$

after subtracting 360° from the angle.   ◆◆◆

◆◆◆ **Example 19:** *An Application:* A certain current **I** is represented by the complex number $1.15\underline{/23.5°}$ amperes, and a complex impedance **Z** is represented by $24.6\underline{/14.8°}$ ohms. Multiply **I** by **Z** to obtain the voltage **V**.

Solution:

$$\mathbf{V} = \mathbf{IZ} = (1.15\underline{/23.5°})(24.6\underline{/14.8°}) = 28.3\underline{/38.3°} \quad (\text{V})$$ ◆◆◆

Quotients

The rule for division of complex numbers in trigonometric form is similar to that for multiplication.

$$\frac{r\angle\theta}{r'\angle\theta'} = \frac{r}{r'}\angle\theta - \theta'$$

Quotients | *The absolute value of the quotient of two complex numbers is the quotient of their absolute values, and the argument is the difference (numerator minus denominator) of their arguments.* | **183**

◆◆◆ **Example 20:** Divide $6\angle70°$ by $2\angle50°$.

Solution: We divide the absolute values, and subtract the angles,

$$\frac{6\angle70°}{2\angle50°} = \left(\frac{6}{2}\right)\angle(70° - 50°)$$
$$= 3\angle20°$$

◆◆◆

Powers

To raise a complex number to a power, we merely have to multiply it by itself the proper number of times, using Eq. 182. For example,

$$(r\angle\theta)^2 = r\angle\theta \cdot r\angle\theta = r \cdot r\angle\theta + \theta = r^2\angle2\theta$$

and
$$(r\angle\theta)^3 = r\angle\theta \cdot r^2\angle2\theta = r \cdot r^2\angle\theta + 2\theta = r^3\angle3\theta$$

Do you see a pattern developing? In general, we get the following equation:

$$(r\angle\theta)^n = r^n\angle n\theta$$

DeMoivre's Theorem | *When a complex number is raised to the nth power, the new absolute value is equal to the original absolute value raised to the nth power, and the new argument is n times the original argument.* | **184**

DeMoivre's theorem is named after Abraham DeMoivre (1667–1754).

◆◆◆ **Example 21:** Raise $2\angle10°$ to the fifth power.

Solution: We raise the absolute value to the fifth power and multiply the angle by 5.

$$(2\angle10°)^5 = 2^5\angle5(10°) = 32\angle50°$$

◆◆◆

Exercise 2 ◆ Complex Numbers in Polar Form

Round to three significant digits, where necessary, in this exercise.

Write each complex number in polar form.

1. $5 + 4i$ **2.** $-3 - 7i$

3. $4 - 3i$ **4.** $8 + 4i$

5. $-5 - 2i$ **6.** $-4 + 7i$

7. $-9 - 5i$ **8.** $7 - 3i$

9. $-4 - 7i$

Write in rectangular form.

10. $5\underline{/28°}$ **11.** $9\underline{/59°}$

12. $4\underline{/63°}$ **13.** $7\underline{/-53°}$

14. $6\underline{/-125°}$

Multiplication and Division

Multiply.

15. $5.82\underline{/44.8°}$ by $2.77\underline{/10.1°}$ **16.** $5\underline{/30°}$ by $2\underline{/10°}$

17. $8\underline{/45°}$ by $7\underline{/15°}$ **18.** $2.86\underline{/38.2°}$ by $1.55\underline{/21.1°}$

Divide.

19. $58.3\underline{/77.4°}$ by $12.4\underline{/27.2°}$ **20.** $24\underline{/50°}$ by $12\underline{/30°}$

21. $50\underline{/72°}$ by $5\underline{/12°}$ **22.** $71.4\underline{/56.4°}$ by $27.7\underline{/15.2°}$

19–3   Complex Numbers on the Calculator

Most graphing calculators can easily handle complex numbers. Here we will show how to enter complex numbers in either form, rectangular or polar, convert that number to either form, and do basic operations with complex numbers having the result displayed in either form. We will do this for both the TI-83/84 and TI-89 calculators.

Entering and Converting Complex Numbers on the Calculator

You can select one of the two complex number modes available on your calculator. You can *enter* a complex number in *any* mode, but it will be *displayed* in the mode you have chosen.

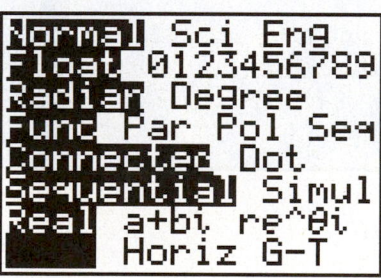

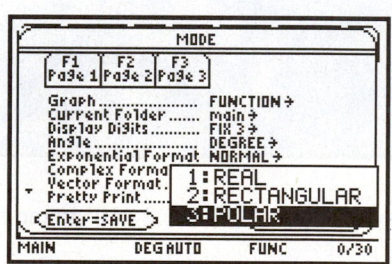

This is the MODE screen on the TI-83/84. The seventh line shows the number modes, **Real, a+bi** (rectangular mode), and **re^θi** (polar mode, which on the TI-83/84 is displayed in exponential form).

MODE screen on the TI-89 showing the complex formats: **REAL, RECTANGULAR,** and **POLAR.**

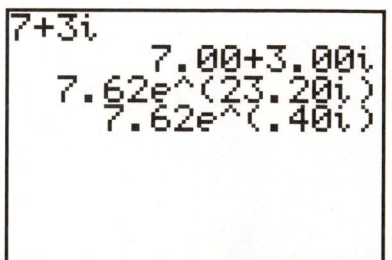

TI-83/84 screen for Example 22.

◆◆◆ **Example 22:** Enter the complex number $7 + 3i$ on the TI-83/84. Have the answer displayed in (a) rectangular form, (b) polar form in degrees, and (c) polar form in radians.

Solution: To enter a complex number, enter the real and the ordinary and the imaginary parts, in any order, separated by a plus or minus sign. Note that i is *not* one of the alphabetic characters but is a special key on the calculator; a second function on the ⋅ key. We enter the number and then

(a) put the calculator into **Rectangular** mode and press ENTER.

(b) put the calculator into **Polar** mode and **Degree** mode, and press ENTER.

(c) put the calculator into **Polar** mode and **Radian** mode, and press ENTER.   ◆◆◆

◆◆◆ Example 23: Repeat the preceding example on the TI-89.

Solution: The steps are the same. On the TI-89 $\boxed{i}$ is a second function on the $\boxed{\text{CATALOG}}$ key. ◆◆◆

◆◆◆ Example 24: Enter the complex number $284\underline{/35.8°}$ in the TI-83/84 and TI-89 calculators. Display the result in rectangular form.

Solution: We enter the complex number, as shown on the accompanying screens. To get the results displayed in rectangular form, we choose that mode from the $\boxed{\text{MODE}}$ menu. Press $\boxed{\text{ENTER}}$ to display the result.

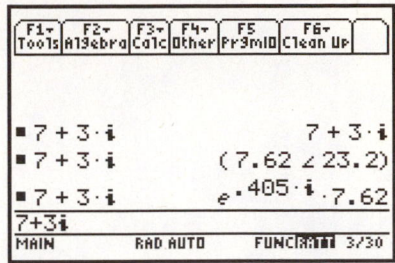

TI-89 screen for Example 23. Note that when in **RADIAN** mode, the calculator displays the result in exponential form.

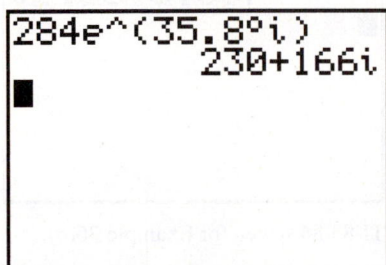

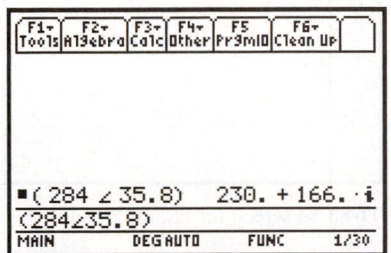

TI-83/84 screen for Example 24. We enter the number in exponential form, which requires that the calculator be in **Radian** mode. The $\boxed{e^x}$ key is a second function on the $\boxed{\text{LN}}$ key. Since our angle is in degrees, we choose the degree symbol from the $\boxed{\text{ANGLE}}$ menu.

TI-89 screen for Example 24. For entering a number in polar form, the $\boxed{L}$ symbol is used, a second function on the $\boxed{\text{EE}}$ key. Note that the complex number entered is enclosed in parentheses. If we put the calculator in **DEGREE** mode, we do not have to specify units for the angle. ◆◆◆

Converting Form Using the ▸Rect and ▸Polar Instructions

We have seen that we can convert between rectangular and polar forms by entering the complex number in one form and having it display in another form. However, there is another way to do these conversions.

◆◆◆ Example 25: Convert $43.4 + 52.5i$ to polar form.

Solution: We enter the complex number in rectangular form, followed by ▸**Polar**. On the TI-83/84 this is found on the $\boxed{\text{MATH}}$ **CPX** menu, and on the TI-89 it is on the $\boxed{\text{MATH}}$ **Matrix/Vector ops** menu. Press $\boxed{\text{ENTER}}$ to get the result.

It does not matter which complex number mode the calculator is in. However, the result will be either in radians or degrees, depending upon which angle mode is chosen.

The ▸**Rect** instruction is used in the same way. ◆◆◆

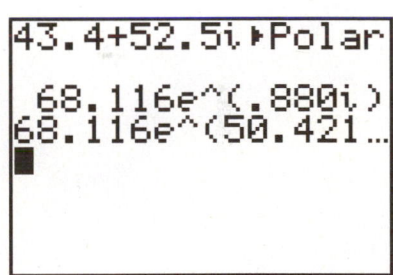

TI-83/84 screen for Example 25. The first result was obtained in **radian** mode and the second in **degree** mode.

Arithmetic Operations on Complex Numbers by Calculator

To perform arithmetic operations on complex numbers by calculator, first select the angular units desired, degrees or radians, and the form, rectangular or polar, in which you want the result displayed. Then simply enter the numbers, in either rectangular or polar form or both, and the indicated operation.

TI-89 screen for Example 25. The first result was obtained in **RADIAN** mode and the second in **DEGREE** mode.

◆◆◆ **Example 26:** Perform the following computations by calculator. Display the results in rectangular form.

(a) $(5.4 + 6.8i) + (7.2 - 8.1i) - (-4.7 - 6.5i)$

(b) $(4.3 - 6.5i)^3$

(c) $(2.94\underline{/25.3°})(1.75\underline{/15.7°})$

Solution: The computations are shown in the following TI-83/84 and TI-89 screens. All are in **RECTANGULAR** mode. If they were instead in **POLAR** mode, the answers would be displayed in polar form.

◆◆◆

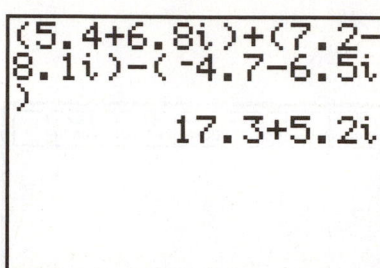

TI-83/84 screen for Example 26(a).

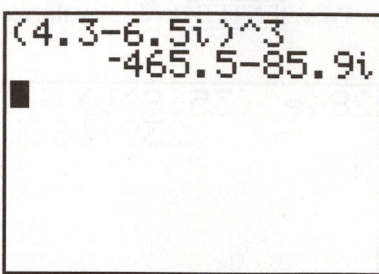

TI-83/84 screen for Example 26(b).

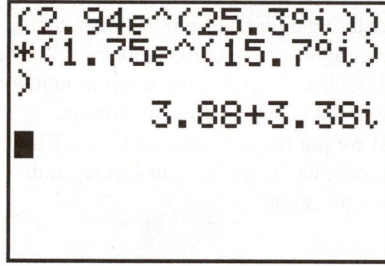

TI-83/84 screen for Example 26(c);
RADIAN mode.

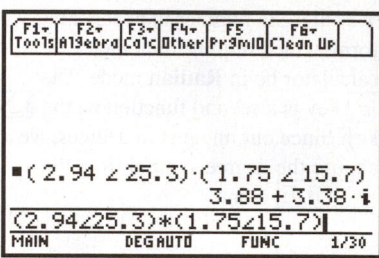

TI-89 screen for Example 26(c);
DEGREE mode.

Exercise 3 ◆ Complex Numbers on the Calculator

Do the following by calculator. Round to three significant digits, where necessary.

Write each complex number in polar form.

1. $5 + 4i$ **2.** $-3 - 7i$

3. $4 - 3i$ **4.** $8 + 4i$

5. $-5 - 2i$ **6.** $-4 + 7i$

Write in rectangular form.

7. $5\underline{/28°}$ **8.** $9\underline{/59°}$

9. $4\underline{/63°}$ **10.** $7\underline{/-53°}$

11. $6\underline{/-125°}$

Combine and simplify.

12. $(3 - 2i) + (-4 + 3i)$ **13.** $(-1 - 2i) - (i + 6)$

14. $(a - 3i) + (a + 5i)$ **15.** $(p + qi) + (q + pi)$

16. $\left(\dfrac{1}{2} + \dfrac{i}{3}\right) + \left(\dfrac{1}{4} - \dfrac{i}{6}\right)$ **17.** $(-84 + 91i) - (28 + 72i)$

Multiply and simplify.

18. $4i(5 - 2i)$
19. $5i(2 + 3i)$
20. $(3 - 5i)(2 + 6i)$
21. $(5 + 4i)(4 + 2i)$
22. $(6 + 3i)(3 - 8i)$
23. $(5 - 3i)(8 + 2i)$
24. $5.82\underline{/44.8°}$ by $2.77\underline{/10.1°}$
25. $5\underline{/30°}$ by $2\underline{/10°}$
26. $8\underline{/45°}$ by $7\underline{/15°}$
27. $2.86\underline{/38.2°}$ by $1.55\underline{/21.1°}$

Divide and simplify.

28. $8i \div 4$
29. $9 \div 3i$
30. $12i \div 6i$
31. $44i \div 2i$
32. $(4 + 2i) \div 2$
33. $8 \div (4 - i)$
34. $58.3\underline{/77.4°}$ by $12.4\underline{/27.2°}$
35. $24\underline{/50°}$ by $12\underline{/30°}$
36. $50\underline{/72°}$ by $5\underline{/12°}$
37. $71.4\underline{/56.4°}$ by $27.7\underline{/15.2°}$

19–4   Vector Operations Using Complex Numbers

An important use of a complex number is that it can represent a vector. Vectors represented by complex numbers can be easily manipulated by calculator to find rectangular components and the resultant of any number of vectors. This is our third look at vectors. It is a good time to glance back at Chapter 7 where we introduced vectors, and Chapter 8 where we manipulated vectors at any angle.

Representing a Vector by a Complex Number

Figure 19–6 shows a plot of the complex number $2 + 3i$. If we connect that point with a line to the origin, we can think of the complex number $2 + 3i$ as *representing a vector* **R** *having a horizontal component of 2 units and a vertical component of 3 units.*

The complex number used to represent a vector can be expressed either in rectangular, polar, or exponential form. In this section we will do all the work by calculator, for which the exponential and polar forms will be most convenient.

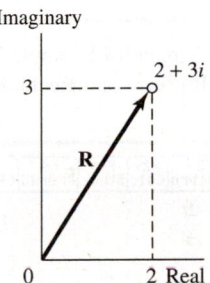

FIGURE 19–6   A vector represented by a complex number.

Components of a Vector

To find the rectangular components of a vector, write it in exponential form and then convert it to rectangular form. The quantities a and b give the horizontal and vertical components, respectively.

◆◆◆ **Example 27:** A vector has a magnitude of 154 and makes an angle of 48.7° with the horizontal, Fig. 19–7. Find the vector's horizontal and vertical components (a) using the TI-83/84 and (b) using the TI-89.

Solution: (a) The TI-83/84 takes complex numbers in exponential form, so we put the calculator into **Radian** mode. We then enter

$$154\,e^{48.7°i}$$

Since we are in radian mode, we affix the degree symbol, found on the $\boxed{\text{ANGLE}}$ menu, to our angle.

(b) On the TI-89 we can enter complex numbers in the simpler polar notation, and our calculator can be in **DEGREE** mode. So we enter the complex number, in parentheses,

$$(154\underline{/48.7°})$$

Since the calculator is in **DEGREE** mode, we do not need the degree symbol on the angle.

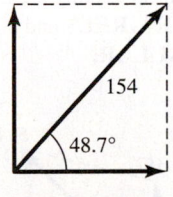

FIGURE 19–7

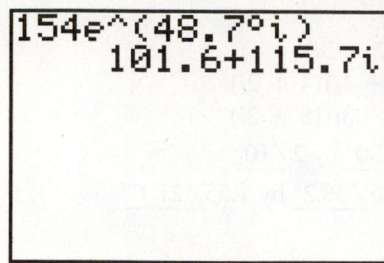

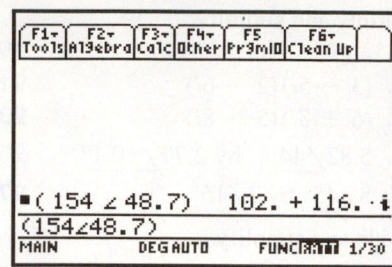

TI-83/84 screen for Example 27.   TI-89 screen for Example 27.

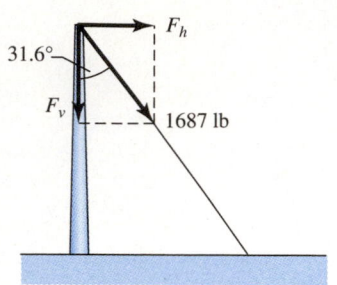

FIGURE 19–8

With either calculator we go into **RECTANGULAR** mode and press ENTER . We see that the horizontal component is 102, and the vertical component is 116.

◆◆◆

◆◆◆ **Example 28:** *An Application:* The tension in the cable, Fig. 19–8, is 1687 lb. Find the vertical and horizontal components of this tension.

Solution: If we take the real axis as vertically downward and the imaginary axis to the right, we can represent the given force by the complex number as

$$1687\,e^{31.6°i} \text{ in exponential form}$$

or

$$1687\underline{/31.6°} \text{ in polar form}$$

We choose the rectangular mode in our calculator, enter the number, and read the values of a and b. We see that the real part of our complex number is 1437 and the imaginary part is 884, so the vertical force is 1437 lb downward, and the horizontal force is 884 lb to the right.

◆◆◆

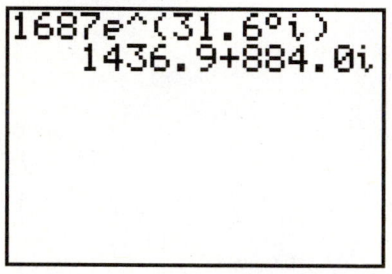

TI-83/84 screen for Example 28. The mode is **Radians** and **Rectangular.**

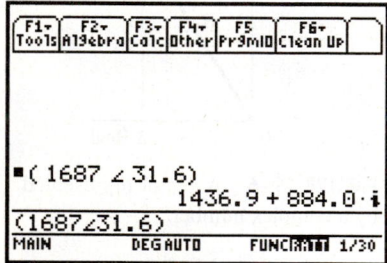

TI-89 screen for Example 28. The modes are **DEGREES** and **RECTANGULAR.**

Resultants of Several Vectors

To find the resultant or vector sum of two vectors, write each as a complex number and then add the two complex numbers. If there are more than two vectors, simply include them in the sum.

◆◆◆ **Example 29:** Find the resultant of two vectors, the first of magnitude 274 at an angle of 14.6°, and the second of magnitude 317 at an angle of 73.8°, Fig. 19–9.

Solution: Our vectors, in exponential form are

$$274\,e^{14.6°i} \quad \text{and} \quad 317\,e^{73.8°i}$$

and in polar form are

$$274\underline{/14.6°} \quad \text{and} \quad 317\underline{/73.8°}$$

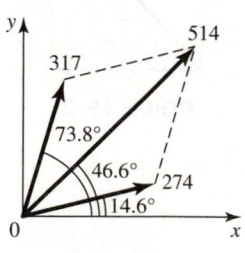

FIGURE 19–9

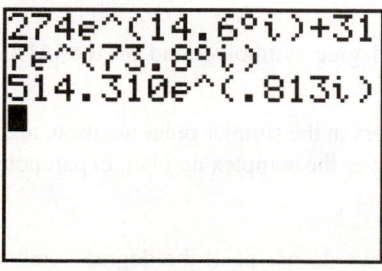

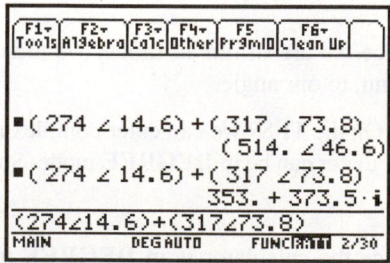

TI-83/84 screen for Example 29.   TI-89 screen for Example 29.

We add these vectors by calculator as shown, with each calculator in **POLAR** mode. Our resultant on the TI-83/84 is 514 and is at an angle of 0.813 radians, which converts to 46.6°.

To find the rectangular components of the resultant, convert the complex number to rectangular form. In the TI-89 screen shown, we have put the calculator into **RECTANGULAR** mode for the second computation. We see that the x component is 353 and the y component is 374.

◆◆◆

◆◆◆ **Example 30:** *An Application:* An aircraft having an airspeed of 475 mi/h encounters a wind of 42.6 mi/h, the directions of each as shown in Fig. 19–10. Find the actual speed and direction of the aircraft by finding the resultant of these two velocities.

Solution: If we take the real axis to the east and the imaginary axis to the north, vectors are

$$475\,e^{71.2°i} \quad \text{and} \quad 42.6\,e^{15.8°i}$$

in exponential form, and

$$475\underline{/71.2°} \quad \text{and} \quad 42.6\underline{/15.8°}$$

in polar form. Adding these by calculator we get a resultant with a magnitude of 500 mi/h and a direction of 67.2° or its radian equivalent of 1.17 radians. Thus the aircrafts travels at a speed of 500 mi/h in a direction of 67.2°, measured northward from the easterly direction.

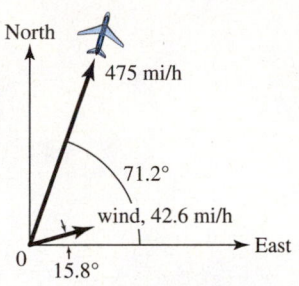

FIGURE 19–10

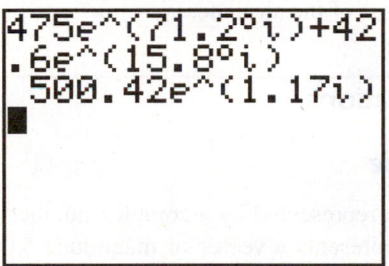

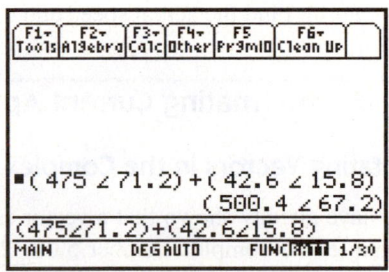

TI-83/84 screen for Example 30, in **radian** and **polar** modes. Recall that the degree symbol is in the ANGLE menu and i is a second function on the • key. To get the angle in degrees, go to degree mode, then use MATH CPX/▶ Polar.

TI-89 screen for Example 30, in **DEGREE** and **POLAR** modes.

◆◆◆

Let us now do the problem from the introduction to this chapter.

◆◆◆ **Example 31:** Find the resultant of the four vectors in Fig. 19–11.

Solution: Rewriting the angles so that they are measured from the positive x direction, our four forces are

$$203\underline{/61.5°}$$
$$225\angle(180° - 47.2°) = 225\underline{/133°}$$
$$238\angle(180° + 41.1°) = 238\underline{/221°}$$
$$221\angle(360° - 39.6°) = 221\underline{/320°}$$

We add these by calculator, as shown, getting a resultant of $80.5\underline{/146°}$. ◆◆◆

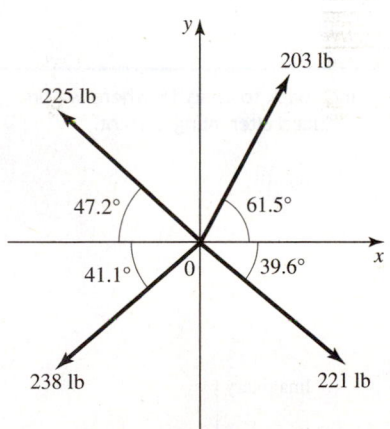

FIGURE 19–11

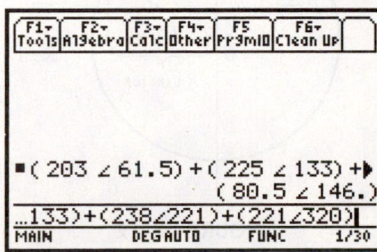

TI-89 screen for Example 31.

Exercise 4 ◆ Vector Operations Using Complex Numbers

Find the rectangular components of each vector.

1. $385\underline{/83.5°}$
2. $1.58\underline{/18.70°}$
3. $28.4\underline{/73.0°}$
4. $8364\underline{/36.20°}$

Find the resultant of each set of vectors.

5. $284\underline{/33.6°}$ and $184\underline{/11.5°}$
6. $1.84\underline{/63.5°}$ and $3.72\underline{/22.6°}$
7. $16.8\underline{/22.5°}$, $38.5\underline{/73.4°}$, and $6.48\underline{/125.5°}$
8. $735\underline{/28.4°}$, $68.4\underline{/22.5°}$, $33.7\underline{/52.4°}$, and $483\underline{/44.8°}$

Applications

9. A crate is dragged by a cable in which the tension is 58.4 lb and which makes an angle of 38.5° with the horizontal. Find the vertical and horizontal components of this tension.
10. A rocket is ascending at a speed of 3356 km/h at an angle of 85.4° with the horizontal. Find the vertical and horizontal components of the rocket's velocity.
11. Two forces of 753 lb and 824 lb act on a body. Their lines of action make an angle of 48.3° with each other. Find the magnitude of their resultant and the angle it makes with the larger force.
12. An airplane is heading 23.6° east of north at 473 mi/h but is being blown off course by a wind that has a speed of 59.6 mi/h and a direction of 68.5° east of north. Find the actual speed and direction of the airplane.

19–5   Alternating Current Applications

Rotating Vectors in the Complex Plane

We have already shown that a vector can be represented by a complex number. For example, the complex number $5.00\underline{/28°}$ represents a vector of magnitude 5.00 at an angle of 28° with the real axis.

A *phasor* (a rotating vector) may also be represented by a complex number $R\underline{/\omega t}$ by replacing the angle θ by ωt, where ω is the angular velocity and t is time.

◆◆◆ **Example 32:** The complex number

$$11\underline{/5t}$$

represents a phasor of magnitude 11 rotating with an angular velocity of 5 rad/s.   ◆◆◆

In Sec. 15–3 we showed that a phasor of magnitude R has a projection on the y axis of $R \sin \omega t$. Similarly, a phasor $R\underline{/\omega t}$ in the complex plane (Fig. 19–12) will have a projection on the imaginary axis of $R \sin \omega t$. Thus

$$R \sin \omega t = \text{imaginary part of } R\underline{/\omega t}$$

Similarly,

$$R \cos \omega t = \text{real part of } R\underline{/\omega t}$$

Thus either a sine or a cosine wave can be represented by a complex number in polar form $R\underline{/\omega t}$, depending upon whether we project onto the imaginary or the real axis. It usually does not matter which we choose, because the sine and cosine functions are identical except for a phase difference of 90°. What does matter is that we

Glance back to Chap. 15 where we first introduced alternating current.

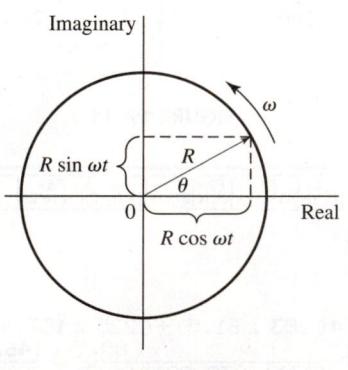

FIGURE 19–12

are consistent. Here we will follow the convention of projecting the phasor onto the imaginary axis, and we will drop the phrase "imaginary part of." Thus we say that

$$R \sin \omega t = R \underline{/\omega t}$$

or, in exponential form, $Re^{i\omega t}$. Similarly, if there is a phase angle ϕ,

$$R \sin(\omega t + \phi) = R \underline{/\omega t + \phi}$$

or $Re^{(\omega t + \phi)i}$.

One final simplification: It is customary to draw phasors at $t = 0$, so that ωt vanishes from the expression. Thus we write

$$R \sin(\omega t + \phi) = R \underline{/\phi}$$

or $Re^{\phi i}$, in exponential form.

Effective or Root Mean Square (rms) Values

Before we write currents and voltages in complex form, we must define the *effective value* of current and voltage. The power P delivered to a resistor by a direct current I is $P = I^2 R$. We now define an effective current I_{eff} that delivers the same power from an alternating current as that delivered by a direct current of the same number of amperes.

$$P = I_{eff}^2 R$$

We get the effective value by taking the square root of the mean value squared. Hence it is also called a *root mean square value*. Thus I_{eff} is often written I_{rms} instead. Figure 19–13(a) shows an alternating current $i = I_m \sin \omega t$ where I_m is the peak current, and Fig. 9-13b shows i^2 the square of that current. The mean value of this alternating quantity is found to deliver the same power in a resistor as a steady quantity of the same value.

$$P = I_{eff}^2 R = (\text{mean } i^2) R$$

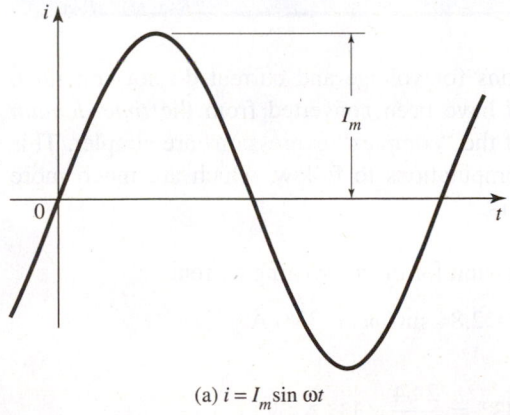
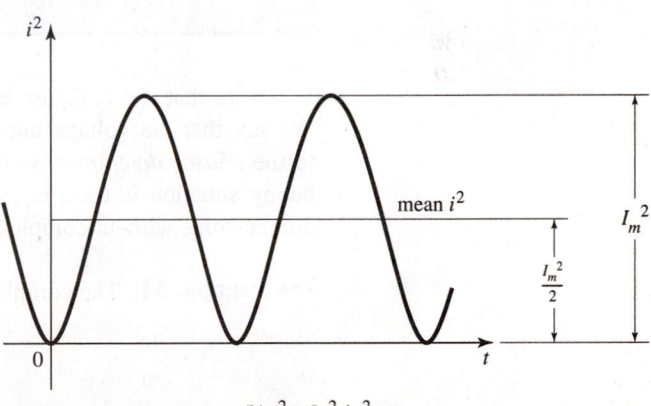

(a) $i = I_m \sin \omega t$

(b) $i^2 = I_m^2 \sin^2 \omega t$

FIGURE 19–13

But the mean value of i^2 is equal to $I_m^2/2$, half the peak value, so

$$I_{eff} = \sqrt{\frac{I_m^2}{2}} = \frac{I_m}{\sqrt{2}}$$

Thus *the effective current I_{eff} is equal to the peak current I_m divided by $\sqrt{2}$.* Similarly, *the effective voltage V_{eff} is equal to the peak voltage V_m divided by $\sqrt{2}$.*

◆◆◆ **Example 33:** An alternating current has a peak value of 2.84 A. What is the effective current?

Solution:

$$I_{\text{eff}} = \frac{I_m}{\sqrt{2}} = \frac{2.84}{\sqrt{2}} = 2.01 \text{ A}$$ ◆◆◆

Alternating Current and Voltage in Complex Form

Next we will write a sinusoidal current or voltage expression in the form of a complex number, because computations are easier in that form. At the same time, we will express the magnitude of the current or voltage in terms of effective value. The effective current I_{eff} is the quantity that is always implied when a current is given, rather than the peak current I_m. It is the quantity that is read when using a meter. The same is true of effective voltage.

To write an alternating current in complex form, we give the effective value of the current and the phase angle. The current is written in boldface type: **I**. Thus we have the following equations:

Complex Voltage and Current	The current is represented by $$i = I_m \sin(\omega t + \phi_2)$$ $$\mathbf{I} = I_{\text{eff}} \underline{/\phi_2} = \frac{I_m}{\sqrt{2}} \underline{/\phi_2}$$	**1075**
	Similarly, the voltage is represented by $$v = V_m \sin(\omega t + \phi_i)$$ $$\mathbf{V} = V_{\text{eff}} \underline{/\phi_1} = \frac{V_m}{\sqrt{2}} \underline{/\phi_1}$$	**1074**

Note that the complex expressions for voltage and current do not contain t. We say that the voltage and current have been converted from the *time domain* to the *phasor domain*. Also note that the "complex" expressions are simpler. This happy situation is used in the ac computations to follow, which are much more cumbersome without complex numbers.

◆◆◆ **Example 34:** The complex expression for the alternating current

$$i = 2.84 \sin(\omega t + 33°) \text{ A}$$

is

$$\mathbf{I} = I_{\text{eff}} \underline{/33°} = \frac{2.84}{\sqrt{2}} \underline{/33°} \text{ A}$$

$$= 2.01 \underline{/33°} \text{ A}$$

or $2.01e^{0.58i}$, where 0.58 is the radian equivalent of 33°.   ◆◆◆

◆◆◆ **Example 35:** The sinusoidal expression for the voltage

$$\mathbf{V} = 84.2 \underline{/-49°} \text{ V}$$

is

$$v = 84.2\sqrt{2} \sin(\omega t - 49°) \text{ V}$$

$$= 119 \sin(\omega t - 49°) \text{ V}$$ ◆◆◆

Complex Impedance

In Sec. 7–5 we drew a vector impedance diagram in which the impedance was the resultant of two perpendicular vectors: the resistance R along the horizontal axis and the reactance X along the vertical axis. If we now use the complex plane and draw the resistance along the *real* axis and the reactance along the *imaginary* axis, we can represent impedance by a complex number.

Complex Impedance	$\mathbf{Z} = R + Xi = Z\underline{/\phi} = Ze^{\phi i}$	1100

where

$$R = \text{circuit resistance}$$
$$X_L = \text{inductive reactance}$$
$$X_C = \text{capacitive reactance}$$
$$X = \text{circuit reactance} = X_L - X_C \tag{1097}$$
$$Z = \text{magnitude of impedance} = \sqrt{R^2 + X^2} \tag{1098}$$
$$\phi = \text{phase angle} = \arctan\frac{X}{R} \tag{1099}$$

◆◆◆ **Example 36:** A circuit has a resistance of 5 Ω in series with an inductive reactance of 55 Ω and a capacitive reactance of 48 Ω, Fig. 19–14(a). Represent the impedance by a complex number.

Solution: The circuit reactance is

$$X = X_L - X_C = 55 - 48 = 7 \ \Omega$$

The impedance, in rectangular form, is then

$$\mathbf{Z} = 5 + 7i$$

as shown in the vector impedance diagram, Fig. 19–14(b). The magnitude of the impedance is

$$\sqrt{5^2 + 7^2} = 8.60$$

The phase angle is

$$\arctan\frac{7}{5} = 54.5°$$

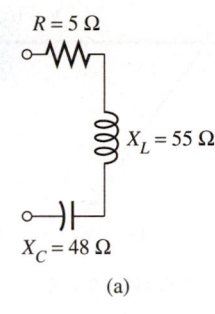

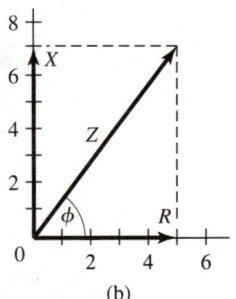

FIGURE 19–14

So we can write the impedance as a complex number in polar form as follows:

$$\mathbf{Z} = 8.60\underline{/54.5°}$$

or $8.60 \ e^{0.951i}$, where 0.951 is the radian equivalent of 54.5°. ◆◆◆

Ohm's Law for Alternating Current

We stated at the beginning of this section that the use of complex numbers would make calculations with alternating current almost as easy as for direct current. We do this by means of the following relationship:

Ohm's Law for ac	$\mathbf{V} = \mathbf{ZI}$	1102

Note the similarity between this equation and Ohm's law, $V = IR$, Eq. 1061. It is used in the same way. Given any two quantities we can find the third.

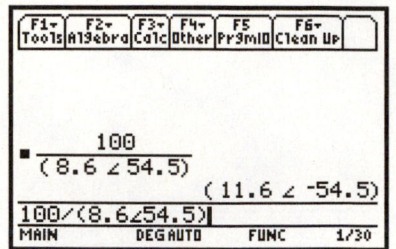

TI-89 screen for Example 37 showing the division of **V** by **Z**.

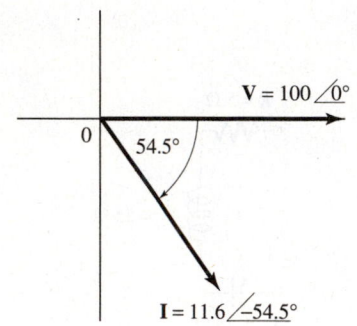

FIGURE 19–15

◆◆◆ **Example 37:** A voltage of 142 sin 200*t* is applied to the circuit of Fig. 19–14(a). Write a sinusoidal expression for the current *i*.

Solution: We first write the voltage in complex form. By Eq. 1074,

$$\mathbf{V} = \frac{142}{\sqrt{2}} \underline{/0°} = 100\underline{/0°}$$

Next we find the complex impedance **Z**. From Example 36,

$$\mathbf{Z} = 8.60\underline{/54.5°}$$

The complex current **I**, by Ohm's law for ac, is

$$\mathbf{I} = \frac{\mathbf{V}}{\mathbf{Z}} = \frac{100\underline{/0°}}{8.60\underline{/54.5°}} = 11.6\underline{/-54.5°}$$

Then the current in sinusoidal form is (Eq. 1075)

$$i = 11.6\sqrt{2} \sin(200t - 54.5°)$$
$$= 16.4 \sin(200t - 54.5°)$$

The current and voltage phasors are plotted in Fig. 19–15, and the instantaneous current and voltage are plotted in Fig. 19–16. Note the phase difference between the voltage and current waves. This phase difference, 54.5°, converted to time, is

$$54.5° = 0.951 \text{ rad} = 200t$$

$$t = \frac{0.951}{200} = 0.00475 \text{ s} = 4.75 \text{ ms}$$

We say that the current lags the voltage by 54.5° or 4.75 ms.

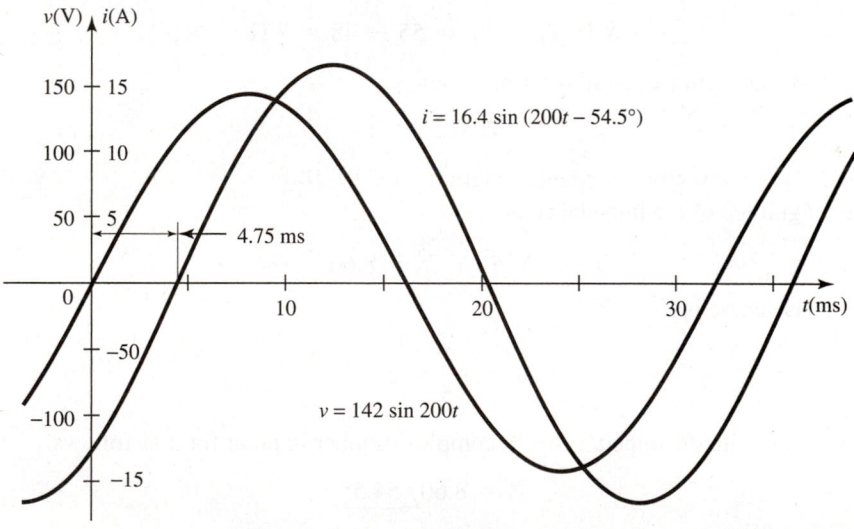

FIGURE 19–16

◆◆◆

Exercise 5 ◆ Alternating Current Applications

1. Find the effective (rms) value of a current that has a peak value of 69.3 A.
2. Find the effective value of a voltage that has a peak value of 155 V.

3. Find the peak value of a voltage that has an effective value of 634 V.

4. Find the peak value of a current that has an rms value of 10.6 A.

Express each current or voltage as a complex number in polar form.

5. $i = 250 \sin(\omega t + 25°)$

6. $v = 1.5 \sin(\omega t - 30°)$

7. $v = 57 \sin(\omega t - 90°)$

8. $i = 2.7 \sin \omega t$

9. $v = 144 \sin \omega t$

10. $i = 2.7 \sin(\omega t - 15°)$

Express each current or voltage in sinusoidal form.

11. $\mathbf{V} = 150 \underline{/0°}$

12. $\mathbf{V} = 1.75 \underline{/70°}$

13. $\mathbf{V} = 300 \underline{/-90°}$

14. $\mathbf{I} = 25 \underline{/30°}$

15. $\mathbf{I} = 7.5 \underline{/0°}$

16. $\mathbf{I} = 15 \underline{/-130°}$

Express the impedance of each circuit as a complex number in rectangular form and in polar form.

17. Fig. 19–17(a)

18. Fig. 19–17(b)

19. Fig. 19–17(c)

20. Fig. 19–17(d)

21. Fig. 19–17(e)

22. Fig. 19–17(f)

23. Fig. 19–17(g)

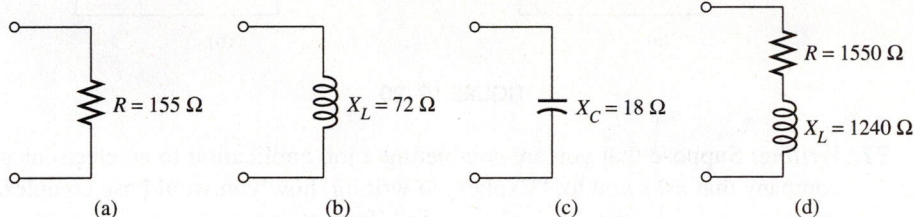

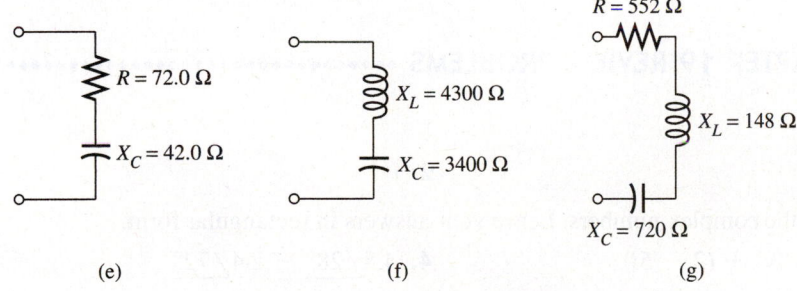

FIGURE 19–17

24. Write a sinusoidal expression for the current i in Fig. 19–18(a) and (b).

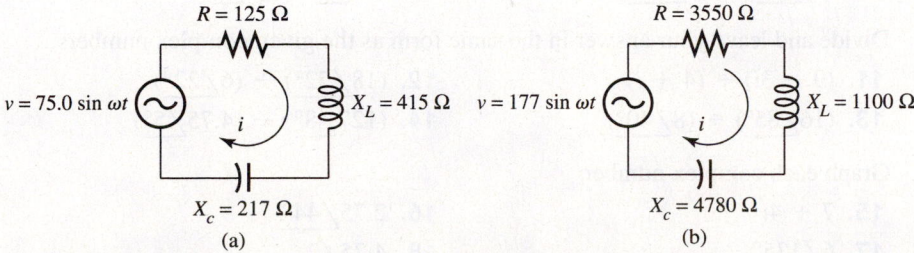

FIGURE 19–18

25. Write a sinusoidal expression for the voltage v in Fig. 19–19(a) and (b).

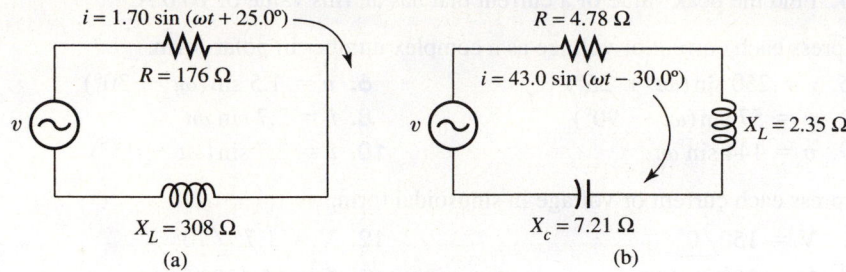

FIGURE 19–19

26. Find the complex impedance **Z** in Fig. 19–20(a) and (b).

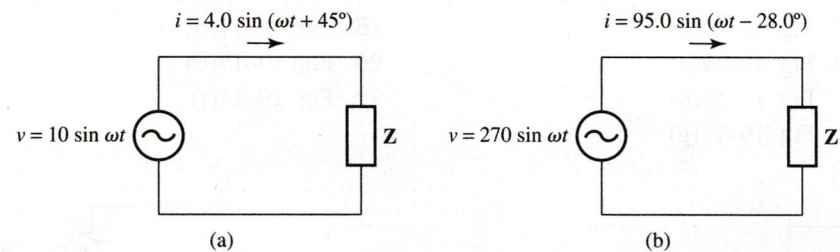

FIGURE 19–20

27. *Writing:* Suppose that you are completing a job application to an electronics company that asks you to, "Explain, in writing, how you would use complex numbers in an electrical calculation, and illustrate your explanation with an example." How would you respond?

◆◆◆ CHAPTER 19 REVIEW PROBLEMS ◆◆◆◆◆◆◆◆◆◆◆◆◆◆◆◆◆◆◆◆◆◆◆◆◆◆◆◆◆

Evaluate.

1. i^{17}

2. i^{25}

Combine the complex numbers. Leave your answers in rectangular form.

3. $(7 - 3i) + (2 + 5i)$

4. $4.8\underline{/28°} - 2.4\underline{/72°}$

5. $52\underline{/50°} + 28\underline{/12°}$

6. $2.7\underline{/7.0} - 4.3\underline{/5.0}$

Multiply. Leave your answer in the same form as the given complex numbers.

7. $(2 - i)(3 + 5i)$

8. $(7.3\underline{/21°})(2.1\underline{/156°})$

9. $(2\underline{/20°}) \times (6\underline{/18°})$

10. $(93\underline{/2}) \times (5\underline{/7})$

Divide and leave your answer in the same form as the given complex numbers.

11. $(9 - 3i) \div (4 + i)$

12. $(18\underline{/72°}) \div (6\underline{/22°})$

13. $(16\underline{/85°}) \div (8\underline{/40°})$

14. $(127\underline{/8°}) \div (4.75\underline{/5°})$

Graph each complex number.

15. $7 + 4i$

16. $2.75\underline{/44°}$

17. $6\underline{/135°}$

18. $4.75\underline{/2}$

Evaluate each power.

19. $(-4 + 3i)^2$

20. $(5\underline{/12°})^3$

21. $(5\underline{/10°})^3$

22. $(2\underline{/3})^5$

23. Express as a complex number in polar form: $i = 45 \sin(\omega t + 32°)$

24. Express in sinusoidal form: $\mathbf{V} = 283 \underline{/-22°}$

25. Write a complex expression for the current i in Fig. 19–21.

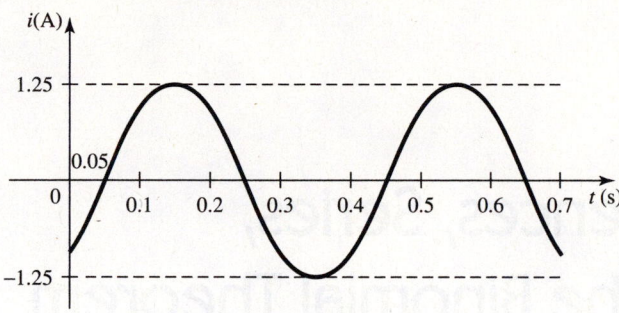

FIGURE 19–21

26. *Project: Quadratics with Complex Roots:* When we solved quadratic equations, we avoided those that resulted in complex roots. These occur when the quantity under the radical sign is negative. Solve each of the following, giving your answer in rectangular form.

(a) $3x^2 - 5x + 7 = 0$

(b) $2x^2 + 3x + 5 = 0$

(c) $x^2 - 2x + 6 = 0$

(d) $4x^2 + x + 8 = 0$

27. *Project: Complex Factors:* Factoring the difference of two squares resulted in real factors. However, factoring the *sum* of two squares will give *complex* factors. Figure out how to factor the following, and give your answer as the product of two complex numbers.

(a) $x^2 + 9$

(b) $b^2 + 25$

(c) $4y^2 + z^2$

(d) $25a^2 + 9b^2$

20

Sequences, Series, and the Binomial Theorem

••• **OBJECTIVES** ••

When you have completed this chapter, you should be able to

- Identify various types of sequences and series.
- Write the general term or a recursion relation for many series.
- Compute any term or the sum of any number of terms of an arithmetic progression or a geometric progression.
- Compute any term of a harmonic progression.
- Insert any number of arithmetic means, harmonic means, or geometric means between two given numbers.
- Compute the sum of an infinite geometric progression.
- Compute, graph, and find sums of sequences by calculator.
- Solve applications problems using series.
- Raise a binomial to a power using the binomial theorem.
- Find any term in a binomial expansion.

••

We start this chapter with a general introduction to *sequences* and *series*. Sequences and series are of great interest because a computer or calculator uses series internally to calculate many functions, such as the sine of an angle, the logarithm of a number, and so on. We then cover some specific sequences, such as the *arithmetic progression* and the *geometric progression*. *Progressions* are used to describe such things as the sequence of heights reached by a swinging pendulum on subsequent swings or the monthly balance on a savings account that is growing with compound interest.

We finish this chapter with an introduction to the *binomial theorem*. The binomial theorem enables us to expand a binomial, such as $(3x^2 - 2y)^5$, without actually having to multiply the terms. It is also useful in deriving formulas in calculus, statistics, and probability.

20–1 Sequences and Series

Before using sequences and series, we must define some terms.

Sequences

A *sequence*

$$u_1, \ u_2, \ u_3, \ \ldots, \ u_n$$

is a set of quantities, called *terms*, which follow each other in a definite order. Each term can be determined either by its position in the sequence or by knowledge of the preceding terms.

◆◆◆ **Example 1:** Here are some different kinds of sequences.

(a) The sequence $1, \frac{1}{2}, \frac{1}{3}, \frac{1}{4}, \frac{1}{5}, \ldots, 1/n$ is called a *finite* sequence because it has a finite number of terms, n.

(b) The sequence $1, \frac{1}{2}, \frac{1}{3}, \frac{1}{4}, \frac{1}{5}, \ldots, 1/n, \ldots$ is called an *infinite* sequence. The three dots at the end indicate that the sequence continues indefinitely.

(c) The sequence $3, 7, 11, 15, \ldots$ is called an *arithmetic sequence*, or *arithmetic progression* (AP), because each term after the first is equal to the sum of the preceding term and a constant. That constant (4, in this example) is called the *common difference*.

(d) The sequence $2, 6, 18, 54, \ldots$ is called a *geometric sequence*, or *geometric progression* (GP), because each term after the first is equal to the *product* of the preceding term and a constant. That constant (3, in this example) is called the *common ratio*.

(e) The sequence $1, 1, 2, 3, 5, 8, \ldots$ is called a *Fibonacci* sequence. Each term after the first two terms is the sum of the two preceding terms. ◆◆◆

Series

A *series* is the indicated sum of a sequence.

$$u_1 + u_2 + u_3 + \cdots + u_n + \cdots \qquad \textbf{189}$$

◆◆◆ **Example 2:** Some different kinds of series are given in this example.

(a) The series $1 + \frac{1}{2} + \frac{1}{3} + \frac{1}{4} + \frac{1}{5}$ is called a *finite series*. It is also called a *positive* series because all of its terms are positive.

(b) The series $2 - 2^2 + 2^3 - 2^4 + \cdots 2^n + \cdots$ is an *infinite series*. It is also called an *alternating series* because the signs of the terms alternate.

(c) The series $6 + 9 + 12 + 15 + \cdots$ is an infinite arithmetic series. The terms of this series form an AP.

(d) $1 - 2 + 4 - 8 + 16 - \cdots$ is an infinite, alternating, geometric series. The terms of this series form a GP. ◆◆◆

General Term

We said that a term can be determined by its position in the sequence. To do this we use an expression called the *general term*.

The *general term* u_n in a sequence or series is an expression involving n (where $n = 1, 2, 3, \ldots$) by which we can obtain any term. The general term is also called the *n*th term. If we have an expression for the general term, we can then find any specific term of the sequence or series.

◆◆◆ **Example 3:** The general term of a certain series is

$$u_n = \frac{n}{2n + 1}$$

Write the first three terms of the series.

Solution: Substituting $n = 1$, $n = 2$, and $n = 3$, in turn, we get

$$u_1 = \frac{1}{2(1) + 1} = \frac{1}{3} \qquad u_2 = \frac{2}{2(2) + 1} = \frac{2}{5} \qquad u_3 = \frac{3}{2(3) + 1} = \frac{3}{7}$$

so our series is

$$\frac{1}{3} + \frac{2}{5} + \frac{3}{7} + \cdots + \frac{n}{2n + 1} + \cdots \qquad \text{◆◆◆}$$

Recursion Relations

We have seen that we can find the terms of a sequence or series given an expression for the nth term. Sometimes we can find each term from one or more immediately preceding terms. The relationship between a term and those preceding it is called a *recursion relation* or *recursion formula*. We have used recursion relations before, for compound interest and for exponential growth in Chap.18.

◆◆◆ **Example 4:** Each term (after the first) in the series $1 + 4 + 13 + 40 + \cdots$ is found by multiplying the preceding term by 3 and adding 1. The recursion relation is then

$$u_n = 3u_{n-1} + 1 \qquad \text{◆◆◆}$$

◆◆◆ **Example 5:** In a *Fibonacci sequence*

$$1, 1, 2, 3, 5, 8, 13, 21, 34, \ldots$$

The Fibonacci sequence is named for Leonardo Fibonacci (ca. 1170– ca. 1250), an Italian number theorist and algebraist who studied this sequence. He is also known as *Leonardo of Pisa*.

each term after the first is the sum of the two preceding terms. Its recursion relation is

$$u_n = u_{n-1} + u_{n-2} \qquad \text{◆◆◆}$$

Sequences and Series by Calculator

We can generate the terms of a sequence or series and graph those terms by calculator. We will show how to do this first by using the general term, and then by means of the recursion formula.

Generating a Sequence or Series Using the General Term

We can generate the terms of a sequence or series by calculator using a built-in operation. On TI calculators it is called the **seq** (for *sequence*) operation. We must enter the general term, the variable, the starting and ending values, and the step size.

◆◆◆ **Example 6:** Find the terms of the sequence whose general term is $2n^2$, starting with $n = 1$ and ending with $n = 5$, in steps of 1 unit.

Solution: On the TI-83/84, we first choose **seq** from the MODE menu. Then we choose **seq** on the LIST OPS menu. We enter

$$2n^2, n, 1, 5, 1$$

Pressing ENTER displays the sequence.

On the TI-89, choose **SEQUENCE** on the MODE menu. Then select **seq** on the MATH List menu. The entry is the same as for the TI-83/84.   ◆◆◆

T1-83/84 screen for Example 6.

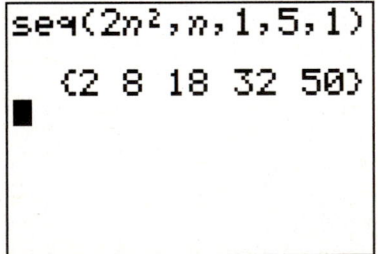

T1-89 screen for Example 6.

Graphing a Sequence or Series Using the General Term

◆◆◆ **Example 7:** Generate and graph the sequence given in Example 3 (a) by TI-83/84 and TI-89.

Solution: We graph the general term

$$u_n = \frac{n}{2n + 1}$$

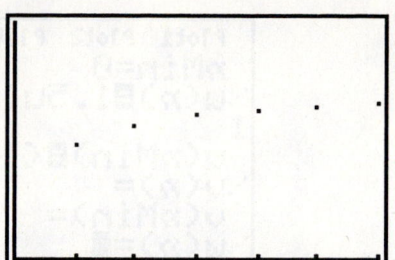

T1-83/84 screen for Example 7. Ticks are spaced 1 unit horizontally and 0.1 unit vertically.

(a) On the TI-83/84, we set the MODE to **SEQ** (for sequence) and choose **DOT** rather than the usual **CONNECTED**. We enter the general term in the Y= screen as $u_n = n/(2n + 1)$. Here the variable name n will appear automatically when pressing the X, T, θ, n key. Choose a suitable WINDOW and press GRAPH. To get a table of values, press TABLE.

(b) On the TI-89, choose **SEQUENCE** from the MODE menu, enter the general term as u_1 in the Y= screen. Here n is an ordinary alpha character. Choose a suitable WINDOW and press GRAPH. To get a table of values, press TABLE. ◆◆◆

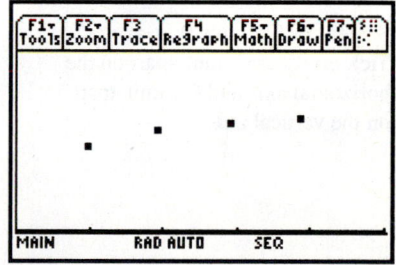

T1-83/84 screen for Example 7, showing the values of the terms for the given sequence.

Graphing a Sequence or Series Using a Recursion Formula

Next we show how to find the values of the terms in a sequence or series and to graph them, using a recursion formula.

◆◆◆ **Example 8:** On the TI-83/84 calculator, graph the first 10 terms of the sequence,

$$u_n = 1.5u_{n-1} + 1$$

starting with a first term of 1.

Solution: In the Mode menu, select **SEQUENTIAL** from the list of graph types.

In the Y= menu, screen 1, set

The starting value for n:	$n\text{Min} = 0$
The recursion formula or general term:	$u(n) = 1.5u(n-1) + 1$
The starting value for u:	$u(n\text{Min}) = 1$

T1-89 screen for Example 7.

In the WINDOW menu we choose

Starting and ending values for n:	$n\text{Min} = 0$	$n\text{Max} = 10$
Number of first term to be plotted:	$\text{PlotStart} = 1$	
Step size for n:	$\text{Plotstep} = 1$	
Smallest and largest x to be displayed:	$X\text{min} = 0$	$X\text{max} = 10$
Horizontal tick mark spacing:	$X\text{scl} = 1$	
Smallest and largest y to be displayed:	$Y\text{min} = 0$	$Y\text{max} = 100$
Vertical tick mark spacing:	$Y\text{scl} = 10$	

T1-89 screen for Example 7, showing the table of values for the sequence.

Press GRAPH to display the graph, screen 2.

To find the value of any term, use TRACE and move the cursor to the desired term, or TABLE to display a table of values for the sequence, screen 3.

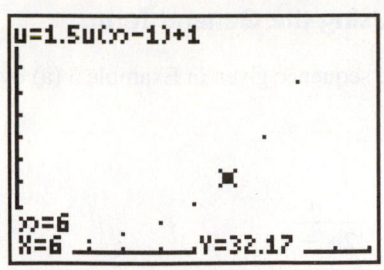

(1) T1-83/84 screens for Example 8. The variable u is the 2nd function on the 7 key and n is obtained by pressing the x, T, θ, n key.

(2) The $\boxed{\text{TRACE}}$ cursor shows that the sixth term is equal to 32.17.

(3) The first six terms in the sequence. Scroll down to get more.

◆◆◆

Sequences and Series by Computer

If we have either an expression for the general term of a sequence or series or the recursion formula, we can use a computer to generate large numbers of terms. By inspecting them, we can get a good idea about how the series behaves.

◆◆◆ **Example 9:** Table 20–1 shows a portion of a computer spreadsheet printout of 80 terms of the series

$$\frac{1}{e} + \frac{2}{e^2} + \frac{3}{e^3} + \cdots + \frac{n}{e^n} + \cdots$$

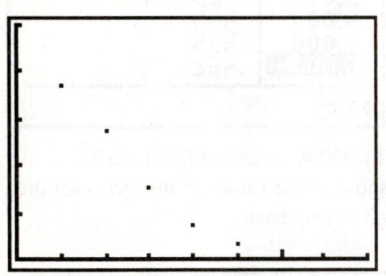

Screen for Example 9: $y = n/e^n$. Tick marks are 1 unit apart on the horizontal axis and 0.1 unit apart on the vertical axis.

TABLE 20–1

n	Term u_n	Partial Sum S_n	Ratio
1	0.367879	0.367879	
2	0.270671	0.638550	0.735759
3	0.149361	0.787911	0.551819
4	0.073263	0.861174	0.490506
5	0.033690	0.894864	0.459849
6	0.014873	0.909736	0.441455
7	0.006383	0.916119	0.429193
8	0.002684	0.918803	0.420434
9	0.001111	0.919914	0.413864
10	0.000454	0.920368	0.408755
.	.	.	.
.	.	.	.
75	0.000000	0.920674	0.372851
76	0.000000	0.920674	0.372785
77	0.000000	0.920674	0.372720
78	0.000000	0.920674	0.372657
79	0.000000	0.920674	0.372596
80	0.000000	0.920674	0.372536

Also shown for each term u_n is the sum of that term and all preceding terms, called the *partial sum*, and the ratio of that term to the one preceding. We notice several things about this series.

1. The terms get smaller and smaller and approach a value of zero in the table and on the screen.
2. The partial sums approach a limit (0.920674).
3. The ratio of two successive terms approaches a limit that is less than 1.

What do those facts tell us about this series? We'll see in the following section. ◆◆◆

Convergence or Divergence of an Infinite Series

When we use a series to do computations, we cannot, of course, work with an infinite number of terms. In fact, for practical computation, we prefer as few terms as possible, as long as we get the needed accuracy. We want a series in which the terms decrease rapidly and approach zero, and for which the sum of the first several terms is not too different from the sum of all of the terms of the series. Such a series is said to *converge* on some limit. A series that does not converge is said to *diverge*.

There are many tests for convergence. A particular test may tell us with certainty if the series converges or diverges. Sometimes, though, a test will fail and we must try another test. Here we consider only three tests.

1. *Magnitude of the terms*. If the terms of an infinite series do not approach zero, the series diverges. However, if the terms do approach zero, we cannot say with certainty that the series converges. Thus having the terms approach zero is a necessary but not a sufficient condition for convergence.

2. *Sum of the terms*. If the sum of the first n terms of a series reaches a limiting value as we take more and more terms, the series converges. Otherwise, the series diverges.

3. *Ratio test*. We compute the ratio of each term to the one preceding it and see how that ratio changes as we go further out in the series. If that ratio approaches a number that is

   (a) less than 1, the series converges.

   (b) greater than 1, the series diverges.

   (c) equal to 1, the test is not conclusive.

◆◆◆ **Example 10:** Does the series of Example 9 converge or diverge?

Solution: We apply the three tests by inspecting the computed values.

(1) The terms approach zero, so this test is not conclusive.

(2) The sum of the terms appears to approach a limiting value (0.920674), indicating convergence.

(3) The ratio of the terms appears to approach a value of less than 1, indicating convergence.

Thus the series appears to converge. Note that we say *appears* to *converge*. The computer run, while convincing, is still not a *proof of convergence*.   ◆◆◆

◆◆◆ **Example 11:** Table 20–2 shows part of a computer run for the series

$$1 + \frac{1}{2} + \frac{1}{3} + \cdots + \frac{1}{n} + \cdots$$

Does the series converge or diverge?

Solution: Again we apply the three tests.

(1) We see that the terms are getting smaller, so this test is not conclusive.

(2) The partial sum S_n does not seem to approach a limit but appears to keep growing even after 495 terms, indicating divergence.

(3) The ratio of the terms appears to approach 1, so this test also is not conclusive.

It thus appears that this series diverges.

TABLE 20–2

n	Term u_n	Partial Sum S_n	Ratio
1	1.000000	1.000000	
2	0.500000	1.500000	0.500000
3	0.333333	1.833333	0.666667
4	0.250000	2.083334	0.750000
5	0.200000	2.283334	0.800000
6	0.166667	2.450000	0.833333
7	0.142857	2.592857	0.857143
8	0.125000	2.717857	0.875000
9	0.111111	2.828969	0.888889
10	0.100000	2.928969	0.900000
.	.	.	.
.	.	.	.
.	.	.	.
75	0.013333	4.901356	0.986667
76	0.013158	4.914514	0.986842
77	0.012987	4.927501	0.987013
78	0.012821	4.940322	0.987180
79	0.012658	4.952980	0.987342
80	0.012500	4.965480	0.987500
.	.	.	.
.	.	.	.
.	.	.	.
495	0.002020	6.782784	0.997980
496	0.002016	6.784800	0.997984
497	0.002012	6.786813	0.997988
498	0.002008	6.788821	0.997992
499	0.002004	6.790825	0.997996
500	0.002000	6.792825	0.998000

◆◆◆

Exercise 1 ◆ Sequences and Series

Do not simplify or give the decimal value of any fractions in this exercise.

Write the first five terms of each series, given the general term.

1. $u_n = 3n$

2. $u_n = 2n + 3$

3. $u_n = \dfrac{n + 1}{n^2}$

4. $u_n = \dfrac{2^n}{n}$

Deduce the general term of each series. Use it to predict the next two terms.

5. $2 + 4 + 6 + \cdots$

6. $1 + 8 + 27 + \cdots$

7. $\dfrac{2}{4} + \dfrac{4}{5} + \dfrac{8}{6} + \dfrac{16}{7} + \cdots$

8. $2 + 5 + 10 + \cdots$

Deduce a recursion relation for each series. Use it to predict the next two terms.

9. $1 + 5 + 9 + \cdots$

10. $5 + 15 + 45 + \cdots$

11. $3 + 9 + 81 + \cdots$

12. $5 + 8 + 14 + 26 + \cdots$

13. *Writing:* State in your own words the difference between an arithmetic progression and a geometric progression. Give a real-world example of each.

14. Write a program or use a spreadsheet to generate the terms of a series, given the general term or a recursion relation. Have the program compute and print each term, the partial sum, and the ratio of that term to the preceding one. Use the program to determine if each of the following series converges or diverges.

(a) $1 + \dfrac{1}{2} + \dfrac{1}{3} + \dfrac{1}{4} + \cdots + \dfrac{1}{n} + \cdots$

(b) $\dfrac{3}{2} + \dfrac{9}{8} + \dfrac{27}{24} + \dfrac{81}{64} + \cdots + \dfrac{3^n}{n \cdot 2^n} + \cdots$

(c) $1 + \dfrac{4}{7} + \dfrac{9}{49} + \dfrac{16}{343} + \cdots + \dfrac{n^2}{7^{n-1}} + \cdots$

20–2   Arithmetic and Harmonic Progressions

Recursion Formula

We stated earlier that an *arithmetic progression* (or AP) is a sequence of terms in which each term after the first equals the sum of the preceding term and a constant, called the *common difference*, d. If a_n is any term of an AP, the recursion formula for an AP is as follows:

AP: Recursion Formula	$a_n = a_{n-1} + d$ *Each term of an AP after the first equals the sum of the preceding term and the common difference.*	190

◆◆◆ **Example 12:** The following sequences are arithmetic progressions. The common difference for each is given in parentheses:

(a) $1, 5, 9, 13, \ldots$ $(d = 4)$

(b) $20, 30, 40, 50, \ldots$ $(d = 10)$

(c) $75, 70, 65, 60, \ldots$ $(d = -5)$

We see that each series is increasing when d is positive and decreasing when d is negative. ◆◆◆

General Term

For an AP whose first term is a and whose common difference is d, the terms are

$$a, \quad a + d, \quad a + 2d, \quad a + 3d, \quad a + 4d, \ldots$$

We see that each term is the sum of the first term and a multiple of d, where the coefficient of d is one less than the number n of the term. So the nth term a_n is given by the following equation:

AP: General Term	$a_n = a + (n - 1)d$ *The nth term of an AP is found by adding the first term and $(n - 1)$ times the common difference.*	191

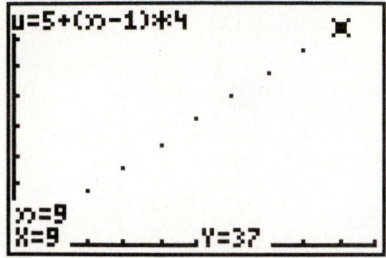

TI-83/84 screen for Example 13. Ticks are spaced 1 unit apart in x and 5 units in y.

◆◆◆ **Example 13:** (a) Write the general term of an AP whose first term is 5 and has a common difference of 4. (b) Find the ninth term. (c) Check by graphing.

Solution:

(a) From Eq. 191, with $a = 5$ and $d = 4$,

$$a_n = 5 + (n - 1)4$$

(b) For the ninth term we set $n = 9$,

$$a_9 = 5 + (9 - 1)4 = 37$$

(c) We graph a_n (using u_n on the TI-83/84) as shown. Then using **value** or $\boxed{\text{TRACE}}$ we find the value $a_n = 37$ at $n = 9$.   ◆◆◆

Of course, Eq. 191 can be used to find any of the four quantities (a, n, d, or a_n) given the other three.

◆◆◆ **Example 14:** Write the AP whose eighth term is 19 and whose fifteenth term is 33.

Solution: Applying Eq. 191 twice gives

$$33 = a + 14d$$
$$19 = a + 7d$$

We now have two equations in two unknowns, which we solve simultaneously. Subtracting the second from the first gives $14 = 7d$, so $d = 2$. Thus, $a = 33 - 14d = 5$, so the general term of our AP is

$$a_n = 5 + (n - 1)2$$

and the AP is then

$$5, \ 7, \ 9, \ldots$$ ◆◆◆

AP: Sum of n Terms

Let us derive a formula for the sum s_n of the first n terms of an AP. Adding term by term gives

$$s_n = a + (a + d) + (a + 2d) + \cdots + (a_n - d) + a_n \tag{1}$$

or, written in reverse order,

$$s_n = a_n + (a_n - d) + (a_n - 2d) + \cdots + (a + d) + a \tag{2}$$

Adding Eqs. (1) and (2) term by term gives

$$2s_n = (a + a_n) + (a + a_n) + (a + a_n) + \cdots$$
$$= n(a + a_n)$$

Dividing both sides by 2 gives the following formula:

AP: Sum of n Terms	$$s_n = \frac{n}{2}(a + a_n)$$ *The sum of n terms of an AP is half the product of n and the sum of the first and nth terms.*

192

◆◆◆ **Example 15:** Find the sum of 10 terms of the AP

$$2, 5, 8, 11, \ldots$$

Solution: From Eq. 191, with $a = 2$ and $d = 3$,

$$a_n = 2 + 3(n - 1)$$

The tenth term is then,

$$a_{10} = 2 + 9(3) = 29$$

Then, from Eq. 192,

$$s_{10} = \frac{10(2 + 29)}{2} = 155 \qquad \blacklozenge\blacklozenge\blacklozenge$$

Sums by Calculator

We earlier showed how to generate a sequence by calculator. Once we have the sequence, we can find its sum using the **sum** operation.

◆◆◆ **Example 16:** Repeat the preceding example by calculator.

Solution: We generate the sequence as we did in Sec. 20–1, using the **seq** operation, from the ⌨LIST⌨ OPS menu, as shown. We then take the sum of that sequence by entering **sum** ⌨Ans⌨. On the TI-83/84, **sum** is found in the ⌨List⌨ Math menu. On the TI-89, **seq** and **sum** are found in the ⌨MATH⌨ List menu. ◆◆◆

We get another form of Eq. 192 by substituting the expression for a_n from Eq. 191, as follows:

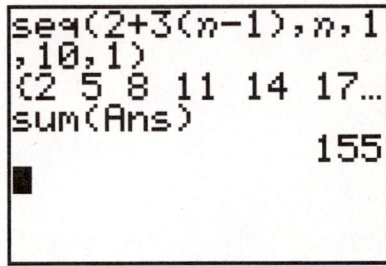

TI-83/84 screen for Example 16.

AP: Sum of n Terms	$s_n = \dfrac{n}{2}[2a + (n - 1)d]$	193

This form is useful for finding the sum without first computing the *n*th term.

◆◆◆ **Example 17:** We repeat Example 15 without first having to find the tenth term. From Eq. 193,

$$s_{10} = \frac{10[2(2) + 9(3)]}{2} = 155 \qquad \blacklozenge\blacklozenge\blacklozenge$$

Sometimes the sum may be given and we must find one of the other quantities in the AP.

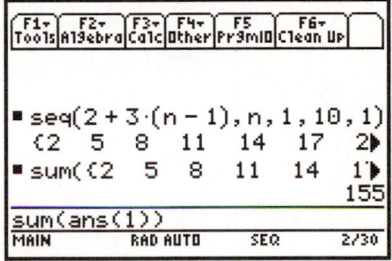

TI-89 screen for Example 16.

◆◆◆ **Example 18:** How many terms of the AP 5, 9, 13, . . . give a sum of 275?

Solution: We seek *n* so that $s_n = 275$. From Eq. 193, with $a = 5$ and $d = 4$,

$$275 = \frac{n}{2}[2(5) + (n - 1)4]$$
$$= 5n + 2n(n - 1)$$

Removing parentheses and collecting terms gives the quadratic equation

$$2n^2 + 3n - 275 = 0$$

From the quadratic formula,

$$n = \frac{-3 \pm \sqrt{9 - 4(2)(-275)}}{2(2)} = 11 \quad \text{or} \quad -12.5$$

We discard the negative root and get 11 terms as our answer. ◆◆◆

◆◆◆ **Example 19:** *An Application.* A person borrows $2450 and agrees to repay the loan in 10 monthly installments. Monthly payments include $245 to reduce the principal, plus interest at a rate of 1.20% per month of the unpaid balance. Find the total amount of interest paid.

Solution:

Interest first month:	$2450 × 0.012 = $29.40
Interest second month:	($2450 − 245) × 0.012 = $26.46
Interest third month:	($2450 − 490) × 0.012 = $23.52

Thus the interest payments form an AP with $a = 29.40$ and $d = -2.94$. The sum of ten terms of this AP is

$$s_{10} = \frac{10}{2}[2(\$29.40) + (10 - 1)(-\$2.94)]$$

$$= \$161.70$$

◆◆◆

Arithmetic Means

The first term a and the last (nth) term a_n of an AP are sometimes called the *extremes*, while the intermediate terms, $a_2, a_3, \ldots, a_{n-1}$, are called *arithmetic means*. We now show, by example, how to insert any number of arithmetic means between two extremes.

◆◆◆ **Example 20:** Insert five arithmetic means between 3 and −9.

Solution: Our AP will have seven terms, with a first term of 3 and a seventh term of −9. From Eq. 191,

$$a_n = a + (n-1)d$$

$$-9 = 3 + 6d$$

from which $d = -2$. The progression is then

$$3, \quad 1, \quad -1, \quad -3, \quad -5, \quad -7, \quad -9$$

and the five arithmetic means are 1, −1, −3, −5, and −7.

◆◆◆

The Arithmetic Mean

Let us insert a single arithmetic mean between two numbers.

◆◆◆ **Example 21:** Find a single arithmetic mean m between the extremes a and b.

Solution: The sequence is a, m, b. The common difference d is $m - a = b - m$. Solving for m gives

$$2m = b + a$$

Dividing by 2 gives

Arithmetic Mean	$m = \dfrac{a + b}{2}$	46

The arithmetic mean which agrees with the common idea of an *average* of two numbers as the sum of those numbers divided by 2.

◆◆◆

Harmonic Progressions

A sequence is called a *harmonic progression* if the reciprocals of its terms form an arithmetic progression.

◆◆◆ **Example 22:** The sequence

$$1, \frac{1}{3}, \frac{1}{5}, \frac{1}{7}, \frac{1}{9}, \frac{1}{11}, \ldots$$

is a harmonic progression because the reciprocals of its terms, which are 1, 3, 5, 7, 9, 11, . . . , form an AP. ◆◆◆

It is not possible to derive an equation for the nth term or for the sum of a harmonic progression. However, we can solve problems involving harmonic progressions by taking the reciprocals of the terms and using the formulas for the AP.

◆◆◆ **Example 23:** Find the tenth term of the harmonic progression

$$2, \frac{2}{3}, \frac{2}{5}, \ldots$$

Solution: We write the reciprocals of the terms,

$$\frac{1}{2}, \frac{3}{2}, \frac{5}{2}, \ldots$$

and note that they form an AP with $a = \frac{1}{2}$ and $d = 1$. The tenth term of the AP is then

$$a_n = a + (n - 1)d$$
$$a_{10} = \frac{1}{2} + (10 - 1)(1)$$
$$= \frac{19}{2}$$

The tenth term of the harmonic progression is the reciprocal of $\frac{19}{2}$, or $\frac{2}{19}$. ◆◆◆

Harmonic Means

To insert *harmonic means* between two terms of a harmonic progression, we simply take the reciprocals of the given terms, insert arithmetic means between those terms, and take reciprocals again.

◆◆◆ **Example 24:** Insert three harmonic means between $\frac{2}{9}$ and 2.

Solution: Taking reciprocals, our AP is

$$\frac{9}{2}, \underline{\quad}, \underline{\quad}, \underline{\quad}, \frac{1}{2}$$

In this AP, $a = \frac{9}{2}$, $n = 5$, and $a_5 = \frac{1}{2}$. We can find the common difference d from the equation

$$a_n = a + (n - 1)d$$
$$\frac{1}{2} = \frac{9}{2} + 4d$$
$$-4 = 4d$$
$$d = -1$$

Having the common difference, we can fill in the missing terms of the AP,

$$\frac{9}{2}, \frac{7}{2}, \frac{5}{2}, \frac{3}{2}, \frac{1}{2}$$

Taking reciprocals again, our harmonic progression is

$$\frac{2}{9}, \frac{2}{7}, \frac{2}{5}, \frac{2}{3}, 2$$

◆◆◆

Exercise 2 ◆ Arithmetic and Harmonic Progressions

General Term

1. Find the fifteenth term of an AP with first term 4 and common difference 3.
2. Find the tenth term of an AP with first term 8 and common difference 2.
3. Find the twelfth term of an AP with first term -1 and common difference 4.
4. Find the ninth term of an AP with first term -5 and common difference -2.
5. Find the eleventh term of the AP

$$9, \quad 13, \quad 17, \ldots$$

6. Find the eighth term of the AP

$$-5, \quad -8, \quad -11, \ldots$$

7. Find the ninth term of the AP

$$x, \quad x + 3y, \quad x + 6y, \ldots$$

8. Find the fourteenth term of the AP

$$1, \quad \frac{6}{7}, \quad \frac{5}{7}, \ldots$$

For problems 9 through 12, write the first five terms of each AP.

9. First term is 3 and thirteenth term is 55.
10. First term is 5 and tenth term is 32.
11. Seventh term is 41 and fifteenth term is 89.
12. Fifth term is 7 and twelfth term is 42.
13. Find the first term of an AP whose common difference is 3 and whose seventh term is 11.
14. Find the first term of an AP whose common difference is 6 and whose tenth term is 77.

Sum of an Arithmetic Progression

15. Find the sum of the first 12 terms of the AP

$$3, \quad 6, \quad 9, \quad 12, \ldots$$

16. Find the sum of the first five terms of the AP

$$1, \quad 5, \quad 9, \quad 13, \ldots$$

17. Find the sum of the first nine terms of the AP

$$5, \quad 10, \quad 15, \quad 20, \ldots$$

18. Find the sum of the first 20 terms of the AP

$$1, \quad 3, \quad 5, \quad 7, \ldots$$

19. How many terms of the AP 4, 7, 10, . . . will give a sum of 375?
20. How many terms of the AP 2, 9, 16, . . . will give a sum of 270?

Arithmetic Means

21. Insert two arithmetic means between 5 and 20.
22. Insert five arithmetic means between 7 and 25.
23. Insert four arithmetic means between -6 and -9.
24. Insert three arithmetic means between 20 and 56.

Harmonic Progressions

25. Find the fourth term of the harmonic progression

$$\frac{3}{5}, \frac{3}{8}, \frac{3}{11}, \ldots$$

26. Find the fifth term of the harmonic progression

$$\frac{4}{19}, \frac{4}{15}, \frac{4}{11}, \ldots$$

Harmonic Means

27. Show that the harmonic mean between two numbers a and b is given by

$$\text{Harmonic Mean} = \frac{2ab}{a + b} \qquad \textbf{48}$$

28. Insert two harmonic means between $\frac{7}{9}$ and $\frac{7}{15}$.

29. Insert three harmonic means between $\frac{6}{21}$ and $\frac{6}{5}$.

Applications

30. *Loan Repayment:* A person agrees to repay a loan of $10,000 with an annual payment of $1000 plus 8% of the unpaid balance.
   (a) Show that the interest payments alone form the AP: $800, $720, $640, ...
   (b) Find the total amount of interest paid.

31. *Simple Interest:* A person deposits $50 in a bank on the first day of each month, at the same time withdrawing all interest earned on the money already in the account.
   (a) If the rate is 1% per month, computed monthly, write an AP whose terms are the amounts withdrawn each month.
   (b) How much interest will have been earned in the 36 months following the first deposit?

32. *Straight-Line Depreciation:* A certain milling machine has an initial value of $150,000 and a scrap value of $10,000 twenty years later. Assuming that the machine depreciates the same amount each year, find its value after 8 years. To find the amount of depreciation for each year, divide the total depreciation (initial value − scrap value) by the number of years of depreciation.

33. *Salary or Price Increase:* A person is hired at a salary of $40,000 and receives a raise of $2500 at the end of each year. Find the total amount earned during 10 years.

34. *Freely Falling Body:* A freely falling body falls $g/2$ feet during the first second, $3g/2$ feet during the next second, $5g/2$ feet during the third second, and so on, where $g \approx 32.2$ ft/s^2. Find the total distance the body falls during the first 10 s.

35. Using the information of problem 34 show that the total distance s fallen in t seconds is $s = \frac{1}{2} gt^2$.

36. *Computer:* Write a program or use a spreadsheet to generate the terms of a series, given the general term or a recursion relation. Have the program compute and print each term, the partial sum, and the ratio of that term to the preceding one. Use the program to determine if each of the following series converges or diverges.

   (a) $1 + \frac{1}{2} + \frac{1}{3} + \frac{1}{4} + \cdots + \frac{1}{n} + \cdots$

(b) $\dfrac{3}{2} + \dfrac{9}{8} + \dfrac{27}{24} + \dfrac{81}{64} + \cdots + \dfrac{3^n}{n \cdot 2^n} + \cdots$

(c) $1 + \dfrac{4}{7} + \dfrac{9}{49} + \dfrac{16}{343} + \cdots + \dfrac{n^2}{7^{n-1}} + \cdots$

37. *Project:* Suppose a car drives one mile at a constant speed of 40 mi/h, another mile at a constant speed of 50 mi/h, a third mile at a constant speed of 60 mi/h, and a fourth mile at a constant speed of 70 mi/h. (a) Is the average speed the average (arithmetic mean) of the four speeds? (b) Can you demonstrate that the arithmetic mean gives the average speed or not? (c) Can you demonstrate that the average speed is in fact given by the *harmonic* mean of the four numbers?

20–3   Geometric Progressions

Recursion Formula

A geometric sequence or *geometric progression* (GP) is one in which each term after the first is formed by multiplying the preceding term by a factor r, called the *common ratio*. Thus if a_n is any term of a GP, the recursion relation is as follows:

GP: Recursion Formula	$a_n = ra_{n-1}$ *Each term of a GP after the first equals the product of the preceding term and the common ratio.*	194

♦♦♦ **Example 25:** Some geometric progressions, with their common ratios given, are as follows:

(a) 2, 4, 8, 16, . . . $(r = 2)$

(b) 27, 9, 3, 1, $\frac{1}{3}$, . . . $\left(r = \frac{1}{3}\right)$

(c) -1, 3, -9, 27, . . . $(r = -3)$ ♦♦♦

General Term

For a GP whose first term is a and whose common ratio is r, the terms are

$$a, \quad ar, \quad ar^2, \quad ar^3, \quad ar^4, \ldots$$

We see that each term after the first is the product of the first term and a power of r, where the power of r is one less than the number n of the term. So the nth term a_n is given by the following equation:

GP: General Term	$a_n = ar^{n-1}$ *The nth term of a GP is found by multiplying the first term by the $n - 1$ power of the common ratio.*	195

♦♦♦ **Example 26:** Find the sixth term of a GP with first term 5 and common ratio 4.

Solution: We substitute into Eq. 195 for the general term of a GP, with $a = 5$, $n = 6$, and $r = 4$.

$$a_6 = 5(4^5) = 5(1024) = 5120$$ ♦♦♦

GP: Sum of *n* Terms

We find a formula for the sum s_n of the first n terms of a GP (also called the sum of n terms of a geometric series) by adding the terms of the GP.

$$s_n = a + ar + ar^2 + ar^3 + \cdots + ar^{n-2} + ar^{n-1} \qquad (1)$$

Multiplying each term in Equation (1) by r gives

$$rs_n = ar + ar^2 + ar^3 + \cdots + ar^{n-1} + ar^n \qquad (2)$$

Subtracting (2) from (1) term by term, we get

$$(1 - r)s_n = a - ar^n$$

Dividing both sides by $(1 - r)$ gives us the following formula:

GP: Sum of *n* Terms	$s_n = \dfrac{a(1 - r^n)}{1 - r}$	196

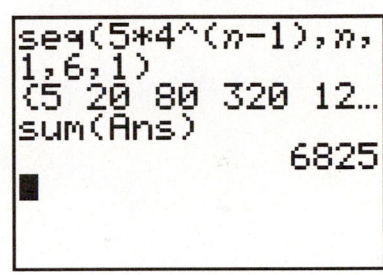

◆◆◆ **Example 27:** Find the sum of the first six terms of the GP in Example 26.

Solution: We substitute into Eq. 196 using $a = 5$, $n = 6$, and $r = 4$.

$$s_n = \frac{a(1 - r^n)}{1 - r} = \frac{5(1 - 4^6)}{1 - 4} = \frac{5(-4095)}{-3} = 6825 \qquad ◆◆◆$$

We can get another equation for the sum of a GP in terms of the nth term a_n. Starting with Eq. 196, $a_n = ar^{n-1}$ we multiply both sides by r and get

$$ra_n = rar^{n-1} = ar^n$$

Substituting ra_n for ar^n in Eq. 196,

$$s_n = \frac{a(1 - r^n)}{1 - r} = \frac{a - ar^n}{1 - r}$$

we get

GP: Sum of *n* Terms	$s_n = \dfrac{a - ra_n}{1 - r}$	197

TI-83/84 solution for Example 27. Here we generate and take the sum of a GP, just as we had done earlier with an AP.

◆◆◆ **Example 28:** Repeat Example 27, given that the sixth term (found in Example 26) is $a_6 = 5120$.

Solution: Substituting, with $a = 5$, $r = 4$, and $a_6 = 5120$, yields

$$s_n = \frac{a - ra_n}{1 - r} = \frac{5 - 4(5120)}{1 - 4} = 6825$$

as in Example 27. ◆◆◆

◆◆◆ **Example 29:** *An Application.* When the shaft in Fig. 20–1 is pulled down 6.50 cm and released, it bobs up and down ten times before the next actuation. To estimate wear in the mechanism over years of use, the total distance traveled for each actuation is needed. Find this total, if the distance moved for each bounce is 85.0% as for the preceding one.

Solution: The distances traveled, in cm, are

$$6.50 \quad 6.50 \times 0.850 \quad 6.50 \times (0.850)^2 \quad 6.50 \times (0.850)^3 \ldots$$

forming a GP with $a = 6.50$ and $r = 0.850$. Using Eq. 196 with $n = 10$,

$$s_{10} = \frac{6.50(1 - 0.850^{10})}{1 - 0.850} = 34.8 \text{ cm} \qquad ◆◆◆$$

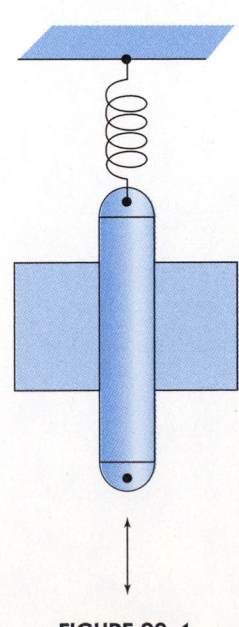

FIGURE 20–1

Geometric Means

As with the AP, the intermediate terms between any two terms are called *means*. A single number inserted between two numbers is called the *geometric mean* between those numbers.

◆◆◆ **Example 30:** Insert a geometric mean b between two numbers a and c.

Solution: Our GP is a, b, c. The common ratio r is then

$$r = \frac{b}{a} = \frac{c}{b}$$

from which $b^2 = ac$, which is rewritten as follows:

Geometric Mean	$b = \pm\sqrt{ac}$ *The geometric mean, or mean proportional, between two numbers is equal to the square root of their product.*	47

This is not new. We studied the mean proportional earlier.   ◆◆◆

◆◆◆ **Example 31:** Find the geometric mean between 3 and 48.

Solution: Letting $a = 3$ and $c = 48$ gives us

$$b = \pm\sqrt{3(48)} = \pm 12$$

Our GP is then

$$3, \quad \mathbf{12}, \quad 48$$

or

$$3, \quad \mathbf{-12}, \quad 48$$

Note that we get *two* solutions.   ◆◆◆

To insert *several* geometric means between two numbers, we first find the common ratio.

◆◆◆ **Example 32:** Insert four geometric means between 2 and $15\frac{3}{16}$.

Solution: Here $a = 2$, $a_6 = 15\frac{3}{16}$, and $n = 6$. Then

$$a_6 = ar^5$$

$$15\frac{3}{16} = 2r^5$$

$$r^5 = \frac{243}{32}$$

$$r = \frac{3}{2}$$

Having r, we can write the terms of the GP. They are

$$2, \quad 3, \quad 4\frac{1}{2}, \quad 6\frac{3}{4}, \quad 10\frac{1}{8}, \quad 15\frac{3}{16}$$ ◆◆◆

Exercise 3 ◆ Geometric Progressions

1. Find the fifth term of a GP with first term 5 and common ratio 2.
2. Find the fourth term of a GP with first term 7 and common ratio −4.
3. Find the sixth term of a GP with first term −3 and common ratio 5.
4. Find the fifth term of a GP with first term −4 and common ratio −2.
5. Find the sum of the first ten terms of the GP in problem 1.
6. Find the sum of the first nine terms of the GP in problem 2.
7. Find the sum of the first eight terms of the GP in problem 3.
8. Find the sum of the first five terms of the GP in problem 4.

Geometric Means

9. Insert a geometric mean between 5 and 45.
10. Insert a geometric mean between 7 and 112.
11. Insert a geometric mean between −10 and −90.
12. Insert a geometric mean between −21 and −84.
13. Insert two geometric means between 8 and 216.
14. Insert two geometric means between 9 and −243.
15. Insert three geometric means between 5 and 1280.
16. Insert three geometric means between 144 and 9.

Applications

17. *Exponential Growth:* Using the equation for exponential growth,

$$y = ae^{nt} \tag{151}$$

    with $a = 1$ and $n = 0.5$, compute values of y for $t = 0, 1, 2, \ldots, 10$. Show that while the values of t form an AP, the values of y form a GP. Find the common ratio.

18. *Exponential Decay:* Repeat problem 17 with the formula for exponential decay:

$$y = ae^{-nt} \tag{153}$$

19. *Cooling:* A certain iron casting is at 1800°F and cools so that its temperature at each minute is 10% less than its temperature the preceding minute. Find its temperature after 1 h.

20. *Light Through an Absorbing Medium:* Sunlight passes through a glass filter. Each millimeter of glass absorbs 20% of the light passing through it. What percentage of the original sunlight will remain after passing through 5.0 mm of the glass?

21. *Radioactive Decay:* A certain radioactive material decays so that after each year the radioactivity is 8% less than at the start of that year. How many years will it take for its radioactivity to be 50% of its original value?

22. *Pendulum:* Each swing of a certain pendulum is 85.0% as long as the one before. If the first swing is 12.0 in., find the entire distance traveled in eight swings.

23. *Bouncing Ball:* A ball dropped from a height of 10.0 ft rebounds to half its height on each bounce. Find the total distance traveled when it hits the ground for the fifth time.

24. *Population Growth:* One of the most famous and controversial references to arithmetic and geometric progressions was made by Thomas Malthus in 1798.

He wrote: *"Population, when unchecked, increases in a geometrical ratio, and subsistence for man in an arithmetical ratio."* Each day the size of a certain colony of bacteria is 25% larger than on the preceding day. If the original size of the colony was 10,000 bacteria, find its size after 5 days.

25. A person has two parents, and each parent has two parents, and so on. We can write a GP for the number of ancestors as 2, 4, 8, ... Find the total number of ancestors in five generations, starting with the parents' generation.

26. *Musical Scale:* The frequency of the "A" note above middle C is, by international agreement, equal to 440 Hz. A note one *octave* higher is at twice that frequency, or 880 Hz. The octave is subdivided into 12 *half-tone* intervals, where each half-tone is higher than the one preceding by a factor equal to the twelfth root of 2. This is called the *equally tempered scale* and is usually attributed to Johann Sebastian Bach (1685–1750). Write a GP showing the frequency of each half-tone, from 440 to 880 Hz. Work to two decimal places.

27. *Chemical Reactions:* Increased temperature usually causes chemicals to react faster. If a certain reaction proceeds 15% faster for each 10°C increase in temperature, by what factor is the reaction speed increased when the temperature rises by 50°C?

28. *Mixtures:* A radiator contains 30% antifreeze and 70% water. One-fourth of the mixture is removed and replaced by pure water. If this procedure is repeated three more times, find the percent antifreeze in the final mixture.

29. *Energy Consumption:* If the U.S. energy consumption is 7.00% higher each year, by what factor will the energy consumption have increased after 10.0 years?

30. *Atmospheric Pressure:* The pressures measured at 1-mi intervals above sea level form a GP, with each value smaller than the preceding by a factor of 0.819. If the pressure at sea level is 29.92 in. Hg, find the pressure at an altitude of 5 mi.

31. *Compound Interest:* A person deposits $10,000 in a bank giving 6% interest, compounded annually. Find to the nearest dollar the value of the deposit after 50 years.

32. *Inflation:* The price of a certain house, now $126,000, is expected to increase by 5% each year. Write a GP whose terms are the value of the house at the end of each year, and find the value of the house after 5 years.

33. *Depreciation:* When calculating depreciation by the *declining-balance method*, a taxpayer claims as a deduction a fixed percentage of the book value of an asset each year. The new book value is then the last book value less the amount of the depreciation. Thus for a machine having an initial book value of $100,000 and a depreciation rate of 40%, the first year's depreciation is 40% of $100,000, or $40,000, and the new book value is $100,000 − $40,000 = $60,000. Thus the book values for each year form the following GP:

$$\$100,000, \quad \$60,000, \quad \$36,000, \ldots$$

Find the book value after 5 years.

20–4 Infinite Geometric Progressions

Sum of an Infinite Geometric Progression

Before we derive a formula for the sum of an infinite geometric progression, let us explore the idea graphically and numerically.

We have already determined that the sum of *n* terms of any geometric progression with a first term *a* and a common ratio *r* is

$$s_n = \frac{a(1 - r^n)}{1 - r} \tag{196}$$

Thus a graph of s_n versus n should tell us about how the sum changes as n increases.

◆◆◆ **Example 33:** Graph the sum s_n versus n for the GP

$$1, \quad 1.2, \quad 1.2^2, \quad 1.2^3, \ldots$$

for $n = 0$ to 20.

Solution: Here, $a = 1$ and $r = 1.2$, so

$$s_n = \frac{1 - 1.2^n}{1 - 1.2} = \frac{1.2^n - 1}{0.2}$$

The sum s_n is graphed as shown. Note that the sum continues to increase. We say that this progression *diverges*. ◆◆◆

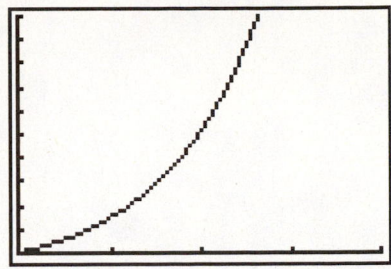

Screen for Example 33. Tick marks are 5 units apart.

Decreasing GP

If the common ratio r in a GP is less than 1, each term in the progression will be less than those preceding it. Such progression is called a *decreasing* GP.

◆◆◆ **Example 34:** Graph the sum s_n versus n for the GP

$$1, \quad 0.8, \quad 0.8^2, \quad 0.8^3, \ldots$$

for $n = 0$ to 20.

Solution: Here, $a = 1$ and $r = 0.8$, so

$$s_n = \frac{1 - 0.8^n}{1 - 0.8} = \frac{1 - 0.8^n}{0.2}$$

The sum is graphed as shown. Note that the sum appears to reach a limiting value, so we say that this progression *converges*. The sum appears to be approaching a value of 5. ◆◆◆

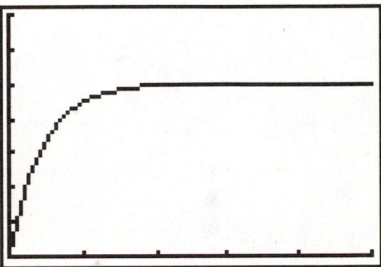

Screen for Example 34. Tick marks are 10 units apart on the horizontal axis and 1 unit apart on the vertical axis.

Our graphs have thus indicated that the sum of an infinite number of terms of a decreasing GP appears to approach a limit. Let us now verify that fact numerically.

◆◆◆ **Example 35:** Find the sum of infinitely many terms of the GP

$$9, \quad 3, \quad 1, \quad \tfrac{1}{3}, \ldots$$

Solution: Knowing that $a = 9$ and $r = \frac{1}{3}$, we can use a computer to compute each term and keep a running sum as shown in Table 20–3. Notice that the terms get smaller and smaller, and the sum appears to approach a value of around 13.5. ◆◆◆

We will confirm the value found in Example 35 after we have derived a formula for the sum of an infinite geometric progression.

Limit Notation

In order to derive a formula for the sum of an infinite geometric progression, we need to introduce *limit notation*. Such notation will also be of great importance in the study of calculus.

First, we use an arrow (→) to indicate that a quantity *approaches* a given value.

◆◆◆ **Example 36:**

(a) $x \rightarrow 5$ means that x gets closer and closer to the value 5.

(b) $n \rightarrow \infty$ means that n grows larger and larger, without bound.

(c) $y \rightarrow 0$ means that y continuously gets closer and closer to zero. ◆◆◆

If some function $f(x)$ approaches some value L as x approaches a, we could write this as

$$f(x) \rightarrow L \quad \text{as} \quad x \rightarrow a$$

TABLE 20–3

	Term	Sum
1	9.0000	9.0000
2	3.0000	12.0000
3	1.0000	13.0000
4	0.3333	13.3333
5	0.1111	13.4444
6	0.0370	13.4815
7	0.0123	13.4938
8	0.0041	13.4979
9	0.0014	13.4993
10	0.0005	13.4998
11	0.0002	13.4999
12	0.0001	13.5000
13	0.0000	13.5000
14	0.0000	13.5000
15	0.0000	13.5000
16	0.0000	13.5000
17	0.0000	13.5000
18	0.0000	13.5000

which is written in more compact notation as follows:

Limit Notation	$$\lim_{x \to a} f(x) = L$$	249

Limit notation lets us compactly write the value to which a function is tending, as in the following example.

◆◆◆ **Example 37:** Here are some examples showing the use of limit notation.

(a) $\lim\limits_{b \to 0} b = 0$

(b) $\lim\limits_{b \to 0} (a + b) = a$

(c) $\lim\limits_{b \to 0} \dfrac{a + b}{c + d} = \dfrac{a}{c + d}$

(d) $\lim\limits_{x \to 0} \dfrac{x + 1}{x + 2} = \dfrac{1}{2}$ ◆◆◆

A Formula for the Sum of an Infinite, Decreasing Geometric Progression

As the number of terms n of an infinite, decreasing geometric progression increases without bound, the term a_n will become smaller and will approach zero. Using our new notation, we may write

$$\lim_{n \to \infty} a_n = 0 \quad \text{when} \quad |r| < 1$$

As before, r is the common ratio. Thus in Eq. 197 for the sum of n terms,

$$s_n = \frac{a - ra_n}{1 - r}$$

the expression ra_n will approach zero as n approaches infinity, so the sum s_∞ of infinitely many terms (called the *sum to infinity*) is

$$s_\infty = \lim_{n \to \infty} \frac{a - ra_n}{1 - r} = \frac{a}{1 - r}$$

Thus letting $S = s_\infty$, the sum to infinity is:

GP: Sum to Infinity	$$S = \frac{a}{1 - r} \quad \text{when} \quad |r| < 1$$	198

◆◆◆ **Example 38:** Find the sum of infinitely many terms of the GP

$$9, \quad 3, \quad 1, \quad \frac{1}{3}, \ldots$$

Solution: Here $a = 9$ and $r = \frac{1}{3}$, so

$$S = \frac{a}{1 - r} = \frac{9}{1 - 1/3} = \frac{9}{2/3} = 13\frac{1}{2}$$

This agrees with the value we found numerically in Example 35.   ◆◆◆

◆◆◆ **Example 39:** *An Application.* For the mechanism in Example 29, find the total distance traveled if the shaft is allowed to come to rest instead of being limited to ten oscillations.

Solution: From Eq. 198, with $a = 6.50$ and $r = 0.850$,

$$S = \frac{6.50}{1 - 0.850} = 43.3 \text{ cm}$$ ◆◆◆

Exercise 4 ◆ Infinite Geometric Progressions

Evaluate each limit.

1. $\lim_{b \to 0} (b - c + 5)$

2. $\lim_{b \to 0} (a + b^2)$

3. $\lim_{b \to 0} \dfrac{3 + b}{c + 4}$

4. $\lim_{b \to 0} \dfrac{a + b + c}{b + c - 5}$

Find the sum of the infinitely many terms of each GP.

5. $144, 72, 36, 18, \ldots$

6. $8, 4, 2, 1, \frac{1}{2}, \ldots$

7. $10, 2, 0.4, 0.08, \ldots$

8. $1, \frac{1}{4}, \frac{1}{16}, \ldots$

Applications

9. Each swing of a certain pendulum is 78% as long as the one before. If the first swing is 10 in., find the entire distance traveled by the pendulum before it comes to rest.

10. A ball dropped from a height of 20 ft rebounds to three-fourths of its height on each bounce. Find the total distance traveled by the ball before it comes to rest.

11. *Team Project, Zeno's Paradox:* "A runner can never reach a finish line 1 mile away because first he would have to run half a mile, and then must run half of the remaining distance, or $\frac{1}{4}$ mile, and then half of that, and so on. Since there are an infinite number of distances that must be run, it will take an infinite length of time, and so the runner will never reach the finish line." Show that the distances to be run form the infinite series;

$$\frac{1}{2}, \frac{1}{4}, \frac{1}{8}, \ldots$$

Then disprove the paradox by actually finding the sum of that series and showing that the sum is *not* infinite.

Zeno's Paradox is named for Zeno of Elea (ca. 490–ca. 435 B.C.E.), a Greek philosopher and mathematician. He wrote several famous paradoxes, including this one.

20–5   The Binomial Theorem

Powers of a Binomial

Recall that a *binomial*, such as $(a + b)$, is a polynomial with two terms. By actual multiplication we can show that

$$(a + b)^1 = a + b$$
$$(a + b)^2 = a^2 + 2ab + b^2$$
$$(a + b)^3 = a^3 + 3a^2b + 3ab^2 + b^3$$
$$(a + b)^4 = a^4 + 4a^3b + 6a^2b^2 + 4ab^3 + b^4$$

We now want a formula for $(a + b)^n$ with which to expand a binomial without actually carrying out the multiplication. In the expansion of $(a + b)^n$, where n is a positive integer, we note the following patterns:

1. There are $n + 1$ terms.
2. The power of a is n in the first term, decreases by 1 in each later term, and reaches 0 in the last term.
3. The power of b is 0 in the first term, increases by 1 in each later term, and reaches n in the last term.
4. Each term has a total degree of n. (That is, the sum of the degrees of the variables is n.)

5. The first coefficient is 1.

6. The product of the coefficient of any term and its power of *a*, divided by the number of the term, gives the coefficient of the next term. (This property gives a *recursion formula* for the coefficients of the binomial expansion.)

These six observations can be expressed as the formula:

$$(a + b)^n = a^n + na^{n-1}b + \frac{n(n-1)}{2} a^{n-2}b^2 + \frac{n(n-1)(n-2)}{2(3)} a^{n-3}b^3 + \cdots + b^n$$

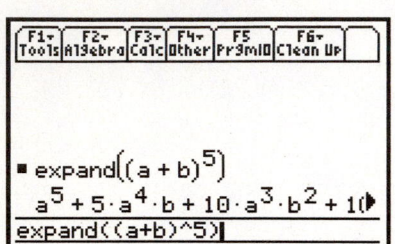

◆◆◆ **Example 40:** Use the binomial theorem to expand $(a + b)^5$.

Solution: From the binomial theorem,

$$(a + b)^5 = a^5 + 5a^4b + \frac{5(4)}{2} a^3b^2 + \frac{5(4)(3)}{2(3)} a^2b^3 + \frac{5(4)(3)(2)}{2(3)(4)} ab^4 + b^5$$
$$= a^5 + 5a^4b + 10a^3b^2 + 10a^2b^3 + 5ab^4 + b^5$$

which can be verified by actual multiplication, as shown in the screen.   ◆◆◆

TI-89 solution for Example 40. Part of the answer is off the screen.

If the expression contains only one variable, we can verify the expansion graphically.

◆◆◆ **Example 41:** Use the result from Example 40 to expand $(x + 1)^5$, and verify graphically.

Solution: Substituting $a = x$ and $b = 1$ into the preceding result gives

$$(x + 1)^5 = x^5 + 5x^4 + 10x^3 + 10x^2 + 5x + 1$$

To verify graphically, we plot the following in the same window.

$$y_1 = (x + 1)^5$$

and

$$y_2 = x^5 + 5x^4 + 10x^3 + 10x^2 + 5x + 1$$

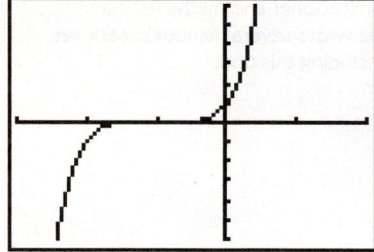

Screen for Example 41.

The two graphs shown are identical. This is not a *proof* that our expansion is correct, but it certainly gives us more confidence in our result.   ◆◆◆

Factorial Notation

We now introduce *factorial notation*. For a positive integer *n*, factorial *n*, written *n*!, is the product of all of the positive integers less than or equal to *n*.

$$n! = 1 \cdot 2 \cdot 3 \cdots (n - 1) \cdot n$$

Of course, it does not matter in which order we multiply the numbers. We may also write

$$n! = n \cdot (n - 1) \cdots 3 \cdot 2 \cdot 1$$

◆◆◆ **Example 42:** Here are some examples showing the use of factorial notation.

(a) $3! = 1 \cdot 2 \cdot 3 = 6$

(b) $5! = 1 \cdot 2 \cdot 3 \cdot 4 \cdot 5 = 120$

(c) $\dfrac{6!3!}{7!} = \dfrac{6!(2)(3)}{7(6!)} = \dfrac{6}{7}$

(d) $0! = 1$ by definition   ◆◆◆

Factorials can be evaluated by most calculators.

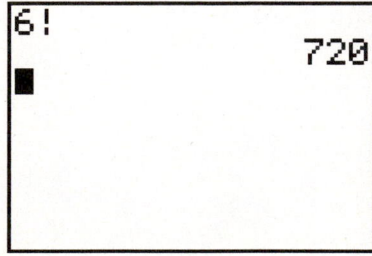

◆◆◆ **Example 43:** Evaluate 6! by calculator.

T1-83/84 screen for Example 43.

Solution: We enter 6 followed by the factorial symbol. On the TI-83/84 and on the TI-89, it is located on the [MATH] PRB (probability) menu. We get

$$6! = 720$$ ◆◆◆

Binomial Theorem Written with Factorial Notation

The use of factorial notation allows us to write the binomial theorem more compactly. Thus:

Binomial Theorem	$(a + b)^n = a^n + na^{n-1}b + \dfrac{n(n-1)}{2!}a^{n-2}b^2 + \dfrac{n(n-1)(n-2)}{3!}a^{n-3}b^3 + \cdots + b^n$	199

◆◆◆ **Example 44:** Using the binomial formula, expand $(a + b)^6$.

Solution: There will be $(n + 1)$ or seven terms. Let us first find the seven coefficients. The first coefficient is always 1, and the second coefficient is always n, or 6. We calculate the remaining coefficients in Table 20–4.

TABLE 20–4

Term	Coefficient
1	1
2	$n = 6$
3	$\dfrac{n(n-1)}{2!} = \dfrac{6(5)}{1(2)} = 15$
4	$\dfrac{n(n-1)(n-2)}{3!} = \dfrac{6(5)(4)}{1(2)(3)} = 20$
5	$\dfrac{n(n-1)(n-2)(n-3)}{4!} = \dfrac{6(5)(4)(3)}{1(2)(3)(4)} = 15$
6	$\dfrac{n(n-1)(n-2)(n-3)(n-4)}{5!} = \dfrac{6(5)(4)(3)(2)}{1(2)(3)(4)(5)} = 6$
7	$\dfrac{n(n-1)(n-2)(n-3)(n-4)(n-5)}{6!} = \dfrac{6(5)(4)(3)(2)(1)}{1(2)(3)(4)(5)(6)} = 1$

The seven terms thus have the coefficients

$$1 \quad 6 \quad 15 \quad 20 \quad 15 \quad 6 \quad 1$$

and our expansion is then

$$(a + b)^6 = a^6 + 6a^5b + 15a^4b^2 + 20a^3b^3 + 15a^2b^4 + 6ab^5 + b^6 \quad ◆◆◆$$

Pascal's Triangle

We now have expansions for $(a + b)^n$, for values of n from 1 to 6, to which we add $(a + b)^0 = 1$. Let us now write the coefficients of the terms of each expansion.

Exponent n

0	1
1	1 1
2	1 2 1
3	1 3 3 1
4	1 4 6 4 1
5	1 5 10 10 5 1
6	1 6 15 20 15 6 1

This array is called *Pascal's triangle*. We may use it to predict the coefficients of expansions with powers higher than 6 by noting that each number in the triangle is equal to the sum of the two numbers above it (thus the 15 in the lowest row is the sum of the 5 and the 10 above it). When expanding a binomial, we may take the coefficients directly from Pascal's triangle, as shown in the following example.

Pascal's triangle is named for Blaise Pascal (1623–62), the French geometer, probabilist, combinatorist, physicist, and philosopher.

◆◆◆ **Example 45:** Expand $(1 - x)^6$, obtaining the binomial coefficients from Pascal's triangle.

Solution: From Pascal's triangle, for $n = 6$, we read the coefficients

$$1 \quad 6 \quad 15 \quad 20 \quad 15 \quad 6 \quad 1$$

We now substitute into the binomial formula with $n = 6$, $a = 1$, and $b = -x$. When one of the terms of the binomial is negative, as here, we must be careful to include the minus sign whenever that term appears in the expansion. Thus

$$\begin{aligned}
(1 - x)^6 &= [1 + (-x)]^6 \\
&= 1^6 + 6(1^5)(-x) + 15(1^4)(-x)^2 + 20(1^3)(-x)^3 \\
&\quad + 15(1^2)(-x)^4 + 6(1)(-x)^5 + (-x)^6 \\
&= 1 - 6x + 15x^2 - 20x^3 + 15x^4 - 6x^5 + x^6 \qquad ◆◆◆
\end{aligned}$$

If either term in the binomial is itself a power, a product, or a fraction, it is a good idea to enclose that entire term in parentheses before substituting.

◆◆◆ **Example 46:** Expand $\left(\dfrac{3}{x^2} + 2y^3\right)^4$.

Solution: The binomial coefficients are

$$1 \quad 4 \quad 6 \quad 4 \quad 1$$

We substitute $\dfrac{3}{x^2}$ for a and $2y^3$ for b.

$$\begin{aligned}
\left(\frac{3}{x^2} + 2y^3\right)^4 &= \left[\left(\frac{3}{x^2}\right) + (2y^3)\right]^4 \\
&= \left(\frac{3}{x^2}\right)^4 + 4\left(\frac{3}{x^2}\right)^3(2y^3) + 6\left(\frac{3}{x^2}\right)^2(2y^3)^2 + 4\left(\frac{3}{x^2}\right)(2y^3)^3 + (2y^3)^4 \\
&= \frac{81}{x^8} + \frac{216y^3}{x^6} + \frac{216y^6}{x^4} + \frac{96y^9}{x^2} + 16y^{12} \qquad ◆◆◆
\end{aligned}$$

Sometimes we may not need the entire expansion but only the first several terms.

◆◆◆ **Example 47:** Find the first four terms of $(x^2 - 2y^3)^{11}$.

Solution: From the binomial formula, with $n = 11$, $a = x^2$, and $b = (-2y^3)$,

$$\begin{aligned}
(x^2 - 2y^3)^{11} &= [(x^2) + (-2y^3)]^{11} \\
&= (x^2)^{11} + 11(x^2)^{10}(-2y^3) + \frac{11(10)}{2}(x^2)^9(-2y^3)^2 \\
&\quad + \frac{11(10)(9)}{2(3)}(x^2)^8(-2y^3)^3 + \cdots \\
&= x^{22} - 22x^{20}y^3 + 220x^{18}y^6 - 1320x^{16}y^9 + \cdots \qquad ◆◆◆
\end{aligned}$$

General Term

We can write the general or rth term of a binomial expansion $(a + b)^n$ by noting the following pattern:

1. The power of b is 1 less than r. Thus a fifth term would contain b^4, and the rth term would contain b^{r-1}.
2. The power of a is n minus the power of b. Thus a fifth term would contain $a^{n-(5-1)}$ or a^{n-5+1}, and the rth term would contain $a^{n-(r-1)}$ or a^{n-r+1}.
3. The coefficient of the rth term is

$$\frac{n!}{(r-1)!(n-r+1)!}$$

These facts are combined into the following formula:

General Term	In the expansion for $(a+b)^n$, rth term $= \dfrac{n!}{(r-1)!(n-r+1)!} a^{n-r+1}b^{r-1}$	**200**

◆◆◆ Example 48: Find the eighth term of the expansion for $(3a+b^5)^{11}$.

Solution: Here, $n=11$ and $r=8$. So $n-r+1=4$. The coefficient of the eighth term is then

$$\frac{n!}{(r-1)!(n-r+1)!} = \frac{11!}{7!\,4!} = 330$$

Then

$$\text{the eighth term} = 330(3a)^{11-8+1}(b^5)^{8-1}$$
$$= 330(81a^4)(b^{35}) = 26{,}730a^4b^{35} \qquad \text{◆◆◆}$$

Fractional and Negative Exponents

So far we have used the binomial theorem only with positive integral exponents. Here we will use it when the exponent is negative or fractional. However, we now obtain an infinite series, called a *binomial series*, with no last term. Further, the binomial series is equal to $(a+b)^n$ only if the series converges.

◆◆◆ Example 49: Write the first four terms of the infinite binomial expansion for $\sqrt{5+x}$.

Solution: We replace the radical with a fractional exponent and substitute into the binomial formula in the usual way.

$$(5+x)^{1/2} = 5^{1/2} + \left(\frac{1}{2}\right)5^{-1/2}x + \frac{\left(\frac{1}{2}\right)\left(-\frac{1}{2}\right)}{2}5^{-3/2}x^2$$

$$+ \frac{\left(\frac{1}{2}\right)\left(-\frac{1}{2}\right)\left(-\frac{3}{2}\right)}{6}5^{-5/2}x^3 + \cdots$$

$$= \sqrt{5} + \frac{x}{2\sqrt{5}} - \frac{x^2}{8(5)^{3/2}} + \frac{x^3}{16(5)^{5/2}} - \cdots$$

Switching to decimal form and working to five decimal places, we get

$$(5+x)^{1/2} \approx 2.23607 + 0.22361x - 0.01118x^2 + 0.00112x^3 - \cdots \qquad \text{◆◆◆}$$

We now ask if this series is valid for *any* x. Let's try a few values. For $x=0$,

$$(5+0)^{1/2} \approx 2.23607$$

which checks within the number of digits retained. For $x = 1$,

$$(5 + 1)^{1/2} = 2.23607 + 0.22361 - 0.01118 + 0.00112 - \cdots$$
$$\approx 2.44962$$

which compares with a calculator value for $(5 + 1)^{1/2}$ of 2.44949. Not bad, considering that we are using only four terms of an infinite series. Of course, we would not use this method to compute the square root of 6. We are just using this value to explore the accuracy of the given series for different values of x.

To determine the effect of larger values of x, we have used a computer to generate eight terms of the binomial series as listed in Table 20–5.

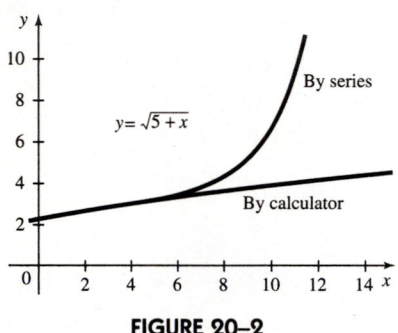

FIGURE 20–2

TABLE 20–5

x	$\sqrt{5 + x}$ by Series	$\sqrt{5 + x}$ by Calculator	Error
0	2.236068	2.236068	0
2	2.645766	2.645751	0.000015
4	3.002957	3	0.002957
6	3.379908	3.316625	0.063283
8	4.148339	3.605551	0.543788
11	9.723964	4	5.723964

Into it we have substituted various x values, from 0 to 11, and compared the value obtained by series with that from a calculator. The values are plotted in Fig. 20–2.

Notice that the accuracy gets worse as x gets larger, with useless values being obtained when x is 6 or greater. This illustrates the following:

> A binomial series is equal to $(a + b)^n$ only if the series converges, and that occurs when $|a| > |b|$.

Our series in Example 49 thus converges when $x < 5$.

The binomial series, then, is given by the following formula:

Binomial Series	$(a + b)^n = a^n + na^{n-1}b + \dfrac{n(n - 1)}{2!}a^{n-2}b^2 + \dfrac{n(n - 1)(n - 2)}{3!}a^{n-3}b^3 + \cdots$ where $\|a\| > \|b\|$	**201**

The binomial series is often written with $a = 1$ and $b = x$, so we give that form here also.

Binomial Series	$(1 + x)^n = 1 + nx + \dfrac{n(n - 1)}{2!}x^2 + \dfrac{n(n - 1)(n - 2)}{3!}x^3 + \cdots$ where $\|x\| < 1$	**202**

◆◆◆ **Example 50:** Expand to four terms:

$$\frac{1}{\sqrt[3]{1/a - 3b^{1/3}}}$$

Solution: We rewrite the given expression without radicals or fractions, and then expand.

$$\frac{1}{\sqrt[3]{1/a - 3b^{1/3}}} = [(a^{-1}) + (-3b^{1/3})]^{-1/3}$$

$$= (a^{-1})^{-1/3} - \frac{1}{3}(a^{-1})^{-4/3}(-3b^{1/3}) + \frac{2}{9}(a^{-1})^{-7/3}(-3b^{1/3})^2 - \frac{14}{81}(a^{-1})^{-10/3}(-3b^{1/3})^3 + \cdots$$

$$= a^{1/3} + a^{4/3}b^{1/3} + 2a^{7/3}b^{2/3} + \frac{14}{3}a^{10/3}b + \cdots \qquad \text{◆◆◆}$$

Exercise 5 ◆ The Binomial Theorem

Factorial Notation

Evaluate each factorial.

1. $6!$

2. $8!$

3. $\dfrac{7!}{5!}$

4. $\dfrac{11!}{9! \, 2!}$

5. $\dfrac{7!}{3! \, 4!}$

6. $\dfrac{8!}{3! \, 5!}$

Binomials Raised to an Integral Power

Verify each expansion. Obtain the binomial coefficients by formula or from Pascal's triangle as directed by your instructor.

7. $(x + y)^7 = x^7 + 7x^6y + 21x^5y^2 + 35x^4y^3 + 35x^3y^4 + 21x^2y^5 + 7xy^6 + y^7$

8. $(4 + 3b)^4 = 256 + 768b + 864b^2 + 432b^3 + 81b^4$

9. $(3a - 2b)^4 = 81a^4 - 216a^3b + 216a^2b^2 - 96ab^3 + 16b^4$

10. $(x^3 + y)^7 = x^{21} + 7x^{18}y + 21x^{15}y^2 + 35x^{12}y^3 + 35x^9y^4 + 21x^6y^5 + 7x^3y^6 + y^7$

11. $(x^{1/2} + y^{2/3})^5 = x^{5/2} + 5x^2y^{2/3} + 10x^{3/2}y^{4/3} + 10xy^2 + 5x^{1/2}y^{8/3} + y^{10/3}$

12. $(1/x + 1/y^2)^3 = 1/x^3 + 3/x^2y^2 + 3/xy^4 + 1/y^6$

13. $(a/b - b/a)^6 = (a/b)^6 - 6(a/b)^4 + 15(a/b)^2 - 20 + 15(b/a)^2 - 6(b/a)^4 + (b/a)^6$

14. $(1/\sqrt{x} + y^2)^4 = x^{-2} + 4x^{-3/2}y^2 + 6x^{-1}y^4 + 4x^{-1/2}y^6 + y^8$

15. $(2a^2 + \sqrt{b})^5 = 32a^{10} + 80a^8b^{1/2} + 80a^6b + 40a^4b^{3/2} + 10a^2b^2 + b^{5/2}$

16. $(\sqrt{x} - c^{2/3})^4 = x^2 - 4x^{3/2}c^{2/3} + 6xc^{4/3} - 4x^{1/2}c^2 + c^{8/3}$

Verify the first four terms of each binomial expansion.

17. $(x^2 + y^3)^8 = x^{16} + 8x^{14}y^3 + 28x^{12}y^6 + 56x^{10}y^9 + \cdots$

18. $(x^2 - 2y^3)^{11} = x^{22} - 22x^{20}y^3 + 220x^{18}y^6 - 1320x^{16}y^9 + \cdots$

19. $(a - b^4)^9 = a^9 - 9a^8b^4 + 36a^7b^8 - 84a^6b^{12} + \cdots$

20. $(a^3 + 2b)^{12} = a^{36} + 24a^{33}b + 264a^{30}b^2 + 1760a^{27}b^3 + \cdots$

General Term

Write the requested term of each binomial expansion, and simplify.

21. Seventh term of $(a^2 - 2b^3)^{12}$

22. Eleventh term of $(2 - x)^{16}$

23. Fourth term of $(2a - 3b)^7$

24. Eighth term of $(x + a)^{11}$

25. Fifth term of $(x - 2\sqrt{y})^{25}$

26. Ninth term of $(x^2 + 1)^{15}$

Binomials Raised to a Fractional or Negative Power

Verify the first four terms of each infinite binomial series.

27. $(1 - a)^{2/3} = 1 - 2a/3 - a^2/9 - 4a^3/81 \ldots$

28. $\sqrt{1 + a} = 1 + a/2 - a^2/8 + a^3/16 \ldots$

29. $(1 + 5a)^{-5} = 1 - 25a + 375a^2 - 4375a^3 \ldots$

30. $(1 + a)^{-3} = 1 - 3a + 6a^2 - 10a^3 \ldots$

31. $1/\sqrt[6]{1 - a} = 1 + a/6 + 7a^2/72 + 91a^3/1296 \ldots$

◆◆◆ CHAPTER 20 REVIEW PROBLEMS ◆◆◆◆◆◆◆◆◆◆◆◆◆◆◆◆◆◆◆◆◆◆◆◆◆◆◆◆◆

1. Find the sum of seven terms of the AP: $-4, -1, 2, \ldots$

2. Find the tenth term of $(1 - y)^{14}$.

3. Insert four arithmetic means between 3 and 18.

4. How many terms of the AP $-5, -2, 1, \ldots$ will give a sum of 63?

5. Insert four harmonic means between 2 and 12.

6. Find the twelfth term of an AP with first term 7 and common difference 5.

7. Find the sum of the first 10 terms of the AP $1, 2\frac{2}{3}, 4\frac{1}{3}, \ldots$.

Expand each binomial.

8. $(2x^2 - \sqrt{x}/2)^5$

9. $(a - 2)^5$

10. $(x - y^3)^7$

11. Find the fifth term of a GP with first term 3 and common ratio 2.

12. Find the fourth term of a GP with first term 8 and common ratio -4.

Evaluate each limit.

13. $\lim\limits_{b \to 0} (2b - x + 9)$

14. $\lim\limits_{b \to 0} (a^2 + b^2)$

Find the sum of infinitely many terms of each GP.

15. $4, 2, 1, \ldots$

16. $\frac{1}{4}, -\frac{1}{16}, \frac{1}{64}, \ldots$

17. $1, -\frac{2}{5}, \frac{4}{25}, \ldots$

18. $27, 9, 3, 1 \ldots$

19. $150, 30, 6, \ldots$

20. Find the seventh term of the harmonic progression $3, 3\frac{3}{7}, 4, \ldots$.

21. Find the sixth term of a GP with first term 5 and common ratio 3.

22. Find the sum of the first seven terms of the GP in problem 21.

23. Find the fifth term of a GP with first term -4 and common ratio 4.

24. Find the sum of the first seven terms of the GP in problem 23.

25. Insert two geometric means between 8 and 125.

26. Insert three geometric means between 14 and 224.

Evaluate each factorial.

27. $\dfrac{4! \, 5!}{2!}$

28. $\dfrac{8!}{4! \, 4!}$

29. $\dfrac{7! \, 3!}{5!}$

30. $\dfrac{7!}{2! \, 5!}$

Find the first four terms of each binomial expansion.

31. $\left(\dfrac{2x}{y^2} - y\sqrt{x} \right)^7$

32. $(1 + a)^{-2}$

33. $\dfrac{1}{\sqrt{1 + a}}$

34. Expand using the binomial theorem:

$$(1 - 3a + 2a^2)^4$$

35. Find the eighth term of $(3a - b)^{11}$.

36. Find the tenth term of an AP with first term 12 if the sum of the first ten terms is 10.

37. *Series Approximations:* It is possible to compute the approximate value of a quantity or function by means of the first several terms of an infinite series. Such a series can be used internally by calculators and computers for this purpose. Several of these well-known series are shown. For each, choose a value of x, write a program or spreadsheet to generate and find the sum of the first ten terms of the series, and compare your results with that obtained by calculator.

$$\sin x = x - \frac{x^3}{3!} + \frac{x^5}{5!} - \frac{x^7}{7!} + \cdots \qquad 167$$

where x is in radians

$$\cos x = 1 - \frac{x^2}{2!} + \frac{x^4}{4!} - \frac{x^6}{6!} + \cdots \qquad 168$$

where x is in radians

$$e = 2 + \frac{1}{2!} + \frac{1}{3!} + \frac{1}{4!} + \cdots \qquad 158$$

$$e^x = 1 + x + \frac{x^2}{2!} + \frac{x^3}{3!} + \frac{x^4}{4!} + \cdots \qquad 159$$

$$\ln x = 2a + \frac{2a^3}{3} + \frac{2a^5}{5} + \frac{2a^7}{7} + \cdots \qquad 161$$

where $a = \dfrac{x-1}{x+1}$

38. *Writing; Fibonacci Sequence:* One famous sequence is the *Fibonacci Sequence* mentioned earlier in this chapter. Write a short paper on this sequence.

39. *On Our Web Site:* We give an entire chapter on infinite series on our Web site: www.wiley.com/college/calter.

Introduction to Statistics and Probability

◆◆◆ **OBJECTIVES** ◆◆

When you have completed this chapter, you should be able to

- Identify data as continuous, discrete, or categorical.
- Represent data as an *x-y* graph, scatter plot, bar graph, pie chart, stem and leaf plot, or boxplot.
- Organize data into frequency distributions, frequency histograms, frequency polygons, and cumulative frequency distributions.
- Calculate the mean, the weighted mean, the median, and the mode.
- Find the range, the quartiles, the deciles, the percentiles, the variance, and the standard deviation.
- Calculate the probabilities for frequency distributions, including the binomial and normal distributions.
- Estimate population means, standard deviations, and proportions within a given confidence interval.
- Make control charts for statistical process control.
- Fit a straight line to a set of data using the method of least squares.

◆◆

Why study statistics? There are several reasons. First, for your work you might have to interpret statistical data or even need to collect such data and make inferences from it. For example, you could be asked to determine if replacing a certain machine on a production line actually reduced the number of defective parts made. Further, a knowledge of statistics can help you to evaluate the claims made in news reports, polls, and advertisements. A toothpaste manufacturer may claim that its product "significantly reduces" tooth decay. A pollster may claim that 52% of the electorate is going to vote for candidate Jones. A newspaper may show a chart that pictures oil consumption going down. It is easy to use statistics to distort the truth, either on purpose or unintentionally. Some knowledge of the subject may help you to separate fact from distortion.

In this chapter we introduce many new terms. We then show how to display statistical data in graphical and numerical forms. How reliable are those statistical data? To answer that question, we include a brief introduction to *probability,* which enables us to describe the *confidence* we can place in those statistics.

21–1   Definitions and Terminology

Populations and Samples

In statistics, an entire group of people or things is called a *population* or *universe*. A population can be *infinite* or *finite*. A small part of the entire group chosen for study is called a *sample*.

♦♦♦ **Example 1:** In a study of the heights of students at Tech College, the entire student body is our *population*. This is a *finite* population. Instead of studying every student in the population, we may choose to work with a *sample* of only 100 students.   ♦♦♦

♦♦♦ **Example 2:** Suppose that we want to determine what percentage of tosses of a coin will be heads. Then all possible tosses of the coin are an example of an *infinite* population. Theoretically, there is no limit to the number of tosses that can be made.   ♦♦♦

Parameters and Statistics

We often use just a few numbers to describe an entire population. Such numbers are called *parameters*. The branch of statistics in which we use parameters to describe a population is referred to as *descriptive statistics*.

♦♦♦ **Example 3:** For the population of all students at Tech College, a computer read the height of each student and computed the average height of all students. It found this *parameter*, average height, to be 68.2 in.   ♦♦♦

Similarly, we use just a few numbers to describe a sample. Such numbers are called *statistics*. When we then use these sample statistics to *infer* things about the entire population, we are engaging in what is called *inductive statistics*.

♦♦♦ **Example 4:** From the population of all students in Tech College, a sample of 100 students was selected. The average of their heights was found to be 67.9 in. This is a sample *statistic*. From it we infer that for the entire population, the average height is 67.9 in. plus or minus some *standard error*. We show how to compute standard error in Sec. 21–6.   ♦♦♦

Thus *parameters* describe *populations,* while *statistics* describe *samples.* The first letters of the words help us to keep track of which goes with which.

Population Parameter

versus

Sample Statistic

Variables

In statistics we distinguish between data that are *continuous* (obtained by measurement), *discrete* (obtained by counting), or *categorical* (belonging to one of several categories).

♦♦♦ **Example 5:** A survey at Tech College asked each student for (a) height, *continuous data;* (b) number of courses taken, *discrete data;* and (c) whether they were in favor (yes or no) of eliminating final exams, *categorical data.*   ♦♦♦

A *variable*, as in algebra, is a quantity that can take on different values within a problem; *constants* do not change value. We call variables *continuous, discrete,* or *categorical,* depending on the type of data they represent.

◆◆◆ **Example 6:** If, in the survey of Example 5, we let

$$H = \text{student's height}$$
$$N = \text{number of courses taken}$$
$$F = \text{opinion on eliminating finals}$$

then H is a continuous variable, N is a discrete variable, and F is a categorical variable.   ◆◆◆

Ways to Display Data

Statistical data are usually presented as a table of values or are displayed in a graph or a chart. Common types of graphs are the x-y graph, scatter plot, bar graph, or pie chart. Here we give examples of the display of discrete and categorical data. We cover display of continuous data in Sec. 21–2.

◆◆◆ **Example 7:** Table 21–1 shows the number of skis made per month by the Ace Ski Company. We show these discrete data as an x-y graph, a scatter plot, a bar chart, and a pie chart in Fig. 21–1.

TABLE 21–1

Month	Pairs Made
Jan.	13,946
Feb.	7,364
Mar.	5,342
Apr.	3,627
May	1,823
June	847
July	354
Aug.	425
Sept.	3,745
Oct.	5,827
Nov.	11,837
Dec.	18,475
Total	73,612

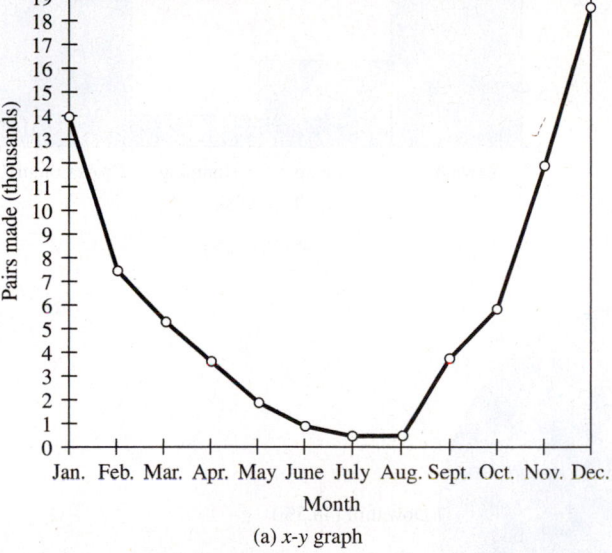

(a) x-y graph

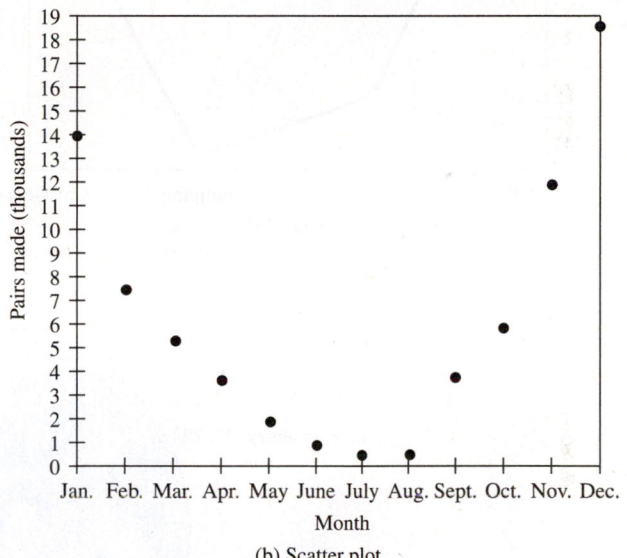

(b) Scatter plot

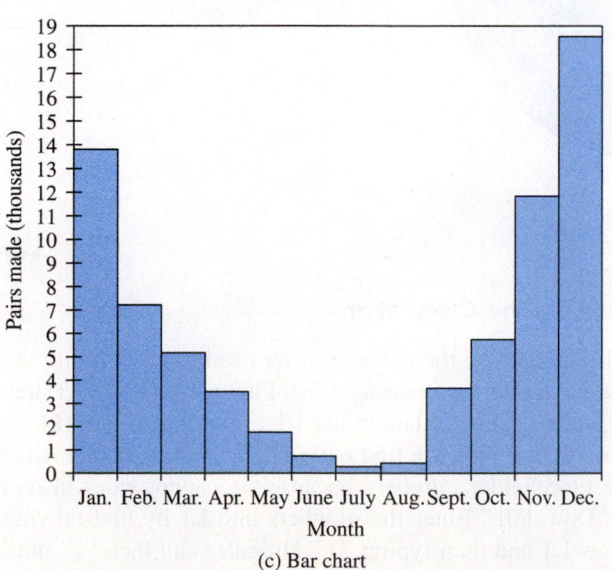

(c) Bar chart

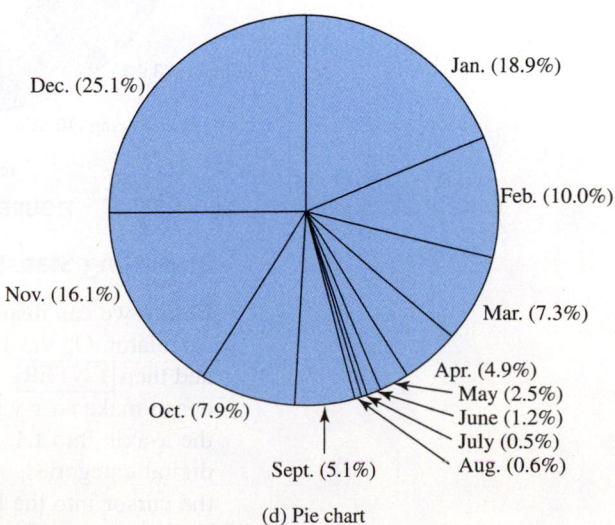

(d) Pie chart

FIGURE 21–1

◆◆◆

TABLE 21–2

Type	Pairs Made
Downhill	35,725
Racing	7,735
Jumping	2,889
Cross-country	27,263
Total	73,612

◆◆◆ **Example 8:** Table 21–2 shows the number of skis of various types made in a certain year by the Ace Ski Company. We show these discrete data as an *x-y* graph, a bar chart, and a pie chart (Fig. 21–2).

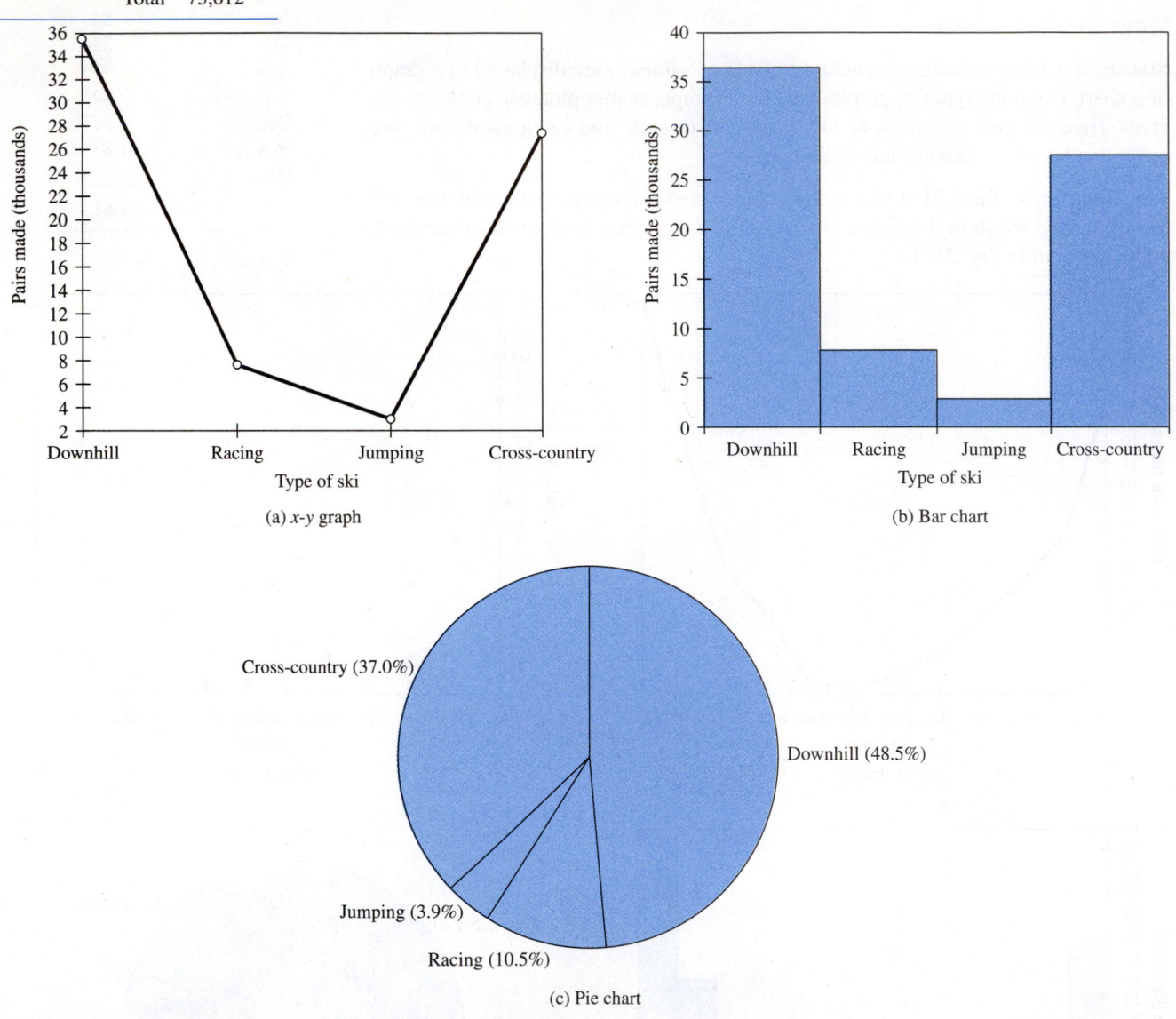

(a) *x-y* graph

(b) Bar chart

(c) Pie chart

FIGURE 21–2   ◆◆◆

Displaying Statistical Data on the Calculator

Before we can display statistical data on the calculator, we need to enter it into the calculator. On the T1-83/84, we do this by creating a list. First press the ⎡STAT⎤ key and then ⎡ENTER⎤. You can enter a table of data in list 1(L1) and higher lists. In order to make an *x-y* graph or a scatter plot, we first enter in placeholder numbers for the *x*-axis into L1. We use placeholder numbers because we cannot enter in non digital categories, such as "Downhill." Enter the numbers into L1 by first moving the cursor into the line below L1 and then typing "1." Hit enter and then "2" until you've entered up to "4." Next, move the cursor over to the first line under L2. Then enter in the values for the sales, in the same way that you entered the list into L1.

When you are done you should have four entries in L1 and four in L2. Next, press, followed by STAT PLOT . Press ENTER to select Plot 1, and then press ENTER again to turn on Plot 1. Move the cursor down and press ENTER to select a scatter plot, or first move the cursor one to the right to select an *x,y*-graph. Then move down to select L1 for Xlist and L2 for Ylist. Finally, press ZOOM and then 9 Zoom Stat to plot.

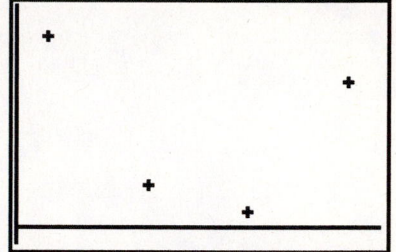

Scatter plot

Exercise 1 ◆ Definitions and Terminology

Types of Data

Label each type of data as continuous, discrete, or categorical.

1. The number of people in each county who voted for Jones.

2. The life of certain 100-W light bulbs.

3. The blood types of patients at a certain hospital.

4. The number of Ford cars sold each day.

5. The models of Ford cars sold each day.

6. The weights of steers in a given herd.

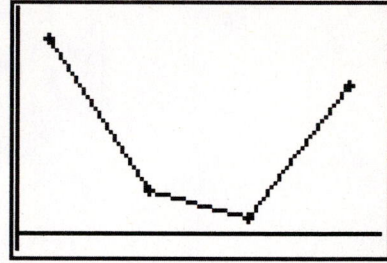

x,y-plot

Screens for Example 8.

Graphical Representation of Data

7. The following table shows the population of a certain town for the years 1920–1930:

Year	Population
1920	5364
1921	5739
1922	6254
1923	7958
1924	7193
1925	6837
1926	7245
1927	7734
1928	8148
1929	8545
1930	8623

Make a scatter plot and an *x-y* graph of the population versus the year.

8. Make a bar graph for the data of problem 7.

9. In a certain election, the tally was as listed in the following table:

Candidate	Number of Votes
Smith	746,000
Jones	623,927
Doe	536,023
Not voting	163,745

Make a bar chart showing the election results.

10. Make a pie chart showing the election results of problem 9. To make the pie chart, simply draw a circle and subdivide the area of the circle into four sectors. Make the central angle (and hence the area) of each sector proportional to the votes obtained by each candidate as a percentage of the total voting population. Label the percent in each sector.

Graphics Calculator

11. Use your graphics calculator to make a scatter plot and an *x-y* graph of the data in problem 7.

Computer

12. Many spreadsheet programs can draw *x-y* graphs, scatter plots and bar graphs. Unlike with the graphics calculator, the data on the *x*-axis can, for certain graphs, be words. Use a spreadsheet to draw any of the graphs in this exercise set.

13. There are a number of computer software packages, such as *Minitab*, designed especially for statistics. These, of course, can draw all of the common statistical graphs. Use one of these programs to draw any of the graphs in this exercise set.

21–2   Frequency Distributions

Raw Data

Data that have not been organized in any way are called *raw data*. Data that have been sorted into ascending or descending order are called an *array*. The *range* of the data is the difference between the largest and the smallest number in that array.

◆◆◆ **Example 9:** Five students were chosen at random and their heights (in inches) were measured. The *raw data* obtained were

$$56.2 \quad 72.8 \quad 58.3 \quad 56.9 \quad 67.5$$

If we sort the numbers into ascending order, we get the *array*

$$56.2 \quad 56.9 \quad 58.3 \quad 67.5 \quad 72.8$$

The *range* of the data is

$$\text{range} = 72.8 - 56.2 = 16.6 \text{ in.} \qquad \text{◆◆◆}$$

Grouped Data

For discrete and categorical data, we have seen that *x-y* graphs and bar charts are effective. These graphs can also be used for *continuous* data, but first we must collect those data into groups. This process is especially helpful when working with a large number of data points. Such groups are usually called *classes*. We create classes by dividing the range into a convenient number of *intervals*. Each interval has an upper and a lower *limit* or *class boundary*. The *class width* is equal to the upper class boundary minus the lower class limit. The intervals are chosen so that a given data point cannot fall directly on a class boundary. We call the center of each interval the *class midpoint*.

Frequency Distribution

Once the classes have been defined, we may tally the number of measurements falling within each class. Such a tally is called a *frequency distribution*. The number of entries in a given class is called the *absolute frequency*. That number, divided by the total number of entries, is called the *relative frequency*. The relative frequency is often expressed as a percent.

To Make a Frequency Distribution:

1. Subtract the smallest number in the data from the largest to find the range of the data.
2. Divide the range into class intervals, usually between 5 and 20. Avoid class limits that coincide with actual data.
3. Get the class frequencies by tallying the number of observations that fall within each class.
4. Get the relative frequency of each class by dividing its class frequency by the total number of observations.

◆◆◆ **Example 10:** The grades of 30 students are as follows:

85.0 62.0 94.6 76.3 77.4 82.3 58.4 86.4 84.4 69.8

91.3 75.0 69.6 86.4 84.1 74.7 65.7 86.4 90.6 69.7

86.7 70.5 78.4 86.4 81.5 60.7 79.8 97.2 85.7 78.5

Group these raw data into convenient classes. Choose the class width, the class limits, and the class midpoints. Find the absolute frequency and the relative frequency of each class.

Solution:

1. The highest value in the data is 97.2 and the lowest is 58.4, so

$$\text{range} = 97.2 - 58.4 = 38.8$$

2. If we divide the range into 5 classes, we get a class width of 38.8/5 or 7.76. If we divide the range into 20 classes, we get a class width of 1.94. Picking a convenient width between those two values, we let

$$\text{class width} = 6$$

Then we choose class limits of

$$56, 62, 68, \ldots, 98$$

Sometimes a given value, such as 62.0 in the preceding grade list, will fall right on a class limit. Although it does not matter much into which of the two adjoining classes we put it, we can easily avoid such ambiguity. One way is to make our class limits the following:

$$56\text{–}61.9, \quad 62\text{–}67.9, \quad 68\text{–}73.9, \ldots, 92\text{–}97.9$$

Thus 62.0 would fall into the second class. Our class midpoints are then

$$59, 65, 71, \ldots, 95$$

3. Our next step is to tally the number of grades falling within each class and get a frequency distribution, shown in Table 21–3.
4. We then divide each frequency by 30 to get the relative frequency.

TABLE 21–3 Frequency distribution of student grades.

Class Limits	Class Midpoint	Tally	Absolute Frequency	Relative Frequency
56–61.9	59	//	2	2/30 = 6.7%
62–67.9	65	//	2	2/30 = 6.7%
68–73.9	71	////	4	4/30 = 13.3%
74–79.9	77	⊮//	7	7/30 = 23.3%
80–85.9	83	⊮/	6	6/30 = 20.0%
86–91.9	89	⊮//	7	7/30 = 23.3%
92–97.9	95	//	2	2/30 = 6.7%
		Total	30	30/30 = 100.00%

◆◆◆

Since all the bars in a frequency histogram have the same width, the *areas* of the bars are proportional to the heights of the bars. The areas are thus proportional to the class frequencies. So if one bar in a relative frequency histogram has a height of 20%, that bar will contain 20% of the area of the whole graph. Thus there is a 20% chance that one grade chosen at random will fall within that class. We will make use of this connection between areas and probabilities in Sec. 21–4.

Frequency Histogram

A *frequency histogram* is a bar chart showing the frequency of occurrence of each class. The width of each bar is equal to the class width, and the height of each bar is equal to the frequency. The horizontal axis can show the class limits, the class midpoints, or both. The vertical axis can show the absolute frequency, the relative frequency, or both.

◆◆◆ **Example 11:** We show a frequency histogram for the data of Example 10 in Fig. 21–3. Note that the horizontal scale shows both the class limits and the class midpoints, and the vertical scales show both absolute frequency and relative frequency.

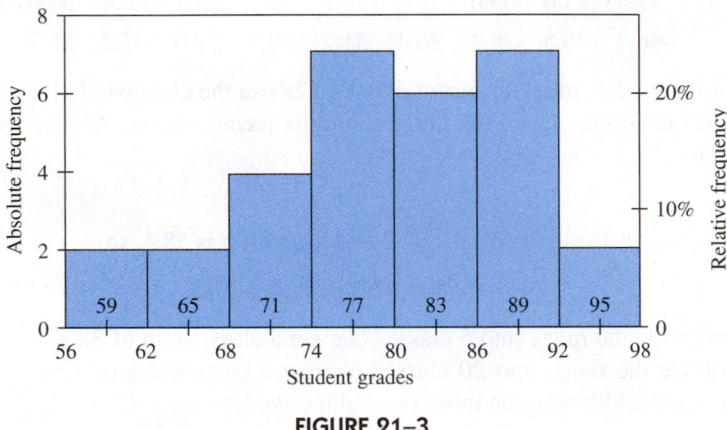

FIGURE 21–3   ◆◆◆

Frequency Polygon

A *frequency polygon* is simply an *x-y* graph in which frequency is plotted against class midpoint. As with the histogram, it can show either absolute or relative frequency, or both.

◆◆◆ **Example 12:** We show a frequency polygon for the data of Example 10 in Fig. 21–4.

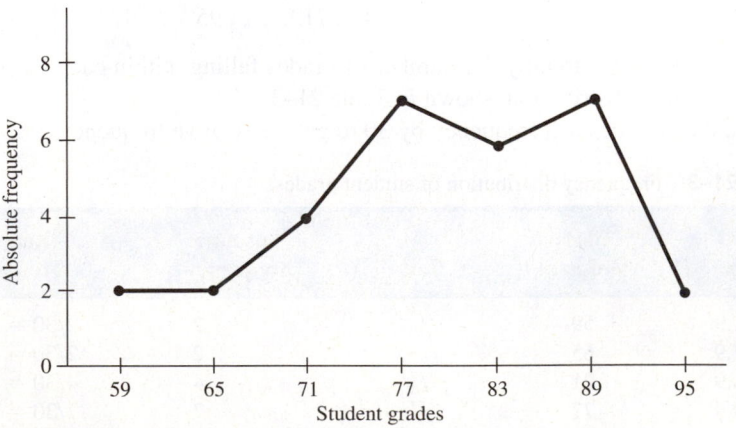

FIGURE 21–4   A frequency polygon for the data of Example 10. Note that this polygon could have been obtained by connecting the midpoints of the tops of the bars in Fig. 21–3.

◆◆◆

Cumulative Frequency Distribution

We obtain a *cumulative* frequency distribution by summing all values *less than* a given class limit. The graph of a cumulative frequency distribution is called a *cumulative frequency polygon,* or *ogive.*

◆◆◆ **Example 13:** Make and graph a cumulative frequency distribution for the data of Example 10.

Solution: We compute the cumulative frequencies by adding the values below the given limits as shown in Table 21–4. Then we plot them in Fig. 21–5.

TABLE 21–4 Cumulative frequency distribution for student grades.

Grade	Cumulative Absolute Frequency	Cumulative Relative Frequency
Under 61.9	2	2/30 (6.7%)
Under 67.9	4	4/30 (13.3%)
Under 73.9	8	8/30 (26.7%)
Under 79.9	15	15/30 (50.0%)
Under 85.9	21	21/30 (70.0%)
Under 91.9	28	28/30 (93.3%)
Under 97.9	30	30/30 (100%)

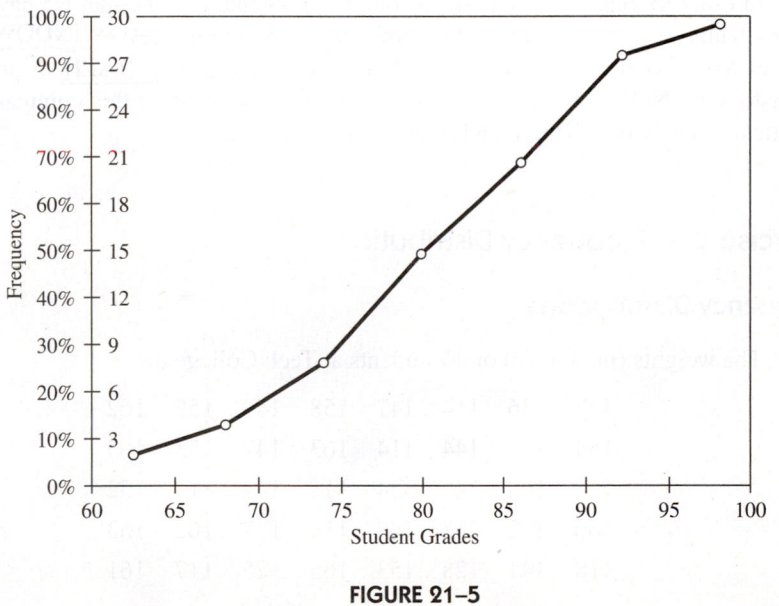

FIGURE 21–5   ◆◆◆

Stem and Leaf Plot

A fast and easy way to get an idea of the distribution of a new set of data is to make a *stem and leaf* plot. We will show how by means of an example.

◆◆◆ **Example 14:** Make a stem and leaf plot for the data in Example 10.

Solution: Let us choose as the *stem* of our plot the first digit of each grade. These range from 5 to 9, representing grades from 50 to 90. Then opposite each number on the stem we write all of the values in the table starting with that number, as shown in Table 21–5.

TABLE 21–5

Stem	Leaves
5	8.4
6	2.0 9.8 9.6 5.7 9.7 0.7
7	6.3 7.4 5.0 4.7 0.5 8.4 9.8 8.5
8	5.0 2.3 6.4 4.4 6.4 4.1 6.4 6.7 6.4 1.5 5.7
9	4.6 1.3 0.6 7.2

We sometimes go on to arrange the leaves in numerical order (not shown here). You may also choose to drop the decimal points.

To see the shape of the distribution, simply turn the page on its side. Notice the similarity in shape to that of Fig. 21–3.

Also note that the stem and leaf plot *preserves the original data.* Thus we could exactly reconstruct the data list of Example 10 from this plot, whereas such a reconstruction is not possible from the grouped data of Table 21–3.

Graphing Frequency Distributions on the Calculator

Screen for Example 11.

Most calculators can create histograms for statistical data. For example, the TI-83/84 will make a histogram using data entered into a list, as described earlier in this chapter. To draw a histogram like the one in Example 11, first use the STAT list editor to enter the grades from Example 10 into L1. Then press STAT PLOT and select Plot 1. On the second item, **Type**, scroll over to select the third type of graph, which is a histogram. Then move down to make sure that you have selected L1 for **Xlist**. In order to make sure that the histogram looks like the one that we have created by hand, we need to set the class limits to be the same. Press WINDOW, and then set **Xmin** to 56, **Xmax** to 98, and **Xscl** to 6. Then press GRAPH to produce the histogram. Note, we don't use ZOOM/9 to display, because this command automatically sets **Xmin**, **Xmax**, and **Xscl** to different values.

Exercise 2 ◆ Frequency Distributions

Frequency Distributions

1. The weights (in pounds) of 40 students at Tech College are

127	136	114	147	158	149	155	162
154	139	144	114	163	147	155	165
172	168	146	154	111	149	117	152
166	172	158	149	116	127	162	153
118	141	128	153	166	125	117	161

(a) Determine the range of the data.
(b) Make an absolute frequency distribution using class widths of 5 lb.
(c) Make a relative frequency distribution.

2. The times (in minutes) for 30 racers to complete a cross-country ski race are

61.4	72.5	88.2	71.2	48.5	48.9	54.8	71.4	99.2	74.5
84.6	73.6	69.3	49.6	59.3	71.4	89.4	92.4	48.4	66.3
85.7	59.3	74.9	59.3	72.7	49.4	83.8	50.3	72.8	69.3

(a) Determine the range of the data.
(b) Make an absolute frequency distribution using class widths of 5 min.
(c) Make a relative frequency distribution.

3. The prices (in dollars) of various computer printers in a distributer's catalog are

850	625	946	763	774	789	584	864	884	698
913	750	696	864	795	747	657	886	906	697
867	705	784	864	946	607	798	972	857	785

   (a) Determine the range of the data.
   (b) Make an absolute frequency distribution using class widths of $50.
   (c) Make a relative frequency distribution.

Frequency Histograms and Frequency Polygons

Draw a histogram, showing both absolute and relative frequency, for the data of

   4. problem 1.
   5. problem 2.
   6. problem 3.

Draw a frequency polygon, showing both absolute and relative frequency, for the data of

   7. problem 1.
   8. problem 2.
   9. problem 3.

Cumulative Frequency Distributions

Make a cumulative frequency distribution showing

   (a) absolute frequency and
   (b) relative frequency, for the data of
   10. problem 1.
   11. problem 2.
   12. problem 3.

Draw a cumulative frequency polygon, showing both absolute and relative frequency, for the data of

   13. problem 10.
   14. problem 11.
   15. problem 12.

Stem and Leaf Plots

Make a stem and leaf plot for the data of
   16. problem 1.
   17. problem 2.
   18. problem 3.

Graphics Calculator

19. Use your graphics calculator to produce histograms of the data in problems 1, 2, and 3.

Computer

20. Some spreadsheets, such as *Excel*, *Quattro Pro*, and *Lotus 123*, have commands for automatically making a frequency distribution, and they can also draw frequency histograms and polygons. If you have such a spreadsheet available, use it to solve any of the problems in this exercise set.

21–3   Numerical Description of Data

We saw in Sec. 21–2 how we can describe a set of data by frequency distribution, a frequency histogram, a frequency polygon, or a cumulative frequency distribution. We can also describe a set of data with just a few numbers, and this more compact description is more convenient for some purposes. For example, if, in a report, you wanted to describe the heights of a group of students and did not want to give the entire frequency distribution, you might simply say:

The mean height is 58 inches, with a standard deviation of 3.5 inches.

The *mean* is a number that shows the *center* of the data; the *standard deviation* is a measure of the *spread* of the data. As mentioned earlier, a number such as the mean or the standard deviation may be found either for an entire population (and are thus population parameters) or for a sample drawn from that population (and so are sample statistics).

Therefore to describe a population or a sample, we need numbers that give both the center of the data and the spread. Further, for sample statistics, we need to give the uncertainty of each figure. This will enable us to make inferences about the larger population.

◆◆◆ **Example 15:** A student recorded the running times for a sample of participants in a race. From that sample she inferred (by methods we'll learn later) that for the entire population of racers

$$\text{mean time} = 23.65 \pm 0.84 \text{ minutes}$$
$$\text{standard deviation} = 5.83 \pm 0.34 \text{ minutes} \qquad ◆◆◆$$

The mean time is called a *measure of central tendency*. We show how to calculate the mean, and other measures of central tendency, later in this section. The standard deviation is called a *measure of dispersion*. We also show how to compute it, and other measures of dispersion, in this section. The "plus-or-minus" values show a degree of uncertainty called the *standard error*. The uncertainty ± 0.84 min is called the *standard error of the mean*, and the uncertainty ± 0.34 min is called the *standard error of the standard deviation*. We show how to calculate standard errors in Sec. 21–6.

Measures of Central Tendency: The Mean

Some common measures of the center of a distribution are the mean, the median, and the mode. The arithmetic mean, or simply the *mean*, of a set of measurements is equal to the sum of the measurements divided by the number of measurements. It is what we commonly call the *average*. It is the most commonly used measure of central tendency. If our n measurements are $x_1, x_2, \ldots, x_n$, then

$$\Sigma x = x_1 + x_2 + \cdots + x_n$$

There is a story, probably untrue, about a statistician who drowned in a lake that had an average depth of 1 ft.

where we use the Greek symbol Σ (sigma) to represent "the sum of." The mean, which we call $\bar{x}$ (read "x bar"), is then given by the following:

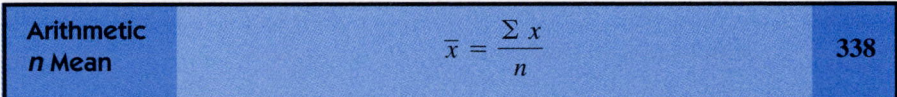

| Arithmetic *n* Mean | $$\bar{x} = \frac{\Sigma x}{n}$$ | 338 |

The arithmetic mean of n measurements is the sum of those measurements divided by n.

We can calculate the mean for a sample or for the entire population. However, we use a *different symbol* in each case.

$\bar{x}$ is the sample mean

μ (mu) is the population mean

◆◆◆ **Example 16:** Find the mean of the following sample:

746   574   645   894   736   695   635   794

Solution: Adding the values gives

$$\Sigma x = 746 + 574 + 645 + 894 + 736 + 695 + 635 + 794 = 5719$$

Then, with $n = 8$,

$$\bar{x} = \frac{5719}{8} = 715$$

rounded to three digits.   ◆◆◆

Weighted Mean

When not all the data are of equal importance, we may compute a *weighted mean,* where the values are weighted according to their importance.

◆◆◆ **Example 17:** A student has grades of 83, 59, and 94 on 3 one-hour exams, a grade of 82 on the final exam, which is equal in weight to 2 one-hour exams, and a grade of 78 on a laboratory report, which is worth 1.5 one-hour exams. Compute the weighted mean.

Solution: If a one-hour exam is assigned a weight of 1, then we have a total of the weights of

$$\Sigma w = 1 + 1 + 1 + 2 + 1.5 = 6.5$$

To get a weighted mean, we add the products of each grade and its weight, and divide by the total weight.

$$\text{weighted mean} = \frac{83(1) + 59(1) + 94(1) + 82(2) + 78(1.5)}{6.5} = 80$$ ◆◆◆

In general, the weighted mean is given by the following:

$$\text{weighted mean} = \frac{\Sigma\,(wx)}{\Sigma\,w}$$

Midrange

We have already noted that the range of a set of data is the difference between the highest and the lowest numbers in the set. The *midrange* is simply the value midway between the two extreme values.

$$\text{midrange} = \frac{\text{highest value} + \text{lowest value}}{2}$$

◆◆◆ **Example 18:** The midrange for the values

$$3, 5, 6, 6, 7, 11, 11, 15$$

is

$$\text{midrange} = \frac{3 + 15}{2} = 9$$

◆◆◆

Mode

Our next measure of central tendency is the *mode.*

Mode	The mode of a set of numbers is the value(s) that occurs most often in the set.	**340**

A set of numbers may have no mode, one mode, or more than one mode.

◆◆◆ **Example 19:**

(a) The set   1   3   3   5   6   7   7   7   9   has the mode 7.
(b) The set   1   1   3   3   5   5   7   7   has no mode.
(c) The set   1   3   3   3   5   6   7   7   7   9   has two modes, 3 and 7. It is called *bimodal.*

◆◆◆

Median

To find the *median,* we simply arrange the data in order of magnitude and pick the *middle value.* For an even number of measurements, we take the mean of the two middle values.

Median	The median of a set of numbers arranged in order of magnitude is the middle value of an odd number of measurements, or the mean of the two middle values of an even number of measurements.	**339**

◆◆◆ **Example 20:** Find the median of the data in Example 16.

Solution: We rewrite the data in order of magnitude.

$$574 \quad 635 \quad 645 \quad 695 \quad 736 \quad 746 \quad 794 \quad 894$$

The two middle values are 695 and 736. Taking their mean gives

$$\text{median} = \frac{695 + 736}{2} = 716$$

after rounding to three significant digits. ◆◆◆

Five-Number Summary

The median of that half of a set of data from the minimum value up to and including the median is called the *lower hinge*. Similarly, the median of the upper half of the data is called the *upper hinge*.

The lowest value in a set of data, together with the highest value, the median, and the two hinges, is called a *five-number summary* of the data.

◆◆◆ **Example 21:** For the data

$$12 \quad 17 \quad 18 \quad 20 \quad 22 \quad 28 \quad 32 \quad 34 \quad 49 \quad 52 \quad 59 \quad 66$$

the median is $(28 + 32)/2 = 30$, so

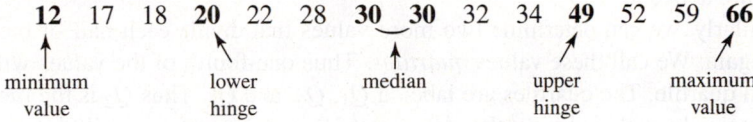

The five-number summary for this set of data may be written

$$[12, 20, 30, 49, 66] \quad\quad ◆◆◆$$

Boxplots

A graph of the values in the five-number summary is one example of a *boxplot*.

◆◆◆ **Example 22:** A boxplot for the data of Example 21 is given in Fig. 21–6. ◆◆◆

Min	Hinge	Median		Hinge	Max
12	20	30		49	66

FIGURE 21–6 A boxplot. It is also called a *box and whisker* diagram.

Measures of Dispersion

We will usually use the mean to describe a set of numbers, but used alone it can be misleading.

◆◆◆ **Example 23:** The set of numbers

$$1 \quad 1 \quad 1 \quad 1 \quad 1 \quad 1 \quad 1 \quad 1 \quad 1 \quad 91$$

and the set

$$10 \quad 10 \quad 10 \quad 10 \quad 10 \quad 10 \quad 10 \quad 10 \quad 10 \quad 10$$

each have a mean value of 10 but are otherwise quite different. Each set has a sum of 100. In the first set most of this sum is concentrated in a single value, but in the second set the sum is *dispersed* among all the values. ◆◆◆

Thus we need some *measure of dispersion* of a set of numbers. Four common ones are the range, the percentile range, the variance, and the standard deviation. We will cover each of these.

Range

We have already introduced the range in Sec. 21–2, and we define it here.

Range	The range of a set of numbers is the difference between the largest and smallest numbers in the set.	341

◆◆◆ **Example 24:** For the set of numbers

$$6 \quad 3 \quad 9 \quad 12 \quad 44 \quad 2 \quad 53 \quad 1 \quad 8$$

the largest value is 53 and the smallest is 1, so the range is

$$\text{range} = 53 - 1 = 52 \qquad \text{◆◆◆}$$

We sometimes give the range by stating the end values themselves. Thus, in Example 24 we might say that the range is from 1 to 53.

Quartiles, Deciles, and Percentiles

We have seen that for a set of data arranged in order of magnitude, the median divides the data into two equal parts. There are as many numbers below the median as there are above it.

Similarly, we can determine two more values that divide each half of the data in half again. We call these values *quartiles.* Thus one-fourth of the values will fall into each quartile. The quartiles are labeled Q_1, Q_2, and Q_3. Thus Q_2 is the median.

Those values that divide the data into 10 equal parts we call *deciles* and label D_1, D_2, ... Those values that divide the data into 100 equal parts are *percentiles,* labelled P_1, P_2, ... Thus the 25th percentile is the same as the first quartile ($P_{25} = Q_1$). The median is the 50th percentile, the fifth decile, and the second quartile ($P_{50} = D_5 = Q_2$). The 70th percentile is the seventh decile ($P_{70} = D_7$).

One measure of dispersion sometimes used is to give the range of values occupied by some given percentiles. Thus we have a *quartile range, decile range,* and *percentile range.* For example, the quartile range is the range from the first to the third quartile.

◆◆◆ **Example 25:** Find the quartile range for the set of data

$$2 \quad 4 \quad 5 \qquad 7 \quad 8 \quad 11 \qquad 15 \quad 18 \quad 19 \qquad 21 \quad 24 \quad 25$$

Solution: The quartiles are

$$Q_1 = \frac{5 + 7}{2} = 6$$

$$Q_2 = \text{the median} = \frac{11 + 15}{2} = 13$$

$$Q_3 = \frac{19 + 21}{2} = 20$$

Then

$$\text{quartile range} = Q_3 - Q_1 = 20 - 6 = 14 \qquad \text{◆◆◆}$$

We see that half the values in Example 25 fall within the quartile range. We may similarly compute the range of any percentiles or deciles. The range from the 10th to the 90th percentile is the one commonly used.

Note that the quartiles are not in the same locations as the upper and lower hinges. The hinges here would be at 7 and 19.

Variance

We now give another measure of dispersion called the *variance,* but we must first mention deviation. We define the *deviation* of any number x in a set of data as the difference between that number and the mean $\bar{x}$ of that set of data. A negative deviation results when the data value is less than the mean. A positive deviation results when the data value is greater than the mean.

◆◆◆ **Example 26:** A certain set of measurements has a mean of 48.3. What are the deviations of the values 24.2 and 69.3 in that set?

Solution: The deviation of 24.2 is

$$24.2 - 48.3 = -24.1$$

and the deviation of 69.3 is

$$69.3 - 48.3 = 21.0 \qquad \text{◆◆◆}$$

To get the *variance* of a population of n numbers, we add up the squares of the deviations of each number in the set and divide by n.

Population Variance	$\sigma^2 = \dfrac{\Sigma(x - \bar{x})^2}{n}$	342

To find the variance of a *sample,* it is more accurate to divide by $n - 1$ rather than n. As with the mean, we use one symbol for the population variance and a different symbol for the sample variance.

$$\sigma^2 \text{ is the population variance}$$
$$s^2 \text{ is the sample variance}$$

Sample Variance	$s^2 = \dfrac{\Sigma(x - \bar{x})^2}{n - 1}$	343

For large samples (over 30 or so), the variance found by either formula is practically the same.

◆◆◆ **Example 27:** Compute the variance for the population

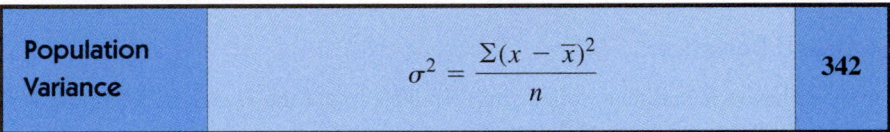

$$1.74 \quad 2.47 \quad 3.66 \quad 4.73 \quad 5.14 \quad 6.23 \quad 7.29 \quad 8.93 \quad 9.56$$

Solution: We first compute the mean, $\bar{x}$.

$$\bar{x} = \frac{1.74 + 2.47 + 3.66 + 4.73 + 5.14 + 6.23 + 7.29 + 8.93 + 9.56}{9}$$

$$= 5.53$$

We then subtract the mean from each of the nine values to obtain deviations. The deviations are then squared and added, as shown in Table 21–6.

TABLE 21–6

Measurement x	Deviation $x - \bar{x}$	Deviation Squared $(x - \bar{x})^2$
1.74	−3.79	14.36
2.47	−3.06	9.36
3.66	−1.87	3.50
4.73	−0.80	0.64
5.14	−0.39	0.15
6.23	0.70	0.49
7.29	1.76	3.10
8.93	3.40	11.56
9.56	4.03	16.24
$\Sigma x = 49.75$	$\Sigma(x - \bar{x}) = -0.02$	$\Sigma(x - \bar{x})^2 = 59.41$

The variance σ^2 is then

$$\sigma^2 = \frac{59.41}{9} = 6.60$$

◆◆◆

Standard Deviation

Once we have the variance, it is a simple matter to get the *standard deviation*. It is the most common measure of dispersion.

Standard Deviation	The standard deviation of a set of numbers is the positive square root of the variance.	344

As before, we use s for the sample standard deviation and σ for the population standard deviation.

◆◆◆ **Example 28:** Find the standard deviation for the data of Example 27.

Solution: We have already found the variance in Example 27.

$$\sigma^2 = 6.60$$

Taking the square root gives the standard deviation.

$$\text{standard deviation} = \sqrt{6.60} = 2.57$$

◆◆◆

To get an intuitive feel for the standard deviation, we have computed it in Fig. 21–7 for several data sets consisting of 12 numbers which can range from 1 to 12. In the first set, all the numbers have the same value, 6, and in the last set every number is different. The data sets in between have differing amounts of repetition. To the right of each data set are a relative frequency histogram and the population standard deviation (computation not shown).

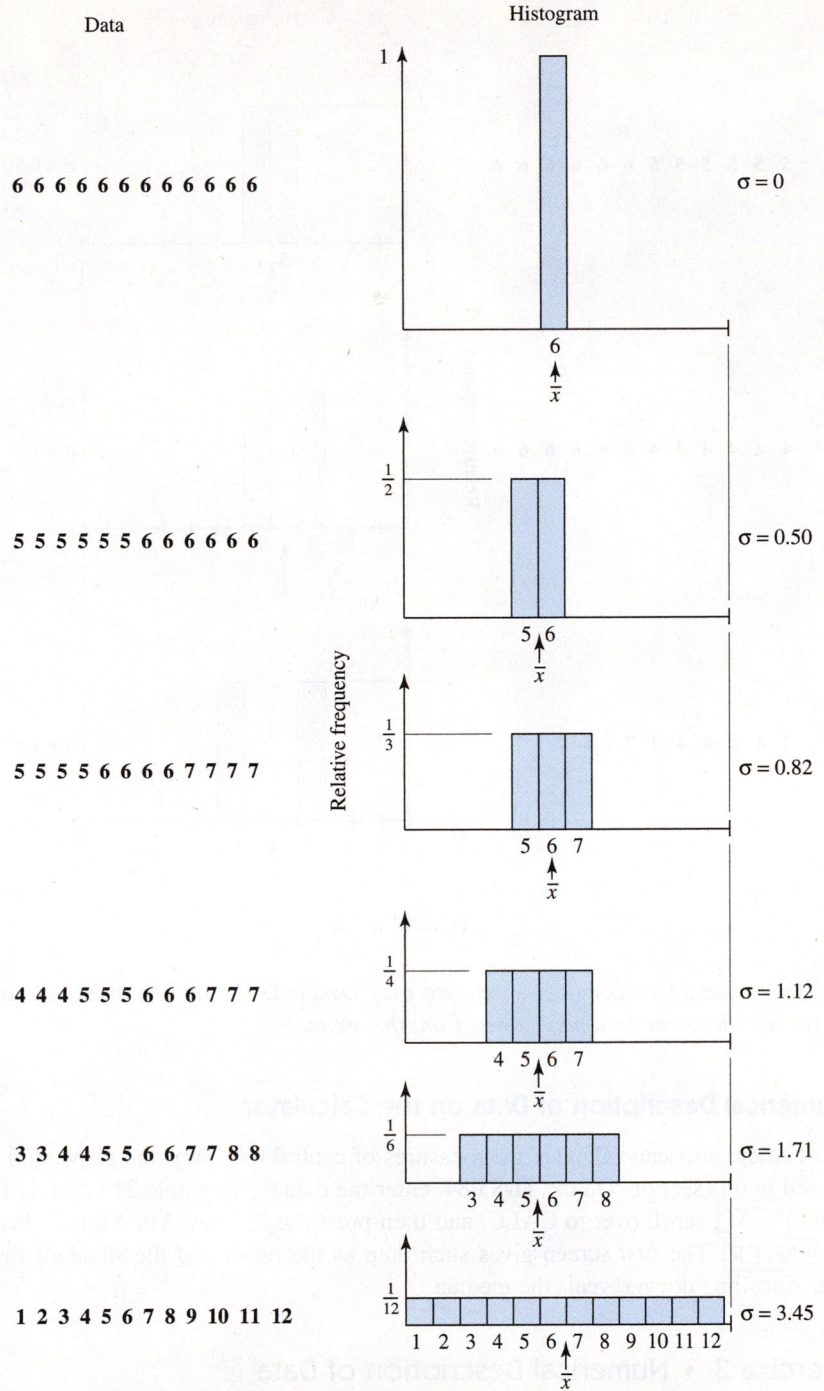

Data

Histogram

6 6 6 6 6 6 6 6 6 6 6 6 $\sigma = 0$

5 5 5 5 5 5 6 6 6 6 6 6 $\sigma = 0.50$

5 5 5 5 6 6 6 6 7 7 7 7 $\sigma = 0.82$

4 4 4 5 5 5 6 6 6 7 7 7 $\sigma = 1.12$

3 3 4 4 5 5 6 6 7 7 8 8 $\sigma = 1.71$

1 2 3 4 5 6 7 8 9 10 11 12 $\sigma = 3.45$

FIGURE 21–7

Note that the most compact distribution has the lowest standard deviation, and as the distribution spreads, the standard deviation increases.

For a final demonstration, let us again take 12 numbers in two groups of six equal values, as shown in Fig. 21–8. Now let us separate the two groups, first by one interval and then by two intervals. Again notice that the standard deviation increases as the data moves further from the mean.

The mean shifts slightly from one distribution to the next, but this does not affect our conclusions.

Data   Histogram

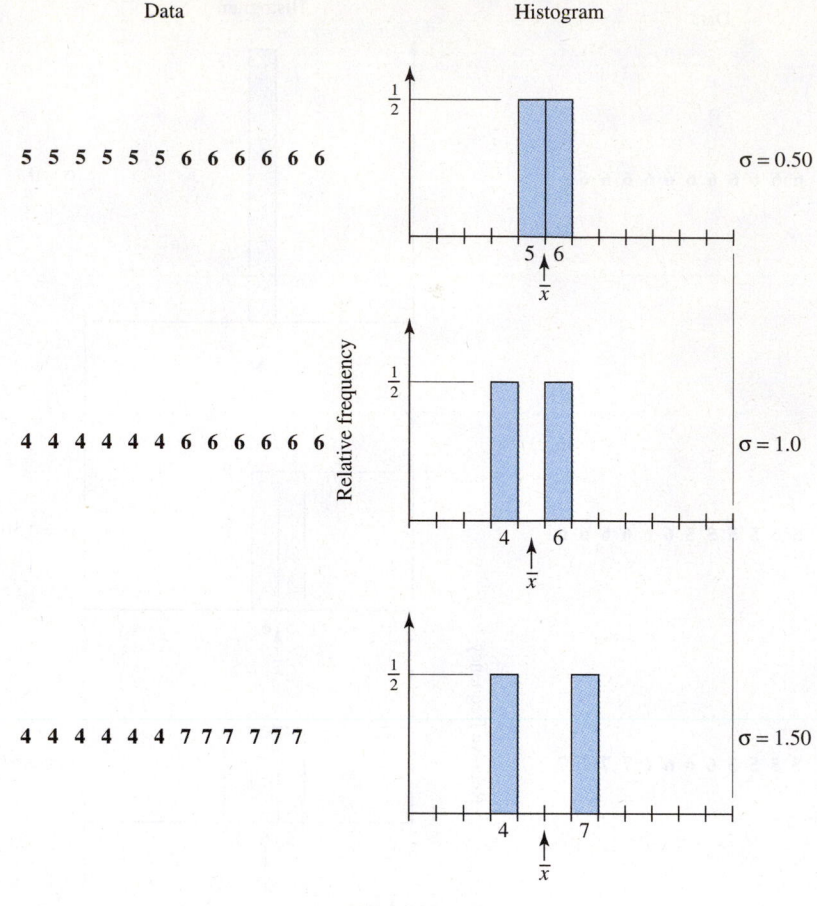

5 5 5 5 5 5 6 6 6 6 6 6   σ = 0.50

4 4 4 4 4 4 6 6 6 6 6 6   σ = 1.0

4 4 4 4 4 4 7 7 7 7 7 7   σ = 1.50

FIGURE 21–8

From these two demonstrations we may conclude that *the standard deviation increases whenever data move away from the mean.*

Numerical Description of Data on the Calculator

Most calculators can calculate the measures of central tendency and dispersion discussed in this section. On the TI-83/84, enter the data for Example 21 into L1. Then press STAT , scroll over to CALC, and then press the "1" key, **Var Stats**, followed by ENTER . The first screen gives such data as the mean and the standard deviation. Scrolling down reveals the median.

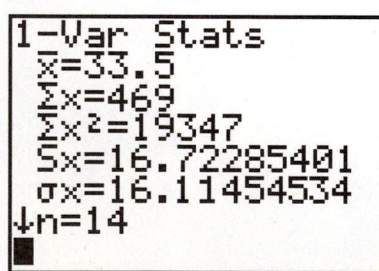

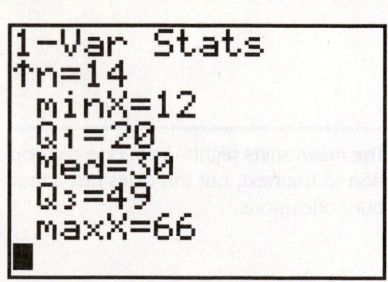

TI-83/84 Screens for Example 21.

Exercise 3 ◆ Numerical Description of Data

Mean

1. Find the mean of the following set of grades:

85   74   69   59   60   96   84   48   89   76   96   68   98   79   76

2. Find the mean of the following set of weights:

173   127   142   164   163   153   116   199

3. Find the mean of the weights in problem 1 of Exercise 2.

4. Find the mean of the times in problem 2 of Exercise 2.

5. Find the mean of the prices in problem 3 of Exercise 2.

Weighted Mean

6. A student's grades and the weight of each grade are given in the following table. Find their weighted mean.

	Grade	Weight
Hour exam	83	5
Hour exam	74	5
Quiz	93	1
Final exam	79	10
Report	88	7

7. A student receives hour-test grades of 86, 92, 68, and 75, a final exam grade of 82, and a project grade of 88. Find the weighted mean if each hour-test counts for 15% of his grade, the final exam counts for 30%, and the project counts for 10%.

Midrange

8. Find the midrange of the grades in problem 1.
9. Find the midrange of the weights in problem 2.

Mode

10. Find the mode of the grades in problem 1.
11. Find the mode of the weights in problem 2.
12. Find the mode of the weights in problem 1 of Exercise 2.
13. Find the mode of the times in problem 2 of Exercise 2.
14. Find the mode of the prices in problem 3 of Exercise 2.

Median

15. Find the median of the grades in problem 1.
16. Find the median of the weights in problem 2.
17. Find the median of the weights in problem 1 of Exercise 2.
18. Find the median of the times in problem 2 of Exercise 2.
19. Find the median of the prices in problem 3 of Exercise 2.

Five-Number Summary

20. Give the five-number summary for the grades in problem 1.
21. Give the five-number summary for the weights in problem 2.

Boxplot

22. Make a boxplot using the results of problem 20.
23. Make a boxplot using the results of problem 21.

Range

24. Find the range of the grades in problem 1.
25. Find the range of the weights in problem 2.

Percentiles

26. Find the quartiles and give the quartile range of the following data:

   28   39   46   53   69   71   83   94   102   117   126

27. Find the quartiles and give the quartile range of the following data:

   1.33   2.28   3.59   4.96   5.23   6.89

   7.91   8.13   9.44   10.6   11.2   12.3

Variance and Standard Deviation

28. Find the variance and standard deviation of the grades in problem 1. Assume that these grades are a sample drawn from a larger population.
29. Find the variance and standard deviation of the weights in problem 2. Assume that these weights are a sample drawn from a larger population.
30. Find the population variance and standard deviation of the weights in problem 1 of Exercise 2.
31. Find the population variance and standard deviation of the times in problem 2 of Exercise 2.
32. Find the population variance and standard deviation of the prices in problem 3 of Exercise 2.

Calculator

33. Use your calculator to find the mean, population variance, and standard deviation for the data in problems 1 and 2.

Computer

34. Most spreadsheet programs can find the mean, the sample or population variance, and standard deviation, counts, and the range. If you have such a program, use it to solve any of the problems in this exercise set.
35. A statistics program for a computer can, of course, do any of the above calculations, and creat boxplots as well. Use such a program to solve any of the problems in this exercise set.
36. Many computer algebra systems can compute the mean, standard deviation, variance, and other statistical measures. Check your manual for the proper instructions, and use CAS to solve any of the problems in this exercise set.

21–4   Introduction to Probability

Why do we need probability to learn statistics? In Sec. 21–3 we learned how to compute certain sample statistics, such as the mean and the standard deviation. Knowing, for example, that the mean height $\bar{x}$ of a sample of students is 69.5 in., we might *infer* that the mean height μ of the entire population is 69.5 in. But how reliable is that number? Is μ equal to 69.5 in. *exactly?* We'll see later that we give population parameters such as μ as a *range* of values, say, 69.5 ± 0.6 in., and we state the *probability* that the true mean lies within that range. We might say, for example, that there is a 68% chance that the true value lies within the stated range.

Further, we may report that the sample standard deviation is, say, 2.6 in. What does that mean? Of 1000 students, for example, how many can be expected to have a height falling within, say, 1 standard deviation of the mean height at that college? We use probability to help us answer questions such as that.

In addition, statistics are often used to "prove" many things, and it takes a knowledge of probability to help decide whether the claims are valid or are merely a result of chance.

◆◆◆ **Example 29:** Suppose that a nurse measures the heights of 14-year-old students in a town located near a chemical plant. Of 100 students measured, she finds 75 students to be shorter than 59 in. Parents then claim that this is caused by chemicals in the air, but the company claims that such heights can occur by chance. We cannot resolve this question without a knowledge of probability. ◆◆◆

We distinguish between *statistical* probability and *mathematical* probability.

◆◆◆ **Example 30:** We can toss a die 6000 times and count the number of times that "one" comes up, say, 1006 times. We conclude that the chance of tossing a "one" is 1006 out of 6000, or about 1 in 6. This is an example of what is called *statistical* probability.

On the other hand, we may reason, without even tossing the die, that each face has an equal chance of coming up. Since there are six faces, the chance of tossing a "one" is 1 in 6. This is an example of *mathematical* probability. ◆◆◆

We first give a brief introduction to mathematical probability and then apply it to statistics.

Probability of a Single Event Occurring

Suppose that an event A can happen in m ways out of a total of n equally likely ways. The probability $P(A)$ of A occurring is defined as follows:

Probability of a Single Event Occurring	$P(A) = \dfrac{m}{n}$	345

Note that the probability $P(A)$ is a number between 0 and 1. $P(A) = 0$ means that there is no chance that A will occur. $P(A) = 1$ means that it is certain that A will occur. Also note that if the probability of A occurring is $P(A)$, the probability of A *not* occurring is $1 - P(A)$.

◆◆◆ **Example 31:** A bag contains 15 balls, 6 of which are red. Assuming that any ball is as likely to be drawn as any other, the probability of drawing a red ball from the bag is $\frac{6}{15}$ or $\frac{2}{5}$. The probability of drawing a ball that is not red is $\frac{9}{15}$ or $\frac{3}{5}$.

◆◆◆

◆◆◆ **Example 32:** A nickel and a dime are each tossed once. Determine (a) the probability $P(A)$ that both coins will show heads and (b) the probability $P(B)$ that not both coins will show heads.

Solution: The possible outcomes of a single toss of each coin are as follows:

Nickel	Dime
Heads	Heads
Heads	Tails
Tails	Heads
Tails	Tails

Thus there are four equally likely outcomes. For event A (both heads), there is only one favorable outcome (heads, heads). Thus

$$P(A) = \frac{1}{4} = 0.25$$

Event B (that not both coins show heads) can happen in three ways (HT, TH, and TT). Thus $P(B) = \frac{3}{4} = 0.75$. Or, alternatively,

$$P(B) = 1 - P(A) = 0.75 \qquad \blacklozenge\blacklozenge\blacklozenge$$

Probability of Two Events Both Occurring

If $P(A)$ is the probability that A will occur, and $P(B)$ is the probability that B will occur, the probability $P(A, B)$ that *both* A and B will occur is as follows:

Probability of Two Events Both Occurring	$P(A, B) = P(A)P(B)$	346

Independent events are those in which the outcome of the first in no way affects the outcome of the second.

> The probability that two independent events will both occur is the product of their individual probabilities.

The rule above can be extended as follows:

Probability of Several Events All Occurring	$P(A, B, C, \dots) = P(A)P(B)P(C)\dots$	347

> The probability that several independent events, $A, B, C, \dots$, will all occur is the product of the probabilities for each.

◆◆◆ **Example 33:** What is the probability that in three tosses of a coin, all will be heads?

Solution: Each toss is independent of the others, and the probability of getting a head on a single toss is $\frac{1}{2}$. Thus the probability of three heads being tossed is

$$\left(\frac{1}{2}\right)\left(\frac{1}{2}\right)\left(\frac{1}{2}\right) = \frac{1}{8}$$

or 1 in 8. If we were to make, say, 800 sets of three tosses, we would expect to toss three heads about 100 times. ◆◆◆

◆◆◆ **Example 34:** Given that the probability of a defective microchip is 0.1, what is the probability that a sample of three chips will have three defectives?

$$(0.1)^3 = 0.001$$

Solution: So there is a 0.1% chance that all three chips in a three-chip sample will be defective. ◆◆◆

Probability of Either of Two Events Occurring

If $P(A)$ and $P(B)$ are the probabilities that events A and B will occur, the probability $P(A + B)$ that *either A or B* or both will occur is as follows:

Probability of Either or Both of Two Events Occurring	$P(A + B) = P(A) + P(B) - P(A, B)$	348

> The probability that either A or B or both will occur is the sum of the individual probabilities for A and B, less the probability that both will occur.

Why subtract $P(A, B)$ in this formula? Since $P(A)$ includes all occurrences of A, it must include $P(A, B)$. Since $P(B)$ includes all occurrences of B, it also includes $P(A, B)$. Thus if we add $P(A)$ and $P(B)$ we will mistakenly be counting $P(A, B)$ twice.

For example, if we count all aces and all spades in a deck of cards, we get 16 cards (13 spades plus 3 additional aces). But if we had added the number of aces (4) to the number of spades (13), we would have gotten the incorrect answer of 17 by counting the ace of spades twice.

◆◆◆ **Example 35:** In a certain cup production line, 6% are chipped, 8% are cracked, and 2% are both chipped and cracked. What is the probability that a cup picked at random will be either chipped or cracked, or both chipped and cracked?

Solution: If

$$P(C) = \text{probability of being chipped} = 0.06$$

and

$$P(S) = \text{probability of being cracked} = 0.08$$

and

$$P(C, S) = \text{probability of being chipped and cracked} = 0.02$$

then

$$P(C + S) = 0.06 + 0.08 - 0.02 = 0.12$$

We would thus predict that 12 cups out of 100 chosen would be chipped or cracked, or both chipped and cracked. ◆◆◆

Probability of Two Mutually Exclusive Events Occurring

Mutually exclusive events A and B are those that cannot both occur at the same time. Thus the probability $P(A, B)$ of both A and B occurring is zero.

◆◆◆ **Example 36:** A tossed coin lands either heads or tails. Thus the probability $P(H, T)$ of a single toss being both heads and tails is

$$P(H, T) = 0 \qquad ◆◆◆$$

Setting $P(A, B) = 0$ in Eq. 348 gives the following equation:

Probability of Two Mutually Exclusive Events Occurring	$P(A + B) = P(A) + P(B)$	349

The probability that either of two mutually exclusive events will occur in a certain trial is the sum of the two individual probabilities.

We can generalize this to n mutually exclusive events. The probability that one of n mutually exclusive events will occur in a certain trial is the sum of the n individual probabilities.

◆◆◆ **Example 37:** In a certain school, 10% of the students have red hair, and 20% have black hair. If we select one student at random, what is the probability that this student will have either red hair or black hair?

Solution: The probability $P(A)$ of having red hair is 0.1, and the probability $P(B)$ of having black hair is 0.2. The two events (having red hair or having black hair) are mutually exclusive. Thus the probability $P(A + B)$ of a student having either red or black hair is

$$P(A + B) = 0.1 + 0.2 = 0.3$$ ◆◆◆

Random Variables

In Sec. 21–1 we defined continuous, discrete, and categorical variables. Now we define a *random* variable.

Suppose that there is a process or an experiment whose outcome is a real number X. If the process is repeated, then X can assume specific values, each determined by chance and each with a certain probability of occurrence. X is then called a *random variable*.

◆◆◆ **Example 38:** The process of rolling a pair of dice once results in a single numerical value X, whose value is determined by chance but whose value has a definite probability of occurrence. Repeated rollings of the dice may result in other values of X, which can take on the values

$$2, 3, 4, 5, 6, 7, 8, 9, 10, 11, 12$$

Here X is a random variable.   ◆◆◆

Discrete and Continuous Random Variables

A *discrete* random variable can take on a finite number of variables. Thus the value X obtained by rolling a pair of dice is a discrete random variable.

A *continuous* random variable may have infinitely many values on a continuous scale, with no gaps.

◆◆◆ **Example 39:** An experiment measuring the temperatures T of the major U.S. cities at any given hour gives values that are determined by chance, depending on the whims of nature, but each value has a definite probability of occurrence. The random variable T is continuous because temperatures are read on a continuous scale.   ◆◆◆

Probability Distributions

If a process or an experiment has X different outcomes, where X is a discrete random variable, we say that the probability of a particular outcome is $P(X)$. If we calculate $P(X)$ for each X, we have what is called a *probability distribution*. Such a distribution may be presented as a table, a graph, or a formula. A probability distribution is sometimes called a *probability density function*, a *relative frequency function*, or a *frequency function*.

◆◆◆ Example 40: Make a probability distribution for a single roll of a single die.

Solution: Let X equal the number on the face that turns up, and $P(X)$ be the probability of that face turning up. That is, $P(5)$ is the probability of rolling a five. Since each number has an equal chance of being rolled, we have the following:

X	$P(X)$
1	$\frac{1}{6}$
2	$\frac{1}{6}$
3	$\frac{1}{6}$
4	$\frac{1}{6}$
5	$\frac{1}{6}$
6	$\frac{1}{6}$

We graph this probability distribution in Fig. 21–9.

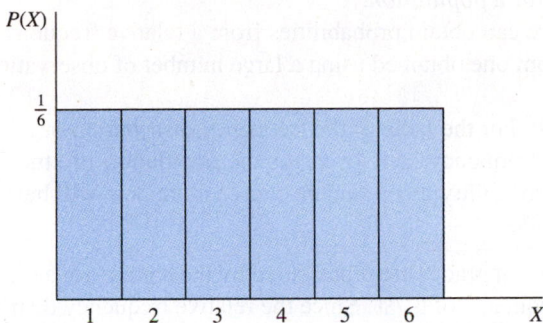

FIGURE 21–9 Probability distribution for the toss of a single die.

◆◆◆

◆◆◆ Example 41: Make a probability distribution for the two-coin toss experiment of Example 32.

Solution: Let X be the number of heads tossed, and $P(X)$ be the probability of X heads being tossed. The experiment has four possible outcomes (HH, HT, TH, TT). The outcome $X = 0$ (no heads) can occur in only one way (TT), so the probability $P(0)$ of not tossing a head is one in four, or $\frac{1}{4}$. The probabilities of the other outcomes are found in a similar way. They are as follows:

X	$P(X)$
0	$\frac{1}{4}$
1	$\frac{1}{2}$
2	$\frac{1}{4}$

They are shown graphically in Fig. 21–10.

◆◆◆

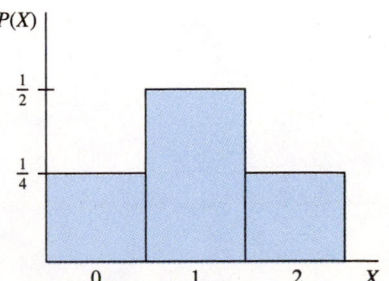

FIGURE 21–10 Probability distribution.

Probabilities as Areas on a Probability Distribution

The probability of an event occurring can be associated with a geometric *area* of the probability distribution or relative frequency distribution.

◆◆◆ **Example 42:** For the distribution in Fig. 21–10, each bar has a width of 1 unit, so the total shaded area is

$$1\left(\frac{1}{4}\right) + 1\left(\frac{1}{2}\right) + 1\left(\frac{1}{4}\right) = 1 \text{ square unit}$$

The bar representing the probability of tossing a "zero" has an area of $\frac{1}{4}$, which is equal to the probability of that event occurring. The areas of the other two bars also give the value of their probabilities.   ◆◆◆

Probabilities from a Relative Frequency Histogram

Suppose we were to repeat the two-coin toss experiment of Example 41 a very large number of times and draw a relative frequency histogram of our results. That relative frequency histogram *would be identical to the probability distribution of Fig. 21–10.* The relative frequency of tossing a "zero" would be $\frac{1}{4}$, of a "one" would be $\frac{1}{2}$, and of a "two" would be $\frac{1}{4}$.

We thus may think of a probability distribution as being the limiting value of a relative frequency distribution as the number of observations increases—in fact, the distribution for a *population*.

Therefore we can obtain probabilities from a relative frequency histogram for a population or from one obtained using a large number of observations.

◆◆◆ **Example 43:** For the table of the frequency distribution of grades (Table 21–3) and the relative frequency histogram for the population of students' grades (Fig. 21–3), find the probability that a student chosen at random will have received a grade between 92 and 98.

Solution: The given grades are represented by the bar having a class midpoint of 95 and a relative frequency of 6.7%. Since the relative frequency distribution for a population probability is no different from the probability distribution, the grade of a student chosen at random has a 6.7% chance of falling within that class.   ◆◆◆

Probability of a Variable Falling between Two Limits

Further, the probability of a variable falling *between two limits* is given by the area of the probability distribution or relative frequency histogram for that variable, between those limits.

◆◆◆ **Example 44:** Using the relative frequency histogram shown in Fig. 21–3, find the chance of a student chosen at random having a grade that falls between 62 and 80.

Solution: This range spans three classes in Fig. 21–3, having relative frequencies of 6.7%, 13.3%, and 23.3%. The chance of falling within one of the three is

$$6.7 + 13.3 + 23.3 = 43.3\%$$

So the probability is 0.433.   ◆◆◆

Thus we have shown that *the area between two limits on a probability distribution or relative frequency histogram for a population is a measure of the probability of a measurement falling within those limits.* This idea will be used again when we compute probabilities from a *continuous* probability distribution such as the normal curve.

Binomial Experiments

A *binomial experiment* is one that can have only two possible outcomes.

◆◆◆ **Example 45:** The following are examples of binomial experiments:

(a) A tossed coin shows either heads or tails.
(b) A manufactured part is either accepted or rejected.
(c) A voter either chooses Smith for Congress or does not choose Smith for Congress. ◆◆◆

Binomial Probability Formula

We call the occurrence of an event a *success* and the nonoccurrence of that event a *failure*. Let us denote the probability of success of a single trial of a binomial experiment by p. Then the probability of failure, q, must be $(1 - p)$.

◆◆◆ **Example 46:** If the probability is 80% ($p = 0.8$) that a single part drawn at random from a production line is accepted, then the probability is 20% ($q = 1 - 0.8$) that a part is rejected. ◆◆◆

If we repeat a binomial experiment n times, we will have x successes, where

$$x = 0, 1, 2, 3, \ldots, n$$

The probability $P(x)$ of having x successes in n trials is given by the following:

Binomial Probability Formula	$P(x) = \dfrac{n!}{(n - x)! \, x!} p^x q^{n-x}$	351

There are tables given in statistics books that can be used instead of this formula. Does this formula look a little familiar? We will soon connect the binomial distribution with what we have already learned about the binomial theorem.

◆◆◆ **Example 47:** For the production line of Example 46, find the probability that of 10 parts taken at random from the line, exactly 6 will be acceptable.

Solution: Here, $n = 10$, $x = 6$, and $p = 0.8$, $q = 0.2$, from before. Substituting into the formula gives

$$P(6) = \frac{10!}{(10 - 6)! \, 6!} (0.8)^6 (0.2)^{10-6}$$

$$= \frac{10!}{4! \, 6!} (0.8)^6 (0.2)^4 = 0.0881$$

or about 9%. ◆◆◆

The Binomial Probability Formula on the Graphing Calculator

It is quite easy to calculate the binomal probability formula on the TI-83/84 and the TI-89. Simply press $\boxed{\text{2ND}}$ and then $\boxed{\text{VARS}}$. Scroll down to binompdf(and hit $\boxed{\text{ENTER}}$. Enter the values for n, p and x, separated by commas, and then close the parentheses and hit $\boxed{\text{ENTER}}$.

Binomial Distribution

In Example 47, we computed a single probability, that 6 out of 10 parts drawn at random will be acceptable. We can also compute the probabilities of having 0, 1, 2, . . . , 10 acceptable parts in a group of 10. If we compute all 10 possible probabilities, we will then have a *probability distribution*.

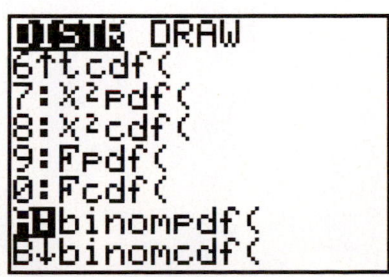

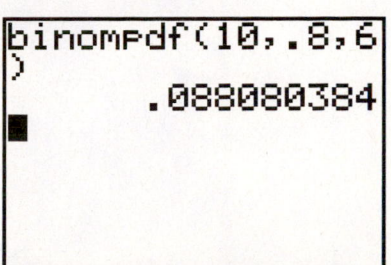

Screens for Example 46.

◆◆◆ **Example 48:** For the production line of Example 46, find the probability that of 10 parts taken at random from the line, 0, 1, 2, . . . , 10 will be acceptable.

Solution: Computing the probabilities in the same way as in the preceding example (work not shown), we get the following binomial distribution:

x	$P(x)$
0	0
1	0
2	0.0001
3	0.0008
4	0.0055
5	0.0264
6	0.0881
7	0.2013
8	0.3020
9	0.2684
10	0.1074
Total	1.0000

This binomial distribution is shown graphically in Fig. 21–11.

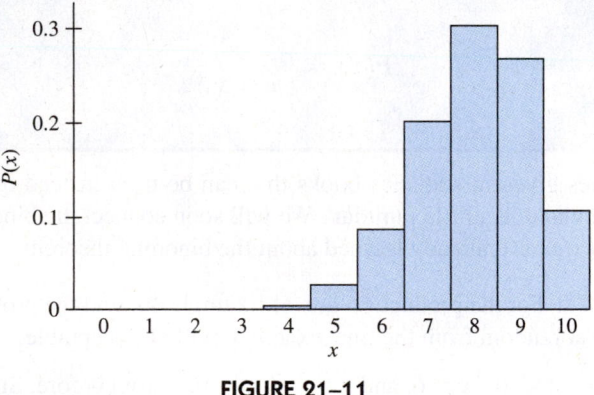

FIGURE 21–11   ◆◆◆

Binomial Distribution Obtained by Binomial Theorem

The n values in a binomial distribution for n trials may be obtained by using the binomial theorem to expand the binomial

$$(q + p)^n$$

where p is the probability of success of a single trial and $q = (1 - p)$ is the probability of failure of a single trial, as before.

◆◆◆ **Example 49:** Use the binomial theorem, Eq. 199, to obtain the binomial distribution of Example 48.

Solution: We expand $(q + p)^n$ using $q = 0.2, p = 0.8$, and $n = 10$.

$$(0.2 + 0.8)^{10}$$

$$= (0.2)^{10}(0.8)^0 + 10(0.2)^9(0.8)^1 + \frac{10(9)}{2!}(0.2)^8(0.8)^2$$

$$+ \frac{10(9)(8)}{3!}(0.2)^7(0.8)^3 + \frac{10(9)(8)(7)}{4!}(0.2)^6(0.8)^4$$

$$+ \frac{10(9)(8)(7)(6)}{5!}(0.2)^5(0.8)^5 + \frac{10(9)(8)(7)(6)(5)}{6!}(0.2)^4(0.8)^6$$

$$+ \frac{10(9)(8)(7)(6)(5)(4)}{7!}(0.2)^3(0.8)^7 + \frac{10(9)(8)(7)(6)(5)(4)(3)}{8!}(0.2)^2(0.8)^8$$

$$+ \frac{10(9)(8)(7)(6)(5)(4)(3)(2)}{9!}(0.2)^1(0.8)^9 + \frac{10!}{10!}(0.2)^0(0.8)^{10}$$

$$= 0.0000 + 0.0000 + 0.0001 + 0.0008 + 0.0055 + 0.0264 + 0.0881$$

$$+ 0.2013 + 0.3020 + 0.2684 + 0.1074$$

As expected, we get the same values as given by binomial probability formula. ◆◆◆

Exercise 4 ◆ Introduction to Probability

Probability of a Single Event

1. We draw a ball from a bag that contains 8 green balls and 7 blue balls. What is the probability that a ball drawn at random will be green?

2. A card is drawn from a deck containing 13 hearts, 13 diamonds, 13 clubs, and 13 spades. What is the chance that a card drawn at random will be a heart?

3. If we toss four coins, what is the probability of getting two heads and two tails? [*Hint:* This experiment has 16 possible outcomes (HHHH, HHHT, . . . , TTTT).]

4. If we toss four coins, what is the probability of getting one head and three tails?

5. If we throw two dice, what is the probability that their sum is 9? [*Hint:* List all possible outcomes, $(1, 1), (1, 2)$, and so on, and count those that have a sum of 9.]

6. If two dice are thrown, what is the probability that their sum is 7?

Probability That Several Events Will All Occur

7. A die is rolled twice. What is the probability that both rolls will give a six?

8. We draw four cards from a deck, replacing each before the next is drawn. What chance is there that all four draws will be a red card?

Probability That Either of Two Events Will Occur

9. At a certain school, 55% of the students have brown hair, 15% have blue eyes, and 7% have both brown hair and blue eyes. What is the probability that a student chosen at random will have either brown hair or blue eyes, or both brown hair and blue eyes?

10. Find the probability that a card drawn from a deck will be either a "picture" card (jack, queen, or king) or a spade card.

Probability of Mutually Exclusive Events

11. What chance is there of tossing a sum of 7 or 9 when two dice are tossed? You may use your results from problems 5 and 6 here.

12. On a certain production line, 5% of the parts are underweight and 8% are overweight. What is the probability that one part selected at random is either underweight or overweight?

Probability Distributions

13. For the relative frequency histogram for the population of students' grades (Fig. 21–3), find the probability that a student chosen at random will have gotten a grade between 68 and 74.

14. Using the relative frequency histogram of Fig. 21–3, find the chance of a student chosen at random having a grade between 74 and 86.

Binomial Distribution

A binomial experiment is repeated n times, with a probability p of success on one trial. Find the probability $P(x)$ of x successes, if

15. $n = 5$, $p = 0.4$, and $x = 3$.

16. $n = 7$, $p = 0.8$, and $x = 4$.

17. $n = 7$, $p = 0.8$, and $x = 6$.

18. $n = 9$, $p = 0.3$, and $x = 6$.

For problems 19 and 20, find, by hand, all of the terms in each probability distribution, and graph.

19. $n = 5$, $p = 0.8$ **20.** $n = 6$, $p = 0.3$

21. What is the probability of tossing 7 heads in 10 tosses of a fair coin?

22. If male and female births are equally likely, what is the probability of five births being all girls?

23. If the probability of producing a defective compact disk is 0.15, what is the probability of getting 10 bad disks in a sample of 15?

24. A certain multiple-choice test has 20 questions, each of which has four choices, only one of which is correct. If a student were to guess every answer, what is the probability of getting 10 correct?

25. Use your graphic calculator to solve any problem from 15–24.

21–5   The Normal Curve

In the preceding section we showed some discrete probability distributions and studied the binomial distribution, one of the most useful of the discrete distributions. Now we introduce *continuous* probabilities and go on to study the most useful of these, the *normal curve*.

Continuous Probability Distributions

Suppose now that we collect data from a "large" population. For example, suppose that we record the weight of every college student in a certain state. Since the population is large, we can use very narrow class widths and still have many observations that fall within each class. If we do so, our relative frequency histogram, Fig. 21–12(a), will have very narrow bars, and our relative frequency polygon, Fig. 21–12(b), will be practically a *smooth curve*.

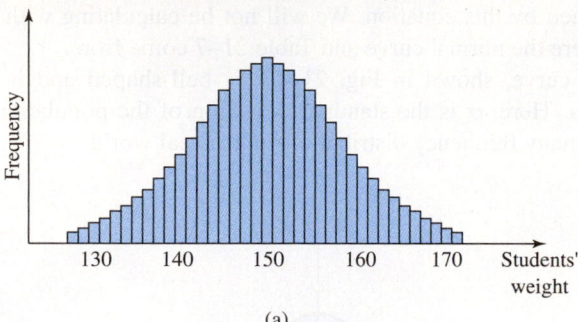

(a)

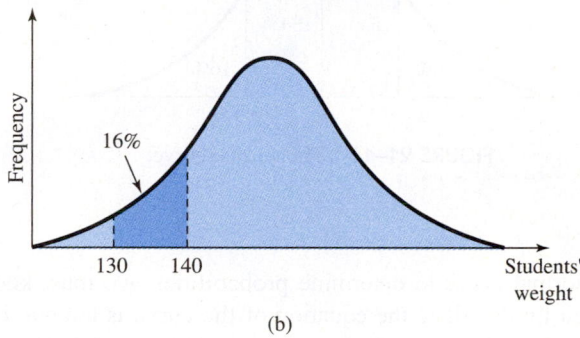

16%

(b)

FIGURE 21–12

We get probabilities from the nearly smooth curve just as we did from the relative frequency histogram. *The probability of an observation falling between two limits* a *and* b *is proportional to the area bounded by the curve and the* x *axis, between those limits.* Let us designate the total area under the curve as 100% (or $P = 1$). Then the probability of an observation falling between two limits is equal to the area between those two limits.

◆◆◆ **Example 50:** Suppose that 16% of the entire area under the curve in Fig. 21–12(b) lies between the limits 130 and 140. That means there is a 16% chance ($P = 0.16$) that one student chosen at random will weigh between 130 and 140 lb.

Normal Distribution

The frequency distribution for many real-life observations, such as a person's height, is found to have a typical "bell" shape. We have seen that the chance of one observation falling between two limits is equal to the area under the relative frequency histogram [such as Fig. 21–12(a)] between those limits. But how are we to get those areas for a curve such as in Fig. 21–12(b)?

One way is to make a *mathematical model* of the probability distribution and from it get the areas of any section of the distribution. One such mathematical model is called the *normal distribution, normal curve,* or *Gaussian distribution.*

The normal distribution is obtained by plotting the following equation:

The Gaussian distribution is named for Carl Friedrich Gauss (1777–1855). Some consider him to be one of the greatest mathematicians of all time. Many theorems and proofs are named after him.

Normal Distribuition	$y = \dfrac{1}{\sigma\sqrt{2\pi}}e^{-(x-\mu)^2/(2\sigma^2)}$	350

Don't be frightened by this equation. We will not be calculating with it but give it only to show where the normal curve and Table 21–7 come from.

The normal curve, shown in Fig. 21–13, is bell-shaped and is symmetrical about its mean, μ. Here, σ is the standard deviation of the population. This curve closely matches many frequency distributions in the real world.

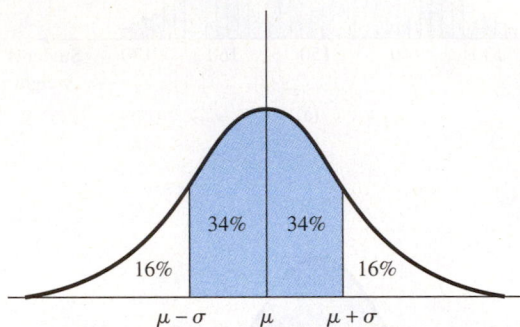

FIGURE 21–13   The normal curve.

To use the normal curve to determine probabilities, we must know the areas between any given limits. Since the equation of the curve is known, it is possible, by methods too advanced to show here, to compute these areas. They are given in Table 21–7. More extensive tables can be found in most statistics books. The areas are given "within z standard deviations of the mean." This means that each area given is that which lies between the mean and z standard deviations *on one side* of the mean, as shown in Fig. 21–14.

TABLE 21–7   Areas under the normal curve within z standard deviations of the mean.

z	Area	z	Area
0.1	0.0398	2.1	0.4821
0.2	0.0793	2.2	0.4861
0.3	0.1179	2.3	0.4893
0.4	0.1554	2.4	0.4918
0.5	0.1915	2.5	0.4938
0.6	0.2258	2.6	0.4952
0.7	0.2580	2.7	0.4965
0.8	0.2881	2.8	0.4974
0.9	0.3159	2.9	0.4981
1.0	0.3413	3.0	0.4987
1.1	0.3643	3.1	0.4990
1.2	0.3849	3.2	0.4993
1.3	0.4032	3.3	0.4995
1.4	0.4192	3.4	0.4997
1.5	0.4332	3.5	0.4998
1.6	0.4452	3.6	0.4998
1.7	0.4554	3.7	0.4999
1.8	0.4641	3.8	0.4999
1.9	0.4713	3.9	0.5000
2.0	0.4772	4.0	0.5000

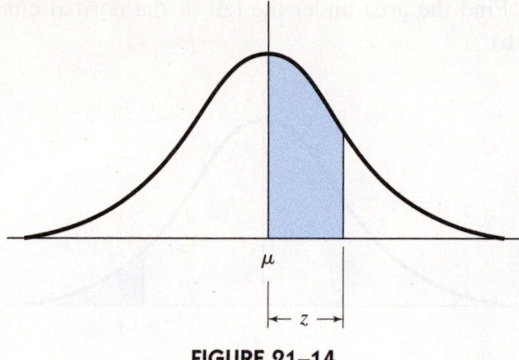

FIGURE 21–14

The number z is called the *standardized variable*, or *normalized variable*. When used for educational testing, it is often called the *standard score*, or *z score*. Since the normal curve is symmetrical about the mean, the given values apply to either the left or the right side.   ◆◆◆

◆◆◆ **Example 51:** Find the area under the normal curve between the following limits: from one standard deviation to the left of the mean, to one standard deviation to the right of the mean.

Solution: From Table 21–7, the area between the mean and one standard deviation is 0.3413. By symmetry, the same area lies to the other side of the mean. The total is thus 2(0.3413), or 0.6826. Thus *about 68% of the area under the normal curve lies within one standard deviation of the mean* (Fig. 21–13). Therefore $100 - 68$ or 32% of the area lies in the "tails," or 16% in each tail.   ◆◆◆

◆◆◆ **Example 52:** Find the area under the normal curve between 0.8 standard deviation to the left of the mean, to 1.1 standard deviations to the right of the mean (Fig. 21–15).

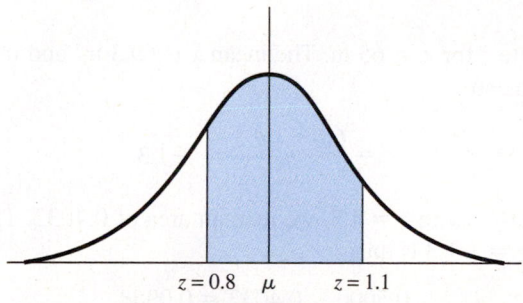

FIGURE 21–15

Solution: From Table 21–7, the area between the mean and 0.8 standard deviation ($z = 0.8$) is 0.2881. Also, the area between the mean and 1.1 standard deviations ($z = 1.1$) is 0.3643. The combined area is thus

$$0.2881 + 0.3643 = 0.6524$$

or 65.24% of the total area.   ◆◆◆

◆◆◆ **Example 53:** Find the area under the tail of the normal curve to the right of $z = 1.4$ (Fig. 21–16).

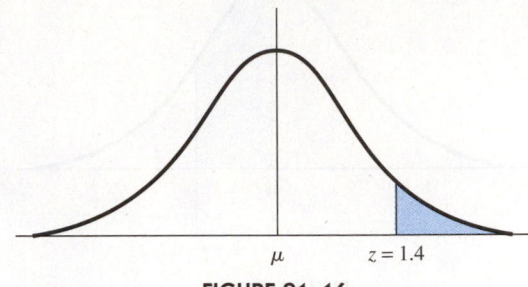

FIGURE 21–16

Solution: From Table 21–7, the area between the mean and $z = 1.4$ is 0.4192. But since the total area to the right of the mean is $\frac{1}{2}$, the area in the tail is

$$0.5000 - 0.4192 = 0.0808$$

or 8.08% of the total area.   ◆◆◆

Since, for a normal distribution, the probability of a measurement falling within a given interval is equal to the area under the normal curve within that interval, we can use the areas to assign probabilities. We must first convert the measurement x to the standard variable z, which tells the number of standard deviations between x and the mean $\bar{x}$. To get the value of z, we simply find the difference between x and $\bar{x}$ and divide that difference by the standard deviation.

$$z = \frac{x - \bar{x}}{s}$$

◆◆◆ **Example 54:** Suppose that the heights of 2000 students have a mean of 69.3 in. with a standard deviation of 3.2 in. Assuming the heights to have a normal distribution, predict the number of students (a) shorter than 65 in., (b) taller than 70 in., and (c) between 65 in. and 70 in.

Solution:

(a) Let us compute z for $x = 65$ in. The mean $\bar{x}$ is 69.3 in. and the standard deviation s is 3.2 in., so

$$z = \frac{65 - 69.3}{3.2} = -1.3$$

From Table 21–7, with $z = 1.3$, we read an area of 0.4032. The area in the tail to the left of $z = -1.3$ is thus

$$0.5000 - 0.4032 = 0.0968$$

Thus there is a 9.68% chance that a single student will be shorter than 65 in. Since there are 2000 students, we predict that

$$9.68\% \text{ of } 2000 \text{ students} = 194 \text{ students}$$

will be within this range.

(b) For a height of 70 in., the value of z is

$$z = \frac{70 - 69.3}{3.2} = 0.2$$

Since the normal curve is symmetrical about the mean, either a positive or a negative value of z will give the same area. Thus for compactness, Table 21–7 shows only positive values of z.

The area between the mean and $z = 0.2$ is 0.0793, so the area to the right of $z = 0.2$ is

$$0.5000 - 0.0793 = 0.4207$$

Since there is a 42.07% chance that a student will be taller than 70 in., we predict that 0.4207(2000) or 841 students will be within that range.

(c) Since 194 + 841 or 1035 students are either shorter than 65 in. or taller than 70 in., we predict that 965 students will be between these two heights. ◆◆◆

Using the Normal Curve on the Calculator

We can graph the normal curve and predict distributions quite easily on the TI-83/84. For example, to solve Example 51 using the calculator, we first press DISTR . We then move the cursor over to DRAW and then press ENTER to select **ShadeNorm(**. Enter -.8 to set the lower bound and then "," and then 1.1 to set the upper bound. Pressing ")" and then ENTER draws the graph and gives us the area under the normal curve between the bounds.

Exercise 5 ◆ The Normal Curve

Use Table 21–7 for the following problems.

1. Find the area under the normal curve between the mean and 1.5 standard deviations in Fig. 21–17.

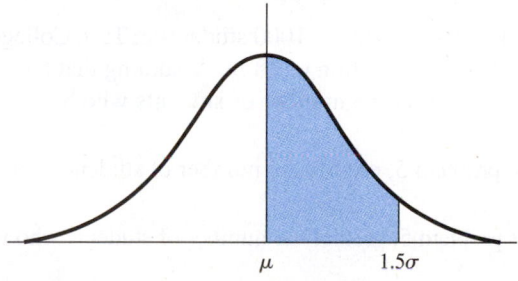

μ   1.5σ

FIGURE 21–17

2. Find the area under the normal curve between 1.6 standard deviations on both sides of the mean in Fig. 21–18.

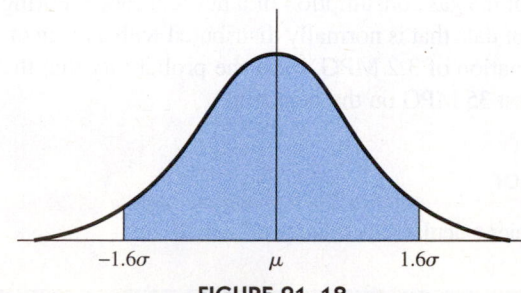

−1.6σ   μ   1.6σ

FIGURE 21–18

3. Find the area in the tail of the normal curve to the left of $z = -0.8$ in Fig. 21–19.

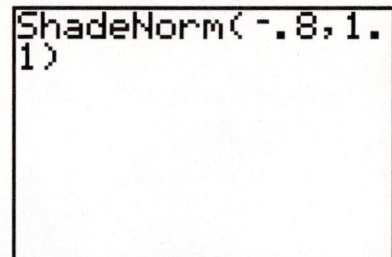

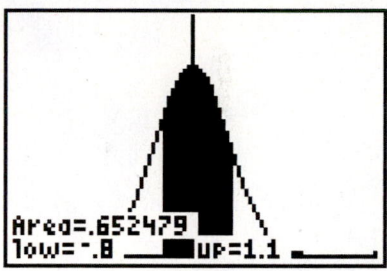

TI-83/84 Screens for Example 52.

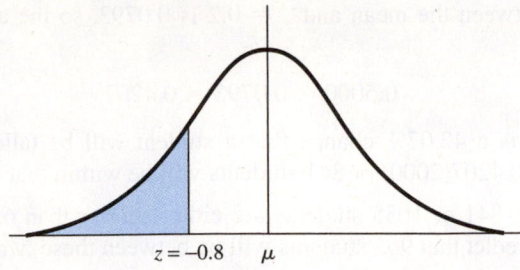

FIGURE 21-19

4. Find the area in the tail of the normal curve to the right of $z = 1.3$ in Fig. 21-20.

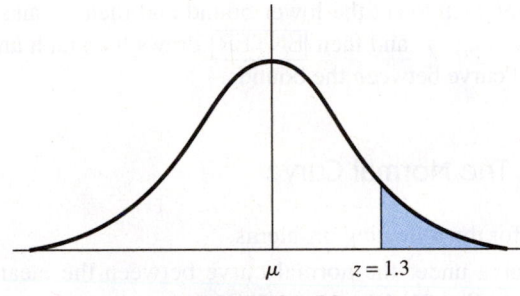

FIGURE 21-20

5. The distribution of the weights of 1000 students at Tech College has a mean of 163 lb. with a standard deviation of 18 lb. Assuming that the weights are normally distributed, predict the number of students who have weights between 130 and 170 lb.

6. For the data of problem 5, predict the number of students who will weigh less than 130 lb.

7. For the data of problem 5, predict the number of students who will weigh more than 195 lb.

8. On a test given to 500 students, the mean grade was 82.6 with a standard deviation of 7.4. Assuming a normal distribution, predict the number of A grades (a grade of 90 or over).

9. For the test of problem 8, predict the number of failing grades (a grade of 60 or less).

10. Computation of the gas consumption of a new car model during a series of trials gives a set of data that is normally distributed with a mean of 32.2 MPG and a standard deviation of 3.2 MPG. Find the probability that the same car will consume at least 35 MPG on the next trial.

Graphics Calculator

11. Use your graphics calculator to solve problems 1–10.

21-6 Standard Errors

When we draw a random sample from a population, it is usually to *infer* something about that population. Typically, from our sample we compute a *statistic* (such as the sample mean $\bar{x}$) and use it to infer a population *parameter* (such as the population mean μ).

◆◆◆ **Example 55:** A researcher measured the heights of a randomly drawn sample of students at Tech College and calculated the mean height of that sample. It was

$$\text{mean height } \bar{x} = 67.50 \text{ in.}$$

From this he inferred that the mean height of the entire population of students at Tech College was

$$\text{mean height } \mu = 67.50 \text{ in.} \qquad ◆◆◆$$

In general,

$$\text{population parameter} \approx \text{sample statistic}$$

How accurate is the population parameter that we find in this way? We know that all measurements are approximate, and we usually give some indication of the accuracy of a measurement. In fact, a population parameter such as the mean height is often given in the form

$$\text{mean height } \mu = 67.50 \pm 0.24 \text{ in.}$$

where ± 0.24 is called the *standard error* of the mean. In this section we show how to compute the standard error of the mean and the standard error of the standard deviation.

Further, when we give a range of values for a statistic, we know that it is possible that another sample can have a mean that lies outside the given range. In fact, the true mean itself can lie outside the given range. That is why a statistician will *give the probability* that the true mean falls within the given range.

◆◆◆ **Example 56:** Using the same example of heights at Tech College, we might say that

$$\text{mean height } \mu = 67.50 \pm 0.24 \text{ in.}$$

with a 68% chance that the mean student height μ falls within the range

$$67.50 - 0.24 \quad \text{to} \quad 67.50 + 0.24$$

or from 67.26 to 67.74 in.   ◆◆◆

We call the interval 67.50 ± 0.24 in Example 56 a *68% confidence interval*. This means that there is a 68% chance that the true population mean falls within the given range. So the complete way to report a population parameter is

$$\text{population parameter} = \text{sample statistic} \pm \text{standard error}$$

within a given confidence interval. We will soon show how to compute this confidence interval, and others as well, but first we must lay some groundwork.

Frequency Distribution of a Statistic

Suppose that you draw a sample of, say, 40 heights from the population of students in your school, and find the mean $\bar{x}$ of those 40 heights, say, 65.6 in. Now measure another 40 heights from the same population and get another $\bar{x}$, say, 69.3 in. Then repeat, drawing one sample of 40 after another, and for each compute the statistic $\bar{x}$. This collection of $\bar{x}$'s now has a frequency distribution of its own (called the $\bar{x}$ distribution) with its own mean and standard deviation.

We now ask, *how is the $\bar{x}$ distribution related to the original distribution?* To answer this, we need a statistical theorem called the *central limit theorem,* which we give here without proof.

Standard Error of the Mean

Suppose that we were to draw *all possible* samples of size n from a given population, and for each sample compute the mean $\bar{x}$, and then make a frequency distribution of the $\bar{x}$'s. The central limit theorem states the following:

1. The mean of the $\bar{x}$ distribution is the same as the mean μ of the population from which the samples were drawn.
2. The standard deviation of the $\bar{x}$ distribution is equal to the standard deviation σ of the population divided by the square root of the sample size n. The standard error or a sampling distribution of a statistic such as $\bar{x}$ is often called its *standard error*. We will denote the standard error of $\bar{x}$ by $SE_{\bar{x}}$. So

$$SE_{\bar{x}} = \frac{\sigma}{\sqrt{n}}$$

3. For a "large" sample (over 30 or so), the $\bar{x}$ distribution is approximately a normal distribution, even if the population from which the samples were drawn is not a normal distribution.

We will illustrate parts of the central limit theorem with an example.

◆◆◆ **Example 57:** Consider the population of 256 integers from 100 to 355 given in Table 21–8. The frequency distribution of these integers [Fig 21–21(a)] shows that each class has a frequency of 32. We computed the population mean μ by adding all of the integers and dividing by 256 and got 227. We also got a population standard deviation σ of 74.

TABLE 21–8

174	230	237	173	214	242	113	177	238	175	303	172	114	179	307	176
190	246	253	189	194	258	129	193	254	191	319	188	130	111	108	103
206	262	269	205	210	274	145	209	270	207	335	204	146	127	124	119
222	278	285	221	226	290	161	225	286	223	351	220	162	143	140	135
300	229	302	228	296	233	298	232	301	239	236	231	304	241	306	240
316	245	318	244	312	249	314	248	317	255	252	247	320	257	322	256
332	261	334	260	328	265	330	264	333	271	268	263	336	273	338	272
348	277	350	276	344	281	346	280	349	287	284	279	352	289	354	288
166	102	109	165	170	234	105	169	110	167	295	164	106	171	299	168
182	118	125	181	186	250	121	185	126	183	311	180	122	187	315	184
198	134	141	197	202	266	137	201	142	199	327	196	138	203	331	200
178	150	157	213	218	282	153	217	158	215	343	212	154	219	347	216
292	101	294	100	297	107	104	235	293	195	323	192	305	115	112	243
308	117	310	116	313	123	120	251	309	211	339	208	321	131	128	259
324	133	326	132	329	139	136	267	325	227	355	224	337	147	144	275
340	149	342	148	345	155	152	283	341	159	156	151	353	163	160	291

The calculations in this example were done by computer, but you will not need a computer to make use of the results of this demonstration.

Then 32 samples of 16 integers each were drawn from the population. For each sample we got the sample mean $\bar{x}$. The 32 values we got for $\bar{x}$ are as follows:

234 229 236 235 227 232 229 224 266 230 218 212 234 226 237 216
170 222 220 218 232 247 262 277 265 252 238 225 191 204 216 223

We now make a frequency distribution for the sample means, which is called a *sampling distribution of the mean* (Fig. 21–21). We further compute the mean of the sample means and get 229, nearly the same value as for the population mean μ.

The standard deviation of the sample means is computed to be 20. Thus we have the following results:

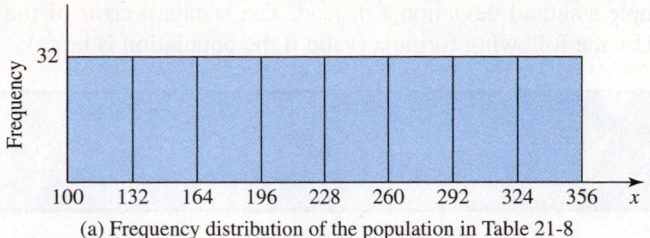

(a) Frequency distribution of the population in Table 21-8

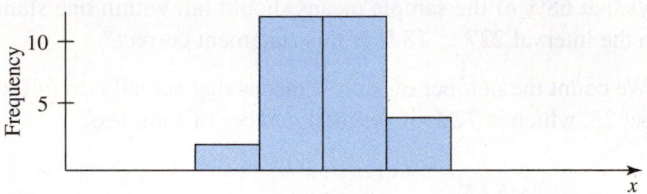

(b) Sampling distribution of the mean (frequency distribution
of the means of 32 samples drawn from the population in
Table 21-8)

FIGURE 21–21

1. The mean of the $\bar{x}$ distribution (229) is approximately equal to the mean μ of the population (227).
2. By the central limit theorem, the standard deviation $SE_{\bar{x}}$ of the $\bar{x}$ distribution is found by dividing σ by the square root of the sample size.

$$SE_{\bar{x}} = \frac{\sigma}{\sqrt{n}} = \frac{74}{\sqrt{16}} = 18.5$$

This compares well with the actual standard deviation of the $\bar{x}$ distribution, which was 20. ◆◆◆

We now take the final step and show how to predict the population mean μ from the mean $\bar{x}$ and standard deviation *s of a single sample*. We also give the probability that our prediction is correct.

Predicting μ from a Single Sample

If the standard deviation of the $\bar{x}$ distribution, $SE_{\bar{x}}$, is small, most of the $\bar{x}$'s will be near the center μ of the population. Thus a particular $\bar{x}$ has a good chance of being close to μ and will hence be a good estimator of μ. Conversely, a large $SE_{\bar{x}}$ means that a given $\bar{x}$ will be a poor estimator of μ.

However, since the frequency distribution of the $\bar{x}$'s is normal, the chance of a single $\bar{x}$ lying within one standard deviation of the mean μ is 68%. Conversely, the mean μ has a 68% chance of lying within one standard deviation of *a single randomly chosen* $\bar{x}$. Thus there is a 68% chance that the true population mean μ falls within the interval

$$\bar{x} \pm SE_{\bar{x}}$$

In this way we can estimate the population mean μ from a *single sample*. Not only that, but we are able to give a range of values within which μ must lie and to give *the probability* that μ will fall within that interval.

There is just one difficulty: To compute the standard error $SE_{\bar{x}}$ by the central limit theorem, we must divide the population standard deviation σ by the square root of the sample size. But σ *is not usually known*. Thus it is common practice to use the sample standard deviation s instead. The standard error of the mean is approximated by the following formula (valid if the population is large):

Standard Error of the Mean $\bar{x}$	$\mathrm{SE}_{\bar{x}} = \dfrac{\sigma}{\sqrt{n}} \approx \dfrac{s}{\sqrt{n}}$	352

◆◆◆ **Example 58:** For the population of 256 integers (Table 21–8), the central limit theorem says that 68% of the sample means should fall within one standard error of μ, or within the interval 227 ± 18.5. Is this statement correct?

Solution: We count the number of sample means that actually do fall within this interval and get 23, which is 72% of the total number of samples.   ◆◆◆

◆◆◆ **Example 59:** One of the 32 samples in Example 57 has a mean of 234 and a standard deviation s of 67. Use these figures to estimate the mean μ of the population.

Solution: Computing the standard error from Eq. 352, with $s = 67$ and $n = 16$, gives

$$\mathrm{SE}_{\bar{x}} \approx \frac{67}{\sqrt{16}} = 17$$

We may then predict that there is a 68% chance that μ will fall within the range

$$\mu = 234 \pm 17$$

We see that the true population mean 227 does fall within that range.   ◆◆◆

◆◆◆ **Example 60:** The heights of 64 randomly chosen students at Tech College were measured. The mean $\bar{x}$ was found to be 68.25 in., and the standard deviation s was 2.21 in. Estimate the mean μ of the entire population of students at Tech College.

Solution: From Eq. 352, with $s = 2.21$ and $n = 64$,

$$\mathrm{SE}_{\bar{x}} \approx \frac{s}{\sqrt{n}} = \frac{2.21}{\sqrt{64}} = 0.28$$

We may then state that there is a 68% chance that the mean height μ of all students at Tech College is

$$\mu = 68.25 \pm 0.28 \text{ in.}$$

In other words, the true mean has a 68% chance of falling within the interval

$$67.97 \text{ to } 68.53 \text{ in.}$$ ◆◆◆

Confidence Intervals

The interval 67.97 to 68.53 in. found in Example 60 is called the *68% confidence interval.* Similarly, there is a 95% chance that the population mean falls within *two* standard errors of the mean.

$$\mu = \bar{x} \pm 2\mathrm{SE}_{\bar{x}}$$

So we call it the *95% confidence interval.*

◆◆◆ **Example 61:** For the data of Example 60, find the 95% confidence interval.

Solution: We had found that $SE_{\bar{x}}$ was 0.28 in. Thus we may state that μ will lie within the interval

$$68.25 \pm 2(0.28)$$

or

$$68.25 \pm 0.56 \text{ in.}$$

This is the 95% confidence interval.　　　　　　　　　　　　　　　　　　　　◆◆◆

We find other confidence intervals in a similar way.

Standard Error of the Standard Deviation

We can compute standard errors for sample statistics other than the mean. The most common is for the standard deviation. The standard error of the standard deviation is called SE_s and is given by the following equation where again, we use the sample standard deviation s if we don't know σ.

Standard Error of the Standard Deviation	$SE_s = \dfrac{\sigma}{\sqrt{2n}}$	353

◆◆◆ **Example 62:** A single sample of size 32 drawn from a population is found to have a standard deviation s of 21.55. Give the population standard deviation σ with a confidence level of 68%.

Solution: From Eq. 353, with $\sigma \approx s = 21.55$ and $n = 32$,

$$SE_s = \frac{21.55}{\sqrt{2(32)}} = 2.69$$

Thus there is a 68% chance that σ falls within the interval

$$21.55 \pm 2.69$$　　　　　　　　　　　　　　　　　　　　◆◆◆

Standard Error of a Proportion

Consider a binomial experiment in which the probability of occurrence (called a *success*) of an event is p and the probability of nonoccurrence (a *failure*) of that event is q (or $1 - p$). For example, for the toss of a die, the probability of success in rolling a three is $p = \frac{1}{6}$, and the probability of failure in rolling a three is $q = \frac{5}{6}$.

If we threw the die n times and recorded the proportion of successes, it would be close to, but probably not equal to, p. Suppose that we then threw the die another n times and got another proportion, and then another n times, and so on. If we did this enough times, the proportion of successes of our samples would form a normal distribution with a mean equal to p and a standard deviation given by the following:

Standard Error of a Proportion	$SE_p = \sqrt{\dfrac{p(1 - p)}{n}}$	354

◆◆◆ **Example 63:** Find the 68% confidence interval for rolling a three for a die tossed 150 times.

Solution: Using Eq. 354, with $p = \frac{1}{6}$ and $n = 150$,

$$SE_p = \sqrt{\frac{\left(\frac{1}{6}\right)\left(\frac{5}{6}\right)}{150}} = 0.030$$

Thus there is a 68% probability that in 150 rolls of a die, the proportion of threes will lie between $\frac{1}{6} + 0.03$ and $\frac{1}{6} - 0.03$.   ◆◆◆

We can use the proportion of success in a single sample to estimate the proportion of success of an entire population. Since we will not usually know p for the population, we use the proportion of successes found from the sample when computing SE_p.

◆◆◆ **Example 64:** In a poll of 152 students at Tech College, 87 said that they would vote for Jones for president of the student union. Estimate the support for Jones among the entire student body.

Solution: In the sample, $\frac{87}{152} = 0.572$ support Jones. So we estimate that her support among the entire student body is 57.2%. To compute the confidence interval, we need the standard error. With $n = 152$, and using 0.572 as an approximation to p, we get

$$SE_p \approx \sqrt{\frac{0.572(1 - 0.572)}{152}} = 0.040$$

We thus expect that there is a 68% chance that the support for Jones is

$$0.572 \pm 0.040$$

or that she will capture between 53.2% and 61.2% of the vote of the entire student body.   ◆◆◆

◆◆◆ **Example 65:** *An Application.* Determine the 99.7% confidence level for measurement of chloroform at sea level in Barbados if a sample measurement yields 136.00 pptv (parts per trillion volume) and a standard error of 0.05.

Solution: To reach the 99.7% confidence level, we need to include measurements within 3 standard errors of the sample measurement. Therefore, there is a 99.7% chance that the concentration of chloroform at sea level is $136.00 +/-3(0.05)$, or, stated another way, that the concentration is between 135.85 and 136.15 pptv.   ◆◆◆

Exercise 6 ◆ Standard Errors

1. The heights of 49 randomly chosen students at Tech College were measured. Their mean $\bar{x}$ was found to be 69.47 in., and their standard deviation s was 2.35 in. Estimate the mean μ of the entire population of students at Tech College with a confidence level of 68%.

2. For the data of problem 1, estimate the population mean with a 95% confidence interval.

3. For the data of problem 1, estimate the population standard deviation with a 68% confidence interval.

4. For the data of problem 1, estimate the population standard deviation with a 95% confidence interval.

5. A single sample of size 32 drawn from a population is found to have a mean of 164.0 and a standard deviation s of 16.31. Give the population mean with a confidence level of 68%.

6. For the data of problem 5, estimate the population mean with a 95% confidence interval.

7. For the data of problem 5, estimate the population standard deviation with a 68% confidence interval.

8. For the data of problem 5, estimate the population standard deviation with a 95% confidence interval.

9. Find the 68% confidence interval for drawing a heart from a deck of cards for 200 draws from the deck, replacing the card each time before the next draw.

10. In a survey of 950 viewers, 274 said that they watched a certain program. Estimate the proportion of viewers in the entire population that watched that program, given the 68% confidence interval.

21–7 Process Control

Statistical Process Control

Statistical process control, or SPC, is perhaps the most important application of statistics for technology students. To cover the subject of statistics usually requires an entire textbook, and SPC is often an entire chapter in such a book or can take an entire textbook by itself. Obviously we can only scratch the surface here. Process control involves continuous testing of samples from a production line. Any manufactured item will have chance variations in weight, dimensions, color, and so forth. As long as the variations are due only to chance, we say that the process is *in control.*

◆◆◆ **Example 66:** If the diameters of steel balls on a certain production line have harmless chance variations between 1.995 mm and 2.005 mm., the production process is said to be in control.   ◆◆◆

However, as soon as there is variation due to causes other than chance, we say that the process is no longer in control.

◆◆◆ **Example 67:** During a certain week, some of the steel balls from the production line of Example 66 were found to have diameters over 2.005. The process was then *out of control.*   ◆◆◆

One problem of goal control is to detect when a process is out of control so that the cause may be eliminated.

◆◆◆ **Example 68:** The period during which the process was out of control in Example 67 was found to occur when the factory air conditioning was out of operation during a heat wave. Now the operators are instructed to stop the production line during similar occurrences.   ◆◆◆

Control Charts

The main tool of SPC is the *control chart.* The idea behind a control chart is simple. We pull samples off a production line at regular time intervals, measure them in some way, and graph those measurements over time. We then establish upper and lower limits between which the sample is acceptable, but outside of which it is not acceptable.

◆◆◆ **Example 69:** Figure 21–22 shows a control chart for the diameters of samples of steel balls of the preceding examples. One horizontal line shows the mean value of the diameter, while the other two give the upper and lower *control limits*. Note that

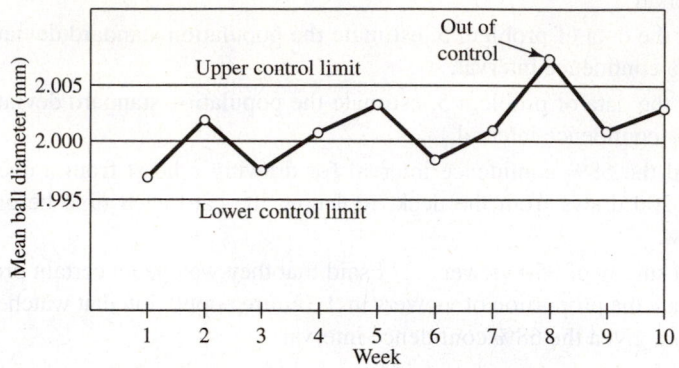

FIGURE 21–22 control chart.

the diameters fluctuate randomly between the control limits until the start of the heat wave, and then go out of control.

◆◆◆

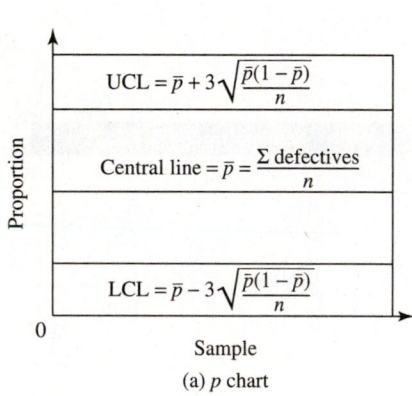

(a) *p* chart

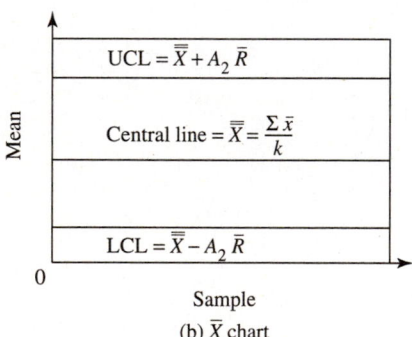

(b) $\overline{X}$ chart

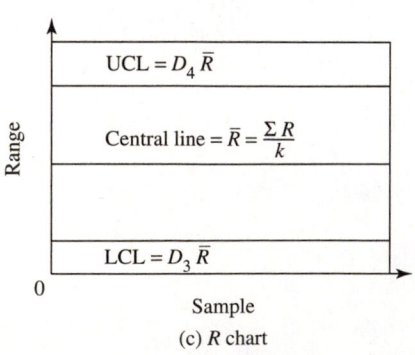

(c) *R* chart

FIGURE 21–23

To draw a control chart, we must decide what variables we want to measure and what statistical quantities we want to compute. Then we must calculate the control limits. A sample can be tested for a categorical variable (such as pass/fail) or for a continuous variable (such as the ball diameter). When testing a categorical variable, we usually compute the *proportion* of those items that pass. When testing a continuous variable, we usually compute the *mean, standard deviation,* or *range.*

We will draw three control charts: one for the proportion of a categorical variable, called a *p chart;* a second for the mean of a continuous variable, called an $\overline{X}$ *chart;* and a third for the range of a continuous variable, called an *R chart.* The formulas for computing the central line and the control limits are summarized in Fig. 21–23. We will illustrate the construction of each chart with an example.

The *p* Chart

◆◆◆ **Example 70:** Make a control chart for the proportion of defective light bulbs coming off a certain production line.

Solution:

1. *Choose a sample size n and the testing frequency.*

Let us choose to test a sample of 1000 bulbs every day for 21 days.

2. *Collect the data. Count the number d of defectives in each sample of 1000. Obtain the proportion defective p for each sample by dividing the number of defectives d by the sample size n.*

The collected data and the proportion defective are given in Table 21–9.

3. *Start the control chart by plotting p versus time.*

TABLE 21–9 Daily samples of 1000 light bulbs per day for 21 days.

Day	Number Defective	Proportion Defective
1	63	0.063
2	47	0.047
3	59	0.059
4	82	0.082
5	120	0.120
6	73	0.073
7	58	0.058
8	55	0.055
9	39	0.039
10	99	0.099
11	85	0.085
12	47	0.047
13	46	0.046
14	87	0.087
15	90	0.090
16	67	0.067
17	85	0.085
18	24	0.024
19	77	0.077
20	58	0.058
21	103	0.103
	Total 1464	

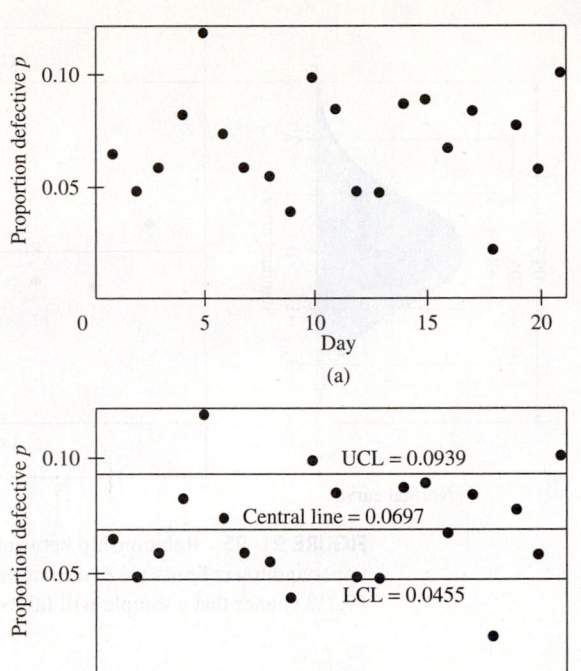

FIGURE 21–24

The daily proportion defective p is plotted in Fig. 21–24(a).

4. *Compute the average proportion defective $\bar{p}$ by dividing the total number of defectives for all samples by the total number of items measured.*

The total number of defectives is 1464, and the total number of bulbs tested is $1000(21) = 21{,}000$. So

$$\bar{p} = \frac{1464}{21{,}000} = 0.0697$$

5. *Compute the standard error SE_p. The standard error for a proportion is in which the probability p of success of a single event is given by*

$$SE_p = \sqrt{\frac{p(1 - p)}{n}} \tag{354}$$

Since we do not know the values for either p or q, we use $\bar{p}$ as an estimator for p.

$$SE_p = \sqrt{\frac{\bar{p}(1 - \bar{p})}{n}} = \sqrt{\frac{0.0697(1 - 0.0697)}{1000}} = 0.00805$$

6. *Choose a confidence interval. Recall from Sec. 21–6 that one standard error will give a 68% confidence interval, two standard errors a 95% confidence interval, three a 99.7% level, and so forth (Fig. 21–25). Set the control limits at these values.*

We will use three-sigma limits, which are the most commonly used. Thus

$$\text{upper control limit} = \bar{p} + 3SE_p = 0.0697 + 3(0.00805) = 0.0939$$
$$\text{lower control limit} = \bar{p} - 3SE_p = 0.0697 - 3(0.00805) = 0.0455$$

We are assuming here that our sampling distribution is approximately normal.

If the calculation of the lower limit gives a negative number, then we have no lower control limit.

The next step would be to *analyze* the finished control chart. We would look for points that are out of control and try to determine their cause. We do not have space here for this discussion, which can be found in any SPC book.

We would also *adjust the control limits* if in our calculation of the limits we used values that we now see are outside those limits. Since those values are not the result of chance but are caused by some production problem, they should not be used in calculating the permissible variations due to chance. Thus we would normally recompute the control limits without using those values. We omit that step here.

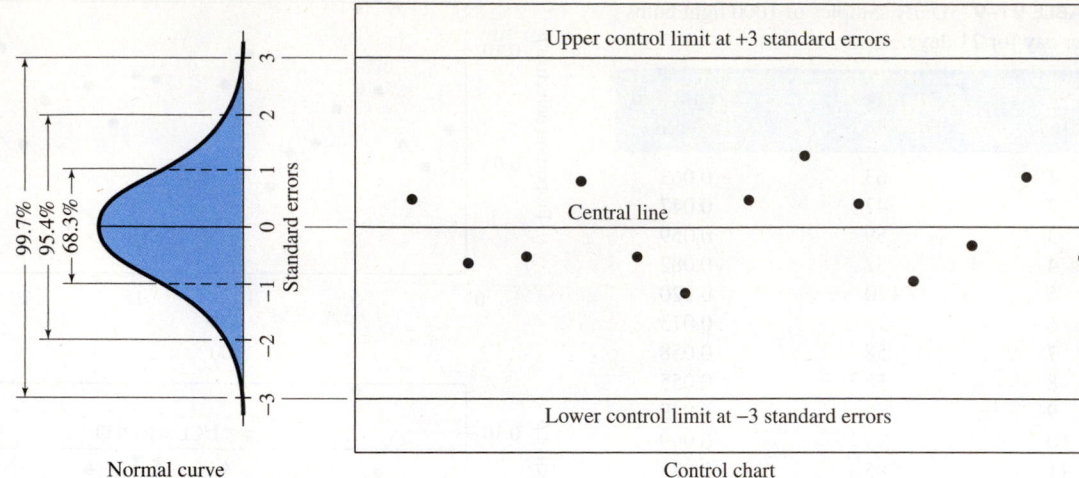

FIGURE 21–25   Relationship between the normal curve and the control chart. Note that the upper and lower limits are placed at three standard errors from the mean, so that there is a 99.7% chance that a sample will fall between those limits.

7. *Draw a horizontal at $\bar{p}$, and upper and lower control limits at $\bar{p} \pm 3SE_p$.*

These are shown in the control chart in Fig. 21–24.   ◆◆◆

Control Chart for a Continuous Variable

We construct a control chart for a continuous variable in much the same way as for a categorical variable. The variables commonly charted are the mean and either the range or the standard deviation. The range is usually preferred over the standard deviation because it is simple to calculate and easier to understand by factory personnel who may not be familiar with statistics. We will now make control charts for the mean and range, commonly called the $\bar{X}$ and *R charts.*

The $\bar{X}$ and *R* Charts

We again illustrate the method with an example. As is usually done, we will construct both charts at the same time.

◆◆◆ **Example 71:** Make $\bar{X}$ and *R* control charts for a sampling of the wattages of the lamps on a production line.

Solution:

1. *Choose a sample size n and the testing frequency.*

Since the measurement of wattage is more time-consuming than just simple counting of defectives, let us choose a smaller sample size than before, say, 5 bulbs every day for 21 days.

2. *Collect the data.*

We measure the wattage in each sample of 5 bulbs. For each sample we compute the sample mean $\bar{X}$ and the sample range *R*. These are given in Table 21–10.

TABLE 21–10 Daily samples of wattages of 5 lamps per day for a period of 21 days.

Day	Wattage of 5 Samples					Mean $\overline{X}$	Range R
1	105.6	92.8	92.6	101.5	102.5	99.0	13.0
2	106.2	100.4	106.6	108.3	109.2	106.1	8.8
3	108.6	103.3	101.8	96.0	98.8	101.7	12.6
4	95.5	106.3	106.1	97.8	101.0	101.3	10.8
5	98.4	91.5	106.3	91.6	93.2	96.2	14.8
6	101.9	102.5	94.5	107.7	108.6	103.0	14.1
7	96.7	94.8	106.9	103.4	96.2	99.6	12.1
8	99.3	107.5	94.8	102.1	108.2	102.4	13.4
9	98.0	108.3	94.8	98.6	102.8	100.5	13.5
10	109.3	98.6	92.2	106.7	96.9	100.7	17.1
11	98.8	95.0	99.4	104.8	96.2	98.8	9.8
12	91.2	103.3	95.3	103.2	108.9	100.4	17.7
13	98.5	102.8	106.4	108.6	94.2	102.1	14.4
14	100.8	108.1	104.7	108.6	97.0	103.8	11.6
15	109.0	102.6	90.5	91.3	90.5	96.8	18.5
16	100.0	104.0	99.5	92.9	93.4	98.0	11.1
17	106.3	106.2	107.6	90.4	102.9	102.7	17.2
18	103.4	91.8	105.7	106.9	106.3	102.8	15.1
19	99.7	106.8	99.6	106.6	90.1	100.6	16.7
20	98.4	104.5	94.5	101.7	96.2	99.1	10.0
21	102.2	103.0	90.3	94.0	103.5	98.6	13.2
					Sums	2114	285.3
					Averages	100.7	13.6

3. *Start the control chart by plotting $\overline{X}$ and R versus time.*

The daily mean and range are plotted as shown in Fig. 21–26.

4. *Compute the average mean $\overline{\overline{X}}$ and the average range $\overline{R}$.*

We take the average of the means (k = number of means = 21).

$$\overline{\overline{X}} = \frac{\Sigma\, \overline{X}}{k} = \frac{2114}{21} = 100.7$$

The average of the ranges is

$$\overline{R} = \frac{\Sigma\, R}{k} = \frac{285.3}{21} = 13.6$$

5. *Compute the control limits.*

When making an $\overline{X}$ and R chart, we normally do not have the standard deviation to use in computing the control limits. The average range $\overline{R}$, multiplied by a suitable constant, is usually used to give the control limits for both the mean and the range charts. For the mean, the three-sigma control limits are

$$\overline{\overline{X}} \pm A_2\overline{R}$$

where A_2 is a constant depending on sample size, given in Table 21–11. The control limits for the range are

$$D_3\overline{R} \quad \text{and} \quad D_4\overline{R}$$

where D_3 and D_4 are also from the table.

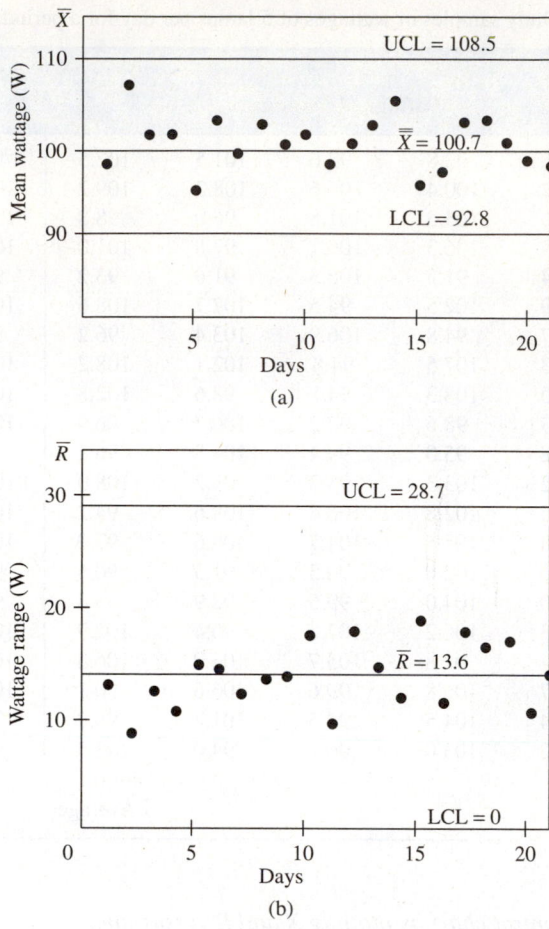

FIGURE 21–26

TABLE 21–11   Factors for computing control chart lines.

n	2	3	4	5	6	7	8	9	10
A_2	1.880	1.023	0.729	0.577	0.483	0.419	0.373	0.337	0.308
D_3	0	0	0	0	0	0.076	0.136	0.184	0.223
D_4	3.268	2.574	2.282	2.114	2.004	1.924	1.864	1.816	1.777

Excerpted from the *ASTM Manual on Presentation of Data and Control Chart Analysis* (1976).
For larger sample sizes, consult a book on SPC.

For a sample size of 5, $A_2 = 0.577$, so the control limits for the mean are

$$\text{UCL} = 100.7 + 0.577(13.6) = 108.5$$

and

$$\text{LCL} = 100.7 - 0.577(13.6) = 92.8$$

Also from the table, $D_3 = 0$ and $D_4 = 2.114$, so the control limits for the range are

$$\text{UCL} = 2.114(13.6) = 28.7$$

and

$$\text{LCL} = 0(13.6) = 0$$

6. *Draw the central line and the upper and lower control limits.*

These are shown in the control chart [Fig. 21–26(b)].   ◆◆◆

Exercise 7 ◆ Process Control

1. The proportion of defectives for samples of 1000 tennis balls per day for 20 days is as follows:

Day	Number Defective	Proportion Defective
1	31	0.031
2	28	0.028
3	26	0.026
4	27	0.027
5	30	0.030
6	26	0.026
7	26	0.026
8	27	0.027
9	29	0.029
10	27	0.027
11	27	0.027
12	26	0.026
13	31	0.031
14	30	0.030
15	27	0.027
16	27	0.027
17	25	0.025
18	26	0.026
19	30	0.030
20	28	0.028
Total	554	

Find the values for the central line, and determine the upper and lower control limits.

2. Draw a p chart for the data of problem 1.

3. The proportion of defectives for samples of 500 calculators per day for 20 days is as follows:

Day	Number Defective	Proportion Defective
1	153	0.306
2	174	0.348
3	139	0.278
4	143	0.286
5	156	0.312
6	135	0.270
7	141	0.282
8	157	0.314
9	125	0.250
10	126	0.252
11	155	0.310
12	174	0.348
13	138	0.276
14	165	0.330
15	144	0.288
16	166	0.332
17	145	0.290
18	153	0.306
19	132	0.264
20	169	0.338
Total	2990	

Find the values for the central line, and determine the upper and lower control limits.

4. Draw a p chart for the data of problem 3.

5. Five pieces of pipe are taken from a production line each day for 20 days, and their lengths are measured as shown in the following table:

Day	Measurements of 5 Samples (in.)				
1	201.7	184.0	201.3	183.9	192.6
2	184.4	184.2	207.2	194.4	193.1
3	217.2	212.9	198.7	185.1	196.9
4	201.4	201.8	182.5	195.0	215.4
5	214.3	197.2	219.8	219.9	194.5
6	190.9	180.6	181.9	203.6	185.1
7	189.5	209.5	207.6	215.2	200.0
8	182.2	216.4	190.7	189.0	204.2
9	184.9	217.3	185.7	183.0	189.7
10	211.2	194.8	208.6	209.4	183.3
11	197.1	189.5	200.1	197.7	187.9
12	207.1	203.1	218.4	199.3	219.1
13	199.1	207.1	205.8	197.4	189.4
14	198.9	196.1	210.3	210.6	217.0
15	215.2	200.8	205.0	186.6	181.6
16	180.2	216.8	182.5	206.3	191.7
17	219.3	195.1	182.2	207.4	191.9
18	204.0	209.9	196.4	202.3	206.9
19	202.2	216.3	217.8	200.7	215.1
20	185.0	194.1	191.2	186.3	201.6
21	181.6	187.8	188.2	191.7	200.1

Find the values for the central line, and determine the upper and lower control limits for the mean.

6. Find the values for the central line for the data of problem 5, and determine the upper and lower control limits for the range.

7. Draw an $\overline{X}$ chart for the data of problem 5.

8. Draw an R chart for the data of problem 5.

9. Five circuit boards are taken from a production line each day for 20 days and are weighed (in grams) as shown in the following table:

Day	Measurements of 5 Samples (g)				
1	16.7	19.5	11.1	18.8	11.8
2	15.0	14.5	14.5	18.4	11.5
3	14.3	16.5	13.3	14.0	14.4
4	19.3	14.0	14.6	15.0	18.0
5	16.2	13.6	16.8	12.4	16.6
6	14.3	16.7	18.4	18.3	16.5
7	13.5	15.5	13.7	19.8	11.6
8	15.6	12.3	16.8	18.3	11.8
9	13.3	15.4	16.1	17.0	12.2
10	17.6	18.6	12.4	19.9	19.3
11	19.0	12.8	14.7	19.6	15.8
12	12.3	12.2	14.9	15.5	12.9
13	16.3	12.3	18.6	14.0	11.3
14	11.4	15.8	13.9	19.1	10.5
15	12.5	11.4	14.1	12.4	14.9
16	18.1	15.1	19.2	15.2	17.6
17	18.7	13.7	11.0	17.0	12.7
18	11.3	11.5	13.4	13.1	13.9
19	16.6	18.7	12.1	14.7	11.2
20	11.2	10.5	16.2	11.0	16.2
21	12.8	18.7	10.4	16.8	17.1

Find the values for the central line, and determine the upper and lower control limits for the mean.

10. Find the values for the central line for the data of problem 9, and determine the upper and lower control limits for the range.

11. Draw an $\overline{X}$ chart for the data of problem 9.

12. Draw an R chart for the data of problem 9.

Computer

13. Some statistics software for the computer, such as *Minitab,* can be used to draw control charts. One simply enters the raw data and specifies which kind of chart is wanted, and the program automatically sets the central line and the control limits and draws the chart. Some software will even detect out-of-control or nonrandom behavior in the data. If you have such software, use it to draw any of the control charts in the exercise set.

21–8   Regression

Curve Fitting

In statistics, *curve fitting,* the fitting of a curve to a set of data points, is called *regression.* The fitting of a straight line to data points is called *linear regression,* while the fitting of some other curve is called *nonlinear regression.* Fitting a curve to *two* sets of data is called *multiple regression.* We will cover linear regression in this section.

The word *regression* was first used in the nineteenth century when these methods were used to determine the extent by which the heights of children of tall or short parents *regressed* or got closer to the mean height of the population.

■ Exploration:

Try this. Graph the following sets of data:

x	0	1	2	3	4	5	6	7	8	9	10
y	2	4	5	4	4	7	5	8	7	6	8

x	0	1	2	3	4	5	6	7	8	9	10
y	8	7	7	6	5	4	7	3	2	2	1

x	0	1	2	3	4	5	6	7	8	9	10
y	4	1	9	7	4	6	2	8	0	3	9

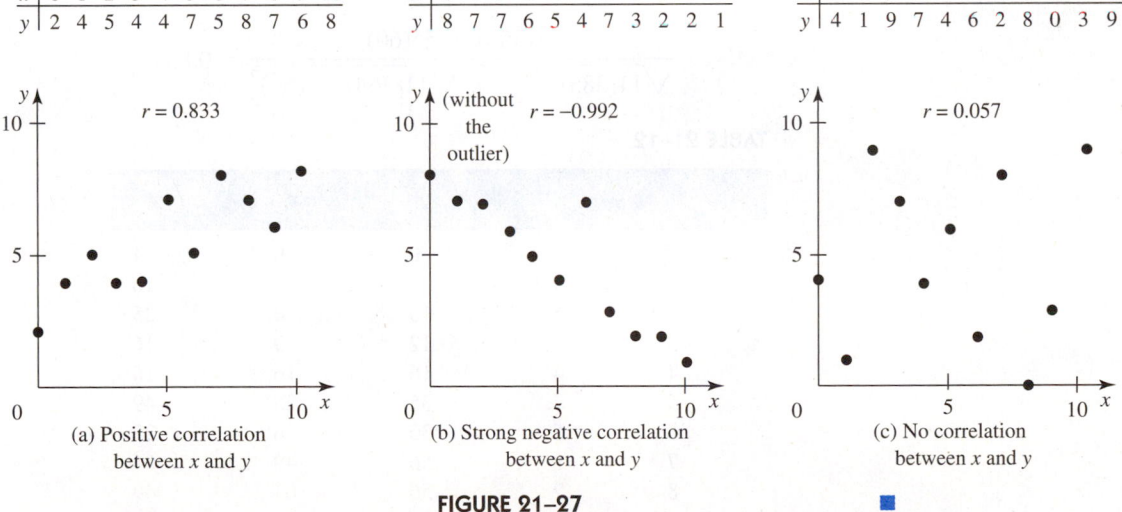

(a) Positive correlation between x and y

(b) Strong negative correlation between x and y

(c) No correlation between x and y

FIGURE 21–27

You should have produced graphs that look like those in Fig. 21-27.

What can you say about the relationship between x and y in these graphs? In (a) we could probably draw a line that would approximate the data points, and most of the points would fall near the line, but not on it. To be more technical, we could say that x and y are moderately *correlated*. We also notice that y increases as x gets larger. Therefore we can say that there is a *positive* correlation between x and y. Another way to think of this is that the slope of the line we draw to approximate the points is positive.

For (b) there is obviously a stronger correlation between x and y, but now this correlation is *negative.* There is also one point, (6, 7), that does not fall anywhere near the line. We call this point an *outlier.* Such points are usually suspected as being the result of an error and are sometimes discarded.

For (c) there is obviously no correlation.

We can now quantify both of the properties we noticed in the graphs. First, we can calculate the correlation coefficient, r, which tells us exactly how strong the correlation between x and y is. This number is very important in determining whether data from a series of measurements actually indicate that two variables have any connection. Second, we can fit a straight line to the data, using *linear regression* and the *least squares method*. In the real world, we can use the equation for this line to calculate what will happen to one quantity when we change the other.

Correlation Coefficient

In Fig. 21–27 we see that some data are more scattered than others. The *correlation coefficient r* gives a numerical measure of this scattering. For a set of n xy pairs, the correlation coefficient is given by the following equation:

Correlation Coefficient	$r = \dfrac{n \sum xy - \sum x \sum y}{\sqrt{n \sum x^2 - (\sum x)^2} \sqrt{n \sum y^2 - (\sum y)^2}}$	355

The formula looks complicated but is actually easy to use when doing the computation in table form.

◆◆◆ **Example 72:** Find the correlation coefficient for the data in Fig. 21–27(a).

Solution: We make a table giving x and y, the product of x and y, and the squares of x and y. We then sum each column as shown in Table 21–12.

Then, from Eq. 355, with $n = 11$,

$$r = \frac{n \sum xy - \sum x \sum y}{\sqrt{n \sum x^2 - (\sum x)^2} \sqrt{n \sum y^2 - (\sum y)^2}}$$

$$= \frac{11(353) - 55(60)}{\sqrt{11(385) - (55)^2} \sqrt{11(364) - (60)^2}} = 0.834$$

TABLE 21–12

x	y	xy	x^2	y^2
0	2	0	0	4
1	4	4	1	16
2	5	10	4	25
3	4	12	9	16
4	4	16	16	16
5	7	35	25	49
6	5	30	36	25
7	8	56	49	64
8	7	56	64	49
9	6	54	81	36
10	8	80	100	64
Sums 55	60	353	385	364
$\sum x$	$\sum y$	$\sum xy$	$\sum x^2$	$\sum y^2$

The correlation coefficient can vary from $+1$ (perfect positive correlation) through 0 (no correlation) to -1 (perfect negative correlation). For comparison with the data in Fig. 21–27(a), the correlation coefficients for the data in parts (b) and (c) are found to be $r = -0.993$ (after discarding the outlier) and $r = 0.057$, respectively (work not shown). ◆◆◆

Linear Regression

In *linear regression,* the object is to find the constants m and b for the equation of a straight line

$$y = mx + b$$

that best fits a given set of data points.

We will first fit the line by eye, and then we will use a more quantitative method to find the equation for the line.

♦♦♦ **Example 73:** Find the equation for the straight line that best fits the data in Fig 21–27(b).

Estimation: Using a straight edge, we draw a straight line through the data, Fig.21–28, trying to balance those points above the line with those below. We then read the y intercept and the rise of the line in a chosen run.

$$y \text{ intercept} = b = 8$$

Our line appears to pass through (10, 1) so in a run of 10 units, the line falls 7 units.

$$\text{slope } m = \frac{\text{rise}}{\text{run}} = \frac{-7}{10} = -0.7$$

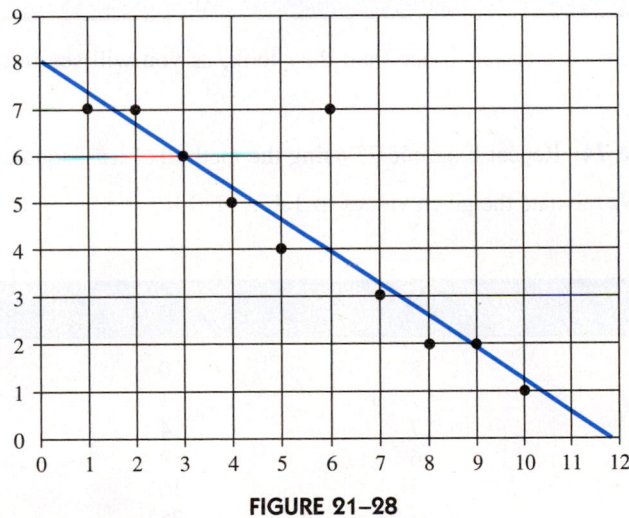

FIGURE 21–28

The equation of the fitting line is then

$$y = 0.7x + 8 \qquad\qquad ♦♦♦$$

Method of Least Squares

Sometimes the points may be too scattered to enable drawing a line with a good fit, or perhaps we desire more accuracy than can be obtained when fitting by eye. Then we may want a method that does not require any manual steps, so that it can be computerized. One such method is the *method of least squares.*

Let us define a *residual* as the vertical distance between a data point and the approximating curve (Fig. 21–29). The method of least squares is a method to fit a straight line to a data set so that *the sum of the squares of the residuals is a minimum*—hence the name *least squares*.

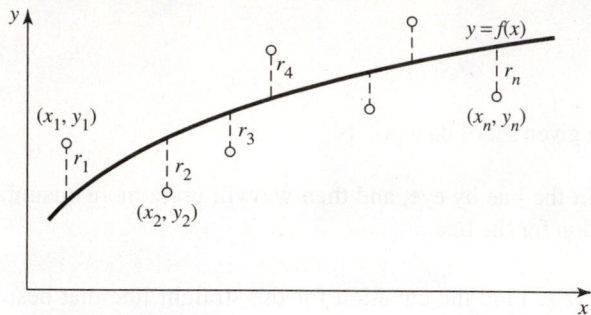

FIGURE 21–29 Definition of a residual.

With this method, the slope and the y intercept of the least squares line are given by the following equations. We give these equations without proof here. They are derived using calculus.

$$\text{slope } m = \frac{n \sum xy - \sum x \sum y}{n \sum x^2 - (\sum x)^2}$$

356

$$y \text{ intercept } b = \frac{\sum x^2 \sum y - \sum x \sum xy}{n \sum x^2 - (\sum x)^2}$$

These equations are easier to use than they look, as you will see in the following example.

◆◆◆ **Example 74:** Repeat Example 73 using the method of least squares.

Solution: We tabulate the given values in Table 21–13.

TABLE 21–13

x	y	x^2	xy
0	8	0	0
1	7	1	7
2	7	4	14
3	6	9	18
4	5	16	20
5	4	25	20
6	7	36	42
7	3	49	21
8	2	64	16
9	2	81	18
10	1	100	10
$\sum x = 55$	$\sum y = 52$	$\sum x^2 = 385$	$\sum xy = 186$

In the third column of Table 21–13, we tabulate the squares of the abscissas given in column 1, and in the fourth column we list the products of x and y. The sums are given below each column.

Substituting these sums into Eq. 356 and letting $n = 11$, we have

$$\text{slope} = \frac{11(186) - 55(52)}{11(385) - (55)^2} = -0.67$$

and

$$y \text{ intercept} = \frac{385(52) - 55(186)}{11(385) - (55)^2} = 8.1$$

which agree well with our graphically obtained values. The equation of our best-fitting line is then

$$y = -0.67x + 8.1$$

Note the close agreement with the values we obtained by eye: slope $= -0.7$ and y intercept $= 8$.   ◆◆◆

This is not the only way to fit a line to a data set, but it is widely regarded as the one that gives the "best fit."

Linear Regression on the Calculator

We can perform a least squares linear regression and calculate the correlation coefficient in a single step on the TI-83/84 and TI-89. For example, after entering the data from Example 73 into L1 and L2, we can press $\boxed{\text{STAT}}$, scroll over to **TESTS**, and then scroll down to select **F:LinRegTTest**. This brings up a menu through which we scroll until the cursor is on **Calculate**. Pressing $\boxed{\text{ENTER}}$ performs the calculation. The value for the y-intercept is stored in a and the slope is stored in b. Scrolling down shows us the value for r, which is the correlation coefficient. Note that we can delete the outlier and recalculate to get the same value for $r = -0.993$ that we calculated earlier (calculations not shown).

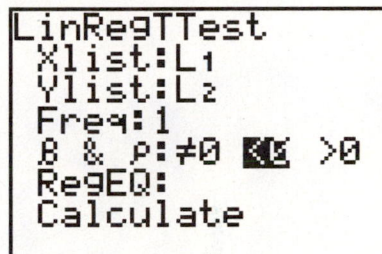

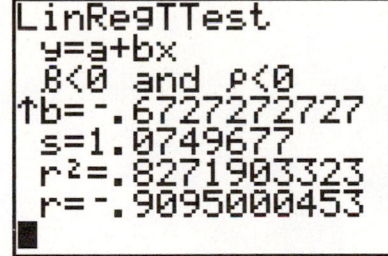

TI-83/84 Screens for Example 74.

Exercise 8 ◆ Regression

Correlation Coefficient

Find the correlation coefficient for each set of data.

1.		2.		3.	
−8.00	−6.238	−20.0	82.29	−11.0	−65.30
−6.66	−3.709	−18.5	73.15	−9.33	−56.78
−5.33	−0.712	−17.0	68.11	−7.66	−47.26
−4.00	1.887	−15.6	59.31	−6.00	−37.21
−2.66	4.628	−14.1	53.65	−4.33	−27.90
−1.33	7.416	−12.6	45.90	−2.66	−18.39
0.00	10.20	−11.2	38.69	−1.00	−9.277
1.33	12.93	−9.73	32.62	0.66	0.081
2.66	15.70	−8.26	24.69	2.33	9.404
4.00	18.47	−6.80	18.03	4.00	18.93
5.33	21.32	−5.33	11.31	5.66	27.86
6.66	23.94	−3.86	3.981	7.33	37.78
8.00	26.70	−2.40	−2.968	9.00	46.64
9.33	29.61	−0.93	−9.986	10.6	56.69
10.60	32.35	0.53	−16.92	12.3	64.74
12.00	35.22	2.00	−23.86	14.0	75.84

Method of Least Squares

Find the least squares line for each set of data.

4. From problem 1.

5. From problem 2.

6. From problem 3.

Graphics Calculator

7. Use your calculator to calculate the correlation coefficient and least squares line for each set of data.

◆◆◆ CHAPTER 21 REVIEW PROBLEMS ◆◆◆◆◆◆◆◆◆◆◆◆◆◆◆◆◆◆◆◆◆◆◆◆◆◆◆◆◆◆◆◆◆◆◆

Label each type of data as continuous, discrete, or categorical.

1. The life of certain radios.

2. The number of houses sold each day.

3. The colors of the cars at a certain dealership.

4. The sales figures for a certain product for the years 1970–80 are as follows:

Year	Number Sold
1970	1344
1971	1739
1972	2354
1973	2958
1974	3153
1975	3857
1976	3245
1977	4736
1978	4158
1979	5545
1980	6493

Make an *x-y* graph of the sales versus the year.

5. Make a bar graph for the data of problem 4.

6. If we toss three coins, what is the probability of getting one head and two tails?

7. If we throw two dice, what chance is there that their sum is 8?

8. A die is rolled twice. What is the probability that both rolls will give a two?

9. At a certain factory, 72% of the workers have brown hair, 6% are left-handed, and 3% both have brown hair and are left-handed. What is the probability that a worker chosen at random will either have brown hair or be left-handed, or both?

10. What is the probability of tossing a five or a nine when two dice are tossed?

11. Find the area under the normal curve between the mean and 0.5 standard deviation to one side of the mean.

12. Find the area under the normal curve between 1.1 standard deviations on both sides of the mean.

13. Find the area in the tail of the normal curve to the left of $z = 0.4$.

14. Find the area in the tail of the normal curve to the right of $z = 0.3$.

Use the following table of data for problems 15 through 29.

146	153	183	148	116	127	162	153
168	161	117	153	116	125	173	131
117	183	193	137	188	159	154	112
174	182	144	144	133	167	192	145
162	138	137	154	141	129	137	152

15. Determine the range of the data.

16. Make a frequency distribution using class widths of 5 starting at 107.5. Show both absolute and relative frequency.

17. Draw a frequency histogram showing both absolute and relative frequency.

18. Draw a frequency polygon showing both absolute and relative frequency.

19. Make a cumulative frequency distribution.

20. Draw a cumulative frequency polygon.

21. Find the mean.

22. Find the median.

23. Find the mode.

24. Find the variance.

25. Find the standard deviation.

26. Predict the population mean with a 68% confidence interval.

27. Predict the population mean with a 95% confidence interval.

28. Predict the population standard deviation with a 68% confidence interval.

29. Predict the population standard deviation with a 95% confidence interval.

30. On a test given to 300 students, the mean grade was 79.3 with a standard deviation of 11.6. Assuming a normal distribution, estimate the number of A grades (a score of 90 or over).

31. For the data of problem 30 estimate the number of failing grades (a score of 60 or less).

32. Determine the quartiles and give the quartile range of the following data:

118   133   135   143   164   173   179   199   212   216   256

33. Determine the quartiles and give the quartile range of the following data:

167   245   327   486   524   639   797   853   974   1136   1162   1183

34. The proportion defective for samples of 700 keyboards per day for 20 days is as follows:

Day	Number Defective	Proportion Defective
1	123	0.176
2	122	0.174
3	125	0.179
4	93	0.133
5	142	0.203
6	110	0.157
7	98	0.140
8	120	0.171
9	139	0.199
10	142	0.203
11	128	0.183
12	132	0.189
13	92	0.131
14	114	0.163
15	92	0.131
16	105	0.150
17	146	0.209
18	113	0.161
19	122	0.174
20	92	0.131
Total	2350	

Find the values of the central line, determine the upper and lower control limits, and make a control chart for the proportion defective.

35. Five castings per day are taken from a production line each day for 21 days and are weighed. Their weights, in grams, are as follows:

Day	Measurements of 5 Samples				
1	1134	1168	995	992	1203
2	718	1160	809	432	650
3	971	638	1195	796	690
4	598	619	942	1009	833
5	374	395	382	318	329
6	737	537	692	562	960
7	763	540	738	969	786
8	777	1021	626	786	472
9	1015	1036	1037	1063	1161
10	797	1028	1102	796	536
11	589	987	765	793	481
12	414	751	1020	524	1100
13	900	613	1187	458	661
14	835	957	680	845	1023
15	832	557	915	934	734
16	353	829	808	626	868
17	472	798	381	723	679
18	916	763	599	338	1026
19	331	372	318	304	354
20	482	649	1133	1022	320
21	752	671	419	715	413

Find the values of the central line, determine the upper and lower control limits, and make a control chart for the mean of the weights.

36. Find the values of the central line, determine the upper and lower control limits, and make a control chart for the range of the weights.

37. Find the correlation coefficient and the least squares fit for the following data:

x	y
5	6.882
11.2	−7.623
17.4	−22.45
23.6	−36.09
29.8	−51.13
36.0	−64.24
42.2	−79.44
48.4	−94.04
54.6	−107.8
60.8	−122.8
67.0	−138.6
73.2	−151.0
79.4	−165.3
85.6	−177.6
91.8	−193.9
98	−208.9

38. For the relative frequency histogram for the population of students' grades (Fig. 21–3), find the probability that a student chosen at random will have received a grade between 74 and 86.

39. Find the 68% confidence interval for drawing a king from a deck of cards, for 200 draws from the deck, replacing the card each time before the next draw.

40. In a survey of 120 shoppers, 71 said that they preferred Brand *A* potato chips over Brand *B*. Estimate the proportion of shoppers in the entire population that would prefer Brand *A*, given the 68% confidence interval.

Writing

41. Give an example of one statistical claim (such as an advertisement, a commercial, or a political message) that you have heard or read lately that has made you skeptical. State your reasons for being suspicious, and point out how the claim could otherwise have been presented to make it more plausible.

Team Project

42. A certain diode has the following characteristics:

Voltage across Diode (V)	Current through Diode (mA)
0	0
5	2.063
10	5.612
15	10.17
20	15.39
25	21.30
30	27.71
35	34.51
40	42.10
45	49.95
50	58.11

Linearize the data using the methods of this chapter. Then apply the method of least squares to find the slope and the y intercept of the straight line obtained. Using those values, write and graph an equation for the current as a function of the voltage. How does your graphed equation compare with the plot of the original points?

Analytic Geometry

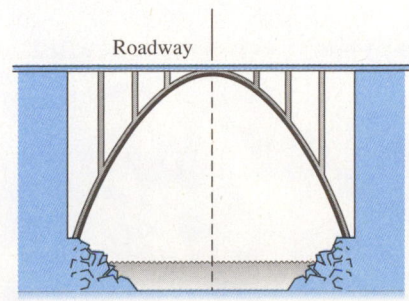

22

⬩⬩⬩ **OBJECTIVES** ⬩⬩⬩⬩⬩⬩⬩⬩⬩⬩⬩⬩⬩⬩⬩⬩⬩⬩⬩⬩⬩⬩⬩⬩⬩⬩⬩⬩⬩⬩⬩⬩⬩⬩⬩

When you have completed this chapter, you should be able to

- Calculate the distance between two points.
- Determine the slope of a line given two points on the line.
- Find the slope of a line given its angle of inclination, and vice versa.
- Determine the slope of a line perpendicular to a given line.
- Calculate the angle between two lines.
- Write the equation of a line using the slope-intercept form, the point-slope form, or the two-point form.
- Solve applied problems involving the straight line.
- Write the equation of a circle, ellipse, parabola, or hyperbola from given information.
- Write an equation in standard form given the equation of any of the previous curves.
- Determine all the features of interest from the standard equation of any of these curves.
- Make a graph of any of these figures.
- Tell, by inspection, whether a given second-degree equation represents a circle, ellipse, parabola, or hyperbola.
- Write a new equation for a curve with the axes shifted when given the equation of that curve.
- Solve applied problems involving any of these figures.

Here we start our study of *analytic geometry*. This is a branch of mathematics in which geometric figures such as points, lines, and circles are placed on the same coordinate axes we introduced in our chapter on graphing. This enables us to write an *equation* for each geometric figure. That equation can then be analyzed and manipulated using algebra to give us more information about the geometric figure than was possible without the equation.

We will first apply analytic geometry to two old and very useful friends, the *straight line* and the *circle*. These are followed by three new curves, the *parabola*, the *ellipse*, and the *hyperbola*. We will see that these curves appear in nature and have an astonishing number of applications, a sampling of which is given here. For example, the parabola is often used for a bridge arch, Fig. 22–1,

Roadway

FIGURE 22–1

because of its strength. Could you compute the length of the vertical members, given their distance from the arch centerline? In this chapter we will show how.

22–1   The Straight Line

Let us begin our study of analytic geometry with the familiar straight line. We already graphed the straight line in an earlier chapter, and here we build upon what we did there. There is no need to point out the usefulness of the straight line in technology, so whatever new tools we can find to work with the straight line should be welcome. We will first learn how to calculate the length of a line segment. Recall that earlier we defined a line segment as the portion of a straight line lying between two endpoints.

Length of a Line Segment Parallel to a Coordinate Axis

The *length of a line segment* is the distance between its endpoints. By *length* we mean the *magnitude* of that length; thus it is always positive.

■ Exploration:

Try this. Without reading further, find the lengths of the three lines in Fig. 22–2. ■

It should be no surprise that to find the length of a line segment lying on or parallel to the *x* axis, we simply subtract the *x* value at one endpoint from that at the other endpoint, and take the absolute value of the result. The procedure is similar for lines on or parallel to the *y* axis.

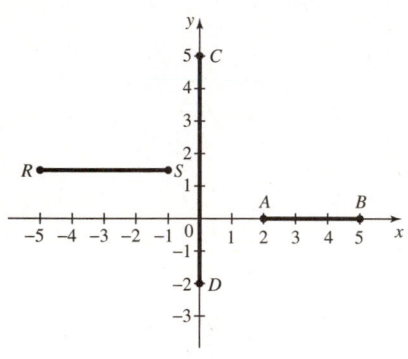

FIGURE 22–2

♦♦♦ **Example 1:** The magnitudes of the lengths of the line segments in Fig. 22–2 are:

(a) $AB = |5 - 2| = 3$

(b) $CD = |5 - (-2)| = 7$

(c) $RS = |-1 - (-5)| = 4$

Note that it doesn't matter if we reverse the order of the endpoints. We get the same result. ♦♦♦

♦♦♦ **Example 2:** Repeating Example 1(a) with the endpoints reversed gives

$$AB = |2 - 5| = |-3| = 3$$

as before. ♦♦♦

Directed Distance

Sometimes when speaking about the length of a line segment or the distance between two points, it is necessary to specify *direction* as well as magnitude. When we specify the *directed distance AB*, for example, we mean the distance *from A to B*, sometimes written $\overline{AB}$. Note that directed distance applies only to a line parallel to a coordinate axis.

♦♦♦ **Example 3:** For the two points $M(5, 2)$ and $N(1, 2)$

(a) the directed distance $NM = 5 - 1 = 4$

(b) and the directed distance $MN = 1 - 5 = -4$ ♦♦♦

Increments

Let us say that a particle is moving along a curve from point P to point Q, as shown in Fig. 22–3. As it moves, its abscissa changes from x_1 to x_2. We call this change an *increment* in x and label it Δx (read "delta x"). Similarly, the ordinate changes from y_1 to y_2 and is labeled Δy. The increments are found simply by subtracting the coordinates at P from those at Q.

| Increments | $\Delta x = x_2 - x_1$ and $\Delta y = y_2 - y_1$ | 203 |

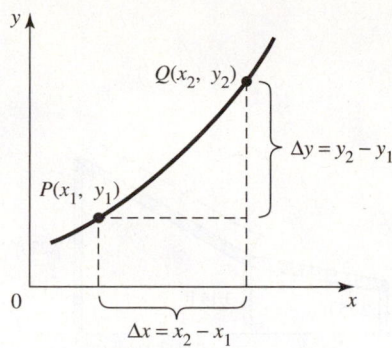

FIGURE 22–3   Increments.

◆◆◆ **Example 4:** A particle moves from $P_1(2, 5)$ to $P_2(7, 3)$. The increments in its coordinates are

$$\Delta x = x_2 - x_1 = 7 - 2 = 5$$

and

$$\Delta y = y_2 - y_1 = 3 - 5 = -2$$ ◆◆◆

Distance Formula

We now wish to find the length of a line segment that is inclined at some angle to the coordinate axes.

■ Exploration:

Try this. Without reading further, see if you can compute the length d of the line segment PQ in Fig. 22–4.   ■

You probably found the length of the line segment in the exploration by applying the Pythagorean theorem. That is exactly what we will now do to derive a formula.

Given the line segment PQ in Fig. 22–4, we first draw a horizontal line through P and drop a perpendicular from Q, forming a right triangle PQR. The sides of the triangle are d, Δx, and Δy. By the Pythagorean theorem,

$$d^2 = (\Delta x)^2 + (\Delta y)^2 = (x_2 - x_1)^2 + (y_2 - y_1)^2$$

Since we want the magnitude of the distance, we take only the positive root and get the following:

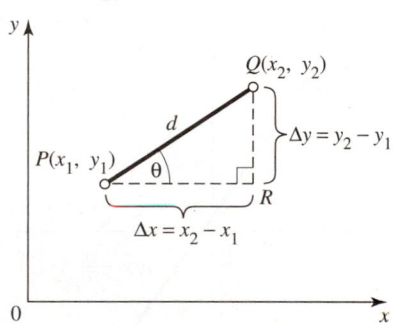

FIGURE 22–4   Length of a line segment.

| Distance Formula | $d = \sqrt{(\Delta x)^2 + (\Delta y)^2}$ $= \sqrt{(x_2 - x_1)^2 + (y_2 - y_1)^2}$ | 204 |

◆◆◆ **Example 5:** Find the length of the line segment, Fig. 22–5, with the endpoints

$$(3, -5) \text{ and } (-1, 6).$$

Solution: Let us give the first point $(3, -5)$ the subscripts 1, and the other point the subscripts 2 (it does not matter which we label 1). So

$$x_1 = 3 \qquad y_1 = -5 \qquad x_2 = -1 \qquad y_2 = 6$$

Substituting into the distance formula, we obtain

$$d = \sqrt{(-1 - 3)^2 + [6 - (-5)]^2} = \sqrt{(-4)^2 + (11)^2} = 11.7 \text{ (rounded)}$$ ◆◆◆

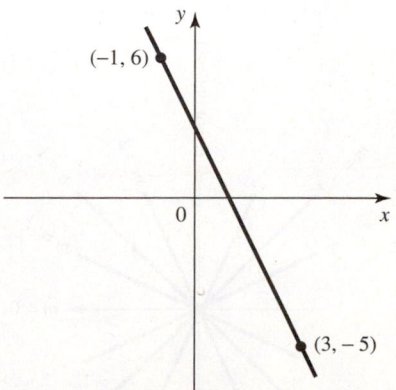

FIGURE 22–5

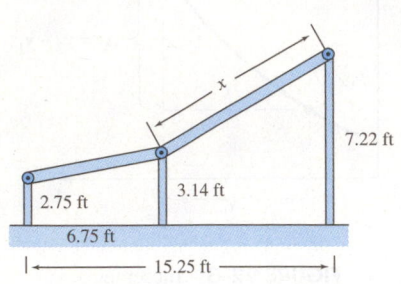

FIGURE 22–6   A conveyor belt.

	Do not take the square root of each term separately.
Common Error	$$d \neq \sqrt{(x_2 - x_1)^2} + \sqrt{(y_2 - y_1)^2}$$

◆◆◆ **Example 6:** *An Application.* Find the length x of the conveyor belt in Fig. 22–6.

Solution: The coordinates of the ends of the belt are (6.75, 3.14) and (15.25, 7.22).

Substituting into the distance formula, we get

$$x = \sqrt{(15.25 - 6.75)^2 + (7.22 - 3.14)^2}$$

$$= 9.43 \text{ ft} \qquad \qquad ◆◆◆$$

Slope

In Fig. 22–4, the increment Δy is called the *rise* from point P to point Q, and the increment Δx is called the *run* from P to Q. The rise between any two points on the line divided by the run between those same points is called the *slope* of the straight line. It is the measure of the *steepness* of a straight line.

Slope	$$m = \frac{\text{rise}}{\text{run}} = \frac{\Delta y}{\Delta x} = \frac{y_2 - y_1}{x_2 - x_1}$$ *The slope is equal to the rise divided by the run.*	**205**

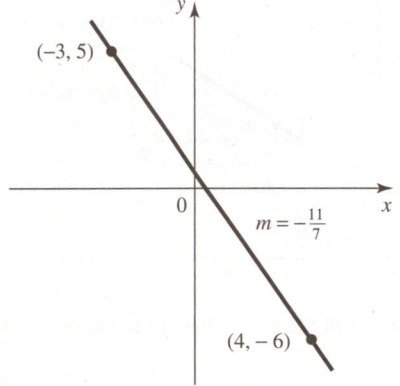

FIGURE 22–7

◆◆◆ **Example 7:** Find the slope of the line connecting the points $(-3, 5)$ and $(4, -6)$ in Fig. 22–7.

Solution: We will see that it does not matter which is called point 1 and which is point 2. Let us choose

$$x_1 = -3, \qquad y_1 = 5, \qquad x_2 = 4, \qquad y_2 = -6$$

Then

$$\text{slope } m = \frac{-6 - 5}{4 - (-3)} = \frac{-11}{7} = -\frac{11}{7}$$

If we had chosen $(x_1 = 4, y_1 = -6)$ and $(x_2 = -3, y_2 = 5)$ in the computation for slope,

$$\text{slope } m = \frac{5 - (-6)}{-3 - 4} = \frac{11}{-7} = -\frac{11}{7}$$

we would have gotten the same result. ◆◆◆

	Be careful not to mix up the subscripts.
Common Error	$$m \neq \frac{y_2 - y_1}{x_1 - x_2}$$

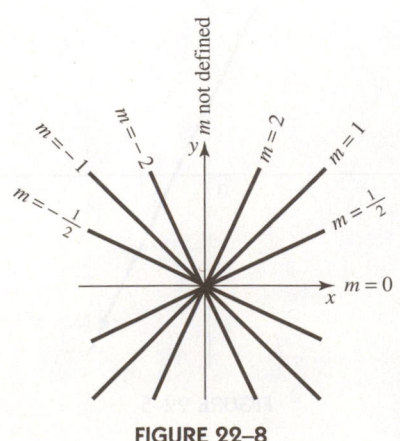

FIGURE 22–8

Slopes of Horizontal and Vertical Lines

For any two points on a horizontal line, the values of y_1 and y_2 in Eq. 205 are equal, making the slope equal to zero. For a vertical line, the values of x_1 and x_2 are equal, giving division by zero. Hence the slope is undefined for a vertical line. The slopes of various lines are shown in Fig. 22–8.

Angle of Inclination

Another measure of the steepness of a line is the *angle of inclination*, θ. It is defined as the smallest positive angle that the line makes with the positive x axis, Fig. 22–9.

From Fig. 22–4,

$$\tan \theta = \frac{\text{opposite side}}{\text{adjacent side}} = \frac{y_2 - y_1}{x_2 - x_1} = \frac{\text{rise}}{\text{run}} = \frac{\Delta y}{\Delta x}$$

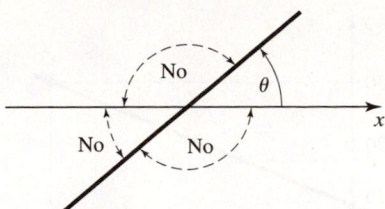

FIGURE 22–9 Angle of inclination, θ.

But this is the definition of the slope m of the line, so we have the following equations:

| Slope | $m = \tan \theta$
 $0° \leq \theta < 180°$

 The slope of a line is equal to the tangent of the angle of inclination θ (except when $\theta = 90°$). | 206 |

The angle of inclination can have values greater than or equal to 0° but less than 180°, with a horizontal line having an angle of inclination of 0° and a vertical line having an angle of inclination of 90°.

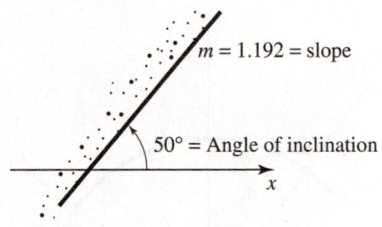

FIGURE 22–10 A coal chute.

◆◆◆ **Example 8:** *An Application.* The slope of a coal chute having an angle of inclination of 50° (Fig. 22–10) is, by Eq. 206,

$$m = \tan 50° = 1.192 \quad \text{(rounded)} \qquad \blacklozenge\blacklozenge\blacklozenge$$

◆◆◆ **Example 9:** Find the angle of inclination of a line having a slope of 3 (Fig. 22–11).

Solution: By Eq. 206, $\tan \theta = 3$, so

$$\theta = \arctan 3 = 71.6° \quad \text{(rounded)} \qquad \blacklozenge\blacklozenge\blacklozenge$$

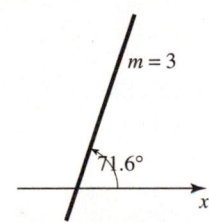

FIGURE 22–11

When the slope is negative, our calculator will give us a negative angle, which we then use to obtain a positive angle of inclination less than 180°.

◆◆◆ **Example 10:** For a line having a slope of -2 (Fig. 22–12), $\arctan(-2) = -63.4°$ (rounded), so

$$\theta = 180° - 63.4° = 116.6° \qquad \blacklozenge\blacklozenge\blacklozenge$$

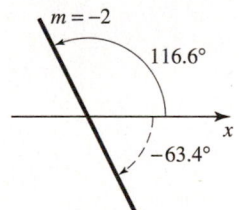

FIGURE 22–12

◆◆◆ **Example 11:** Find the angle of inclination of the line passing through $(-2.47, 1.74)$ and $(3.63, -4.26)$ (Fig. 22–13).

Solution: The slope, from Eq. 205, is

$$m = \frac{-4.26 - 1.74}{3.63 - (-2.47)} = -0.984$$

from which $\theta = 135.5°$. $\qquad \blacklozenge\blacklozenge\blacklozenge$

When *different scales* are used for the x and y axes, the angle of inclination *will appear distorted.*

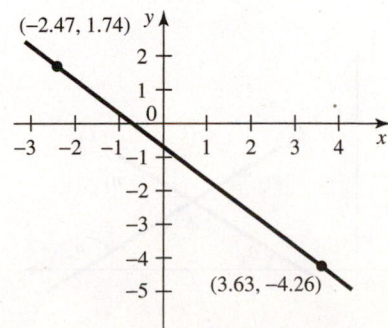

FIGURE 22–13

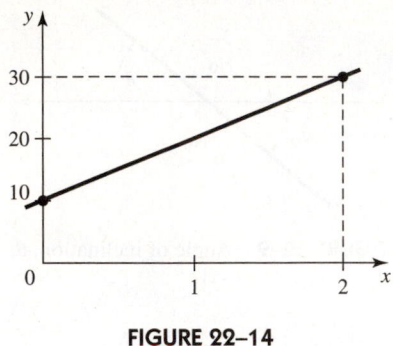

FIGURE 22–14

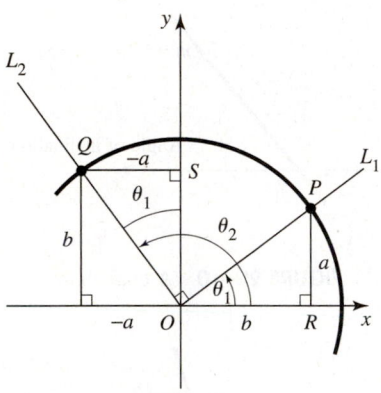

FIGURE 22–15 Slopes of perpendicular lines.

◆◆◆ Example 12: The angle of inclination of the line in Fig. 22–14 is $\theta = \arctan 10 = 84.3°$. Notice that the angle in the graph appears much smaller than 84.3° because of the different scales on each axis.   **◆◆◆**

Slopes of Parallel and Perpendicular Lines

Parallel lines have, of course, *equal* slopes.

Slopes of Parallel Lines	$m_1 = m_2$ *Parallel lines have equal slopes.*	213

Now let us draw a line L_1 having a slope of m_1 and an angle of inclination θ_1, as in Fig. 22–15. We add a second line L_2 perpendicular to L_1. Let its slope be m_2 and its angle of inclination be θ_2, where $\theta_2 = \theta_1 + 90°$ since the two lines are perpendicular. We choose a point P on L_1, draw a circle of radius OP, and mark point Q on L_2. Thus $OQ = OP$. Note that $\angle QOS = \theta_2 - 90° = (\theta_1 + 90°) - 90° = \theta_1$, so right triangles OPR and OQS are congruent. Thus the magnitudes of their corresponding sides a and b are equal. The slope m_1 of L_1 is a/b, and the slope m_2 of L_2 is $-b/a$, so we have the following equation:

Slopes of Perpendicular Lines	$m_1 = -\dfrac{1}{m_2}$ *The slope of a line is the negative reciprocal of the slope of a perpendicular to that line.*	214

◆◆◆ Example 13: Any line perpendicular to a line whose slope is 5 has a slope of $-\frac{1}{5}$.   **◆◆◆**

Common Error	The minus sign in Eq. 214 is often forgotten. $m_1 \neq \dfrac{1}{m_2}$

◆◆◆ Example 14: Find the slope of the line (a) parallel to and (b) perpendicular to a line having a slope of $-2/3$.

Solution:

(a) The parallel line has the same slope, or $-2/3$.
(b) The perpendicular line has a slope that is the negative reciprocal of $-2/3$, or $+3/2$.   **◆◆◆**

Angle of Intersection Between Two Lines

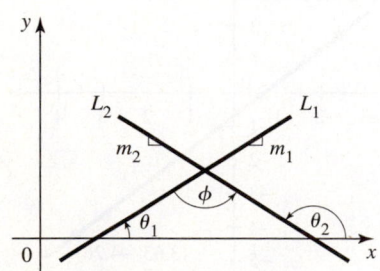

FIGURE 22–16 Angle of intersection between two lines.

Figure 22–16 shows two lines L_1 and L_2 intersecting at an angle ϕ, measured counterclockwise from line 1 to line 2. We want a formula for ϕ in terms of the slopes m_1 and m_2 of the lines. Since an exterior angle equals the sum of the two opposite interior angles, Eq. 107, θ_2 equals the sum of the angle of inclination θ_1 and the angle of intersection ϕ. So $\phi = \theta_2 - \theta_1$. Taking the tangent of both sides gives $\tan \phi = \tan(\theta_2 - \theta_1)$, or

$$\tan \phi = \frac{\tan \theta_2 - \tan \theta_1}{1 + \tan \theta_1 \tan \theta_2}$$

by the trigonometric identity for the tangent of the difference of two angles (Eq. 130). But the tangent of an angle of inclination is the slope, so we have

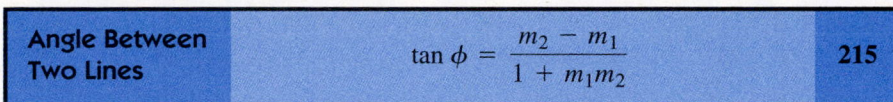

Angle Between Two Lines	$\tan \phi = \dfrac{m_2 - m_1}{1 + m_1 m_2}$	215

◆◆◆ **Example 15:** The tangent of the angle of intersection between line L_1, having a slope of 4, and line L_2, having a slope of -1, is

$$\tan \phi = \frac{-1 - 4}{1 + 4(-1)} = \frac{-5}{-3} = \frac{5}{3}$$

from which $\phi = \arctan \frac{5}{3} = 59.0°$ (rounded). This angle is measured counterclockwise from line 1 to line 2, as shown in Fig. 22–17. ◆◆◆

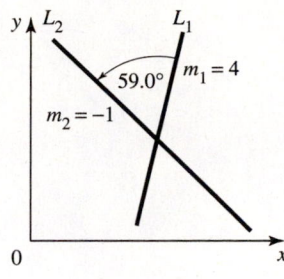

FIGURE 22–17

Exercise 1 ◆ The Straight Line

Directed Distance

Find the directed distance AB.

1. $A(3, 0)$; $B(5, 0)$
2. $B(-6, -6)$; $A(-6, -7)$
3. $B(-8, -2)$; $A(-8, -5)$
4. $A(-9, -2)$; $B(17, -2)$
5. $A(3.95, -2.07)$; $B(-3.95, -2.07)$
6. $B(11.5, 3.68)$; $A(11.5, -5.38)$

Increments

A particle moves from point A to point B. Find the increments Δx and Δy in its coordinates.

7. $A(2, 4)$, $B(5, 7)$
8. $A(3, 6)$, $B(6, 4)$
9. $A(-4, 4)$, $B(5, -8)$
10. $A(-9, -5)$, $B(-3, -8)$

Length of a Line Segment

Find the length of the line segment with the given endpoints.

11. $(5, 0)$ and $(2, 0)$
12. $(0, 3)$ and $(0, -5)$
13. $(-2, 0)$ and $(7, 0)$
14. $(-4, 0)$ and $(-6, 0)$
15. $(0, -2.74)$ and $(0, 3.86)$
16. $(55.34, 0)$ and $(25.38, 0)$
17. $(5.59, 3.25)$ and $(8.93, 3.25)$
18. $(-2.06, -5.83)$ and $(-2.06, -8.34)$
19. $(8.38, -3.95)$ and $(2.25, -4.99)$

Slope

Find the slope of each straight line.

20. Rise $= 4$; run $= 2$
21. Rise $= 6$; run $= 4$
22. Rise $= -4$; run $= 4$
23. Rise $= -9$; run $= -3$
24. Connecting $(2, 4)$ and $(5, 7)$
25. Connecting $(5, 2)$ and $(3, 6)$
26. Connecting $(-2, 5)$ and $(5, -6)$
27. Connecting $(3, -3)$ and $(-6, 2)$

Angle of Inclination

Find the slope of the line having the given angle of inclination.

28. $38.2°$
29. $77.9°$
30. 1.83 rad
31. $58°14'$
32. $156.3°$
33. $132.8°$

Find the angle of inclination, in decimal degrees to three significant digits, of a line having the given slope.

34. $m = 3$ **35.** $m = 1.84$ **36.** $m = -4$

37. $m = -2.75$ **38.** $m = 0$ **39.** $m = -15$

Find the angle of inclination, in decimal degrees to three significant digits, of a line passing through the given points.

40. $(5, 2)$ and $(-3, 4)$ **41.** $(-2.5, -3.1)$ and $(5.8, 4.2)$

42. $(6, 3)$ and $(-1, 5)$ **43.** $(x, 3)$ and $(x + 5, 8)$

Slopes of Parallel and Perpendicular Lines

Find the slopes of the lines parallel to, and perpendicular to, each line with the given slope.

44. $m = 5$ **45.** $m = 2$ **46.** $m = 4.8$

47. $m = -1.85$ **48.** $m = -2.85$ **49.** $m = -5.372$

Angle Between Two Lines

50. Find the angle of intersection between line L_1 having a slope of 1 and line L_2 having a slope of 6.

51. Find the angle of intersection between line L_1 having a slope of 3 and line L_2 having a slope of -2.

52. Find the angle of intersection between line L_1 having an angle of inclination of 35° and line L_2 having an angle of inclination of 160°.

53. Find the angle of intersection between line L_1 having an angle of inclination of 22° and line L_2 having an angle of inclination of 86°.

Applications

54. Find the length of girder AB in Fig. 22–18.
55. Find the distance between the centers of the holes in Fig. 22–19.
56. A triangle has vertices at $(3, 5)$, $(-2, 4)$, and $(4, -3)$. Find the length of each side. Then compute the area using Hero's formula (Eq. 103). Work to three significant digits.

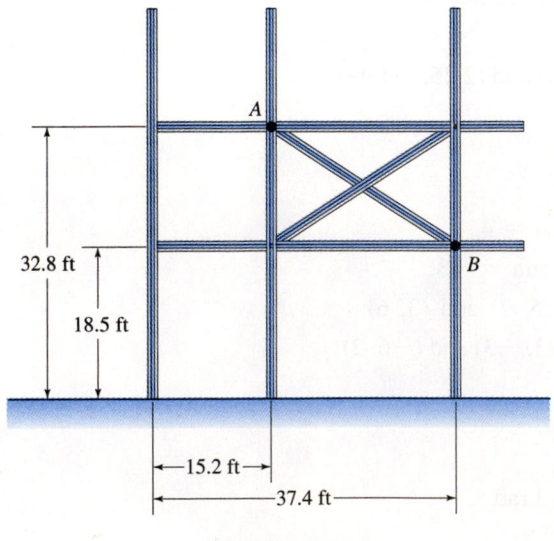

FIGURE 22–18

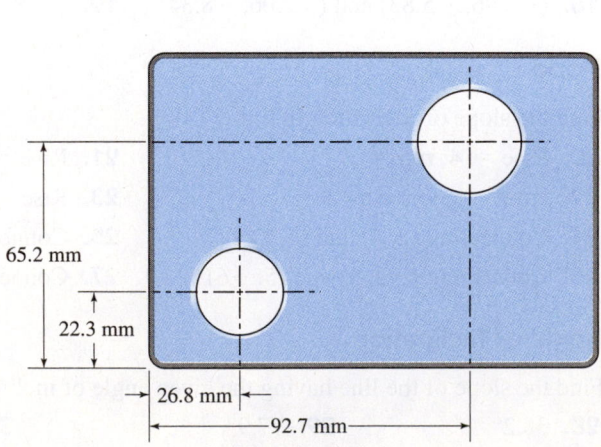

FIGURE 22–19

57. The distance between two stakes on a slope is taped at 2055 ft, and the angle of the slope with the horizontal is 12.3°. Find the horizontal distance between the stakes.

58. What is the angle of inclination with the horizontal of a roadbed that rises 15.0 ft in each 250 ft, measured horizontally?

59. How far apart must two stakes on a 7° slope be placed so that the horizontal distance between them is 1250 m?

60. In some fields such as highway design, the word *grade* is used instead of slope. To find the grade, multiply the slope by 100 and affix the % sign. Thus a 5% grade rises 5 units for every 100 units of run. On a 5% road grade, at what angle is the road inclined to the horizontal? How far does one rise in traveling uphill 500 ft, measured along the road?

61. A straight tunnel under a river is 755 ft long and descends 12.0 ft in this distance. What angle does the tunnel make with the horizontal?

62. A straight driveway slopes downward from a house to a road and is 28.0 m in length. If the angle of inclination from the road to the house is 3.60°, find the height of the house above the road.

63. An escalator is built so as to rise 2.00 m for each 3.00 m of horizontal travel. Find its angle of inclination.

64. A straight highway makes an angle of 4.50° with the horizontal. How much does the highway rise in a distance of 2500 ft, measured along the road?

22–2 Equation of a Straight Line

We said earlier that one value of analytic geometry was the ability to write an equation for a geometric figure. Then we could analyze that figure using algebra. Here we will write an equation for a geometric figure, the straight line. Such an equation would relate y to x for any point on the figure, so that given one coordinate, we could find the other from the equation.

Slope-Intercept Form

We covered the slope-intercept form of the equation of a straight line in our chapter on graphing, and we will repeat some of that material here. We will also show a few other forms of the straight line equation.

One way we can get such an equation is from our definition of slope. Let $P(x, y)$ be any point on a line, Fig. 22–20, and let the y intercept, $(0, b)$ be a second point on the line. Between P and the y intercept, the rise is $(y - b)$ and the run is $(x - 0)$. The slope m of the line is then

$$m = \frac{y - b}{x - 0}$$

Simplifying, we have $mx = y - b$, or

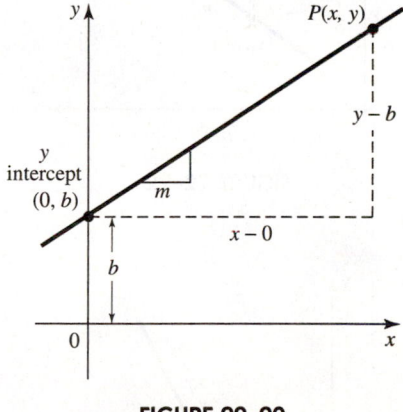

FIGURE 22–20

Straight Line Slope-Intercept Form	$y = mx + b$	210

This is called the *slope-intercept form* of the equation of a straight line because the slope m and the y intercept b are easily identified once the equation is in this form. For example, in the equation $y = 2x + 1$,

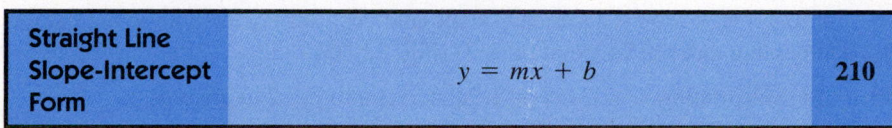

Here m, the slope, is 2 and b, the y intercept, is 1.

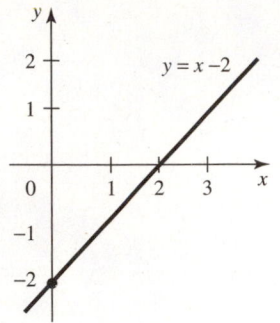

FIGURE 22-21

♦♦♦ **Example 16:** Write the equation, in slope-intercept form, of the straight line that has a slope of 1 and a y intercept of –2.

Solution: Substituting into Eq. 210 with $m = 1$ and $b = -2$

$$y = x - 2$$

For our graph we first plot the y intercept 2 units down from the origin. From there we lay out a rise of 1 unit in a run of 1 unit to get another point, and then connect the two points to get our graph, Fig. 22–21.   ♦♦♦

■ **Exploration:**

Try this. In the same viewing window,

(a) graph several lines $y = mx + b$, taking $m = 1$ and values of b of $-2, -1, 0, 1, 2$. What do you conclude is the effect of the magnitude and algebraic sign of the constant b?

(b) graph several lines $y = mx + b$, taking $b = 0$ and values of m of $-2, -1, 0, 1, 2$. What do you conclude is the effect of the magnitude and algebraic sign of the constant m?   ■

Your explorations should have clearly shown that

• the value of the y intercept b determines how far above or below the origin the line cuts the y axis, and

• the value of the slope m determines the steepness of the line, either in the positive or negative direction.

Point-Slope Form

We can quickly write the equation of a straight line, as we have seen, when we know its slope and y intercept. But what if we have *other* information instead? Here we will show how to write the equation of a line if we know one point on the line and the slope of the line.

Let us first write an equation for a line that has slope m, but that passes through a given point (x_1, y_1) which is *not*, in general, on either axis, as in Fig. 22–22. Again using the definition of slope (Eq. 205), with a general point (x, y), we get the following form:

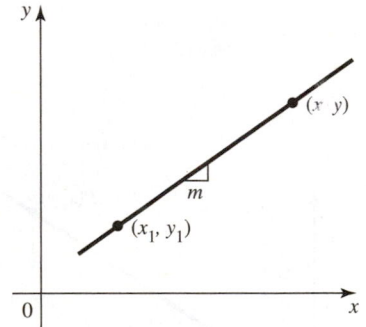

FIGURE 22-22

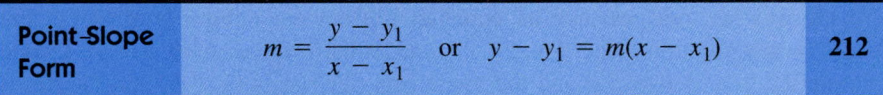

Point-Slope Form	$m = \dfrac{y - y_1}{x - x_1}$ or $y - y_1 = m(x - x_1)$	212

While this *point-slope* form will enable us to quickly write the equation, it is not the best for graphing by calculator. For that we prefer the equation in explicit form. So we will usually convert our equation to slope-intercept form.

♦♦♦ **Example 17:** Write the equation in slope-intercept form of the line having a slope of 2 and passing through the point $(1, -3)$ (Fig. 22–23).

Solution: Substituting $m = 2$, $x_1 = 1$, $y_1 = -3$ into Eq. 212 gives us

$$2 = \frac{y - (-3)}{x - 1}$$

Multiplying by $x - 1$ yields

$$2x - 2 = y + 3$$

or, in slope-intercept form,

$$y = 2x - 5$$ ♦♦♦

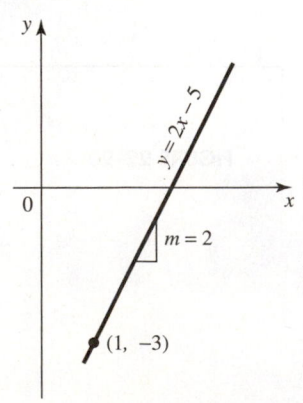

FIGURE 22-23

✦✦✦ Example 18: Write the equation in slope-intercept form of the line passing through the point (3, 2) and perpendicular to the line $y = 3x - 7$ (Fig. 22–24).

Solution: The slope of the given line is 3, so the slope of our perpendicular line is $-\frac{1}{3}$. Using the point-slope form, we obtain

$$-\frac{1}{3} = \frac{y - 2}{x - 3}$$

Going to slope-intercept form, $x - 3 = -3y + 6$, or

$$y = -\frac{x}{3} + 3 \qquad \text{✦✦✦}$$

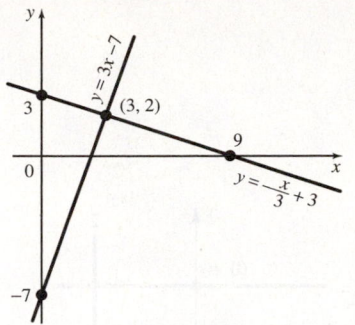

FIGURE 22–24

Two-Point Form

If two points on a line are known, the equation of the line is easily written using the two-point form, which we now derive. If we call the points P_1 and P_2 in Fig. 22–25, the slope of the line is

$$m = \frac{y_2 - y_1}{x_2 - x_1}$$

The slope of the line segment connecting P_1 with any other point P on the same line is

$$m = \frac{y - y_1}{x - x_1}$$

Since these slopes must be equal, we get the following equation:

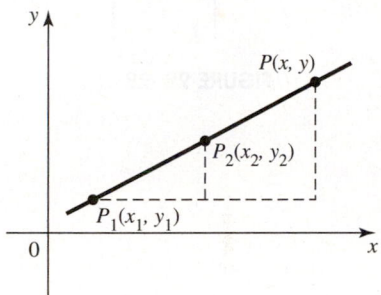

FIGURE 22–25

Two-Point Form	$\dfrac{y - y_1}{x - x_1} = \dfrac{y_2 - y_1}{x_2 - x_1}$	**211**

✦✦✦ Example 19: Write the equation in slope-intercept form of the line passing through the points $(1, -3)$ and $(-2, 5)$ (Fig. 22–26).

Solution: Calling the first given point P_1 and the second P_2 and substituting into Eq. 211, we have

$$\frac{y - (-3)}{x - 1} = \frac{5 - (-3)}{-2 - 1} = \frac{8}{-3}$$

$$\frac{y + 3}{x - 1} = -\frac{8}{3}$$

Putting the equation into slope-intercept form, we have $3y + 9 = -8x + 8$, or

$$y = -\frac{8}{3}x - \frac{1}{3} \qquad \text{✦✦✦}$$

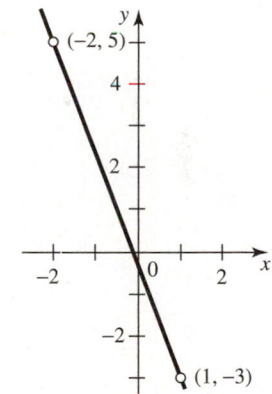

FIGURE 22–26

✦✦✦ Example 20: *An Application.* The *inrun,* the straight portion of a certain ski jump is 125 m long and has an angle of inclination of 31.0°, Fig. 22–27. (a) Write the equation of the inrun, taking axes as shown, and (b) find the heights h_1 and h_2 of the support posts.

Solution: (a) The slope is

$$m = \tan 31.0° = 0.601$$

and the y intercept is

$$b = 28.2$$

so our equation is

$$y = 0.601x + 28.2$$

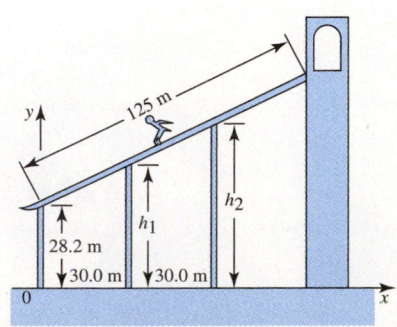

FIGURE 22–27 A ski jump.

(b) Substituting into this equation to get the heights at the given x values,

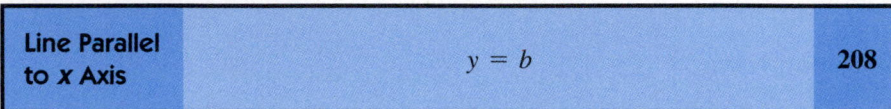

$$h_1 = y(30.0) = 0.601(30.0) + 28.2 = 46.2 \text{ m}$$
$$h_2 = y(60.0) = 0.601(60.0) + 28.2 = 64.3 \text{ m}$$ ◆◆◆

Lines Parallel to the Coordinate Axes

Line 1 in Fig. 22–28 is parallel to the x axis. Its slope is 0 and it cuts the y axis at $(0, b)$. From the point-slope form, $y - y_1 = m(x - x_1)$, we get $y - b = 0(x - 0)$, or the following:

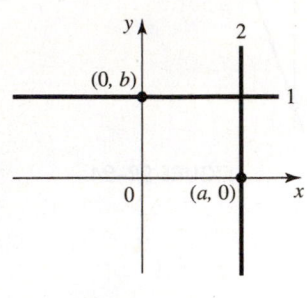

FIGURE 22–28

Line Parallel to *x* Axis	$y = b$	208

Line 2 has an undefined slope, but we get its equation by noting that $x = a$ at every point on the line, *regardless of the value of y*. Thus

Line Parallel to *y* Axis	$x = a$	209

◆◆◆ **Example 21:**

(a) A line that passes through the point $(5, -2)$ and is parallel to the x axis has the equation $y = -2$.
(b) A line that passes through the point $(5, -2)$ and is parallel to the y axis has the equation $x = 5$.   ◆◆◆

Equation of a Straight Line in General Form

The slope-intercept form is the equation of a straight line in *explicit* form, $y = f(x)$. We can also write the equation of a straight line in *implicit* form, $f(x, y) = 0$, simply by transposing all terms to one side of the equal sign and simplifying. We usually write the x term first, then the y term, and finally the constant term. This form is referred to as the *general form* of the equation of a straight line.

General Form	$Ax + By + C = 0$	207

◆◆◆ **Example 22:** Change the equation $y = \frac{6}{7}x - 5$ from slope-intercept form to general form.

Solution: Subtracting y from both sides, we have

$$0 = \frac{6}{7}x - 5 - y$$

Multiplying by 7 and rearranging gives $6x - 7y - 35 = 0$.   ◆◆◆

◆◆◆ **Example 23:** Rewrite this general equation in slope-intercept form.

$$2x - 3y + 5 = 0$$

Solution: Solving for y,

$$3y = 2x + 5$$

$$y = \frac{2}{3}x + \frac{5}{3}$$ ◆◆◆

Exercise 2 ◆ Equation of a Straight Line

Write the equation of each straight line in slope-intercept form, and make a graph.

1. Slope $= 4$; y intercept $= -3$
2. Slope $= -1$; y intercept $= 2$
3. Slope $= 3$; y intercept $= -1$
4. Slope $= -2$; y intercept $= 3$
5. Slope $= 2.3$; y intercept $= -1.5$
6. Slope $= -1.5$; y intercept $= 3.7$

Write the equation of each line in Fig. 22–29 in general form.

7. line A **8.** line B
9. line C **10.** line D

Find the slope and the y intercept for each equation, and make a graph.

11. $y = 3x - 5$ **12.** $y = 7x + 2$
13. $y = -\frac{1}{2}x - \frac{1}{4}$ **14.** $y = -3x + 2$

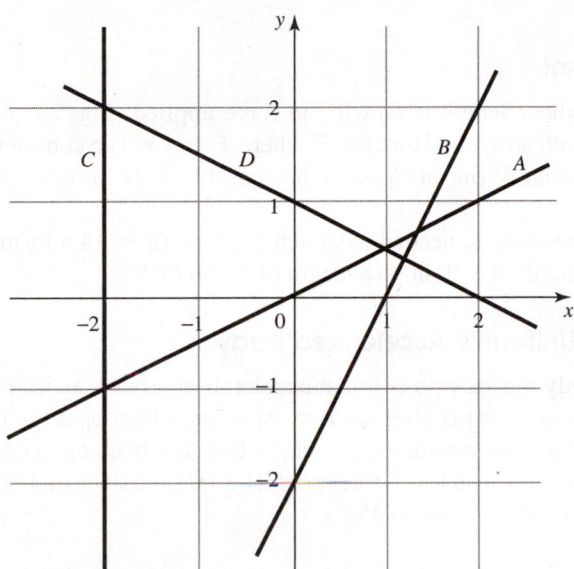

FIGURE 22–29

Write the equation of each line in slope-intercept form.

15. Slope $= 2$; passes through $(3, 4)$
16. Slope $= 3$; passes through $(2, 5)$

Write the equation of each line in general form.

17. Slope $= -4$; passes through $(-2, 5)$
18. Slope $= -2$; passes through $(-2, -3)$
19. y intercept $= 3$; parallel to $y = 5x - 2$
20. y intercept $= -2.3$; parallel to $2x - 3y + 1 = 0$
21. y intercept $= -5$; perpendicular to $y = 3x - 4$
22. y intercept $= 2$; perpendicular to $4x - 3y = 7$
23. passes through $(-2, 5)$; parallel to $y = 5x - 1$
24. passes through $(4, -1)$; parallel to $4x - y = -3$

25. passes through $(-4, 2)$; perpendicular to $y = 5x - 3$

26. passes through $(6, 1)$; perpendicular to $6y - 2x = 3$

27. passes through $(5, 2)$; is parallel to the x axis.

28. passes through $(-3, 6)$; is parallel to the y axis.

29. passes through points $(3, 5)$ and $(-1, 2)$

30. passes through points $(4.24, -1.25)$ and $(3.85, 4.27)$

31. x intercept $= 5$; y intercept $= -3$

32. x intercept $= -2$; y intercept $= 6$

Rewrite each equation in general form.

33. $y = 3x + 2$ **34.** $y = 4x - 5$

35. $y = -2x + 6$ **36.** $y = -4x - 3$

Rewrite each equation in slope-intercept form.

37. $x + 2y + 5 = 0$ **38.** $3x + 4y - 6 = 0$

39. $-7x + 3y + 8 = 0$ **40.** $-2x + 2y - 9 = 0$

Applications

Spring Constant

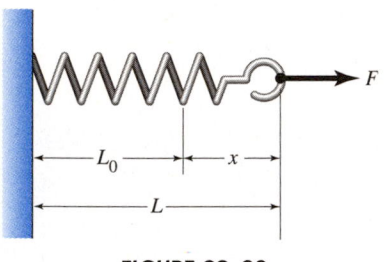

FIGURE 22–30

41. A spring whose length is L_0 with no force applied (Fig. 22–30) stretches an amount x with an applied force of F, where $F = kx$. The constant k is called the *spring constant*. Write an equation, in slope-intercept form, for F in terms of k, L, and L_0.

42. What force would be needed to stretch a spring ($k = 14.5$ lb/in.) from an unstretched length of 8.50 in. to a length of 12.50 in.?

Velocity of a Uniformly Accelerated Body

43. When a body moves with constant acceleration a (such as in free fall), its velocity v at any time t is given by $v = v_0 + at$, where v_0 is the initial velocity. Note that this is the equation of a straight line. If a body has a constant acceleration of 2.15 m/s^2 and has a velocity of 21.8 m/s at 5.25 s, find (a) the initial velocity and (b) the velocity at 25.0 s.

Resistance Change with Temperature

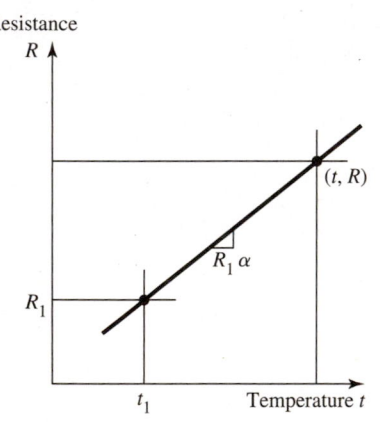

FIGURE 22–31   Resistance change with temperature.

44. The resistance of metal is a linear function of the temperature (Fig. 22–31) for certain ranges of temperature. The slope of the line is $R_1\alpha$, where α is the temperature coefficient of resistance at temperature t_1. If R is the resistance of any temperature t, write an equation for R as a function of t.

45. Using a value for α of $\alpha = 1/(234.5t_1)$ (for copper), find the resistance of a copper conductor at 75.0°C if its resistance at 20.0°C is 148.4 Ω.

46. If the resistance of the copper conductor in problem 45 is 1255 Ω at 20.0°C, at what temperature will the resistance be 1265 Ω?

Thermal Expansion

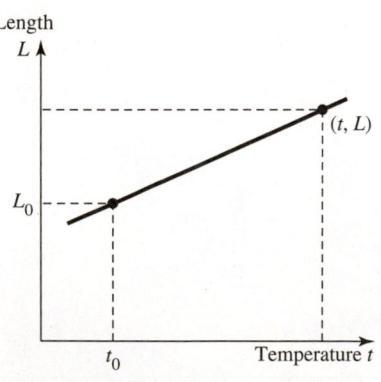

FIGURE 22–32   Thermal expansion.

47. When a bar is heated, its length will increase from an initial length L_0 at temperature t_0 to a new length L at temperature t. The plot of L versus t is a straight line (Fig. 22–32) with a slope of $L_0\alpha$, where α is the coefficient of thermal expansion. Derive the equation $L = L_0(1 + \alpha \, \Delta t)$, where Δt is the change in temperature, $t - t_0$.

48. A steel pipe is 21.50 m long at 0°C. Find its length at 75.0°C if α for steel is 12.0×10^{-6} per Celsius degree.

Fluid Pressure

49. The pressure at a point located at a depth x ft from the surface of a liquid varies directly as the depth. If the pressure at the surface is 20.6 lb/in.^{2} and increases by 0.432 lb/in.^{2} for every foot of depth, write an equation for P as a function of the depth x (in feet). At what depth will the pressure be 30.0 lb/in.^{2}?

50. A straight pipe slopes downward from a reservoir to a water turbine (Fig. 22–33). The pressure head at any point in the pipe, expressed in feet, is equal to the vertical distance between the point and the surface of the reservoir. If the reservoir surface is 25 ft above the upper end of the pipe, write an expression for the head H as a function of the horizontal distance x. At what distance x will the head be 35 ft?

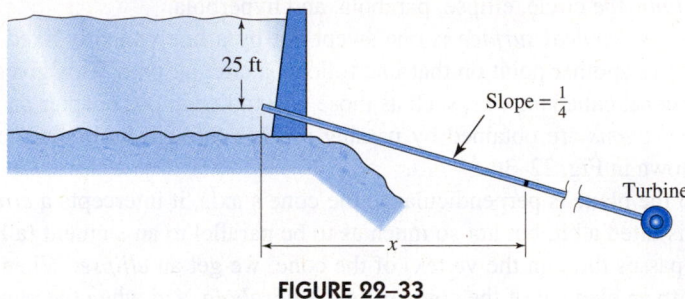

FIGURE 22–33

Temperature Gradient

51. Figure 22–34 shows a uniform wall whose inside face is at temperature t_i and whose outside face is at t_o. The temperatures within the wall plot as a straight line connecting t_i and t_o. Write the equation $t = f(x)$ of that line, taking $x = 0$ at the inside face, if $t_i = 25.0°C$ and $t_o = -5.0°C$. At what x will the temperature be 0°C? What is the slope of the line?

Hooke's Law

52. The increase in length of a wire in tension is directly proportional to the applied load P. Write an equation for the length L of a wire that has an initial length of 3.00 m and that stretches 1.00 mm for each 12.5 N. Find the length of the wire with a load of 750 N.

Straight-Line Depreciation

53. The straight-line method is often used to depreciate a piece of equipment for tax purposes. Starting with the purchase price P, the item is assumed to drop in value the same amount each year (the annual depreciation) until the salvage value S is reached (Fig. 22–35). Write an expression for the book value y (the value at any time t) as a function of the number of years t. For a lathe that cost $15,428 and has a salvage value of $2264, find the book value after 15 years if it is depreciated over a period of 20 years.

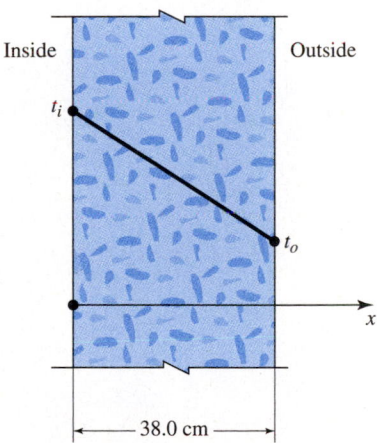

FIGURE 22–34 Temperature drop in a wall. The slope of the line (in °C/cm) is called the *temperature gradient*. The amount of heat flowing through the wall is proportional to the temperature gradient.

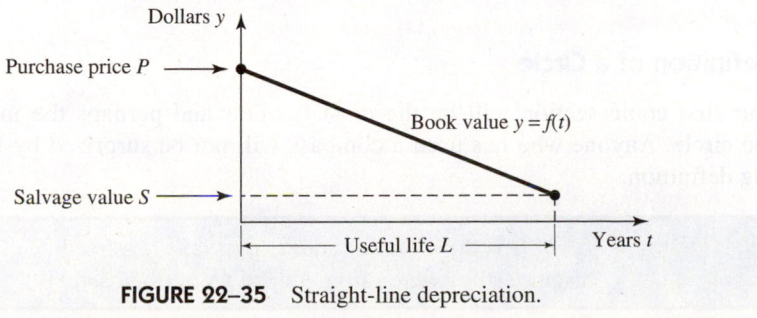

FIGURE 22–35 Straight-line depreciation.

54. *Temperature Conversions*: The freezing point of water is 32° Fahrenheit (F), or 0° Celsius (C). The boiling point of water is 212°F or 100°C. The curve connecting these two point pairs is a straight line. Use the two-point form of the equation of a straight line to derive an equation connecting degrees Fahrenheit and degrees Celsius. (We use this equation to convert between Fahrenheit and Celsius.)

22–3   The Circle

The Conic Sections

We turn now from the straight line to a very interesting group of curves called the *conic sections:* the circle, ellipse, parabola, and hyperbola.

A *circular conical surface* is one swept out by a line which is fixed at a point (the vertex) as another point on that line follows a circular path. This gives an upper and lower cone, called *nappes*, such as those we studied in our chapter on geometry. The *conic sections* are obtained by passing a plane through a circular conical surface, as shown in Fig. 22–36.

When the plane is perpendicular to the cone's axis, it intercepts a *circle*. When the plane is tilted a bit, but not so much as to be parallel to an element (a line on the cone that passes through the vertex) of the cone, we get an *ellipse*. When the plane is parallel to an element of the cone, we get a *parabola*, and when the plane is steep enough to cut the upper cone, we get the two-branched *hyperbola*.

> The great Greek geometer Apollonius of Perga (c. 255–170 B.C.E.) is credited as the first to define the conic sections in this way.

■ Exploration:

Try this. Make a model of the conic sections. Construct a cone (from clay, styrofoam, plaster poured into a conical cardboard mold, a wood turning, etc.) and cut it as shown to form any of the four curves.   ■

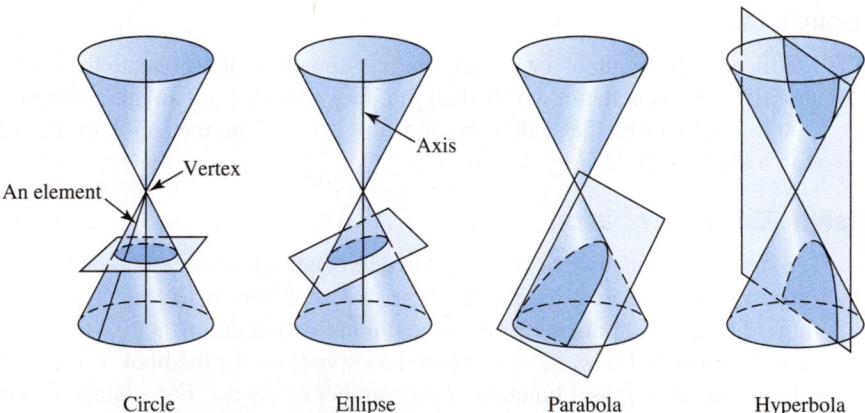

| Circle | Ellipse | Parabola | Hyperbola |

FIGURE 22–36   Conic sections.

Definition of a Circle

Our first conic section will be the most familiar and perhaps the most useful: the circle. Anyone who has used a compass will not be surprised by the following definition.

| **Definition of a Circle** | A *circle* is the set of all points in a plane at a fixed distance (the *radius*) from a fixed point (the *center*). | **217** |

Standard Equation of a Circle: Center at Origin

We will now derive an equation for a circle of radius r, starting with the simplest case of a circle whose center is at the origin (Fig. 22–37). Let x and y be the coordinates of any point P on the circle. The equation we develop will give a relationship between x and y that will have two meanings: geometrically, (x, y) will represent a point on the circle; and algebraically, the numbers corresponding to those points that satisfy that equation.

We observe that, by the definition of a circle, the distance OP must be constant and equal to r. But, by the distance formula (Eq. 204),

$$OP = \sqrt{x^2 + y^2} = r$$

Squaring, we get a *standard equation of a circle* (also called the *standard form* of the equation).

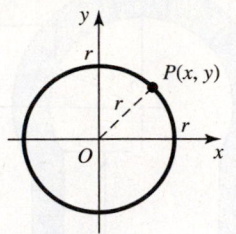

FIGURE 22–37 Circle with center at origin.

Standard Equation, Circle of Radius r: Center at Origin	$x^2 + y^2 = r^2$	218

Note that both x^2 and y^2 have the same coefficient. Otherwise, the graph is not a circle.

◆◆◆ **Example 24:** Write, in standard form, the equation of a circle of radius 3, whose center is at the origin.

Solution:
$$x^2 + y^2 = 3^2 = 9 \qquad ◆◆◆$$

◆◆◆ **Example 25:** Find the center and radius of the circle with the equation

$$x^2 + y^2 = 64$$

Solution: We recognize this as the equation of a circle whose center is at the origin. Finding the radius,

$$r = \sqrt{64} = 8 \qquad ◆◆◆$$

◆◆◆ **Example 26:** *An Application. Circular Arches.* Portions of a circle are commonly used for arches, as shown in Fig. 22–38. (a) Write the equation for the circle in the

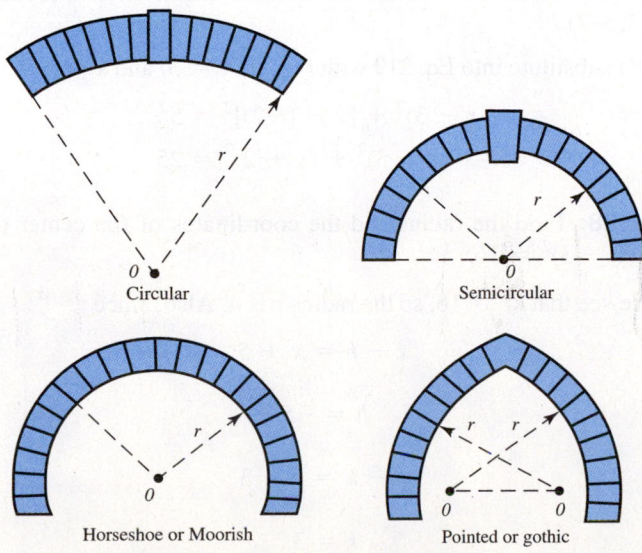

Circular

Semicircular

Horseshoe or Moorish

Pointed or gothic

FIGURE 22–38 Arches based on the circle.

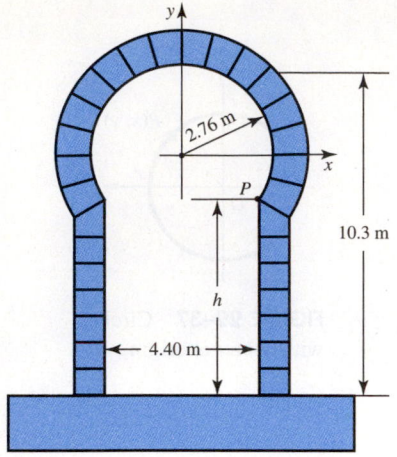

FIGURE 22–39 Horseshoe arch over a doorway.

horseshoe arch (also called a *Moorish* arch), Fig. 22–39. Take the origin at the center of the circle. (b) Use your equation to find h.

Solution: (a) The circle has a radius of 2.76 m, so substituting into the equation for a circle with center at the origin, Eq. 218, gives

$$x^2 + y^2 = (2.76)^2$$

(b) The x coordinate of point P is half the width of the doorway, or 2.20 m.

Substituting into our equation gives

$$(2.20)^2 + y^2 = (2.76)^2$$

Solving for y,

$$y^2 = (2.76)^2 - (2.20)^2$$

$$y^2 = 2.78$$

$$y = \pm 1.67 \text{ m}$$

We take the negative value, since P is below the center of the circle. Finally,

$$h = 10.3 - 2.76 - 1.67 = 5.87 \text{ m}$$ ◆◆◆

Standard Equation of a Circle: Center Not at the Origin

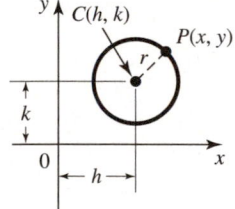

FIGURE 22–40 Circle with center at (h, k).

Figure 22–40 shows a circle whose center has the coordinates (h, k). We can think of the difference between this circle and the one in Fig. 22–37 as having its center moved or *translated* h units to the right and k units upward. Our derivation is similar to the preceding one.

$$CP = r = \sqrt{(x - h)^2 + (y - k)^2}$$

Squaring, we get the following equation:

Standard Equation, Circle of Radius *r*: Center at *(h, k)*	$(x - h)^2 + (y - k)^2 = r^2$	219

◆◆◆ **Example 27:** Write, in standard form, the equation of a circle of radius 5 whose center is at $(3, -2)$.

Solution: We substitute into Eq. 219 with $r = 5$, $h = 3$, and $k = -2$.

$$(x - 3)^2 + [y - (-2)]^2 = 5^2$$
$$(x - 3)^2 + (y + 2)^2 = 25$$ ◆◆◆

◆◆◆ **Example 28:** Find the radius and the coordinates of the center of the circle $(x + 5)^2 + (y - 3)^2 = 16$.

Solution: We see that $r^2 = 16$, so the radius r is 4. Also, since

$$x - h = x + 5$$

then

$$h = -5$$

and since

$$y - k = y - 3$$

then

$$k = 3$$

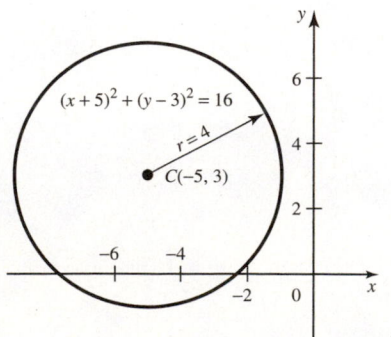

FIGURE 22–41

So the center is at $(-5, 3)$ as shown in Fig. 22–41.   ◆◆◆

Common Error	It is easy to get the signs of h and k wrong. In Example 28, do *not* take $$h = +5 \quad \text{and} \quad k = -3$$

Graphing a Circle by Calculator

A computer algebra system may be able to graph an equation entered in implicit form. However, to graph a circle with a graphics calculator and most graphics utilities, we must put an equation in explicit form by solving for y. When we do that we get *two* functions, one for the upper half of the circle and another for the lower half. To get the complete circle, we must graph *both*.

◆◆◆ **Example 29:** Graph the circle

$$(x - 2)^2 + (y + 1)^2 = 15$$

Solution: We first solve for y.

$$(y + 1)^2 = 15 - (x - 2)^2$$

$$y + 1 = \pm\sqrt{15 - (x - 2)^2}$$

$$y = -1 \pm \sqrt{15 - (x - 2)^2}$$

$$y_1 = -1 + \sqrt{15 - (x - 2)^2}$$

$$y_2 = -1 - \sqrt{15 - (x - 2)^2}$$

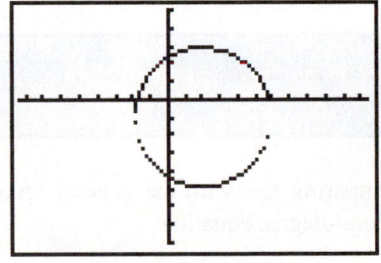

TI-83/84 Screen for Example 29.

Our graph will sometimes show a gap because of the spacing of pixels on the calculator screen. ◆◆◆

Screens for Example 29 show the $\boxed{Y =}$ screen and the graph. Note that the graph may not appear circular on a particular grapher, but some have a feature to adjust the horizontal and vertical scales so that circles do appear circular. On TI calculators that feature is $\boxed{\text{ZOOM}}$ Square.

If we know in advance that a curve is symmetrical about the x axis, we could simply enter **Y2 = –Y1** instead of entering the entire equation (Y1 is found in the $\boxed{\text{VARS}}$ menu). We will show this method later.

Translation of Axes

We see that Eq. 219 for a circle with its center at (h, k) is almost identical to Eq. 218 for a circle with center at the origin, *except that x has been replaced by $(x - h)$ and y has been replaced by $(y - k)$*. This same substitution will, of course, work for curves other than the circle, and we will use it to translate or *shift* the axes for the other conic sections.

Translation of Axes	To translate or shift the *axes* of a curve to the left by a distance h and downward by a distance k, replace x by $(x - h)$ and y by $(y - k)$ in the equation of the curve.

General Equation of a Circle

A second-degree equation in x and y which has all possible terms would have an x^2 term, a y^2 term, an xy term, and all terms of lesser degree as well. The general second-degree equation is usually written with terms in the following order, where A, B, C, D, E, and F are constants:

General Second-Degree Equation	$Ax^2 + Bxy + Cy^2 + Dx + Ey + F = 0$	216

There are six constants in this equation, but only five are independent. We can divide through by any constant and thus make it equal to 1.

To get the general equation of a circle we expand Eq. 219 and get

$$(x - h)^2 + (y - k)^2 = r^2$$
$$x^2 - 2hx + h^2 + y^2 - 2ky + k^2 = r^2$$

Rearranging gives

$$x^2 + y^2 - 2hx - 2ky + (h^2 + k^2 - r^2) = 0$$

If we now replace the constants $(-2h)$ by D, $(-2k)$ by E, and $(h^2 + k^2 - r^2)$ by F, we get.

General Equation of a Circle	$x^2 + y^2 + Dx + Ey + F = 0$	220

Comparing this with the general second-degree equation, we see that the general second-degree equation

$$Ax^2 + Bxy + Cy^2 + Dx + Ey + F = 0$$

represents a circle if $B = 0$ and $A = C$.

◆◆◆ **Example 30:** Write this standard equation in general form.

$$(x + 5)^2 + (y - 3)^2 = 16$$

Solution: Expanding, we obtain

$$x^2 + 10x + 25 + y^2 - 6y + 9 = 16$$

or

$$x^2 + y^2 + 10x - 6y + 18 = 0$$ ◆◆◆

Changing from General to Standard Form

When we want to go from general form to standard form, we must *complete the square*, both for x and for y. We learned how to complete the square in our chapter on quadratic equations. Glance back there if you need to refresh your memory.

◆◆◆ **Example 31:** Write the equation $2x^2 + 2y^2 - 18x + 16y + 60 = 0$ in standard form. Find the radius and center, and plot the curve.

Solution: We first divide by 2:

$$x^2 + y^2 - 9x + 8y + 30 = 0$$

and then separate the x and y terms:

$$(x^2 - 9x) + (y^2 + 8y) = -30$$

We complete the square for the x terms with $\frac{81}{4}$ because we take $\frac{1}{2}$ of (-9) and square it. Similarly, for completing the square for the y terms, $\left(\frac{1}{2} \times 8\right)^2 = 16$. Completing the square on the left and compensating on the right gives

$$\left(x^2 - 9x + \frac{81}{4}\right) + (y^2 + 8y + 16) = -30 + \frac{81}{4} + 16$$

Factoring each group of terms yields

$$\left(x - \frac{9}{2}\right)^2 + (y + 4)^2 = \left(\frac{5}{2}\right)^2$$

So

$$r = \frac{5}{2} \qquad h = \frac{9}{2} \qquad k = -4$$

The circle is shown in Fig. 22–42.   ◆◆◆

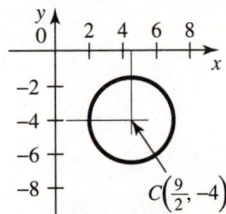

FIGURE 22–42

An Application

◆◆◆ **Example 32:** Write an equation for the circle shown in Fig. 22–43, and find the dimensions A and B produced by the circular cutting tool.

Estimate: The easiest way to get an estimate here is to make a sketch and measure the distances. We get

$$A \approx 2.6 \text{ in.} \quad \text{and} \quad B \approx 2.7 \text{ in.}$$

Solution: In order to write an equation, we must have coordinate axes. Otherwise the quantities x and y in the equation will have no meaning. Since no axes are given, we are free to draw them anywhere we please. Our first impulse might be to place the origin at the corner p, since most of the dimensions are referenced from that point, but

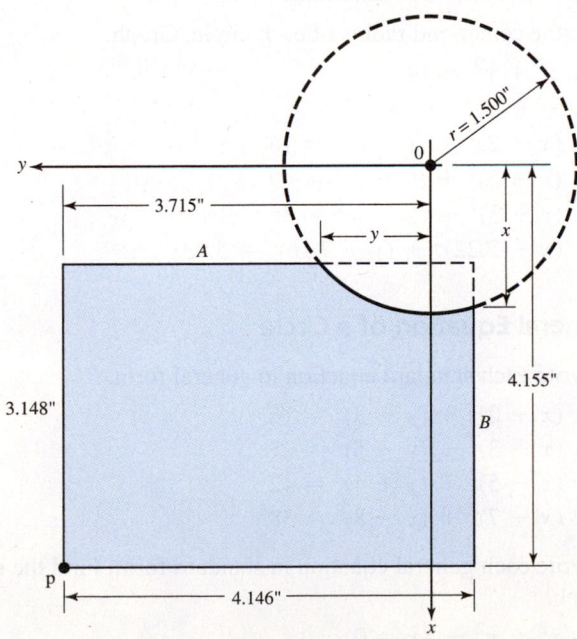

FIGURE 22–43

our equation will be simpler if we place the origin at the center of the circle. Let us draw axes reversed from our usual direction, as shown, so that not all our numbers will be negative. The equation of the circle is then

$$x^2 + y^2 = (1.500)^2$$
$$= 2.2500$$

The distance from the y axis to the top edge of the block is $4.155 - 3.148 = 1.007$ in. Substituting this value for x into the equation of the circle will give the corresponding value for y. When $x = 1.007$,

$$y^2 = 2.250 - (1.007)^2 = 1.236$$
$$y = 1.112 \text{ in.}$$

Similarly, the distance from the x axis to the right edge of the block is $4.146 - 3.715 = 0.431$ in. When $y = -0.431$,

$$x^2 = 2.250 - (-0.431)^2 = 2.064$$
$$x = 1.437$$

Finally,

$$A = 3.715 - 1.112 = 2.603 \text{ in.}$$

and

$$B = 4.155 - 1.437 = 2.718 \text{ in.} \quad \blacklozenge\blacklozenge\blacklozenge$$

Exercise 3 ◆ The Circle

Standard Equation of a Circle

Write the equation of each circle in standard form. Graph.

1. center at $(0, 0)$; radius $= 7$
2. center at $(0, 0)$; radius $= 4.82$
3. center at $(2, 3)$; radius $= 5$
4. center at $(5, 2)$; radius $= 10$
5. center at $(5, -3)$; radius $= 4$
6. center at $(-3, -2)$; radius $= 11$

Find the center and radius of each circle. Graph.

7. $x^2 + y^2 = 49$
8. $x^2 + y^2 = 64.8$
9. $(x - 2)^2 + (y + 4)^2 = 16$
10. $(x + 5)^2 + (y - 2)^2 = 49$
11. $(y + 5)^2 + (x - 3)^2 = 36$
12. $(x - 2.22)^2 + (y + 7.16)^2 = 5.93$

General Equation of a Circle

Rewrite each standard equation in general form.

13. $(x + 2)^2 + (y + 3)^2 = 16$
14. $(x + 3)^2 + (y - 5)^2 = 25$
15. $(x - 5)^2 + (y + 4)^2 = 42$
16. $(x - 7)^2 + (y - 8)^2 = 38$

Rewrite each general equation in standard form. Find the center and radius. Graph.

17. $x^2 + y^2 - 8x = 0$
18. $x^2 + y^2 - 2x - 4y = 0$
19. $x^2 + y^2 - 10x + 12y + 25 = 0$

20. $x^2 + y^2 - 4x + 2y = 36$
21. $x^2 + y^2 + 6x - 2y = 15$
22. $x^2 + y^2 - 2x - 6y = 39$

Applications

Even though you may be able to solve some of these with only the Pythagorean theorem, we suggest that you use analytic geometry for the practice.

23. Prove that any angle inscribed in a semicircle is a right angle.
24. Write the equation of the circle in Fig. 22–44, taking the axes as shown. Use your equation to find A and B.
25. Write the equations for each of the circular arches in Fig. 22–45, taking the axes as shown. Solve simultaneously to get the point of intersection P, and compute the height h of the column.
26. Write the equation of the centerline of the circular street shown in Fig. 22–46, taking the origin at the intersection O. Use your equation to find the distance y.
27. Each side of a Gothic arch is a portion of a circle. Write the equation of one side of the Gothic arch shown in Fig. 22–47. Use the equation to find the width w of the arch at a height of 3.00 ft.

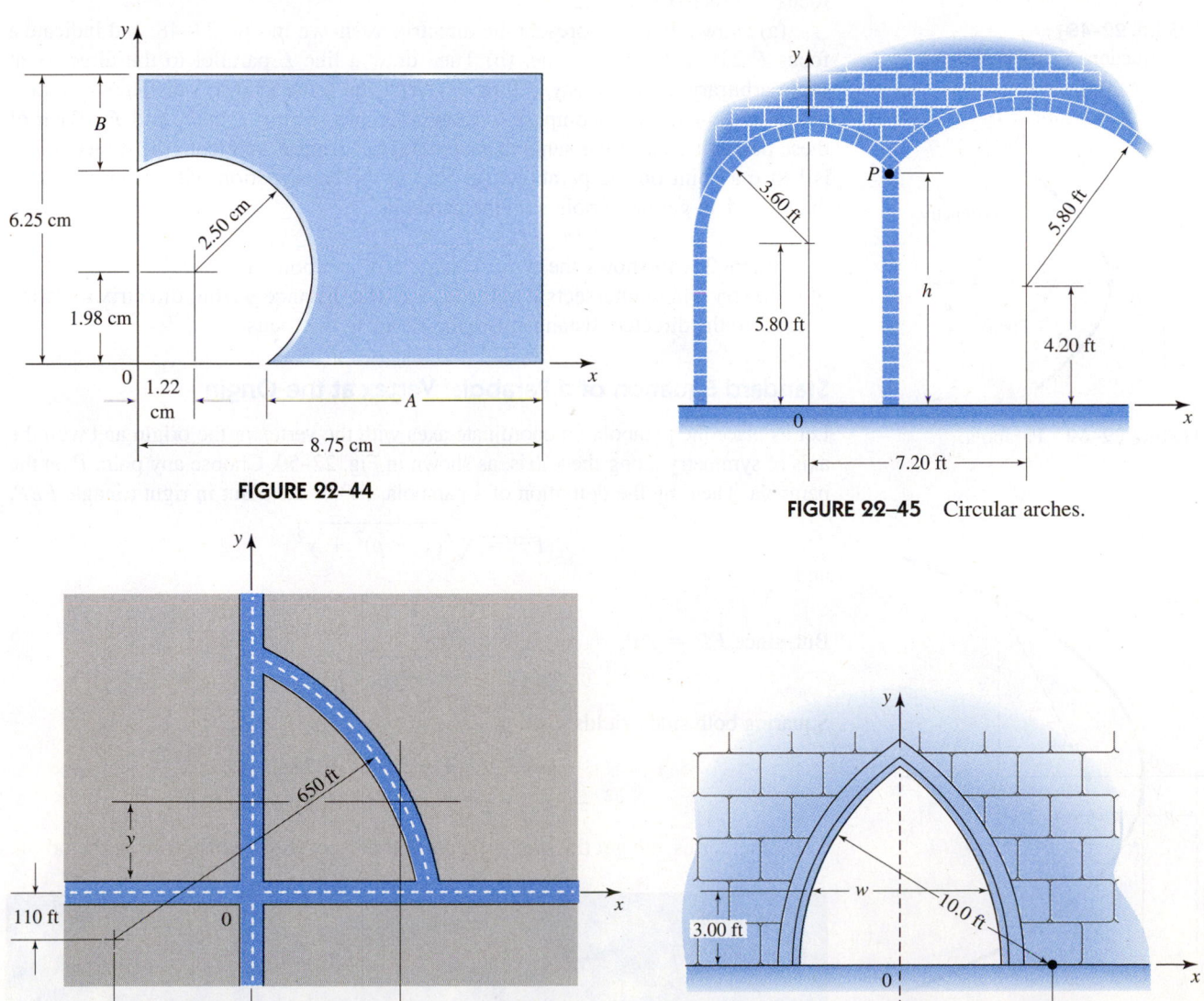

FIGURE 22–44

FIGURE 22–45   Circular arches.

FIGURE 22–46   Circular street.

FIGURE 22–47   Gothic arch.

22–4   The Parabola

Our second conic section is the *parabola*. It is another amazing curve, describing the deflection of beams, the path of objects thrown or dropped, and the orbits of some comets. It is used in the design of optical devices, highway curves, and much more.

Definition of a Parabola

We said that the parabola results when we cut a cone by a plane that is parallel to an element of the cone. Here is another definition.

Definition of a Parabola	A *parabola* is the set of points in a plane, each of which is equidistant from a fixed point, the *focus*, and a fixed line, the *directrix*.	221

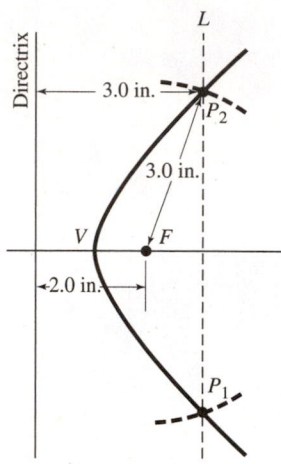

FIGURE 22–48
Construction of a parabola.

■ **Exploration:**

Try this. Use the preceding definition to construct a parabola whose distance from focus to directrix is 2.0 in.

(a) Draw a line to represent the directrix as shown in Fig. 22–48, and indicate a focus F 2.0 in. from that line. (b) Then draw a line L parallel to the directrix at some arbitrary distance, say, 3.0 in. (c) With the same (3.0-in.) distance as radius and F as center, use a compass to draw arcs intersecting L at P_1 and P_2. Each of these points is now at the same distance (3.0 in.) from F and from the directrix and is hence a point on the parabola. (d) Repeat the construction with distances other than 3.0 in. to get more points on the parabola.   ■

Figure 22–49 shows the typical shape of a parabola. The parabola has an *axis of symmetry* which intersects it at the *vertex*. The distance p from directrix to vertex is equal to the directed distance from the vertex to the focus.

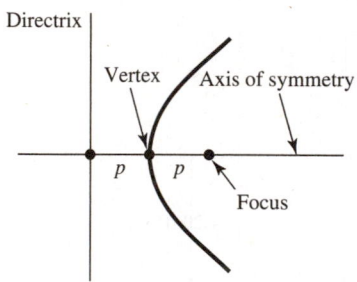

FIGURE 22–49   Parabola.

Standard Equation of a Parabola: Vertex at the Origin

Let us place the parabola on coordinate axes with the vertex at the origin and with the axis of symmetry along the x axis, as shown in Fig. 22–50. Choose any point P on the parabola. Then, by the definition of a parabola, $FP = AP$. But in right triangle FBP,

$$FP = \sqrt{(x - p)^2 + y^2}$$

and

$$AP = p + x$$

But, since $FP = AP$,

$$\sqrt{(x - p)^2 + y^2} = p + x$$

Squaring both sides yields

$$(x - p)^2 + y^2 = p^2 + 2px + x^2$$
$$x^2 - 2px + p^2 + y^2 = p^2 + 2px + x^2$$

Collecting terms, we get the standard equation of a parabola with vertex at the origin.

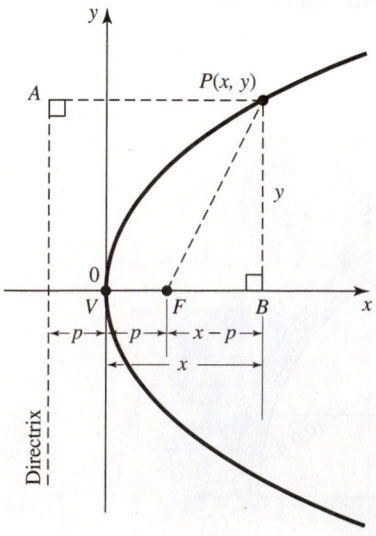

FIGURE 22–50

Standard Equation of a Parabola: Vertex at Origin, Axis Horizontal		$y^2 = 4px$	222

(b) When $x = 18.6$ ft,

$$(18.6)^2 = 50.1y$$
$$h = 6.91 \text{ ft}$$ ◆◆◆

Focal Width of a Parabola

The *latus rectum* of a parabola is a line through the focus which is perpendicular to the axis of symmetry, such as line AB in Fig. 22–55. The length of the latus rectum is also called the *focal width*. We will find the focal width, or length L, of the latus rectum, by substituting the coordinates (p, h) of point A into Eq. 222.

$$h^2 = 4p(p) = 4p^2$$
$$h = \pm 2p$$

The focal width is twice h, so we have the following equation:

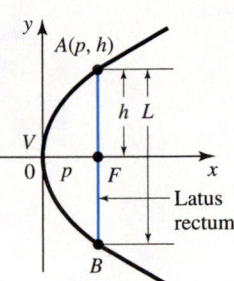

FIGURE 22–55

Focal Width of a Parabola	$L = \lvert 4p \rvert$ *The focal width (length of the latus rectum) of a parabola is four times the distance from vertex to focus.*	227

The focal width is useful for making a quick sketch of the parabola.

◆◆◆ **Example 38:** A parabola opening upward has its vertex at the origin and its focus 2 units from the vertex. Make a quick sketch.

Solution: We're given $p = 2$, so the focal width is

$$L = \lvert 4(2) \rvert = 8 \text{ units}$$

For a quick sketch we plot the vertex and the ends of the latus rectum and connect these points, Fig. 22–56.   ◆◆◆

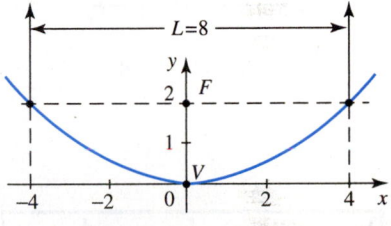

FIGURE 22–56

Standard Equation of a Parabola: Vertex Not at the Origin

As with the circle, when the vertex of the parabola is not at the origin but at (h, k), our equations will be similar to Eqs. 222 and 223, except that x is replaced with $x - h$ and y is replaced with $y - k$.

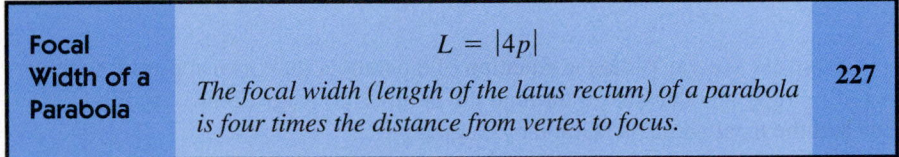

Standard Equations of a Parabola: Vertex at (h, k)	Axis Horizontal		$(y - k)^2 = 4p(x - h)$	224
	Axis Vertical		$(x - h)^2 = 4p(y - k)$	225

◆◆◆ **Example 39:** Find the vertex, focus, focal width, and equation of the axis for the parabola $(y - 3)^2 = 8(x + 2)$.

Solution: The given equation is of the same form as Eq. 224, so the axis is horizontal. Also, $h = -2$, $k = 3$, and $4p = 8$. So the vertex is at $V(-2, 3)$ (Fig. 22–57).

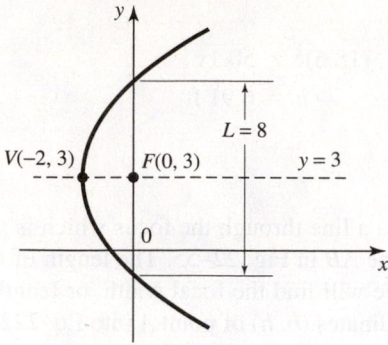

FIGURE 22–57

Since $p = \frac{8}{4} = 2$, the focus is 2 units to the right of the vertex, at $F(0, 3)$. The focal width is $4p$, so $L = 8$. The axis is horizontal and 3 units from the x axis, so its equation is $y = 3$.   ◆◆◆

◆◆◆ **Example 40:** (a) Write the equation of a parabola that opens upward, with vertex at $(-1, 2)$, and that passes through the point $(1, 3)$ (Fig. 22–58). (b) Find the focus and the focal width. (c) Graph by calculator.

Solution: (a) We substitute $h = -1$ and $k = 2$ into Eq. 225.

$$(x + 1)^2 = 4p(y - 2)$$

Now, since $(1, 3)$ is on the parabola, these coordinates must satisfy our equation. Substituting, we get

$$(1 + 1)^2 = 4p(3 - 2)$$

Solving for p, we obtain $2^2 = 4p$, or $p = 1$. So the equation is

$$(x + 1)^2 = 4(y - 2)$$

(b) The focus is p units above the vertex, at $(-1, 3)$. The focal width is, by Eq. 227,

$$L = |4p| = 4(1) = 4 \text{ units}$$

(c) To graph the curve by calculator we solve for y, as usual.

$$y - 2 = \frac{(x + 1)^2}{4}$$

$$y = \frac{(x + 1)^2}{4} + 2$$

We enter this equation as **Y1**, select the viewing window, $x = -5$ to 5 and $y = -1$ to 7, and get the graph as shown.   ◆◆◆

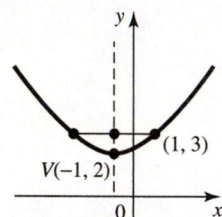

FIGURE 22–58

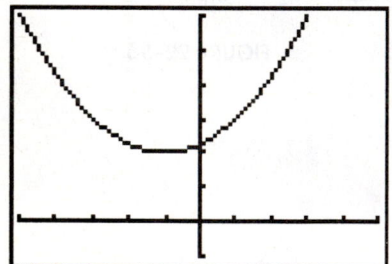

Screen for Example 40. Tick marks are one unit apart on both axes.

General Equation of a Parabola

We get the general equation of the parabola by expanding the standard equation (Eq. 224) as follows:

$$(y - k)^2 = 4p(x - h)$$
$$y^2 - 2ky + k^2 = 4px - 4ph$$

or

$$y^2 - 4px - 2ky + (k^2 + 4ph) = 0$$

which is of the following general form (where C, D, E, and F are constants):

General Equation of a Parabola with Horizontal Axis	$Cy^2 + Dx + Ey + F = 0$	226

◆◆◆◆◆ **Example 41:** Rewrite this standard equation in general form.

$$(x - 1)^2 = 3(y + 4)$$

Solution: Removing parentheses and rearranging gives

$$x^2 - 2x + 1 = 3y + 12$$

$$x^2 - 2x - 3y - 11 = 0 \qquad ◆◆◆$$

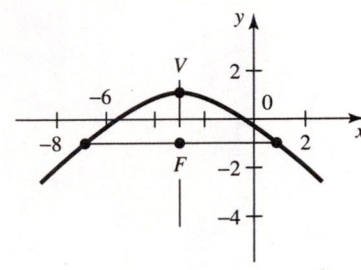

FIGURE 22–59

Completing the Square

As with the circle, we go from general to standard form by completing the square.

◆◆◆ **Example 42:** Find the vertex, focus, and focal width for the parabola

$$x^2 + 6x + 8y + 1 = 0$$

Solution: Separating the x and y terms, we have

$$x^2 + 6x = -8y - 1$$

Completing the square by adding 9 to both sides, we obtain

$$x^2 + 6x + 9 = -8y - 1 + 9$$

Factoring,

$$(x + 3)^2 = -8(y - 1)$$

which is the form of Eq. 225, with $h = -3$, $k = 1$, and $p = -2$.

The vertex is $(-3, 1)$. Since the parabola opens downward (Fig. 22–59), the focus is 2 units below the vertex, at $(-3, -1)$. The focal width is $|4p|$, or 8 units. ◆◆◆

We see that the equation of a parabola having a horizontal axis of symmetry has a y^2 term but no x^2 term. Conversely, the equation for a parabola with vertical axis has an x^2 term but not a y^2 term. The parabola is the only conic for which there is only one variable squared.

If the coefficient B of the xy term in the general second-degree equation (Eq. 216) were not zero, it would indicate that the axis of symmetry was *rotated* by some amount and was no longer parallel to a coordinate axis. The presence of an xy term indicates rotation of the ellipse and hyperbola as well.

Exercise 4 ◆ The Parabola

Standard Equation of a Parabola: Vertex at the Origin

Find the coordinates of the focal point and the focal width for each parabola. Graph.

1. $y^2 = 8x$ **2.** $x^2 = 16y$

3. $7x^2 + 12y = 0$ **4.** $3y^2 + 5x = 0$

Write the standard equation of each parabola. Find the focus and make a graph.

5. passes through $(6, 4)$; axis vertical.

6. passes through $(25, 20)$; axis horizontal

7. passes through $(3, 2)$ and $(3, -2)$

8. passes through $(3, 4)$ and $(-3, 4)$

Standard Equation of a Parabola: Vertex Not at the Origin

Find the vertex, focus, focal width, and equation of the axis for each parabola. Make a graph.

9. $(y - 5)^2 = 12(x - 3)$

10. $(x + 2)^2 = 16(y - 6)$

11. $(x - 3)^2 = 24(y + 1)$

12. $(y + 3)^2 = 4(x + 5)$

13. $3x + 2y^2 + 4y - 4 = 0$

14. $y^2 + 8y + 4 - 6x = 0$

15. $y - 3x + x^2 + 1 = 0$

16. $x^2 + 4x - y - 6 = 0$

Write the equation of each parabola in standard form. Find all missing features and graph.

17. vertex at $(1, 2)$; $L = 8$; axis is $y = 2$; opens to the right

18. axis is $y = 3$; passes through $(6, -1)$ and $(3, 1)$

19. vertex $(0, 2)$; axis is $x = 0$; passes through $(-4, -2)$

General Equation of a Parabola

Rewrite each standard equation in general form.

20. $(x + 3)^2 = 2(y + 2)$

21. $(x + 6)^2 = 4(y - 5)$

22. $(y + 7)^2 = 3(x - 2)$

23. $(y - 5)^2 = 9(x + 8)$

Write each general equation in standard form. Find the vertex, focus, and focal width, and make a graph.

24. $x^2 + 4x + 6y + 2 = 0$

25. $x^2 - 3x + 7y + 4 = 0$

26. $y^2 + 6y - 2x + 9 = 0$

Trajectories

27. A ball thrown into the air will, neglecting air resistance, follow a parabolic path, as shown in Fig. 22–60. Write the equation of the path, taking axes as shown. Use your equation to find the height of the ball when it is at a horizontal distance of 95.0 ft from O.

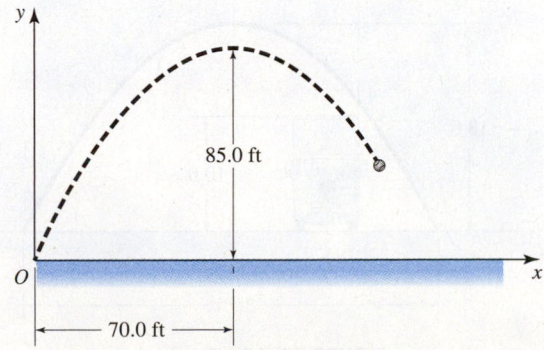

85.0 ft

70.0 ft

FIGURE 22–60 Ball thrown into the air.

28. Some comets follow a parabolic orbit with the sun at the focal point (Fig. 22–61). Taking axes as shown, write the equation of the path if the distance p is 75 million kilometers.

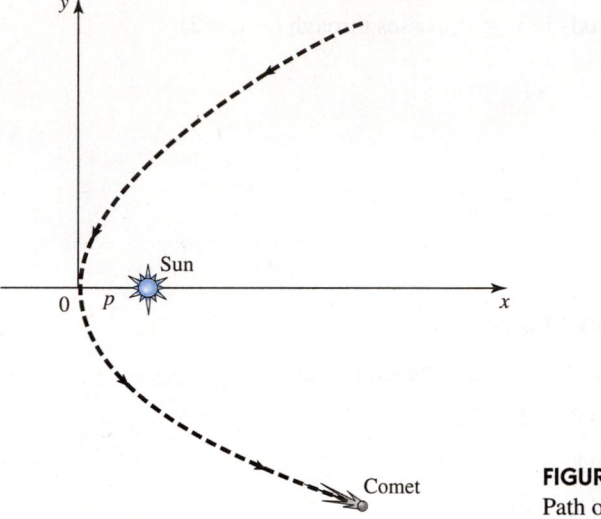

FIGURE 22–61
Path of a comet.

29. An object dropped from a moving aircraft (Fig. 22–62) will follow a parabolic path if air resistance is negligible. A weather instrument released at a height of 3520 m is observed to strike the water at a distance of 2150 m from the point of release. Write the equation of the path, taking axes as shown. Find the height of the instrument when x is 1000 m.

Parabolic Arch

30. A 10-ft-high truck passes under a parabolic arch, as shown in Fig. 22–63. Find the maximum distance x that the side of the truck can be from the center of the road.

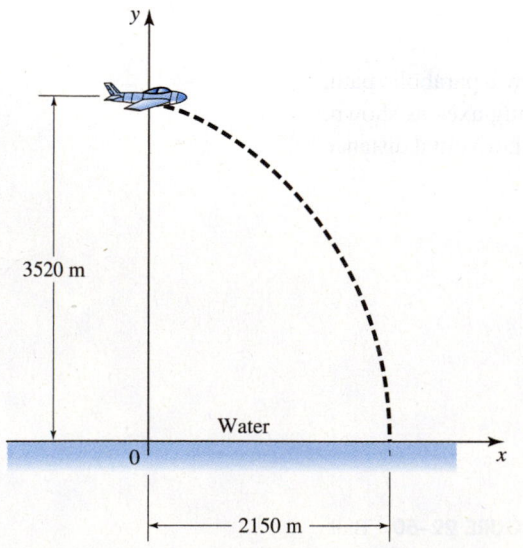

FIGURE 22–62   Object dropped from an aircraft.

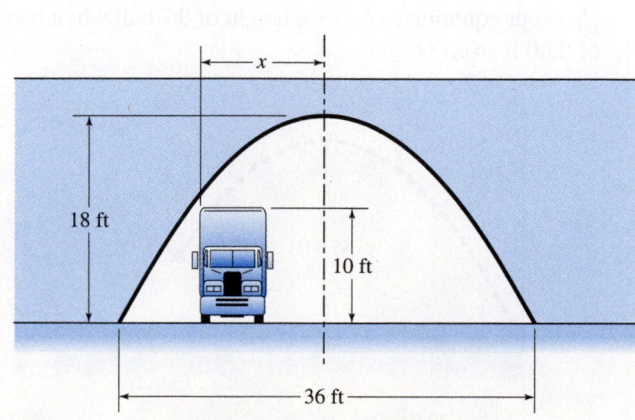

FIGURE 22–63   Parabolic arch.

31. Assuming the bridge cable *AB* of Fig. 22–64 to be a parabola, write its equation, taking axes as shown.

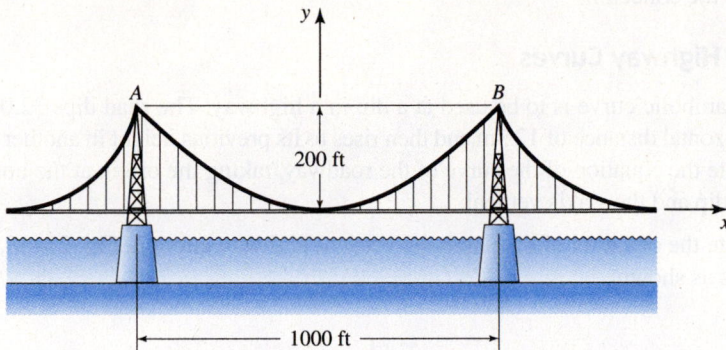

FIGURE 22–64 Parabolic bridge cable. A cable will hang in the shape of a parabola if the vertical load per horizontal foot is constant.

32. A parabolic arch supports a roadway as shown in Fig. 22–65. Write the equation of the arch, taking axes as shown. Use your equation to find the vertical distance from the roadway to the arch at a horizontal distance of 50.0 m from the center.

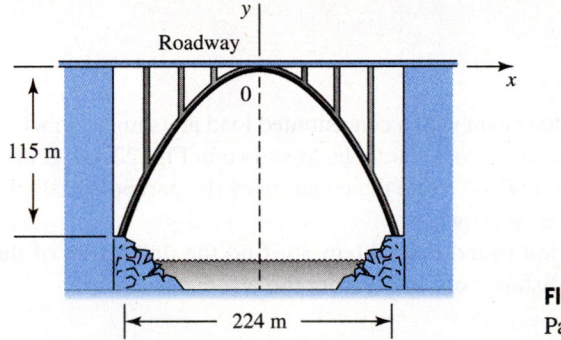

FIGURE 22–65
Parabolic arch.

Parabolic Reflector

33. A certain solar collector consists of a long panel of polished steel bent into a parabolic shape (Fig. 22–66), which focuses sunlight onto a pipe *P* at the focal point of the parabola. At what distance *x* should the pipe be placed?

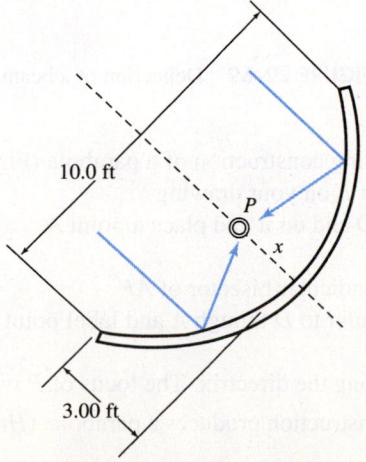

FIGURE 22–66 Parabolic solar collector.

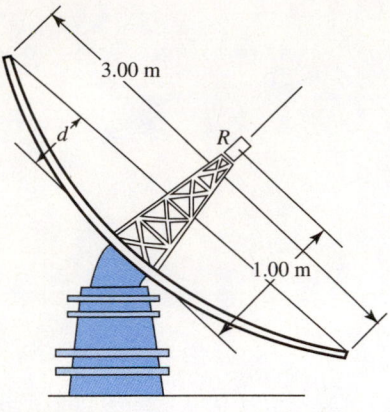

FIGURE 22–67   Parabolic antenna.

34. A parabolic collector for receiving television signals from a satellite is shown in Fig. 22–67. The receiver *R* is at the focus, 1.00 m from the vertex. Find the depth *d* of the collector.

Vertical Highway Curves

35. A parabolic curve is to be used at a dip in a highway. The road dips 32.0 m in a horizontal distance of 125 m and then rises to its previous height in another 125 m. Write the equation of the curve of the roadway, taking the origin at the bottom of the dip and the *y* axis vertical.

36. Write the equation of the parabolic vertical highway curve in Fig. 22–68, taking axes as shown.

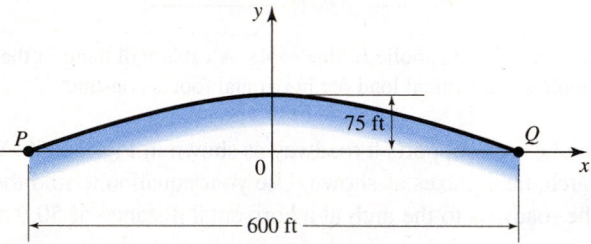

FIGURE 22–68   Road over a hill.

Beams

37. A simply supported beam with a concentrated load at its midspan will deflect approximately in the shape of a parabola, as shown in Fig. 22–69. If the deflection at the midspan is 1.00 in., write the equation of the parabola (called the *elastic curve*), taking axes as shown.

38. Using the equation found in problem 36, find the deflection of the beam in Fig. 22–69 at a distance of 10.0 ft from the left end.

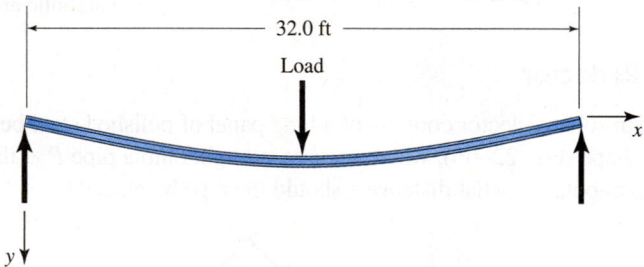

FIGURE 22–69   Deflection of a beam.

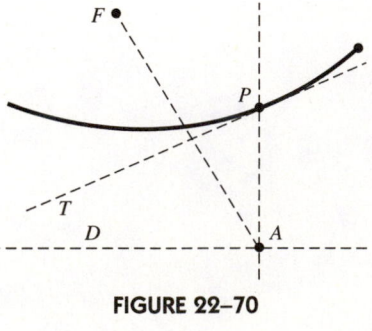

FIGURE 22–70

39. *CAD:* Do the following construction of a parabola (Fig. 22–70).
  • Place a focal point *F* on your drawing
  • Draw a directrix *D* and on it and place a point *A*
  • Draw *AF*
  • Draw *T*, the perpendicular bisector of *AF*
  • Draw a perpendicular to *D* though *A* and label point *P* where this line intersects *T*
  • Finally, drag *A* along the directrix. The locus of *P* will be a parabola.

Explain why this construction produces a parabola. (*Hint:* What type of triangle is *AFP*?)

Demonstrate this construction to your class.

40. *Project:* If a projectile is launched with a horizontal velocity v_x and vertical velocity v_y, its position after t seconds is given by

$$x = v_x t$$

$$y = v_y t - (g/2)t^2$$

where g is the acceleration due to gravity. Eliminate t from this pair of equations to get $y = f(x)$, and show that this equation represents a parabola.

22–5 The Ellipse

Definition of the Ellipse

Our third conic is the ellipse, another curve that has many applications. As we said, it is formed when a cone is cut by a plane that is not perpendicular to the cone's axis but is not tilted so much as to be parallel to an element of the cone. We also define an ellipse as a set of points meeting the following conditions:

Definition of an Ellipse	An *ellipse* is the set of all points in a plane such that the sum of the distances from each point on the ellipse to two fixed points (called the *foci*) is constant.	**229**

■ Exploration:

Try this. Push two tacks into a drafting board, spaced by about 6 inches apart, Fig. 22–71. Pass a loop of string around the tacks, pull it taut with a pencil, and trace a curve.

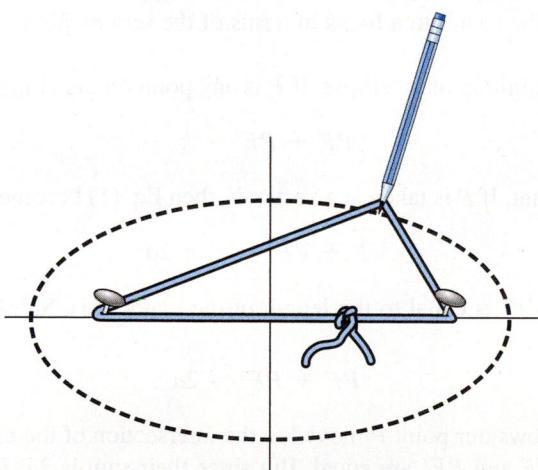

FIGURE 22–71 Construction of an ellipse.

The curve you got was apparently an ellipse. Comparing your construction with the definition above, can you say whether your ellipse is exact or approximate?

Now measure half the distance between the tacks, half the long dimension of your ellipse, and half the short dimension. Can you see how those three half-dimensions are related? Change the distance between the tacks, retie the string, and draw another ellipse. Does your relationship between the three half-dimensions still hold? ■

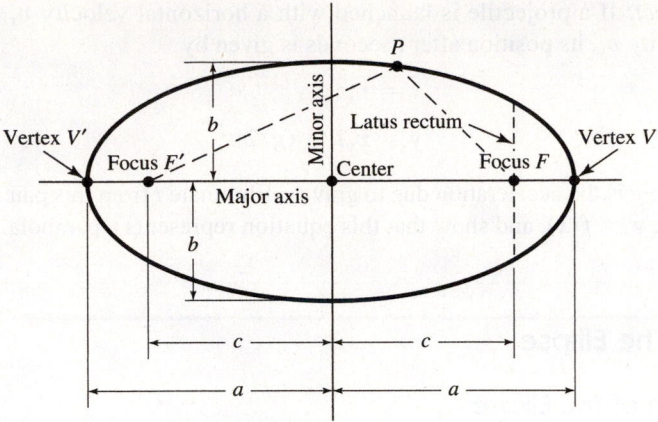

FIGURE 22–72   Ellipse.

Figure 22–72 shows the typical shape of the ellipse. An ellipse has two axes of symmetry: the *major axis* and *minor axis*, which intersect at the *center* of the ellipse. A *vertex* is a point where the ellipse crosses the major axis.

It is often convenient to speak of *half* the lengths of the major and minor axes, and these are called the *semimajor* and *semiminor* axes, whose length we label *a* and *b*, respectively. The distance from either focus to the center is labeled *c*.

A line from a point *P* on the ellipse to a focus is called a *focal radius*. Thus *PF* and *PF′* are focal radii.

Distance to Focus

Before deriving an equation for the ellipse, let us first write an expression for the distance *c* from the center to a focus in terms of the semimajor axis *a* and the semiminor axis *b*.

From our definition of an ellipse, if *P* is any point on the ellipse, then

$$PF + PF' = k \tag{1}$$

where *k* is constant. If *P* is taken at a vertex *V*, then Eq. (1) becomes

$$VF + VF' = k = 2a \tag{2}$$

because $VF + VF'$ is equal to the length of the major axis. Substituting back into Eq. (1) gives us

$$PF + PF' = 2a \tag{3}$$

Figure 22–73 shows our point *P* moved to the intersection of the ellipse and the minor axis. Here *PF* and *PF′* are equal. But since their sum is 2*a*, *PF* and *PF′* must each equal *a*. By the Pythagorean theorem, $c^2 + b^2 = a^2$, or

$$c^2 = a^2 - b^2$$

or

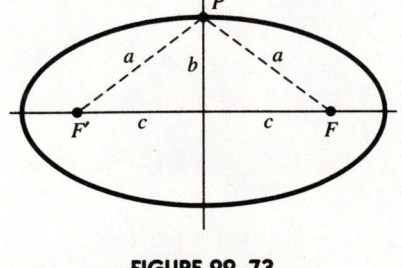

FIGURE 22–73

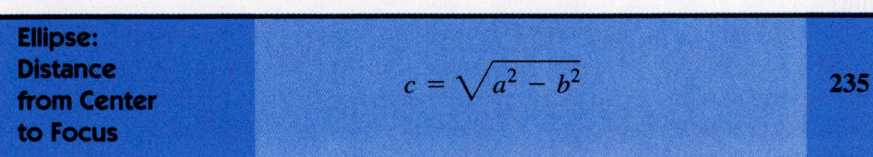

Ellipse: Distance from Center to Focus	$c = \sqrt{a^2 - b^2}$	235

Did you get this result from your exploration?

Standard Equation of Ellipse: Center at Origin, Major Axis Horizontal

Let us place an ellipse on coordinate axes with its center at the origin and major axis along the x axis, as shown in Fig. 22–74. If $P(x, y)$ is any point on the ellipse, then by the definition of the ellipse,

$$PF + PF' = 2a$$

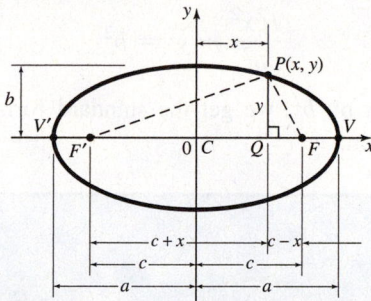

FIGURE 22–74   Ellipse with center at origin.

To get PF and PF' in terms of x and y, we first drop a perpendicular from P to the x axis. Then in triangle PQF,

$$PF = \sqrt{(c - x)^2 + y^2}$$

and in triangle PQF',

$$PF' = \sqrt{(c + x)^2 + y^2}$$

Substituting yields

$$PF + PF' = \sqrt{(c - x)^2 + y^2} + \sqrt{(c + x)^2 + y^2} = 2a$$

Rearranging, we obtain

$$\sqrt{(c + x)^2 + y^2} = 2a - \sqrt{(c - x)^2 + y^2}$$

Squaring, and then expanding the binomials, we have

$$x^2 + 2cx + c^2 + y^2 = 4a^2 - 4a\sqrt{(c - x)^2 + y^2} + c^2 - 2cx + x^2 + y^2$$

Collecting terms, we get

$$cx = a^2 - a\sqrt{(c - x)^2 + y^2}$$

Dividing by a and rearranging yields

$$a - \frac{cx}{a} = \sqrt{(c - x)^2 + y^2}$$

Squaring both sides again gives

$$a^2 - 2cx + \frac{c^2 x^2}{a^2} = c^2 - 2cx + x^2 + y^2$$

Collecting terms, we have

$$a^2 + \frac{c^2 x^2}{a^2} = c^2 + x^2 + y^2$$

But $c = \sqrt{a^2 - b^2}$. Substituting, we get

$$a^2 + \frac{x^2}{a^2}(a^2 - b^2) = a^2 - b^2 + x^2 + y^2$$

or

$$a^2 + x^2 - \frac{b^2 x^2}{a^2} = a^2 - b^2 + x^2 + y^2$$

Collecting terms and rearranging gives us

$$\frac{b^2 x^2}{a^2} + y^2 = b^2$$

Finally, dividing through by b^2, we get the standard form of the equation of an ellipse with center at origin.

Standard Equation of an Ellipse: Center at Origin, Major Axis Horizontal	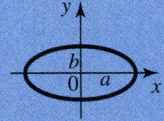	$\dfrac{x^2}{a^2} + \dfrac{y^2}{b^2} = 1$ $a > b$	**230**

Here a is half the length of the major axis, and b is half the length of the minor axis.

Note that if $a = b$, this equation reduces to the equation of a circle.

◆◆◆ **Example 43:** Find the vertices and foci for the ellipse

$$16x^2 + 36y^2 = 576$$

Solution: To be in standard form, our equation must have 1 (unity) on the right side. Dividing by 576 and simplifying gives us

$$\frac{x^2}{36} + \frac{y^2}{16} = 1$$

from which $a = 6$ and $b = 4$. The vertices are then $V(6, 0)$ and $V'(-6, 0)$, as shown in Fig. 22–75. The distance c from the center to a focus is

$$c = \sqrt{6^2 - 4^2} = \sqrt{20} \approx 4.47$$

So the foci are $F(4.47, 0)$ and $F'(-4.47, 0)$.   ◆◆◆

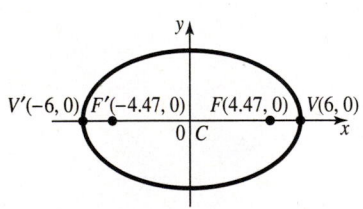

FIGURE 22–75

◆◆◆ **Example 44:** *An Application.* The deck sorrounding an elliptical hot tub is to have an elliptical opening with a length of 6.00 ft and a width of 4.00 ft, Fig. 22–76. Taking the origin at the center of the ellipse, (a) write the equation of the ellipse so that a pattern can be made for cutting the deck from stone and (b) find the width of the opening 2.00 ft from its center.

Solution: (a) Taking the x axis along the major axis of the ellipse, we can use Eq. 230 with $a = 3.00$ and $b = 2.00$. We get

$$\frac{x^2}{9.00} + \frac{y^2}{4.00} = 1$$

FIGURE 22–76   An elliptical hot tub.

(b) When $x = 2.00$ ft,

$$\frac{y^2}{4.00} = 1 - \frac{(2.00)^2}{9.00} = 0.556$$

Solving for y,

$$y^2 = 4.00(0.556) = 2.22$$

$$y = 1.49 \text{ ft}$$

So the width of the opening 2.00 ft from its center is $2(1.49) = 2.98$ ft. ◆◆◆

Graphing an Ellipse with a Graphics Utility

Unless your grapher can accept equations in implicit form, we must write the equation in explicit form by solving for y, as we did with our other curves.

◆◆◆ **Example 45:** Graph the ellipse

$$16x^2 + 36y^2 = 576$$

Solution: Solving for y we get

$$36y^2 = 576 - 16x^2$$

$$y^2 = \frac{576 - 16x^2}{36}$$

$$y = \pm\frac{1}{6}\sqrt{576 - 16x^2}$$

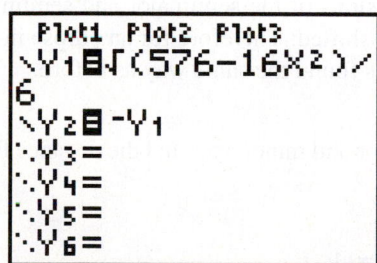
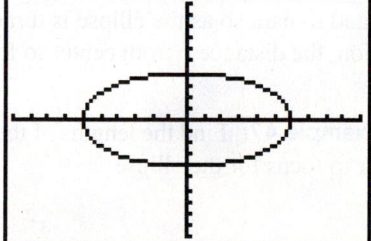

TI-83/84 screens for Example 45.

As before, we must graph both an upper and a lower portion of the curve. Here we take advantage of symmetry about the x axis, and for the lower portion set **Y2 = −Y1.** ◆◆◆

◆◆◆ **Example 46:** An ellipse whose center is at the origin and whose major axis is on the x axis has a minor axis of 10 units and passes through the point (6, 4), as shown in Fig. 22–77. Write the equation of the ellipse in standard form.

Solution: Since the major axis is on the x axis, our equation will have the form of Eq. 230. Substituting, with $b = 5$, we have

$$\frac{x^2}{a^2} + \frac{y^2}{25} = 1$$

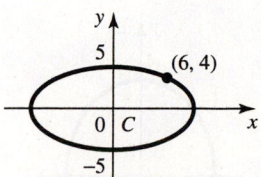

FIGURE 22–77

Since the ellipse passes through (6, 4), these coordinates must satisfy our equation. Substituting gives

$$\frac{36}{a^2} + \frac{16}{25} = 1$$

Solving for a^2, we multiply by the LCD, $25a^2$

$$36(25) + 16a^2 = 25a^2$$
$$9a^2 = 36(25)$$
$$a^2 = 100$$

So our final equation is

$$\frac{x^2}{100} + \frac{y^2}{25} = 1$$ ◆◆◆

Standard Equation of Ellipse: Center at Origin, Major Axis Vertical

When the major axis is vertical rather than horizontal, the only effect on the standard equation is to interchange the positions of x and y.

Standard Equation of an Ellipse: Center at Origin, Major Axis Vertical	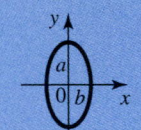 $\dfrac{y^2}{a^2} + \dfrac{x^2}{b^2} = 1$ $a > b$	**231**

Notice that the quantities a and b are dimensions of the semimajor and semiminor axes and remain so as the ellipse is turned or shifted. Therefore for an ellipse in any position, the distance c from center to focus is found the same way as before.

◆◆◆ **Example 47:** Find the lengths of the major and minor axes and the distance from center to focus for the ellipse

$$\frac{x^2}{16} + \frac{y^2}{49} = 1$$

Solution: How can we tell which denominator is a^2 and which is b^2? It is easy: a is always greater than b. So $a = 7$ and $b = 4$. Thus the major and minor axes are 14 units and 8 units long (Fig. 22–78). From Eq. 235,

$$c = \sqrt{a^2 - b^2} = \sqrt{49 - 16} = \sqrt{33}$$ ◆◆◆

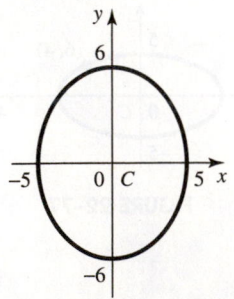

FIGURE 22–78

◆◆◆ **Example 48:** Write the equation of an ellipse with center at the origin, whose major axis is 12 units on the y axis and whose minor axis is 10 units (Fig. 22–79).

Solution: Substituting into Eq. 231, with $a = 6$ and $b = 5$, we obtain

$$\frac{y^2}{36} + \frac{x^2}{25} = 1$$ ◆◆◆

FIGURE 22–79

Common Error	Do not confuse a and b in the ellipse equations. The *larger* denominator is always a^2. Also, the variable (x or y) in the same term with a^2 tells the direction of the major axis.

Standard Equation of Ellipse: Center Not at Origin

Now consider an ellipse whose center is not at the origin, but at (h, k). The equation for such an ellipse will be the same as before, except that x is replaced by $x - h$ and y is replaced by $y - k$.

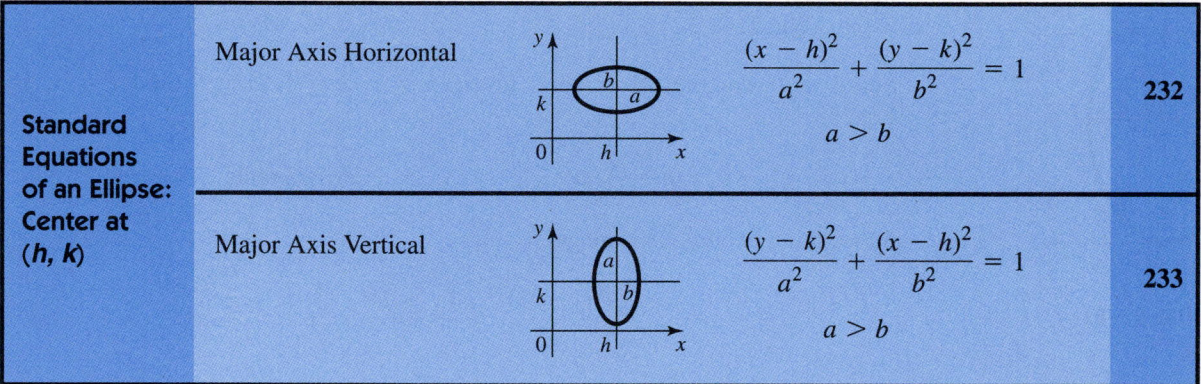

| Standard Equations of an Ellipse: Center at (h, k) | Major Axis Horizontal | $\dfrac{(x-h)^2}{a^2} + \dfrac{(y-k)^2}{b^2} = 1$ $a > b$ | 232 |
| | Major Axis Vertical | $\dfrac{(y-k)^2}{a^2} + \dfrac{(x-h)^2}{b^2} = 1$ $a > b$ | 233 |

◆◆◆ **Example 49:** Find the center, vertices, and foci for the ellipse

$$\frac{(x-5)^2}{9} + \frac{(y+3)^2}{16} = 1$$

Graph by calculator.

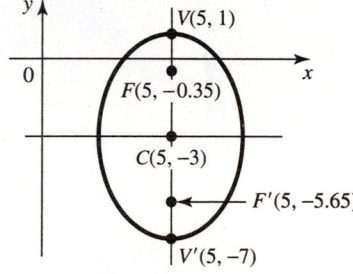

Solution: From the given equation, $h = 5$ and $k = -3$, so the center is $C(5, -3)$, as shown in Fig. 22–80. Also, $a = 4$ and $b = 3$, and the major axis is vertical because a is with the y term. By setting $x = 5$ in the equation of the ellipse, we find that the vertices are $V(5, 1)$ and $V'(5, -7)$. Or we could locate the vertices simply by noting that they are a distance $a = 4$ above and below the center.

The distance c to the foci is

$$c = \sqrt{4^2 - 3^2} = \sqrt{7} \approx 2.65$$

so the foci are $F(5, -0.35)$ and $F'(5, -5.65)$.

To graph this ellipse by calculator, we solve for y.

$$\frac{(y+3)^2}{16} = 1 - \frac{(x-5)^2}{9}$$

$$= \frac{9 - (x-5)^2}{9}$$

$$(y+3)^2 = \frac{16}{9}[9 - (x-5)^2]$$

$$y + 3 = \pm\frac{4}{3}\sqrt{9 - (x-5)^2}$$

$$y = -3 \pm \frac{4}{3}\sqrt{9 - (x-5)^2}$$

FIGURE 22–80 Problems such as these have many quantities to keep track of. It's a good idea to make a sketch and on it place each quantity as it is known.

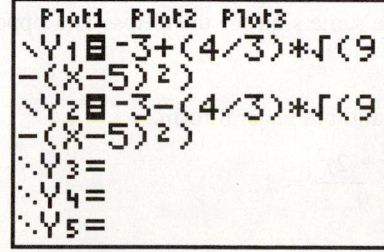

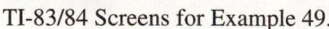

TI-83/84 Screens for Example 49.

The screens for the equation entry and the graph are shown. As before, we must graph the upper and lower portions separately.   ◆◆◆

◆◆◆ **Example 50:** Write the equation in standard form of an ellipse with vertical major axis 10 units long, center at (3, 5), and whose distance between focal points is 8 units (Fig. 22–81).

Solution: From the information given, $h = 3$, $k = 5$, $a = 5$, and $c = 4$. By Eq. 235,

$$b = \sqrt{a^2 - c^2} = \sqrt{25 - 16} = 3 \text{ units}$$

Substituting into Eq. 233, we get

$$\frac{(y - 5)^2}{25} + \frac{(x - 3)^2}{9} = 1$$ ◆◆◆

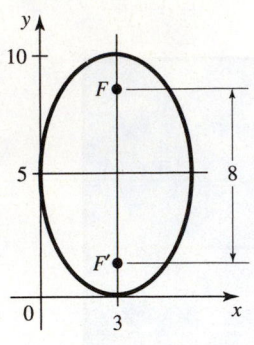

FIGURE 22–81

General Equation of an Ellipse

As we did with the circle, we now expand the standard equation for the ellipse to get the general equation. Starting with Eq. 232,

$$\frac{(x - h)^2}{a^2} + \frac{(y - k)^2}{b^2} = 1$$

we multiply through by $a^2 b^2$ and expand the binomials:

$$b^2(x^2 - 2hx + h^2) + a^2(y^2 - 2ky + k^2) = a^2 b^2$$

$$b^2 x^2 - 2b^2 hx + b^2 h^2 + a^2 y^2 - 2a^2 ky + a^2 k^2 - a^2 b^2 = 0$$

or

$$b^2 x^2 + a^2 y^2 - 2b^2 hx - 2a^2 ky + (b^2 h^2 + a^2 k^2 - a^2 b^2) = 0$$

which is of the general form, again combining constants into A, C, D, E, and F,

General Equation of an Ellipse	$Ax^2 + Cy^2 + Dx + Ey + F = 0$	234

Comparing this with the general second-degree equation, we see that $B = 0$. As for the parabola (and the hyperbola, as we will see), this indicates that the axes of the curve are parallel to the coordinate axes. In other words, the curve is not rotated.

Also note that neither A nor C is zero. That tells us that the curve is not a parabola. Further, A and C have different values, telling us that the curve cannot be a circle. We'll see that A and C will have the same sign for the ellipse and opposite signs for the hyperbola.

◆◆◆ **Example 51:** Rewrite this standard equation in general form.

$$\frac{(x - 1)^2}{4} + \frac{(y + 2)^2}{9} = 1$$

Solution: We multiply through by the LCD (36) and expand each binomial.

$$9(x^2 - 2x + 1) + 4(y^2 + 4y + 4) = 36$$

$$9x^2 - 18x + 9 + 4y^2 + 16y + 16 = 36$$

$$9x^2 + 4y^2 - 18x + 16y - 11 = 0$$

which is our equation in general form.   ◆◆◆

Completing the Square

As before, we go from general to standard form by completing the square.

◆◆◆ **Example 52:** Find the center, foci, vertices, and major and minor axes for the ellipse $9x^2 + 25y^2 + 18x - 50y - 191 = 0$.

Solution: Grouping the x terms and the y terms gives us

$$(9x^2 + 18x) + (25y^2 - 50y) = 191$$

Factoring gives

$$9(x^2 + 2x) + 25(y^2 - 2y) = 191$$

Completing the square, we obtain

$$9(x^2 + 2x + 1) + 25(y^2 - 2y + 1) = 191 + 9 + 25$$

Factoring gives us

$$9(x + 1)^2 + 25(y - 1)^2 = 225$$

Finally, dividing by 225, we have

$$\frac{(x + 1)^2}{25} + \frac{(y - 1)^2}{9} = 1$$

We see that $h = -1$ and $k = 1$, so the center is at $(-1, 1)$. Also, $a = 5$, so the major axis is 10 units and is horizontal, and $b = 3$, so the minor axis is 6 units. A vertex is located 5 units to the right of the center, at $(4, 1)$, and 5 units to the left, at $(-6, 1)$. From Eq. 235,

$$c = \sqrt{a^2 - b^2}$$

$$= \sqrt{25 - 9}$$

$$= 4$$

So the foci are at $(3, 1)$ and $(-5, 1)$. This ellipse is shown in Fig. 22–82.   ◆◆◆

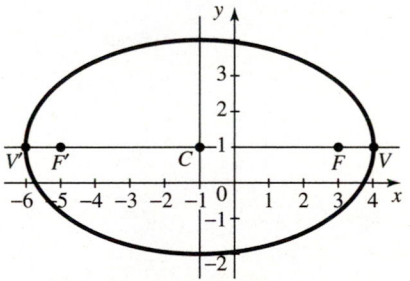

FIGURE 22–82

	In Example 52 we needed 1 to complete the square on x, but we added 9 to the right side because the expression containing the 1 was multiplied by a factor of 9. Similarly, for y, we needed 1 on the left but added 25 to the right. It is very easy to forget to multiply by those factors.
Common Error	

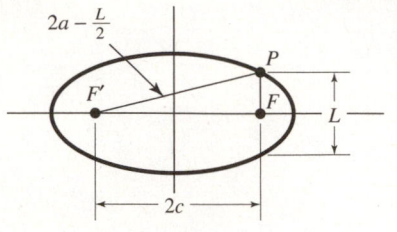

FIGURE 22–83

As with the parabola, the main use for L is for quick sketching of the ellipse.

Focal Width of an Ellipse

The focal width L, or length of the latus rectum, is the width of the ellipse through the focus (Fig. 22–83). The sum of the focal radii PF and PF' from a point P at one end of the latus rectum must equal $2a$, by the definition of an ellipse. So $PF' = 2a - PF$, or, since $L/2 = PF$, $PF' = 2a - L/2$. Squaring both sides, we obtain

$$(PF')^2 = 4a^2 - 2aL + \frac{L^2}{4} \tag{1}$$

But in right triangle PFF',

$$(PF')^2 = \left(\frac{L}{2}\right)^2 + (2c)^2 = \frac{L^2}{4} + 4c^2 = \frac{L^2}{4} + 4a^2 - 4b^2 \tag{2}$$

since $c^2 = a^2 - b^2$. Equating Eqs. (1) and (2) and collecting terms gives $2aL = 4b^2$, or the following:

Focal Width of an Ellipse	$L = \dfrac{2b^2}{a}$ *The focal width of an ellipse is twice the square of the semiminor axis, divided by the semimajor axis.*	236

◆◆◆ **Example 53:** The focal width of an ellipse that is 25 m long and 10 m wide is

$$L = \frac{2(5^2)}{12.5} = \frac{50}{12.5} = 4 \text{ m}$$

◆◆◆

Exercise 5 ◆ The Ellipse

Standard Equation, Center at Origin

Find the coordinates of the vertices and foci for each ellipse.

1. $\dfrac{x^2}{25} + \dfrac{y^2}{16} = 1$

2. $\dfrac{x^2}{49} + \dfrac{y^2}{36} = 1$

3. $3x^2 + 4y^2 = 12$

4. $64x^2 + 15y^2 = 960$

5. $4x^2 + 3y^2 = 48$

6. $8x^2 + 25y^2 = 200$

Write the equation of each ellipse in standard form.

7. vertices at $(\pm 5, 0)$; foci at $(\pm 4, 0)$

8. vertical major axis 8 units long; a focus at $(0, 2)$

9. horizontal major axis 12 units long; passes through $(3, \sqrt{3})$

10. passes through $(4, 6)$ and $(2, 3\sqrt{5})$

11. horizontal major axis 26 units long; distance between foci $= 24$

12. vertical minor axis 10 units long; distance from focus to vertex $= 1$

13. passes through $(1, 4)$ and $(-6, 1)$

14. distance between foci $= 18$; sum of axes $= 54$; horizontal major axis

Standard Equation, Center Not at Origin

Find the coordinates of the center, vertices, and foci for each ellipse. Round to three significant digits where needed.

15. $\dfrac{(x-2)^2}{16} + \dfrac{(y+2)^2}{9} = 1$

16. $\dfrac{(x+5)^2}{25} + \dfrac{(y-3)^2}{49} = 1$

17. $5x^2 + 20x + 9y^2 - 54y + 56 = 0$

18. $16x^2 - 128x + 7y^2 + 42y = 129$

19. $7x^2 - 14x + 16y^2 + 32y = 89$

20. $3x^2 - 6x + 4y^2 + 32y + 55 = 0$

21. $25x^2 + 150x + 9y^2 - 36y + 36 = 0$

Write the equation of each ellipse.

22. minor axis $= 10$; foci at $(13, 2)$ and $(-11, 2)$

23. center at $(0, 3)$; vertical major axis $= 12$; length of minor axis $= 6$

24. center at $(2, -1)$; a vertex at $(2, 5)$; length of minor axis $= 3$

25. center at $(-2, -3)$; a vertex at $(-2, 1)$; a focus halfway between vertex and center

General Equation

Write each standard equation in general form.

26. $\dfrac{x^2}{25} + \dfrac{y^2}{9} = 1$

27. $\dfrac{x^2}{16} + \dfrac{y^2}{36} = 4$

28. $\dfrac{(x-3)^2}{16} + \dfrac{(y-4)^2}{25} = 12$

29. $\dfrac{(x+5)^2}{9} + \dfrac{(y+7)^2}{4} = 22$

Write each general equation in standard form. Find the center, foci, vertices, and the lengths a and b of the semiaxes, and make a graph.

30. $4x^2 + y^2 - 16x + 6y + 21 = 0$

31. $x^2 + 9y^2 - 4x + 18y + 4 = 0$

32. $16x^2 + 25y^2 + 96x - 50y - 231 = 0$

33. $49x^2 + 81y^2 + 294x + 810y - 1503 = 0$

Focal Width

34. Find the focal width of an ellipse that is 4 units wide and 8 units long.

35. Find the focal width of an ellipse that is 44 units wide and 22 units long.

Applications

For problems 36 through 39, first write the equation for each ellipse.

36. A certain bridge arch is in the shape of half an ellipse 120 ft wide and 30.0 ft high. At what horizontal distance from the center of the arch is the height equal to 15.0 ft?

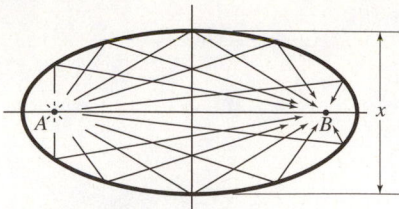

FIGURE 22–84   Focusing property of the ellipse.

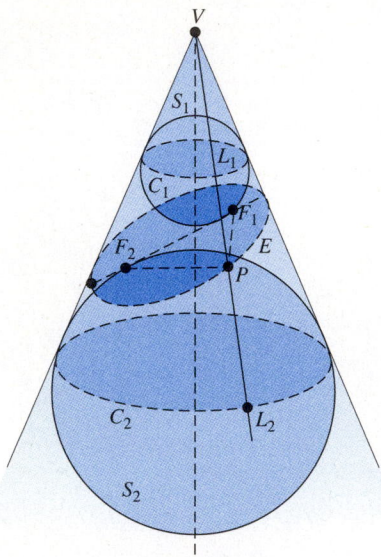

FIGURE 22–86

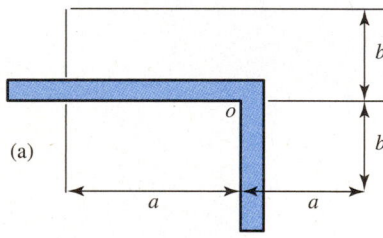

(a)

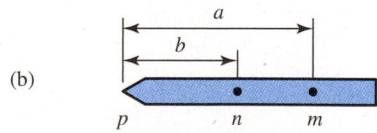

(b)

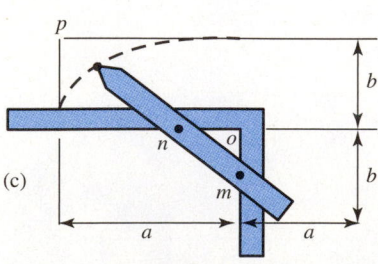

(c)

FIGURE 22–87

37. A curved mirror in the shape of an ellipse will reflect all rays of light coming from one focus onto the other focus (Fig. 22–84). A certain spot heater is to be made with a heating element at A and the part to be heated at B, contained within an ellipsoid (a solid obtained by rotating an ellipse about one axis). Find the width x of the chamber if its length is 25 cm and the distance from A to B is 15 cm.

38. The paths of the planets and certain comets are ellipses, with the sun at one focal point. The path of Halley's comet is an ellipse with a major axis of 36.18 AU and a minor axis of 9.12 AU. An *astronomical unit,* AU, is the distance between the earth and the sun, about 92.6 million miles. What is the greatest distance that Halley's comet gets from the sun?

39. An elliptical culvert (Fig. 22–85) is filled with water to a depth of 1.0 ft. Find the width w of the stream.

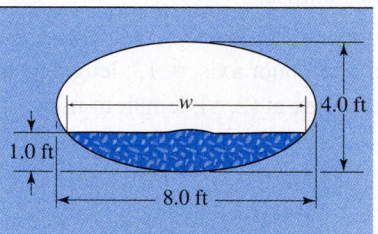

FIGURE 22–85

40. *Project: Another Definition of the Ellipse.* We have defined an ellipse as the set of all points in a plane such that the sum of the distances from two fixed points is constant. We have also defined it as the curve obtained by intersecting a plane with a cone. Referring to Fig. 22–86, show that these two definitions are consistent.

    (a) Let E be the curve in which a plane intersects a cone, and P be any point on that curve.

    (b) Insert a sphere S_1 into the cone so that it touches the cone along circle C_1 and is tangent to the cutting plane at point F_1. Similarly insert sphere S_2 touching the cone along circle C_2 and the plane at point F_2.

    (c) Show that $PF_1 + PF_2$ is a constant, regardless of where on E the point P is chosen.

    (*Hint:* Draw element VP and extend it to where it intersects C_2 at L_2. Then use the fact that two tangents drawn to a sphere from a common point are equal.)

41. *Project:* If the area of an ellipse is twice the area of the inscribed circle, find the length of the semimajor axis of the ellipse.

Workshop Methods for Constructing an Ellipse

42. Draw an ellipse using tacks and strings, as shown in Fig. 22–71, with major axis of 84.0 cm and minor axis of 58.0 cm. How far apart must the tacks be placed?

43. *Project: Drawing an Ellipse Using the Framing Square.* Draw an ellipse by this method and demonstrate it to your class. (a) Draw the major and minor axes of the ellipse and tape down a framing square with the inner edges of its legs along these axes, as shown in Fig. 22–87(a). (b) On a pointed stick place pins m and n at distances a and b from the pointed end P, Fig. 22–86(b). (c) Move the stick while keeping the two pins in contact with inner edges of the square, and P will describe one quadrant of an ellipse. Find the other quadrants by symmetry.

44. *Project: Drawing an Approximate Ellipse.* Sometimes an exact ellipse is not needed and an approximate one, an *oval*, will do. Draw an oval by this method. Then draw an exact ellipse having the same length axes and compare. (a) Draw a rectangle whose dimensions are half the major axes of the ellipse. (b) Subdivide one side into an equal number of parts, says 8 as in Fig. 22–88. Number each point from left to right. (c) Subdivide the other side into the same number of equal parts and number each point from top to bottom. (d) The lines connecting points having the same number will form the tangents to an approximate ellipse. Use symmetry to complete the ellipse.

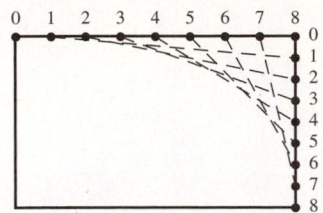

FIGURE 22–88 Oval by tangent lines.

22–6 The Hyperbola

Our final conic will be the hyperbola. Recall that this is the two-branched curve we get when a plane that is parallel to the axis of a cone intercepts both nappes of that cone. It is also defined as follows:

Definition of a Hyperbola	A *hyperbola* is the set of all points in a plane such that the distances from each point to two fixed points, the *foci*, have a constant difference.	238

Figure 22–89 shows the typical shape of the hyperbola. The point midway between the foci is called the *center* of the hyperbola. The line passing through the foci and the center is one *axis* of the hyperbola. The hyperbola crosses that axis at points called the *vertices*. The line segment connecting the vertices is called the *transverse axis*. A second axis of the hyperbola passes through the center and is perpendicular to the transverse axis. The segment of this axis shown in bold in Fig. 22–89 is called the *conjugate axis*. Half the lengths of the transverse and conjugate axes are the *semitransverse* and *semiconjugate* axes, respectively. They are also referred to as *semiaxes*.

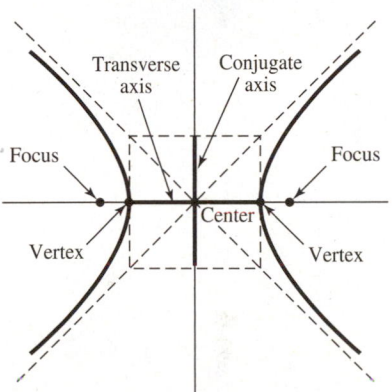

FIGURE 22–89 Hyperbola.

Standard Equations of a Hyperbola with Center at Origin

We place the hyperbola on coordinate axes, with its center at the origin and its transverse axis on the *x* axis (Fig. 22–90). Let *a* be half the transverse axis, and let *c* be half the distance between foci. Now take any point *P* on the hyperbola and draw the focal radii *PF* and *PF'*. Then, by the definition of the hyperbola, $|PF' - PF| = $ constant. An expression for the constant can be found by moving *P* to the vertex *V*, where

$$|PF' - PF| = VF' - VF$$
$$= 2a + V'F' - VF$$
$$= 2a$$

since $V'F'$ and VF are equal. So $|PF' - PF| = 2a$. But in right triangle $PF'D$,

$$PF' = \sqrt{(x + c)^2 + y^2}$$

and in right triangle *PFD*,

$$PF = \sqrt{(x - c)^2 + y^2}$$

so

$$\sqrt{(x + c)^2 + y^2} - \sqrt{(x - c)^2 + y^2} = 2a$$

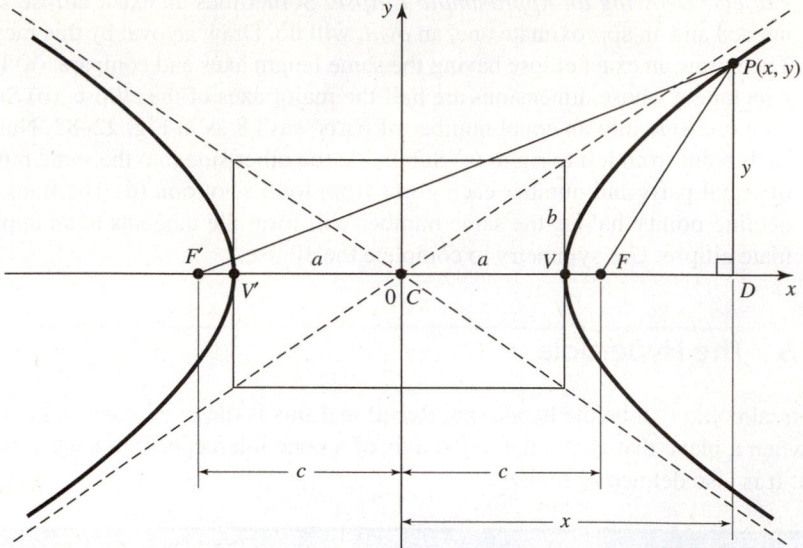

FIGURE 22–90   Hyperbola with center at origin.

Notice that this equation is almost identical to the one we had when deriving the equation for the ellipse and the derivation is almost identical as well. However, to eliminate c from our equation, we define a new quantity b, such that $b^2 = c^2 - a^2$. We'll soon give a geometric meaning to the quantity b.

The remainder of the derivation is left as an exercise. Follow the same steps we used for the ellipse. The resulting equations are very similar to those for the ellipse.

Standard Equations of a Hyperbola: Center at Origin	Transverse Axis Horizontal		$\dfrac{x^2}{a^2} - \dfrac{y^2}{b^2} = 1$	**239**
	Transverse Axis Vertical		$\dfrac{y^2}{a^2} - \dfrac{x^2}{b^2} = 1$	**240**

Asymptotes of a Hyperbola

In general, if the distance from a point P on a curve to some line L approaches zero as the distance from P to the origin increases without bound, then L is called an *asymptote*. In Fig. 22–90, the dashed diagonal lines are asymptotes of the hyperbola. The two branches of the hyperbola approach, but do not intersect, the asymptotes.

We find the slope of these asymptotes from the equation of the hyperbola. Solving Eq. 239 for y gives $y^2 = b^2(x^2/a^2 - 1)$, or

$$y = \pm b \sqrt{\frac{x^2}{a^2} - 1}$$

As x gets large, the 1 under the radical sign becomes insignificant in comparison with x^2/a^2, and the equation for y becomes

$$y = \pm \frac{b}{a}x$$

This is the equation of a straight line having a slope of b/a. Thus as x increases, the branches of the hyperbola more closely approach straight lines of slopes $\pm b/a$.

For a hyperbola whose transverse axis is *vertical*, the slopes of the asymptotes can be shown to be $\pm a/b$.

Slope of the Asymptotes of a Hyperbola	Transverse Axis Horizontal	slope $= \pm \dfrac{b}{a}$	245
	Transverse Axis Vertical	slope $= \pm \dfrac{a}{b}$	246

Common Error	The asymptotes are not (usually) perpendicular to each other. The slope of one is **not** the negative reciprocal of the other.

Looking back at Fig. 22–90, we can now give meaning to the quantity b. If an asymptote has a slope b/a, it must have a rise of b in a run equal to a. Thus b is the distance, perpendicular to the transverse axis, from vertex to asymptote. It is half the length of the conjugate axis. Here again, for emphasis, is the relationship between a, b, and c.

Hyperbola: Distance from Center to Focus	$c = \sqrt{a^2 + b^2}$	244

♦♦♦ **Example 54:** Find (a) the coordinates of the center, (b) the lengths a and b of the semiaxes, (c) the coordinates of the vertices, (d) the coordinates of the foci, and (e) the slope of the asymptotes for the hyperbola

$$\frac{x^2}{25} - \frac{y^2}{36} = 1$$

Solution:

(a) This equation is of the same form as Eq. 239, so we know that the center is at the origin and that the transverse axis is on the x axis.

(b) Also, $a^2 = 25$ and $b^2 = 36$, so

$$a = 5 \quad \text{and} \quad b = 6$$

(c) The vertices have the coordinates

$$V(5, 0) \quad \text{and} \quad V'(-5, 0)$$

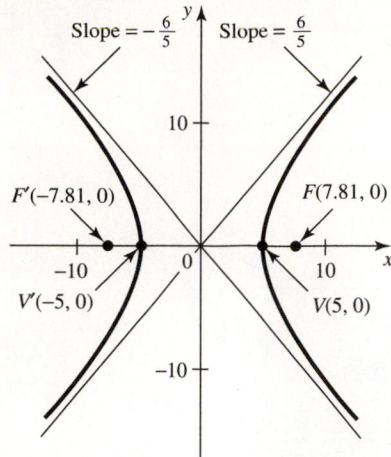

FIGURE 22–91 Graph of $\dfrac{x^2}{25} - \dfrac{y^2}{36} = 1$.

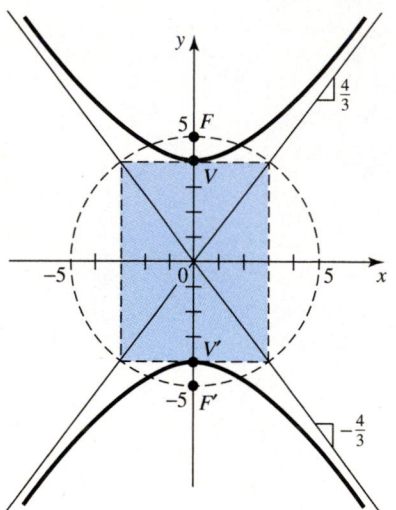

FIGURE 22–92

(d) Then, from Eq. 244,

$$c^2 = a^2 + b^2 = 25 + 36 = 61$$

so $c = \sqrt{61} \approx 7.81$. The coordinates of the foci are then

$$F(7.81, 0) \quad \text{and} \quad F'(-7.81, 0)$$

(e) The slopes of the asymptotes are

$$\pm \frac{b}{a} = \pm \frac{6}{5}$$

This hyperbola is shown in Fig. 22–91. In the following section we show how to make such a graph.   ◆◆◆

◆◆◆ **Example 55:** A hyperbola whose center is at the origin has a focus at $(0, -5)$ and a transverse axis 8 units long, as shown in Fig. 22–92. (a) Write the standard equation of the hyperbola, and (b) find the slope of the asymptotes.

Solution: (a) The transverse axis is 8, so $a = 4$. A focus is 5 units below the origin, so the transverse axis must be vertical, and $c = 5$. From Eq. 244,

$$b = \sqrt{c^2 - a^2} = \sqrt{25 - 16} = 3$$

Substituting into Eq. 240 gives us

$$\frac{y^2}{16} - \frac{x^2}{9} = 1$$

(b) From Eq. 246,

$$\text{slope of asymptotes} = \pm \frac{a}{b} = \pm \frac{4}{3}$$ ◆◆◆

Common Error	Be sure that you know the direction of the transverse axis before computing the slopes of the asymptotes, which are $\pm b/a$ when the axis is horizontal, but $\pm a/b$ when the axis is vertical.

Manually Graphing a Hyperbola

A good way to start a sketch of the hyperbola is to draw a rectangle whose dimension along the transverse axis is $2a$ and whose dimension along the conjugate axis is $2b$. The asymptotes are then drawn along the diagonals of this rectangle. Half the diagonal of the rectangle has a length $\sqrt{a^2 + b^2}$, which is equal to c, the distance to the foci. Thus an arc of radius c will cut extensions of the transverse axis at the focal points.

◆◆◆ **Example 56:** Graph the hyperbola $64x^2 - 49y^2 = 3136$.

Solution: We write the equation in standard form by dividing by 3136.

$$\frac{x^2}{49} - \frac{y^2}{64} = 1$$

This is the form of Eq. 239, so the transverse axis is horizontal, with $a = 7$ and $b = 8$. We draw a rectangle of width 14 and height 16 (shown shaded in Fig. 22–93), thus locating the vertices at $(\pm 7, 0)$. Diagonals through the rectangle give us the asymptotes of slopes $\pm \frac{8}{7}$. We locate the foci by swinging an arc of radius c, equal to half the diagonal of the rectangle. Thus

$$c = \sqrt{7^2 + 8^2} = \sqrt{113} \approx 10.6$$

The foci then are at $(\pm 10.6, 0)$. We obtain a few more points by computing the focal width. As with the parabola and ellipse, a perpendicular through a focus con-

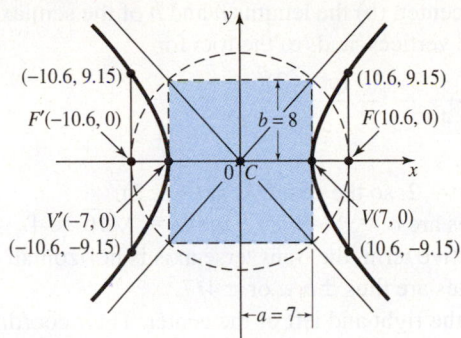

FIGURE 22–93

necting two points on the hyperbola is called a *latus rectum*. Its length, called the *focal width*, is $2b^2/a$, the same as for the ellipse.

$$L = \frac{2b^2}{a} = \frac{2(64)}{7} \approx 18.3$$

This gives us the additional points $(10.6, 9.15)$, $(10.6, -9.15)$, $(-10.6, 9.15)$ and $(-10.6, -9.15)$. ◆◆◆

Graphing a Hyperbola by Graphics Calculator

We solve the given equation for y, as before, unless your grapher can accept equations in implicit form.

◆◆◆ **Example 57:** The equation for Example 56, in explicit form, is

$$y = \pm 8\sqrt{\frac{x^2}{49} - 1}$$

(work not shown) and the calculator screen is shown. We must again graph both the upper and lower portions of the curve. Here, symmetry about the x axis has allowed us to use $\mathbf{Y2 = -Y1}$ as the second equation. ◆◆◆

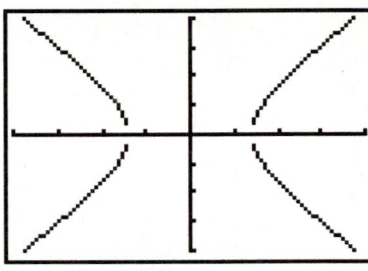

Screen for Example 57.

Standard Equation of Hyperbola: Center Not at Origin

As with the other conics, we shift the axes by replacing x by $(x - h)$ and y by $(y - k)$ in Eqs. 239 and 240.

Standard Equations of a Hyperbola: Center at (h, k)	Transverse Axis Horizontal		$\dfrac{(x - h)^2}{a^2} - \dfrac{(y - k)^2}{b^2} = 1$	**241**
	Transverse Axis Vertical		$\dfrac{(y - k)^2}{a^2} - \dfrac{(x - h)^2}{b^2} = 1$	**242**

The equations for c and for the slope of the asymptotes are still valid for these cases.

Common Error	Do not confuse these equations with those for the ellipse. Here the terms have **opposite signs**. Also, a^2 is always the denominator of the **positive** term, even though it may be smaller than b^2. As with the ellipse, the variable in the same term as a^2 tells the direction of the transverse axis.

◆◆◆ **Example 58:** Find (a) the center, (b) the lengths a and b of the semiaxes, (c) the slope of the asymptotes, (d) the vertices, and (e) the foci for

$$\frac{(x+3)^2}{49} - \frac{(y-2)^2}{16} = 1$$

Solution:

(a) We see that $h = -3$ and $k = 2$, so the center is at $(-3, 2)$.
(b) The lengths of the semiaxes are $a = \sqrt{49} = 7$ and $b = \sqrt{16} = 4$.
(c) Since the x term is the positive term, the transverse axis is horizontal. The slopes of the asymptotes are thus $\pm b/a$ or $\pm 4/7$.
(d) The vertices are 7 units to the right and left of the center. Their coordinates are thus $(-10, 2)$ and $(4, 2)$.
(e) We compute c to find the focal points.

$$c^2 = a^2 + b^2 = 49 + 16 = 65$$
$$c = 8.06$$

The foci are then 8.06 units to the left and right of the center so their coordinates are $(-11.06, 2)$ and $(5.06, 2)$.

These features are shown in Fig. 22–94.

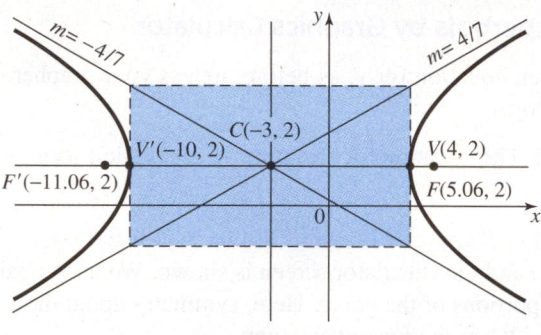

FIGURE 22–94   ◆◆◆

◆◆◆ **Example 59:** A certain hyperbola has a focus at $(1, 0)$, passes through the origin, has its transverse axis on the x axis, and has a distance of 10 between its focal points. Write its standard equation.

Solution: In Fig. 22–95 we plot the given focus $F(1, 0)$. Since the hyperbola passes through the origin, and the transverse axis also passes through the origin, a vertex must be at the origin as well. Then, since $c = 5$, we go 5 units to the left of F and plot the center, $(-4, 0)$. Thus $a = 4$ units, from center to vertex. Then, by Eq. 244,

$$b = \sqrt{c^2 - a^2}$$
$$= \sqrt{25 - 16} = 3$$

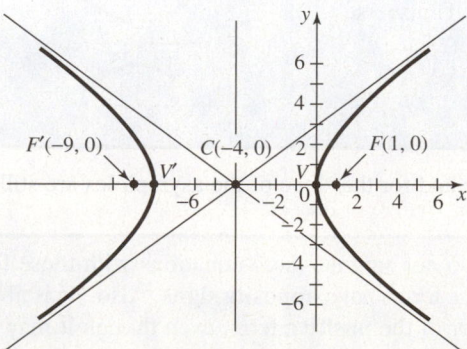

FIGURE 22–95

Substituting into Eq. 241 with $h = -4$, $k = 0$, $a = 4$, and $b = 3$, we get

$$\frac{(x + 4)^2}{16} - \frac{y^2}{9} = 1$$

as the required equation.   ◆◆◆

General Equation of a Hyperbola

Since the standard equations for the ellipse and hyperbola are identical except for the minus sign, it is not surprising that their general equations should be identical except for a minus sign. We showed how to change between standard and general forms in the section on the ellipse. The general equation we got for the ellipse is the same for the hyperbola. But here we will see that A and C have *opposite* signs, while for the ellipse, A and C have the *same* sign.

General Equation of a Hyperbola	$Ax^2 + Cy^2 + Dx + Ey + F = 0$	243

Comparing this with the general second-degree equation, we see that $B = 0$, indicating that the axes of the curve are parallel to the coordinate axes.

Also note that neither A nor C is zero, which tells us that the curve is not a parabola. Further, A and C have different values, telling us that the curve cannot be a circle.

The methods used to convert between standard and general forms are the same as for the ellipse, and are not repeated here.

Hyperbola Whose Asymptotes Are the Coordinate Axes

The graph of an equation of the form

$$xy = k$$

where k is a constant, is a hyperbola similar to those we have studied in this section but rotated so that the coordinate axes are now the asymptotes, and the transverse and conjugate axes are at $45°$.

Hyperbola: Axes Rotated 45°	$xy = k$	248

◆◆◆ **Example 60:** Manually plot the equation $xy = -4$.

Solution: We select values for x and evaluate $y = -4/x$.

x	-4	-3	-2	-1	0	1	2	3	4
y	1	$\frac{4}{3}$	2	4	—	-4	-2	$-\frac{4}{3}$	-1

The graph of Fig. 22–96 shows vertices at $(-2, 2)$ and $(2, -2)$. We see from the graph that a and b are equal and that each is the hypotenuse of a right triangle of side 2. Therefore

$$a = b = \sqrt{2^2 + 2^2} = \sqrt{8}$$

and from Eq. 244,

$$c = \sqrt{a^2 + b^2} = \sqrt{16} = 4$$ ◆◆◆

When the constant k in Eq. 248 is negative, the hyperbola lies in the second and fourth quadrants, as in Example 60. When k is positive, the hyperbola lies in the first and third quadrants.

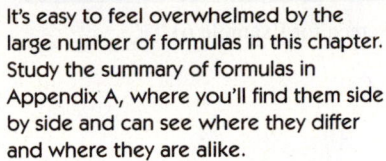

It's easy to feel overwhelmed by the large number of formulas in this chapter. Study the summary of formulas in Appendix A, where you'll find them side by side and can see where they differ and where they are alike.

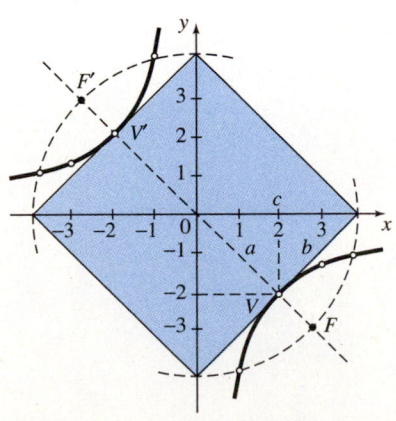

FIGURE 22–96 Hyperbola whose asymptotes are the coordinate axes.

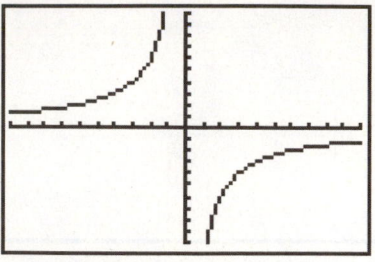

Screen for Example 61.

◆◆◆ **Example 61:** Write an equation of the hyperbola with center at the origin and asymptotes on the coordinate axes, which passes through the point $(3, -4)$. Graph by calculator.

Solution: This hyperbola has an equation of the form $xy = k$. Substituting the known point gives

$$k = xy = 3(-4) = -12$$

Our equation is then $xy = -12$. Entering the equation $y = -12/x$ into a graphics calculator, setting a suitable range, and graphing gives the curve shown.   ◆◆◆

Exercise 6 ◆ The Hyperbola

Standard Equation of Hyperbola with Center at Origin

Find the vertices, the foci, the lengths a and b of the semiaxes, and the slope of the asymptotes for each hyperbola. Graph some.

1. $\dfrac{x^2}{16} - \dfrac{y^2}{25} = 1$ **2.** $\dfrac{y^2}{25} - \dfrac{x^2}{49} = 1$

3. $16x^2 - 9y^2 = 144$ **4.** $9x^2 - 16y^2 = 144$

5. $x^2 - 4y^2 = 16$ **6.** $3x^2 - y^2 + 3 = 0$

Write the equation of each hyperbola in standard form.

7. vertices at $(\pm 5, 0)$; foci at $(\pm 13, 0)$

8. vertices at $(0, \pm 7)$; foci at $(0, \pm 10)$

9. distance between foci $= 8$; transverse axis $= 6$ and is horizontal

10. conjugate axis $= 4$; foci at $(\pm 2.5, 0)$

11. transverse and conjugate axes equal; passes through $(3, 5)$; transverse axis vertical

12. transverse axis $= 16$ and is horizontal; conjugate axis $= 14$

13. transverse axis $= 10$ and is vertical; curve passes through $(8, 10)$

14. transverse axis $= 8$ and is horizontal; curve passes through $(5, 6)$

Standard Equation of Hyperbola with Center Not at Origin

Find the center, lengths of the semiaxes, foci, slope of the asymptotes, and vertices for each hyperbola. Graph some

15. $\dfrac{(x - 2)^2}{25} - \dfrac{(y + 1)^2}{16} = 1$

16. $\dfrac{(y + 3)^2}{25} - \dfrac{(x - 2)^2}{49} = 1$

Write the equation of each hyperbola in standard form.

17. center at $(3, 2)$; length of the transverse axis $= 8$ and is vertical; length of the conjugate axis $= 4$

18. foci at $(-1, -1)$ and $(-5, -1)$; $a = b$

19. foci at $(5, -2)$ and $(-3, -2)$; vertex halfway between center and focus

20. length of the conjugate axis $= 8$ and is horizontal; center at $(-1, -1)$; length of the transverse axis $= 16$

Write each standard equation in general form.

21. $\dfrac{x^2}{25} - \dfrac{y^2}{9} = 1$ **22.** $\dfrac{x^2}{16} - \dfrac{y^2}{36} = 4$

23. $\dfrac{(x - 3)^2}{16} - \dfrac{(y - 4)^2}{25} = 12$ **24.** $\dfrac{(x + 5)^2}{9} - \dfrac{(y + 7)^2}{4} = 22$

Write each general equation in standard form by completing the square. Find the center, lengths of the semiaxes, foci, slope of the asymptotes, and vertices.

25. $16x^2 - 64x - 9y^2 - 54y = 161$

26. $16x^2 + 32x - 9y^2 + 592 = 0$

27. $4x^2 + 8x - 5y^2 - 10y + 19 = 0$

28. $y^2 - 6y - 3x^2 - 12x = 12$

Hyperbola Whose Asymptotes Are Coordinate Axes

29. Write the equation of a hyperbola with center at the origin and with asymptotes on the coordinate axes, whose vertex is at $(6, 6)$.

30. Write the equation of a hyperbola with center at the origin and with asymptotes on the coordinate axes, which passes through the point $(-9, 2)$.

Applications

31. A hyperbolic mirror (Fig. 22–97) has the property that a ray of light directed at one focus will be reflected so as to pass through the other focus. Write the equation of the mirror shown, taking the axes as indicated.

32. A ship (Fig. 22–98) receives simultaneous radio signals from stations P and Q, on the shore. The signal from station P is found to arrive 375 microseconds (μs) later than that from Q. Assuming that the signals travel at a rate of 0.186 mi/μs, find the distance from the ship to each station (*Hint:* The ship will be on one branch of a hyperbola H whose foci are at P and Q. Write the equation of this hyperbola, taking axes as shown, and then substitute 75.0 mi for y to obtain the distance x.)

33. Boyle's law states that under certain conditions, the product of the pressure and volume of a gas is constant, or $pv = c$. This equation has the same form as the hyperbola (Eq. 248). If a certain gas has a pressure of 25.0 lb/in.^{2} at a volume of 1000 in.^{3}, write Boyle's law for this gas, and make a graph of pressure versus volume.

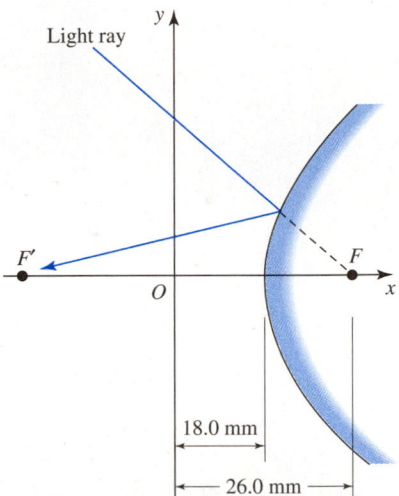

FIGURE 22–97 Hyperbolic mirror. This type of mirror is used in the *Cassegrain* form of reflecting telescope.

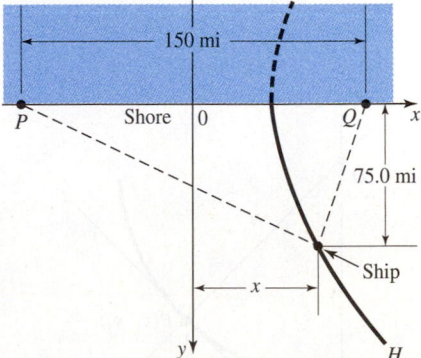

FIGURE 22–98

••• CHAPTER 22 REVIEW PROBLEMS •••••••••••••••••••••••••••••••••

1. Find the distance between the points $(3, 0)$ and $(7, 0)$.

2. Find the distance between the points $(4, -4)$ and $(1, -7)$ to 3 significant digits.

3. Find the slope of the line perpendicular to a line that has an angle of inclination of $34.8°$.

4. Find the angle of inclination in degrees of a line with a slope of -3.

5. Find the angle of inclination of a line perpendicular to a line having a slope of 1.55.

6. Find the angle of inclination of a line passing through $(-3, 5)$ and $(5, -6)$.

7. Find the slope of a line perpendicular to a line having a slope of $a/2b$.

8. Write the equation of a line having a slope of -2 and a y intercept of 5.

9. Find the slope and y intercept of the line $2y - 5 = 3(x - 4)$.

10. Write the equation of a line having a slope of $2p$ and a y intercept of $p - 3q$.

11. Write the equation of the line passing through $(-5, -1)$ and $(-2, 6)$.

12. Write the equation of the line passing through $(-r, s)$ and $(2r, -s)$.

13. Write the equation of the line having a slope of 5 and passing through the point $(-4, 7)$.

14. Write the equation of the line having a slope of $3c$ and passing through the point $(2c, c - 1)$.

15. Write the equation of the line having an x intercept of -3 and a y intercept of 7.

16. Find the acute angle between two lines if one line has a slope of 1.50 and the other has a slope of 3.40.

17. Write the equation of the line that passes through $(2, 5)$ and is parallel to the x axis.

18. Find the angle of intersection between line L_1 having a slope of 2 and line L_2 having a slope of 7.

19. Find the directed distance AB between the points $A(-2, 0)$ and $B(-5, 0)$.

20. Find the angle of intersection between line L_1 having an angle of inclination of $18°$ and line L_2 having an angle of inclination of $75°$.

21. Write the equation of the line that passes through $(-3, 6)$ and is parallel to the y axis.

22. Find the increments in the coordinates of a particle that moves along a curve from $(3, 4)$ to $(5, 5)$.

23. Find the area of a triangle with vertices at $(6, 4)$, $(5, -2)$, and $(-3, -4)$ to 3 significant digits.

Identify the curve represented by each equation. Find, where applicable, the center, vertices, foci, radius, semiaxes, and so on.

24. $x^2 - 2x - 4y^2 + 16y = 19$

25. $x^2 + 6x + 4y = 3$

26. $x^2 + y^2 = 8y$

27. $25x^2 - 200x + 9y^2 - 90y = 275$

28. $16x^2 - 9y^2 = 144$

29. $x^2 + y^2 = 9$

Tangents and Normals. In problems 50 through 59, a tangent T, of slope m, and a normal N are drawn to a curve at the point $P(x_1, y_1)$, as shown in Fig. 22–99. Show the following.

30. The equation of the tangent is
$$y - y_1 = m(x - x_1)$$

31. The equation of the normal is
$$x - x_1 + m(y - y_1) = 0$$

32. The x intercept A of the tangent is
$$x_1 - y_1/m$$

33. The y intercept B of the tangent is
$$y_1 - mx_1$$

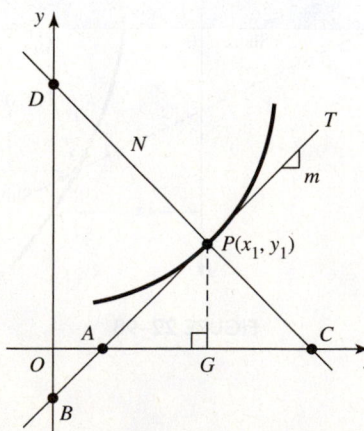

FIGURE 22–99 Tangent and Normal to a curve.

34. The length of the tangent from P to the x axis is

$$PA = \frac{y_1}{m}\sqrt{1 + m^2}$$

35. The length of the tangent from P to the y axis is

$$PB = x_1\sqrt{1 + m^2}$$

36. The x intercept C of the normal is

$$x_1 + my_1$$

37. The y intercept D of the normal is

$$y_1 + x_1/m$$

38. The length of the normal from P to the x axis is

$$PC = y_1\sqrt{1 + m^2}$$

39. The length of the normal from P to the y axis is

$$PD = \frac{x_1}{m}\sqrt{1 + m^2}$$

40. *Project:* We have already shown how to graph each of the conic sections. Now graph, in the same viewing window,

the ellipse

$$\frac{x^2}{a^2} + \frac{y^2}{b^2} = 1$$

the hyperbola

$$\frac{x^2}{a^2} - \frac{y^2}{b^2} = 1$$

and the hyperbola

$$\frac{y^2}{a^2} - \frac{x^2}{b^2} = 1$$

using the same values of a and b for each. Also graph the asymptotes of the hyperbolas. Describe your results. Do you see anything similar about the three curves?

41. Write the equation for an ellipse whose center is at the origin, whose major axis ($2a$) is 20 and is horizontal, and whose minor axis ($2b$) equals the distance ($2c$) between the foci.

42. Write an equation for the circle passing through $(0, 0)$, $(8, 0)$, and $(0, -6)$.

43. Write the equation for a parabola whose vertex is at the origin and whose focus is $(-4.25, 0)$.

44. Write an equation for a hyperbola whose transverse axis is horizontal, with center at $(1, 1)$ passing through $(6, 2)$ and $(-3, 1)$.

45. Write the equation of a circle whose center is $(-5, 0)$ and whose radius is 5.

46. Write the equation of a hyperbola whose center is at the origin, whose transverse axis $= 8$ and is horizontal, and which passes through $(10, 25)$.

47. Write the equation of an ellipse whose foci are $(2, 1)$ and $(-6, 1)$, and the sum of the focal radii is 10.

48. Write the equation of a parabola whose axis is the line $y = -7$, whose vertex is 3 units to the right of the y axis, and which passes through $(4, -5)$.

49. Graphically find the intercepts of the curve $y^2 + 4x - 6y = 16$.

50. Find the points of intersection $x^2 + y^2 + 2x + 2y = 2$ and $3x^2 + 3y^2 + 5x + 5y = 10$.

51. A stone bridge arch is in the shape of half an ellipse and is 15.0 m wide and 5.00 m high. Find the height of the arch at a distance of 6.00 m from its center.

52. Write the equation of a hyperbola centered at the origin, where the conjugate axis = 12 and is vertical, and the distance between foci is 13.

53. A parabolic arch is 5.00 m high and 6.00 m wide at the base. Find the width of the arch at a height of 2.00 m above the base.

54. A stone propelled into the air follows a parabolic path and reaches a maximum height of 56.0 ft in a horizontal distance of 48.0 ft. At what horizontal distances from the launch point will the height be 25.0 ft?

55. *Writing:* Suppose that the day before your visit to your former high school math class, the teacher unexpectedly asks you to explain how that day's topic, the conic sections, got that name. Write a paragraph on what you will tell the class about how the circle, ellipse, parabola, and hyperbola (and the point and straight line, too) can be formed by intersecting a plane and a cone. You may plan to illustrate your talk with a clay model or a piece of cardboard rolled into a cone. You will probably be asked what the conic sections are good for, so write out at least one use for each curve.

Derivatives of Algebraic Functions

••• OBJECTIVES ••

When you have completed this chapter, you should be able to

- Evaluate limits of algebraic functions.
- Find the derivative of an algebraic expression using the delta method, by using the rules for derivatives, or by calculator.
- Find the slope of the tangent line to a curve at a given point.
- Find the derivative of an implicit algebraic function.
- Write the differential of an algebraic function.
- Find higher derivatives of an algebraic function.

Here we begin our study of *calculus*, a topic we will pursue for the remainder of this text. We will start with the *limit concept*, our main use for which will be for the definition of the *derivative*. We will learn how to take a derivative by a process called the *delta method*, and also by using our graphing calculators. We will then use the delta method to derive some rules to let us find derivatives more easily. At each step we will use our calculators to check and verify our results.

We will first find derivatives of explicit functions and then go on to implicit functions, derivatives with respect to other variables, and second derivatives. Our purpose in each case will be to provide the tools needed for the applications to follow. We will define a *differential* to lay some groundwork for integration and for differential equations. We will cover only algebraic functions here, saving trigonometric, logarithmic, and exponential functions for a later chapter.

This chapter is followed by two chapters devoted entirely to applications of the derivative. You will see there that calculus will let us solve problems that are not possible by other means. Here, however, we will give just a few applications to give you an idea of what is to follow.

For example, suppose the position y of a point P in a mechanism, Fig. 23–1, is given by $y = 2.35t^2 - 3.93t + 5.26$ cm. How would you find the velocity and acceleration of that point? We will show how to find these quantities by means of the derivative.

The development of calculus is often considered one of the greatest achievements of science. It was conceived independently in the second half of the seventeenth century by the English mathematician, physicist, and astronomer Sir Isaac Newton (1642–1727) and by the German philosopher and analyst Gottfried Wilhelm von Leibniz (1646–1716).

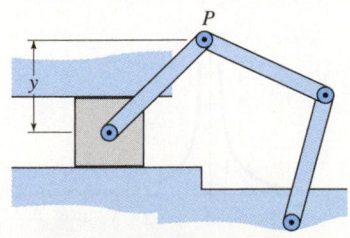

FIGURE 23–1

23–1 Limits

When we learn about derivatives in the next section, we will be finding *limits* of functions, so in this section we will learn what a limit is and how to find one. This is not our first use of the limit idea. We used it to find the value of *e*, the base of natural logarithms in our chapter on logarithms.

■ Exploration:

Try this. Graph the function $y = 1/x^2$, for $x = -10$ to 20.

• What happens to the value of *y* when *x* equals 0?

• What happens to the value of *y* when *x* gets close to 0? In other words, what happens to the value of *y* as *x approaches* 0?

• Which do you think is more useful? To say that

  (a) the given function is not defined at $x = 0$, or

  (b) the given function approaches infinity as *x* approaches 0.   ■

You may have concluded from your exploration that more information is conveyed by (b), and in fact, much of what we do in calculus is based on statements of that sort.

Limit Notation

Suppose that *x* and *y* are related by a function, such as $y = 3x$. Then if we are given a value of *x*, we can find a corresponding value for *y*. For example, when *x is* 2, then *y is* 6.

  We now want to extend our mathematical language to be able to say what will happen to *y not* when *x is* a certain value, but when *x approaches* a certain value. For example, when *x approaches* 2, then *y approaches* 6. The *notation* we use to say the same thing is

$$\lim_{x \to 2} (3x) = 6$$

We read this as "the limit of 3*x*, as *x* approaches 2, is equal to 6." In general, if the function $f(x)$ approaches some value *L* as *x* approaches *a*, we convey that idea with the notation:

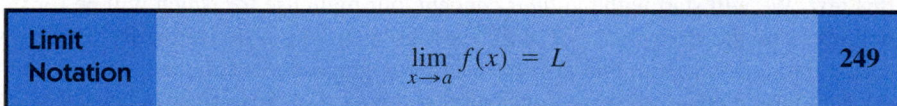

Limit Notation	$\lim_{x \to a} f(x) = L$	249

  Why bother with new notation? Why not just say, in the preceding example, that *y is* 6 when *x is* 2? Why is it necessary to creep up on the answer like that? It is true that limit notation offers no advantage in an example such as the last one. But we really need it when our function is *not even defined* at the limit.

◆◆◆ **Example 1:** The function $y = 1/x^2$ from our exploration is graphed in Fig. 23–2. Even though our function is not even defined at $x = 0$, because it results in division by zero, we can still write

$$\lim_{x \to 0} \frac{1}{x^2} = \infty$$

Further, even though *y* never reaches 0, we can still write

$$\lim_{x \to \infty} \frac{1}{x^2} = 0$$ ◆◆◆

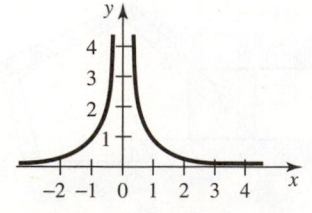

FIGURE 23–2 Graph of $y = 1/x^2$.

Finding Simple Limits

We first try to evaluate a limit by substituting into the function the value which x is approaching.

◆◆◆ **Example 2:** Evaluate

$$\lim_{x \to 3} (x^2 + 4x - 5)$$

Solution: Substituting 3 for x gives

$$\lim_{x \to 3} (x^2 + 4x - 5) = 9 + 12 - 5 = 16 \qquad ◆◆◆$$

Sometimes substitution will result in division by zero. We can usually avoid this by first simplifying the expression.

◆◆◆ **Example 3:** Evaluate

$$\lim_{x \to 3} \frac{x^2 + 2x - 15}{x - 3}$$

Solution: We see that substituting 3 for x will give division by zero. So let us first factor the numerator and simplify.

$$\lim_{x \to 3} \frac{x^2 + 2x - 15}{x - 3} = \lim_{x \to 3} \frac{(x - 3)(x + 5)}{x - 3} = \lim_{x \to 3} (x + 5)$$

Now substituting 3 for x gives

$$\lim_{x \to 3} \frac{x^2 + 2x - 15}{x - 3} = \lim_{x \to 3} (x + 5) = 8 \qquad ◆◆◆$$

Limits Found Graphically

As we saw in the exploration, we can find the approximate limit of a function simply by graphing the function and observing what happens at the value that x is approaching. The calculator is also useful for *visualizing* a limit found algebraically, or approximately *verifying* that limit.

◆◆◆ **Example 4:** Graphically evaluate

$$\lim_{x \to 0} \frac{(x + 1)^2 - 1}{x}$$

Solution: We graph the function $f(x) = \dfrac{(x + 1)^2 - 1}{x}$ as shown. We get no value for $x = 0$ as that results in division by zero. But we note that the value of the expression approaches 2 as x approaches zero, so we write

$$\lim_{x \to 0} \frac{(x + 1)^2 - 1}{x} = 2$$

Do not expect to *see* a gap in the graph at $x = 0$. The gap is infinitesimally small and will not show. ◆◆◆

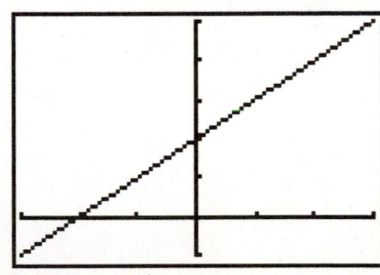

Screen for Example 4.

Limits Found Numerically

A simple numerical way to evaluate a limit is to substitute values of x closer and closer to the value that it is to approach, and see if the expression approaches a limit. We will use such a numerical method in the next example.

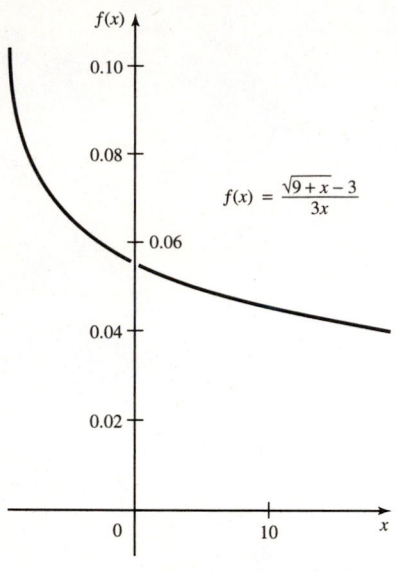

FIGURE 23–3

♦♦♦ **Example 5:** Evaluate $\lim\limits_{x \to 0} \dfrac{\sqrt{9 + x} - 3}{3x}$.

Solution: We see that when x *equals* zero, we get

$$\frac{\sqrt{9 + 0} - 3}{3(0)} = \frac{3 - 3}{0}$$

$$= \frac{0}{0}$$

a result that is indeterminate. But what happens when x *approaches* zero? Let us use the calculator to substitute smaller and smaller values of x. Working to five decimal places, we get the following values:

x	10	1	0.1	0.01	0.001	0.0001
	0.04530	0.05409	0.05540	0.05554	0.05555	0.05556

So as x approaches zero, the given expression appears to approach 0.05556 as a limit (Fig. 23–3). This, of course, is not a *proof* that the limit found is the correct one or even that a limit exists. ♦♦♦

Common Error	Be sure to distinguish between a denominator that *is* zero and one that *is approaching* zero. In the first case we have division by zero, but in the second case we might get a useful answer.

Limits by Computer Algebra System

A calculator or computer that can do symbolic processing can be used to find limits. On the TI-89, for example, we select **limit** from the **Calc** menu. We then enter the function, the variable, and the value which the variable is approaching. Pressing $\boxed{\text{ENTER}}$ will display the limit. It can either be a numerical value, as in our examples so far, or an expression, as we will see a little later.

♦♦♦ **Example 6:** Use the TI-89 calculator to evaluate the limits in Examples 2–5.

Solution: The calculator screens are as follows:

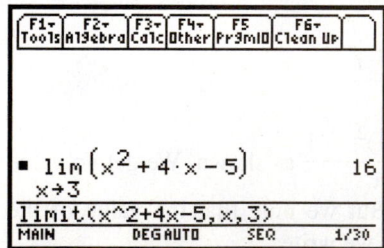

Screen for Example 2.

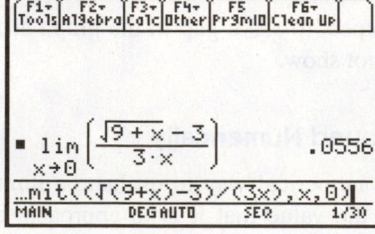

Screen for Example 3.

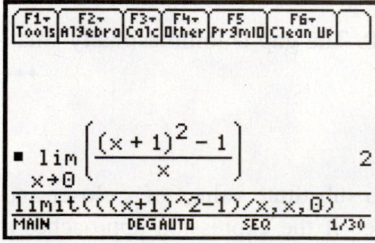

Screen for Example 4.

Screen for Example 5. ♦♦♦

Limit of a Constant

Now suppose we want the limit of a constant C as x approaches some value a. This is written

$$\lim_{x \to a} C$$

If we graph the function $f(x) = C$, Fig. 23–4, we see that its value is C for any value of x. Thus *the limit of a constant is the constant itself.*

$$\lim_{x \to a} C = C$$

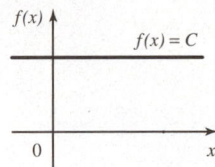

FIGURE 23–4   Graph of $f(x) = C$.

◆◆◆ **Example 7:** Here are some limits of constants.

(a) $\displaystyle\lim_{x \to \infty} 5 = 5$ (b) $\displaystyle\lim_{x \to 0} 5 = 5$ (c) $\displaystyle\lim_{x \to 0} 3\pi = 3\pi$ ◆◆◆

Limits Involving Zero or Infinity

Limits involving zero or infinity can usually be evaluated using the following facts. (Here, C is a nonzero constant.)

Limits Involving Zero or Infinity	(1) $\displaystyle\lim_{x \to 0} Cx = 0$	(4) $\displaystyle\lim_{x \to \infty} Cx = \infty$	
	(2) $\displaystyle\lim_{x \to 0} \frac{x}{C} = 0$	(5) $\displaystyle\lim_{x \to \infty} \frac{x}{C} = \infty$	**250**
	(3) $\displaystyle\lim_{x \to 0} \frac{C}{x} = \infty$	(6) $\displaystyle\lim_{x \to \infty} \frac{C}{x} = 0$	

◆◆◆ **Example 8:** Here are some examples of the use of these rules.

(a) $\displaystyle\lim_{x \to 0} 5x^3 = 0$

(b) $\displaystyle\lim_{x \to 0} \left(7 + \frac{x}{2}\right) = 7$

(c) $\displaystyle\lim_{x \to 0} \left(3 + \frac{25}{x^2}\right) = \infty$

(d) $\displaystyle\lim_{x \to \infty} (3x - 2) = \infty$

(e) $\displaystyle\lim_{x \to \infty} \left(8 + \frac{x}{4}\right) = \infty$

(f) $\displaystyle\lim_{x \to \infty} \left(\frac{5}{x} - 3\right) = -3$

Each function is graphed in Fig. 23–5. ◆◆◆

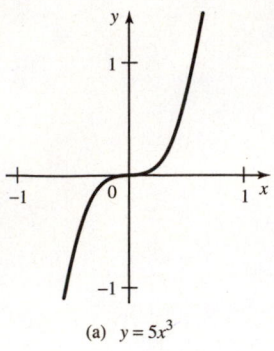

(a) $y = 5x^3$

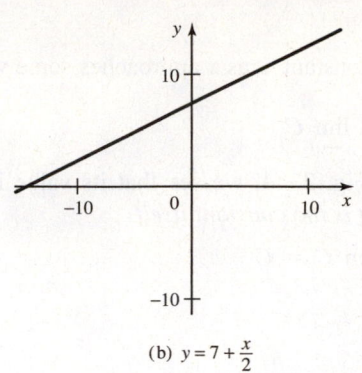

(b) $y = 7 + \dfrac{x}{2}$

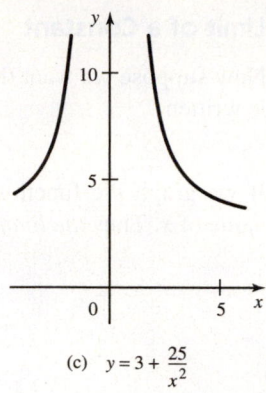

(c) $y = 3 + \dfrac{25}{x^2}$

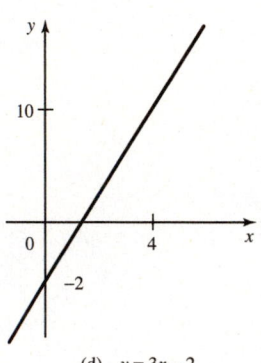

(d) $y = 3x - 2$

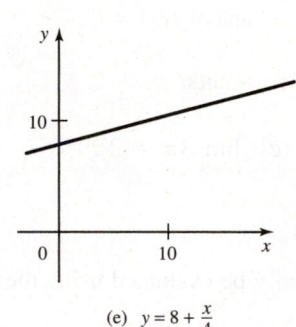

(e) $y = 8 + \dfrac{x}{4}$

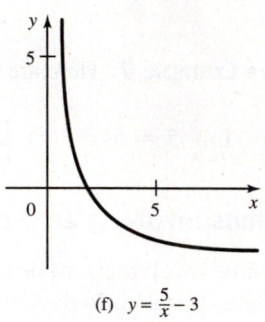

(f) $y = \dfrac{5}{x} - 3$

FIGURE 23–5

When we want the limit, as x becomes infinite, of the quotient of two polynomials, such as

$$\lim_{x \to \infty} \frac{4x^3 - 3x^2 + 5}{3x - x^2 - 5x^3}$$

we see that both numerator and denominator become infinite so the limit is indeterminate. However, the limit of such an expression can often be found if *we divide both numerator and denominator by the highest power of x occurring in either.*

◆◆◆ **Example 9:** Evaluate $\lim\limits_{x \to \infty} \dfrac{4x^3 - 3x^2 + 5}{3x - x^2 - 5x^3}$.

Solution: Dividing both numerator and denominator by x^3 gives us

$$\lim_{x \to \infty} \frac{4x^3 - 3x^2 + 5}{3x - x^2 - 5x^3} = \lim_{x \to \infty} \frac{4 - \dfrac{3}{x} + \dfrac{5}{x^3}}{\dfrac{3}{x^2} - \dfrac{1}{x} - 5}$$

$$= \frac{4 - 0 + 0}{0 - 0 - 5} = -\frac{4}{5}$$

as shown in Fig. 23–6.

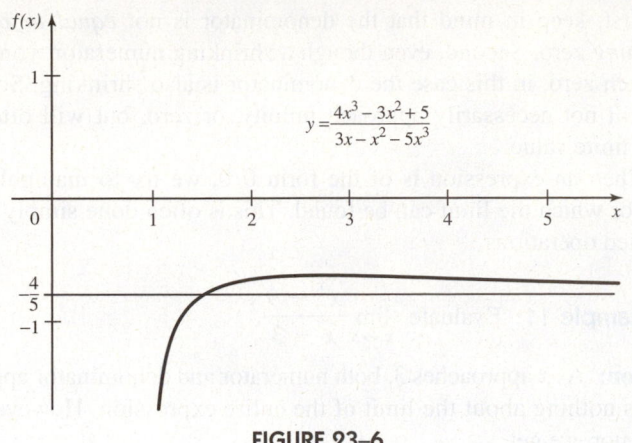

$$y = \frac{4x^3 - 3x^2 + 5}{3x - x^2 - 5x^3}$$

FIGURE 23–6 ◆◆◆

Limits Depending on Direction of Approach

Rule 3 in Eq. 250 needs a bit more explanation. At the start of this section we found that

$$\lim_{x \to 2} (3x) = 6$$

In this case it does not matter whether 2 is approached *from the right* (from values of x greater than 2) or *from the left* (from values of x less than 2). Now we will indicate that x is approaching a limit, say 2, from the *right* by

$$x \to 2^+$$

and from the *left* by

$$x \to 2^-$$

But sometimes the direction of approach *does* matter, as in the following example.

◆◆◆ **Example 10:** The function $y = 1/x$ is graphed in Fig. 23–7. When x approaches zero from "above" (that is, from the *right* in Fig. 23–7), which we write $x \to 0^+$, we get

$$\lim_{x \to 0^+} \frac{1}{x} = +\infty$$

which means that y grows without bound in the positive direction. But when x approaches zero from "below" (from the *left* in Fig. 27–7), we write

$$\lim_{x \to 0^-} \frac{1}{x} = -\infty$$

which means that y grows without bound in the negative direction. ◆◆◆

A limit that depends on the direction of approach is sometimes called a *left-hand limit* or a *right-hand limit*.

Limits of the Form 0/0

The purpose of this entire discussion of limits is to prepare us for the idea of the *derivative*. There we will have to find the limit of a fraction in which *both the numerator and the denominator approach zero*.

At first glance, when we see a numerator approaching zero, we expect the entire fraction to approach zero. But when we see a denominator approaching zero, we throw up our hands and cry "division by zero." What then do we make of a fraction in which *both* numerator and denominator approach zero?

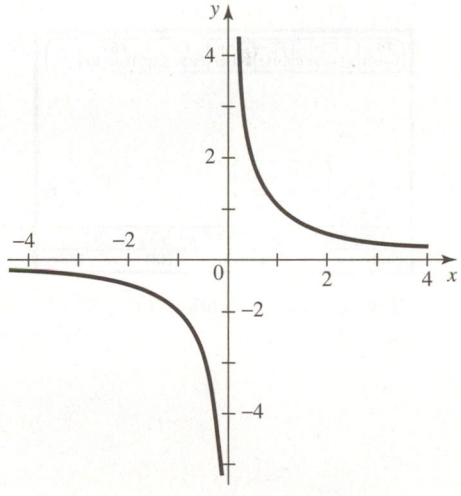

FIGURE 23–7 Graph of $y = 1/x$.

First, keep in mind that the denominator is not *equal to* zero; it is only *approaching* zero. Second, even though a shrinking numerator would make a fraction approach zero, in this case the denominator is also shrinking. So, in fact, our fraction will not necessarily approach infinity, or zero, but will often approach some useful finite value.

When an expression is of the form $0/0$, we try to manipulate it into another form for which the limit can be found. This is often done simply by performing the indicated operations.

◆◆◆ **Example 11:** Evaluate $\lim\limits_{x \to 3} \dfrac{x^2 - 9}{x - 3}$.

Solution: As x approaches 3, both numerator and denominator approach zero, which tells us nothing about the limit of the entire expression. However, if we factor the numerator, we get

$$\lim_{x \to 3} \frac{x^2 - 9}{x - 3} = \lim_{x \to 3} \frac{(x - 3)(x + 3)}{x - 3}$$
$$= \lim_{x \to 3} \left(\frac{x - 3}{x - 3}\right)(x + 3)$$

Now as x approaches 3, both the numerator and the denominator of the fraction $(x - 3)/(x - 3)$ approach zero. But since the numerator and denominator are equal, the fraction will equal 1 for any nonzero value of x. So

$$\lim_{x \to 3} \frac{x^2 - 9}{x - 3} = \lim_{x \to 3} (1)(x + 3) = 6$$

as shown in Fig. 23–8.   ◆◆◆

If an expression *cannot* be factored or otherwise manipulated into a form where the limit can be found, we can often find the limit graphically, numerically, or by a computer algebra system, as shown in the screen.

When the Limit Is an Expression

Our main use for limits will be for finding derivatives in the following sections of this chapter. There we will evaluate limits in which both the numerator and the denominator approach zero, and the resulting limit is an *expression* rather than a single number. A limit typical of the sort we will have to evaluate later is given in the following example.

◆◆◆ **Example 12:** Evaluate $\lim\limits_{d \to 0} \dfrac{(x + d)^2 + 5(x + d) - x^2 - 5x}{d}$.

Solution: If we try to set d equal to zero,

$$\frac{(x + 0)^2 + 5(x + 0) - x^2 - 5x}{0} = \frac{x^2 + 5x - x^2 - 5x}{0} = \frac{0}{0}$$

we get the indeterminate expression $0/0$. Instead, let us remove parentheses in the original expression. We get

$$\lim_{d \to 0} \frac{(x + d)^2 + 5(x + d) - x^2 - 5x}{d} = \lim_{d \to 0} \frac{x^2 + 2dx + d^2 + 5x + 5d - x^2 - 5x}{d}$$

Combining similar terms causes the x^2 and $5x$ terms to drop out.

$$= \lim_{d \to 0} \frac{2dx + d^2 + 5d}{d}$$

(left margin figure)

$f(x)$

6

4

2

$f(x) = \dfrac{x^2 - 9}{x - 3}$

0   1   2   3 x

FIGURE 23–8

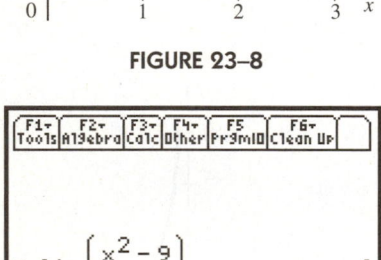

TI-89 calculator solution for Example 11.

We can now divide each term in the numerator by the d in the denominator.

$$= \lim_{d \to 0} (2x + d + 5)$$

Finally, letting d approach zero gives

$$\lim_{d \to 0} \frac{(x + d)^2 + 5(x + d) - x^2 - 5x}{d} = 2x + 5$$

◆◆◆

TI-89 calculator solution for Example 12. Here our limit is an expression rather than a number.

Exercise 1 ◆ Limits

Evaluate each limit.

1. $\displaystyle\lim_{x \to 2} (x^2 + 2x - 7)$

2. $\displaystyle\lim_{x \to -1} (x^3 - 3x^2 - 5x - 5)$

3. $\displaystyle\lim_{x \to 2} \frac{x^2 - x - 1}{x + 3}$

4. $\displaystyle\lim_{x \to 5} \frac{5 + 4x - x^2}{5 - x}$

5. $\displaystyle\lim_{x \to 5} \frac{x^2 - 25}{x - 5}$

6. $\displaystyle\lim_{x \to -7} \frac{49 - x^2}{x + 7}$

7. $\displaystyle\lim_{x \to 1} \frac{x^2 + 2x - 3}{x - 1}$

8. $\displaystyle\lim_{x \to 5} \frac{x^2 - 12x + 35}{5 - x}$

9. $\displaystyle\lim_{x \to 0} \frac{\sin x}{\tan x}$

10. $\displaystyle\lim_{x \to 4} \frac{x^2 - 16}{x - 4}$

11. $\displaystyle\lim_{x \to -3} \frac{x^2 + 2x - 3}{x + 3}$

12. $\displaystyle\lim_{x \to 1} \frac{x^3 - x^2 + 2x - 2}{x - 1}$

Limits Involving Zero or Infinity

13. $\displaystyle\lim_{x \to 0} (4x^2 - 5x - 8)$

14. $\displaystyle\lim_{x \to 0} \frac{3 - 2x}{x + 4}$

15. $\displaystyle\lim_{x \to 0} \frac{\sqrt{x} - 4}{\sqrt[3]{x} + 5}$

16. $\displaystyle\lim_{x \to 0} \frac{3 + x - x^2}{(x + 3)(5 - x)}$

17. $\displaystyle\lim_{x \to 0} \left(\frac{1}{2 + x} - \frac{1}{2} \right) \cdot \frac{1}{x}$

18. $\displaystyle\lim_{x \to 0} x \cos x$

19. $\displaystyle\lim_{x \to \infty} \frac{2x + 5}{x - 4}$

20. $\displaystyle\lim_{x \to \infty} \frac{5x - x^2}{2x^2 - 3x}$

21. $\displaystyle\lim_{x \to \infty} \frac{x^2 + x - 3}{5x^2 + 10}$

22. $\displaystyle\lim_{x \to \infty} \frac{4 + 2^x}{3 + 2^x}$

Limits Depending on Direction of Approach

23. $\displaystyle\lim_{x \to 0^+} \frac{7}{x}$

24. $\displaystyle\lim_{x \to 0^+} \frac{e^x}{x}$

25. $\displaystyle\lim_{x \to 0^+} \frac{x + 1}{x}$

26. $\displaystyle\lim_{x \to 0^-} \frac{x + 1}{x}$

27. $\displaystyle\lim_{x \to 2^-} \frac{5 + x}{x - 2}$

28. $\displaystyle\lim_{x \to 1^+} \frac{2x + 3}{1 - x}$

When the Limit Is an Expression

29. $\displaystyle\lim_{d \to 0} x^2 + 2d + d^2$

30. $\displaystyle\lim_{d \to 0} 2x + d$

31. $\displaystyle\lim_{d\to0}\frac{(x+d)^2-x^2}{x^2(x+d)}$ **32.** $\displaystyle\lim_{d\to0}\frac{(x+d)^2-x^2}{d}$

33. $\displaystyle\lim_{d\to0}\frac{3(x+d)-3x}{d}$ **34.** $\displaystyle\lim_{d\to0}\frac{[2(x+d)+5]-(2x+5)}{d}$

35. $\displaystyle\lim_{d\to0}3x+d-\frac{1}{(x+d+2)(x-2)}$

36. $\displaystyle\lim_{d\to0}\frac{[(x+d)^2+1]-(x^2+1)}{d}$

37. $\displaystyle\lim_{d\to0}\frac{(x+d)^3-x^3}{d}$ **38.** $\displaystyle\lim_{d\to0}\frac{\sqrt{x+d}-\sqrt{x}}{d}$

39. $\displaystyle\lim_{d\to0}\frac{(x+d)^2-2(x+d)-x^2+2x}{d}$

40. $\displaystyle\lim_{d\to0}\frac{\dfrac{7}{x+d}-\dfrac{7}{x}}{d}$

23–2   Rate of Change and the Tangent

Many technical applications require that we find the *rate of change* of a function. For example, velocity is the rate of change of displacement and current is the rate of change of charge. In this section we will show how rate of change is related to the tangent to a curve and later apply this idea to technical applications. But first let us define the tangent to a curve.

Tangent to a Curve

■ Exploration:

Try this.

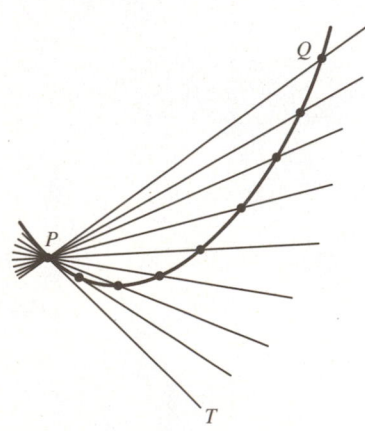

- Draw any curve, Fig. 23–9, and on it place any point *P*.
- At *P* draw, by eye, what you think is meant by the tangent *T* to the curve at *P*, keeping in mind what you know about the tangent to a circle.
- Then place another point *Q* on the curve, to either side of *P*. Connect *P* and *Q* with a straight line. Line *PQ* is called a *secant line*.
- Next move *Q* closer to *P* and draw another secant line.
- Repeat, moving *Q* ever closer to *P*, drawing a new secant line each time.

FIGURE 23–9

What is happening to the slope of the secant as *Q* approaches *P*? How does that slope compare with the tangent you drew? From this exploration, can you venture a definition of the slope of the tangent line? Can you state it using the idea of a *limit* from the preceding section?   ■

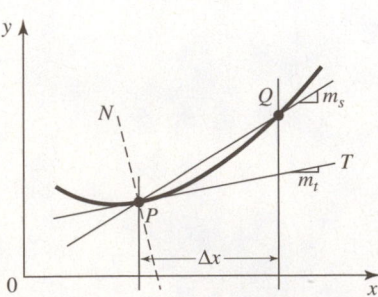

Your exploration may have suggested to you that *the tangent to a curve is the limiting position of the secant PQ, as Q approaches P.*

Let's now state this idea in a more precise way. If Δx is the horizontal distance between *P* and *Q* (Fig. 23–10), we say that *Q* approaches *P* as Δx approaches zero. We can thus say that the slope m_t of the tangent line *T* is the limit of the slope m_s of the secant line *PQ* as Δx approaches zero. We can now restate this idea in compact form using our new limit notation, as follows:

$$m_t = \lim_{\Delta x\to0} m_s$$

FIGURE 23–10   Tangent to a curve as the limiting position of the secant. This figure will have an important role in our introduction of the derivative in the following section.

The *normal* to a curve at a point *P* is the line perpendicular to the tangent at that point. A normal *N* is shown dashed in Fig. 23–10. Recall that the slope of a line is the *negative reciprocal* of the slope of a perpendicular to that line.

Rate of Change Equal to Slope of Tangent

We will now relate *slope of the tangent* to *rate of change*. To do this let us use some familiar ideas about uniform motion.

Recall that in *uniform motion*, the rate of change of displacement with respect to time (the velocity) is constant. The graph of displacement versus time, shown in Fig. 23–11(a), is a straight line whose *slope is equal to the rate of change of displacement with respect to time*.

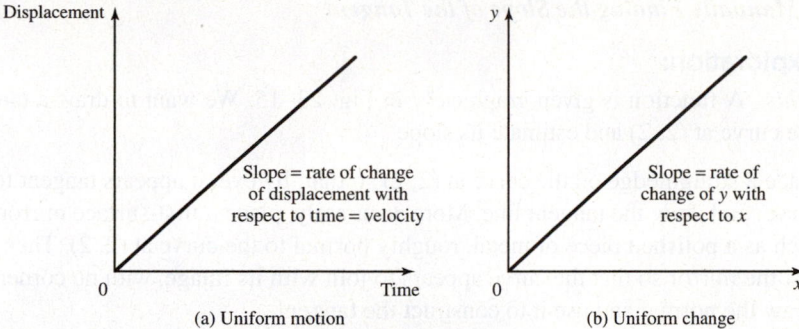

FIGURE 23–11

In general, for a function whose graph is a straight line, Fig. 23–11(b), *its slope gives the rate of change of y with respect to x*. For a function $y = f(x)$ whose graph is *not* a straight line (Fig. 23–12), we cannot give the rate of change for the entire function. However, we can choose an interval on the curve and give the *average rate of change over that interval*. For the interval PQ (Fig. 23–13), the average rate of change is equal to the change in y divided by the change in x over that interval. Thus *the slope of the secant line PQ gives us the average rate of change from P to Q*.

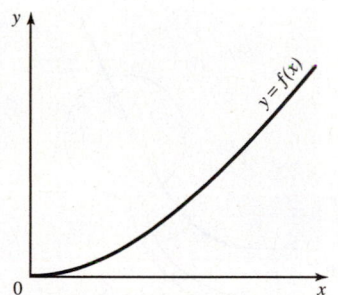

FIGURE 23–12   Nonuniform change.

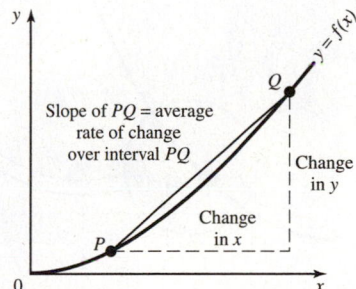

FIGURE 23–13   Average rate of change. We introduced the idea of an increment earlier, and the symbols Δx and Δy to stand for small increments in x and y.

Finally, let us move Q ever closer to P, Fig. 23–14. In doing so, the average rate of change for the interval PQ approaches the *instantaneous rate of change* at P. At the same time the slope of PQ approaches the slope of the tangent T at P. In the limit, *the instantaneous rate of change at P is equal to the slope of the tangent line at P*.

Finding the Slope of the Tangent

So we have seen that the slope of a tangent to a curve at some point will give us the instantaneous rate of change at that point. This important quantity will enable us to solve many problems in technology. But how do we find that slope?

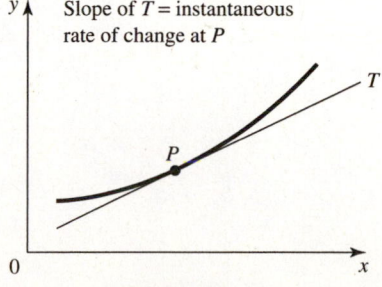

FIGURE 23–14

Here we will show four methods:

(a) manually
(b) by zooming in on a calculator graph of the curve.
(c) by using the **Tangent** function from the DRAW menu on the TI-83/84.
(d) numerically

(In the following section we will find the slope of the tangent by a new method—by finding the *derivative*.)

(a) *Manually Finding the Slope of the Tangent*

■ **Exploration:**

Try this. A function is given graphically in Fig. 23–15. We want to draw a tangent to the curve at (2, 2) and estimate its slope.

- Place a straightedge on the curve at (2, 2) so that, by eye, it appears tangent to the curve, and draw the tangent line. More accurately, place a first-surface mirror, such as a polished piece of metal, roughly normal to the curve at (2, 2). Then adjust the mirror so that the curve appears to join with its image, with no corner. Draw the normal and use it to construct the tangent.
- Measure the rise of the tangent over some chosen run to compute the slope, and hence the rate of change of the function at that point.   ■

(b) *Finding the Slope by Zooming In*

■ **Exploration:**

Try this:

- Use your graphing calculator to graph $y = x^2$. Set the viewing window from -10 to 10 on both axes.
- Keeping some point, say $P(1, 1)$, approximately in the center of the viewing window, repeatedly zoom in, as in Fig. 23–16.

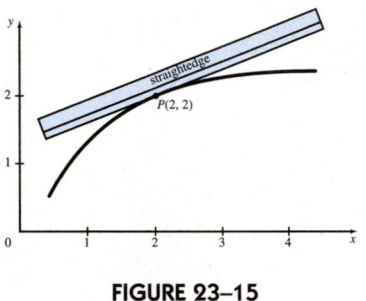

FIGURE 23–15

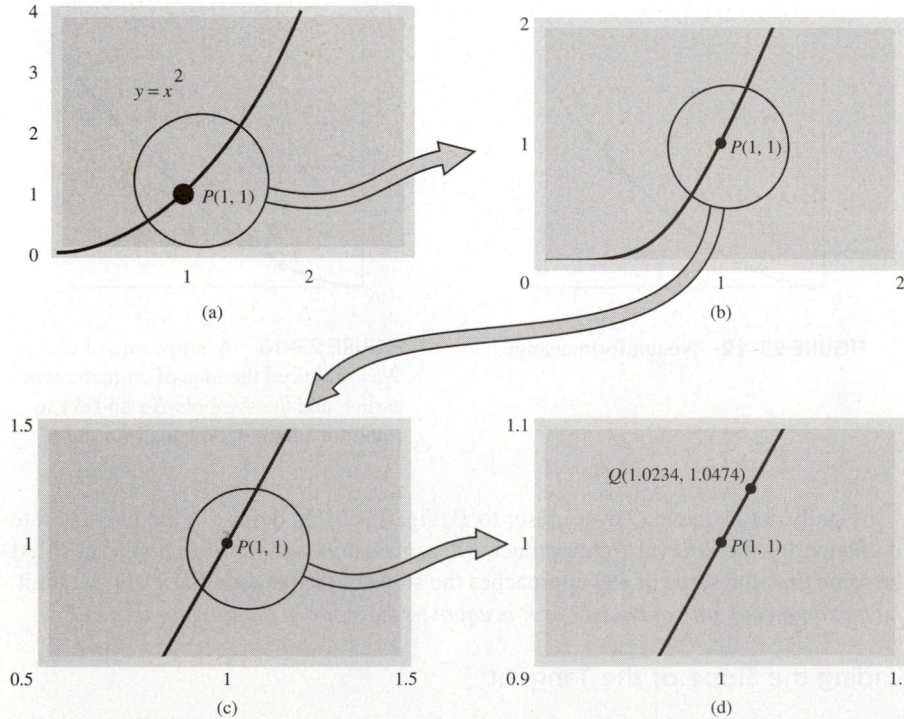

FIGURE 23–16   Zooming in on $y = x^2$ at $P(1, 1)$. The circles show the portion of the curve that is magnified in each succeeding screen.

What can you say happens to the shape of the curve as you zoom in? Do you see why a curve is sometimes referred to as *locally straight*?

• In the viewing window in which the curve appears as a straight line, use TRACE to locate a second point Q on the curve. Read its coordinates and use them to compute the slope of PQ. ■

(c) *Finding the Slope by Using the Tangent Function on a Graphics Calculator*

■ **Exploration:**

Try this. Using your graphics calculator, graph $y = x^2$. Then

• Select **Tangent** from the DRAW menu.
• Enter the value of x at which you want the tangent drawn to the curve. Choose $x = 1$ and press ENTER.

   You should get a screen as shown at right. How do you interpret this screen? What is the equation in the lower left corner? How would you find the slope of the tangent line? How does this slope compare with what you found by zooming in, in the preceding exploration? ■

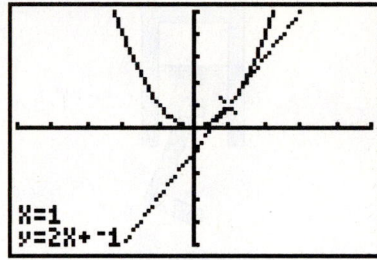

Screen for the exploration.

(d) *Finding the Slope of the Tangent Numerically*

If we have an equation for the given function, we can compute the approximate slope of the tangent line at some point P by the following numerical method.

   We place a point Q at a horizontal distance Δx from P. Then we let Q approach P in steps, at each step computing the slope m_s of the secant line PQ.

   By watching how the slope changes as we proceed, we can get a measure of the accuracy of our answer. We continue reducing Δx and repeating the computation of the slope until that computed value does not change, within the accuracy we want.

◆◆◆ **Example 13:** Use the numerical method to find the instantaneous rate of change for the function $y = x^2$ at $P(1, 1)$, approximate to three significant digits.

Solution: Let us place Q to the right of P at a distance of $\Delta x = 1$ as shown in Fig. 23–17. The abscissa of Q is then 2, and the ordinate is 2^2 or 4. The slope of the secant PQ from $P(1, 1)$ to $Q(2, 4)$ is

$$\text{slope} = \frac{\Delta y}{\Delta x} = \frac{4 - 1}{2 - 1} = 3$$

Let us then repeat the computation by computer, halving Δx each time. We get the values shown in Table 23–1.

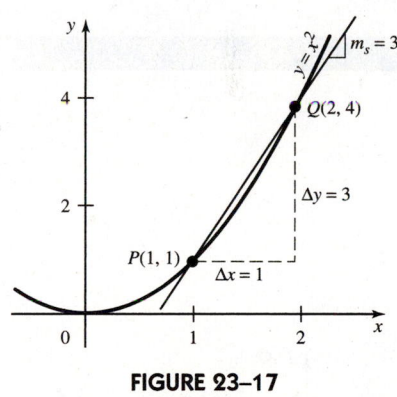

FIGURE 23–17

TABLE 23–1

Δx	x	y	Δy	Slope
1.0000	2.0000	4.0000	3.0000	3.0000
0.5000	1.5000	2.2500	1.2500	2.5000
0.2500	1.2500	1.5625	0.5625	2.2500
0.1250	1.1250	1.2656	0.2656	2.1250
0.0625	1.0625	1.1289	0.1289	2.0625
0.0313	1.0313	1.0635	0.0635	2.0313
0.0156	1.0156	1.0315	0.0315	2.0156
0.0078	1.0078	1.0157	0.0157	2.0078
0.0039	1.0039	1.0078	0.0078	2.0039
0.0020	1.0020	1.0039	0.0039	2.0020
0.0010	1.0010	1.0020	0.0020	2.0010
0.0005	1.0005	1.0010	0.0010	2.0005
0.0002	1.0002	1.0005	0.0005	2.0002
0.0001	1.0001	1.0002	0.0002	2.0001

Notice that the value of the slope appears to be reaching a limiting value of 2. Also, if we wanted only three significant digits, we could have stopped at $\Delta x = 0.0039$ because the second decimal place in the slope no longer changes beyond that value. ◆◆◆

Exercise 2 ◆ Rate of Change and the Tangent

Graph the given function. Then find the slope or rate of change of the curve at the given value of x, either manually, by zooming in, by using the TANGENT feature on your calculator, or numerically, as directed by your instructor.

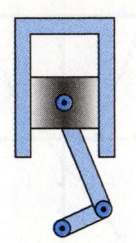

FIGURE 23–18

1. $y = x^2$ at $x = 2$
2. $y = -x^2$ at $x = 3$
3. $y = \sqrt{x}$ at $x = 1$
4. $y = \sqrt{x} + x^2$ at $x = 2$
5. $y = x - x^2$ at $x = 3$
6. $y = x^2 - 3$ at $x = 5$
7. $y = 3 - \sqrt{x}$ at $x = 1$
8. $y = \sqrt{x} - x$ at $x = 4$

9. *An Application:* The pressure p (lb/in.^{2}) and the volume v (in.^{3}) in the cylinder, Fig. 23–18, are related by the equation

$$pv = 3650$$

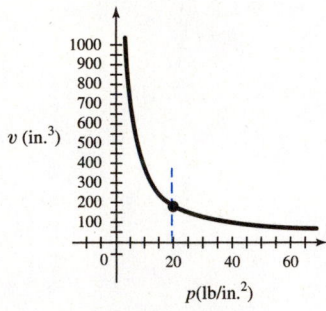

FIGURE 23–19

which is graphed in Fig. 23–19. Use any means to find the rate of change of volume with respect to pressure, at $p = 20.0$ lb/in.^{2}.

23–3 The Derivative

We see that we can get the approximate slope of the tangent line, and hence the rate of change of a function, at any point that we wish. Our next step is to derive a *formula* for finding the same thing. The advantages of having a formula are that (1) it will give the rate of change *everywhere* on the curve, not just at one point, (2) it will give the *exact* value, and (3) it is faster and easier to use than numerical or graphical methods. We will derive such a formula in the same way we found the slope of the tangent earlier. But now we will work with *symbols* instead of numbers.

We place two points P and Q on a graph of our function, Fig. 23–20. Let their horizontal spacing be Δx. We see that in a run of Δx the rise of the secant line PQ is

$$\text{rise of } PQ = f(x + \Delta x) - f(x)$$

The slope of the secant line is thus

$$\text{slope of } PQ = \frac{f(x + \Delta x) - f(x)}{\Delta x}$$

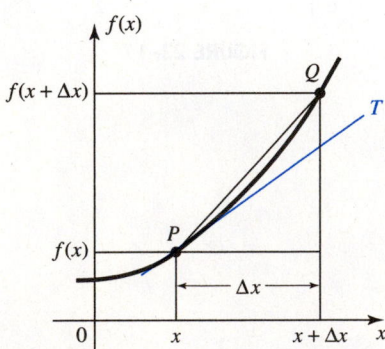

FIGURE 23–20 Derivation of the derivative.

We now let Δx approach zero, so that point Q will approach point P along the curve. Since the tangent T is the limiting position of the secant PQ,

$$\text{slope of } T = \lim_{\Delta x \to 0} \frac{f(x + \Delta x) - f(x)}{\Delta x}$$

This important quantity is called the *derivative* of $f(x)$. It is given the symbol $f'(x)$, read "*eff prime of x.*"

The Derivative	$$f'(x) = \lim_{\Delta x \to 0} \frac{f(x + \Delta x) - f(x)}{\Delta x}$$	**251**

◆◆◆ **Example 14:** Find the derivative of $f(x) = x^2$.

Solution: We substitute into Eq. 251, with $f(x) = x^2$ and $f(x + \Delta x) = (x + \Delta x)^2$

$$
\begin{aligned}
f'(x) &= \lim_{\Delta x \to 0} \frac{f(x + \Delta x) - f(x)}{\Delta x} \\
&= \lim_{\Delta x \to 0} \frac{(x + \Delta x)^2 - x^2}{\Delta x} \\
&= \lim_{\Delta x \to 0} \frac{x^2 + 2x\,\Delta x + (\Delta x)^2 - x^2}{\Delta x} \\
&= \lim_{\Delta x \to 0} \frac{2x\,\Delta x + (\Delta x)^2}{\Delta x} \\
&= \lim_{\Delta x \to 0} (2x + \Delta x) = 2x
\end{aligned}
$$

◆◆◆

Common Error	The symbol Δx is a *single symbol.* It is not Δ *times x.* Thus $$x \cdot \Delta x \neq \Delta x^2$$

Another Symbol for the Derivative

The numerator in our definition of the derivative

$$f'(x) = \lim_{\Delta x \to 0} \frac{f(x + \Delta x) - f(x)}{\Delta x}$$

is simply the value of the function at Q minus the value of the function at P. It is the *rise* from P to Q. If we now introduce the variable y by letting $y = f(x)$, we can call that rise Δy, Fig. 23–21. Our definition of the derivative then becomes

$$f'(x) = \lim_{\Delta x \to 0} \frac{\Delta y}{\Delta x}$$

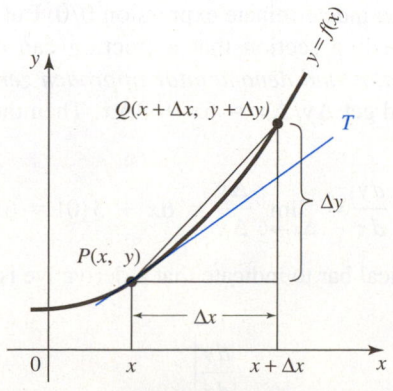

FIGURE 23–21

This leads us to a new symbol for the derivative. By definition,

$$\frac{dy}{dx} = f'(x) = \lim_{\Delta x \to 0} \frac{\Delta y}{\Delta x}$$

We will see later that dy and dx in the symbol dy/dx can be given separate meanings of their own. But for now we will treat dy/dx as a single symbol.

Derivatives by the Delta Method

Instead of substituting directly into Eq. 251 as we did in Example 14, some prefer to apply this equation in a series of steps. This is sometimes referred to as the *delta method, delta process*, or *four-step rule*. Our main use for the delta method will be to derive *rules* with which we can quickly find derivatives.

◆◆◆ Example 15:

(a) Find the derivative of the function $y = 3x^2$ by the delta method.
(b) Evaluate the derivative at $x = 1$.

Solution:

(a)

(1) Starting from $P(x, y)$, Fig. 23–22(a), locate a second point Q, spaced from P by a horizontal distance Δx and by a vertical distance Δy. Since the coordinates of $Q(x + \Delta x, y + \Delta y)$ must satisfy the given function, we may substitute $x + \Delta x$ for x and $y + \Delta y$ for y in the original equation.

$$\begin{aligned} y + \Delta y &= 3(x + \Delta x)^2 \\ &= 3[x^2 + 2x\,\Delta x + (\Delta x)^2] \\ &= 3x^2 + 6x\,\Delta x + 3(\Delta x)^2 \end{aligned}$$

(2) Find the rise Δy from P to Q, Fig. 23–22(b), by subtracting the original function.

$$(y + \Delta y) - y = 3x^2 + 6x\,\Delta x + 3(\Delta x)^2 - 3x^2$$
$$\Delta y = 6x\,\Delta x + 3(\Delta x)^2$$

(3) Find the slope of the secant line PQ, Fig. 23–22(c), by dividing the rise Δy by the run Δx.

$$\frac{\Delta y}{\Delta x} = \frac{6x\,\Delta x + 3(\Delta x)^2}{\Delta x}$$

(4) Let Δx approach zero. This causes Δy also to approach zero and appears to make $\Delta y/\Delta x$ equal to the indeterminate expression $0/0$. But recall from our study of limits in the preceding section that a fraction can often have a limit *even when both numerator and denominator approach zero*. To find it, we divide through by Δx and get $\Delta y/\Delta x = 6x + 3\Delta x$. Then the slope of the tangent T, Fig. 23–22(d), is

$$\frac{dy}{dx} = \lim_{\Delta x \to 0} \frac{\Delta y}{\Delta x} = 6x + 3(0) = 6x$$

(b) We will use a vertical bar to indicate that a derivative is to be evaluated at a given value. Thus

$$\frac{dy}{dx}\bigg|_{x=a}$$

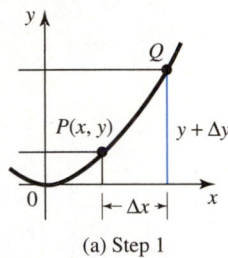

(a) Step 1

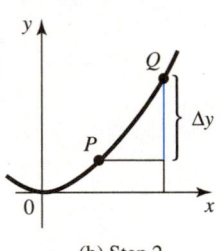

(b) Step 2

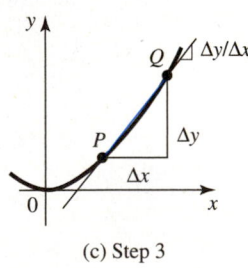

(c) Step 3

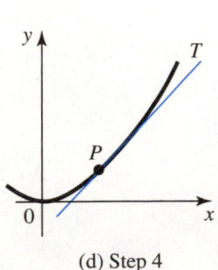

(d) Step 4

FIGURE 23–22 Derivatives by the delta method.

means to evaluate the derivative dy/dx at $x = a$. When $x = 1$, we write

$$\left.\frac{dy}{dx}\right|_{x=1} = 6(1) = 6$$

The curve $y = 3x^2$ is graphed in Fig. 23–23 with a tangent line of slope 6 drawn at the point $(1, 3)$. ♦♦♦

If our expression is a fraction with x in the denominator, step 2 will require us to find a common denominator, as in the following example.

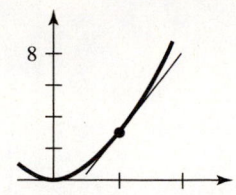

FIGURE 23–23 Graph of $y = 3x^2$.

♦♦♦ **Example 16:** Use the delta method to find the derivative of

$$y = \frac{3}{x^2 + 1}$$

Solution: Following the four steps, we have

(1) Substitute $x + \Delta x$ for x and $y + \Delta y$ for y.

$$y + \Delta y = \frac{3}{(x + \Delta x)^2 + 1}$$

(2) Subtracting the original function gives us

$$y + \Delta y - y = \frac{3}{(x + \Delta x)^2 + 1} - \frac{3}{x^2 + 1}$$

Combining the fractions over a common denominator gives

$$\Delta y = \frac{3(x^2 + 1) - 3[(x + \Delta x)^2 + 1]}{[(x + \Delta x)^2 + 1](x^2 + 1)}$$

which simplifies to

$$\Delta y = \frac{-6x\,\Delta x - 3(\Delta x)^2}{[(x + \Delta x)^2 + 1](x^2 + 1)}$$

(3) Dividing by Δx, we obtain

$$\frac{\Delta y}{\Delta x} = \frac{-6x - 3\,\Delta x}{[(x + \Delta x)^2 + 1](x^2 + 1)}$$

(4) Letting Δx approach zero gives

$$\frac{dy}{dx} = \frac{-6x}{(x^2 + 1)^2}$$ ♦♦♦

The following example shows how to use the delta method to differentiate (that is, to find the derivative of) an expression containing a radical.

♦♦♦ **Example 17:** Find the slope of the tangent to the curve $y = \sqrt{x}$ at $x = 4$.

Solution: We first find the derivative in four steps.

(1) $y + \Delta y = \sqrt{x + \Delta x}$

(2) $\Delta y = \sqrt{x + \Delta x} - \sqrt{x}$

The later steps will be easier if we now write this expression as a fraction with no radicals in the numerator. When simplifying radicals, we used to rationalize the denominator. Here we rationalize the *numerator* instead. To do this, we multiply (and divide) the binomial obtained in step (2) by its conjugate and get

$$\Delta y = (\sqrt{x + \Delta x} - \sqrt{x}) \cdot \frac{\sqrt{x + \Delta x} + \sqrt{x}}{\sqrt{x + \Delta x} + \sqrt{x}}$$

$$= \frac{(x + \Delta x) - x}{\sqrt{x + \Delta x} + \sqrt{x}}$$

$$= \frac{\Delta x}{\sqrt{x + \Delta x} + \sqrt{x}}$$

(3) We now have a fraction with no radicals in the numerator. Dividing by Δx gives us

$$\frac{\Delta y}{\Delta x} = \frac{1}{\sqrt{x + \Delta x} + \sqrt{x}}$$

(4) Letting Δx approach zero gives

$$\frac{dy}{dx} = \frac{1}{2\sqrt{x}}$$

When $x = 4$,

$$\left.\frac{dy}{dx}\right|_{x=4} = \frac{1}{4}$$

which is the slope of the tangent to $y = \sqrt{x}$ at $x = 4$. ◆◆◆

Other Variables

Mathematical ideas do not, of course, depend upon which symbols we use to express them. So instead of the variables x and y used up to now, we can use any letters we choose.

◆◆◆ **Example 18:** *An Application.* The displacement s of a point in a certain mechanism is given by $s = 3t^2$, where t is the elapsed time. The *velocity v* of the point, we will soon see, is given by the derivative of the displacement with respect to time, or

$$v = \frac{ds}{dt}$$

Find this derivative for the given function.

Solution: For Example 15, we found the derivative of $y = 3x^2$ to be

$$\frac{dy}{dx} = 6x$$

Simply by switching variables, we get

$$v = \frac{ds}{dt} = 6t$$ ◆◆◆

More Symbols for the Derivative

In addition to $f'(x)$ and dy/dx, the symbol $y'(x)$ is sometimes used. The $f'(x)$ and $y'(x)$ symbols are handy for specifying the derivative at a particular value of x, instead of using the vertical bar.

◆◆◆ **Example 19:** To specify a derivative evaluated at $x = 2$, we can write

$$y'(2), \text{ or } f'(2),$$

instead of the clumsier

$$\left.\frac{dy}{dx}\right|_{x=2} \qquad \text{◆◆◆}$$

The Derivative as an Operator

We can think of the derivative as an *operator:* one that operates on a function to produce the derivative of that function. The symbol d/dx or D in front of an expression indicates that the expression is to be differentiated. For example, the symbols

$$\frac{d}{dx}(\ ) \quad \text{or} \quad D_x(\ ) \quad \text{or} \quad D(\ )$$

mean to find the derivative of the expression enclosed in parentheses. $D(\ )$ means to differentiate the function with respect to its independent variable. Keep in mind that even though the notation is different, we find the derivative in *exactly the same way.*

◆◆◆ **Example 20:** We saw in Example 15 that if $y = 3x^2$, then $dy/dx = 6x$. This same result can also be written

$$\frac{d(3x^2)}{dx} = 6x$$

or

$$D_x(3x^2) = 6x$$

or

$$D(3x^2) = 6x \qquad \text{◆◆◆}$$

Continuity and Discontinuity

A curve is called *continuous* if it contains no breaks or gaps, and it is said to be *discontinuous* at a value of x where there is a break or gap. The derivative does not exist at such points. It also does not exist where the curve has a jump, corner, cusp, or any other feature at which it is not possible to draw a unique tangent line, as at the points shown in Fig. 23–24. At such points, we say that the function is not *differentiable.*

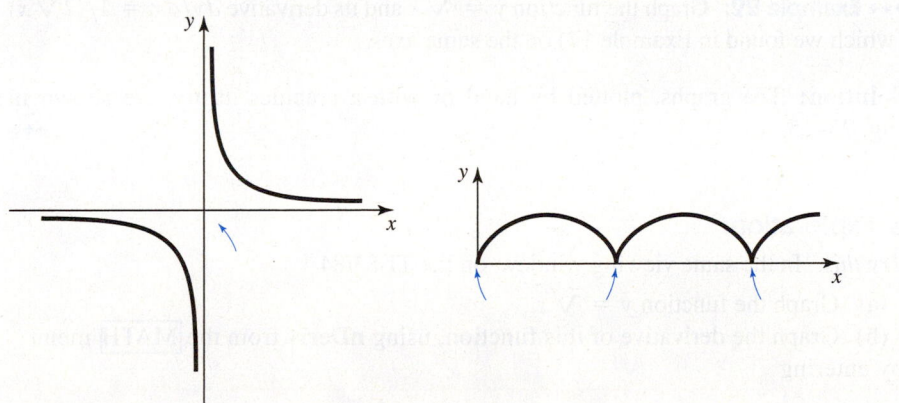
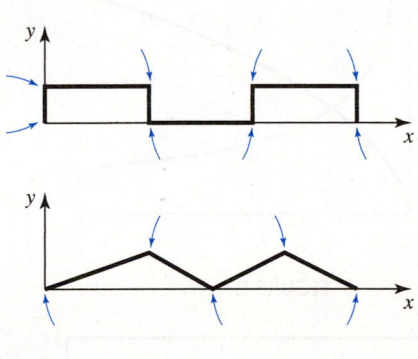

FIGURE 23–24 The arrows show the points at which the derivative does not exist.

Approximate Derivatives by Calculator

Many calculators can find the derivative at a point by an approximate numerical method, similar to the way we numerically found the slope of a tangent in the preceding section. On the TI-83/84 calculator, there are two ways to find derivatives:

(a) By using **nDeriv** from the MATH menu
(b) By using **dy/dx** from the CALC menu.

♦♦♦ **Example 21:** Use the TI-83/84 to find the derivative of $\sqrt{x}$ at $x = 4$ (a) using **nDeriv** and (b) using *dy/dx*.

Solution:

(a) We select **nDeriv** from the $\boxed{\text{MATH}}$ menu (**nDeriv** stands for the *numerical* derivative). Following that command we enter the function, the variable with respect to which we are taking the derivative, and the value of x at which we want the derivative. So we enter

$$\text{nDeriv}(\sqrt{x},x,4)$$

We press $\boxed{\text{ENTER}}$ and get 0.25 for a result, as shown. This agrees with our result from Example 17.

(b) We first graph the function $Y1 = \sqrt{x}$. We then select *dy/dx* from the $\boxed{\text{CALC}}$ menu, enter 4 for the value of x at which we want the derivative, and press $\boxed{\text{ENTER}}$. The derivative at that point is displayed below the graph.

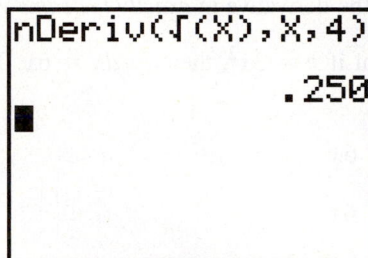
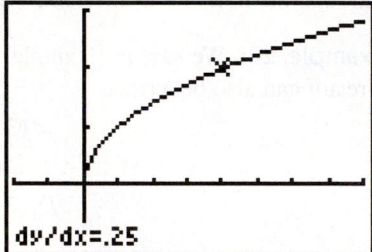

TI-83/84 screens for Example 21.   ♦♦♦

Graph of a Derivative

A derivative can be graphed just like any other function.

♦♦♦ **Example 22:** Graph the function $y = \sqrt{x}$ and its derivative $dy/dx = 1/(2\sqrt{x})$ (which we found in Example 17) on the same axes.

Solution: The graphs, plotted by hand or with a graphics utility, are shown in Fig. 23–25.   ♦♦♦

■ **Exploration:**

Try this. In the same viewing window, on the TI-83/84

(a) Graph the function $y = \sqrt{x}$.

(b) Graph the derivative of this function, using **nDeriv** from the $\boxed{\text{MATH}}$ menu by entering

$$Y_1 = \text{nDeriv}(\sqrt{(X)}, X, X)$$

(c) Graph the tangent to the given function at $x = 1$, using **Tangent** from the $\boxed{\text{DRAW}}$ menu.

What is the slope of the given curve at $x = 1$? Using $\boxed{\text{TRACE}}$, find the ordinate of the derivative curve, also at $x = 1$. What do you observe? Repeat, drawing the tangent line at some other value of x and comparing the slope at that x to the ordinate of the derivative curve. What conclusion can you draw?   ■

FIGURE 23–25

TI-83/84 screen for the exploration.
Ticks are 1 unit apart.

◆◆◆ **Example 23:** Graph the function from Example 15, $y = 3x^2$ and its derivative, in the same viewing window.

Solution: For Y1 we enter the function itself. For Y2 we enter either

$$\text{nDeriv}(3X^2, X, X)$$

or

$$\text{nDeriv}(Y1, X, X)$$

where **nDeriv** is from the MATH menu and Y1 is found in the VARS menu. Note that instead of entering a particular value of x at which to find the derivative, we enter x to indicate all values of x. The screens are shown.

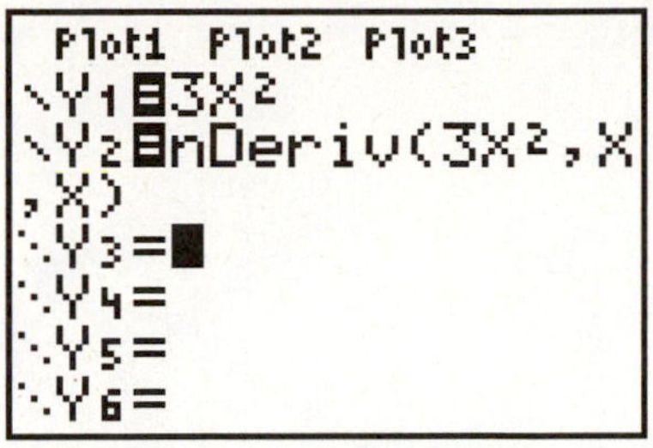

TI-83/84 screens for Example 23.

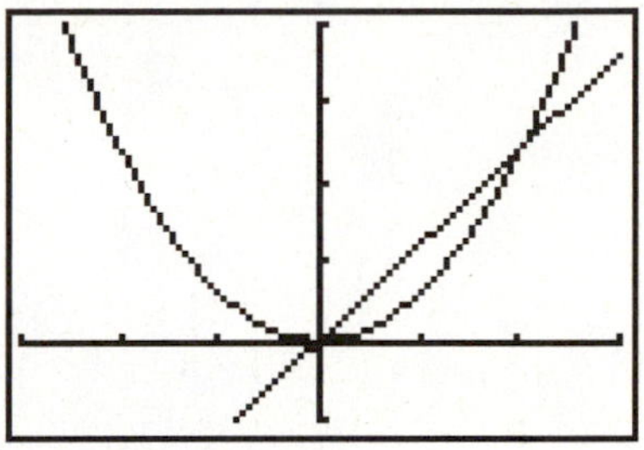
Tick spacing is 1 unit in x and 5 units in y.

Note that the graph of the derivative is a straight line with the equation $dy/dx = 6x$, as was found in Example 15. ◆◆◆

We can also use our calculator to check a derivative that was found by other methods. We can graph that derivative, as well as the **nDeriv** of the given function, in the same viewing window. One curve should exactly overlay the other. Perhaps a more convincing check is to display a table of values, as in the following example.

◆◆◆ **Example 24:** In the first screen we enter for Y1 the derivative of $y = 3x^2$ that we had found earlier, and for Y2 we ask for the **nDeriv** of $y = 3x^2$. Note in the second screen that the values for Y1 and Y2 are identical.

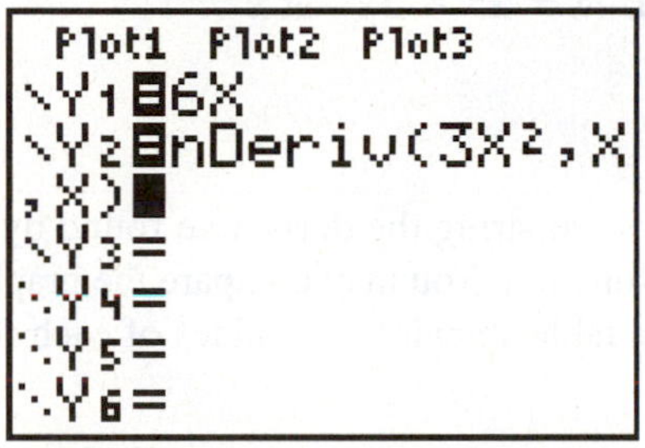

TI-83/84 screens for Example 23.

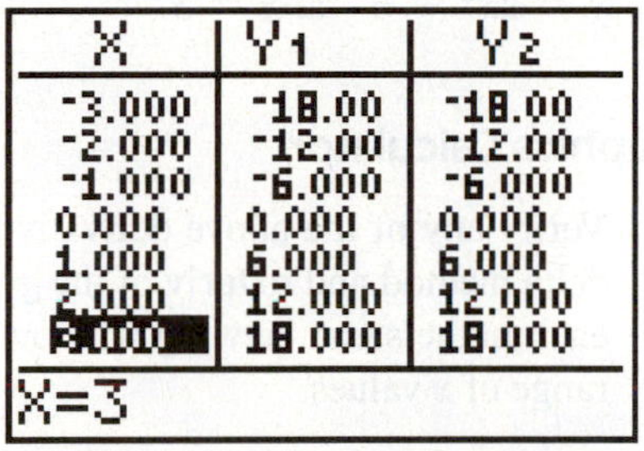

This display is obtained by pressing TABLE. ◆◆◆

Symbolic Differentiation by Calculator or Computer

We can find derivatives with a calculator or computer that can do symbolic algebra. On the TI-89, for example, the operation ***d***(from the **Calc** menu will find derivatives with respect to a given variable. We must enter the function followed by that variable.

◆◆◆ **Example 25:** Find the derivative of $3x^2$ on the TI-89.

Solution: We select ***d***(from the **Calc** menu, enter **3*x*^2, *x***, and press ENTER to obtain the derivative, $6x$. ◆◆◆

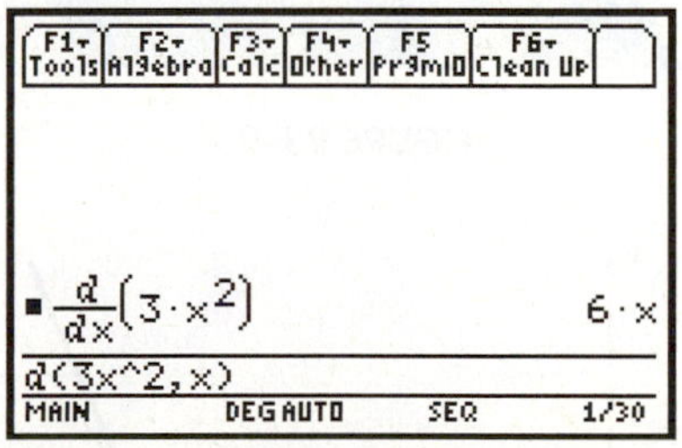

TI-89 screen for Example 25.

To find a derivative evaluated at a given point, in a single step, include a vertical bar | and the value at which the derivative is to be evaluated.

◆◆◆ **Example 26:** Find the derivative of $3x^2$, evaluated at $x = 4$, on the TI-89.

Solution: We enter the derivative as in the preceding example, followed by | $x = 4$.

$$d(3x^\wedge 2,x)|x = 4$$

and press ENTER to obtain the value 24. ◆◆◆

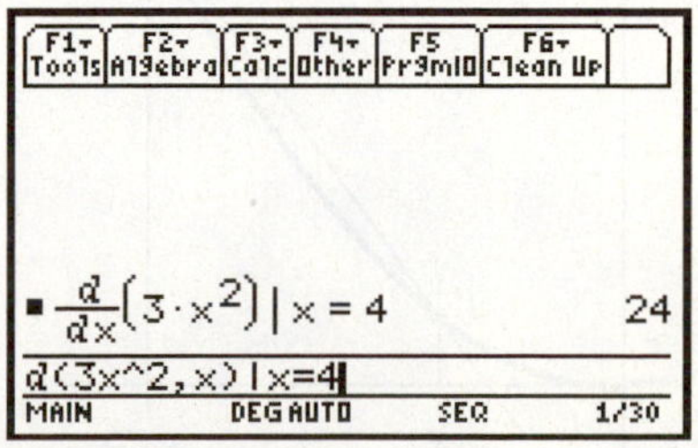

TI-89 screen for Example 26.

Exercise 3 ◆ The Derivative

Delta Method

Find the derivative by the delta method.

1. $y = 3x - 2$
2. $y = 4x - 3$
3. $y = 2x + 5$
4. $y = 7 - 4x$
5. $y = x^2 + 1$
6. $y = x^2 - 3x + 5$
7. $y = x^3$
8. $y = x^3 - x^2$
9. $y = \dfrac{3}{x}$
10. $y = \dfrac{x}{(x-1)^2}$
11. $y = \sqrt{3 - x}$
12. $y = \dfrac{1}{\sqrt{x}}$

Find the slope of the tangent or the rate of change at the given value of x.

13. $y = \dfrac{1}{x^2}$ at $x = 1$
14. $y = \dfrac{1}{x+1}$ at $x = 2$
15. $y = x + \dfrac{1}{x}$ at $x = 2$
16. $y = \dfrac{1}{x}$ at $x = 3$
17. $y = 2x - 3$ at $x = 3$
18. $y = 16x^2$ at $x = 1$
19. $y = 2x^2 - 6$ at $x = 3$
20. $y = x^2 + 2x$ at $x = 1$

Graphics Calculator

21. Verify any of the above derivatives by comparing the derivative found by the delta method and **nDeriv** of the given function. You may compare the graph of each in the same viewing window, or a table showing the values of each for a range of x values.

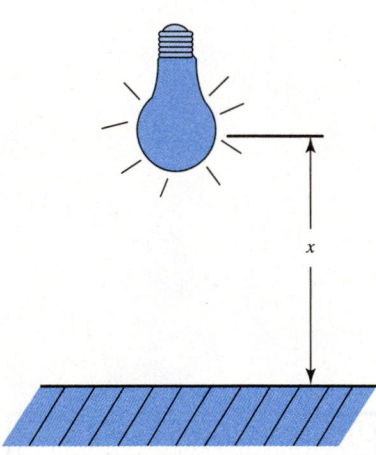

Other Symbols for the Derivative

22. If $y = 2x^2 - 3$, find y'.
23. In problem 22, find $y'(3)$.
24. In problem 22, find $y'(-1)$.
25. If $f(x) = 7 - 4x^2$, find $f'(x)$.
26. In problem 25, find $f'(2)$.
27. In problem 25, find $f'(-3)$.

FIGURE 23–26

Operator Notation

Find the derivative.

28. $\dfrac{d}{dx}(3x + 2)$
29. $\dfrac{d}{dx}(x^2 - 1)$
30. $D_x(7 - 5x)$
31. $D_x(x^2)$
32. $D(3x + 2)$
33. $D(x^2 - 1)$

34. *An application:* A certain light source produces an illumination of I lux on a surface at a distance of x m, Fig. 23–26, where $I = 258/x^2$. Find dI/dx, the rate of change of illumination with respect to distance.

35. *Writing:* We find the derivative by the delta method by first finding Δy divided by Δx and then letting Δx approach zero. Explain in your own words why this doesn't give division by zero, causing us to junk the whole calculation.

36. *CAD:* Plot a curve, Fig. 23–27, and on the x axis locate points at x and at $x + \Delta x$.

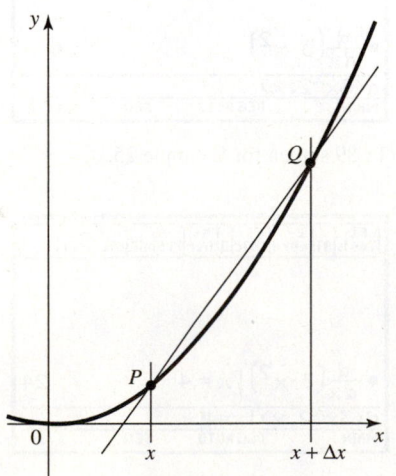

FIGURE 23–27

From these points draw lines perpendicular to the x axis. Label their intersections with the curve as P and Q. Draw secant line PQ.

Now drag Q closer to P. What happens to secant line PQ as Δx shrinks to zero?

Demonstrate your construction in class using a computer projector.

23–4   Rules for Derivatives

In the preceding section we have shown several ways to find the derivative, and hence, the rate of change of a function. We made use of the calculator and computer, and also the delta method. These are fine, but for many functions the fastest and easiest way to find a derivative is by applying a *rule*. Here we will use the delta method to derive a few such rules.

The Derivative of a Constant

For the function $y = c$, if x is increased by an amount Δx, then y (being constant) remains unchanged. Thus $\Delta y = 0$, and so $\Delta x/\Delta y = 0$. The limit, as Δx approaches zero, is also zero. So we get the rule,

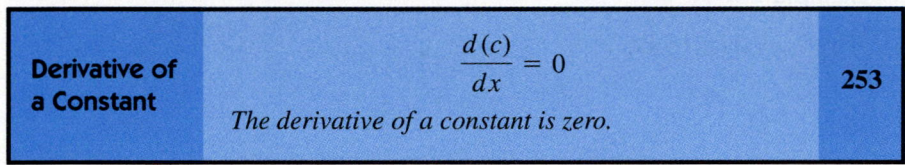

| Derivative of a Constant | $$\dfrac{d(c)}{dx} = 0$$ *The derivative of a constant is zero.* | 253 |

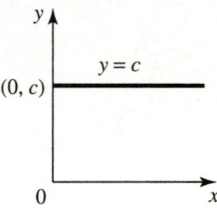

FIGURE 23–28

This is not surprising, because the graph of the function $y = c$ is a straight line parallel to the x axis (Fig. 23–28) whose slope is, of course, zero.

◆◆◆ **Example 27:** If $y = 2\pi^2$, then

$$\frac{d(2\pi^2)}{dx} = 0$$

◆◆◆

Derivative of a Constant Times a Power Function

We let $y = cx^n$, where n is a positive integer and c is any constant. Using the delta method:

1. We substitute $x + \Delta x$ for x and $y + \Delta y$ for y.

$$y + \Delta y = c(x + \Delta x)^n$$

Here we have a binomial, $(x + \Delta x)$, raised to a power n. Recall, from Chapter 20 that we can expand such a binomial by using the *binomial formula*, Eq. 199,

$$(a + b)^n = a^n + na^{n-1}b + \frac{n(n-1)}{2!}a^{n-2}b^2 + \frac{n(n-1)(n-2)}{3!}a^{n-3}b^3 + \cdots + b^n$$

Substituting into the binomial formula with $a = x$ and $b = \Delta x$ we get

$$y + \Delta y = c\left[x^n + nx^{n-1}(\Delta x) + \frac{n(n-1)}{2}x^{n-2}(\Delta x)^2 + \cdots + (\Delta x)^n\right]$$

2. Subtracting $y = cx^n$ we get Δy on the left side, and on the right side the first term x^n is eliminated.

$$\Delta y = c\left[nx^{n-1}(\Delta x) + \frac{n(n-1)}{2}x^{n-2}(\Delta x)^2 + \cdots + (\Delta x)^n\right]$$

3. Every term on the right contains the factor Δx. We can therefore divide both sides by Δx and we get

$$\frac{\Delta y}{\Delta x} = c\left[nx^{n-1} + \frac{n(n-1)}{2}x^{n-2}\Delta x + \cdots + (\Delta x)^{n-1}\right]$$

4. Finally, letting Δx approach zero, all terms but the first vanish. Thus

$$\frac{dy}{dx} = cnx^{n-1}$$

or

Derivative of a Power Function	$$\frac{d}{dx}cx^n = cnx^{n-1}$$ *The derivative of a constant times a power of x is equal to the product of the constant, the exponent, and x raised to the exponent reduced by 1.*	**256**

◆◆◆ **Example 28:**

(a) If $y = x^3$, then, by Eq. 256,

$$\frac{dy}{dx} = 3x^{3-1} = 3x^2$$

(b) If $y = 5x^2$, then

$$\frac{dy}{dx} = 5(2)x^{2-1} = 10x$$ ◆◆◆

Power Function with Negative Exponent

In our derivative of Eq. 256, we had required that the exponent n be a positive integer. We'll now show that the rule is also valid when n is a *negative* integer.

If n is a negative integer, then $m = -n$ is a positive integer. So

$$y = cx^n = cx^{-m} = \frac{c}{x^m}$$

We again use the delta method.

1. We substitute $x + \Delta x$ for x, and $y + \Delta y$ for y.

$$y + \Delta y = \frac{c}{(x + \Delta x)^m}$$

2. Subtracting gives us

$$\Delta y = \frac{c}{(x + \Delta x)^m} - \frac{c}{x^m} = c\frac{x^m - (x + \Delta x)^m}{x^m(x + \Delta x)^m}$$

$$= c\frac{x^m - (x^m + mx^{m-1}\Delta x + kx^{m-2}\Delta x^2 + \cdots + \Delta x^m)}{x^m(x + \Delta x)^m}$$

3. Dividing by Δx yields

$$\frac{\Delta y}{\Delta x} = -c\frac{mx^{m-1} + kx^{m-2}\Delta x + \cdots + \Delta x^{m-1}}{x^m(x + \Delta x)^m}$$

4. Letting Δx go to zero, we get

$$\frac{dy}{dx} = -c\,\frac{mx^{m-1}}{x^{2m}} = -cmx^{m-1-2m} = -cmx^{-m-1} = cnx^{n-1}$$

We get the same result as for a positive integral exponent. This shows that Eq. 256 is valid when the exponent n is a negative integer as well as when it is positive.

◆◆◆ Example 29:

(a) If $y = x^{-4}$, then

$$\frac{dy}{dx} = -4x^{-5} = -\frac{4}{x^5}$$

(b) $\dfrac{d(3x^{-2})}{dx} = 3(-2)x^{-3} = -\dfrac{6}{x^3}$

(c) If $y = -3/x^2$, then $y = -3x^{-2}$ and

$$y' = -3(-2)x^{-3} = \frac{6}{x^3} \qquad \text{◆◆◆}$$

Power Function with Fractional Exponent

We have shown that the exponent n in Eq. 256 can be a positive or a negative integer. The rule is also valid when n is a positive or a negative rational number. We'll prove it later in this chapter.

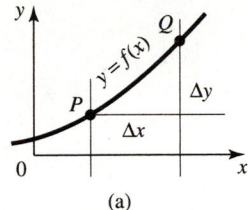

(a)

◆◆◆ Example 30: If $y = x^{-5/3}$, then

$$\frac{dy}{dx} = -\frac{5}{3}x^{-8/3} \qquad \text{◆◆◆}$$

To find the derivative of a radical, write it in exponential form, and use Eq. 256.

◆◆◆ Example 31: If $y = \sqrt[3]{x^2}$, then

$$\frac{dy}{dx} = \frac{d}{dx}x^{2/3} = \frac{2}{3}x^{-1/3} = \frac{2}{3\sqrt[3]{x}} \qquad \text{◆◆◆}$$

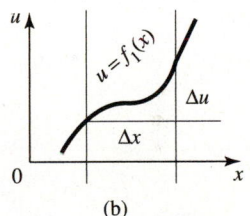

(b)

Derivative of a Sum

We want the derivative of the function

$$y = u + v + w$$

where u, v, and w are all functions of x. These may be visualized as shown in Fig. 23–29. We use the delta method as before. Starting from $P(x, y)$ on the curve $y = f(x)$, we locate a second point Q spaced from P by a horizontal increment Δx. In a run of Δx, the graph of $y = f(x)$ rises by an amount that we call Δy, Fig. 23–29(a).

But in a run of Δx, the graph of u also has a rise, and we call this rise Δu. Similarly, the graphs of v and w will rise by amounts that we call Δv and Δw.

Thus at $(x + \Delta x)$, the values of u, v, w, and y are $(u + \Delta u)$, $(v + \Delta v)$, $(w + \Delta w)$, and $(y + \Delta y)$. Substituting these values into the original function $y = u + v + w$ gives

$$y + \Delta y = (u + \Delta u) + (v + \Delta v) + (w + \Delta w)$$

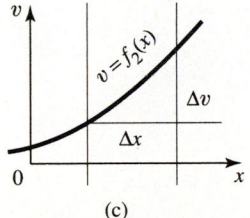

(c)

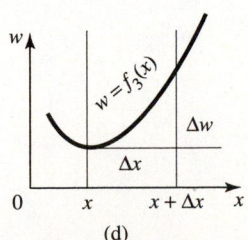

(d)

FIGURE 23–29

Subtracting the original function we get

$$y + \Delta y - y = u + \Delta u + v + \Delta v + w + \Delta w - u - v - w$$
$$\Delta y = \Delta u + \Delta v + \Delta w$$

Dividing by Δx gives us

$$\frac{\Delta y}{\Delta x} = \frac{\Delta u + \Delta v + \Delta w}{\Delta x}$$

So

$$\frac{dy}{dx} = \lim_{\Delta x \to 0}\left(\frac{\Delta u}{\Delta x} + \frac{\Delta v}{\Delta x} + \frac{\Delta w}{\Delta x}\right)$$

$$= \lim_{\Delta x \to 0}\frac{\Delta u}{\Delta x} + \lim_{\Delta x \to 0}\frac{\Delta v}{\Delta x} + \lim_{\Delta x \to 0}\frac{\Delta w}{\Delta x}$$

Here we used, without proof, the idea that the limit of a sum of several functions is equal to the sum of the individual limits. This leads us to the following sum rule:

Derivative of a Sum	$$\frac{d}{dx}(u + v + w) = \frac{du}{dx} + \frac{dv}{dx} + \frac{dw}{dx}$$ *The derivative of the sum of several functions is equal to the sum of the derivatives of those functions.*	**257**

◆◆◆ **Example 32:** Differentiate.

$$\frac{d}{dx}(2x^3 - 3x^2 + 5x + 4)$$

Solution: We often have to apply several rules for derivatives in one problem. Here we need, in addition to the rule for sums, the rules for a power function and a constant.

$$\frac{d}{dx}(2x^3 - 3x^2 + 5x + 4) = 6x^2 - 6x + 5 \qquad\qquad ◆◆◆$$

◆◆◆ **Example 33:** Find the derivative of $y = \dfrac{x^2 + 3}{x}$ and verify by calculator.

Solution: At first glance it looks as if none of the rules learned so far apply here. But if we divide by x we get

$$y = \frac{x^2 + 3}{x} = x + \frac{3}{x} = x + 3x^{-1}$$

Taking the derivative gives

$$\frac{dy}{dx} = 1 + 3(-1)x^{-2}$$

$$= 1 - \frac{3}{x^2}$$

Check: We approximately verify this by the TI-83/84 calculator by entering for Y1 the derivative we just found, and for Y2 the **nDeriv** of the given function, screen (1). We then display a table showing Y1 and Y2 for a range of x values, and check that they are equal screen (2).

As another check, we have taken the derivative symbolically, TI-89 screen (3).

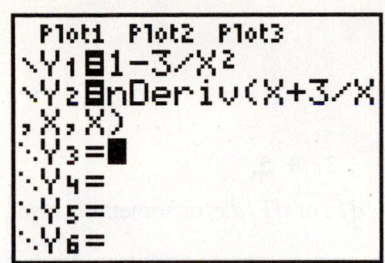

(1) TI-83/84 checks for Example 33.   (2)

(3) After taking the derivative on the TI-89, we have used the **expand** operation from the **Algebra** menu to put the derivative into the same form as that found manually.

◆◆◆

Other Symbols for the Derivative

Let us not forget the other symbols that are used to indicate a derivative.

◆◆◆ Example 34:

(a) if $y = 3x^2$, then $y' = 6x$

(b) if $f(x) = 3x^2$, then $f'(x) = 6x$

(c) $D_x(3x^2) = 6x$

(d) $D(3x^2) = 6x$ ◆◆◆

Derivative at a Given Point

As before, we can use the y' and f' notation to denote the derivative evaluated at *a specific point*.

◆◆◆ Example 35: If $y = x^2 - 2x + 3$, find $y'(1)$.

Solution: The derivative is

$$y' = 2x - 2$$

At $x = 1$,

$$y'(1) = 2(1) - 2 = 0$$ ◆◆◆

◆◆◆ Example 36: If $f(x) = 2x^3 + 4x$, find $f'(3)$.

Solution: Differentiating gives $f'(x) = 6x^2 + 4$, so

$$f'(3) = 6(3)^2 + 4 = 58$$ ◆◆◆

Functions with Other Variables

So far we have mostly been using x for the independent variable and y for the dependent variable, but now let us get some practice using *other* variables.

◆◆◆ Example 37:

(a) If $s = 3t^2 - 4t + 5$, then

$$\frac{ds}{dt} = 6t - 4$$

(b) If $y = 7u - 5u^3$, then

$$\frac{dy}{du} = 7 - 15u^2$$

(c) If $u = 9 + x^2 - 2x^4$, then

$$\frac{du}{dx} = 2x - 8x^3$$

◆◆◆

◆◆◆ **Example 38:** Find the derivative of $T = 4z^2 - 2z + 5$.

Solution: What are we to find here, dT/dz, or dz/dT, or dT/dx, or something else?

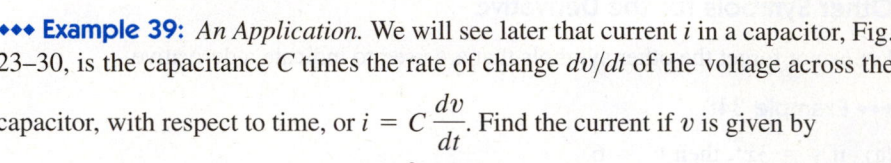

> When we say "derivative," we mean the derivative of the given function with respect to the **independent variable,** unless otherwise noted.

Here we find the derivative of T with respect to the independent variable z.

$$\frac{dT}{dz} = 8z - 2$$

◆◆◆

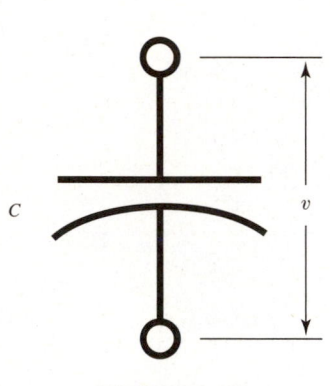

FIGURE 23–30

◆◆◆ **Example 39:** *An Application.* We will see later that current i in a capacitor, Fig. 23–30, is the capacitance C times the rate of change dv/dt of the voltage across the capacitor, with respect to time, or $i = C\dfrac{dv}{dt}$. Find the current if v is given by

$$v = 2.95t^2 + 1.44t - 3.85$$

Solution: Taking the derivative we get

$$\frac{dv}{dt} = 2(2.95)t + 1.44$$

So,

$$i = C\frac{dv}{dt} = C(5.90t + 1.44)$$

◆◆◆

Exercise 4 ◆ Rules for Derivatives

Find the derivative of each function. Verify some of your results by calculator. As usual, the letters $a, b, c, \ldots$ represent constants.

Derivative of a Constant

1. $y = 8$ **2.** $y = \pi$
3. $y = a^2$ **4.** $y = 3b + 7c$

Derivative of a Constant Times a Power Function

5. $y = x$ **6.** $y = 3x$
7. $y = x^7$ **8.** $y = x^4$
9. $y = 3x^2$ **10.** $y = 5.4x^3$

Power Function with Negative Exponent

11. $y = x^{-5}$ **12.** $y = 2x^{-3}$
13. $y = \dfrac{1}{x}$ **14.** $y = \dfrac{1}{x^2}$
15. $y = \dfrac{3}{x^3}$ **16.** $y = \dfrac{3}{2x^2}$

Power Function with Fractional Exponent

17. $y = 7.5x^{1/3}$

18. $y = 4x^{5/3}$

19. $y = 4\sqrt{x}$

20. $y = 3\sqrt[3]{x}$

21. $y = -17\sqrt{x^3}$

22. $y = -2\sqrt[5]{x^4}$

Derivative of a Sum

23. $y = 3 - 2x$

24. $y = 4x^2 + 2x^3$

25. $y = 3x - x^3$

26. $y = x^4 + 3x^2 + 2$

27. $y = 3x^3 + 7x^2 - 2x + 5$

28. $y = x^3 - x^{3/2} + 3x$

29. $y = ax + b$

30. $y = ax^5 - 5bx^3$

31. $y = \dfrac{x^2}{2} - \dfrac{x^7}{7}$

32. $y = \dfrac{x^3}{1.75} + \dfrac{x^2}{2.84}$

33. $y = 2x^{3/4} + 4x^{-1/4}$

34. $y = \dfrac{2}{x} - \dfrac{3}{x^2}$

35. $y = 2x^{4/3} - 3x^{2/3}$

36. $y = x^{2/3} - a^{2/3}$

Other Symbols for the Derivative

37. If $y = 2x^3 - 3$, find y'.

38. If $f(x) = 7 - 4x^2$, find $f'(x)$.

Evaluate each expression.

39. $\dfrac{d}{dx}(3x^5 + 2x)$

40. $\dfrac{d}{dx}(2.5x^2 - 1)$

41. $D_x(7.8 - 5.2x^{-2})$

42. $D_x(4x^2 - 1)$

43. $D(3x^2 + 2x)$

44. $D(1.75x^{-2} - 1)$

Derivative at a Given Point

45. If $y = x^3 - 5$, find $y'(1)$.

46. If $f(x) = 1/x^2$, find $f'(2)$.

47. If $f(x) = 2.75x^2 - 5.02x$, find $f'(3.36)$.

48. If $y = \sqrt{83.2x^3}$, find $y'(1.74)$.

49. Find the slope of the tangent to the curve $y = x^2 - 2$, where x equals 2.

50. Find the slope of the tangent to the function $y = x - x^2$ at $x = 2$.

51. Find the rate of change of the function $y = 5x^3$ at $x = 0.500$.

52. Find the rate of change of the function $y = 3.45x^2 - 2.74x$ at $x = 1.34$.

Functions with Other Variables

Find the derivative with respect to the independent variable.

53. $v = 5t^2 - 3t + 4$

54. $z = 9 - 8w + w^2$

55. $s = 58.3t^3 - 63.8t$

56. $x = 3.82y + 6.25y^4$

57. $y = \sqrt{5w^3}$

58. $w = \dfrac{5}{x} - \dfrac{3}{x^2}$

59. $v = \dfrac{85.3}{t^4}$

60. $T = 3.55\sqrt{1.06w^5}$

An Application

Remember that we have two entire chapters of applications following this one, and that we are only giving a few in this chapter as a sample of things to come.

61. The position s of a point in a mechanism is given by

$$s = 5t^2 + 3t \quad \text{in.}$$

where t is the time in seconds. The velocity of the point is found by taking the first derivative of the displacement, or ds/dt. Take the derivative and evaluate it at $t = 3.55$ s.

62. *Writing:* Do you feel it is reasonable to learn rules for derivatives, now that we can use calculators or computers to find derivatives? Can you think of any reasons for continuing to learn them? Write a memo to the head of your mathematics department stating whether or not you think the present method of learning derivatives should be changed, and give your reasons.

23–5   Derivative of a Function Raised to a Power

In the preceding section we derived a few rules for taking derivatives. However, there are many simple functions not covered by those rules, so we add to them here and in the following section.

Composite Functions

We earlier derived Rule 256 for finding the derivative of x raised to a power. But Rule 256 does not apply when we have an *expression* raised to a power, such as

$$y = (2x + 7)^5 \tag{1}$$

We may consider this function as being made up of two parts. One part is the quantity, which we shall call u, that is being raised to the power. The other part is the power itself .

$$u = 2x + 7 \tag{2}$$

Then y can be written

$$y = u^5 \tag{3}$$

Our original function [Eq. (1)], which can be obtained by combining Eqs. (2) and (3), is called a *composite function*. We introduced composite functions in our earlier chapter on functions.

The Chain Rule

The *chain rule* will enable us to take the derivative of a composite function, such as Eq. (1). Consider the situation where y is a function of u,

$$y = g(u)$$

and u, in turn, is a function of x,

$$u = h(x)$$

so y is the composite function of $y = g[h(x)]$, which we will now call $f(x)$,

$$y = g[h(x)] = f(x)$$

The graphs of $y = f(x)$ and $u = h(x)$ are shown in Fig. 23–31. We now start our derivation of the chain rule by recalling our definition of the derivative as the limit of the quotient $\Delta y / \Delta x$ as Δx approaches zero

$$\frac{dy}{dx} = \lim_{\Delta x \to 0} \frac{\Delta y}{\Delta x} \tag{251}$$

where Δy is the rise of the graph of $y = f(x)$ in a run of Δx, Fig. 23–31(a). But in a run of Δx, the graph of u also has a rise, and we call this rise Δu, Fig. 23–31(b). Before taking the limit in Eq. 251, let us multiply the quotient $\Delta y / \Delta x$ by $\Delta u / \Delta u$, assuming here that Δu is not zero. (The chain rule is true even for those rare cases where Δu may be zero, but it takes a more complicated proof to show this.)

$$\frac{\Delta y}{\Delta x} = \frac{\Delta y}{\Delta x} \cdot \frac{\Delta u}{\Delta u}$$

Rearranging gives us

$$\frac{\Delta y}{\Delta x} = \frac{\Delta y}{\Delta u} \cdot \frac{\Delta u}{\Delta x}$$

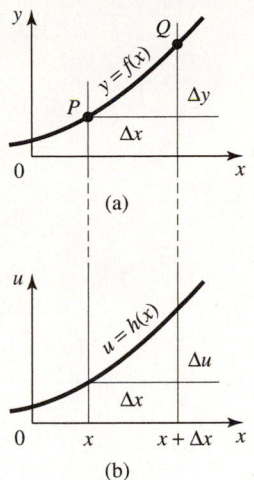

We now let Δx approach zero and apply a theorem (which we'll use without proof) that the limit of a product of two functions is equal to the product of the limits of the two functions.

$$\lim_{\Delta x \to 0} \frac{\Delta y}{\Delta x} = \lim_{\Delta x \to 0} \frac{\Delta y}{\Delta u} \cdot \frac{\Delta u}{\Delta x} = \left(\lim_{\Delta x \to 0} \frac{\Delta y}{\Delta u} \right) \left(\lim_{\Delta x \to 0} \frac{\Delta u}{\Delta x} \right)$$

FIGURE 23–31

But Δu also approaches zero as Δx approaches zero, so we may write

$$\lim_{\Delta x \to 0} \frac{\Delta y}{\Delta x} = \lim_{\Delta u \to 0} \frac{\Delta y}{\Delta u} \cdot \lim_{\Delta x \to 0} \frac{\Delta u}{\Delta x}$$

Then, by our definition of the derivative, Eq. 251,

$$\lim_{\Delta x \to 0} \frac{\Delta y}{\Delta x} = \frac{dy}{dx}$$

Similarly,

$$\lim_{\Delta u \to 0} \frac{\Delta y}{\Delta u} = \frac{dy}{du} \quad \text{and} \quad \lim_{\Delta x \to 0} \frac{\Delta u}{\Delta x} = \frac{du}{dx}$$

So we get the following rule:

Chain Rule	$$\frac{dy}{dx} = \frac{dy}{du} \cdot \frac{du}{dx}$$ *If y is a function of u, and u is a function of x, then the derivative of y with respect to x is the product of the derivative of y with respect to u and the derivative of u with respect to x.* **252**

The Power Rule

We now use the chain rule to find the derivative of a function raised to a power, $y = cu^n$. If, in Eq. 256, $\dfrac{d}{dx}(cx^n) = cnx^{n-1}$, we replace x by u, we get

$$\frac{dy}{du} = \frac{d}{du}(cu^n) = cnu^{n-1}$$

Then, by the chain rule, we get dy/dx by multiplying by du/dx.

$$\frac{dy}{dx} = \frac{dy}{du} \cdot \frac{du}{dx}$$

and obtain the following rule:

Derivative of a Function Raised to a Power, the Power Rule	$$\dfrac{d\,(cu^{n})}{dx} = cnu^{n-1}\dfrac{du}{dx}$$ *The derivative (with respect to x) of a constant times a function raised to a power is equal to the product of the constant, the power, the function raised to the power less 1, and the derivative of the function (with respect to x).*	**258**

◆◆◆ **Example 40:** Take the derivative of $y = (x^3 + 1)^5$. Check by calculator.

Solution: We use the rule for a function raised to a power, with

$$n = 5 \quad \text{and} \quad u = x^3 + 1$$

Then

$$\frac{du}{dx} = 3x^2$$

So

$$\frac{dy}{dx} = nu^{n-1}\frac{du}{dx}$$
$$= 5(x^3 + 1)^4(3x^2)$$

Common Error	Don't forget $\dfrac{du}{dx}$!

Now simplifying our answer, we get

$$\frac{dy}{dx} = 15x^2(x^3 + 1)^4$$

Check: We approximately verify this by the TI-83/84 calculator by entering for Y1 the derivative we just found, and for Y2 the **nDeriv** of the given function, screen (1). We then display a table showing Y1 and Y2 for a range of x values, and check that they are approximately equal, screen (2). Also shown is the derivative found symbolically on the TI-89, screen (3).

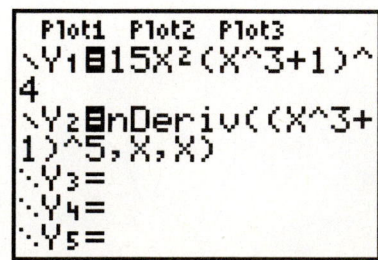

(1) TI-83/84 screens for Example 40.

(2)

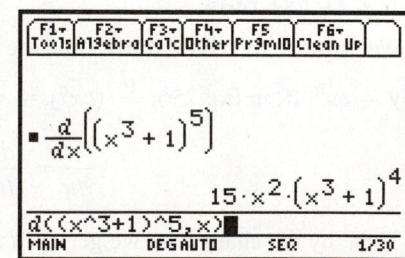

(3) TI-89 screen for Example 40.

◆◆◆

Our next example shows the use of the power rule when *the exponent is negative.*

◆◆◆ **Example 41:** Take the derivative of $y = \dfrac{3}{x^2 + 2}$.

Solution: Rewriting our function as $y = 3(x^2 + 2)^{-1}$ and applying the power rule, Rule 258, with $u = x^2 + 2$, we get

$$\frac{dy}{dx} = 3(-1)(x^2 + 2)^{-2}(2x)$$

$$= -\frac{6x}{(x^2 + 2)^2} \qquad ◆◆◆$$

We'll now use the power rule for *fractional exponents*, even though we will not prove that the rule is valid for them until later in this chapter.

◆◆◆ **Example 42:** Differentiate $y = \sqrt[3]{1 + x^2}$.

Solution: We rewrite the radical in exponential form.

$$y = (1 + x^2)^{1/3}$$

Then, by Rule 258, with $u = 1 + x^2$, and $du/dx = 2x$,

$$\frac{dy}{dx} = \frac{1}{3}(1 + x^2)^{-2/3}(2x)$$

Or, returning to radical form,

$$\frac{dy}{dx} = -\frac{2x}{3\sqrt[3]{(1 + x^2)^2}} \qquad ◆◆◆$$

◆◆◆ **Example 43:** If $f(x) = \dfrac{5}{\sqrt{x^2 + 3}}$, find $f'(1)$.

Solution: Rewriting our function without radicals gives us

$$f(x) = 5(x^2 + 3)^{-1/2}$$

Taking the derivative yields

$$f'(x) = 5\left(-\frac{1}{2}\right)(x^2 + 3)^{-3/2}(2x)$$

$$= -\frac{5x}{(x^2 + 3)^{3/2}}$$

Substituting $x = 1$, we obtain

$$f'(1) = -\frac{5}{(1 + 3)^{3/2}} = -\frac{5}{8} \qquad ◆◆◆$$

Exercise 5 ◆ Derivative of a Function Raised to a Power

Find the derivative of each function. Check some by calculator.

1. $y = (2x + 1)^5$

2. $y = (2 - 3x^2)^3$

3. $y = (3x^2 + 2)^4 - 2x$

4. $y = (x^3 + 5x^2 + 7)^2$

5. $y = (2 - 5x)^{3/5}$

6. $y = -\dfrac{2}{x + 1}$

7. $y = \dfrac{2.15}{x^2 + a^2}$

8. $y = \dfrac{31.6}{1 - 2x}$

9. $y = \dfrac{3}{x^2 + 2}$ **10.** $y = \dfrac{5}{(x^2 - 1)^2}$

11. $y = \left(a - \dfrac{b}{x}\right)^2$ **12.** $y = \left(a + \dfrac{b}{x}\right)^3$

13. $y = \sqrt{1 - 3x^2}$ **14.** $y = \sqrt{2x^2 - 7x}$

15. $y = \sqrt{1 - 2x}$ **16.** $y = \dfrac{b}{a}\sqrt{a^2 - x^2}$

17. $y = \sqrt[3]{4 - 9x}$ **18.** $y = \sqrt[3]{a^3 - x^3}$

19. $y = \dfrac{1}{\sqrt{x + 1}}$ **20.** $y = \dfrac{1}{\sqrt{x^2 - x}}$

Find the derivative.

21. $\dfrac{d}{dx}(3x^5 + 2x)^2$ **22.** $\dfrac{d}{dx}(1.5x^2 - 3)^3$

23. $D_x(4.8 - 7.2x^{-2})^2$ **24.** $D(3x^3 - 5)^3$

Find the derivative with respect to the independent variable.

25. $v = (5t^2 - 3t + 4)^2$ **26.** $z = (9 - 8w)^3$

27. $s = (8.3t^3 - 3.8t)^{-2}$ **28.** $x = \sqrt{3.2y + 6.2y^4}$

29. Find the derivative of the function $y = (4.82x^2 - 8.25x)^3$ when $x = 3.77$.

30. Find the slope of the tangent to the curve $y = 1/(x + 1)$ at $x = 2$.

31. If $y = (x^2 - x)^3$, find $y'(3)$.

32. If $f(x) = \sqrt[3]{2x} + (2x)^{2/3}$, find $f'(4)$.

An Application

33. We will see in a later chapter that the *acceleration* of a point is the *rate of change of the velocity* of the point. If the velocity of the arm of an industrial robot is given by $v = 3.45(t^2 + 2)^2$ ft/s, where t is the time in seconds, take the derivative of this velocity to find the acceleration, and evaluate it at $t = 1.00$ s.

23–6   Derivatives of Products and Quotients

We are nearly done adding rules now. Here we add a few more that will enable us to find derivatives of products and quotients. We will not need any others until we later cover logarithmic, exponential, and trigonometric functions.

Derivative of a Product

We often need the derivative of the product of two expressions, such as $y = (x^2 + 2)\sqrt{x-5}$, where each of the expressions is itself a function of x. Let us label these expressions u and v. So our function is

$$y = uv$$

where u and v are functions of x. These may be visualized in Fig. 23–32.

We use the delta method as before. Starting from $P(x, y)$ on the curve $y = f(x)$, we locate a second point Q spaced from P by a horizontal increment Δx. In a run of Δx, the graph of $y = f(x)$ is seen to rise by an amount that we call Δy, Fig. 23–32(a).

But in a run of Δx, the graph of u also has a rise, and we call this rise Δu, Fig. 23–32(b). Similarly, the graph of v will rise by an amount that we call Δv, Fig. 23–32(c).

Thus at $(x + \Delta x)$, the values of u, v, and y are $(u + \Delta u)$, $(v + \Delta v)$, and $(y + \Delta y)$. Substituting these values into the original function $y = uv$ gives

$$y + \Delta y = (u + \Delta u)(v + \Delta v)$$
$$= uv + u\,\Delta v + v\,\Delta u + \Delta u\,\Delta v$$

Subtracting $y = uv$ gives us

$$\Delta y = u\,\Delta v + v\,\Delta u + \Delta u\,\Delta v$$

Dividing by Δx yields

$$\frac{\Delta y}{\Delta x} = u\frac{\Delta v}{\Delta x} + v\frac{\Delta u}{\Delta x} + \Delta u\frac{\Delta v}{\Delta x}$$

As Δx now approaches 0, Δy, Δu, and Δv also approach 0, and

$$\lim_{\Delta x \to 0}\frac{\Delta y}{\Delta x} = \frac{dy}{dx}, \quad \lim_{\Delta x \to 0}\frac{\Delta u}{\Delta x} = \frac{du}{dx}, \quad \lim_{\Delta x \to 0}\frac{\Delta v}{\Delta x} = \frac{dv}{dx}$$

so

$$\frac{dy}{dx} = u\frac{dv}{dx} + v\frac{du}{dx} + 0\frac{dv}{dx}$$

which can be rewritten as follows:

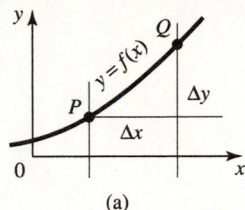

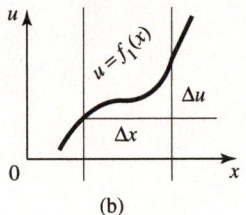

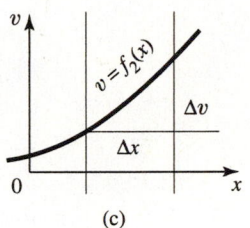

FIGURE 23–32

Derivative of a Product of Two Factors	$$\dfrac{d(uv)}{dx} = u\dfrac{dv}{dx} + v\dfrac{du}{dx}$$ *The derivative of a product of two factors is equal to the first factor times the derivative of the second factor, plus the second factor times the derivative of the first.*	**259**

◆◆◆ **Example 44:** Find the derivative of $y = (x^2 + 2)(x - 5)$.

Solution: We let the first factor be u, and the second be v.

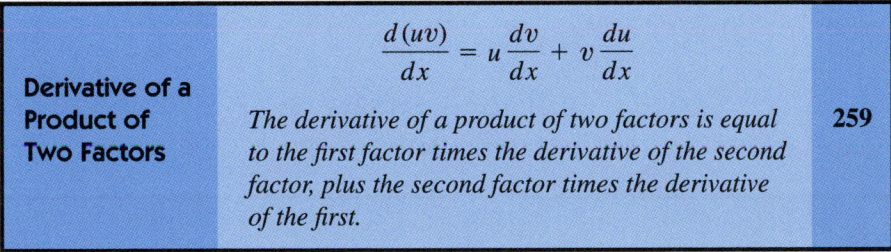

$$y = (x^2 + 2)(x - 5)$$
$$\underbrace{\hspace{2cm}}_{u}\ \underbrace{\hspace{1.5cm}}_{v}$$

So

$$\frac{du}{dx} = 2x \quad \text{and} \quad \frac{dv}{dx} = 1$$

Using the product rule,

$$\frac{dy}{dx} = (x^2 + 2)\frac{d}{dx}(x - 5) + (x - 5)\frac{d}{dx}(x^2 + 2)$$
$$= (x^2 + 2)(1) + (x - 5)(2x)$$
$$= x^2 + 2 + 2x^2 - 10x$$
$$= 3x^2 - 10x + 2$$

◆◆◆

In Example 44 we could have multiplied the two factors together and taken the derivative term by term. Try it and see if you get the same result.

♦♦♦ **Example 45:** Differentiate $y = (x + 5)\sqrt{x - 3}$.

Solution: By the product rule,

$$\frac{dy}{dx} = (x + 5) \cdot \frac{1}{2}(x - 3)^{-1/2}(1) + \sqrt{x - 3}\,(1)$$

$$= \frac{x + 5}{2\sqrt{x - 3}} + \sqrt{x - 3} \qquad\qquad\qquad ♦♦♦$$

Derivative of a Constant Times a Function

Let us use the product rule for the product cu, where c is a constant and u is a function of x.

$$\frac{d}{dx}cu = c\frac{du}{dx} + u\frac{dc}{dx} = c\frac{du}{dx}$$

since the derivative dc/dx of a constant is zero. Thus:

Derivative of a Constant Times a Function	$$\dfrac{d(cu)}{dx} = c\dfrac{du}{dx}$$ *The derivative of the product of a constant and a function is equal to the constant times the derivative of the function.*	255

♦♦♦ **Example 46:** If $y = 3(x^2 - 3x)^5$, then

$$\frac{dy}{dx} = 3\frac{d}{dx}(x^2 - 3x)^5$$

$$= 3(5)(x^2 - 3x)^4(2x - 3)$$

$$= 15(x^2 - 3x)^4(2x - 3) \qquad\qquad ♦♦♦$$

Tip	If one of the factors is a constant, it is much easier to use Rule 255 for a constant times a function, rather than the product rule.

Products with More Than Two Factors

Our rule for the derivative of a product having two factors can easily be extended. Take an expression with three factors, for example, which can be written as the product of *two* factors as follows:

$$y = uvw = (uv)w$$

Then using the product rule, twice,

$$\frac{dy}{dx} = uv\frac{dw}{dx} + w\frac{d(uv)}{dx}$$

$$= uv\frac{dw}{dx} + w\left(u\frac{dv}{dx} + v\frac{du}{dx}\right)$$

Thus,

Derivative of a Product of Three Factors	$$\frac{d(uvw)}{dx} = uv\frac{dw}{dx} + uw\frac{dv}{dx} + vw\frac{du}{dx}$$ *The derivative of the product of three factors is an expression of three terms, each term being the product of two of the factors and the derivative of the third factor.*	**260**

◆◆◆ **Example 47:** Differentiate $y = x^2(x - 2)^5\sqrt{x + 3}$.

Solution: By Eq. 260,

$$\frac{dy}{dx} = x^2(x - 2)^5\left[\frac{1}{2}(x + 3)^{-1/2}\right] + x^2(x + 3)^{1/2}[5(x - 2)^4]$$

$$+ (x - 2)^5(x + 3)^{1/2}(2x)$$

$$= \frac{x^2(x - 2)^5}{2\sqrt{x + 3}} + 5x^2(x - 2)^4\sqrt{x + 3} + 2x(x - 2)^5\sqrt{x + 3} \quad ◆◆◆$$

We now generalize this result (without proof) to cover any number of factors:

Derivative of a Product of *n* Factors	*The derivative of the product of n factors is an expression of n terms, each term being the product of n − 1 of the factors and the derivative of the other factor.*	**261**

◆◆◆ **Example 48:** *An Application.* The shape of a deflected beam is called its elastic curve and is given by the deflection y at a distance x. For the beam of Fig. 23–33 the elastic curve is given by

$$y = kx(l - x)(2l - x)$$

where l is the length of the beam and k is a constant depending on the properties of the beam. Write an equation giving the slope of the elastic curve by taking the derivative dy/dx of the given equation.

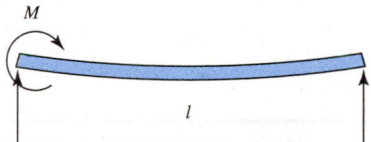

FIGURE 23–33 A beam simply supported at its ends, with a turning moment M at the left end.

Solution: Instead of multiplying out, we will use Eq. 261 to find the derivative.

$$\frac{dy}{dx} = kx(l - x)(-1) + kx(-1)(2l - x) + k(l - x)(2l - x)$$

$$= -kx(l - x) - kx(2l - x) + k(l - x)(2l - x)$$

$$= -6kxl + kx^2 + kl^2 \quad ◆◆◆$$

Derivative of a Quotient

To find the derivative of the function

$$y = \frac{u}{v}$$

where u and v are functions of x, we first rewrite it as a *product*,

$$y = uv^{-1}$$

Now, using the rule for products and the rule for a power function,

$$\frac{dy}{dx} = u(-1)v^{-2}\frac{dv}{dx} + v^{-1}\frac{du}{dx}$$

$$= -\frac{u}{v^2}\frac{dv}{dx} + \frac{1}{v}\frac{du}{dx}$$

We combine the two fractions over the LCD, v^2, and rearrange.

$$\frac{dy}{dx} = \frac{v}{v^2}\frac{du}{dx} - \frac{u}{v^2}\frac{dv}{dx}$$

which can be rewritten as follows:

| **Derivative of a Quotient** | $$\frac{d}{dx}\left(\frac{u}{v}\right) = \frac{v\dfrac{du}{dx} - u\dfrac{dv}{dx}}{v^2}$$
 The derivative of a quotient equals the denominator times the derivative of the numerator minus the numerator times the derivative of the denominator, all divided by the square of the denominator. | **262** |

◆◆◆ **Example 49:** Take the derivative of $y = \dfrac{2x^3}{(4x + 1)}$. Verify by calculator.

Solution: The numerator is $u = 2x^3$, so

$$\frac{du}{dx} = 6x^2$$

and the denominator is $v = 4x + 1$, so

$$\frac{dv}{dx} = 4$$

| **Common Error** | It is very easy, in Eq. 262, to interchange u and v, by mistake. |

Applying the quotient rule yields

$$\frac{dy}{dx} = \frac{(4x + 1)(6x^2) - (2x^3)(4)}{(4x + 1)^2}$$

Simplifying, we get

$$\frac{dy}{dx} = \frac{24x^3 + 6x^2 - 8x^3}{(4x + 1)^2} = \frac{16x^3 + 6x^2}{(4x + 1)^2} = \frac{2x^2(8x + 3)}{(4x + 1)^2}$$

Some prefer to use the product rule to do quotients, treating the quotient u/v as the product uv^{-1}.

Check: For an approximate verification by the TI-83/84 calculator, we enter for Y1 the derivative we just found, and for Y2 the **nDeriv** of the given function, screen (1). We then display a table showing Y1 and Y2 for a range of x values and check that they are approximately equal, screen (2). We also show the derivative taken symbolically on the TI-89, screen (3).

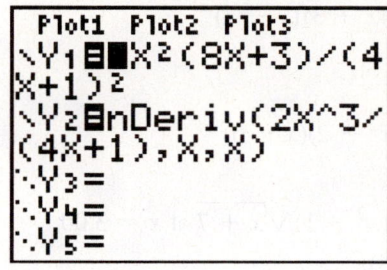

Plot1 Plot2 Plot3
\Y1◼X²(8X+3)/(4X+1)²
\Y2◼nDeriv(2X^3/(4X+1),X,X)
∙∙Y3=
∙∙Y4=
∙∙Y5=

X	Y₁	Y₂
0.000	0.00	2.0E⁻⁶
1.00	.88	.88
2.00	1.88	1.88
3.00	2.88	2.88
4.00	3.88	3.88
5.00	4.88	4.88
6.00	5.88	5.88
X=0		

F1▾ F2▾ F3▾ F4▾ F5 F6▾
Tools Algebra Calc Other PrgmIO Clean Up

$$\cdot \frac{d}{dx}\left[\frac{2 \cdot x^3}{4 \cdot x + 1}\right]$$

$$\frac{2 \cdot x^2 \cdot (8 \cdot x + 3)}{(4 \cdot x + 1)^2}$$

d((2x^3)/(4x+1),x)■

MAIN   DEG AUTO   SEQ   1/30

(1) TI-83/84 screens for Example 49.   (2)   (3) TI-89 screen for Example 49.

◆◆◆

◆◆◆ **Example 50:** Find $s'(3)$ if $s = \dfrac{(t^3 - 3)^2}{\sqrt{t + 1}}$.

Solution: By the quotient rule,

$$s' = \frac{\sqrt{t + 1}(2)(t^3 - 3)(3t^2) - (t^3 - 3)^2\left(\frac{1}{2}\right)(t + 1)^{-1/2}}{t + 1}$$

$$= \frac{6t^2(t^3 - 3)\sqrt{t + 1} - \dfrac{(t^3 - 3)^2}{2\sqrt{t + 1}}}{t + 1}$$

We could simplify now, but it will be easier to just substitute into the unsimplified expression. Letting $t = 3$ gives

$$s'(3) = \frac{6(9)(27 - 3)\sqrt{3 + 1} - \dfrac{(27 - 3)^2}{2\sqrt{3 + 1}}}{3 + 1} = 612$$

◆◆◆

Exercise 6 ◆ Derivatives of Products and Quotients

Find the derivative. Verify some by calculator. The answers to some of these problems, especially the quotients, may need a lot of simplification to match the book answer. Don't be discouraged if your answer does not appear to check at first.

Products

Some of these can be multiplied out. For a few of these, take the derivative both before and after multiplying out, and compare the two.

1. $y = x(x^2 - 3)$ **2.** $y = x^3(5 - 2x)$

3. $y = x(x^2 - 2)^2$ **4.** $y = x(x - 9)^3$

5. $y = (5 + 3x)(3 + 7x)$ **6.** $y = (7 - 2x)(x + 4)$

7. $y = (x + 3)(5x - 6)$ **8.** $y = (4x - 1)(3x + 3)$

9. $y = x^3(8.24x - 6.24x^3)$ **10.** $y = x\sqrt{1 + 2x}$

11. $y = 3x\sqrt{5 + x^2}$ **12.** $y = x^2\sqrt{3 - 4x}$

13. $y = \sqrt{x}(3x^2 + 2x - 3)$ **14.** $y = (3x + 1)^3\sqrt{4x - 2}$

15. $y = (2x^2 - 3)\sqrt[3]{3x + 5}$ **16.** $\dfrac{d}{dx}(4x + 3)(x - 7)$

17. $\dfrac{d}{dx}(2x^5 + 5x)^2(x - 3)$

18. $D_x(4x - 9)(x + 5)$ **19.** $D_x(x^2 - 2)(x - 6)$

20. If $y = (x + 2)\sqrt{x + 5}$, find $y'(2.34)$.

21. Find the rate of change of the function $y = (x^2 - 1)\sqrt{x + 7}$ at $x = 3.00$.

22. If $f(x) = (2x + 3)\sqrt{3x + 1}$, find $f'(2.88)$.

Constant Times a Function

23. $y = 6(x - 9)$ **24.** $y = 8(x^2 + 1)$

25. $y = \pi(2x - 4)^3$ **26.** $y = 3\pi(x - 1)^3$

Products with More Than Two Factors

27. $y = x(x - 7)(x + 1)$ **28.** $y = x(x + 2)(x - 9)^2$

29. $y = x(x + 1)^2(x - 2)^3$ **30.** $y = x\sqrt{x + 1}\sqrt[3]{x}$

Quotients

Find the derivative of each function.

31. $y = \dfrac{x}{x + 2}$ **32.** $y = \dfrac{x}{x^2 + 1}$

33. $y = \dfrac{x^2}{4 - x^2}$ **34.** $y = \dfrac{x - 1}{x + 1}$

35. $y = \dfrac{x + 2}{x - 3}$ **36.** $y = \dfrac{2x - 1}{(x - 1)^2}$

37. $y = \dfrac{x^{1/2}}{x^{1/2} + 1}$ **38.** $s = \sqrt{\dfrac{t - 1}{t + 1}}$

39. $w = \dfrac{z}{\sqrt{z^2 - a^2}}$ **40.** $v = \sqrt{\dfrac{1 + 2t}{1 - 2t}}$

41. Find the slope of the tangent to the curve $y = \sqrt{16 + 3x}/x$ at $x = 3$.

42. Find the derivative of the function $y = x/(7.42x^2 - 2.75x)$ when $x = 1.47$.

43. If $y = x/\sqrt{8 - x^2}$, find $y'(2)$.

44. If $f(x) = x^2/\sqrt{1 + x^3}$, find $f'(2)$.

An Application

45. The temperature T inside a certain furnace varies with the time t according to the function $T = (t + 3)\sqrt{t + 1}$ °F, where t is in hours. Find the rate of change of T when $t = 2.35$ h.

23–7 Other Variables, Implicit Relations, and Differentials

We now know how to find the derivative of a great many different functions. But so far

- our derivative has always been with respect to the variable in the function
- our function has always been in explicit form

Here we will learn how to take derivatives without having these limitations, enabling us to handle various applications later. We will also show how to put a derivative into *differential form*, a prerequisite to finding an integral and also to solving a differential equation, both of which we do in later chapters.

Derivatives with Respect to Other Variables

Do you recall our "Power Rule" from earlier in this chapter? Here it is again.

$$\frac{d(cu^n)}{dx} = cnu^{n-1}\frac{du}{dx} \qquad 258$$

Up to now, the variable in the function u has been the *same variable* that we take the derivative with respect to, as in the following examples.

◆◆◆ **Example 51:**

(a) $\dfrac{d}{dx}(x^3) = 3x^2\dfrac{dx}{dx} = 3x^2$ $\qquad$ (b) $\dfrac{d}{dt}(t^5) = 5t^4\dfrac{dt}{dt} = 5t^4$

$\qquad\qquad$ same $\qquad\qquad\qquad\qquad\qquad\qquad$ same $\qquad\qquad\qquad\qquad$ ◆◆◆

Since $dx/dx = dt/dt = 1$, we have not bothered to write these in. Of course, Rule 258 is just as valid when our independent variable is *different* from the variable that we are taking the derivative with respect to. The following examples show the power rule being applied to such cases.

◆◆◆ **Example 52:**

(a) $\dfrac{d}{dx}(u^5) = 5u^4\dfrac{du}{dx}$ $\qquad\qquad$ (b) $\dfrac{d}{dt}(w^4) = 4w^3\dfrac{dw}{dt}$

$\qquad\qquad$ different $\qquad\qquad\qquad\qquad\qquad\qquad$ different

(c) $\dfrac{d}{dx}(y^6) = 6y^5\dfrac{dy}{dx}$ $\qquad\qquad$ (d) $\dfrac{d}{dx}(y) = 1y^0\dfrac{dy}{dx} = \dfrac{dy}{dx}$

$\qquad\qquad$ different $\qquad\qquad\qquad\qquad\qquad\qquad$ different $\qquad\qquad\qquad\qquad$ ◆◆◆

As we said before, mathematical ideas do not, of course, depend upon which letters of the alphabet we happen to have chosen when doing a derivation. Thus in any of our rules you can replace *any* letter, say, x, with *any other* letter, such as z or t, as long as we do it throughout.

Common Error	It is very easy to forget to include the dy/dx in problems such as the following: $$\frac{d}{dx}(y^6) = 6y^5\frac{dy}{dx}$$ ⤷Don't forget!

Our other rules for derivatives (for sums, products, quotients, etc.) also work when the independent variable(s) of the function is different from the variable we are taking the derivative with respect to.

♦♦♦ **Example 53:** Here are some derivative of sums and products

(a) $\dfrac{d}{dx}(2y + z^3) = 2\dfrac{dy}{dx} + 3z^2\dfrac{dz}{dx}$ (b) $\dfrac{d}{dt}(wz) = w\dfrac{dz}{dt} + z\dfrac{dw}{dt}$

(c) $\dfrac{d}{dx}(x^2y^3) = x^2(3y^2)\dfrac{dy}{dx} + y^3(2x)\dfrac{dx}{dx}$

$\qquad\qquad = 3x^2y^2\dfrac{dy}{dx} + 2xy^3$ ♦♦♦

We usually think of x as being the independent variable and y the dependent variable, and we have been taking the derivative dy/dx of y with respect to x. Sometimes, however, the positions of x and y will be reversed, and we will want to find dx/dy, as in the following example.

> We will have to find dx/dy when later solving arc-length problems.

♦♦♦ **Example 54:** Find the derivative of x with respect to y, (dx/dy), if

$$x = y^2 - 3y + 2$$

Solution: We take the derivative exactly as before, except that x is where y usually is, and vice versa.

$$\dfrac{dx}{dy} = 2y\dfrac{dy}{dy} - 3\dfrac{dy}{dy} = 2y - 3 \qquad ♦♦♦$$

Derivatives of Implicit Relations

Recall that in an *implicit relation*, neither variable is isolated on one side of the equals sign.

♦♦♦ **Example 55:** $x^2 + y^2 = y^3 - x$ is an implicit relation.   ♦♦♦

We need to be able to differentiate implicit relations because we cannot always solve for one of the variables before differentiating.

To find the derivative dy/dx of an implicit relation between x and y,

(a) Take the derivative of both sides of the equation, with respect to x.
(b) Rearrange so that all dy/dx terms are on one side of the equation.
(c) Factor out dy/dx.
(d) Solve for dy/dx by dividing.

When taking derivatives, keep in mind that the derivative of x with respect to x is 1, and that the derivative of y with respect to x is dy/dx.

♦♦♦ **Example 56:** Given the implicit relation in Example 55, find dy/dx.

Solution: Given

$$x^2 + y^2 = y^3 - x$$

(a) We take the derivative term by term.

$$2x\dfrac{dx}{dx} + 2y\dfrac{dy}{dx} = 3y^2\dfrac{dy}{dx} - \dfrac{dx}{dx}$$

or

$$2x(1) + 2y\dfrac{dy}{dx} = 3y^2\dfrac{dy}{dx} - 1$$

(b) Collecting the dy/dx terms on one side,

$$2y\dfrac{dy}{dx} - 3y^2\dfrac{dy}{dx} = -2x - 1$$

(c) factoring

$$(2y - 3y^2)\frac{dy}{dx} = -2x - 1$$

(d) dividing,

$$\frac{dy}{dx} = \frac{2x + 1}{3y^2 - 2y}$$

Note that the derivative, unlike those for explicit functions, contains both x and y.

◆◆◆

When taking implicit derivatives, it is convenient to use the y' notation instead of dy/dx.

◆◆◆ **Example 57:** Find the derivative dy/dx for the relation

$$x^2 y^3 = 5$$

Solution: Using the product rule, we obtain

$$x^2 (3y^2)y' + y^3 (2x) = 0$$

$$3x^2 y^2 y' = -2xy^3$$

$$y' = -\frac{2xy^3}{3x^2 y^2} = -\frac{2y}{3x}$$

◆◆◆

◆◆◆ **Example 58:** Find dy/dx, given $x^2 + 3x^3 y + y^2 = 4xy$.

Solution:

(a) Taking the derivative of each term, we obtain

$$2x + 3x^3 y' + y(9x^2) + 2yy' = 4xy' + 4y$$

(b) Moving all terms containing y' to the left side and the other terms to the right side yields

$$3x^3 y' + 2yy' - 4xy' = 4y - 2x - 9x^2 y$$

(c) Factoring gives us

$$(3x^3 + 2y - 4x)y' = 4y - 2x - 9x^2 y$$

(d) Dividing, we get y' alone on the left side.

$$y' = \frac{4y - 2x - 9x^2 y}{3x^3 + 2y - 4x}$$

◆◆◆

◆◆◆ **Example 59:** *An Application.* The equation of the elliptical arch, Fig. 23–34, is

$$\frac{x^2}{(8.75)^2} + \frac{y^2}{(6.25)^2} = 1$$

(a) Write an expression for the slope at any point on the arch, and (b) evaluate the slope where the arch touches the beam at the point (4.00, 5.56).

Solution: (a) We find the slope by finding y'. We will do this implicitly.

$$\frac{2x}{(8.75)^2} + \frac{2yy'}{(6.25)^2} = 0$$

Solving for y',

$$2yy' = -(6.25)^2 \frac{2x}{(8.75)^2}$$

$$y' = -\frac{0.510x}{y}$$

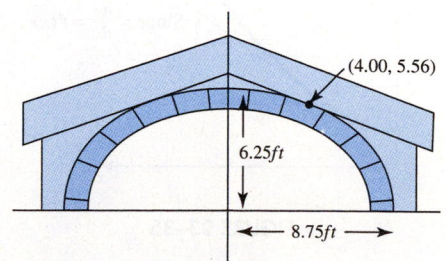

FIGURE 23–34 An elliptical arch.

(b) At (4.00, 5.56),

$$y' = \frac{0.510(4.00)}{5.56} = -0.367$$

◆◆◆

Power Function with Fractional Exponent

Now that we are able to take derivatives implicitly, we can show that the power rule, Eq. 254

$$\frac{d}{dx}x^n = nx^{n-1}$$

and hence Eq. 258

$$\frac{d(cu^n)}{dx} = cnu^{n-1}\frac{du}{dx}$$

are both valid when the exponent n is a fraction.

Let $n = p/q$, where p and q are both integers, positive or negative. Then

$$y = x^n = x^{p/q}$$

Raising both sides to the qth power, we have

$$y^q = x^p$$

Using the power rule, we take the derivative of each side.

$$qy^{q-1}\frac{dy}{dx} = px^{p-1}\frac{dx}{dx}$$

Solving for dy/dx we get

$$\frac{dy}{dx} = \frac{p}{q}\frac{x^{p-1}}{y^{q-1}} = n\frac{x^{p-1}}{[x^{(p/q)}]^{(q-1)}}$$

since $p/q = n$ and $y = x^{p/q}$. Applying the laws of exponents gives

$$\frac{dy}{dx} = n\frac{x^{p-1}}{x^{p-p/q}} = nx^{p-1-p+n} = nx^{n-1}$$

We have now shown that the power rule works for any rational exponent, positive or negative. It is also valid for an irrational exponent (such as π), as we will show later.

Differentials

Up to now, we have treated the symbol dy/dx as a *whole*, and not as the quotient of two quantities dy and dx. Here, we give dy and dx *separate* names and meanings of their own. The quantity dy is called the *differential of y*, and dx is called the *differential of x*.

These two differentials, dx and dy, have a simple geometric interpretation. Figure 23–35 shows a tangent drawn to a curve $y = f(x)$ at some point P. The slope of the curve is found by evaluating dy/dx at P. *The differential dy is then the rise of the tangent line, in some arbitrary run dx.* Since the rise of a line is equal to the slope of the line times the run, we get the following equation:

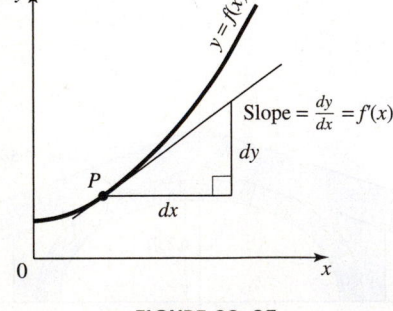

FIGURE 23–35

Differential of y	$dy = f'(x)dx$	**279**

where we have represented the slope by $f'(x)$ instead of dy/dx, to avoid confusion. If we take the derivative of some function, say, $y = x^3$, we get

$$\frac{dy}{dx} = 3x^2$$

Since we may now think of dy and dx as separate quantities (differentials), we can multiply both sides by dx.

$$dy = 3x^2 \, dx$$

This expression is said to be in *differential form*. We will see later that an equation containing a derivative (called a *differential equation*) is often written in differential form before it is solved. Thus to find dy, the differential of y, given some function of x, simply take the derivative of the function and multiply by dx.

◆◆◆ **Example 60:** If $y = 3x^2 - 2x + 5$, find the differential dy.

Solution: Taking the derivative gives us

$$\frac{dy}{dx} = 6x - 2$$

Multiplying by dx, we get the differential of y.

$$dy = (6x - 2) \, dx \qquad\qquad ◆◆◆$$

To find the differential of an *implicit* function, simply find the derivative as in earlier examples, and then multiply both sides by dx.

◆◆◆ **Example 61:** Find the differential of the implicit function

$$x^3 + 3xy - 2y^2 = 8$$

Solution: Differentiating term by term gives us

$$3x^2 + 3x\frac{dy}{dx} + 3y - 4y\frac{dy}{dx} = 0$$

$$(3x - 4y)\frac{dy}{dx} = -3x^2 - 3y$$

$$\frac{dy}{dx} = \frac{-3x^2 - 3y}{3x - 4y} = \frac{3x^2 + 3y}{4y - 3x}$$

$$dy = \frac{3x^2 + 3y}{4y - 3x}dx \qquad\qquad ◆◆◆$$

Exercise 7 ◆ Other Variables, Implicit Relations, and Differentials

Derivatives with Respect to Other Variables

1. If $y = 2u^3$, find dy/dw.

2. If $z = (w + 3)^2$, find dz/dy.

3. If $w = y^2 + u^3$, find dw/du.

4. If $y = 3x^2$, find dy/du.

Find the derivative.

5. $\dfrac{d}{dx}(x^3 y^2)$

6. $\dfrac{d}{dx}(w^2 - 3w - 1)$

7. $\dfrac{d}{dt}\sqrt{3z^2 + 5}$

8. $\dfrac{d}{dz}(y - 3)\sqrt{y - 2}$

Find the derivative dx/dy of x with respect to y.

9. $x = y^2 - 7y$

10. $x = (y - 3)^2$

11. $x = (y - 2)(y + 3)^5$

12. $x = \dfrac{y^2}{(4 - y)^3}$

Derivatives of Implicit Relations

Find dy/dx. (Treat a and r as constants.)

13. $5x - 2y = 7$

14. $2x + 3y^2 = 4$

15. $xy = 5$

16. $x^2 + 3xy = 2y$

17. $y^2 = 4ax$

18. $y^2 - 2xy = a^2$

19. $x^3 + y^3 - 3axy = 0$

20. $x^2 + y^2 = r^2$

21. $y + y^3 = x + x^3$

22. $x + 2x^2 y = 7$

23. $y^3 - 4x^2 y^2 + y^4 = 9$

24. $y^{3/2} + x^{3/2} = 16$

Find the slope of the tangent to each curve at the given point.

25. $x^2 + y^2 = 25$ at $x = 2$ in the first quadrant

26. $x^2 + y^2 = 25$ at $(3, 4)$

27. $2x^2 + 2y^3 - 9xy = 0$ at $(1, 2)$

28. $x^2 + xy + y^2 - 3 = 0$ at $(1, 1)$

Differentials

Write the differential dy for each function.

29. $y = x^3$

30. $y = x^2 + 2x$

31. $y = \dfrac{x - 1}{x + 1}$

32. $y = (2 - 3x^2)^3$

33. $y = x^3 + 3x$

34. $y = \sqrt{1 - 2x}$

Write the differential dy in terms of x, y, and dx for each implicit relation.

35. $3x^2 - 2xy + 2y^2 = 3$

36. $x^3 + 2y^3 = 5$

37. $2x^2 + 3xy + 4y^2 = 20$

38. $2\sqrt{x} + 3\sqrt{y} = 4$

An Application

39. A semicircular arch, Fig. 23–36, touches roof member AB at a distance of 6.25 ft from the center of the circle. Find the slope of AB. *Hint:* The equation of the semicircle is $x^2 + y^2 = (8.25)^2$. Take its derivative implicitly, solve for dy/dx, and evaluate it at $x = 6.25$.

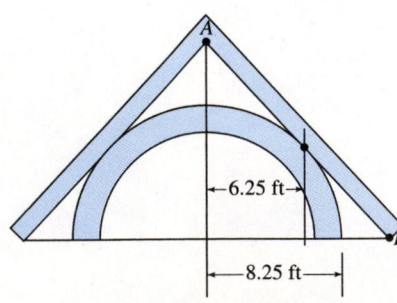

FIGURE 23–36

Remember that the bulk of our applications of the derivative are in the two following chapters.

23–8 Higher-Order Derivatives

■ Exploration:

Try this:

(a) Take the derivative of $y = 2x^3$.
(b) Now take the derivative of your derivative.
(c) Finally, take the derivative of the result of step (b).

In your exploration *you have just taken the first, second, and third derivatives* of the given function. That's all there is to it! ■

After taking the derivative of a function, we may then take the derivative of the derivative. That is called the *second derivative*. Our original derivative we now call the *first derivative*. The symbols used for the second derivative are

$$\frac{d^2y}{dx^2} \quad \text{or} \quad y'' \quad \text{or} \quad f''(x) \quad \text{or} \quad D^2y$$

We can then go on to find third, fourth, and higher derivatives. We will have many uses for the second derivative in the next few chapters, but we will seldom need derivatives higher than second order.

◆◆◆ **Example 62:** Given the function

$$y = x^4 + 2x^3 - 3x^2 + x - 5$$

Differentiating we get

$$y' = 4x^3 + 6x^2 - 6x + 1$$

Taking higher derivatives gives

$$y'' = 12x^2 + 12x - 6$$
$$y''' = 24x + 12$$
$$y^{(iv)} = 24$$
$$y^{(v)} = 0$$

and all higher derivatives will also be zero. ◆◆◆

◆◆◆ **Example 63:** Find the second derivative of $y = (x + 2)\sqrt{x - 3}$.

Solution: Using the product rule, we obtain the first derivative

$$y' = (x + 2)\left(\frac{1}{2}\right)(x - 3)^{-1/2} + \sqrt{x - 3}\,(1)$$

Now taking the second derivative,

$$y'' = \frac{1}{2}\left[(x + 2)\left(-\frac{1}{2}\right)(x - 3)^{-3/2} + (x - 3)^{-1/2)}(1)\right] + \frac{1}{2}(x - 3)^{-1/2}$$

$$= -\frac{x + 2}{4(x - 3)^{3/2}} + \frac{1}{\sqrt{x - 3}} \qquad ◆◆◆$$

Tip	You can usually save a lot of work if you *simplify the first derivative* before taking the second.

Higher Derivatives by Calculator

■ Exploration:

Try this. In the $\boxed{Y=}$ window, screen (1), enter the function $y = x^3$ for Y1. For Y2 enter the derivative of Y1. For Y3 enter the derivative of Y2. Graph the three functions, screen (2). (Y1 and Y2 are found in the $\boxed{VARS}$ menu.)

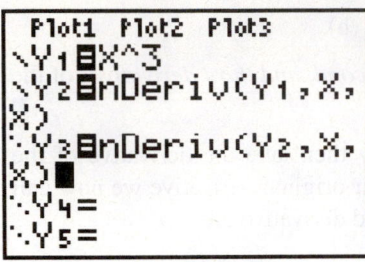

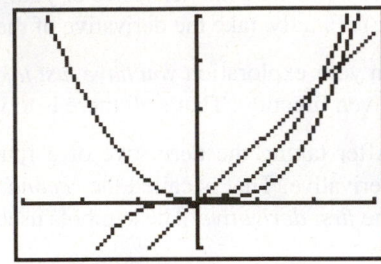

(1) TI-83/84 screens.   (2) Tick marks are 1 unit apart in x and 5 units apart in y.

What was the degree of the original function? Of the first derivative? The second? Can you generalize about what happens to the degree of a polynomial as you take successive derivatives?   ■

Higher-order derivatives can be found directly on a calculator that does symbolic algebra. On the TI-89, for example, after selecting $d(\ )$ from the **Calc** menu and entering the function and the variable, as usual, we add another integer indicating the order of the derivative to be taken.

◆◆◆ **Example 64:** Find the second derivative of $f(x) = x^3$ using the TI-89.

Solution: Following the function and the variable, we enter the number **2** to indicate the second derivative. The screen is shown.   ◆◆◆

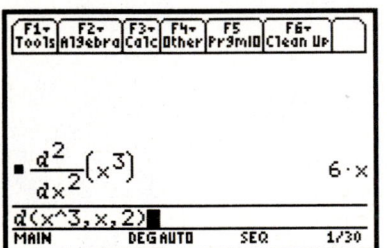

TI-89 screen for Example 64.

Exercise 8 ◆ Higher-Order Derivatives

Find the second derivative of each function. Verify some by calculator.

1. $y = 2x^3$ **2.** $y = 5x^4$

3. $y = 4x^3 + 3x^2$ **4.** $y = 5x^4 - 9x$

5. $y = 3x^4 - x^3 + 5x$ **6.** $y = x^3 - 3x^2 + 6$

7. $y = \dfrac{x^2}{x+2}$ **8.** $y = \dfrac{3+x}{3-x}$

9. $y = \sqrt{5 - 4x^2}$ **10.** $y = \sqrt{x+2}$

11. If $y = 3x^3 + 2x^2$, find $y''(2)$.   **12.** If $f(x) = x^2 - 4x^4$, find $f''(3)$.

An Application

13. The velocity of a moving point is given by the first derivative of the displacement, and the acceleration is given by the second derivative of the displacement. Find the velocity and acceleration at $t = 1.55$ s, of a point whose displacement is given by

$$s = 4.55t^3 + 2.85t^2 + 5.22 \qquad \text{cm}$$

where t is the elapsed time, in seconds.

••• CHAPTER 23: REVIEW PROBLEMS •••••••••••••••••••••••••••••••

1. Evaluate $\lim\limits_{x \to 1} \dfrac{x^2 + 2 - 3x}{x - 1}$.

2. Find dy/dx if $x^{2/3} + y^{2/3} = 9$.

3. If $f(x) = \sqrt{4x^2 + 9}$, find $f'(2)$.

4. Evaluate $\lim\limits_{x \to \infty} \dfrac{5x + 3x^2}{x^2 - 1 - 3x}$.

5. Find dy/dx if $\dfrac{x^2}{4} + \dfrac{y^2}{9} = 1$.

6. Find dy/dx if $y = \dfrac{x^2 + 5}{3 - x^2}$.

7. Find dy/dx by the delta method if $y = 5x - 3x^2$.

8. Find the slope of the tangent to the curve $y = \dfrac{1}{\sqrt{25 - x^2}}$ at $x = 3$.

9. Evaluate $\lim\limits_{x \to 0} \dfrac{e^x}{x}$.

10. If $y = 2.15x^3 - 6.23$, find $y'(5.25)$.

11. Evaluate $\lim\limits_{x \to -7} \dfrac{x^2 + 6x - 7}{x + 7}$.

12. Find the derivative $\dfrac{d}{dx}(3x + 2)$.

13. Find the derivative $D(3x^4 + 2)^2$.

14. Find the derivative $D_x(x^2 - 1)(x + 3)^{-4}$.

15. If $v = 5t^2 - 3t + 4$, find dv/dt.

16. If $z = 9 - 8w + w^2$, find dz/dx.

17. Evaluate $\lim\limits_{x \to 5} \dfrac{25 - x^2}{x - 5}$.

18. If $f(x) = 7x - 4x^3$, find $f''(x)$.

19. Find the following derivative: $D_x(21.7x + 19.1)(64.2 - 17.9x^{-2})^2$

20. If $s = 58.3t^3 - 63.8t$, find ds/dt.

21. Find $\dfrac{d}{dx}(4x^3 - 3x + 2)$.

Find the derivative.

22. $y = 6x^2 - 2x + 7$

23. $y = (3x + 2)(x^2 - 7)$

24. $y = 16x^3 + 4x^2 - x - 4$

25. $y = \dfrac{2x}{x^2 - 9}$

26. $y = 2x^{17} + 4x^{12} - 7x + 1/(2x^2)$

27. $y = (2x^3 - 4)^2$

28. $y = \dfrac{x^2 + 3x}{x - 1}$

29. $y = x^{2/5} + 2x^{1/3}$

Find $f''(x)$.

30. $f(x) = 6x^4 + 4x^3 - 7x^2 + 2x - 17$

31. $f(x) = (2x + 1)(5x^2 - 2)$

Write the differential dy for each function.

32. $y = 3x^4$

33. $y = (2x - 5)^2$

34. $x^2 + 3y^2 = 36$

35. $x^3 - y - 2x = 5$

36. *Project:* For a function assigned by your instructor, find the derivative in as many ways as you can: analytically, graphically, numerically, using a calculator or a computer.

24

Graphical Applications of the Derivative

✦✦✦ **OBJECTIVES** ✦✦

When you have completed this chapter, you should be able to

- Write the equation of the tangent or the normal to a curve.
- Find the angle of intersection between two curves.
- Find the values of *x* for which a given curve is increasing or decreasing.
- Determine the concavity of a curve.
- Find maximum and minimum points on a curve.
- Test whether a point is a maximum or a minimum point.
- Find points of inflection on a curve.
- Graph and interpret curves and regions with the aid of the techniques of this chapter.

◆◆◆

Now that we are able to find the derivative by several methods, we turn to applications. In this chapter we cover graphical applications, and in the next chapter applications from technology.

In the preceding chapter we showed several methods for finding the slope of the tangent to a curve at a given point. Here we build on that, writing the equation of such a tangent, as well as of the normal.

We then use derivatives to determine whether a curve is rising or falling at a particular place, find its concavity, and locate maximum, minimum, and inflection points. We finish by showing how calculus can help us to analyze a graph of a function. For example, could you say whether the curve at the left has maximum and minimum points, and give their locations? In this chapter we will show how to do that.

All the applications in this chapter are graphical, but they will give us the needed background to tackle the many applications from technology in the following chapter.

It is not surprising that this chapter requires heavy use of the graphics calculator.

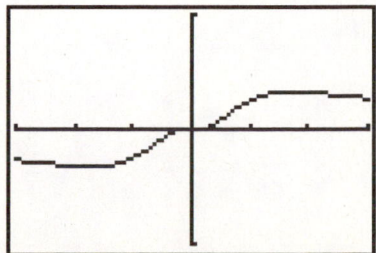

787

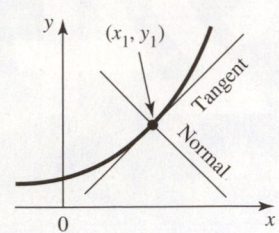

FIGURE 24–1   Tangent and normal to a curve.

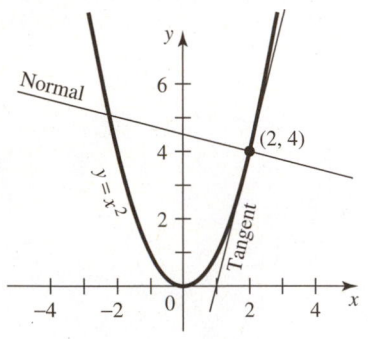

FIGURE 24–2

24–1   Equations of Tangents and Normals

In the preceding chapter we defined tangents and normals to a curve, Fig. 24–1. We also gave several methods for finding the slope of the tangent at a particular point on the curve: manually, by zooming in, by using the **Tangent** function on the calculator, and numerically. We then showed that the derivative, evaluated at a point on a curve, also gives the slope of the tangent at that point.

$$\text{slope of tangent at } x_1 = m_t = y'(x_1)$$

Having the slope of the tangent and a point through which it passes, we now use the point-slope form for the equation of a straight line to write the equation of that tangent. For the normal, we recall that its slope is the negative reciprocal of the slope of the tangent, enabling us to write its equation.

◆◆◆ **Example 1:** Write the equation of (a) the tangent and of (b) the normal to the curve $y = x^2$ at $x = 2$, Fig. 24–2. Check graphically.

Solution:

(a) Of our several ways of finding the slope of the tangent, let us choose the derivative. The derivative of $y = x^2$ is

$$y' = 2x$$

So

$$y'(2) = 2(2) = 4 = m_t.$$

Also, when $x = 2$, $y = 2^2 = 4$. So the slope is 4 at the point (2, 4). Using the point-slope form of the straight-line equation, we obtain

$$\frac{y - 4}{x - 2} = 4$$

or

$$y - 4 = 4x - 8$$

So $y = 4x - 4$ is the equation of the tangent.

(b) The slope of the normal is the negative reciprocal of the slope of the tangent (Eq. 214) so,

$$m_n = -\frac{1}{m_t} = -\frac{1}{4}$$

Again using the point-slope form, we have

$$\frac{y - 4}{x - 2} = -\frac{1}{4}$$

$$4y - 16 = -x + 2$$

So $y = \frac{9}{2} - \frac{x}{4}$ is the equation of the normal.

Graphical Check: In the same viewing window we graph the given function, the tangent, and the normal. We check that the tangent and the normal pass through the given point on the curve, screens (1) and (2).

Another graphical check is by means of the **Tangent** feature on the calculator, as described in the preceding chapter.

• Graph $y = x^2$.
• Select **Tangent** from the DRAW menu.

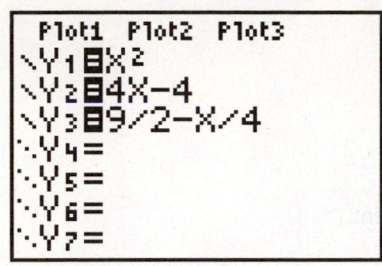

(1) TI-83/84 screens for Example 1.

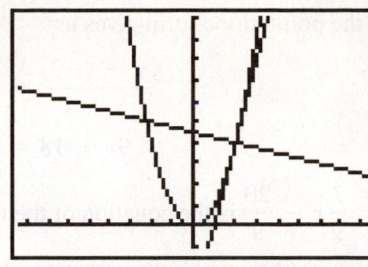

(2) Tick marks are 1 unit apart. Choose **Zsquare** to make the tangent and normal appear perpendicular.

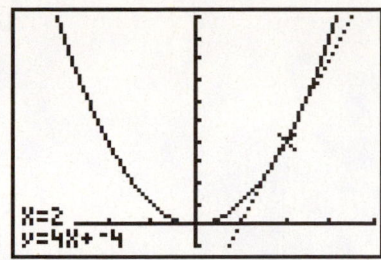

(3) Here the tangent is drawn by the calculator by selecting **Tangent** and specifying a point on the curve.

- Enter the value of x at which you want the tangent drawn to the curve. Choose $x = 2$ and press ENTER.

The tangent is drawn, and on the TI-83/84, the equation is also displayed, screen (3). ◆◆◆

Implicit Relations

When the equation of the curve is an implicit relation, you may choose to solve for y, when possible, before taking the derivative. Often, though, it is easier to take the derivative implicitly, as in the following example.

◆◆◆ **Example 2:** Find (a) the equation of the tangent to the ellipse $4x^2 + 9y^2 = 40$ at the point $(1, -2)$ and (b) the x intercept of the tangent (Fig. 24–3). Check by graphing.

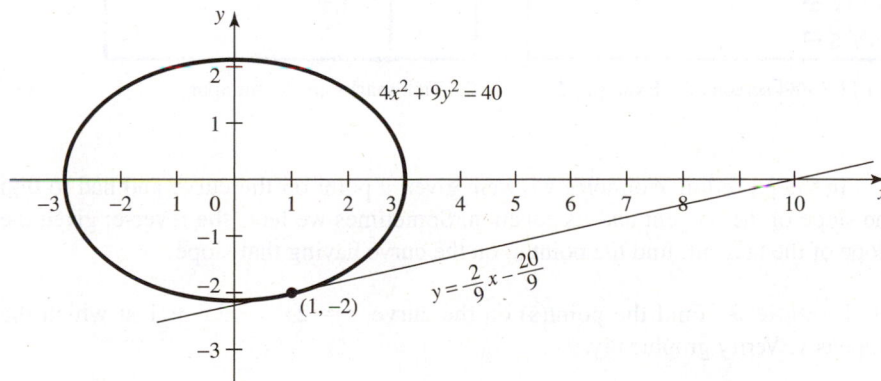

FIGURE 24–3

Solution:

(a) Taking the derivative implicitly, we have $8x + 18yy' = 0$, or

$$18yy' = -8x$$

$$y' = -\frac{8x}{18y} = -\frac{4x}{9y}$$

At $(1, -2)$,

$$y' = -\frac{4(1)}{9(-2)} = \frac{2}{9} = m_t$$

Using the point-slope form gives us

$$\frac{y - (-2)}{x - 1} = \frac{2}{9}$$

$$9y + 18 = 2x - 2$$

So $y = \dfrac{2}{9}x - \dfrac{20}{9}$ is the equation of the tangent.

(b) Setting y equal to zero in the equation of the tangent gives $2x - 20 = 0$, or an x intercept of

$$x = 10$$

Graphical Check: To graph the ellipse we first solve for y as we did in Chapter 22, and get

$$y = \pm\frac{2}{3}\sqrt{10 - x^2}$$

(work not shown). We graph the upper and lower parts of the ellipse, as well as the equation of the tangent, screen (1), and check the point of intersection and the x intercept screen (2) using **zero** from the $\boxed{\text{CALC}}$ menu.

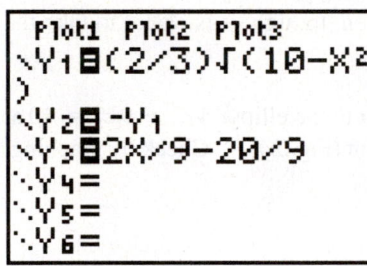

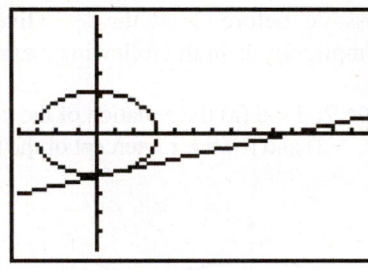

(1) TI-83/84 screens for Example 2.   (2) Tick marks are 1 unit apart.   ◆◆◆

In the preceding examples we were given a point on the curve and had to find the slope of the tangent and its equation. Sometimes we have the reverse: given the slope of the tangent, find the point(s) on the curve having that slope.

◆◆◆ **Example 3:** Find the point(s) on the curve $y = 2x^2 - 3x + 1$ at which the slope is 1. Verify graphically.

Solution: We find the slope everywhere on the parabola by taking the derivative.

$$y' = 4x - 3$$

To find the value of x where the slope is 1 we set the derivative equal to 1 and solve for x.

$$4x - 3 = 1$$

$$x = 1$$

Substituting back into the given equation gives

$$y(1) = 2(1)^2 - 3(1) + 1 = 0$$

So the slope is 1 at the point $(1, 0)$.

Graphical Check: We graph the original function and draw the tangent at $x = 1$, and check that its slope is 1.   ◆◆◆

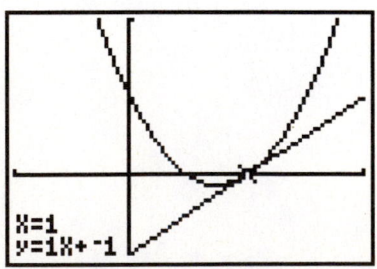

Graphical check for Example 3.
Tick marks are 1 unit apart.

Angle of Intersection of Two Curves

The angle between two curves is defined as the angle between their tangents at the point of intersection, which is found from Eq. 215,

$$\tan \phi = \frac{m_2 - m_1}{1 + m_1 m_2}$$

◆◆◆ **Example 4:** Find the angle of intersection between the parabolas (a) $y^2 = x$ and (b) $y = x^2$ at the point of intersection $(1, 1)$ (Fig. 24–4). Check graphically.

Solution:

(a) Taking the derivative of $y^2 = x$, we have $2yy' = 1$, or

$$y' = \frac{1}{2y}$$

At $(1, 1)$,

$$y' = \frac{1}{2} = m_1$$

(b) Taking the derivative of $y = x^2$ gives us $y' = 2x$. At $(1, 1)$,

$$y' = 2 = m_2$$

Then, from Eq. 215

$$\tan \phi = \frac{2 - \frac{1}{2}}{1 + \frac{1}{2}(2)} = 0.75$$

$$\phi = 36.9°$$

Graphical Check: We graph each parabola. Then using the **Tangent** feature from the [DRAW] menu, we draw the tangent to each curve at the point $(1, 1)$, screen (2). From the displayed equation for each tangent, we read the slope of each, getting $\frac{1}{2}$ and 2 as before.

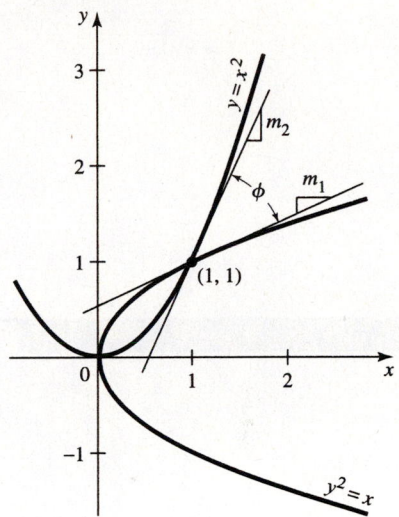

FIGURE 24–4 Angle of intersection of two curves. If the points of intersection are not known, solve the two equations simultaneously.

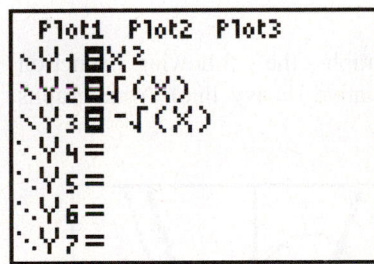

(1) TI-83/84 screens for Example 4.

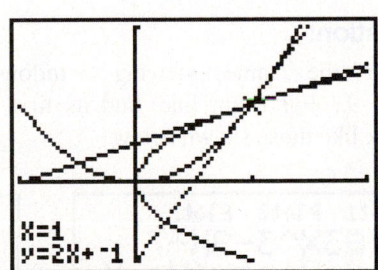

(2) Tick marks are 1 unit apart. While in the [DRAW] menu you can switch between curves using the up and down arrows. ◆◆◆

Exercise 1 ◆ Equations of Tangents and Normals

For problems 1 through 6, write the equations of the tangent and normal at the given point. Check some by calculator.

1. $y = x^2 + 2$ at $x = 1$
2. $y = x^3 - 3x$ at $(2, 2)$
3. $y = 3x^2 - 1$ at $x = 2$
4. $y = x^2 - 4x + 5$ at $(1, 2)$
5. $x^2 + y^2 = 25$ at $(3, 4)$
6. $16x^2 + 9y^2 = 144$ at $(2, 2.98)$

7. Find the first quadrant point on the curve $y = x^3 - 3x^2$ at which the slope $= 9$.
8. Write the equation of the tangent to the parabola $y^2 = 4x$ that makes an angle of $45°$ with the x axis.
9. Each of two tangents to the circle $x^2 + y^2 = 25$ has a slope $\frac{3}{4}$. Find the points of contact.
10. Find the equation of the line tangent to the parabola $y = x^2 - 3x + 2$ which has a slope of $-2/5$.

Angles Between Curves

Find the angle(s) of intersection, to nearest tenth of a degree, between the given curves.

11. $y = 2 - x$ and $y = x^2$ at $(1, 1)$
12. $y = 2x$ and $y = 2 - x^2$ at $(0.732, 1.46)$
13. $y = x^2 + x - 2$ and $y = x^2 - 5x + 4$ at $(1, 0)$
14. $y = -2x$ and $y = x^2(1 - x)$ at $(0, 0)$, $(2, -4)$, and $(-1, 2)$

24–2   Maximum, Minimum, and Inflection Points

Here we will learn how to find some features of a curve that will help us later to analyze a function, and also lead to useful technical applications in the following chapter. We will find these features using the derivative from the preceding chapter, and also by graphics calculator. We will often use one method to find a feature, and another to check. Let us start by showing how to find sections of a curve that are increasing, and those that are decreasing, without actually graphing the curve. As usual, we limit this discussion to *smooth curves*, without cusps gaps, or corners at which there are no derivatives.

Increasing and Decreasing Functions

■ Exploration:

Try this. In the same viewing window, graph the following function $y = 3x^3 - 9x + 7$ (light line) and its first derivative (heavy line). Your graphs should look like those shown below.

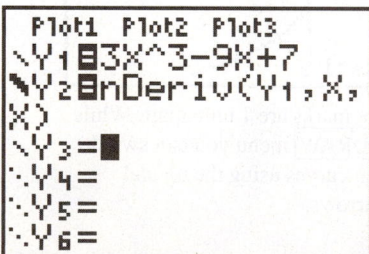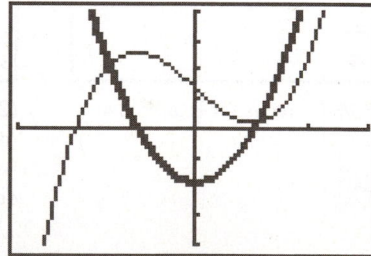

TI-83/84 screens for the exploration. The function is graphed with a lighter line than its derivative. Tick marks are 1 unit apart in x and 5 units apart in y.

- At what values of x is the derivative curve zero? What is happening to the function itself at those x values?
- For what values of x is the derivative curve positive? What is happening to the function at those x values?
- At what values of x is the derivative curve negative? What is happening to the function at those x values?
- What can you generalize about your findings?

When we talk about an increasing or decreasing function, we really mean the *interval(s)* where the function is increasing or decreasing. Few functions increase or decrease everywhere.

You may have concluded from your exploration that the first derivative is *positive* in an interval where the function itself is *increasing*, and *negative* in an interval where the function is *decreasing*. This, of course, gives us a way to determine if a function is increasing or decreasing at a particular place on a curve or to locate increasing and decreasing intervals of a curve, without actually graphing it.

◆◆◆ **Example 5:** Is the function $y = 2x^3 - 5x + 7$ increasing or decreasing at $x = 2$?

Solution: The derivative is $y' = 6x^2 - 5$. At $x = 2$,

$$y'(2) = 6(4) - 5 = +19$$

A *positive* derivative means that the given function is *increasing* at $x = 2$.   ◆◆◆

◆◆◆ **Example 6:** For what values of x is the curve $y = 3x^2 - 12x - 2$ rising, and for what values of x is it falling? Solve by using the first derivative and check by graphing.

Solution: The first derivative is $y' = 6x - 12$. This derivative will be equal to zero when $6x - 12 = 0$, or $x = 2$. We see that y' is negative for values of x less than 2. A *negative* derivative tells us that our original function is *falling* in that region. Further, the derivative is positive for $x > 2$, so the given function is rising in that region, as shown.   ◆◆◆

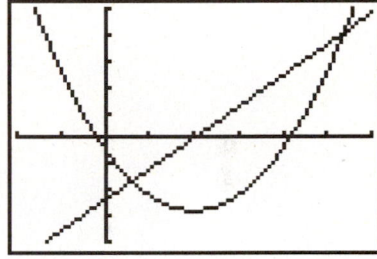

Graphical check for Example 6. Graph of the given function and the first derivative. Tick marks are 1 unit apart on the x axis and 5 units apart on the y axis.

Concavity

A curve may be either *concave upward*, Fig. 24–5(a), or concave *downward*, Fig. 24–5(b).

■ **Exploration:**

Try this. In the same window graph the function $y = 3x^3 - 3x$, and its second derivative.

- For what interval of x is the second derivative positive? What can you say about the concavity of the given function in that interval?

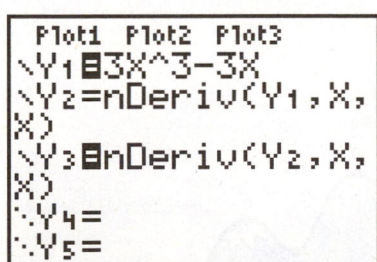

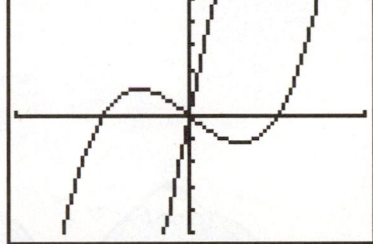

TI-83/84 screens for the exploration. Notice that we have *deselected* Y2 so that it does not graph by moving the cursor to the equals sign and pressing ENTER.

Graph of the given function and its second derivative. Tick marks are 1 unit apart.

(a) Concave upward

(b) Concave downward

FIGURE 24–5 It might help to think of the curve as a bowl. When it is concave upward, it will hold (+) water, and concave downward it will spill (−) water.

- For what interval of x is the second derivative negative? What can you say about the concavity of the given function in that interval?
- For what value of x is the second derivative zero? What can you say about the concavity of the given function at that value?
- What conclusions can you draw from this exploration?   ■

You may have concluded that

(a) The second derivative is positive where a curve is concave upward.
(b) The second derivative is negative where a curve is concave downward.
(c) The second derivative is zero where a curve is neither concave upward nor downward.

We may use these ideas to determine the concavity of curves at particular points or regions without making a graph.

◆◆◆ **Example 7:** Is the curve $y = 3x^4 - 7x^2 - 2$ concave upward or concave downward at (a) $x = 0$ and (b) $x = 1$?

Solution: The first derivative is $y' = 12x^3 - 14x$, and the second derivative is $y'' = 36x^2 - 14$.

(a) At $x = 0$,

$$y''(0) = -14$$

A *negative* second derivative tells that the given curve is concave *downward* at that point.

(b) At $x = 1$,

$$y''(1) = 36 - 14 = 22$$

or *positive*, so the curve is concave *upward* at that point.   ◆◆◆

Maximum and Minimum Points

Figure 24–6 shows a path over the mountains from *A* to *H*. It goes over three peaks, *B*, *D*, and *F*. These are called *maximum points* on the curve from *A* to *H*. The highest, peak *D*, is called the *absolute maximum* on the curve from *A* to *H*, while peaks *B* and *F* are called *relative maximum* points. Similarly, valley *G* is called an *absolute minimum*, while valleys *C* and *E* are *relative minimums*. All the peaks and valleys are referred to as *extreme values*. Figure 24–7 shows maximum and minimum points on a graph of $y = f(x)$.

FIGURE 24–6   Path over the mountains.

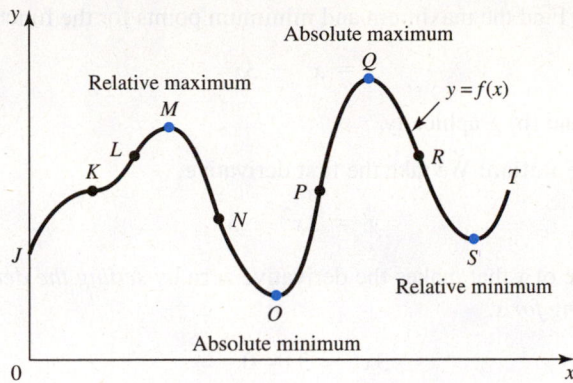

FIGURE 24–7 Maximum and minimum points.

■ Exploration:

Try this. In the same window graph the function $y = x^4 - 3x^2$ and its first derivative.

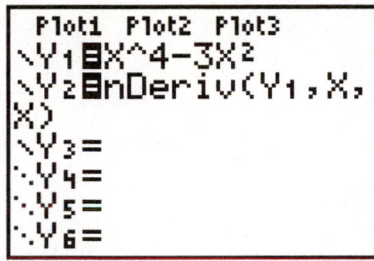

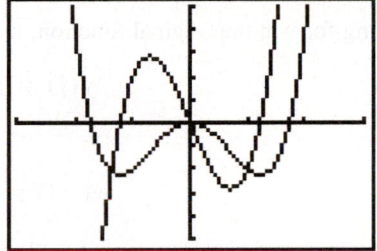

TI-83/84 screens for the exploration. Tick marks are 1 unit apart.

- At what value(s) of x is the first derivative zero?
- What do you see on the function itself at those point(s)?
- What does a zero first derivative tell you about the slope of the tangent on the function itself?
- What conclusions can you draw from this exploration? ■

You may have concluded that the firs t derivative (and hence the slope of the tangent and the rate of change) is zero at maximum and minimum points. Hence we can use the first derivative to find such points, which are called *stationary* points. The slope is also zero at a rare point, such as K in Fig. 24–7, that is neither a maximum nor a minimum point.

To find maximum and minimum points (and other stationary points), find the values of x for which the first derivative is zero.	**280**

We can find such points graphically or analytically, as in the following example.

◆◆◆ **Example 8:** Find the maximum and minimum points for the function

$$y = x^3 - 3x$$

(a) analytically and (b) graphically.

(a) **Analytical Solution:** We take the first derivative,

$$y' = 3x^2 - 3$$

We find the value of x that makes the derivative zero by *setting the derivative equal to zero and solving for x.*

$$3x^2 - 3 = 0$$

Factoring, we get

$$3(x^2 - 1) = 0$$

Multiplying both sides by 1/3 gives

$$x^2 - 1 = 0$$
$$x^2 = 1$$
$$x = \pm 1$$

Solving for y in the original function, we have

$$y(1) = 1 - 3 = -2$$

and

$$y(-1) = -1 + 3 = 2$$

So the points of zero slope are $(1, -2)$ and $(-1, 2)$.

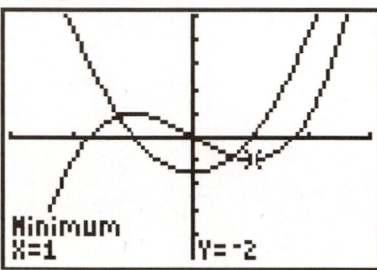

TI-83/84 screen for Example 8. Graph of the given function and the first derivative. The calculator was asked to locate a minimum point. Tick marks are 1 unit apart on the x axis and 2 units apart on the y axis.

(b) **Graphical Solutions**

Using the Graph of the Derivative: We graph the given function and the first derivative, as shown. We then locate the values of x at which the first derivative is zero using **zero** from the CALC menu. We substitute back to find the corresponding y values.

Using the Built-in Features for Finding Maximum and Minimum Points: We graph the function itself and have the calculator locate the maximum and minimum points using **maximum** and **minimum** from the CALC menu.   ◆◆◆

Testing for Maximum or Minimum

The simplest way to tell whether a stationary point is a maximum, a minimum, or neither, is to *graph the curve.* There are also a few tests that we can use to identify maximum and minimum points without having to graph the function. They are (a) the first-derivative test, (b) the second-derivative test, and (c) the ordinate test.

(a) In the *first-derivative test*, we look at the slope of the tangent on either side of a stationary point. At the minimum point B, shown in Fig. 24–8, for example, we see that the slope is negative to the left (at A) and positive to the right (at C) of that point. The reverse is true for a maximum point. Since the slope is given by the first derivative, we have the following:

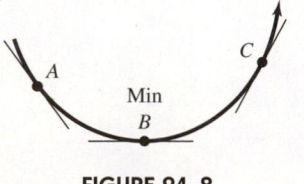

FIGURE 24–8

| **First-Derivative Test** | The first derivative is negative to the left of, and positive to the right of, a minimum point. The reverse is true for a maximum point. | **281** |

We use this test only on points close to the suspected maximum or minimum. If our test points are too far away, the curve might have already changed direction.

(b) The *second-derivative test* uses the fact that a minimum point occurs in a region of a curve that is concave upward (y'' is positive), while a maximum point occurs where a curve is concave downward (y'' is negative).

| **Second-Derivative Test** | If the first derivative at some point is zero, then, if the second derivative is

1. positive, the point is a minimum.
2. negative, the point is a maximum.
3. zero, the test fails. | **282** |

This test will not work on rare occasions. For example, the function $y = x^4$ has a minimum point at the origin, but the second derivative there is zero, not a positive number as expected.

| **Common Error** | It is tempting to group *maximum* with *positive,* and *minimum* with *negative*. Remember that they are just the *reverse* of this. |

(c) The *ordinate* test can distinguish between a maximum point and a minimum point, by checking the height of the curve to either side of the point.

| **Ordinate Test** | Find y a small distance to either side of the point to be tested. If y is greater there, we have a minimum; if it is less, we have a maximum. | **283** |

◆◆◆ **Example 9:** The function $y = x^3 - 3x$ was found in Example 8 to have stationary points at $(1, -2)$ and $(-1, 2)$. Use all three tests to show if $(1, -2)$ is a maximum or a minimum.

Solution:

(a) *First-derivative test:* We evaluate the first derivative a small distance to either side of the point, say, at $x = 0.9$ and $x = 1.1$.

$$y' = 3x^2 - 3$$

$$y'(0.9) = 3(0.9)^2 - 3 = -0.57$$

$$y'(1.1) = 3(1.1)^2 - 3 = 0.63$$

Slopes that are negative to the left and positive to the right of the point indicate a minimum point.

(b) *Second-derivative test:* Evaluating y'' at the point gives

$$y'' = 6x$$
$$y''(1) = 6$$

A positive second derivative also indicates a minimum point.

(c) *Ordinate test:* We compute y at $x = 0.9$ and $x = 1.1$.

$$y(0.9) = (0.9)^3 - 3(0.9) = -1.97$$
$$y(1.1) = (1.1)^3 - 3(1.1) = -1.97$$

Thus the curve is higher a small distance to either side of the extreme value $(1, -2)$, again indicating a minimum point. ◆◆◆

Implicit Relations

The procedure for finding maximum and minimum points is no different for an implicit relation, although it usually takes more work.

◆◆◆ **Example 10:** Find any maximum and minimum points on the curve

$$x^2 + 4y^2 - 6x + 2y + 3 = 0.$$

Solution: Taking the derivative implicitly gives

$$2x + 8yy' - 6 + 2y' = 0$$
$$y'(8y + 2) = 6 - 2x$$
$$y' = \frac{3 - x}{4y + 1}$$

Setting this derivative equal to zero gives $3 - x = 0$, or

$$x = 3$$

Substituting $x = 3$ into the original equation, we get

$$9 + 4y^2 - 18 + 2y + 3 = 0$$

Collecting terms and dividing by 2 yields

$$2y^2 + y - 3 = 0$$

Factoring gives us

$$(y - 1)(2y + 3) = 0$$
$$y = 1 \quad \text{and} \quad y = -\frac{3}{2}$$

So the points of zero slope are $(3, 1)$ and $(3, -\frac{3}{2})$.

We now apply the second-derivative test. Using the quotient rule gives

$$y'' = \frac{(4y + 1)(-1) - (3 - x)4y'}{(4y + 1)^2}$$
$$= -\frac{(4y + 1) + 4(3 - x)y'}{(4y + 1)^2}$$

Replacing y' by $(3 - x)/(4y + 1)$ and simplifying, we have

$$y'' = -\frac{(4y + 1) + 4(3 - x)\dfrac{3 - x}{4y + 1}}{(4y + 1)^2}$$
$$= -\frac{(4y + 1)^2 + 4(3 - x)^2}{(4y + 1)^3}$$

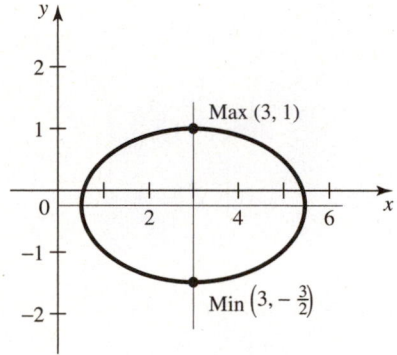

FIGURE 24–9 Graph of $x^2 + 4y^2 - 6x + 2y + 3 = 0$.

Max (3, 1)

Min $\left(3, -\frac{3}{2}\right)$

At (3, 1), $y'' = -0.2$. The negative second derivative tells that (3, 1) is a maximum point. At $(3, -\frac{3}{2})$, $y'' = 0.2$, telling that we have a minimum, as shown in Fig. 24–9. ◆◆◆

Inflection Points

A point where the curvature changes from concave upward to concave downward (or vice versa) is called an *inflection point*, or *point of inflection*. In Fig. 24–7 they are points K, L, N, P, R.

■ Exploration:

Try this. Graph the function $y = x^4 - 4x^2$ and its second derivative in the same window.

- At what values of x is the second derivative zero?
- At these values of x what appears to be happening on the function itself?
- What conclusions can you draw? ■

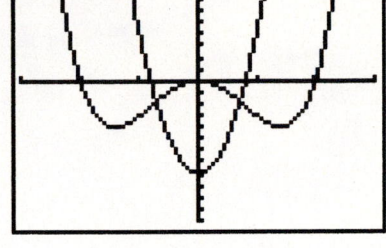

Screen for the exploration. Tick marks are one unit apart.

We saw in our exploration that the second derivative was positive where a curve was concave upward and negative where concave downward. In going from positive to negative, the second derivative must somewhere be zero, and this is at the point of inflection.

| Inflection Points | To find points of inflection, set the second derivative to zero and solve for x. Test by seeing if the second derivative changes sign a small distance to either side of the point. | 284 |

Unfortunately, a curve may have a point of inflection where the second derivative does not exist, such as the curve $y = x^{1/3}$ at $x = 0$. Fortunately, these cases are rare.

As with finding maxima and minima, we can use either an analytical or a graphical method to find inflection points.

◆◆◆ **Example 11:** Find any points of inflection on the curve

$$y = x^3 - 3x^2 - 5x + 7$$

Analytical Solution: We take the derivative twice.

$$y' = 3x^2 - 6x - 5$$

$$y'' = 6x - 6$$

We now set y'' to 0 and solve for x.

$$6x - 6 = 0$$

$$x = 1$$

and

$$y(1) = 1 - 3 - 5 + 7 = 0$$

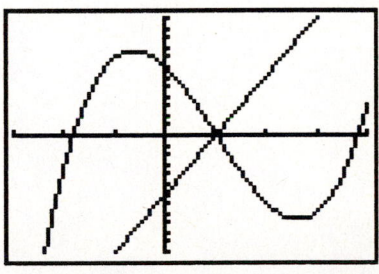

Screen for Example 11. Tick marks are 1 unit apart.

A graph of the given function shows the point (1, 0) clearly to be a point of inflection. If there were any doubt, we would test it by seeing if the second derivative has opposite signs on either side of the point.

Graphical Solution: We graph the function and its second derivative as shown. Using $\boxed{\text{TRACE}}$ and $\boxed{\text{ZOOM}}$, we find that the second derivative has a zero at $x = 1$, as for the analytical solution.   ◆◆◆

Summary

Figure 24–10 summarizes the ideas of this section.

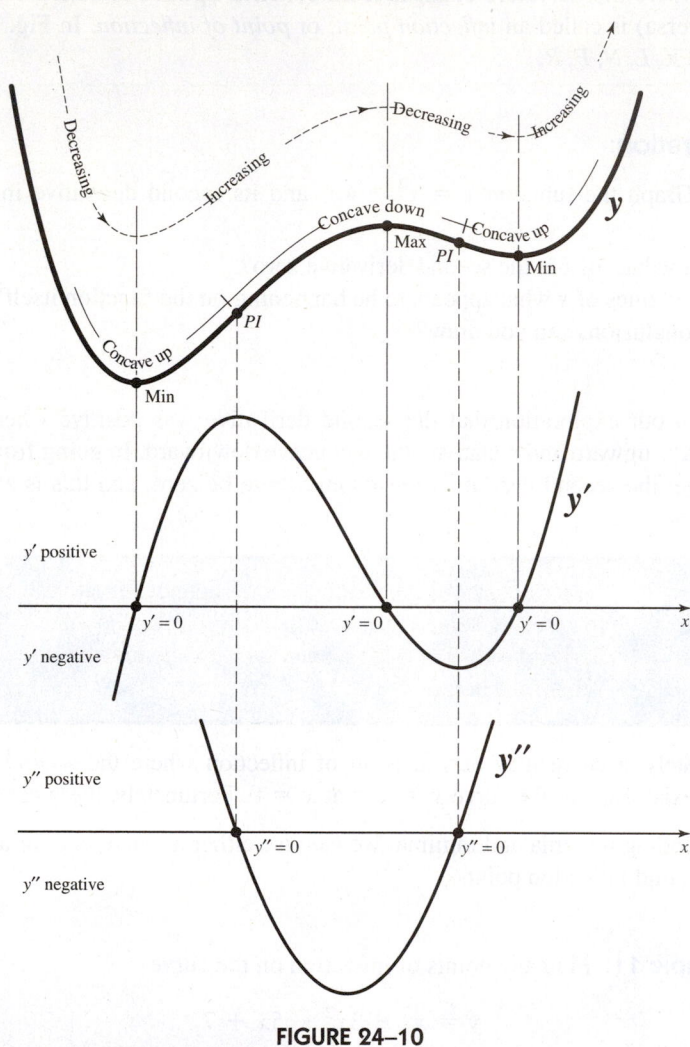

FIGURE 24–10

The function y is

> higher on both sides near a minimum point.
> lower on both sides near a maximum point.

The first derivative y' is

> positive when y is increasing and negative when y is decreasing.
> zero at a maximum or minimum point.

The second derivative y'' is

> positive when the y curve is concave upward.
> negative when the y curve is concave downward.
> positive at a minimum point.
> negative at a maximum point.
> zero at an inflection point.

Exercise 2 ◆ Maximum, Minimum, and Inflection Points

Increasing and Decreasing Functions

Use the derivative to say whether each function is increasing or decreasing at the value indicated. Check by graphing.

1. $y = 3x^2 - 4$ at $x = 2$

2. $y = x^2 - x - 3$ at $x = 0$

3. $y = 4x^2 - x$ at $x = -2$

4. $y = x^3 + 2x - 4$ at $x = -1$

5. $y = 3 - 2x + x^2$ at $x = 0$

6. $y = x^3 - 4x$ at $x = 2$

7. $y = x^3 + x^2$ at $x = -2$

8. $y = x^4 + x - 3$ at $x = 0$

Use the derivative to find the values of x for which each function is increasing, and for which it is decreasing. Check by graphing.

9. $y = 3x + 5$

10. $y = 4x^2 + 16x - 7$

11. $y = x^3 - 3$

12. $y = 2x^3 + 4x$

13. $y = 2x + x^2$

14. $y = 4x - x^3$

15. $y = 5x + x^5$

16. $y = x^4 + 3$

Concavity

Use the second derivative to state whether each curve is concave upward or concave downward at the given value of x. Check by graphing.

17. $y = x^4 + x^2$ at $x = 2$

18. $y = 4x^5 - 5x^4$ at $x = 1$

19. $y = -2x^3 - 2\sqrt{x + 2}$ at $x = \frac{1}{4}$

20. $y = \sqrt{x^2 + 3x}$ at $x = 2$

21. $y = 2x + x^2$ at $x = 1$

22. $y = x^3 - x$ at $x = 2$

23. $y = x + x^3$ at $x = -1$

24. $y = x^4 + x$ at $x = 1$

Round all approximate answers in this exercise to three significant digits.

Maximum and Minimum Points

Use derivatives to find any maximum and minimum points for each function. Distinguish between maximum and minimum points by graphing calculator, by the first-derivative test, the second-derivative test, or the ordinate test. Check by graphing.

25. $y = x^2$

26. $y = x^3 + 3x^2 - 2$

27. $y = x^3 - 7x^2 + 36$

28. $y = x^4 - 4x^3$

29. $y = 2x^2 - x^4$

30. $16y = x^2 - 32x$

31. $y = x^4 - 4x$

32. $2y = x^2 - 4x + 6$

33. $y = 3x^4 - 4x^3 - 12x^2$

34. $y = 2x^3 - 9x^2 + 12x - 3$

35. $y = x^3 + 3x^2 - 9x + 5$

36. $y = (x - 2)^2(2x + 1)$

Implicit Relations

Use derivatives to find and check for any maximum and minimum points for each relation. Check by graphing.

37. $4x^2 + 9y^2 = 36$

38. $x^2 + y^2 - 2x + 4y = 4$

39. $x^2 - x - 2y^2 + 36 = 0$

40. $x^2 + y^2 - 8x - 6y = 0$

41. $y^2 + 2y = 2x^4 + 2x$

42. $x^2 + y^5 + x = 16$

Inflection Points

Use the second derivative to find any inflection points for each function. Check by graphing.

43. $y = x^3 + 3$　　　　　　　　　　**44.** $y = 2x^3 + x^2 - 3$

45. $y = x^4 - x^3 + 1$　　　　　　　**46.** $y = x^4 + 2x^3 + 2x - 3$

47. $y = x^3 - x$　　　　　　　　　**48.** $y = 4x - 3x^3$

49. $y = 5x^3 - 2x^2 + 1$　　　　　　**50.** $y = x^3 - 5x - 1$

24–3　Sketching, Verifying, and Interpreting Graphs

We already know how to graph a function by hand and to quickly get a graph of any function by using a graphics calculator. What is there left to learn about graphing?

■ Exploration:

Try this. Graph the function $y = \dfrac{8}{8 + x^2}$, with the viewing window set to -4 to 4 on the x axis and -1 to 2 on the y axis, screen (1).

　　The curve appears to be symmetrical about the y axis. Is it really, or will the apparent symmetry disappear if we zoom out? There appear to be inflection points. If there are, where exactly are they? The curve appears to be flattening out as it goes farther from the origin. Does it approach an asymptote? If so, what is its value? Or can the curve turn and reach higher y values?　　　■

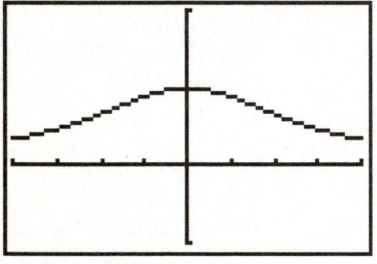

(1) Tick marks are 1 unit apart.

■ Exploration:

Try this. With a viewing window of $x = -3$ to 3 and $y = -1$ to 1, graph the function

$$y = \frac{x^3}{(1 + x^2)^2}$$

　　The graph, screen (2), seems to have maximum and minimum points. Where exactly are they? Is that a point of inflection near the origin? Are there others? Is this curve symmetrical about the origin? How can you know for sure? What about end behavior? Are there asymptotes, and, if so, what are their values? Will the curve turn and recross the x axis, or will it turn away from the x axis?　　■

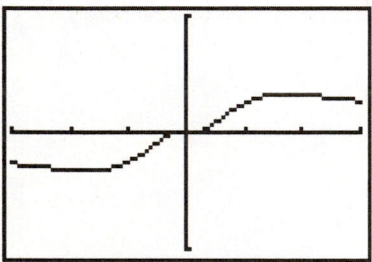

(2) Tick marks are 1 unit apart.

■ Exploration:

Try this. With a viewing window of $x = -2$ to 8 and $y = -2$ to 2, graph the function

$$y = \frac{\sqrt{x}}{x - 2}$$

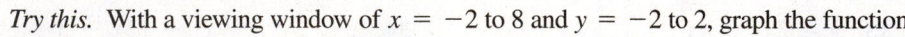

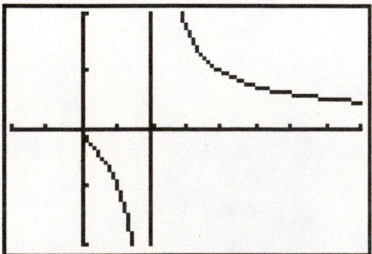

(3) Tick marks are 1 unit apart.

　　Can you explain why the graph, screen (3), does not extend to the left of the origin? Is it correct as shown, or is this an error? Is there an inflection point in the left branch? Where? What happens when each branch approaches $x = 2$? What happens far from the origin? Does the right branch have an asymptote? A minimum point?　　■

■ **Exploration:**

Try this. With a viewing window of $x = -3$ to 3 and $y = -10$ to 30, graph the function

$$y = 2x^2 + \frac{4}{x}$$

What if your supervisor asked you to explain the strange shape of this curve, screen (4). Could you? Why does it break and swing wildly away from the origin? Is there an inflection point in the left branch? Where? There is clearly a minimum point on the right branch. Can you say for sure that there are no others? ■

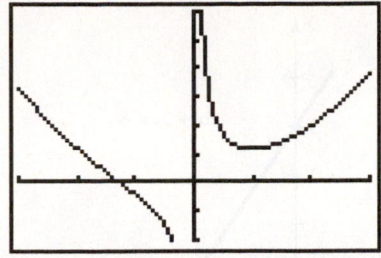

(4) Tick marks are 1 unit apart on the x axis and 5 units apart on the y axis.

■ **Exploration:**

Try this. Photocopy the four screens from the preceding four explorations, and screen (5) from this one. Cut them apart, cut off the titles, and scramble them. Then, without regraphing, see if you can pair each graph with the appropriate equation. Can you tell which graph has no corresponding equation? ■

It is not enough to be able to make a graph—anyone with a graphics calculator can now do it. We must be able to *interpret* a graph, to discover hidden behavior, to explain it to others, and to confirm that it is correct. We must also be sure that we have a *complete* graph, with no features of interest outside the viewing window.

With our new calculus tools we are now able to do all these things. We will review them here, along with some that we've had earlier.

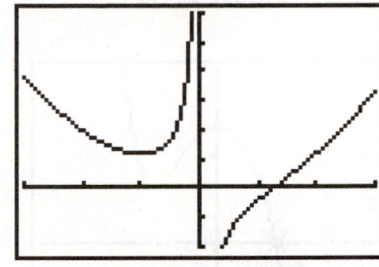

(5) Tick marks are 1 unit apart.

> ### Curve-Sketching Tips
>
> **A.** Identify the **type of equation**. This will often save you time.
> **B.** Find any **intercepts**.
> **C.** Take advantage of **symmetry**.
> **D.** Determine the **extent** of the function to locate any places where the curve does not exist.
> **E.** Sketch in any **asymptotes**.
> **F.** Locate the regions in which the function is **increasing** or **decreasing.**
> **G.** Locate any **maximum, minimum,** and **inflection** points.
> **H.** Determine the **end behavior** as x gets very large.
> **I.** Be sure to make **a complete graph**. Plot extra points if needed.

We will now give examples of each.

A. Type of Equation: Does the equation look like any we have studied before—a linear function, power function, quadratic function, trigonometric function, exponential function, logarithmic function, or equation of a conic? For a polynomial, the *degree* will tell the maximum number of extreme points you can expect.

◆◆◆ **Example 12:** The function $y = x^5 - 3x^2 + 4$ is a polynomial of fifth degree, so its graph (Fig. 24–11) may have up to four extreme points and three inflection points. There may, however, be fewer than these. ◆◆◆

B. Intercepts: Find the x intercept(s), or root(s), by setting y equal to zero and solving the resulting equation for x. Find the y intercept(s) by setting x equal to zero and solving for y.

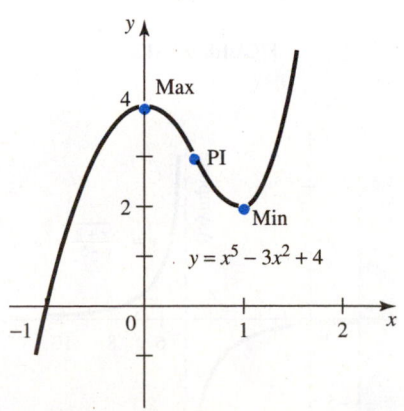

FIGURE 24–11

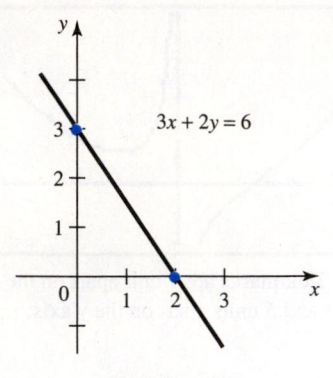

FIGURE 24–12

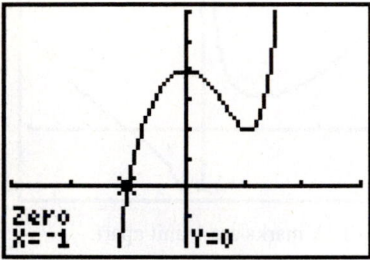

Screen for Example 13 (b) with tick marks 1 unit apart. We have used the **zero** operation to locate the root at $x = -1$.

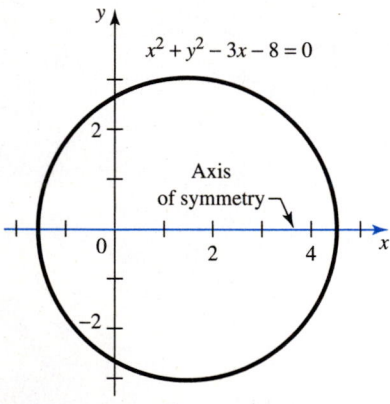

FIGURE 24–13

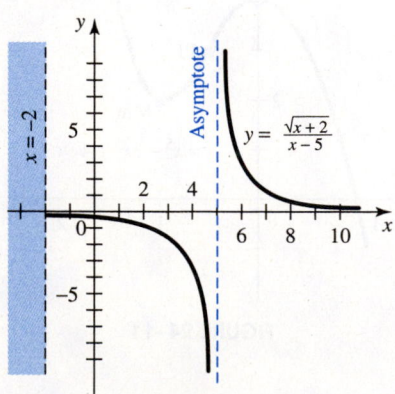

FIGURE 24–14

◆◆◆ Example 13:

(a) The function $3x + 2y = 6$ has a y intercept at

$$3(0) + 2y = 6$$

or at $y = 3$, and an x intercept at

$$3x + 2(0) = 6$$

or at $x = 2$ (Fig. 24–12).

(b) The function of Example 12 has a y intercept at

$$y = 0 - 0 + 4 = 4$$

We find the x intercept approximately using the **zero** operation from the $\boxed{\text{CALC}}$ menu on the TI-83/84, getting $x = -1$. ◆◆◆

C. Symmetry: A curve is symmetrical about the

(a) x axis, if the equation does not change when we substitute $-y$ for y.
(b) y axis, if the equation does not change when we substitute $-x$ for x.
(c) origin, if the equation does not change when we substitute $-x$ for x and $-y$ for y.

A function whose graph is symmetrical about the y axis is called an *even function*; one symmetrical about the origin is called an *odd function*.

◆◆◆ Example 14: Given the function $x^2 + y^2 - 3x - 8 = 0$:

(a) Substituting $-y$ for y gives $x^2 + (-y)^2 - 3x - 8 = 0$, or $x^2 + y^2 - 3x - 8 = 0$. This is the same as the original, indicating symmetry about the x axis.

(b) Substituting $-x$ for x gives $(-x)^2 + y^2 - 3(-x) - 8 = 0$, or $x^2 + y^2 + 3x - 8 = 0$. This equation is different from the original, indicating *no* symmetry about the y axis.

(c) Substituting $-x$ for x and $-y$ for y gives $(-x)^2 + (-y)^2 - 3(-x) - 8 = 0$ or $x^2 + y^2 + 3x - 8 = 0$. The equation is different from the original, indicating no symmetry about the origin, as shown in Fig. 24–13. ◆◆◆

D. Extent: Look for values of the variables that give division by zero or that result in negative numbers under a radical sign. The curve will not exist at these values.

◆◆◆ Example 15: For the function

$$y = \frac{\sqrt{x + 2}}{x - 5}$$

y is not real for $x < -2$ or for $x = 5$ (Fig. 24–14). ◆◆◆

E. Asymptotes: Look for some value of x that when *approached* from above or from below, will cause y to become infinite. You will then have found a *vertical asymptote*. Then if y approaches some particular value as x becomes infinite, we will have found a *horizontal asymptote*. Similarly, check what happens to y when x becomes infinite in the negative direction.

◆◆◆ Example 16: The function in Example 15 becomes infinite as x approaches 5 from above or below, so we expect a vertical asymptote at $x = 5$ (Fig. 24–14). ◆◆◆

As x gets very large, the numbers 2 and -5 become insignificant compared to x, so we have

$$y \approx \frac{\sqrt{x}}{x} \approx \frac{1}{\sqrt{x}}$$

Thus when x approaches $+\infty$, y approaches zero, giving us a horizontal asymptote at $y = 0$. The curve approaches the same asymptote in the negative direction, but as we saw in Example 15, the curve does not exist for $x < -2$.

F. Increasing or Decreasing Function: We saw that the first derivative is positive in regions where the function is increasing, and negative where the function is decreasing. Thus inspection of the first derivative can show where the curve is rising and where it is falling.

◆◆◆ **Example 17:** For the function $y = x^2 - 4x + 2$, the first derivative $y' = 2x - 4$ is positive for $x > 2$ and negative for $x < 2$. Thus we expect the curve to fall in the region to the left of $x = 2$ and rise in the region to the right of $x = 2$ (Fig. 24–15).

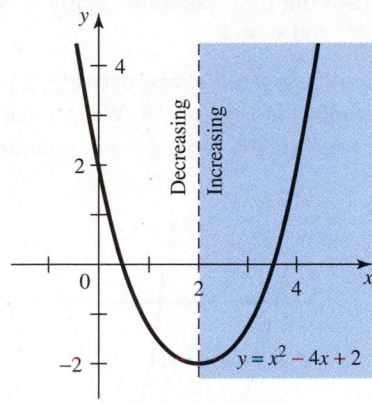

FIGURE 24–15

G. Maximum, Minimum, and Inflection Points: Maxima and minima are found where y' is zero. Inflection points are found where y'' is zero. These points may be found analytically or graphically.

◆◆◆ **Example 18:** To find the maximum and minimum points on the curve $y = x^3 - 2x^2 + x + 1$, we set the first derivative

$$y' = 3x^2 - 4x + 1$$

to zero and solve for x, getting

$$x = \tfrac{1}{3} \quad \text{and} \quad x = 1$$

Solving for the corresponding values of y we get, $y(1/3) = (1/3)^3 - 2(1/3)^2 + (1/3) + 1 = 1.15$ and $y(1) = (1)^3 - 2(1)^2 + (1) + 1 = 1$, to three significant digits, giving a maximum at $\left(\tfrac{1}{3}, 1.15\right)$ and a minimum at $(1, 1)$. To get the inflection point, we set the second derivative

$$y'' = 6x - 4$$

to zero, getting $x = \tfrac{2}{3}$. Solving for y gives,

$$y(2/3) = (2/3)^3 - 2(2/3)^2 + (2/3) + 1 = 1.07.$$

The inflection point is then $\left(\tfrac{2}{3}, 1.07\right)$. For a graphical solution, we graph y, y', and y'' on the same axes, as shown in Fig. 24–16. The zeros on the y' graph locate the maxima and minima, and the zero on the y'' graph locates the inflection point.   ◆◆◆

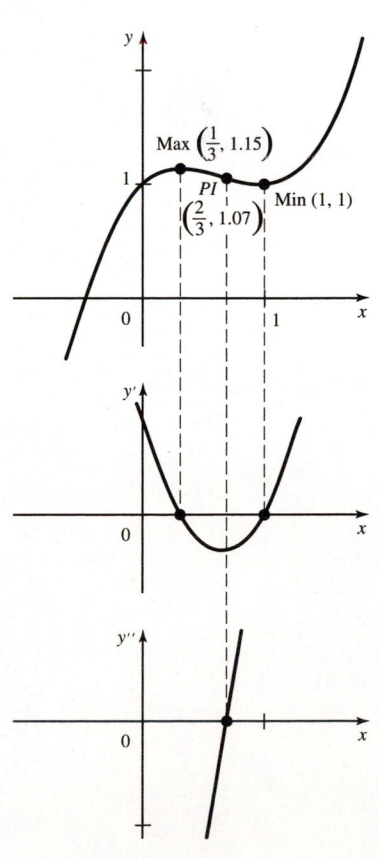

FIGURE 24–16

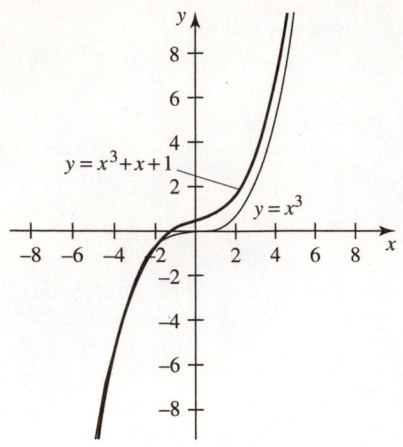

FIGURE 24–17

H. End Behavior: What happens to y as x gets very large in both the positive and the negative directions? Does y continue to grow without bound or approach some asymptote, or is it possible that the curve will turn and perhaps cross the x axis again? Similarly, can you say what happens to x as y gets very large?

◆◆◆ **Example 19:** As x grows, for the function $y = x^3 + x + 1$, the second and third terms on the right side become less significant compared to the x^3 term. So, far from the origin, the function will have the appearance of the function $y = x^3$ (Fig. 24–17). ◆◆◆

I. Complete Graph: Once we have found all of the zeros, maxima, minima, inflection points, and asymptotes, and have investigated the end behavior, we can be quite sure that we have not missed any features of interest.

Graphing Regions

In later applications we will have to identify *regions* that are bounded by two or more curves. Here we get some practice in doing that.

◆◆◆ **Example 20:** Locate the first-quadrant region bounded by the y axis and the curves $y = x^2$, $y = 1/x$, and $y = 4$.

Solution: The three curves given are, respectively, a parabola, a hyperbola, and a straight line, and are graphed in Fig. 24–18. We see that three different closed areas are formed, but the shaded area is the only one *bounded by each one of the given curves* and the y axis.

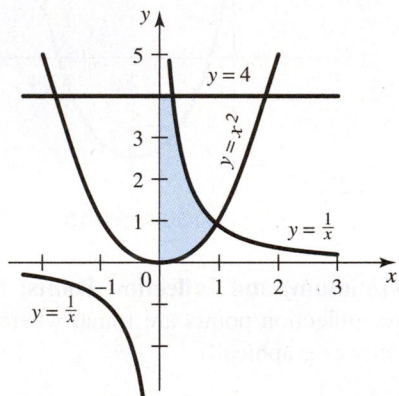

FIGURE 24–18 Graphing a region bounded by several curves. ◆◆◆

Exercise 3 ◆ Sketching, Verifying, and Interpreting Graphs

Make a complete graph of each function. Locate all features of interest.

1. $y = 4x^2 - 5$ 2. $y = 3x - 2x^2$

3. $y = 5 - \dfrac{1}{x}$ 4. $y = \dfrac{3}{x} + x^2$

5. $y = x^4 - 8x^2$ 6. $y = \dfrac{1}{x^2 - 1}$

7. $y = x^3 - 9x^2 + 24x - 7$ 8. $y = x\sqrt{1 - x}$

9. $y = 5x - x^5$

10. $y = \dfrac{9}{x^2 + 9}$

11. $y = \dfrac{6x}{3 + x^2}$

12. $y = \dfrac{x}{\sqrt{4 - x^2}}$

13. $y = x^3 - 6x^2 + 9x + 3$

14. $y = x^2\sqrt{6 - x^2}$

15. $y = \dfrac{96x - 288}{x^2 + 2x + 1}$

16. $y = \dfrac{\sqrt{x}}{x - 1}$

17. $y = \dfrac{x}{\sqrt{x^2 + 1}}$

18. $y = \dfrac{x^3}{(1 + x^2)^2}$

19. $y = x^2 + 2x$

20. $y = x^2 - 3x + 2$

21. $y = x^3 + 4x^2 - 5$

22. $y = x^4 - x^2$

Graph the region bounded by the given curves.

23. $y = 3x^2$ and $y = 2x$

24. $y^2 = 4x$, $x = 5$, and the x axis, in the first quadrant

25. $y = 5x^2 - 2x$, the y axis, and $y = 4$, in the second quadrant

26. $y = 4/x$, $y = x$, $x = 6$, and the x axis

••• CHAPTER 24 REVIEW PROBLEMS •••••••••••••••••••••••••••••••

1. Find any maximum and minimum points on the curve $y = 4x^2 - 2x + 5$.

2. Find any points of inflection on the curve $y = x^3(1 + x^2)$.

3. Find the maximum and minimum points on the curve $y = 3\sqrt[3]{x} - x$.

4. Find any maximum points, minimum points, and points of inflection for the function $3y = x^3 - 3x^2 - 9x + 11$.

5. Write the equations of the tangent and normal to the curve $y = \dfrac{1 + 2x}{3 - x}$ at the point $(2, 5)$.

6. Find the coordinates of the point on the curve $y = \sqrt{13 - x^2}$ at which the slope of the tangent is $-\frac{2}{3}$.

7. Find the x intercept of the tangent to the curve $y = \sqrt{x^2 + 7}$ at the point $(3, 4)$.

8. Graph the function $y = \dfrac{4 - x^2}{\sqrt{1 - x^2}}$, and locate any features of interest (roots, asymptotes, maximum/minimum points, and points of inflection).

9. For which values of x is the curve $y = \sqrt{4x}$ rising, and for what values is it falling?

10. Is the function $y = \sqrt{5 - 3x}$ increasing or decreasing at $x = 1$?

11. Is the curve $y = x^2 - x^5$ concave upward or concave downward at $x = 1$?

12. Graph the region bounded by the curves $y = 3x^3$, $x = 1$, and the x axis.

13. Write the equations of the tangent and of the normal to the curve $y = 3x^3 - 2x + 4$ at $x = 2$.

14. Find the x intercepts of the tangent and of the normal in problem 13.

15. Find the angle of intersection between the curves $y = x^2/4$ and $y = 2/x$.

16. Graph the function $y = 3x^3\sqrt{9 - x^2}$, and locate any features of interest.

17. *Writing:* In this chapter we gave nine things to look for when graphing a function. Can you list at least seven from memory? Write a paragraph explaining how just one of the nine is useful in graphing.

More Applications
of the Derivative

◆◆◆ OBJECTIVES ◆◆◆◆◆◆◆◆◆◆◆◆◆◆◆◆◆◆◆◆◆◆◆◆◆◆◆◆◆◆◆◆◆◆◆◆◆◆◆

When you have completed this chapter, you should be able to

- Solve applied problems involving instantaneous rate of change.
- Solve for currents and voltages in capacitors and inductors.
- Compute velocities and accelerations for straight-line or curvilinear motion.
- Solve motion problems given by parametric equations.
- Solve related rate applications.
- Solve applied maximum-minimum problems.

◆◆

Following our purely graphical applications we move on to more physical applications of the derivative. What is this stuff good for? Here we give the answer to that question, at least for the derivative. We will see an amazing assortment of technical applications that can be attacked using the derivative. They fall into three main groups:

- Finding a rate of change, including motion of a point and rate of change in electrical applications.
- Solving applications where *two* things are changing at the same time, with the rate of change of one related to the rate of change of the other.
- Optimization, where we want to find the lowest cost, or the strongest beam, or the shortest time. Here we make use of our ability to find maximum and minimum points on a curve, as learned in the preceding chapter. For example, if you were designing a cylindrical storage container to contain a given volume, Fig. 25–1, how would you choose the radius and the length so that the container needed the least amount of material? We will do problems of this sort here.

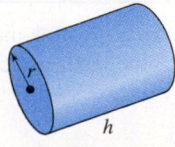

FIGURE 25–1

809

25–1 Rate of Change

Our first set of applications deals with the *rate of change* of a physical quantity. We have already found rates of change of various functions; here our functions will be equations from technology.

Rate of Change Given by the Derivative

We already established that the first derivative of a function gives the *rate of change* of that function.

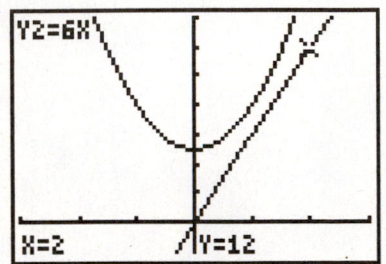

Screen for Example 1. Tick marks are 1 unit apart on the *x* axis and 2 units apart on the *y* axis.

♦♦♦ **Example 1:** Find the instantaneous rate of change of the function $y = 3x^2 + 5$ when $x = 2$.

Solution: Taking the derivative,

$$y' = 6x$$

When $x = 2$,

$$y'(2) = 6(2) = 12 = \text{instantaneous rate of change when } x = 2$$

Graphical Solution: Graph the derivative of the given function and determine its value at the required point. The screen shows graphs of $y = 3x^2 + 5$, and its derivative $y' = 6x$. Notice that the derivative has a value of 12 at $x = 2$.   ♦♦♦

Of course, rates of change can involve variables other than x and y. When two or more related quantities are changing, we often speak about the *rate of change* of one quantity with respect to one of the other quantities. For example, if a steel rod is placed in a furnace, its temperature and its length both increase. Since the length varies with the temperature, we can speak about the *rate of change of length with respect to temperature*. But the length of the bar is also varying with time, so we can speak about the *rate of change of length with respect to time.*

Rate of Change with Respect to Time

Time rates are the most common rates of change we need to calculate. Given a function $y = f(t)$, where t is time, we find the time rate of change of y simply by taking the derivative with respect to t; that is, we find dy/dt. We then usually have to evaluate that derivative at a given value of t.

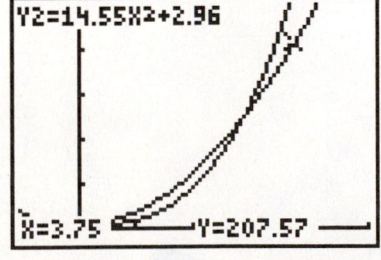

Screen for Example 2. Tick marks are 1 unit apart on the *x* axis and 50 units apart on the *y* axis.

♦♦♦ **Example 2:** The temperature T (°F) in a certain furnace varies with time t (s) according to the function $T = 4.85t^3 + 2.96t$. Find the rate of change of temperature at $t = 3.75$ s.

Solution: Taking the derivative of T with respect to t,

$$\frac{dT}{dt} = 14.6t^2 + 2.96 \quad \text{°F/s}$$

At $t = 3.75$ s,

$$\left. \frac{dT}{dt} \right|_{t=3.75\,\text{s}} = 14.6(3.75)^2 + 2.96 = 208\text{°F/s}$$

Graphical Solution: We graph the original function and the derivative as shown and determine the value of dT/dt when $t = 3.75$ s.   ♦♦♦

Electric Current

The idea of *time rate of change* finds many applications in electrical technology, a few of which we introduce here. We will return to this topic once more, after we learn how to take derivatives of exponential and trigonometric functions, as currents and voltages are often expressed by those functions. In this introductory section we will limit ourselves to algebraic functions.

The *coulomb* (C) is the unit of electrical *charge*. The *current* in amperes (A) is the number of coulombs passing a point in a circuit in 1 second. Charge is usually denoted by the letter q and current by the letter i. If the current varies with time, then the *instantaneous current* is given by the following equation:

Current	$$i = \frac{dq}{dt}$$	**1078**
	Current is the rate of change of charge, with respect to time.	

◆◆◆ **Example 3:** The charge through a 2.85-Ω resistor is given by

$$q = 1.08t^3 - 3.82t \quad \text{C}$$

Write an expression for (a) the instantaneous current through the resistor, (b) the instantaneous voltage across the resistor, and (c) the instantaneous power in the resistor. (d) Evaluate each at 2.00 s.

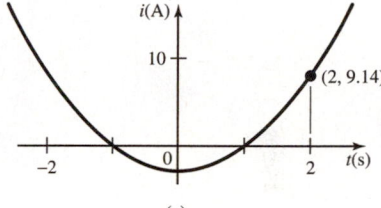

(a)

Solution:

(a) $\qquad i = dq/dt = 3.24t^2 - 3.82 \text{ A}$

(b) By Ohm's law,

$$v = Ri = 2.85(3.24t^2 - 3.82)$$
$$= 9.23t^2 - 10.9 \text{ V}$$

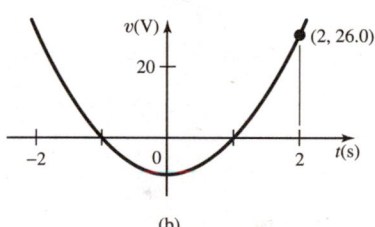

(b)

(c) Since $P = vi$,

$$P = (9.23t^2 - 10.9)(3.24t^2 - 3.82)$$
$$= 29.9t^4 - 70.6t^2 + 41.6 \text{ W}$$

(d) At $t = 2.00$ s,

$$i = 3.24(4.00) - 3.82 = 9.14 \text{ A}$$
$$v = 9.23(4.00) - 10.9 = 26.0 \text{ V}$$
$$P = 29.9(16.0) - 70.6(4.00) + 41.6 = 238 \text{ W}$$

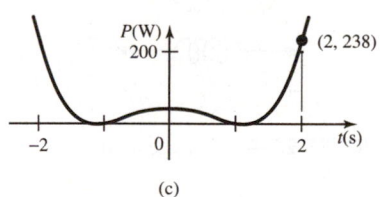

(c)

Graphical Check: Graphs of i, v, and P, given in Fig. 25–2, show these values at $t = 2.00$ s. ◆◆◆

FIGURE 25–2

Current in a Capacitor

If a steady voltage is applied across a capacitor, no current will flow into the capacitor (after the initial transient currents have died down). But if the applied voltage varies with time, the instantaneous current i to the capacitor will be proportional to the rate of change of the voltage. The constant of proportionality is called the *capacitance C*.

Current in a Capacitor	$$i = C\frac{dv}{dt}$$	**1080**
	The current to a capacitor equals the capacitance times the rate of change of the voltage, with respect to time.	

The units here are volts for v, seconds for t, farads for C, and amperes for i.

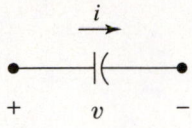

FIGURE 25–3 Current in a capacitor. A microfarad equals 10^{-6} farad.

The sign convention is that the current is assumed to flow in the direction of the voltage drop, as shown in Fig. 25–3. If the current is assumed to be in the direction of the voltage *rise,* then one of the sides in Eq. 1080 is taken as negative.

◆◆◆ **Example 4:** The voltage applied to a 2.85-microfarad (μF) capacitor is $v = 1.47t^2 + 48.3t - 38.2$ V. Find the current at $t = 2.50$ s.

Solution: The derivative of the voltage equation is $dv/dt = 2.94t + 48.3$. Then, from Eq. 1080,

$$i = C\frac{dv}{dt} = (2.85 \times 10^{-6})(2.94t + 48.3) \text{ A}$$

At $t = 2.50$ s,

$$i = (2.85 \times 10^{-6})[2.94(2.50) + 48.3] = 159 \times 10^{-6} \text{ A}$$
$$= 0.159 \text{ mA} \qquad\qquad ◆◆◆$$

Voltage Across an Inductor

If the current through an inductor (such as a coil of wire) is steady, there will be no voltage drop across the inductor. But if the current varies, a voltage will be induced that is proportional to the rate of change of the current. The constant of proportionality L is called the *inductance* and is measured in henrys (H).

Voltage Across an Inductor	$$v = L\frac{di}{dt}$$ *The voltage across an inductor equals the inductance times the rate of change of the current with respect to time.*	**1086**

The sign convention is similar to that for the capacitor: The current is assumed to flow in the direction of the voltage drop, as shown in Fig. 25–4. Otherwise, one term in Eq. 1086 is taken as negative.

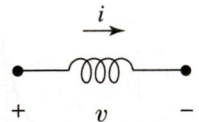

FIGURE 25–4 Voltage across an inductor.

◆◆◆ **Example 5:** The current in a 8.75-H inductor is given by

$$i = \sqrt{t^2 + 5.83t}$$

Find the voltage across the inductor at $t = 5.00$ s.

Solution: By Eq. 1086,

$$v = 8.75\frac{di}{dt}$$
$$= 8.75\left(\frac{1}{2}\right)(t^2 + 5.83t)^{-1/2}(2t + 5.83)$$

At $t = 5.00$ s,

$$v = 8.75\left(\frac{1}{2}\right)[25.0 + 5.83(5.00)]^{-1/2}(10.0 + 5.83)$$
$$= 9.41 \text{ V}$$

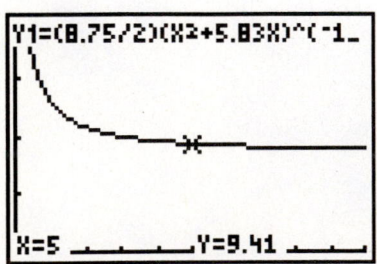

Graphical check for Example 5. Graph of the voltage, showing a point at (5, 9.41). Tick marks are 1 unit apart on the x axis and 5 units apart on the y axis.

Graphical Check: This result is verified graphically as shown.   ◆◆◆

Rate of Change with Respect to Another Variable

While time rates of change are important in technology, so are rates of change with respect to other variables. For example, while the rate of water flowing from a tank can vary with time, it can also vary with the depth of water in the tank.

In problems of this sort, we usually have one variable related to another (other quantities being assumed to be constant) and we want the rate of change of one variable with respect to the other. As usual, we find this by taking the derivative.

◆◆◆ **Example 6:** The pressure p of air in a cylinder is related to the volume v of that air by Boyle's law,

$$p = \frac{k}{v}$$

where k is a constant. (a) Find the rate of change of p with respect to v. (b) Evaluate it when $p = 164$ lb/in.^{2} and $v = 274$ in.^{3}.

Solution: Let us first find k by substituting the given values.

$$k = pv = (164)(274) = 44{,}940$$

(a) Now taking the derivative of p with respect to v,

$$p = kv^{-1}$$

$$\frac{dp}{dv} = -kv^{-2} = -\frac{k}{v^2}$$

(b) Substituting the given values we get

$$\frac{dp}{dv} = -\frac{k}{v^2} = -\frac{44{,}940}{(274)^2}$$

$$= -0.599 \quad (\text{lb/inch}^2)/\text{inch}^3 \qquad \text{◆◆◆}$$

Beam Deflection

◆◆◆ **Example 7:** For the cantilever beam, Fig. 25–5, the deflection y at a distance x from the built-in end is given by

$$y = \frac{Px^2}{6EI}(3L - x)$$

where P is the applied load, E is the modulus of elasticity of the beam material, I is the moment of inertia of the beam's cross-section, and L is the length of the beam. This equation describes the shape of the deflected beam, or the *elastic curve*. A useful quantity for beam analysis is the slope dy/dx of the elastic curve. Find it, taking P, E, I, and L as constants.

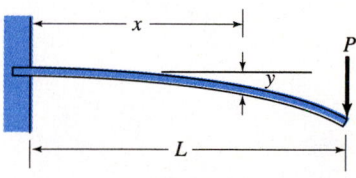

FIGURE 25–5

Solution: Let us remove parentheses and take the derivative term by term.

$$y = \frac{PL}{2EI}x^2 - \frac{P}{6EI}x^3$$

$$\frac{dy}{dx} = \frac{PL}{EI}x - \frac{P}{2EI}x^2$$

$$= \frac{P}{EI}\left(Lx - \frac{x^2}{2}\right)$$

This equation gives us the slope of the elastic curve at any x. ◆◆◆

Exercise 1 ◆ Rate of Change

Rate of Change with Respect to Time

1. The temperature T inside a certain furnace is described by the equation $T = 55.6t^2 + 28.2t + 44.8°$F, where t is the elapsed time in hours. Find the time rate of change of the temperature at $t = 2.00$ h.

2. The pressure p in a tank varies with time according to the function $p = 34.6t^3 - 44.5t$ lb/in.^{2}, where t is in minutes. What is the time rate of change of pressure at $t = 5.50$ min?

3. The quantity q of water flowing in a pipe is given by $q = \dfrac{44.5}{\sqrt{t^2 - 26.4}}$ ft^3/min, where t is the time, in minutes. Find the time rate of change of q at $t = 15.3$ min.

Electric Current

4. The charge q (in coulombs) through a 4.82-Ω resistor varies with time according to the function $q = 3.48t^2 - 1.64t$. Write an expression for the instantaneous current through the resistor.

5. Evaluate the current in problem 4 at $t = 5.92$ s.

6. Find the voltage across the resistor of problem 4. Evaluate it at $t = 1.75$ s.

7. Find the instantaneous power in the resistor of problem 4. Evaluate it at $t = 4.88$ s.

8. The charge q (in coulombs) at a resistor varies with time according to the function $q = 22.4t + 41.6t^3$. Write an expression for the instantaneous current through the resistor, and evaluate it at 2.50 s.

Current in a Capacitor

9. The voltage applied to a 33.5-μF capacitor is $v = 6.27t^2 - 15.3t + 52.2$ V. Find the current at $t = 5.50$ s.

10. The voltage applied to a 1.25-μF capacitor is $v = 3.17 + 28.3t + 29.4t^2$ V. Find the current at $t = 33.2$ s.

Voltage Across an Inductor

11. The current in a 1.44-H inductor is given by $i = 5.22t^2 - 4.02t$. Find the voltage across the inductor at $t = 2.00$ s.

12. The current in a 8.75-H inductor is given by $i = 8.22 + 5.83t^3$. Find the voltage across the inductor at $t = 25.0$ s.

Rate of Change with Respect to Another Variable

13. The air in a certain cylinder is at a pressure of 25.5 lb/in.^{2} when its volume is 146 in.^{3}. Find the rate of change of the pressure with respect to volume as the piston descends farther. Use Boyle's law, $pv = k$.

14. A certain light source produces an illumination of 655 lux on a surface at a distance of 2.75 m. Find the rate of change of illumination with respect to distance, and evaluate it at 2.75 m. Use the inverse square law, $I = k/d^2$.

15. A spherical balloon starts to shrink as the gas escapes. Find the rate of change of its volume with respect to its radius when the radius is 1.00 m. ($V = \dfrac{4}{3}\pi r^3$)

16. The power dissipated in a certain resistor is 865 W at a current of 2.48 A. What is the rate of change of the power with respect to the current as the current starts to increase? Use Eq. 1066, $P = I^2R$.

17. The period (in seconds) for a pendulum of length L in. to complete one oscillation is equal to $P = 0.324\sqrt{L}$. Find the rate of change of the period with respect to length when the length is 9.00 in.

18. The temperature T at a distance x in. from the end of a certain heated bar is given by $T = 2.24x^3 + 1.85x + 95.4$ (°F). Find the rate of change of temperature with respect to distance, which is called the *temperature gradient,* at a point 3.75 in. from the end.

Beam Deflection

19. The cantilever beam in Fig. 25–6 has a deflection y at a distance x from the built-in end of

$$y = \frac{wx^2}{24EI}(x^2 + 6L^2 - 4Lx)$$

where E is the modulus of elasticity and I is the moment of inertia. Write an expression for the rate of change of deflection with respect to the distance x. Regard E, I, w, and L as constants.

20. The equation of the elastic curve for the beam of Fig. 25–7 is

$$y = \frac{wx}{24EI}(L^3 - 2Lx^2 + x^3)$$

Write an expression for the rate of change of deflection (the slope) of the elastic curve at $x = L/4$. Regard E, I, w, and L as constants.

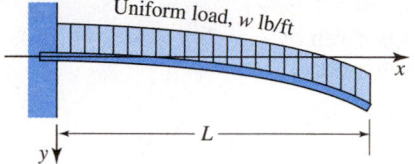

FIGURE 25–6 Cantilever beam with uniform load.

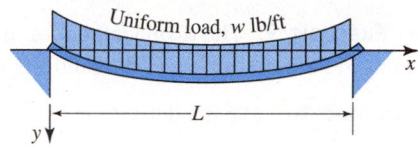

FIGURE 25–7 Simply supported beam with uniform load.

25–2 Motion of a Point

We saw in the preceding section that the rates of change that we often have to find are *with respect to time.* We continue with time rates of change here as we consider the motion of a point. We will find velocity and acceleration of a moving point.

Displacement and Velocity in Straight-Line Motion

Let us first consider *straight-line motion;* later we will study curvilinear motion.

A particle moving along a straight line is said to be in *rectilinear motion.* To define the position P of the particle in Fig. 25–8, we first choose an origin O on the straight line. Then the distance s from O to P is called the *displacement* of the particle. Further, the displacement of the particle at any instant of time t is determined if we have a function $s = f(t)$.

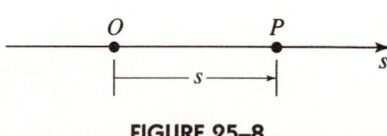

FIGURE 25–8

◆◆◆ **Example 8:** The displacement of a certain particle is given by

$$s = f(t) = 3t^2 + 4t \quad \text{m}$$

where t is in seconds. Find the displacement at 2.15 s.

Solution: Substituting 2.15 for t gives

$$s = 3(2.15)^2 + 4(2.15) = 22.5 \text{ m}$$

◆◆◆

In this section let us assume that the integers in the given equations are exact numbers.

Next we distinguish between speed and velocity. As an object moves along some path, the distance traveled *along the path* per unit time is called the *speed*. No account is taken of any change in direction; hence speed is a *scalar* quantity.

Velocity, on the other hand, is a *vector* quantity, having both *magnitude* and *direction*. The sentence "I drove my car 60 miles per hour on the interstate" is referring to the *speed* of the car. The statement "I drove my car 60 miles per hour and in the due north direction" is referring to the *velocity* of the car.

For an object moving along a curved path, the magnitude of the velocity along the path is equal to the speed, and the *direction* of the velocity is the same as that of the *tangent to the curve* at that point. We will also speak of the components of that velocity in directions other than along the path, usually in the x and y directions.

As with average and instantaneous rates of change, we also can have average speed and average velocity, or instantaneous speed or instantaneous velocity.

Velocity is the rate of change of displacement and hence is given by the derivative of the displacement. If we give displacement the symbol s, then we have the following equation:

Instantaneous Velocity	$v = \dfrac{ds}{dt}$	1023
	The velocity is the rate of change of the displacement.	

◆◆◆ **Example 9:** The displacement of an object is given by $s = 2t^2 + 5t + 4$ (in.), where t is the time in seconds. Find the velocity at 1.00 s.

Solution: We take the derivative

$$v = \frac{ds}{dt} = 4t + 5$$

At $t = 1.00$ s,

$$v(1.00) = \left.\frac{ds}{dt}\right|_{t=1.00} = 4(1.00) + 5 = 9.00 \text{ in./s}$$

Graphical Solution: As with other rate of change problems, we can graph the first derivative and determine its value at the required point. Thus the graph for velocity shows a value of 9.00 in./s at $t = 1.00$ s.   ◆◆◆

Acceleration in Straight-Line Motion

Acceleration is defined as the time rate of change of velocity. Like velocity, it is also a vector quantity. Since the velocity is itself the derivative of the displacement, the acceleration is the derivative of the derivative of the displacement, or the *second derivative* of displacement, with respect to time. We write the second derivative of s with respect to t as follows:

$$\frac{d^2s}{dt^2}$$

Instantaneous Acceleration	$a = \dfrac{dv}{dt} = \dfrac{d^2s}{dt^2}$	1025
	The acceleration is the rate of change of the velocity.	

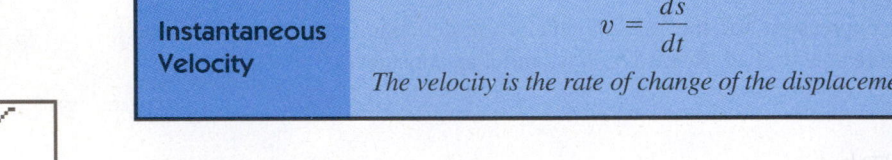

Screen for Example 9. Graph of the given function and the first derivative. At $x = 1.00$ the derivative is 9.00. Tick marks are one-half unit apart on the x axis and 5 units apart on the y axis.

◆◆◆ **Example 10:** One point in a certain mechanism moves according to the equation $s = 3t^3 + 5t - 3$ cm, where t is in seconds. Find the instantaneous velocity and acceleration at $t = 2.00$ s.

Solution: We take the derivative twice with respect to t.

$$v = \frac{ds}{dt} = 9t^2 + 5$$

and

$$a = \frac{dv}{dt} = 18t$$

At $t = 2.00$ s,

$$v(2.00) = 9(2.00)^2 + 5 = 41.0 \text{ cm/s}$$

and

$$a(2.00) = 18(2.00) = 36.0 \text{ cm/s}^2$$

Graphical Check: These values are verified graphically in Fig. 25–9. ◆◆◆

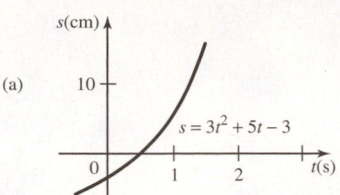

(a)

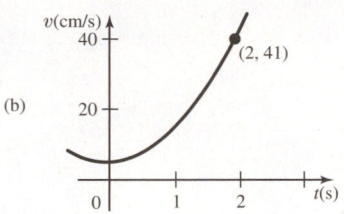

(b)

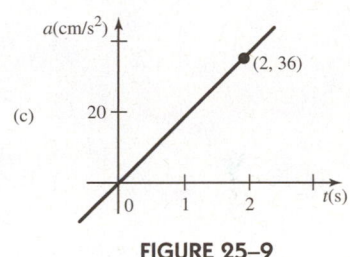

(c)

FIGURE 25–9

Velocity in Curvilinear Motion

Figure 25–10 shows a point moving along a curved path. At any instant we may think of the direction of the point as being *tangent* to the curve. Thus if the speed is known and the direction of the tangent can be found, the *instantaneous velocity* (a vector having both magnitude and direction) can also be found.

A more useful way of giving the instantaneous velocity, however, is by its x and y components (Fig. 25–11). If the magnitude and direction of the velocity are known, the components can be found by resolving the velocity vector into its x and y components.

FIGURE 25–10

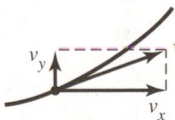

FIGURE 25–11 x and y components of velocity.

◆◆◆ **Example 11:** A point moves along the curve $y = 2x^3 - 5x^2 - 1$ cm.

(a) Find the direction of travel at $x = 2.00$ cm.

(b) If the speed v of the point along the curve is 3.00 cm/s, find the x and y components of the velocity when $x = 2.00$ cm.

Solution:

(a) Taking the derivative of the given function gives

$$dy/dx = 6x^2 - 10x$$

When $x = 2.00$,

$$y'(2.00) = 6(2.00)^2 - 10(2.00) = 4.00$$

The slope of the curve at that point is thus 4.00 and the direction of travel is $\tan^{-1} 4.00 = 76.0°$, as shown in Fig. 25–12.

(b) Resolving the velocity vector v into x and y component gives us

$$v_x = 3.00 \cos 76.0° = 0.726 \text{ cm/s}$$

$$v_y = 3.00 \sin 76.0° = 2.91 \text{ cm/s}$$ ◆◆◆

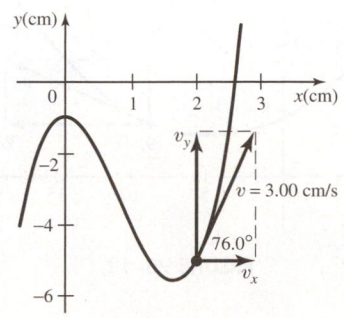

FIGURE 25–12

Displacement Given by Parametric Equations

If the x displacement and y displacement are each given by a separate function of time (parametric equations), we may find the x and y components directly by taking the derivative of each equation.

	(a)	(b)	
x and y Components of Velocity	$v_x = \dfrac{dx}{dt}$	$v_y = \dfrac{dy}{dt}$	**1033**

Once we have expressions for the x and y components of velocity, we simply have to take the derivative again to get the x and y components of acceleration.

	(a)	(b)	
x and y Components of Acceleration	$a_x = \dfrac{dv_x}{dt} = \dfrac{d^2x}{dt^2}$	$a_y = \dfrac{dv_y}{dt} = \dfrac{d^2y}{dt^2}$	**1035**

◆◆◆ **Example 12:** A point moves along a curve such that its horizontal displacement is

$$x = 4t + 5 \quad \text{cm}$$

and its vertical displacement is

$$y = 3 + t^2 \quad \text{cm}$$

Find (a) the horizontal and vertical components of the instantaneous velocity at $t = 1.00$ s and (b) the magnitude and direction of the instantaneous velocity.

Solution:

(a) Using Eqs. 1033, we take derivatives.

$$v_x = 4 \quad \text{and} \quad v_y = 2t$$

At $t = 1.00$ s,

$$v_x = 4.00 \text{ cm/s} \quad \text{and} \quad v_y = 2(1.00) = 2.00 \text{ cm/s}$$

(b) We get the resultant of v_x and v_y by vector addition.

$$v = \sqrt{v_x^2 + v_y^2} = \sqrt{16.0 + 4.00} = 4.47 \text{ cm/s}$$

Finding the direction of the resultant, we have

$$\tan \theta = v_y/v_x = 2.00/4.00 = 1/2$$
$$\theta = 26.6°$$

Figure 25–13 shows a parametric plot of the given equations. At $t = 1.00$ s, the moving point is at (9, 4). The figure also shows the velocity tangent to the curve and the x and y components of the velocity.

The screens for obtaining the parametric plot by calculator are shown. The calculator must be put into PARAMETRIC mode before entering the functions in the $\boxed{Y=}$ screen.

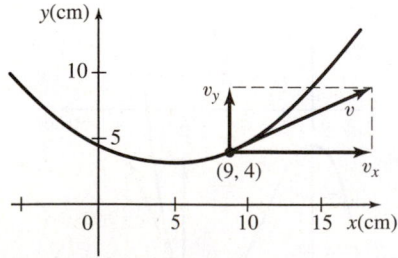

FIGURE 25–13

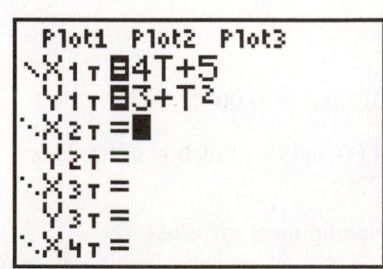

(a) TI-83/84 check for Example 12.

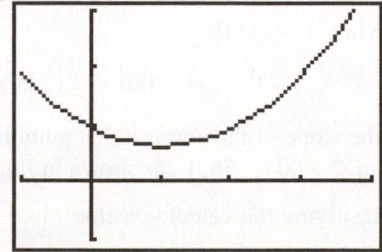

(b) Parametric plot of the given functions. Tick marks are 5 units apart.   ◆◆◆

Trajectories

A *trajectory* is the path followed by a projectile, such as a ball thrown into the air. Trajectories are usually described by parametric equations, with the horizontal and vertical motions considered separately.

If air resistance is neglected, a projectile will move horizontally with constant velocity, and will fall with constant acceleration, just like any other freely falling body. Thus if the initial velocities are v_{0x} and v_{0y} in the horizontal and vertical directions, the parametric equations of motion are then

$$x = v_{0x}t \quad \text{and} \quad y = v_{0y}t - \frac{gt^2}{2}$$

Here, the term $-gt^2/2$ accounts for the downward pull of gravity, without which the object would continue traveling upward.

◆◆◆ **Example 13:** A projectile is launched with initial velocity v_0 at an angle θ (Fig. 25–14). The horizontal and vertical components of the vector v_0 are

$$v_{0x} = v_0 \cos \theta$$

$$v_{0y} = v_0 \sin \theta$$

so the parametric equations of motion are

$$x = (v_0 \cos \theta)t \quad \text{and} \quad y = (v_0 \sin \theta)t - \frac{gt^2}{2}$$

Find the horizontal and vertical velocities and accelerations.

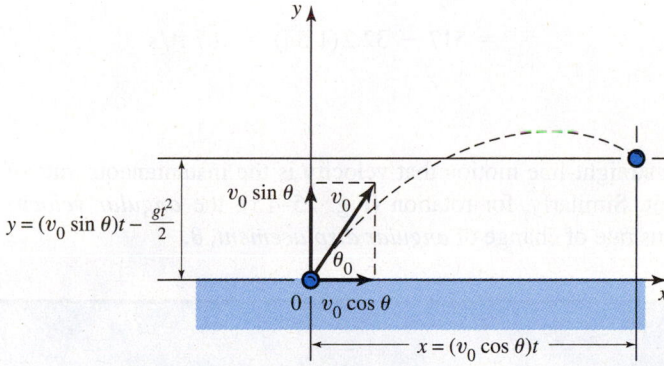

FIGURE 25–14 A trajectory.

Solution: We take derivatives, remembering that v_0, $\cos \theta$, and $\sin \theta$ are constants.

$$v_x = \frac{dx}{dt} = v_0 \cos \theta \quad \text{and} \quad v_y = \frac{dy}{dt} = v_0 \sin \theta - gt$$

We take derivatives again to get the accelerations.

$$a_x = \frac{dv_x}{dt} = 0 \quad \text{and} \quad a_y = \frac{dv_y}{dt} = -g$$

As expected, we get a horizontal motion with constant velocity and a vertical motion with constant acceleration. ◆◆◆

◆◆◆ **Example 14:** A projectile is launched at an angle of 62.0° to the horizontal with an initial velocity of 585 ft/s. Find (a) the horizontal and vertical positions of the projectile and (b) the horizontal and vertical velocities after 1.55 s.

Solution:

(a) We first resolve the initial velocity into horizontal and vertical components.

$$v_{0x} = v_0 \cos \theta = 585 \cos 62.0° = 275 \text{ ft/s}$$
$$v_{0y} = v_0 \sin \theta = 585 \sin 62.0° = 517 \text{ ft/s}$$

The horizontal and vertical displacements are then

$$x = v_{0x}t = 275t \quad \text{ft}$$

and

$$y = v_{0y}t - gt^2/2 = 517t - gt^2/2 \quad \text{ft}$$

At $t = 1.55$ s, and with $g = 32.2 \text{ ft/s}^2$,

$$x = 275(1.55) = 426 \text{ ft}$$

and

$$y = 517(1.55) - 32.2(1.55)^2/2 = 763 \text{ ft}$$

(b) The horizontal velocity is constant, so

$$v_x = v_{0x} = 275 \text{ ft/s}$$

and the vertical velocity is

$$v_y = v_{0y} - gt$$
$$= 517 - 32.2(1.55) = 467 \text{ ft/s} \qquad \text{◆◆◆}$$

Rotation

We saw for straight-line motion that velocity is the instantaneous rate of change of displacement. Similarly, for rotation (Fig. 25–15), the *angular velocity, ω*, is the instantaneous rate of change of *angular displacement, θ*.

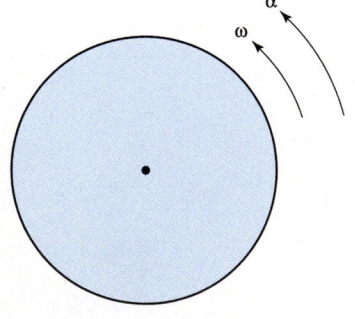

FIGURE 25–15

Angular Velocity	$\omega = \dfrac{d\theta}{dt}$ *The angular velocity is the instantaneous rate of change of the angular displacement with respect to time.*	1029

Similarly for acceleration:

Angular Acceleration	$\alpha = \dfrac{d\omega}{dt} = \dfrac{d^2\theta}{dt^2}$ *The angular acceleration is the instantaneous rate of change of the angular velocity with respect to time.*	1031

◆◆◆ **Example 15:** The angular displacement of a rotating body is given by $\theta = 1.75t^3 + 2.88t^2 + 4.88$ rad. Find (a) the angular velocity and (b) the angular acceleration at $t = 2.00$ s.

Solution:

(a) From Eq. 1029,

$$\omega = \frac{d\theta}{dt} = 5.25t^2 + 5.76t$$

At 2.00 s, $\omega = 5.25\,(4.00) + 5.76\,(2.00) = 32.5$ rad/s.

(b) From Eq. 1031,

$$\alpha = \frac{d\omega}{dt} = 10.5t + 5.76$$

At 2.00 s, $\alpha = 10.5\,(2.00) + 5.76 = 26.8$ rad/s^2.

Graphical Check: These results are verified graphically in Fig. 25–16. ◆◆◆

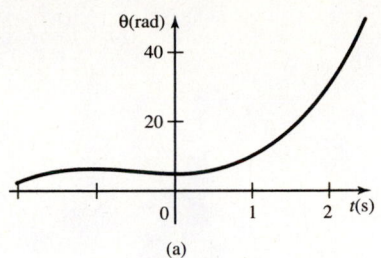

(a)

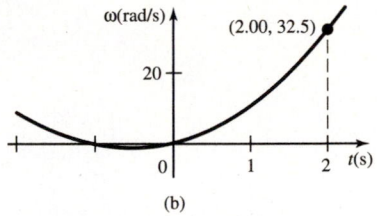

(b)

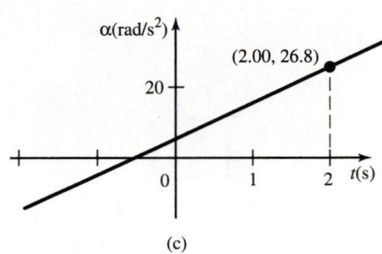

(c)

FIGURE 25–16

Solve or verify some of these problems graphically.

Exercise 2 ◆ Motion of a Point

Straight-Line Motion

Find the instantaneous velocity and acceleration at the given time for the straight-line motion described by each equation, where s is in centimeters and t is in seconds. In this exercise assume that the integers in the given equations are exact numbers and give approximate answers to three significant digits.

1. $s = 32t - 8t^2$ at $t = 2.00$
2. $s = 6t^2 - 2t^3$ at $t = 1.00$
3. $s = t^2 + t^{-1} + 3$ at $t = 0.500$
4. $s = (t + 1)^4 - 3(t + 1)^3$ at $t = -1.00$
5. $s = 120t - 16t^2$ at $t = 4.00$
6. $s = 3t - t^4 - 8$ at $t = 1.00$

7. The distance in feet traveled in time t seconds by a point moving in a straight line is given by the formula $s = 40t + 16t^2$. Find the velocity and the acceleration at the end of 2.00 s.

8. A car moves according to the equation $s = 250t^2 - \frac{5}{4}t^4$, where t is measured in minutes and s in feet.
   (a) How far does the car go in the first 10.0 min?
   (b) What is the maximum speed?
   (c) How far has the car moved when its maximum speed is reached?

9. If the distance traveled by a ball rolling down an incline in t seconds is s feet, where $s = 6t^2$, find its speed when $t = 5.00$ s.

10. The height s in feet reached by a ball t seconds after being thrown vertically upward at 320 ft/s is given by $s = 320t - 16t^2$. Find (a) the greatest height reached by the ball and (b) the velocity with which it reaches the ground.

11. A rocket was fired straight upward so that its height in feet after t seconds was $s = 2000t - 16t^2$.
    (a) What was its initial velocity?
    (b) What was its greatest height?
    (c) What was its velocity at the end of 10.0 s?

12. The height h in kilometers to which a balloon will rise in t minutes is given by the formula

$$h = \frac{10t}{\sqrt{4000 + t^2}}$$

At what rate is the balloon rising at the end of 30.0 min?

13. If the equation of motion of a point is $s = 16t^2 - 64t + 64$, find the position and acceleration at which the point first comes to rest.

Curvilinear Motion

14. A point moves along the curve $y = 2x^3 - 3x^2 - 2$ cm.
   (a) Find the direction of travel at $x = 1.50$ cm.
   (b) If the speed of the point along the curve is 3.75 cm/s, find the x and y components of the velocity when $x = 1.50$ cm.

15. A point moves along the curve $y = x^4 + x^2$ in.
   (a) Find the direction of travel at $x = 2.55$ in.
   (b) If the speed of the point along the curve is 1.25 in./s, find the x and y components of the velocity when $x = 2.55$ in.

Equations Given in Parametric Form

16. A point moves along a curve such that its horizontal displacement is $x = 3t^2 + 5t$ cm and its vertical displacement is $y = 13 - 3t^2$ cm. Find the horizontal and vertical components of the instantaneous velocity and acceleration at $t = 4.55$ s.

17. Find the magnitude and direction of the resultant velocity in problem 16.

18. A point has horizontal and vertical displacements (in cm) of $x = 4 - 2t^2$ and $y = 5t^2 + 3$, respectively.
   (a) Find the x and y components of the velocity and acceleration at $t = 2.75$ s.
   (b) Find the magnitude and direction of the resultant velocity.

Trajectories

19. A projectile is launched at an angle of 43.0° to the horizontal with an initial velocity of 6350 ft/s. Find (a) the horizontal and vertical positions of the projectile and (b) the horizontal and vertical velocities, after 7.00 s.

20. A projectile is launched at an angle of 27.0° to the horizontal with an initial velocity of 1260 ft/s. Find (a) the horizontal and vertical positions of the projectile and (b) the horizontal and vertical velocities, after 3.25 s.

Rotation

21. The angular displacement of a rotating body is given by $\theta = 44.8t^3 + 29.3t^2 + 81.5$ rad. Find the angular velocity at $t = 4.25$ s.

22. Find the angular acceleration in problem 21 at $t = 22.4$ s.

23. The angular displacement of a rotating body is given by $\theta = 184 + 271t^3$ rad. Find (a) the angular velocity and (b) the angular acceleration, at $t = 1.25$ s.

24. The angular displacement of a rotating body is given by $\theta = 2.84t^3 - 7.25$ rad. Find (a) the angular velocity and (b) the angular acceleration, at $t = 4.82$ s.

25–3 Related Rates

In *related rate* applications, there are *two* quantities changing with time. The rate of change of one of the quantities is given, and the other must be found. A procedure that can be followed is

> 1. Locate the *given rate*. Since it is a rate, it can be *expressed as a derivative* with respect to time.
> 2. Determine the *unknown* rate. Express it also as a derivative with respect to time.
> 3. Find an *equation* linking the variable in the given rate with that in the unknown rate. If there are other variables in the equation, they must be eliminated by means of other relationships.
> 4. Take the derivative of the equation *with respect to time*.
> 5. Substitute the given values and solve for the unknown rate.

You would, of course, study the problem statement and make a diagram, as you would for other applications.

◆◆◆ **Example 16:** A 20.0-ft ladder leans against a building (Fig. 25–17). The foot of the ladder is pulled away from the building at a rate of 2.00 ft/s. How fast is the top of the ladder falling when its foot is 10.0 ft from the building?

Estimate: Note that when the foot of the ladder is 10 ft from the wall, the ladder's angle is arccos (10/20) or 60°. If the ladder is at 45°, the top and foot should move at the same speed; when the angle is steeper, the top should move more slowly than the foot. Thus we expect the top to move at a speed less than 2 ft/s. Also, since y is decreasing, we expect the rate of change of y to be negative.

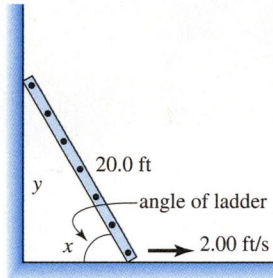

FIGURE 25–17 The ladder problem.

Solution:

1. If we let x be the distance from the foot of the ladder to the building, we have

$$\frac{dx}{dt} = 2.00 \text{ ft/s}$$

2. If y is the distance from the ground to the top of the ladder, we are looking for dy/dt.
3. The equation linking x and y is the Pythagorean theorem.

$$x^2 + y^2 = 20.0^2$$

Solving for y,

$$y = (400 - x^2)^{1/2}$$

4. Taking the derivative with respect to t,

$$\frac{dy}{dt} = \frac{1}{2}(400 - x^2)^{-1/2}(-2x)\frac{dx}{dt}$$

5. Substituting $x = 10.0$ ft and $dx/dt = 2.00$ ft/s gives

$$\frac{dy}{dt} = \frac{1}{2}(400 - 100)^{-1/2}(-20.0)(2.00) = -1.15 \text{ ft/s}$$

The negative sign indicates that y is decreasing.

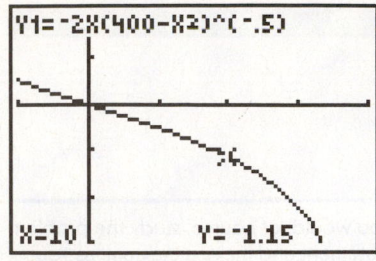

TI-83/84 screen for Example 16. Tick marks are 5 units apart on the x axis and 1 unit apart on the vertical axis.

Graphical Solution: The derivative dy/dt, after substituting 2.00 for dx/dt, is

$$dy/dt = -2x(400 - x^2)^{-1/2}$$

We plot this derivative as shown, and we graphically determine that the value of dy/dt at $x = 10.0$ ft is -1.15 ft/s.

Alternate Solution: Steps 1 through 3 are the same as above, but instead of solving for y as we did in step 3, we now take the derivative *implicitly*. It will often be easier to take the derivative implicitly rather than first to solve for one of the variables.

$$2x\frac{dx}{dt} + 2y\frac{dy}{dt} = 0$$

When $x = 10.0$ ft, the height of the top of the ladder is

$$y = \sqrt{400 - (10.0)^2} = 17.3 \text{ ft}$$

We now substitute 2.00 for dx/dt, 10.0 for x, and 17.3 ft for y.

$$2(10.0)(2.00) + 2(17.3)\frac{dy}{dt} = 0$$

$$\frac{dy}{dt} = -1.15 \text{ ft/s} \qquad \blacklozenge\blacklozenge\blacklozenge$$

Common Errors

In step 4 of Example 16, when taking the derivative of x^2, it is tempting to take the derivative with respect to x, rather than t. Also, don't forget the dx/dt in the derivative.

$$\frac{d}{dt}x^2 = 2x\boxed{\frac{dx}{dt}}$$

↑ — Don't forget!

Students often substitute the given values too soon. For example, if we had substituted $x = 10.0$ and $y = 17.3$ before taking the derivative, we would have gotten

$$(10.0)^2 + (17.3)^2 = (20.0)^2$$

Taking the derivative now gives us

$$0 = 0!$$

Do not substitute the given values until *after* you have taken the derivative.

In the next example when we find an equation linking the variables, it contains *three* variables. In such a case we need a *second equation* with which to eliminate one variable.

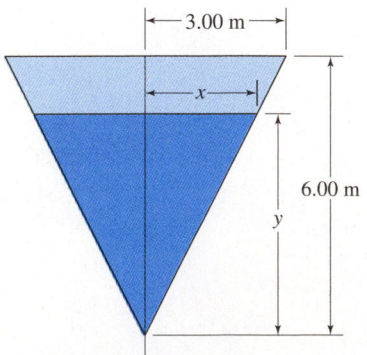

FIGURE 25–18   Conical tank.

◆◆◆ **Example 17:** A conical tank with vertex down has a base radius of 3.00 m and a height of 6.00 m (Fig. 25–18). Water flows in at a rate of 2.00 m³/h. How fast is the water level rising when the depth y is 3.00 m?

Estimate: Suppose that our tank was *cylindrical* with a radius of 1.5 m and cross-sectional area of $\pi(1.5)^2 \approx 7 \text{ m}^2$. The water would rise at a constant rate, equal to the incoming flow rate divided by the cross-sectional area of the tank, or

$$2 \text{ m}^3/\text{h} \div 7 \text{ m}^2 \approx 0.3 \text{ m/h}$$

The conical tank should have about this value when half full, as in our example, less where the tank is wider and greater where narrower.

Solution: Let x equal the base radius and V the volume of water when the tank is partially filled. Then

1. Given: $dV/dt = 2.00 \text{ m}^3/\text{h}$.

2. Unknown: dy/dt when $y = 3.00$ m.

3. The equation linking V and y is that for the volume of a cone, $V = (\pi/3)x^2 y$. But, in addition to the two variables in our derivatives, V and y, we have the third variable, x. We must eliminate x by means of another equation. By similar triangles,

$$\frac{x}{y} = \frac{3}{6}$$

from which $x = \dfrac{y}{2}$. Substituting yields

$$V = \frac{\pi}{3} \cdot \frac{y^2}{4} \cdot y = \frac{\pi}{12} y^3$$

4. Now we have V as a function of y only. Taking the derivative gives us

$$\frac{dV}{dt} = \frac{\pi}{4} y^2 \frac{dy}{dt}$$

5. Substituting 3.00 for y and 2.00 for dV/dt, we obtain

$$2.00 = \frac{\pi}{4} (9.00) \frac{dy}{dt}$$

$$\frac{dy}{dt} = \frac{8.00}{9.00\pi} = 0.283 \text{ m/h}$$

This agrees well with our estimate.

Graphical Solution: The derivative dy/dt, after substituting 2.00 for dV/dt, is

$$\frac{dy}{dt} = 2\left(\frac{4}{\pi y^2}\right) = \frac{8}{\pi y^2}$$

We plot dy/dt as shown and graphically find that dy/dt is 0.283 when y is 3.00 m. ◆◆◆

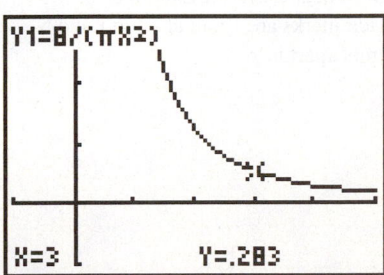

Screen for Example 17. We need to redefine our axes for this graph. The horizontal axis, usually x, is now y. The vertical axis, usually y, is now dy/dt. For our equation, instead of $dy/dt = 8/\pi y^2$, we enter $y = 8/\pi x^2$.

Common Error	You cannot write an equation linking the variables until you have *defined* those variables. Indicate them right on your diagram. Draw axes if needed.

Two Moving Objects

Some applications have two *independently* moving objects, as in the following example.

◆◆◆ **Example 18:** Ship A leaves a port P and travels west at 11.5 mi/h. After 2.25 h, ship B leaves P and travels north at 19.4 mi/h. How fast are the ships separating 5.00 h after the departure of A?

Solution: Figure 25–19 shows the ships t hours after A has left. Ship A has gone $11.5t$ mi and B has gone $19.4(t - 2.25)$ mi. The distance S between them is given by

$$S^2 = (11.5t)^2 + [19.4(t - 2.25)]^2$$
$$= 132t^2 + 376(t - 2.25)^2$$

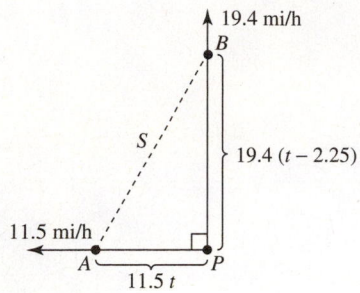

FIGURE 25–19 Note that we give the position of each ship after t hours have elapsed.

We could take the square root of both sides before taking the derivative, but it is sometimes easier to take the derivative first. We then have

$$2S\frac{dS}{dt} = 2(132)t + 2(376)(t - 2.25)$$

$$\frac{dS}{dt} = \frac{132t + 376t - 846}{S} = \frac{508t - 846}{S}$$

At $t = 5.00$ h, A has gone $11.5(5.00) = 57.5$ mi, and B has gone

$$19.4(5.00 - 2.25) = 53.35 \text{ mi}$$

The distance S between them is then

$$S = \sqrt{(57.5)^2 + (53.35)^2} = 78.4 \text{ mi}$$

Substituting gives

$$\frac{dS}{dt} = \frac{508(5.00) - 846}{78.4} = 21.6 \text{ mi/h}$$

Graphical Check: After substituting $S = 78.4$ mi/h, our derivative is

$$\frac{dS}{dt} = 6.48t - 10.8$$

A graph of dS/dt shows a value of 21.6 at $t = 5.00$ s. ◆◆◆

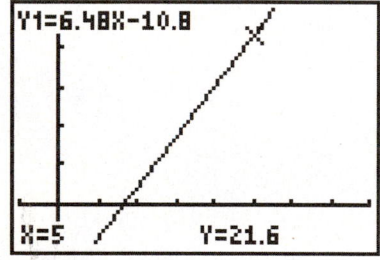

Graphical check for Example 18. Tick marks are 1 unit apart in x and 5 units apart in y.

Common Error	Be sure to distinguish which quantities in a problem are *constants* and which are *variables*. Represent each variable by a letter, and do not substitute a given numerical value for a variable *until the very last step*.

Exercise 3 ◆ Related Rates

One Moving Object

1. An airplane flying horizontally at a height of 8000 m and at a rate of 100 m/s passes directly over a pond. How fast is its straight-line distance from the pond increasing 1 min later?

2. A ship moving 30.0 mi/h is 6.00 mi from a straight beach and is moving parallel to the beach. How fast is the ship approaching a lighthouse on the beach when 10.0 mi (straight-line distance) from it?

3. A person is running at the rate of 8.00 mi/h on a horizontal street directly toward the foot of a tower 100 ft high. How fast is the person approaching the top of the tower when 50.0 ft from the foot?

Ropes and Cables

4. A boat is fastened to a rope that is wound around a winch 20.0 ft above the level at which the rope is attached to the boat. The boat is drifting away at the horizontal rate of 8.00 ft/s. How fast is the rope increasing in length when 30.0 feet of rope is out?

5. A boat with its anchor on the bottom at a depth of 40.0 m is drifting away from the anchor at 4.00 m/s, while the anchor cable slips from the boat at water level. At what rate is the cable leaving the boat when 50.0 meters of cable is out? Assume that the cable is straight.

6. A kite is at a constant height of 120 ft and moves horizontally, at 4.00 mi/h, in a straight line away from the person holding the cord. Assuming that the cord remains straight, how fast is the cord being paid out when its length is 130 ft?

7. A rope runs over a pulley at A and is attached at B as shown in Fig. 25–20. The rope is being wound in at the rate of 4.00 ft/s. How fast is B rising when AB is horizontal?

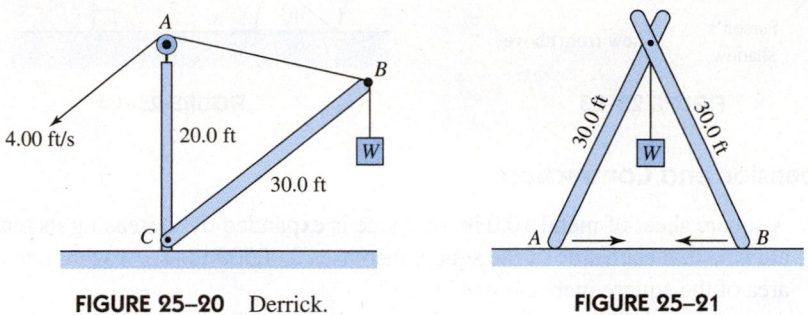

FIGURE 25–20 Derrick.   **FIGURE 25–21**

8. A weight W is being lifted between two poles as shown in Fig. 25–21. How fast is W being raised when A and B are 20.0 ft apart if they are being drawn together, each moving at the rate of 9.00 in./s?

9. A bucket is raised by a person who walks away from the building at 12.0 in./s (Fig. 25–22). How fast is the bucket rising when $x = 80.0$ in.?

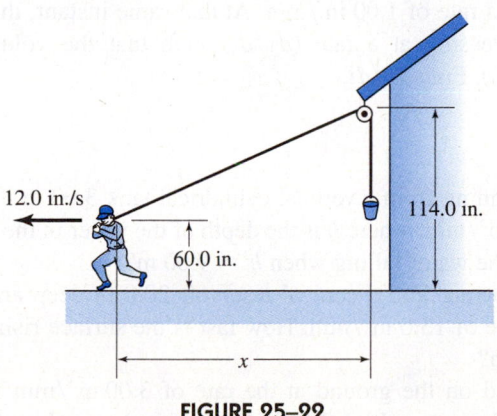

FIGURE 25–22

Moving Shadows

10. A light is 100 ft from a wall (Fig. 25–23). A person runs at 13.0 ft/s away from the wall. Find the speed of the shadow on the wall when the person's distance from the wall is 50.0 ft.

11. A lamp is located on the ground 30.0 ft from a building. A person 6.00 ft tall walks from the light toward the building at a rate of 5.00 ft/s. Find the rate at which the person's shadow on the wall is shortening when the person is 15.0 ft from the building.

12. A ball dropped from a height of 100 ft is at height s at t seconds, where $s = 100 - gt^2/2$ ft (Fig. 25–24). The sun, at an altitude of 40°, casts a shadow of the ball on the ground. Find the rate dx/dt at which the shadow is traveling along the ground when the ball has fallen 50.0 ft. Remember, the sun is so far away that the angle of the ball to the sun will always be 40°.

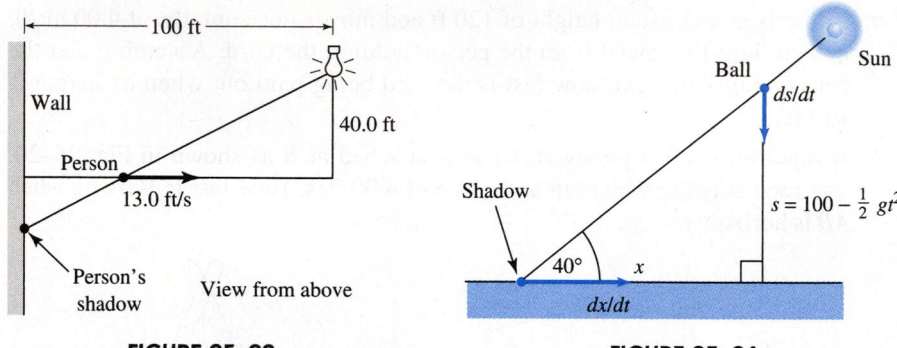

FIGURE 25–23   **FIGURE 25–24**

Expansion and Contraction

13. A square sheet of metal 10.0 in. on a side is expanded by increasing its temperature so that each side of the square increases 0.00500 in./s. At what rate is the area of the square increasing at 20.0 s?

14. A circular plate in a furnace is expanding so that its radius is changing 0.010 cm/s. How fast is the area of one face changing when the radius is 5.00 cm?

15. The volume of a cube is increasing at 10.0 in.³/min. At the instant when its volume is 125 in.³, what is the rate of change of its edge?

16. The edge of an expanding cube is changing at the rate of 0.00300 in./s. Find the rate of change of its volume when its edge is 5.00 in. long.

17. At some instant the diameter x of a cylinder (Fig. 25–25) is 10.0 in. and is increasing at a rate of 1.00 in./min. At that same instant, the height y is 20.0 in. and is decreasing at a rate (dy/dt) such that the volume is not changing $(dV/dt = 0)$. Find dy/dt.

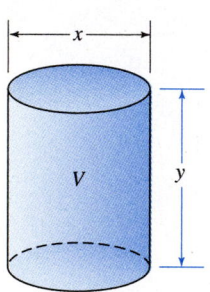

FIGURE 25–25

Fluid Flow

18. Water is running from a vertical cylindrical tank 3.00 m in diameter at the rate of $3\pi\sqrt{h}$ m³/min, where h is the depth of the water in the tank. How fast is the surface of the water falling when $h = 9.00$ m?

19. Water is flowing into a conical reservoir 20.0 m deep and 10.0 m across the top, at a rate of 15.0 m³/min. How fast is the surface rising when the water is 8.00 m deep?

20. Sand poured on the ground at the rate of 3.00 m³/min forms a conical pile whose height is one-third the diameter of its base. How fast is the altitude of the pile increasing when the radius of its base is 2.00 m?

21. A horizontal trough 10.0 ft long has ends in the shape of an isosceles right triangle (Fig. 25–26). If water is poured into it at the rate of 8.00 ft³/min, at what rate is the surface of the water rising when the water is 2.00 ft deep?

Gas Laws

22. An expanding tank contains 734,000 cubic inches of gas at a particular instant, with a pressure of 5.25 lb/in.^{2}. As the volume of the tank increases the pressure decreases at the rate of 0.575 lb/in.^{2} per hour. Find the rate of increase of the volume. Use Boyle's law, which says that the pressure times the volume is constant.

23. The adiabatic law for the expansion of air is $pv^{1.4} = C$. If at a given time the volume is observed to be 10.0 ft^3, and the pressure is 50.0 lb/in.^{2}, at what rate is the pressure changing if the volume is decreasing 1.00 ft^3/s?

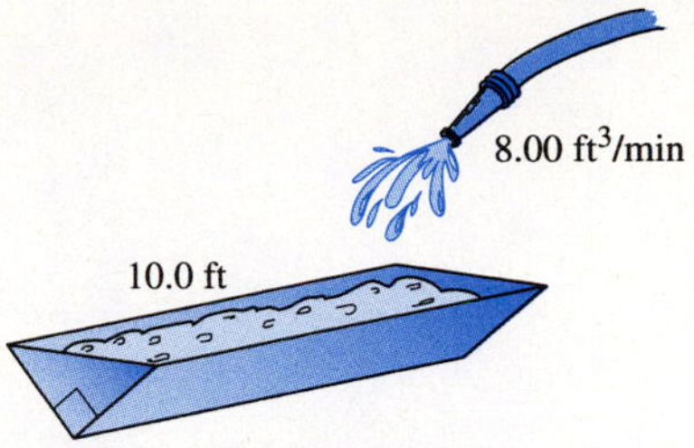

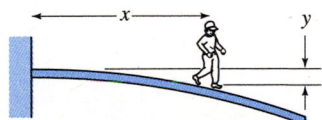

FIGURE 25–26

Two Moving Objects

24. Two trains start from the same point at the same time, one going east at a rate of 40.0 mi/h and the other going south at 60.0 mi/h. Find the rate at which they are separating after 1.00 h of travel.

25. An airplane leaves a field at noon and flies east at 100 km/h. A second airplane leaves the same field at 1 P.M. and flies south at 150 km/h. How fast are the airplanes separating at 2 P.M.?

26. An elevated train on a track 30.0 m above the ground crosses a street (which is at right angles to the track) at the rate of 20.0 m/s. At that instant, an automobile, approaching at the rate of 30.0 m/s, is 40.0 m from a point directly beneath the track. Find how fast the train and the automobile are separating 2.00 s later.

27. As a person is jogging over a bridge at 5.00 ft/s, a boat, 30.0 ft beneath the jogger and moving perpendicular to the bridge, passes directly underneath at 10.0 ft/s. How fast are the person and the boat separating 3.00 s later?

Miscellaneous

28. The speed v ft/s of a certain bullet passing through wood is given by $v = 500\sqrt{1 - 3x}$, where x is the depth in feet. Find the rate at which the speed is decreasing after the bullet has penetrated 3.00 in. (*Hint:* When substituting for dx/dt, simply use the given expression for v.)

29. As a man walks a distance x along a board (Fig. 25–27), he sinks a distance of y in., where

$$y = \frac{Px^3}{3EI}$$

Here, P is the person's weight, 165 lb; E is the modulus of elasticity of the material in the board, 1,320,000 lb/in.^{2}; and I is the modulus of elasticity of the cross section, 10.9 in.^{4}. If he moves at the rate of 25.0 in./s, how fast is he sinking when $x = 75.0$ in.?

30. A stone dropped into a calm lake causes a series of circular ripples. The radius of the outer one increases at 2.00 ft/s. How rapidly is the disturbed area changing at the end of 3.00 s?

FIGURE 25–27

25–4   Optimization

In an earlier chapter, we found *extreme points*, the peaks and valleys on a curve. We did this by finding the points where the slope (and hence the first derivative) was zero. We now apply the same idea to problems in which we find, for example, the point of minimum cost, or the point of maximum efficiency, or the point of maximum carrying capacity.

> ### Suggested Steps for Optimization Problems
>
> 1. Locate the quantity (which we will call Q) to be maximized or minimized, and locate the independent variable, say, x, which is to be varied in order to maximize or minimize Q.
> 2. Write an equation linking Q and x. If this equation contains another variable as well, that variable must be eliminated by means of a second equation. A graph of $Q = f(x)$ will show any maximum or minimum points.
> 3. Take the derivative dQ/dx.
> 4. Locate the values of x for which the derivative is zero. This may be done graphically by finding the zeros for the graph of the derivative or analytically by setting the derivative to zero and solving for x.
> 5. Check any extreme points found to see if they are maxima or minima. This can be done simply by looking at your graph of $Q = f(x)$, by applying your knowledge of the physical problem, or by using the first- or second-derivative tests. Also check whether the maximum or minimum value you seek is at one of the endpoints. An endpoint can be a maximum or minimum point in the given interval, even though the slope there is not zero.

A list of general suggestions such as these is usually so vague as to be useless without examples. Our first example is a number puzzle in which the relation between the variables is given verbally in the problem statement.

◆◆◆ **Example 19:** What two positive numbers whose product is 100 have the least possible sum?

Solution:

(1) We want to minimize the sum S of the two numbers by varying one of them, which we call x. Then since their product is 100,

$$\frac{100}{x} = \text{other number}$$

(2) The sum of the two numbers is

$$S = x + \frac{100}{x}$$

(3) Taking the derivative yields

$$\frac{dS}{dx} = 1 - \frac{100}{x^2}$$

(4) Setting the derivative to zero and solving for x gives us

$$x^2 = 100$$
$$x = \pm 10$$

Since we are asked for positive numbers, we discard the -10. The other number is $100/10 = 10$.

(5) But have we found those numbers that will give a *minimum* sum, as requested, or a *maximum* sum? We can check this by means of the second-derivative test. Taking the second derivative, we have

$$S'' = \frac{200}{x^3}$$

When $x = 10$,

$$S''(10) = \frac{200}{1000} = 0.2$$

which is *positive*, indicating a *minimum* point.

Graphical Solution: A graph of S versus x, shown in screen (1), shows a minimum at $x = 10$. Another way to locate the minimum is to graph the derivative, screen (2) and note where it crosses the x axis.

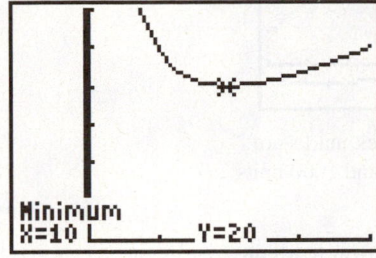

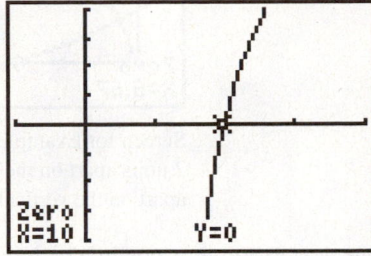

(1) Screen for Example 19. Tick marks are 5 units apart on the x axis and 5 units apart on the vertical axis.

(2) Tick marks are 5 units apart on the x axis and 0.1 unit apart on the vertical axis.

◆◆◆

In the next example, the equation linking the variables is easily written from the geometrical relationships in the problem.

◆◆◆ **Example 20:** An open-top box is to be made from a square of sheet metal 40 cm on a side by cutting a square from each corner and bending up the sides along the dashed lines (Fig. 25–28). Find the dimension x of the cutout that will result in a box of the greatest volume.

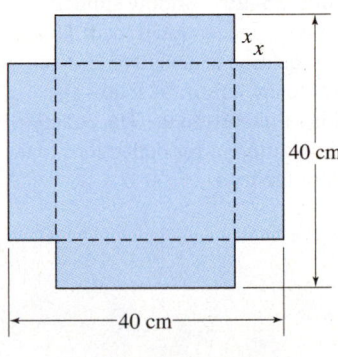

FIGURE 25–28

Solution:

(1) We want to maximize the volume V by varying x.

(2) The equation is

$$V = \text{length} \cdot \text{width} \cdot \text{depth}$$
$$= (40 - 2x)(40 - 2x)x$$
$$= x(40 - 2x)^2$$

(3) Taking the derivative, we obtain

$$\frac{dV}{dx} = x(2)(40 - 2x)(-2) + (40 - 2x)^2$$
$$= -4x(40 - 2x) + (40 - 2x)^2$$
$$= (40 - 2x)(-4x + 40 - 2x)$$
$$= (40 - 2x)(40 - 6x)$$

(4) Setting the derivative to zero and solving for x, we have

$$40 - 2x = 0 \qquad \Big| \qquad 40 - 6x = 0$$
$$x = 20 \text{ cm} \qquad \Big| \qquad x = \frac{20}{3} = 6.67 \text{ cm}$$

We discard $x = 20$ cm, because it is a minimum value and results in the entire sheet of metal being cut away. Thus we keep $x = 6.67$ cm as our answer.

Graphical Check: The graph of the volume equation, shown in screen (1), shows a maximum at $x \approx 6.67$, so we don't need a further test of that point. The graph of the derivative has a zero at $x = 6.67$, verifying our solution.

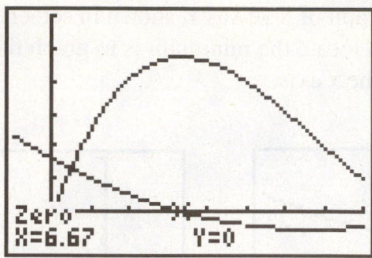

Screen for Example 20. Tick marks are
2 units apart on the *x* axis and 1000 units
apart on the vertical axis.

◆◆◆

Our next example is one in which the equation is given.

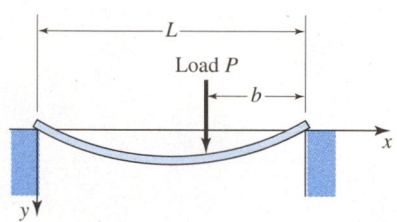

FIGURE 25–29 Simply supported
beam with concentrated load. The
given equation is actually valid only for
points to the left of the load—the
region of interest to us. The equation is
slightly different for deflections to the
right of the load.

◆◆◆ **Example 21:** The deflection *y* of a simply supported beam with a concentrated
load *P* (Fig. 25–29) at a distance *x* from the left end of the beam is given by

$$y = \frac{Pbx}{6LEI}(L^2 - x^2 - b^2)$$

where *E* is the modulus of elasticity, and *I* is the moment of inertia of the
beam's cross section. Find the value of *x* at which the deflection is a maximum for a
20.0-ft-long beam with a concentrated load 5.00 ft from the right end.

Estimate: It seems reasonable that the maximum deflection would be at the center,
at $x = 10$ ft from the left end of the beam. But it is equally reasonable that the maximum deflection would be at the concentrated load, at $x = 15$. So our best guess
might be that the maximum deflection is somewhere between 10 and 15 ft.

Solution:

(1) We want to find the distance *x* from the left end at which the deflection is a
     maximum.
(2) The equation is given in the problem statement.
(3) We take the derivative using the product rule, noting that every quantity but *x*
     or *y* is a constant.

$$\frac{dy}{dx} = \frac{Pb}{6LEI}[x(-2x) + (L^2 - x^2 - b^2)(1)]$$

(4) We set this derivative equal to zero and solve for *x*.

$$2x^2 = L^2 - x^2 - b^2$$
$$3x^2 = L^2 - b^2$$
$$x = \pm\sqrt{\frac{L^2 - b^2}{3}}$$

We drop the negative value, since *x* cannot be negative in this problem. Now
substituting $L = 20.0$ ft and $b = 5.00$ ft gives us

$$x = \sqrt{\frac{400 - 25}{3}} = 11.2 \text{ ft}$$

Thus the maximum deflection occurs between the load and the midpoint of the
beam, as expected.

(5) It is clear from the physical problem that our point is a maximum, so no test
     is needed.   ◆◆◆

The equation linking the variables in Example 21 had only two variables, x and y. In the following example, our equation has *three* variables, one of which must be eliminated before we take the derivative. We eliminate the third variable by means of a second equation.

◆◆◆ Example 22: Assume that the strength of a rectangular beam varies directly as its width and the square of its depth. Find the dimensions of the strongest beam that can be cut from a round log 12.0 inches in diameter (Fig. 25–30).

Estimate: A *square* beam cut from a 12-in. log would have a width of about $8\frac{1}{2}$ in. But we are informed, and know from experience, that depth contributes more to the strength than width, so we expect a width less than $8\frac{1}{2}$ in. But from experience we would be suspicious of a very narrow beam, say, 2 in. or less. So we expect a width between 2 and $8\frac{1}{2}$ in. The depth has to be greater than $8\frac{1}{2}$ in. but cannot exceed 12 in. These numbers bracket our answer.

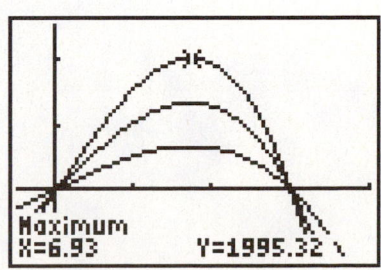

FIGURE 25–30 Beam cut from a log.

Solution:

(1) We want to maximize the strength S by varying the width x.
(2) The strength is $S = kxy^2$, where k is a constant of proportionality. Note that we have three variables, S, x, and y. We must eliminate x or y. By the Pythagorean theorem,

$$x^2 + y^2 = 12.0^2$$

so

$$y^2 = 144 - x^2$$

Substituting, we obtain

$$S = kx(144 - x^2)$$

(3) The derivative is

$$\frac{dS}{dx} = kx(-2x) + k(144 - x^2)$$

$$= -3kx^2 + 144k$$

(4) Setting the derivative equal to zero and solving for x gives us $3x^2 = 144$, so

$$x = \pm 6.93 \text{ in.}$$

We discard the negative value, of course. The depth is

$$y = \sqrt{144 - 48} = 9.80 \text{ in.}$$

(5) But have we found the dimensions of the beam with *maximum* strength, or *minimum* strength? Let us use the second-derivative test to tell us. The second derivative is

$$\frac{d^2S}{dx^2} = -6kx$$

When $x = 6.93$,

$$\frac{d^2S}{dx^2} = -41.6k$$

Since k is positive, the second derivative is negative, which tells us that we have found a *maximum*.

Graphical Solution or Check: As before, we can solve the problem or check the analytical solution graphically by graphing the function and its derivative. But here the equations contain an unknown constant k!

The screen shows the function graphed with assumed values of $k = 1$, 2, and 3. Notice that *changing k does not change the horizontal location of the maximum point.* Thus we can do a graphical solution or check with any value of k, with $k = 1$ being the simplest choice.

◆◆◆

Maximum
X=6.93   Y=1995.32

Screen for Example 22. Tick marks are 4 units apart on the x axis and 1000 units apart on the vertical axis.

Sometimes a graphical solution is the only practical one, as in the following example.

◆◆◆ **Example 23:** Find the dimensions x and y and the minimum cost for a 10.0-ft^3-capacity, open-top box with square bottom (Fig. 25–31). The sides are aluminum at \$1.28/ft^2, and the bottom is copper at \$2.13/ft^2. An aluminum-to-aluminum joint costs \$3.95/ft, and a copper-to-aluminum joint costs \$2.08/ft.

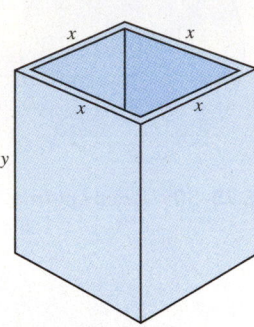

FIGURE 25–31

Solution: The costs are as follows:

an aluminum side	\$1.28 xy
the copper bottom	\$2.13x^2
one alum-to-alum joint	\$3.95y
one alum-to-copper joint	\$2.08x

The total cost C is then

$$C = 4(1.28xy) + 2.13x^2 + 4(3.95y) + 4(2.08x)$$

We can eliminate y from this equation by noting that the volume, 10.0 ft^3, is equal to x^2y, so

$$y = \frac{10.0}{x^2}$$

Substituting gives

$$C = 4(1.28x)\frac{10.0}{x^2} + 2.13x^2 + 4(3.95)\frac{10.0}{x^2} + 4(2.08x)$$

This simplifies to

$$C = 2.13x^2 + 8.32x + 51.2x^{-1} + 158x^{-2}$$

Taking the derivative,

$$C' = 4.26x + 8.32 - 51.2x^{-2} - 316x^{-3}$$

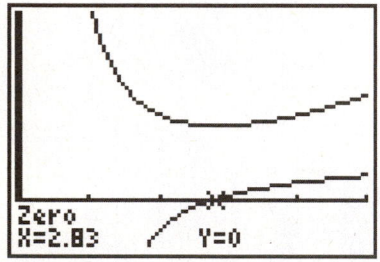

Screen for Example 23. Tick marks are 1 unit apart on the x axis and 50 units apart on the vertical axis.

For an *analytical solution*, we would set C' to zero and attempt to solve for x. Instead we will do a *graphical solution*. As in the preceding examples, we graph the function and its derivative as shown and find a minimum point at $x = 2.83$ ft. Substituting back to get y,

$$y = \frac{10.0}{x^2} = \frac{10.0}{(2.83)^2} = 1.25 \text{ ft}$$

We can get the total cost C by substituting back, or we can read it off the graph. Either way we get $C = \$78.47$.   ◆◆◆

Exercise 4 ◆ Optimization

Number Puzzles

1. What number added to half the square of its reciprocal gives the smallest sum?

2. Separate the number 10 into two parts such that their product will be a maximum.

3. Separate the number 20 into two parts such that the product of one part and the square of the other part is a maximum.

4. Separate the number 5 into two parts such that the square of one part times the cube of the other part shall be a maximum.

Minimum Perimeter

5. A rectangular garden (Fig. 25–32) laid out along your neighbor's lot contains 432 m². It is to be fenced on all sides. If the neighbor pays for half the shared fence, what should be the dimensions of the garden so that your cost is a minimum?

6. It is required to enclose a rectangular field by a fence (Fig. 25–33) and then divide it into two lots by a fence parallel to the short sides. If the area of the field is 25,000 ft², find the lengths of the sides so that the total length of fence will be a minimum.

7. A rectangular pasture 162 yd² in area is built so that a long, straight wall serves as one side of it. If the length of the fence along the remaining three sides is the least possible, find the dimensions of the pasture.

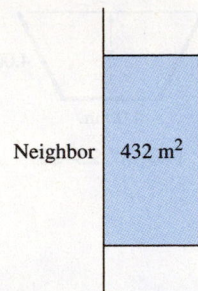

FIGURE 25–32 Rectangular garden.

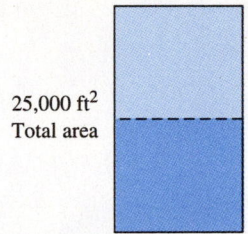

FIGURE 25–33 Rectangular field.

Maximum Volume of Containers

8. Find the volume of the largest open-top box that can be made from a rectangular sheet of metal 6.00 in. by 16.0 in. (Fig. 25–34) by cutting a square from each corner and turning up the sides.

9. Find the height and base diameter of a cylindrical, topless tin cup of maximum volume if its area (sides and bottom) is 100 cm².

10. The slant height of a certain cone is 50.0 cm. What cone height will make the volume a maximum?

Maximum Area of Plane Figures

11. Find the area of the greatest rectangle that has a perimeter of 20.0 in.

12. A window composed of a rectangle surmounted by an equilateral triangle is 15.0 ft in perimeter (Fig. 25–35). Find the dimensions that will make its total area a maximum.

13. Two corners of a rectangle are on the x axis between $x = 0$ and 10 (Fig. 25–36). The other two corners are on the lines whose equations are $y = 2x$ and $3x + y = 30$. For what value of y will the area of the rectangle be a maximum?

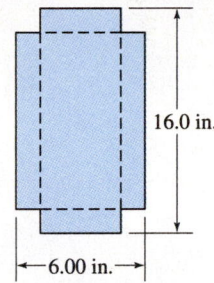

FIGURE 25–34 Sheet metal for box.

Maximum Cross-Sectional Area

14. A trough is to be made of a long rectangular piece of metal by bending up two edges so as to give a rectangular cross section. If the width of the original piece is 14.0 in., how deep should the trough be made in order that its cross-sectional area be a maximum?

15. A gutter is to be made of a strip of metal 12.0 in. wide, with the cross section having the form shown in Fig. 25–37. What depth x gives a maximum cross-sectional area?

FIGURE 25–35 Window.

Minimum Distance

16. Ship A is traveling due south at 40.0 mi/h, and ship B is traveling due west at the same speed. Ship A is now 10.0 mi from the point at which their paths will eventually cross, and ship B is now 20.0 mi from that point. What is the closest that the two ships will get to each other?

17. Find the point Q on the curve $y = x^2/2$ that is nearest the point $(4, 1)$ (Fig. 25–38).

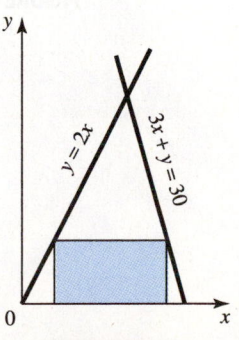

FIGURE 25–36

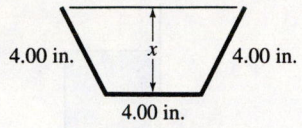

FIGURE 25–37

18. Given the branch of the parabola $y^2 = 8x$ in the 1st quadrant and the point $P(6, 0)$ on the x axis (Fig. 25–39), find the coordinates of point Q so that PQ is a minimum.

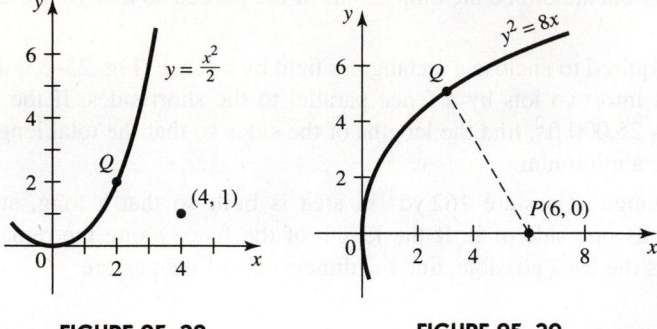

FIGURE 25–38   **FIGURE 25–39**

Inscribed Plane Figures

19. Find the dimensions of the rectangle of greatest area that can be inscribed in an equilateral triangle, each of whose sides is 10.0 in., if one of the sides of the rectangle is on a side of the triangle (Fig. 25–40). (*Hint:* Let the independent variable be the height x of the rectangle.)

20. Find the area of the largest rectangle with sides parallel to the coordinate axes which can be inscribed in the figure bounded by the two parabolas $3y = 12 - x^2$ and $6y = x^2 - 12$ (Fig. 25–41).

21. Find the dimensions of the largest rectangle that can be inscribed in an ellipse whose major axis is 20.0 units and whose minor axis is 14.0 units (Fig. 25–42).

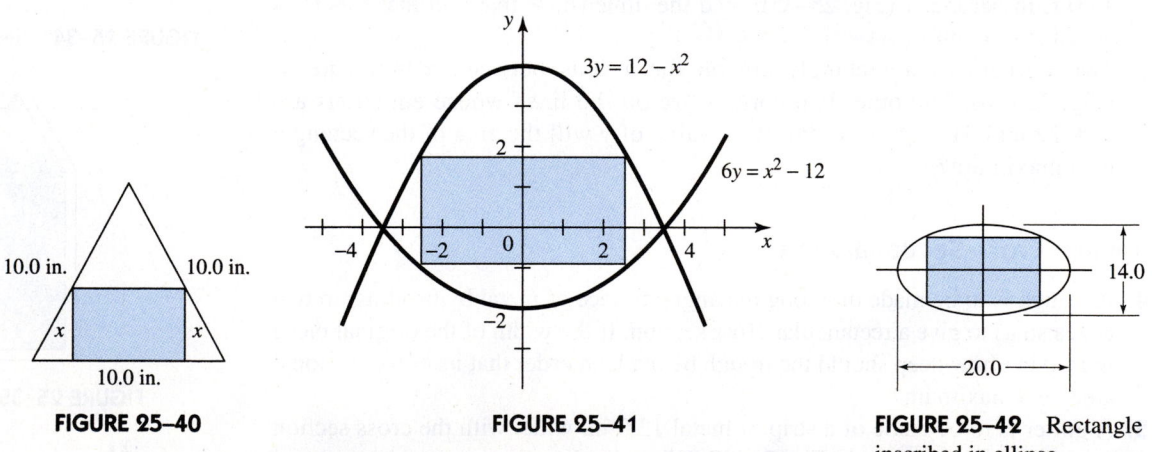

FIGURE 25–40   **FIGURE 25–41**   **FIGURE 25–42** Rectangle inscribed in ellipse.

Inscribed Volumes

22. Find the dimensions of the rectangular parallelepiped of greatest volume and with a square base that can be cut from a solid sphere 18.0 inches in diameter (Fig. 25–43).

23. Find the dimensions of the right circular cylinder of greatest volume that can be inscribed in a sphere with a diameter of 10.0 cm (Fig. 25–44).

24. Find the height of the cone of minimum volume circumscribed about a sphere of radius 10.0 m (Fig. 25–45).

25. Find the altitude of the cone of maximum volume that can be inscribed in a sphere of radius 9.00 ft.

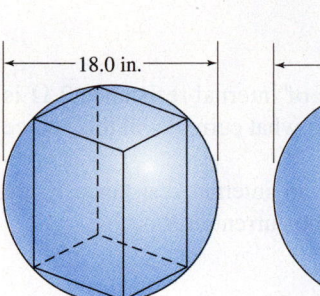

FIGURE 25–43
Parallelepiped
inscribed in sphere.

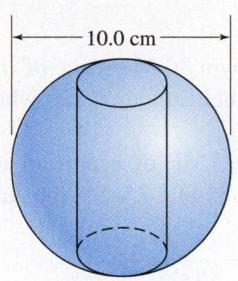

FIGURE 25–44
Cylinder inscribed
in sphere.

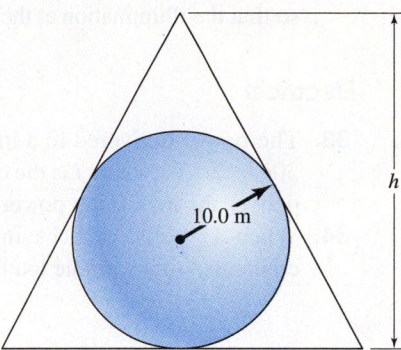

FIGURE 25–45 Cone circumscribed
about sphere.

Most Economical Dimensions of Containers

26. What should be the diameter of a can holding 1 qt (58 in.³) and requiring the least amount of metal, if the can is open at the top?

27. A silo (Fig. 25–46) has a hemispherical roof, cylindrical sides, and circular floor, all made of steel. Find the dimensions for a silo having a volume of 755 m³ (including the dome) that needs the least steel.

Minimum Travel Time

28. A man in a rowboat at P (Fig. 25–47) 6.00 mi from shore desires to reach point Q on the shore at a straight-line distance of 10.0 mi from his present position. If he can walk 4.00 mi/h and row 3.00 mi/h, at what point L should he land in order to reach Q in the shortest time?

Beam Problems

29. The strength S of the beam in Fig. 25–30 is given by $S = kxy^2$, where k is a constant. Find x and y for the strongest rectangular beam that can be cut from an 18.0-in.-diameter cylindrical log.

30. The stiffness Q of the beam in Fig. 25–30 is given by $Q = kxy^3$, where k is a constant. Find x and y for the stiffest rectangular beam that can be cut from an 18.0-in.-diameter cylindrical log.

Light

31. The intensity E of illumination at a point due to a light at a distance x from the point is given by $E = kI/x^2$, where k is a constant and I is the intensity of the source. A light M has an intensity three times that of N (Fig. 25–48). At what distance from M is the illumination a minimum?

32. The intensity of illumination I from a given light source is given by

$$I = \frac{k \sin \phi}{d^2}$$

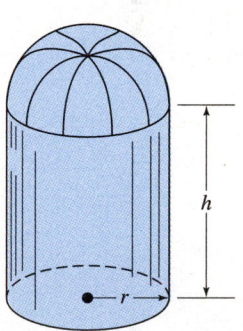

FIGURE 25–46 Silo.

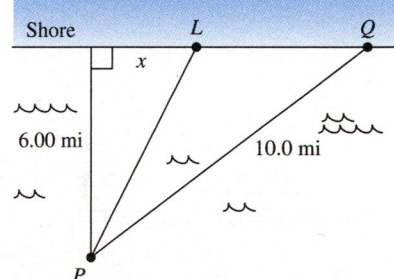

FIGURE 25–47 Find the fastest route from boat P to shore at Q.

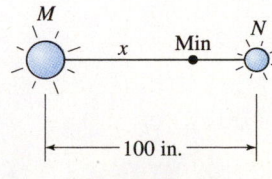

FIGURE 25–48

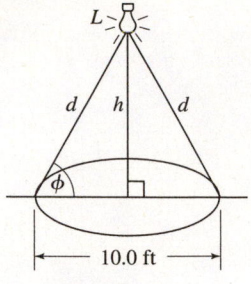

FIGURE 25–49

where k is a constant, ϕ is the angle at which the rays strike the surface, and d is the distance between the surface and the light (Fig. 25–49). At what height h should a light be suspended directly over the center of a circle 10.0 ft in diameter so that the illumination at the circumference will be a maximum?

Electrical

33. The power delivered to a load by a 30-V source of internal resistance 2 Ω is $30i - 2i^2$ W, where i is the current in amperes. For what current will this source deliver the maximum power?

34. When 12 cells, each having an EMF of e and an internal resistance r, are connected to a variable load R as in Fig. 25–50, the current in R is

$$i = \frac{3e}{\dfrac{3r}{4} + R}$$

Show that the maximum power (i^2R) delivered to the load is a maximum when the load R is equal to the equivalent internal resistance of the source, $3r/4$.

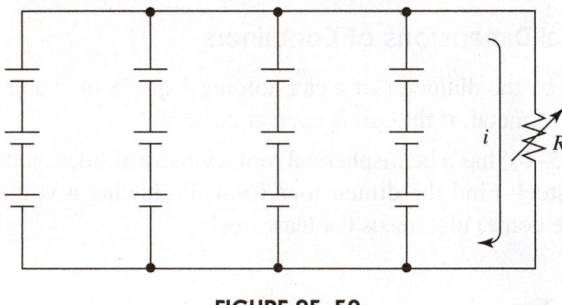

FIGURE 25–50

35. A certain transformer has an efficiency E when delivering a current i, where

$$E = \frac{115i - 25 - i^2}{115i}$$

At what current is the efficiency of the transformer a maximum?

Mechanisms

36. If the lever in Fig. 25–51 weighs 12.0 lb per foot, find its length so as to make the lifting force F a minimum.

37. The efficiency E of a screw is given by

$$E = \frac{x - \mu x^2}{x + \mu}$$

where μ is the coefficient of friction and x the tangent of the pitch angle of the screw. Find x for maximum efficiency if $\mu = 0.45$.

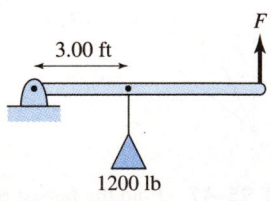

FIGURE 25–51

Graphical Solution

The following problems require a graphical solution.

38. Find the dimensions x and y and the minimum cost for a 28.7 m³-capacity box with square bottom (Fig. 25–52). The material for the sides costs \$5.87/m², and the joints cost \$4.29/m. The box has both a top and a bottom, at \$7.42/m².

39. A cylindrical tank (Fig. 25–53) with capacity 10,000 ft^3 has ends costing $4.23/ft^2 and cylindrical side costing $3.81/ft^2. The welds cost $5.85/ft. Find the radius, height, and total cost, for a tank of minimum cost.

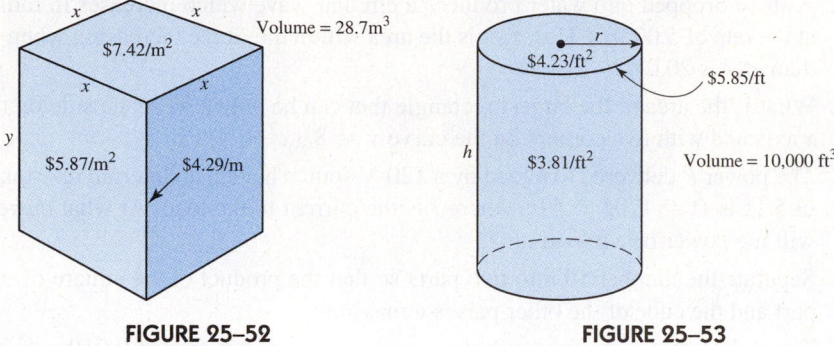

FIGURE 25–52   **FIGURE 25–53**

••• CHAPTER 25 REVIEW PROBLEMS •••••••••••••••••••••••••••••••

1. Airplane A is flying south at a speed of 120 ft/s. It passes over a bridge 12 min before another airplane, B, which is flying east at the same height at a speed of 160 ft/s. How fast are the airplanes separating 12 min after B passes over the bridge?

2. Find the velocity at 3.55 s of a point moving in a straight line according to the equation, $s = t^3 + 4$.

3. A person walks toward the base of a 60.0-m-high tower at the rate of 5.0 km/h. At what rate does the person approach the top of the tower when 80.0 m from the base?

4. A point moves along the hyperbola $x^2 - y^2 = 144$ with a horizontal velocity $v_x = 15.0$ cm/s. Find the total velocity when the point is at (13.0, 5.00).

5. A conical tank with vertex down has a vertex angle of 60.0°. Water flows from the tank at a rate of 5.00 cm^3/min. At what rate is the inner surface of the tank being exposed when the water is 6.00 cm deep?

6. Find the instantaneous velocity and acceleration at $t = 2.0$ s for a point moving in a straight line according to the equation $s = 4t^2 - 6t$.

7. A turbine blade (Fig. 25–54) is driven by a jet of water having a speed s. The power output from the turbine is given by $P = k(sv - v^2)$, where v is the blade speed and k is a constant. Find the blade speed v for maximum power output.

8. A pole (Fig. 25–55) is braced by a cable 24.0 ft long. Find the distance x from the foot of the pole to the cable anchor so that the moment produced by the tension in the cable, about the foot of the pole, is a maximum. Assume that the tension in the cable does not change as the anchor point is changed.

9. Find the height of a right circular cylinder of maximum volume that can be inscribed in a sphere of radius 6.

10. Three sides of a trapezoid each have a length of 10 units. What length must the fourth side be to make the area a maximum?

11. The air in a certain balloon has a pressure of 40.0 lb/in.^{2} and a volume of 5.0 ft^3 and is expanding at the rate of 0.20 ft^3/s. If the pressure and volume are related by the equation $pv^{1.41}$ = constant, find the rate at which the pressure is changing.

12. A certain item costs $10 to make, and the number that can be sold is estimated to be inversely proportional to the cube of the selling price. What selling price will give the greatest net profit?

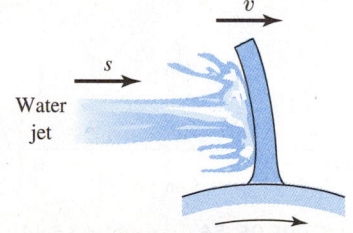

FIGURE 25–54   Turbine blade.

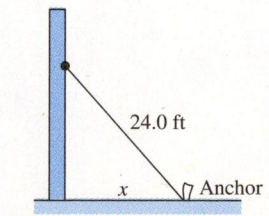

FIGURE 25–55   Pole braced by a cable.

13. The distance s of a point moving in a straight line is given by

$$s = -t^3 + 3t^2 + 24t + 28$$

At what times and at what distances is the point at rest?

14. A stone dropped into water produces a circular wave which increases in radius at the rate of 5.00 ft/s. How fast is the area within the ripple increasing when its diameter is 20.0 ft?

15. What is the area of the largest rectangle that can be drawn with one side on the x axis and with two corners on the curve $y = 8/(x^2 + 4)$?

16. The power P delivered to a load by a 120-V source having an internal resistance of 5 Ω is $P = 120I - 5I^2$, where I is the current to the load. At what current will the power be a maximum?

17. Separate the number 10 into two parts so that the product of the square of one part and the cube of the other part is a maximum.

18. The radius of a circular metal plate is increasing at the rate of 0.010 m/s. At what rate is the area increasing when the radius is 2.00 m?

19. Find the dimensions of the largest rectangular box with square base and open top that can be made from 300 in.^{2} of metal.

20. The voltage applied to a 3.25-μF capacitor is $v = 1.03t^2 + 1.33t + 2.52$ V. Find the current at $t = 15.0$ s.

21. The angular displacement of a rotating body is given by $\theta = 18.5t^2 + 12.8t + 14.8$ rad. Find (a) the angular velocity and (b) the angular acceleration at $t = 3.50$ s.

22. The charge through a 8.24-Ω resistor varies with time according to the function $q = 2.26t^3 - 8.28$ C. Write an expression for the instantaneous current through the resistor. Remember, $i = C\dfrac{dv}{dt}$.

23. The charge at a resistor varies with time according to the function $q = 2.84t^2 + 6.25t^3$ C. Write an expression for the instantaneous current through the resistor, and evaluate it at 1.25 s.

24. *Writing:* Once again, our writing assignment is to make up an application, but now one that leads to a max/min or a related rate problem. As before, swap with a classmate; solve each other's problems; note anything unclear, unrealistic, or ambiguous; and then rewrite your problem if needed.

25. *Team Project:* Lay out a race course (Fig. 25–56) on your school athletic field. The object is to get from A to C in the shortest time. You may take any path, but you may run only on line AB and must walk anywhere else.

Each member of your team should clock his or her rate of running and of walking. Then, using the ideas from problem 28 of Exercise 4 as a guide, compute for each of you the point P at which you should leave line AB for minimum time. Mark these points.

When ready, challenge another class to a race. Be careful to avoid students who have taken calculus or who are members of the track team.

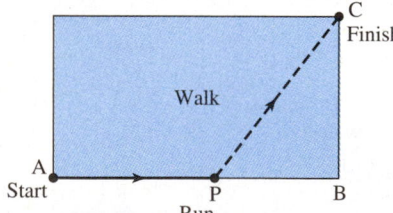

FIGURE 25–56 An athletic field.

Integration

◆◆◆ OBJECTIVES ◆◆◆

When you have completed this chapter, you should be able to

- Find the integral of certain algebraic functions.
- Check an integral by differentiating.
- Check an integral by calculator.
- Solve simple differential equations.
- Evaluate the constant of integration, given boundary conditions.
- Perform successive integration.
- Use a table of integrals.
- Evaluate definite integrals.
- Determine approximate areas under curves by the midpoint method.
- Find the approximate area under a curve by calculator.
- Apply the fundamental theorem of calculus to determine exact areas under curves.

◆◆

Every operation in mathematics has its inverse. For example, we reverse the squaring operation by taking the square root; the arcsin is the inverse of the sine, and so on. In this chapter we learn how to reverse the process of differentiation with the process of *integration*.

The material in this and the following three chapters usually falls under the heading of *integral calculus*, in contrast to the differential calculus already introduced. We define the *definite integral* and show how to evaluate it. Next we discuss the problem of finding the area bounded by a curve and the x axis between two given values of x. We find such areas, first approximately by the *midpoint method* and then exactly by means of the definite integral. In the process we develop the *fundamental theorem of calculus*, which ties together the derivative, the integral, and the area under a curve.

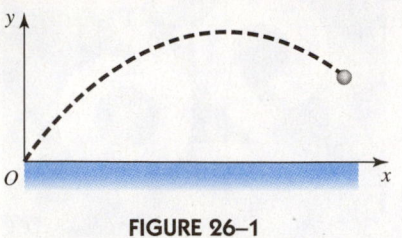

FIGURE 26-1

Integration has a great many practical applications in technical work. As we did with the derivative, we will give the mathematics in this chapter, followed by two chapters of applications. We will, however, have an occasional application here to give an idea of what is to come. For example, if we had an equation for the displacement of a point on a body, say the projectile of Fig. 26–1, we saw that we could take the derivative of that equation to get the velocity of that point. Now we will be able to do the reverse; given the velocity, we can take the *integral* to get the displacement. This is just one of a long list of applications of the integral.

26–1　The Indefinite Integral

As with any new mathematical idea, we must first have new definitions, symbols, and some theory. We begin with the idea that finding an integral is *the inverse of finding a derivative*.

Reversing the Process of Differentiation

An *integral* of an expression is a new expression which, if differentiated, gives the original expression.

◆◆◆ **Example 1:** The derivative of x^3 is $3x^2$, so an integral of $3x^2$ is x^3.

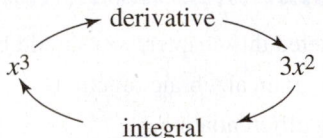

◆◆◆

The derivative of x^3 is $3x^2$, but the derivative of $x^3 + 6$ is also $3x^2$. The derivatives of $x^3 - 99$ and of $x^3 + $ *any constant* are also $3x^2$. This constant, called the *constant of integration*, must be included when we find the integral of a function. We will learn how to evaluate the constant of integration later.

◆◆◆ **Example 2:** The derivative of $x^3 + $ any constant is $3x^2$, so the integral of $3x^2$ is $x^3 + $ a constant. We use C to stand for the constant.

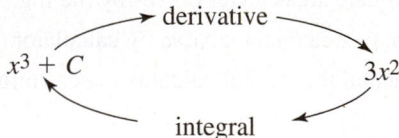

◆◆◆

The process of finding the integral is called *integration*. The variable, x in this case, is called the *variable of integration*. Another name for the integral is the *antiderivative*.

The Integral Sign

We have seen above that the derivative of $x^3 + C$ is $3x^2$, so the integral of $3x^2$ is $x^3 + C$. Let us now state this same idea more formally.
Let there be a function

$$F(x) + C$$

Its derivative is

$$\frac{d}{dx}[F(x) + C] = F'(x) + 0 = F'(x)$$

where we have used the familiar prime (′) notation to indicate a derivative. Going to differential form, we have

$$d[F(x) + C] = F'(x)\, dx$$

Thus the differential of $F(x) + C$ is $F'(x)\,dx$. Conversely,

Integral of $F'(x)\,dx = F(x) + C$

Instead of writing "Integral of," we use the *integral sign* $\int$ to indicate that operation. It is no accident that the integral sign looks like the letter S. We will see later that it stands for *summation*.

Indefinite Integral	$$\int F'(x)\,dx = F(x) + C$$ *The integral of the differential of a function is equal to the function itself, plus a constant.*	286

The integral is called *indefinite* because of the unknown constant. In a later section we will study, we will study the *definite* integral, which has no unknown constant.

We read Eq. 286 as "the integral of $F'(x)\,dx$ with respect to x is $F(x) + C$." The expression $F'(x)\,dx$ to be integrated is called the *integrand*, and C, as we said earlier, is the *constant of integration*.

◆◆◆ **Example 3:** What is the indefinite integral of $3x^2\,dx$?

Solution: The function x^3 has $3x^2\,dx$ as a differential, so the integral of $3x^2\,dx$ is $x^3 + C$.

$$\int 3x^2\,dx = x^3 + C$$

◆◆◆

If we let $F'(x)$ be denoted by $f(x)$, Equation 286 becomes the following:

Indefinite Integral or Antiderivative	$$\int f(x)\,dx = F(x) + C$$ where $f(x) = F'(x)$.	287

Common Error	We use both capital and lowercase F in this section. Be careful not to mix them up.

Some Rules for Finding Integrals

We can obtain rules for integration by reversing the rules we previously derived for differentiation. Such a list of rules, called a *table of integrals*, is given in Appendix C. This is a very short table of integrals; some fill entire books.

Let there be a function $u = F(x)$ whose derivative is

$$\frac{du}{dx} = F'(x)$$

or

$$du = F'(x)\,dx$$

Just as we can take the derivative of both sides of an equation, we can take the integral of both sides. This gives

$$\int du = \int F'(x)\,dx = F(x) + C$$

by Eq. 286. Substituting $F(x) = u$, we get our first rule for finding integrals.

$$\int du = u + C$$

Rule 1

The integral of the differential of a function is equal to the function itself, plus a constant of integration.

We will often say "integrate" instead of the longer "find the integral of."

◆◆◆ **Example 4:** Integrate $\int d(3x - x^2)$.

Solution: By Rule 1, the integral of the differential of a function is the function itself, plus a constant, so

$$\int d(3x - x^2) = 3x - x^2 + C$$ ◆◆◆

◆◆◆ **Example 5:** Here are some more examples of the use of Rule 1.

(a) $\displaystyle\int dy = y + C$ (b) $\displaystyle\int dz = z + C$ (c) $\displaystyle\int d(x^3) = x^3 + C$

(d) $\displaystyle\int d(x^2 + 2x) = x^2 + 2x + C$ (e) $\displaystyle\int d\left(\frac{u^{n+1}}{n+1}\right) = \frac{u^{n+1}}{n+1} + C$ ◆◆◆

For our second rule, we use the fact that the derivative of a constant times a function equals the constant times the derivative of the function. Reversing the process, we have the following:

$$\int a\,f(x)\,dx = a\int f(x)\,dx = a\,F(x) + C$$

Rule 2

The integral of a constant times a function is equal to the constant times the integral of the function, plus a constant of integration.

Rule 2 says that we may move *constants* (*not* variables!) to the left of the integral sign, as in the following example:

◆◆◆ **Example 6:** Find $\int 5\,dx$.

Solution: By Rule 2,

$$\int 5\,dx = 5\int dx$$

Then, by Rule 1,

$$5\int dx = 5(x + C_1)$$
$$= 5x + C$$

where we have replaced the constant $5C_1$ by the constant C.   ◆◆◆

We get our third rule by noting that the derivative of a function with several terms is the sum of the derivatives of each term. Reversing gives the following:

$$\int [f(x) + g(x) + h(x) + \cdots]\,dx$$

Rule 3

$$= \int f(x)\,dx + \int g(x)\,dx + \int h(x)\,dx + \cdots + C$$

The integral of a function with several terms equals the sum of the integrals of those terms, plus a constant.

Rule 3 says that when integrating an expression having several terms, we may *integrate each term separately*. The constant C comes from combining the individual constants, $C_1, C_2, \dots$ into a single constant.

◆◆◆ **Example 7:** Integrate $\int (3x^2 + 5)\, dx$.

Solution:

By Rule 3,

$$\int (3x^2 + 5)\, dx = \int 3x^2\, dx + \int 5\, dx$$

By Rule 2,

$$= \int 3x^2\, dx + 5 \int dx$$

$$= x^3 + C_1 + 5x + C_2$$

$$= x^3 + 5x + C$$

where we have combined C_1 and C_2 into a single constant C. From now on, we will not bother writing C_1 and C_2 but will combine them immediately into a single constant C. ◆◆◆

Our fourth rule is for a power of x.

Rule 4

$$\int x^n\, dx = \frac{x^{n+1}}{n+1} + C \qquad (n \neq -1)$$

The integral of x raised to a power is x raised to that power increased by 1, divided by the new power, plus a constant.

We can prove this rule simply by taking the derivative.

$$\frac{d}{dx}\left(\frac{x^{n+1}}{n+1} + C\right) = (n+1)\frac{x^n}{n+1} + 0$$

$$= x^n$$

Note that Rule 4 is not valid for $n = -1$, for this would give division by zero. We'll derive a later rule for when n is -1.

◆◆◆ **Example 8:** Integrate $\int x^3\, dx$.

Solution: We use Rule 4 with $n = 3$.

$$\int x^3\, dx = \frac{x^{3+1}}{3+1} + C = \frac{x^4}{4} + C \qquad\qquad ◆◆◆$$

◆◆◆ **Example 9:** Integrate $\int x^5\, dx$.

Solution: This is similar to Example 8, except that the exponent is 5. By Rule 4,

$$\int x^5\, dx = \frac{x^{5+1}}{5+1} + C = \frac{x^6}{6} + C \qquad\qquad ◆◆◆$$

◆◆◆ **Example 10:** Integrate $\int 7x^3\, dx$.

Solution: This is similar to Example 8, except that here our function is multiplied by 7. By Rule 2, we may move the constant factor 7 to the left of the integral sign.

$$\int 7x^3\, dx = 7 \int x^3\, dx$$

$$= 7\left(\frac{x^4}{4} + C_1\right)$$

$$= \frac{7x^4}{4} + 7C_1$$

$$= \frac{7x^4}{4} + C$$

Where $C = 7C_1$. ◆◆◆

◆◆◆ **Example 11:** Integrate $\int (x^3 + x^5)\, dx$.

Solution: By Rule 3, we can integrate each term separately.

$$\int (x^3 + x^5)\, dx = \int x^3\, dx + \int x^5\, dx$$

$$= \frac{x^4}{4} + \frac{x^6}{6} + C$$

Even though each of the two integrals has produced its own constant of integration, we have combined them immediately into the single constant C.   ◆◆◆

◆◆◆ **Example 12:** Integrate $\int (5x^2 + 2x - 3)\, dx$.

Solution: Integrating term by term yields

$$\int (5x^2 + 2x - 3)\, dx = \frac{5x^3}{3} + x^2 - 3x + C$$ ◆◆◆

Rule 4 is also used when the exponent n is not an integer.

◆◆◆ **Example 13:** Integrate $\int x^{2/3}\, dx$.

Solution: We apply Rule 4 with $n = 2/3$ and get

$$\int x^{\frac{2}{3}} = \frac{x^{\frac{5}{3}}}{\frac{5}{3}} + C$$

$$= \frac{3}{5} x^{\frac{5}{3}} + C$$ ◆◆◆

To find the integral of a *radical*, change the radical to exponential form and proceed as in the last example.

◆◆◆ **Example 14:** Integrate $\int \sqrt[3]{x}\, dx$.

Solution: We write the radical in exponential form with $n = \frac{1}{3}$.

$$\int \sqrt[3]{x}\, dx = \int x^{1/3}\, dx$$

By Rule 4,

$$= \frac{x^{4/3}}{\frac{4}{3}} + C$$

$$= \frac{3x^{4/3}}{4} + C$$ ◆◆◆

The exponent n can also be *negative* (with the exception of -1, which would result in division by zero).

◆◆◆ **Example 15:** Integrate $\int \frac{1}{x^3}\, dx$.

Solution:

$$\int \frac{1}{x^3}\, dx = \int x^{-3}\, dx$$

$$= \frac{x^{-2}}{-2} + C = -\frac{1}{2x^2} + C$$ ◆◆◆

Simplify Before Integrating

If an expression does not seem to fit any given rule at first, try changing its form by performing the indicated operations (squaring, removing parentheses, and so on).

◆◆◆ **Example 16:** Integrate $\int (x + 2)(x^2 - 1)\, dx$.

Solution: None of our rules seems to fit, so we try to rewrite the given function in a different form. Let us multiply out.

$$\int (x + 2)(x^2 - 1)\, dx = \int (x^3 + 2x^2 - x - 2)\, dx$$

We can now use Rule 2 to integrate term by term, and Rule 4 to integrate each term.

$$\int (x^3 + 2x^2 - x - 2)\, dx = \frac{x^4}{4} + \frac{2x^3}{3} - \frac{x^2}{2} - 2x + C \qquad ◆◆◆$$

◆◆◆ **Example 17:** Integrate $\int (x^2 + 3)^2\, dx$.

Solution: None of our rules (so far) seem to fit. Rule 3, for example, is for x raised to a power, *not* for $(x^2 + 3)$ raised to a power. However, if we square $x^2 + 3$ before integrating we get

$$\int (x^2 + 3)^2\, dx = \int (x^4 + 6x^2 + 9)\, dx$$

$$= \frac{x^5}{5} + \frac{6x^3}{3} + 9x + C$$

$$= \frac{x^5}{5} + 2x^3 + 9x + C \qquad ◆◆◆$$

◆◆◆ **Example 18:** Integrate $\displaystyle\int \frac{x^5 - 2x^3 + 5x}{x}\, dx$.

Solution: This problem looks complicated at first, but let us perform the indicated division

$$\int \frac{x^5 - 2x^3 + 5x}{x}\, dx = \int (x^4 - 2x^2 + 5)\, dx$$

Now Rules 3 and 4 can be used.

$$= \frac{x^5}{5} - \frac{2x^3}{3} + 5x + C \qquad ◆◆◆$$

Checking an Integral by Differentiating

Many rules for integration are presented without derivation or proof, so you would be correct in being suspicious of them. However, you can convince yourself that a rule works (and that you have used it correctly) simply by taking the derivative of your result. You should get back the original expression.

◆◆◆ **Example 19:** Taking the derivative of the expression obtained in Example 15, we have

$$\frac{d}{dx}\left(-\frac{1}{2x^2} + C\right) = \frac{d}{dx}\left(-\frac{x^{-2}}{2} + C\right) = -\frac{1}{2}(-2x^{-3}) + 0 = x^{-3} = \frac{1}{x^3}$$

which is the expression we started with, so our integration was correct. ◆◆◆

Graphing an Indefinite Integral by Calculator

We can use our calculators to compute and graph the approximate integral of a given function. The ability to find integrals by calculator is extremely important because most functions cannot be integrated analytically.

On the TI-83/84 we use the **fnInt** operation (standing for *function integral*) found in the MATH menu. On the TI-89, we use **nInt** from the CALC menu.

◆◆◆ Example 20: Graph the integral $\int x^2 \, dx$ by calculator.

Solution: In screen (1) we enter

$$Y1 = \text{fnInt}(X^2, X, 0, X)$$

selecting **fnInt** from the MATH menu. Here the first quantity in the parentheses is the function, x^2, the second is the variable of integration, x, and the last two are limits of integration that we will learn about later. For now, to make a graph we enter 0 and x. The graph is shown in screen (2).

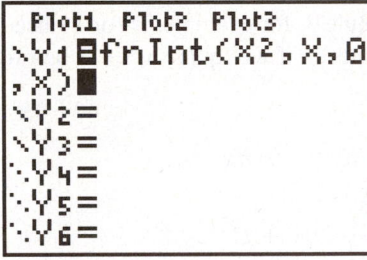 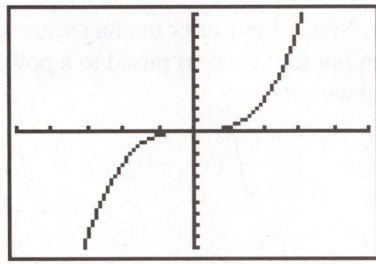

(1) TI-83/84 screens for Example 20.

(2) Tick marks are 1 unit apart. You can speed up the drawing of the graph by setting **Xres** to a larger value.   ◆◆◆

The work is even easier if the function is already defined in the Y = screen.

◆◆◆ Example 21: Graph the function $y = x^2$ and its integral, in the same viewing window.

Solution: We enter the function itself as Y1 and the integral as Y2.

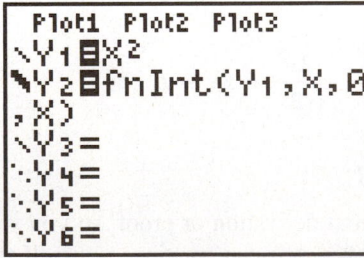 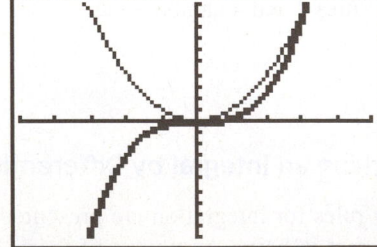

TI-83/84 screens for Example 21.

Graph of the given function, shown light, and its integral, shown heavy. Tick marks are 1 unit apart.   ◆◆◆

Checking an Integral by Calculator

For a calculator check of an integral that was found analytically, (a) graph that integral, with $C = 0$, and (b) have the calculator compute and graph the integral of the given function. The graphs should overlay.

◆◆◆ **Example 22:** Evaluate $\int 4x^3 \, dx$, and check graphically.

Solution: We take the integral using Rule 4.

$$\int 4x^3 \, dx = 4\left(\frac{x^4}{4}\right) + C$$
$$= x^4 + C$$

We graph that function (with $C = 0$) as Y1. In the same viewing window we graph the **fnInt** of $4x^3$, as Y2. We graph Y2 with a heavier line than Y1. The two graphs are identical, showing that our integral is correct.

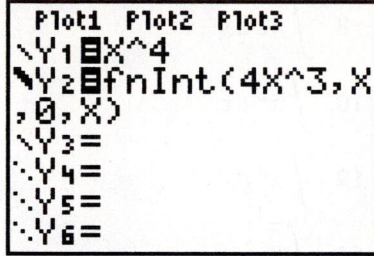

TI-83/84 screen for Example 22.

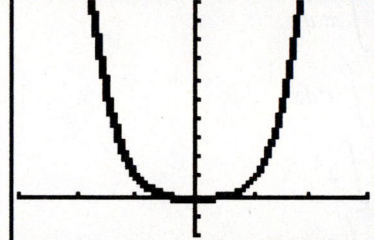

Graph of x^4 and $\int 4x^3 \, dx$. It helps to have the first curve plotted lightly or dashed, and the second heavy, so that you may see the second overlaying the first. Tick marks are 1 unit apart. ◆◆◆

Symbolic Integration by Calculator

We can find integrals on a calculator that does symbolic processing. Here, instead of getting a graph, we will get an *expression*.

◆◆◆ **Example 23:** Evaluate $\int 4x^3 \, dx$ on the TI-89 calculator.

Solution: We press $\boxed{\int}$ or select $\int(\,)$ from the **Calc** menu. We follow this by the expression to be integrated ($4x^3$) and the variable of integration (x). Do not enter "dx." Pressing $\boxed{\text{ENTER}}$ gives the integral, as shown. ◆◆◆

Of course, a calculator that does symbolic mathematics can also be used to graph an integral.

◆◆◆ **Example 24:** Graph the integral $\int 4x^3 \, dx$ on the TI-89 calculator, for $x = 0$ to 3.

Solution: We enter the integral into the $\boxed{Y=}$ editor just as we did in Example 23, set the viewing window, and press $\boxed{\text{ENTER}}$.

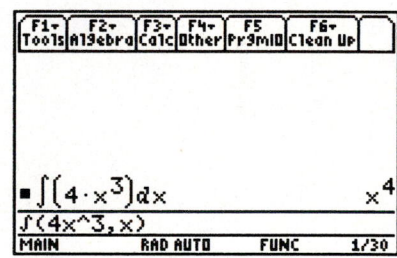

TI-89 screen for Example 23. Notice that the constant of integration is not given. We will use this same instruction later in this chapter to evaluate *definite* integrals.

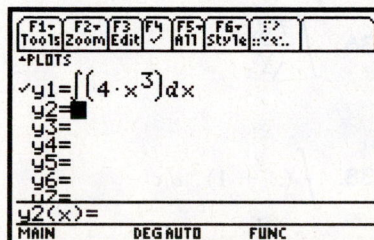

TI-89 screens for Example 24. The $\boxed{Y=}$ editor on the TI-89, showing the integral entered as y1.

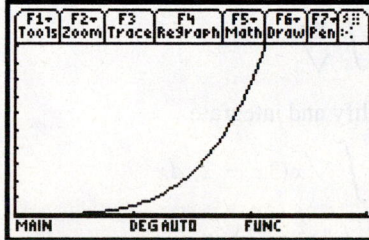

Graph of the integral. Tick marks are 1 unit apart on the x axis and 2 units apart on the y axis. ◆◆◆

Exercise 1 ◆ The Indefinite Integral

Find each indefinite integral. Check some by calculator.

1. $\int dx$

2. $\int dy$

3. $\int 6\,dx$

4. $\int -2\,dx$

5. $\int 5dx$

6. $\int 7dy$

7. $\int \pi\,dx$

8. $\int 42\,dx$

9. $\int x\,dx$

10. $\int x^2\,dx$

11. $\int x^4\,dx$

12. $\int x^3\,dx$

13. $\int 5x\,dx$

14. $\int 9x^2\,dx$

15. $\int 8x^4\,dx$

16. $\int \pi x^3\,dx$

17. $\int (x + 4)\,dx$

18. $\int (9 - x^2)\,dx$

19. $\int (x^2 + 4x)\,dx$

20. $\int (x + 6 - x^2)\,dx$

21. $\int (3x^2 - 24x + 4)\,dx$

22. $\int (x^3 + 7 - 2x^2)\,dx$

23. $\int x^{1/2}\,dx$

24. $\int \frac{7}{2}x^{5/2}\,dx$

25. $\int x^{4/3}\,dx$

26. $\int \frac{4}{3}x^{1/3}\,dx$

27. $\int \sqrt{5x}\,dx$

28. $\int \sqrt[3]{4x}\,dx$

29. $\int 9\sqrt[5]{2x}\,dx$

30. $\int 7\sqrt[3]{8x}\,dx$

31. $\int \frac{1}{x^2}\,dx$

32. $\int \frac{3}{x^3}\,dx$

33. $\int \frac{dx}{x^4}$

34. $\int \frac{5\,dx}{x^2}$

35. $\int \frac{dx}{\sqrt{x}}$

36. $\int \frac{7\,dx}{\sqrt[3]{x}}$

Simplify and integrate.

37. $\int \sqrt{x}(3x - 2)\,dx$

38. $\int (x + 1)^2\,dx$

39. $\int \frac{4x^2 - 2\sqrt{x}}{x}\,dx$

40. $\int (x + 2)(x - 3)\,dx$

41. $\int (1 - x)^3\,dx$

42. $\int \frac{x^3 + 2x^2 - 3x - 6}{x + 2}\,dx$

43. ■ Exploration:

Try this. In the same viewing window, graph

(a) For Y1, the function $x^3 + 2x$, light.
(b) For Y2, the integral of Y1, light.
(c) For Y3, the derivative of Y2, heavy.

What do you find? Why are there only two curves in our viewing window, when we had graphed *three* functions? What can you conclude from this exploration?

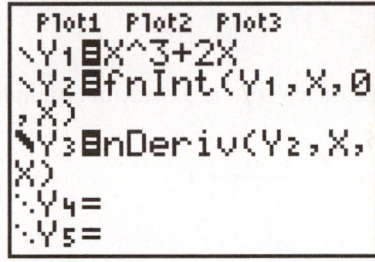

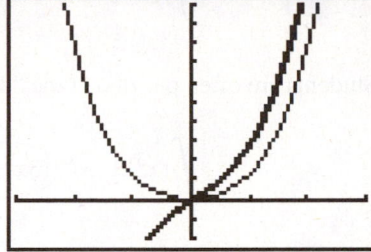

TI-83/84 screens for problem 43.   Tick marks are 1 unit apart.

■

26–2   Rules for Finding Integrals

Our small but growing table of integrals has four rules so far. To these we will now add rules for integrating a power function and for integrating an expression in the form du/u.

Integral of a Power Function

The integral of u^n, where u is some function x, is given by

Rule 5	$\displaystyle \int u^n\, du = \frac{u^{n+1}}{n+1} + C \quad (n \neq -1)$

Again, we prove this rule by taking the derivative.

$$\frac{d}{du}\left(\frac{u^{n+1}}{n+1} + C\right) = (n+1)\frac{u^n}{n+1} + 0$$
$$= u^n$$

The expression u in Rule 5 can be any function of x, say, $x^3 + 3$. However, in order for us to use Rule 5, *the quantity u^n must be followed by the derivative of u.* Thus if $u = x^3 + 3$, then $du = 3x^2\, dx$, as in the following example.

◆◆◆ **Example 25:** Find the integral $\int (x^3 + 3)^6 (3x^2)\, dx$.

Solution: If we let

$$u = x^3 + 3$$

we see that the derivative of u is

$$\frac{du}{dx} = 3x^2$$

or, in differential form,

$$du = 3x^2\, dx$$

Notice now that our given integral exactly matches Rule 5.

$$\int \underbrace{(x^3 + 3)^6}_{u} \overset{\overset{n}{\frown}}{} \underbrace{(3x^2)dx}_{du}$$

We now apply Rule 5.

$$\int (x^3 + 3)^6 (3x^2)dx = \frac{(x^3 + 3)^7}{7} + C \qquad \text{◆◆◆}$$

Students are often puzzled at the "disappearance" of the $3x^2\,dx$ in Example 25.

$$\int (x^3 + 3)^6 \boxed{(3x^2)dx} = \frac{(x^3 + 3)^7}{7} + C$$

↑————————— Where did this go?

The $3x^2\,dx$ is the differential of x^3 and does not remain after integration. Do not be alarmed when it vanishes. As before, we may check our integration by taking its derivative. We should get back the original expression, including the part that seemed to "vanish."

◆◆◆ **Example 26:** Check the integration in Example 25 by taking the derivative.

Solution: If

$$y = \frac{1}{7}(x^3 + 3)^7 + C$$

then

$$\frac{dy}{dx} = \frac{1}{7}(7)(x^3 + 3)^6(3x^2) + 0$$

or

$$dy = (x^3 + 3)^6(3x^2)\,dx$$

This is the same expression that we started with. ◆◆◆

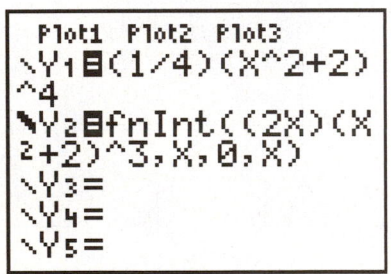

(a) TI-83/84 screens for Example 27.

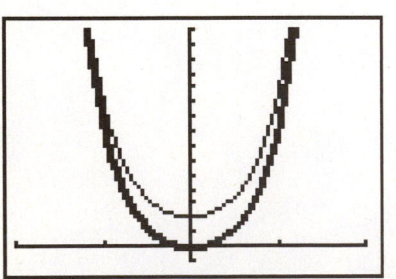

(b) Tick marks are 1 unit apart on the x axis and 2 units apart on the y axis.

◆◆◆ **Example 27:** Evaluate $\int (x^2 + 2)^3\,(2x\,dx)$. Check by calculator.

Solution: Our integral is already in the form of Rule 5, so

$$\int \underbrace{(x^2 + 2)^3}_{u} \overset{\overset{n}{\frown}}{} \underbrace{(2x\,dx)}_{du} = \frac{(x^2 + 2)^4}{4} + C$$

Graphical Check: In the same window we graph as Y1 the integral found above (light line) and as Y2 the **fnInt** (**nInt** on the TI-89) of the given function (heavy line). Note that the two curves have the same shape but do *not* overlay. However, they seem to differ by a constant amount, being 4 units apart in the y direction. This is accounted for by the constant of integration. In fact, if we subtract 4 units from Y1, the curves exactly overlay. ◆◆◆

◆◆◆ Example 33: If it is true that

$$\int 4x^3 \, dx = x^4 + C$$

it is also true that

$$\int 4z^3 \, dz = z^4 + C$$

and that

$$\int 4t^3 \, dt = t^4 + C \qquad \text{◆◆◆}$$

Integral of bu/u

Our next rule applies when the function u appears in the *denominator*.

Rule 6	$$\int \frac{du}{u} = \ln|u| + C \quad (u \neq 0)$$

We cannot prove this rule by taking the derivative, as we did with other rules, because we have not yet learned how to take the derivative of a logarithmic function. Instead, we can gain some confidence that this rule is true by graphing the integral $\int du/u$ as well as $\ln|u|$ in the same viewing window.

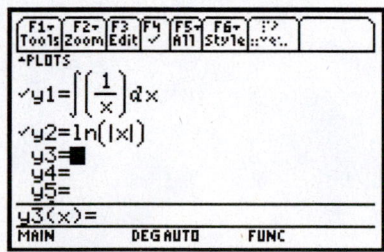

(1) TI-89 screen showing the two functions entered in the $\boxed{Y=}$ editor.

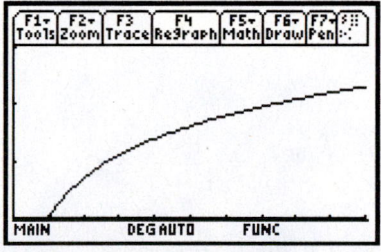

(2) Graph of the two functions. We see that y2 exactly overlays y1.

◆◆◆ Example 34: Integrate $\displaystyle\int \frac{5}{x} \, dx$.

Solution: This matches Rule 6 if we let $x = u$. So du then equals dx. Then

$$\int \frac{5}{x} \, dx = 5 \int \frac{dx}{x} = 5 \ln|x| + C \qquad \text{◆◆◆}$$

◆◆◆ Example 35: Integrate $\displaystyle\int \frac{x \, dx}{3 - x^2}$.

Solution: Let $u = 3 - x^2$. Then du is $-2x \, dx$. Our integral does not have the -2 factor, so we insert a factor of -2 and compensate with a factor of $-\frac{1}{2}$.

$$\int \frac{x \, dx}{3 - x^2} = -\frac{1}{2} \int \frac{-2x \, dx}{3 - x^2} = -\frac{1}{2} \ln|3 - x^2| + C$$

by Rule 6. ◆◆◆

As before, if the expression does not appear to match any rule, try performing the indicated operations.

◆◆◆ Example 36: Evaluate $\displaystyle\int \frac{7 - x^2}{x} \, dx$.

Solution: This does not seem to match any of our rules, so let us try to divide through by x.

$$\int \frac{7 - x^2}{x}\, dx = \int \left(\frac{7}{x} - x \right) dx$$

We can now integrate term by term.

$$\int \left(\frac{7}{x} - x \right) dx = 7 \int \frac{dx}{x} - \int x\, dx$$

$$= 7 \ln|x| - \frac{x^2}{2} + C \qquad \blacklozenge\blacklozenge\blacklozenge$$

Summary of Integration Rules So Far

Our table of integrals now contains six rules, summarized in the following table. These will be enough for us to do the applications in the following chapter. We will learn more rules later.

1. $\displaystyle \int du = u + C$

2. $\displaystyle \int a\, f(x)\, dx = a \int f(x)\, dx = a\, F(x) + C$

3. $\displaystyle \int [f(x) + g(x) + h(x) + \cdots]\, dx = \int f(x)\, dx + \int g(x)\, dx$
$$+ \int h(x)\, dx + \cdots + C$$

4. $\displaystyle \int x^n\, dx = \frac{x^{n+1}}{n+1} + C \qquad (n \neq -1)$

5. $\displaystyle \int u^n\, du = \frac{u^{n+1}}{n+1} + C \qquad (n \neq -1)$

6. $\displaystyle \int \frac{du}{u} = \ln|u| + C \qquad (u \neq 0)$

Miscellaneous Rules from a Table of Integrals

Now that you can use Rules 1 through 6, you should be able to use many of the rules from the table of integrals in Appendix C. We will restrict ourselves here to finding integrals of algebraic expressions, for which you should be able to identify the du portion in the rule. We will do transcendental functions in a later chapter.

◆◆◆ **Example 37:** Integrate $\displaystyle \int \frac{dx}{4x^2 + 25}$.

Solution: From the table we find the following:

Rule 56	$\displaystyle \int \frac{du}{a^2 + b^2 u^2} = \frac{1}{ab} \tan^{-1} \frac{bu}{a} + C$

Letting $a = 5$, $b = 2$, $u = x$, and $du = dx$, our integral matches Rule 56 if we rearrange the denominator.

$$\int \frac{dx}{25 + 4x^2} = \frac{1}{10} \tan^{-1} \frac{2x}{5} + C$$ ◆◆◆

◆◆◆ **Example 38:** Integrate $\displaystyle\int \frac{dx}{(4x^2 + 9)2x}$.

Solution: We match this with the following rule:

Rule 60	$\displaystyle\int \frac{du}{u(u^2 + a^2)} = \frac{1}{2a^2} \ln\left| \frac{u^2}{u^2 + a^2} \right| + C$

with $a = 3$, $u = 2x$, and $du = 2\,dx$. Thus

$$\int \frac{dx}{(4x^2 + 9)2x} = \frac{1}{2} \int \frac{2\,dx}{(2x)[(2x)^2 + 3^2]} = \frac{1}{36} \ln\left| \frac{4x^2}{4x^2 + 9} \right| + C$$ ◆◆◆

Common Error	When using the table of integrals, be sure that the integral chosen completely matches the given integral, and be sure especially that *all factors of du are present.*

◆◆◆ **Example 39:** Integrate $\displaystyle\int \frac{dx}{x\sqrt{x^2 + 16}}$.

Solution: We use Rule 64.

Rule 64	$\displaystyle\int \frac{du}{u\sqrt{u^2 + a^2}} = \frac{1}{a} \ln\left| \frac{u}{a + \sqrt{u^2 + a^2}} \right| + C$

with $u = x$, $a = 4$, and $du = dx$. Thus

$$\int \frac{dx}{x\sqrt{x^2 + 16}} = \frac{1}{4} \ln\left| \frac{x}{4 + \sqrt{x^2 + 16}} \right| + C$$ ◆◆◆

Exercise 2 ◆ Rules for Finding Integrals

Evaluate each integral. Check some by calculator.

Integral of a Power Function

1. $\displaystyle\int (x^4 + 1)^3 4x^3 \, dx$

2. $\displaystyle\int (2x^2 - 6)^3 4x \, dx$

3. $\displaystyle\int 9(x^3 + 1)^2 x^2 \, dx$

4. $\displaystyle\int 6(x^2 - 1)^2 x \, dx$

5. $\displaystyle\int 2(x^2 + 2x)^3 (2x + 2) \, dx$

6. $\displaystyle\int (x + 1)(x^2 + 2x + 6)^2 \, dx$

7. $\displaystyle\int \frac{dx}{(1-x)^2}$

8. $\displaystyle\int \frac{4x\,dx}{(9-x^2)^2}$

9. $\displaystyle\int x\sqrt{x^2-2}\,dx$

10. $\displaystyle\int 3x\sqrt{x^2-1}\,dx$

11. $\displaystyle\int \frac{x^2\,dx}{\sqrt{1-x^3}}$

12. $\displaystyle\int x\sqrt{x^2-5}\,dx$

Other Variables

13. $\displaystyle\int 5t^4\,dt$

14. $\displaystyle\int (w^2+2)^3\,(w\,dw)$

15. $\displaystyle\int (z^3+3z)^3\,(z^2+1)\,dz$

16. $\displaystyle\int y\sqrt{y^2-7}\,dy$

17. $\displaystyle\int \frac{t^2+9}{t}\,dt$

18. $\displaystyle\int (w^3-3w)^4\,(w^2-1)\,dw$

Integral of du/u

19. $\displaystyle\int \frac{3}{x}\,dx$

20. $\displaystyle\int \frac{dx}{x-1}$

21. $\displaystyle\int \frac{5z^2}{z^3-3}\,dz$

22. $\displaystyle\int \frac{t\,dt}{6-t^2}$

23. $\displaystyle\int \frac{x+1}{x}\,dx$

24. $\displaystyle\int \frac{w^2+5}{w}\,dw$

Miscellaneous Rules from a Table of Integrals

Integrate, using the rule from Appendix C whose number is given.

25. $\displaystyle\int \frac{dt}{4-9t^2}$ Rule 57.

26. $\displaystyle\int \frac{ds}{\sqrt{s^2-16}}$ Rule 62.

27. $\displaystyle\int \sqrt{25-9x^2}\,dx$ Rule 69.

28. $\displaystyle\int \sqrt{4x^2+9}\,dx$ Rule 66.

29. $\displaystyle\int \frac{dx}{x^2+9}$ Rule 56.

30. $\displaystyle\int \frac{dx}{x^2+2x}$ Rule 49.

31. $\displaystyle\int \frac{dx}{16x^2+9}$ Rule 56.

32. $\displaystyle\int x\sqrt{1+3x}\,dx$ Rule 52.

Integrate, finding an appropriate rule from Appendix C.

33. $\displaystyle\int \frac{dx}{9-4x^2}$

34. $\displaystyle\int \sqrt{1+9x^2}\,dx$

35. $\displaystyle\int \frac{dx}{x^2-4}$

36. $\displaystyle\int \frac{dy}{\sqrt{25-y^2}}$

37. $\displaystyle\int \sqrt{\frac{x^2}{4}-1}\,dx$

38. $\displaystyle\int \frac{x^2\,dx}{\sqrt{3x+5}}$

39. $\displaystyle\int \frac{5x\,dx}{\sqrt{1-x^4}}$

26–3 Simple Differential Equations

To tackle the applications that follow, mostly in the next two chapters, we have to solve equations that contain derivatives, and evaluate all constants of integration. Here we show how to do that, for simple cases.

Solving a Simple Differential Equation

Suppose we have the derivative of a function, say,

$$\frac{dy}{dx} = 3x^2$$

Any equation that contains a derivative, such as this one, is called a *differential equation*. Since it contains only a *first* derivative, it is called a *first-order* differential equation. To *solve* this differential equation means to find the equation $y = F(x)$ of which $3x^2$ is the derivative. The steps are

(a) Write the equation in differential form by multiplying both sides by dx.
(b) Take the integral of both sides.

◆◆◆ **Example 40:** Solve the differential equation $\dfrac{dy}{dx} = 3x^2$.

Solution:

(a) Multiplying by dx gives

$$dy = 3x^2\, dx$$

(b) Now taking the integral of both sides of the equation

$$\int dy = \int 3x^2\, dx$$

so

$$y + C_1 = x^3 + C_2$$

Next we combine the two arbitrary constants, C_1 and C_2.

$$y = x^3 + C_2 - C_1 = x^3 + C$$

This is the function whose derivative is $3x^2$. This function is also called the *solution* to the differential equation $dy/dx = 3x^2$. ◆◆◆

Note that our solution to the preceding problem contains a constant of integration. But if our aim is to use the integral to solve practical problems from technology, of what use is an equation that contains an unknown constant? What we need is a way to evaluate such a constant.

Finding the Constant of Integration

To find the constant of integration we need another piece of information. Such additional information is called a *boundary condition*, or if our variable is time, an *initial condition*. A boundary condition is often in the form of a point through which the curve passes, as in the following example.

◆◆◆ **Example 41:** Evaluate the constant of integration in the preceding example if it is known that the curve passes through the point (1, 2).

Solution: We had found that

$$y = x^3 + C$$

Since $y = 2$ when $x = 1$, we have

$$2 = (1)^3 + C$$

from which $C = 1$. Our solution, with no unknown constant, is then

$$y = x^3 + 1$$ ◆◆◆

An Application

In our applications of the derivative, we saw that velocity is the rate of change of displacement. So we took the derivative of displacement to get velocity. The reverse of this is that we take the *integral* of velocity to get displacement.

◆◆◆ **Example 42:** A certain body thrown downward has a velocity of

$$v = 18.2 + 32.2t \qquad \text{ft/s}$$

and its displacement is 55.6 ft at $t = 2.00$ s. Write an equation for its displacement s.

Solution: Since velocity is the first derivative of displacement, we write

$$v = \frac{ds}{dt} = 18.2 + 32.2t$$

Going to differential form and integrating

$$\int ds = \int (18.2 + 32.2t) \, dt$$

Integrating,

$$s = 18.2t + \frac{32.2t^2}{2} + C$$

To evaluate C we let $s = 55.6$ when $t = 2.00$.

$$55.6 = 18.2\,(2.00) + 16.1\,(2.00)^2 + C$$

Solving, we get $C = -45.2$. Our complete equation for s, with no unknown constant, is

$$s = 18.2t + 16.1t^2 - 45.2 \text{ ft}$$ ◆◆◆

> Keep in mind that our main applications of the integral will come in the two following chapters.

Solving a Simple Second-Order Differential Equation

An equation that contains a *second* derivative, such as

$$y'' = 4x - 2$$

is called a *second-order* differential equation. To solve such an equation we must *integrate twice*. However, each time we integrate we get another constant of integration. To evaluate both we need *two* additional pieces of information, as in the following example.

◆◆◆ **Example 43:** Find the equation that has a second derivative $y'' = 4x - 2$ and that has a slope of 1 at the point $(2, 9)$.

Solution: We can write the second derivative as

$$\frac{d(y')}{dx} = 4x - 2$$

or, in differential form,

$$d(y') = (4x - 2)\, dx$$

Integrating gives

$$y' = \int (4x - 2)\, dx = 2x^2 - 2x + C_1$$

But the slope, and hence y', is 1 when $x = 2$, so

$$C_1 = 1 - 2(2)^2 + 2(2) = -3$$

This gives $y' = 2x^2 - 2x - 3$ or, in differential form,

$$dy = (2x^2 - 2x - 3)\, dx$$

Integrating again gives

$$y = \int (2x^2 - 2x - 3)\, dx = \frac{2x^3}{3} - x^2 - 3x + C_2$$

But $y = 9$ when $x = 2$, so

$$C_2 = 9 - \frac{2(2^3)}{3} + 2^2 + 3(2) = \frac{41}{3}$$

Our final equation, with all of the constants evaluated, is thus

$$y = \frac{2x^3}{3} - x^2 - 3x + \frac{41}{3} \qquad\qquad ◆◆◆$$

Another Application

We found in the preceding chapter that *acceleration* is the rate of change of velocity, which is itself the rate of change of displacement. Thus acceleration is therefore the rate of change of the rate of change of displacement. Since rates of change are found by taking the derivative, we see that acceleration is the *second derivative* of displacement. Thus if we want to find displacement from the acceleration, we have to *integrate twice*.

◆◆◆ **Example 44:** A body moves with an acceleration of 5.86 m/s^2. It has an initial velocity of 4.55 m/s and an initial displacement of 3.94 m. Write the equations for velocity and displacement, with all constants evaluated.

Solution: Since the acceleration is dv/dt, we write

$$\frac{dv}{dt} = 5.86$$

$$dv = 5.86\, dt$$

Integrating gives

$$\int dv = 5.86 \int dt$$

$$v = 5.86t + C_1$$

Since $v = 4.55$ when $t = 0$, we get

$$C_1 = 4.55 - 5.86(0) = 4.55$$

so

$$v = 5.86t + 4.55 \text{ m/s}$$

Now since $v = ds/dt$,

$$\frac{ds}{dt} = 5.86t + 4.55$$

$$ds = (5.86t + 4.55)\,dt$$

Integrating again,

$$\int ds = \int (5.86t + 4.55)\,dt$$

$$s = 5.86\frac{t^2}{2} + 4.55t + C_2$$

Since $s = 3.94$ when $t = 0$, we get $C_2 = 3.94$. Our final equation is then

$$s = 2.93\,t^2 + 4.55t + 3.94 \quad \text{m}$$

◆◆◆

Exercise 3 ◆ Simple Differential Equations

Simple First-Order Differential Equations

Solve each differential equation.

1. $\dfrac{dy}{dx} = 4x^2$ **2.** $\dfrac{dy}{dx} = 2x(x^2 + 6)$

3. $\dfrac{dy}{dx} = x^{-3}$ **4.** $\dfrac{ds}{dt} = 10t^{-6}$

5. $\dfrac{ds}{dt} = \dfrac{1}{2}t^{-2/3}$ **6.** $\dfrac{dv}{dt} = 6t^3 - 3t^{-2}$

Finding the Constant of Integration

Solve each differential equation, including evaluation of the constant of integration.

7. $y' = 3x$, passes through $(2, 6)$

8. $y' = x^2$, passes through $(1, 1)$

9. $y' = \sqrt{x}$, passes through $(2, 4)$

10. If $dy/dx = 2x + 1$, and $y = 7$ when $x = 1$, find the value of y when $x = 3$.

11. If $dy/dx = \sqrt{2x}$, and $y = \frac{1}{3}$ when $x = \frac{1}{2}$, find the value of y when $x = 2$.

Simple Second-Order Differential Equations

12. Find the equation of a curve that has a second derivative $y'' = x$ if it has a slope of $7/2$ at the point $(3, 0)$.

13. Find the equation of a curve that has a second derivative $y'' = 4$ if it has a slope of 3 at the point $(2, 6)$.

14. Find the equation of a curve that has a second derivative $y'' = 12/x^3$ if it has a slope of -6 at the point $(1, 0)$.

15. Find the equation of a curve that has a second derivative $y'' = 12x^2 - 6$ if it has a slope of 20 at the point $(2, 4)$.

Applications

16. A certain body thrown downward has a velocity of $v = 32.2t + 43.4$ m/s and its displacement is 28.5 m at $t = 1.00$ s. Write an equation for its displacement s.

17. A body moves with an acceleration of 21.5 m/s². It has an initial velocity of 27.6 m/s and an initial displacement of 44.3 m. Write the equations for velocity and displacement, with all constants evaluated.

26–4   The Definite Integral

While the indefinite integral will enable us to solve a great many problems from technology, there are others that are best tackled with what is called a *definite* integral. We will use the definite integral to find the area of an airplane rudder, the surface area and volume of a rocket, the length of a bridge cable, the centroid of a wind generator vane, the fluid pressure on a dam, the work needed to compress a spring, the moment of inertia of a flywheel, and much more.

We will explain what definite integrals are in this section, and how to evaluate them. We will see that all our rules for integrals can still be used. Most of our applications will come soon afterwards.

The Fundamental Theorem of Calculus

Earlier we learned how to find the indefinite integral of a function. For example,

$$\int x^2 \, dx = \frac{x^3}{3} + C \tag{1}$$

We can, of course, evaluate the integral at some particular value, say, $x = 6$. Substituting into Eq. (1) gives us

$$\int x^2 \, dx \bigg|_{x=6} = \frac{6^3}{3} + C = 72 + C$$

Similarly, we can evaluate the same integral at, say, $x = 3$. Again substituting in to Eq. (1) yields

$$\int x^2 \, dx \bigg|_{x=3} = \frac{3^3}{3} + C = 9 + C$$

Suppose, now, that we subtract the second integral from the first. We get

$$72 + C - (9 + C) = 63$$

Although we do not know the value of the constant C, we do know that it has the same value in both expressions since both were obtained from Eq. (1), so C will drop out when we subtract.

We now introduce *new notation*. We let

$$\int x^2 \, dx \bigg|_{x=6} - \int x^2 \, dx \bigg|_{x=3} = \int_3^6 x^2 \, dx$$

The expression on the right is called a *definite integral*. Here 6 is called the *upper limit,* and 3 is the *lower limit.* This notation tells us to evaluate the integral at the upper limit and, from that, subtract the integral evaluated at the lower limit. Notice that a definite integral (unlike the indefinite integral) has a *numerical* value, in this case, 63. But what is this number, and what is it good for? We'll soon see that the definite integral gives us the area under the curve $y = x^2$, from $x = 3$ to 6, and that it has many applications. In general, if

$$\int f(x) \, dx = F(x) + C$$

We require, as usual, that the function $f(x)$ be *continuous* in the interval under consideration.

then

$$\int f(x)\, dx \bigg|_{x=a} = F(a) + C$$

$$\int f(x)\, dx \bigg|_{x=b} = F(b) + C$$

and

$$\int_a^b f(x)\, dx = F(b) + C - F(a) - C$$

The constants drop out, leaving the following equation:

Definite Integral

$$\int_a^b f(x)\, dx = F(b) - F(a)$$

The definite integral of a function is equal to the integral of that function evaluated at the upper limit b minus the integral evaluated at the lower limit a.

288

The equation defining the definite integral is called *the fundamental theorem of calculus*, because it connects the processes of differentiation and of integration.

We will now show how to evaluate a definite integral. Unlike with an indefinite integral, which gives us an equation, here we will get a *numerical value*.

Evaluating a Definite Integral

To *evaluate* a definite integral, first integrate the expression (omitting the constant of integration), and write the upper and lower limits on a vertical bar or bracket to the right of the integral. Next substitute the upper limit and then the lower limit, and subtract.

◆◆◆ **Example 45:** Evaluate $\int_2^4 x^2\, dx$.

Solution:

$$\int_2^4 x^2\, dx = \frac{x^3}{3}\bigg|_2^4$$

$$= \frac{4^3}{3} - \frac{2^3}{3} = \frac{56}{3} \qquad ◆◆◆$$

◆◆◆ **Example 46:** Evaluate $\int_0^3 (2x - 5)^3\, dx$.

Solution: We insert a factor of 2 in the integral and compensate with a factor of $\frac{1}{2}$ in front.

$$\int_0^3 (2x - 5)^3\, dx = \frac{1}{2}\int_0^3 (2x - 5)^3 (2\, dx)$$

$$= \frac{1}{2}\left(\frac{1}{4}\right)(2x - 5)^4 \bigg|_0^3$$

$$= (1/8)[(6 - 5)^4 - (0 - 5)^4]$$

$$= (1/8)[1 - 625] = -78 \qquad ◆◆◆$$

We have seen that integrals will sometimes produce expressions with absolute value signs, as in the following example.

◆◆◆ **Example 47:**

$$\int_{-3}^{-2} \frac{dx}{x} = \ln|x| \Big|_{-3}^{-2} = \ln|-2| - \ln|-3|$$

The logarithm of a negative number is not defined. But here we are taking the logarithm of the *absolute value* of a negative number. Thus

$$\ln|-2| - \ln|-3| = \ln 2 - \ln 3 \cong -0.405 \qquad \text{◆◆◆}$$

Discontinuity

Recall that a function is called *discontinuous* if there is a break, jump, cusp, corner or gap in the curve, and is *nondifferentiable* at such points. The definite integral is *not defined* over an interval containing any of these features.

◆◆◆ **Example 48:** The integral $\int_{-3}^{2} \dfrac{dx}{x}$ is not defined because the function $y = 1/x$ is discontinuous at $x = 0$. ◆◆◆

Common Error	Be sure that your function is continuous and differentiable between the given limits before evaluating a definite integral.

A simple but rough way to check for corners, cusps, and jumps is with a graph. This will show many discontinuities, but small gaps may go undetected.

◆◆◆ **Example 49:** Is the function $y = \sqrt{x^2 - 4}$ continuous from $x = 3$ to 6?

Solution: Our graph shows continuity between the given limits. However, the function is discontinuous between $x = -2$ and 2. ◆◆◆

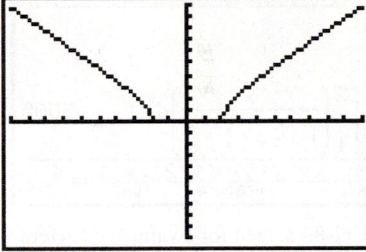

Screen for Example 49. Tick marks are 1 unit apart.

Finding a Definite Integral by Calculator

In an earlier section, we used

$$\mathbf{fnInt}\,(u, x, o, x)$$

to designate the approximate indefinite integral, obtained by calculator, of a function u whose variable of integration is x. We could graph the indefinite integral but could not, of course, obtain a numerical value for it. We will now use

$$\mathbf{fnInt}\,(u, x, a, b)$$

to designate the approximate *definite* integral, obtained by calculator, of a function u whose variable of integration is x, between the limits a and b. Here we will obtain a numerical value for the expression.

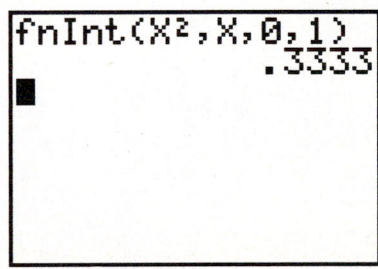

(1) TI-83/84 screen for Example 50.

◆◆◆ **Example 50:** (a) Evaluate $\int_{0}^{1} x^2\,dx$ analytically, (b) check by TI-83/84 calculator, and (c) check by TI-89 calculator.

Solution:

(a) $\displaystyle\int_{0}^{1} x^2\,dx = (1/3)x^3 \Big|_{0}^{1} = 1/3[1^3 - 0^3] = 1/3$

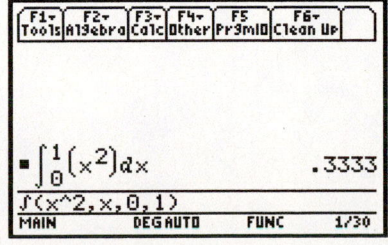

(2) TI-89 screen for Example 50.

(b) On the TI-83/84, we enter **fnInt(X^2, X, 0, 1)** and get the rounded value 0.3333, which agrees with our analytical result, screen (1).

(c) On the TI-89, we press $\boxed{\int}$ or select ∫( ) from the **Calc** menu. We follow this by the expression to be integrated (x^2), the variable of integration (x), the lower limit (0), and the upper limit (1). Pressing $\boxed{=}$ gives the integral, screen (2). ◆◆◆

(1) TI-83/84 screen for Example 51. As usual, you must check your calculator manual to see how these operations are performed on your own device.

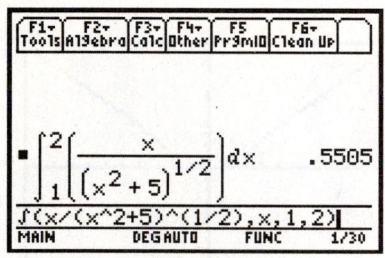

(2) TI-89 screen for Example 51. Here we press the $\boxed{\approx}$ key to get the approximate decimal value.

◆◆◆ **Example 51:** Evaluate $\displaystyle\int_1^2 \frac{x\,dx}{\sqrt{x^2+5}}$ and verify by calculator.

Solution: We write the radical in exponential form,

$$\int_1^2 \frac{x\,dx}{\sqrt{x^2+5}} = \int_1^2 (x^2+5)^{-\frac{1}{2}}(x\,dx)$$

We match Rule 5 by multiplying $x\,dx$ by 2, and compensate with $\frac{1}{2}$ in front of the integral. We then integrate and substitute in the limits.

$$\int_1^2 (x^2+5)^{-\frac{1}{2}}(x\,dx) = \frac{1}{2}\int_1^2 (x^2+5)^{-\frac{1}{2}}(2x\,dx)$$

$$= (1/2)\frac{(x^2+5)^{(1/2)}}{(1/2)}\Big|_1^2$$

$$= \sqrt{x^2+5}\,\Big|_1^2$$

$$= \sqrt{9} - \sqrt{6} = 0.5505$$

This is verified by screens (1) and (2). ◆◆◆

Exercise 4 ◆ The Definite Integral

Evaluate each definite integral to three significant digits. Check some by calculator.

1. $\displaystyle\int_1^2 x\,dx$

2. $\displaystyle\int_{-2}^2 x^2\,dx$

3. $\displaystyle\int_1^3 7x^2\,dx$

4. $\displaystyle\int_{-2}^2 3x^4\,dx$

5. $\displaystyle\int_0^4 (x^2+2x)\,dx$

6. $\displaystyle\int_{-2}^2 x^2(x+2)\,dx$

7. $\displaystyle\int_2^4 (x+3)^2\,dx$

8. $\displaystyle\int_0^a (x-x^3)\,dx$

9. $\displaystyle\int_1^{10} \frac{dx}{x}$

10. $\displaystyle\int_1^e \frac{dx}{x}$

11. $\displaystyle\int_0^1 \frac{x\,dx}{\sqrt{4+x^2}}$

12. $\displaystyle\int_{-2}^{-3} \frac{2x\,dx}{1+x^2}$

13. $\displaystyle\int_0^1 \frac{dx}{\sqrt{3-2x}}$

14. $\displaystyle\int_0^1 \frac{x\,dx}{\sqrt{2-x^2}}$

An Application

15. The volume of the rocket nose cone, Fig. 26–2, is given by

$$\int_0^{48.0} 12\pi x \, dx$$

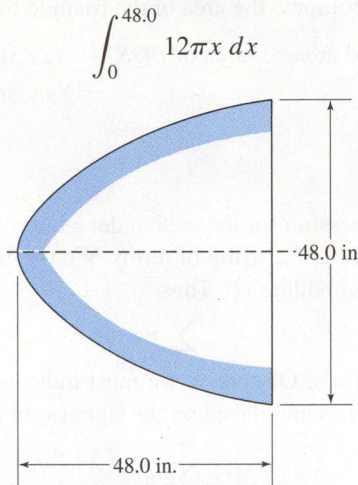

48.0 in.

48.0 in.

FIGURE 26–2 Rocket nose cone.

Evaluate this integral.

26–5 Approximate Area Under a Curve

In this section we will find the approximate area under a curve both graphically and numerically. These methods are valuable not only as a lead-in to finding exact areas by integration, but also for finding areas under the many functions that cannot be integrated.

■ Exploration:

Try this. On a sheet of graph paper, draw coordinate axes, draw any smooth curve, and draw vertical lines at upper and lower limits *a* and *b*, as in Fig. 26–3.

- Can you devise a way to get an approximate value for the shaded area shown?
- Can you think of more than one way to do it?   ■

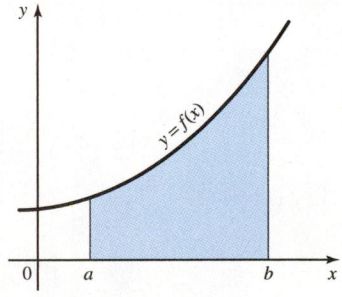

FIGURE 26–3

Estimating Areas

We want to find the approximate value of the area under a curve, such as $y = f(x)$ in Fig. 26–3, between two limits *a* and *b*. Our graphical approach is simple. After plotting the curve, we will subdivide the required area into rectangles, find the area of each rectangle, and add them.

◆◆◆ Example 52: Find the approximate area under the curve $y = x^2/3$ between the limits $x = 1$ and $x = 3$.

Solution: We draw the curve as shown in Fig. 26–4 between the upper and lower limits. We subdivide the required area into squares $\frac{1}{2}$ unit on a side, and we count them, estimating the fractional part of those that are incomplete. We count around 12 squares, each with an area of $\frac{1}{4}$ square units, getting

$$\text{area} \approx 3 \text{ square units}$$ ◆◆◆

Another way to estimate the area under a curve is to simply sketch a rectangle or triangle of roughly the same area right on the graph of the given curve and compute its area.

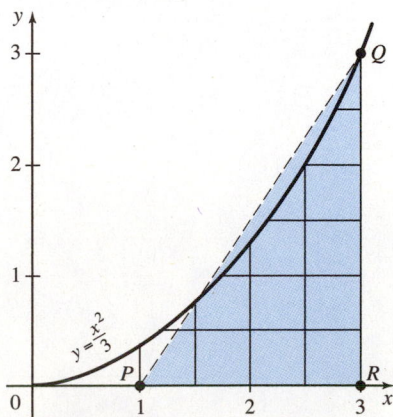

FIGURE 26–4 We have chosen very large squares for this illustration. Of course, smaller squares would give greater accuracy.

◆◆◆ **Example 53:** Make a quick estimate of the area in Example 52.

Solution: We draw line PQ as shown in Fig. 26–4, trying to balance the excluded and included areas, and compute the area of the triangle formed.

$$\text{shaded area} \cong \text{area of } PQR = \tfrac{1}{2}(2)(3)$$
$$= 3 \text{ square units} \qquad \text{◆◆◆}$$

Summation Notation

Before we derive an expression for the area under a curve, we must learn some new notation to express the sum of a string of terms. We use the Greek capital sigma Σ to stand for summation, or adding up. Thus

$$\sum n$$

means to sum a string of n's. Of course, we must indicate a starting and an ending value for n, and these values are placed on the sigma symbol. Thus

$$\sum_{n=1}^{5} n$$

means to add up the n's starting with $n = 1$ and ending with $n = 5$.

$$\sum_{n=1}^{5} n = 1 + 2 + 3 + 4 + 5 = 15$$

◆◆◆ **Example 54:** Evaluate $\displaystyle\sum_{n=1}^{4} (n^2 - 1)$.

Solution:

$$\sum_{n=1}^{4} (n^2 - 1) = (1^2 - 1) + (2^2 - 1) + (3^2 - 1) + (4^2 - 1)$$
$$= 0 + 3 + 8 + 15 = 26 \qquad \text{◆◆◆}$$

◆◆◆ **Example 55:**

(a) $\displaystyle\sum_{k=2}^{5} k^2 = 2^2 + 3^2 + 4^2 + 5^2 = 54$

(b) $\displaystyle\sum_{x=1}^{4} f(x) = f(1) + f(2) + f(3) + f(4)$

(c) $\displaystyle\sum_{i=1}^{n} f(x_i) = f(x_1) + f(x_2) + f(x_3) + \cdots + f(x_n)$ ◆◆◆

In the following section, we use the summation notation for expressions similar to that of Example 55(c).

A Numerical Technique: The Midpoint Method

We will now show a method that is easily computerized and that will lead us to a method for finding exact areas later. Also, the notation that we introduce now will be used again later.

Figure 26–5 shows a graph of some function $f(x)$. Our problem is to find the area (shown lightly shaded) bounded by that curve, the x axis, and the lines $x = a$ and $x = b$.

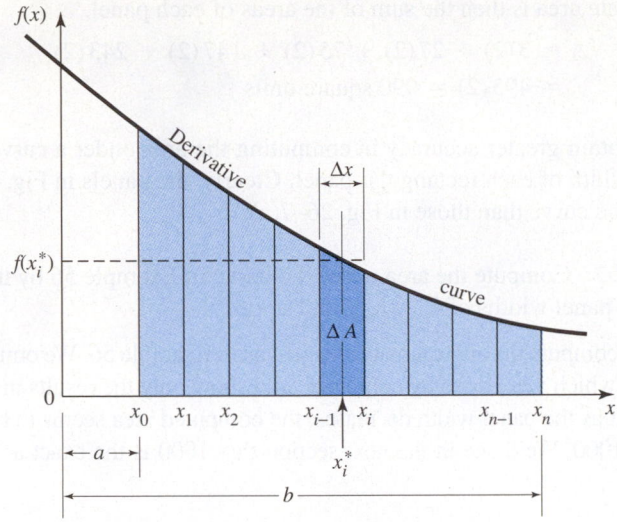

FIGURE 26–5

We start by subdividing that area into n vertical strips, called *panels*, by drawing vertical lines at $x_0, x_1, x_2, \ldots, x_n$. The panels do not have to be of equal width, but we make them equal for simplicity. Let the width of each panel be Δx.

Now look at one particular panel, the one lying between x_{i-1} and x_i (shown shaded darkly). At the midpoint of this panel we choose a point x_i^*. The height of the curve at this value of x is then $f(x_i^*)$. The area ΔA of the dark panel is then *approximately* equal to the area of a rectangle of width Δx and height $f(x_i^*)$.

$$\Delta A \approx f(x_i^*)\, \Delta x$$

The area of the first panel is, similarly, $f(x_1^*)\, \Delta x$; of the second panel, $f(x_2^*)\, \Delta x$; and so on. To get an approximate value for the total area, we add up the areas of each panel.

$$A \approx f(x_1^*)\, \Delta x + f(x_2^*)\, \Delta x + f(x_3^*)\, \Delta x + \cdots + f(x_n^*)\, \Delta x$$

Rewriting this expression using our sigma notation gives

$$A \approx \sum_{i=1}^{n} f(x_i^*)\, \Delta x \tag{289}$$

These are called *Riemann sums,* after Georg Friedrich Bernhard Riemann (1826–66).

Midpoint Method	$A \approx \sum_{i=1}^{n} f(x_i^*)\, \Delta x$	**289**
	where $f(x_i^*)$ is the height of the *i*th panel at its midpoint.	

◆◆◆ **Example 56:** Use the midpoint method to calculate the approximate area under the curve $f(x) = 3x^2$ from $x = 0$ to 10, taking panels of width 2.

Solution: Our graph (Fig. 26–6) shows the panels, with midpoints at 1, 3, 5, 7, and 9. At each midpoint x^*, we compute the height $f(x^*)$ of the curve.

x^*	1	3	5	7	9
$f(x^*)$	3	27	75	147	243

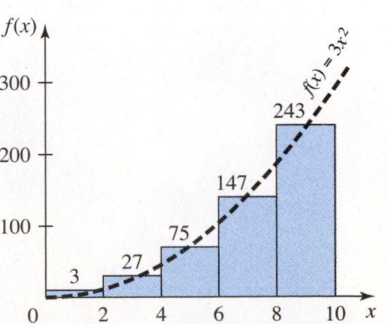

FIGURE 26–6 Area by midpoint method.

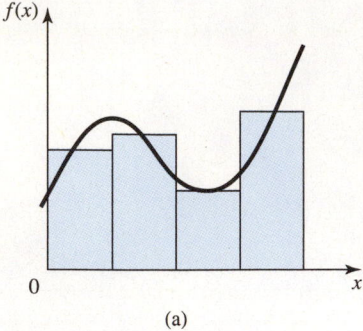

(a)

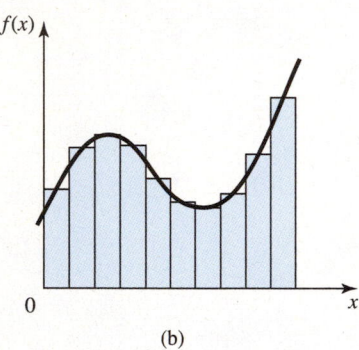

(b)

FIGURE 26–7 More panels give greater accuracy.

TABLE 26–1

Panel Width	Area
2.0000	990.0000
1.0000	997.5000
0.5000	999.3750
0.2500	999.8438
0.1250	999.9609
0.0625	999.9902
0.0313	999.9968
0.0156	999.9996

The approximate area is then the sum of the areas of each panel.

$$A \approx 3(2) + 27(2) + 75(2) + 147(2) + 243(2)$$
$$= 495(2) = 990 \text{ square units}$$ ◆◆◆

We can obtain greater accuracy in computing the area under a curve simply by reducing the width of each rectangular panel. Clearly, the panels in Fig. 26–7(b) are a better fit to the curve than those in Fig. 26–7(a).

◆◆◆ **Example 57:** Compute the area under the curve in Example 56 by the midpoint method, using panel widths of 2, 1, $\frac{1}{2}$, $\frac{1}{4}$, and so on.

Solution: We compute the approximate area just as in Example 56. We omit the tedious computations (which were done by computer) and show only the results in Table 26–1.

Notice that as the panel width decreases, the computed area seems to be approaching a limit of 1000. We'll see in the next section that 1000 is the exact area under the curve.   ◆◆◆

Exercise 5 ◆ Approximate Area Under a Curve

Estimation of Areas

Estimate the approximate area of each shaded region, in square units.

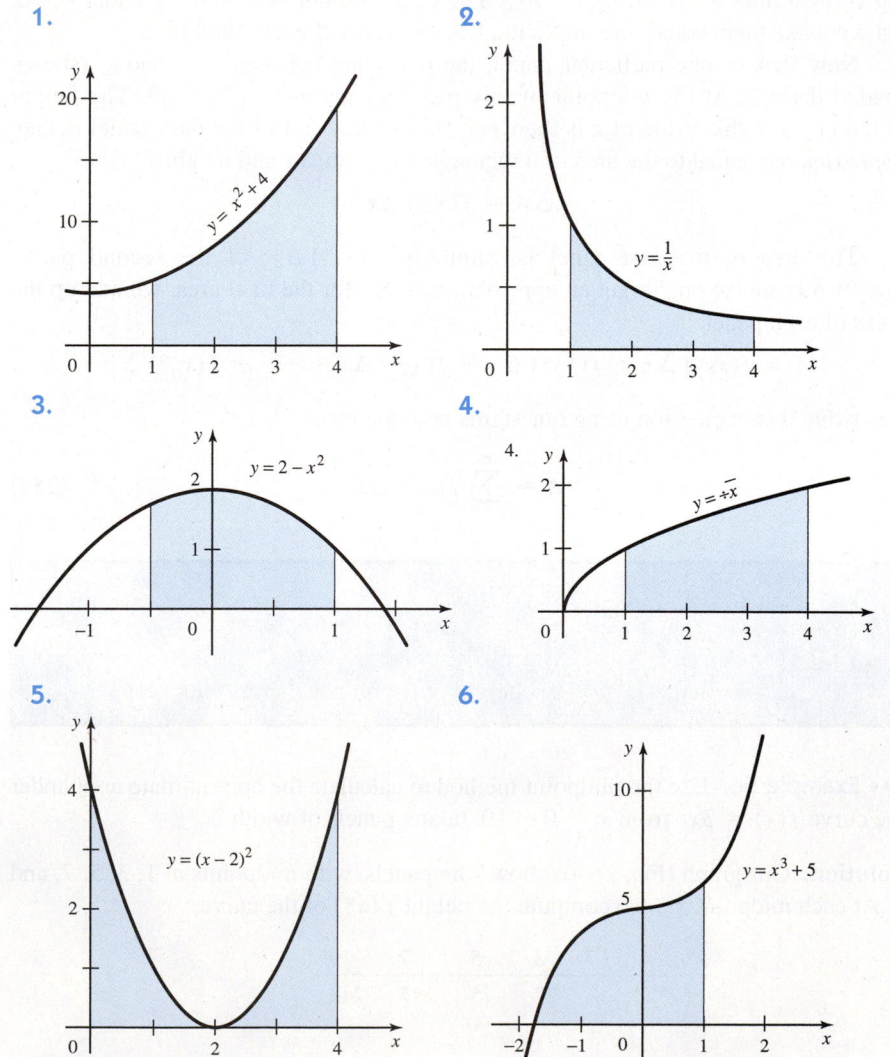

Graph each region. Make a quick estimate of the indicated area, and then use a graphical method to find its approximate value.

7. $y = x^2 + 1$ from $x = 0$ to 8

8. $y = x^2 + 3$ from $x = -4$ to 4

9. $y = 1/x$ from $x = 2$ to 10

10. $y = 2 + x^4$ from $x = -2$ to 0

Sigma Notation

Evaluate each expression.

11. $\displaystyle\sum_{n=1}^{5} n$ **12.** $\displaystyle\sum_{r=1}^{9} r^2$ **13.** $\displaystyle\sum_{n=1}^{7} 3n$

14. $\displaystyle\sum_{m=1}^{4} \frac{1}{m}$ **15.** $\displaystyle\sum_{n=1}^{5} n(n-1)$ **16.** $\displaystyle\sum_{q=1}^{6} \frac{q}{q+1}$

Approximate Areas by Midpoint Method

Using panels 2 units wide, find the approximate area (in square units) under each curve by the midpoint method.

17. $y = x^2 + 1$ from $x = 0$ to 8

18. $y = x^2 + 3$ from $x = -4$ to 4

19. $y = \dfrac{1}{x}$ from $x = 2$ to 10

20. $y = 2 + x^4$ from $x = -10$ to 0

26–6   Exact Area Under a Curve

We saw with the midpoint method that we could obtain a more accurate area by using a greater number of narrower panels. In the limit, we find the *exact* area by using an *infinite* number of panels, each of infinitesimal width.

Starting with Eq. 289,

$$A \approx \sum_{i=1}^{n} f(x_i{}^*)\,\Delta x$$

we let the panel width Δx approach zero. As it does so, the number of panels approaches infinity, and the sum of the areas of the panels approaches the exact area A under the curve.

Exact Area Under a Curve	$A = \displaystyle\lim_{\Delta x \to 0} \sum_{i=1}^{n} f(x_i{}^*)\,\Delta x$	290

Exact Area by Integration

Equation 290 gives us the exact area under a curve, but not a practical way to find it. We will now derive a formula that will give us the exact area. The derivation will be long, but the formula itself will be very simple. In fact, it is one that you have already used.

Suppose we want to find the exact area under some curve $f(x)$, between the values a and b, Fig. 26–8(b). Directly above the graph of $f(x)$, let us graph its integral $F(x)$, Fig. 26–8(b). Thus, the upper curve $F(x)$ is the integral of the lower curve $f(x)$; conversely the lower curve is the derivative of the upper curve.

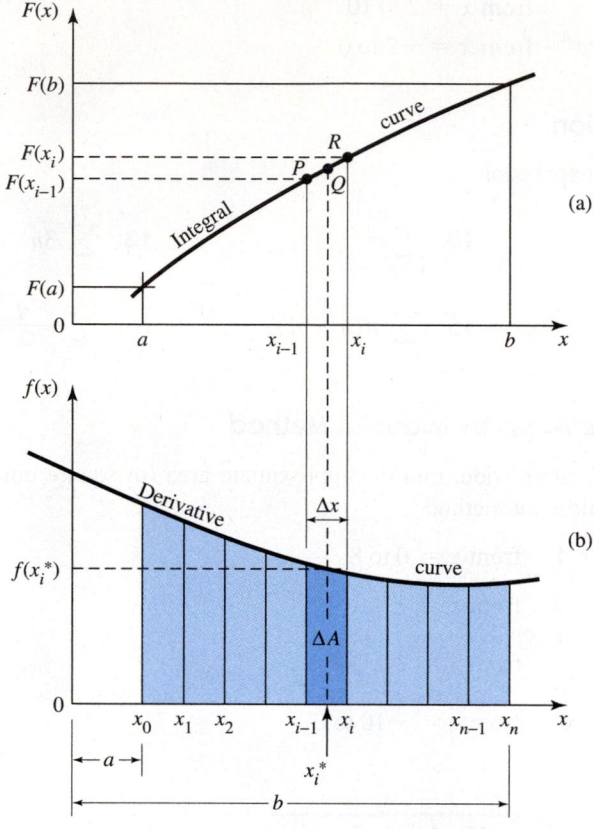

FIGURE 26–8   Area as the limit of a sum.

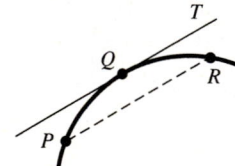

FIGURE 26–9   There is a theorem, called the *mean value theorem*, that says there must be at least one point Q between P and R at which the slope is equal to the slope of PR. We won't prove it, but can you see intuitively that it must be so?

For the midpoint method, we arbitrarily selected x_i^* at the midpoint of each panel. We now do it differently. *We select x_i^* so that the slope at point Q on the integral (upper) curve is equal to the slope of the straight line PR,* as shown in Fig. 26–9.

The slope at Q is equal to $f(x_i^*)$, and the slope of PR is equal to

$$f(x_i^*) = \frac{\text{rise}}{\text{run}} = \frac{F(x_i) - F(x_{i-1})}{\Delta x}$$

or

$$f(x_i^*)\Delta x = F(x_i) - F(x_{i-1})$$

If we write this expression for each panel, we get

$$f(x_1^*)\,\Delta x = F(x_1) - F(a)$$
$$f(x_2^*)\,\Delta x = F(x_2) - F(x_1)$$
$$f(x_3^*)\,\Delta x = F(x_3) - F(x_2)$$
$$\vdots$$
$$f(x_{n-1}^*)\,\Delta x = F(x_{n-1}) - F(x_{n-2})$$
$$f(x_n^*)\,\Delta x = F(b) - F(x_{n-1})$$

If we add all of these equations, every term on the right drops out except $F(a)$ and $F(b)$.

$$f(x_1^*)\,\Delta x + f(x_2^*)\,\Delta x + f(x_3^*)\,\Delta x + \cdots + f(x_n^*)\,\Delta x = F(b) - F(a)$$

$$\sum_{i=1}^{n} f(x_i^*)\,\Delta x = F(b) - F(a)$$

As before, we let Δx approach zero.

$$\lim_{\Delta x \to 0} \sum_{i=1}^{n} f(x_i^*)\,\Delta x = F(b) - F(a)$$

The left side of this equation is equal to the exact area A under the curve (Eq. 290). The right side is equal to the definite integral from a to b of the function $f(x)$ that we had earlier derived as Eq. 288. Thus:

Exact Area Under a Curve	$$A = \int_a^b f(x)\,dx = F(b) - F(a)$$ *The exact area under a curve is given by the definite integral of the given fuction between the given limits.*	**291**

We get the amazingly simple result that the area under a curve between two limits is equal to the change in the integral between the same limits, as shown graphically in Fig. 26–10.

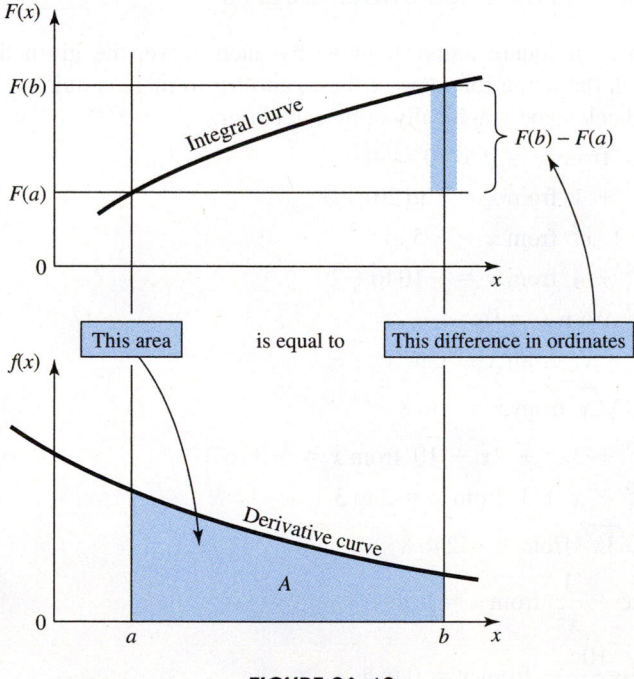

FIGURE 26–10

So, *to find the exact area under a curve, simply evaluate the definite integral of the given function between the given limits.*

◆◆◆ **Example 58:** Find the area bounded by the curve $y = 3x^2 + x + 1$, the x axis, and the lines $x = 2$ and $x = 5$.

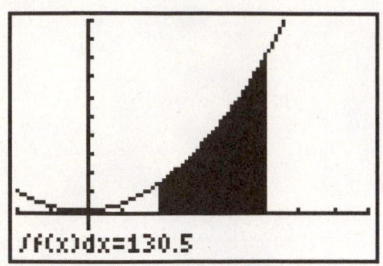

(1) TI-83/84 screen for Example 59. Tick marks are 1 unit apart on the *x* axis and 10 units apart on the *y* axis.

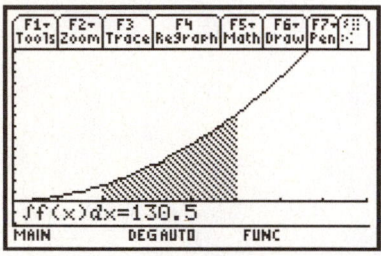

(2) TI-89 screen for Example 59.

Solution: By Eq. 291,

$$A = \int_2^5 (3x^2 + x + 1)\, dx = x^3 + \frac{x^2}{2} + x \Big|_2^5$$

$$= \left(5^3 + \frac{5^2}{2} + 5 \right) - \left(2^3 + \frac{2^2}{2} + 2 \right)$$

$$= 130.5 \text{ square units} \qquad \text{◆◆◆}$$

Area Under a Curve by Calculator

To find the area under a curve that has already been graphed,

- select $\int f(x)\, dx$ from the appropriate menu
- select the upper and lower limits

The calculator will shade in the area under the curve between the selected limits and give the numerical value of that area.

◆◆◆ **Example 59:** Verify the result of the preceding problem by calculator.

Solution: We enter the function itself in the $\boxed{Y =}$ screen and have it graphed. We then select $\int f(x)\, dx$. This is found in the $\boxed{\text{CALC}}$ menu on the TI-83/84 and on the $\boxed{\text{MATH}}$ **Calculus** menu on the TI-89. Then enter 2 for the lower limit and 5 for the upper limit. The result, 130.5, is displayed below the graph. ◆◆◆

Exercise 6 ◆ Exact Area Under a Curve

Find the area (in square units) bounded by each curve, the given lines, and the *x* axis. Sketch the curve for some of these, and try to make a quick estimate of the area. Also check some graphically or by calculator.

1. $y = 2x$ from $x = 0$ to 10

2. $y = x^2 + 1$ from $x = 1$ to 20

3. $y = 3 + x^2$ from $x = -5$ to 5

4. $y = x^4 + 4$ from $x = -10$ to -2

5. $y = x^3$ from $x = 0$ to 4

6. $y = 9 - x^2$ from $x = 0$ to 3

7. $y = 1/\sqrt{x}$ from $x = \frac{1}{2}$ to 8

8. $y = x^3 + 3x^2 + 2x + 10$ from $x = -3$ to 3

9. $y = x^2 + x + 1$ from $x = 2$ to 3

10. $y = \sqrt{3x}$ from $x = 2$ to 8

11. $y = 2x + \dfrac{1}{x^2}$ from $x = 1$ to 4

12. $y = \dfrac{10}{\sqrt{x + 4}}$ from $x = 0$ to 5

We are doing only simple area problems here. In the following chapter we will find areas between curves, areas below the *x* axis, and other more complicated types of areas. Remember that the bulk of our applications will come in the next two chapters. There we will show how to set up an integral such as this one.

An Application

13. A ship's deck, Fig. 26–11, has the shape of two intersecting parabolas. Its area in square feet is given by

$$2 \int_{-30}^{23} \left(6.5 - \frac{13x^2}{1800} \right) dx$$

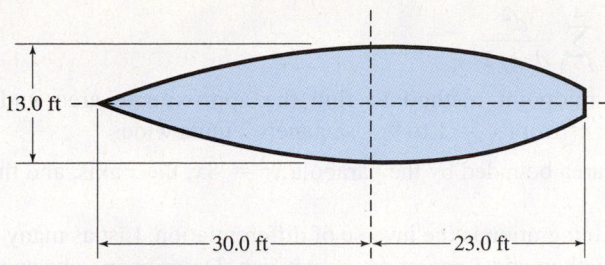

FIGURE 26–11   Top view of a ship's deck.

Find this area.

••• CHAPTER 26 REVIEW PROBLEMS •••••••••••••••••••••••••••••••

Perform each integration.

1. $\displaystyle\int \frac{dx}{\sqrt[3]{x}}$

2. $\displaystyle\int \frac{x^4 + x^3 + 1}{x^3} \, dx$

3. $\displaystyle\int 3.1 \, y^2 \, dy$

4. $\displaystyle\int \frac{2dt}{t^2}$

5. $\displaystyle\int \sqrt{4x} \, dx$

6. $\displaystyle\int x^2 (x^3 - 4)^2 \, dx$

7. $\displaystyle\int (x^4 - 2x^3)(2x^3 - 3x^2) \, dx$

8. $\displaystyle\int \frac{x \, dx}{x^2 + 3}$

9. $\displaystyle\int \frac{dx}{x + 5}$

10. $\displaystyle\int_{-1}^{0} \frac{dx}{1 - x}$

11. $\displaystyle\int_{2}^{7} (x^2 - 2x + 3) \, dx$

12. $\displaystyle\int_{0}^{2} \frac{x \, dx}{4 + x^2}$

13. $\displaystyle\int_{0}^{a} (\sqrt{a} - \sqrt{x})^2 \, dx$

14. $\displaystyle\int_{2}^{1} \sqrt{7 - 3x} \, dx$

15. $\displaystyle\int_{0}^{3} \sqrt{9 + 25x^2} \, dx$

16. $\displaystyle\int_{0}^{2} \sqrt{x^2 + 25} \, dx$

17. $\displaystyle\int_{1}^{4} x\sqrt{1 + 5x} \, dx$

18. $\displaystyle\int_{0}^{2} \sqrt{1 + 9x^2} \, dx$

19. $\displaystyle\int \frac{dx}{9 - 4x^2}$

20. $\displaystyle\int \frac{dx}{\sqrt{x^2 - 25}}$

21. $\displaystyle\int \frac{dx}{9 + 4x^2}$

22. $\displaystyle\int_{1}^{4} \frac{dx}{x^2 + 4x}$

23. $\displaystyle\int_{2}^{6} \frac{dx}{\sqrt{4 + 9x^2}}$

24. Find the equation of a curve that passes through the point $(3, 0)$, has a slope of 9 at that point, and has a second derivative $y'' = x$.

25. The rate of growth of the number N of bacteria in a culture is $dN/dt = 0.5N$. If $N = 100$ when $t = 0$, derive the formula for N at any time.

26. Evaluate $\displaystyle\sum_{n=1}^{5} n^2(n - 1)$.

27. Evaluate $\displaystyle\sum_{d=1}^{4} \frac{d^2}{d+1}$.

28. Use the midpoint method to find the approximate area under the curve $y = 5 + x^2$ from $x = 1$ to 9. Use panels 2 units wide.

29. Find the area bounded by the parabola $y^2 = 8x$, the x axis, and the line $x = 2$.

30. *Writing:* Integration is the inverse of differentiation. List as many other pairs of inverse mathematical operations as you can. Describe in your own words when the inverse of each operation gives an indefinite result.

31. *Project:* Given a function by your instructor that can be integrated, and an upper and lower limit, (a) make an accurate graph, and estimate the area of the given region by counting boxes on the graph paper, (b) use the midpoint method with different panel widths to calculate the same area, and (c) find the exact area by integrating. Compare the results obtained by the various methods.

32. *On Our Web Site:* There are a great many methods for finding integrals that we have not shown here, but that may be found under *Methods of Integration* at our Web site: www.wiley.com/college/calter

Applications of the Integral

OBJECTIVES

When you have completed this chapter, you should be able to

- Use the integral to solve motion problems.
- Use the integral to solve problems involving electric circuits.
- Find exact areas under curves and between curves.
- Find the volume of a solid of revolution using the disk or shell method.
- Determine a volume of a solid of revolution rotated about the x axis, the y axis, or a noncoordinate axis.

While we had a few applications in the last chapter, here we start with two chapters completely devoted to applications of the integral. Our first applications concern the motion of a point. In our chapter on the derivative, we found velocity and acceleration by taking derivatives. Here we do the reverse; we find velocity and displacement by taking the *integral*. Motion is followed by the application of the integral to electric circuits, a topic we will take up again after we learn how to take derivatives and integrals of the logarithmic, exponential, and trigonometric functions.

Next we present a fast way to set up the integral and use it to find the exact area under a curve. We found areas in the preceding chapter, and here do more advanced problems, such as the area between curves and the area between a curve and the y axis. This leads to applications such as verifying the areas of familiar plane figures, finding areas of structural members, culverts, windows, and so forth.

Finally, we learn how to compute volumes of solids of revolution. We use this to verify the formulas for volumes of common solids, finding volumes of rocket nose cones, structural members, tanks for liquids, and so on. For example, how would you find the capacity of the aircraft wing tank, Fig. 27–1, given its dimensions and the shape of the curved surface? We will do such a calculation here.

Throughout this chapter we support our analytical methods with the graphics calculator.

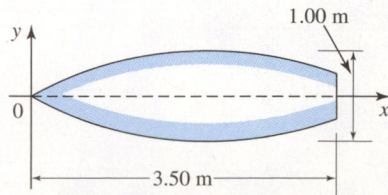

FIGURE 27–1 Airplane wing tank.

27–1   Applications to Motion

Displacement

In an earlier chapter we saw that the velocity v of a moving point was defined as the rate of change of the displacement s of the point. The velocity was thus equal to the derivative of the displacement, or $v = ds/dt$. We now reverse the process and find the displacement when given the velocity. Since $ds = v\,dt$, integrating gives the following equation:

Displacement	$$s = \int v\,dt$$	1022

Displacement is the integral of velocity.

Thus if given an expression for velocity, we can get one for displacement by integrating. We use the initial conditions to evaluate the constant of integration, as in the following example.

◆◆◆ **Example 1:** A point in a mechanism has an initial displacement of 2.75 ft, and has a velocity given by

$$v = 3.74t + 5.85 \qquad \text{ft/s}$$

(a) Write an equation for the displacement s and (b) evaluate it at $t = 5.00$ s.

Solution:

(a) Since $v = ds/dt$, we write

$$\frac{ds}{dt} = 3.74t + 5.85$$

$$ds = (3.74t + 5.85)\,dt$$

Integrating gives

$$s = \int (3.74t + 5.85)\,dt = \frac{3.74t^2}{2} + 5.85t + C$$

We find the constant by noting that $s = 2.75$ ft at $t = 0$. That gives $C = 2.75$, so our complete equation for displacement is

$$s = 1.87t^2 + 5.85t + 2.75 \qquad \text{ft}$$

(b) At $t = 5.00$ s,

$$s = 1.87(5.00)^2 + 5.85(5.00) + 2.75 = 78.8 \text{ ft} \qquad\qquad ◆◆◆$$

Velocity

In our applications of the derivative, we saw that acceleration was the derivative of velocity. The reverse is also true, that the velocity is the integral of the acceleration.

Instantaneous Velocity	$$v = \int a\,dt$$	1024

Velocity is the integral of the acceleration.

Given an expression for acceleration, we can get one for velocity by integrating. As before, we use the initial conditions to obtain the constant of integration.

◆◆◆ **Example 2:** A point on an industrial robot has an initial velocity of 9.85 m/s and has an acceleration given by

$$a = 11.1 + 15.6t^2 \qquad m/s^2$$

(a) Write an equation for the velocity v and (b) evaluate it at $t = 2.00$ s.

Solution:

(a) Since $a = dv/dt$, we write

$$\frac{dv}{dt} = 11.1 + 15.6t^2$$

$$dv = (11.1 + 15.6t^2)\, dt$$

Integrating gives

$$v = \int (11.1 + 15.6t^2)\, dt = 11.1t + \frac{15.6t^3}{3} + C$$

We find the constant by noting that $v = 9.85$ m/s at $t = 0$. That gives $C = 9.85$, so our complete equation for velocity is

$$v = 11.1t + 5.20t^3 + 9.85 \qquad m/s$$

(b) At $t = 2.00$ s,

$$v = 11.1(2.00) + 5.20(2.00)^3 + 9.85 = 73.6 \text{ m/s} \qquad ◆◆◆$$

Freely Falling Body

Integration provides us with a slick way to derive the equations for the displacement and velocity of a freely falling body.

◆◆◆ **Example 3:** An object falls with constant acceleration, g, due to gravity. Write the equations for the displacement and velocity of the body at any time.

Solution: We are given that $a = dv/dt = g$, so

$$dv = g\, dt$$

Integrating, we find that

$$v = \int g\, dt = gt + C_1$$

When $t = 0$, $v = C_1$, so C_1 is the initial velocity. Let us relabel the constant as v_0.

$$v = v_0 + gt \qquad \boxed{1019}$$

But $v = ds/dt$, so

$$ds = (v_0 + gt)\, dt$$

$$s = \int (v_0 + gt)\, dt = v_0 t + \frac{gt^2}{2} + C_2$$

When $t = 0$, $s = C_2$, so we interpret C_2 as the initial displacement. Let us call it s_0. So the displacement is:

$$s = s_0 + v_0 t + \frac{gt^2}{2} \qquad \boxed{1018}$$

◆◆◆

♦♦♦ **Example 4:** A body is thrown downward with an initial velocity of 7.55 ft/s. Write an equation for its (a) acceleration, (b) velocity, and (c) displacement. (d) Evaluate each at $t = 4.00$ s.

Solution:

(a) The acceleration of a freely falling body (even one thrown downward) is

$$a = 32.2 \text{ ft/s}^2$$

(b) The velocity is, from Eq. 1019,

$$v = v_0 + gt$$

$$= 7.55 + 32.2t \qquad \text{ft/s}$$

since $v_0 = 7.55$ ft/s.

(c) The displacement, from Eq. 1018, is

$$s = s_0 + v_0 t + \frac{gt^2}{2}$$

$$= 0 + 7.55t + 16.1t^2 \qquad \text{ft}$$

where we have taken the initial displacement s_0 equal to 0.

(d) At $t = 4.00$ s,

$$v = 7.55 + 32.2(4.00) = 136 \text{ ft/s}$$

and

$$s = 7.55(4.00) + 16.1(4.00)^2 = 288 \text{ ft} \qquad\qquad ♦♦♦$$

Motion Along a Curve

In an earlier chapter the motion of a point along a curve was described by parametric equations, with the x and y displacements each given by a separate function of time. We saw that dx/dt gave the velocity v_x in the x direction and that dy/dt gave the velocity v_y in the y direction. Now, given the velocities, we integrate to get the displacements.

	(a) $\qquad\qquad$ (b)	
Displacement in x **and** y **Directions**	$x = \displaystyle\int v_x \, dt \qquad\qquad y = \displaystyle\int v_y \, dt$	**1032**

Similarly, if we have parametric equations for the accelerations in the x and y directions, we integrate to get the velocities.

	(a) $\qquad\qquad$ (b)	
Velocity in x **and** y **Directions**	$v_x = \displaystyle\int a_x \, dt \qquad\qquad v_y = \displaystyle\int a_y \, dt$	**1034**

◆◆◆ **Example 5:** An object starts from the point (2, 4) with initial velocities $v_x = 7$ cm/s and $v_y = 5$ cm/s, and it moves along a curved path. It has x and y accelerations of $a_x = 3t$ cm/s^2 and $a_y = 5$ cm/s^2. Write expressions for the x and y components of (a) velocity and (b) displacement. (c) Find x and y when $t = 4$ s and (d) graph the parametric equations of displacement.

Solution:

(a) We integrate to find the velocities.

$$v_x = \int 3t\, dt = \frac{3t^2}{2} + C_1 \quad \text{and} \quad v_y = \int 5\, dt = 5t + C_2$$

At $t = 0$, $v_x = 7$ and $v_y = 5$, so

$$v_x = \frac{3t^2}{2} + 7 \quad \text{cm/s} \quad \text{and} \quad v_y = 5t + 5 \quad \text{cm/s}$$

(b) Integrating again gives the displacements.

$$x = \int \left(\frac{3t^2}{2} + 7\right) dt \quad \text{and} \quad y = \int (5t + 5)\, dt$$

$$x = \frac{t^3}{2} + 7t + C_3 \qquad y = \frac{5t^2}{2} + 5t + C_4$$

At $t = 0$, $x = 2$ and $y = 4$, so our complete equations for the displacements are

$$x = \frac{t^3}{2} + 7t + 2 \text{ cm} \quad \text{and} \quad y = \frac{5t^2}{2} + 5t + 4 \text{ cm}$$

(c) At $t = 4$ s,

$$x = \frac{4^3}{2} + 7(4) + 2 = 62 \text{ cm}$$

and

$$y = \frac{5(4^2)}{2} + 5(4) + 4 = 64 \text{ cm}$$

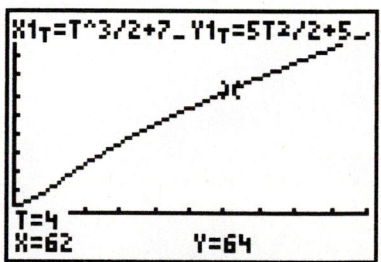

(d) A parametric plot of the displacements is given, showing the values of x and y at $t = 4$ s.

◆◆◆

Screen for Example 5. Glance back at Chap. 15 if you've forgotten how to make a parametric plot.

Rotation

In an earlier chapter we saw that the angular velocity ω of a rotating body was given by the derivative $d\theta/dt$ of the angular displacement θ. Thus θ is the integral of the angular *velocity*.

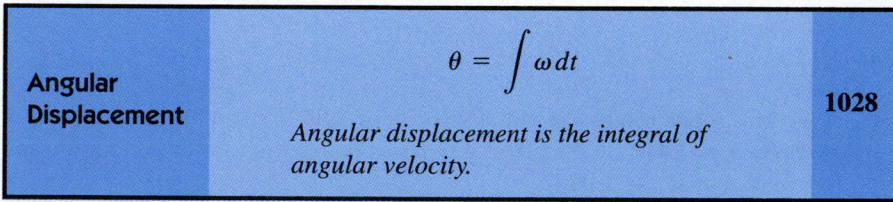

| Angular Displacement | $$\theta = \int \omega\, dt$$ | 1028 |

Angular displacement is the integral of angular velocity.

Similarly,

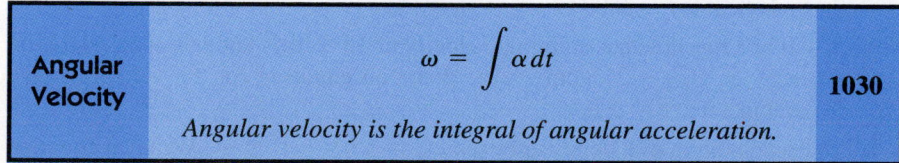

| Angular Velocity | $$\omega = \int \alpha\, dt$$ | 1030 |

Angular velocity is the integral of angular acceleration.

♦♦♦ **Example 6:** A flywheel in a machine starts from rest and accelerates at $3.85t$ rad/s^2. Find the angular velocity and the total number of revolutions after 10.0 s.

Solution: We integrate to get the angular velocity.

$$\omega = \int 3.85t \, dt = \frac{3.85t^2}{2} + C_1 \qquad \text{rad/s}$$

Since the flywheel starts from rest, $\omega = 0$ at $t = 0$, so $C_1 = 0$. Integrating again gives the angular displacement.

$$\theta = \int \frac{3.85t^2}{2} \, dt = \frac{3.85t^3}{6} + C_2 \qquad \text{rad}$$

Since θ is 0 at $t = 0$, we get $C_2 = 0$. Now evaluating ω and θ at $t = 10.0$ s yields

$$\omega = \frac{3.85(10.0)^2}{2} = 192 \text{ rad/s}$$

and, recalling that 2π radians equals 1 revolution,

$$\theta = \frac{3.85(10.0)^3}{6} = 642 \text{ rad} = 102 \text{ revolutions} \qquad \text{♦♦♦}$$

Exercise 1 ♦ Applications to Motion

Displacement

1. A point in a machine has an initial displacement of 12.6 cm and has a velocity given by $v = 11.6t + 21.4$ cm/s. (a) Write an equation for the displacement s and (b) evaluate it at $t = 7.00$ s.

2. At a particular location in a mechanism, the initial displacement is 6.48 in. and the velocity is given by $v = 1.83 + 2.28t^2$ in./s. (a) Write an equation for the displacement s and (b) evaluate it at $t = 4.00$ s.

3. A car starts from rest and continues at a rate of $v = \frac{1}{8}t^2$ ft/s. Find the function that relates the distance s the car has traveled to the time t in seconds. How far will the car go in 4 s?

4. A body is moving at the rate $v = \frac{3}{2}t^2$ m/s. Find the distance that it will move in t seconds if $s = 0$ when $t = 0$.

Velocity

5. A pin on a robot arm has an initial velocity of 2.58 ft/s and has an acceleration given by $a = 1.41t^2 + 5.28$ ft/s^2. (a) Write an equation for the velocity v and (b) evaluate it at $t = 1.00$ s.

6. A point in a mechanism has an initial velocity of 44.3 in./s and has an acceleration given by $a = 52.6t^2 - 41.1t$ in./s^2. (a) Write an equation for the velocity v and (b) evaluate it at $t = 2.00$ s.

7. A part in a machine has an initial velocity of 15.8 cm/s and has an acceleration given by $a = t^3 - 25.8$ cm/s^2. (a) Write an equation for the velocity v and (b) evaluate it at $t = 5.00$ s.

8. The acceleration of a point is given by $a = 4.00 - t^2$ m/s^2. Write an equation for the velocity if $v = 2.00$ m/s when $t = 3.00$ s.

Freely Falling Body

9. A body is thrown downward with an initial velocity of 1.77 ft/s. Write an equation for its (a) acceleration, (b) velocity, and (c) displacement. (d) Evaluate each at $t = 3.00$ s.

10. A ball is thrown downward with an initial velocity of 21.5 ft/s. Write an equation for the ball's (a) acceleration, (b) velocity, and (c) displacement. (d) Evaluate each at $t = 4.00$ s.

11. The acceleration of a falling body is $a = -32.2$ ft/s². Find the relation between s and t if $s = 0$ and $v = 20$ ft/s when $t = 0$.

Motion Along a Curve

12. A point starts from rest at the origin and moves along a curved path with x and y accelerations of $a_x = 2.00$ cm/s² and $a_y = 8.00t$ cm/s². Write expressions for the x and y components of velocity.

13. A point starts from rest at the origin and moves along a curved path with x and y accelerations of $a_x = 5.00t^2$ cm/s² and $a_y = 2.00t$ cm/s². Find the x and y components of velocity at $t = 10.0$ s.

14. A point starts from $(5, 2)$ with initial velocities of $v_x = 2.00$ cm/s and $v_y = 4.00$ cm/s and moves along a curved path. It has x and y accelerations of $a_x = 7t$ and $a_y = 2$. Find the x and y displacements at $t = 5.00$ s.

15. A point starts from $(9, 1)$ with initial velocities of $v_x = 6.00$ cm/s and $v_y = 2.00$ cm/s and moves along a curved path. It has x and y accelerations of $a_x = 3t$ and $a_y = 2t^2$. Find the x and y components of velocity at $t = 15.0$ s.

Rotation

16. A wheel starts from rest and accelerates at 3.00 rad/s². Find the angular velocity after 12.0 s.

17. A certain gear starts from rest and accelerates at $8.5t^2$ rad/s². Find the total number of revolutions after 20.0 s.

18. A link in a mechanism rotating with an angular velocity of 3.00 rad/s is given an acceleration of 5.00 rad/s² at $t = 0$. Find the angular velocity after 20.0 s.

19. A pulley in a magnetic tape drive is rotating at 1.25 rad/s when it is given an acceleration of 7.24 rad/s² at $t = 0$. Find the angular velocity at 2.00 s.

20. *Project:* Trajectories. A projectile, Fig. 27–2, is launched at an angle θ with initial velocity v. The horizontal acceleration x'' is zero and the vertical acceleration y'' is $-g$.

$$x'' = 0$$
$$y'' = -g$$

(a) Integrate each of these expressions twice to get expressions for the horizontal and vertical displacements, x and y.

(b) Evaluate the constants of integration, noting that

$$x(0) = 0 \qquad\qquad y(0) = 0$$

$$x'(0) = vt\cos\theta \qquad\qquad y'(0) = vt\sin\theta - \frac{1}{2}gt^2$$

(c) Compute the range by setting to 0, solving for the nonzero value of t, and substituting back to get x. You should get

$$\text{Range} = \frac{2v^2}{g}\sin\theta\cos\theta$$

(d) Show that this is equivalent to

$$\text{Range} = \frac{v^2}{g}\sin 2\theta$$

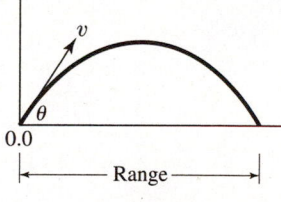

FIGURE 27–2

27–2   Applications to Electric Circuits

Recall that we had electrical applications of the *derivative* in Chap. 25. There we stated that it was only an *introduction*, because currents and voltages are often expressed by exponential or trigonometric functions, which we have not yet covered. The same is true here, and we will revisit this topic in Chapter 29, where we learn how to differentiate and integrate the logarithmic, exponential, and trigonometric functions. For now we will limit ourselves to simple algebraic expressions.

Charge

We stated in Chap. 25 that the current i (amperes, A) at some point in a conductor was equal to the time rate of change of the charge q (coulombs, C) passing that point, or

$$i = \frac{dq}{dt}$$

We can now solve this equation for q. Multiplying by dt gives

$$dq = i\,dt$$

Integrating, we get the following:

Charge	$q = \displaystyle\int i\,dt$ coulombs	1079

✦✦✦ **Example 7:** The current to a certain capacitor is given by $i = 2t^3 + t^2 + 3$. The initial charge on the capacitor is 6.83 C. Find (a) an expression for the charge on the capacitor and (b) the charge when $t = 5.00$ s.

Solution:

(a) Integrating the expression for current, we obtain

$$q = \int i\,dt = \int (2t^3 + t^2 + 3)\,dt = \frac{t^4}{2} + \frac{t^3}{3} + 3t + k \qquad \text{Coulombs}$$

We will use the letter k for the constant of integration in electrical problems to avoid confusion with C used for capacitance, or Coulombs. We find the constant of integration k by substituting the initial conditions, $q = 6.83$ C at $t = 0$. So $k = 6.83$. Our complete equation is then

$$q = \frac{t^4}{2} + \frac{t^3}{3} + 3t + 6.83 \qquad \text{Coulombs}$$

(b) When $t = 5.00$ s,

$$q = \frac{(5.00)^4}{2} + \frac{(5.00)^3}{3} + 3(5.00) + 6.83 = 376 \text{ C} \qquad \text{✦✦✦}$$

Voltage Across a Capacitor

The current in a capacitor has already been given by Eq. 1080, $i = C\,dv/dt$, where i is in amperes (A), C in farads (F), v in volts (V), and t in seconds (s). We now integrate to find the voltage across the capacitor.

$$dv = \frac{1}{C}i\,dt$$

Voltage Across a Capacitor	$v = \dfrac{1}{C}\displaystyle\int i\,dt$ volts	1081

◆◆◆ **Example 8:** A 1.25-F capacitor that has an initial voltage of 25.0 V is charged with a current that varies with time according to the equation $i = t\sqrt{t^2 + 6.83}$ A. Find the voltage across the capacitor at 1.00 s.

Solution: By Eq. 1081

$$v = \frac{1}{1.25}\int t\sqrt{t^2 + 6.83}\, dt = 0.80\left(\frac{1}{2}\right)\int (t^2 + 6.38)^{1/2}(2t\, dt)$$

$$= \frac{0.40(t^2 + 6.83)^{3/2}}{3/2} + k = 0.267(t^2 + 6.83)^{3/2} + k$$

where we have again used k for the constant of integration to avoid confusion with the symbol for capacitance. Since $v = 25.0$ V when $t = 0$, we get

$$k = 25.0 - 0.267(6.83)^{3/2} = 20.2 \text{ V}$$

When $t = 1.00$ s,

$$v = 0.267(1.00^2 + 6.83)^{3/2} + 20.2 = 26.0 \text{ V} \qquad \text{◆◆◆}$$

Current in an Inductor

The voltage across an inductor was given by Eq. 1086 as $v = L\, di/dt$, where L is the inductance in henrys. Integrating this equation we get:

Current in an Inductor	$i = \dfrac{1}{L}\displaystyle\int v\, dt$ amperes	1085

◆◆◆ **Example 9:** The voltage across a 10.6-H inductor is $v = \sqrt{3t + 25.4}$ V. Find the current in the inductor at 5.25 s if the initial current is 6.15 A.

Solution: From Eq. 1085,

$$i = \frac{1}{10.6}\int \sqrt{3t + 25.4}\, dt = 0.0943\left(\frac{1}{3}\right)\int (3t + 25.4)^{1/2}(3dt)$$

$$= \frac{0.0314(3t + 25.4)^{3/2}}{3/2} + k = 0.0210(3t + 25.4)^{3/2} + k$$

When $t = 0$, $i = 6.15$ A, so

$$k = 6.15 - 0.0210(25.4)^{3/2} = 3.46$$

When $t = 5.25$ s,

$$i = 0.0210[3(5.25) + 25.4]^{3/2} + 3.46 = 9.00 \text{ A} \qquad \text{◆◆◆}$$

Exercise 2 ◆ Applications to Electric Circuits

Charge

1. The current to a capacitor is given by $i = 2t + 3$. The initial charge on the capacitor is 8.13 C. Find the charge when $t = 1.00$ s.
2. The current to a certain circuit is given by $i = t^2 + 4$. If the initial charge is zero, find the charge at 2.50 s.
3. The current to a certain capacitor is $i = 3.25 + t^3$. If the initial charge on the capacitor is 16.8 C, find the charge when $t = 3.75$ s.

Voltage Across a Capacitor

4. A 21.5-F capacitor with zero initial voltage has a charging current of $i = \sqrt{t}$. Find the voltage across the capacitor at 2.00 s.

5. A 15.2-F capacitor has an initial voltage of 2.00 V. It is charged with a current given by $i = t\sqrt{5 + t^2}$. Find the voltage across the capacitor at 1.75 s.

6. A 75.0-μF capacitor has an initial voltage of 125 V and is charged with a current equal to $i = \sqrt{t + 16.3}$ mA. Find the voltage across the capacitor at 4.00 s.

Current in an Inductor

7. The voltage across a 1.05-H inductor is $v = \sqrt{23t}$ V. Find the current in the inductor at 1.25 s if the initial current is zero.

8. The voltage across a 52.0-H inductor is $v = t^2 - 3t$ V. If the initial current is 2.00 A, find the current in the inductor at 1.00 s.

9. The voltage across a 15.0-H inductor is given by $v = 28.5 + \sqrt{6t}$ V. Find the current in the inductor at 2.50 s if the initial current is 15.0 A.

27–3   Finding Areas by Integration

Another Way to Set Up an Integral

In the preceding chapter we found the area under a curve and the x axis, between two given limits. We did this by setting up and evaluating a definite integral. Now we will do somewhat different area problems: areas between curves and areas bounded by the y axis, and later we will find volumes, length of arc, and so forth.

For each of these we need to write a slightly different definite integral. To avoid doing a long derivation for each, we now give *an intuitive shortcut*.

Think of the integral sign as the letter S, standing for *sum*. It indicates that we are to *add up* the elements that are written after the integral sign. Thus

$$A = \int_a^b f(x)\,dx$$

can be read, "The area A is the sum of all of the elements having a height $f(x)$ and a width dx, between the limits of a and b," as shown in Fig. 27–3.

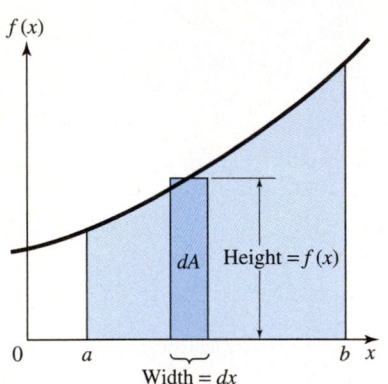

FIGURE 27–3

♦♦♦ **Example 10:** Find the area bounded by the curve $y = x^2 + 3$, the x axis, and the lines $x = 1$ and $x = 4$ (Fig. 27–4).

Estimate: Let us enclose the given area in a rectangle of width 3 and height 19, shown dashed in the figure. We see that the given area occupies more than half the area of the rectangle, say, about 60%. Thus a reasonable estimate of the required area would then be 60% of (3) (19), or 34 square units.

Solution: The usual steps are as follows:

(1) Make a sketch showing the bounded area, as in Fig. 27–4. Locate a point (x, y) on the curve.

(2) Through (x, y) draw a rectangular element of area, which we call dA. Next give the rectangle dimensions. We call the width dx because it is measured in the x direction, and we call the height y. The area of the element is thus

$$dA = y\,dx = (x^2 + 3)\,dx$$

(3) We think of A as the sum of all the small dA's. We accomplish the summation by integration.

$$A = \int dA = \int (x^2 + 3)\,dx$$

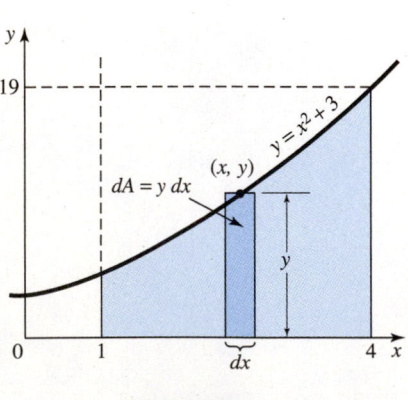

FIGURE 27–4

(4) We locate the limits of integration from the figure. Since we are summing the elements in the x direction, our limits must be on x. It is clear that we start the summing at $x = 1$ and end it at $x = 4$. So

$$A = \int_1^4 (x^2 + 3)\,dx$$

(5) Check that all parts of the integral, including the integrand, the differential, and the limits of integration, are in terms of the *same variable*. In our example, everything is in terms of x, and the limits are on x, so we can proceed. If, however, our integral contained both x and y, one of the variables would have to be eliminated. We show how to do this in a later example.

(6) Our integral is now set up. Integrating gives

$$A = \int_1^4 (x^2 + 3)\,dx = \frac{x^3}{3} + 3x \Big|_1^4$$

$$= \frac{4^3}{3} + 3(4) - \left[\frac{1^3}{3} + 3(1)\right] = 30 \text{ square units} \quad \blacklozenge\blacklozenge\blacklozenge$$

This looks like a long procedure, but it will save us a great deal of time later.

Areas Bounded by the *x* Axis

Let us now use the definite integral to find a few more areas under curves.

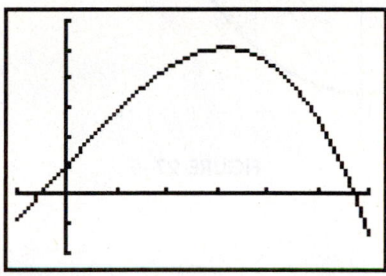

Screen for Example 11. It is possible to draw vertical lines on some calculators. On the TI-83/84 select **Vertical** from the $\boxed{\text{DRAW}}$ menu. Tick marks are 1 unit apart on the x axis and 5 units apart on the y axis.

◆◆◆ **Example 11:** Find the area under the curve $y = 9.51x - \dfrac{x^3}{3.14} + 5.25$ from $x = 1$ to 4.

Estimate:

Our sketch is made by calculator, as shown. We can imagine our area as being enclosed within a rectangle 3 units wide and about 25 units high, with an area of about 75 square units. Our area occupies about 90% of that rectangle, so we expect an answer of about 70 square units.

Solution: Let us take vertical elements of area, each of width dx, height y, and area dA.

$$dA = y\,dx = \left(9.51x - \frac{x^3}{3.14} + 5.25\right) dx$$

We now integrate to get the area.

$$A = \int_1^4 \left(9.51x - \frac{x^3}{3.14} + 5.25\right) dx = \frac{9.51x^2}{2} - \frac{x^4}{12.6} + 5.25x \Big|_1^4$$

$$= 76.76 - 9.936 = 66.8 \text{ square units}$$

which agrees with our estimate.   ◆◆◆

Sometimes the limits are not given and must be found by some other means.

◆◆◆ **Example 12:** Find the first-quadrant area bounded by the coordinate axes and the curve $y = 4x - x^3 + 2$.

Solution: Our graph shows that the curve has an x and a y intercept. We could try to find the x intercept analytically by setting y equal to zero and solving for x. A simpler way is to use the graphics calculator. We would use the $\boxed{\text{ZERO}}$ operation, as shown, or $\boxed{\text{TRACE}}$ and $\boxed{\text{ZOOM}}$.

We now integrate, between limits of 0 and 2.214.

$$A = \int_0^{2.214} (4x - x^3 + 2)\,dx = \frac{4x^2}{2} - \frac{x^4}{4} + 2x \Big|_0^{2.214}$$

$$= \frac{4(2.214)^2}{2} - \frac{(2.214)^4}{4} + 2(2.214) = 8.22 \text{ square units} \quad \blacklozenge\blacklozenge\blacklozenge$$

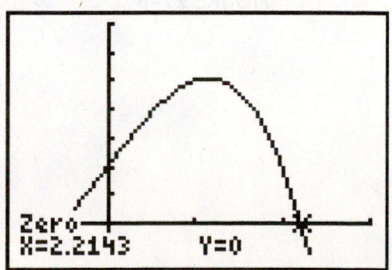

Screen for Example 12. Tick marks are 1 unit apart.

If a curve crosses the x axis somewhere between the lower and upper limits, as in Fig. 27–5, part of the required area will be below the x axis. If we set up our integral in the usual way, that area will come out *negative*. However, if we want the total area between the curve and the x axis, simply find each area separately and add their absolute values.

◆◆◆ **Example 13:** Find the total area bounded by the curve $y = x^2 - 4$ and the x axis between $x = 1$ and 3.

Solution: Our sketch (Fig. 27–5) shows two regions. If we integrate from $x = 1$ to $x = 3$, then A_1 will come out negative and A_2 will be positive, and our result will be A_2 *minus* A_1. If we want the *sum* of A_1 and A_2 we must set up two separate integrals.

$$A_1 = \int_1^2 (x^2 - 4)\, dx = \left. \frac{x^3}{3} - 4x \right|_1^2 = -\frac{5}{3}$$

$$A_2 = \int_2^3 (x^2 - 4)\, dx = \left. \frac{x^3}{3} - 4x \right|_2^3 = \frac{7}{3}$$

Adding absolute values gives

$$A = |A_1| + |A_2| = \left| -\frac{5}{3} \right| + \left| \frac{7}{3} \right| = 4 \text{ square units} \qquad ◆◆◆$$

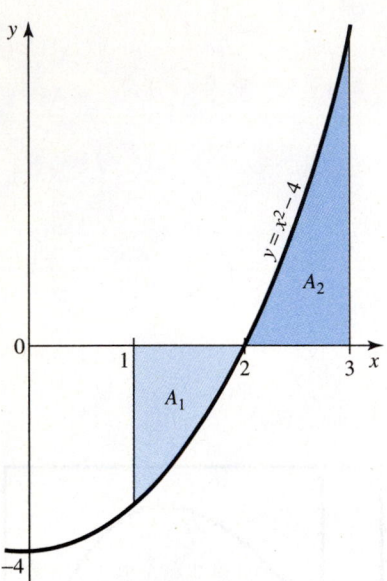

FIGURE 27–5

In our next example we find the area under a curve that has a *discontinuity*.

◆◆◆ **Example 14:** Find the area bounded by the curve $y = 1/x$ and the x axis (a) from $x = 1$ to 4, (b) from $x = -1$ to 4, and (c) from $x = -4$ to -1.

Solution: Integrating, we obtain

$$A = \int_a^b \frac{dx}{x} = \left. \ln|x| \right|_a^b = \ln|b| - \ln|a|$$

(a) For the limits 1 to 4,

$$A = \ln 4 - \ln 1 = 1.386 \text{ square units}$$

(b) For the limits -1 to 4,

$$A = \ln|4| - \ln|-1| = \ln 4 - \ln 1 = 1.386 \text{ square units (?)}$$

We appear to get the same area between the limits -1 and 4 as we did for the limits 1 and 4. However, a graph (Fig. 27–6) shows that the curve $y = 1/x$ is *discontinuous* at $x = 0$, so we cannot integrate over the interval -1 to 4.

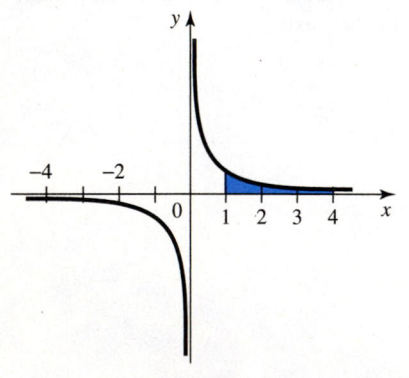

FIGURE 27–6

Common Errors	Don't try to set up these area problems without making a sketch. Don't try to integrate across a discontinuity.

A definite integral between limits a and b is called an *improper* integral if the integrand is discontinuous at some x within the interval $a \leq x \leq b$.

(c) For the limits -4 to -1,

$$A = \ln|-1| - \ln|-4|$$
$$= \ln 1 - \ln 4 = -1.386 \text{ square units}$$

This area lies below the x axis and has a negative value, as expected. ◆◆◆

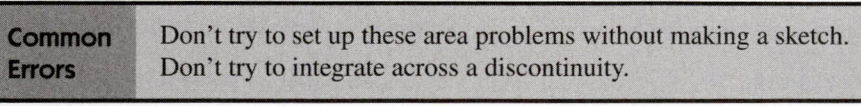

Areas by Graphics Calculator

In the preceding chapter we showed three ways to find the approximate area under a curve. On the TI-83/84 or similar calculators,

1. use the $\int f(x)\,dx$ operation from the $\boxed{\text{CALC}}$ menu on a formerly plotted curve, or
2. use the **fnInt** operation to evaluate a definite integral.

On the TI-89 or similar calculators,

3. use the $\boxed{\int}$ key or select $\int(\ )$ from the **Calc** menu.

We will now refresh our memories of these operations to find the area under a curve, where other methods may fail.

◆◆◆ **Example 15:** Find the approximate area under the curve

$$y = x^2 \ln(2.65x - 1.73)$$

from $x = 2$ to $x = 4$, by the three calculator methods.

Solution: We have not yet learned to integrate expressions containing logarithms and are not sure They are integrable even if we knew the rules. This is an ideal place to use the calculator.

(1) On the TI-83/84, we enter the function, screen (1). Then on the graph screen, screen (2), we select $\int(x)\,dx$ from the $\boxed{\text{CALC}}$ menu. We enter the lower and upper limits when prompted. The calculator then shades in the required area and displays its value below.

(2) On the TI-83/84 we select **fnInt** from the $\boxed{\text{MATH}}$ menu, screen (3), enter the function, the variable of integration x, the lower limit 2, and the upper limit 4. Note that we get the same result as in Method 1, as expected.

(3) On the TI-89, we press $\boxed{\int}$ or select $\int(\ )$ from the **Calc** menu. We follow this by the expression to be integrated, the variable of integration, and the lower and upper limits. Pressing $\boxed{\approx}$ gives the integral screen (4).

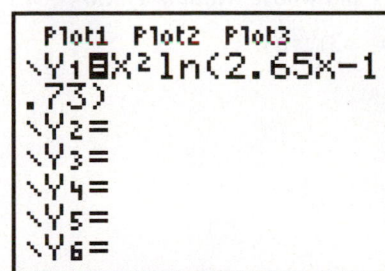

(1) The $\boxed{\text{Y} =}$ editor on the TI-83/84.

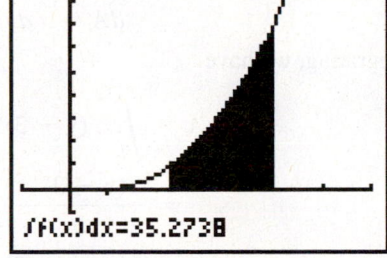

(2) Tick marks are 1 unit apart on the x axis and 5 units apart on the y axis.

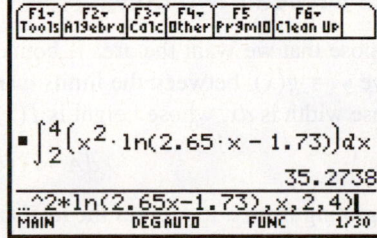

(3) Using **fnInt** on the TI-83/84.

(4) Using the TI-89.   ◆◆◆

Areas Bounded by the *y* Axis

So far we have only used *vertical* elements when setting up our integral. Sometimes, however, *horizontal* elements are more convenient. This is true when our area is bounded by the *y* axis, rather than the *x* axis, as in the following example.

◆◆◆ **Example 16:** Find the first-quadrant area bounded by the curve $y = x^2 + 3$, the *y* axis, and the lines $y = 7$ and $y = 12$ (Fig. 27–7).

Estimate: We enclose the given area in a 5×3 rectangle, shown dashed, of area 15. From this we subtract a roughly triangular portion whose area is $\frac{1}{2}(1)(5)$, or 2.5, getting an estimate of 12.5 square units.

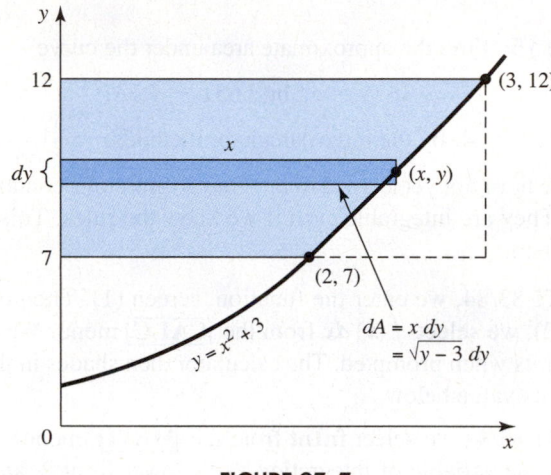

FIGURE 27–7

Solution: We locate a point (x, y) on the curve. If we were to draw a vertical element through (x, y), its height would be 5 units in the region to the left of the point $(2, 7)$, and $12 - y$ units in the region to the right of $(2, 7)$. Thus we would need two different expressions for the height of the element. To avoid this complication we choose a *horizontal* element whose length is simply x and whose width is dy. So

$$dA = x\, dy = (y - 3)^{1/2} dy$$

Integrating, we have

$$A = \int_7^{12} (y - 3)^{1/2} dy$$

$$= \frac{2(y - 3)^{3/2}}{3} \Big|_7^{12}$$

$$= \frac{2}{3}(9)^{3/2} - \frac{2}{3}(4)^{3/2} = \frac{38}{3} \text{ square units}$$ ◆◆◆

Area Between Two Curves

Suppose that we want the area *A* bounded by an upper curve $y = f(x)$ and a lower curve $y = g(x)$, between the limits *a* and *b* (Fig. 27–8). We draw a vertical element whose width is dx, whose height is $f(x) - g(x)$, and whose area dA is

$$dA = [f(x) - g(x)]\, dx$$

Integrating from *a* to *b* gives the total area.

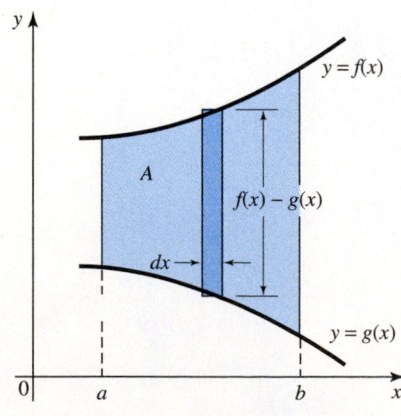

FIGURE 27–8 Area between two curves.

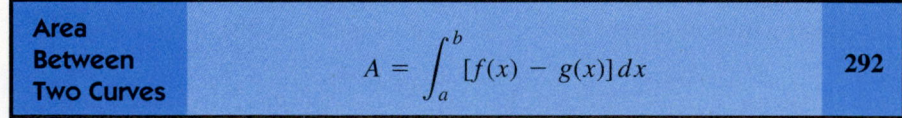

| Area Between Two Curves | $A = \int_a^b [f(x) - g(x)]\, dx$ | 292 |

It is important to get *positive* lengths for the elements. To do this, be sure to properly identify the "upper curve" for the region and subtract from it the values on the "lower curve."

The steps we follow are then

(a) Graph both curves in the same window.
(b) Write an equation for $f(x) - g(x)$. Simplify.
(c) Get the area by integrating the equation from the preceding step between the given limits.

◆◆◆ **Example 17:** Find the first-quadrant area bounded by the curves

$$y = x^2 + 3 \text{ and } y = 3x - x^2 \text{ between } x = 0 \text{ and } x = 3$$

Estimate: We make a sketch as shown in Fig. 27–9, and we "box in" the given area in a 3×12 rectangle, shown dashed, whose area is 36 square units. Let us estimate that the given area is less than half of that, or less than 18 square units.

Solution: Let the upper curve be $\qquad f(x) = x^2 + 3$

and the lower curve be $\qquad g(x) = 3x - x^2$

Subtracting the lower from the upper gives

$$f(x) - g(x) = (x^2 + 3) - (3x - x^2) = 2x^2 - 3x + 3$$

The area bounded by the curves is then

$$A = \int_0^3 (2x^2 - 3x + 3)\, dx = \frac{2x^3}{3} - \frac{3x^2}{2} + 3x \Big|_0^3$$

$$= 18 - \frac{27}{2} + 9 = 13.5 \text{ square units}$$

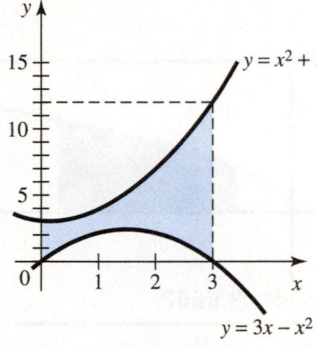

FIGURE 27–9

Alternate Solution: Another way of finding the area between two curves is to find the area between each and the x axis, and subtract one from the other. This method is useful for the calculator. Screen (1) shows the area under the upper curve, while screen (2) shows the area under the lower curve. For each we have the calculator find the area, as we showed earlier in this section. Their difference is 13.5, as found above.

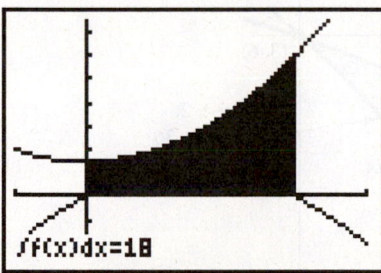 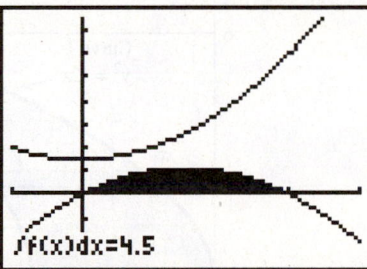

(1) Screen for Example 17. Tick marks are 1 unit apart.

(2) Tick marks are 1 unit apart.

◆◆◆

Don't worry if all or part of the desired area is below the x axis. Just follow the same procedure as when the area is above the axis, and the signs will work out by themselves. Be sure, however, that the lengths of the elements are *positive* by subtracting the lower curve from the upper curve.

◆◆◆ **Example 18:** Find the area bounded by the curves $y = \sqrt{x}$ and $y = x - 3$ between $x = 1$ and 4 (Fig. 27–10).

Estimate: Here our enclosing rectangle, shown dashed, is 3 by 4 units, with an area of 12 square units. The shaded area is about half of that, or about 6 square units.

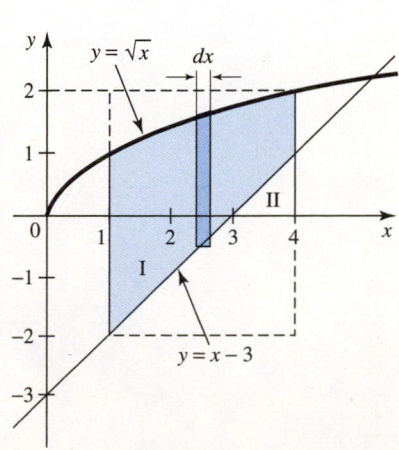

FIGURE 27–10

Solution: Letting $f(x) = \sqrt{x}$ and $g(x) = x - 3$, we get

$$A = \int_a^b [f(x) - g(x)]\,dx$$

$$= \int_1^4 [\sqrt{x} - x + 3]\,dx$$

$$= \frac{2x^{3/2}}{3} - \frac{x^2}{2} + 3x \Big|_1^4$$

$$= 6\tfrac{1}{6} \text{ square units}$$

Alternate Solution: If one of the bounding curves is just a straight line, we can often find our area by adding or subtracting the areas of triangles from the area between the curve and the x axis. This enables use of the calculator.

By calculator, as shown, we get an area between the curve and the x axis equal to 4.6667. We then add the area of triangle I, Fig. 27–10, and subtract the area of triangle II.

$$A = 4.6667 + \frac{1}{2}(2)(2) - \frac{1}{2}(1)(1) = 6.1667 \text{ square units}$$

which agrees with our previous answer and our estimate.   ◆◆◆

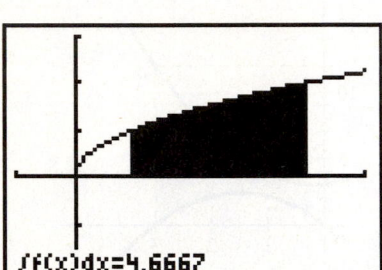

$\int f(x)dx=4.6667$

TI-83/84 screen for the Alternate solution to Example 18. Tick marks are 1 unit apart.

Sometimes we are asked to find the area *enclosed* by two curves, where the crossing points of two curves are the limits of integration. We find the points of intersection by solving the equations simultaneously, or by calculator using TRACE and ZOOM or **intersect**. These will be the limits of integration a and b.

In our next example we will also find it easier to take *horizontal elements* when setting up our integral.

◆◆◆ **Example 19:** Find the area bounded by the curves $y^2 = 12x$ and $y^2 = 24x - 36$.

Solution: We first plot the two curves as shown in Fig. 27–11. We recognize from their equations that they are parabolas opening to the right. We find their points of intersection by solving simultaneously.

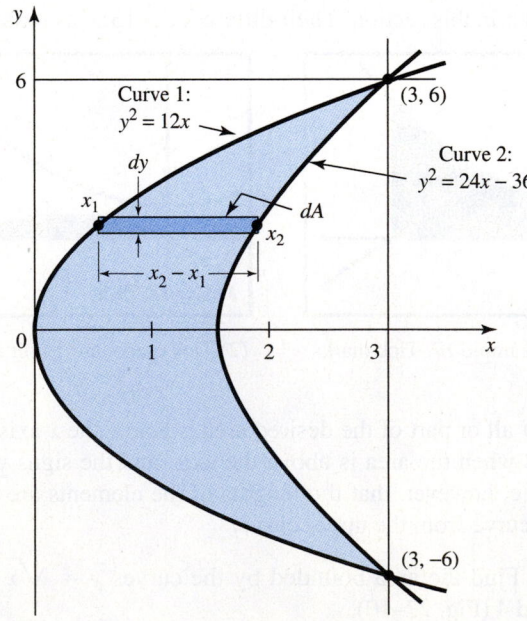

FIGURE 27–11

$$24x - 36 = 12x$$

$$x = 3$$

$$y = \pm\sqrt{12x} = \pm 6$$

Since we have symmetry about the x axis, let us solve for the first-quadrant area and later double it. We draw a horizontal strip of width dy and length $x_2 - x_1$ where x_2 is on the rightmost curve, and x_1 is on the left. The area dA is then $dA = (x_2 - x_1)\,dy$, where x_1 and x_2 are found by solving each given equation for x.

$$x_1 = \frac{y^2}{12} \quad \text{and} \quad x_2 = \frac{y^2 + 36}{24}$$

Integrating, we get

$$A = \int_0^6 (x_2 - x_1)\,dy$$

$$= \int_0^6 \left(\frac{y^2}{24} + \frac{36}{24} - \frac{y^2}{12} \right) dy$$

$$= \int_0^6 \left(\frac{3}{2} - \frac{y^2}{24} \right) dy$$

$$= \frac{3y}{2} - \frac{y^3}{72} \Big|_0^6$$

$$= \frac{3(6)}{2} - \frac{(6)^3}{72} = 6 \text{ square units}$$

By symmetry, the total area between the two curves is twice this, or 12 square units. ◆◆◆

Common Error	Don't always assume that a vertical element is the best choice. Try setting up the integral in Example 19 using a vertical element. What problems arise?

◆◆◆ **Example 20:** Find the area bounded by the parabola $y = 2 - x^2$ and the straight line $y = x$.

Solution: Our sketch, shown in Fig. 27–12 and in the calculator screen, shows two points of intersection. To find them, let us solve the equations simultaneously by setting one equation equal to the other.

$$2 - x^2 = x$$

or $x^2 + x - 2 = 0$. We can solve this quadratic by factoring.

$$(x + 2)(x - 1) = 0$$

so $x = -2$ and $x = 1$. Substituting back gives the points of intersection $(1, 1)$ and $(-2, -2)$, as we also see in the calculator plot.

We draw a vertical element whose width is dx and whose height is the upper curve minus the lower, or

$$2 - x^2 - x$$

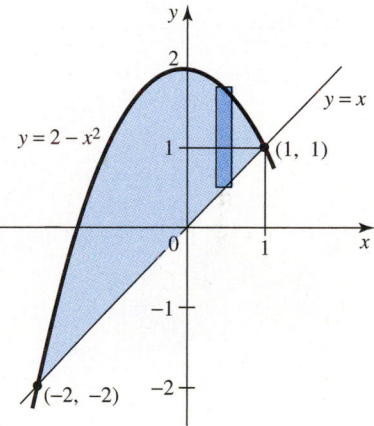

FIGURE 27–12

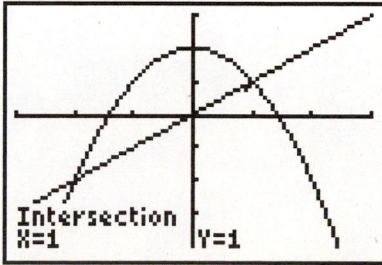

Screen for Example 20. Calculator plot of the two curves. Here **intersect** was used to locate the points of intersection, one of which is shown. Tick marks are 1 unit apart.

The area dA of the strip is then

$$dA = (2 - x^2 - x)\,dx$$

We integrate, taking as limits the values of x (-2 and 1) found earlier by simultaneous solution of the given equations.

$$A = \int_{-2}^{1} (2 - x^2 - x)\,dx$$

$$= 2x - \frac{x^3}{3} - \frac{x^2}{2}\bigg|_{-2}^{1} = 4\frac{1}{2} \text{ square units}$$ ◆◆◆

◆◆◆ **Example 21:** *An Application.* A triangular glass prism for a periscope is shown in Fig. 27–13. (a) Use integration to verify the formula $A = bh/2$ for the triangular face, and find (b) the volume of the prism, and (c) the weight of the prism, taking the density of the glass as 2.55 gm/cm^3.

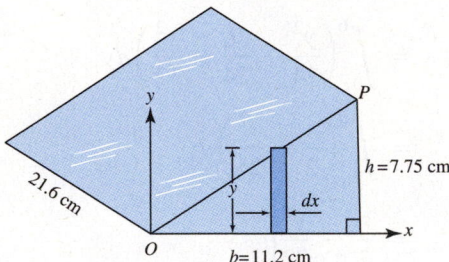

FIGURE 27–13   A triangular prism.

Solution: (a) Taking axes as shown, the straight line OP has the equation

$$y = \frac{h}{b}x + 0$$

We draw a vertical element of area of height y and width dx. The area of the triangular end is then

$$A = \int y\,dx = \int_{0}^{b} \frac{h}{b}x\,dx$$

$$= \frac{h}{b} \cdot \frac{x^2}{2}\bigg|_{0}^{b} = \frac{h}{b} \cdot \frac{b^2}{2} - 0$$

$$= \frac{bh}{2}$$

We have thus verified our familiar formula for the area of a triangle as one-half its base times its height.

(b)   Volume = area of face × length

$$= \frac{(11.2)(7.75)}{2}(21.6) = 937 \text{ cm}^3$$

(c)   Weight = volume × density

$$= 937 \text{ cm}^3\left(\frac{2.55 \text{ g}}{\text{cm}^3}\right) = 2390 \text{ g}$$

or 2.39 kg.   ◆◆◆

Exercise 3 ◆ Finding Areas by Integration

Find or check some of these by calculator. Give any approximate answers to at least three significant digits.

Area Bounded by the x Axis

Find the area bounded by each curve, the x axis, and the given limits.

1. $y = 3x^2 + 2x$ from $x = 1$ to 3

2. $y = 4 + 2x^2 + 2x$ from $x = 2$ to 4

3. $y = 3\sqrt{x}$ from $x = 1$ to 5

4. $y = x + \sqrt{x}$ from $x = 1$ to 2

Find the first-quadrant area bounded by each curve and both coordinate axes.

5. $y^2 = 16 - x$ **6.** $y = x^3 - 8x^2 + 15x$

7. $x + y + y^2 = 2$ **8.** $\sqrt{x} + \sqrt{y} = 1$

9. Find the area bounded by the curve $10y = x^2 - 80$, the x axis, and the lines $x = 1$ and $x = 6$.

10. Find the area bounded by the curve $y = x^3$, the x axis, and the lines $x = -3$ and $x = 0$.

Find only the portion of the area below the x axis.

11. $y = x^3 - 4x^2 + 3x$ **12.** $y = x^2 - 4x + 3$

Areas Bounded by the y Axis

Find the first-quadrant area bounded by the given curve, the y axis, and the given lines.

13. $y = x^2 + 2$ from $y = 3$ to 5 **14.** $8y^2 = x$ from $y = 0$ to 10

15. $y^3 = 4x$ from $y = 0$ to 4 **16.** $y = 4 - x^2$ from $y = 0$ to 3

Area Between Two Curves

17. Find the area bounded by the parabola $y = 6 + 4x - x^2$ and the line $y = 2x - 2$ between $x = 0$ and $x = 4$.

18. Find the area bounded by the curve $y^2 = x^3$ and the line $x = 4$.

19. Find the area bounded by the curve $y^3 = x^2$ and the line $y = \frac{1}{3}x + \frac{4}{3}$ between $x = -1$ and $x = 8$.

20. Find the area bounded by the parabola $y = 3 - x^2$ and the line $y = x + 1$.

21. Find the area bounded by the curves $y^2 = 4x$ and $2x - y = 4$.

22. Find the area between the parabolas $y^2 = 4x$ and $x^2 = 4y$.

23. Find the area between the parabolas $y^2 = 2x$ and $x^2 = 2y$.

24. Find the area bounded by the curves $y = 2 - x$ and $y^2 = x - 1$.

Areas of Geometric Figures

Use integration to verify the formula for the area of each figure. (*Hint:* For problem 29 and several of those to follow, integrate using Rule 69.)

25. square of side a **26.** rectangle with sides a and b

27. right triangle [Fig. 27–14(a)]

28. triangle [Fig. 27–14 (b)]

29. circle of radius r

30. segment of circle [Fig. 27–14(c)]

31. ellipse [Fig. 27–14(d)]

32. parabola [Fig. 27–14(e)]

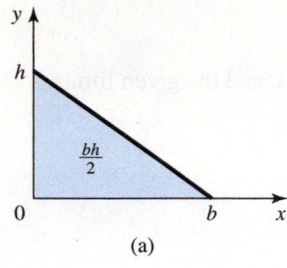

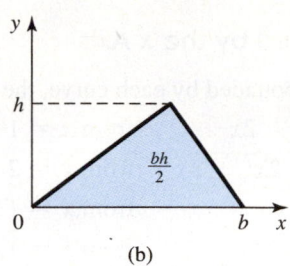

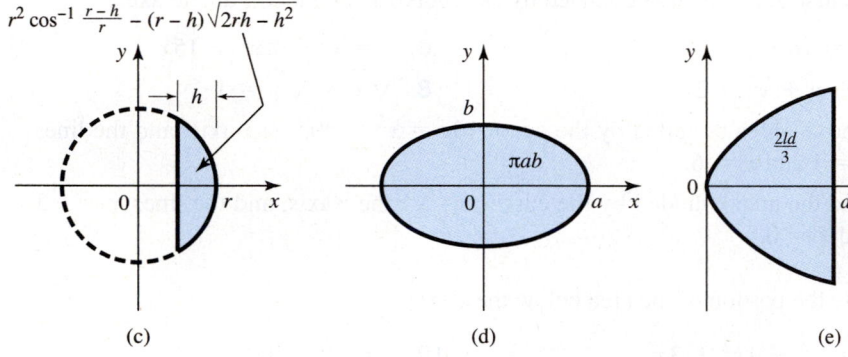

FIGURE 27–14   Areas of some geometric figures.

Applications

Note that no equations are given for the curves in these applications. Thus before setting up each integral, you must draw axes so as to obtain the simplest equation, and then write the equation of the curve.

33. An elliptical culvert is partly full of water (Fig. 27–15). Find, by integration, the cross-sectional area of the water.

34. A mirror (Fig. 27–16) has a parabolic face. Find the volume of glass in the mirror.

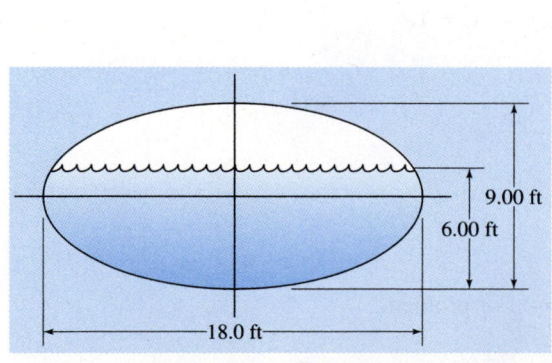

FIGURE 27–15   Elliptical culvert.

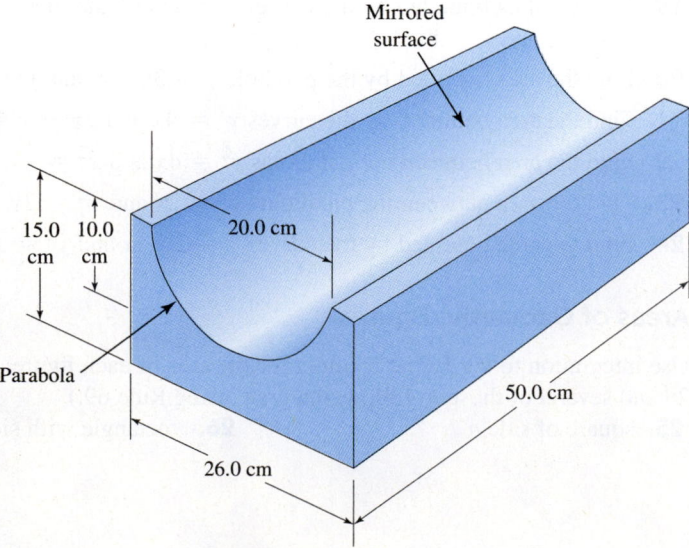

FIGURE 27–16   Parabolic mirror.

35. Figure 27–17 shows a concrete column that has an elliptical cross section. Find the volume of concrete in the column.

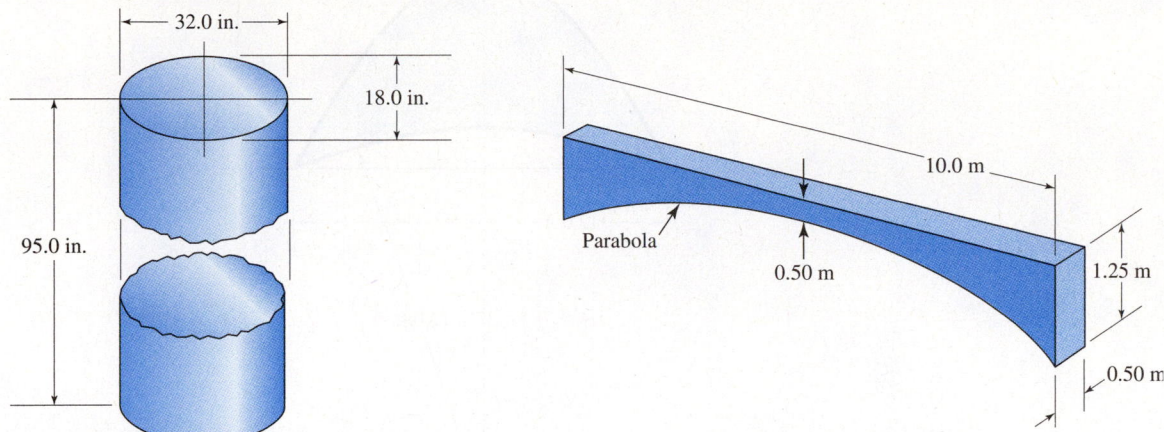

FIGURE 27–17   Concrete column.

FIGURE 27–18   Roof beam.

36. A concrete roof beam for an auditorium has a straight top edge and a parabolic lower edge (Fig. 27–18). Find the volume of concrete in the beam.

37. The deck of a certain ship has the shape of two intersecting parabolic curves (Fig. 27–19). Find the area of the deck.

38. A lens (Fig. 27–20) has a cross section formed by two intersecting circular arcs. Find by integration the cross-sectional area of the lens.

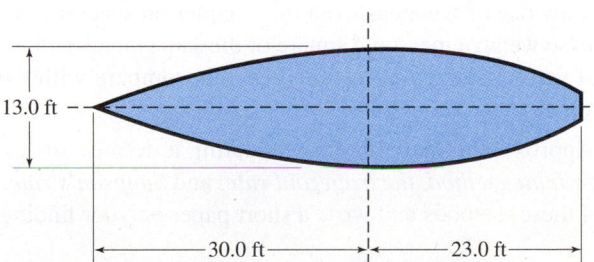

FIGURE 27–19   Top view of a ship's deck.

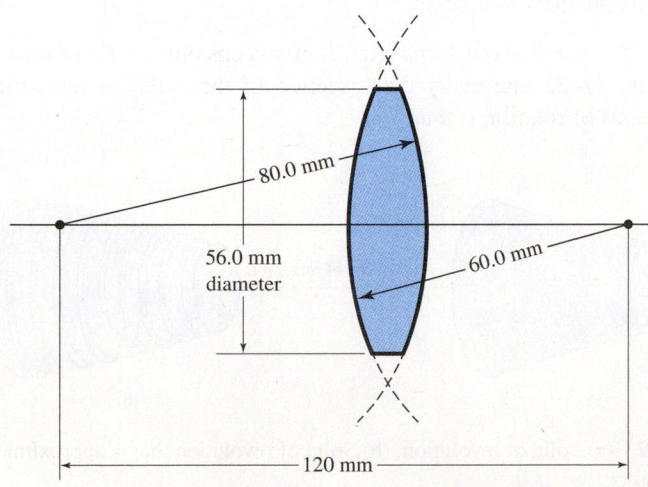

FIGURE 27–20   Cross section of a lens.

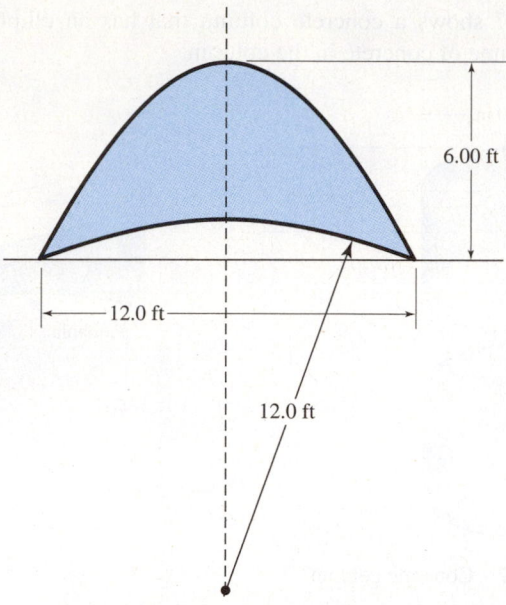

FIGURE 27–21   Window.

39. A window (Fig. 27–21) has the shape of a parabola above and a circular arc below. Find the area of the window.

40. *Project:* Draw one of the areas from this chapter on graph paper and estimate its area by counting boxes. How does this compare with the area found by integration?

41. *Project:* Draw one of the areas from this chapter on sheet metal, cut it out, and weigh it. Also weigh a measured square of the same metal, and use it to estimate the area of the original cutout. How does this compare with the result gotten by integration?

42. *Writing:* Approximate methods for evaluating a definite integral include the *average ordinate method, the trapezoid rule,* and *Simpson's rule.* Research one or more of these methods and write a short paper on your findings.

27–4   Volumes by Integration

Solids of Revolution

When an area is rotated about some axis, *L,* it sweeps out a *solid of revolution.* It is clear from Fig. 27–22 that every cross section of that solid of revolution at right angles to the axis of rotation *is a circle.*

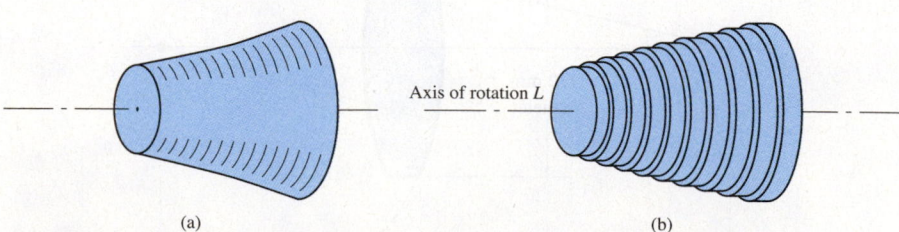

FIGURE 27–22   (a) Solid of revolution. (b) Solid of revolution that is approximated by a stack of thin disks.

When the area A is rotated about an axis L located at some fixed distance from the area, we get a solid of revolution with a cylindrical hole down its center, Fig. 27–23(b). The cross section of this solid at right angles to the axis of rotation consists of a ring bounded by an outer circle and an inner concentric circle.

When the area B is rotated about axis L, Fig. 27–23(c), we get a solid of revolution with a hole of varying diameter down its center. We first learn how to calculate the volume of a solid with no hole, and then we cover "hollow" solids of revolution.

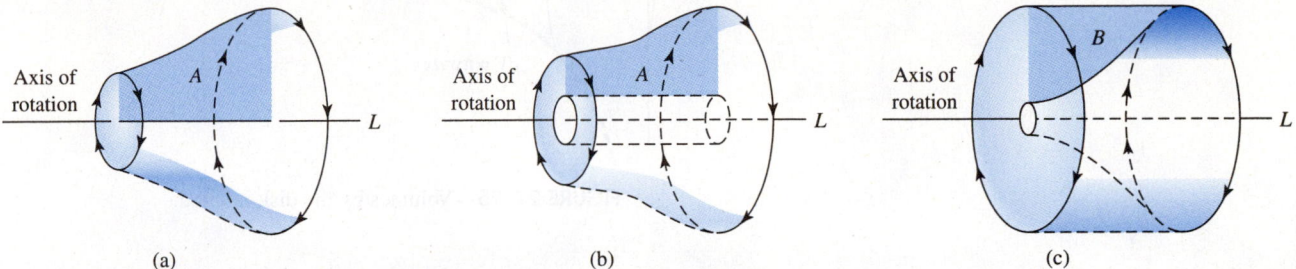

(a) (b) (c)

FIGURE 27–23 (a) Solid of revolution. (b) Solid of revolution with an axial hole. (c) Solid of revolution with axial hole of varying diameter.

Volumes by the Disk Method: Rotation About the *x* Axis

We may think of a solid of revolution, Fig. 27–22(a), as being made up of a stack of thin disks, like a stack of coins of different sizes, Fig. 27–22(b). Each disk is called an *element* of the total volume. We let the radius of one such disk be r (which varies with the disk's position in the stack) and let the thickness be equal to dh. Since a disk (Fig. 27–24) is a cylinder, we calculate its volume dV by Eq. 293.

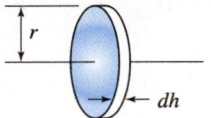

FIGURE 27–24 One disk.

$$dV = \pi r^2 \, dh \qquad\qquad 293$$

Now using the shortcut method for setting up a definite integral, we "sum" the volumes of all such disk-shaped elements by integrating from one end of the solid to the other.

Volumes by the Disk Method	$V = \pi \displaystyle\int_a^b r^2 \, dh$	294

In an actual problem, we must express r and h in terms of x and y, as in the following example.

◆◆◆ **Example 22:** The area bounded by the curve $y = 8/x$, the x axis, and the lines $x = 1$ and $x = 8$ is rotated about the x axis. Find the volume generated.

Estimate: We sketch the solid as shown in Fig. 27–25 and try to visualize a right circular cone having roughly the same volume, say, with height 7 and base radius 5. Such a cone would have a volume of $\left(\frac{1}{3}\right) \pi \, (5^2)(7)$, or $58\,\pi$.

Solution: On our figure we sketch in a typical disk touching the curve at some point (x, y). The radius r of the disk is equal to y, and the thickness dh of the disk is dx. So, by Eq. 294,

$$V = \pi \int_a^b y^2 \, dx$$

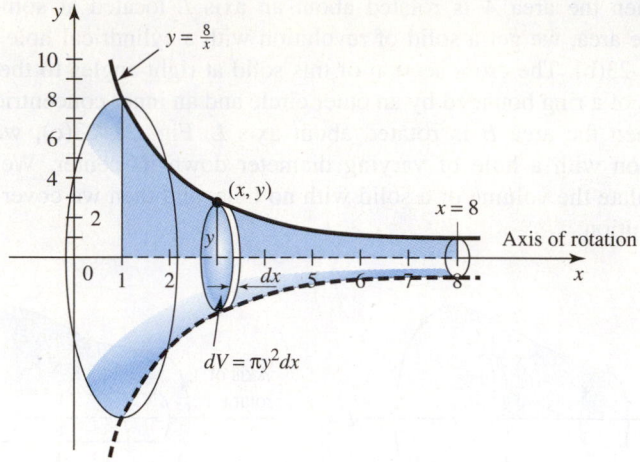

FIGURE 27–25   Volumes by the disk method.

For y we substitute $8/x$, and for a and b, the limits 1 and 8.

$$V = \pi \int_1^8 \left(\frac{8}{x}\right)^2 dx = 64\pi \int_1^8 x^{-2} dx$$

$$= -64\pi \left[x^{-1} \right]_1^8$$

$$= -64\pi \left(\frac{1}{8} - 1 \right) = 56\pi \text{ cubic units} \qquad \blacklozenge\blacklozenge\blacklozenge$$

Common Error	Remember that all parts of the integral, including the limits, must be expressed in terms of the *same variable*.

Volumes by the Disk Method: Rotation About the *y* Axis

When our area is rotated about the y axis rather than the x axis, we take our element of area as a horizontal disk rather than a vertical one.

◆◆◆ **Example 23:** Find the volume generated when the first-quadrant area bounded by $y = x^2$, the y axis, and the lines $y = 0$ and $y = 4$ is rotated about the y axis.

Estimate: The volume of the given solid must be less than that of the circumscribed cylinder, $\pi(2^2)(4)$ or 16π, and greater than the volume of the inscribed cone, $16\pi/3$. Thus we have bracketed our answer between $5\frac{1}{3}\pi$ and 16π.

Solution: Our disk-shaped element of volume, shown in Fig. 27–26, now has a radius x and a thickness dy.

So, by Eq. 294,

$$V = \pi \int_a^b x^2 dy$$

Substituting y for x^2 and inserting the limits 0 and 4, we have

$$V = \pi \int_0^4 y \, dy = \pi \left[\frac{y^2}{2} \right]_0^4$$

$$= \frac{\pi}{2}(4^2 - 0) = 8\pi \text{ cubic units}$$

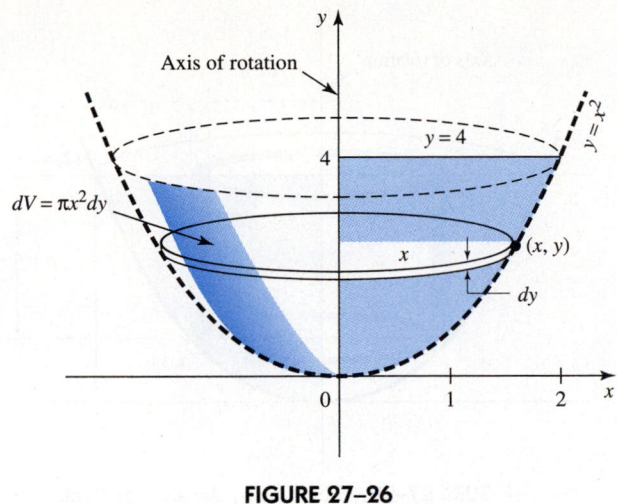

FIGURE 27–26

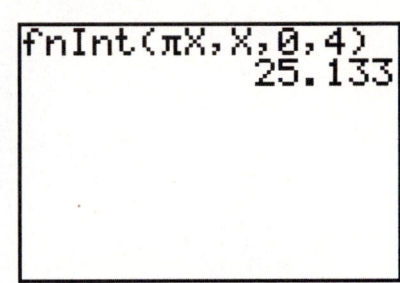

Check by Calculator: Let us check the integration by using **fnInt**. We get 25.133, which is equivalent to 8π. ◆◆◆

Volumes by the Shell Method

Instead of using a thin disk for our element of volume, it is sometimes easier to use a *thin-walled shells* (Fig. 27–27). To visualize such shells, imagine a solid of revolution to be turned from a log, with the axis of revolution along the centerline of the log. Each annual growth ring would have the shape of a thin-walled shell, with the solid of revolution being made up of many such shells nested one inside the other, Fig. 27–28.

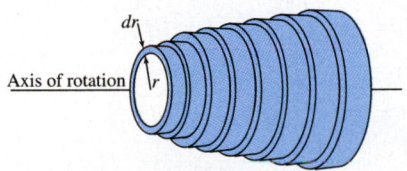

FIGURE 27–28 Solid of revolution
approximated by nested thin-walled shells.

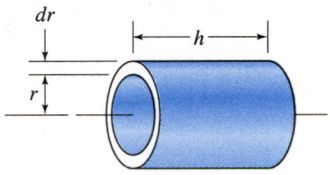

FIGURE 27–27 Thin-walled shell.

The volume dV of one thin-walled cylindrical shell of radius r, height h, and wall thickness dr is

dV = circumference × height × wall thickness $= 2\pi r h \, dr$	**297**

Integrating gives the following:

Volumes by the Shell Method	$V = 2\pi \displaystyle\int_a^b r h \, dr$	**298**

Notice that here the integration limits are in terms of r. As with the disk method, r and h must be expressed in terms of x and y in a particular problem.

◆◆◆ **Example 24:** Repeat Example 23 using the shell method to find the volume generated.

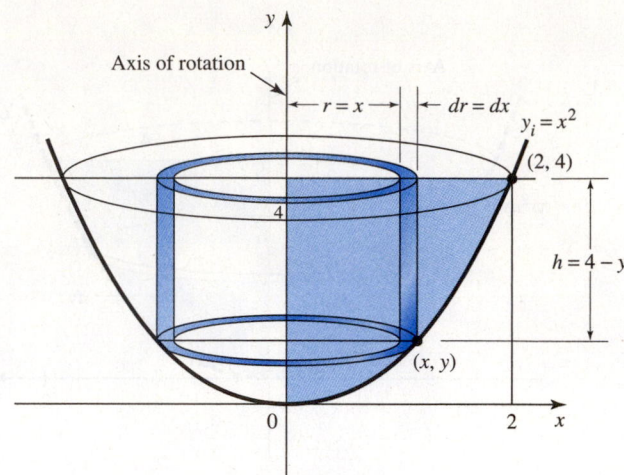

FIGURE 27–29   Volumes by the shell method.

Solution: Through a point (x, y) on the curve, Fig. 27–29, we sketch an element of area of thickness dx. The upper end of that element is 4 units from the x axis and its lower end is y units from the x axis, so the element's height is $4 - y$. As the first-quadrant area rotates about the y axis, generating a solid of revolution, our element of area sweeps out a shell of radius $r = x$ and thickness $dr = dx$. The volume dV of the shell is then

$$dV = 2\pi r h\, dr \;=\; 2\pi x(4 - y)\,dx$$
$$= 2\pi x(4 - x^2)\,dx$$

since $y = x^2$. Integrating gives the total volume.

$$V = 2\pi \int_0^2 x(4 - x^2)\,dx$$

$$= 2\pi \int_0^2 (4x - x^3)\,dx$$

$$= 2\pi \left[2x^2 - \frac{x^4}{4} \right]_0^2 \;=\; 2\pi \left[2(2)^2 - \frac{2^4}{4} \right]$$

$$= 8\pi \text{ cubic units}$$

as we got for Example 23.   ◆◆◆

Solid of Revolution with Hole

To find the volume of a solid of revolution with an axial hole, such as in Fig. 27–23(b) and (c), we can first find the volume of hole and solid separately, and then subtract. Or we can find the volume of the solid of revolution directly by either the washer or the shell method, as in the following example.

◆◆◆ **Example 25:** The first-quadrant area bounded by the curve $y^2 = 4x$, the x axis, and the line $x = 4$ is rotated about the y axis. Find the volume generated (a) by the washer method and (b) by the shell method.

Solution:

(a) *Washer method*: The volume dV of a thin washer, which is actually a cylinder with a hole (Fig. 27–30) is given by

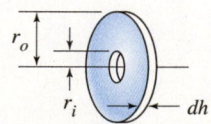

FIGURE 27–30   Washer or ring.

$$dV = \pi(r_o^2 - r_i^2)\,dh \qquad\qquad \textbf{295}$$

where r_o is the outer radius and r_i is the inner radius. Integrating gives the following.

Volume by the Washer Method	$$V = \pi \int_a^b (r_o^2 - r_i^2)\,dh$$	**296**

On our given solid, Fig. 27–31(a), we show an element of volume in the shape of a washer centered on the y axis. Its outer and inner radii are x_o and x_i, respectively, and its thickness dh is here dy. Then, by Eq. 296,

$$V = \pi \int_a^b (x_o^2 - x_i^2)\,dy$$

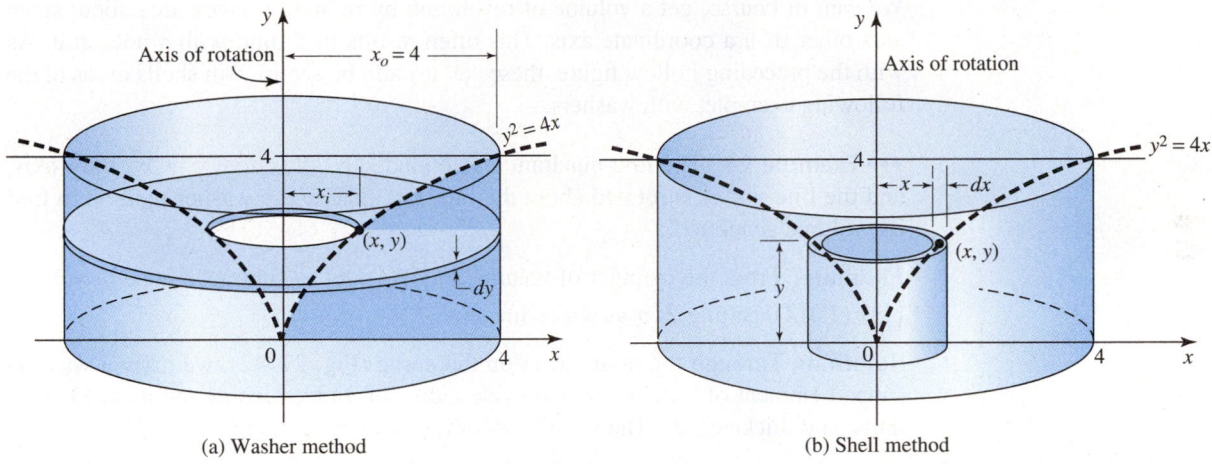

(a) Washer method   (b) Shell method

FIGURE 27–31

In our problem, $x_o = 4$ and $x_i = y^2/4$. Substituting these values and placing the limits of 0 and 4 on y gives

$$V = \pi \int_0^4 \left(16 - \frac{y^4}{16} \right) dy$$

Integrating, we obtain

$$V = \pi \left[16y - \frac{y^5}{80} \right]_0^4 = \pi \left[16(4) - \frac{4^5}{80} \right] = \frac{256}{5}\pi \text{ cubic units}$$

(b) *Shell method:* On the given solid we indicate an element of volume in the shape of a shell, Fig. 27–31(b). Its inner radius r is x, its thickness dr is dx, and its height h is y. Then, by Eq. 297,

$$dV = 2\pi xy\,dx$$

So we get the total volume by integrating.

$$V = 2\pi \int_a^b xy\,dx$$

Replacing y with $2\sqrt{x}$ and placing limits of 0 and 4 on x, we have

$$V = 2\pi \int_0^4 x(2x^{1/2})\,dx$$

$$= 4\pi \int_0^4 x^{3/2}\,dx$$

$$= 4\pi \left[\frac{x^{5/2}}{5/2} \right]_0^4$$

$$= \frac{8\pi}{5} (4)^{5/2} = \frac{256}{5} \pi \text{ cubic units}$$

as by the washer method.   ♦♦♦

Common Error	In Example 25 the radius x in the shell method varies from 0 to 4. The limits of integration are therefore 0 to 4, not -4 to 4.

Rotation About a Noncoordinate Axis

We can, of course, get a volume of revolution by rotating a given area about some axis other than a coordinate axis. This often results in a solid with a hole in it. As with the preceding hollow figure, these can usually be set up with shells or, as in the following example, with washers.

♦♦♦ **Example 26:** The first-quadrant area bounded by the curve $y = x^2$, the y axis, and the line $y = 4$ is rotated about the line $x = 3$. Use the washer method to find the volume generated.

Estimate: From the cylinder of volume $\pi(3^2)(4)$, let us subtract a cone of volume $\left(\frac{1}{3}\right)\pi(3^2)(4)$, getting 24π as our estimate.

Solution: Through the point (x, y) on the curve (Fig. 27–32), we draw a washer-shaped element of volume, with outside radius of 3 units, inside radius of $(3 - x)$ units, and thickness dy. The volume dV of the elements is then, by Eq. 295,

$$dV = \pi[3^2 - (3 - x)^2] \, dy$$
$$= \pi(9 - 9 + 6x - x^2) \, dy$$
$$= \pi(6x - x^2) \, dy$$

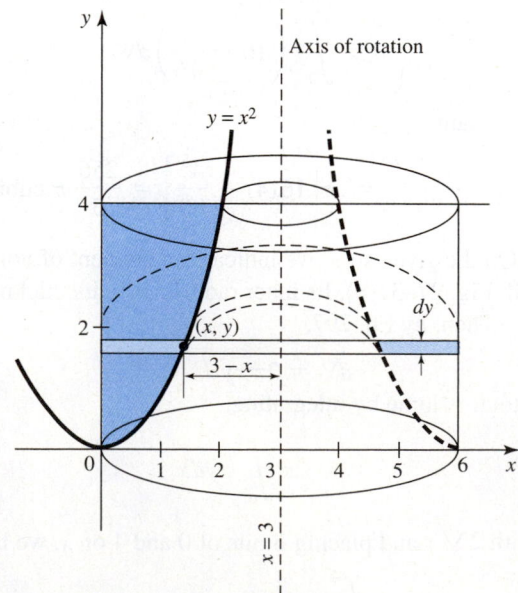

FIGURE 27–32

Substituting $\sqrt{y}$ for x gives

$$dV = \pi(6\sqrt{y} - y)\,dy$$

Integrating, we get

$$V = \pi \int_0^4 (6y^{1/2} - y)\,dy = \pi\left[\frac{12y^{3/2}}{3} - \frac{y^2}{2}\right]_0^4$$

$$= \pi\left[4(4)^{3/2} - \frac{16}{2}\right] = 24\pi \text{ cubic units} \qquad \bullet\bullet\bullet$$

Exercise 4 ◆ Volumes by Integration

Perform or check some of your integrals by calculator. Give any approximate answers to three significant digits.

Rotation About the x Axis

Find the volume generated by rotating the first-quadrant area bounded by each set of curves and the x axis about the x axis. Use either the disk or the shell method.

1. $y = x^3$ and $x = 2$

2. $y = \dfrac{x^2}{4}$ and $x = 4$

3. $y = \dfrac{x^{3/2}}{2}$ and $x = 2$

4. $y^2 = x^3 - 3x^2 + 2x$ and $x = 1$

5. $y^2(2 - x) = x^3$ and $x = 1$

6. $\sqrt{x} + \sqrt{y} = 1$

7. $x^{2/3} + y^{2/3} = 1$ from $x = 0$ to 1

8. $y^2 = x\left(\dfrac{x - 3}{x - 4}\right)$

Rotation About the y Axis

Find the volume generated by rotating about the y axis the first-quadrant area bounded by each set of curves.

9. $y = x^3$, the y axis, and $y = 8$

10. $2y^2 = x^3$, $x = 0$, and $y = 2$

11. $9x^2 + 16y^2 = 144$

12. $\left(\dfrac{x}{2}\right)^2 + \left(\dfrac{y}{3}\right)^{2/3} = 1$

13. $y^2 = 4x$ and $y = 4$

Solid of Revolution with a Hole

Find the volume generated by rotating about the indicated axis the first-quadrant area bounded by the given pair of curves.

14. $y = 3x^2$ and $x = 2$, about the y axis.

15. $y = 2\sqrt{x}$ and $x = 3$, about the y axis.

16. $y = 2x^3$, the y axis, and $y = 7$, about the x axis.

17. $y = 3\sqrt{x}$, and $y = 2$, about the x axis.

Rotation About a Noncoordinate Axis

Find the volume generated by rotating about the indicated axis the first-quadrant area bounded by each set of curves.

18. $x = 4$ and $y^2 = x^3$, about $x = 4$
19. above $y = 3$ and below $y = 4x - x^2$, about $y = 3$
20. $y^2 = x^3$ and $y = 8$, about $y = 9$
21. $y = 4$ and $y = 4 + 6x - 2x^2$, about $y = 4$

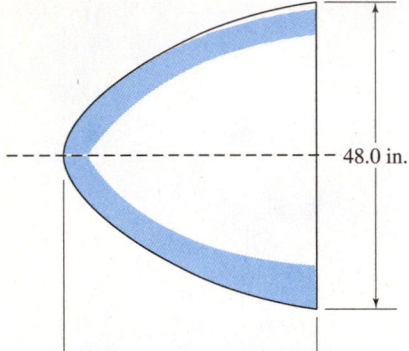

FIGURE 27–33   Rocket nose cone.

Volumes of Familiar Solids

22. *Cylinder:* Derive the formula for the volume of a cylinder by rotating the area bounded by the x axis, the line $x = h$, and the line $y = r$ about the x axis.
23. *Cone:* Derive the formula for the volume of a right circular cone. Rotate the area bounded by the x axis, the line $x = h$, and the line $y = rx/h$ about the x axis.
24. *Sphere:* Derive the formula for the volume of a hemisphere by rotating the area bounded by the x and y axes and the curve $x^2 + y^2 = r^2$ about the x axis. Double it to get the volume of a sphere.

Applications

25. The nose cone of a certain rocket is a paraboloid of revolution, the figure formed by revolving a parabola about its axis (Fig. 27–33). Find its volume.
26. A wing tank for an airplane is a solid of revolution formed by rotating the curve , from   to 3.50, about the x axis (Fig. 27–34). Find the volume of the tank.

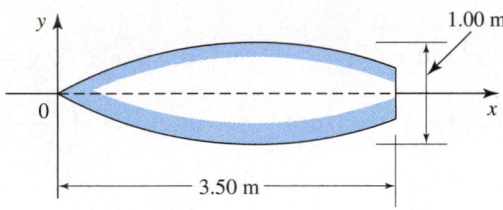

FIGURE 27–34   Airplane wing tank.

27. A bullet (Fig. 27–35) consists of a cylinder and a paraboloid of revolution and is made of lead having a density of 11.3 g/cm^3. Find its weight.
28. A telescope mirror, shown in cross section in Fig. 27–36, is formed by rotating the area under the hyperbola $y^2/100 - x^2/1225 = 1$ about the y axis, and has a 20.0-cm-diameter hole at its center. Find the volume of glass in the mirror.

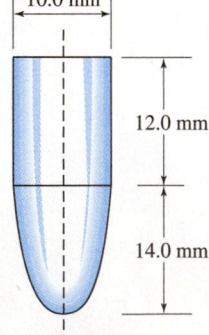

FIGURE 27–35   Bullet.

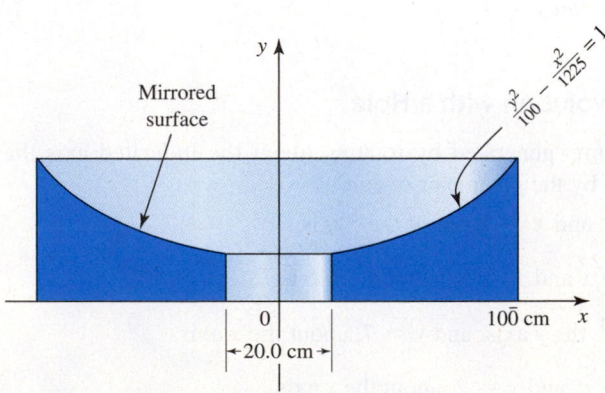

FIGURE 27–36   Telescope mirror.

29. *Project:* Use a wood-turning lathe to make a model of one of the solids of revolution given in this chapter. Find its volume by immersing it in water, and compare it to the value obtained by integration.

30. *Project:* Make an approximate model of one of the solids of revolution by stacking disks of equal thickness, cut from cardboard or other thin material. Add up the volumes of the disks and compare the volume you get with that obtained by integration.

••• **CHAPTER 27 REVIEW PROBLEMS** ••••••••••••••••••••••••••••••••••••••

Give any approximate answers to at least three significant digits.

1. Find the volume generated when the area bounded by the parabolas $y^2 = 4x$ and $y^2 = 5 - x$ is rotated about the x axis.

2. Find the area bounded by the coordinate axes and the curve $x^{1/2} + y^{1/2} = 2$.

3. Find the area between the curve $x^2 = 8y$ and $y = \dfrac{64}{x^2 + 16}$. (Use Rule 56.)

4. The area bounded by the parabola $y^2 = 4x$, from $x = 0$ to 8, is rotated about the x axis. Find the volume generated.

5. Find the area bounded by the curve $y^2 = x^3$ and the line $x = 4$.

6. Find the area bounded by the curve $y = 1/x$ and the x axis, between the limits $x = 1$ and $x = 3$.

7. Find the area bounded by the curve $y^2 = x^3 - x^2$ and the line $x = 2$.

8. Find the area bounded by $xy = 6$, the lines $x = 1$ and $x = 6$, and the x axis.

9. A flywheel starts from rest and accelerates at $7.25t^2$ rad/s^2. Find the angular velocity and the total number of revolutions after 20.0 s.

10. The current to a certain capacitor is $i = t^3 + 18.5$ A. If the initial charge on the capacitor is 6.84 C, find the charge when $t = 5.25$ s.

11. A point starts from (1, 1) with initial velocities of $v_x = 4$ cm/s and $v_y = 15$ cm/s and moves along a curved path. It has x and y components of acceleration of $a_x = t$ and $a_y = 5t$. Write expressions for the x and y components of velocity and displacement.

12. A 15.0-F capacitor has an initial voltage of 25.0 V and is charged with a current equal to $i = \sqrt{4t + 21.6}$ A. Find the voltage across the capacitor at 14.0 s.

13. Find the volume of the solid generated by rotating the ellipse $x^2/16 + y^2/9 = 1$ about the x axis.

14. Find the area bounded by the curves $y^2 = 8x$ and $x^2 = 8y$.

15. Find the volume generated when the area bounded by $y^2 = x^3$, $x = 4$, and the x axis is rotated about the line $y = 8$.

16. Find the volume generated when the area bounded by the curve $y^2 = 16x$, from $x = 0$ to 4, is rotated about the x axis.

17. Find the entire area of the ellipse

$$\frac{x^2}{16} + \frac{y^2}{9} = 1$$

(Use Rule 69.)

18. The acceleration of an object that starts from rest is given by $a = 3t$. Write equations for the velocity and displacement of the object.

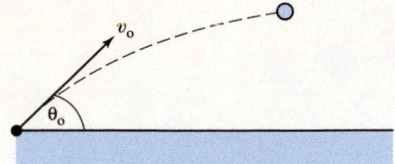

FIGURE 27–37   A projectile.

19. The voltage across a 25.0-H inductor is given by $v = 8.9 + \sqrt{3t}$ V. Find the current in the inductor at 5.00 s if the initial current is 1.00 A.

20. Find the volume generated when the area bounded by $y^2 = x^3$, the y axis, and $y = 8$ is rotated about the line $x = 4$.

21. *Writing:* List the steps needed to find the area between two curves, and give a short description of each step.

22. *Project:* A projectile is launched with initial velocity v_0 at an angle θ_0 with the horizontal, as shown in Fig. 27–37. Derive the equations for the displacement of the projectile.

$$x = (v_0 \cos \theta_0)t + x_0$$

and

$$y = (v_0 \sin \theta_0)t - gt^2/2 + y_0$$

More Applications
of the Integral

28

In this, our second chapter on applications of the integral, we will compute several quantities that are important in technical work: *arc length*, such as the length of a suspension bridge cable; *area of a surface of revolution*, perhaps the outer surface of a rocket; a *centroid of an area*, such as a vane in a wind generator; *fluid pressure*, to compute the force, for example, on a hatch in a submerged research vehicle; *work*, such as that needed to move an arm in a mechanism; and *moment of inertia*, a quantity needed to compute the strength of beams and the dynamics of moving bodies.

For example, to determine the drag and the heat transfer to or from the rocket nose cone, Fig. 28–1, you would need to know its surface area. Could you find it, given its dimensions and the shape of the curve, with the tools you already have? You can, with a little more help, and we will show you how.

We will learn how to set up an integral for each. We will see that this is a great place to use the calculator to evaluate integrals, as some are long and difficult to evaluate analytically. Here we will use both calculator and analytical methods.

We will have more applications of the integral later, after we have learned to integrate logarithmic, exponential, and trigonometric functions.

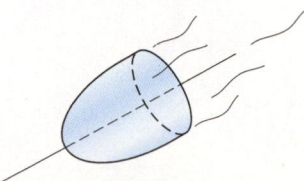

FIGURE 28–1

28–1   Length of Arc

Building on our applications from the last chapter, we now show how to find the length of a curve, such as the distance around an elliptical courtyard. It is the length we would get if the curve were stretched out straight and then measured.

■ Exploration:

Try this. Draw a smooth, continuous curve between two points, Fig. 28–2. Then

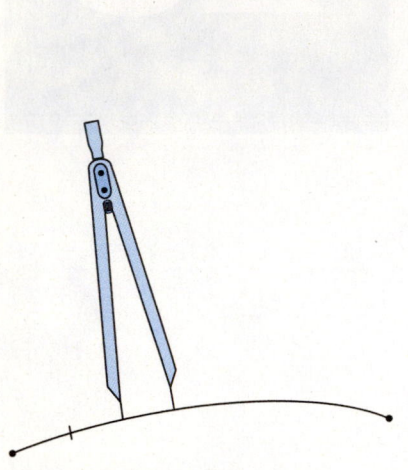

FIGURE 28–2   Measuring a curve by stepping off with dividers.

- With dividers, step along the curve from one endpoint to the other, just as you would measure a distance on a road map. Count the number of steps, and estimate the fraction of a step at the end of the curve.
- Measure the opening between the points of the dividers.
- Multiply the number of steps by the opening to get the approximate length of the curve.
- Repeat a few times, each time with a smaller divider opening.

What do you conclude? Does this exploration suggest a way to find the *exact* length using calculus?   ■

You may have concluded that the smaller the step, the greater the accuracy. Here we will use steps that are vanishingly small, and use integration to add them up.

Again we use our intuitive method to set up the integral. We think of the curve, Fig. 28–3, as being made up of many short sections, each of length Δs. By the Pythagorean theorem,

$$(\Delta s)^2 \approx (\Delta x)^2 + (\Delta y)^2$$

Dividing by $(\Delta x)^2$ gives us

$$\frac{(\Delta s)^2}{(\Delta x)^2} \approx 1 + \frac{(\Delta y)^2}{(\Delta x)^2}$$

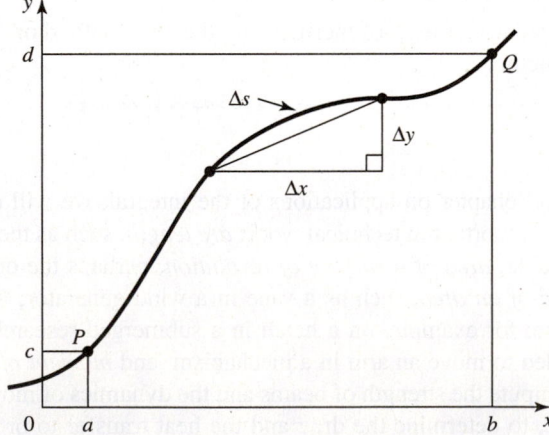

FIGURE 28–3   Length of arc. We are considering here (as usual) only smooth, continuous curves.

Taking the square root yields

$$\frac{\Delta s}{\Delta x} \approx \sqrt{1 + \left(\frac{\Delta y}{\Delta x}\right)^2}$$

We now let the number of these short sections of curve approach infinity as we let Δx approach zero.

$$\frac{ds}{dx} = \lim_{\Delta x \to 0} \frac{\Delta s}{\Delta x} = \sqrt{1 + \left(\frac{dy}{dx}\right)^2}$$

or, in a differential form

$$ds = \sqrt{1 + \left(\frac{dy}{dx}\right)^2}\, dx$$

Now thinking of integration as a summing process, we use it to add up all of the small segments of length ds as follows:

Length of Arc	$s = \displaystyle\int_a^b \sqrt{1 + \left(\dfrac{dy}{dx}\right)^2}\, dx$	299

◆◆◆ **Example 1:** Find the first-quadrant length of the curve $y = x^{3/2}$ from $x = 0$ to 4 (Fig. 28–4).

Estimate: Computing the straight-line distance between the two endpoints $(0, 0)$ and $(4, 8)$ of our arc gives $\sqrt{4^2 + 8^2}$, or about 8.94. Thus we expect the curve connecting those points to be slightly longer than 8.94.

Solution: Let us first find $(dy/dx)^2$.

$$y = x^{3/2}$$

$$\frac{dy}{dx} = \frac{3}{2}x^{1/2}$$

$$\left(\frac{dy}{dx}\right)^2 = \frac{9x}{4}$$

Then, by Eq. 299,

$$s = \int_0^4 \sqrt{1 + \left(\frac{dy}{dx}\right)^2}\, dx = \int_0^4 \sqrt{1 + \frac{9x}{4}}\, dx$$

$$= \frac{4}{9}\int_0^4 \left(1 + \frac{9x}{4}\right)^{1/2}\left(\frac{9}{4}\, dx\right) = \frac{8}{27}\left[\left(1 + \frac{9x}{4}\right)^{3/2}\right]_0^4$$

$$= \frac{8}{27}(10^{3/2} - 1) \approx 9.07$$

Integrating by Calculator: Every integral in this chapter may be evaluated using a rule (often Rule 66) from the Table of Integrals, Appendix C. They are, however, long and difficult, so this is an ideal place to use our calculators. We will, therefore, show a calculator screen for each integral. ◆◆◆

Another Form of the Arc Length Equation

Another form of the equation for arc length, which can be derived in a similar way to Eq. 299, follows.

Length of Arc	$s = \displaystyle\int_c^d \sqrt{1 + \left(\dfrac{dx}{dy}\right)^2}\, dy$	300

This equation is more useful when the equation of the curve is given in the form $x = f(y)$, instead of the more usual form $y = f(x)$.

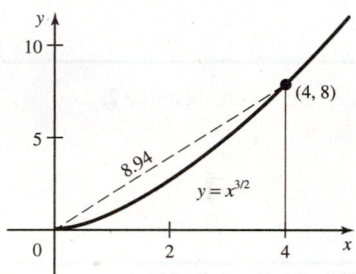

FIGURE 28–4

```
fnInt(√(1+9X/4),
X,0,4)
              9.0734
```

TI-83 screen for Example 1. The TI-84 displays an integral sign for the same operation.

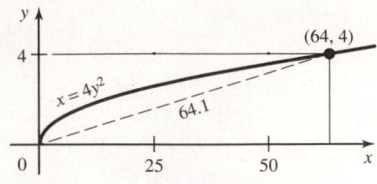

FIGURE 28–5

TI-83 screen for Example 2.

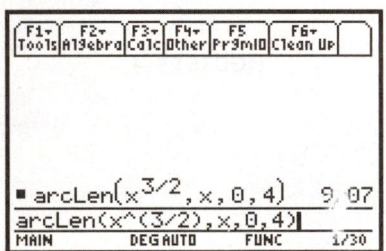

(1) TI-89 screen for Example 3.

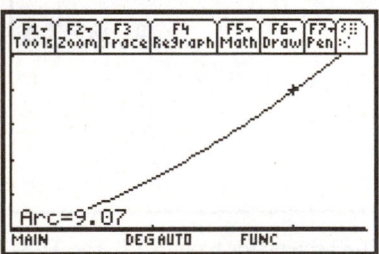

(2) Tick marks are 2 units apart. Note that the **MATH** menu we are using is not the one from the keyboard, as in (a), but is the function key F5.

◆◆◆ **Example 2:** Find the length of the curve $x = 4y^2$ between $y = 0$ and 4.

Estimate: The straight-line distance between the endpoints $(0, 0)$ and $(64, 4)$ is $\sqrt{64^2 + 4^2}$, or about 64.1 (Fig. 28–5). As before, we expect the curve to be slightly longer than that.

Solution: Taking the derivative, we have

$$\frac{dx}{dy} = 8y$$

so $(dx/dy)^2 = 64y^2$. Substituting into Eq. 300, we have

$$s = \int_0^4 \sqrt{1 + 64y^2} \, dy = \frac{1}{8} \int_0^4 \sqrt{1 + (8y)^2} \, (8 \, dy)$$

Using Rule 66, with $u = 8y$ and $a = 1$, yields

$$s = \frac{1}{8}\left[\frac{8y}{2}\sqrt{1 + 64y^2} + \frac{1}{2}\ln\left|8y + \sqrt{1 + 64y^2}\right| \right]_0^4$$

$$= \frac{1}{8}\left[4(4)\sqrt{1 + 64(16)} + \frac{1}{2}\ln\left|32 + \sqrt{1 + 64(16)}\right| - 0 \right] = 64.3$$

Integrating by Calculator: The calculator screen for the integral in this example is as shown.　　　　◆◆◆

Arc Length by Calculator

Some calculators can find arc lengths directly. On the TI-89, for example, one can either find the arc length of (a) a given function between two points or (b) a section of a previously graphed curve.

◆◆◆ **Example 3:** Find the arc length of the curve in Example 1 by both methods, on the TI-89.

Solution: The given function was $y = x^{3/2}$ and the limits were 0 and 4.

(a) We select **arcLen()** from the **MATH** Calculus menu. We then enter the function, the variable x, and the limits 0 and 4, and press ⌐≈. The screen is shown in screen (1).

(b) We graph the curve in the usual way. Then in the **MATH** menu we select **Arc,** enter the lower and upper values of x when prompted, and press **ENTER**. The graph, with the selected points and the arc length, is then displayed, screen (2).　　　　◆◆◆

Exercise 1 ◆ Length of Arc

Find the length of each curve. Evaluate or check each integral by calculator or by rule. Round approximate answers to three significant digits.

1. $y = x^{3/2}$　　　　　　in the first quadrant from $x = 0$ to 5/9
2. $y = (1/6)x^2$　　　　　from the origin to the point (4, 8/3)
3. $y = (36 - x^2)^{1/2}$　　in the first quadrant (*Hint:* Use Rule 61.)
4. $y = (1 - x^{2/3})^{3/2}$　　from $x = 0$ to 1

5. $y = (1/4)x^2$ from $x = 0$ to 4

6. $y = \dfrac{x^{3/2}}{\sqrt{2}}$ from $x = 0$ to 10

7. The arch of the parabola $y = 4x - x^2$ that lies above the x axis

8. $x = 2y^2$ from $y = 0$ to 2

9. $x = (1/8)y^2$ from $y = 0$ to 4

10. $x = 2y^{2/3}$ from $y = 0$ to 8

Applications

11. Find the length of the cable AB in Fig. 28–6 that is in the shape of a parabola.

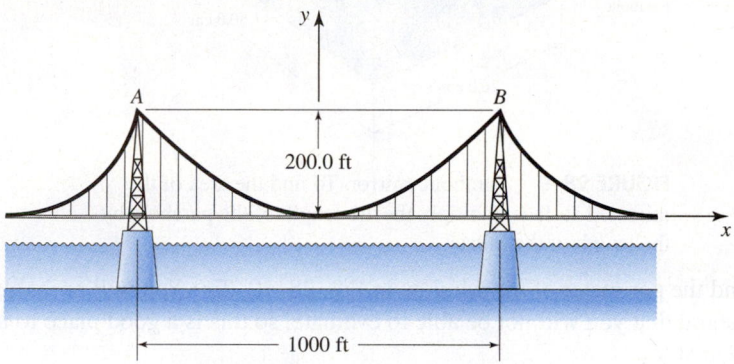

FIGURE 28–6   Suspension bridge.

12. A roadway has a parabolic shape at the top of a hill (Fig. 28–7). If the road is 30 ft wide, find the cost of paving from P to Q at the rate of \$35 per square foot.

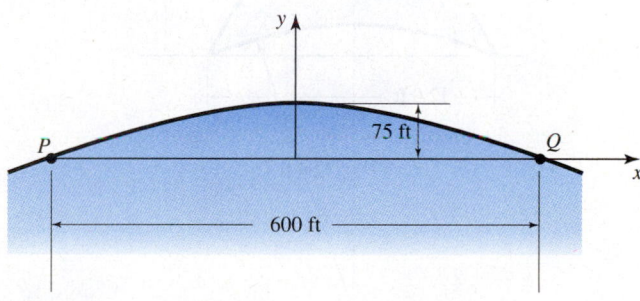

FIGURE 28–7   Road over a hill.

13. The equation of the bridge arch in Fig. 28–8 is $y = 0.0625x^2 - 5x + 100$. Find its length.

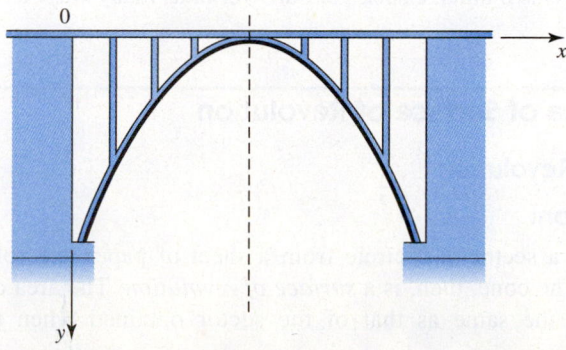

FIGURE 28–8   Parabolic bridge arch. Note that the y axis is positive downward.

14. Find the surface area of the curved portion of the mirror in Fig. 28–9.

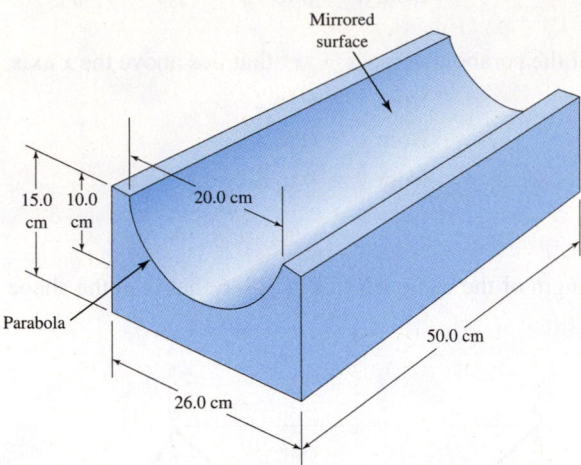

FIGURE 28–9 Parabolic mirror. To find the area of the mirrored surface, multiply the arc length of the parabola by the length of the mirror.

15. Find the perimeter of the window in Fig. 28–10. *Tip:* You will probably get an integral that you will not be able to evaluate, so this is a good place to use your calculator.

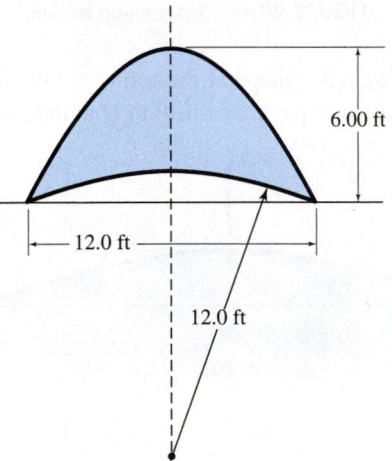

FIGURE 28–10 Window.

16. *Project:* Find the perimeter of an ellipse whose major axis is 10 units and whose minor axis is 6 units. Check your answer in as many ways as you can think of.

28–2   Area of Surface of Revolution

Surface of Revolution

■ Exploration:

Try this. Cut a sector of a circle from a sheet of paper and roll it into a cone, Fig. 28–11. The cone, then, is a *surface of revolution*. The area of that surface of revolution is the same as that of the sector obtained when the cone is laid out flat.   ■

In the exploration you could easily find the area of the surface of revolution by geometry. Here we will use calculus to find areas of more complex surfaces.

We showed earlier that a solid of revolution is the figure we get by rotating an area about some axis. A *surface of revolution* is simply the surface of that solid of revolution. Alternately, we can think of a surface of revolution as being generated by a *curve* rotating about some axis.

Rotation About the *x* Axis

Let us take the curve *PQ* in Fig. 28–12 and rotate it about the *x* axis. It sweeps out a surface of revolution, while the small section *ds* sweeps out a hoop-shaped element of that surface.

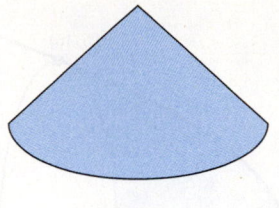

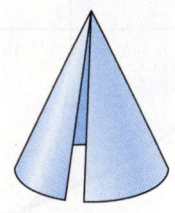

FIGURE 28–11

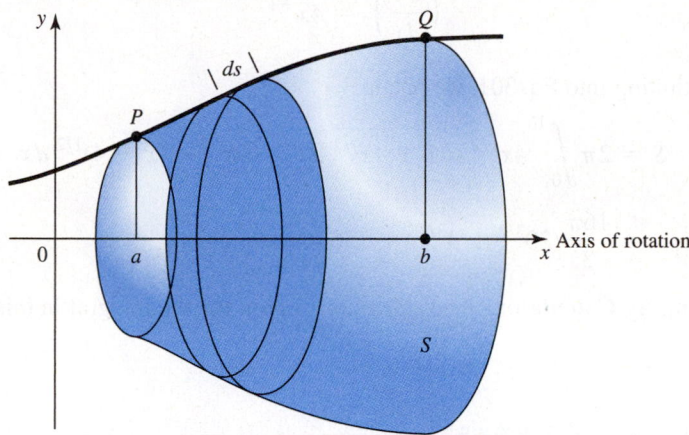

FIGURE 28–12 Surface of revolution.

The area of a hoop-shaped or circular band is equal to the length of the edge times the average circumference of the hoop. Our element *ds* is at a radius *y* from the *x* axis, so the area *dS* of the hoop is

$$dS = 2\pi y\, ds$$

But from Sec. 28–1, we saw that

$$ds = \sqrt{1 + \left(\frac{dy}{dx}\right)^2}\, dx$$

So

$$dS = 2\pi y \sqrt{1 + \left(\frac{dy}{dx}\right)^2}\, dx$$

We then integrate from *a* to *b* to sum up the areas of all such elements.

Area of Surface of Revolution About *x* Axis	$$s = 2\pi \int_a^b y\sqrt{1 + \left(\frac{dy}{dx}\right)^2}\, dx$$	301

Common Error	We are using the capital letter *S* for surface area and lowercase *s* for arc length. Be careful not to confuse the two.

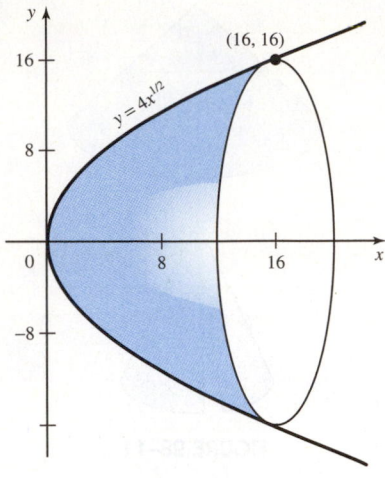

FIGURE 28–13

$$8\pi\int_0^{16}(\sqrt{(X+4)})dX$$
$$1364.5859$$

TI-84 screen for Example 4. Note how the display differs from that of the TI-83.

••• **Example 4:** The portion of the parabola $y = 4\sqrt{x}$ between $x = 0$ and 16 is rotated about the x axis (Fig. 28–13). Find the area of the surface of revolution generated.

Estimate: Let us approximate the given surface by a hemisphere of radius 16. Its area is $\frac{1}{2}(4\pi)\,16^2$, or about 1600 square units.

Solution: We first find the derivative.

$$y = 4x^{1/2}$$
$$\frac{dy}{dx} = 2x^{-1/2}$$
$$\left(\frac{dy}{dx}\right)^2 = 4x^{-1}$$

Substituting into Eq. 301, we obtain

$$S = 2\pi\int_0^{16} 4x^{1/2}\sqrt{1 + 4x^{-1}}\,dx = 8\pi\int_0^{16}(x+4)^{1/2}\,dx$$
$$= \frac{16\pi}{3}(20^{3/2} - 4^{3/2}) \approx 1365 \text{ square units}$$

Integrating by Calculator: The calculator screen for the integral in this example is shown. ◆◆◆

Rotation About the y Axis

The equation for the area of a surface of revolution whose axis of revolution is the y axis can be derived in a similar way.

Area of Surface of Revolution About y Axis	$S = 2\pi\int_a^b x\sqrt{1 + \left(\dfrac{dy}{dx}\right)^2}\,dx$	302

••• **Example 5:** The portion of the curve $y = x^2$ lying between the points $(0, 0)$ and $(2, 4)$ is rotated about the y axis (Fig. 28–14). Find the area of the surface generated.

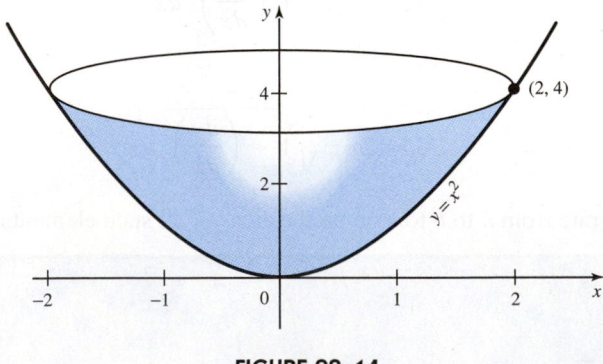

FIGURE 28–14

Solution: Taking the derivative gives $dy/dx = 2x$, so

$$\left(\frac{dy}{dx}\right)^2 = 4x^2$$

Substituting into Eq. 302, we have

$$S = 2\pi \int_0^2 x\sqrt{1 + 4x^2}\,dx = \frac{2\pi}{8}\int_0^2 (1 + 4x^2)^{1/2}(8x\,dx)$$

$$= \frac{\pi}{4} \cdot \frac{2(1 + 4x^2)^{3/2}}{3}\Bigg|_0^2 = \frac{\pi}{6}(17^{3/2} - 1^{3/2}) = 36.2 \text{ square units}$$

```
2πfnInt(X√(1+4X²
),X,0,2)
            36.1769
■
```

TI-83 screen for Example 5.

Integrating by Calculator: The calculator screen for the integral in this example is shown. ◆◆◆

Exercise 2 ◆ Area of Surface of Revolution

Find the area of each surface of revolution. Perform or check some of the integrals by calculator. Give all approximate answers to three significant digits.

Rotation About the *x* Axis

1. $y = \dfrac{x^3}{9}$ from $x = 0$ to 3

2. $y = \sqrt{2x}$ from $x = 0$ to 4

3. $y = 3\sqrt{x}$ from $x = 0$ to 4

4. $y = 2\sqrt{x}$ from $x = 0$ to 1

5. $y = \sqrt{4 - x}$ in the first quadrant

6. $y = \sqrt{24 - 4x}$ from $x = 3$ to 6

7. $y = (1 - x^{2/3})^{3/2}$ in the first quadrant

8. $y = \dfrac{x^3}{6} + \dfrac{1}{2x}$ from $x = 1$ to 3

Rotation About the *y* Axis

9. $y = 3x^2$ from $x = 0$ to 5

10. $y = 6x^2$ from $x = 2$ to 4

11. $y = 4 - x^2$ from $x = 0$ to 2

12. $y = 24 - x^2$ from $x = 2$ to 4

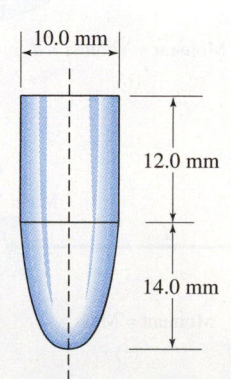

FIGURE 28–15 Rocket nose cone.

48.0 in.

48.0 in.

Geometric Figures

13. Find the surface area of a sphere by rotating the curve $x^2 + y^2 = r^2$ about a diameter.

14. Find the area of the curved surface of a cone by rotating about the *x* axis the line connecting the origin and the point (a, b).

Applications

15. Find the surface area of the nose cone shown in Fig. 28–15.

16. Find the cost of copper plating 10,000 bullets (Fig. 28–16) at the rate of \$15 per square meter.

10.0 mm

12.0 mm

14.0 mm

FIGURE 28–16 Bullet.

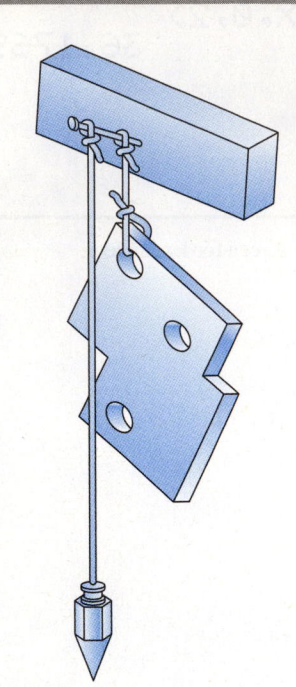

FIGURE 28–17 Any object hung from a point will swing to where its center of gravity is directly below the point of suspension.

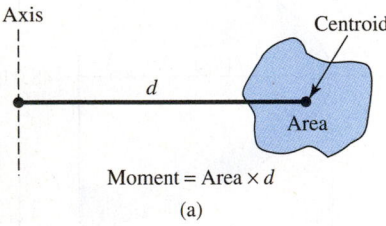

(a)

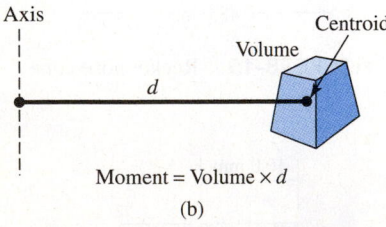

(b)

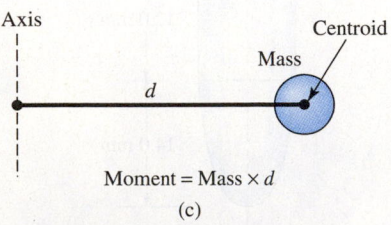

(c)

FIGURE 28–18 Moment of (a) an area, (b) a volume, and (c) a mass.

28–3   Centroids

For a variety of applications we can greatly simplify a problem by treating an entire plane area as though it was concentrated at a single point, called the *centroid*. We can do the same for many applications involving a three-dimensional object, such as a cube. If that object has weight or mass, we refer to its centroid as its *center of gravity*.

■ Exploration:

Try this. Cut a plane area from cardboard, and suspend it so that it can swing freely, Fig. 28–17. Hang a string and a weight from the same suspension point and trace the position of the string on the cutout. Repeat a few times, hanging the cutout and the weight from a new point each time.

What do you observe? Do your traced lines pass through a single point? If so, what is the significance of that point?   ■

■ Exploration:

Try this. With the cutout from the preceding exploration horizontal, try to balance it on the point of a pencil. Mark the balance point. How does it compare with the point found above?   ■

First Moment

To calculate the position of the centroid in figures of various shapes, we need the idea of the *first moment*, often referred to simply as *the moment*. It is not new to us. We know from our previous work that the *moment of a force* about some point *a* is the product of the force F and the perpendicular distance d from the point to the line of action of the force. In a similar way, we speak of the *moment of an area* about some axis

$$\text{moment of an area} = \text{area} \times \text{distance to the centroid}$$

or *moment of a volume*

$$\text{moment of a volume} = \text{volume} \times \text{distance to the centroid}$$

or *moment of a mass*

$$\text{moment of a mass} = \text{mass} \times \text{distance to the centroid}$$

In each case, Fig. 28–18, the distance is that from the axis about which we take the moment, measured to some point on the area, volume, or mass. But to which point on the figure shall we measure? To the *centroid* or *center of gravity*.

In contrast to the first moments that we are finding now, we will learn about the *second moment* or *moment of inertia* later in this chapter.

Centroids of Simple Shapes

For simple plane figures (square, rectangle, circle, etc.) and simple solids (cube, sphere, cylinder, and so forth) the centroid is exactly where you would expect it to be, at the center of the figure.

If a figure has an axis of symmetry, the centroid always lies on that axis. If there are two axes of symmetry, the centroid is at their intersection. The centroid of any triangle is located at the intersection of the three *medians*, the lines drawn from each vertex to the midpoint of the opposite side.

We can easily find the centroid of an area that can be subdivided into simple regions, each of whose centroid location is known by symmetry. We first find the moment of each region about some convenient axis by multiplying the area of that region by the distance of its centroid from the axis. We then use the following fact: For an area subdivided into smaller regions, *the moment of that area about a given axis is equal to the sum of the moments of the individual regions about that same axis.*

◆◆◆ **Example 6:** Locate the centroid of the shape in Fig. 28–19.

Solution: We subdivide the area into simple shapes as in Fig. 28–19(b) and compute its area.

$$\text{Area} = 40 + 16 + 20 = 76 \text{ in.}^2$$

Next we choose x and y axes from which we will measure the coordinates $\bar{x}$ and $\bar{y}$ of the centroid, and locate, by inspection, the centroid of each rectangular area.

The upper rectangle, for example, has an area of 40 in.² The centroid *of that rectangle only* is at its center, which is at a distance of 5 in. from the y axis. The moment about the y axis of that rectangle only is thus 40(5). Similarly, the moments of the middle and lower rectangles about the y axis are 16(6) and 20(5). The moment M_y about the y axis of the *entire* region is the sum of the moments of the individual regions.

$$M_y = 40(5) + 16(6) + 20(5) = 396 \text{ in.}^3$$

Since

$$\text{area} \times \text{distance to centroid} = \text{moment}$$

then the distance $\bar{x}$ to the centroid is

$$\bar{x} = \frac{\text{moment about } y \text{ axis, } M_y}{\text{area}}$$

$$= \frac{396 \text{ in.}^3}{76 \text{ in.}^2} = 5.21 \text{ in.}$$

Similarly, the moment M_x about the x axis is

$$M_x = 40(8) + 16(4) + 20(1) = 404 \text{ in.}^3$$

So the distance $\bar{y}$ to the centroid is

$$\bar{y} = \frac{\text{moment about } x \text{ axis, } M_x}{\text{area}}$$

$$= \frac{404 \text{ in.}^3}{76 \text{ in.}^2} = 5.32 \text{ in.}$$

The centroid is shown in Fig. 28–19(c). ◆◆◆

Centroids of Areas by Integration

If an area does not have axes of symmetry whose intersection gives us the location of the centroid, we can often find it by integration. We subdivide the area into thin strips, compute the first moment of each, sum these moments by integration, and then divide by the total area to get the distance to the centroid.

Consider the area bounded by the curves $y_1 = f_1(x)$ and $y_2 = f_2(x)$ and the lines $x = a$ and $x = b$ (Fig. 28–20). We draw a vertical element of area of width dx and height $(y_2 - y_1)$. Since the strip is narrow, all points on it may be

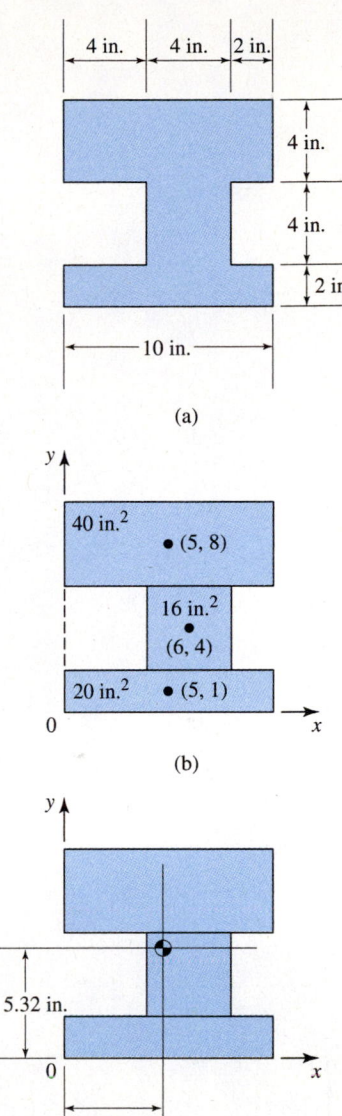

(a)

(b)

$\bar{y} = 5.32$ in.

$\bar{x} = 5.21$ in.

(c)

FIGURE 28–19

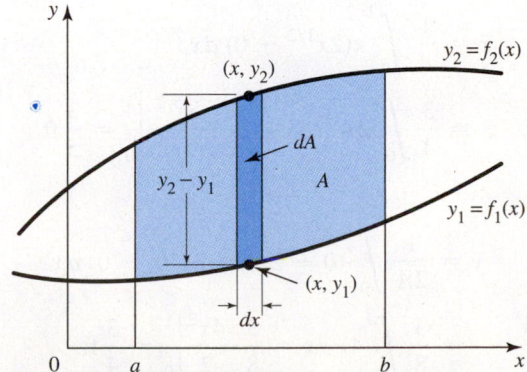

FIGURE 28–20 Centroid of irregular area found by integration.

considered to be the same distance x from the y axis. The moment dM_y of that strip about the y axis is then

$$dM_y = x\, dA = x(y_2 - y_1)\, dx$$

since $dA = (y_2 - y_1)\, dx$. We get the total moment M_y by integrating.

$$M_y = \int_a^b x(y_2 - y_1)\, dx$$

But since the moment M_y is equal to the area A times the distance $\bar{x}$ to the centroid, we get $\bar{x}$ by dividing the moment by the area. Thus:

Horizontal Distance to Centroid	$\bar{x} = \dfrac{1}{A}\displaystyle\int_a^b x(y_2 - y_1)\, dx$	303

To find $\bar{x}$, we must have the area A. It can be found by integration.

We now find the moment about the x axis. The centroid of the vertical element is at its midpoint, which is at a distance of $(y_1 + y_2)/2$ from the x axis. Thus the moment of the element about the x axis is

$$dM_x = \frac{y_1 + y_2}{2}(y_2 - y_1)\, dx$$

Integrating and dividing by the area gives us $\bar{y}$.

Vertical Distance to Centroid	$\bar{y} = \dfrac{1}{2A}\displaystyle\int_a^b (y_1 + y_2)(y_2 - y_1)\, dx$	304

Equations 303 and 304 apply only for *vertical elements*. When using horizontal elements, interchange x and y in these equations.

Our first example is for an area bounded by one curve and the x axis.

◆◆◆ **Example 7:** Find the centroid of the area bounded by the parabola $y^2 = 4x$, the x axis, and the line $x = 1$ (Fig. 28–21).

Solution: We will need the area, so we find that first. We draw a vertical strip having an area $y\, dx$ and integrate.

$$A = \int y\, dx = 2\int_0^1 x^{1/2}\, dx = 2\left[\frac{x^{3/2}}{3/2}\right]_0^1 = \frac{4}{3}\ \text{ft}^2$$

Then, by Eq. 303,

$$\bar{x} = \frac{1}{A}\int_0^1 x(2x^{1/2} - 0)\, dx$$

$$= \frac{3}{4}\int_0^1 2x^{3/2}\, dx = \frac{3}{2}\cdot\frac{x^{5/2}}{(5/2)}\Big|_0^1 = \frac{3}{5}\ \text{ft}$$

and, by Eq. 304,

$$\bar{y} = \frac{1}{2A}\int_0^1 (0 + 2x^{1/2})(2x^{1/2} - 0)\, dx$$

$$= \frac{3}{8}\int_0^1 4x\, dx = \frac{3}{8}\cdot\frac{4x^2}{2}\Big|_0^1 = \frac{3}{4}\ \text{ft}$$

Integrating by Calculator: The TI-83 calculator screens for the three integrals in this example are given in screens (1)–(3).

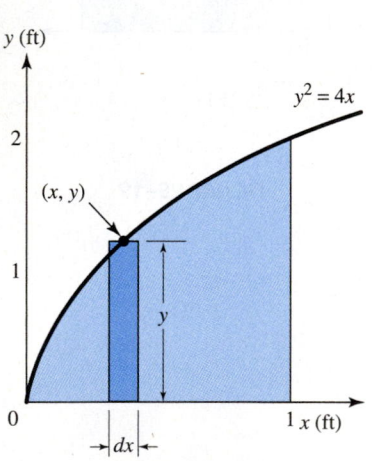

FIGURE 28–21

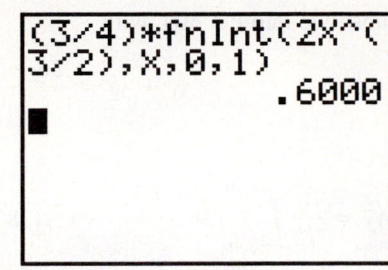 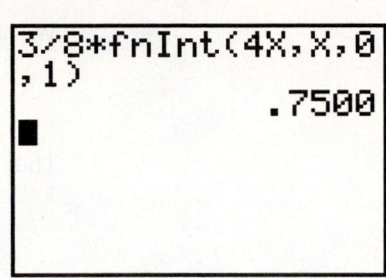

| (1) Calculation of A. | (2) Calculation of $\bar{x}$. | (3) Calculation of $\bar{y}$. |

Check: Does the answer seem reasonable? The given area extends from $x = 0$ to 1, with more area lying to the right of $x = \frac{1}{2}$ than to the left. Thus we would expect $\bar{x}$ to be between $x = \frac{1}{2}$ and 1, which it is. Similarly, we would expect the centroid to be located between $y = 0$ and 1, which it is. ◆◆◆

■ **Exploration:**

Try this. Make a graph as in Example 7, being sure that you use the same scale for the x and y axes, and paste the graph to cardboard or thin metal. Cut out the shaded area and locate the centroid by suspending it as in Fig. 28–17 or by balancing it on the point of a pencil. Compare your experimental result with the calculated one. ■

Areas Bounded by Two Curves

For areas that are not bounded by a curve and a coordinate axis but are instead bounded by two curves, the work is only slightly more complicated, as shown in the following example.

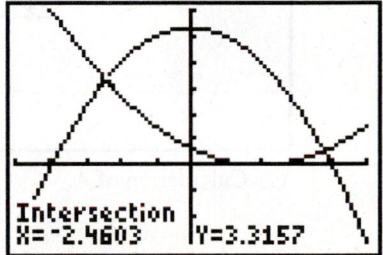

◆◆◆ **Example 8:** Find the coordinates of the centroid of the area bounded by the curves $6y = x^2 - 4x + 4$ and $3y = 16 - x^2$.

Solution: We plot the curves as shown and find their points of intersection graphically or by solving simultaneously, as follows. Multiplying the second equation by -2 and adding the resulting equation to the first gives

$$3x^2 - 4x - 28 = 0$$

Solving by quadratic formula (work not shown), we find that the points of intersection are at $x = -2.46$ and $x = 3.79$. We then take a vertical strip whose width is dx and whose height is $y_2 - y_1$, where

Screen for Example 8. Graph of the given curves showing one of the two points of intersection. Here, y_1 is the parabola open upward. Tick marks are 1 unit apart.

$$y_2 - y_1 = \left(\frac{16 - x^2}{3}\right) - (x^2 - 4x + 4) = \frac{16}{3} - \frac{x^2}{3} - \frac{x^2}{6} + \frac{4x}{6} - \frac{4}{6} = \frac{28 + 4x - 3x^2}{6}$$

The area is then

$$A = \int_{-2.46}^{3.79} (y_2 - y_1)\,dx$$

$$= \frac{1}{6}\int_{-2.46}^{3.79} (28 + 4x - 3x^2)\,dx = 20.4$$

Then, from Eq. 303,

$$A\bar{x} = \int_{-2.46}^{3.79} x(y_2 - y_1)\,dx$$

$$= \frac{1}{6}\int_{-2.46}^{3.79} (28x + 4x^2 - 3x^3)\,dx = 13.6$$

$$\bar{x} = \frac{M_y}{A} = \frac{13.6}{20.4} = 0.667$$

We now substitute into Eq. 304, with

$$y_1 + y_2 = \frac{36 - 4x - x^2}{6}$$

Thus

$$A\bar{y} = \frac{1}{72} \int_{-2.46}^{3.79} (36 - 4x - x^2)(28 + 4x - 3x^2)\,dx$$

$$= \frac{1}{72} \int_{-2.46}^{3.79} (3x^4 + 8x^3 - 152x^2 + 32x + 1008)\,dx$$

$$= \frac{1}{72}\left[\frac{3x^5}{5} + \frac{8x^4}{4} - \frac{152x^3}{3} + \frac{32x^2}{2} + 1008x\right]_{-2.46}^{3.79} = 52.5$$

$$\bar{y} = \frac{M_x}{A} = \frac{52.5}{20.4} = 2.57$$

Integrating by Calculator: The TI-83 calculator screens for the three integrals in this sample are given in screens (1)–(3).

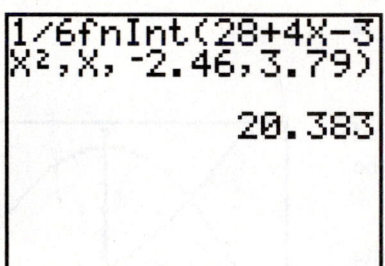

(1) Calculation of A.

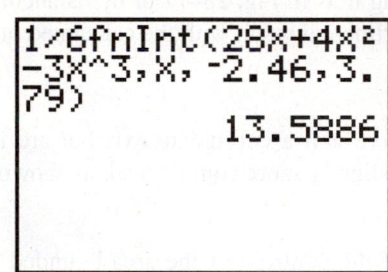

(2) Calculation of $\bar{x}$.

(3) Calculation of $\bar{y}$.

◆◆◆

Centroids of Solids of Revolution

A solid of revolution is, of course, symmetrical about the axis of revolution, so the centroid must be on that axis. We only have to find the position of the centroid along that axis. The procedure is similar to that for an area. We think of the solid as being subdivided into many small elements of volume, find the sum of the moments for each element by integration, and set this equal to the moment of the entire solid (the product of its total volume and the distance to the centroid). We then divide by the volume to obtain the distance to the centroid.

Of course, these methods work only for a solid that is *homogeneous*, that is, one whose density is the same throughout.

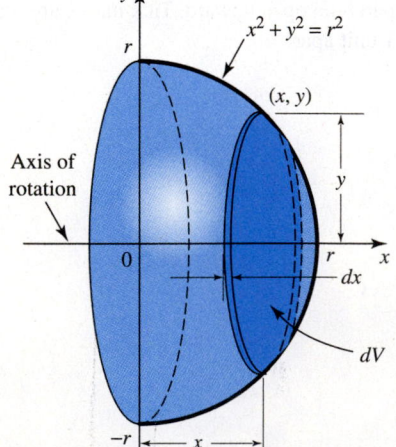

FIGURE 28–22   Finding the centroid of a hemisphere.

◆◆◆ **Example 9:** Find the centroid of a hemisphere of radius r.

Solution: We place the hemisphere on coordinate axes as shown in Fig. 28–22 and consider it as the solid obtained by rotating the first-quadrant portion of the curve $x^2 + y^2 = r^2$ about the x axis. Through the point (x, y) we draw an element of volume of radius y and thickness dx, at a distance x from the base of the hemisphere. Its volume is thus

$$dV = \pi y^2\,dx$$

and its moment about the base of the hemisphere (the y axis) is

$$dM_y = \pi xy^2\,dx = \pi x(r^2 - x^2)\,dx$$

Integrating gives us the total moment.

$$M_y = \pi \int_0^r x(r^2 - x^2)\,dx = \pi \int_0^r (r^2 x - x^3)\,dx$$

$$= \pi \left[\frac{r^2 x^2}{2} - \frac{x^4}{4} \right]_0^r = \frac{\pi r^4}{4}$$

The total moment also equals the volume ($\frac{2}{3}\pi r^3$ for a hemisphere) times the distance $\bar{x}$ to the centroid, so

$$\bar{x} = \frac{\pi r^4/4}{2\pi r^3/3} = \frac{3r}{8}$$

Thus the centroid is located at $(3r/8, 0)$.

Integration by Calculator: To check the integration by calculator we need to assign some value to r. The integration for M_y is shown in the calculator screen, with $r = 2$. This value compares with our previous value, with $r = 2$, of

$$M_y = \frac{\pi r^4}{4} = \frac{\pi(2^4)}{4} = 12.5664 \qquad \text{◆◆◆}$$

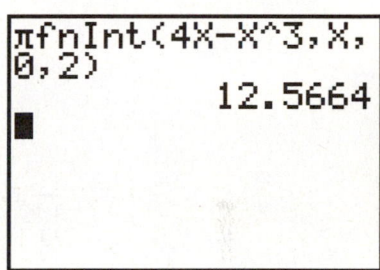

TI-83 screen for Example 9.

As with centroids of areas, we can state the method we have just used for finding the centroid of a solid of revolution as a formula. For the volume V (Fig. 28–23) formed by rotating the curve $y = f(x)$ about the x axis, the distance to the centroid is the following:

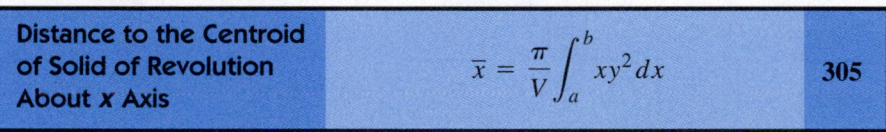

Distance to the Centroid of Solid of Revolution About x Axis	$\bar{x} = \dfrac{\pi}{V} \displaystyle\int_a^b xy^2\,dx$	305

For volumes formed by rotation about the y axis, we simply interchange x and y in Eq. 305 and get

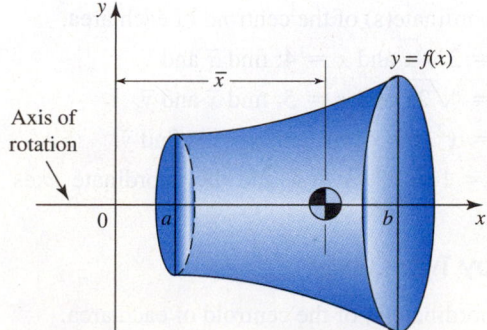

FIGURE 28–23 Centroid of a solid of revolution.

Distance to the Centroid of Solid of Revolution About y Axis	$\bar{y} = \dfrac{\pi}{V} \displaystyle\int_c^d yx^2\, dy$	306

These formulas work for solid figures only, not for those that have holes down their center.

Exercise 3 ◆ Centroids

Centroids of Simple Shapes

1. Find the centroid of four particles of equal mass located at $(0, 0)$, $(4, 2)$, $(3, -5)$, and $(-2, -3)$.

2. Find the centroid in Fig. 28–24(a).

3. Find the centroid in Fig. 28–24(b).

4. Find the centroid in Fig. 28–24(c).

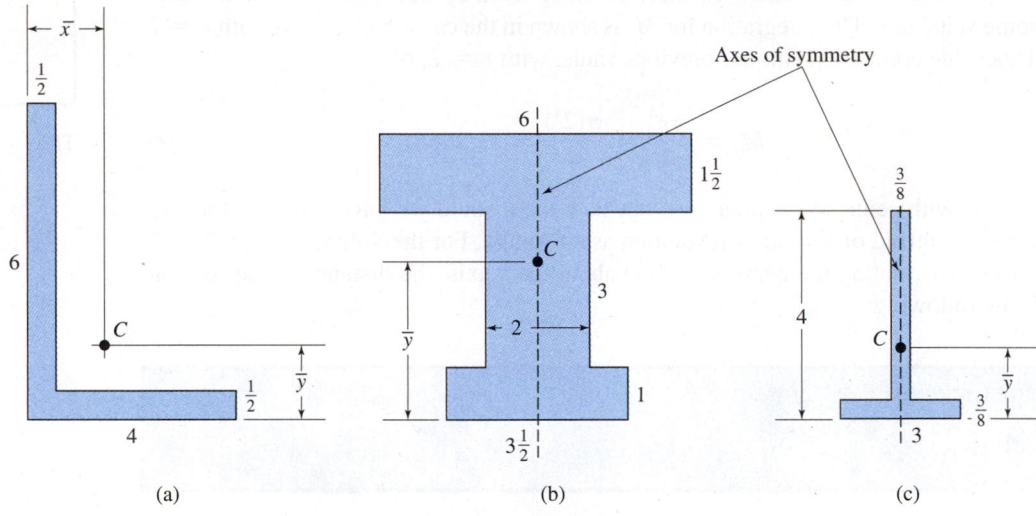

(a)   (b)   (c)

FIGURE 28–24

Centroids of Areas by Integration

Find the specified coordinate(s) of the centroid of each area.

5. bounded by $y = 2\sqrt{x}$ and $x = 4$; find $\bar{x}$ and $\bar{y}$.

6. bounded by $y = \sqrt{2x}$ and $x = 5$; find $\bar{x}$ and $\bar{y}$.

7. bounded by $y = x^2$, the x axis, and $x = 3$; find $\bar{y}$.

8. bounded by $y = 1 - 2\sqrt{x} + x$ and the coordinate axes (area $= \frac{1}{6}$); find $\bar{x}$ and $\bar{y}$.

Areas Bounded by Two Curves

Find the specified coordinate(s) of the centroid of each area.

9. bounded by $y = x^3$ and $y = 4x$ in the first quadrant; find $\bar{x}$.

10. bounded by $x = 4y - y^2$ and $y = x$; find $\bar{y}$.

11. bounded by $y^2 = x$ and $x^2 = y$; find $\bar{x}$ and $\bar{y}$.

12. bounded by $y^2 = 4x$ and $y = 2x - 4$; find $\bar{y}$.

13. bounded by $2y = x^2$ and $y = x^3$; find $\bar{x}$.

14. bounded by $y = x^2 - 2x - 3$ and $y = 6x - x^2 - 3$ (area = 21.33); find $\bar{x}$ and $\bar{y}$.

15. bounded by $y = x^2$ and $y = 2x + 3$ in the first quadrant; find $\bar{x}$.

Centroids of Volumes of Revolution

Find the distance from the origin to the centroid of each volume.

16. formed by rotating the area bounded by $x^2 + y^2 = 4$, $x = 0$, $x = 1$, and the x axis about the x axis.

17. formed by rotating the area bounded by $6y = x^2$, the line $x = 6$, and the x axis about the x axis.

18. formed by rotating the first-quadrant area under the curve $y^2 = 4x$, from $x = 0$ to 1, about the x axis.

19. formed by rotating the area bounded by $y^2 = 4x$, $y = 6$, and the y axis about the y axis.

20. formed by rotating the first-quadrant portion of the ellipse $x^2/64 + y^2/36 = 1$ about the x axis.

21. a right circular cone of height h, measured from its base.

Applications

22. A certain airplane rudder (Fig. 28–25) consists of one quadrant of an ellipse and a quadrant of a circle. Find the coordinates of the centroid.

23. The vane on a certain wind generator has the shape of a semicircle attached to a trapezoid (Fig. 28–26). Find the distance $\bar{x}$ to the centroid.

24. A certain rocket (Fig. 28–27) c onsists of a cylinder attached to a paraboloid of revolution. Find the distance from the nose to the centroid of the total volume of the rocket.

25. An optical instrument contains a mirror in the shape of a paraboloid of revolution (Fig. 28–28) hollowed out of a cylindrical block of glass. Find the distance from the flat bottom of the mirror to the centroid of the mirror.

26. *Writing:* Suppose that you have designed a tank in the shape of a solid of revolution and have found its centroid by integration. Your manager, a practical person, insists that it is impossible to find the center of gravity of something that isn't even built yet. Write a memo to your manager explaining how it is done.

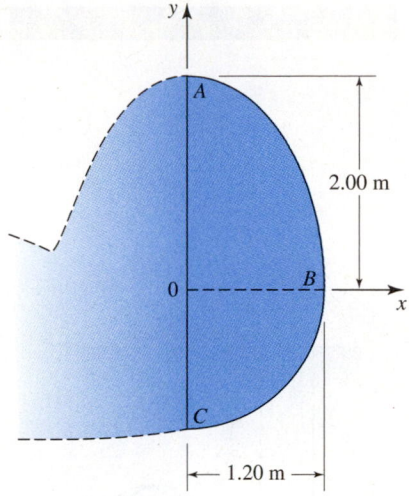

FIGURE 28–25   Airplane rudder.

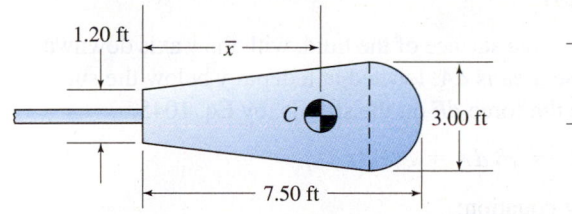

FIGURE 28–26   Wind vane.

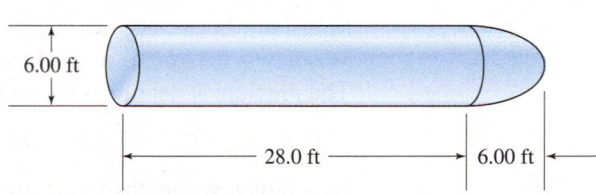

FIGURE 28–27   Rocket.

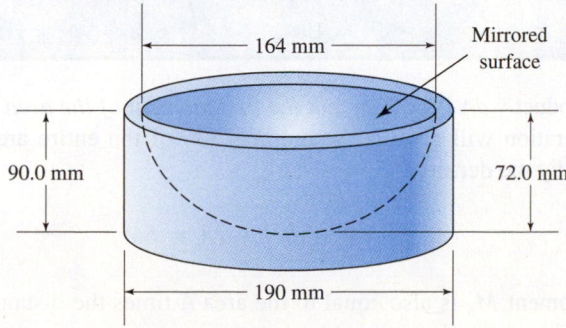

FIGURE 28–28   Paraboloidal mirror.

27. *Project:* The gasoline tank in a certain truck has the shape of a horizontal cylinder. When the fuel gage stopped working the trucker inserted a stick all the way into the tank and marked "*full*" where the stick met the top of the tank. The she marked the halfway points as "*half full*." At what point on the stick should she mark "*quarter full*?" *Hint:* Find the centroid of a semicircle.

28–4   Fluid Pressure

The pressure at any point on a submerged surface varies directly with the depth of that point below the surface. Thus the pressure on a diver at a depth of 50 ft will be twice that at 25 ft.

The pressure on a submerged *area* is equal to the weight of the column of fluid above that area. Thus a square foot of area at a depth of 20 ft supports a column of water having a volume of $20 \times 1^2 = 20$ ft^3. Since the density of water is 62.4 lb/ft^3, the weight of this column is $20 \times 62.4 = 1248$ lb, so the pressure is 1248 lb/ft^2 at a depth of 20 ft. Recall that weight = volume × density (Eq. 1042). Further, Pascal's law says that the pressure is the same in *all directions*, so the same 1248 lb/ft^2 will be felt by a surface that is horizontal, vertical, or at any angle.

The *force* exerted by the fluid (*fluid pressure*) can be found by multiplying the pressure per unit area at a given depth by the total area.

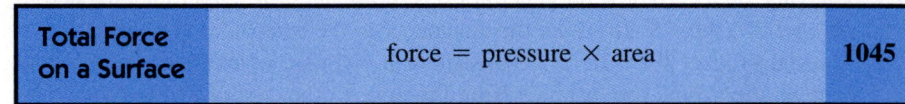

| Total Force on a Surface | force = pressure × area | 1045 |

A complication arises from an area that has points at various depths and hence has different pressures over its surface. To compute the force on such a surface, we first compute the force on a narrow horizontal strip of area, assuming that the pressure is the same everywhere on that strip, and then add up the forces on all such strips by integration.

◆◆◆ **Example 10:** Find an expression for the force on the vertical area A submerged in a fluid of density δ (Fig. 28–29).

Solution: Let us take our origin at the surface of the fluid, with the y axis downward. We draw a horizontal strip whose area is dA, located at a depth y below the surface. The pressure at depth y is $y\delta$, so the force dF on the strip is, by Eq. 1045,

$$dF = y\delta\, dA = \delta(y\, dA)$$

Integrating, we get the following equation:

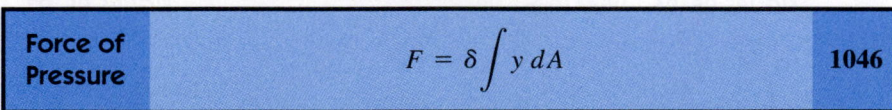

| Force of Pressure | $F = \delta \displaystyle\int y\, dA$ | 1046 |

But the product $y\, dA$ is nothing but *the first moment of the area dA about the x axis*. Thus integration will give us the moment M_x of the entire area, about the x axis, multiplied by the density.

$$F = \delta \int y\, dA = \delta M_x$$

But the moment M_x is also equal to the area A times the distance $\overline{y}$ to the centroid, so we have the following:

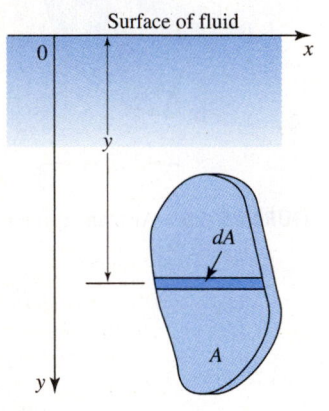

Surface of fluid

FIGURE 28–29 Force on a submerged vertical surface. Sometimes it is more convenient to take the origin at the surface of the liquid.

	$F = \delta \bar{y} A$	
Force of Pressure	*The force of pressure on a submerged, vertical surface is equal to the product of its area, the distance to its centroid, and the density of the fluid.*	**1047**

◆◆◆

We can compute the force on a submerged surface by using either Eq. 1046 or 1047. We give now an example of each method.

◆◆◆ **Example 11:** A vertical triangular wall in a dam (Fig. 28–30) holds back water whose level is at the top of the wall. Use integration to find the total force of pressure on the wall.

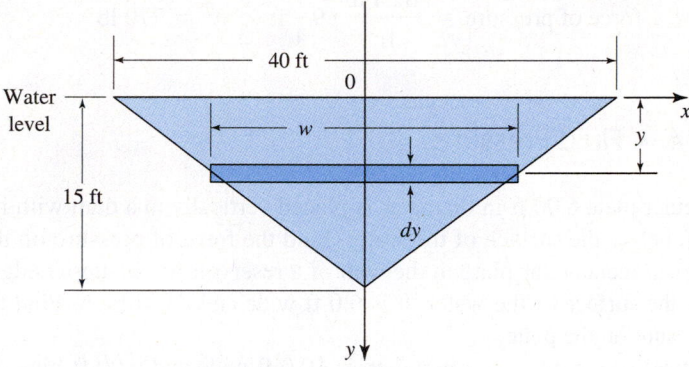

FIGURE 28–30

Estimate: We can sketch in the medians and their point of intersection at the centroid of the triangle, and estimate it to be five feet below the surface. Then

force = (density) × (distance to centroid) × (area), or $62.4(5) \left(\frac{1}{2}\right)(40)(15) = 93{,}600$ lb.

Solution: We sketch an element of area with width w and height dy. So

$$dA = w \, dy \qquad (1)$$

We wish to express w in terms of y. By similar triangles,

$$\frac{w}{40} = \frac{15 - y}{15}$$

$$w = \frac{8}{3}(15 - y)$$

Substituting into Eq. (1) yields

$$dA = \frac{8}{3}(15 - y) \, dy$$

Then, by Eq. 1046,

$$F = \delta \int y \, dA = \frac{8\,\delta}{3} \int_0^{15} y(15 - y) \, dy$$

$$= \frac{8\,\delta}{3} \int_0^{15} (15y - y^2) \, dy$$

$$= \frac{8\,\delta}{3} \left[\frac{15y^2}{2} - \frac{y^3}{3} \right]_0^{15}$$

$$= \frac{8\,(62.4)}{3} \left[\frac{15\,(15)^2}{2} - \frac{(15)^3}{3} \right] = 93{,}600 \text{ lb}$$

TI-83 screen for Example 11.

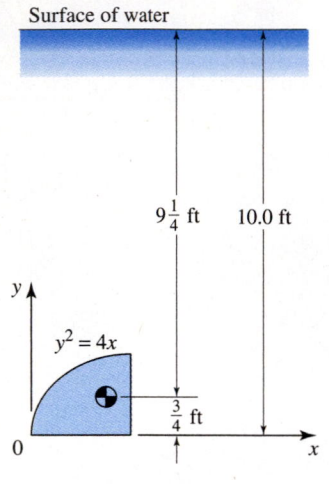

FIGURE 28–31

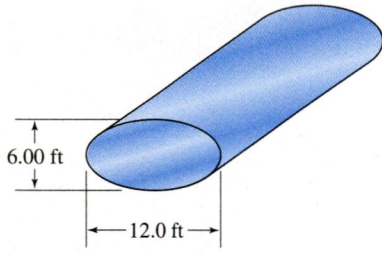

FIGURE 28–32   Oil tank.

Integration by Calculator: The screen for evaluating the integral by calculator is shown.
◆◆◆

◆◆◆ **Example 12:** The area shown in Fig. 28–31 is submerged in water (density $= 62.4$ lb/ft^3) so that the origin is 10.0 ft below the surface. Find the force on the area using Eq. 1047.

Solution: The area and the distance to the centroid have already been found in Example 7 of this chapter: area $= \frac{4}{3}$ ft^2 and $\bar{y} = \frac{3}{4}$ ft up from the origin, as shown in Fig. 28–31. The depth of the centroid below the surface is then

$$10.0 - \frac{3}{4} = 9\frac{1}{4} \text{ ft}$$

Thus by Eq. 1047,

$$\text{force of pressure} = \frac{62.4 \text{ lb}}{\text{ft}^3} \cdot 9\frac{1}{4} \text{ ft} \cdot \frac{4}{3} \text{ ft}^2 = 770 \text{ lb}$$
◆◆◆

Exercise 4 ◆ Fluid Pressure

1. A circular plate 6.00 ft in diameter is placed vertically in a dam with its center 50.0 ft below the surface of the water. Find the force of pressure on the plate.
2. A vertical rectangular plate in the wall of a reservoir has its upper edge 20.0 ft below the surface of the water. It is 6.0 ft wide and 4.0 ft high. Find the force of pressure on the plate.
3. A vertical rectangular gate in a dam is 10.0 ft wide and 6.00 ft high. Find the force on the gate when the water level is 8.00 ft above the top of the gate.
4. A vertical cylindrical tank has a diameter of 30.0 ft and a height of 50.0 ft. Find the total force on the curved surface when the tank is full of water.
5. A trough, whose cross section is an equilateral triangle with a vertex down, has sides 2.00 ft long. Find the total force on one end when the trough is full of water.
6. A horizontal cylindrical boiler 4.00 ft in diameter is half full of water. Find the force on one end.
7. A horizontal tank of oil (density $= 60.0$ lb/ft^3) has ends in the shape of an ellipse with horizontal axis 12.0 ft long and vertical axis 6.00 ft long (Fig. 28–32). Find the force on one end when the tank is half full.
8. The cross section of a certain trough is a parabola with vertex down. It is 2.00 ft deep and 4.00 ft wide at the top. Find the force on one end when the trough is full of water.

28–5   Work

It is important to be able to compute the work necessary to perform certain mechanical tasks. This enables us, with other information, to select the right size motor or engine, the speed of operation, the rating for a transmission or clutch, and so forth. Here we will compute the work required to perform a few simple tasks.

Definition

When a constant force acts on an object that moves in the direction of the force, the *work* done by the force is defined as the product of the force and the distance moved by the object.

Work Done by a Constant Force	work = force × distance	1005

◆◆◆ **Example 13:** The work needed to lift a 100-lb weight a distance of 2 ft is 200 ft · lb.   ◆◆◆

Variable Force

Equation 1005 applies when the force is *constant*, but this is not always the case. For example, the force needed to stretch a spring increases as the spring gets extended (Fig. 28–33). Or, as another example, the expanding gases in an automobile cylinder exert a variable force on the piston.

If we let the variable force be represented by $F(x)$, acting in the x direction from $x = a$ to $x = b$, the work done by this force may be defined as follows:

Work Done by a Variable Force	$$W = \int_a^b F(x)\,dx$$	1006

We first apply this formula to find the work done in stretching or compressing a spring.

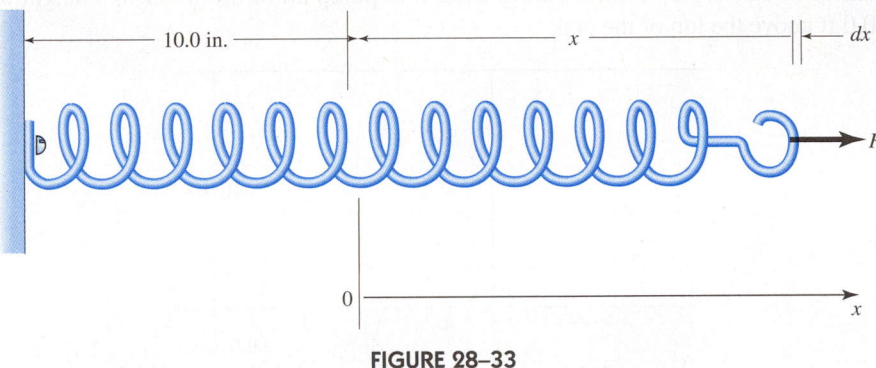

FIGURE 28–33

◆◆◆ **Example 14:** A certain spring (Fig. 28–33) has a free length (when no force is applied) of 10.0 in. The spring constant is 12.0 lb/in. Find the work needed to stretch the spring from a length of 12.0 in. to a length of 14.0 in.

Estimate: The amount of stretch starts at 2 in., requiring a force of 2(12), or 24 lb, and ends at 4 in. with a force of 48 lb. Assuming an average force of 36 lb over the 2-in. travel gives an estimate of 72 in. · lb.

Solution: We draw the spring partly stretched, as shown, taking our x axis in the direction of movement of the force, with zero at the free position of the end of the spring. The force needed to hold the spring in this position is equal to the spring constant k times the deflection x.

	$F = kx$	1060

If we assume that the force does not change when stretching the spring an additional small amount dx, the work done is

$$dW = F\,dx = kx\,dx$$

We get the total work by integrating.

$$W = \int F\,dx = k \int_{2}^{4} x\,dx = \left.\frac{kx^2}{2}\right|_{2}^{4}$$

$$= \frac{12.0}{2}(4^2 - 2^2) = 72.0 \text{ in.} \cdot \text{lb}$$

Integration by Calculator: The screen for evaluating the integral by calculator is shown.   ◆◆◆

```
12fnInt(X,X,2,4)
          72.0000
```

TI-83 screen for Example 14.

Common Error	Be sure to measure spring deflections from the *free* end of the spring, not from the fixed end.

Another typical problem is that of finding the work needed to pump fluid out of a tank. Such problems may be solved by noting that the work required is equal to the weight of the fluid times the distance which the centroid of the fluid (when still in the tank) must be raised.

◆◆◆ **Example 15:** A hemispherical tank (Fig. 28–34) is filled with water having a density of 62.4 lb/ft^3. Find the work needed to pump all of the water to a height of 10.0 ft above the top of the tank.

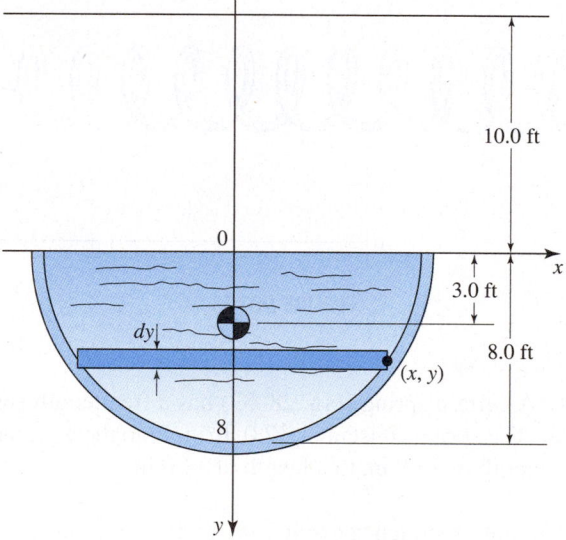

FIGURE 28–34   Hemispherical tank.

Solution: The distance to the centroid of a hemisphere of radius r was found in Example 9 to be $3r/8$, so the centroid of our hemisphere is at a distance

$$\bar{y} = \frac{3}{8} \cdot 8 = 3.0 \text{ ft}$$

as shown. It must therefore be raised a distance of 13.0 ft. We find the weight of the tankful of water by multiplying its volume times its density.

$$\text{weight} = \text{volume} \times \text{density}$$

$$= \frac{2}{3}\pi(8^3)(62.4) = 66,900 \text{ lb}$$

The work done is then

$$\text{work} = 66,900 \text{ lb}(13.0 \text{ ft}) = 870,000 \text{ ft} \cdot \text{lb}$$ ◆◆◆

If we do not know the location of the centroid, we can find the work directly by integration.

◆◆◆ **Example 16:** Repeat Example 15 by integration, assuming that the location of the centroid is not known.

Solution: We choose coordinate axes as shown in Fig. 28–34. Through some point (x, y) on the curve, we draw an element of volume whose volume dV is

$$dV = \pi x^2 \, dy$$

and whose weight is

$$62.4 \, dV = 62.4 \pi x^2 \, dy$$

Since this element must be lifted a distance of $(10 + y)$ ft, the work required is

$$dW = (10 + y)(62.4 \pi x^2 \, dy)$$

Integrating gives us

$$W = 62.4 \pi \int (10 + y) x^2 \, dy$$

But using the equation for a circle (Eq. 218), we have

$$x^2 = r^2 - y^2 = 64 - y^2$$

We substitute to get the expression in terms of y.

$$W = 62.4 \pi \int_0^8 (10 + y)(64 - y^2) \, dy$$

$$= 62.4 \pi \int_0^8 (640 + 64y - 10y^2 - y^3) \, dy$$

$$= 62.4 \pi \left[640y + 32y^2 - \frac{10y^3}{3} - \frac{y^4}{4} \right]_0^8 = 870{,}000 \text{ ft} \cdot \text{lb}$$

as by the method of Example 15.

Integration by Calculator: The screen for evaluating the integral by calculator is shown. ◆◆◆

$$62.4\pi \int_0^8 ((10+X)(6\blacktriangleright$$

$$869874$$

TI-84 screen for Example 16.

Exercise 5 ◆ Work

Springs

1. A certain spring has a free length of 12.0 in. and a spring constant of 50.0 lb/in. How much work is required to stretch the spring from a length of 14.0 in. to 16.0 in.?

2. A spring whose free length is 10.0 in. has a spring constant of 12.0 lb/in. Find the work needed to stretch this spring from 12.0 in. to 15.0 in.

3. A spring has a spring constant of 8.0 lb/in. and a free length of 5.0 in. Find the work required to stretch it from 6.0 in. to 8.0 in.

Tanks

4. Find the work required to pump all of the water to the top and out of a vertical cylindrical tank that is 16.0 ft in diameter, 20.0 ft deep, and completely filled at the start.

5. A hemispherical tank 12.0 ft in diameter is filled with water to a depth of 4.00 ft. How much work is needed to pump the water to the top of the tank?

6. A conical tank 20.0 ft deep and 20.0 ft across the top is full of water. Find the work needed to pump the water to a height of 15.0 ft above the top of the tank.

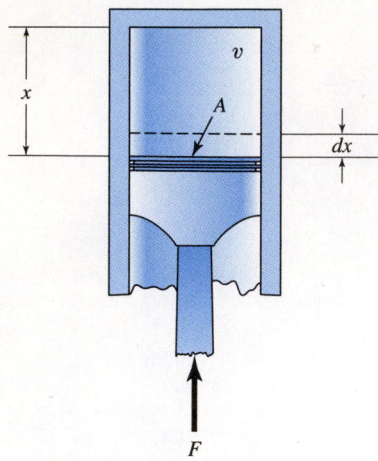

FIGURE 28–35 Piston and cylinder. Assume that the pressure and volume are related by the equation pv = constant.

7. A tank has the shape of a frustum of a cone, with a bottom diameter of 8.00 ft, a top diameter of 12.0 ft, and a height of 10.0 ft. How much work is needed to pump the contents to a height of 10.0 ft above the tank, if it is filled with oil of density 50.0 lb/ft^3?

Gas Laws

8. Find the work needed to compress air initially at a pressure of 15.0 lb/in.^{2} from a volume of 200 ft^3 to 50.0 ft^3. [*Hint:* The work dW done in moving the piston (Fig. 28–35) is $dW = F\,dx$. Express both F and dx in terms of v by means of Eqs. 91 for the volume of a cylinder, and 1045 (Force = pressure), and integrate.]

9. Air is compressed (Fig. 28–35) from an initial pressure of 15.0 lb/in.^{2} and volume of 200 ft^3 to a pressure of 80.0 lb/in.^{2}. How much work was needed to compress the air?

10. If the pressure and volume of air are related by the equation $pv^{1.4} = k$, find the work needed to compress air initially at 14.5 lb/in.^{2} and 10.0 ft^3 to a pressure of 100 lb/in.^{2}.

Miscellaneous

11. The force of attraction (in pounds) between two masses separated by a distance d is equal to k/d^2, where k is a constant. If two masses are 50 ft apart, find the work needed to separate them another 50 ft. Express your answer in terms of k.

12. Find the work needed to wind up a vertical cable 100 ft long, weighing 3.0 lb/ft.

13. A 500-ft-long cable weighs 1.00 lb/ft and is hanging from a tower with a 200-lb weight at its end. How much work is needed to raise the weight and the cable a distance of 20.0 ft?

28–6   Moment of Inertia

Earlier in this chapter we defined the *first* moment; here we learn about the *second* moment, or *moment of inertia*. This is an important quantity in technology.

The moment of inertia is very important in beam design. It is not only the cross-sectional area of a beam that determines how well the beam will resist bending, but also *how far* that area is located from the axis of the beam. The moment of inertia, taking into account both the area and its location with respect to the beam axis, is a measure of the resistance to bending. Further, the polar moment of inertia of a rotating body determines how much torque is needed to accelerate that body, or how much torsion a shaft can take before breaking.

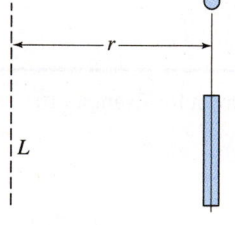

FIGURE 28–36

Moment of Inertia of an Area

Figure 28–36 shows two small areas at a distance r from a line L in the same plane. In each case, the dimensions of the area are such that we may consider all points on the area as being at the same distance r from the line L.

We earlier defined the *first moment* of the area about L as being the product of the area times the distance to the line. We now define the *second moment*, or *moment of inertia I*, as the product of the area times the *square* of the distance to the line.

Moment of Inertia of an Area	$I = Ar^2$	307

The distance r, which is the distance to the line, is called the *radius of gyration*. We write I_x to denote the moment of inertia about the x axis, and I_y for the moment of inertia about the y axis.

◆◆◆ **Example 17:** Find the moment of inertia of the thin strip in Fig. 28–36 if it has a length of 8.00 cm and a width of 0.200 cm and is 7.00 cm from axis L.

Solution: The area of the strip is $8.00(0.200) = 1.60$ cm^2, so the moment of inertia is, by Eq. 307,

$$I = 1.60(7.00)^2 = 78.4 \text{ cm}^4 \qquad ◆◆◆$$

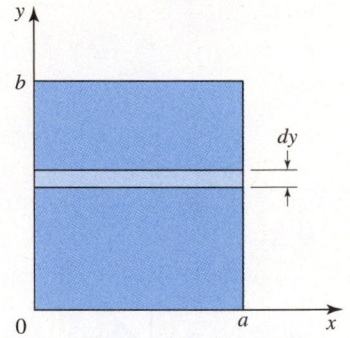

Moment of Inertia of a Rectangle

In Example 17, our area was a thin strip parallel to the axis, with all points on the area at the same distance r from the axis. But what shall we use for r when dealing with an extended area, such as the rectangle in Fig. 28–37? Again calculus comes to our aid. Since we can easily compute the moment of inertia of a thin strip, we slice our area into many thin strips and add up their individual moments of inertia by integration.

◆◆◆ **Example 18:** Compute the moment of inertia of a rectangle about its base (Fig. 28–37).

FIGURE 28–37 Moment of inertia of a rectangle.

Solution: We draw a single strip of area parallel to the axis about which we are taking the moment, here the x axis. This strip has a width dy and a length a. Its area is then

$$dA = a \, dy$$

All points on the strip are at a distance y from the x axis, so the moment of inertia, dI_x, of the single strip is

$$dI_x = y^2 (a \, dy)$$

We add up the moments of all such strips by integrating from $y = 0$ to b.

$$I_x = a \int_0^b y^2 dy = \frac{ay^3}{3}\Big|_0^b = \frac{ab^3}{3} \qquad ◆◆◆$$

Radius of Gyration

If all of the area in a plane figure were squeezed into a single thin strip of equal area and placed parallel to the x axis at such a distance that it had the *same moment of inertia* as the original rectangle, it would be at a distance r that we call the *radius of gyration*. If I is the moment of inertia of some area A about some axis, then

$$Ar^2 = I$$

Thus:

Radius of Gyration	$r = \sqrt{\dfrac{I}{A}}$	311

We use r_x and r_y to denote the radius of gyration about the x and y axes, respectively.

◆◆◆ **Example 19:** Find the radius of gyration for the rectangle in Example 18.

Solution: The area of the rectangle is ab, so, by Eq. 311,

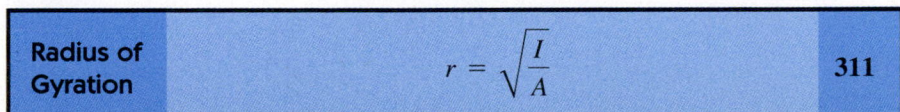

$$r_x = \sqrt{\frac{I_x}{A}} = \sqrt{\frac{ab^3}{3ab}} = \frac{b}{\sqrt{3}} \approx 0.577b$$

Note that the *centroid* is at a distance of $0.5b$ from the edge of the rectangle. We have thus shown that *the radius of gyration is not equal to the distance to the centroid.* ◆◆◆

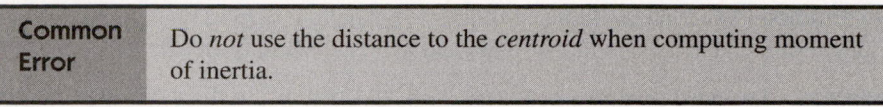

| Common Error | Do *not* use the distance to the *centroid* when computing moment of inertia. |

Moment of Inertia of an Area by Integration

Since we are now able to write the moment of inertia of a rectangular area, we can derive formulas for the moment of inertia of other areas, such as in Fig. 28–38. The area of the vertical strip shown in Fig. 28–38 is dA, and its distance from the y axis is x, so its moment of inertia about the y axis is, by Eq. 307,

$$dI_y = x^2\, dA = x^2 y\, dx$$

since $dA = y\, dx$. The total moment is then found by integrating.

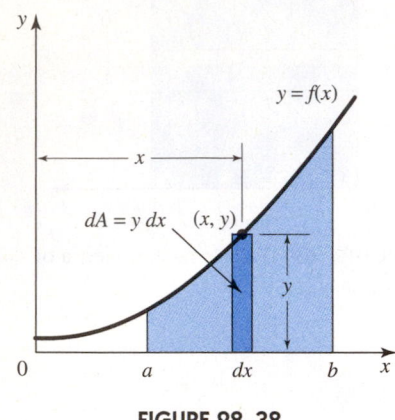

FIGURE 28–38

| Moment of Inertia of an Area About y Axis | $I_y = \displaystyle\int x^2 y\, dx$ | 309 |

To find the moment of inertia about the x axis, we use the result of Example 18: that the moment of inertia of a rectangle about its base is $ab^3/3$. Thus the moment of inertia of the rectangle in Fig. 28–38 is

$$dI_x = \frac{1}{3} y^3\, dx$$

As before, the total moment of inertia is found by integrating.

| Moment of Inertia of an Area About x Axis | $I_x = \dfrac{1}{3}\displaystyle\int y^3\, dx$ | 308 |

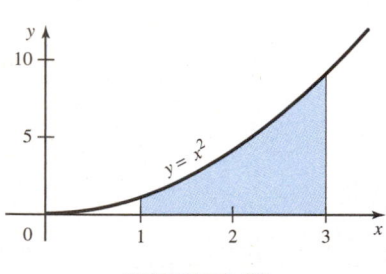

FIGURE 28–39

◆◆◆ **Example 20:** Find the moment of inertia about the x and y axes for the area under the curve $y = x^2$ from $x = 1$ to 3 (Fig. 28–39).

Solution: From Eq. 309,

$$I_y = \int_1^3 x^2 (x^2)\, dx$$

$$= \int_1^3 x^4\, dx$$

$$= \left.\frac{x^5}{5}\right|_1^3 \approx 48.4$$

Now using Eq. 308, with $y^3 = (x^2)^3 = x^6$,

$$I_x = \frac{1}{3}\int_1^3 x^6\, dx = \frac{1}{3}\left[\frac{x^7}{7}\right]_1^3 = \frac{1}{21}(3^7 - 1^7) = \frac{2186}{21} \approx 104.1$$

Integration by Calculator: The screens for the two integrations in this example are shown on the left.

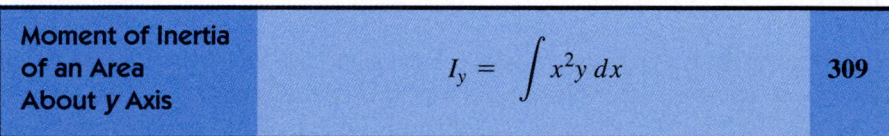

```
fnInt(X^4,X,1,3)

          48.4000
■
```

(a) TI-83 screens for Example 20.

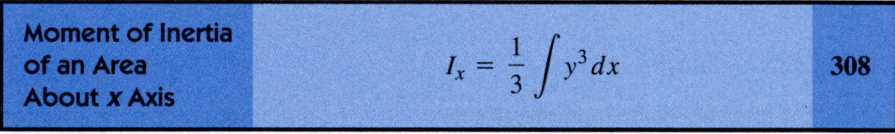

```
(1/3)fnInt(X^6,X
,1,3)
          104.0952
```

(b)

◆◆◆ **Example 21:** Find the radius of gyration about the x axis of the area in Example 20.

Solution: We first find the area of the figure.

$$A = \int_1^3 x^2 \, dx = \frac{x^3}{3}\bigg|_1^3 = \frac{3^3}{3} - \frac{1^3}{3} \approx 8.667$$

From Example 20, $I_x = 104.1$. So, by Eq. 311,

$$r_x = \sqrt{\frac{I_x}{A}} = \sqrt{\frac{104.1}{8.667}} = 3.466$$

◆◆◆

Polar Moment of Inertia

In the preceding sections we found the moment of inertia of an area about some line in the plane of that area. Now we find the moment of inertia of a *solid of revolution* about its axis of revolution. We call this the *polar moment of inertia*. The polar moment of inertia is needed when studying rotation of rigid bodies and torsion of shafts. We first find the polar moment of inertia for a thin-walled shell and then use that result to find the polar moment of inertia for any solid of revolution.

Polar Moment of Inertia by the Shell Method

A thin-walled cylindrical shell (Fig. 28–40) has a volume equal to the product of its circumference, wall thickness, and height.

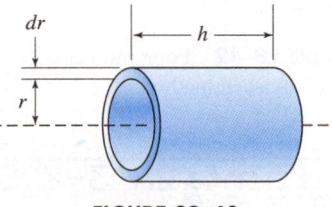

FIGURE 28–40

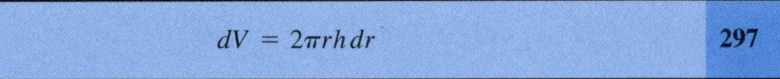

$dV = 2\pi rh \, dr$	297

If we let m represent the mass per unit volume, the mass of the shell is then

$$dM = 2\pi mrh \, dr$$

Since we may consider all particles in this shell to be at a distance r from the axis of revolution, we obtain the moment of inertia of the shell by multiplying the mass by r^2.

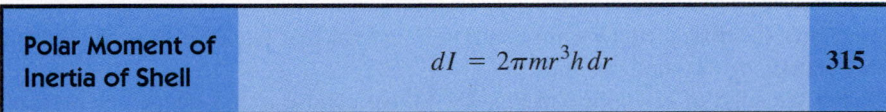

Polar Moment of Inertia of Shell	$dI = 2\pi mr^3h \, dr$	315

We then think of the entire solid of revolution as being formed by concentric shells. The polar moment of inertia of an entire solid of revolution is then found by adding up the shells by integration.

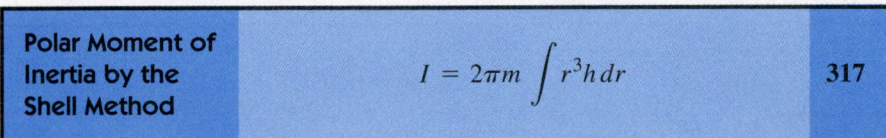

Polar Moment of Inertia by the Shell Method	$I = 2\pi m \int r^3h \, dr$	317

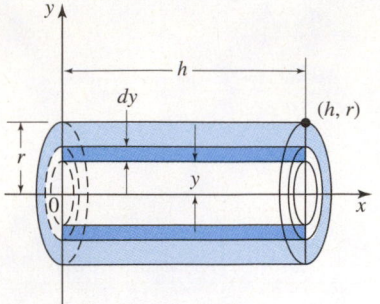

FIGURE 28–41 Polar moment of inertia of a solid cylinder.

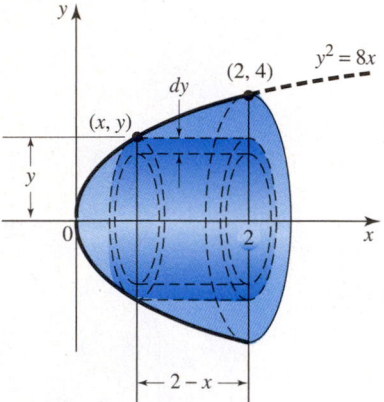

FIGURE 28–42 Polar moment of inertia by shell method.

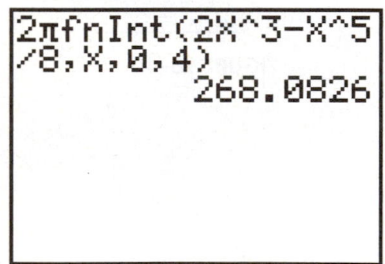

TI-83 screen for Example 23.

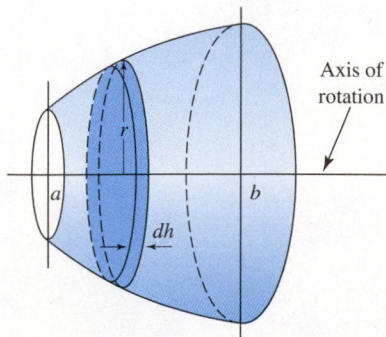

FIGURE 28–43 Polar moment of inertia by disk method.

♦♦♦ **Example 22:** Find the polar moment of inertia of a solid cylinder (Fig. 28–41) about its axis.

Solution: We draw elements of volume in the form of concentric shells. The moment of inertia of each is

$$dI_x = 2\pi m y^3 h \, dy$$

Integrating yields

$$I_x = 2\pi m h \int_0^r y^3 \, dy = 2\pi m h \left[\frac{y^4}{4}\right]_0^r$$

$$= \frac{\pi m h r^4}{2}$$

But since the volume of the cylinder is $\pi r^2 h$, we get

$$I_x = \frac{1}{2} m V r^2$$

In other words, *the polar moment of inertia of a cylindrical solid is equal to half the product of its density, its volume, and the square of its radius.* ♦♦♦

♦♦♦ **Example 23:** The first-quadrant area under the curve $y^2 = 8x$, from $x = 0$ to 2, is rotated about the x axis. Use the shell method to find the polar moment of inertia of the paraboloid generated (Fig. 28–42).

Solution: Our shell element has a radius y, a length $2 - x$, and a thickness dy. By Eq. 315,

$$dI_x = 2\pi m y^3 (2 - x) \, dy$$

Replacing x by $y^2/8$ gives

$$dI_x = 2\pi m y^3 \left(2 - \frac{y^2}{8}\right) dy$$

Integrating gives us

$$I_x = 2\pi m \int_0^4 \left(2y^3 - \frac{y^5}{8}\right) dy = 2\pi m \left[\frac{2y^4}{4} - \frac{y^6}{48}\right]_0^4 \approx 268m$$

Integration by Calculator: We can check our integration by choosing a value for m, say $m = 1$. The screen is shown. ♦♦♦

Polar Moment of Inertia by the Disk Method

Sometimes the disk method will result in an integral that is easier to evaluate than that obtained by the shell method.

For the solid of revolution in Fig. 28–43, we choose a disk-shaped element of volume of radius r and thickness dh. Since it is a cylinder, we use the moment of inertia of a cylinder found in Example 22.

$$dI = \frac{m\pi r^4 \, dh}{2}$$

Integrating gives the following:

Polar Moment of Inertia by the Disk Method	$I = \dfrac{m\pi}{2} \displaystyle\int_a^b r^4 \, dh$	**316**

◆◆◆ **Example 24:** Repeat Example 23 by the disk method.

Solution: We use Eq. 316 with $r = y$ and $dh = dx$.

$$I_x = \frac{m\pi}{2} \int y^4 \, dx$$

But $y^2 = 8x$, so $y^4 = 64x^2$. Substituting yields

$$I_x = 32m\pi \int_0^2 x^2 \, dx$$

$$= 32m\pi \left[\frac{x^3}{3} \right]_0^2$$

$$= 32m\pi \left(\frac{2^3}{3} \right) \approx 268m$$

as before. Note how the disk method has resulted in a simpler integral in this case.

◆◆◆

Exercise 6 ◆ Moment of Inertia

Moment of Inertia of an Area by Integration

Perform or check some of the integrations in this set by calculator.

1. Find the moment of inertia about the x axis of the area bounded by $y = x$, $y = 0$, and $x = 1$.

2. Find the radius of gyration about the x axis of the first-quadrant area bounded by $y^2 = 4x$, $x = 4$, and $y = 0$.

3. Find the radius of gyration about the y axis of the area in problem 2.

4. Find the moment of inertia about the y axis of the area bounded by $x + y = 3$, $x = 0$, and $y = 0$.

5. Find the moment of inertia about the x axis of the first-quadrant area bounded by the curve $y = 4 - x^2$ and the coordinate axes.

6. Find the moment of inertia about the y axis of the area in problem 5.

7. Find the moment of inertia about the x axis of the first-quadrant area bounded by the curve $y^3 = 1 - x^2$.

8. Find the moment of inertia about the y axis of the area in problem 1.

Polar Moment of Inertia

Find the polar moment of inertia of the volume formed when a first-quadrant area with the following boundaries is rotated about the x axis.

9. bounded by $y = x$, $x = 2$, and the x axis

10. bounded by $y = x + 1$, from $x = 1$ to 2, and the x axis

11. bounded by the curve $y = x^2$, the line $x = 2$, and the x axis

12. bounded by $\sqrt{x} + \sqrt{y} = 2$ and the coordinate axes

Find the polar moment of inertia of each solid with respect to its axis in terms of the total mass M of the solid.

13. a right circular cone of height h and base radius r

14. a sphere of radius r

15. a paraboloid of revolution bounded by a plane through the focus perpendicular to the axis of symmetry

16. *Project:* Draw two or more shapes having the same area but different moments of inertia, such as a rectangle and a circle. Using each shape as the cross section of a beam, make a small model of each out of wood. Load each beam with similar weights. Write a paragraph describing how the moment of inertia seems to affect the stiffness of the beam.

17. *Writing:* Find and tabulate the moments of inertia of various shapes from a structural engineering handbook or by surfing the Web. Include simple shapes as well as structural members like Ells, I-beams, and so forth. Which shapes have the greatest moment of inertia in the vertical direction? Summarize your findings in a short report.

18. *Project:* Find formulas for the strength and stiffness of simple beams. Where does the moment of inertia appear in these formulas? Can you predict how the moment of inertia will affect the beam's strength? Its stiffness?

◆◆◆ CHAPTER 28 REVIEW PROBLEMS ◆◆◆◆◆◆◆◆◆◆◆◆◆◆◆◆◆◆◆◆◆◆◆◆◆◆◆◆◆

1. Find the distance from center to the centroid of a semicircle of radius 10.0.

2. A bucket that weighs 3.00 lb and has a volume of 2.00 ft^3 is filled with water. It is being raised at a rate of 5.0 ft/s while water leaks from the bucket at a rate of 0.0100 ft^3/s. Find the work done in raising the bucket 100 ft.

3. Find the centroid of the area bounded by one quadrant of the ellipse $x^2/16 + y^2/9 = 1$.

4. Find the radius of gyration of the area bounded by the parabola $y^2 = 16x$ and its latus rectum, with respect to its latus rectum.

5. A conical tank is 8.00 ft in diameter at the top and is 12.0 ft deep. It is filled with a liquid having a density of 80.0 lb/ft^3. How much work is required to pump all of the liquid to the top of the tank?

6. Find the moment of inertia of a right circular cone of base radius r, height h, and mass M with respect to its axis.

7. A cylindrical, horizontal tank is 6.00 ft in diameter and is half full of water. Find the force on one end of the tank.

8. Find the moment of inertia of a sphere of radius r and density m with respect to a diameter.

9. A spring has a free length of 12.00 in. and a spring constant of 5.45 lb/in. Find the work needed to compress it from a length of 11.00 in. to 9.00 in.

10. Find the coordinates of the centroid of the area bounded by the curve $y^2 = 2x$ and the line $y = x$.

11. A horizontal cylindrical tank 8.00 ft in diameter is half full of oil (60.0 lb/ft^3). Find the force on one end.

12. Find the length of the curve $y = \frac{4}{5}x^2$ from the origin to $x = 4$.

13. The area bounded by the parabolas $y^2 = 4x$ and $y^2 = x + 3$ is rotated about the x axis. Find the surface area of the solid generated.

14. Find the length of the parabola $y^2 = 8x$ from the vertex to one end of the latus rectum.

15. The area bounded by the curves $x^2 = 4y$ and $x - 2y + 4 = 0$ and by the y axis is rotated about the y axis. Find the surface area of the volume generated.

16. The cables on a certain suspension bridge hang in the shape of a parabola. The towers are 100 m apart, and the cables dip 10.0 m below the tops of the towers. Find the length of the cables.

17. The first-quadrant area bounded by the curves $y = x^3$ and $y = 4x$ is rotated about the x axis. Find the surface area of the solid generated.

18. Find the centroid of a paraboloid of revolution bounded by a plane through the focus perpendicular to the axis of symmetry.

Trigonometric, Logarithmic, and Exponential Functions

Up to now we have found derivatives and integrals of algebraic functions only. Here we extend our rules to cover trigonometric, logarithmic, and exponential functions. With these we will be able to solve a much larger range of technical applications than before. We will do problems similar to those in earlier chapters, tangents, related rates, maximum-minimum, and so forth, but now with these new functions.

We previously applied the derivative and the integral to electric circuits, finding current, charge, voltage, and so forth. But as these quantities are often expressed as trigonometric or exponential functions, we will revisit them here. For example, we earlier learned that the current in a capacitor was the derivative of the voltage across that capacitor divided by the capacitance. If you were given that the voltage across the capacitor, Fig. 29–1, was

$$v = 115 \left(1 - e^{-\frac{t}{374}}\right) \text{V}$$

would you then be able to find the current in that capacitor? You will, by the end of this chapter.

If you have not worked with the trigonometric, logarithmic, and exponential functions lately, this would be a good time to review them before starting this chapter.

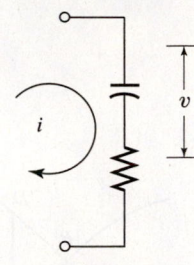

FIGURE 29–1

29–1 Derivatives of the Sine and Cosine Functions

Derivative of sin *u* Approximated Graphically

Before deriving a formula for the derivative of the sine function, let us get a clue as to its derivative graphically.

◆◆◆ **Example 1:** Graph $y = \sin x$, with x in radians. Use the slopes at points on that graph to sketch the graph of the derivative.

Solution: We graph $y = \sin x$ as shown in Fig. 29–2(a) and the slopes as shown in Fig. 29–2(b). Note that the slope of the sine curve is zero at points A, B, C, and D, so the derivative curve must cross the x axis at points A′, B′, C′, and D′. We estimate the slope to be 1 at points E and F, which gives us points E′ and F′ on the derivative curve. Similarly, the slope is −1 at G, giving us point G′ on the derivative curve. We then note that the sine curve is rising from A to B and from C to D, so the derivative curve must be positive in those intervals. Similarly, the sine curve is falling from B to C, so the derivative curve is negative in this interval. Using all of this information, we sketch in the derivative curve.

◆◆◆

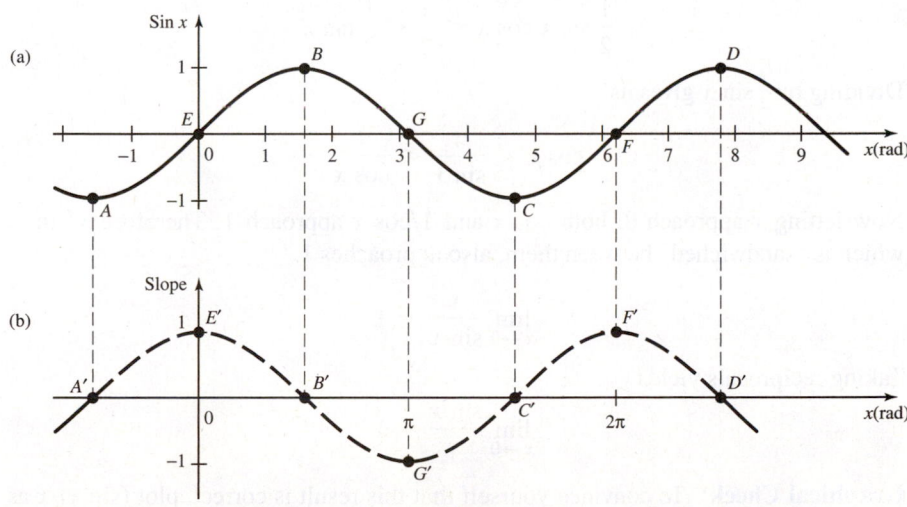

FIGURE 29–2

We see that the derivative of the sine curve seems to be another sine curve, but shifted by $\pi/2$ radians. We might recognize it as a cosine curve. So from our graphs it appears that

$$\frac{d(\sin x)}{dx} = \cos x$$

We'll soon see that this is, in fact, correct.

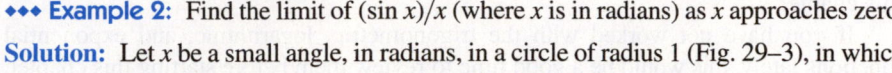

FIGURE 29–3

Limit of $(\sin x)/x$

We will soon derive a formula for the derivative of $\sin x$. In that derivation we will need the limit of $(\sin x)/x$ as x approaches zero.

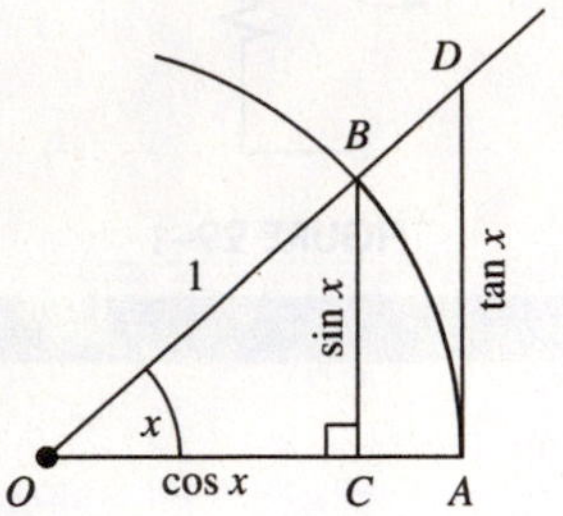

♦♦♦ **Example 2:** Find the limit of $(\sin x)/x$ (where x is in radians) as x approaches zero.

Solution: Let x be a small angle, in radians, in a circle of radius 1 (Fig. 29–3), in which

$$\sin x = \frac{BC}{1} = BC \quad \cos x = \frac{OC}{1} = OC \quad \tan x = \frac{AD}{1} = AD$$

We see that the area of the sector OAB is greater than the area of triangle OBC but less than the area of triangle OAD. But, by Eq. 102,

$$\text{area of triangle } OBC = \frac{1}{2} BC \cdot OC$$

$$= \frac{1}{2} \sin x \cos x$$

and

$$\text{area of triangle } OAD = \frac{1}{2} \cdot OA \cdot AD$$

$$= \frac{1}{2} \tan x$$

Also, by Eq. 77,

$$\text{area of sector } OAB = \frac{1}{2} r^2 x$$

$$= \frac{x}{2}$$

So

$$\frac{1}{2} \sin x \cos x < \frac{x}{2} < \frac{1}{2} \tan x$$

Dividing by $\frac{1}{2} \sin x$ gives us

$$\cos x < \frac{x}{\sin x} < \frac{1}{\cos x}$$

Now letting x approach 0, both $\cos x$ and $1/\cos x$ approach 1. Therefore $x/\sin x$, which is "sandwiched" between them, also approaches 1.

$$\lim_{x \to 0} \frac{x}{\sin x} = 1$$

Taking reciprocals yields

$$\lim_{x \to 0} \frac{\sin x}{x} = 1$$

Graphical Check: To convince yourself that this result is correct, plot $(\sin x)/x$ as shown in Fig. 29–4. Then zoom in on the y intercept as much as you like to see that its coordinates are $(0, 1)$.

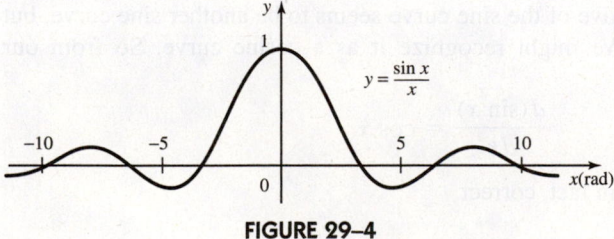

$$y = \frac{\sin x}{x}$$

FIGURE 29–4   ♦♦♦

Derivative of sin *u* Found Analytically

We want to be able to find derivatives of expressions such as

$$y = \sin 2x \quad \text{or} \quad y = \sin(x^2 + 3x)$$

or expressions of the form $y = \sin u$, where u is a function of x. We will use the delta method to derive a rule for finding this derivative. You might want to glance back at the delta method.

Recall that we start the delta method by giving an increment Δu to u, thus causing y to change by an amount Δy.

$$y = \sin u$$
$$y + \Delta y = \sin(u + \Delta u)$$

Subtracting the original equation gives

$$\Delta y = \sin(u + \Delta u) - \sin u$$

We now make use of the identity

$$\sin \alpha - \sin \beta = 2 \cos \frac{\alpha + \beta}{2} \sin \frac{\alpha - \beta}{2}$$

to transform our equation for Δy into a more useful form. To use this identity, we let

$$\alpha = u + \Delta u \quad \text{and} \quad \beta = u$$

so

$$\Delta y = 2 \cos\left(\frac{u + \Delta u + u}{2}\right) \sin\left(\frac{u + \Delta u - u}{2}\right)$$
$$= 2[\cos(u + \Delta u/2)] \sin(\Delta u/2)$$

Now dividing by Δu, we have

$$\frac{\Delta y}{\Delta u} = \frac{2[\cos(u + \Delta u/2)] \sin(\Delta u/2)}{\Delta u}$$

Dividing numerator and denominator by 2, we get

$$\frac{\Delta y}{\Delta u} = \frac{[\cos(u + \Delta u/2)] \sin(\Delta u/2)}{\Delta u/2}$$
$$= [\cos(u + \Delta u/2)] \frac{\sin(\Delta u/2)}{\Delta u/2}$$

If we now let Δu approach zero, the quantity

$$\frac{\sin(\Delta u/2)}{\Delta u/2}$$

approaches 1, as we saw in Example 2. When we evaluated this limit in the preceding section, we required the angle ($\Delta u/2$ in this case) to be *in radians*. Thus the formula we derive here also requires the angle to be in radians. Also, the quantity $\Delta u/2$ will approach 0, leaving us with

$$\frac{dy}{du} = \cos u$$

Now, by the chain rule,

$$\frac{dy}{dx} = \frac{dy}{du} \cdot \frac{du}{dx}$$

Do you remember the chain rule? Glance back at Chap. 23.

Thus:

Derivative of the Sine	$$\frac{d(\sin u)}{dx} = \cos u \frac{du}{dx}$$ *The derivative of the sine of some function is the cosine of that function, multiplied by the derivative of that function.*	**263**

◆◆◆ **Example 3:**

(a) If $y = \sin 3x$, then

$$\frac{dy}{dx} = (\cos 3x) \cdot \frac{d}{dx} 3x$$

$$= 3 \cos 3x$$

(b) If $y = \sin(x^3 + 2x^2)$, then

$$y' = \cos(x^3 + 2x^2) \cdot \frac{d}{dx}(x^3 + 2x^2)$$

$$= \cos(x^3 + 2x^2)(3x^2 + 4x)$$

$$= (3x^2 + 4x)\cos(x^3 + 2x^2)$$

(c) If $y = \sin^3 x$, then by the power rule

$$y' = 3(\sin x)^2 \cdot \frac{d}{dx}(\sin x)$$

$$= 3 \sin^2 x \cos x \qquad\qquad ◆◆◆$$

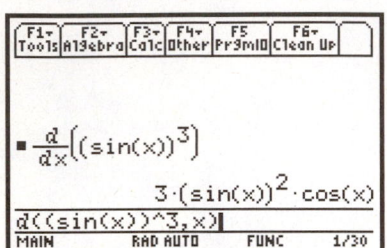

TI-89 calculator check for Example 3(c), with the calculator in **Radian** mode. Recall that $\sin^3 x$ is the same as $(\sin x)^3$.

◆◆◆ **Example 4:** Find the slope of the tangent to the curve $y = x^2 - \sin^2 x$ at $x = 2$ (Fig. 29–5).

Solution: Taking the derivative yields

$$y' = 2x - 2 \sin x \cos x$$

At $x = 2$ rad,

$$y'(2) = 4 - 2 \sin 2 \cos 2$$

$$= 4 - 2(0.909)(-0.416)$$

$$= 4.757 \qquad\qquad ◆◆◆$$

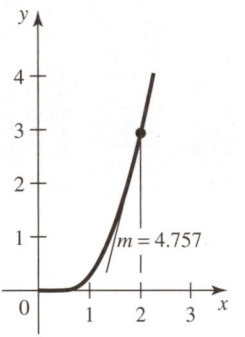

FIGURE 29–5   Graph of $y = x^2 - \sin^2 x$.

Common Error	Remember that x is in *radians* unless otherwise specified. Be sure that your calculator is in radian mode.

Derivative of $\cos u$

We now take the derivative of $y = \cos u$.

We do not need the delta method again because we can relate the cosine to the sine with Eq. 118b, $\cos A = \sin B$, where A and B are complementary angles ($B = \pi/2 - A$), Fig. 29–6. So

$$y = \cos u = \sin\left(\frac{\pi}{2} - u\right)$$

Then, by Eq. 263,

$$\frac{dy}{dx} = \cos\left(\frac{\pi}{2} - u\right)\left(-\frac{du}{dx}\right)$$

But $\cos(\pi/2 - u) = \sin u$, so we have the following equation.

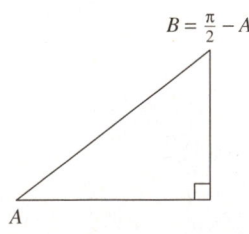

FIGURE 29–6

Derivative of the Cosine	$$\dfrac{d(\cos u)}{dx} = -\sin u \dfrac{du}{dx}$$ *The derivative of the cosine of some function is the negative of the sine of that function, multiplied by the derivative of that function.*	264

♦♦♦ **Example 5:** If $y = \cos 3x^2$, then

$$y' = (-\sin 3x^2)\frac{d}{dx}(3x^2)$$

$$= (-\sin 3x^2)6x = -6x \sin 3x^2 \qquad ♦♦♦$$

Common Errors	Recall that $$\cos 3x^2 \text{ is not the same as } (\cos 3x)^2$$ and $$\cos 3x^2 \text{ is not the same as } \cos(3x)^2$$ but that $$\cos 3x^2 \text{ is the same as } \cos(3x^2)$$

Knowing the derivative of both $\sin x$ and $\cos x$ allows us to take second and higher derivatives of these trigonometric functions.

♦♦♦ **Example 6:** Find the second derivative of $y = \sin x$.

Solution:
$$y' = \cos x$$
$$y'' = -\sin x \qquad ♦♦♦$$

Of course, our former rules for products, quotients, powers, and so forth, work for trigonometric functions as well.

♦♦♦ **Example 7:** Differentiate $y = \sin 3x \cos 5x$.

Solution: Using the product rule, we have

$$\frac{dy}{dx} = (\sin 3x)(-\sin 5x)(5) + (\cos 5x)(\cos 3x)(3)$$

$$= -5 \sin 3x \sin 5x + 3 \cos 5x \cos 3x \qquad ♦♦♦$$

♦♦♦ **Example 8:** Differentiate $y = \dfrac{2 \cos x}{\sin 3x}$.

Solution: Using the quotient rule gives us

$$y' = \frac{(\sin 3x)(-2 \sin x) - (2 \cos x)(3 \cos 3x)}{(\sin 3x)^2}$$

$$= \frac{-2 \sin 3x \sin x - 6 \cos x \cos 3x}{\sin^2 3x} \qquad ♦♦♦$$

♦♦♦ **Example 9:** Find the maximum and minimum points on the curve $y = 3 \cos x$.

Solution: We take the derivative, set it equal to zero, and solve for x.

$$y' = -3 \sin x = 0$$
$$\sin x = 0$$
$$x = 0, \pm\pi, \pm2\pi, \pm3\pi, \ldots$$

The second derivative is

$$y'' = -3 \cos x$$

which is negative when x equals $0, \pm2\pi, \pm4\pi, \ldots$, so these are the locations of the maximum points. The others, where the second derivative is positive, are minimum points.

Graphical Check: This agrees with our plot of the cosine curve in Fig. 29–7. ♦♦♦

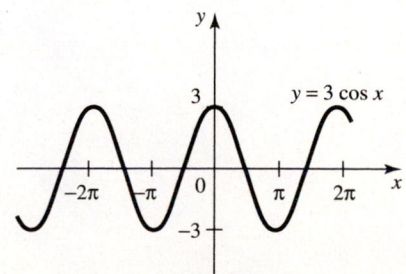

FIGURE 29–7

Implicit derivatives involving trigonometric functions are handled the same way as shown in Chapter 23.

◆◆◆ **Example 10:** Find dy/dx if $\sin xy = x^2 y$.

Solution:

$$(\cos xy)(xy' + y) = x^2 y' + 2xy$$
$$xy'\cos xy + y \cos xy = x^2 y' + 2xy$$
$$xy'\cos xy - x^2 y' = 2xy - y \cos xy$$
$$(x \cos xy - x^2)y' = 2xy - y \cos xy$$
$$y' = \frac{2xy - y \cos xy}{x \cos xy - x^2}$$

◆◆◆

Electrical Applications

Here we will do applications similar to those given in earlier chapters, but now with the trigonometric functions. As we noted in an earlier chapter, we do not intend to teach electrical technology here, but simply to reinforce material learned in other courses. There we introduced the following formulas, where i is the current in amperes (A), q is the charge in coulombs (C), v is voltage in volts (V), L is inductance in henrys (H), and t is time in seconds.

Current	$$i = \frac{dq}{dt}$$ *Current is the rate of change of charge, with respect to time.*	**1078**
Instantaneous Current in a Capacitor	$$i = C\frac{dv}{dt}$$ *The current in a capacitor equals the capacitance times the rate of change of the voltage, with respect to time.*	**1080**
Instantaneous Voltage Across an Inductor	$$v = L\frac{di}{dt}$$ *The voltage across an inductor equals the inductance times the rate of change of the current with respect to time.*	**1086**

◆◆◆ **Example 11:** The charge through a 4.85-Ω resistor is given by

$$q = 3.74 \sin(44.6t + 1.44) \qquad C$$

Write an expression for (a) the instantaneous current through the resistor and (b) the instantaneous voltage across the resistor.

Solution:

(a) Taking the derivative we get, by Eq. 1078,

$$i = \frac{dq}{dt} = 3.74[\cos(44.6t + 1.44)](44.6)$$
$$= 167 \cos(44.6t + 1.44) \qquad A$$

(b) By Ohm's law,

$$v = ri = (4.85)(167)\cos(44.6t + 1.44)$$
$$= 810 \cos(44.6t + 1.44) \qquad V$$

◆◆◆

♦♦♦ **Example 12:** The voltage applied to a 4.82-microfarad (μF) capacitor, Fig. 29–8, is

$$v = 2.85 \cos(26.4t + 0.56) \quad \text{V}$$

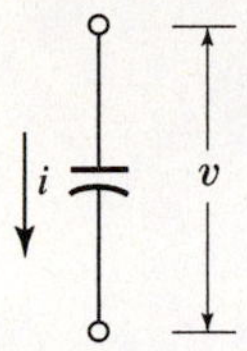

(a) Write an expression for the current in the capacitor and (b) evaluate the current at $t = 0.1$ s.

Solution:

(a) Taking the derivative,

$$\frac{dv}{dt} = [-2.85 \sin(26.4t + 0.56)](26.4)$$

$$= -75.2 \sin(26.4t + 0.56)$$

Then by Eq. 1080

$$i = C\frac{dv}{dt} = (4.82 \times 10^{-6})[-75.2 \sin(26.4t + 0.56)]$$

$$= (-362 \times 10^{-6}) \sin(26.4t + 0.56) \quad \text{A}$$

(b) At $t = 0.1$ s,

$$i = (-362 \times 10^{-6}) \sin(2.64 + 0.56)$$

$$= 21.1 \times 10^{-6} \quad \text{A}$$

$$= 21.1 \quad \mu\text{A}$$

♦♦♦

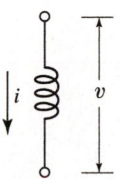

FIGURE 29–8

♦♦♦ **Example 13:** The current in a 11.6-H inductor, Fig. 29–9, is

$$i = 4.37 \sin(6.83t + 0.55) \quad \text{A}$$

(a) Write an expression for the voltage across that inductor and (b) evaluate it at $t = 1.60$ s.

Solution:

(a) Taking the derivative we get

$$\frac{di}{dt} = [4.37 \cos(6.83t + 0.55)](6.83)$$

$$= 29.8 \cos(6.83t + 0.55)$$

Then by Eq. 1086,

$$v = L\frac{di}{dt} = 11.6[29.8 \cos(6.83t + 0.55)]$$

$$= 346 \cos(6.83t + 0.55) \quad \text{V}$$

(b) At $t = 1.60$ s,

$$v = 346 \cos(6.83 \times 1.60 + 0.55)$$

$$= 161\text{V}$$

♦♦♦

FIGURE 29–9

Exercise 1 ♦ Derivatives of the Sine and Cosine Functions

First Derivatives

Find the derivative.

1. $y = \sin x$

2. $y = 3 \cos 2x$

3. $y = \cos^3 x$

4. $y = \sin x^2$

5. $y = \sin 3x$

6. $y = \cos 6x$

7. $y = \sin x \cos x$

8. $y = 15.4 \cos^5 x$

9. $y = 3.75 \, x \cos x$

10. $y = x^2 \cos x^2$

11. $y = \sin^2(\pi - x)$

12. $y = \dfrac{\sin \theta}{\theta}$

13. $y = \sin 2\theta \cos \theta$

14. $y = \sin^5\theta$

15. $y = \sin^2 x \cos x$

16. $y = \sin 2x \cos 3x$

17. $y = 1.23 \sin^2 x \cos 3x$

18. $y = \frac{1}{2} \sin^2 x$

19. $y = \sqrt{\cos 2t}$

20. $y = 42.7 \sin^2 t$

Second Derivatives

For problems 21 through 23, find the second derivative of each function.

21. $y = \cos x$

22. $y = \frac{1}{4} \cos 2\theta$

23. $y = x \cos x$

24. If $f(x) = x^2 \cos^3 x$, find $f''(0)$.

25. If $f(x) = x \sin(\pi/2)x$, find $f''(1)$.

Implicit Functions

Find dy/dx for each implicit function.

26. $y \sin x = 1$

27. $xy - y \sin x - x \cos y = 0$

28. $y = \cos(x - y)$

29. $x = \sin(x + y)$

30. $x \sin y - y \sin x = 0$

Tangents

Find the slope of the tangent to four significant digits at the given value of x.

31. $y = \sin x$ at $x = 2$ rad

32. $y = x - \cos x$ at $x = 1$ rad

33. $y = x \sin \dfrac{x}{2}$ at $x = 2$ rad

34. $y = \sin x \cos 2x$ at $x = 1$ rad

Extreme Values and Inflection Points

For each curve, find the maximum, minimum, and inflection points between $x = 0$ and 2π.

35. $y = \sin x$

36. $y = \dfrac{x}{2} - \sin x$

37. $y = 3 \sin x - 4 \cos x$

Electrical Applications

38. The charge through a 59.3-Ω resistor is given by

$$q = 224 \sin(83.4t + 3.83) \qquad \text{C}$$

Write an expression for the instantaneous current through the resistor.

39. The voltage applied to a 22.5-microfarad (μF) capacitor is

$$v = 11.5 \cos(2.84t + 0.75) \qquad \text{V}$$

(a) Write an expression for the current in the capacitor and (b) evaluate the current at $t = 0.2$ s.

40. The current in a 38.3-H inductor is

$$i = 33.5 \sin(82.4t + 0.77) \qquad \text{A}$$

(a) Write an expression for the voltage across that inductor and (b) evaluate it at $t = 2.60$ s.

29–2 Derivatives of the Other Trigonometric Functions

Derivative of tan u

To find the derivative of $y = \tan u$, we first use the identity

$$y = \frac{\sin u}{\cos u}$$

Using the quotient rule (Eq. 262), we have

$$\frac{dy}{dx} = \frac{\cos^2 u \dfrac{du}{dx} + \sin^2 u \dfrac{du}{dx}}{\cos^2 u}$$

$$= \frac{(\sin^2 u + \cos^2 u) \dfrac{du}{dx}}{\cos^2 u}$$

and, since $\sin^2 u + \cos^2 u = 1$,

$$\frac{dy}{dx} = \frac{1}{\cos^2 u} \frac{du}{dx}$$

and, since $1/\cos^2 u = \sec^2 u$,

$$\frac{dy}{dx} = \sec^2 u \frac{du}{dx}$$

Thus,

Derivative of the Tangent	$$\dfrac{d(\tan u)}{dx} = \sec^2 u \dfrac{du}{dx}$$ *The derivative of the tangent of some function is the secant squared of that function, multiplied by the derivative of that function.*	**265**

◆◆◆ **Example 14:**

(a) If $y = 3 \tan x^2$, then

$$y' = 3(\sec^2 x^2)(2x) = 6x \sec^2 x^2$$

(b) If $y = 2 \sin 3x \tan 3x$, then, by the product rule,

$$y' = 2[\sin 3x (\sec^2 3x)(3) + \tan 3x (\cos 3x)(3)]$$

$$= 6 \sin 3x \sec^2 3x + 6 \cos 3x \tan 3x \qquad \text{◆◆◆}$$

Derivatives of cot *u*, sec *u*, and csc *u*

Each of these derivatives can be obtained using the rules for the sine, cosine, and tangent already derived and using the following identities:

$$\cot u = \frac{\cos u}{\sin u} \quad \sec u = \frac{1}{\cos u} \quad \csc u = \frac{1}{\sin u}$$

We list those derivatives here, together with those already found.

Derivatives of the Trigonometric Functions	$\dfrac{d(\sin u)}{dx} = \cos u \dfrac{du}{dx}$	263
	$\dfrac{d(\cos u)}{dx} = -\sin u \dfrac{du}{dx}$	264
	$\dfrac{d(\tan u)}{dx} = \sec^2 u \dfrac{du}{dx}$	265
	$\dfrac{d(\cot u)}{dx} = -\csc^2 u \dfrac{du}{dx}$	266
	$\dfrac{d(\sec u)}{dx} = \sec u \tan u \dfrac{du}{dx}$	267
	$\dfrac{d(\csc u)}{dx} = -\csc u \cot u \dfrac{du}{dx}$	268

You should memorize at least the first three of these. Notice how the signs alternate. The derivative of each cofunction is negative. Again, the rules learned for derivatives earlier apply to all the trigonometric functions.

◆◆◆ **Example 15:**

(a) If $y = \sec(x^3 - 2x)$, then

$$y' = [\sec(x^3 - 2x)\tan(x^3 - 2x)](3x^2 - 2)$$
$$= (3x^2 - 2)\sec(x^3 - 2x)\tan(x^3 - 2x)$$

(b) If $y = \cot^3 5x$, then, by the power rule,

$$y' = 3(\cot 5x)^2(-\csc^2 5x)(5)$$
$$= -15 \cot^2 5x \csc^2 5x$$ ◆◆◆

The derivatives of the trigonometric functions, in combination with the basic definitions of the functions themselves, allow us to handle many applications.

◆◆◆ **Example 16:** In Fig. 29–10, link L is pivoted at B but slides along the fixed pin P. As slider C moves in the slot at a constant rate of 4.26 cm/s, the angle θ changes. Find the rate at which θ is changing when $S = 6.00$ cm.

Estimate: When $S = 6.00$ cm, θ is equal to arctan $(8.63/6.00) = 55.2°$. After, say, 0.1 s, S has decreased by 0.426 cm, so then θ is equal to arctan $(8.63/5.57) = 57.2°$. Thus θ has increased by about 2° in 0.1 s, or about 20 deg/s.

Solution: S and θ are related by the tangent function, so

$$\tan \theta = \frac{8.63}{S}$$

or

$$S = \frac{8.63}{\tan \theta} = 8.63 \cot \theta$$

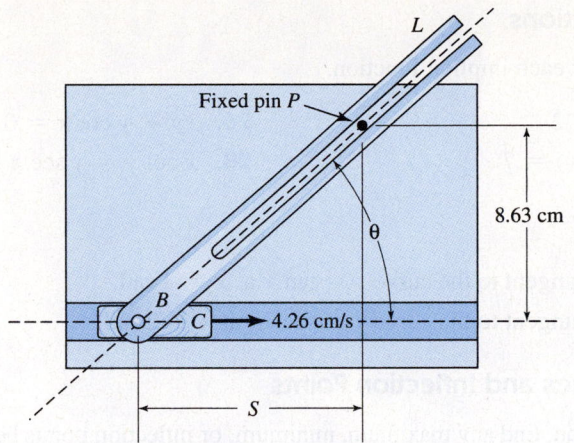

FIGURE 29–10   Pivoted link and slider.

Taking the derivative with respect to *time* gives

$$\frac{dS}{dt} = 8.63(-\csc^2\theta)\frac{d\theta}{dt}$$

Solving for $d\theta/dt$ gives

$$\frac{d\theta}{dt} = -\frac{1}{8.63\csc^2\theta}\frac{dS}{dt} = -\frac{\sin^2\theta}{8.63}\frac{dS}{dt}$$

When $S = 6.00$ cm, $dS/dt = -4.26$ cm/s, and $\theta = 55.2°$ (from our estimate). Substituting, we get

$$\frac{d\theta}{dt} = -\frac{(\sin 55.2°)^2}{8.63}(-4.26) = 0.333 \text{ rad/s}$$

or 19.1 deg/s, which agrees well with our estimate.   ◆◆◆

Exercise 2 ◆ Derivatives of the Other Trigonometric Functions

First Derivative

Find the derivative in problems 1–16.

1. $y = \tan 2x$ **2.** $y = \sec 4x$

3. $y = 5\csc 3x$ **4.** $y = 9\cot 8x$

5. $y = 3.25\tan x^2$ **6.** $y = 5.14\sec 2.11x^2$

7. $y = 7\csc x^3$ **8.** $y = 9\cot 3x^3$

9. $y = x\tan x$ **10.** $y = x\sec x^2$

11. $y = 5x\csc 6x$ **12.** $y = 9x^2\cot 2x$

13. $w = \sin\theta\tan 2\theta$ **14.** $s = \cos t\sec 4t$

15. $v = 5\tan t\csc 3t$ **16.** $z = 2\sin 2\theta\cot 8\theta$

17. If $y = 5.83\tan^2 2x$, find $y'(1)$.   **18.** If $f(x) = \sec^3 x$, find $f'(3)$.

19. If $f(x) = 3\csc^3 3x$, find $f'(3)$.   **20.** If $y = 9.55x\cot^2 8x$, find $y'(1)$.

Second Derivative

Find the second derivative.

21. $y = 3\tan x$ **22.** $y = 2\sec 5\theta$

23. If $y = 3\csc 2\theta$, find $y''(1)$.   **24.** If $f(x) = 6\cot 4x$, find $f''(3)$.

Implicit Functions

Find dy/dx for each implicit function.

25. $y \tan x = 2$

26. $xy + y \cot x = 0$

27. $\sec(x + y) = 7$

28. $x \cot y = y \sec x$

Tangents

29. Find the tangent to the curve $y = \tan x$ at $x = 1$ rad.

30. Find the tangent to the curve $y = \sec 2x$ at $x = 2$ rad.

Extreme Values and Inflection Points

For each function, find any maximum, minimum, or inflection points between 0 and π.

31. $y = 2x - \tan x$

32. $y = \tan x - 4x$

Rate of Change

33. An object moves with simple harmonic motion so that its displacement y at time t is $y = 6 \sin 4t$ cm. Find the velocity and acceleration of the object when $t = 0.0500$ s.

Related Rates

34. A ship is sailing at 10.0 km/h in a straight line (Fig. 29–11). It keeps its searchlight trained on a reef that is at a perpendicular distance of 3.00 km from the path of the ship. How fast (rad/h) is the light turning when the distance d is 5.00 km?

35. Two cables pass over fixed pulleys A and B (Fig. 29–12), forming an isosceles triangle ABP. Point P is being raised at the rate of 3.00 in./min. How fast is θ changing when h is 4.00 ft?

36. The illumination at a point P on the ground (Fig. 29–13) due to a flare F is $I = k \sin \theta / d^2$ lux, where k is a constant. Find the rate of change of I when the flare is 100.0 ft above the ground and falling at a rate of 1.0 ft/s if the illumination at P at that instant is 65 lux.

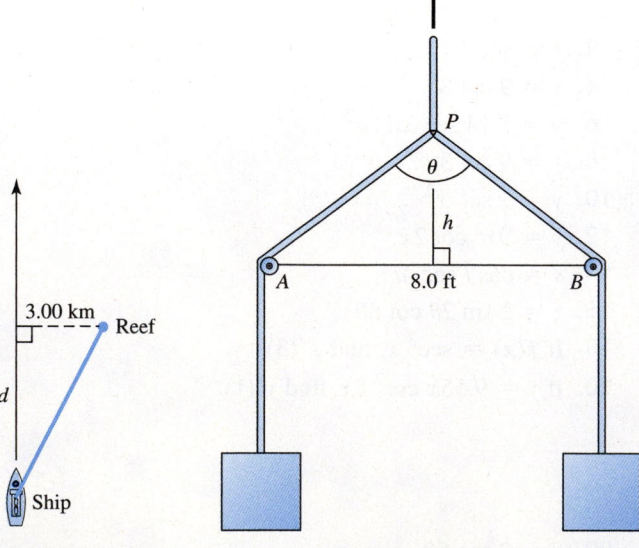

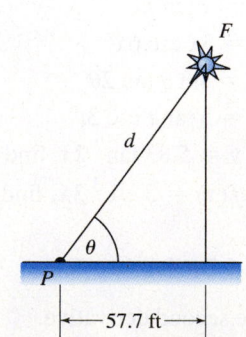

FIGURE 29–11   **FIGURE 29–12**   **FIGURE 29–13** Falling flare.

Optimization

37. Find the length of the shortest ladder that will touch the ground, the wall, and the house in Fig. 29–14.

38. The range x of a projectile fired at an angle θ with the horizontal at a velocity v (Fig. 29–15) is $x = (v^2/g) \sin 2\theta$, where g is the acceleration due to gravity. Find θ for the maximum range.

39. A force F (Fig. 29–16) pulls the weight along a horizontal surface. If f is the coefficient of friction, then

$$F = \frac{fW}{f \sin \theta + \cos \theta}$$

Find θ for a minimum force when $f = 0.60$.

40. A 20.0-ft-long steel girder is dragged along a corridor 10.0 ft wide and then around a corner into another corridor at right angles to the first (Fig. 29–17). Neglecting the thickness of the girder, what must the width of the second corridor be to allow the girder to turn the corner?

41. If the girder in Fig. 29–17 is to be dragged from a 12.8-ft-wide corridor into another 5.4-ft-wide corridor, find the length of the longest girder that will fit around the corner. (Neglect the thickness of the girder.)

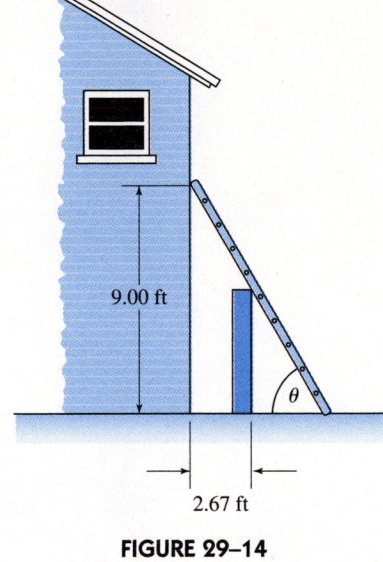

FIGURE 29–14

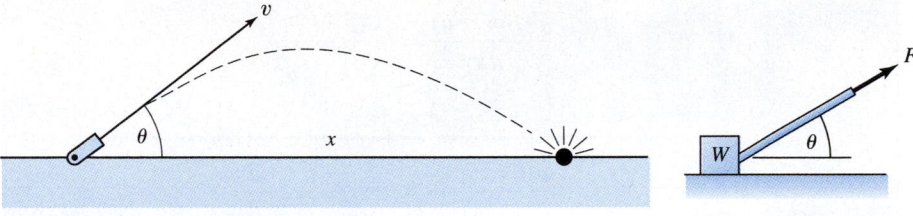

FIGURE 29–15   **FIGURE 29–16**

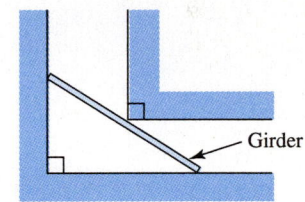

FIGURE 29–17 Top view of a girder dragged along a corridor.

29–3   Derivatives of the Inverse Trigonometric Functions

We will next use our ability to take derivatives of the trigonometric functions to find the derivatives of the *inverse* trigonometric functions. We start with the derivative of $y = \mathrm{Sin}^{-1} u$ where y is some angle whose sine is u, as in Fig. 29–18, whose value we restrict to the range $-\pi/2$ to $\pi/2$. We can then write

$$\sin y = u$$

Taking the derivative yields

$$\cos y \frac{dy}{dx} = \frac{du}{dx}$$

so

$$\frac{dy}{dx} = \frac{1}{\cos y} \frac{du}{dx} \tag{1}$$

However, from Eq. 125,

$$\cos^2 y = 1 - \sin^2 y$$

$$\cos y = \pm\sqrt{1 - \sin^2 y}$$

But since y is restricted to values between $-\pi/2$ and $\pi/2$, $\cos y$ cannot be negative. So

$$\cos y = +\sqrt{1 - \sin^2 y} = \sqrt{1 - u^2}$$

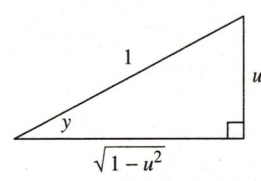

FIGURE 29–18

Substituting into Eq. (1), we have the following equation:

Derivative of the Arcsin	$$\frac{d(\mathrm{Sin}^{-1} u)}{dx} = \frac{1}{\sqrt{1 - u^2}}\frac{du}{dx}$$ $$-1 < u < 1$$	**269**

♦♦♦ **Example 17:** If $y = \mathrm{Sin}^{-1} 3x$, then

$$y' = \frac{1}{\sqrt{1 - (3x)^2}} \quad (3)$$

$$= \frac{3}{\sqrt{1 - 9x^2}}$$

♦♦♦

Derivatives of Arccos, Arctan, Arccot, Arcsec, and Arccsc

The rules for taking derivatives of the remaining inverse trigonometric functions, and the Arcsin as well, are as follows:

	$$\frac{d(\mathrm{Sin}^{-1} u)}{dx} = \frac{1}{\sqrt{1 - u^2}}\frac{du}{dx}$$ $$-1 < u < 1$$	**269**
	$$\frac{d(\mathrm{Cos}^{-1} u)}{dx} = \frac{-1}{\sqrt{1 - u^2}}\frac{du}{dx}$$ $$-1 < u < 1$$	**270**
Derivatives of the Inverse Trigonometric Functions	$$\frac{d(\mathrm{Tan}^{-1} u)}{dx} = \frac{1}{1 + u^2}\frac{du}{dx}$$	**271**
	$$\frac{d(\mathrm{Cot}^{-1} u)}{dx} = \frac{-1}{1 + u^2}\frac{du}{dx}$$	**272**
	$$\frac{d(\mathrm{Sec}^{-1} u)}{dx} = \frac{1}{u\sqrt{u^2 - 1}}\frac{du}{dx}$$ $$|u| > 1$$	**273**
	$$\frac{d(\mathrm{Csc}^{-1} u)}{dx} = \frac{-1}{u\sqrt{u^2 - 1}}\frac{du}{dx}$$ $$|u| > 1$$	**274**

Try to derive one or more of these equations. Follow steps similar to those we used for the derivative of Arcsin.

♦♦♦ **Example 18:**

(a) If $y = \mathrm{Cot}^{-1}(x^2 + 1)$, then

$$y' = \frac{-1}{1 + (x^2 + 1)^2}(2x)$$

$$= \frac{-2x}{2 + 2x^2 + x^4}$$

(b) If $y = \text{Cos}^{-1} \sqrt{1 - x}$, then

$$y' = \frac{-1}{\sqrt{1 - (1 - x)}} \cdot \frac{-1}{2\sqrt{1 - x}}$$

$$= \frac{1}{2\sqrt{x - x^2}}$$

◆◆◆

Exercise 3 ◆ Derivatives of the Inverse Trigonometric Functions

Find the derivative.

1. $y = x \, \text{Sin}^{-1} x$

2. $y = \text{Sin}^{-1} \dfrac{x}{a}$

3. $y = \text{Cos}^{-1} \dfrac{x}{a}$

4. $y = \text{Tan}^{-1}(\sec x + \tan x)$

5. $y = \text{Sin}^{-1} \dfrac{\sin x - \cos x}{\sqrt{2}}$

6. $y = \sqrt{2ax - x^2} + a \, \text{Cos}^{-1} \dfrac{\sqrt{2ax - x^2}}{a}$

7. $y = t^2 \, \text{Cos}^{-1} t$

8. $y = \text{Arcsin } 2x$

9. $y = \text{Arctan}(1 + 2x)$

10. $y = \text{Arccot}(2x + 5)^2$

11. $y = \text{Arccot } \dfrac{x}{a}$

12. $y = \text{Arcsec } \dfrac{1}{x}$

13. $y = \text{Arccsc } 2x$

14. $y = \text{Arcsin } \sqrt{x}$

15. $y = t^2 \, \text{Arcsin } \dfrac{t}{2}$

16. $y = \text{Sin}^{-1} \dfrac{x}{\sqrt{1 + x^2}}$

17. $y = \text{Sec}^{-1} \dfrac{a}{\sqrt{a^2 - x^2}}$

Find the slope of the tangent to each curve.

18. $y = x \, \text{Arcsin } x$ at $x = \frac{1}{2}$

19. $y = \dfrac{\text{Arctan } x}{x}$ at $x = 1$

20. $y = x^2 \, \text{Arccsc } \sqrt{x}$ at $x = 2$

21. $y = \sqrt{x} \, \text{Arccot } \dfrac{x}{4}$ at $x = 4$

22. Find the equations of the tangents to the curve $y = \text{Arctan } x$ having a slope of $\frac{1}{4}$.

29–4 Derivatives of Logarithmic Functions

Derivative of $\log_b u$

Let us now use the delta method again to find the derivative of the logarithmic function $y = \log_b u$. We first let u take on an increment Δu and y an increment Δy.

$$y = \log_b u$$
$$y + \Delta y = \log_b(u + \Delta u)$$

Subtracting gives us

$$\Delta y = \log_b(u + \Delta u) - \log_b u = \log_b \frac{u + \Delta u}{u}$$

by the law of logarithms for quotients. Now dividing by Δu yields

$$\frac{\Delta y}{\Delta u} = \frac{1}{\Delta u} \log_b \frac{u + \Delta u}{u}$$

We now do some manipulation to get our expression into a form that will be easier to evaluate. We start by multiplying the right side by u/u.

$$\frac{\Delta y}{\Delta u} = \frac{u}{u} \cdot \frac{1}{\Delta u} \log_b \frac{u + \Delta u}{u}$$

$$= \frac{1}{u} \cdot \frac{u}{\Delta u} \log_b \frac{u + \Delta u}{u}$$

Then, using the law of logarithms for powers (Eq. 141), we have

$$\frac{\Delta y}{\Delta u} = \frac{1}{u} \log_b \left(\frac{u + \Delta u}{u} \right)^{u/\Delta u}$$

We now let Δu approach zero.

$$\lim_{\Delta u \to 0} \frac{\Delta y}{\Delta u} = \frac{dy}{du} = \lim_{\Delta u \to 0} \frac{1}{u} \log_b \left(\frac{u + \Delta u}{u} \right)^{u/\Delta u}$$

$$= \frac{1}{u} \log_b \left[\lim_{\Delta u \to 0} \left(1 + \frac{\Delta u}{u} \right)^{u/\Delta u} \right]$$

Let us simplify the expression inside the brackets by making the substitution

$$k = \frac{u}{\Delta u}$$

Then k will approach infinity as Δu approaches zero, and the expression inside the brackets becomes

$$\lim_{\Delta u \to 0} \left(1 + \frac{\Delta u}{u} \right)^{u/\Delta u} = \lim_{k \to \infty} \left(1 + \frac{1}{k} \right)^k$$

This limit defines the number e, the familiar base of natural logarithms.

Definition of *e*	$e \equiv \lim\limits_{k \to \infty} \left(1 + \dfrac{1}{k} \right)^k$	137

Glance back at Chap. 18 where we derived this expression. Our derivative thus becomes

$$\frac{dy}{du} = \frac{1}{u} \log_b e$$

We are nearly finished now. Using the chain rule, we get dy/dx by multiplying dy/du by du/dx. Thus,

Derivative of $\log_b u$	$\dfrac{d(\log_b u)}{dx} = \dfrac{1}{u} \log_b e \dfrac{du}{dx}$	275a

Or, since $\log_b e = 1/\ln b$,

Derivative of $\log_b u$	$\dfrac{d(\log_b u)}{dx} = \dfrac{1}{u \ln b} \dfrac{du}{dx}$	275b

This form is more useful when the base b is a number other than 10.

◆◆◆ **Example 19:** Take the derivative of $y = \log(x^2 - 3x)$. (Recall that when the base is not specified, as here, we have a *common* logarithm, with a base of 10.)

Solution:

$$\frac{dy}{dx} = \frac{1}{x^2 - 3x}(\log e)(2x - 3)$$

$$= \frac{2x - 3}{x^2 - 3x} \log e$$

Since $\log e = 1/\ln 10$, we can also write the result as

$$\frac{dy}{dx} = \frac{2x - 3}{\ln(10)x(x - 3)}$$

◆◆◆

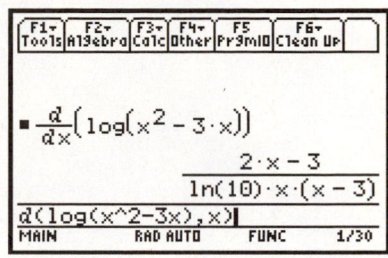

TI-89 calculator check for Example 19.

◆◆◆ **Example 20:** Find the derivative of $y = x \log_3 x^2$.

Solution: By the product rule,

$$y' = x\left(\frac{1}{x^2 \ln 3}\right)(2x) + \log_3 x^2$$

$$= \frac{2}{\ln 3} + \log_3 x^2 \approx 1.82 + \log_3 x^2$$

◆◆◆

Derivative of ln *u*

Our efforts in deriving Eqs. 275a and 275b will now pay off, because we can use those results to find the derivative of the natural logarithm of a function, as well as the derivatives of exponential functions in the following sections, without having to use the delta method again. To find the derivative of $y = \ln u$ we use Eq. 275a.

$$y = \ln u$$

$$\frac{dy}{dx} = \frac{1}{u}\ln e \frac{du}{dx}$$

But, by Eq. 144, $\ln e = 1$. Thus,

Derivative of ln *u*	$$\dfrac{d(\ln u)}{dx} = \dfrac{1}{u}\dfrac{du}{dx}$$	**276**
	The derivative of the natural logarithm of a function is the reciprocal of that function multiplied by its derivative.	

◆◆◆ **Example 21:** Differentiate $y = \ln(2x^3 + 5x)$.

Solution: By Eq. 276,

$$\frac{dy}{dx} = \frac{1}{2x^3 + 5x}(6x^2 + 5)$$

$$= \frac{6x^2 + 5}{2x^3 + 5x}$$

◆◆◆

The rule for derivatives of the logarithmic function is often used with our former rules for derivatives.

◆◆◆ **Example 22:** Take the derivative of $y = x^3 \ln(5x + 2)$.

Solution: Using the product rule together with our rule for logarithms gives us

$$\frac{dy}{dx} = x^3\left(\frac{1}{5x+2}\right)5 + [\ln(5x+2)]3x^2 = \frac{5x^3}{5x+2} + 3x^2\ln(5x+2) \quad ◆◆◆$$

Our work is sometimes made easier if we first use the laws of logarithms to simplify a given expression.

◆◆◆ **Example 23:** Take the derivative of $y = \ln\dfrac{x\sqrt{2x-3}}{\sqrt[3]{4x+1}}$.

Solution: We first rewrite the given expression using the laws of logarithms.

$$y = \ln x + \frac{1}{2}\ln(2x-3) - \frac{1}{3}\ln(4x+1)$$

We now take the derivative term by term.

$$\frac{dy}{dx} = \frac{1}{x} + \frac{1}{2}\left(\frac{1}{2x-3}\right)2 - \frac{1}{3}\left(\frac{1}{4x+1}\right)4$$

$$= \frac{1}{x} + \frac{1}{2x-3} - \frac{4}{3(4x+1)} \quad ◆◆◆$$

Logarithmic Differentiation

Here we use logarithms to aid in differentiating nonlogarithmic expressions. Derivatives of some complicated expressions can be found more easily if we first take the logarithm of both sides of the given expression and simplify by means of the laws of logarithms. (These operations will not change the meaning of the original expression.) We then take the derivative.

◆◆◆ **Example 24:** Differentiate $y = \dfrac{\sqrt{x-2}\sqrt[3]{x+3}}{\sqrt[4]{x+1}}$.

Solution: We will use logarithmic differentiation here. Instead of proceeding in the usual way, we first take the natural log of both sides of the equation and apply the laws of logarithms. We could instead take the common log of both sides, but the natural log has a simpler derivative.

$$\ln y = \frac{1}{2}\ln(x-2) + \frac{1}{3}\ln(x+3) - \frac{1}{4}\ln(x+1)$$

Taking the derivative, we have

$$\frac{1}{y}\frac{dy}{dx} = \frac{1}{2}\left(\frac{1}{x-2}\right) + \frac{1}{3}\left(\frac{1}{x+3}\right) - \frac{1}{4}\left(\frac{1}{x+1}\right)$$

Finally, multiplying by y to solve for dy/dx gives us

$$\frac{dy}{dx} = y\left[\frac{1}{2(x-2)} + \frac{1}{3(x+3)} - \frac{1}{4(x+1)}\right] \quad ◆◆◆$$

If desired, we could now use the original expression to replace the y in the answer. In other cases, this method will allow us to take derivatives not possible with our other rules.

◆◆◆ **Example 25:** Find the derivative of $y = x^{2x}$.

Solution: This is not a power function because the exponent is not a constant. Nor is it an exponential function because the base is not a constant. So neither Eq. 258 nor

277 applies. But let us use logarithmic differentiation by first taking the natural logarithm of both sides.

$$\ln y = \ln x^{2x} = 2x \ln x$$

Now taking the derivative by means of the product rule yields

$$\frac{1}{y}\frac{dy}{dx} = 2x\left(\frac{1}{x}\right) + (\ln x)2$$

$$= 2 + 2\ln x$$

Multiplying by y, we get

$$\frac{dy}{dx} = 2(1 + \ln x)y$$

Replacing y by x^{2x} gives us

$$\frac{dy}{dx} = 2(1 + \ln x)x^{2x} \qquad \blacklozenge\blacklozenge\blacklozenge$$

Exercise 4 ◆ Derivatives of Logarithmic Functions

Derivative of $\log_b u$

Differentiate.

1. $y = \log 7x$

2. $y = \log x^{-2}$

3. $y = \log_b x^3$

4. $y = \log_a(x^2 - 3x)$

5. $y = \log(x\sqrt{5 + 6x})$

6. $y = \log_a\left(\dfrac{1}{2x + 5}\right)$

7. $y = x \log \dfrac{2}{x}$

8. $y = \log \dfrac{(1 + 3x)}{x^2}$

Derivative of $\ln u$

Differentiate.

9. $y = \ln 3x$

10. $y = \ln x^3$

11. $y = \ln(x^2 - 3x)$

12. $y = \ln(4x - x^3)$

13. $y = 2.75x \ln 1.02x^3$

14. $w = z^2 \ln(1 - z^2)$

15. $y = \dfrac{\ln(x + 5)}{x^2}$

16. $y = \dfrac{\ln x^2}{3 \ln(x - 4)}$

17. $s = \ln \sqrt{t - 5}$

18. $y = 5.06 \ln \sqrt{x^2 - 3.25x}$

With Trigonometric Functions

Differentiate.

19. $y = \ln \sin x$

20. $y = \ln \sec x$

21. $y = \sin x \ln \sin x$

22. $y = \ln(\sec x + \tan x)$

Implicit Relations

Find dy/dx.

23. $y \ln y + \cos x = 0$ **24.** $\ln x^2 - 2x \sin y = 0$

25. $x - y = \ln(x + y)$ **26.** $xy = a^2 \ln \dfrac{x}{a}$

27. $\ln y + x = 10$

Logarithmic Differentiation

Differentiate. Remember to start these by taking the logarithm of both sides.

28. $y = \dfrac{\sqrt{x + 2}}{\sqrt[3]{2 - x}}$ **29.** $y = \dfrac{\sqrt{a^2 - x^2}}{x}$

30. $y = x^x$ **31.** $y = x^{\sin x}$

32. $y = (\cot x)^{\sin x}$ **33.** $y = (\text{Cos}^{-1} x)^x$

Tangent to a Curve

Find the slope of the tangent at the point indicated.

34. $y = \log x$ at $x = 1$ **35.** $y = \ln x$ where $y = 0$

36. $y = \ln(x^2 + 2)$ at $x = 4$ **37.** $y = \log(4x - 3)$ at $x = 2$

38. Find the equation of the tangent to the curve $y = \ln x$ at $y = 0$.

Angle of Intersection

Find the angle of intersection of each pair of curves.

39. $y = \ln(x + 1)$ and $y = \ln(7 - 2x)$ at $x = 2$

40. $y = x \ln x$ and $y = x \ln(1 - x)$ at $x = \frac{1}{2}$

Extreme Values and Points of Inflection

Find the maximum, minimum, and inflection points for each curve.

41. $y = x \ln x$ **42.** $y = x^3 \ln x$

43. $y = \dfrac{x}{\ln x}$ **44.** $y = \ln(8x - x^2)$

Roots

Find the smallest positive root between $x = 0$ and 10 by any approximate method, to two decimal places.

45. $x - 10 \log x = 0$ **46.** $\tan x - \log x = 0$

Applications

47. A certain underwater cable has a core of copper wires covered by insulation. The speed of transmission of a signal along the cable is

$$S = x^2 \ln \frac{1}{x}$$

where x is the ratio of the radius of the core to the thickness of the insulation. What value of x gives the greatest signal speed?

48. The heat loss q per foot of cylindrical pipe insulation (Fig. 29–19) having an inside radius r_1 and outside radius r_2 is given by the logarithmic equation

$$q = \frac{2\pi k(t_1 - t_2)}{\ln(r_2/r_1)} \text{ Btu/h}$$

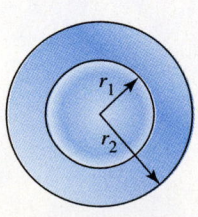

FIGURE 29–19   Insulated pipe.

where t_1 and t_2 are the inside and outside temperatures (°F) and k is the conductivity of the insulation. A 4-in.-thick insulation having a conductivity of 0.036 is wrapped around a 9-in.-diameter pipe at 550°F, and the surroundings are at 90°F. Find the rate of change of heat loss q if the insulation thickness is decreasing at the rate of 0.1 in./h.

49. The pH value of a solution having a concentration C of hydrogen ions is pH $= -\log_{10} C$. Find the rate at which the pH is changing when the concentration is 20×10^{-5} moles/liter and decreasing at the rate of 5.5×10^{-5} per minute.

50. The difference in elevation, in feet, between two locations having barometric readings of B_1 and B_2 in. of mercury is given by $h = 60{,}470 \log B_2/B_1$, where B_1 is the pressure at the upper location. At what rate is an airplane changing in elevation when the barometric pressure outside the airplane is 21.5 in. of mercury and decreasing at the rate of 0.500 in. per minute? (*Hint*: Treat B_2 as a *constant*.)

51. The power input to a certain amplifier is 2.0 W; the power output is normally 400 W. But because of a defective component is dropping at the rate of 0.50 W per day. Use Eq. 1103 to find the rate (decibels per day) at which the decibels are decreasing.

29–5 Derivative of the Exponential Function

Knowing how to find derivatives of the logarithmic functions will now enable us to find derivatives of the exponential functions. Glance back at Chapter 18 for a review of the exponential functions. You will see there how extremely useful they are in technology.

Derivative of b^u

We start with the derivative of the exponential function $y = b^u$, where b is a constant and u is a function of x. We can get the derivative without having to use the delta method by using the rule we derived for the logarithmic function (Eq. 276). We first take the natural logarithm of both sides.

$$y = b^u$$
$$\ln y = \ln b^u = u \ln b$$

by Eq. 141. We now take the derivative of both sides, remembering that $\ln b$ is a constant.

$$\frac{1}{y}\frac{dy}{dx} = \frac{du}{dx}\ln b$$

Multiplying by y, we have

$$\frac{dy}{dx} = y\frac{du}{dx}\ln b$$

Finally, replacing y by b^u gives us the following:

Derivative of b^u	$$\frac{d(b^u)}{dx} = b^u\frac{du}{dx}\ln b$$ *The derivative of a base b raised to a function u is the product of b^u, the derivative of the function, and the natural log of the base.*	277

Note that the derivative of b^u has in it the same function, b^u. This repetition forms the basis of many applications.

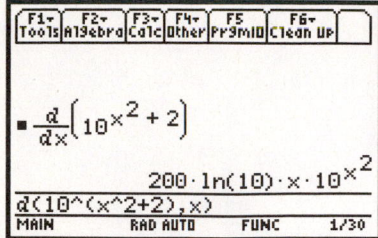

TI-89 calculator check for Example 26.

◆◆◆ **Example 26:** Find the derivative of $y = 10^{x^2+2}$.

Solution: By Eq. 277,

$$\frac{dy}{dx} = 10^{x^2+2}(2x) \ln 10 = 2x(\ln 10)10^{x^2+2}$$

$$= 2x \ln 10 (10^{x^2})(10^2)$$

$$= 200x (\ln 10)10^{x^2}$$

◆◆◆

Common Error	Do not confuse the exponential function $y = b^x$ with the power function $y = x^n$. The derivative of b^x is not xb^{x-1}! The derivative of b^x is $b^x \ln b$.

Derivative of e^u

We will use Eq. 277 mostly when the base b is the base of natural logarithms, e.

$$y = e^u$$

Taking the derivative by Eq. 277 gives us

$$\frac{dy}{dx} = e^u \frac{du}{dx} \ln e$$

But since $\ln e = 1$, we get the following:

Derivative of e^u	$$\frac{d}{dx} e^u = e^u \frac{du}{dx}$$ *The derivative of an exponential function e^u is the same exponential function, multiplied by the derivative of the exponent.*	278

Note that $y = c_1 e^{mx}$ and its derivatives $y' = c_2 e^{mx}$, $y'' = c_3 e^{mx}$, etc., are all *like terms*, since c_1, c_2, and c_3 are constants. We use this fact later when solving differential equations.

◆◆◆ **Example 27:** Find the first, second, and third derivatives of $y = 2e^{3x}$.

Solution: By Rule 278,

$$y' = 6e^{3x} \quad y'' = 18e^{3x} \quad y''' = 54e^{3x}$$

◆◆◆

◆◆◆ **Example 28:** Find the derivative of $y = e^{x^3+5x^2}$.

Solution: By Rule 278,

$$\frac{dy}{dx} = e^{x^3+5x^2}(3x^2 + 10x)$$

◆◆◆

◆◆◆ **Example 29:** Find the derivative of $y = x^2 e^{x^2}$.

Solution: Using the product rule together with Eq. 278, we have

$$\frac{dy}{dx} = x^2(e^{x^2})(2x) + e^{x^2}(2x)$$

$$= (2x^3 + 2x)e^{x^2}$$

$$= 2xe^{x^2}(x^2 + 1)$$

◆◆◆

TI-89 calculator check for Example 29.

Derivative of $y = x^n$, Where n Is Any Real Number

We now return to some unfinished business regarding the power rule. We have already shown that the derivative of the power function x^n is nx^{n-1}, when n is any rational number, positive or negative. Now we show that n can also be an irrational number, such as e or π. Using the fact that

$$x = e^{\ln x}$$

raising to the nth power gives

$$x^n = (e^{\ln x})^n = e^{n \ln x}$$

Then

$$\frac{d}{dx} x^n = \frac{d}{dx} e^{n \ln x}$$

By Eqs. 278 and 276,

$$\frac{d}{dx} x^n = \frac{n}{x} e^{n \ln x}$$

Substituting x^n for $e^{n \ln x}$ gives

$$\frac{d}{dx} x^n = \frac{n}{x} x^n = nx^{n-1}$$

Thus the power rule holds when the exponent n is any real number.

◆◆◆ **Example 30:** The derivative of $3x^\pi$ is

$$\frac{d}{dx} 3x^\pi = 3\pi x^{\pi - 1}$$

◆◆◆

Electrical Applications

Now let us repeat some of the electrical applications we did with trigonometric functions, using the same formulas given there, but now those involving the exponential function. We will start by finding the current in a capacitor.

◆◆◆ **Example 31:** The voltage applied to a certain 14.8-microfarad (μF) capacitor, Fig. 29–20, is

$$v = 115\left(1 - e^{-\frac{t}{374}}\right) \qquad \text{V}$$

(a) Write an expression for the current in the capacitor and (b) evaluate the current at $t = 300$ s.

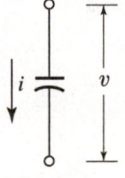

FIGURE 29–20

Solution:

(a) Taking the derivative,

$$\frac{dv}{dt} = 0 - 115e^{-t/374}\left(-\frac{1}{374}\right)$$

$$= 0.307e^{-t/374}$$

Then by Eq. 1080,

$$i = C\frac{dv}{dt} = (14.8 \times 10^{-6})(0.307)e^{-t/374}$$

$$= 4.54e^{-t/374} \,\mu\text{A}$$

(b) At $t = 300$ s,

$$i = 4.54e^{-300/374} = 2.04 \qquad \mu\text{A}$$

◆◆◆

Next we will find the voltage across an inductor.

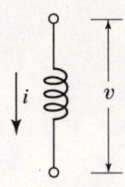

FIGURE 29–21

◆◆◆ **Example 32:** The current in a 58.3-H inductor, Fig. 29–21, is

$$i = 458e^{-t/295} \quad \text{A}$$

(a) Write an expression for the voltage across that inductor and (b) evaluate it at
$t = 125$ s.

Solution: (a) Taking the derivative we get

$$\frac{di}{dt} = 458e^{-t/295}\left(-\frac{1}{295}\right)$$

$$= -1.55e^{-t/295}$$

Then by Eq. 1086,

$$v = L\frac{di}{dt} = -(58.3)(1.55)e^{-t/295}$$

$$= -90.4e^{-t/295} \quad \text{V}$$

(b) At $t = 125$ s,

$$v = 90.4e^{-125/295} = -59.2 \quad \text{V}$$

◆◆◆

Exercise 5 ◆ Derivative of the Exponential Function

Derivative of b^u

Differentiate.

1. $y = 3^{2x}$ 2. $y = 10^{2x+3}$

3. $y = (x)(10^{2x+3})$ 4. $y = 10^{3x}$

5. $y = 2^{x^2}$ 6. $y = 7^{2x}$

Derivative of e^u

Differentiate.

7. $y = e^{2x}$ 8. $y = e^{x^2}$

9. $y = e^{e^x}$ 10. $y = e^{3x^2+4}$

11. $y = e^{\sqrt{1-x^2}}$ 12. $y = xe^x$

13. $y = \dfrac{2}{e^x}$ 14. $y = (3x+2)e^{-x^2}$

15. $y = x^2e^{3x}$ 16. $y = xe^{-x}$

17. $y = \dfrac{e^x}{x}$ 18. $y = \dfrac{x^2-2}{e^{3x}}$

19. $y = \dfrac{e^x - e^{-x}}{x^2}$ 20. $y = \dfrac{e^x - x}{e^{-x} + x^2}$

21. $y = (x + e^x)^2$ 22. $y = (e^x + 2x)^3$

23. $y = \dfrac{(1 + e^x)^2}{x}$ 24. $y = \left(\dfrac{e^x + 1}{e^x - 1}\right)^2$

With Trigonometric Functions

Differentiate.

25. $y = \sin^3 e^x$

26. $y = e^x \sin x$

27. $y = e^\theta \cos 2\theta$

28. $y = e^x(\cos bx + \sin bx)$

Implicit Relations

Find dy/dx.

29. $e^x + e^y = 1$

30. $e^x \sin y = 0$

31. $e^y = \sin(x + y)$

Evaluate each expression.

32. $f'(2)$ where $f(x) = e^{\sin(\pi x/2)}$

33. $f''(0)$ where $f(t) = e^{\sin t} \cos t$

34. $f'(1)$ and $f''(1)$ where $f(x) = e^{x-1} \sin \pi x$

With Logarithmic Functions

Differentiate.

35. $y = e^x \ln x$

36. $y = \ln(x^2 e^x)$

37. $y = \ln x^{e^x}$

38. $y = \ln e^{2x}$

Find the second derivative.

39. $y = e^t \cos t$

40. $y = e^{-t} \sin 2t$

41. $y = e^x \sin x$

42. $y = \frac{1}{2}(e^x + e^{-x})$

Approximate Solution

Find the smallest root that is greater than zero to two decimal places using any method.

43. $e^x + x - 3 = 0$

44. $xe^{-0.02x} = 1$

45. $5e^{-x} + x - 5 = 0$

46. $e^x = \tan x$

Maximum, Minimum, and Inflection Points

47. Find the minimum point of $y = e^{2x} + 5e^{-2x}$.

48. Find the maximum point and the points of inflection of $y = e^{-x^2}$.

49. Find the maximum and minimum points for one cycle of $y = 10e^{-x} \sin x$.

Applications

50. If \$10,000 is invested for t years at an annual interest rate of 10% compounded continuously, it will accumulate to an amount y, where $y = 10{,}000e^{0.1t}$. At what rate, in dollars per year, is the balance growing when (a) $t = 0$ years and (b) $t = 10$ years?

51. If we assume that the price of an automobile is increasing or "inflating" exponentially at an annual rate of 8%, at what rate in dollars per year is the price of a car that initially cost \$9000 increasing after 3 years?

52. When a certain object is placed in an oven at 1000°F, its temperature T rises according to the equation $T = 1000(1 - e^{-0.1t})$, where t is the elapsed time (minutes). How fast is the temperature rising (in degrees per minute) when (a) $t = 0$ and (b) $t = 10.0$ min?

53. A catenary has the equation $y = \frac{1}{2}(e^x + e^{-x})$. We have seen the catenary before. It is the shape taken by a rope or chain suspended from both ends. Find the slope of the catenary when $x = 5$.

54. Verify that the minimum point on the catenary described in problem 53 occurs at $x = 0$.

55. The speed N of a certain flywheel is decaying exponentially according to the equation $N = 1855e^{-0.5t}$ (rev/min), where t is the time (min) after the power is disconnected. Find the angular acceleration (the rate of change of N) when $t = 1$ min.

56. The height y of a certain pendulum released from a height of 50.0 cm is $y = 50.0e^{-0.5t}$ cm, where t is the time after release in seconds. Find the vertical component of the velocity of the pendulum when $t = 1.00$ s.

57. The number of bacteria in a certain culture is growing exponentially. The number N of bacteria at any time t (h) is $N = 10,000\, e^{0.1t}$. At what rate (number of bacteria per hour) is the population increasing when (a) $t = 0$ and (b) $t = 100$ h?

58. The force F needed to hold a weight W (Fig. 29–22) is $F = We^{-\mu\theta}$ where $\mu = $ the coefficient of friction. For a certain beam with $\mu = 0.15$, an angle wrap of 4.62 rad is needed to hold a weight of 200 lb with a force of 100 lb. Find the rate of change of F if the rope is unwrapping at a rate of 15°/s.

59. The atmospheric pressure at a height of h miles above the earth's surface is given by $p = 29.92e^{-h/5}$ in. of mercury. Find the rate of change of the pressure on a rocket that is at 18.0 mi and climbing at a rate of 1500 mi/h.

60. The equation in problem 59 becomes $p = 2121e^{-0.000037h}$ when h is in feet and p is in pounds per square foot. Find the rate of change of pressure on an aircraft at 5000 ft climbing at a rate of 10 ft/s.

61. The approximate density of seawater at a depth of h miles is $d = 64.0e^{0.00676h}$ lb/ft³. Find the rate of change of density, with respect to depth, at a depth of 1.00 mile.

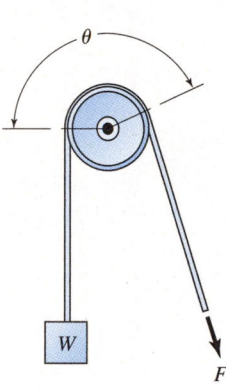

FIGURE 29–22

Electrical Applications

62. The voltage applied to a certain 218-microfarad (μF) capacitor is

$$v = 25.4(1 - e^{-t/285}) \quad \text{V}$$

(a) Write an expression for the current in the capacitor and (b) evaluate the current at $t = 200$ s.

63. The voltage applied to a certain 185-microfarad (μF) capacitor is

$$v = 448(1 - e^{-t/122}) \quad \text{V}$$

(a) Write an expression for the current in the capacitor and (b) evaluate the current at $t = 150$ s.

64. The current in a 88.3-H inductor is

$$i = 115e^{-t/624} \text{ A}$$

(a) Write an expression for the voltage across that inductor and (b) evaluate it at $t = 250$ s.

65. The current in a 37.2-H inductor is

$$i = 225e^{-t/128} \text{ A}$$

(a) Write an expression for the voltage across that inductor and (b) evaluate it at $t = 155$ s.

29–6 Integral of the Exponential and Logarithmic Functions

Let us leave the derivative and turn our attention to the *integral*. We will present rules for integrating the exponential and logarithmic functions here, and then in the next section, the trigonometric functions.

Integral of $e^u \, du$

Since the derivative of e^u is

$$\frac{d(e^u)}{dx} = e^u \frac{du}{dx}$$

or $d(e^u) = e^u \, du$, then integrating gives

$$\int e^u \, du = \int d(e^u) = e^u + C$$

or the following, from Appendix C:

Rule 8	$\displaystyle \int e^u \, du = e^u + C$

◆◆◆ **Example 33:** Integrate $\int e^{6x} \, dx$.

Solution: To match the form $\int e^u \, du$, let

$$u = 6x$$
$$du = 6 \, dx$$

We insert a factor of 6 and compensate with $\frac{1}{6}$; then we use Rule 8.

$$\int e^{6x} \, dx = \frac{1}{6} \int e^{6x} (6 \, dx) = \frac{1}{6} e^{6x} + C \qquad \text{◆◆◆}$$

We now do a definite integral. Simply substitute the limits, as before.

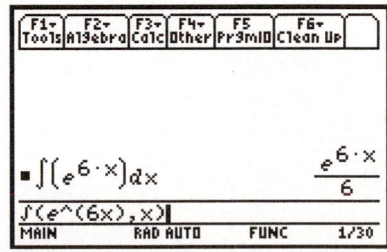

TI-89 calculator check for Example 33. Remember that the calculator does not show the constant of integration.

◆◆◆ **Example 34:** Integrate $\displaystyle \int_0^3 \frac{6e^{\sqrt{3x}}}{\sqrt{3x}} \, dx$.

Solution: Since the derivative of $\sqrt{3x}$ is $\frac{3}{2}(3x)^{-1/2}$, we insert a factor of $\frac{3}{2}$ and compensate.

$$\int_0^3 \frac{6e^{\sqrt{3x}}}{\sqrt{3x}} \, dx = 6 \int_0^3 e^{\sqrt{3x}} (3x)^{-1/2} \, dx$$

$$= 6 \left(\frac{2}{3}\right) \int_0^3 e^{\sqrt{3x}} \left[\frac{3}{2}(3x)^{-1/2} \, dx\right]$$

$$= 4 e^{\sqrt{3x}} \Big|_0^3$$

$$= 4e^3 - 4e^0 \approx 76.3 \qquad \text{◆◆◆}$$

Integral of $b^u\ du$

The derivative of $\dfrac{b^u}{\ln b}$ is

$$\frac{d}{dx}\left(\frac{b^u}{\ln b}\right) = \frac{1}{\ln b}(b^u)(\ln b)\frac{du}{dx} = b^u\frac{du}{dx}$$

or, in differential form, $d\left(\dfrac{b^u}{\ln b}\right) = b^u\ du$. Thus the integral of $b^u\ du$ is as follows:

Rule 9	$\displaystyle\int b^u\ du = \frac{b^u}{\ln b} + C \quad (b > 0, b \neq 1)$

◆◆◆ **Example 35:** Integrate $\displaystyle\int 3xa^{2x^2}\ dx$.

Solution:

$$\int 3xa^{2x^2}\ dx = 3\int a^{2x^2}x\ dx$$

$$= 3\left(\frac{1}{4}\right)\int a^{2x^2}(4x\ dx)$$

$$= \frac{3a^{2x^2}}{4\ln a} + C \qquad\qquad ◆◆◆$$

Integral of $\ln u$

To integrate the natural logarithm of a function, we use Rule 43, which we give here without proof.

Rule 43	$\displaystyle\int \ln u\ du = u(\ln u - 1) + C$

◆◆◆ **Example 36:** Integrate $\displaystyle\int x\ln(3x^2)\ dx$.

Solution: We put our integral into the form of Rule 43 and integrate.

$$\int x\ln(3x^2)\ dx = \left(\tfrac{1}{6}\right)\int \ln(3x^2)(6x\ dx)$$

$$= \left(\tfrac{1}{6}\right)(3x^2)(\ln 3x^2 - 1) + C$$

$$= \tfrac{1}{2}x^2(\ln 3x^2 - 1) + C \qquad\qquad ◆◆◆$$

Integral of $\log u$

To integrate the *common* logarithm of a function, we first convert it to a natural logarithm using Eq. 147.

	$\log N = \dfrac{\ln N}{\ln 10} \approx \dfrac{\ln N}{2.3026}$	**147**

◆◆◆ **Example 37:** Integrate $\displaystyle\int_3^4 \log(3x - 7)\,dx$.

Solution: We convert the common log to a natural log and apply Rule 43.

$$\int_3^4 \log(3x - 7)\,dx = \int_3^4 \frac{\ln(3x - 7)}{\ln 10}\,dx$$

$$= \frac{1}{3\ln 10}\int_3^4 \ln(3x - 7)(3\,dx)$$

$$= \frac{1}{3\ln 10}(3x - 7)\Big[\ln(3x - 7) - 1\Big]_3^4$$

$$= \frac{1}{3\ln 10}[5(\ln 5 - 1) - 2(\ln 2 - 1)]$$

$$\approx 0.530 \qquad\qquad ◆◆◆$$

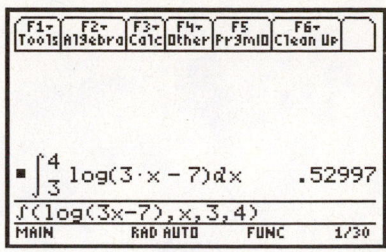

TI-89 calculator check for Example 37.

Electrical Applications

In an earlier chapter we had some applications of the integral to electric circuits. There we presented the following formulas:

Charge	$q = \displaystyle\int i\,dt$	1079
Voltage Across a Capacitor	$v = \dfrac{1}{C}\displaystyle\int i\,dt$	1081
Current in an Inductor	$i = \dfrac{1}{L}\displaystyle\int v\,dt$	1085

As before, i is the current in amperes (A), q is the charge in coulombs (C), v is voltage in volts (V), L is inductance in henrys (H), and t is time in seconds.

Now let us apply them to cases involving the exponential function, and in the following section, the trigonometric functions.

◆◆◆ **Example 38:** The current in a certain 25 F capacitor is given by

$$i = 18.5e^{-t/27.5}\ \text{A}$$

Write an expression for the voltage across the capacitor when charging, Fig. 29–23.

Solution: Integrating gives

$$\int i\,dt = 18.5\int e^{-t/27.5}\,dt$$

$$= (18.5)(-27.5)\int e^{-t/27.5}\left(-\frac{dt}{27.5}\right)$$

$$= -509e^{-t/27.5} + k_1$$

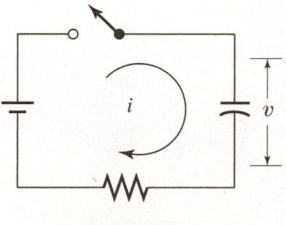

FIGURE 29–23

Multiplying by $1/C$ now gives us the voltage.

$$v = \frac{1}{C} \int i \, dt = -\frac{509}{25} e^{-t/27.5} + k$$

$$= -20.4 e^{-t/27.5} + k \quad \text{V}$$

where we have replaced $k_1/25$ by k. For a charging capacitor, the voltage is 0 at $t = 0$, so

$$0 = -20.4 e^0 + k$$

from which $k = 20.4$. Our complete equation is then

$$v = -20.4 e^{-t/27.5} + 20.4$$

$$= 20.4 (1 - e^{-t/27.5}) \quad \text{V} \qquad \bullet\bullet\bullet$$

Exercise 6 ◆ Integrals of Exponential and Logarithmic Functions

Integrate.

Exponential Functions

1. $\displaystyle\int a^{5x} \, dx$ **2.** $\displaystyle\int a^{9x} \, dx$

3. $\displaystyle\int 5^{7x} \, dx$ **4.** $\displaystyle\int 10^x \, dx$

5. $\displaystyle\int a^{3y} \, dy$ **6.** $\displaystyle\int x a^{3x^2} \, dx$

7. $\displaystyle\int 4 e^x \, dx$ **8.** $\displaystyle\int e^{2x} \, dx$

9. $\displaystyle\int x e^{x^2} \, dx$ **10.** $\displaystyle\int x^2 e^{x^3} \, dx$

11. $\displaystyle\int_1^2 x e^{3x^2} \, dx$ **12.** $\displaystyle\int_3^4 \sqrt{e^t} \, dt$

13. $\displaystyle\int \frac{e^{\sqrt{x}} \, dx}{\sqrt{x}}$ **14.** $\displaystyle\int (x + 3) e^{x^2 + 6x - 2} \, dx$

15. $\displaystyle\int_0^1 (e^x - 1)^2 \, dx$ **16.** $\displaystyle\int_2^3 x e^{-x^2} \, dx$

17. $\displaystyle\int \frac{e^{\sqrt{x-2}}}{\sqrt{x - 2}} \, dx$ **18.** $\displaystyle\int \frac{(e^{x/2} - e^{-x/2})^2}{4} \, dx$

19. $\displaystyle\int (e^{x/a} + e^{-x/a}) \, dx$ **20.** $\displaystyle\int (e^{x/a} - e^{-x/a})^2 \, dx$

Logarithmic Functions

21. $\displaystyle\int \ln 3x \, dx$ **22.** $\displaystyle\int \ln 7x \, dx$

23. $\displaystyle\int_1^2 x \ln x^2 \, dx$

24. $\displaystyle\int \log(5x - 3) \, dx$

25. $\displaystyle\int_2^4 x \log(x^2 + 1) \, dx$

26. $\displaystyle\int_1^3 x^2 \log(2 + 3x^3) \, dx$

27. Find the area under the curve $y = e^{2x}$ from $x = 1$ to 3.

28. The first-quadrant area bounded by the catenary $y = \frac{1}{2}(e^x + e^{-x})$ from $x = 0$ to 1 is rotated about the x axis. Find the volume generated.

29. The first-quadrant area bounded by $y = e^x$ and $x = 1$ is rotated about the line $x = 1$. Find the volume generated.

30. Find the length of the catenary $y = (a/2)(e^{x/a} + e^{-x/a})$ from $x = 0$ to 6. Use $a = 3$.

31. The curve $y = e^{-x}$ is rotated about the x axis. Find the area of the surface generated, from $x = 0$ to 100.

32. Find the horizontal distance $\bar{x}$ to the centroid of the area formed by the curve $y = \frac{1}{2}(e^x + e^{-x})$, the coordinate axes, and the line $x = 1$.

33. Find the vertical distance $\bar{y}$ to the centroid of the area formed by the curve $y = e^x$ between $x = 0$ and 1.

34. A volume of revolution is formed by rotating the curve $y = e^x$ between $x = 0$ and 1 about the x axis. Find the distance from the origin to the centroid.

35. Find the moment of inertia of the area bounded by the curve $y = e^x$, the line $x = 1$, and the coordinate axes, with respect to the x axis.

Electrical Applications

36. The current in a certain 228 F capacitor is given by

$$i = 25.5e^{-t/83.5} \text{ A}$$

Write an expression for the voltage across the capacitor when charging from an initial voltage of zero.

37. The current in a certain 3.85 F capacitor is given by

$$i = 84.6e^{-t/127} \text{ A}$$

Write an expression for the voltage across the capacitor when charging from an initial voltage of zero.

29–7   Integrals of the Trigonometric Functions

To our growing list of rules we add those for the six trigonometric functions. By Eq. 264,

$$\frac{d(-\cos u)}{dx} = \sin u \, \frac{du}{dx}$$

or $d(-\cos u) = \sin u \, du$. Taking the integral of both sides gives

$$\int \sin u \, du = -\cos u + C$$

The integrals of the other trigonometric functions are found in the same way. We will not do the derivation for each but you can convince yourself that each of the following rules is correct by taking the derivative, as we did earlier.

Rule 10	$\int \sin u \, du = -\cos u + C$
Rule 11	$\int \cos u \, du = \sin u + C$
Rule 12	$\int \tan u \, du = -\ln|\cos u| + C$
Rule 13	$\int \cot u \, du = \ln|\sin u| + C$
Rule 14	$\int \sec u \, du = \ln|\sec u + \tan u| + C$
Rule 15	$\int \csc u \, du = \ln|\csc u - \cot u| + C$

We use these rules just as we did the preceding ones: Match the given integral *exactly* with one of the rules, inserting a factor and compensating when necessary, and then copy off the integral in the rule.

◆◆◆ **Example 39:** Integrate $\int_0^1 x \sin x^2 \, dx$.

Solution:

$$\int_0^1 x \sin x^2 \, dx = \int_0^1 \sin x^2 (x \, dx)$$

$$= \frac{1}{2} \int_0^1 \sin x^2 (2x \, dx) = -\frac{1}{2} \cos x^2 \Big|_0^1$$

$$= -\frac{1}{2}(\cos 1 - \cos 0) \approx 0.2298 \qquad ◆◆◆$$

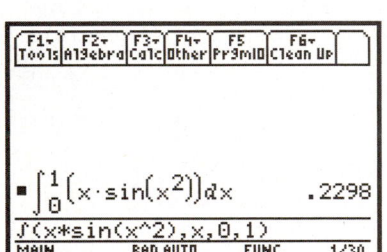

TI-89 calculator check for Example 39.

Sometimes the trigonometric identities can be used to simplify an expression before integrating.

◆◆◆ **Example 40:** Integrate $\int \dfrac{\cot 5x}{\cos 5x} \, dx$.

Solution: We replace $\cot 5x$ by $\cos 5x / \sin 5x$.

$$\int \frac{\cot 5x}{\cos 5x} \, dx = \int \frac{\cos 5x}{\sin 5x \cos 5x} \, dx = \int \frac{1}{\sin 5x} \, dx$$

$$= \int \csc 5x \, dx$$

$$= \frac{1}{5} \int \csc 5x \, (5 \, dx) = \frac{1}{5} \ln|\csc 5x - \cot 5x| + C$$

by Rule 15. ◆◆◆

Miscellaneous Rules from the Table

Now that you can use Rules 1 through 15, you should find it no harder to use any rule from the table of integrals.

◆◆◆ **Example 41:** Integrate $\int e^{3x} \cos 2x \, dx$.

Solution: We search the table for a similar form and find the following:

Rule 42	$\int e^{au} \cos bu \, du = \dfrac{e^{au}}{a^2 + b^2}(a \cos bu + b \sin bu) + C$

This matches our integral if we set

$$a = 3 \quad b = 2 \quad u = x \quad du = dx$$

so

$$\int e^{3x} \cos 2x \, dx = \frac{e^{3x}}{3^2 + 2^2} (3 \cos 2x + 2 \sin 2x) + C$$

$$= \frac{e^{3x}}{13} (3 \cos 2x + 2 \sin 2x) + C \qquad ◆◆◆$$

An Electrical Application

◆◆◆ **Example 42:** The current in a point in a certain circuit is given by

$$i = 6.84 \sin(5.83t + 1.46) \qquad A$$

(a) Write an expression for the charge at that point, assuming an initial charge of 0, and (b) evaluate it at $t = 1.00$ s.

Solution:

(a) The charge is the integral of the current (Eq. 1079) so,

$$q = \int i \, dt = 6.84 \int \sin(5.83t + 1.46) \, dt$$

$$= \frac{6.84}{5.83} \int [\sin(5.83t + 1.46)]5.83 \, dt$$

$$= -1.17 \cos(5.83t + 1.46) + k \qquad C$$

Letting $q = 0$ at $t = 0$ gives

$$0 = -1.17 \cos(1.46) + k$$

from which $k = 0.129$. Our complete equation is then

$$q = -1.17 \cos(5.83t + 1.46) + 0.129 \qquad C$$

(b) At $t = 1.00$ s,

$$q = -1.17 \cos(5.83 + 1.46) + 0.129$$

$$= -0.496 \qquad C \qquad ◆◆◆$$

Exercise 7 ◆ Integrals of the Trigonometric Fuctions

Integrate

1. $\int \sin 3x \, dx$

2. $\int \cos 7x \, dx$

3. $\int \tan 5\theta \, d\theta$

4. $\int \sec 2\theta \, d\theta$

5. $\displaystyle\int \sec 4x \, dx$ **6.** $\displaystyle\int \cot 8x \, dx$

7. $\displaystyle\int 3 \tan 9\theta \, d\theta$ **8.** $\displaystyle\int 7 \sec 3\theta \, d\theta$

9. $\displaystyle\int x \sin x^2 \, dx$ **10.** $\displaystyle\int 5x \cos 2x^2 \, dx$

11. $\displaystyle\int \theta^2 \tan \theta^3 \, d\theta$ **12.** $\displaystyle\int \theta \sec 2\theta^2 \, d\theta$

13. $\displaystyle\int \sin (x + 1) \, dx$ **14.** $\displaystyle\int \cos (7x - 3) \, dx$

15. $\displaystyle\int \tan (4 - 5\theta) \, d\theta$ **16.** $\displaystyle\int \sec (2\theta + 3) \, d\theta$

17. $\displaystyle\int x \sec (4x^2 - 3) \, dx$ **18.** $\displaystyle\int 3x^2 \cot (8x^3 + 3) \, dx$

19. $\displaystyle\int x \cos x^2 \, dx$ **20.** $\displaystyle\int 3x^2 \cos x^3 \, dx$

21. $\displaystyle\int_0^\pi \sin \phi \, d\phi$ **22.** $\displaystyle\int_0^{\pi/2} \cos \phi \, d\phi$

23. $\displaystyle\int_0^\pi \cos \frac{\theta}{2} \, d\theta$ **24.** $\displaystyle\int_{\pi/3}^{\pi/2} \sin^2 x \cos x \, dx$

Find the area under each curve.

25. $y = \sin x$ from $x = 0$ to π

26. $y = 2 \cos x$ from $x = -\pi/2$ to $\pi/2$

27. $y = 2 \sin \frac{1}{2} \pi x$ from $x = 0$ to 2 rad

28. Find the area between the curve $y = \sin x$ and the x axis from $x = 1$ rad to 3 rad.

29. Find the area between the curve $y = \cos x$ and the x axis from $x = 0$ to $\frac{3}{2} \pi$.

30. The area bounded by one arch of the sine curve $y = \sin x$ and the x axis is rotated about the x axis. Find the volume of the solid generated.

31. Find the surface area of the volume of revolution of problem 30.

32. The area bounded by one arch of the sine curve $y = \sin x$ and the x axis is rotated about the y axis. Find the volume generated.

33. Find the coordinates of the centroid of the area bounded by the x axis and a half-cycle of the sine curve $y = \sin x$.

34. Find the radius of gyration of the area under one arch of the sine curve $y = \sin x$ with respect to the x axis.

Electrical Applications

35. The current at a point in a certain circuit is given by

$$i = 84.3 \sin (11.5t + 5.48) \text{ A}$$

(a) Write an expression for the charge at that point, assuming an initial charge of 0, and (b) evaluate it at $t = 2.00$ s.

36. An expression for the current at a point in a certain circuit is

$$i = 273 \sin (382t + 0.573) \text{ A}$$

(a) Assuming an initial charge of 0, write an expression for the charge at that point and (b) evaluate it at $t = 3.50$ s.

29–8 Average and Root Mean Square Values

We are now able to do two applications that usually require integration of a trigonometric function.

Average Value of a Function

The area A under the curve $y = f(x)$ (Fig. 29–24) between $x = a$ and b is, by
Eq. 291,

$$A = \int_a^b f(x)\, dx$$

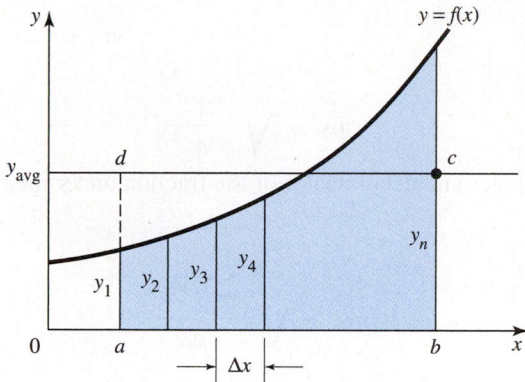

FIGURE 29–24   Average value, or average ordinate.

Within the same interval, the *average* value of that function, y_{avg}, is that value of y
which will cause the rectangle $abcd$ to have the same area as that under the curve, or

$$(b - a)\, y_{avg} = A = \int_a^b f(x)\, dx$$

Thus,

Average Value	$y_{avg} = \dfrac{1}{b - a} \displaystyle\int_a^b f(x)\, dx$	318

◆◆◆ **Example 43:** Find the average value of a half-cycle of the sinusoidal voltage

$$v = V \sin \theta \qquad \text{(volts)}$$

Solution: By Eq. 318, with $a = 0$ and $b = \pi$,

$$V_{avg} = \frac{V}{\pi - 0} \int_0^\pi \sin \theta\, d\theta$$

$$= \frac{V}{\pi} \Big[-\cos \theta \Big]_0^\pi$$

$$= \frac{V}{\pi} (-\cos \pi + \cos 0) = \frac{2}{\pi} V \approx 0.637\, V \qquad \text{(volts)} \qquad \text{◆◆◆}$$

Root Mean Square (RMS) Value of a Function

While the average value of a function is useful for many applications, it is not for others. For example, the average value of a sine wave is zero, because any part of the wave above the x axis is exactly matched by portions below the axis. Thus, electrical power expressed as a sine wave *would have an average value of zero*! That, of course, does not give a measure of its ability to turn a motor or heat an element. For such applications, a more useful measure is the root mean square value.

The *root mean square* (rms) value of a function is the square root of the average of the squares of the ordinates. In Fig. 29–24, if we take n values of y spaced apart by a distance Δx, the rms value is approximately

$$\text{rms} \cong \sqrt{\frac{y_1^2 + y_2^2 + y_3^2 + \cdots + y_n^2}{n}}$$

or, using summation notation,

$$\text{rms} \cong \sqrt{\frac{\sum_{i=1}^{n} y_i^2}{n}}$$

Multiplying numerator and denominator of the fraction under the radical by Δx, we obtain

$$\text{rms} \cong \sqrt{\frac{\sum_{i=1}^{n} y_i^2 \Delta x}{n \, \Delta x}}$$

But $n\Delta x$ is simply the width $(b - a)$ of the interval. If we now let n approach infinity, we get

$$\lim_{n \to \infty} \sum_{i=1}^{n} y_i^2 \Delta x = \int_a^b [f(x)]^2 \, dx$$

Therefore:

Root Mean Square Value	$\text{rms} = \sqrt{\dfrac{1}{b-a} \displaystyle\int_a^b [f(x)]^2 \, dx}$	**319**

◆◆◆ **Example 44:** Find the rms value for the sinusoidal voltage of Example 43.

Solution: We substitute into Eq. 319 with $a = 0$ and $b = \pi$.

$$\text{rms} = \sqrt{\frac{1}{\pi - 0} \int_0^\pi V^2 \sin^2 \theta \, d\theta} = \sqrt{\frac{V^2}{\pi} \int_0^\pi \sin^2 \theta \, d\theta}$$

But, by Rule 16,

$$\int_0^\pi \sin^2 \theta \, d\theta = \frac{\theta}{2} - \frac{\sin 2\theta}{4} \Big|_0^\pi$$

$$= \frac{\pi}{2} - \frac{\sin 2\pi}{4} = \frac{\pi}{2}$$

So

$$\text{rms} = \sqrt{\frac{V^2}{\pi} \cdot \frac{\pi}{2}} = \frac{V}{\sqrt{2}} \approx 0.707V \qquad \text{Volts}$$

In electrical work, the rms value of an alternating current or voltage is also called the *effective* value. We see then that the effective value is 0.707 of the peak value.

◆◆◆

Exercise 8 ◆ Average and Root Mean Square Values

Find the average ordinate for each function in the given interval.

1. $y = x^2$ from 0 to 6

2. $y = x^3$ from -5 to 5

3. $y = \sqrt{1 + 2x}$ from 4 to 12

4. $y = \dfrac{x}{\sqrt{9 + x^2}}$ from 0 to 4

5. $y = \sin^2 x$ from 0 to $\pi/2$

6. $2y = \cos 2x + 1$ from 0 to π

Find the rms value for each function in the given interval.

7. $y = 2x + 1$ from 0 to 6

8. $y = \sin 2x$ from 0 to $\pi/2$

9. $y = x + 2x^2$ from 1 to 4

10. $y = 3 \tan x$ from 0 to $\pi/4$

11. $y = 2 \cos x$ from $\pi/6$ to $\pi/2$

12. $y = 5 \sin 2x$ from 0 to $\pi/6$

••• CHAPTER 29 REVIEW PROBLEMS •••••••••••••••••••••••••••••••

Find dy/dx.

1. $y = \dfrac{a}{2}(e^{x/a} - e^{-x/a})$

2. $y = 5^{2x+3}$

3. $y = 8 \tan \sqrt{x}$

4. $y = \sec^2 x$

5. $y = x \operatorname{Arctan} 4x$

6. $y = \dfrac{1}{\sqrt{\operatorname{Arcsin} 2x}}$

7. $y^2 = \sin 2x$

8. $y = xe^{2x}$

9. $y = x \sin x$

10. $y = x^2 \sin x$

11. $y = x^3 \cos x$

12. $y = \ln \sin(x^2 + 3x)$

13. $y = \dfrac{\sin x}{x}$

14. $y = (\log x)^2$

15. $y = \log x(1 + x^2)$

16. $\cos(x - y) = 2x$

17. $y = \ln(x + \sqrt{x^2 + a^2})$

18. $y = \ln(x + 10)$

19. $y = \csc 3x$

20. $y = \ln(x^2 + 3x)$

21. $y = \dfrac{\sin x}{\cos x}$

22. $y = \dfrac{1}{\cos^2 x}$

23. $y = \ln(2x^3 + x)$

24. $y = x \operatorname{Arcsin} 2x$

25. $y = x^2 \operatorname{Arccos} x$

26. Find the minimum point of the curve $y = \ln(x^2 - 2x + 3)$.

27. Find the points of inflection of the curve $xy = 4 \log(x/2)$.

28. Find a minimum point and a point of inflection on the curve $y \ln x = x$. Write the equation of the tangent at the point of inflection.

Integrate.

29. $\displaystyle\int_0^1 3x \sin x^2 \, dx$

30. $\displaystyle\int \sin 7x \, dx$

31. $\displaystyle\int 6x \cos 2x^2 \, dx$

32. $\displaystyle\int \sin(5x - 4) \, dx$

33. $\displaystyle\int_0^{\pi/2} 3 \cos 5x \, dx$ **34.** $\displaystyle\int_0^{\pi/2} 7x \sin 2x^2 \, dx$

35. $\displaystyle\int x \ln x^2 \, dx$ **36.** $\displaystyle\int e^{3x+4} \, dx$

37. $\displaystyle\int x \ln(3x^2 - 2) \, dx$ **38.** $\displaystyle\int_1^4 5x^2 \log(3x^3 + 7) \, dx$

39. $\displaystyle\int_0^{\pi} x^2 \cos(x^3 + 5) \, dx$ **40.** $\displaystyle\int e^{4-2x} \, dx$

41. Find the average value for the function $y = \sin^2 x$ for $x = 0$ to 2π.

42. Find the rms value for the function $y = 2x + x^2$ for the interval $x = -1$ to 3.

43. At what x between $-\pi/2$ and $\pi/2$ is there a maximum on the curve $y = 2 \tan x - \tan^2 x$?

Find the value of dy/dx for the given value of x.

44. $y = x \operatorname{Arccos} x$ at $x = -\frac{1}{2}$ **45.** $y = \dfrac{\operatorname{Arcsec} 2x}{\sqrt{x}}$ at $x = 1$

46. If $x^2 + y^2 = \ln y + 2$, find y' and y'' at the point $(1, 1)$.

Find the equation of the tangent to each curve.

47. $y = \sin x$ at $x = \pi/6$

48. $y = x \ln x$ parallel to the line $3x - 2y = 5$

49. At what x is the tangent to the curve $y = \tan x$ parallel to the line $y = 2x + 5$?

Find the smallest positive root between $x = 0$ and 10 to three decimal places by any method.

50. $\sin 3x - \cos 2x = 0$ **51.** $2 \sin \frac{1}{2}x - \cos 2x = 0$

52. Find the angle of intersection between $y = \ln(x^3/8 - 1)$ and $y = \ln(3x - x^2/4 - 1)$ at $x = 4$.

53. A casting is taken from one oven at 1500°F and placed in another oven whose temperature is 0°F and rising. The temperature T of the casting after t h is given by $T = 100t + 1500e^{-0.2t}$. Find the minimum temperature reached by the casting and the time at which it occurs.

54. A statue 11 ft tall is on a pedestal so that the bottom of the statue is 25 ft above eye level. How far from the statue (measured horizontally) should an observer stand so that the statue will subtend the greatest angle at the observer's eye?

55. The voltage applied to a 184-microfarad (μF) capacitor is

$$v = 22.5 \cos(48.3t + 0.95) \quad \text{V}$$

(a) Write an expression for the current in the capacitor and (b) evaluate the current at $t = 0.1$ s.

56. The current in a certain 84.6 F capacitor is given by

$$i = 39.4e^{-t/237} \text{ A}$$

Write an expression for the voltage across the capacitor when charging, assuming an initial voltage of 0.

57. The current at a point in a certain circuit is given by

$$i = 735 \sin(33.6t + 0.73) \text{ A}$$

(a) Write an expression for the charge at that point, assuming an initial charge of 0, and (b) evaluate it at $t = 2.00$ s.

58. An expression for the current at a point in a certain circuit is

$$i = 24.6 \sin(24.6t + 0.663) \text{ A}$$

    (a) Assuming an initial charge of 0, write an expression for the charge at that point and (b) evaluate it at $t = 1.00$ s.

59. The voltage applied to a certain 482-microfarad (μF) capacitor is

$$v = 274(1 - e^{-t/335}) \text{ V}$$

    (a) Write an expression for the current in the capacitor and (b) evaluate the current at $t = 200$ s.

60. The charge through a 63.5-Ω resistor is given by

$$q = 184 \sin(39.6t + 0.383) \text{ C}$$

Write an expression for (a) the instantaneous current through the resistor and (b) the instantaneous voltage across the resistor.

61. The current in a certain 834 F capacitor is given by

$$i = 63.5e^{-t/77.3} \text{ A}$$

Write an expression for the voltage across the capacitor when charging, assuming an initial voltage of 0.

62. *Project:* A capacitor (Fig. 29–25) is charged to 300 V. When the switch is closed, the voltage across R is initially 300 V but then drops according to the equation $V_1 = 300e^{-t/RC}$, where t is the time (in seconds), R is the resistance, and C is the capacitance. If the voltage V_2 also starts to rise at the instant of switch closure so that $V_2 = 100t$:

    (a) Graph V_1, V_2, and V.
    (b) Show that the total voltage V is $300e^{-t/RC} + 100t$.
    (c) Graphically find t when V is a minimum.
    (d) Find t when V is a minimum by setting the derivative dV/dt equal to zero and solving for t.

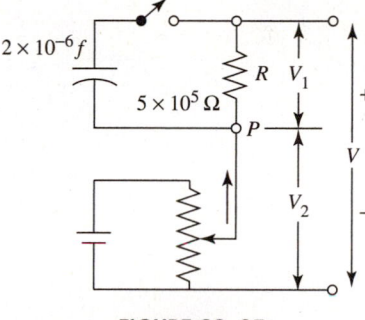

FIGURE 29–25

30

First-Order Differential Equations

••• **OBJECTIVES** ••

When you have completed this chapter, you should be able to

- Solve simple differential equations by calculator, graphically or numerically.
- Solve first-order differential equations that have separable variables or that are exact.
- Solve homogeneous first-order differential equations.
- Solve first-order linear differential equations and Bernoulli's equation.
- Solve applications of first-order differential equations.

••

A *differential equation* is simply an equation that contains one or more derivatives. We have already solved simple types in an earlier chapter and here we go on to more difficult types.

After some definitions, we show how to solve differential equations by calculator, and by graphical and numerical methods. These are useful for cases where a differential equation cannot be solved by other means. We then show how to solve several types of first-order differential equations analytically.

Our main applications will be in exponential growth and decay, motion, and electric circuits. For these, we will write a differential equation from the problem statement and go on to solve it. For example, suppose the temperature of a part in a heat-treating furnace increases with time so that its rate of increase is proportional to the difference between its final temperature T_f and its present temperature T. We can express this relationship as

$$\frac{dT}{dt} = n(T_f - T)$$

where n is a constant. How would you now solve that simple differential equation to arrive at a formula for exponential growth? We will show how in this chapter.

Here we limit ourselves to first-order differential equations and will cover those of *second order* in the following chapter.

30–1 Definitions

A *differential equation* is one that contains one or more derivatives. We sometimes refer to differential equations by the abbreviation DE.

◆◆◆ **Example 1:** Some differential equations, using different notation for the derivative, are

(a) $\dfrac{dy}{dx} + 5 = 2xy$ (b) $y'' - 4y' + xy = 0$

(c) $Dy + 5 = 2xy$ (d) $D^2y - 4Dy + xy = 0$ ◆◆◆

Differential Form

A differential equation containing the derivative dy/dx is put into *differential form* simply by multiplying through by dx.

◆◆◆ **Example 2:** Example 1(a) in differential form is

$$dy + 5\, dx = 2xy\, dx$$ ◆◆◆

Ordinary Versus Partial Differential Equations

We get an *ordinary* differential equation when our differential equation contains only two variables and "ordinary" derivatives, as in Examples 1 and 2. When the equation contains *partial* derivatives, because of the presence of three or more variables, we have a *partial differential equation*.

◆◆◆ **Example 3:** $\dfrac{\partial x}{\partial t} = 5\dfrac{\partial y}{\partial t}$ is a partial differential equation.

The symbol ∂ is used for partial derivatives.   ◆◆◆

We cover only *ordinary* differential equations in this book.

Order

The *order* of a differential equation is the order of the highest-order derivative in the equation.

◆◆◆ **Example 4:**

(a) $\dfrac{dy}{dx} - x = 2y$ is of *first* order.   (b) $\dfrac{d^2y}{dx^2} - \dfrac{dy}{dx} = 3x$ is of *second* order.

(c) $5y''' - 3y'' = xy$ is of *third* order.   ◆◆◆

Degree

The *degree of a derivative* is the power to which that derivative is raised.

◆◆◆ **Example 5:** $(y')^2$ is of second degree.   ◆◆◆

The *degree of a differential equation* is the degree of the highest-order derivative in the equation. We will cover first-order DEs in this chapter, an second-order in the next.

The equation must be rationalized and cleared of fractions before its degree can be determined.

◆◆◆ **Example 6:**

(a) $(y'')^3 - 5(y')^4 = 7$ is a second-order equation of *third* degree.

(b) To find the degree of the differential equation

$$\frac{x}{\sqrt{y'-2}} = 1$$

we clear fractions and square both sides, getting

$$\sqrt{y'-2} = x$$

or

$$y' - 2 = x^2$$

which is of the first degree.   ◆◆◆

It's important to recognize the type of DE here for the same reason as for other equations. It lets us pick the right method of solution.

Common Error	Don't confuse order and degree. The symbols $$\frac{d^2y}{dx^2} \text{ and } \left(\frac{dy}{dx}\right)^2$$ have different meanings.

Solving a Simple Differential Equation

We have already solved some simple differential equations in an earlier chapter by multiplying both sides of the equation by dx and integrating.

◆◆◆ **Example 7:** Solve the differential equation $dy/dx = x^2 - 3$.

Solution: We first put our equation into differential form, getting $dy = (x^2 - 3)dx$. Integrating, we get

$$y = \int (x^2 - 3)dx = \frac{x^3}{3} - 3x + C$$ ◆◆◆

Checking a Solution

Any *function* that satisfies a differential equation is called a *solution* of that equation. Thus to check a solution, we substitute it and its derivatives into the original equation and see if an identity is obtained.

◆◆◆ **Example 8:** Is the function $y = e^{2x}$ a solution of the differential equation $y'' - 3y' + 2y = 0$?

Solution: Taking the first and second derivatives of the function gives

$$y' = 2e^{2x} \quad \text{and} \quad y'' = 4e^{2x}$$

Substituting the function and its derivatives into the differential equation gives

$$4e^{2x} - 6e^{2x} + 2e^{2x} = 0$$

which is an identity. Thus $y = e^{2x}$ is a solution to the differential equation. We will see shortly that it is one of many solutions to the given equation.   ◆◆◆

General and Particular Solutions

■ **Exploration:** We have just seen that the function $y = e^{2x}$ is a solution of the differential equation $y'' - 3y' + 2y = 0$. But is it the *only* solution? *Try this.*

Substitute the following functions yourself, and you will see that they are also solutions to the given equation.

$$y = 4e^{2x} \tag{1}$$
$$y = Ce^{2x} \tag{2}$$
$$y = C_1 e^{2x} + C_2 e^x \tag{3}$$

where C, C_1, and C_2 are *arbitrary constants*.

There is a simple relation, which we state without proof, between the order of a differential equation and the number of constants in the solution.

> The solution of an *n*th-order differential equation can have at most *n* arbitrary constants. A solution having the maximum number of constants is called the *general solution* or *complete solution*.

The differential equation in Example 8 is of second order, so the solution can have up to two arbitrary constants. Thus Eq. (3) is the general solution whereas Eqs. (1) and (2) are called *particular solutions*. When we later solve a differential equation, we will first obtain the general solution. Then, by using other given information, we will evaluate the arbitrary constants to obtain a particular solution. The "other given information" is referred to as *boundary conditions* or *initial conditions*.

Exercise 1 ◆ Definitions

Give the order and degree of each equation, and state whether it is an ordinary or partial differential equation.

1. $\dfrac{dy}{dx} + 3xy = 5$ **2.** $y'' + 3y' = 5x$

3. $D^3 y - 4Dy = 2xy$ **4.** $\dfrac{\partial^2 y}{\partial x^2} + 4y = 7$

5. $3(y'')^4 - 5y' = 3y$ **6.** $4\dfrac{dy}{dx} - 3\left(\dfrac{d^2 y}{dx^2}\right)^3 = x^2 y$

Solve each differential equation.

7. $\dfrac{dy}{dx} = 7x$ **8.** $2y' = x^2$

9. $4x - 3y' = 5$ **10.** $3Dy = 5x + 2$

11. $dy = x^2\, dx$ **12.** $dy - 4x\, dx = 0$

Show that each function is a solution to the given differential equation.

13. $y' = \dfrac{2y}{x}, \; y = Cx^2$ **14.** $\dfrac{dy}{dx} = \dfrac{x^2}{y^3}, \; 4x^3 - 3y^4 = C$

15. $Dy = \dfrac{2y}{x}, \; y = Cx^2$

16. $y' \cot x + 3 + y = 0, \; y = C \cos x - 3$

30–2   Solving a DE by Calculator, Graphically, and Numerically

We have said that any function that satisfies a differential equation is a solution to that equation. That function can be given

(a) analytically, as an equation
(b) graphically, as a plotted curve
(c) numerically, as a table of point pairs

In this section we will find all three kinds of solutions, with our analytical solution found by symbolic math on a calculator. We will do analytical solutions by hand later in this chapter.

The methods of this section are especially important because analytical solutions cannot be found by hand for many differential equations.

Solution by Calculator

Many calculators that can do symbolic processing can solve a differential equation. We must enter the DE, the independent variable, and the dependent variable. If boundary conditions are also entered, we will get a particular solution. Otherwise we will get a general solution with unknown constants. We will first do an example of a general solution.

◆◆◆ Example 9: Use the TI-89 to solve the DE

$$\frac{dy}{dx} = \frac{x}{y}$$

Solution:

(a) We select **deSolve** from the MATH Calculus menu. (b) Then enter the DE, $y' = x/y$. Indicate a derivative using the prime (′) symbol. (c) Enter the variables x and y. (d) Press ENTER to display the general solution

$$y^2 = x^2 + C$$

Notice that the constant is displayed as **@1** on the calculator.   **◆◆◆**

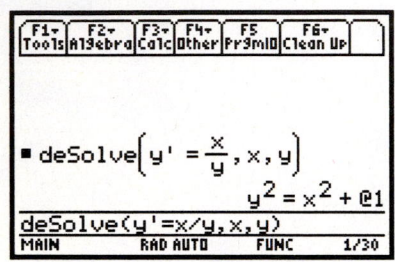

TI-89 screen for Example 9.

Now let us repeat the preceding example with boundary conditions.

◆◆◆ Example 10: Solve the DE from the preceding example with the boundary conditions

$$y = 1 \text{ when } x = 0$$

Solution: The steps are similar, but now immediately following the DE, we enter boundary conditions in the form

and $y(0) = 1$

where **and** is from the MATH Test menu. The complete entry is then

deSolve $(y' = x/y$ and $y(0) = 1, x, y)$

Pressing ENTER displays the particular solution

$$y^2 = x^2 + 1$$ **◆◆◆**

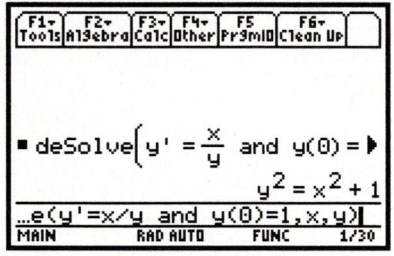

TI-89 screen for Example 10.

Graphical Solution by Slope Fields

A differential equation, such as

$$dy/dx = x - 2y$$

relates the variables x and y to the derivative dy/dx. But, remember that the derivative is the *slope m* of the curve. Replacing dy/dx with m gives

$$m = x - 2y$$

Using this equation, we can compute the slope at any point. The graph of the slopes is called a *slope field* (also called a *direction field* or *tangent field*). To get the solutions to a differential equation, we simply sketch in the curves that have the slopes of the surrounding slope field.

◆◆◆ Example 11:

(a) Construct a slope field for the differential equation

$$dy/dx = x - 2y$$

for $x = 0$ to 5 and $y = 0$ to 5.

(b) Sketch the solution that has the boundary conditions $y = 2$ when $x = 0$.

Solution:

(a) Computing slopes gives the following:

$$\text{At } (0, 0) \ m = 0 - 2(0) = 0$$

$$\text{At } (0, 1) \ m = 0 - 2(1) = -2$$

$$\vdots \qquad \vdots$$

$$\text{At } (5, 5) \ m = 5 - 2(5) = -5$$

It takes 25 computations to get all the points. We save time by using a computer and get the following:

	5	-10	-9	-8	-7	-6	-5
	4	-8	-7	-6	-5	-4	-3
y	3	-6	-5	-4	-3	-2	-1
	2	-4	-3	-2	-1	0	1
	1	-2	-1	0	1	2	3
	0	0	1	2	3	4	5
		0	1	2	3	4	5
				x			

We make our slope field by drawing a short line with the required slope at each point, as shown in Fig. 30–1.

(b) Several solutions are shown dashed in Fig. 30–1. The solution that has the boundary conditions $y = 2$ when $x = 0$ is shown as a solid line.

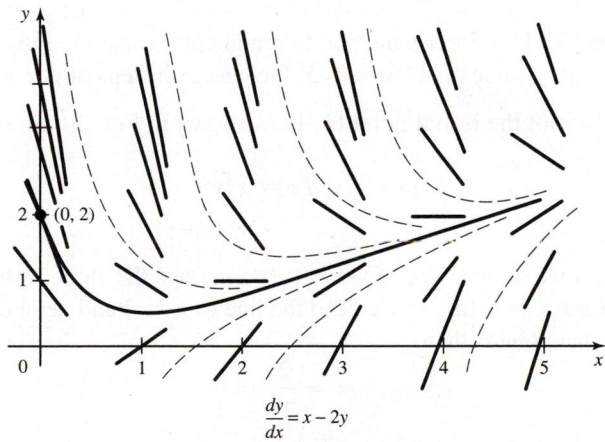

$$\frac{dy}{dx} = x - 2y$$

FIGURE 30–1 This is obviously a lot of work, which usually makes it an impractical method of solution. It is, however, a good way to *visualize* a solution to a DE. ◆◆◆

Euler's Method: Graphical Solution

We will now describe a technique, called *Euler's method*, for solving a differential equation approximately. We use it here for a graphical solution of a DE, and in the next section we use it for a numerical solution.

Suppose that we have a first-order differential equation, which we write in the form

$$y' = f(x, y) \tag{1}$$

and a boundary condition that $x = x_p$ when $y = y_p$. We seek a solution $y = F(x)$ such that the graph of this function (Fig. 30–2) passes through $P(x_p, y_p)$ and has a slope at P given by Eq. (1), $m_p = f(x_p, y_p)$.

Euler's method is named after the Swiss mathematician Leonhard Euler (1707–83).

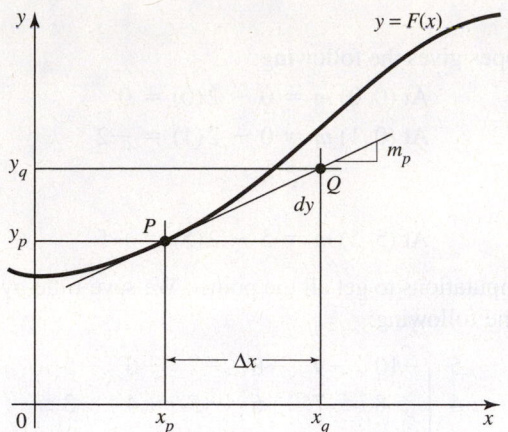

FIGURE 30–2 Euler's method.

Having the slope at P, we then step a distance Δx to the right. The rise dy of the tangent line is then

$$\text{rise} = (\text{slope})(\text{run}) = m_p \, \Delta x$$

Thus the point Q has the coordinates $(x_p + \Delta x, y_p + m_p \, \Delta x)$. *Point Q is probably not on the curve $y = F(x)$ but, if Δx is small enough, may be close enough to use as an approximation to the curve.*

From Q, we repeat the process enough times to reconstruct as much of the curve $y = F(x)$ as is needed.

◆◆◆ **Example 12:** Use Euler's method to graphically solve the DE, $dy/dx = x/y^2$, from the boundary value $(1, 1)$ to $x = 5$. Increase x in steps of $\Delta x = 1$ unit.

Solution: We plot the initial point $(1, 1)$ as shown in Fig. 30–3. The slope at that point is then

$$m = dy/dx = x/y^2$$
$$= 1/1^2 = 1$$

Through $(1, 1)$ we draw a line of slope 1. We *assume* the slope to be constant over the interval from $x = 1$ to 2. We extend the line to $x = 2$ and get a new point $(2, 2)$. The slope at that point is then

$$m = \frac{2}{2^2} = \frac{1}{2}$$

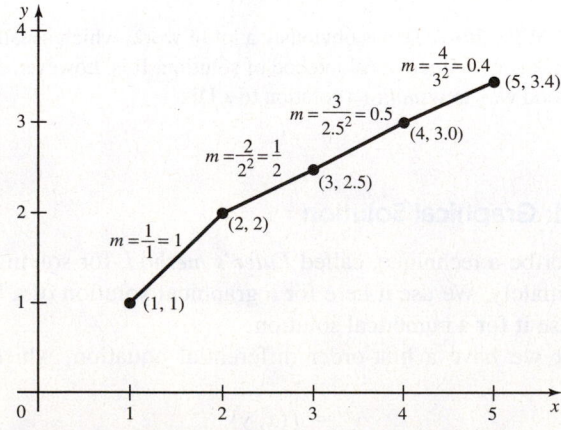

FIGURE 30–3 Graphical solution of $\dfrac{dy}{dx} = \dfrac{x}{y^2}$.

Through (2, 2) we draw a line of slope $\frac{1}{2}$ and extend it to get $(3, 2\frac{1}{2})$. We continue in this manner to get the final point (5, 3.4). ◆◆◆

Note that our solution in Example 12 was in the form of a graph. With the numerical method to follow, our solution will be in the form of a table of point pairs.

Euler's Method: Numerical Solution

We can express Euler's method by means of two iteration formulas. If we have the coordinates (x_p, y_p) at any point P, and the slope m_p at P, we find the coordinates (x_q, y_q) of Q by the following iteration formulas:

Euler's Method	$x_q = x_p + \Delta x$ $y_q = y_p + m_p \Delta x$	320

◆◆◆ **Example 13:** Find an approximate solution to $y' = x^2/y$, with the boundary condition that $y = 2$ when $x = 3$. Calculate y for $x = 3$ to 10 in steps of 1.

Solution: The slope at (3, 2) is

$$m = y'(3, 2) = \frac{9}{2} = 4.5$$

If $\Delta x = 1$, the rise is

$$dy = m \, \Delta x = 4.5(1) = 4.5$$

The ordinate of our next point is then $2 + 4.5 = 6.5$. The abscissa of the next point is $3 + 1 = 4$. So the coordinates of our next point are (4, 6.5). Repeating the process using (4, 6.5) as (x_p, y_p), we calculate the next point by first getting m and dy.

$$m = y'(4, 6.5) = \frac{16}{6.5} = 2.462$$

$$dy = 2.462(1) = 2.462$$

So the next point is (5, 8.962). The remaining values are given in Table 30–1, which was computer generated.

TABLE 30–1

x	Approximate y	Exact y	Error
3	2.00000	2.00000	0.00000
4	6.50000	5.35413	1.14587
5	8.96154	8.32666	0.63487
6	11.75124	11.40176	0.34948
7	14.81475	14.65151	0.16324
8	18.12226	18.09236	0.02991
9	21.65383	21.72556	−0.07173
10	25.39451	25.54734	−0.15283

The solution to our differential equation then is in the form of a set of (x, y) pairs, the first two columns in Table 30–1. We do not get an equation for our solution.

We normally use numerical methods for a differential equation whose exact solution cannot be found. The exact solution, which we'll find later, is $3y^2 = 2x^3 - 42$, which we'll now use to compute the values shown in the third column of the table, with the difference between exact and approximate values in the fourth column. Note that the error at $x = 10$ is about 0.6%. The exact and the approximate values are graphed in Fig. 30–4.

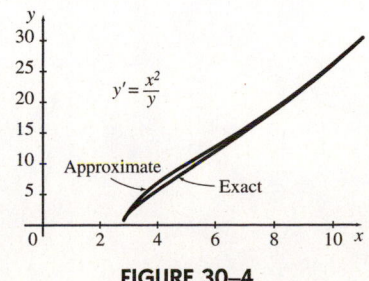

FIGURE 30–4

Reducing the step size will result in better accuracy (and more work). However, more practical numerical methods for solving DEs are given on our companion web site. ◆◆◆

Exercise 2 ◆ Solving a DE by Calculator, Graphically, and Numerically

Calculator Solution

Solve each differential equation by calculator.

1. $y' = x/y$
2. $y' = 2y/x$
3. $y' = x^2/y^3$
4. $y' = x/4y$ $y(5) = 2$
5. $y' = x^2/y^2$ $y(0) = 1$
6. $y' = 3x/y^2$ $y(4) = 1$

Graphical and Numerical Solution

Solve each differential equation and find the approximate value of y requested. Start at the given boundary value and use a slope field or Euler's graphical or numerical method, as directed by your instructor.

7. $y' = x$ Start at $(0, 1)$. Find $y(2)$.
8. $y' = y$ Start at $(0, 1)$. Find $y(3)$.
9. $y' = x - 2y$ Start at $(0, 4)$. Find $y(3)$.
10. $y' = x^2 - y^2 - 1$ Start at $(0, 2)$. Find $y(2)$.

11. *Computer:* Use a spreadsheet to compute the values of the slopes for a slope field for any of the DEs in this chapter.

12. *Computer:* Use a spreadsheet for the numerical solution by Euler's method of any of the DEs in this chapter.

30–3   First-Order DE: Variables Separable

Given a differential equation of first order, $dy/dx = f(x, y)$, it is sometimes possible to *separate* the variables x and y. That is, when we multiply both sides by dx, the resulting equation can be put into a form that has dy multiplied by a function of y only and dx multiplied by a function of x only.

Form of First-Order DE, Variables Separable	$f(y)\,dy = g(x)\,dx$	321

If this is possible, we can obtain a solution simply by integrating term by term.

◆◆◆ **Example 14:** Solve the differential equation $y' = x^2/y$.

Solution:

(1) Rewrite the equation in differential form: We do this by replacing y' with dy/dx and multiplying by dx.

$$dy = \frac{x^2\,dx}{y}$$

(2) Separate the variables: We do this here by multiplying both sides by y, and we get $y\,dy = x^2\,dx$. The variables are now separated, and our equation is in the form of Eq. 321.

(3) Integrate:

$$\int y \, dy = \int x^2 \, dx$$

$$\frac{y^2}{2} = \frac{x^3}{3} + C_1$$

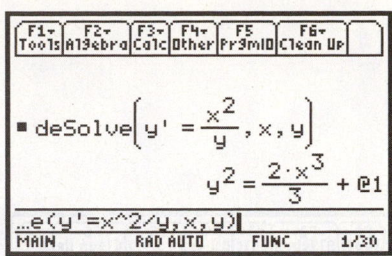

TI-89 calculator check for Example 14.

where C_1 is an arbitrary constant.

(4) Simplify the answer: We can do this by multiplying by the LCD 6.

$$3y^2 = 2x^3 + C$$

where we have replaced $6C_1$ by C. We'll often leave our answer in implicit form, as we have done here.   ◆◆◆

Simplifying the Solution

Often the simplification of a solution to a differential equation will involve several steps.

◆◆◆ **Example 15:** Solve the differential equation $dy/dx = 4xy$.

Solution: Multiplying both sides by dx/y and then integrating gives us

$$\frac{dy}{y} = 4x \, dx$$

$$\ln|y| = 2x^2 + C$$

We can leave our solution in this implicit form or solve for y.

$$|y| = e^{2x^2 + C} = e^{2x^2} e^C$$

Let us replace e^C by another constant k. Since k can be positive or negative, we can remove the absolute value symbols from y, getting

$$y = ke^{2x^2}$$ ◆◆◆

The solution to a DE can take on different forms, depending on how you simplify it. Don't be discouraged if your solution does not at first match the one in the answer key or on the calculator.

> We will often interchange one arbitrary constant with another of different form, such as replacing C_1 by $8C$, or by $\ln C$, or by $\sin C$, or by e^C, or by any other form that will help us to simplify an expression.

The laws of exponents or of logarithms will often be helpful in simplifying an answer, as in the following example.

◆◆◆ **Example 16:** Solve the differential equation $dy/dx = y/(5 - x)$.

Solution:

(1) Going to differential form, we have

$$dy = \frac{y \, dx}{5 - x}$$

(2) Separating the variables yields

$$\frac{dy}{y} = \frac{dx}{5 - x}$$

(3) Integrating gives

$$\ln|y| = -\ln|5 - x| + C_1$$

(4) Simplifying gives

$$\ln|y| + \ln|5 - x| = C_1$$

Then using our laws of logarithms we get

$$\ln|y(5 - x)| = C_1$$

Going to exponential form,

$$y(5 - x) = e^{C_1}$$

$$y = \frac{e^{C_1}}{5 - x}$$

where $x \neq 5$. We can leave this expression as it is or simplify it by letting $C = e^{C_1}$, getting

$$y = \frac{C}{5 - x} \qquad \blacklozenge\blacklozenge\blacklozenge$$

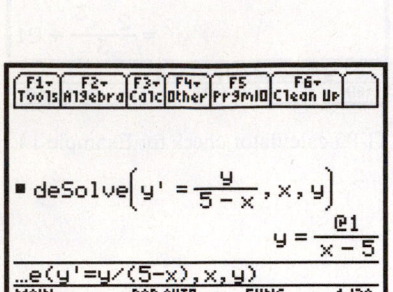

TI-89 calculator check for Example 16. Here the constant @1 is equal to $-C$ in our example.

Logarithmic, Exponential, or Trigonometric Equations

To separate the variables in certain equations, it may be necessary to use our laws of exponents or logarithms, or the trigonometric identities.

◆◆◆ **Example 17:** Solve the equation $4y' \, e^{4y} \cos x = e^{2y} \sin x$.

Solution: We replace y' with dy/dx and multiply through by dx, getting $4e^{4y} \cos x \, dy = e^{2y} \sin x \, dx$. We can eliminate x on the left side by dividing through by $\cos x$. Similarly, we can eliminate y from the right by dividing by e^{2y}.

$$\frac{4e^{4y}}{e^{2y}} \, dy = \frac{\sin x}{\cos x} \, dx$$

or $4e^{2y} \, dy = \tan x \, dx$. Integrating gives the solution

$$2e^{2y} = -\ln|\cos x| + C \qquad \blacklozenge\blacklozenge\blacklozenge$$

Particular Solution

We can evaluate the constant in our solution to a differential equation when given suitable *boundary conditions*, as in the following example.

◆◆◆ **Example 18:** Solve the equation $2y(1 + x^2)y' + x(1 + y^2) = 0$, subject to the condition that $y = 2$ when $x = 0$.

Solution: Separating variables and integrating, we obtain

$$\frac{2y \, dy}{1 + y^2} + \frac{x \, dx}{1 + x^2} = 0$$

$$\int \frac{2y \, dy}{1 + y^2} + \frac{1}{2} \int \frac{2x \, dx}{1 + x^2} = 0$$

$$\ln|1 + y^2| + \frac{1}{2} \ln|1 + x^2| = C$$

Simplifying, we drop the absolute value signs since $(1 + x^2)$ and $(1 + y^2)$ cannot be negative. Then we multiply by 2 and apply the laws of logarithms.

$$2 \ln(1 + y^2) + \ln(1 + x^2) = 2C$$

$$\ln(1 + x^2)(1 + y^2)^2 = 2C$$

$$(1 + x^2)(1 + y^2)^2 = e^{2C}$$

Applying the boundary conditions that $y = 2$ when $x = 0$ gives

$$e^{2C} = (1 + 0)(1 + 2^2)^2 = 25$$

Our particular solution is then $(1 + x^2)(1 + y^2)^2 = 25$.   ◆◆◆

Exercise 3 ◆ First-Order DE: Variables Separable

General Solution

Find the general solution to each differential equation. Try some by calculator.

We'll have applications later in this chapter.

1. $y' = \dfrac{x}{y}$

2. $\dfrac{dy}{dx} = \dfrac{2y}{x}$

3. $dy = x^2 y\, dx$

4. $y' = xy^3$

5. $y' = \dfrac{x^2}{y^3}$

6. $y' = \dfrac{x^2 + x}{y - y^2}$

7. $xy\, dx - (x^2 + 1)\, dy = 0$

8. $y' = x^3 y^5$

9. $(1 + x^2)\, dy + (y^2 + 1)\, dx = 0$

10. $y' = x^2 e^{-3y}$

11. $\sqrt{1 + x^2}\, dy + xy\, dx = 0$

12. $y^2\, dx = (1 - x)\, dy$

13. $(y^2 + 1)\, dx = (x^2 + 1)\, dy$

14. $y^3\, dx = x^3\, dy$

15. $(2 + y)\, dx + (x - 2)\, dy = 0$

16. $y' = \dfrac{e^{x-y}}{e^x + 1}$

17. $(x - xy^2)\, dx = -(x^2 y + y)\, dy$

With Exponential Functions

18. $dy = e^{-x}\, dx$

19. $ye^{2x} = (1 + e^{2x})y'$

20. $e^y(y' + 1) = 1$

21. $e^{x-y}\, dx + e^{y-x}\, dy = 0$

With Trigonometric Functions

22. $(3 + y)\, dx + \cot x\, dy = 0$

23. $\tan y\, dx + (1 + x)\, dy = 0$

24. $\tan y\, dx + \tan x\, dy = 0$

25. $\cos x \sin y\, dy + \sin x \cos y\, dx = 0$

26. $\sin x \cos^2 y\, dx + \cos^2 x\, dy = 0$

27. $4 \sin x \sec y\, dx = \sec x\, dy$

Particular Solution

Using the given boundary condition, find the particular solution to each differential equation. Try some by calculator.

28. $x\, dx = 4y\, dy$, $x = 5$ when $y = 2$

29. $y^2 y' = x^2$, $x = 0$ when $y = 1$

30. $\sqrt{x^2 + 1}\, y' + 3xy^2 = 0$, $x = 1$ when $y = 1$

31. $y' \sin y = \cos x$, $x = \pi/2$ when $y = 0$

32. $x(y + 1)y' = y(1 + x)$, $x = 1$ when $y = 1$

30–4 Exact First-Order DE

When the left side of a first-order differential equation is the exact differential of some function, we call that equation an *exact differential equation*. Even if we cannot separate the variables in such a differential equation, we might still solve it by integrating a *combination* of terms.

◆◆◆ **Example 19:** Solve $y\,dx + x\,dy = x\,dx$.

Solution: The variables are not now separated, and we see that no amount of manipulation will separate them. However, the combination of terms $y\,dx + x\,dy$ on the left side may ring a bell. In fact, it is the derivative of the product of x and y.

$$\frac{d(xy)}{dx} = x\frac{dy}{dx} + y\frac{dx}{dx} \qquad \text{(Eq. 259)}$$

or

$$d(xy) = x\,dy + y\,dx$$

This, then, is the left side of our given equation. The right side contains only x's, so we integrate.

$$\int (y\,dx + x\,dy) = \int x\,dx$$

$$\int d(xy) = \int x\,dx$$

$$xy = \frac{x^2}{2} + C$$

or

$$y = \frac{x}{2} + \frac{C}{x} \qquad\qquad ◆◆◆$$

Integrable Combinations

The expression $y\,dx + x\,dy$ from Example 19 is called an *integrable combination*. Some of the most frequently used combinations are as follows:

Integrable Combinations		
$x\,dy + y\,dx = d(xy)$		322
$\dfrac{x\,dy - y\,dx}{x^2} = d\left(\dfrac{y}{x}\right)$		323
$\dfrac{y\,dx - x\,dy}{y^2} = d\left(\dfrac{x}{y}\right)$		324
$\dfrac{x\,dy - y\,dx}{x^2 + y^2} = d\left(\tan^{-1}\dfrac{y}{x}\right)$		325

◆◆◆ **Example 20:** Solve $dy/dx = y(1 - xy)/x$.

Solution: We go to differential form and clear denominators by multiplying through by $x\,dx$, getting $x\,dy = y(1 - xy)\,dx$. Removing parentheses, we obtain

$$x\,dy = y\,dx - xy^2\,dx$$

On the lookout for an integrable combination, we move the $y\,dx$ term to the left side.

$$x\,dy - y\,dx = -xy^2\,dx$$

We see now that the left side will be the differential of x/y if we multiply by $-1/y^2$. Any expression that we multiply by (such as $-1/y^2$ here) to make our equation exact is called an *integrating factor.*

$$\frac{y\,dx - x\,dy}{y^2} = x\,dx$$

Integrating gives $x/y = x^2/2 + C_1$, or

$$y = \frac{2x}{x^2 + C} \qquad \text{◆◆◆}$$

A certain amount of trial-and-error work may be needed to get the DE into a suitable form.

Particular Solution

As before, we substitute boundary conditions to evaluate the constant of integration.

◆◆◆ Example 21: Solve $2xy\,dy - 4x\,dx + y^2\,dx = 0$ such that $x = 1$ when $y = 2$.

Solution: It looks as though the first and third terms might form an integrable combination, so we transpose the $-4x\,dx$.

$$2xy\,dy + y^2\,dx = 4x\,dx$$

The left side is the derivative of the product xy^2.

$$d(xy^2) = 4x\,dx$$

Integrating gives $xy^2 = 2x^2 + C$. Substituting the boundary conditions, we get $C = xy^2 - 2x^2 = 1(2)^2 - 2(1)^2 = 2$, so our particular solution is

$$xy^2 = 2x^2 + 2 \qquad \text{◆◆◆}$$

Exercise 4 ◆ Exact First-Order DE

Integrable Combinations

Find the general solution of each differential equation. Try some by calculator.

1. $y\,dx + x\,dy = 7\,dx$

2. $x\,dy = (4 - y)\,dx$

3. $x\dfrac{dy}{dx} = 3 - y$

4. $y + xy' = 9$

5. $2xy' = x - 2y$

6. $x\dfrac{dy}{dx} = 2x - y$

7. $x\,dy = (3x^2 + y)\,dx$

8. $(x + y)\,dx + x\,dy = 0$

9. $3x^2 + 2y + 2xy' = 0$

10. $(1 - 2x^2y)\dfrac{dy}{dx} = 2xy^2$

11. $(2x - y)y' = x - 2y$

12. $\dfrac{y\,dx - x\,dy}{y^2} = x\,dx$

13. $y\,dx - x\,dy = 2y^2\,dx$

14. $(x - 2x^2y)\,dy = y\,dx$

15. $(4y^3 + x)\dfrac{dy}{dx} = y$

16. $(y - x)y' + 2xy^2 + y = 0$

17. $3x - 2y^2 - 4xyy' = 0$

18. $\dfrac{x\,dy - y\,dx}{x^2 + y^2} = 5\,dy$

Using the given boundary condition, find the particular solution to each differential equation.

19. $4x = y + xy'$, $x = 3$ when $y = 1$

20. $y\,dx = (x - 2x^2y)\,dy$, $x = 1$ when $y = 2$

21. $y = (3y^3 + x)\dfrac{dy}{dx}$, $x = 1$ when $y = 1$

22. $4x^2 = -2y - 2xy'$, $x = 5$ when $y = 2$

23. $3x - 2y = (2x - 3y)\dfrac{dy}{dx}$, $x = 2$ when $y = 2$

24. $5x = 2y^2 + 4xyy'$, $x = 1$ when $y = 4$

30–5 First-Order Homogeneous DE

Recognizing a Homogeneous Differential Equation

If each variable in a function is replaced by t times the variable, and a power of t can be completely factored out, we say that the function is *homogeneous*. The power of t that can be factored out of the function is the *degree of homogeneity* of the function.

In other words, a function $f(x, y)$ is said to be homogeneous of degree n if

$$f(tx, ty) = t^n f(x, y)$$

◆◆◆ **Example 22:** Is $\sqrt{x^4 + xy^3}$ a homogeneous function?

Solution: We replace x with tx and y with ty and get

$$\sqrt{(tx)^4 + (tx)(ty)^3} = \sqrt{t^4x^4 + t^4xy^3} = t^2\sqrt{x^4 + xy^3}$$

Since we were able to completely factor out a t^2, we say that our function is homogeneous to the second degree. ◆◆◆

A homogeneous *polynomial* is one in which every term is of the same degree.

◆◆◆ **Example 23:**

(a) $x^2 + xy - y^2$ is a homogeneous polynomial of degree 2.

(b) $x^2 + x - y^2$ is not homogeneous. ◆◆◆

A first-order *homogeneous differential equation* is one of the following form:

Form of First-Order Homogeneous DE	$M\,dx + N\,dy = 0$	326

Here M and N are functions of x and y and are homogeneous functions of the same degree.

◆◆◆ **Example 24:**

(a) $(x^2 + y^2)\,dx + xy\,dy = 0$ is a first-order homogeneous differential equation.

(b) $(x^2 + y^2)\,dx + x\,dy = 0$ is not homogeneous. ◆◆◆

Identifying which equations are homogeneous and which are not allows us to pick the correct method for solving a given equation.

Solving a First-Order Homogeneous Differential Equation

Sometimes we can transform a homogeneous differential equation whose variables cannot be separated into one whose variables can be separated by making the substitution

$$y = vx$$

as explained in the following example.

◆◆◆ Example 25: Solve

$$x\frac{dy}{dx} - y = \sqrt{x^2 + y^2}$$

Solution: We first check if the given equation is homogeneous of first degree. We put the equation into the form of Eq. 326 by multiplying by dx and rearranging.

$$x\,dy - y\,dx = \sqrt{x^2 + y^2}\,dx$$

$$(\sqrt{x^2 + y^2} + y)\,dx - x\,dy = 0$$

This is now in the form of Eq. 326,

$$M\,dx + N\,dy = 0$$

with $M = \sqrt{x^2 + y^2} + y$ and $N = -x$. To test if M is homogeneous, we replace x by tx and y by ty.

$$\sqrt{(tx)^2 + (ty)^2} + ty = \sqrt{t^2x^2 + t^2y^2} + ty$$

$$= t\sqrt{x^2 + y^2} + ty$$

$$= t[\sqrt{x^2 + y^2} + y]$$

We see that t can be completely factored out and is of first degree. Thus M is homogeneous of first degree, as is N, so the given differential equation is homogeneous.

To solve, we make the substitution $y = vx$ to transform the given equation into one whose variables can be separated. However, when we substitute for y, we must also substitute for dy/dx. Let

$$y = vx$$

Then by the product rule

$$\frac{dy}{dx} = v + x\frac{dv}{dx}$$

Substituting into the given equation yields

$$x\left(v + x\frac{dv}{dx}\right) - vx = \sqrt{x^2 + v^2x^2}$$

which simplifies to

$$x\frac{dv}{dx} = \sqrt{1 + v^2}$$

Separating variables, we have

$$\frac{dv}{\sqrt{1 + v^2}} = \frac{dx}{x}$$

Integrating by Rule 62, we obtain

$$\ln|v + \sqrt{1 + v^2}| = \ln|x| + C_1$$

$$\ln\left|\frac{v + \sqrt{1 + v^2}}{x}\right| = C_1$$

$$\frac{v + \sqrt{1 + v^2}}{x} = e^{C_1}$$

$$v + \sqrt{1 + v^2} = Cx$$

where $C = e^{C_1}$. Now subtracting v from both sides, squaring, and simplifying gives

$$C^2x^2 - 2Cvx = 1$$

Finally, substituting back, $v = y/x$, we get

$$C^2x^2 - 2Cy = 1 \qquad \text{◆◆◆}$$

Exercise 5 ♦ First-Order Homogeneous DE

Find the general solution to each differential equation.

1. $(x - y)\,dx - 2x\,dy = 0$
2. $(3y - x)\,dx = (x + y)\,dy$
3. $(x^2 - xy)y' + y^2 = 0$
4. $(x^2 - xy)\,dy + (x^2 - xy + y^2)\,dx = 0$
5. $xy^2\,dy - (x^3 + y^3)\,dx = 0$
6. $2x^3y' + y^3 - x^2y = 0$

Using the given boundary condition, find the particular solution.

7. $x - y = 2xy'$, $x = 1$, $y = 1$
8. $3xy^2\,dy = (3y^3 - x^3)\,dx$, $x = 1$, $y = 2$
9. $(x^3 + y^3)\,dx - xy^2\,dy = 0$, $x = 1$, $y = 0$

30–6   First-Order Linear DE

When describing the degree of a term, we usually add the degrees of each variable in the term. Thus x^2y^3 is of fifth degree.

Sometimes, however, we want to describe the degree of a term with regard to just one of the variables. Thus we say that x^2y^3 is of the second degree in x and of the third degree in y.

In determining the degree of a term, we must also consider any derivatives in that term. We consider dy/dx to contribute one "degree" to y in this computation, d^2y/dx^2 to contribute two, and so on. Thus the term $xy\,dy/dx$ is of first degree in x and of second degree in y.

A first-order differential equation is called *linear* if each term is of first degree or less *in the dependent variable y*.

◆◆◆ Example 26:

(a) $y' + x^2y = e^x$ is linear.

(b) $y' + xy^2 = e^x$ is not linear because y is squared in the second term.

(c) $y\,dy/dx - xy = 5$ is not linear because we must add the exponents of y and dy/dx, making the first term of second degree.   ◆◆◆

A first-order linear differential equation can always be written in the following standard form:

Form of First-Order Linear DE	$\dfrac{dy}{dx} + Py = Q$	327

where P and Q are functions of x only.

◆◆◆ **Example 27:** Write the equation $xy' - e^x + y = xy$ in the form of Eq. 327.

Solution: Rearranging gives $xy' + y - xy = e^x$. Factoring, we have

$$xy' + (1 - x)y = e^x$$

Dividing by x gives us the standard form,

$$y' + \frac{1-x}{x}y = \frac{e^x}{x}$$

where $P = (1 - x)/x$ and $Q = e^x/x$. ◆◆◆

Integrating Factor

The left side of Eq. 327, a first-order linear differential equation, can always be made into an integrable combination by multiplying by an *integrating factor R*. We now find such a factor.

Multiplying Eq. 327 by R, the left side becomes

$$R\frac{dy}{dx} + yRP \tag{1}$$

Let us try to make the left side *the exact derivative of the product Ry of y and the integrating factor R*. The derivative of Ry is

$$R\frac{dy}{dx} + y\frac{dR}{dx} \tag{2}$$

But Eqs. (1) and (2) will be equal if $dR/dx = RP$. Since P is a function of x only, we can separate variables.

$$\frac{dR}{R} = P\,dx$$

Integrating, $\ln R = \int P\,dx$, which can be rewritten as follows:

Integrating Factor	$R = e^{\int P\,dx}$	328

We omit the constant of integration because we seek only one integrating factor.

Thus, *multiplying a given first-order linear equation by the integrating factor* $R = e^{\int P\,dx}$ *will make the left side of Eq. 327 the exact derivative of Ry*. We can then proceed to integrate to get the solution of the given differential equation.

◆◆◆ **Example 28:** Solve $\dfrac{dy}{dx} + \dfrac{4y}{x} = 3$.

Solution: Our equation is in standard form with $P = 4/x$ and $Q = 3$. Then

$$\int P\,dx = 4\int\frac{dx}{x} = 4\ln|x| = \ln x^4$$

Our integrating factor R is then

$$R = e^{\int P\,dx} = e^{\ln x^4} = x^4$$

Can you show why $e^{\ln x^4} = x^4$?

Multiplying our given equation by x^4 and going to differential form gives

$$x^4\,dy + 4x^3y\,dx = 3x^4\,dx$$

Notice that the left side is now the derivative of y times the integrating factor, $d(x^4y)$. Integrating, we obtain

$$x^4y = \frac{3x^5}{5} + C$$

or

$$y = \frac{3x}{5} + \frac{C}{x^4}$$ ◆◆◆

In summary, to solve the first-order linear equation

$$y' + Py = Q$$

multiply by an integrating factor $R = e^{\int P\,dx}$ and the solution to the equation becomes

$$yR = \int QR\,dx$$

which can be expressed as follows:

Solution to First-Order Linear DE	$ye^{\int P\,dx} = \int Qe^{\int P\,dx}dx$	329

Next we try an equation having trigonometric functions.

◆◆◆ Example 29: Solve $y' + y \cot x = \csc x$.

Solution: This is a first-order differential equation in standard form, with $P = \cot x$ and $Q = \csc x$. Thus $\int P\,dx = \int \cot x\,dx = \ln|\sin x|$. The integrating factor R is e raised to the $\ln|\sin x|$, or simply $\sin x$. Then, by Eq. 329, the solution to the differential equation is

$$y \sin x = \int \csc x \sin x\,dx$$

Since $\csc x \sin x = 1$, this simplifies to

$$y \sin x = \int dx$$

so

$$y \sin x = x + C \qquad\qquad ◆◆◆$$

Particular Solution

As before, we find the general solution and then substitute the boundary conditions to evaluate C.

◆◆◆ Example 30: Solve $y' + \left(\dfrac{1 - 2x}{x^2}\right)y = 1$ given that $y = 2$ when $x = 1$.

Solution: We are in standard form with $P = (1 - 2x)/x^2$. We find the integrating factor by first integrating $\int P\,dx$. This gives

$$P = x^{-2} - 2x^{-1}$$

$$\int P\,dx = -\frac{1}{x} - 2\ln|x|$$

$$= -\frac{1}{x} - \ln x^2$$

Our integrating factor is then

$$R = e^{\int P\,dx} = e^{-1/x - \ln x^2} = \frac{e^{-1/x}}{e^{\ln x^2}} = \frac{1}{x^2 e^{1/x}}$$

Substituting into Eq. 329, the general solution is

$$\frac{y}{x^2 \, e^{1/x}} = \int \frac{dx}{x^2 \, e^{1/x}}$$

$$= \int e^{-1/x} x^{-2} \, dx = e^{-1/x} + C$$

When $x = 1$ and $y = 2$, $C = 2/e - 1/e = 1/e$, so the particular solution is

$$\frac{y}{x^2 e^{1/x}} = e^{-1/x} + \frac{1}{e}$$

or

$$y = x^2 \left(1 + \frac{e^{1/x}}{e} \right) \qquad \blacklozenge\blacklozenge\blacklozenge$$

Equations Reducible to Linear Form

Sometimes a first-order differential equation that is *nonlinear* can be expressed in linear form. One type of nonlinear equation easily reducible to linear form is one in which the right side contains a power of y as a factor, an equation known as a *Bernoulli's equation*.

Bernoulli's Equation	$\dfrac{dy}{dx} - Gy = Hy^n$	330

Here G and H are functions of x only. We put a Bernoulli's equation into the form of a first-order linear differential equation (Eq. 327) by making the substitution

$$z = y^{1-n}$$

◆◆◆ **Example 31:** Solve the equation $y' + y/x = x^2 y^6$.

Solution: This is a Bernoulli's equation with

$$G = -1/x \qquad H = x^2 \qquad n = 6$$

If we let $z = y^{1-n} = y^{-5}$, the derivative of z with respect to x is then

$$\frac{dz}{dx} = -5y^{-6} \frac{dy}{dx}$$

$$= -5z^{6/5} \frac{dy}{dx}$$

since $y = z^{-1/5}$. Solving for dy/dx gives

$$\frac{dy}{dx} = -\frac{z^{-6/5}}{5} \frac{dz}{dx}$$

Substituting for y and dy/dx in the given equation we get

$$-\frac{z^{-6/5}}{5} \frac{dz}{dx} + \frac{z^{-1/5}}{x} = x^2 z^{-6/5}$$

Multiplying by -5 and dividing through by $z^{-6/5}$ gives

$$\frac{dz}{dx} - \frac{5}{x} z = -5x^2 \qquad\qquad (1)$$

This is now a first-order linear differential equation in z and matches the form of Eq. 327. We proceed to solve it as before, with $P = -5/x$ and $Q = -5x^2$. We now find the integrating factor. The integral of P is

$$\int \left(\frac{-5}{x} \right) dx = -5 \ln x = \ln x^{-5},$$

so the integrating factor is

$$R = e^{\int P dx} = x^{-5} = \frac{1}{x^5}$$

From Eq. 329,

$$\frac{z}{x^5} = -5 \int x^{-3} \, dx = \frac{-5x^{-2}}{-2} + C$$

This is the solution of the differential equation, Eq. (1). We now solve for z and then substitute back $z = 1/y^5$,

$$z = \frac{5}{2} x^3 + Cx^5 = \frac{1}{y^5}$$

Our solution of the given DE is then

$$y^5 = \frac{1}{5x^3/2 + Cx^5}$$

or

$$y^5 = \frac{2}{5x^3 + C_1 x^5}$$ ◆◆◆

Summary

To solve a first-order differential equation,

1. If you can *separate the variables*, integrate term by term.
2. If the DE is exact (contains an *integrable combination*), isolate that combination on one side of the equation and take the integral of both sides.
3. If the DE is *homogeneous,* start by substituting $y = vx$ and $dy/dx = v + x \, dv/dx$.
4. If the DE is *linear,* $y' + Py = Q$, multiply by the integrating factor $R = e^{\int P \, dx}$, and the solution will be $y = (1/R) \int QR \, dx$.
5. If the DE is a *Bernoulli's equation,* $y' + Py = Qy^n$, substitute $z = y^{1-n}$ to make it first-order linear, and proceed as in step 4.
6. If *boundary conditions* are given, substitute them to evaluate the constant of integration.

Exercise 6 ◆ First-Order Linear DE

Find the general solution to each differential equation.

1. $y' + \dfrac{y}{x} = 4$

2. $y' + \dfrac{y}{x} = 3x$

3. $xy' = 4x^3 - y$

4. $\dfrac{dy}{dx} + xy = 2x$

5. $y' = x^2 - x^2 y$

6. $y' - \dfrac{y}{x} = \dfrac{-1}{x^2}$

7. $y' = \dfrac{3 - xy}{2x^2}$

8. $y' = x + \dfrac{2y}{x}$

9. $xy' = 2y - x$

10. $(x + 1)y' - 2y = (x + 1)^4$

11. $y' = \dfrac{2 - 4x^2 y}{x + x^3}$

12. $(x + 1)y' = 2(x + y + 1)$

13. $xy' + x^2 y + y = 0$

14. $(1 + x^3) \, dy = (1 - 3x^2 y) \, dx$

15. $y' + y = e^x$

16. $y' = e^{2x} + y$

17. $y' = 2y + 4e^{2x}$

18. $xy' - e^x + y + xy = 0$

19. $y' = \dfrac{4 \ln x - 2x^2 y}{x^3}$

With Trigonometric Expressions

20. $y' + y \sin x = 3 \sin x$

21. $y' + y = \sin x$

22. $y' + 2xy = 2x \cos x^2$

23. $y' = 2 \cos x - y$

24. $y' = \sec x - y \cot x$

Bernoulli's Equation

25. $y' + \dfrac{y}{x} = 3x^2 y^2$

26. $xy' + x^2 y^2 + y = 0$

27. $y' = y - xy^2(x + 2)$

28. $y' + 2xy = xe^{-x^2} y^3$

Particular Solution

Using the given boundary condition, find the particular solution to each differential equation.

29. $xy' + y = 4x$, $x = 1$ when $y = 5$

30. $\dfrac{dy}{dx} + 5x = x - xy$, $x = 2$ when $y = 1$

31. $y' + \dfrac{y}{x} = 5$, $x = 1$ when $y = 2$

32. $y' = 2 + \dfrac{3y}{x}$, $x = 2$ when $y = 6$

33. $y' = \tan^2 x + y \cot x$, $x = \dfrac{\pi}{4}$ when $y = 2$

34. $\dfrac{dy}{dx} + 5y = 3e^x$, $x = 1$ when $y = 1$

30–7 Geometric Applications of First-Order DEs

Now that we are able to solve some simple differential equations of first order, we turn to applications. Here we not only must solve the equation but also must first *set up* the differential equation. The geometric problems we do first will help to prepare us for the physical applications that follow.

Setting Up a Differential Equation

When reading the problem statement, look for the words "slope" or "rate of change." Each of these can be represented by the first derivative.

◆◆◆ Example 32:

(a) The statement "the slope of a curve at every point is equal to twice the ordinate" is represented by the differential equation

$$\frac{dy}{dx} = 2y$$

(b) The statement "the ratio of abscissa to ordinate at each point on a curve is proportional to the rate of change at that point" can be written

$$\frac{x}{y} = k\frac{dy}{dx}$$

(c) The statement "the slope of a curve at every point is inversely proportional to the square of the ordinate at that point" can be described by the equation

$$\frac{dy}{dx} = \frac{k}{y^2} \qquad \qquad \blacklozenge\blacklozenge\blacklozenge$$

Finding an Equation Whose Slope Is Specified

Once the equation is written, it is solved by the methods of the preceding sections.

♦♦♦ **Example 33:** The slope of a curve at each point is one-tenth the product of the ordinate and the square of the abscissa, and the curve passes through the point (2, 3). Find the equation of the curve.

Solution: The differential equation is, from the problem statement,

$$\frac{dy}{dx} = \frac{x^2 y}{10}$$

In solving a differential equation, we first see if the variables can be separated. In this case they can be.

$$\frac{10\, dy}{y} = x^2 dx$$

Integrating gives us

$$10 \ln |y| = \frac{x^3}{3} + C$$

At (2, 3),

$$C = 10 \ln 3 - \frac{8}{3} = 8.32$$

Our curve thus has the equation $\ln |y| = x^3/30 + 0.832$. ♦♦♦

Tangents and Normals to Curves

In setting problems involving tangents and normals to curves, recall that

$$\text{slope of the tangent} = \frac{dy}{dx} = y'$$

$$\text{slope of the normal} = -\frac{1}{y'}$$

♦♦♦ **Example 34:** A curve passes through the point (4, 2), as shown in Fig. 30–5. If from any point P on the curve, the line OP and the tangent PT are drawn, the triangle OPT is isosceles. Find the equation of the curve.

Solution: The slope of the tangent is dy/dx, and the slope of OP is y/x. Since the triangle is isosceles, these slopes must be equal but of opposite signs.

$$\frac{dy}{dx} = -\frac{y}{x}$$

We can solve this equation by separation of variables or as the integrable combination $x\, dy + y\, dx = 0$. Either way gives the hyperbola

$$xy = C$$

(see Eq. 248). At the point (4, 2), $C = 4(2) = 8$. So our equation is $xy = 8$. ♦♦♦

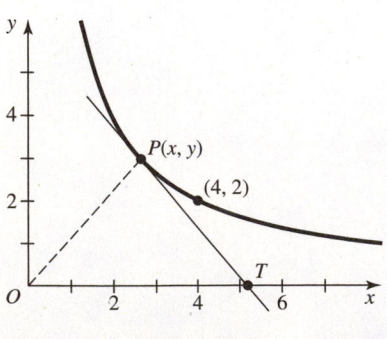

FIGURE 30–5

Orthogonal Trajectories

If we graph a relation that has an arbitrary constant, we get a *family of curves*. For example, the relation $x^2 + y^2 = C^2$ represents a family of circles of radius C, whose center is at the origin. Another curve that cuts each curve of the family at right angles is called an *orthogonal trajectory* to that family.

To find the orthogonal trajectory to a family of curves,

(1) Differentiate the equation of the family to get the slope.
(2) Eliminate the constant contained in the original equation.
(3) Take the negative reciprocal of the slope to get the slope of the orthogonal trajectory.
(4) Solve the resulting differential equation to get the equation of the orthogonal trajectory.

◆◆◆ **Example 35:** Find the equation of the orthogonal trajectories to the parabolas $y^2 = px$.

Solution:

(1) The derivative is $2yy' = p$, so

$$y' = \frac{p}{2y} \tag{1}$$

(2) The constant p, from the original equation, is y^2/x. Substituting y^2/x for p in Equation (1) gives $y' = y/2x$.

Common Error	Be sure to eliminate the constant (p in this example) before continuing.

(3) The slope of the orthogonal trajectory is, by Eq. 214, the negative reciprocal of the slope of the given family.

$$y' = \frac{-2x}{y}$$

(4) Separating variables yields

$$y \, dy = -2x \, dx$$

Integrating gives the solution,

$$\frac{y^2}{2} = \frac{-2x^2}{2} + C_1$$

which is a family of ellipses, $2x^2 + y^2 = C$ (Fig. 30–6).   ◆◆◆

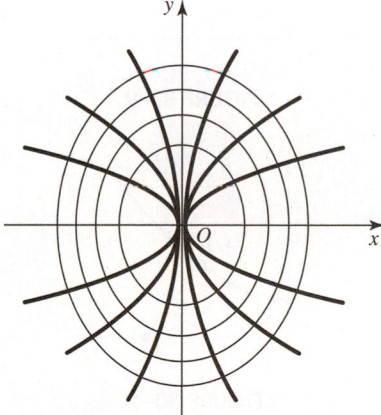

FIGURE 30–6 Orthogonal ellipses and parabolas. Each intersection is at 90°.

Exercise 7 ◆ Geometric Applications of First-Order DEs

Slope of Curves

1. Find the equation of the curve that passes through the point $(2, 9)$ and whose slope is $y' = x + 1/x + y/x$.
2. The slope of a certain curve at any point is equal to the reciprocal of the ordinate at the point. Write the equation of the curve if it passes through the point $(1, 3)$.

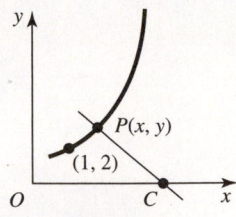

FIGURE 30–7

The equations for the various distances associated with tangents and normals were given in the Review Problems for our chapter on analytic geometry.

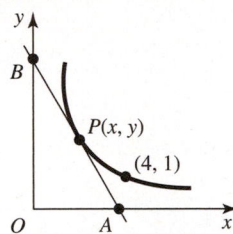

FIGURE 30–8

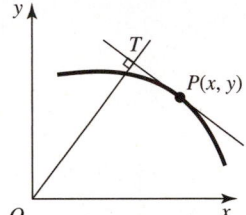

FIGURE 30–9

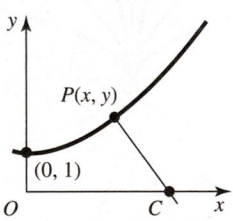

FIGURE 30–10

3. Find the equation of a curve whose slope at any point is equal to the abscissa of that point divided by the ordinate and which passes through the point (3, 4).

4. Find the equation of a curve that passes through (1, 1) and whose slope at any point is equal to the product of the ordinate and abscissa.

5. A curve passes through the point (2, 3) and has a slope equal to the sum of the abscissa and ordinate at each point. Find its equation.

Tangents and Normals

6. The distance to the x intercept C of the normal (Fig. 30–7) is given by $OC = x + y'y$. Write the equation of the curve passing through (1, 2) for which OC is equal to three times the abscissa of P.

7. A certain first-quadrant curve (Fig. 30–8) passes through the point (4, 1). If a tangent is drawn through any point P, the portion AB of the tangent that lies between the coordinate axes is bisected by P. Find the equation of the curve, given that

$$AP = -\left(\frac{y}{y'}\right)\sqrt{1 + (y')^2} \quad \text{and} \quad BP = x\sqrt{1 + (y')^2}$$

8. A tangent PT is drawn to a curve at a point P (Fig. 30–9). The distance OT from the origin to a tangent through P is given by

$$OT = \frac{xy' - y}{\sqrt{1 + (y')^2}}$$

Find the equation of the curve passing through the point (2, 4) so that OT is equal to the abscissa of P.

9. Find the equation of the curve passing through (0, 1) for which the length PC of a normal through P equals the square of the ordinate of P (Fig. 30–10), where

$$PC = y\sqrt{1 + (y')^2}$$

10. Find the equation of the curve passing through (4, 4) such that the distance OT (Fig. 30–9) is equal to the ordinate of P.

Orthogonal Trajectories

Write the equation of the orthogonal trajectories to each family of curves.

11. the circles, $x^2 + y^2 = r^2$
12. the parabolas, $x^2 = ay$
13. the hyperbolas, $x^2 - y^2 = Cy$

30–8 Exponential Growth and Decay

In our chapter on the exponential function we derived Eqs. 151, 153, and 154 for exponential growth and decay and exponential growth to an upper limit by means of compound interest formulas. Here we derive the equation for exponential growth to an upper limit by solving the differential equation that describes such growth. The derivations of equations for exponential growth and decay are left as an exercise.

◆◆◆ **Example 36:** A quantity starts from zero and grows with time such that its rate of growth is proportional to the difference between the final amount a and the present amount y. Find an equation for y as a function of time.

Solution: The amount present at time t is y, and the rate of growth of y we write as dy/dt. Since the rate of growth is proportional to $(a - y)$, we write the differential equation

$$\frac{dy}{dt} = n(a - y)$$

where n is a constant of proportionality. Separating variables, we obtain

$$\frac{dy}{a - y} = n\,dt$$

Integrating gives us

$$-\ln(a - y) = nt + C$$

Going to exponential form and simplifying yields

$$a - y = e^{-nt-C} = e^{-nt}e^{-C} = C_1 e^{-nt}$$

where $C_1 = e^{-C}$. Applying the initial condition that $y = 0$ when $t = 0$ gives $C_1 = a$, so our equation becomes $a - y = ae^{-nt}$, which can be rewritten as follows:

Exponential Growth to an Upper Limit	$y = a(1 - e^{-nt})$	154

So we have just verified, by setting up and solving a differential equation, the expression for exponential growth to an upper limit that we got in an earlier chapter from the compound interest formula   ◆◆◆

Motion in a Resisting Fluid

Here we continue our study of motion. We earlier showed that the instantaneous velocity is given by the derivative of the displacement, and that the instantaneous acceleration is given by the derivative of the velocity (or by the second derivative of the displacement). Later we solved simple differential equations to find displacement given the velocity or acceleration. Here we do a type of problem that we were not able to solve then.

In this type of problem, an object falls through a fluid (usually air or water) which exerts a *resisting force* that is proportional to the velocity of the object and in the opposite direction. We set up these problems using Newton's second law, $F = ma$, and we'll see that the motion follows the law for exponential growth described by Eq. 154.

◆◆◆ **Example 37:** A crate falls from rest from an airplane. The air resistance is proportional to the crate's velocity, and the crate reaches a limiting speed of 218 ft/s.

(a) Write an equation for the crate's velocity.
(b) Find the crate's velocity after 0.75 s.

Solution:

(a) By Newton's second law,

$$F = ma = \frac{W}{g}\frac{dv}{dt}$$

where W is the weight of the crate, dv/dt is the acceleration, and $g = 32.2$ ft/s^2. Taking the downward direction as positive, the resultant force F is equal to $W - kv$, where k is a constant of proportionality. So

$$W - kv = \frac{W}{32.2}\frac{dv}{dt}$$

We can find k by noting that the acceleration must be zero when the limiting speed (218 ft/s) is reached. Thus

$$W - 218k = 0$$

So $k = W/218$. Our differential equation, after multiplying by $218/W$, is then $218 - v = 6.77\,dv/dt$. Separating variables and integrating, we have

$$\frac{dv}{218 - v} = 0.148 \, dt$$

$$\ln|218 - v| = -0.148t + C$$

$$218 - v = e^{-0.148t + C} = e^{-0.148t}e^{C}$$

$$= C_1 e^{-0.148t}$$

Notice that the weight W has dropped out. Since $v = 0$ when $t = 0$, $C_1 = 218 - 0 = 218$. Then

$$v = 218(1 - e^{-0.148t})$$

We'll have more problems involving exponential growth and decay in the following section on electric circuits.

(b) Note that our equation for v is of the same form as Eq. 154 for exponential growth to an upper limit. When $t = 0.75$ s,

$$v = 218(1 - e^{-0.111}) = 22.9 \text{ ft/s} \qquad \blacklozenge\blacklozenge\blacklozenge$$

Exercise 8 ◆ Exponential Growth and Decay

Exponential Growth

1. A quantity grows with time such that its rate of growth dy/dt is proportional to the present amount y. Use this statement to derive the equation for exponential growth, $y = ae^{nt}$.

2. A biomedical company finds that a certain bacterium used for crop insect control will grow exponentially at the rate of 12.0% per hour. Starting with 1000 bacteria, how many will the company have after 10.0 h?

3. If the U.S. energy consumption in 2000 was 158 million barrels (bbl) per day oil equivalent and is growing exponentially at a rate of 6.9% per year, estimate the daily oil consumption in the year 2020.

Exponential Decay

4. A quantity decreases with time such that its rate of decrease dy/dt is proportional to the present amount y. Use this statement to derive the equation for exponential decay, $y = ae^{-nt}$.

5. An iron ingot is 1850°F above room temperature. If it cools exponentially at 3.50% per minute, find its temperature (above room temperature) after 2.50 h.

6. A certain pulley in a tape drive is rotating at 2550 rev/min. After the power is shut off, its speed decreases exponentially at a rate of 12.5% per second. Find the pulley's speed after 5.00 s.

Exponential Growth to an Upper Limit

7. A forging, initially at 0°F, is placed in a furnace at 1550°F, where its temperature rises exponentially at the rate of 6.50% per minute. Find its temperature after 25.0 min.

8. If we assume that the compressive strength of concrete increases exponentially with time to an upper limit of 4000 lb/in.^{2}, and that the rate of increase is 52.5% per week, find the strength after 2 weeks.

Motion in a Resisting Medium

9. A 45.5-lb carton is initially at rest. It is then pulled horizontally by a 19.4-lb force in the direction of motion and is resisted by a frictional force which is equal (in pounds) to four times the carton's velocity (in ft/s). Show that the differential equation of motion is $dv/dt = 13.7 - 2.83v$.

10. For the carton in problem 9, find the velocity after 1.25 s.

11. An instrument package is dropped from an airplane. It falls from rest through air whose resisting force is proportional to the speed of the package. The terminal speed is 155 ft/s. Show that the acceleration is given by the differential equation $a = dv/dt = g - gv/155$.

12. Find the speed of the instrument package in problem 11 after 0.50 s.

13. Find the displacement of the instrument package in problem 11 after 1.00 s.

14. A 157-lb stone falls from rest from a cliff. If the air resistance is proportional to the square of the stone's speed, and the limiting speed of the stone is 125 ft/s, show that the differential equation of motion is $dv/dt = g - gv^2/15{,}625$.

15. Find the time for the velocity of the stone in problem 14 to be 60.0 ft/s.

16. A 15.0-lb ball is thrown downward from an airplane with a speed of 21.0 ft/s. If we assume the air resistance to be proportional to the ball's speed, and the limiting speed is 135 ft/s, show that the velocity of the ball is given by $v = 135 - 114e^{-t/4.19}$ ft/s.

17. Find the time at which the ball in problem 16 is going at a speed of 70.0 ft/s.

18. A box falls from rest and encounters air resistance proportional to the cube of the speed. The limiting speed is 12.5 ft/s. Show that the acceleration is given by the differential equation $60.7\, dv/dt = 1953 - v^3$.

30–9 Series *RL* and *RC* Circuits

Series *RL* Circuit

Figure 30–11 shows a resistance of R ohms in series with an inductance of L henrys. The switch can connect these elements either to a battery of voltage E (position 1, charge) or to a short circuit (position 2, discharge). In either case, our objective is to find the current i in the circuit. We will see that it is composed of two parts: a *steady-state* current that flows long after the switch has been thrown, and a *transient* current that dies down shortly after the switch is thrown.

◆◆◆ **Example 38:** *Inductor Charging:* After being in position 2 for a long time, the switch in Fig. 30–11 is thrown into position 1 at $t = 0$. Write an expression for (a) the current i and (b) the voltage across the inductor.

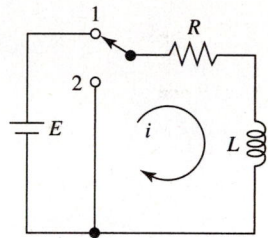

FIGURE 30–11 *RL* circuit.

Solution:

(a) The voltage v_L across an inductance L is given by Eq. 1086, $v_L = L\, di/dt$. Using Kirchhoff's voltage law (Eq. 1067) gives

$$Ri + L\frac{di}{dt} = E \tag{1}$$

We separate variables,

$$L\, di = (E - Ri)\, dt$$
$$dt = \frac{L\, di}{E - Ri}$$

Taking the integral of both sides, with an initial current of 0 at $t = 0$,

$$\int_0^t dt = \int_0^i \frac{L\, di}{E - Ri} = -\frac{L}{R}\int_0^i \frac{di}{i - E/R}$$

Note that we have used a definite integral here. We could also use an indefinite integral and later evaluate the constant of integration by substituting the initial conditions.

The right side is of the form $\int du/u$ (Integral No. 7). Integrating,

$$t = -\frac{L}{R}\Big[\ln(i - E/R)\Big]_0^i = -\frac{L}{R}\big[\ln(i - E/R) - \ln(0 - E/R)\big] - \frac{Rt}{L}$$
$$= \ln\left(\frac{i - E/R}{-E/R}\right) = \ln\left(1 - \frac{Ri}{E}\right)$$

Switching from logarithmic to exponential form gives

$$1 - \frac{Ri}{E} = e^{-Rt/L}$$

From which we get

| Current in a Charging Inductor | $i = \dfrac{E}{R}(1 - e^{-Rt/L})$ | 1087 |

The first term in this expression (E/R) is the steady-state current, and the second term $(E/R)\,(e^{-Rt/L})$ is the transient current.

(b) From Eq. (1), the voltage across the inductor is

$$v_L = L\frac{di}{dt} = E - Ri$$

Using Eq. 1087, we see that $Ri = E - Ee^{-Rt/L}$. Then

$$v_L = E - E + Ee^{-Rt/L}$$

Thus:

| Voltage Across a Charging Inductor | $v_L = Ee^{-Rt/L}$ | 1089 |

Note that the equation for the current is the same as that for exponential growth to an upper limit (Eq. 154), and that the equation for the voltage across the inductor is of the same form as for exponential decay (Eq. 153).

◆◆◆

Series *RC* Circuit

We now analyze the *RC* circuit as we did the *RL* circuit.

◆◆◆ **Example 39:** *Capacitor Discharging:* A fully charged capacitor (Fig. 30–12) is discharged by throwing the switch from 1 to 2 at $t = 0$. Write an expression for (a) the voltage across the capacitor and (b) the current i.

Solution:

(a) If the voltage across the capacitor is v, then the voltage across the resistor must be $-v$, since the sum of the voltages around the loop must be zero. Since the current $(-v/R)$ through the resistor must equal the current $(C\,dv/dt)$ through the capacitor we write

$$-\frac{v}{R} = C\frac{dv}{dt}$$

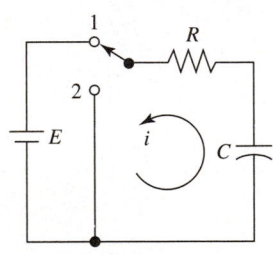

FIGURE 30–12   *RC* circuit.

Separating variables and integrating gives

$$\frac{dv}{v} = -\frac{dt}{RC}$$

$$\ln v = -\frac{t}{RC} + k$$

Recall that in electrical problems, we will use k for the constant of integration, saving C for capacitance. Similarly, R here is resistance, not the integrating constant.

At $t = 0$ the voltage across the capacitor is the battery voltage E, so $k = \ln E$. Substituting, we obtain

$$\ln v - \ln E = \ln\frac{v}{E} = -\frac{t}{RC}$$

Or, in exponential form, $v/E = e^{-t/RC}$, which is also expressed as follows:

Voltage Across a Discharging Capacitor	$v = Ee^{-t/RC}$	**1084**

(b) We get the current through the resistor (and the capacitor) by dividing the voltage v by R.

Current in a Discharging Capacitor	$i = \dfrac{E}{R} e^{-t/RC}$	**1082**

Equations 1082 and 1084 are both for exponential decay.

◆◆◆

◆◆◆ **Example 40:** For the circuit of Fig. 30–12, $R = 1540\ \Omega$, $C = 125\ \mu\text{F}$, and $E = 115$ V. If the switch is thrown from position 1 to position 2 at $t = 0$, find the current and the voltage across the capacitor at $t = 60$ ms.

Solution: We first compute $1/RC$.

$$\frac{1}{RC} = \frac{1}{1540 \times 125 \times 10^{-6}} = 5.19$$

Then, from Eqs. 1082 and 1084,

$$i = \frac{E}{R} e^{-t/RC} = \frac{115}{1540} e^{-5.19t}$$

and

$$v = Ri = Ee^{-t/RC} = 115e^{-5.19t}$$

At $t = 0.060$ s, $e^{-5.19t} = 0.732$, so

$$i = \frac{115}{1540}(0.732) = 0.0547\ \text{A} = 54.7\ \text{mA}$$

and

$$v = 115(0.732) = 84.2\ \text{V}$$

◆◆◆

Alternating Source

We now consider the case where the *RL* circuit or *RC* circuit is connected to an alternating rather than a direct source of voltage.

◆◆◆ **Example 41:** A switch (Fig. 30–13) is closed at $t = 0$, thus applying an alternating voltage of amplitude E to a resistor and an inductor in series. Write an expression for the current.

Solution: By Kirchhoff's voltage law, $Ri + L\,di/dt = E \sin \omega t$, or

$$\frac{di}{dt} + \frac{R}{L} i = \frac{E}{L} \sin \omega t$$

This is a first-order linear differential equation. Our integrating factor is $e^{\int R/L\, dt} = e^{Rt/L}$. Thus

$$ie^{Rt/L} = \frac{E}{L} \int e^{Rt/L} \sin \omega t \, dt$$

$$= \frac{E}{L} \frac{e^{Rt/L}}{(R^2/L^2 + \omega^2)} \left(\frac{R}{L} \sin \omega t - \omega \cos \omega t \right) + k$$

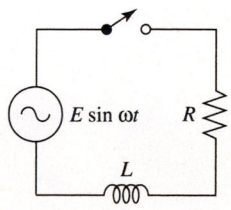

FIGURE 30–13 *RL* circuit with ac source.

by Rule 41. We now divide through by $e^{Rt/L}$ and after some manipulation get

$$i = E \cdot \frac{R \sin \omega t - \omega L \cos \omega t}{R^2 + \omega^2 L^2} + ke^{-Rt/L}$$

From the impedance triangle (Fig. 30–14), we see that $R^2 + \omega^2 L^2 = Z^2$, the square of the impedance. Further, by Ohm's law for ac, $E/Z = I$, the amplitude of the current wave. Thus

$$i = \frac{E}{Z}\left(\frac{R}{Z} \sin \omega t - \frac{\omega L}{Z} \cos \omega t\right) + ke^{-Rt/L}$$

$$= I\left(\frac{R}{Z} \sin \omega t - \frac{\omega L}{Z} \cos \omega t\right) + ke^{-Rt/L}$$

Again from the impedance triangle, $R/Z = \cos \phi$ and $\omega L/Z = \sin \phi$. Substituting yields

$$i = I(\sin \omega t \cos \phi - \cos \omega t \sin \phi) + ke^{-Rt/L}$$

$$= I \sin(\omega t - \phi) + ke^{-Rt/L}$$

which we get by means of the trigonometric identity for the sine of the difference of two angles (Eq. 128). Evaluating k, we note that $i = 0$ when $t = 0$, so

$$k = -I \sin(-\phi) = I \sin \phi = \frac{IX_L}{Z}$$

where, from the impedance triangle, $\sin \phi = X_L/Z$. Substituting, we obtain

$$i = \underbrace{I \sin(\omega t - \phi)}_{\substack{\text{steady-state} \\ \text{current}}} + \underbrace{\frac{IX_L}{Z}e^{-Rt/L}}_{\substack{\text{transient} \\ \text{current}}}$$

Our current thus has two parts: (1) a steady-state alternating current of magnitude I, out of phase with the applied voltage by an angle ϕ; and (2) a transient current with an initial value of IX_L/Z, which decays exponentially.   ◆◆◆

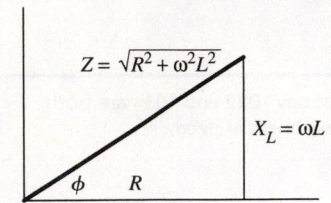

FIGURE 30–14   Impedance triangle.

$Z = \sqrt{R^2 + \omega^2 L^2}$

$X_L = \omega L$

ϕ R

Exercise 9 ◆ Series RL and RC Circuits

Series RL Circuit

1. If the inductor in Fig. 30–11 is discharged by throwing the switch from position 1 to 2, show that the current decays exponentially according to the function $i = (E/R)e^{-Rt/L}$.

2. The voltage across an inductor is equal to $L\, di/dt$. Show that the magnitude of the voltage across the inductance in problem 1 decays exponentially according to the function $v = Ee^{-Rt/L}$.

3. We showed that when the switch in Fig. 30–11 is thrown from 2 to 1 (charging), the current grows exponentially to an upper limit and is given by $i = (E/R)(1 - e^{-Rt/L})$. Show that the voltage across the inductance ($L\, di/dt$) decays exponentially and is given by $v = Ee^{-Rt/L}$.

4. For the circuit of Fig. 30–11, $R = 382\ \Omega$, $L = 4.75\ H$, and $E = 125\ V$. If the switch is thrown from 2 to 1 (charging), find the current and the voltage across the inductance at $t = 2.00\ ms$.

Series RC Circuit

5. The voltage v across the capacitor in Fig. 30–12 during charging is described by the differential equation $(E - v)/R = C\, dv/dt$. Solve this differential equation to show that the voltage is given by $v = E(1 - e^{-t/RC})$. (*Hint:* Write the given equation in the form of a first-order linear differential equation, and solve using Eq. 329.)

6. Show that in problem 5 the current through the resistor (and hence through the capacitor) is given by $i = (E/R)e^{-t/RC}$.

7. For the circuit of Fig. 30–12, $R = 538\ \Omega$, $C = 525\ \mu\text{F}$, and $E = 125$ V. If the switch is thrown from 2 to 1 (charging), find the current and the voltage across the capacitor at $t = 2.00$ ms.

Circuits in Which *R*, *L*, or *C* Is Not Constant

8. For the circuit of Fig. 30–11, $L = 2.00$ H, $E = 60.0$ V, and the resistance decreases with time according to the expression

$$R = 4.00/(t + 1)$$

Show that the current i is given by

$$i = 10(t + 1) - 10(t + 1)^{-2}$$

9. For the circuit in problem 8, find the current at $t = 1.55$ ms.

10. For the circuit of Fig. 30–11, $R = 10.00\ \Omega$, $E = 100$ V, and the inductance varies with time according to the expression $L = 5.00t + 2.00$. Show that the current i is given by the expression $i = 10.0 - 40/(5t + 2)^2$.

11. For the circuit in problem 10, find the current at $t = 4.82$ ms.

12. For the circuit of Fig. 30–11, $E = 300$ V, the resistance varies with time according to the expression $R = 4.00t$, and the inductance varies with time according to the expression $L = t^2 + 4.00$. Show that the current i as a function of time is given by $i = 100t(t^2 + 12)/(t^2 + 4)^2$.

13. For the circuit in problem 12, find the current at $t = 1.85$ ms.

14. For the circuit of Fig. 30–12, $C = 2.55\ \mu\text{F}$, $E = 625$ V, and the resistance varies with current according to the expression $R = 493 + 372i$. Show that the differential equation for current is $di/dt + 1.51i\ di/dt + 795i = 0$.

Series *RL* or *RC* Circuit with Alternating Current

15. For the circuit of Fig. 30–13, $R = 233\ \Omega$, $L = 5.82$ H, and $E = 58.0 \sin 377t$ V. If the switch is closed when E is zero and increasing, show that the current is given by $i = 26.3 \sin(377t - 83.9°) + 26.1e^{-40t}$ mA.

16. For the circuit in problem 15, find the current at $t = 2.00$ ms.

17. For the circuit in Fig. 30–12, the applied voltage is alternating and is given by $E = E_{max} \sin \omega t$. If the switch is thrown from 2 to 1 (charging) when e is zero and increasing, show that the current is given by

$$i = (E_{max}/Z)[\sin(\omega t + \phi) - e^{-t/RC}\sin \phi]$$

In your derivation, follow the steps used for the *RL* circuit with an ac source.

18. For the circuit in problem 17, $R = 837\ \Omega$, $C = 2.96\ \mu\text{F}$, and $E = 58.0 \sin 377t$. Find the current at $t = 1.00$ ms.

19. *Project:* Using the methods of Sec. 30–9 as a guide, derive the formula for the voltage across a capacitor in a series *RC* circuit, when charging.

Voltage Across a Capacitor When Charging	$v = E(1 - e^{-t/RC})$	**1083**

20. *Project:* Using the methods of Sec. 30–9 as a guide, derive the formula for the current across an inductor in a series *RL* circuit, when discharging.

Current in an Inductor When Discharging	$i = \dfrac{E}{R}e^{-Rt/L}$	**1088**

••• CHAPTER 30 REVIEW PROBLEMS •••••••••••••••••••••••••••••

Find the general solution to each first-order differential equation.

1. $xy + y + xy' = e^x$

2. $y + xy' = 4x^3$

3. $y' + y - 2\cos x = 0$

4. $y + x^2y^2 + xy' = 0$

5. $2y + 3x^2 + 2xy' = 0$

6. $y' - 3x^2y^2 + \dfrac{y}{x} = 0$

7. $y^2 + (x^2 - xy)y' = 0$

8. $y' = e^{-y} - 1$

9. $(1 - x)\dfrac{dy}{dx} = y^2$

10. $y' \tan x + \tan y = 0$

11. $y + 2xy^2 + (y - x)y' = 0$

Using the given boundary conditions, find the particular solution to each differential equation.

12. $x\,dx = 2y\,dy$, $x = 3$ when $y = 1$

13. $y' \sin y = \cos x$, $x = \dfrac{\pi}{4}$ when $y = 0$

14. $xy' + y = 4x$, $x = 2$ when $y = 1$

15. $y\,dx = (x - 2x^2y)\,dy$, $x = 2$ when $y = 1$

16. $3xy^2\,dy = (3y^3 - x^3)\,dx$, $x = 3$ when $y = 1$

17. A gear is rotating at 1550 rev/min. Its speed decreases exponentially at a rate of 9.50% per second after the power is shut off. Find the gear's speed after 6.00 s.

18. For the circuit of Fig. 30–11, $R = 1350\ \Omega$, $L = 7.25$ H, and $E = 225$ V. If the switch is thrown from position 2 to position 1, find the current and the voltage across the inductance at $t = 3.00$ ms.

19. Write the equation of the orthogonal trajectories to each family of parabolas, $x^2 = 4y$.

20. For the circuit of Fig. 30–12, $R = 2550\ \Omega$, $C = 145\ \mu$F, and $E = 95.0$ V. If the switch is thrown from position 2 to position 1, find the current and the voltage across the capacitor at $t = 5.00$ ms.

21. Find the equation of the curve that passes through the point $(1, 2)$ and whose slope is $y' = 2 + y/x$.

22. An object is dropped and falls from rest through air whose resisting force is proportional to the speed of the package. The terminal speed is 275 ft/s. Show that the acceleration is given by the differential equation $a = dv/dt = g - gv/275$.

23. A certain yeast is found to grow exponentially at the rate of 15.0% per hour. Starting with 500 g of yeast, how many grams will there be after 15.0 h?

24. *Writing:* State in words what a differential equation is. Explain the difference between a first-order DE and a second-order DE.

25. *Team Project:* Take a differential equation from Exercise 2. Solve it

by slope field

by Euler's graphical method

by Euler's numerical method

analytically

with each team member using a different method. Compare your results.

26. *On Our Web Site:* Another method for solving differential equations is by use of the *Laplace transform*. A complete treatment of the Laplace transform, with electrical applications, is given in our text Web site. Also in that section are given more advanced numerical methods than are shown here.

See www.wiley.com/college/calter

31

Second-Order Differential Equations

♦♦♦ **OBJECTIVES** ♦♦♦

When you have completed this chapter, you should be able to

- Solve second-order differential equations that have separable variables.
- Use the auxiliary equation to determine the general solution to second-order differential equations with right side equal to zero.
- Use the method of undetermined coefficients to solve a second-order differential equation with right side not equal to zero.
- Use second-order differential equations to solve problems involving mechanical vibrations.
- Use second-order differential equations to solve *RLC* circuits.

♦♦♦

We conclude our study of differential equations with second-order equations. These, you recall, will have second derivatives. They may, of course, also have first derivatives, but no third or higher derivatives. Here we solve types of second-order equations that are fairly simple, but of great practical importance just the same.

Certain applications cannot be described by first-order differential equations. For mechanical vibrations, for example, we must include acceleration in the equation. As we've seen in earlier units, acceleration is found by taking the *second derivative* of the displacement. This gives rise to a second-order differential equation. Thus an equation describing the motion of the block suspended from a spring, Fig. 31–1, has the form

$$\frac{d^2y}{dt^2} + a\frac{dy}{dt} + bt = 0$$

where a and b are constants. To analyze the motion of the block we must solve this second-order differential equation, and we will do just that in this chapter.

Equations describing circuits that have both capacitance and inductance, in addition to resistance, also contain the second derivative and lead to second-order differential equations.

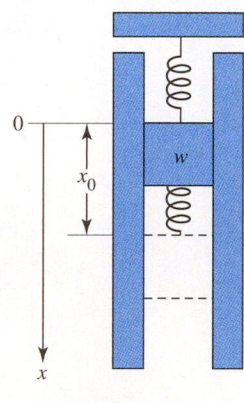

FIGURE 31–1

31–1 Second-Order DE

The General Second-Order Linear DE

A linear differential equation of second order can be written in the form

$$Py'' + Qy' + Ry = S$$

where P, Q, R, and S are constants or functions of x.

A second-order linear differential equation *with constant coefficients* is one where P, Q, and R *are constants*, although S can be a function of x, such as in the following equation:

Form of Second-Order Linear DE, Right Side Not Zero	$ay'' + by' + cy = f(x)$	336

where a, b, and c are constants. This is the type of equation we will solve in this chapter.

Equation 336 is sometimes called *homogeneous* if $f(x)$ is zero, but we will not use this misleading term.

Operator Notation

Differential equations are often written using the D operator that we have already introduced, where

$$Dy = y' \quad D^2y = y'' \quad D^3y = y''' \quad \text{etc.}$$

Thus Eq. 336 can be written

$$aD^2y + bDy + cy = f(x)$$

We'll usually use the more familiar y' notation rather than the D operator.

Second-Order Differential Equations with Separable Variables

We will develop methods for solving the general second-order equation later. However, simple differential equations of second order that are lacking a first derivative term can be solved by separation of variables, as in the following example.

◆◆◆ **Example 1:** Solve the equation $y'' = 3 \cos x$ (where x is in radians) if $y' = 1$ at the point $(2, 1)$.

Solution: Replacing y'' by $d(y')/dx$ and multiplying both sides by dx to separate variables, we have

$$d(y') = 3 \cos x \, dx$$

Integrating gives us

$$y' = 3 \sin x + C_1$$

Since $y' = 1$ when $x = 2$ rad, $C_1 = 1 - 3 \sin 2 = -1.73$, so

$$y' = 3 \sin x - 1.73$$

or $dy = 3 \sin x \, dx - 1.73 \, dx$. Integrating again, we have

$$y = -3 \cos x - 1.73x + C_2$$

At the point $(2, 1)$, $C_2 = 1 + 3 \cos 2 + 1.73(2) = 3.21$. Our solution is then

$$y = -3 \cos x - 1.73x + 3.21 \qquad \text{◆◆◆}$$

Notice that we had to integrate *twice* to solve a second-order DE and that *two constants* of integration had to be evaluated.

Solution by Calculator

Many calculators that can do symbolic processing can solve a second-order differential equation. We must enter the DE, the independent variable, and the dependent variable. If boundary conditions are also entered, we will get a particular solution. Otherwise we will get a general solution with unknown constants. We will repeat Example 1, first getting a general solution, and then a particular solution.

Example 2: Use the TI-89 to solve the DE

$$\frac{d^2y}{dx^2} = 3 \cos x$$

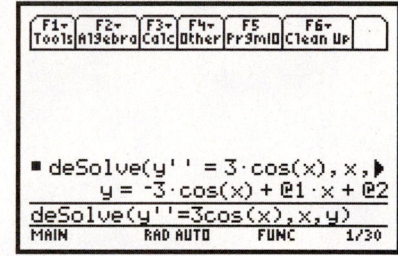

Solution:

(a) We select **deSolve** from the $\boxed{\text{MATH}}$ **Calculus** menu.

(b) Then enter the DE, $y'' = 3 \cos x$. Indicate a second derivative using the prime (′) symbol twice.

(c) Enter the independent variables x and y.

(d) Press $\boxed{\text{ENTER}}$ to display the general solution

TI-89 screen for Example 2.

$$y = -3 \cos x + C_1 x + C_2$$

Notice that the constants are displayed as @1 and @2 on the calculator. ◆◆◆

Now let us repeat the preceding example with boundary conditions.

Example 3: Solve the DE from the preceding example with the boundary conditions

$$y' = 1 \text{ at the point } (2, 1)$$

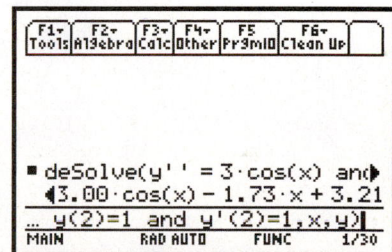

Solution: The steps are similar, but now immediately following the DE, we enter the boundary conditions in the form

$$\textbf{and } y(2) = 1 \text{ and } y'(2) = 1$$

where **and** is from the $\boxed{\text{MATH}}$ **Test** menu. The complete entry is then

$$\textbf{deSolve}(y'' = 3 \cos(x) \textbf{ and } y(2) = 1 \textbf{ and } y'(2) = 1, x, y)$$

TI-89 screen for Example 3. Unfortunately, the entire equations do not fit on the screen.

Pressing $\boxed{\approx}$ displays the particular solution

$$y = -3.00 \cos x - 1.73x + 3.21$$ ◆◆◆

Exercise 1 ◆ Second-Order DE

Solve each equation for y. Try some by calculator.

1. $y'' = 5$ **2.** $y'' = x$

3. $y'' = 3e^x$ **4.** $y'' = \sin 2x$

5. $y'' - x^2 = 0$ where $y' = 1$ at the point $(0, 0)$

31–2   Constant Coefficients and Right Side Zero

Solving a Second-Order Equation with Right Side Equal to Zero

If the right side, $f(x)$, in Eq. 336 is zero, and a, b, and c are constants, we have the following equation:

Form of Second-Order DE, Right Side Zero	$ay'' + by' + cy = 0$	331

To solve this equation, we note that the sum of the three terms on the left side must equal zero. Thus a solution must be a value of y that will make these terms alike, so they may be combined to give a sum of zero. Also note that each term on the left contains y or its first or second derivative. Thus *a possible solution is a function such that it and its derivatives are like terms.* Recall from the chapter on the integration of exponential functions that one such function was the exponential function

$$y = e^{mx}$$

It has derivatives $y' = me^{mx}$ and $y'' = m^2e^{mx}$, which are all like terms. We thus try this for our solution. Substituting $y = e^{mx}$ and its derivatives into Eq. 331 gives

$$am^2e^{mx} + bme^{mx} + ce^{mx} = 0$$

Factoring, we obtain

$$e^{mx}(am^2 + bm + c) = 0$$

Since e^{mx} can never be zero, this equation is satisfied only when $am^2 + bm + c = 0$. This is called the *auxiliary* or *characteristic equation.*

Auxiliary Equation of Second-Order DE, Right Side Zero	$am^2 + bm + c = 0$	332

The auxiliary equation is a quadratic and has two roots which we call m_1 and m_2. Either of the two values of m makes the auxiliary equation equal to zero. Thus we get *two solutions* to Eq. 331: $y_1 = e^{m_1x}$ and $y_2 = e^{m_2x}$.

Note that the coefficients in the auxiliary equation are the same as those in the original DE. We can thus get the auxiliary equation by inspection of the DE.

General Solution

We now show that if y_1 and y_2 are each solutions to $ay'' + by' + cy = 0$, then $y = c_1y_1 + c_2y_2$ is also a solution.

We first observe that if $y = c_1y_1 + c_2y_2$, then

$$y' = c_1y'_1 + c_2y'_2 \quad \text{and} \quad y'' = c_1y''_1 + c_2y''_2$$

Substituting into the differential equation (331), we have

$$a(c_1y''_1 + c_2y''_2) + b(c_1y'_1 + c_2y'_2) + c(c_1y_1 + c_2y_2) = 0$$

This simplifies to

$$c_1(ay''_1 + by'_1 + cy_1) + c_2(ay''_2 + by'_2 + cy_2) = 0$$

But $ay''_1 + by'_1 + cy_1 = 0$ and $ay''_2 + by'_2 + cy_2 = 0$ from Eq. 331, so

$$c_1(0) + c_2(0) = 0$$

showing that if y_1 and y_2 are solutions to the differential equation, then $y = c_1y_1 + c_2y_2$ is also a solution to the differential equation. Since $y = e^{m_1x}$ and $y = e^{m_2x}$ are solutions to Eq. 331, the complete solution is then of the form $y = c_1e^{m_1x} + c_2e^{m_2x}$.

General Solution of Second-Order DE, Right Side Zero	$y = c_1e^{m_1x} + c_2e^{m_2x}$	333

We see that the solution to a second-order differential equation depends on the nature of the roots of the auxiliary equation. We thus look at the roots of the auxiliary equation in order to quickly write the solution to the differential equation.

The roots of a quadratic can be real and unequal, real and equal, or nonreal. We now give examples of each case.

Roots Real and Unequal

◆◆◆ **Example 4:** Solve the equation $y'' - 3y' + 2y = 0$.

Solution: We get the auxiliary equation by inspection.

$$m^2 - 3m + 2 = 0$$

It factors into $(m - 1)$ and $(m - 2)$. Setting each factor equal to zero gives $m = 1$ and $m = 2$. These roots are real and unequal; our solution, by Eq. 333, is then

$$y = c_1 e^x + c_2 e^{2x}$$

◆◆◆

Sometimes one root of the auxiliary equation will be zero, as in the next example.

◆◆◆ **Example 5:** Solve the equation $y'' - 5y' = 0$.

Solution: The auxiliary equation is $m^2 - 5m = 0$, which factors into $m(m - 5) = 0$. Setting each factor equal to zero gives

$$m = 0 \quad \text{and} \quad m = 5$$

Our solution is then

$$y = c_1 + c_2 e^{5x}$$

◆◆◆

If the auxiliary equation cannot be factored, we use the quadratic formula to find its roots.

◆◆◆ **Example 6:** Solve $4.82y'' + 5.85y' - 7.26y = 0$.

Solution: The auxiliary equation is $4.82m^2 + 5.85m - 7.26 = 0$. By the quadratic formula,

$$m = \frac{-5.85 \pm \sqrt{(5.85)^2 - 4(4.82)(-7.26)}}{2(4.82)} = 0.762 \quad \text{and} \quad -1.98$$

Our solution is then

$$y = c_1 e^{0.762x} + c_2 e^{-1.98x}$$

◆◆◆

Roots Real and Equal

If $b^2 - 4ac$ is zero, the auxiliary equation has the double root, where $m_1 = m_2 = m$

$$m = \frac{-b \pm \sqrt{b^2 - 4ac}}{2a} = -\frac{b}{2a}$$

Our solution $y = c_1 e^{mx} + c_2 e^{mx}$ seems to contain two arbitrary constants, but it actually does not. Factoring gives

$$y = (c_1 + c_2) e^{mx} = c_3 e^{mx}$$

where $c_3 = c_1 + c_2$. A solution with one constant cannot be a complete solution for a second-order equation.

Let us *assume* that there is *a second solution* $y = ue^{mx}$, where u is some function of x that we are free to choose. Differentiating, we have

$$y' = mue^{mx} + u'e^{mx}$$

and

$$y'' = m^2ue^{mx} + mu'e^{mx} + mu'e^{mx} + u''e^{mx}$$

Substituting into $ay'' + by' + cy = 0$ gives

$$a(m^2ue^{mx} + 2mu'e^{mx} + u''e^{mx}) + b(mue^{mx} + u'e^{mx}) + cue^{mx} = 0$$

which simplifies to

$$e^{mx}[u(am^2 + bm + c) + u'(2am + b) + au''] = 0$$

But $am^2 + bm + c = 0$. Also, $m = -b/2a$, so $2am + b = 0$. Our equation then becomes $e^{mx}(au'') = 0$. Since e^{mx} cannot be zero, we have $u'' = 0$. Thus any u that has a zero second derivative will make ue^{mx} a solution to the differential equation. The simplest u (not a constant) for which $u'' = 0$ is $u = x$. Thus xe^{mx} is a solution to the differential equation, and the complete solution to Eq. 331 is as follows:

Equal Roots	$y = c_1e^{mx} + c_2xe^{mx}$	334

◆◆◆ **Example 7:** Solve $y'' - 6y' + 9y = 0$.

Solution: The auxiliary equation $m^2 - 6m + 9 = 0$ has the double root $m = 3$. Our solution is then

$$y = c_1e^{3x} + c_2xe^{3x} \qquad \text{◆◆◆}$$

Euler's Formula

When the roots of the auxiliary equation are nonreal, our solution will contain expressions of the form e^{bix}. In the following section we will want to simplify such expressions using *Euler's formula*, which we derive here.

Let $z = \cos\theta + i\sin\theta$, where $i = \sqrt{-1}$ and θ is in radians. Then

$$\frac{dz}{d\theta} = -\sin\theta + i\cos\theta$$

Multiplying by i (and recalling that $i^2 = -1$) gives

$$i\frac{dz}{d\theta} = -i\sin\theta - \cos\theta = -z$$

Multiplying by $-i$, we get $dz/d\theta = iz$. We now separate variables and integrate.

$$\frac{dz}{z} = i\,d\theta$$

Integrating, we obtain

$$\ln z = i\theta + c$$

When $\theta = 0$, $z = \cos 0 + i\sin 0 = 1$. So $c = \ln z - i\theta = \ln 1 - 0 = 0$. Thus $\ln z = i\theta$, or, in exponential form, $z = e^{i\theta}$. But $z = \cos\theta + i\sin\theta$, so we arrive at Euler's formula.

Euler's Formula	$e^{i\theta} = \cos\theta + i\sin\theta$	185

We can get two more useful forms of Eq. 185 for e^{bix} and e^{-bix}. First we set $\theta = bx$. Thus

$$e^{bix} = \cos bx + i\sin bx \tag{1}$$

Further,

$$e^{-bix} = \cos(-bx) + i\sin(-bx) = \cos bx - i\sin bx \tag{2}$$

since $\cos(-A) = \cos A$, and $\sin(-A) = -\sin A$.

Roots Not Real

We return to our second-order differential equation whose solution we are finding by means of the auxiliary equation. We now see that if the auxiliary equation has the nonreal roots $a + bi$ and $a - bi$, our solution becomes

$$y = c_1e^{(a+bi)x} + c_2e^{(a-bi)x}$$
$$= c_1e^{ax}e^{bix} + c_2e^{ax}e^{-bix} = e^{ax}(c_1e^{bix} + c_2e^{-bix})$$

Using Euler's formula gives

$$y = e^{ax}[c_1(\cos bx + i \sin bx) + c_2(\cos bx - i \sin bx)]$$
$$= e^{ax}[(c_1 + c_2)\cos bx + i(c_1 - c_2)\sin bx]$$

Replacing $c_1 + c_2$ by C_1, and $i(c_1 - c_2)$ by C_2 gives the following:

Nonreal Roots	$y = e^{ax}(C_1 \cos bx + C_2 \sin bx)$	335a

A more compact form of the solution may be obtained by using the equation for the sum of a sine wave and a cosine wave of the same frequency. In our chapter on trigonometric graphs we derived the formula

$$A \sin \omega t + B \cos \omega t = R \sin(\omega t + \phi) \qquad (166)$$

where

$$R = \sqrt{A^2 + B^2} \quad \text{and} \quad \phi = \arctan \frac{B}{A}$$

Thus the solution to the differential equation can take the following alternative form:

Nonreal Roots	$y = Ce^{ax} \sin(bx + \phi)$	335b

We'll find this form handy for applications.

where $C = \sqrt{C_1^2 + C_2^2}$ and $\phi = \arctan C_1/C_2$.

◆◆◆ **Example 8:** Solve $y'' - 4y' + 13y = 0$.

Solution: The auxiliary equation $m^2 - 4m + 13 = 0$ has the roots $m = 2 \pm 3i$, two nonreal roots. Substituting into Eq. 335a with $a = 2$ and $b = 3$ gives

$$y = e^{2x}(C_1 \cos 3x + C_2 \sin 3x)$$

or

$$y = Ce^{2x} \sin(3x + \phi)$$

in the alternative form of Eq. 335b.   ◆◆◆

Summary

The types of solutions to a second-order differential equation with right side zero and with constant coefficients are summarized here.

Second-Order DE with Right Side Zero $ay'' + by' + cy = 0$		
If the roots of auxiliary equation $am^2 + bm + c = 0$ are	**Then the solution to $ay'' + by' + cy = 0$ is**	
Real and unequal	$y = c_1 e^{m_1 x} + c_2 e^{m_2 x}$	333
Real and equal	$y = c_1 e^{mx} + c_2 x e^{mx}$	334
Nonreal	$y = e^{ax}(C_1 \cos bx + C_2 \sin bx)$	335a
	$y = Ce^{ax} \sin(bx + \phi)$	335b

Particular Solution

As before, we use the boundary conditions to find the two constants in the solution.

◆◆◆ **Example 9:** Solve $y'' - 4y' + 3y = 0$ if $y' = 5$ at $(1, 2)$.

Solution: The auxiliary equation $m^2 - 4m + 3 = 0$ has roots $m = 1$ and $m = 3$. Our solution is then

$$y = c_1 e^x + c_2 e^{3x}$$

At $(1, 2)$ we get

$$2 = c_1 e + c_2 e^3 \tag{1}$$

Here we have one equation and two unknowns. We get a second equation by taking the derivative of y.

$$y' = c_1 e^x + 3c_2 e^{3x}$$

Substituting the boundary condition $y' = 5$ when $x = 1$ gives

$$5 = c_1 e + 3c_2 e^3 \tag{2}$$

We solve Eqs. (1) and (2) simultaneously. Subtracting Eq. (1) from Eq. (2) gives $2c_2 e^3 = 3$, or

$$c_2 = \frac{3}{2e^3}$$

$$= 0.0747$$

Then, from Eq. (1),

$$c_1 = \frac{2 - (0.0747)e^3}{e}$$

$$= 0.184$$

Our particular solution is then

$$y = 0.184 e^x + 0.0747 e^{3x} \qquad \text{◆◆◆}$$

Third-Order Differential Equations

We now show (without proof) how the method of the preceding sections can be extended to simple third-order equations that can be easily factored.

◆◆◆ Example 10: Solve $y''' - 4y'' - 11y' + 30y = 0$.

Solution: We write the auxiliary equation by inspection.

$$m^3 - 4m^2 - 11m + 30 = 0$$

which factors, by trial and error, into

$$(m - 2)(m - 5)(m + 3) = 0$$

giving roots of 2, 5, and -3. The solution to the given equation is then

$$y = C_1 e^{2x} + C_2 e^{5x} + C_3 e^{-3x} \qquad \text{◆◆◆}$$

Exercise 2 ◆ Constant Coefficients and Right Side Zero

Find the general solution to each differential equation.

Second-Order DE, Roots of Auxiliary Equation Real and Unequal

1. $y'' - 6y' + 5y = 0$ **2.** $2y'' - 5y' - 3y = 0$

3. $y'' - 3y' + 2y = 0$ **4.** $y'' + 4y' - 5y = 0$

5. $y'' - y' - 6y = 0$ **6.** $y'' + 5y' + 6y = 0$

7. $5y'' - 2y' = 0$ **8.** $y'' + 4y' + 3y = 0$

9. $6y'' + 5y' - 6y = 0$ **10.** $y'' - 4y' + y = 0$

Second-Order DE, Roots of Auxiliary Equation Real and Equal

11. $y'' - 4y' + 4y = 0$ **12.** $y'' - 6y' + 9y = 0$

13. $y'' - 2y' + y = 0$ **14.** $9y'' - 6y' + y = 0$

15. $y'' + 4y' + 4y = 0$ **16.** $4y'' + 4y' + y = 0$

17. $y'' + 2y' + y = 0$ **18.** $y'' - 10y' + 25y = 0$

Second-Order DE, Roots of Auxiliary Equation Not Real

19. $y'' + 4y' + 13y = 0$ **20.** $y'' - 2y' + 2y = 0$

21. $y'' - 6y' + 25y = 0$ **22.** $y'' + 2y' + 2y = 0$

23. $y'' + 4y = 0$ **24.** $y'' + 2y = 0$

25. $y'' - 4y' + 5y = 0$ **26.** $y'' + 10y' + 425y = 0$

Particular Solution

Solve each differential equation. Use the given boundary conditions to find the constants of integration.

27. $y'' + 6y' + 9y = 0,$ $y = 0$ and $y' = 3$ when $x = 0$

28. $y'' + 3y' + 2y = 0,$ $y = 0$ and $y' = 1$ when $x = 0$

29. $y'' - 2y' + y = 0,$ $y = 5$ and $y' = -9$ when $x = 0$

30. $y'' + 3y' - 4y = 0,$ $y = 4$ and $y' = -2$ when $x = 0$

31. $y'' - 2y' = 0,$ $y = 1 + e^2$ and $y' = 2e^2$ when $x = 1$

32. $y'' + 2y' + y = 0,$ $y = 1$ and $y' = -1$ when $x = 0$

33. $y'' - 4y = 0,$ $y = 1$ and $y' = -1$ when $x = 0$

34. $y'' + 9y = 0,$ $y = 2$ and $y' = 0$ when $x = \pi/6$

35. $y'' + 2y' + 2y = 0,$ $y = 0$ and $y' = 1$ when $x = 0$

36. $y'' + 4y' + 13y = 0,$ $y = 0$ and $y' = 12$ when $x = 0$

Third-Order DE

Solve each differential equation.

37. $y''' - 2y'' - y' + 2y = 0$ **38.** $y''' - y' = 0$

39. $y''' - 6y'' + 11y' - 6y = 0$ **40.** $y''' + y'' - 4y' - 4y = 0$

41. $y''' - 3y'' - y' + 3y = 0$ **42.** $y''' - 7y' + 6y = 0$

43. $4y''' - 3y' + y = 0$ **44.** $y''' - y'' = 0$

31–3   Right Side Not Zero

Complementary Function and Particular Integral

We'll now see that the solution to a differential equation is made up of *two parts:* the *complementary function* and the *particular integral*. We show this now for a first-order equation and later for a second-order equation.

    The solution to a first-order differential equation, say,

$$y' + \frac{y}{x} = 4 \tag{1}$$

can be found by the methods of the preceding chapter. The solution to Eq. (1) is

$$y = \frac{c}{x} + 2x$$

Note that the solution has two parts. Let us label one part y_c and the other y_p. Thus $y = y_c + y_p$, where $y_c = c/x$ and $y_p = 2x$.

Let us substitute for y only $y_c = c/x$ into the left side of Eq. (1), and for y' the derivative $-c/x^2$.

$$-\frac{c}{x^2} + \frac{c}{x^2} = 0$$

We get 0 instead of the required 4 on the right-hand side. Thus y_c *does not satisfy Eq. (1).* It does, however, satisfy what we call the *reduced equation,* obtained by setting the right side equal to zero. We call y_c the *complementary function.*

We now substitute only $y_p = 2x$ as y in the left side of Eq. (1). Similarly, for y', we substitute 2. We now have

$$2 + \frac{2x}{x} = 4$$

We see that y_p *does* satisfy Eq. (1) and is hence a solution. But it cannot be a complete solution because it has no arbitrary constant. We call y_p a *particular integral.* The quantity y_c had the required constant but did not, by itself, satisfy Eq. (1).

However, *the sum of y_c and y_p satisfies Eq. (1) and has the required number of constants, and it is hence the complete solution.*

Complete Solution	$y = \quad y_c \quad + \quad y_p$ $= \begin{pmatrix} \text{complementary} \\ \text{function} \end{pmatrix} + \begin{pmatrix} \text{particular} \\ \text{integral} \end{pmatrix}$	**337**

Common Error	Don't confuse *particular integral* with *particular solution.*

Second-Order Differential Equations

We have seen that the solution to a first-order equation is made up of a complementary function and a particular integral. But is the same true of the second-order equation?

$$ay'' + by' + cy = f(x) \tag{336}$$

If a particular integral y_p is a solution to Eq. 336, we get, on substituting,

$$ay_p'' + by_p' + cy_p = f(x) \tag{2}$$

If the complementary function y_c is a complete solution to the reduced equation

$$ay'' + by' + cy = 0 \tag{331}$$

we get

$$ay_c'' + by_c' + cy_c = 0 \tag{3}$$

Adding Eqs. (2) and (3) gives us

$$a(y_p'' + y_c'') + b(y_p' + y_c') + c(y_p + y_c) = f(x)$$

Since the sum of the two derivatives is the derivative of the sum, we get

$$a(y_p + y_c)'' + b(y_p + y_c)' + c(y_p + y_c) = f(x)$$

This shows that $y_p + y_c$ is a solution to Eq. 336.

◆◆◆ **Example 11:** Given that the solution to $y'' - 5y' + 6y = 3x$ is

$$y = \underbrace{c_1 e^{3x} + c_2 e^{2x}}_{\substack{\text{complementary} \\ \text{function}}} + \underbrace{\frac{x}{2} + \frac{5}{12}}_{\substack{\text{particular} \\ \text{integral}}}$$

prove by substitution that (a) the complementary function will make the left side of the given equation equal to zero and that (b) the particular integral will make the left side equal to $3x$.

Solution: Given $y'' - 5y' + 6y = 3x$,

(a) Let
$$y_c = c_1 e^{3x} + c_2 e^{2x}$$
$$y_c' = 3c_1 e^{3x} + 2c_2 e^{2x}$$
$$y_c'' = 9c_1 e^{3x} + 4c_2 e^{2x}$$

Substituting on the left gives

$$9c_1 e^{3x} + 4c_2 e^{2x} - 5(3c_1 e^{3x} + 2c_2 e^{2x}) + 6(c_1 e^{3x} + c_2 e^{2x})$$

which equals zero.

(b) Let
$$y_p = \frac{x}{2} + \frac{5}{12}$$
$$y_p' = \frac{1}{2}$$
$$y_p'' = 0$$

Substituting gives

$$0 - 5\left(\frac{1}{2}\right) + 6\left(\frac{x}{2} + \frac{5}{12}\right)$$

or $3x$.   ◆◆◆

Finding the Particular Integral

We already know how to find the complementary function y_c. We set the right side of the given equation to zero and then solve that (reduced) equation just as we did in the preceding section.

To see how to find the particular integral y_p, we start with Eq. 336 and isolate y. We get

$$y = \frac{1}{c} [f(x) - ay'' - by']$$

For this equation to balance, y must contain terms similar to those in $f(x)$. Further, y must contain terms similar to those in its own first and second derivatives. Thus it seems reasonable to try a solution consisting of the sum of $f(x)$, $f'(x)$, and $f''(x)$, each with an (as yet) undetermined (constant) coefficient.

◆◆◆ **Example 12:** Find y_p for the equation $y'' - 5y' + 6y = 3x$.

Solution: Here $f(x) = 3x$, $f'(x) = 3$, and $f''(x) = 0$. Then

$$y_p = Ax + B$$

where A and B are constants yet to be found. This is sometimes called the trial function.   ◆◆◆

Finding the Constants in a Particular Integral

To find the constants A and B in the particular integral y_p,

(a) Take the first and second derivatives of y_p.
(b) Substitute y_p and its derivatives into the given differential equation.
(c) Equate coefficients of like terms and solve for A and B.

This is called the *method of undetermined coefficients*. It is best shown by example.

◆◆◆ **Example 13:** Find the constants A and B in Example 12. Write the complete solution to the differential equation, given that the complementary function is

$$y_c = c_1 e^{3x} + c_2 e^{2x}$$

Solution: From Example 12,

$$y_p = Ax + B$$

Taking derivatives,

$$y_p' = A$$
$$y_p'' = 0$$

Substituting into the original differential equation,

$$y'' - 5y' + 6y = 3x$$
$$0 - 5A + 6(Ax + B) = 3x$$

or

$$6Ax + (6B - 5A) = 3x$$

In order for an equation to be true, the coefficients of like powers of x on both sides of the equation must be equal. Thus we equate the coefficient of x on the left with that on the right,

$$6A = 3$$

or $A = 1/2$. Equating the constant term on the left with that on the right gives

$$6B - 5A = 0$$

From which

$$B = \frac{5A}{6} = \frac{5}{12}$$

Our particular integral is then

$$y_p = Ax + B = \frac{x}{2} + \frac{5}{12}$$

The complete solution is then

$$y = y_c + y_p = c_1 e^{3x} + c_2 e^{2x} + \frac{x}{2} + \frac{5}{12} \qquad ◆◆◆$$

General Procedure

Thus to solve a second-order linear differential equation with constant coefficients (right side not zero):

$ay'' + by' + cy = f(x)$	**336**

1. Find the *complementary function* y_c by solving the auxiliary equation.
2. Write the *particular integral* y_p. It should contain each term from the right side $f(x)$ (less coefficients) as well as the first and higher derivatives of each term of $f(x)$ (less coefficients). Discard any duplicates. When we say "duplicate," we mean terms that are alike, regardless of numerical coefficient. These are discussed in the following part of this section.
3. If a term in y_p is a duplicate of one in y_c, multiply that term in y_p by x^n, using the lowest n that *eliminates that duplication* and any new duplication with other terms in y_p.
4. *Write* y_p, each term with an undetermined coefficient.
5. *Substitute* y_p and its first and second derivatives into the differential equation.
6. *Evaluate the coefficients* by the method of undetermined coefficients.
7. *Combine* y_c and y_p to obtain the complete solution.

We illustrate these steps in the following example.

◆◆◆ **Example 14:** Solve $y'' - y' - 6y = 36x + 50 \sin x$.

Solution:

(1) The auxiliary equation $m^2 - m - 6 = 0$ has the roots $m = 3$ and $m = -2$, so the complementary function is $y_c = c_1 e^{3x} + c_2 e^{-2x}$.

(2) The terms in $f(x)$ and their derivatives (less coefficients) are

Term in $f(x)$	x	$\sin x$
First derivative	constant	$\cos x$
Second derivative	0	$\sin x$
Third derivative	0	$\cos x$
⋮	⋮	⋮

We see that the sine and cosine terms will keep repeating, so eliminating duplicates, we retain an x term, a constant term, a $\sin x$ term, and a $\cos x$ term.

(3) Comparing the terms from y_c and those possible terms that will form y_p, we see that no term in y_p is a duplicate of one in y_c.

(4) Our particular integral, y_p, is then

$$y_p = A + Bx + C \sin x + D \cos x$$

(5) The derivatives of y_p are

$$y_p' = B + C \cos x - D \sin x$$

and

$$y_p'' = -C \sin x - D \cos x$$

Substituting into the differential equation gives

$$y_p'' - y_p' - 6y_p = 36x + 50 \sin x$$

$$(-C \sin x - D \cos x) - (B + C \cos x - D \sin x)$$

$$-6(A + Bx + C \sin x + D \cos x) = 36x + 50 \sin x$$

(6) Collecting terms and equating coefficients of like terms from the left and the right sides of this equation gives the equations

$$-6B = 36 \qquad -B - 6A = 0 \qquad D - 7C = 50 \qquad -7D - C = 0$$

from which $A = 1$, $B = -6$, $C = -7$, and $D = 1$. Our particular integral is then

$$y_p = 1 - 6x - 7 \sin x + \cos x$$

(7) The complete solution is thus $y_c + y_p$, or

$$y = c_1 e^{3x} + c_2 e^{-2x} + 1 - 6x - 7 \sin x + \cos x \qquad ◆◆◆$$

Duplicate Terms in the Solution

The terms in the particular integral y_p must be *independent*. If y_p has duplicate terms, only one should be kept. However, if a term in y_p is a duplicate of one in the complementary function y_c (except for the coefficient), *multiply that term in y_p by x^n, using the lowest n that will eliminate that duplication* and any new duplication with other terms in y_p.

◆◆◆ **Example 15:** Solve the equation $y'' - 4y' + 4y = e^{2x}$.

Solution:

(1) The complementary function (work not shown) is

$$y_c = c_1 e^{2x} + c_2 x e^{2x}$$

(2) Our particular integral should contain e^{2x} and its derivatives, which are also of the form e^{2x}. But since these are duplicates, we need e^{2x} only once.

(3) But e^{2x} is a duplicate of the first term, $c_1 e^{2x}$, in y_c. If we multiply by x, we see that xe^{2x} is now a duplicate of the *second* term in y_c. We thus need $x^2 e^{2x}$.

We now see that in the process of eliminating duplicates with y_c, we may have created a new duplicate within y_p. If so, we again multiply by x^n, using the lowest n that would eliminate that new duplication as well. Here, $x^2 e^{2x}$ does not duplicate any term in y_p, so we proceed.

(4) Our particular integral is thus $y_p = Ax^2 e^{2x}$.

(5) Taking derivatives

$$y_p' = 2Ax^2 e^{2x} + 2Axe^{2x}$$

and

$$y_p'' = 4Ax^2 e^{2x} + 8Axe^{2x} + 2Ae^{2x}$$

Substituting into the differential equation gives

$$4Ax^2 e^{2x} + 8Axe^{2x} + 2Ae^{2x} - 4(2Ax^2 e^{2x} + 2Axe^{2x}) + 4(Ax^2 e^{2x}) = e^{2x}$$

(6) Collecting terms and solving for A gives $A = \frac{1}{2}$.

(7) Our complete solution is then

$$y = c_1 e^{2x} + c_2 xe^{2x} + \frac{1}{2} x^2 e^{2x}$$

◆◆◆

Exercise 3 ◆ Right Side Not Zero

Solve each second-order differential equation.

With Algebraic Expressions

1. $y'' - 4y = 12$
3. $y'' - y' - 2y = 4x$
5. $y'' - 4y = x^3 + x$

2. $y'' + y' - 2y = 3 - 6x$
4. $y'' + y' = x + 2$

With Exponential Expressions

6. $y'' + 2y' - 3y = 42e^{4x}$
8. $y'' + y' = 6e^x + 3$
10. $y'' - y = e^x + 2e^{2x}$
12. $y'' - y = xe^x$

7. $y'' - y' - 2y = 6e^x$
9. $y'' - 4y = 4x - 3e^x$
11. $y'' + 4y' + 4y = 8e^{2x} + x$

With Trigonometric Expressions

13. $y'' + 4y = \sin 2x$
15. $y'' + 2y' + y = \cos x$
17. $y'' + y = 2\cos x - 3\cos 2x$

14. $y'' + y' = 6\sin 2x$
16. $y'' + 4y' + 4y = \cos x$
18. $y'' + y = \sin x + 1$

With Exponential and Trigonometric Expressions

19. $y'' + y = e^x \sin x$
21. $y'' - 4y' + 5y = e^{2x} \sin x$

20. $y'' + y = 10e^x \cos x$
22. $y'' + 2y' + 5y = 3e^{-x} \sin x - 10$

Particular Solution

Find the particular solution to each differential equation, using the given boundary conditions.

23. $y'' - 4y' = 8$, $\quad y = y' = 0$ when $x = 0$

24. $y'' + 2y' - 3y = 6$, $\quad y = 0$ and $y' = 2$ when $x = 0$

25. $y'' + 4y = 2$, $\quad y = 0$ when $x = 0$ and $y = \frac{1}{2}$ when $x = \pi/4$

26. $y'' + 4y' + 3y = 4e^{-x}$, $\quad y = 0$ and $y' = 2$ when $x = 0$

27. $y'' - 2y' + y = 2e^x$, $\quad y' = 2e$ at $(1, 0)$

28. $y'' - 9y = 18 \cos 3x + 9$, $\quad y = -1$ and $y' = 3$ when $x = 0$

29. $y'' + y = -2 \sin x$, $\quad y = 0$ at $x = 0$ and $\pi/2$

31–4 Mechanical Vibrations

Free Vibrations

An important use for second-order differential equations is the analysis of mechanical vibrations. We first consider *free vibrations*, such as a vibrating spring, and later study *forced vibrations*, such as those caused by an unbalanced motor.

A block of weight W hangs from a spring with spring constant k (Fig. 31–2). The block is pulled down a distance x_0 from its rest position and released, its motion retarded by a frictional force proportional to the velocity dv/dt of the block. By Newton's second law of motion, the force F on the body equals the product of the mass m and its acceleration a.

$$ma = F$$

But $m = W/g$ (where g is the acceleration due to gravity) and a is the second derivative of the displacement x. Further, the force on the block is equal to the spring force kx acting upward, plus the frictional force $c \, dx/dt$, where c is called the coefficient of friction. So if the block is moving in the positive (downward) direction, we have

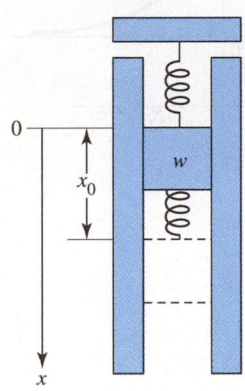

FIGURE 31–2 A block hanging from a spring.

$$\frac{W}{g} \frac{d^2x}{dt^2} = -kx - c \frac{dx}{dt}$$

Rearranging gives

$$\frac{d^2x}{dt^2} + \frac{cg}{W} \frac{dx}{dt} + \frac{kg}{W} x = 0 \tag{1}$$

Making the substitutions

$$2a = \frac{cg}{W} \quad \text{and} \quad \omega_n^2 = \frac{kg}{W}$$

our equation becomes

$$x'' + 2ax' + \omega_n^2 x = 0 \tag{2}$$

This is a second-order linear differential equation with a right side of zero, which we solve as before. The auxiliary equation $m^2 + 2am + \omega_n^2 = 0$ has the roots

$$m = \frac{-2a \pm \sqrt{4a^2 - 4\omega_n^2}}{2} = -a \pm \sqrt{a^2 - \omega_n^2} \tag{3}$$

We saw that the solution to a second-order differential equation depends on the nature of the roots of the auxiliary equation. Now we will see that each of these cases corresponds to a particular type of motion. These are

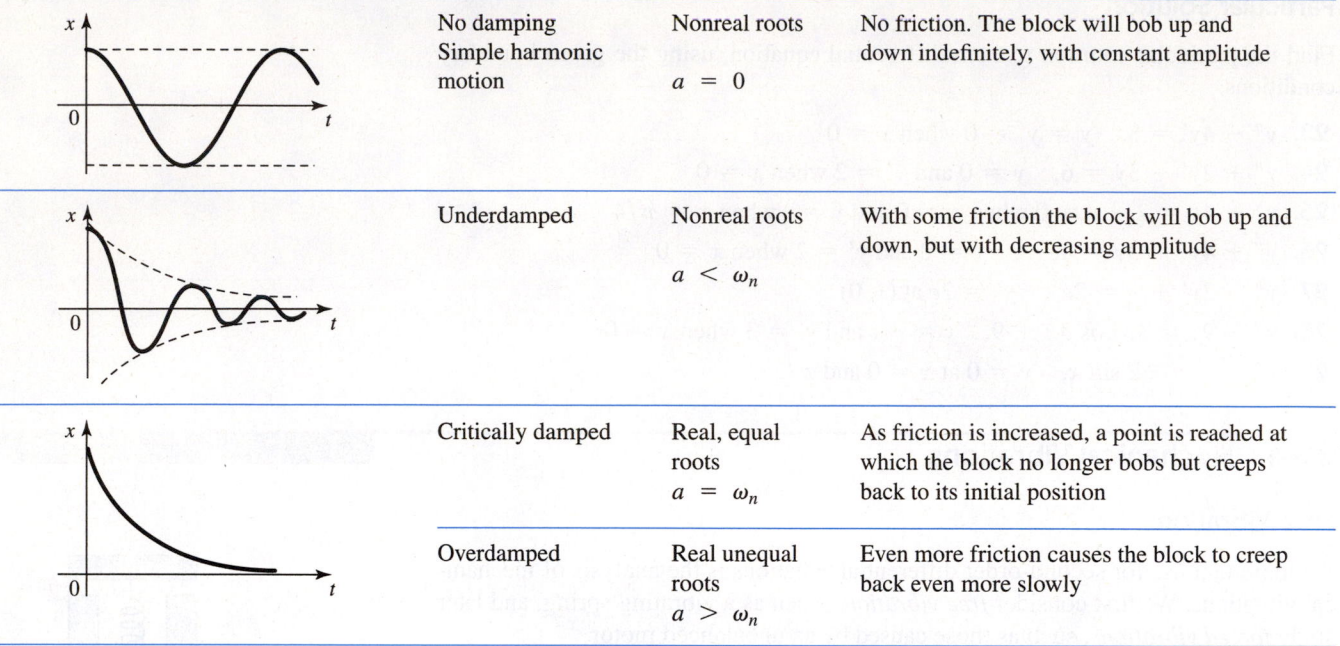

	No damping Simple harmonic motion	Nonreal roots $a = 0$	No friction. The block will bob up and down indefinitely, with constant amplitude
	Underdamped	Nonreal roots $a < \omega_n$	With some friction the block will bob up and down, but with decreasing amplitude
	Critically damped	Real, equal roots $a = \omega_n$	As friction is increased, a point is reached at which the block no longer bobs but creeps back to its initial position
	Overdamped	Real unequal roots $a > \omega_n$	Even more friction causes the block to creep back even more slowly

Underdamped Free Vibrations ($a < \omega_n$)

When a is less than ω_n, the auxiliary equation, Eq. (3), gives the nonreal roots

$$m = -a \pm i\sqrt{\omega_n^2 - a^2}$$
$$= -a \pm i\omega_d$$

if we make the substitution $\omega_d^2 = \omega_n^2 - a^2$. The solution to Eq. (2) is then (using the alternative form)

$$x = Ce^{-at}\sin(\omega_d t + \phi)$$

a damped sine wave of maximum amplitude C. From the physical problem, we know that the maximum amplitude is the initial displacement x_0, So

$$x = x_0 e^{-at}\sin(\omega_d t + \phi)$$

Also, at $t = 0$,

$$\frac{x}{x_0} = 1 = \sin\phi$$

from which $\phi = \pi/2$. Then $\sin(\omega_d t + \pi/2) = \cos\omega_d t$, so

Underdamped Free Vibrations	$x = x_0 e^{-at}\cos\omega_d t$	1039

The motion is thus a cosine wave whose amplitude decreases exponentially. The angular velocity ω_d is called the *damped angular velocity*. From our earlier substitution,

Damped Angular Velocity	$\omega_d = \sqrt{\omega_n^2 - a^2}$ where $a = \dfrac{cg}{2W}$	1040

◆◆◆ **Example 16:** The block in Fig. 31–2 weighs 25.9 lb, the spring constant is 110 lb/in., and $c = 1.26$ lb/(in./s). It is pulled down 0.625 in. from the rest position and released with zero velocity at $t = 0$. Find (a) the damped angular velocity, (b) the frequency, and (c) the period. (d) Write an equation for the displacement. Take $g = 386$ in./s^2.

Solution:

(a) We first find a and ω_n.

$$a = \frac{cg}{2W} = \frac{(1.26\ \text{lb}/(\text{in.}/\text{s}))(386\ \text{in.}/\text{s}^2)}{2(25.9\ \text{lb})} = 9.39\ \text{rad/s}$$

and

$$\omega_n = \sqrt{\frac{kg}{W}} = \sqrt{\frac{110\ \text{lb/in.}\ (386\ \text{in.}/\text{s}^2)}{25.9\ \text{lb}}} = 40.5\ \text{rad/s}$$

Since $a < \omega_n$, we have underdamping. Then by Eq. 1040,

$$\omega_d = \sqrt{(40.5)^2 - (9.39)^2} = 39.4\ \text{rad/s}$$

(b) The frequency f is then

$$f = \frac{\omega_d}{2\pi} = \frac{39.4}{2\pi} = 6.27\ \text{Hz (cycles/s)}$$

(c) The period is the reciprocal of the frequency, so

$$Period = \frac{1}{f} = \frac{1}{6.27} = 0.159\ \text{s}$$

(d) By Eq. 1039,

$$x = x_0 e^{-at} \cos \omega_d t = 0.625 e^{-9.39t} \cos 39.4t\ \text{in.}$$

The displacement is graphed in Fig. 31–3, showing a cosine wave enclosed within an "envelope" which decreases exponentially.   ◆◆◆

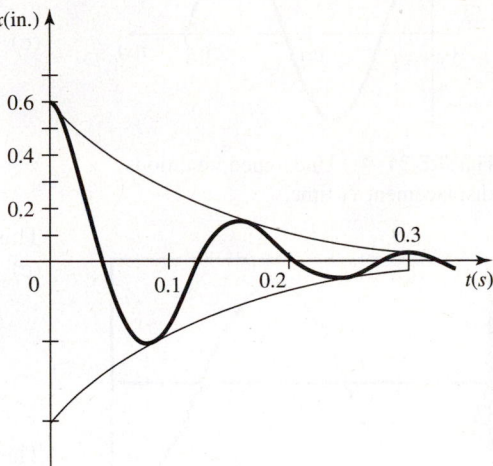

FIGURE 31–3
Underdamped vibrations.

Simple Harmonic Motion ($a = 0$)

When there is no damping, we have a special case of underdamped motion called *simple harmonic motion*. The coefficient of friction c is 0 and hence a is zero, and Eq. 1039 reduces to

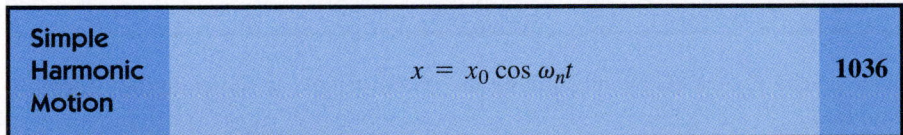

Simple Harmonic Motion	$x = x_0 \cos \omega_n t$	1036

We see that the displacement is a cosine function of amplitude x_0 (it would be a sine function if we had chosen $x = 0$, rather than $x = x_0$, at $t = 0$). The quantity ω_n is called the *undamped angular velocity*. From before,

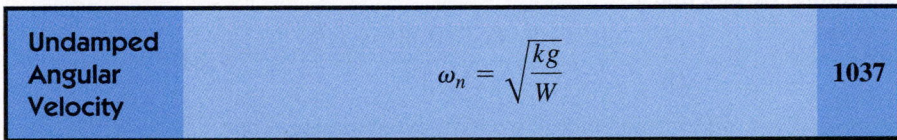

Undamped Angular Velocity	$\omega_n = \sqrt{\frac{kg}{W}}$	1037

The frequency obtained by dividing ω_n by 2π is called the *natural frequency* f_n.

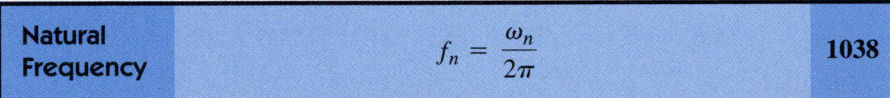

Natural Frequency	$f_n = \frac{\omega_n}{2\pi}$	1038

◆◆◆ **Example 17:** If the frictional coefficient c in Example 16 is zero, find

(a) the undamped angular velocity,
(b) the natural frequency, and
(c) the period.
Write equations for
(d) the displacement and
(e) the velocity.

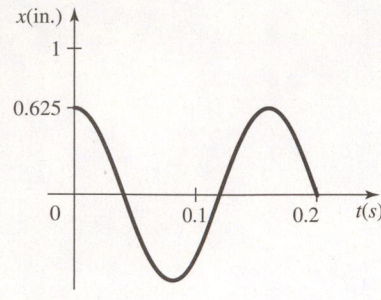

FIGURE 31–4 Undamped vibrations, displacement vs time.

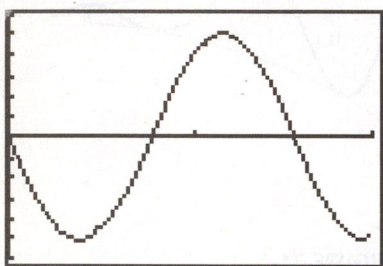

Screen for Example 17 showing the velocity vs. time. Ticks are spaced 0.1 s horizontally and 5 in./s vertically

Solution:

(a) The undamped angular velocity ω_n is from (Example 14) 40.5 rad/s.

(b) By Eq. 1038,

$$f_n = \frac{\omega_n}{2\pi} = \frac{40.5}{2\pi} = 6.44 \text{ Hz}$$

(c) The period is the reciprocal of the frequency, so

$$Period = \frac{1}{f_n} = \frac{1}{6.44} = 0.155 \text{ s}$$

(d) By Eq. 1036,

$$x = x_0 \cos \omega_n t = 0.625 \cos 40.5t \text{ in.}$$

This is graphed in Fig. 31–4.

(e) Taking the derivative,

$$v = \frac{dx}{dt} = -\omega_n x_0 \sin \omega_n t = -(40.5)(0.625)\sin 40.5t$$

$$= -25.3 \sin 40.5t \text{ in./s} \qquad \text{◆◆◆}$$

The velocity curve is shown in the screen.

Overdamped Free Vibrations ($a > \omega_n$)

When a is greater than ω_n, the auxiliary equation has the real and unequal roots

$$m = -a \pm \sqrt{a^2 - \omega_n^2}$$

The solution to the differential equation of motion is then

Overdamped Free Vibrations	$x = C_1 e^{m_1 t} + C_2 e^{m_2 t}$	**1041**

We evaluate C_1 and C_2 by substituting the initial values, as shown in the following example.

◆◆◆ Example 18: If $c = 7.46$ lb/ (in./s) for the block in Examples 16 and 17, write an equation for the displacement and the velocity.

Solution: We first find a.

$$a = \frac{cg}{2W} = \frac{(7.46 \text{ lb/(in./s)})(386 \text{ in./s}^2)}{2(25.9 \text{ lb})} = 55.6 \text{ rad/s}$$

Since $a > \omega_n$ (40.5 from before), we have overdamping. Then

$$\sqrt{a^2 - \omega_n^2} = \sqrt{(55.6)^2 - (40.5)^2} = 38.1 \text{ rad/s}$$

So

$$m_1 = -55.6 - 38.1 = -93.7 \quad \text{and} \quad m_2 = -55.6 + 38.1 = -17.5$$

By Eq. 1041,

$$x = C_1 e^{-93.7t} + C_2 e^{-17.5t}$$

Taking the derivative gives the velocity,

$$v = -93.7C_1 e^{-93.7t} - 17.5C_2 e^{-17.5t}$$

Substituting $x = 0.625$ and $v = 0$ at $t = 0$ in the equations for x and v gives

$$C_1 + C_2 = 0.625$$

and

$$93.7C_1 + 17.5C_2 = 0$$

Solving simultaneously gives $C_1 = -0.144$ and $C_2 = 0.769$. Substituting back, we get

$$x = -0.144e^{-93.7t} + 0.769e^{-17.5t} \quad \text{in.}$$

and

$$v = 13.5e^{-93.7t} - 13.5e^{-17.5t} \quad \text{in./s}$$

The displacement is graphed in Fig. 31–5 and the velocity is shown in the screen.

◆◆◆

FIGURE 31–5 Overdamped vibrations, displacement vs time.

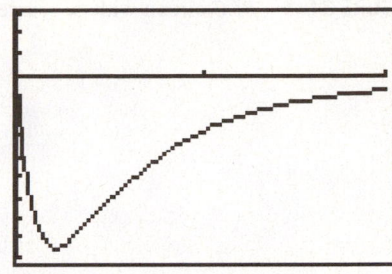

Screen for Example 18 showing the velocity vs. time. Ticks are spaced 0.1 s horizontally and 1 in./s vertically

Exercise 4 ◆ Mechanical Vibrations

Simple Harmonic Motion

1. The displacement x of an object at t seconds is given by $x = 3.75 \cos 182t$ in. Find (a) the period and (b) the amplitude of this motion.

2. What is the earliest time at which the displacement of the object in problem 1 is -3.00 in.?

3. The equation of motion of a certain wood block bobbing in water is $x'' + 225x = 0$, where x is in centimeters and t is in seconds. The initial conditions are $x = 0$ and $v = 15.0 \text{ cm/s}$ at $t = 0$. Write an equation for x as a function of time.

4. In problem 3, find (a) the maximum displacement and (b) the earliest time at which it occurs.

5. A 2.00-lb weight hangs motionless from a spring and is seen to stretch the spring 8.00 in. from its free length. It is pulled down an additional 1.00 in. and released. Write the equation of motion of the weight, taking zero at the original (motionless) position.

6. The maximum velocity in simple harmonic motion occurs when an object passes through its zero position. Find the maximum velocity and its time of occurrence for the weight in problem 5.

7. The motion of a pendulum hanging from a long string and swinging through small angles approximates simple harmonic motion. If the initial conditions are $x_0 = 1.50 \text{ cm}$ and $a_0 = -3.00 \text{ cm/s}^2$, write an equation for the displacement as a function of time.

8. Find the period for the motion of the pendulum in problem 7.

Damped Vibrations

9. The equation of motion of a certain car shock absorber is given by $x = 2.50e^{-3t} \cos 55t$ in., where t is in seconds. Make a graph of x versus t.

10. In problem 9, is the motion underdamped, overdamped, or critically damped?

11. For the shock absorber in problem 9, find x when t is 0.30 second.

12. For the shock absorber in problem 9, use any approximate method to find the earliest time at which x is half its initial value.

13. An object has a differential equation of motion given by $x'' + 5x' + 4x = 0$, with the initial conditions $x_0 = 1.50 \text{ cm}$ and $v_0 = 15.3 \text{ cm/s}$. Write an equation for x as a function of time.

14. For the weight of problem 5, assume a resisting force numerically equal to 3.00 times the velocity, in ft/s. Write an equation for the displacement.

15. In problem 14, (a) what type of damping do we have and (b) what weight W will produce critical damping?

31–5 *RLC* Circuits

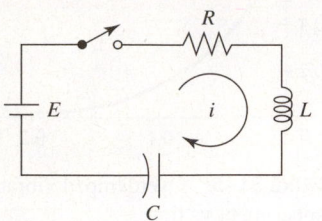

FIGURE 31–6 *RLC* circuit with dc source.

In the preceding chapter we studied the *RL* circuit and the *RC* circuit. Each gave rise to a first-order differential equation. We'll now see that the *RLC* circuit will result in a *second-order* differential equation.

A switch (Fig. 31–6) is closed at $t = 0$. The sum of the voltage drops must equal the applied voltage, so

$$Ri + L\frac{di}{dt} + \frac{q}{C} = E \tag{1}$$

Replacing q by $\int i\,dt$ and differentiating gives

$$R\frac{di}{dt} + L\frac{d^2i}{dt^2} + \frac{i}{C} = 0$$

or

$$Li'' + Ri' + \left(\frac{1}{C}\right)i = 0 \tag{2}$$

This is a second-order linear differential equation, which we now solve as we did before. The auxiliary equation $Lm^2 + Rm + 1/C = 0$ has the roots

$$m = \frac{-R \pm \sqrt{R^2 - 4L/C}}{2L}$$

$$= -\frac{R}{2L} \pm \sqrt{\frac{R^2}{4L^2} - \frac{1}{LC}}$$

We now let

$$a^2 = \frac{R^2}{4L^2} \quad \text{and} \quad \omega_n^2 = \frac{1}{LC}$$

We define the resonant frequency as ω_n. We'll say why it is called the resonant frequency later in this section:

Resonant Frequency of *RLC* Circuit with dc Source	$\omega_n = \dfrac{1}{\sqrt{LC}}$	1090

and our roots become

$$m = -a \pm \sqrt{a^2 - \omega_n^2}$$

or

$$m = -a \pm j\omega_d$$

where $\omega_d^2 = \omega_n^2 - a^2$. In this section we will use j rather than i for the imaginary unit. This is common practice in electrical work, where i is reserved for current.

We have three possible cases, listed in Table 31–1.

TABLE 31–1 *RLC* Circuit with dc Source: Second-Order Differential Equation

ROOTS OF AUXILIARY EQUATION	TYPE OF SOLUTION	
Nonreal	Underdamped	$a < \omega_n$
	No damping (series *LC* circuit)	$a = 0$
Real, equal	Critically damped	$a = \omega_n$
Real, unequal	Overdamped	$a > \omega_n$

Of these, we will consider the underdamped and overdamped cases with a dc source, and the underdamped case with an ac source.

Underdamped with dc Source ($a < \omega_n$)

We first consider the case where R is not zero but is low enough so that $a < \omega_n$. The roots of the auxiliary equation are then nonreal, and the current is

$$i = e^{-at}(k_1 \sin \omega_d t + k_2 \cos \omega_d t)$$

Since i is zero at $t = 0$, we get $k_2 = 0$. The current is then

$$i = k_1 e^{-at} \sin \omega_d t \tag{1}$$

Taking the derivative yields

$$\frac{di}{dt} = k_1 \omega_d e^{-at} \cos \omega_d t - ak_1 e^{-at} \sin \omega_d t$$

At $t = 0$, the capacitor behaves as a short circuit, and there is also no voltage drop across the resistor (since $i = 0$). The entire voltage E then appears across the inductor. Since $E = L \, di/dt$, then $di/dt = E/L$, so

$$k_1 = \frac{E}{\omega_d L}$$

Substituting into (1) gives the current:

Underdamped **RLC Circuit**	$i = \dfrac{E}{\omega_d L} e^{-at} \sin \omega_d t$	**1092**

where, from our previous substitution,

Underdamped **RLC Circuit**	$\omega_d = \sqrt{\omega_n^2 - a^2} = \sqrt{\omega_n^2 - \dfrac{R^2}{4L^2}}$	**1093**

We get a damped sine wave whose amplitude decreases exponentially with time.

◆◆◆ **Example 19:** A switch (Fig. 31–6) is closed at $t = 0$. If $R = 225 \ \Omega$, $L = 1.50$ H, $C = 4.75 \ \mu$F, and $E = 75.4$ V, write an expression for the instantaneous current.

Solution: We first compute LC.

$$LC = 1.50(4.75 \times 10^{-6}) = 7.13 \times 10^{-6}$$

Then, by Eq. 1090,

$$\omega_n = \sqrt{\frac{1}{LC}} = \sqrt{\frac{10^6}{7.13}} = 375 \text{ rad/s}$$

and

$$a = \frac{R}{2L} = \frac{225}{2(1.50)} = 75.0 \text{ rad/s}$$

Then, by Eq. 1093,

$$\omega_d = \sqrt{\omega_n^2 - a^2} = \sqrt{(375)^2 - (75.0)^2} = 367 \text{ rad/s}$$

The instantaneous current is then

$$i = \frac{E}{\omega_d L} e^{-at} \sin \omega_d t = \frac{75.4}{367(1.50)} e^{-75t} \sin 367t \text{ A}$$

$$= 137 e^{-75t} \sin 367t \text{ mA}$$

This curve is plotted in Fig. 31–7 showing the damped sine wave. ◆◆◆

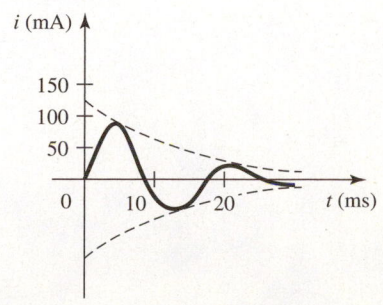

FIGURE 31–7

No Damping: The Series *LC* Circuit

When the resistance R is zero, $a = 0$ and $\omega_d = \omega_n$. From Eq. 1092 we get the following:

Series *LC* Circuit	$i = \dfrac{E}{\omega_n L} \sin \omega_n t$	**1091**

which represents a sine wave with amplitude $E/\omega_n L$. This, of course, is a theoretical case, because a real circuit always has some resistance.

◆◆◆ Example 20: Repeat Example 19 with $R = 0$.

Solution: The value of ω_n, from before, is 375 rad/s. The amplitude of the current wave is

$$\frac{E}{\omega_n L} = \frac{75.4}{375\,(1.50)} = 0.134 \text{ A}$$

so the instantaneous current is

$$i = 134 \sin 375t \text{ mA} \qquad\qquad ◆◆◆$$

Overdamped with dc Source ($a > \omega_n$)

If the resistance is relatively large, so that $a > \omega_n$, the auxiliary equation has the real and unequal roots

$$m_1 = -a + j\omega_d \quad \text{and} \quad m_2 = -a - j\omega_d$$

The current is then

$$i = k_1 e^{m_1 t} + k_2 e^{m_2 t} \qquad\qquad (1)$$

Since $i(0) = 0$, we have

$$k_1 + k_2 = 0 \qquad\qquad (2)$$

Taking the derivative of Eq. (1), we obtain

$$\frac{di}{dt} = m_1 k_1 e^{m_1 t} + m_2 k_2 e^{m_2 t}$$

Since $di/dt = E/L$ at $t = 0$,

$$\frac{E}{L} = m_1 k_1 + m_2 k_2 \qquad\qquad (3)$$

Solving Eqs. (2) and (3) simultaneously gives

$$k_1 = \frac{E}{(m_1 - m_2)\,L} \quad \text{and} \quad k_2 = -\frac{E}{(m_1 - m_2)\,L}$$

where $m_1 - m_2 = -a + j\omega_d + a + j\omega_d = 2j\omega_d$. The current is then given by the following equation:

Overdamped *RLC* Circuit	$i = \dfrac{E}{2j\omega_d L}\left[e^{(-a + j\omega_d)t} - e^{(-a - j\omega_d)t}\right]$	**1094**

◆◆◆ Example 21: For the circuit of Examples 19 and 20, let $R = 2550\ \Omega$, and compute the instantaneous current.

Solution: $a = \dfrac{R}{2L} = \dfrac{2550}{2\,(1.50)} = 850$ rad/s. Since $\omega_n = 375$ rad/s, we have

$$\omega_d = \sqrt{(375)^2 - (850)^2} = j\,763 \text{ rad/s}$$

Then $-a - j\omega_d = -87.0$ and $-a + j\omega_d = -1613$. From Eq. 1092,

$$i = \frac{75.4}{2(-763)(1.50)} (e^{-1613t} - e^{-87.0t})$$

$$= 32.9 (e^{-87.0t} - e^{-1613t}) \text{ mA}$$

This equation is graphed in Fig. 31–8.

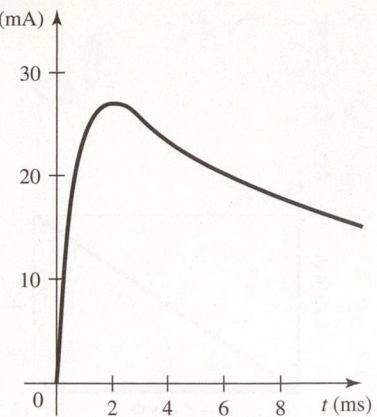

FIGURE 31–8 Current in an overdamped circuit.

Underdamped with ac Source

Up to now we have considered only a dc source. We now repeat the underdamped (low-resistance) case with an alternating voltage $E \sin \omega t$.

A switch (Fig. 31–9) is closed at $t = 0$. The sum of the voltage drops must equal the applied voltage, so

$$Ri + L\frac{di}{dt} + \frac{q}{C} = E \sin \omega t \tag{1}$$

Replacing q by $\int i\, dt$ and differentiating gives

$$R\frac{di}{dt} + L\frac{d^2i}{dt^2} + \frac{i}{C} = \omega E \cos \omega t$$

or

$$Li'' + Ri' + \left(\frac{1}{C}\right)i = \omega E \cos \omega t \tag{2}$$

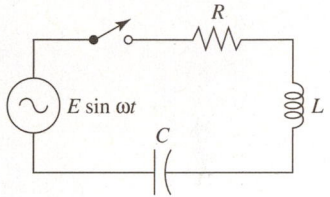

FIGURE 31–9 *RLC* circuit with ac source.

The complementary function is the same as was calculated for the dc case.

$$i_c = e^{-at}(k_1 \sin \omega_d t + k_2 \cos \omega_d t)$$

The particular integral i_p will have a sine term and a cosine term.

$$i_p = A \sin \omega t + B \cos \omega t$$

Taking first and second derivatives yields

$$i' = \omega A \cos \omega t - \omega B \sin \omega t$$
$$i'' = -\omega^2 A \sin \omega t - \omega^2 B \cos \omega t$$

Substituting into Eq. (2), we get

$$-L\omega^2 A \sin \omega t - L\omega^2 B \cos \omega t - R\omega B \sin \omega t + R\omega A \cos \omega t + \frac{A}{C} \sin \omega t$$

$$+ \frac{B}{C} \cos \omega t = \omega E \cos \omega t$$

Equating the coefficients of the sine terms gives

$$-L\omega^2 A - R\omega B + \frac{A}{C} = 0$$

from which

$$-RB = A\left(\omega L - \frac{1}{\omega C}\right) = AX \tag{3}$$

where X is the reactance of the circuit. Equating the coefficients of the cosine terms gives

$$-L\omega^2 B + R\omega A + \frac{B}{C} = \omega E$$

or

$$RA - E = B\left(\omega L - \frac{1}{\omega C}\right) = BX \tag{4}$$

Solving Eqs. (3) and (4) simultaneously gives

$$A = \frac{RE}{R^2 + X^2} = \frac{RE}{Z^2}$$

where Z is the impedance of the circuit, and

$$B = -\frac{EX}{R^2 + X^2} = -\frac{EX}{Z^2}$$

Our particular integral thus becomes

$$i_p = \frac{RE}{Z^2} \sin \omega t - \frac{EX}{Z^2} \cos \omega t$$

$$= \frac{E}{Z^2}(R \sin \omega t - X \cos \omega t)$$

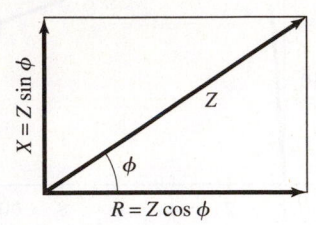

FIGURE 31–10   Impedance triangle.

From the impedance triangle (Fig. 31–10),

$$R = Z \cos \phi \quad \text{and} \quad X = Z \sin \phi$$

where ϕ is the phase angle. Thus

$$R \sin \omega t - X \cos \omega t = Z \sin \omega t \cos \phi - Z \cos \omega t \sin \phi$$

$$= Z \sin(\omega t - \phi)$$

by the trigonometric identity (Eq. 128). Thus i_p becomes $E/Z \sin(\omega t - \phi)$, or

$$i_p = I_{max} \sin(\omega t - \phi)$$

since E/Z gives the maximum current I_{max}. The total current is the sum of the complementary function i_c and the particular integral i_p.

$$i = \underbrace{e^{-at}(k_1 \sin \omega_d t + k_2 \cos \omega_d t)}_{\substack{\text{transient} \\ \text{current}}} + \underbrace{I_{max} \sin(\omega t - \phi)}_{\substack{\text{steady-state} \\ \text{current}}}$$

Thus the current is made up of two parts: a *transient* part that dies quickly with time and a *steady-state* part that continues as long as the ac source is connected. We are usually interested only in the steady-state current.

Steady-State Current for *RLC* Circuit, ac Source, Underdamped Case	$i_{ss} = \dfrac{E}{Z} \sin(\omega t - \phi)$	**1101**

◆◆◆ **Example 22:** Find the steady-state current for an *RLC* circuit if $R = 345 \ \Omega$, $L = 0.726$ H, $C = 41.4 \ \mu$F, and $E = 155 \sin 285t$.

Solution: By Eqs. 1095 and 1096,

$$X_L = \omega L = 285(0.726) = 207 \ \Omega$$

and

$$X_c = \frac{1}{\omega C} = \frac{1}{285(41.4 \times 10^{-6})} = 84.8 \ \Omega$$

The total reactance is, by Eq. 1097,

$$X = X_L - X_c = 207 - 84.8 = 122 \ \Omega$$

The impedance Z is found from Eq. 1098,

$$Z = \sqrt{R^2 + X^2} = \sqrt{(345)^2 + (122)^2} = 366 \ \Omega$$

The phase angle, from Eq. 1099, is

$$\phi = \tan^{-1} \frac{X}{R} = \tan^{-1} \frac{122}{345} = 19.5°$$

The steady-state current is then

$$
\begin{aligned}
i_{ss} &= \frac{E}{Z} \sin(\omega t - \phi) \\
&= \frac{155}{366} \sin(285t - 19.5°) \\
&= 0.423 \sin(285t - 19.5°) \text{ A}
\end{aligned}
$$

Figure 31–11 shows the applied voltage and the steady-state current, with a phase difference of 19.5°, or 0.239 ms.

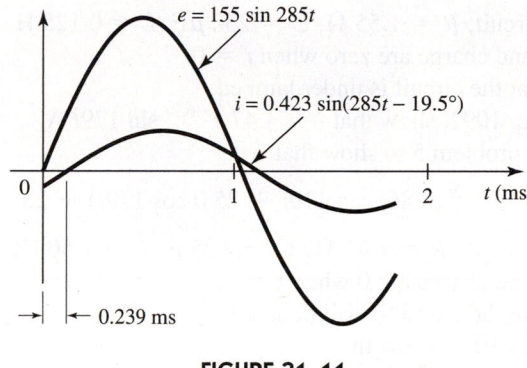

FIGURE 31–11 ◆◆◆

Resonance

The current in a series *RLC* circuit will be a maximum when the impedance Z is zero. This will occur when the reactance X is zero, so

$$X = \omega L - \frac{1}{\omega C} = 0$$

Solving for ω, we get $\omega^2 = 1/LC$ or ω_n^2. Thus,

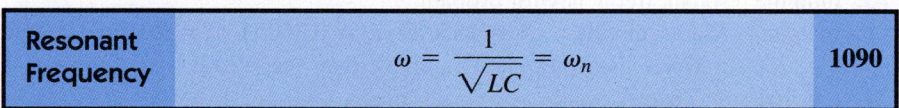

| Resonant Frequency | $\omega = \dfrac{1}{\sqrt{LC}} = \omega_n$ | **1090** |

◆◆◆ **Example 23:** Find the resonant frequency for the circuit of Example 22, and write an expression for the steady-state current at that frequency.

Solution: The resonant frequency is

$$\omega_n = \frac{1}{\sqrt{LC}} = \frac{1}{\sqrt{0.726 \, (41.4 \times 10^{-6})}} = 182 \text{ rad/s}$$

Since $X_L = X_C$, then $X = 0$, $Z = R$, and $\phi = 0$. Thus I_{max} is $155/345 = 0.449$ A, and the steady-state current is then

$$i_{ss} = 449 \sin 182t \text{ mA}$$

◆◆◆

Exercise 5 ♦ RLC Circuits

Series LC Circuit with dc Source

1. In an LC circuit, $C = 1.00 \ \mu F$, $L = 1.00$ H, and $E = 100$ V. At $t = 0$, the charge and the current are both zero. Using Eq. 1091, show that $i = 0.1 \sin 1000t$ A.

2. For problem 1, take the integral of i to show that

$$q = 10^{-4}(1 - \cos 1000t) \ \text{C}$$

3. For the circuit of problem 1, with $E = 0$ and an initial charge of 255×10^{-6} C (i_0 is still 0), the differential equation, in terms of charge, is $Lq'' + q/C = 0$. Solve this DE for q, and show that $q = 255 \cos 1000t \ \mu C$.

4. By differentiating the expression for charge in problem 3, show that the current is $i = -255 \sin 1000t$ mA.

Series RLC Circuit with dc Source

5. In an RLC circuit, $R = 1.55 \ \Omega$, $C = 250 \ \mu F$, $L = 0.125$ H, and $E = 100$ V. The current and charge are zero when $t = 0$.
   (a) Show that the circuit is underdamped.
   (b) Using Eq. 1092, show that $i = 4.47 \ e^{-6.2t} \sin 179t$ A.

6. Integrate i in problem 5 to show that

$$q = -e^{-6.2t}(0.866 \sin 179t + 25.0 \cos 179t) + 25.0 \ \text{mC}$$

7. In an RLC circuit, $R = 1.75 \ \Omega$, $C = 4.25$ F, $L = 1.50$ H, and $E = 100$ V. The current and charge are 0 when $t = 0$.
   (a) Show that the circuit is overdamped.
   (b) Using Eq. 1094, show that

$$i = -77.8e^{-1.01t} + 77.8e^{-0.155t} \qquad \text{A}$$

8. Integrate the expression for i for the circuit in problem 7 to show that

$$q = 77.0e^{-1.01t} - 502e^{-0.155t} + 425 \qquad \text{C}$$

Series RLC Circuit with ac Source

9. For an RLC circuit, $R = 10.5 \ \Omega$, $L = 0.125$ H, $C = 225 \ \mu F$, and $e = 100 \sin 175t$. Using Eq. 1101, show that the steady-state current is $i = 9.03 \sin(175t + 18.5°)$.

10. Find the resonant frequency for problem 9.

11. In an RLC circuit, $R = 1550 \ \Omega$, $L = 0.350$ H, $C = 20.0 \ \mu F$, and $e = 250 \sin 377t$. Using Eq. 1101, show that the steady-state current is $i = 161 \sin 377t$ mA.

12. If $R = 10.5 \ \Omega$, $L = 0.175$ H, $C = 1.50 \times 10^{-3}$ F, and $e = 175 \sin 55t$, find the amplitude of the steady-state current.

13. Find the resonant frequency for the circuit in problem 12.

14. Find the amplitude of the steady-state current at resonance for the circuit in problem 12.

15. For the circuit of Fig. 31–9, $R = 110 \ \Omega$, $L = 5.60 \times 10^{-6}$ H, and $\omega = 6.00 \times 10^5$ rad/s. What value of C will produce resonance?

16. If an RLC circuit has the values $R = 10 \ \Omega$, $L = 0.200$ H, $C = 500 \ \mu F$, and $e = 100 \sin \omega t$, find the resonant frequency.

17. For the circuit in problem 16, find the amplitude of the steady-state current at resonance.

18. Project: In the electrical laboratory construct one of the circuits given in this section (or create one of your own). With an oscilloscope, measure the voltage and/or current at the specified place in the circuit. Do you get what is predicted by the differential equations? Make a presentation to your class to convince yourself and your classmates that the convoluted computations of this section really do describe what happens in the real world.

••• CHAPTER 31 REVIEW PROBLEMS •••••••••••••••••••••••••••••••

Solve each differential equation.

1. $y'' - 2y = 2x^3 + 3x$

2. $y'' - 4y' - 5y = e^{4x}$

3. $y''' - 2y' = 0$

4. $y'' - 5y' = 6$, $y = y' = 1$ when $x = 0$

5. $y'' + 2y' - 3y = 0$

6. $y'' - 7y' - 18y = 0$

7. $y'' + 6y' + 9y = 0$, $y = 0$ and $y' = 2$ when $x = 0$

8. $y''' - 6y'' + 11y' - 6y = 0$

9. $y'' - 2y = 3x^3 e^x$

10. $y'' + 2y = 3\sin 2x$

11. $y'' + y' = 5\sin 3x$

12. $y'' + 3y' + 2y = 0$, $y = 1$ and $y' = 2$ when $x = 0$

13. $y'' + 9y = 0$

14. $y'' + 3y' - 4y = 0$

15. $y'' - 2y' + y = 0$, $y = 0$ and $y' = 2$ when $x = 0$

16. $y'' + 5y' - y = 0$

17. $y'' - 4y' + 4y = 0$

18. $y'' + 3y' - 4y = 0$, $y = 0$ and $y' = 2$ when $x = 0$

19. $y'' + 4y' + 4y = 0$

20. $y'' + 6y' + 9y = 0$

21. $y'' - 2y' + y = 0$

22. $9y'' - 6y' + y = 0$

23. A 5.00-lb weight hangs motionless from a spring, which is seen to stretch 7.00 in. from its free length. The weight is pulled down another 1.50 in. and released. Write the equation of motion of the weight, taking zero at the original (motionless) position.

24. In problem 23, assume a resisting force numerically equal to 2.50 times the velocity, in ft/s. Write an equation for the displacement.

25. For an LC circuit, $C = 1.75\ \mu F$, $L = 2.20$ H, and $E = 110$ V. At $t = 0$ the charge and the current are both zero. Find i as a function of time.

26. In problem 25, find the charge as a function of time.

27. For an RLC circuit, $R = 3.75\ \Omega$, $C = 150\ \mu F$, $L = 0.100$ H, and $E = 120$ V. The current and charge are 0 when $t = 0$. Find i as a function of time.

28. In problem 27, find an expression for the charge.

29. *Writing:* We have given seven steps for the solution of a second-order linear differential equation with constant coefficients. List as many of these steps as you can and write a one-sentence explanation of each.

30. The deflection curve for a beam is given by

$$\frac{d^2y}{dx^2} = \frac{M}{EI}$$

where E is the modulus of elasticity of the material, I is the moment of inertia of the beam cross section, y is the deflection of the beam, and M is the bending moment at a distance x from one end.

(a) For the cantilever beam (Fig. 31–12) show that the deflection curve is

$$\frac{d^2y}{dx^2} = \frac{Px - PL}{EI}$$

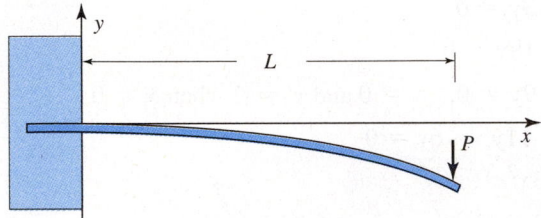

FIGURE 31–12   A cantilever beam.

(b) Solve this DE for the slope dy/dx and for the deflection y. Evaluate any constants of integration.

31. *On Our Web Site:* Another method for solving second-order differential equations is by use of the *Laplace transform*. A complete treatment of the Laplace transform, with electrical applications, is given on our text Web site. Also given in that section are more advanced numerical methods than are shown here.

See www.wiley.com/college/calter

A

Summary of Facts and Formulas

Many mathematics courses cover only a fraction of the topics in this text. Further, some of these formulas are included for reference even though they may not appear elsewhere in the text. We hope that this Summary will provide a handy reference, not only for your current course but for others, and for your technical work after graduation.

	No.			Page
ADDITION & SUBTRACTION	1	Rules of Signs	$a + (-b) = a - (+b) = a - b$	10
	2		$a + (+b) = a - (-b) = a + b$	11
	3	Commutative Law	$a + b = b + a$	12, 68
	4	Associative Law	$a + (b + c) = (a + b) + c = (a + c) + b$	12
	5	Approximate Numbers	When adding or subtracting approximate numbers, keep as many decimal places in your answer as contained in the number having the fewest decimal places.	12
MULTIPLICATION	6	Rules of Signs (a and b are positive numbers)	$(+a)(+b) = (-a)(-b) = +ab$	16, 80
	7		$(+a)(-b) = (-a)(+b) = -(+a)(+b) = -ab$	16, 80
	8	Commutative Law	$ab = ba$	81
	9	Associative Law	$a(bc) = (ab)c = (ac)b = abc$	81
	10	Distributive Law	$a(b + c) = ab + ac$	83, 320
	11	Approximate Numbers	When multiplying two or more approximate numbers, round the result to as many digits as in the factor having the fewest significant digits.	17
DIVISION	12	Rules of Signs (a and b are positive numbers)	$\dfrac{+a}{+b} = \dfrac{-a}{-b} = -\dfrac{-a}{+b} = -\dfrac{+a}{-b} = \dfrac{a}{b}$	20, 93
	13		$\dfrac{+a}{-b} = \dfrac{-a}{+b} = -\dfrac{-a}{-b} = -\dfrac{a}{b}$	20, 93
	14	Approximate Numbers	After dividing one approximate number by another, round the quotient to as many digits as there are in the original number having the fewest significant digits.	21
	15	Division Involving Zero	Zero divided by any quantity (except zero) is zero. Division by zero is not defined. It is an illegal operation in mathematics.	22
PERCENTAGE	16		Amount = base × rate $\quad A = BP$	52
	17		Percent change $= \dfrac{\text{new value} - \text{original value}}{\text{original value}} \times 100$	54
	18		Percent error $= \dfrac{\text{measured value} - \text{known value}}{\text{known value}} \times 100$	56
	19		Percent concentration of ingredient $A = \dfrac{\text{amount of } A}{\text{amount of mixture}} \times 100$	56, 124
	20		Percent efficiency $= \dfrac{\text{output}}{\text{input}} \times 100$	55

	No.				Page
EXPONENTS	21	Positive Integral Exponent		$x^n = \underbrace{x \cdot x \cdot x \cdot \ldots x}_{n \text{ factors}}$	24, 72, 77, 90
	22	Law of Exponents	Products	$x^a \cdot x^b = x^{a+b}$	72, 77, 381
	23		Quotients	$\dfrac{x^a}{x^b} = x^{a-b} \quad (x \neq 0)$	73, 77, 93, 382
	24		Power Raised to a Power	$(x^a)^b = x^{ab} = (x^b)^a$	74, 77, 381
	25		Product Raised to a Power	$(xy)^n = x^n \cdot y^n$	74, 77, 381
	26		Quotient Raised to a Power	$\left(\dfrac{x}{y}\right)^n = \dfrac{x^n}{y^n} \quad (y \neq 0)$	75, 77, 382
	27		Zero Exponent	$x^0 = 1 \quad (x \neq 0)$	75, 77, 381
	28		Negative Exponent	$x^{-a} = \dfrac{1}{x^a} \quad (x \neq 0)$	32, 76, 77, 380
	29	Fractional Exponents		$a^{1/n} = \sqrt[n]{a}$	29, 385
	30			$a^{m/n} = \sqrt[n]{a^m} = (\sqrt[n]{a})^m$	386
RADICALS	31	Rules of Radicals	Root of a Product	$\sqrt[n]{ab} = \sqrt[n]{a}\,\sqrt[n]{b}$	387
	32		Root of a Quotient	$\sqrt[n]{\dfrac{a}{b}} = \dfrac{\sqrt[n]{a}}{\sqrt[n]{b}}$	387
SPECIAL PRODUCTS AND FACTORING	33	Trinomials	Difference of Two Squares	$a^2 - b^2 = (a - b)(a + b)$	324
	34		Sum of Two Cubes	$a^3 + b^3 = (a + b)(a^2 - ab + b^2)$	333
	35		Difference of Two Cubes	$a^3 - b^3 = (a - b)(a^2 + ab + b^2)$	333
	36	Binomials	Test for Factorability	$ax^2 + bx + c$ is factorable if $b^2 - 4ac$ is a perfect square	327
	37		Leading Coefficient = 1	$x^2 + (a + b)x + ab = (x + a)(x + b)$	
	38		General Quadratic Trinomial	$acx^2 + (ad + bc)x + bd = (ax + b)(cx + d)$	327
	39		Perfect Square Trinomials	$a^2 + 2ab + b^2 = (a + b)^2$	330
	40			$a^2 - 2ab + b^2 = (a - b)^2$	330

No.				Page
FRACTIONS	41	Simplifying	$\dfrac{ad}{bd} = \dfrac{a}{b}$	337
	42	Multiplication	$\dfrac{a}{b} \cdot \dfrac{c}{d} = \dfrac{ac}{bd}$	340
	43	Division	$\dfrac{a}{b} \div \dfrac{c}{d} = \dfrac{a}{b} \cdot \dfrac{d}{c} = \dfrac{ad}{bc}$	341
	44	Addition and Subtraction — Same Denominators	$\dfrac{a}{b} \pm \dfrac{c}{b} = \dfrac{a \pm c}{b}$	344
	45	Addition and Subtraction — Different Denominators	$\dfrac{a}{b} \pm \dfrac{c}{d} = \dfrac{ad}{bd} \pm \dfrac{bc}{bd} = \dfrac{ad \pm bc}{bd}$	345
THREE MEANS	46	Arithmetic Mean between a and b.	$\dfrac{a + b}{2}$	596
	47	Geometric Mean between a and c.	$\pm\sqrt{ac}$	494, 602
	48	Harmonic Mean between a and b.	$\dfrac{2ab}{a + b}$	599
VARIATION	49	$k = $ Constant of Proportionality — Direct	$y \propto x$ or $y = kx$	502
	50	$k = $ Constant of Proportionality — Inverse	$y \propto \dfrac{1}{x}$ or $y = \dfrac{k}{x}$	509
	51	$k = $ Constant of Proportionality — Joint	$y \propto xw$ or $y = kxw$	513

SYSTEMS OF LINEAR EQUATIONS

No.				Page
52	Algebraic Solution	$\begin{aligned} a_1x + b_1y &= c_1 \\ a_2x + b_2y &= c_2 \end{aligned}$	$x = \dfrac{b_2c_1 - b_1c_2}{a_1b_2 - a_2b_1}, \quad \text{and} \quad y = \dfrac{a_1c_2 - a_2c_1}{a_1b_2 - a_2b_1}$ $\qquad$ where $a_1b_2 - a_2b_1 \neq 0$	282
53	Algebraic Solution	$\begin{aligned} a_1x + b_1y + c_1z &= k_1 \\ a_2x + b_2y + c_2z &= k_2 \\ a_3x + b_3y + c_3z &= k_3 \end{aligned}$	$x = \dfrac{b_2c_3k_1 + b_1c_2k_3 + b_3c_1k_2 - b_2c_1k_3 - b_3c_2k_1 - b_1c_3k_2}{a_1b_2c_3 + a_3b_1c_2 + a_2b_3c_1 - a_3b_2c_1 - a_1b_3c_2 - a_2b_1c_3}$ $y = \dfrac{a_1c_3k_2 + a_3c_2k_1 + a_2c_1k_3 - a_3c_1k_2 - a_1c_2k_3 - a_2c_3k_1}{a_1b_2c_3 + a_3b_1c_2 + a_2b_3c_1 - a_3b_2c_1 - a_1b_3c_2 - a_2b_1c_3}$ $z = \dfrac{a_1b_2k_3 + a_3b_1k_2 + a_2b_3k_1 - a_3b_2k_1 - a_1b_3k_2 - a_2b_1k_3}{a_1b_2c_3 + a_3b_1c_2 + a_2b_3c_1 - a_3b_2c_1 - a_1b_3c_2 - a_2b_1c_3}$	311
54	Value of a Determinant — Second Order		$\begin{vmatrix} a_1 & b_1 \\ a_2 & b_2 \end{vmatrix} = a_1b_2 - a_2b_1$	302
55	Value of a Determinant — Third Order		$\begin{vmatrix} a_1 & b_1 & c_1 \\ a_2 & b_2 & c_2 \\ a_3 & b_3 & c_3 \end{vmatrix} = a_1b_2c_3 + a_3b_1c_2 + a_2b_3c_1 - (a_3b_2c_1 + a_1b_3c_2 + a_2b_1c_3)$	311
56	Value of a Determinant — Minors		The signed minor of element b in the determinant $\begin{vmatrix} a & b & c \\ d & e & f \\ g & h & i \end{vmatrix}$ is $-\begin{vmatrix} d & f \\ g & i \end{vmatrix}$	310
57	Value of a Determinant — Minors		To find the value of a determinant: 1. Choose any row or any column to develop by minors. 2. Write the product of every element in that row or column and its signed minor. 3. Add these products to get the value of the determinant.	310
58	Determinants — Cramer's Rule		The solution for any variable is a fraction whose denominator is the determinant of the coefficients, and whose numerator is also the determinant of the coefficients, except that the column of coefficients of the variable being solved for is replaced by the column of constants.	312
59	Cramer's Rule — Two Equations		$x = \dfrac{\begin{vmatrix} c_1 & b_1 \\ c_2 & b_2 \end{vmatrix}}{\begin{vmatrix} a_1 & b_1 \\ a_2 & b_2 \end{vmatrix}} \quad \text{and} \quad y = \dfrac{\begin{vmatrix} a_1 & c_1 \\ a_2 & c_2 \end{vmatrix}}{\begin{vmatrix} a_1 & b_1 \\ a_2 & b_2 \end{vmatrix}}$	304
60	Cramer's Rule — Three Equations		$x = \dfrac{\begin{vmatrix} k_1 & b_1 & c_1 \\ k_2 & b_2 & c_2 \\ k_3 & b_3 & c_3 \end{vmatrix}}{\Delta}, \quad y = \dfrac{\begin{vmatrix} a_1 & k_1 & c_1 \\ a_2 & k_2 & c_2 \\ a_3 & k_3 & c_3 \end{vmatrix}}{\Delta}, \quad z = \dfrac{\begin{vmatrix} a_1 & b_1 & k_1 \\ a_2 & b_2 & k_2 \\ a_3 & b_3 & k_3 \end{vmatrix}}{\Delta}$ Where $\Delta = \begin{vmatrix} a_1 & b_1 & c_1 \\ a_2 & b_2 & c_2 \\ a_3 & b_3 & c_3 \end{vmatrix} \neq 0$	312

QUADRATICS

No.			Page
61	General Form	$ax^2 + bx + c = 0$	368
62	Quadratic Formula	$x = \dfrac{-b \pm \sqrt{b^2 - 4ac}}{2a}$	369, 371
63	The Discriminant	If a, b, and c are real, and $\quad$ if $b^2 - 4ac > 0$ the roots are real and unequal $\quad$ if $b^2 - 4ac = 0$ the roots are real and equal $\quad$ if $b^2 - 4ac < 0$ the roots are not real	

	No.				Page
INTERSECTING LINES	64		Opposite angles of two intersecting straight lines are equal.		177
	65		If two parallel straight lines are cut by a transversal, corresponding angles are equal and alternate interior angles are equal.		178
	66		If two lines are cut by a number of parallels, the corresponding segments are proportional.		178
QUADRILATERALS	67		Square	$\text{Area} = a^2$	
	68		Rectangle	$\text{Area} = ab$	
	69		Parallelogram: Diagonals bisect each other	$\text{Area} = bh$	
	70		Rhombus: Diagonals intersect at right angles	$\text{Area} = ah$	
	71		Trapezoid	$\text{Area} = \dfrac{(a+b)h}{2}$	
POLYGON	72	n sides	Sum of the interior angles $= (n-2)\,180°$		189
CIRCLE	73	Definition of π	$\pi = \dfrac{\text{circumference}}{\text{diameter}} \cong 3.1416$		191
	74		Circumference $= 2\pi r = \pi d$		191
	75		$\text{Area} = \pi r^2 = \dfrac{\pi d^2}{4}$		47, 191
	76		Central angle θ (radians) $= \dfrac{s}{r}$		413
	77		Area of sector $= \dfrac{rs}{2} = \dfrac{r^2\theta}{2}$		410
	78		1 revolution $= 2\pi$ radians $= 360°$ $1° = 60$ minute 1 minute $= 60$ seconds		406
	79	Area of Segment	$A = r^2 \arccos \dfrac{r-h}{r} - (r-h)\sqrt{2rh - h^2}$ where $\arccos \dfrac{r-h}{r}$ is in radians		410
	80		$A = \dfrac{r^2}{2}(\theta - \sin\theta)$ where θ is in radians		411, 490

	No.				Page
CIRCLE *(Continued)*	81	Inscribed and Central Angle	If an inscribed angle ϕ and a central angle θ subtend the same arc, the central angle is twice the inscribed angle. $\quad \theta = 2\phi$		192
	82		Any angle inscribed in a semicircle is a right angle.		193
	83		Tangents to a Circle	Tangent AP is perpendicular to radius OA.	192
	84			Tangent AP = tangent BP OP bisects angle APB	193
	85		Intersecting Chords	$ab = cd$	193
	86	Perpendicular Bisector of a Chord	The perpendicular bisector of a chord passes through the center of the circle.		192
SOLIDS	87		Cube	Volume = a^3	198
	88			Surface area = $6a^2$	198
	89		Rectangular Parallelepiped	Volume = lwh	197
	90			Surface area = $2(lw + hw + lh)$	197
	91		Any Cylinder or Prism	Volume = (area of base)(altitude)	197, 201
	92		Right Cylinder or Prism	Lateral area (not incl. bases) = (perimeter of base)(altitude)	197, 201
	93		Sphere	Volume = $\frac{4}{3}\pi r^3$	203
	94			Surface area = $4\pi r^2$	203
	95		Any Cone or Pyramid	Volume = $\frac{1}{3}$(area of base)(altitude)	198, 202
	96		Right Circular Cone or Regular Pyramid	Lateral area = $\frac{1}{2}$(perimeter of base) $\times$ (slant height) Does not include the base.	198, 202
	97		Any Cone or Pyramid	Volume = $\frac{h}{3}\left(A_1 + A_2 + \sqrt{A_1 A_2}\right)$	198, 202
	98		Right Circular Cone or Regular Pyramid	Lateral area = $\frac{s}{2}$(sum of base perimeters) = $\frac{s}{2}(P_1 + P_2)$	198, 202
SIMILAR FIGURES	99		Corresponding dimensions of plane or solid similar figures are in proportion.		498
	100		Areas of similar plane or solid figures are proportional to the squares of any two corresponding dimensions.		499
	101		Volumes of similar solid figures are proportional to the cubes of any two corresponding dimensions.		499

	No.				Page
ANY TRIANGLE	102	Areas		Area $= \frac{1}{2}bh$	181
	103		Hero's Formula: $\text{Area} = \sqrt{s(s-a)(s-b)(s-c)}$ where $s = \frac{1}{2}(a+b+c)$		181
	104		Sum of the Angles	$A + B + C = 180°$	182, 212
	105		Law of Sines	$\dfrac{a}{\sin A} = \dfrac{b}{\sin B} = \dfrac{c}{\sin C}$	241
	106		Law of Cosines	$a^2 = b^2 + c^2 - 2bc \cos A$ $b^2 = a^2 + c^2 - 2ac \cos B$ $c^2 = a^2 + b^2 - 2ab \cos C$	247
	107		Exterior Angle	$\theta = A + B$	182
SIMILAR TRIANGLES	108	If two angles of a triangle equal two angles of another triangle, the triangles are similar.			183
	109	Corresponding sides of similar triangles are in proportion.			183
RIGHT TRIANGLES	110		Pythagorean Theorem	$a^2 + b^2 = c^2$	183, 212
	111	Trigonometric Functions	Sine	$\sin \theta = \dfrac{y}{r} = \dfrac{\text{opposite side}}{\text{hypotenuse}}$	208, 212, 221, 232
	112		Cosine	$\cos \theta = \dfrac{x}{r} = \dfrac{\text{adjacent side}}{\text{hypotenuse}}$	208, 212, 221, 232
	113		Tangent	$\tan \theta = \dfrac{y}{x} = \dfrac{\text{opposite side}}{\text{adjacent side}}$	208, 212, 221, 232
	114		Cotangent	$\cot \theta = \dfrac{x}{y} = \dfrac{\text{adjacent side}}{\text{opposite side}}$	232
	115		Secant	$\sec \theta = \dfrac{r}{x} = \dfrac{\text{hypotenuse}}{\text{adjacent side}}$	232
	116		Cosecant	$\csc \theta = \dfrac{r}{y} = \dfrac{\text{hypotenuse}}{\text{opposite side}}$	232
	117	Reciprocal Relations	(a) $\csc \theta = \dfrac{1}{\sin \theta}$	(b) $\sec \theta = \dfrac{1}{\cos \theta}$ (c) $\cot \theta = \dfrac{1}{\tan \theta}$	234, 462
	118	A and B are Complementary Angles	Cofunctions	(a) $\sin A = \cos B$ (d) $\cot A = \tan B$ (b) $\cos A = \sin B$ (e) $\sec A = \csc B$ (c) $\tan A = \cot B$ (f) $\csc A = \sec B$	235
COORDINATE SYSTEMS	119		Rectangular	$x = r \cos \theta$	455
	120			$y = r \sin \theta$	455
	121		Polar	$r = \sqrt{x^2 + y^2}$	455
	122			$\theta = \arctan \dfrac{y}{x}$	455

No.				Page
123	Quotient Relations	$\tan\theta = \dfrac{\sin\theta}{\cos\theta}$		463
124		$\cot\theta = \dfrac{\cos\theta}{\sin\theta}$		463
125	Pythagorean Relations	$\sin^2\theta + \cos^2\theta = 1$		463
126		$1 + \tan^2\theta = \sec^2\theta$		464
127		$1 + \cot^2\theta = \csc^2\theta$		464
128	Sum or Difference of Two Angles	$\sin(\alpha \pm \beta) = \sin\alpha\cos\beta \pm \cos\alpha\sin\beta$		470
129		$\cos(\alpha \pm \beta) = \cos\alpha\cos\beta \mp \sin\alpha\sin\beta$		471
130		$\tan(\alpha \pm \beta) = \dfrac{\tan\alpha \pm \tan\beta}{1 \mp \tan\alpha\tan\beta}$		472
131	Double-Angle Relations	$\sin 2\alpha = 2\sin\alpha\cos\alpha$		474
132		(a) $\cos 2\alpha = \cos^2\alpha - \sin^2\alpha$ (b) $\cos 2\alpha = 1 - 2\sin^2\alpha$ (c) $\cos 2\alpha = 2\cos^2\alpha - 1$		475
133		$\tan 2\alpha = \dfrac{2\tan\alpha}{1 - \tan^2\alpha}$		476
134	Half-Angle Relations	$\sin\dfrac{\alpha}{2} = \pm\sqrt{\dfrac{1 - \cos\alpha}{2}}$		477
135		$\cos\dfrac{\alpha}{2} = \pm\sqrt{\dfrac{1 + \cos\alpha}{2}}$		478
136		(a) $\tan\dfrac{\alpha}{2} = \dfrac{1 - \cos\alpha}{\sin\alpha}$ (b) $\tan\dfrac{\alpha}{2} = \dfrac{\sin\alpha}{1 + \cos\alpha}$ (c) $\tan\dfrac{\alpha}{2} = \pm\sqrt{\dfrac{1 - \cos\alpha}{1 + \cos\alpha}}$		479
137	Definition of e	$\displaystyle\lim_{k\to\infty}\left(1 + \dfrac{1}{k}\right)^k = e$		525, 956
138	Exponential to Logarithmic Form	If $b^x = y$ then $x = \log_b y$ $(y > 0, b > 0, b \ne 1)$		533
139	Laws of Logarithms Products	Products	$\log_b MN = \log_b M + \log_b N$	540
140		Quotients	$\log_b \dfrac{M}{N} = \log_b M - \log_b N$	540
141		Powers	$\log_b M^P = p\log_b M$	541
142		Roots	$\log_b \sqrt[q]{M} = \dfrac{1}{q}\log_b M$	543
143		Log of 1	$\log_b 1 = 0$	544
144		Log of the Base	$\log_b b = 1$	544
145	Log of the Base Raised to a power		$\log_b b^n = n$	544
146	Base Raised to a Logarithm of the Same Base		$b^{\log_b x} = x$	545
147	Change of Base	$\log N = \dfrac{\ln N}{\ln 10} \approx \dfrac{\ln N}{2.3026}$		545, 968

TRIGONOMETRIC IDENTITIES

LOGARITHMS

No.				Page
148	Power function		$y = ax^n$	505
149	Exponential Function		$y = a(b)^{nx}$	
150	Series Approximation	$b^x = 1 + x \ln b + \dfrac{(x \ln b)^2}{2!} + \dfrac{(x \ln b)^3}{3!} + \cdots (b > 0)$		532
151	Exponential Growth		$y = ae^{nt}$	525
152			Doubling Time or Half-Life $\qquad t = \dfrac{\ln 2}{n}$	551
153	Exponential Decay		$y = ae^{-nt}$	526
154	Exponential Growth to an Upper Limit		$y = a(1 - e^{-nt})$	526, 1005
155	Time Constant	$T = \dfrac{1}{\text{growth rate } n}$		527
156	Recursion Relation for Exponential Growth	$y_t = By_{t-1}$		
157	Nonlinear Growth Equation	$y_t = By_{t-1}(1 - y_{t-1})$		561

SOME USEFUL FUNCTIONS

	No.			Page

<table>
<tr><td rowspan="22" style="writing-mode: vertical">SOME USEFUL FUNCTIONS (Continued)</td></tr>
</table>

No.			Page
158	Series Approximations	$e = 2 + \dfrac{1}{2!} + \dfrac{1}{3!} + \dfrac{1}{4!} + \cdots$	531, 616
159		$e^x = 1 + x + \dfrac{x^2}{2!} + \dfrac{x^3}{3!} + \dfrac{x^4}{4!} + \cdots$	532, 616
160	Logarithmic Function	$y = \log_b x$ $(x > 0, b > 0, b \neq 1)$	
161	Series Approximation	$\ln x = 2a + \dfrac{2a^3}{3} + \dfrac{2a^5}{5} + \dfrac{2a^7}{7} + \cdots$ where $a = \dfrac{x-1}{x+1}$	539, 616
162	Sine Wave of Amplitude a	$y = a \sin(bx + c)$	
163		Period $P = \dfrac{360}{b}$ deg/cycle $= \dfrac{2\pi}{b}$ rad/cycle	424
164		Frequency $= \dfrac{1}{P} = \dfrac{b}{360}$ cycle/deg $= \dfrac{b}{2\pi}$ cycle/rad	424
165		Phase shift $= -\dfrac{c}{b}$	426
166	Addition of a Sine Wave and a Cosine Wave	$A \sin \omega t + B \cos \omega t = R \sin(\omega t + \phi)$ where $R = \sqrt{A^2 + B^2}$ and $\phi = \arctan \dfrac{B}{A}$	447, 474
167	Series Approximations	$\sin x = x - \dfrac{x^3}{3!} + \dfrac{x^5}{5!} - \dfrac{x^7}{7!} + \cdots$	616
168	x is in radians	$\cos x = 1 - \dfrac{x^2}{2!} + \dfrac{x^4}{4!} - \dfrac{x^6}{6!} + \cdots$	616

No.				Page
169		Rectangular	$a + bi$	563, 569
170		Polar	$r \angle \theta$	569
171		Trigonometric	$r(\cos \theta + i \sin \theta)$	569
172		Exponential	$re^{i\theta}$	569
173	Forms of a Complex Number		where $r = \sqrt{a^2 + b^2}$	569
174			$\theta = \arctan \dfrac{b}{a}$	
175			$a = r \cos \theta$	569
176			$b = r \sin \theta$	
177	Powers of i		$i = \sqrt{-1}, \quad i^2 = -1, \quad i^3 = -i, \quad i^4 = 1, \quad i^5 = i, \text{etc.}$	564
178		Sums	$(a + bi) + (c + id) = (a + c) + i(b + d)$	563
179		Differences	$(a + bi) - (c + id) = (a - c) + i(b - d)$	564
180	Rectangular Form	Products	$(a + bi)(c + id) = (ac - bd) + i(ad + bc)$	565
181		Quotients	$\dfrac{a + bi}{c + id} = \dfrac{ac + bd}{c^2 + d^2} + i\dfrac{bc - ad}{c^2 + a^2}$	567
182		Products	$r \angle\theta \cdot r' \angle\theta' = rr' \angle{\theta + \theta'}$	570
183	Polar Form	Quotients	$r\angle\theta \div r' \angle\theta' = \dfrac{r}{r'} \angle{\theta - \theta'}$	571
184		Roots and Powers	DeMoivre's Theorem: $(r\angle\theta)^n = r^n \angle{n\theta}$	571
185		Euler's Formula	$e^{i\theta} = \cos \theta + i \sin \theta$	1018
186		Products	$r_1 e^{i\theta_1} \cdot r_2 e^{i\theta_2} = r_1 r_2 e^{i(\theta_1 + \theta_2)}$	
187	Exponential Form	Quotients	$\dfrac{r_1 e^{i\theta_1}}{r_2 e^{i\theta_2}} = \dfrac{r_1}{r_2} e^{i(\theta_1 - \theta_2)}$	
188		Powers and Roots	$(re^{i\theta})^n = r^n e^{in\theta}$	

COMPLEX NUMBERS

	No.				Page
PROGRESSIONS	189	Series Notation		$u_1 + u_2 + u_3 + \cdots + u_n + \cdots$	587
	190	Arithmetic Progression	Recursion Formula	$a_n = a_{n-1} + d$	593
	191		General Term	$a_n = a + (n-1)d$	593
	192	Common Difference $= d$	Sum of n Terms	$s_n = \dfrac{n(a + a_n)}{2}$	594
	193			$s_n = \dfrac{n}{2}[2a + (n-1)d]$	595
	194	Geometric Progression	Recursion Formula	$a_n = r a_{n-1}$	600
	195		General Term	$a_n = ar^{n-1}$	600
	196		Sum of n Terms	$s_n = \dfrac{a(1 - r^n)}{1 - r}$	601
	197	Common Ratio $= r$		$s_n = \dfrac{a - r a_n}{1 - r}$	601
	198		Sum to Infinity	$S = \dfrac{a}{1 - r}$ where $|r| < 1$	606
BINOMIAL THEOREM	199	Binomial Theorem		$(a+b)^n = a^n + na^{n-1}b + \dfrac{n(n-1)}{2!}a^{n-2}b^2 + \dfrac{n(n-1)(n-2)}{3!}a^{n-3}b^3 + \cdots + b^n$	609
	200	General Term		$r\text{th term} = \dfrac{n!}{(r-1)!(n-r+1)!}a^{n-r+1}b^{r-1}$	611
	201	Binomial Series		$(a+b)^n = a^n + na^{n-1}b + \dfrac{n(n-1)}{2!}a^{n-2}b^2 + \dfrac{n(n-1)(n-2)}{3!}a^{n-3}b^3 + \cdots$ where $|a| > |b|$	612
	202			$(1+x)^n = 1 + nx + \dfrac{n(n-1)}{2!}x^2 + \dfrac{n(n-1)(n-2)}{3!}x^3 + \cdots n$ where $|x| < 1$	612
THE STRAIGHT LINE	203	Increments		$\Delta x = x_2 - x_1, \quad \Delta y = y_2 - y_1$	681
	204		Distance Formula	$d = \sqrt{(\Delta x)^2 + (\Delta y)^2} = \sqrt{(x_2 - x_1)^2 + (y_2 - y_1)^2}$	681
	205		Slope	$m = \dfrac{\text{rise}}{\text{run}} = \dfrac{\Delta y}{\Delta x} = \dfrac{y_2 - y_1}{x_2 - x_1}$	167, 682
	206			$m = \tan(\text{angle of inclination}) = \tan\theta \quad 0 \le \theta < 180°$	683
	207			General Form	$Ax + By + C = 0$	690
	208			Parallel to x axis	$y = b$	690
	209		Equation of Straight Line	Parallel to y axis	$x = a$	690
	210			Slope-Intercept Form	$y = mx + b$	168, 687
	211			Two-Point Form	$\dfrac{y - y_1}{x - x_1} = \dfrac{y_2 - y_1}{x_2 - x_1}$	689
	212			Point-Slope Form	$m = \dfrac{y - y_1}{x - x_1}$	688
	213		If L_1 and L_2 are parallel, then		$m_1 = m_2$	684
	214		If L_1 and L_2 are perpendicular, then		$m_1 = -\dfrac{1}{m_2}$	684
	215		Angle of Intersection $\tan\phi = \dfrac{m_2 - m_1}{1 + m_1 m_2}$			685

	No.				Page

CONIC SECTIONS	216		General Second-Degree Equation	$Ax^2 + Bxy + Cy^2 + Dx + Ey + F = 0$
				Circle: $A = C$ and $B = O$
				Parabola: Either $A = 0$ or $C = 0$
				Ellipse: $A \neq C$ but have the same sign
				Hyperbola: $A \neq C$ and have opposite signs
				For the parabola, ellipse, and hyperbola:
				                                  $B = 0$ when an axis of the curve is
				                                  parallel to a coordinate axis
					698

	217	**Circle of Radius r**		The set of points in a plane equidistant from a fixed point	694
	218		**Standard Form**	$x^2 + y^2 = r^2$	695
	219			$(x - h)^2 + (y - k)^2 = r^2$	696
	220		General Form	$x^2 + y^2 + Dx + Ey + F = 0$	698
	221	**Parabola**		The set of points in a plane such that the distance PF from each point P to a fixed point F (the focus) is equal to the distance PD to a fixed line (the directrix) $$PF = PD$$	702
	222		**Standard Form**	$y^2 = 4px$	702
	223			$x^2 = 4py$	704
	224			$(y - k)^2 = 4p(x - h)$	705
	225			$(x - h)^2 = 4p(y - k)$	705
	226		General Form	$Cy^2 + Dx + Ey + F = 0$ or $Ax^2 + Dx + Ey + F = 0$	707
	227		Focal Width or length of latus rectum	$L = |4p|$	705
	228		*Area*	$Area = \dfrac{2}{3}ab$	

No.				Page
229		The set of points in a plane such that the sum of the distances PF and PF' from each point P to two fixed points F and F' (the foci) is constant, and equal to the length $2a$ of the major axis $$PF + PF' = 2a$$		713
230			$$\frac{x^2}{a^2} + \frac{y^2}{b^2} = 1$$ $$a > b$$	716
231			$$\frac{y^2}{a^2} + \frac{x^2}{b^2} = 1$$ $$a > b$$	718
232			$$\frac{(x-h)^2}{a^2} + \frac{(y-k)^2}{b^2} = 1$$ $$a > b$$	719
233			$$\frac{(y-k)^2}{a^2} + \frac{(x-h)^2}{b^2} = 1$$ $$a > b$$	719
234		General Form	$$Ax^2 + Cy^2 + Dx + Ey + F = 0$$ $A \neq C$, but have same signs	720
235		Distance from Center to Focus	$$c = \sqrt{a^2 - b^2}$$	714
236		Focal Width or length of latus rectum	$$L = \frac{2b^2}{a}$$	722
237		Area $= \pi ab$		

CONIC SECTIONS *(Continued)*   Ellipse   Standard Form

No.					Page
238			The set of points in a plane such that the difference of the distances PF and PF' from each point P to two fixed points F and F' (the foci) is constant, and equal to the distance $2a$ between the vertices $$PF' - PF = 2a$$		725
239		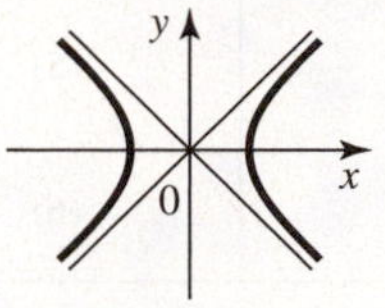		$$\dfrac{x^2}{a^2} - \dfrac{y^2}{b^2} = 1$$	726
240		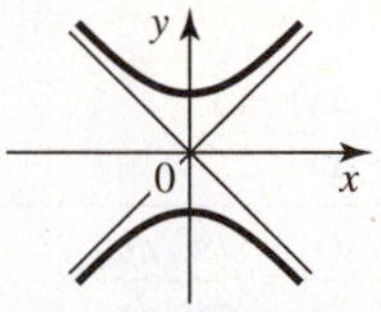		$$\dfrac{y^2}{a^2} - \dfrac{x^2}{b^2} = 1$$	726
241		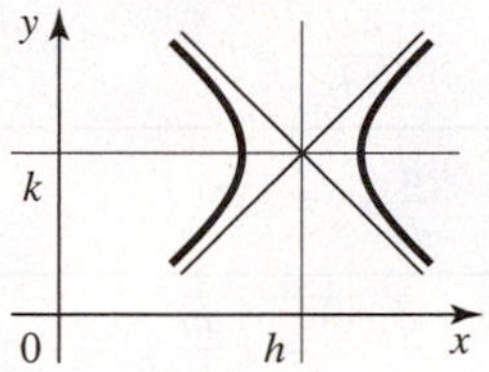		$$\dfrac{(x-h)^2}{a^2} - \dfrac{(y-k)^2}{b^2} = 1$$	729
242		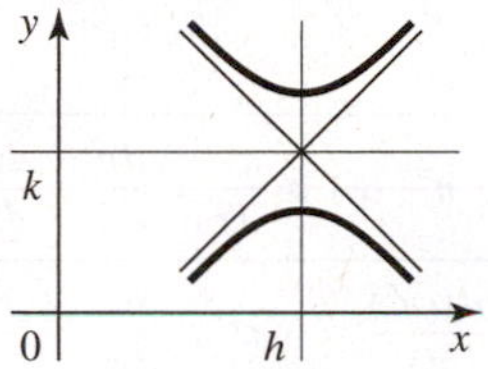		$$\dfrac{(y-k)^2}{a^2} - \dfrac{(x-h)^2}{b^2} = 1$$	729
243		General Form		$$Ax^2 + Cy^2 + Dx + Ey + F = 0$$ $A \neq C$, and have opposite signs	731
244		Distance from Center to Focus		$$c = \sqrt{a^2 + b^2}$$	727
245		Slope of Asymptotes	Transverse Axis Horizontal	$$\text{Slope} = \pm \dfrac{b}{a}$$	727
246			Transverse Axis Vertical	$$\text{Slope} = \pm \dfrac{a}{b}$$	727
247		Length of Latus Rectum		$$L = \dfrac{2b^2}{a}$$	
248		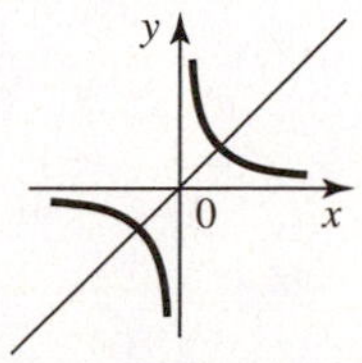		Axes Rotated 45° $$xy = k$$	731

CONIC SECTIONS (Continued)

Hyperbola

Standard Form

	No.				Page
LIMITS	249			Limit Notation	$\lim\limits_{x\to a} f(x) = L$	738
	250		Limits involving zero or infinity	(1) $\lim\limits_{x\to 0} Cx = 0$ (4) $\lim\limits_{x\to \infty} Cx = \infty$ (2) $\lim\limits_{x\to 0} \dfrac{x}{C} = 0$ (5) $\lim\limits_{x\to \infty} \dfrac{x}{C} = \infty$ (3) $\lim\limits_{x\to 0} \dfrac{C}{x} = +\infty$ (6) $\lim\limits_{x\to \infty} \dfrac{C}{x} = 0$	741
DERIVATIVES	251		Definition of the Derivative	$f'(x) = \dfrac{dy}{dx} = \lim\limits_{\Delta x\to 0} \dfrac{\Delta y}{\Delta x} = \lim\limits_{\Delta x\to 0} \dfrac{f(x + \Delta x) - f(x)}{\Delta x}$ $= \lim\limits_{\Delta x\to 0} \dfrac{(y + \Delta y) - y}{\Delta x}$	751
	252		The Chain Rule	$\dfrac{dy}{dx} = \dfrac{dy}{du}\cdot\dfrac{du}{dx}$	767
	253	**Rules for Derivatives**	Of a Constant	$\dfrac{d(c)}{dx} = 0$	759
	254		Of a Power Function	$\dfrac{d}{dx}x^n = nx^{n-1}$	
	255		Of a Constant Times a Function	$\dfrac{d(cu)}{dx} = c\dfrac{du}{dx}$	772
	256		Of a Constant Times a Power of x	$\dfrac{d}{dx}cx^n = cnx^{n-1}$	760
	257		Of a Sum	$\dfrac{d}{dx}(u + v + w) = \dfrac{du}{dx} + \dfrac{dv}{dx} + \dfrac{dw}{dx}$	762
	258		Of a Function Raised to a Power	$\dfrac{d(cu^n)}{dx} = cnu^{n-1}\dfrac{du}{dx}$	768, 777
	259		Of a Product	$\dfrac{d(uv)}{dx} = u\dfrac{dv}{dx} + v\dfrac{du}{dx}$	771
	260		Of a Product of Three Factors	$\dfrac{d(uvw)}{dx} = uv\dfrac{dw}{dx} + uw\dfrac{dv}{dx} + vw\dfrac{du}{dx}$	773
	261		Of a Product of n Factors	The derivative is an expression of n terms, each term being the product of $n - 1$ of the factors and the derivative of the other factor.	773
	262		Of a Quotient	$\dfrac{d}{dx}\left(\dfrac{u}{v}\right) = \dfrac{v\dfrac{du}{dx} - u\dfrac{dv}{dx}}{v^2}$	774

	No.				Page
DERIVATIVES (*Continued*)	263	Rules for Derivatives (*cont.*)	Of the Trigonometric Functions	$\frac{d(\sin u)}{dx} = \cos u \frac{du}{dx}$	943, 950
	264			$\frac{d(\cos u)}{dx} = -\sin u \frac{du}{dx}$	944, 950
	265			$\frac{d(\tan u)}{dx} = \sec^2 u \frac{du}{dx}$	949, 950
	266			$\frac{d(\cot u)}{dx} = -\csc^2 u \frac{du}{dx}$	950
	267			$\frac{d(\sec u)}{dx} = \sec u \tan u \frac{du}{dx}$	950
	268			$\frac{d(\csc u)}{dx} = -\csc u \cot u \frac{du}{dx}$	950
	269		Of the Inverse Trigonometric Functions	$\frac{d(\mathrm{Sin}^{-1} u)}{dx} = \frac{1}{\sqrt{1-u^2}} \frac{du}{dx} \quad -1 < u < 1$	954
	270			$\frac{d(\mathrm{Cos}^{-1} u)}{dx} = \frac{-1}{\sqrt{1-u^2}} \frac{du}{dx} \quad -1 < u < 1$	954
	271			$\frac{d(\mathrm{Tan}^{-1} u)}{dx} = \frac{1}{1+u^2} \frac{du}{dx}$	954
	272			$\frac{d(\mathrm{Cot}^{-1} u)}{dx} = \frac{-1}{1+u^2} \frac{du}{dx}$	954
	273			$\frac{d(\mathrm{Sec}^{-1} u)}{dx} = \frac{1}{u\sqrt{u^2-1}} \frac{du}{dx} \quad \lvert u \rvert > 1$	954
	274			$\frac{d(\mathrm{Csc}^{-1} u)}{dx} = \frac{-1}{u\sqrt{u^2-1}} \frac{du}{dx} \quad \lvert u \rvert > 1$	954
	275		Of Logarithmic and Exponential Functions	(a) $\frac{d}{dx}(\log_b u) = \frac{1}{u} \log_b e \frac{du}{dx}$ (b) $\frac{d}{dx}(\log_b u) = \frac{1}{u \ln b} \frac{du}{dx}$	956
	276			$\frac{d}{dx}(\ln u) = \frac{1}{u} \frac{du}{dx}$	957
	277			$\frac{d}{dx} b^u = b^u \frac{du}{dx} \ln b$	961
	278			$\frac{d}{dx} e^u = e^u \frac{du}{dx}$	962
	279		Differential of y	$dy = f'(x)\,dx$	780
	280	Graphical Applications	Maximum and Minimum Points	To find maximum and minimum points (and other stationary points) set the first derivative equal to zero and solve for x.	795
	281		First-Derivative Test	The first derivative is negative to the left of, and positive to the right of, a minimum point. The reverse is true for a maximum point.	797
	282		Second-Derivative Test	If the first derivative at some point is zero, then, if second derivative is 1. Positive, the point is a minimum. 2. Negative, the point is a maximum. 3. Zero, the test fails.	797
	283		Ordinate Test	Find y a small distance to either side of the point to be tested. If y is greater there, we have a minimum; if less, we have a maximum.	797
	284		Inflection Points	To find points of inflection, set the second derivative to zero and solve for x. Test by seeing if the second derivative changes sign a small distance to either side of the point.	799
	285		Newton's Method	$x_{n+1} = x_n - \frac{f(x_n)}{f'(x_n)}$	

	No.				Page
INTEGRALS	286		The Indefinite Integral	$$\int F'(x)\,dx = F(x) + C$$	843
	287			$$\int f(x)\,dx = F(x) + C \quad \text{where} \quad f(x) = F'(x)$$	843
	288		The Fundamental Theorem	$$\int_a^b f(x)\,dx = F(b) - F(a)$$	864
	289		Midpoint Method	$$A \cong \sum_{i=1}^n f(x_i{}^*)\Delta x$$ where $f(x_i{}^*)$ is the height of the ith panel at its midpoint	869
	290		Defined by Riemann Sums	$$A = \lim_{\Delta x \to 0} \sum_{i=1}^n f(x_i{}^*)\Delta x = \int_a^b f(x)\,dx$$	871
	291	Exact Area under a Curve	By Integration	$$A = \int_a^b f(x)\,dx = F(b) - F(a)$$	873
	292		Areas between Two Curves	$$A = \int_a^b [f(x) - g(x)]\,dx$$	890
APPLICATIONS OF THE DEFINITE INTEGRAL	293	Volumes of Solids of Revolution — Disk Method		$$\text{Volume} = dV = \pi r^2\,dh$$	899
	294			$$V = \pi \int_a^b r^2\,dh$$	899
	295	Washer Method		$$dV = \pi(r_o^2 - r_i^2)\,dh$$	902
	296			$$V = \pi \int_a^b (r_o^2 - r_i^2)\,dh$$	903
	297	Shell Method		$$dV = 2\pi r h\,dr$$	901, 935
	298			$$V = 2\pi \int_a^b r h\,dr$$	901

No.				Page

APPLICATIONS OF THE DEFINITE INTEGRAL *(Continued)*

Length of Arc

299	$s = \int_a^b \sqrt{1 + \left(\dfrac{dy}{dx}\right)^2}\, dx$	911
300	$s = \int_c^d \sqrt{1 + \left(\dfrac{dx}{dy}\right)^2}\, dy$	911

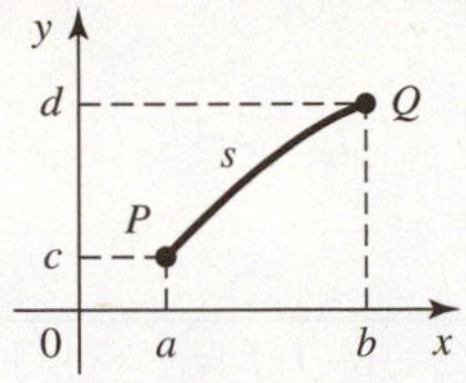

Surface Area

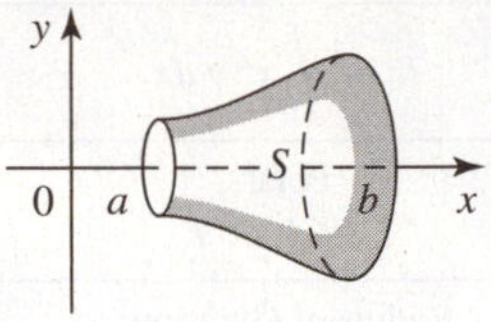 301	About x Axis: $S = 2\pi \int_a^b y \sqrt{1 + \left(\dfrac{dy}{dx}\right)^2}\, dx$	915
302	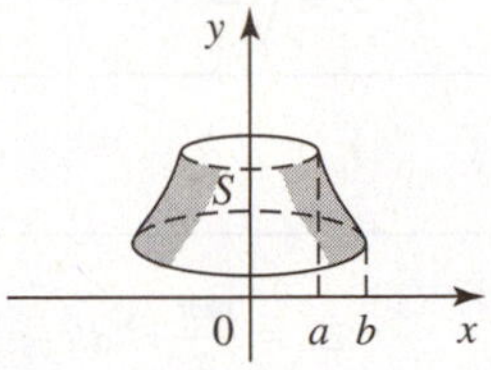 About y Axis: $S = 2\pi \int_a^b x \sqrt{1 + \left(\dfrac{dy}{dx}\right)^2}\, dx$	916

Centroids

Of Plane Area:

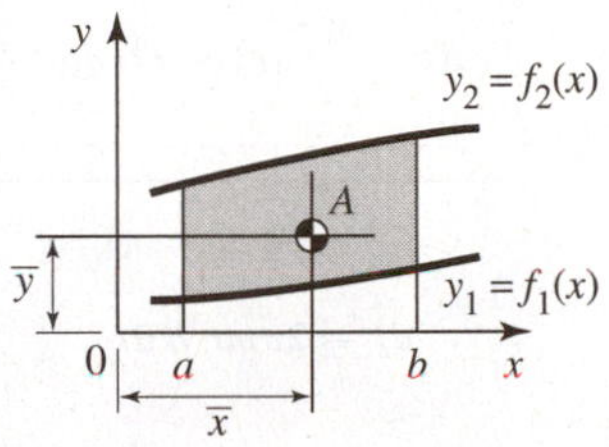

303	$\bar{x} = \dfrac{1}{A} \int_a^b x(y_2 - y_1)\, dx$	920
304	$\bar{y} = \dfrac{1}{2A} \int_a^b (y_1 + y_2)(y_2 - y_1)\, dx$	920

Of Solid of Revolution of Volume V:

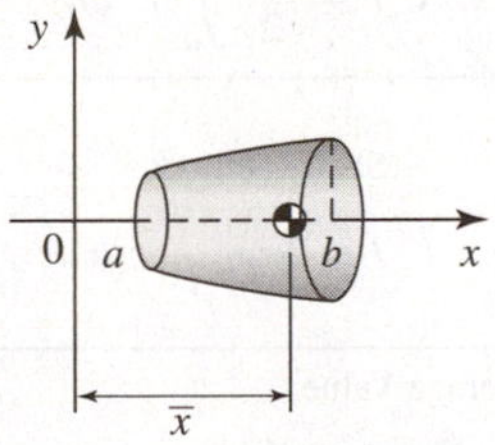

305	About x Axis: $\bar{x} = \dfrac{\pi}{V} \int_a^b xy^2\, dx$	923
306	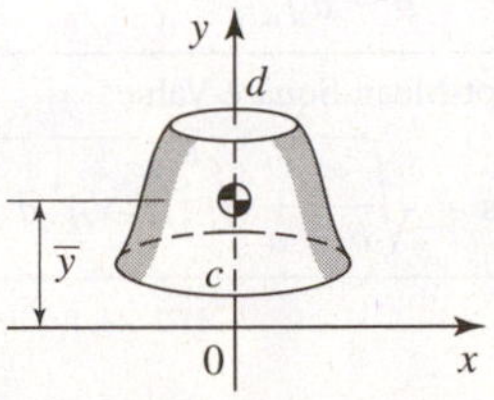 About y Axis: $\bar{y} = \dfrac{\pi}{V} \int_c^d yx^2\, dy$	924

No.						Page
			Thin Strip			
307		Of Areas			$I_p = Ar^2$	932
308			Extended Area		$I_x = \dfrac{1}{3}\int y^3\,dx$	934
309					$I_y = \int x^2 y\,dx$	934
310					Polar $I_o = I_x + I_y$	
311					Radius of Gyration: $r = \sqrt{\dfrac{I}{A}}$	933
312	Moment of Intertia				$I_p = Ar^2$	
313		Of Masses	Disk:	About Axis of Revolution (Polar Moment of Inertia)	$dI = \dfrac{m\pi}{2}\,r^4\,dh$	
314			Ring:		$dI = \dfrac{m\pi}{2}\,(r_o^4 - r_i^4)\,dh$	
315			Shell:		$dI = 2\pi m r^3 h\,dr$	935
316			Solid of Revolution:		Disk Method: $I = \dfrac{m\pi}{2}\displaystyle\int_a^b r^4\,dh$	936
317					Shell Method: $I = 2\pi m\displaystyle\int r^3 h\,dr$	935
318	Average and rms Values				Average Value: $y_{\text{avg}} = \dfrac{1}{b-a}\displaystyle\int_a^b f(x)\,dx$	975
319					Root-Mean-Square Value: $\text{rms} = \sqrt{\dfrac{1}{b-a}\displaystyle\int_a^b [f(x)]^2\,dx}$	976

APPLICATIONS OF THE DEFINITE INTEGRAL (Continued)

	No.						Page
DIFFERENTIAL EQUATIONS	320	First-Order	Euler's Method		$x_q = x_p + \Delta x$ $y_q = y_p + m_p \Delta x$		987
	321		Variables Separable		$f(y)\, dy = g(x)\, dx$		988
	322		Integrable Combinations		$x\, dy + y\, dx = d(xy)$		992
	323				$\dfrac{x\, dy - y\, dx}{x^2} = d\left(\dfrac{y}{x}\right)$		992
	324				$\dfrac{y\, dx - x\, dy}{y^2} = d\left(\dfrac{x}{y}\right)$		992
	325				$\dfrac{x\, dy - y\, dx}{x^2 + y^2} = d\left(\tan^{-1}\dfrac{y}{x}\right)$		992
	326		Homogeneous		$M\, dx + N\, dy = 0$ (Substitute $y = ux$)		994
	327		First-Order Linear	Form	$y' + Py = Q$		996
	328			Integrating Factor	$R = e^{\int P dx}$		997
	329			Solution	$ye^{\int P dx} = \displaystyle\int Q e^{\int P dx}$		998
	330		Bernoulli's Equation		$\dfrac{dy}{dx} - Gy = Hy^n$ (Substitute $z = y^{1-n}$)		999
	331	Second-Order	Right Side Zero	Form	$ay'' + by' + cy = 0$		1016
	332			Auxiliary Equation	$am^2 + bm + c = 0$		1016
				Form of Solution	Roots of Auxiliary Equation	Solution	1016
	333				Real and Unequal	$y = c_1 e^{m_1 x} + c_2 e^{m_2 x}$	1018, 1019
	334				Real and Equal	$y = c_1 e^{mx} + c_2 x e^{mx}$	1019
	335				Nonreal	(a) $y = e^{ax}(C_1 \cos bx + C_2 \sin bx)$ or (b) $y = C e^{ax} \sin(bx + \phi)$	1019
	336		Right Side Not Zero	Form	$ay'' + by' + cy = f(x)$		1014, 1024
	337			Complete Solution	$y = \underset{\substack{\uparrow \\ \text{complementary} \\ \text{function}}}{y_c} + \underset{\substack{\uparrow \\ \text{particular} \\ \text{integral}}}{y_p}$		1022

STATISTICS AND PROBABILITY	338	Measures of Central Tendency	Arithmetic Mean	$\bar{x} = \dfrac{\sum x}{n}$	629

No.	Category	Name	Formula	Page
338	Measures of Central Tendency	Arithmetic Mean	$\bar{x} = \dfrac{\sum x}{n}$	629
339		Median	The median of a set of numbers arranged in order of magnitude is the middle value, or the mean of the two middle values.	630
340		Mode	The mode of a set of numbers is the measurement(s) that occurs most often in the set.	630
341	Measures of Dispersion	Range	The range of a set of numbers is the difference between the largest and the smallest number in the set.	632
342		Population Variance σ^2	$\sigma^2 = \dfrac{\sum(x - \bar{x})^2}{n}$	633
343		Sample Variance s^2	$s^2 = \dfrac{\sum(x - \bar{x})^2}{n - 1}$	633
344		Standard Deviation s	The standard deviation of a set of numbers is the positive square root of the variance.	634
345	Probability	Of a Single Event	$P(A) = \dfrac{\text{number of ways in which } A \text{ can happen}}{\text{number of equally likely ways}}$	639
346		Of Two Events Both Occurring	$P(A, B) = P(A)P(B)$	640
347		Of Several Events All Occurring	$P(A, B, C, \ldots) = P(A)P(B)P(C)\ldots$	640
348		Of Either of Two Events Occurring	$P(A + B) = P(A) + P(B) - P(A, B)$	641
349		Of Two Mutually Exclusive Events Occurring	$P(A + B) = P(A) + P(B)$	641
350	Gaussian Distribution		$y = \dfrac{1}{\sigma\sqrt{2\pi}} e^{-(x-\mu)^2/2\sigma^2}$ where μ = population mean and σ = population standard deviation	649
351	Binomial Probability Formula		$P(x) = \dfrac{n!}{(n - x)! \, x!} p^x q^{n-x}$	645
352	Standard Error	Of the Mean	$SE_{\bar{x}} = \dfrac{\sigma}{\sqrt{n}} = \dfrac{s}{\sqrt{n}}$	658
353		Of the Standard Deviation	$SE_s = \dfrac{\sigma}{\sqrt{2n}}$	659
354		Of a Proportion	$SE_p = \sqrt{\dfrac{p(1 - p)}{n}}$	659
355	Correlation Coefficient		$r = \dfrac{n\sum xy - \sum x \sum y}{\sqrt{n\sum x^2 - (\sum x)^2}\,\sqrt{n\sum y^2 - (\sum y)^2}}$	670
356	Least Squares Line	Slope m	$m = \dfrac{n\sum xy - \sum x \sum y}{n\sum x^2 - (\sum x)^2}$	672
		y intercept b	$b = \dfrac{\sum x^2 \sum y - \sum x \sum xy}{n\sum x^2 - (\sum x)^2}$	672

Applications

Note that the applications numbers start with 1000.

MIXTURES	1000	Mixture containing Ingredients A, B, C, ...	Total amount of mixture = amount of A + amount of B + ...		123
	1001		Final amount of each ingredient = initial amount + amount added − amount removed		123
	1002	Combination of Two Mixtures	Final amount of A = amount of A in first mixture + amount of A in second mixture		124
	1003	Fluid Flow	Amount of flow = flow rate × elapsed time $A = QT$		
WORK	1004		Amount done = rate of work × time worked		130
	1005		Constant Force	Work = force × distance = Fd	928
	1006		Variable Force	Work = $\int_a^b F(x)\, dx$	929
FINANCIAL	1007	Unit Cost		Unit cost = $\dfrac{\text{total cost}}{\text{number of units}}$	
	1008	Interest: Principal a Invested at Rate n for t years Accumulates to Amount y	Simple	$y = a(1 + nt)$	
	1009		Compounded Annually	$y = a(1 + n)^t$	524
	1010			Recursion Relation $y_t = y_{t-1}(1 + n)$	523
	1011		Compounded m times/yr	$y = a\left(1 + \dfrac{n}{m}\right)^{mt}$	524
STATICS	1012		Moment about Point a	$M_a = Fd$	127
	1013	Equations of Equilibrium (Newton's First Law)	The sum of all horizontal forces = 0		127
	1014		The sum of all vertical forces = 0		127
	1015		The sum of all moments about any point = 0		127
	1016		Coefficient of Friction	$\mu = \dfrac{f}{N}$	

No.					Page
1017			Uniform Motion (Constant Speed)	Distance = rate × time $D = Rt$	
1018			Uniformly Accelerated (Constant Acceleration a, Initial Velocity v_0) For free fall, $a = g = 9.807 \text{ m/s}^2 = 32.2 \text{ ft/s}^2$	Displacement at Time t $\quad s = v_0 t + \dfrac{at^2}{2}$	359, 879
1019				Velocity at Time t $\quad v = v_0 + at$	879
1020				Newton's Second Law $\quad F = ma$	
1021		Linear Motion		Average Speed $\quad$ Average speed $= \dfrac{\text{total distance traveled}}{\text{total time elapsed}}$	
1022			Nonuniform Motion	Displacement $\quad s = \displaystyle\int v\, dt$	878
1023				Instantaneous Velocity $\quad v = \dfrac{ds}{dt}$	816
1024				$\quad v = \displaystyle\int a\, dt$	878
1025				Instantaneous Acceleration $\quad a = \dfrac{dv}{dt} = \dfrac{d^2 s}{dt^2}$	816
1026	MOTION	Rotation	Uniform Motion	Angular Displacement $\quad \theta = \omega t$	416
1027				Linear Speed of Point at Radius r $\quad v = \omega r$	417
1028			Nonuniform Motion	Angular Displacement $\quad \theta = \displaystyle\int \omega\, dt$	881
1029				Angular Velocity $\quad \omega = \dfrac{d\theta}{dt}$	820
1030				$\quad \omega = \displaystyle\int \alpha\, dt$	881
1031				Angular Acceleration $\quad \alpha = \dfrac{d\omega}{dt} = \dfrac{d^2\theta}{dt^2}$	820
1032		Curvilinear Motion	x and y Components	Displacement $\quad$ (a) $x = \displaystyle\int v_x\, dt$ $\quad$ (b) $y = \displaystyle\int v_y\, dt$	880
1033				Velocity $\quad$ (a) $v_x = \dfrac{dx}{dt}$ $\quad$ (b) $v_y = \dfrac{dy}{dt}$	818
1034				(a) $v_x = \displaystyle\int a_x\, dt$ $\quad$ (b) $v_y = \displaystyle\int a_y\, dt$	880
1035				Acceleration $\quad$ (a) $a_x = \dfrac{dv_x}{dt}$ $\quad$ (b) $a_y = \dfrac{dv_y}{dt}$ $\qquad = \dfrac{d^2 x}{dt^2} \qquad = \dfrac{d^2 y}{dt^2}$	818

	No.				Page
MECHANICAL VIBRATIONS	1036		Simple Harmonic Motion (No Damping)	$x = x_0 \cos \omega_n t$	1029
	1037			Undamped Angular Velocity $\quad \omega_n = \sqrt{\dfrac{kg}{W}}$	1029
	1038			Natural Frequency $\quad f_n = \dfrac{\omega_n}{2\pi}$	1029
	1039		Underdamped	$x = x_0 e^{-at} \cos \omega_d t$	1028
	1040			Damped Angular Velocity $\quad \omega_d = \sqrt{\omega_n^2 - \dfrac{c^2 g^2}{W^2}}$	1028
	1041	Coefficient of friction = c	Overdamped	$x = C_1 e^{m_1 t} + C_2 e^{m_2 t}$	1030
MATERIAL PROPERTIES	1042	Density		Density $= \dfrac{\text{weight}}{\text{volume}} \quad$ or $\quad \dfrac{\text{mass}}{\text{volume}}$	
	1043	Mass		Mass $= \dfrac{\text{weight}}{\text{acceleration due to gravity}}$	
	1044	Specific Gravity		SG $= \dfrac{\text{density of substance}}{\text{density of water}}$	
	1045	Pressure	Total Force on a Surface	Force = pressure $\times$ area	926
	1046		Force on a Submerged Surface	$F = \delta \displaystyle\int y\, dA$	926
	1047			$F = \delta \bar{y} A$	927
	1048	pH		pH $= -10 \log$ concentration	559
TEMPERATURE	1049	Conversions between Degrees Celsius (C) and Degrees Fahrenheit (F)		$C = \frac{5}{9}(F - 32)$	
	1050			$F = \frac{9}{5}C + 32$	
STRENGTH OF MATERIALS	1051		Normal Stress	$\sigma = \dfrac{P}{a}$	
	1052		Strain	$\epsilon = \dfrac{e}{L}$	
	1053		Modulus of Elasticity and Hooke's Law	$E = \dfrac{PL}{ae}$	359
	1054	Tension or Compression		$E = \dfrac{\sigma}{\epsilon}$	
	1055	Thermal Expansion	Elongation	$e = \alpha L\, \Delta t$	
	1056		New Length	$L = L_0 (1 + \alpha \Delta t)$	359
	1057		Strain	$\epsilon = \dfrac{e}{L} = \alpha \Delta t$	
	1058		Stress, if Restrained	$\sigma = E\epsilon = E\alpha\, \Delta t$	
	1059	Temperature change $= \Delta t$ Coefficient of thermal expansion $= \alpha$	Force, if Restrained	$P = a\sigma = aE\alpha\, \Delta t$	
	1060		Force needed to Deform a Spring	$F =$ spring constant $\times$ distance $= kx$	929

	No.				Page
ELECTRICAL TECHNOLOGY	1061	Ohm's Law		Current $=\frac{\text{voltage}}{\text{resistance}}$ $I=\frac{V}{R}$	
	1062	Combinations of Resistors	In Series	$R = R_1 + R_2 + R_3 + \cdots$	
	1063		In Parallel	$\frac{1}{R} = \frac{1}{R_1} + \frac{1}{R_2} + \frac{1}{R_3} + \cdots$	359
	1064	Power Dissipated in a Resistor 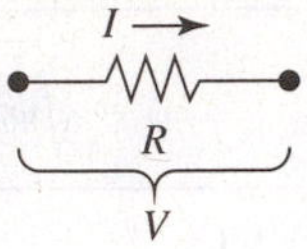		Power $= P = VI$	
	1065			$P = \frac{V^2}{R}$	
	1066			$P = I^2R$	
	1067	Kirchhoff's Laws	Loops	The sum of the voltage rises and drops around any closed loop is zero.	
	1068		Nodes	The sum of the currents entering and leaving any node is zero.	
	1069	Resistance Change with Temperature		$R = R_1[1 + \alpha(t - t_1)]$	360
	1070	Resistance of a wire 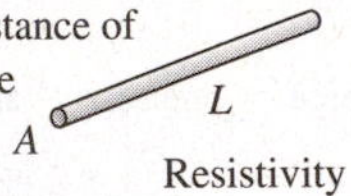 Resistivity ρ		$R = \frac{\rho L}{A}$	
	1071	Combinations of Capacitors	In Series	$\frac{1}{C} = \frac{1}{C_1} + \frac{1}{C_2} + \frac{1}{C_3} + \cdots$	
	1072		In Parallel	$C = C_1 + C_2 + C_3 + \cdots$	
	1073	Charge on a Capacitor at Voltage V		$Q = CV$	
			Sinusoidal Form	Complex Form	
	1074	Alternating Voltage	$v = V_m \sin(\omega t + \phi_1)$	$\mathbf{V} = V_{eff} \angle \phi_1$	580
	1075	Alternating Current	$i = I_m \sin(\omega t + \phi_2)$	$\mathbf{I} = I_{eff} \angle \phi_2$	580
	1076	Period		$P = \frac{2\pi}{\omega}$ seconds	435
	1077	Frequency		$f = \frac{1}{P} = \frac{\omega}{2\pi}$ hertz (cycles/s)	436
	1078	Current		$i = \frac{dq}{dt}$	811, 946
	1079	Charge		$q = \int i\,dt$ coulombs	884, 969
	1080	Capacitor: Instantaneous Current		$i = C\frac{dv}{dt}$	811, 946
	1081	Capacitor: Instantaneous Voltage		$v = \frac{1}{C}\int i\,dt$ volts	884, 969
	1082	Capacitor: Current when Charging or Discharging	Series RC Circuit 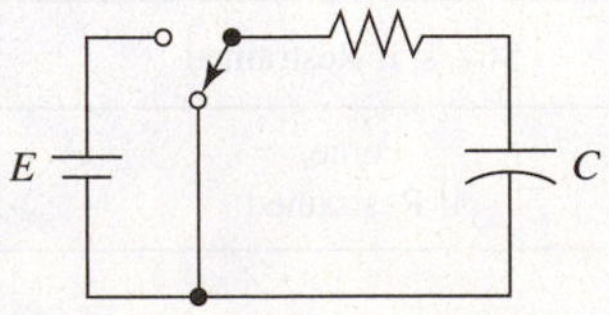	$i = \frac{E}{R}e^{-t/RC}$	531, 1009
	1083	Capacitor: Voltage when Charging		$v = E(1 - e^{-t/RC})$	1011
	1084	Capacitor: Voltage when Discharging		$v = Ee^{-t/RC}$	1009

No.					Page
1085	Inductor		Instantaneous Current	$i = \dfrac{1}{L}\displaystyle\int v\,dt \quad$ amperes	885, 969
1086			Instantaneous Voltage	$v = L\dfrac{di}{dt}$	812, 946
1087			Current when Charging	**Series RL Circuit** $i = \dfrac{E}{R}(1 - e^{-Rt/L})$	1008
1088			Current when Discharging	$i = \dfrac{E}{R}e^{-Rt/L}$	1011
1089			Voltage when Charging or Discharging	$v = Ee^{-Rt/L}$	1008
1090		DC Source	Resonant Frequency	$\omega_n = \dfrac{1}{\sqrt{LC}}$	1032, 1037
1091			No Resistance: The Series LC Circuit:	$i = \dfrac{E}{\omega_n L}\sin \omega_n t$	1034
1092			Underdamped	$i = \dfrac{E}{\omega_d L}e^{-at}\sin \omega_d t$	1033
1093				$\omega_d = \sqrt{\omega_n^2 - \dfrac{R^2}{4L^2}}$	1033
1094			Overdamped	$i = \dfrac{E}{2j\omega_d L}\left[e^{(-a+j\omega_d)t} - e^{(-a-j\omega_d)t}\right]$	1034
1095		AC Source	Inductive Reactance	$X_L = \omega L$	
1096			Capacitive Reactance	$X_C = \dfrac{1}{\omega C}$	
1097			Total Reactance	$X = X_L - X_C$	227
1098			Impedance	$|Z| = \sqrt{R^2 + X^2} = \sqrt{R^2 + \left(\omega L - \dfrac{1}{\omega C}\right)^2}$	228
1099			Phase Angle	$\phi = \arctan\dfrac{X}{R}$	228
1100			Complex Impedance	$Z = R + jX = Z\underline{\diagup \phi} = Ze^{j\phi}$	581
1101			Steady-State Current	$i_{ss} = \dfrac{E}{Z}\sin(\omega t - \phi)$	1036
1102	Ohm's Law for AC			$V = ZI$	581
1103	Power Gain or Loss			$G_p = 10\log_{10}\dfrac{P_2}{P_1}\quad$ dB	556
1104	Voltage Gain or Loss			$G_v = 20\log_{10}\dfrac{V_2}{V_1}\quad$ dB	557
1105	Sound Level Gain or Loss			$G_s = 10\log_{10}\dfrac{I_2}{I_1}\quad$ dB	558

ELECTRICAL TECHNOLOGY (Continued)

Series RLC Circuit

R

Source

L

C

Conversion Factors

UNIT	EQUALS
LENGTH	
1 angstrom	1×10^{-10} meter
	1×10^{-4} micrometer (micron)
1 centimeter	10^{-2} meter
	0.3937 inch
1 foot	12 inches
	0.3048 meter
1 inch	25.4 millimeters
	2.54 centimeters
1 kilometer	3281 feet
	0.5400 nautical mile
	0.6214 statute mile
	1094 yards
1 light-year	9.461×10^{12} kilometers
	5.879×10^{12} statute miles
1 meter	10^{10} angstroms
	3.281 feet
	39.37 inches
	1.094 yards
1 micron	10^{4} angstroms
	10^{-4} centimeter
	10^{-6} meter
1 nautical mile (International)	8.439 cables
	6076 feet
	1852 meters
	1.151 statute miles
1 statute mile	5280 feet
	8 furlongs
	1.609 kilometers
	0.8690 nautical mile
1 yard	3 feet
	0.9144 meter

UNIT	EQUALS
ANGLES	
1 degree	60 minutes
	0.01745 radian
	3600 seconds
	2.778×10^{-3} revolution
1 minute of arc	0.01667 degree
	2.909×10^{-4} radian
	60 seconds
1 radian	0.1592 revolution
	57.296 degrees
	3438 minutes
1 second of arc	2.778×10^{-4} degree
	0.01667 minute
AREA	
1 acre	4047 square meters
	43560 square feet
1 are	0.02471 acre
	1 square dekameter
	100 square meters
1 hectare	2.471 acres
	100 ares
	10000 square meters
1 square foot	144 square inches
	0.09290 square meter
1 square inch	6.452 square centimeters
1 square kilometer	247.1 acres
1 square meter	10.76 square feet
1 square mile	640 acres
	2.788×10^{7} square feet
	2.590 square kilometers
VOLUME	
1 board-foot	144 cubic inches
1 bushel (U.S.)	1.244 cubic feet
	35.24 liters
1 cord	128 cubic feet
	3.625 cubic meters
1 cubic foot	7.481 gallons (U.S. liquid)
	28.32 liters
1 cubic inch	0.01639 liter
	16.39 milliliters
1 cubic meter	35.31 cubic feet
	10^{6} cubic centimeter
1 cubic millimeter	6.102×10^{-5} cubic inch
1 cubic yard	27 cubic feet
	0.7646 cubic meter
1 gallon (imperial)	277.4 cubic inches
	4.546 liters
1 gallon (U.S. liquid)	231 cubic inches
	3.785 liters

UNIT	EQUALS
VOLUME (Continued)	
1 kiloliter	35.31 cubic feet
	1.308 cubic yards
	220 imperial gallons
1 liter	10^3 cubic centimeters
	10^6 cubic millimeters
	10^{-3} cubic meter
	61.02 cubic inches
MASS	
1 gram	10^{-3} kilogram
	6.854×10^{-5} slug
1 kilogram	1000 grams
	0.06854 slug
1 slug	14.59 kilograms
	14,590 grams
1 metric ton	1000 kilograms
FORCE	
1 dyne	10^{-5} newton
1 newton	10^5 dynes
	0.2248 pound
	3.597 ounces
1 pound	4.448 newtons
	16 ounces
1 ton	2000 pounds
AT SEA LEVEL	
1 kilogram	2.205 pounds
VELOCITY	
1 foot/minute	0.3048 meter/minute
	0.011364 mile/hour
1 foot/second	1097 kilometers/hour
	18.29 meters/minute
	0.6818 mile/hour
1 kilometer/hour	3281 feet/hour
	54.68 feet/minute
	0.6214 mile/hour
1 kilometer/minute	3281 feet/minute
	37.28 miles/hour
1 knot	6076 feet/hour
	101.3 feet/minute
	1.852 kilometers/hour
	30.87 meters/minute
	1.151 miles/hour
1 meter/hour	3.281 feet/hour
1 mile/hour	1.467 feet/second
	1.609 kilometers/hour

UNIT	EQUALS
POWER	
1 British thermal unit/hour	0.2929 watt
1 Btu/pound	2.324 joules/gram
1 Btu-second	1.414 horsepower
	1.054 kilowatts
	1054 watts
1 horsepower	42.44 Btu/minute
	550 footpounds/second
	746 watts
1 kilowatt	3414 Btu/hour
	737.6 footpounds/second
	1.341 horsepower
	10^3 joules/second
	999.8 international watt
1 watt	44.25 footpounds/minute
	1 joule/second
PRESSURE	
1 atmosphere	1.013 bars
	14.70 pounds/square inch
	760 torrs
	101 kilopascals
1 bar	10^6 baryes
	14.50 pounds-force/square inch
1 barye	10^{-6} bar
1 inch of mercury	0.03386 bar
	70.73 pounds/square foot
1 pascal	1 newton/square meter
1 pound/square inch	0.06803 atmosphere
ENERGY	
1 British thermal unit	1054 joules
	1054 wattseconds
1 foot-pound	1.356 joules
	1.356 newtonmeters
1 joule	0.7376 foot-pound
	1 wattsecond
	0.2391 calories
1 kilowatthour	3410 British thermal units
	1.341 horsepowerhours
1 newtonmeter	0.7376 footpounds
1 watthour	3.414 British thermal units
	2655 footpounds
	3600 joules

Source: Adapted from P. Calter, *Schaum's Outline of Technical Mathematics*, McGraw-Hill Book Company, New York, 1979.

Table of Integrals

Note: Many integrals have alternate forms that are not shown here. Don't be surprised if another table of integrals gives an expression that looks very different than one listed here. Further, a computer algebra system may give integrals that look much different than these, but that will result in the same numerical answers after substituting limits.

Basic forms

1. $\int du = u + C$

2. $\int af(x)\,dx = a\int f(x)\,dx = aF(x) + C$

3. $\int [f(x) + g(x) + h(x) + \cdots]\,dx = \int f(x)\,dx + \int g(x)\,dx + \int h(x)\,dx + \cdots + C$

4. $\int x^n\,dx = \dfrac{x^{n-1}}{n+1} + C \ (n \neq -1)$

5. $\int u^n\,du = \dfrac{u^{n-1}}{n+1} + C \ (n \neq -1)$

6. $\int u\,dv = uv - \int v\,du$

7. $\int \dfrac{du}{u} = \ln|u| + C$

8. $\int e^u\,du = e^u + C$

9. $\int b^u\,du = \dfrac{b^u}{\ln b} + C \ (b > 0, b \neq 1)$

Trigonometric functions

10. $\int \sin u\,du = -\cos u + C$

11. $\int \cos u\,du = \sin u + C$

12. $\int \tan u\,du = -\ln|\cos u| + C$

13. $\int \cot u\,du = \ln|\sin u| + C$

14. $\int \sec u\,du = \ln|\sec u + \tan u| + C$

15. $\int \csc u\,du = \ln|\csc u - \cot u| + C$

Squares of the *trigonometric functions*	16. $\displaystyle\int \sin^2 u\, du = \frac{u}{2} - \frac{\sin 2u}{4} + C$
	17. $\displaystyle\int \cos^2 u\, du = \frac{u}{2} + \frac{\sin 2u}{4} + C$
	18. $\displaystyle\int \tan^2 u\, du = \tan u - u + C$
	19. $\displaystyle\int \cot^2 u\, du = -\cot u - u + C$
	20. $\displaystyle\int \sec^2 u\, du = \tan u + C$
	21. $\displaystyle\int \csc^2 u\, du = -\cot u + C$

Cubes of the *trigonometric functions*	22. $\displaystyle\int \sin^3 u\, du = \frac{\cos^3 u}{3} - \cos u + C$
	23. $\displaystyle\int \cos^3 u\, du = \sin u - \frac{\sin^3 u}{3} + C$
	24. $\displaystyle\int \tan^3 u\, dx = \frac{1}{2}\tan^2 u + \ln|\cos u| + C$
	25. $\displaystyle\int \cot^3 u\, dx = -\frac{1}{2}\cot^2 u - \ln|\sin u| + C$
	26. $\displaystyle\int \sec^3 u\, du = \frac{1}{2}\sec u \tan u + \frac{1}{2}\ln|\sec u + \tan u| + C$
	27. $\displaystyle\int \csc^3 u\, du = -\frac{1}{2}\csc u \cot u + \frac{1}{2}\ln|\csc u - \cot u| + C$

Miscellaneous *trigonometric forms*	28. $\displaystyle\int \sec u \tan u\, du = \sec u + C$
	29. $\displaystyle\int \csc u \cot u\, du = -\csc u + C$
	30. $\displaystyle\int \sin^2 u \cos^2 u\, du = \frac{u}{8} - \frac{1}{32}\sin 4u + C$
	31. $\displaystyle\int u \sin u\, du = \sin u - u \cos u + C$
	32. $\displaystyle\int u \cos u\, du = \cos u + u \sin u + C$
	33. $\displaystyle\int u^2 \sin u\, du = 2u \sin u + (u^2 - 2)\cos u + C$
	34. $\displaystyle\int u^2 \cos u\, du = 2u \cos u + (u^2 - 2)\sin u + C$
	35. $\displaystyle\int \mathrm{Sin}^{-1} u\, du = u\,\mathrm{Sin}^{-1} u + \sqrt{1 - u^2} + C$
	36. $\displaystyle\int \mathrm{Tan}^{-1} u\, du = u\,\mathrm{Tan}^{-1} u - \ln\sqrt{1 + u^2} + C$

Exponential and *logarithmic forms*	37. $\displaystyle\int u e^{au}\, du = \frac{e^{au}}{a^2}(au - 1) + C$
	38. $\displaystyle\int u^2 e^{au}\, du = \frac{e^{au}}{a^3}(a^2 u^2 - 2au + 2) + C$
	39. $\displaystyle\int u^n e^u\, du = u^n e^u - n \int u^{n-1} e^u\, du$
	40. $\displaystyle\int \frac{e^u\, du}{u^n} = \frac{-e^u}{(n-1)u^{n-1}} - \frac{1}{n-1}\int \frac{e^u\, du}{u^{n-1}} \quad (n \neq 1)$

Exponential and
logarithmic forms

41. $\int e^{au} \sin bu \, du = \dfrac{e^{au}}{a^2 + b^2} (a \sin bu - b \cos bu) + C$

42. $\int e^{au} \cos bu \, du = \dfrac{e^{au}}{a^2 + b^2} (a \cos bu + b \sin bu) + C$

43. $\int \ln u \, du = u(\ln u - 1) + C$

44. $\int u^n \ln |u| \, du = u^{n+1} \left[\dfrac{\ln |u|}{n+1} - \dfrac{1}{(n+1)^2} \right] + C \ \ (n \neq -1)$

Forms involving $a + bu$

45. $\int \dfrac{u \, du}{a + bu} = \dfrac{1}{b^2}[a + bu - a \ln |a + bu|] + C$

46. $\int \dfrac{u^2 \, du}{a + bu} = \dfrac{1}{b^3} \left[\dfrac{1}{2}(a + bu)^2 - 2a(a + bu) + a^2 \ln |a + bu| \right] + C$

47. $\int \dfrac{u \, du}{(a + bu)^2} = \dfrac{1}{b^2} \left[\dfrac{a}{a + bu} + \ln |a + bu| \right] + C$

48. $\int \dfrac{u^2 \, du}{(a + bu)^2} = \dfrac{1}{b^3} \left[a + bu - \dfrac{a^2}{a + bu} - 2a \ln |a + bu| \right] + C$

49. $\int \dfrac{du}{u(a + bu)} = \dfrac{-1}{a} \ln \left| \dfrac{a + bu}{u} \right| + C$

50. $\int \dfrac{du}{u^2(a + bu)} = \dfrac{-1}{au} + \dfrac{b}{a^2} \ln \left| \dfrac{a + bu}{u} \right| + C$

51. $\int \dfrac{du}{u(a + bu)^2} = \dfrac{1}{a(a + bu)} - \dfrac{1}{a^2} \ln \left| \dfrac{a + bu}{u} \right| + C$

52. $\int u \sqrt{a + bu} \, du = \dfrac{2(3bu - 2a)}{15b^2} (a + bu)^{3/2} + C$

53. $\int u^2 \sqrt{a + bu} \, du = \dfrac{2(15b^2u^2 - 12abu + 8a^2)}{105b^3} (a + bu)^{3/2} + C$

54. $\int \dfrac{u \, du}{\sqrt{a + bu}} = \dfrac{2(bu - 2a)}{3b^2} \sqrt{a + bu} + C$

55. $\int \dfrac{u^2 \, du}{\sqrt{a + bu}} = \dfrac{2(3b^2u^2 - 4abu + 8a^2)}{15b^3} \sqrt{a + bu} + C$

Forms involving $u^2 \pm a^2$
($a > 0$)

56. $\int \dfrac{du}{a^2 + b^2u^2} = \dfrac{1}{ab} \operatorname{Tan}^{-1} \dfrac{bu}{a} + C$

57. $\int \dfrac{du}{u^2 - a^2} = \dfrac{1}{2a} \ln \left| \dfrac{u - a}{u + a} \right| + C$

58. $\int \dfrac{u^2 \, du}{u^2 - a^2} = u + \dfrac{a}{2} \ln \left| \dfrac{u - a}{u + a} \right| + C$

59. $\int \dfrac{u^2 \, du}{u^2 + a^2} = u - a \operatorname{Tan}^{-1} \dfrac{u}{a} + C$

60. $\int \dfrac{du}{u(u^2 \pm a^2)} = \dfrac{\pm 1}{2a^2} \ln \left| \dfrac{u^2}{u^2 \pm a^2} \right| + C$

Forms involving
$\sqrt{a^2 \pm u^2}$
and
$\sqrt{u^2 \pm a^2}$

($a > 0$)

61. $\int \dfrac{du}{\sqrt{a^2 - u^2}} = \operatorname{Sin}^{-1} \dfrac{u}{a} + C$

62. $\int \dfrac{du}{\sqrt{u^2 \pm a^2}} = \ln |u + \sqrt{u^2 \pm a^2}| + C$

63. $\int \dfrac{u^2 \, du}{\sqrt{u^2 \pm a^2}} = \dfrac{u}{2} \sqrt{u^2 \pm a^2} \mp \dfrac{a^2}{2} \ln |u + \sqrt{u^2 \pm a^2}| + C$

64. $\int \dfrac{du}{u\sqrt{u^2 + a^2}} = \dfrac{1}{a} \ln \left| \dfrac{u}{a + \sqrt{u^2 + a^2}} \right| + C$

Forms involving

$\sqrt{a^2 \pm b^2}$

and

$\sqrt{b^2 \pm a^2}$

$(a > 0)$

65. $\displaystyle\int \frac{du}{u\sqrt{u^2 - a^2}} = \frac{1}{a}\,\text{Sec}^{-1}\,\frac{u}{a} + C$

66. $\displaystyle\int \sqrt{u^2 \pm a^2}\,du = \frac{u}{2}\sqrt{u^2 \pm a^2} \pm \frac{a^2}{2}\ln|u + \sqrt{u^2 \pm a^2}| + C$

67. $\displaystyle\int \frac{\sqrt{u^2 + a^2}}{u}\,du = \sqrt{u^2 + a^2} - a\ln\left|\frac{a + \sqrt{u^2 + a^2}}{u}\right| + C$

68. $\displaystyle\int \frac{\sqrt{u^2 - a^2}}{u}\,du = \sqrt{u^2 - a^2} - a\,\text{Sec}^{-1}\,\frac{u}{a} + C$

69. $\displaystyle\int \sqrt{a^2 - u^2}\,du = \frac{u}{2}\sqrt{a^2 - u^2} + \frac{a^2}{2}\,\text{Sin}^{-1}\,\frac{u}{a} + C$

70. $\displaystyle\int u^2\sqrt{a^2 - u^2}\,du = \frac{-u}{4}(a^2 - u^2)^{3/2} + \frac{a^2 u}{8}\sqrt{a^2 - u^2} + \frac{a^4}{8}\,\text{Sin}^{-1}\,\frac{u}{a} + C$

71. $\displaystyle\int \frac{\sqrt{a^2 - u^2}}{u}\,du = \sqrt{a^2 - u^2} - a\ln\left|\frac{a + \sqrt{a^2 - u^2}}{u}\right| + C$

72. $\displaystyle\int \frac{\sqrt{a^2 - u^2}}{u^2}\,du = \frac{-\sqrt{a^2 - u^2}}{u} - \text{Sin}^{-1}\,\frac{u}{a} + C$

73. $\displaystyle\int \frac{u^2\,du}{\sqrt{a^2 - u^2}} = \frac{-u}{2}\sqrt{a^2 - u^2} + \frac{a^2}{2}\,\text{Sin}^{-1}\,\frac{u}{a} + C$

Answers to Selected Problems

Note: The graphs in this appendix are so tiny that it is difficult to convey highly accurate information with them. Please look at these graphs for general trends only. For larger graphs, please see the SSM and ISM.

••• CHAPTER 1 •••

Exercise 1

1. $7 < 10$ **3.** $-3 < 4$ **5.** $\dfrac{3}{4} = 0.75$ **7.** 4   **9.** -6 **11.** 24   **13.** 3   **15.** 4   **17.** 1

19. 4   **21.** 1   **23.** 3   **25.** 38.47   **27.** 96.84   **29.** 398.37   **31.** 14.0   **33.** 5.7

35. 398.4   **37.** 28,600   **39.** 3,845,200   **41.** 9.28   **43.** 0.0482   **45.** 0.0838   **47.** 34.927

49. 4.0373   **51.** 5.9373

Exercise 2

1. 1789   **3.** -1129 **5.** -850 **7.** 1827   **9.** 4931   **11.** 593.44   **13.** -0.00031 **15.** 78,388 mi^2

17. 35.0 cm   **19.** 41.1 Ω

Exercise 3

1. -8 **3.** 120   **5.** 11.3   **7.** 0.525   **9.** $-17,800$ **11.** 22.9   **13.** 10.2   **15.** 21.36

17. $3320   **19.** $1440   **21.** 540 W   **23.** 2375 g   **25.** 1751°

Exercise 4

1. -7 **3.** 6   **5.** 163   **7.** -0.347 **9.** 0.7062   **11.** 70,840   **13.** 0.00144   **15.** -0.00253

17. -175 **19.** 0.2003   **21.** 371.708 m   **23.** 18.1 ft   **25.** 314 Ω   **27.** 0.279

Exercise 5

1. 8   **3.** 81   **5.** 1000   **7.** 100   **9.** 625   **11.** 871   **13.** 61.6   **15.** 7.41   **17.** 156

19. 133   **21.** -27 **23.** -64 **25.** -514 **27.** -151 **29.** -2.35 **31.** -22.9

33. 1   **35.** 0.01   **37.** 0.0675   **39.** 0.0177   **41.** -0.00646 **43.** -0.158 **45.** 2.04   **47.** 2.12

49. 2.01   **51.** 1.50   **53.** 105   **55.** 48.6   **57.** 470 ft   **59.** 45,900 cm^3 **61.** $3151.35   **63.** 3

65. -3 **67.** -2 **69.** 1.365   **71.** 27.8   **73.** 19.39   **75.** -1.79 **77.** 1.77 s   **79.** 6.07

Exercise 6

1. 3340   **3.** -5940 **5.** 5   **7.** 3   **9.** 121   **11.** 27   **13.** 27   **15.** 24   **17.** 12   **19.** 2

21. 30   **23.** 46.2   **25.** 978   **27.** 2.28   **29.** 0.160   **31.** 59.8   **33.** 55.8   **35.** 3.51   **37.** 7.17

39. 3.23   **41.** 0.871   **43.** 7.93

Exercise 7

1. 100,000   **3.** 0.00001   **5.** 10,000   **7.** 10^6 **9.** 10^{-3} **11.** 1.86×10^5
13. 2.5742×10^4 **15.** 9.83×10^4 **17.** 2850   **19.** 90,000   **21.** 0.003667
23. 358   **25.** 134×10^{-3} **27.** 3.74×10^{-3} **29.** 18.64   **31.** 7.739   **33.** 2,660,000
35. 1.07×10^4 **37.** 3.1×10^{-2} **39.** 10^7 **41.** 10^{-9} **43.** 10^{-5} **45.** 4×10^2
47. 6×10^8 **49.** 10^{-2} **51.** 10^{-8} **53.** 4×10^2 **55.** 5×10^{-3} **57.** 3×10^6
59. 1.55×10^6 **61.** 2.11×10^4 **63.** 1.7×10^2 **65.** $9.79 \times 10^6 \, \Omega$ **67.** 2.72×10^{-6} W
69. 6.77×10^{-5} F

Exercise 8

1. 12.7 ft   **3.** 9144 in.   **5.** 58,000 lb   **7.** 44.8 ton   **9.** 364 km   **11.** 735.9 kg
13. 6.2×10^3 megohms   **15.** $9.348 \times 10^{-3} \, \mu$F   **17.** 1194 ft   **19.** 32.7 N   **21.** 17.6 L   **23.** 3.60 m
25. 0.587 acre   **27.** 2.30 m^2 **29.** 243 acres   **31.** 12,720 in.^{3} **33.** 3.31 mi/h   **35.** 107 km/h
37. 18.3 births/week   **39.** 19.8 ¢/m^2 **41.** 236 ¢/lb   **43.** 87.42°   **45.** 72.2061°   **47.** 275.3097°
49. 61°20′20″   **51.** 177°20′38″   **53.** 128°15′32″   **55.** 42.9 ft   **57.** 9,790,000 N/cm^2 **59.** 958 Ω
61. 117 gal   **63.** 14.0 lb   **65.** 8.00 cm, 2.72 cm, 3.15 cm, 3.62 cm, 13.3 cm   **67.** 5.699 m^2 **69.** 2.450 gal

Exercise 9

1. 372%   **3.** 0.55%   **5.** 40.0%   **7.** 70.0%   **9.** 0.23   **11.** 2.875   **13.** $\frac{3}{8}$ **15.** $1\frac{1}{2}$ **17.** 105 tons
19. 220 kg   **21.** 120 L   **23.** 1090 Ω **25.** $1562   **27.** 518   **29.** 65.6   **31.** 100   **33.** 108 km
35. $4457   **37.** 45.9%   **39.** 18.5%   **41.** 33.4%   **43.** 11.6%   **45.** 1.5%   **47.** 96.6%   **49.** 31.3%
51. 10.5%   **53.** 5.99%   **55.** 67.0%   **57.** 73%   **59.** 2.3%   **61.** 112.5 and 312.5 V   **63.** 37.5%
65. 25 L

Review Problems

1. 83.35   **3.** 88.1   **5.** 0.346   **7.** 94.7   **9.** 5.46   **11.** 6.8   **13.** 30.6   **15.** 17.4   **17.** (a) 179
(b) 1.08 (c) 4.86 (d) 45,700   **19.** 70.2%   **21.** 3.16×10^6 **23.** 12,500 kW   **25.** 3.4%

27. 7×10^{11} bbl   **29.** 10,200   **31.** 4.16×10^{-10} **33.** 2.42   **35.** $-\frac{2}{3} < -0.660$ **37.** 2370

39. −1.72   **41.** 64.5%   **43.** 219 N   **45.** 525 ft   **47.** 22.0%   **49.** 7216   **51.** 4.939   **53.** 109
55. 0.207   **57.** 14.7   **59.** 6.20   **61.** 2   **63.** 83.4   **65.** 93.52 cm   **67.** 9.07   **69.** 75.2%
71. 0.0737   **73.** 7.239   **75.** 121   **77.** 1.21

◆◆◆ CHAPTER 2 ◆◆◆

Exercise 1

1., 3. 1. and 4.   **5., 7.** 6. and 7.   **9.** 2   **11.** 3   **13.** 3, a, x **15.** 7, x, x, y, y, y **17.** 6   **19.** −1
21. 2a   **23.** second   **25.** fifth   **27.** second   **29.** third

Exercise 2

1. $10x$ **3.** $7a$ **5.** $6ab$ **7.** $6.0x$ **9.** $75.1a$ **11.** $-12.7ab$ **13.** $-x$ **15.** $7a$
17. $-8ab$ **19.** $3.6x$ **21.** $72.7m$ **23.** $62.6xy$ **25.** $6x$ **27.** $-5a$
29. $13.9x$ **31.** $-32.7m$ **33.** $8x$ **35.** $2ab$ **37.** $-23.8m$ **39.** $4x - 2$
41. $-25.5ab + 104.7$ **43.** $1.49y + 5.47$ **45.** $2b$ **47.** $12b - 2a - 16c - 8d$
49. $5by^5 - 72bx^5 - 8bx^4 + 23by^4$ **51.** $7.71a + 1.59y - 3.6b$ **53.** $2x - 2w$
55. $43.9xy + 14.3x - 44.7y$ **57.** $0.04x + 400$ **59.** $2\pi r^2 + 2\pi rh$

Exercise 3

1. 8　　**3.** 64　　**5.** 243　　**7.** -32　　**9.** -125　　**11.** m^7　　**13.** x^6　　**15.** a^9　　**17.** 10^8

19. z^2　　**21.** 2　　**23.** 10　　**25.** b　　**27.** 10^2　　**29.** w^6　　**31.** p^{12}　　**33.** 2^{10}　　**35.** y^3　　**37.** $4x^2$

39. $a^3x^3y^3$　　**41.** a^3c^3　　**43.** $256a^{12}c^8$　　**45.** $\dfrac{x^2}{y^2}$　　**47.** $-\dfrac{8}{125}$　　**49.** $\dfrac{16x^4}{9y^4}$　　**51.** 1　　**53.** $108a^3$　　**55.** c

57. 4　　**59.** $\dfrac{1}{x}$　　**61.** $\dfrac{x^4}{16}$　　**63.** $4/w^4 - 3/z^2$　　**65.** $\dfrac{27y^6}{64x^9}$　　**67.** $5x^{-3}$　　**69.** $x^2w^4z^{-2}$　　**71.** $b^2c^2d^{-4}$

73. $\dfrac{x^9y^6w^{15}}{z^6}$　　**75.** $\dfrac{a^{15}m^{15}}{b^{18}n^{18}}$　　**77.** $64.4t^2$ ft　　**79.** $R^{-1} = R_1^{-1} + R_2^{-1}$

Exercise 4

1. x^6　　**3.** $-x^5$　　**5.** $6ab^2$　　**7.** $15m^3n^3$　　**9.** $40.8a^3$　　**11.** $6.60x^2y^3$　　**13.** $6ab^{n+1}$

15. $3.66a^{m+n}$　　**17.** $30w^5$　　**19.** $24a^3b^{2+n}$　　**21.** $17.3w^6x^7y^6$　　**23.** $5.92a^{x+4}b^6c^{x+4}$　　**25.** $10.6x^2$

Exercise 5

1. $-x - 2$　　**3.** $x + 2$　　**5.** $2a + b$　　**7.** $2x - y$　　**9.** $bx + 2x$　　**11.** $x^2 - 5x$　　**13.** $2.58x^2 - 4.79x$

15. $b^6 + 8b^4$　　**17.** $0.22 - 2a$　　**19.** $3x + 3$　　**21.** $7a - 2x$　　**23.** $4418x + 9.49a$　　**25.** $-2b - 2z$

27. $6z + 3c - 9a$　　**29.** $-5x - 11$　　**31.** $18a^3b + 12a^2b^2 - 6ab^3$

33. $-22.0x^3y^3 - 6.35x^3y^2 + 30.1x^2y^3 + 15.2x^2y^2$　　**35.** $14p + 2q + 2$　　**37.** $-7y^3 + 31y^2 - y - 12$

39. $12y^2 - 13xy + a$　　**41.** $-12x + 5y + 2z$　　**43.** $R_1 + R_2 - R_3 + R_4 - R_5 - R_6 + R_7$

Exercise 6

1. $x^2 + xy + xz + yz$　　**3.** $8m^3 + 2m^2n - 4mn - n^2$　　**5.** $2x^2 + xy - y^2$　　**7.** $12x^2y^4 + 7a^3bxy^2 - 12a^6b^2$

9. $2a^2 - 21x^2 - 11ax$　　**11.** $a^2x^2 - 25b^2$　　**13.** $2.93x^2 + 1.82xy - 1.11y^2$

15. $13.1y^4 + 6.28a^3by^2 - 18.5a^6b^2$　　**17.** $LW - 3L + 2W - 6$

Exercise 7

1. $x^2 - xy + x + 3y - 12$　　**3.** $4w^3 + 2w^2 - 22w + 10$

5. $x^4 + 3.88x^3 - 2.15x^2 - 14.4x + 23.4$　　**7.** $6x^3 - 12x^2y - 9xy^2 + 9y^3$　　**9.** $a^3 + 3a^2 - 2$

11. $c^3 - 2c^2m + c^2n + cm^2 - m^2n$　　**13.** $x^2 - y^2 - 2yz - z^2$

15. $28x^2 - 11xy + 42x + y^2 - 6y$　　**17.** $a^4 - 11.3a^3 + 67.3a^2 - 193a + 127$

19. $-x^2y^2 + 2mxy^2 - m^2y^2 + a^2m^2$　　**21.** $xy - bx + lx + ay + wy - ab + al - bw + lw$

Exercise 8

1. $x^2 + 2xy + y^2$　　**3.** $a^2 - 2ad + d^2$　　**5.** $B^2 + 2BD + D^2$　　**7.** $24.2y^2 + 30.7yz + 9.73z^2$

9. $36x^2 + 60nx + 25n^2$　　**11.** $w^2 - 2w + 1$　　**13.** $b^6 - 26b^3 + 169$

15. $15.1x^4 - 10.3x + 1.77$　　**17.** $x^2 + 2xy + 2xz + y^2 + 2yz + z^2$　　**19.** $a^2 + 2ab - 2a + b^2 - 2b + 1$

21. $c^4 - 2c^3d + 3c^2d^2 - 2cd^3 + d^4$　　**23.** $x^3 - 3x^2y + 3xy^2 - y^3$

25. $27.5m^3 + 59.1m^2n + 42.3mn^2 + 10.1n^3$　　**27.** $c^3 + 3c^2d + 3cd^2 + d^3$

29. $8x^6y^3 + 36x^5y^4 + 54x^4y^5 + 27x^3y^6$　　**31.** $x^2 + 4x + 4$

33. $4.19r^3 - 25.1r^2 + 50.3r - 33.5$

Exercise 9

1. x^3　　**3.** $5z$　　**5.** $-2a$　　**7.** $11.8b$　　**9.** $2.55d$　　**11.** $6p^2q^3r$　　**13.** $-8mn$　　**15.** $-4a^3bc$

17. $-6n^3$　　**19.** $5a^2y$　　**21.** $-4ac$　　**23.** $-3by$　　**25.** $8xy$　　**27.** y^6　　**29.** a^{x-y}　　**31.** $-5a^2z$

33. $\dfrac{19}{ab^2}$　　**35.** $r/3$

Exercise 10

1. $15x^2 + 3x$ **3.** $12d^4 - 2d$ **5.** $4c^2 + 3c^3$ **7.** $-3 - 4p$ **9.** $-5x^2 - 3x$ **11.** $7.95bmn + 3.10b$

13. $1.45a^2 - 2.11ab$ **15.** $y^2z - xy$ **17.** $mn^2 - mn - m^2n$ **19.** $x^3 - x^2y^2 - y^3$ **21.** $\dfrac{c^2}{d^2} - 4c + \dfrac{d}{c}$

23. $\dfrac{a^2}{b^2} + 2 - \dfrac{b^2}{a^2}$ **25.** $\dfrac{3z^3}{x} - 4x^2 - 2z$ **27.** $r^3 - \dfrac{pq^3}{r} - \dfrac{q^2r^2}{p}$ **29.** $d^3 - \dfrac{4c^3}{d} - 3cd$

31. $12c - \dfrac{8c^2}{b} - \dfrac{4}{bc}$

Exercise 11

1. $a + 8$ **3.** $a + 8$ **5.** $x + 5$ **7.** $a - 8$ **9.** $-3x - 2$ **11.** $a^2 + 3a + 1$ **13.** $x - 6 + \dfrac{15}{(x + 2)}$

15. $5x + 13 - \dfrac{35}{(3 - x)}$

Review Problems

1. $b^6 - b^4x^2 + b^4x^3 - b^2x^5 + b^2x^4 - x^6$ **3.** 3.86×10^{14} **5.** $9x^4 - 6mx^3 - 6m^3x + 10m^2x^2 + m^4$

7. $8x^3 + 12x^2 + 6x + 1$ **9.** $6a^2x^5$ **11.** $a - b - c$ **13.** $16a^2 - 24ab + 9b^2$ **15.** $x^2y^2 - 6xy + 8$

17. $9x^2 + 12xy + 4y^2$ **19.** $ab - b^4 - a^2b$ **21.** $16m^4 - c^4$ **23.** $2a^2$ **25.** $24 - 8y$

27. $6x^3 + 9x^2y - 3xy^2 - 6y^3$ **29.** $13w - 6$ **31.** $a^5 + 32c^5$ **33.** $(a - c)^{m-2}$ **35.** $x + y$

37. $b^3 - 9b^2 + 27b - 27$ **39.** $x^2/2y^3$ **41.** $x^2 + 2x - 8$ **43.** $ab - 2 - 3b^2$ **45.** $5a - 10x - 2$

47. 8.01×10^6 **49.** $10x^3y^4$ **51.** $x^3 + 3x^2 - 4x$ **53.** 1.77×10^8 **55.** $6a^3b^3$

57. $x^4 + x + 1 + R(x^2 + x + 1)$ **59.** $0.13x + 540$ **61.** $4.02t^2$ ft

◆◆◆ CHAPTER 3

Exercise 1

1. 7 **3.** 2 **5.** $\dfrac{9}{4}$ **7.** -4 **9.** 2 **11.** $\dfrac{9}{4}$ **13.** 11 **15.** 2 **17.** 4 **19.** $-\dfrac{1}{2}$ **21.** 5

23. 3 **25.** 3 **27.** 7 **29.** $-\dfrac{66}{5}$ **31.** $-\dfrac{5}{7}$ **33.** 6 **35.** 0 **37.** 5 **39.** $\dfrac{2}{3}$ **41.** 22 **43.** 8

45. $\dfrac{59}{3}$ **47.** -6 **49.** $-\dfrac{12}{7}$ **51.** 2 **53.** 26 **55.** $\dfrac{1}{21}$ **57.** 56 **59.** 4 **61.** $\dfrac{15}{11}$ **63.** 1.15

65. -7.77 **67.** -0.164 **69.** -3.28 **71.** -1.97 **73.** $\dfrac{b - 9}{b}$ **75.** $\dfrac{b - 7}{a}$ **77.** $-18/7$

79. $-1/2$ **81.** $-7/2$ **83.** $-7/2$ **85.** 0 **87.** 1.77 tons   **89.** 5.3 months

Exercise 2

1. x and $3x + 10$ **3.** x and $x + 42$ or $x - 42$ **5.** $\dfrac{x}{6x + 4}$ **7.** $0.11x$ gal   **9.** (a) $32 - x$ (b) $\dfrac{320}{x}$

11. 8 **13.** 9 **15.** 5

Exercise 3

1. 3.38 h   **3.** 130 km, 1.52 h   **5.** 4.95 days   **7.** 327 mi   **9.** 3:17 P.M., 902 km

Exercise 4

1. 8 technicians   **3.** \$1,226,027   **5.** 57,000 gal   **7.** \$42 for skis, \$168 for boots   **9.** \$60,000   **11.** \$4608

13. \$34,027 @ 6.75%, \$139,897 @ 8.24%   **15.** \$398 for computer, \$597 for printer

Exercise 5

1. 336 gal   **3.** 489 kg of 18% alloy, 217 kg of 31% alloy   **5.** 1400 kg   **7.** 1.29 liters   **9.** 63.0 lb
11. 59.3 lb

Exercise 6

1. (a) 4.97 ft; (b) 4000 lb   **3.** 172 in   **5.** 4.61 ft   **7.** 11.5 in.; 37.7 lb

Exercise 7

1. 2 days   **3.** $5\frac{5}{6}$ days   **5.** 2.4 h   **7.** 4.2 h   **9.** 12.4 h   **11.** 7.8 weeks   **13.** 5.6 winters   **15.** 6.1 h

Review Problems

1. 12   **3.** 5   **5.** 0   **7.** $10\frac{1}{2}$ **9.** 2   **11.** 4   **13.** 9   **15.** 4/3   **17.** 1.77   **19.** -0.578
21. 25/4   **23.** 6   **25.** $-13/5$ **27.** 13   **29.** $-5/7$ **31.** $-5/2$ **33.** $-7/3$ **35.** $-6/b$
37. $\dfrac{a+10}{2}$ **39.** $\dfrac{5a-c+8}{a}$ **41.** 14,600 tons, 10,400 tons   **43.** \$30   **45.** 13, 15, 17, 19
47. $R_1 = 608$ lb, $R_2 = 605$ lb   **49.** 18.3 L   **51.** \$46,764   **53.** \$80,000   **55.** 5.71 tons   **57.** 0.180 ton
59. 139 lb   **61.** 6.44 L   **63.** 7 technicians   **65.** 3060 kg   **67.** 149 km   **69.** 58,100 km
71. 27.1 h; 665 km

◆◆◆ CHAPTER 4 ◆◆

Exercise 1

1. Is a function   **3.** Not a function   **5.** Not a function   **7.** Yes   **9.** Explicit   **11.** Implicit
13. x is independent; y is dependent.   **15.** x and y are independent; w is dependent.
17. x and y are independent; z is dependent.   **19.** $y = 2x + 4$ **21.** $y = (2 - 5x)/2$ **23.** 6   **25.** -21
27. 12.5   **29.** -5 **31.** 2160 ft; 4450 ft; 7540 ft   **33.** 0.21 in.; 0.44 in.   **35.** 64.7 W; 144 W; 257 W

Exercise 2

1. $y = x^3$ **3.** $y = x + 2x^2$ **5.** $y = (2/3)(x - 4)$ **7.** $c = \sqrt{a^2 + b^2}$ **9.** $P = RI^2$
11. $s = 2.25 + 0.65w$ dollars   **13.** $2a^2 + 4$ **15.** $5a + 5b + 1$ **17.** 20   **19.** -52 **21.** $x = (y - 3)/5$
23. $x = 5/(5y + 1)$ **25.** $q = \dfrac{p^2 + 5p}{2}$ **27.** $e = PL/aE$ **29.** $x^2 + 2$ **31.** $(x - 4)^2$ **33.** -77
35. -125 **37.** $y = (x + 15)/14$ **39.** $y = \dfrac{x + 1}{8}$ **41.** $y = \dfrac{x + 6}{5}$ **43.** $y = -x - 18$
45. Domain $= -10, -7, 0, 5, 10$; Range $= 3, 7, 10, 20$ **47.** $x \geq 7$; $y \geq 0$ **49.** $x \neq 0$; all y **51.** $x < 1$; $y > 0$
53. $x \geq 1$; $y \geq 0$

Review Problems

1. (a) Is a function   (b) Not a function   (c) Not a function   **3.** $S = 4\pi r^2$
5. (a) Explicit; y independent, w dependent. (b) Implicit.   **7.** $w = (3 - x^2 - y^2)/2$ **9.** $7x^2$ **11.** 6   **13.** 28
15. 13/90   **17.** y is 7 less than 5 times the cube of x.   **19.** $x = \pm\sqrt{(6 - y)/3}$ **21.** $y = 8 - x$
23. $g[f(x)] = 25x^2 + 5x$ **25.** 15.1   **27.** 9.50   **29.** 934; 1400; 1540 in   **31.** 40,800; 76,100; 154,000
33. (a) $f[g(x)] = 7 - 10x^2$ (b) $g[f(x)] = 20x^2 - 140x + 245$ (c) $f[g(5)] = -243$ (d) $g[f(5)] = 45$

◆◆◆ CHAPTER 5 ◆◆

Exercise 1

1. Fourth   **3.** Second   **5.** Fourth   **7.** First and fourth   **9.** $x = 7$

11. $E(-1.8, -0.7)$; $F(-1.4, -1.4)$; $G(1.4, -0.6)$; $H(2.5, -1.9)$

13. 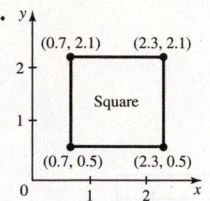   **15.** 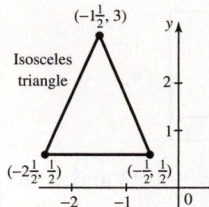   **17.** $(4, -5)$ **19.** 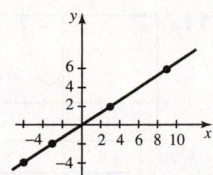   **21.**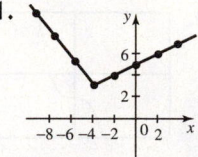

23. (a) 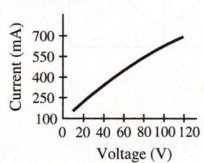   (b) 67 V   **25.** (a) 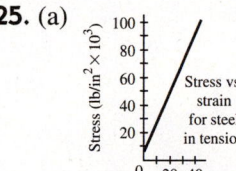   (b) 0.0008 in./in.

Exercise 2

1. 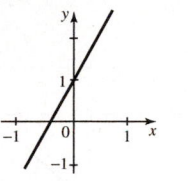   **3.** 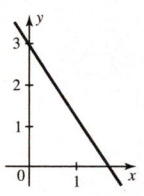   **5.** 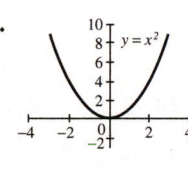   **7.** 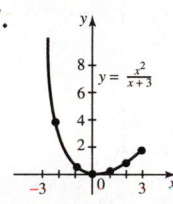   **9.**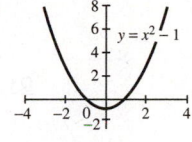

11. 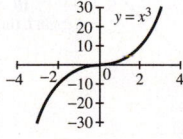   **13.** 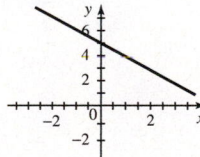   **15.**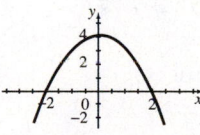

17. 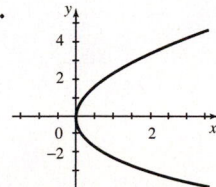   **19.** 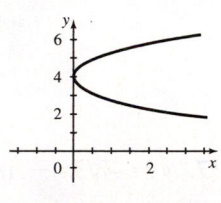   **21.**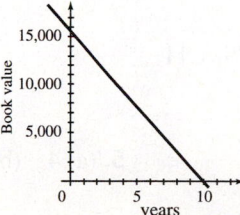

23. 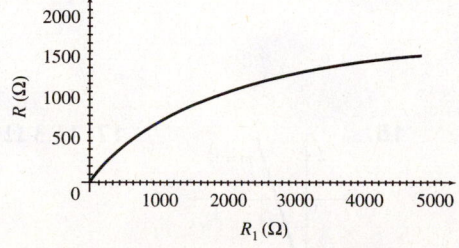   **25.**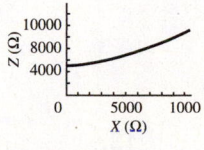

Exercise 3

1. 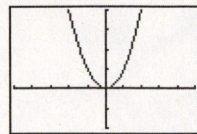 **3.** 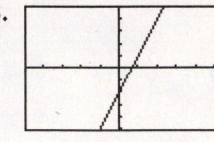 **5.** 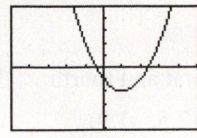 **7.**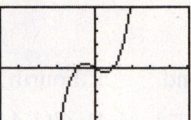

9. 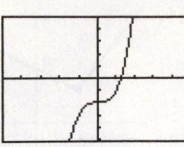 **11.** 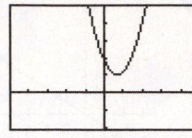 **13.**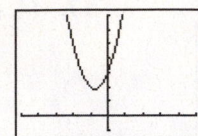

15. Zeros at $x = -2.20$ and 0.91. Min at $(-0.64, -16.9)$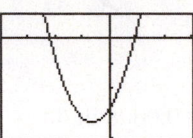

17. Zeros at $x = -1.23$ and 1.23. Min at $(0, -31)$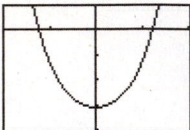

Exercise 4

1. 2   **3.** −0.797   **5.** 1   **7.** −1.40

9. 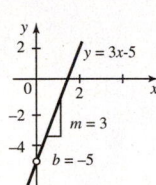 **11.** 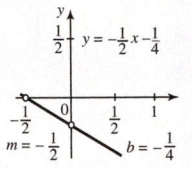 **13.** 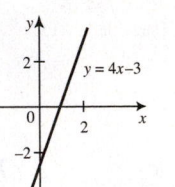 **15.** 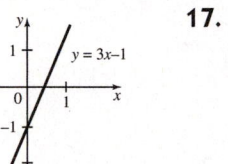 **17.**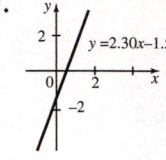

19. 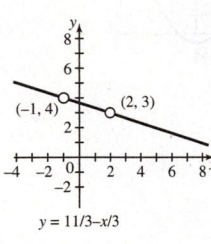 **21.** 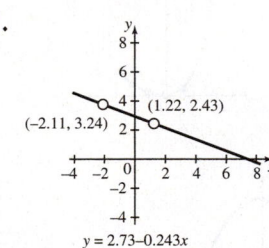 **23.** 3/5   **25.** 105 lb

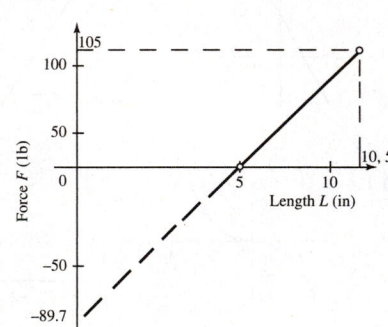

Exercise 5

1. 3.14   **3.** −0.34, 2.47   **5.** −0.79, 1.11

Review Problems

1. 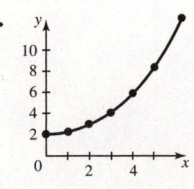 **3.** 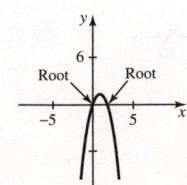 **5.** (a) 4   (b) 2/5   **7.** $y = -8/3x - 1/3$

9. 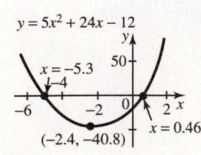 **11.** 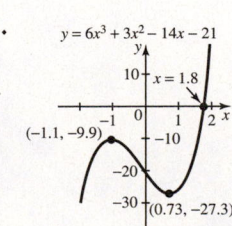 **13.** 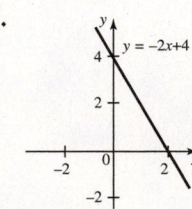 **15.** 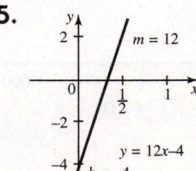 **17.** 11.3 Ω

••• CHAPTER 6 ••

Exercise 1

1. (a) 62.8°; (b) 64.6°; (c) 37.7°   **3.** 5.05   **5.** $A = C = 46.3°$; $B = D = 134°$

Exercise 2

1. $\theta = 77°$, $\phi = 103°$ **3.** 235 sq. units   **5.** $a = 27.6$ units, $b = 58.2$ units   **7.** 57.5 units   **9.** $4405
11. 36.0 ft   **13.** 28.3 ft   **15.** 264 m   **17.** 19.8 in.   **19.** 43.39 acres   **21.** 5.2 mm   **23.** 6.62 in.
25. 2.30 in.   **27.** 2.62 m

Exercise 3

1. (a) 34.0 in.^{2}; 23.3 in.   (b) 23.2 m^2; 19.3 m   (c) 282,000 cm^2; 2240 cm   (d) 4070 in.^{2}; 258 in.
3. $3042   **5.** $348   **7.** $3220   **9.** $861   **11.** 1000 bricks   **13.** (a) 30°   (b) 45°   (c) 60°   (d) 67.5°

Exercise 4

1. 30.3 cm; 73.0 cm^2 **3.** 11.9 in.; 445 in.^{2} **5.** 3.55 ft; 22.3 ft   **7.** 45.4 cm   **9.** 0.738 unit
11. 44.7 units   **13.** 15.8 in.   **15.** 104 m   **17.** 15.1 in.   **19.** 218 cm   **21.** 1.200 in.
23. 247 cm   **25.** 0.820 m   **27.** 3.52 ft   **29.** 7.500 in.

Exercise 5

1. 26.6 in.^{3} **3.** 6.34×10^7 in.^{3} **5.** 40 ft^2, 12 ft^3 **7.** 1800 ft^2, 8830 ft^3 **9.** (a) 84.4 in.^{2}, 52.7 in.^{3}
(b) 4150 cm^2, 18,200 cm^3 (c) 30.1 ft^2, 11.2 ft^3 **11.** 100 cuts   **13.** 428 loads   **15.** $56\frac{1}{4}$ board ft
17. 770 yd^3 **19.** 128 ft^3 **21.** $3\frac{1}{2}$ loads   **23.** 4.6 yd^3 **27.** 123 in.^{3}, 118 in.^{2} **29.** 104 in.^{2}
31. (a) 3950 ft^3 (b) 1180 ft^2 **33.** 35.4 ft

Exercise 6

1. (a) 2860 in.^{2}, 4730 in.^{2}, 24,700 in.^{3} (b) 95,100 cm^2, 123,000 cm^2, 3,190,000 cm^3 (c) 31.7 m^2, 52.6 m^2, 28.9 m^3
3. 177 in.^{3} **5.** 3530 mm^3 **7.** 2.01 liters   **9.** 14.8 m^3 **11.** 857 gal   **13.** (a) 208 in.^{2}, 284 in.^{2}, 317 in.^{3}
(b) 4.52 m^2, 6.79 m^2, 1.28 m^3 **15.** 5650 cm^2 **17.** 660 ft^3 **19.** 12.0 ft^3 **23.** $A = 15.8$; $r = 1.12$
25. $r = 1.91$ cm; $V = 29.3$ cm^3 **27.** 4   **29.** (a) 12.9 m   (b) 4180 kg

Review Problems

1. 1440 mi/h   **3.** 2.88 m   **5.** 13 m   **7.** 175,000 square units   **9.** 43.1 in.   **11.** 17.5 m   **13.** 161°
15. 213 cm^3 **17.** 1030 in.^{2} **19.** 1150 m^2 **21.** 4830 in.^{2} **23.** 98,300 in.^{3} **25.** 7.91 cm

••• CHAPTER 7 ••

Exercise 1

1. (a) $\sin \theta = 0.6325$ $\cos \theta = 0.7750$ $\tan \theta = 0.8162$ (b) $\sin \theta = 0.5607$ $\cos \theta = 0.8295$ $\tan \theta = 0.6759$
    (c) $\sin \theta = 0.5321$ $\cos \theta = 0.8486$ $\tan \theta = 0.6270$ (d) $\sin \theta = 0.5109$ $\cos \theta = 0.8607$ $\tan \theta = 0.5937$

3.

	sin	cos	tan
(a)	0.7581	0.6521	1.1626
(b)	0.6280	0.7782	0.8069
(c)	0.3140	0.9494	0.3307
(d)	0.0361	0.9993	0.0361
(e)	0.9966	0.0819	12.1632
(f)	0.4802	0.8771	0.5475
(g)	0.9598	0.2807	3.4197
(h)	0.6934	0.7206	0.9623

(i) 0.0583   0.9983   0.0584
(j) 0.8525   0.5226   1.6312
(k) 0.9795   0.2016   4.8587
(l) 0.3754   0.9269   0.4050
(m) 0.9940   0.1094   9.0821
(n) 0.9798   0.2002   4.8933
(o) 0.5561   0.8311   0.6690
(p) 0.8948   0.4465   2.0042

5. 56.8°   **7.** 39.6°   **9.** 32.1°   **11.** 30.5°   **13.** 10.1°   **15.** 76.8°   **17.** 36.9°   **19.** 31.3°

Exercise 2

1. $B = 47.1°$; $b = 167$; $c = 228$ **3.** $b = 1.08$; $A = 58.1°$; $c = 2.05$ **5.** $B = 25.3°$; $b = 134$; $c = 314$
7. $a = 6.44$; $c = 11.3$; $A = 34.8°$ **9.** $a = 50.3$; $c = 96.5$; $B = 58.6°$ **11.** $c = 470$; $A = 54.3°$; $B = 35.7°$
13. $a = 2.80$; $A = 35.2°$; $B = 54.8°$ **15.** $b = 25.6$; $A = 46.9°$; $B = 43.1°$ **17.** $b = 48.5$; $A = 40.4°$; $B = 49.6°$
19. $a = 414$; $A = 61.2°$; $B = 28.8°$

Exercise 3

1. 402 m   **3.** 285 ft   **5.** 64.9 m   **7.** 39.9 yd   **9.** 128 m   **11.** 21.4 km; 14.7 km
13. 432 mi; S 58°31′ W   **15.** 156 mi north; 162 mi east   **17.** 30.2°; 20.5 ft   **19.** 37.6°
21. $A = 52.4°$, $B = 37.6°$; $AB = 1.23$ m; Area $= 0.366$ m^2 **23.** 35.3°   **25.** 19.5°   **27.** 122,000 square units
29. 355 in.   **31.** 77.5 mm; 134 mm; 134 mm; 77.5 mm   **33.** 1.550 in.   **35.** 5.53 in.   **37.** 0.866 cm; 0.433 cm
39. (a) 2.160 in.   (b) 18.10°   (c) 26.30°

Exercise 4

	r	sin	cos	tan	θ
1.	5.32	0.906	0.423	2.14	65.0°
3.	6.63	0.828	0.561	1.48	55.9°
5.	5.20	0.929	0.371	2.50	68.2°

Exercise 5

1. 3.28; 3.68   **3.** 0.9917; 1.602   **5.** 589; 593   **7.** 9.63; 20.6   **9.** 4.05; 17.9   **11.** 616; 51.7°
13. 8811; 56.70°   **15.** 2.42; 31.1°   **17.** 8.36; 54.7°   **19.** 4.57; 29.8°

Exercise 6

1. 12.3 N   **3.** 1520 N   **5.** 25.7°   **7.** 4.94 tons   **9.** 119 km/h   **11.** 5.70 m/min; 1.76 min
13. 115 mi/h   **15.** $X = 354$ Ω; $Z = 372$ Ω   **17.** 7.13 Ω; 53.7°

Review Problems

1. 0.9558, 0.2940, 3.2506   **3.** 0.8674, 0.4976, 1.7433   **5.** 34.5°   **7.** 22.1°   **9.** 55.0°
11. $B = 61.5°$; $a = 2.02$; $c = 4.23$ **13.** 356; 810   **15.** 473; 35.5°   **17.** 7.27 ft   **19.** 0.5120   **21.** 1.3175
23. 0.9085   **25.** 60.1°   **27.** 46.5°   **29.** 18.6°   **31.** $AC = 74.98$ ft; N 28°18′ E

◆◆◆ CHAPTER 8 ◆◆

Exercise 1

	r	sin	cos	tan	cot	sec	csc
1.	5.83	−0.858	0.514	−1.67	−0.600	1.95	1.17
3.	25.0	−0.280	0.960	−0.292	−3.43	1.04	3.57
5.	3.49	−0.890	0.455	−1.96	−0.511	2.20	−1.12
7.		0.9816	−0.1908	−5.145			
9.		−0.4848	0.8746	−0.5543			
11.		−0.8898	0.4563	−1.950			
13.		0.8090	−0.5878	−1.376			
15.		0.9108	−0.4128	−2.206			
17.		0.7880	0.6157	1.280			

19. 1.3602   **21.** 1.3598   **23.** −2.0145   **25.** −2.1730   **27.** 0.8785   **29.** sin 17°   **31.** csc 4.4°
33. sec 7.3°   **35.** −0.574, −0.919

Exercise 2

1. 17°   **3.** 55°   **5.** 69.3°   **7.** IV   **9.** II   **11.** IV   **13.** I or II
15. I or IV   **17.** neg   **19.** pos   **21.** neg

	sin	cos	tan
23.	−	−	+
25.	−	+	−

27. 50.9°; 129.1°   **29.** 33.2°; 326.8°   **31.** 81.1°; 261.1°   **33.** 219.5°; 320.5°
35. 54.8°; 305.2°   **37.** 195.0°; 345.0°

Exercise 3

1. (a) $B = 71.1°$; $b = 8.20$; $c = 6.34$ (b) $C = 27.0°$; $a = 212$; $c = 119$ (c) $B = 103.6°$; $b = 21.7$; $c = 15.7$
(d) $B = 93.0°$; $a = 102$; $b = 394$ **3.** $C = 117.59°$; $b = 8423$; $c = 12,050$ **5.** $B = 71.65°$; $a = 0.8309$; $b = 1.126$
7. (a) $B = 32.8°$; $C = 100°$; $c = 413$ (b) $B = 57.1°$; $C = 57.0°$; $b = 1.46$ (c) $C = 107.8°$; $B = 29.2°$; $c = 29.3$
9. $B = 40.9$; $A = 77.4°$; $a = 423$ **11.** $B = 26.0°$; $C = 108.4°$; $b = 4.80$ $B' = 62.8°$; $C' = 71.6°$; $b' = 9.75$
13. 358 ft; 225 ft

Exercise 4

1. (a) $A = 44.2°$; $B = 29.8°$; $c = 21.7$ (b) $A = 80.9°$; $C = 47.7°$; $b = 1.54$ (c) $B = 50.3°$; $C = 65.9°$; $a = 21.3$
(d) $A = 52.6°$; $B = 81.1°$; $c = 663$ **3.** $B = 30.8°$; $C = 34.2°$; $a = 82.8$ **5.** $A = 26.1°$; $C = 24.9°$; $b = 329$
7. (a) $A = 56.9°$; $B = 95.8°$; $C = 27.3°$ (b) $A = 18.9°$; $B = 128°$; $C = 33.1°$ (c) $A = 44.2°$; $B = 29.8°$; $C = 106°$
(d) $A = 67.7°$; $B = 77.1°$; $C = 35.2°$ **9.** $A = 44.71°$; $B = 61.13°$; $C = 74.16°$
11. $A = 26.2°$; $B = 129.0°$; $C = 25.1°$ **13.** 756 ft

Exercise 5

1. 30.8 m; 85.6 m   **3.** 32.3°; 60.3°; 87.4°   **5.** N 48.8° W   **7.** S 59.4° E   **9.** 598 km   **11.** 28.3 m
13. 77.3 m; 131 m   **15.** 107 ft   **17.** 337 m   **19.** 33.7 cm   **21.** 73.4 in.   **23.** 53.8 mm; 78.2 mm
25. 419   **27.** $A = 45.0°$; $B = 60.0°$; $C = 75.0°$; $AC = 1220$; $AB = 1360$ **29.** 21.9 ft

Exercise 6

	RESULTANT	ANGLE
1.	521	10.0°
3.	87.1	31.9°
5.	6708	41.16°

7. 9.14; 54.8°   **9.** 1090; 34.0°   **11.** 39.8 at 26.2°   **13.** 37.3 N at 20.5°   **15.** 121 N at N 59.8° W

17. 1720 N at 29.8° from larger force   **19.** 44.2° and 20.7°   **21.** Wind: S 37.4° E; plane: S 84.8° W

23. 413 km/h at N 41.2° E   **25.** 632 km/h; 3.09°   **27.** 26.9 A; 32.3°

Review Problems

1. $A = 24.1°; B = 20.9°; c = 77.6$ **3.** $A = 61.6°; C = 80.0°; b = 1.30$ **5.** $B = 20.2°; C = 27.8°; a = 82.4$
7. IV   **9.** II   **11.** neg   **13.** neg   **15.** neg   **17.** sin -0.800, cos -0.600, tan 1.33
19. 1200 at 36.3°   **21.** 22.1 at 121°   **23.** 1.11 km   **25.** sin 0.0872, cos -0.9962, tan -0.0875
27. sin 0.7948, cos -0.6069, tan -1.3095 **29.** 0.4698   **31.** 1.469   **33.** S 3.5° E   **35.** 130.8°; 310.8°
37. 47.5°; 132.5°   **39.** 80.0°; 280.0°   **41.** 695 lb; 17.0°   **43.** -224 cm/s; 472 cm/s   **45.** 148 mm

••• CHAPTER 9 •••

Exercise 1

1. $(2, -1)$ **3.** $(1, 2)$ **5.** $(-0.24, 0.90)$ **7.** $(3, 5)$ **9.** $\left(-\dfrac{3}{4}, 3\right)$ **11.** $(-3, 3)$ **13.** $(1, 2)$ **15.** $(3, 2)$

17. $(15, 6)$ **19.** $(3, 4)$ **21.** $(2, 3)$ **23.** $(-3, 5)$ **25.** $(1.63, 0.0970)$
27. $m = 2, n = 3$ **29.** $w = 6, z = 1$ **31.** $(9.36, 4.69)$ **33.** $v = 1.06, w = 2.58$

Exercise 2

1. (a) 9.85 mi/h (b) 4.27 mi/h   **3.** 5.3 mi/h and 1.3 mi/h   **5.** \$2500 at 4%   **7.** \$2684 at 6.2% and \$1716 at 9.7%
9. \$6000 for 2 years   **11.** 2740 lb mixture; 271 lb sand   **13.** 5.57 lb peat; 11.6 lb vermiculite
15. $T_1 = 490$ lb; $T_2 = 185$ lb   **17.** Carpenter: 25.0 days; helper: 37.5 days   **19.** 18,000 gal/h and 12,000 gal/h
21. 92,700 people and 162,000 people   **23.** $l_1 = 13.1$ mA; $l_2 = 22.4$ mA   **25.** $R_1 = 27.6 \ \Omega; \alpha = 0.00511$
27. $h = 352$ ft; $d = 899$ ft   **29.** $v_0 = 0.381$ cm/s; $a = 3.62$ cm/s^2

Exercise 3

1. $(60, 36)$ **3.** $(87/7, 108/7)$ **5.** $(15, 12)$ **7.** $m = 4, n = 3$ **9.** $r = 3.77, s = 1.23$ **11.** $(1/2, 1/3)$

13. $(1/3, 1/2)$ **15.** $(1/10, 1/12)$ **17.** $w = \dfrac{1}{36}, z = \dfrac{1}{60}$ **19.** $x = \dfrac{3}{5a}, y = \dfrac{1}{5b}$ **21.** $x = \dfrac{6}{5p}, y = \dfrac{1}{5q}$

23. $\left(\dfrac{a + 2b}{7}, \dfrac{3b - 2a}{7}\right)$ **25.** $\left(\dfrac{c(n - d)}{an - dm}, \dfrac{c(m - a)}{an - dm}\right)$

Exercise 4

1. $(15, 20, 25)$ **3.** $(1, 2, 3)$ **5.** $(5, 6, 7)$ **7.** $(1, 2, 3)$ **9.** $(3, 4, 5)$ **11.** $a = -3, b = 4, c = -3/2$
13. $(3, 6, 9)$ **15.** $(2, -1, 5)$ **17.** $(3a, 2a, a)$ **19.** $(c, c, ab/c)$ **21.** 2/3A, 7/3A, 2/3A
23. $F_1 = 9890$ lb, $F_2 = 9860$ lb, $F_3 = 9360$ lb   **25.** 86.0 lb zinc; 0512 lb tin; 4.62 lb lead; 2.25 lb nickel; 9.05 lb manganese.

Review Problems

1. $(3, 5)$ **3.** $(5, -2)$ **5.** $(2, -1, 1)$ **7.** $\left(\dfrac{-9}{5}, \dfrac{54}{5}, \dfrac{-21}{5}\right)$ **9.** $(8, 10)$ **11.** $(2, 3)$

13. $[(a + b - c)/2, (a - b + c)/2, (b - a + c)/2]$ **15.** $(7, 5)$ **17.** $(1, -5, -4)$ **19.** $(2, 3, 1)$
21. $(13, 17)$ **23.** \$450 for A; \$270 for B **25.** 3/7   **27.** 9×16 units

◆◆◆ CHAPTER 10 ◆◆

Exercise 1

1. A, B, D, E, F, I, J, K   **3.** C, H   **5.** I   **7.** B, I   **9.** F   **11.** 6   **13.** 4×3 **15.** 2×4

Exercise 2

1. $(3, 5)$ **3.** $(-3, 3)$ **5.** $(1, 2)$ **7.** $(3, 2)$ **9.** $(15, 6)$ **11.** $(3, 4)$ **13.** $(2, 3)$ **15.** $(1.63, 0.0971)$
17. $m = 2, n = 3$ **19.** $w = 6, z = 1$ **21.** $(-0.462, 2.31)$ **23.** $(5, 6, 7)$ **25.** $(15, 20, 25)$ **27.** $(1, 2, 3)$
29. $(3, 4, 5)$ **31.** $(6, 8, 10)$ **33.** $(-2.30, 4.80, 3.09)$ **35.** $(3, 6, 9)$ **37.** $x = 2, y = 3, z = 4, w = 5$
39. $x = -4, y = -3, z = 2, w = 5$ **41.** $x = a - c, y = b + c, z = 0, w = a - b$
43. $x = 4, y = 5, z = 6, w = 7, u = 8$ **45.** $v = 3, w = 2, x = 4, y = 5, z = 6$ **47.** $t = 53.2$ min; $d = 200$ mi
49. $1.54, -0.377, -1.15$ **51.** $-1.01, 1.69, 2.01, 1.20$ **53.** $1597, 774, 453, 121$

Exercise 3

1. -14 **3.** 15 **5.** -27 **7.** 17.3 **9.** $-2/5$ **11.** $ad - bc$ **13.** $\left(-\dfrac{3}{4}, 3\right)$

15. $(-3, 3)$ **17.** $(1, 2)$ **19.** $(3, 2)$ **21.** $(87/7, 108/7)$ **23.** $(15, 12)$ **25.** $(15, 6)$ **27.** $(3, 4)$

29. $(2, 3)$ **31.** $(-3, 5)$ **33.** $m = 4, n = 3$ **35.** $m = 2, n = 3$ **37.** $w = 6, z = 1$

39. $v = 1.05, w = 2.58$ **41.** $\left(\dfrac{dp - bq}{ad - bc}, \dfrac{aq - cp}{ad - bc}\right)$ **43.** $x = \dfrac{12 - bd}{8 - bc}, y = \dfrac{2d - 3c}{8 - bc}$

Exercise 4

1. 11 **3.** 45 **5.** 48 **7.** -28 **9.** 2 **11.** 18 **13.** -66 **15.** $(5, 6, 7)$ **17.** $(15, 20, 25)$
19. $(1, 2, 3)$ **21.** $(3, 4, 5)$ **23.** $(6, 8, 10)$ **25.** $(-2.30, 4.80, 3.09)$ **27.** $(3, 6, 9)$
29. $x = 2, y = 3, z = 4, w = 5$ **31.** $x = -4, y = -3, z = 2, w = 5$ **33.** $x = a - c, y = b + c, z = 0, w = a - b$
35. $x = 4, y = 5, z = 6, w = 7, u = 8$ **37.** $v = 3, w = 2, x = 4, y = 5, z = 6$
39. $x = -0.927, y = 2.28, z = 1.38, u = -0.385, v = 2.48$

Review Problems

1. 20 **3.** 0 **5.** -3 **7.** 0 **9.** 18 **11.** 15 **13.** 0 **15.** -29 **17.** 133
19. $x = -4, y = -3, z = 2, w = 1$ **21.** $(3, 5)$ **23.** $(4, 3)$ **25.** $(5, 1)$ **27.** $(2, 5)$ **29.** $(2, 1)$
31. $(3, 2)$ **33.** $(9/7, 37/7)$ **35.** $(3.19, 1.55)$ **37.** $(-15.1, 3.52, 11.4)$

◆◆◆ CHAPTER 11 ◆◆

Exercise 1

1. $y^2(3 + y)$ **3.** $x^3(x^2 - 2x + 3)$ **5.** $a(3 + a - 3a^2)$ **7.** $2q(2p + 3q + q^2)$
9. $(1/x)(3 + 2/x - 5/x^2)$ **11.** $(5m/2n)(1 + 3m/2n - 5m^2/4)$ **13.** $a^2(5b + 6c)$ **15.** $xy(4x + cy + 3y^2)$
17. $3ay(a^2 - 2ay + 3y^2)$ **19.** $cd(5a - 2cd + b)$ **21.** $4x^2(2y^2 + 3z^2)$ **23.** $ab(3a + c - d)$
25. $L_0(1 + \alpha t)$ **27.** $R_1[1 + \alpha(t - t_1)]$ **29.** $t(v_0 + at/2)$

Exercise 2

1. $(2 - x)(2 + x)$ **3.** $(3a - x)(3a + x)$ **5.** $4(x - y)(x + y)$ **7.** $(x - 3y)(x + 3y)$
9. $(3c - 4d)(3c + 4d)$ **11.** $(3y - 1)(3y + 1)$ **13.** $(m - n)(m + n)(m^2 + n^2)$ **15.** $(2m - 3n^2)(2m + 3n^2)$
17. $(a^2 - b)(a^2 + b)(a^4 + b^2)(a^8 + b^4)$ **19.** $(5x^2 - 4y^3)(5x^2 + 4y^3)$ **21.** $(4a^2 - 11)(4a^2 + 11)$

23. $(5a^2b^2 - 3)(5a^2b^2 + 3)$ **25.** $\left(\dfrac{1}{a} + \dfrac{1}{b}\right)\left(\dfrac{1}{a} - \dfrac{1}{b}\right)$ **27.** $\left(\dfrac{a}{x} + \dfrac{b}{y}\right)\left(\dfrac{a}{x} - \dfrac{b}{y}\right)$ **29.** $\pi(r_2 - r_1)(r_2 + r_1)$

31. $4\pi(r_1 - r_2)(r_1 + r_2)$ **33.** $m(v_1 - v_2)(v_1 + v_2)/2$ **35.** $\pi h(R - r)(R + r)$

Exercise 3

1. $(x - 7)(x - 3)$ **3.** $(x - 9)(x - 1)$ **5.** $(x + 10)(x - 3)$ **7.** $(x + 4)(x + 3)$ **9.** $(x - 7)(x + 3)$
11. $(x + 4)(x + 2)$ **13.** $(b - 5)(b - 3)$ **15.** $(b - 4)(b + 3)$ **17.** $2(y - 10)(y - 3)$
19. $(4x - 1)(x - 3)$ **21.** $(5x + 1)(x + 2)$ **23.** $(3b + 2)(4b - 3)$ **25.** $(2a - 3)(a + 2)$
27. $(x - 7)(5x - 3)$ **29.** $3(x + 1)(x + 1)$ **31.** $(3x + 2)(x - 1)$ **33.** $2(2x - 3)(x - 1)$
35. $(2a - 1)(2a + 3)$ **37.** $(3a - 7)(3a + 2)$ **39.** $(x + 2)^2$ **41.** $(y - 1)^2$ **43.** $2(y - 3)^2$
45. $(3 + x)^2$ **47.** $(3x + 1)^2$ **49.** $9(y - 1)^2$ **51.** $4(2 + a)^2$ **53.** $(x - 15)(x - 20)$
55. $(R - 300)(R - 100)$ **57.** $(2t - 9)(8t - 5)$ **59.** $(m - 500)^2$

Exercise 4

1. $(a^2 + 4)(a + 3)$ **3.** $(x - 1)(x^2 + 1)$ **5.** $(x - b)(x + 3)$ **7.** $(x - 2)(3 + y)$
9. $(x + y - 2)(x + y + 2)$ **11.** $(m - n + 2)(m + n - 2)$ **13.** $(4 + x)(16 - 4x + x^2)$
15. $2(a - 2)(a^2 + 2a + 4)$ **17.** $(x - 1)(x^2 + x + 1)$ **19.** $(x + 1)(x^2 - x + 1)$
21. $(a + 4)(a^2 - 4a + 16)$ **23.** $(x + 5)(x^2 - 5x + 25)$ **25.** $8(3 - a)(9 + 3a + a^2)$
27. $(4\pi/3)(r_2 - r_1)(r_2^2 + r_2 r_1 + r_1^2)$

Exercise 5

1. $x \neq 0$ **3.** $x \neq 5$ **5.** $x \neq 1; x \neq 2$ **7.** -1 **9.** $d - c$ **11.** $\dfrac{2}{3}$ **13.** $\dfrac{15}{7}$ **15.** $a/3$

17. $3m/4p^2$ **19.** $\dfrac{x + 2}{x^2 + 2x + 4}$ **21.** $\dfrac{n(m - 4)}{3(m - 2)}$ **23.** $\dfrac{2(a + 1)}{a - 1}$ **25.** $\dfrac{x + z}{x^2 + xz + z^2}$

27. $\dfrac{a - 2}{a - 3}$ **29.** $(x - 1)/2y$ **31.** $\dfrac{2}{3(x^4 y^4 - 1)}$

Exercise 6

1. $\dfrac{2}{15}$ **3.** $\dfrac{6}{7}$ **5.** $2\dfrac{2}{15}$ **7.** $9\dfrac{3}{8}$ **9.** $1\dfrac{1}{2}$ **11.** $a^4 b^4/2y^{2n}$ **13.** $x - a$ **15.** $ax/30$ **17.** $7/15$

19. $7/32$ **21.** $38\dfrac{2}{5}$ **23.** $1\dfrac{11}{12}$ **25.** $19\dfrac{3}{13}$ **27.** $7ax/2y$ **29.** $2cx/3az$ **31.** $(a + 2)/(d + c)$

33. $5(x + y)/(x - y)$ **35.** $(a + x)/2$ **37.** $(a + 1)(2a + 3)$ **39.** $7\dfrac{1}{2}$ in.   **41.** $\dfrac{pq + qq_1}{pq_1 + qq_1}$ **43.** $4F/\pi d^2$

45. $3F/4\pi r^3 D$ **47.** $\pi d/s$

Exercise 7

1. 1 **3.** $\dfrac{1}{7}$ **5.** $5/3$ **7.** $7/6$ **9.** $19/16$ **11.** $7/18$ **13.** $13/5$ **15.** $6/a$ **17.** $(a + 3)/y$

19. x **21.** $x/(a - b)$ **23.** $19a/10x$ **25.** $(5b - a)/6$ **27.** $\dfrac{9 - x}{x^2 - 1}$

29. $97\dfrac{11}{16}$ in.   **31.** $16\dfrac{3}{4}$ in.   **33.** $4\dfrac{7}{12}$ mi   **35.** $2/25$ min   **37.** $\dfrac{11}{8}$ machines   **39.** $\dfrac{2h(a + b) - \pi d^2}{4}$

41. $\dfrac{V_1 V_2 d + V V_2 d_1 + V V_1 d_2}{V V_1 V_2}$

Exercise 8

1. $\dfrac{85}{12}$ 3. $\dfrac{65}{132}$ 5. 69/95   7. $\dfrac{3(4x + y)}{4(3x - y)}$ 9. $y/(y - x)$ 11. $\dfrac{5(3a^2 + x)}{3(20 + x)}$ 13. $\dfrac{2(3acx + 2d)}{3(2acx + 3d)}$

15. $\dfrac{2x^2 - y^2}{x - 3y}$ 17. $\dfrac{(x + 2)(x - 1)}{(x + 1)(x - 2)}$ 19. $\dfrac{(d_1 + d_2)V_1V_2}{d_1V_2 + d_2V_1}$ 21. $-1/[x(x + h)]$

Exercise 9

1. 12   3. 20   5. 14   7. 72   9. 24   11. 7   13. 24   15. 24   17. 5/2   19. 12
21. 17   23. 3/13   25. $-2/3$ 27. $-11/4$ 29. 1/2   31. No solution   33. -2
35. 15.0 cm/s   37. 12, 24, and 18 days   39. 510 ft/min   41. 5.6 winters

Exercise 10

1. $bc/2a$ 3. $\dfrac{bz - ay}{a - b}$ 5. $\dfrac{a^2d + 3d^2}{4ac + d^2}$ 7. $\dfrac{b + cd}{a^2 + a - d}$ 9. $2z/a + 3w$ 11. $(b - m)/3$

13. $(c - m)/(a - b - d)$ 15. b 17. $w = \dfrac{ab}{5d}$ 19. $z = \dfrac{5mn + 2}{m}$ 21. $y = \dfrac{3 - m^2n}{m + 1}$

23. $\dfrac{w}{w(w + y) - 1}$ 25. $(p - q)/3p$ 27. $(5b - 2a)/3$ 29. $\dfrac{a^2(c + 1)(c - a)}{c^2}$ 31. $\dfrac{ab}{a + b}$

33. $\sqrt[3]{24CP^2/w^2}$ 35. $L/(1 + \alpha\Delta t)$ 37. $2(s - v_0t)/t^2$ 39. $y/(1 + nt)$ 41. $(E + I_2R_2)/(R_1 + R_2)$

43. $\dfrac{R}{1 + \alpha(t - t_1)}$ 45. $F/(m_1 + m_2 + m_3)$ 47. $RT/(v^2 - Rg)$

Review Problems

1. $(x - 5)(x + 3)$ 3. $(x^3 - y^2)(x^3 + y^2)$ 5. $(2x - 1)(x + 2)$ 7. $(2x - y/3)(4x^2 + 2xy/3 + y^2/9)$
9. $2a(xy - 3)(xy + 3)$ 11. $(3a + 4)(a - 2)$ 13. $(2/3)(a/2 + 2b/3)(a/2 - 2b/3)$ 15. $2(x - 5a)^2$
17. $(5a - 1)(1 - a)$ 19. $(a - 4)(a + 2)$ 21. $(x - 10)(x - 11)$ 23. $3(x + 3)(x - 5)$
25. $(4m - 3n)(16m^2 + 12mn + 9n^2)$ 27. $(3a - 4)(5a + 3)$ 29. $(a - b)(x + y)$
31. $(3a + 2b + c)/(a - b)$ 33. 4   35. -16 37. $(np - m)(mp - n)$ 39. $\dfrac{am + bn + cp}{m + n + p}$ 41. 47/7

43. 6   45. 1   47. $\dfrac{5x^2 - 4xy + 18y^2}{30x^2y^2}$ 49. $\dfrac{-10}{9}$ 51. $\dfrac{4xy}{x^2 - y^2}$ 53. $\dfrac{2ax^2y}{7w}$ 55. $\dfrac{b}{b + 1}$

57. $5(a - c)$ 59. $\dfrac{2a + 3}{3a + 5}$ 61. $\dfrac{\pi}{3}(r_1^2h - r_2^2d)$

⬩⬩⬩ CHAPTER 12 ⬩⬩⬩

Exercise 1

1. 3.17, 8.83   3. 3.89, -4.89 5. 1.96, -5.96 7. 0.401, -0.485 9. 0.170, -0.599 11. 2.87, 0.465
13. 0.907, -2.57 15. 5.87, -1.87 17. 12.0, -14.0 19. 5.08, 10.9   21. 2.81, -1.61 23. 8.37, -2.10
25. 0.132, -15.1 27. 2.31, 0.623   29. 4.26 in.

Exercise 2

1. $[1, -6]$ 3. $[11.7, 0.255]$ 5. $[6.84, 0.658]$ 7. $[4.83, 0.165]$ 9. $[8.47, 0.957]$ 11. $[11.1, 0.123]$
13. $[12.1, 0.371]$ 15. $[1.39, -1.55]$ 17. $[1.74, -0.608]$

Exercise 3

1. 2/3 or 3/2   **3.** 4 and 11   **5.** 5 and 15   **7.** 4 m × 6 m   **9.** 15 cm × 30 cm   **11.** 162 m × 200 m
13. 2.26 in.   **15.** 44.9 mi/h   **17.** 40 mi/h; 50 mi/h   **19.** 5 km/h   **21.** 15.6 h   **23.** 21.8 days
25. 9.07 ft and 15.9 ft   **27.** 24.5 s   **29.** 0.381 A and 0.769 A   **31.** −0.5 A and 0.1 A   **33.** 6.47 in.

Review Problems

1. 6, −1   **3.** 0, 5   **5.** 0, −2   **7.** 5, −2/3   **9.** 1, 1/2   **11.** ±3   **13.** 1.79, −2.79
15. 0.692, −0.803   **17.** $\pm\sqrt{10}$ **19.** 1/2, −2   **21.** −0.182, 9.18   **23.** ±5   **25.** 3.54, −2.64
27. 0.777, −2.49   **29.** 0.716, −2.32   **31.** 202 bags   **33.** 15 ft × 30 ft   **35.** 3 mi/h   **37.** $1\frac{1}{2}$ s
39. 16 in. wide × 5.0 in. deep or 10 in. wide × 8.0 in. deep   **41.** 3.56 km/h   **43.** 12 ft × 12 ft

••• CHAPTER 13 •••

Exercise 1

1. $3/x$ **3.** 2   **5.** $a/4b^2$ **7.** a^3/b^2 **9.** p^3/q **11.** 1   **13.** $1/(16a^6b^4c^{12})$ **15.** $1/(x+y)$
17. $m^4/(1-6m^2n)^2$ **19.** $2/x + 1/y^2$ **21.** $b/3$ **23.** $9y^2/4x^2$ **25.** $27q^3y^3/8p^3x^3$ **27.** $9a^8b^6/25x^4y^2$
29. $1/(3m)^3 - 2/n^2$ **31.** $x^{2n} + 2x^n y^m + y^{2m}$ **33.** $x/2y^7$ **35.** p **37.** $4^n a^{6n} x^{6n}/9^n b^{4n} y^{2n}$
39. $3p^2/2q^2x^4z^3$ **41.** $5^p w^{2p}/2^p z^p$ **43.** $25n^4x^8/9m^6y^6$ **45.** $R^{-1} = R_1^{-1} + R_2^{-1}$ **47.** $2I^2R$

Exercise 2

1. $\sqrt[4]{a}$ **3.** $\sqrt[4]{z^3}$ **5.** $\sqrt{m-n}$ **7.** $\sqrt[3]{y/x}$ **9.** $b^{1/2}$ **11.** y **13.** $(a+b)^{1/n}$
15. xy **17.** $3\sqrt{2}$ **19.** $3\sqrt{7}$ **21.** $2\sqrt[3]{7}$ **23.** $a\sqrt{a}$ **25.** $6x\sqrt{y}$ **27.** $\sqrt{21}/7$
29. $\sqrt[3]{2}/2$ **31.** $\sqrt[3]{6}/3$ **33.** $\sqrt{2x}/2x$ **35.** $2x^2\sqrt[3]{2y}$ **37.** $6y^2\sqrt[5]{xy}$ **39.** $a\sqrt{a-b}$
41. $3\sqrt{m^3 + 2n}$ **43.** $a\sqrt{15ab}/(5b)$ **45.** $\sqrt[3]{x/x}$ **47.** $x\sqrt[3]{18/3}$ **49.** $\omega_n = (kg/W)^{1/2}$
51. $Z = (R^2 + X^2)^{1/2}$ **53.** $a\sqrt{10}$

Exercise 3

1. $\sqrt{6}$ **3.** $\sqrt{6}$ **5.** $25\sqrt{2}$ **7.** $-\sqrt[3]{2}$ **9.** $-5\sqrt[3]{5}$ **11.** $-\sqrt[4]{3}$ **13.** $10x\sqrt{2y}$
15. $11\sqrt{6}/10$ **17.** $(a+b)\sqrt{x}$ **19.** $12\sqrt{6}$ **21.** $\sqrt{30}/8$ **23.** $6\sqrt[6]{108}$ **25.** $2\sqrt[4]{45}$
27. $4\sqrt[3]{2}$ **29.** $9a\sqrt[3]{bc}$ **31.** $\sqrt{abcd}/(bd)$ **33.** $\sqrt[6]{x^2y^3}$ **35.** $a^2b^2\sqrt[6]{108a^5}$
37. $32x\sqrt[3]{2x}$ **39.** $25 + 40\sqrt{x} + 16x$ **41.** $a + 10a\sqrt{ab} + 25a^2b$ **43.** 3/4   **45.** 3/2
47. $\dfrac{2\sqrt[6]{a^5b^2c^3}}{ac}$ **49.** $\dfrac{8 + 5\sqrt{2}}{2}$ **51.** $4\sqrt{ax}/a$ **53.** $6\sqrt[3]{2x}/x$ **55.** $(x-2a)\sqrt{y}$ **57.** $-4a\sqrt{3x}$
59. $(5a + 3c)\sqrt[3]{10b}$ **61.** $(a^2b^2c^2 - 2a + bc)\sqrt[5]{a^3bc^2}$ **63.** $2\sqrt{15} - 6$ **65.** $x^2\sqrt{1-xy}$ **67.** $a^2 - b$
69. $16x - 12\sqrt{xy} - 10y$ **71.** $\dfrac{x - \sqrt{xy}}{x-y}$ **73.** $\dfrac{a^2 + 2a\sqrt{b} + b}{a^2 - b}$ **75.** $\dfrac{3\sqrt{3mn} - 3m - \sqrt{6n} + \sqrt{2mn}}{3n - m}$
77. $24x^4\sqrt{3x}$

Exercise 4

1. 36   **3.** 4   **5.** 13.8   **7.** 8   **9.** 8   **11.** 7   **13.** 4.79   **15.** 3   **17.** 14.1 cm; 29.0 cm
19. 3.91 m; 4.77 m   **21.** $C = 1/(\omega^2 L \pm \omega\sqrt{Z^2 - R^2})$

Review Problems

1. $2\sqrt{13}$ **3.** $3\sqrt[3]{6}$ **5.** $\sqrt[3]{2}$ **7.** $9x\sqrt[4]{x}$ **9.** $(a-b)x\sqrt[3]{(a-b)^2 x}$ **11.** $x^2\sqrt{1-xy}$

13. $-\dfrac{1}{46}(6 + 15\sqrt{2} - 4\sqrt{3} - 10\sqrt{6})$ **15.** $-xy\sqrt{2}$ **17.** $\sqrt[6]{a^3 b^2}$ **19.** $9 + 12\sqrt{x} + 4x$ **21.** $7\sqrt{2}$

23. $\sqrt{2b}/(2b)$ **25.** $72\sqrt{2}$ **27.** 10   **29.** 14   **31.** 15.2   **33.** $x^{2n-1} + (xy)^{n-1} + x^n y^{n-2} + y^{2n-3}$

35. $27x^3 y^6/8$ **37.** $8x^9 y^6$ **39.** $3/w^2$ **41.** $1/(3x)$ **43.** $1/x - 2/y^2$ **45.** $1/(9x^2) + (y^8/4x^4)$

47. $p^{2a-1} + (pq)^{a-1} + p^a q^{a-2} + q^{2a-3}$ **49.** p^2/q **51.** r^2/s^3 **53.** 1   **55.** $V = 36\pi r^3$

57. $B = \sqrt{A}\sqrt{C}$

◆◆◆ CHAPTER 14 ◆◆

Exercise 1

1. 0.834 rad   **3.** 0.6152 rad   **5.** 9.74 rad   **7.** 0.279 rev   **9.** 0.497 rev

11. 0.178 rev   **13.** 162°   **15.** 21.3°   **17.** 65.3°   **19.** $\pi/3$

21. $11\pi/30$ **23.** $7\pi/10$ **25.** $13\pi/30$ **27.** $20\pi/9$ **29.** $9\pi/20$

31. $22\frac{1}{2}°$ **33.** 147.3°   **35.** 20°   **37.** 157.5°   **39.** 24°

41. 15°   **43.** 0.8660   **45.** 0.4863   **47.** -0.3090 **49.** 1.067

51. -2.747 **53.** -0.8090 **55.** -0.2582 **57.** 0.5854   **59.** 0.8129

61. 1.337   **63.** 0.2681   **65.** 1.309   **67.** 1.116   **69.** 0.5000

71. 0.1585   **73.** 19.1 in.^{2} **75.** 485 cm^2 **77.** 1130 cm^2 **79.** 2280 cm^2

81. 3.86 in.

Exercise 2

1. 6.07 in.   **3.** 230 ft   **5.** 43.5 in.   **7.** 2.21 rad   **9.** 1.24 rad

11. 0.824 cm   **13.** 125 ft   **15.** 3790 mi   **17.** 6210 mi   **19.** 448 mm

21. 55 mm   **23.** 40,600 bits   **25.** 0.251 m   **27.** $r = 355$ mm; $R = 710$ mm; $\theta = 60.8°$

29. 14.1 *cm*   **31.** 87.591 ft

Exercise 3

1. 194 rad/s; 11,100 deg/s   **3.** 12.9 rev/min; 1.35 rad/s   **5.** 8.02 rev/min; 0.840 rad/s   **7.** 621 ft/min

9. 7.86 rev/min   **11.** 4350°   **13.** 0.00749 s   **15.** 25.0 mm   **17.** 5.39 rev   **19.** 66,700 mi/h

21. 2790 ft/min   **23.** 85.7 rev/min

Review Problems

1. 77.1°   **3.** 20°   **5.** 165°   **7.** 2.42 rad/s   **9.** $5\pi/3$

11. $23\pi/18$ **13.** 0.3420   **15.** 1.000   **17.** -0.01827 **19.** 155 rev/min

21. 336 mi/h   **23.** 1.1089   **25.** -0.8812 **27.** -1.0291 **29.** 0.9777

31. 0.9097   **33.** 222 mm   **35.** 2830 cm^3

••• CHAPTER 15 •••

Exercise 1

1. amplitude = 2; period = 360°; phase shift = 0°
3. amplitude = 1; period = 180°; phase shift = 0°
5. amplitude = 3; period = 180°; phase shift = 0°
7. amplitude = 1; period = 360°; phase shift = $-15°$
9. amplitude = 1; period = 2π; phase shift = $\pi/2$
11. amplitude = 3; period = 360°; phase shift = $-45°$
13. amplitude = 4; period = 2π; phase shift = $\pi/4$
15. amplitude = 1; period = 180°; phase shift = $-27.5°$
17. amplitude = 1; period = $2\pi/3$; phase shift = $\pi/9$
19. amplitude = 3; period = 180°; phase shift = $-27.5°$
21. amplitude = 2; period = $2\pi/3$; phase shift = $\pi/6$
23. amplitude = 3.73; period = 83.3°; phase shift = $-12.8°$

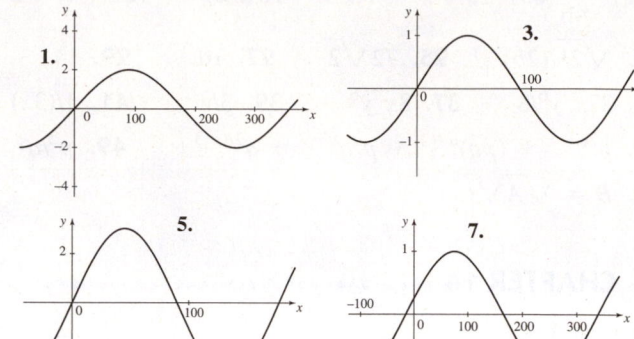

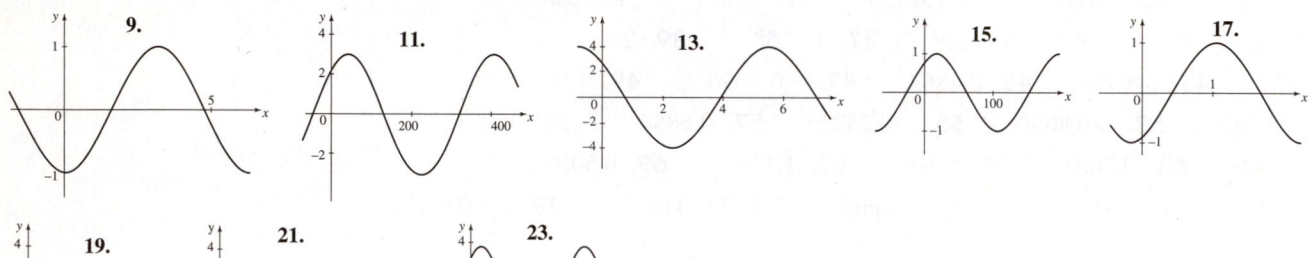

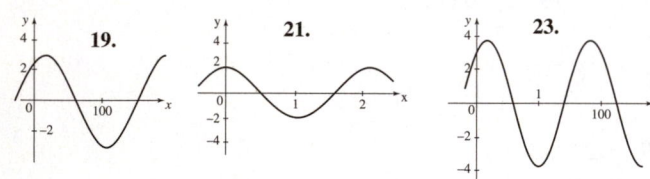

25. zeros at $x = 82.5°$ and 172.5°; $y = 0.7071$ at $x = 15°$ **27.** zeros at $x = 1$ and 2.05; $y = 0$ at $x = 1$

29. 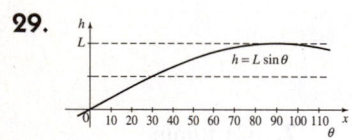   **31.** 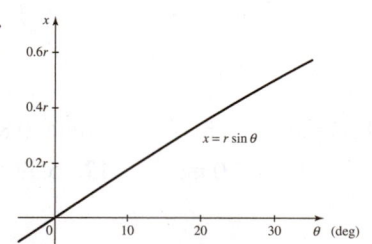   **33.** $r \sin(\theta + 38.6°)$

Exercise 2

3. $y = -2 \sin\left(\dfrac{x}{3} + \dfrac{\pi}{12}\right)$

Exercise 3

1. $P = 0.0147$ s; $\omega = 427$ rad/s   **3.** $P = 0.0002$ s; $\omega = 31{,}400$ rad/s   **5.** $f = 8$ Hz; $\omega = 50.3$ rad/s
7. 3.33 s   **9.** $P = 0.0138$ s; $f = 72.4$ Hz   **11.** $P = 0.0126$ s; $f = 79.6$ Hz
13. $P = 400$ ms; amplitude = 10; $\phi = 1.1$ rad   **15.** $y = 5 \sin(750t + 15°)$

17.
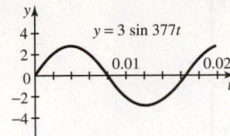
$y = 3 \sin 377t$

19.
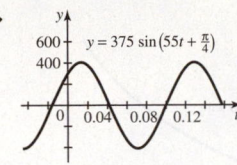
$y = 375 \sin\left(55t + \frac{\pi}{4}\right)$

21.
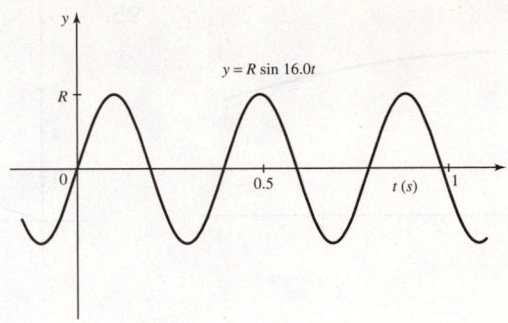
$y = R \sin 16.0t$

23. $V_{max} = 4.27$ V; $P = 13.6$ ms; $f = 73.7$ Hz; $\phi = 27° = 0.471$ rad; $v(0.12) = -2.11$ V

25. $i = 49.2 \sin(220t + 63.2°)$ mA

29.
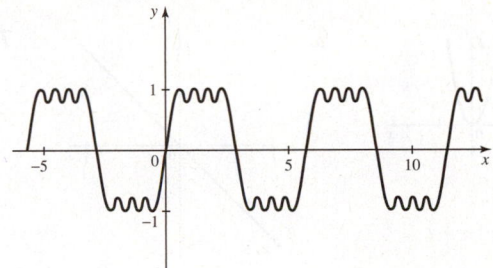

Exercise 4

1.
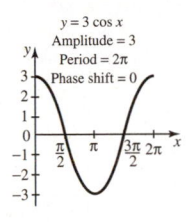
$y = 3 \cos x$
Amplitude = 3
Period = 2π
Phase shift = 0

3.
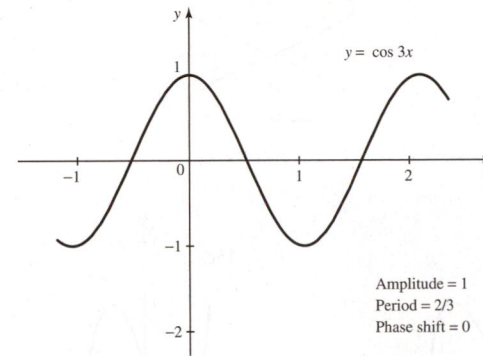
$y = \cos 3x$
Amplitude = 1
Period = 2/3
Phase shift = 0

5.
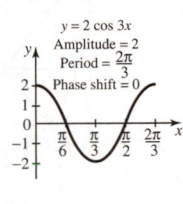
$y = 2 \cos 3x$
Amplitude = 2
Period = $\frac{2\pi}{3}$
Phase shift = 0

7.
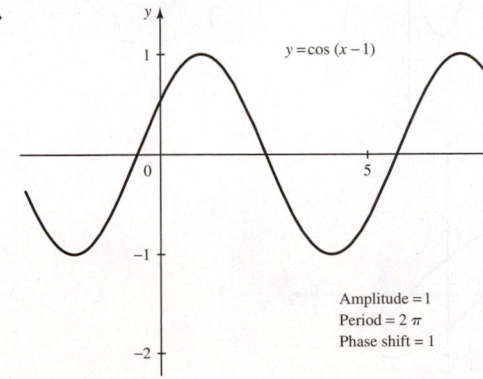
$y = \cos(x - 1)$
Amplitude = 1
Period = 2π
Phase shift = 1

9.
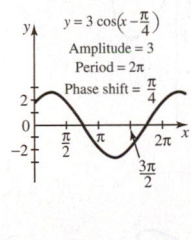
$y = 3 \cos\left(x - \frac{\pi}{4}\right)$
Amplitude = 3
Period = 2π
Phase shift = $\frac{\pi}{4}$

11.
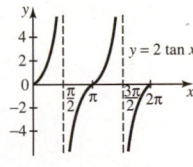
$y = 2 \tan x$

13.
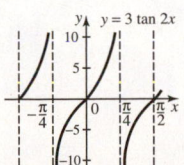
$y = 3 \tan 2x$

15.
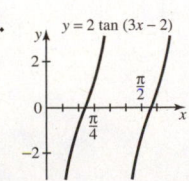
$y = 2 \tan(3x - 2)$

17.
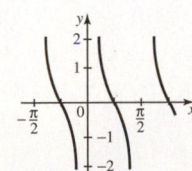

19.
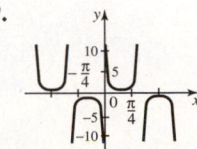

21.
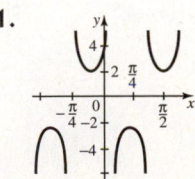

23.

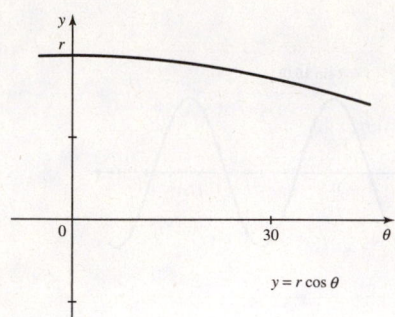

$y = r \cos \theta$

25.

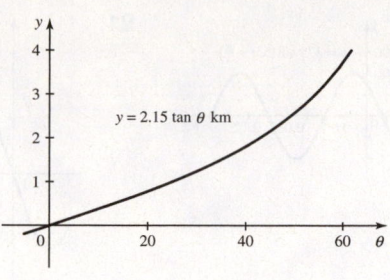

$y = 2.15 \tan \theta$ km

27.

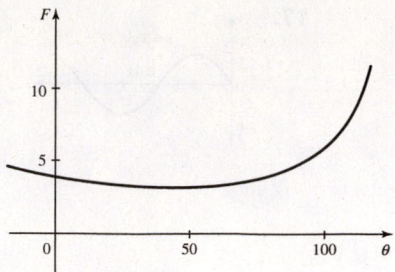

Exercise 5

1.

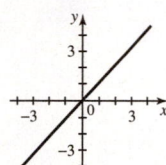

3.

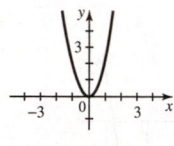

5.

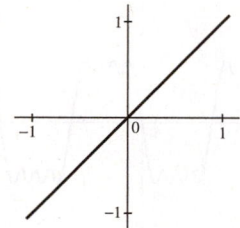

7.

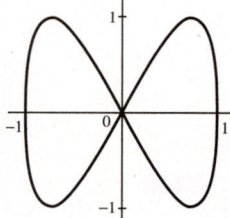

9.

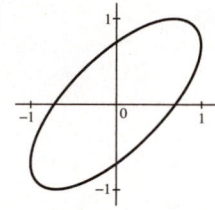

11.

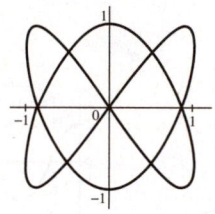

13.

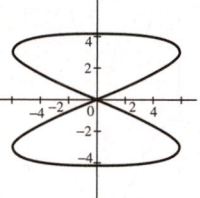

15.

15a.

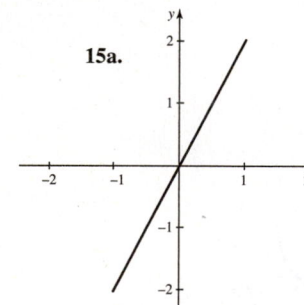

15b.

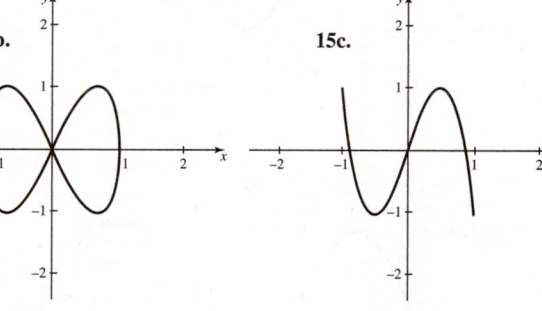

15c.

15d.

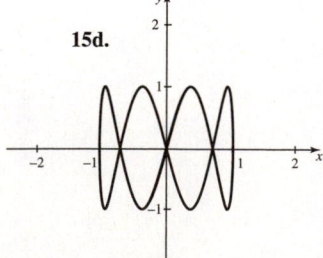

15e.

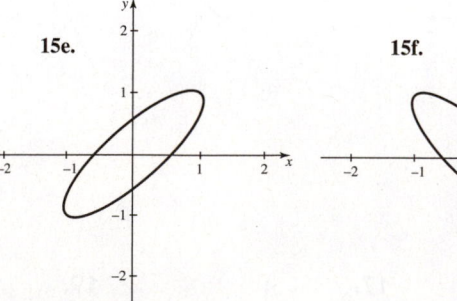

15f.

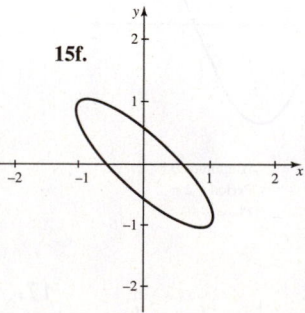

15g. 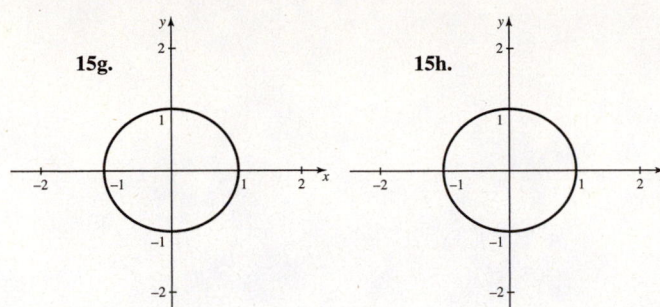 **15h.**

17. (a) 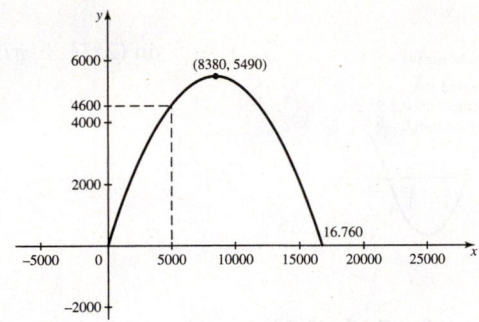   **(b)** 5490 ft   **(c)** 8380 ft   **(d)** 16,800 ft   **(e)** 4600 ft

Exercise 6

1., 3., 5., 7., 9., 11.

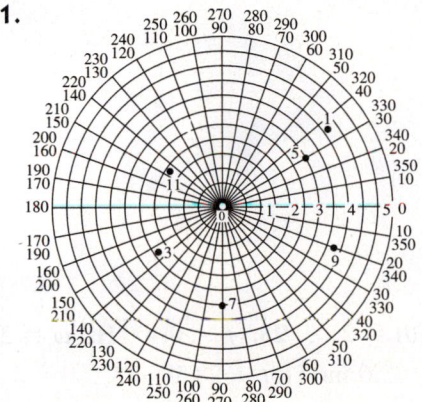

13. 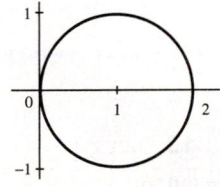   **15.** 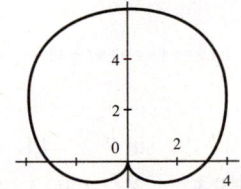   **17.** 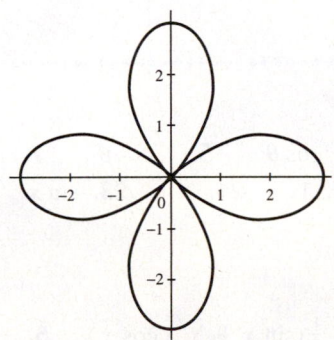   **19.**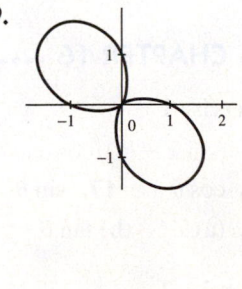

23. $(6.71, 63.4°)$ **25.** $(5.00, 36.9°)$ **27.** $(7.61, 231°)$ **29.** $(597, 238°)$

31. $(3.41, 3.66)$ **33.** $(298, -331)$ **35.** $(2.83, -2.83)$ **37.** $(12.3, -8.60)$

39. $(-7.93, 5.76)$ **41.** $x^2 + y^2 = 2y$ **43.** $x^2 + y^2 = 1 - y/x$

45. $x^2 + y^2 = 4 - x$ **47.** $r \sin \theta + 3 = 0$ **49.** $r = 1$ **51.** $\sin \theta = r \cos^2 \theta$

53.

r (in.)	θ (deg)	x (in.)	y (in.)
4.25	0	4.25	0.00
4.25	15	4.11	1.10
4.25	30	3.68	2.13
4.25	45	3.01	3.01
4.25	60	2.13	3.68
4.25	75	1.10	4.11
4.25	90	0.00	4.25

Review Problems

1.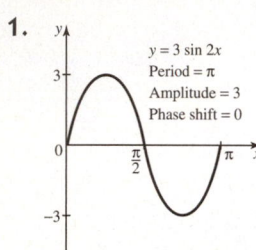

3.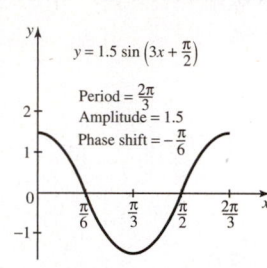

5.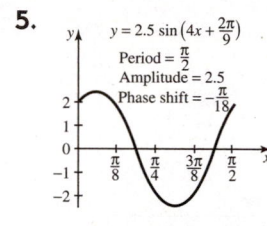

7. $y = 5 \sin(2x/3 + \pi/9)$

9.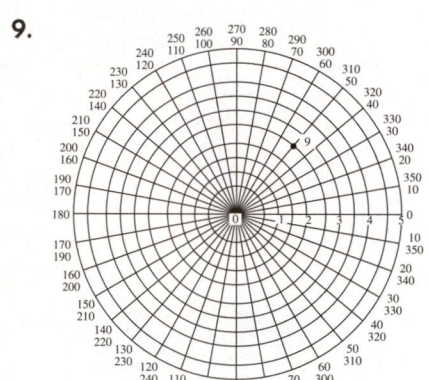

11.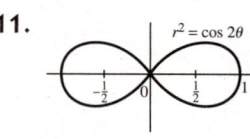

13. $(7.62, 23.2°)$ **15.** $(57.0, 245°)$

17. $(-61, -22)$ **19.** $y^2 = 4x + 4$ **21.** $r(\cos\theta - 3\sin\theta) = 2$ **23.** $f = 0.400$ Hz; $\omega = 2.51$ rad/s

25. $P = 0.140$ s; $f = 7.13$ Hz   **27.** 250 Hz   **29.** $i = 92.6 \sin(515t + 28.3°)$ mA

31. $v_{max} = 27.4$ V; $P = 8.54$ ms; $f = 117$ Hz; $\phi = 37° = 0.646$ rad; $v(0.250) = 17.8$ V

◆◆◆ CHAPTER 16 ◆◆

Exercise 1

1. $(\sin x - 1)/\cos x$ **3.** $1/\cos\theta$ **5.** $1/\cos\theta$ **7.** $-\tan^2 x$ **9.** $\sin\theta$ **11.** $\sec\theta$ **13.** $\tan x$

15. $\cos\theta$ **17.** $\sin\theta$ **19.** 1   **21.** 1   **23.** $\sin x$ **25.** $\sec x$ **27.** $\tan^2 x$ **29.** $-\tan^2\theta$

55. (a) Z^2 (b) $\tan\theta$

Exercise 2

1. $\frac{1}{2}(\sqrt{3}\sin\theta + \cos\theta)$ **3.** $\frac{1}{2}(\sin x + \sqrt{3}\cos x)$ **5.** $-\sin x$ **7.** $\sin\theta\cos 2\phi + \cos\theta\sin 2\phi$

9. $\cos 7x$ **11.** $\sin\theta$

Exercise 3

1. 1   **3.** $\sin 2x$ **23.** (a) $2v_0 \sin\dfrac{\theta}{g} = t$

Exercise 4

1. 2.05   **3.** 2.89   **5.** −10.9   **7.** 1.39   **9.** −0.500   **11.** 0.933   **13.** 0.137   **15.** 86.5 ft
17. 1.24 s   **19.** (a) miter = 35.3°; bevel = 30.0° (b) miter = 53.6°; bevel = 38.8°
21. miter = 58.7°; bevel = 15.5°

Exercise 5

1. 30°, 150°   **3.** 45°, 225°   **5.** 60°, 120°, 240°, 300°   **7.** 90°   **9.** 30°, 150°, 210°, 330°
11. 45°, 135°, 225°, 315°   **13.** 0°, 45°, 180°, 225°   **15.** 60°, 120°, 240°, 300°   **17.** 60°, 180°, 300°
19. 120°, 240°   **21.** 0°, 60°, 120°, 180°, 240°, 300°   **23.** 135°, 315°   **25.** 0°, 60°, 180°, 300°
27. 0.858 s   **29.** 13.7°, 76.3°

Review Problems

15. 30°, 90°, 150°   **17.** 45°, 90°, 135°, 225°, 270°, 315°   **19.** 90°, 306.9°   **21.** 60°, 300°
23. 3.82   **25.** 0.183   **29.** 4.2 ft

◆◆◆ CHAPTER 17 ◆◆◆

Exercise 1

1. 9/2   **3.** 8/3   **5.** 8   **7.** 8   **9.** $3x$ **11.** $-5a$ **13.** $(x + 2)/x$ **15.** ±12
17. ±15   **19.** 300 turns   **21.** 2.63   **23.** (a) 2.15; (b) 2.24; (c) 96.0%

Exercise 2

1. 0.951 m   **3.** 639 lb   **5.** 74.2 mm   **7.** 2.02 ft^2 **9.** $338   **11.** 43.8 acres
13. 161 acres   **15.** 7570 ft^3

Exercise 3

1. 197   **3.** 71.5

5.

x	9	11	15
y	45	55	75

7.

x	115	125	138	154
y	140	152	167	187

9. 6850   **11.** 418 mi   **13.** 4470 Ω   **15.** 15,136 parts   **17.** 36.6 MW

Exercise 4

1. 2050   **3.** 79.3   **5.** 2799   **7.** 66.7

9.

x	18.2	75.6	27.5
y	29.7	8840	154

11.

x	315	122	782
y	203	148	275

13.

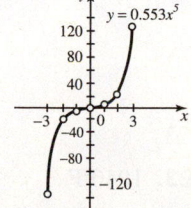

15. 396 m   **17.** 4.03 s   **19.** 765 W   **21.** 1.73   **23.** f8   **25.** f13

Exercise 5

1. 1420   **3.** y is halved

5.

x	306	622	91.5
y	125	61.5	418

7.

x	9236	3567	5725
y	1136	1828	1443

9. 41.4% increase

11.

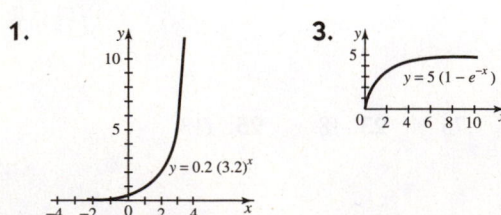

13. 103 lb/in.^{2} **15.** 4.49×10^{-6} dyne   **17.** 79 lb

19. 200 lux   **21.** 5.31 m   **23.** 0.909   **25.** 33.3% increase

Exercise 6

1. 352   **3.** 4.2% increase

5.

w	x	y
46.2	18.3	127
19.5	41.2	121
116	8.86	155
12.2	43.5	79.8

7. 13.9   **9.** $y = 4.08x^{3/2}/w$ **11.** $13\frac{3}{4}$ % decrease   **13.** New resistance $= \frac{1}{3}$ original   **15.** 9/4

17. 169 W   **19.** 3.58 m^3 **21.** 3.0 weeks   **23.** 12,600 lb   **25.** 394 times/s   **27.** 2.21

Review Problems

1. 1040   **3.** 10.4% increase   **5.** 121 L/min   **7.** 1.73   **9.** 29 yr   **11.** Shortened by 0.3125 in.

13. 8.5   **15.** 22 h 52 m   **17.** 215,000 mi   **19.** 42 kg   **21.** 2.8   **23.** 45% increase   **25.** 469 gal

27. 102 m^2 **29.** 1.1 in.   **31.** 17.6 m^2 **33.** 12.7 tons   **35.** 144 oz   **37.** $2162   **39.** 1.5

41. 761 cm^2

◆◆◆ CHAPTER 18 ◆◆

Exercise 1

1.

3.

5. (a) 7.55 m (b) 3.55 m (c) 2.99 m   **7.** $4852   **9.** (a) $ 6.73 (b) $ 7.33 (c) $7.39   **11.** $1823

13. 285 units   **15.** 12.4 million   **17.** 18.2 million barrels   **19.** 506°C   **21.** 0.100 A   **23.** 1040°F

25. 9.568 in. Hg   **27.** 72 mA   **29.** 30%

Exercise 2

1. $\log_3 81 = 4$ **3.** $\log_4 4096 = 6$ **5.** $\log_x 995 = 5$ **7.** $10^2 = 100$ **9.** $5^3 = 125$
11. $3^{57} = x$ **13.** 1.441 **15.** 0.773 **17.** 1.684 **19.** 2.922 **21.** 1.438 **23.** 906 **25.** 7.69
27. 2.32 **29.** 94,200 **31.** 3.877 **33.** 7.7685 **35.** 0.6125 **37.** 0.3177 **39.** 0.3104 **41.** 71.1
43. 6.33 **45.** 379 **47.** 144.2 **49.** 14.3 yr **51.** 14 yr **53.** 2773 ft **55.** 17.5 yr

Exercise 3

1. $\log 2 - \log 3$ **3.** $\log a + \log b$ **5.** $\log x + \log y + \log z$ **7.** $\log 3 + \log x - \log 4$ **9.** $-\log 2 - \log x$
11. $\log a + \log b + \log c - \log d$ **13.** $\log 12$ **15.** $\log(3/2)$ **17.** $\log 27$ **19.** $\log(a^3 c^4 / b^2)$
21. $\log(xy^2 z^3 / ab^2 c^3)$ **23.** $2^x = xy^2$ **25.** $p - q = 100$ **27.** 1 **29.** 2 **31.** x **33.** $3y$ **35.** 3.63
37. 1.639 **39.** 2.28 **41.** 195 **43.** −8.80 **45.** 5.46

Exercise 4

1. 2.81 **3.** 16.1 **5.** 2.46 **7.** 1.17 **9.** 5.10 **11.** 1.49 **13.** 0.239 **15.** −3.44 **17.** 1.39
19. 0.0416 s **21.** 22.4 s **23.** 55 yr **25.** 1.15 mi **27.** 9.7 yr **29.** 19.8 yr **31.** 42.3 yr

Exercise 5

1. 2 **3.** 1/3 **5.** 2/3 **7.** 2/3 **9.** 2 **11.** 3 **13.** 25 **15.** 6 **17.** 3/2 **19.** 47.5
21. −11.0, 9.05 **23.** 10/3 **25.** 22 **27.** 0.928 **29.** 0.916 **31.** 12 **33.** 101 **35.** 25.56 in. Hg
37. 1300 kW **39.** 11.2 in. **41.** 10^{-7} **43.** 40 **45.** −3.01 dB **47.** 13 dB **49.** 39 dB
51. −0.915 dB **53.** 6.02 dB

Review Problems

1. $\log_x 352 = 5.2$ **3.** $\log_{24} x = 1.4$ **5.** $x^{124} = 5.2$ **7.** 3/4 **9.** 1/128 **11.** $\log 3 + \log x - \log z$
13. $\log 10$ **15.** $\log(\sqrt{p}/\sqrt[4]{q})$ **17.** 2.5611 **19.** 1.2695 **21.** 701.5 **23.** 0.337 **25.** 4.4394
27. −4.7410 **29.** 4.362 **31.** 2.101 **33.** 13.44 **35.** 3.17 **37.** 1.00×10^{-3} **39.** $2071
41. 828 rev/min **43.** 23 yr **47.** 2600 V

••• CHAPTER 19 •••

Exercise 1

1. $-1 + i$ **3.** $2a + 2i$ **5.** $3/4 + i/6$ **7.** $4.03 + 1.20i$ **9.** i **11.** i **13.** $14i$ **15.** -15 **17.** $-96i$
19. -25 **21.** $6 - 8i$ **23.** $8 + 20i$ **25.** $36 + 8i$ **27.** $42 - 39i$ **29.** $21 - 20i$
31, 33, 35. **37.** $2 + 3i$ **39.** $p - qi$ **41.** $n + mi$ **43.** $2i$ **45.** 2

47. $2 + i$ **49.** $-5/2 + i/2$ **51.** $\dfrac{11}{34} - \dfrac{41}{34}i$

Exercise 2

1. $6.40\underline{/38.7°}$ **3.** $5\underline{/323°}$ **5.** $5.39\underline{/202°}$ **7.** $10.3\underline{/209°}$ **9.** $8.06\underline{/240°}$ **11.** $4.64 + 7.71i$
13. $4.21 - 5.59i$ **15.** $16.1\underline{/54.9°}$ **17.** $56\underline{/60°}$ **19.** $4.70\underline{/50.2°}$ **21.** $10\underline{/60°}$

Exercise 3

1. $6.40\underline{/38.7°}$ **3.** $5\underline{/323°}$ **5.** $5.39\underline{/202°}$ **7.** $4.41 + 2.35i$ **9.** $1.82 + 3.56i$ **11.** $-3.44 - 4.91i$
13. $-7-3i$ **15.** $(p + q) + (p + q)i$ **17.** $-112 + 19i$ **19.** $-15 + 10i$ **21.** $12 + 26i$ **23.** $46 - 14i$
25. $10\underline{/40°}$ **27.** $4.43\underline{/59.3°}$ **29.** $-3i$ **31.** 22 **33.** $(32 + 8i)/17$ **35.** $2\underline{/20°}$ **37.** $2.58\underline{/41.2°}$

Exercise 4

1. $43.6, 383$ **3.** $8.30, 27.2$ **5.** $460\underline{/24.9°}$ **7.** $53.7\underline{/64.9°}$ **9.** 36.4 lb vertical; 45.7 lb horizontal
11. 1440 lb; $23.0°$

Exercise 5

1. 49.0 A   **3.** 897 V   **5.** $177\underline{/25°}$ **7.** $40\underline{/-90°}$ **9.** $102\underline{/0°}$ **11.** $212 \sin \omega t$ **13.** $424 \sin(\omega t - 90°)$
15. $11 \sin \omega t$ **17.** $155 + 0i; 155\underline{/0°}$ **19.** $0 - 18i; 18\underline{/270°}$ **21.** $72.0 - 42i; 83.4\underline{/-30.3°}$
23. $552 - 572i; 795\underline{/-46.0°}$ **25.** (a) $v = 603 \sin(\omega t + 85.3°)$; (b) $v = 293 \sin(\omega t - 75.5°)$

Review Problems

1. i **3.** $9 + 2i$ **5.** $60.8 + 45.7i$ **7.** $11 + 7i$ **9.** $12\underline{/38°}$ **11.** $33/17 - 21i/17$ **13.** $2\underline{/45°}$
15, 17.   **19.** $7 - 24i$ **21.** $125\underline{/30°}$ **23.** $32\underline{/32°}$ **25.** $0.884\underline{/-45°}$

15, 17

27. (a) $(x + 3i)(x - 3i)$ (b) $(b + 5i)(b - 5i)$ (c) $(2y + zi)(2y - zi)$ (d) $(5a + 3bi)(5a - 3bi)$

◆◆◆◆◆ CHAPTER 20 ◆◆

Exercise 1

1. $3 + 6 + 9 + 12 + 15 + \cdots + 3n + \cdots$ **3.** $2 + 3/4 + 4/9 + 5/16 + 6/25 + \cdots + (n + 1)/n^2 + \cdots$
5. $u_n = 2n$; 8, 10   **7.** $u_n = 2^n/(n + 3)$; 32/8, 64/9   **9.** $u_n = u_{n-1} + 4$; 13, 17
11. $u_n = (u_{n-1})^2$; 6561, 43,046,721

Exercise 2

1. 46   **3.** 43   **5.** 49   **7.** $x + 24y$ **9.** $3, 7\frac{1}{3}, 11\frac{2}{3}, 16, 20\frac{1}{3}, \ldots$ **11.** $5, 11, 17, 23, 29, \ldots$ **13.** -7
15. 234   **17.** 225   **19.** 15   **21.** 10, 15   **23.** $-6\frac{3}{5}, -7\frac{1}{5}, -7\frac{4}{5}, -8\frac{2}{5}$ **25.** 3/14   **29.** 6/17, 6/13, 6/9
31. (a) $0.50, 1.00, 1.50, \ldots$ (b) $333   **33.** $512,500

Exercise 3

1. 80   **3.** -9375 **5.** 5115   **7.** $-292,968$ **9.** ± 15 **11.** ± 30 **13.** 24, 72   **15.** $\pm 20, 80, \pm 320$
17. $e^{1/2}$ or 1.649   **19.** $3.2°F$ **21.** 8.3 yr   **23.** 28.8 ft   **25.** 62   **27.** 2.0   **29.** 1.97
31. $184,202   **33.** $7776

Exercise 4

1. $5 - c$ **3.** $3/(c + 4)$ **5.** 288   **7.** 12.5   **9.** 45.5 in.   **11.** sum $= 1$

Exercise 5

1. 720   **3.** 42   **5.** 35   **21.** $59136a^{12}b^{18}$ **23.** $-15120a^4b^3$ **25.** $202400x^{21}y^2$

Review Problems

1. 35 **3.** 6, 9, 12, 15 **5.** $2\frac{2}{5}$, 3, 4, 6 **7.** 85 **9.** $a^5 - 10a^4 + 40a^3 - 80a^2 + 80a - 32$ **11.** 48

13. $9 - x$ **15.** 8 **17.** 5/7 **19.** $187\frac{1}{2}$ **21.** 1215 **23.** -1024 **25.** 20, 50 **27.** 1440 **29.** 252

31. $128x^7/y^{14} - 448x^{13/2}/y^{11} + 672x^6/y^8 - 560x^{11/2}/y^5 + \cdots$ **33.** $1 - a/2 + 3a^2/8 - 5a^3/16 + \cdots$

35. $-26{,}730a^4b^7$

◆ CHAPTER 21

Exercise 1

1. Discrete **3.** Categorical **5.** Categorical **7.** 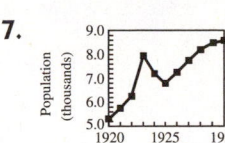 **9.**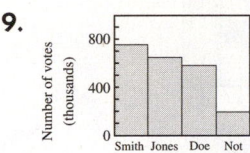

Exercise 2

1. (a) Range = 172 − 111 = 61

Class Midpt.	Class	Limits	1(b) Abs. Freq.	1(c) Rel. Freq. (%)	10(a) Cumulative Freq. Abs.	10(b) Cumulative Freq. Rel. (%)
113	110.5	115.5	3	7.5	3	7.5
118	115.5	120.5	4	10.0	7	17.5
123	120.5	125.5	1	2.5	8	20.0
128	125.5	130.5	3	7.5	11	27.5
133	130.5	135.5	0	0.0	11	27.5
138	135.5	140.5	2	5.0	13	32.5
143	140.5	145.5	2	5.0	15	37.5
148	145.5	150.5	6	15.0	21	52.5
153	150.5	155.5	7	17.5	28	70.0
158	155.5	160.5	2	5.0	30	75.0
163	160.5	165.5	5	12.5	35	87.5
168	165.5	170.5	3	7.5	38	95.0
173	170.5	175.5	2	5.0	40	100.0

3. (a) Range = 972 − 584 = 388

Class Midpt.	Class	Limits	3(b) Abs. Freq.	3(c) Rel. Freq. (%)	12(a) Cumulative Freq. Abs.	12(b) Cumulative Freq. Rel. (%)
525	500.1	550	0	0.0	0	0.0
575	550.1	600	1	3.3	1	3.3
625	600.1	650	2	6.7	3	10.0
675	650.1	700	4	13.3	7	23.3
725	700.1	750	3	10.0	10	33.3
775	750.1	800	7	23.3	17	56.7
825	800.1	850	1	3.3	18	60.0
875	850.1	900	7	23.3	25	83.3
925	900.1	950	4	13.3	29	96.7
975	950.1	1000	1	3.3	30	100.0

5. 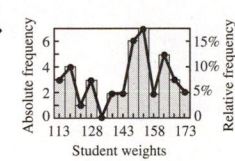 **7.** 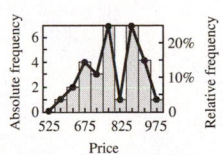 **9.**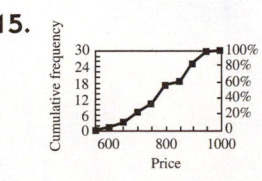

11. Range = 99.2 − 48.4 = 50.8

Class Midpt.	Class	Limits	2(b) Abs. Freq.	2(c) Rel. Freq. (%)	11(a) Cumulative Freq. Abs.	11(b) Cumulative Freq. Rel. (%)
47.5	45.05	50.05	5	16.7	5	16.7
52.5	50.05	55.05	2	6.7	7	23.3
57.5	55.05	60.05	3	10.0	10	33.3
62.5	60.05	65.05	1	3.3	11	36.7
67.5	65.05	70.05	3	10.0	14	46.7
72.5	70.05	75.05	9	30.0	23	76.7
77.5	75.05	80.05	0	0.0	23	76.7
82.5	80.05	85.05	2	6.7	25	83.3
87.5	85.05	90.05	3	10.0	28	93.3
92.5	90.05	95.05	1	3.3	29	96.7
97.5	95.05	100.05	1	3.3	30	100.0

13.

15.

17. 3
   4 8.5 8.9 9.6 8.4 9.4
   5 4.8 9.3 9.3 9.3 0.3
   6 1.4 9.3 6.3 9.3
   7 2.5 1.2 1.4 4.5 3.6 1.4 4.9 2.7 2.8
   8 8.2 4.6 9.4 5.7 3.8
   9 9.2 2.4
  10

Exercise 3

1. 77   **3.** 145 lb   **5.** $796   **7.** 81.6   **9.** 157.5   **11.** None   **13.** 59.3 min   **15.** 76   **17.** 149 lb
19. $792   **21.** 116, 142, 158, 164, 199   **23.** 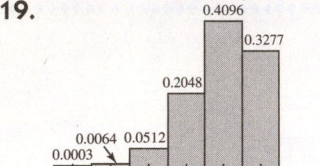   **25.** 83   **27.** 4.28, 7.40 and 10.02; 5.74
29. 697, 26.4   **31.** 201, 14.2

Exercise 4

1. 8/15   **3.** 0.375   **5.** 1/9   **7.** 1/36   **9.** 0.63   **11.** 5/18   **13.** 0.133   **15.** 0.230   **17.** 0.367
19.

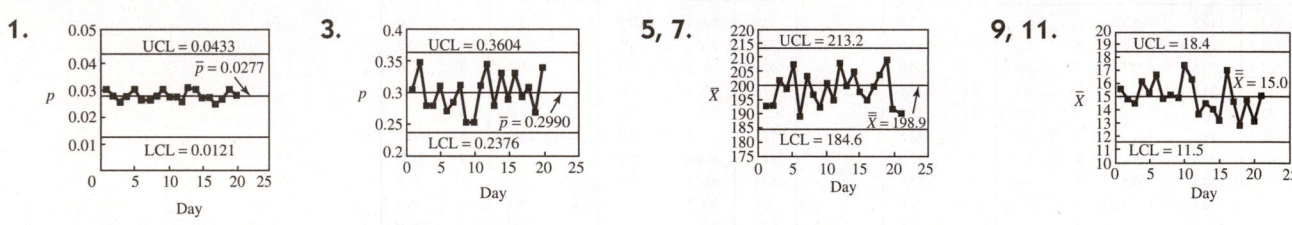

21. 0.117   **23.** 7.68×10^{-6}

Exercise 5

1. 0.4332   **3.** 0.2119   **5.** 620 students   **7.** 36 students   **9.** 1

Exercise 6

1. 69.47 ± 0.34 in.   **3.** 2.35 ± 0.24 in.   **5.** 164.0 ± 2.88   **7.** 16.31 ± 2.04   **9.** 0.250 ± 0.031

Exercise 7

1.

3.

5, 7.

9, 11.

Exercise 8

1. 1.00   **3.** 1.00   **5.** $y = -4.83x - 15.0$

Review Problems

1. Continuous   **3.** Categorical

5.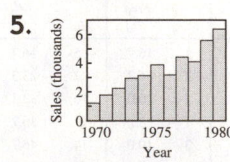

7. 5/36   **9.** 0.75   **11.** 0.1915   **13.** 0.3446   **15.** 81

17.

19.

Data	Cumulative Absolute Frequency	Cumulative Relative Frequency
Under 112.5	1	1/40 (2.5%)
Under 117.5	5	5/40 (12.5%)
Under 122.5	5	5/40 (12.5%)
Under 127.5	7	7/40 (17.5%)
Under 132.5	9	9/40 (22.5%)
Under 137.5	13	13/40 (32.5%)
Under 142.5	15	15/40 (37.5%)
Under 147.5	19	19/40 (47.5%)
Under 152.5	21	21/40 (52.5%)
Under 157.5	26	26/40 (65%)
Under 162.5	30	30/40 (75%)
Under 167.5	31	31/40 (77.5%)
Under 172.5	32	32/40 (80%)
Under 177.5	34	34/40 (85%)
Under 182.5	35	35/40 (87.5%)
Under 187.5	37	37/40 (92.5%)
Under 192.5	39	39/40 (97.5%)
Under 197.5	40	40/40 (100%)

21. $\bar{x} = 150$ **23.** 137 and 153   **25.** $s = 22.1$ **27.** 150 ± 7.0 **29.** 22.1 ± 4.94 **31.** 13

33. $Q_1 = 407, Q_2 = 718, Q_3 = 1055$; quartile range $= 648$

35.

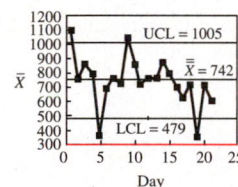

37. $-1.00; y = -2.31x + 18.1$ **39.** 0.0769 ± 0.0188

◆◆◆◆◆ **CHAPTER 22** ◆◆

Exercise 1

1. 2   **3.** 3   **5.** -7.90 **7.** $\Delta x = 3, \Delta y = 3$ **9.** $\Delta x = 9, \Delta y = -12$ **11.** 3   **13.** 9   **15.** 6.60

17. 3.34   **19.** 6.22   **21.** 3/2   **23.** 3   **25.** -2 **27.** $-5/9$ **29.** 4.66   **31.** 1.615   **33.** -1.080

35. $61.5°$ **37.** $110°$ **39.** $93.8°$ **41.** $41.3°$ **43.** $45.0°$ **45.** $2, -\dfrac{1}{2}$ **47.** $-1.85, 0.541$

49. $-5.372, 0.1862$ **51.** $45°$ **53.** $64°$ **55.** 78.6 mm   **57.** 2008 ft   **59.** 1260 m   **61.** $0.911°$

63. $33.7°$

Exercise 2

1.   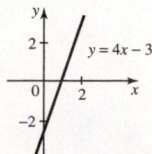   **3.**   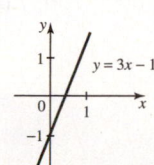   **5.**   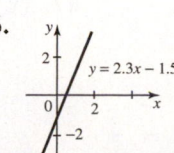   **7.** $x - 2y = 0$ **9.** $x + 2 = 0$

11. 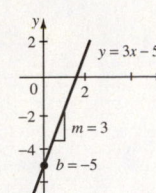 **13.**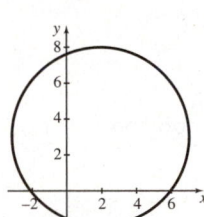

15. $y = 2x - 2$ **17.** $4x + y + 3 = 0$ **19.** $5x - y + 3 = 0$ **21.** $x + 3y + 15 = 0$ **23.** $5x - y + 15 = 0$

25. $x + 5y - 6 = 0$ **27.** $y - 2 = 0$ **29.** $3x - 4y + 11 = 0$ **31.** $3x - 5y - 15 = 0$ **33.** $3x - y + 2 = 0$

35. $2x + y - 6 = 0$ **37.** $y = -\dfrac{x}{2} - \dfrac{5}{2}$ **39.** $y = \dfrac{7x}{3} - \dfrac{8}{3}$ **41.** $F = kL - kL_0$

43. (a) 10.5 m/s; (b) 64.3 m/s   **45.** 150.1 Ω   **49.** $P = 20.6 + 0.432x$; 21.8 ft

51. $t = -0.789x + 25.0$; $x = 31.7$ cm; $m = -0.789$ **53.** $y = P + t(S - P)/L$; \$5555

Exercise 3

1. $x^2 + y^2 = 49$ **3.** $(x - 2)^2 + (y - 3)^2 = 25$ **5.** $(x - 5)^2 + (y + 3)^2 = 16$

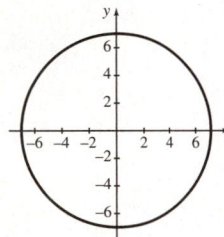   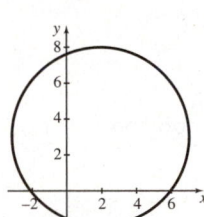   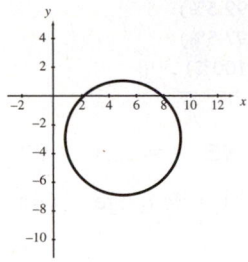

7. $C(0, 0)$; $r = 7$ **9.** $C(2, -4)$; $r = 4$ **11.** $C(3, -5)$; $r = 6$

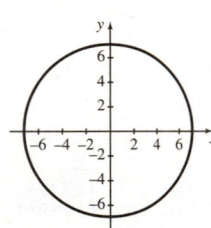   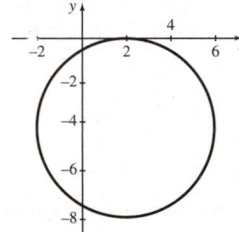   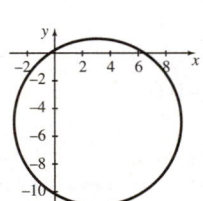

13. $x^2 + y^2 + 4x + 6y - 3 = 0$ **15.** $x^2 + y^2 - 10x + 8y - 1 = 0$

17. $(x - 4)^2 + y^2 = 16$; $C(4, 0)$; $r = 4$ **19.** $(x - 5)^2 + (y + 6)^2 = 36$; $C(5, -6)$; $r = 6$

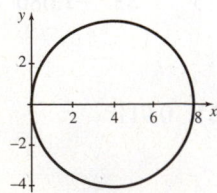   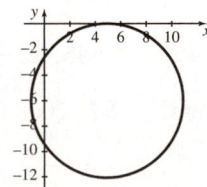

21. $(x + 3)^2 + (y - 1)^2 = 25$; $C(-3, 1)$; $r = 5$ **25.** 8.02 ft   **27.** $(x - 6)^2 + y^2 = 100$; 7.08 ft

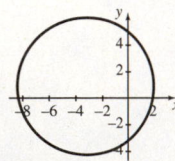

Exercise 4

1. $F(2, 0)$; $L = 8$

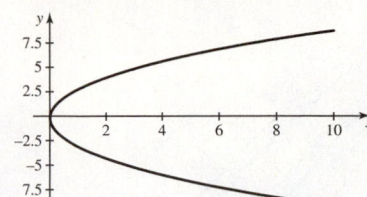

3. $F(0, -3/7)$; $L = 12/7$

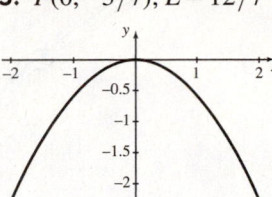

5. $x^2 = 9y$; $F(0, 9/4)$

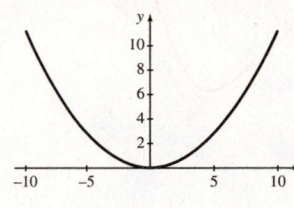

7. $3y^2 = 4x$; $F(1/3, 0)$

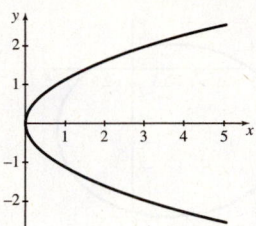

9. $V(3, 5)$; $F(6, 5)$; $L = 12$; $y = 5$

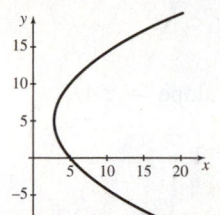

11. $V(3, -1)$; $F(3, 5)$; $L = 24$; $x = 3$

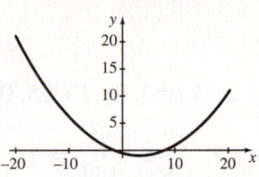

13. $V(2, -1)$; $F(13/8, -1)$; $L = 3/2$; $y = -1$

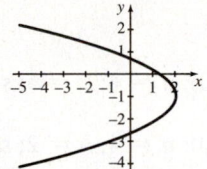

15. $V(3/2, 5/4)$; $F(3/2, 1)$; $L = 1$; $x = 3/2$

17. $(y - 2)^2 = 8(x - 1)$; $F(3, 2)$

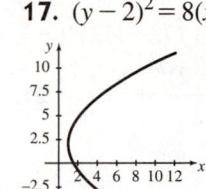

19. $x^2 = -4(y - 2)$; $F(0, 1)$; $L = 4$

21. $x^2 + 12x - 4y + 56 = 0$

23. $y^2 - 9x - 10y - 47 = 0$

25. $\left(x - \frac{3}{2}\right)^2 = -7\left(y + \frac{1}{4}\right)$; $V\left(\frac{3}{2}, -\frac{1}{4}\right)$; $F\left(\frac{3}{2}, -2\right)$; $L = 7$

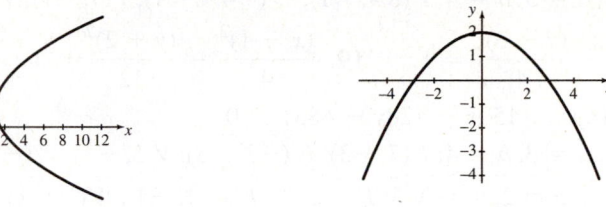

27. $(x - 70.0)^2 = -57.6(y - 85.0)$; 74.1 ft

29. $x^2 = -1310(y - 3520)$; 2760 m

31. $x^2 = 1250y$

33. 2.08 ft

35. $x^2 = 488y$

37. $(x - 16.0)^2 = -3070\left(y - \frac{1}{12}\right)$

Exercise 5

1. $V(\pm 5, 0)$; $F(\pm 3, 0)$

3. $V(\pm 2, 0)$; $F(\pm 1, 0)$

5. $V(0, \pm 4)$; $F(0, \pm 2)$

7. $\dfrac{x^2}{25} + \dfrac{y^2}{9} = 1$

9. $\dfrac{x^2}{36} + \dfrac{y^2}{4} = 1$

11. $\dfrac{x^2}{169} + \dfrac{y^2}{25} = 1$

13. $\dfrac{3x^2}{115} + \dfrac{7y^2}{115} = 1$

15. $C(2, -2)$; $V(6, -2), (-2, -2)$; $F(4.65, -2), (-0.65, -2)$

17. $C(-2, 3)$; $V(1, 3), (-5, 3)$; $F(0, 3), (-4, 3)$

19. $C(1, -1)$; $V(5, -1), (-3, -1)$; $F(4, -1), (-2, -1)$

21. $C(-3, 2)$; $V(-3, 7), (-3, -3)$; $F(-3, 6), (-3, -2)$

23. $\dfrac{x^2}{9} + \dfrac{(y - 3)^2}{36} = 1$

25. $\dfrac{(x + 2)^2}{12} + \dfrac{(y + 3)^2}{16} = 1$

27. $9x^2 + 4y^2 - 576 = 0$

29. $4x^2 + 9y^2 + 40x + 126y - 251 = 0$

31. $\dfrac{(x - 2)^2}{9} + (y + 1)^2 = 1$ $C(2, -1)$; $a = 3, b = 1$; $V(5, -1), (-1, -1)$; $F(4.83, -1), (-0.838, -1)$

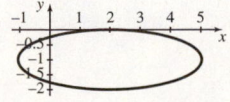

33. $\dfrac{(x+3)^2}{81} + \dfrac{(y+5)^2}{49} = 1$ $C(-3, -5)$; $a = 9$, $b = 7$; $V(6, -5)$, $(-12, -5)$; $F(2.66, -5)$, $(-8.66, -5)$

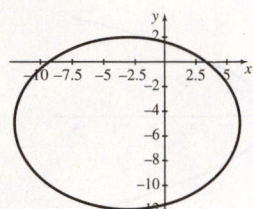

35. 11   **37.** 20 cm   **39.** 6.9 ft

Exercise 6

1. $V(\pm 4, 0)$; $F(\pm\sqrt{41}, 0)$; $a = 4$; $b = 5$; slope $= \pm 5/4$ **3.** $V(\pm 3, 0)$; $F(\pm 5, 0)$; $a = 3$; $b = 4$; slope $= \pm 4/3$

5. $V(\pm 4, 0)$; $F(\pm 2\sqrt{5}, 0)$; $a = 4$; $b = 2$; slope $= \pm 1/2$ **7.** $\dfrac{x^2}{25} - \dfrac{y^2}{144} = 1$ **9.** $\dfrac{x^2}{9} - \dfrac{y^2}{7} = 1$

11. $\dfrac{y^2}{16} - \dfrac{x^2}{16} = 1$ **13.** $\dfrac{y^2}{25} - \dfrac{3x^2}{64} = 1$

15. $C(2, -1)$; $a = 5$, $b = 4$; $F(8.4, -1)$, $F'(-4.4, -1)$; $V(7, -1)$; $V'(-3, -1)$; slope $= \pm 4/5$

17. $\dfrac{(y-2)^2}{16} - \dfrac{(x-3)^2}{4} = 1$ **19.** $\dfrac{(x-1)^2}{4} - \dfrac{(y+2)^2}{12} = 1$ **21.** $9x^2 - 25y^2 - 225 = 0$

23. $25x^2 - 16y^2 - 150x + 128y - 4831 = 0$

25. $C(2, -3)$; $a = 3$, $b = 4$; $F(7, -3)$, $F'(-3, -3)$; $V(5, -3)$, $V'(-1, -3)$; slope $= \pm 4/3$

27. $C(-1, -1)$; $a = 2$, $b = \sqrt{5}$; $F(-1, 2)$, $F'(-1, -4)$; $V(-1, 1)$, $V'(-1, -3)$; slope $= \pm 2/\sqrt{5}$ **29.** $xy = 36$

31. $\dfrac{x^2}{324} - \dfrac{y^2}{352} = 1$ **33.** $pv = 25{,}000$

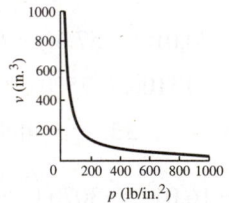

Review Problems

1. 4   **3.** -1.44 **5.** 147°   **7.** $-2b/a$ **9.** $m = 3/2$; $b = -7/2$ **11.** $7x - 3y + 32 = 0$ **13.** $5x - y + 27 = 0$

15. $7x - 3y + 21 = 0$ **17.** $y = 5$ **19.** -3 **21.** $x + 3 = 0$ **23.** 23.0   **25.** Parabola; $V(-3, 3)$; $F(-3, 2)$

27. Ellipse; $C(4, 5)$; $a = 10$, $b = 6$; $F(4, 13)$, $(4, -3)$; $V(4, 15)$, $(4, -5)$ **29.** Circle; $C(0, 0)$; $r = 3$ **41.** $x^2 + 2y^2 = 100$

43. $y^2 = -17x$ **45.** $(x+5)^2 + y^2 = 25$ **47.** $\dfrac{(x+2)^2}{25} + \dfrac{(y-1)^2}{9} = 1$ **49.** x int. $= 4$; y int. $= 8$ and -2

51. 3.00 m   **53.** 4.65 m

◆◆◆◆◆◆◆ CHAPTER 23 ◆◆◆

Exercise 1

1. 1   **3.** 1/5   **5.** 10   **7.** 4   **9.** 1   **11.** -4 **13.** -8 **15.** $-4/5$

17. $-1/4$ **19.** 2   **21.** 1/5   **23.** $+\infty$ **25.** $+\infty$ **27.** $-\infty$ **29.** x^2

31. 0   **33.** 3   **35.** $3x - \dfrac{1}{x^2 - 4}$ **37.** $3x^2$ **39.** $2x - 2$

Exercise 2

1. 4   **3.** 1/2   **5.** –5   **7.** –1/2   **9.** –9.13

Exercise 3

1. 3   **3.** 2   **5.** $2x$ **7.** $3x^2$ **9.** $-3/x^2$ **11.** $-\dfrac{1}{2\sqrt{3-x}}$ **13.** –2   **15.** 3/4   **17.** 2   **19.** 12

23. 12   **25.** $-8x$ **27.** 24   **29.** $2x$ **31.** $2x$ **33.** $2x$

Exercise 4

1. 0   **3.** 0   **5.** 1   **7.** $7x^6$ **9.** $6x$ **11.** $-5x^{-6}$ **13.** $-1/x^2$ **15.** $-9x^{-4}$ **17.** $2.5x^{-2/3}$

19. $2x^{-1/2}$ **21.** $\dfrac{-51\sqrt{x}}{2}$ **23.** –2   **25.** $3-3x^2$ **27.** $9x^2+14x-2$ **29.** a **31.** $x-x^6$

33. $(3/2)x^{-1/4}-x^{-5/4}$ **35.** $(8/3)x^{1/3}-2x^{-1/3}$ **37.** $6x^2$ **39.** $15x^4+2$ **41.** $10.4/x^3$ **43.** $6x+2$

45. 3   **47.** 13.5   **49.** 4   **51.** 3.75   **53.** $10t-3$ **55.** $175t^2-63.8$ **57.** $\dfrac{3\sqrt{5w}}{2}$ **59.** $-341/t^5$

61. $\dfrac{ds}{dt}=10t+3$, at $t=3.55s$, $\dfrac{ds}{dt}=38.5$ in./s

Exercise 5

1. $10(2x+1)^4$ **3.** $24x(3x^2+2)^3-2$ **5.** $\dfrac{-3}{(2-5x)^{2/5}}$ **7.** $\dfrac{-4.30x}{(x^2+a^2)^2}$ **9.** $-\dfrac{6x}{(x^2+2)^2}$ **11.** $\dfrac{2b}{x^2}\left(a-\dfrac{b}{x}\right)$

13. $\dfrac{-3x}{\sqrt{1-3x^2}}$ **15.** $\dfrac{-1}{\sqrt{1-2x}}$ **17.** $\dfrac{-3}{(4-9x)^{2/3}}$ **19.** $\dfrac{-1}{2\sqrt{(x+1)^3}}$ **21.** $2(3x^5+2x)(15x^4+2)$

23. $\dfrac{28.8(4.8-7.2x^{-2})}{x^3}$ **25.** $2(5t^2-3t+4)(10t-3)$ **27.** $\dfrac{7.6-49.8t^2}{(8.3t^3-3.8t)^3}$ **29.** 118,000   **31.** 540

33. $v'=a=13.8\,t(t^2+2)$; at $t=1.00$ s; $a=41.4$ ft/s^2

Exercise 6

1. $3x^2-3$ **3.** $5x^4-12x^2+4$ **5.** $42x+44$ **7.** $10x+9$ **9.** $33.0x^3-37.4x^5$

11. $\dfrac{3x^2}{\sqrt{5+x^2}}+3\sqrt{5+x^2}$ **13.** $\dfrac{15x^2+6x-3}{2\sqrt{x}}$ **15.** $\dfrac{2x^2-3}{(3x+5)^{2/3}}+4x\sqrt[3]{3x+5}$

17. $(2x^5+5x)^2+(2x-6)(2x^5+5x)(10x^4+5)$ **19.** $3x^2-12x-2$ **21.** 20.2   **23.** 6

25. $6\pi(2x-4)^2$ **27.** $3x^2-12x-7$ **29.** $3x(x+1)^2(x-2)^2+(2x^2+2x)(x-2)^3+(x+1)^2(x-2)^3$

31. $\dfrac{2}{(x+2)^2}$ **33.** $\dfrac{8x}{(4-x^2)^2}$ **35.** $\dfrac{-5}{(x-3)^2}$ **37.** $\dfrac{1}{2\sqrt{x}(\sqrt{x}+1)^2}$ **39.** $\dfrac{-a^2}{(z^2-a^2)^{3/2}}$ **41.** −0.456

43. 1   **45.** $T'=3.29$ °F/h at $t=2.35$ h

Exercise 7

1. $6u^2\dfrac{du}{dw}$ **3.** $2y\dfrac{dy}{du}+3u^2$ **5.** $2x^3y\dfrac{dy}{dx}+3x^2y^2$ **7.** $\dfrac{3z}{\sqrt{3z^2+5}}\dfrac{dz}{dt}$ **9.** $2y-7$ **11.** $(6y-7)(y+3)^4$

13. 5/2   **15.** $-y/x$ **17.** $2a/y$ **19.** $\dfrac{ay-x^2}{y^2-ax}$ **21.** $\dfrac{1+3x^2}{1+3y^2}$ **23.** $\dfrac{8xy}{3y-8x^2+4y^2}$ **25.** −0.436

27. 14/15   **29.** $3x^2\,dx$ **31.** $\dfrac{2\,dx}{(x+1)^2}$ **33.** $(3x^2+3)\,dx$ **35.** $\dfrac{y-3x}{2y-x}\,dx$ **37.** $-\dfrac{4x+3y}{3x+8y}\,dx$

39. $dy/dx=-1.16$ at $x=6.25$

Exercise 8

1. $12x$ **3.** $24x + 6$ **5.** $36x^2 - 6x$ **7.** $\dfrac{8}{(x+2)^3}$ **9.** $\dfrac{-20}{(5-4x^2)^{3/2}}$ **11.** 40

13. $v = 41.6$ cm/s; $a = 48.0$ cm/s^2

Review Problems

1. -1 **3.** $8/5$ **5.** $-\dfrac{9x}{4y}$ **7.** $5 - 6x$ **9.** $+\infty$ **11.** -8 **13.** $72x^7 + 48x^3$ **15.** $10t - 3$

17. -10 **19.** $71.6x^{-3}(21.7x + 19.1)(64.2 - 17.9x^{-2}) + 21.7(64.2 - 17.9x^{-2})^2$ **21.** $12x^2 - 3$

23. $9x^2 + 4x - 21$ **25.** $\dfrac{-2x^2 - 18}{(x^2 - 9)^2}$ **27.** $24x^5 - 48x^2$ **29.** $(2/5)x^{-3/5} + (2/3)x^{-2/3}$ **31.** $60x + 10$

33. $(8x - 20)\, dx$ **35.** $(3x^2 - 2)\, dx$

◆◆◆◆◆ CHAPTER 24 ◆◆◆

Exercise 1

1. $2x - y + 1 = 0$; $x + 2y - 7 = 0$ **3.** $12x - y - 13 = 0$; $x + 12y - 134 = 0$

5. $3x + 4y = 25$; $4x - 3y = 0$ **7.** $(3, 0)$ **9.** $(3, -4)$ and $(-3, 4)$ **11.** $71.6°$ **13.** $36.9°$

Exercise 2

1. Increasing   **3.** Decreasing   **5.** Decreasing   **7.** Increasing   **9.** Increasing for all x
11. Increasing for all x **13.** Increasing for $x > -1$ **15.** Increasing $-1 < x < 1$; Decreasing for $x < -1$, $x > 1$ **17.** Upward
19. Downward   **21.** Upward   **23.** Downward   **25.** min $(0, 0)$ **27.** max $(0, 36)$; min $(4.67, -14.8)$
29. max $(1, 1)$, $(-1, 1)$; min $(0, 0)$ **31.** min $(1, -3)$ **33.** max $(0, 0)$; min $(-1, -5)$, $(2, -32)$
35. max $(-3, 32)$; min $(1, 0)$ **37.** max $(0, 2)$; min $(0, -2)$ **39.** max $(0.500, -4.23)$; min $(0.500, 4.23)$
41. min $(-0.630, 0.766)$ max $(-0.630, -1.23)$ **43.** PI $(0, 3)$

45. PI$(0, 1)$ $\left(\dfrac{1}{2}, \dfrac{15}{16}\right)$ **47.** PI $(0, 0)$ **49.** PI $(0.133, 0.976)$

Exercise 3

1. 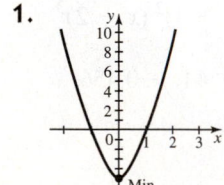   **3.** 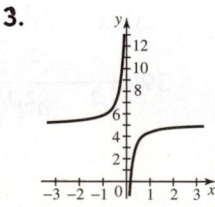   **5.** 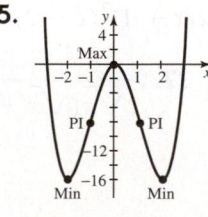   **7.** 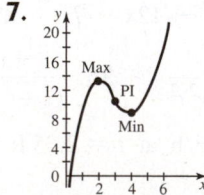   **9.**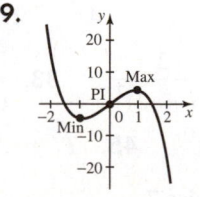

11. 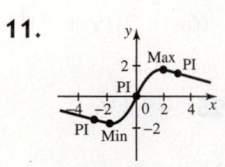   **13.** 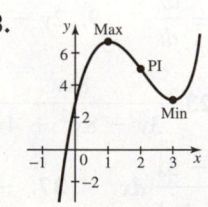   **15.** 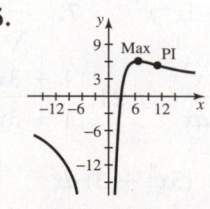   **17.**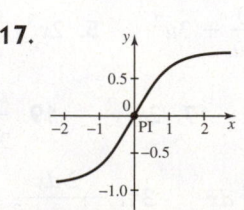

19. 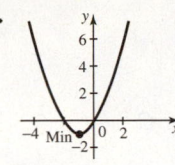   **21.** 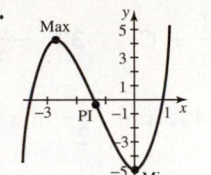   **23.** 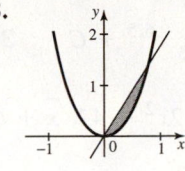   **25.**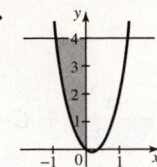

Review Problems

1. min $(0.250, 4.75)$ **3.** max $(1, 2)$; min $(-1, -2)$ **5.** $7x - y - 9 = 0$; $x + 7y - 37 = 0$ **7.** $-7/3$
9. Rising for $x > 0$; never falls   **11.** Downward   **13.** $34x - y = 44$; $x + 34y = 818$ **15.** $71.6°$

◆◆◆◆◆ CHAPTER 25 ◆◆◆

Exercise 1

1. $251 °F/h$ **3.** $-0.227 \text{ ft}^3/\text{min}^2$ **5.** 39.6 A **7.** 5040 W **9.** 1.80 mA **11.** 24.3 V
13. $-0.175 \text{ lb/in.}^2/\text{in.}^3$ **15.** $12.6 \text{ m}^3/\text{m}$ **17.** 0.0540 s/in. **19.** $\dfrac{dy}{dx} = \dfrac{wx}{6EI}(x^2 + 3L^2 - 3Lx)$

Exercise 2

1. $v = 0$; $a = -16.0$ **3.** $v = -3.00$; $a = 18.0$ **5.** $v = -8.00$; $a = -32.0$
7. $v = 104 \text{ ft/s}$; $a = 32.0 \text{ ft/s}^2$ **9.** 60.0 ft/s **11.** a) 2000 ft/s b) $62,500 \text{ ft}$ c) 1680 ft/s
13. $s = 0$; $a = 32 \text{ units/s}^2$ **15.** a) $89.2°$ b) $v_x = 0.0175 \text{ in./s}$; $v_y = 1.25 \text{ in./s}$ **17.** $42.3 \text{ cm/s at } 320°$
19. a) $x = 32,500 \text{ ft}$; $y = 29,500 \text{ ft}$ b) $v_x = 4640 \text{ ft/s}$; $v_y = 4110 \text{ ft/s}$ **21.** 2680 rad/s **23.** 1270 rad/s; 2030 rad/s^2

Exercise 3

1. 60 m/s **3.** 3.58 mi/h **5.** 2.40 m/s **7.** 4.47 ft/s **9.** 9.95 in./s **11.** 4.00 ft/s **13.** $0.101 \text{ in.}^2/\text{s}$
15. 0.133 in./min **17.** -4.00 in./min **19.** 1.19 m/min **21.** 0.200 ft/min **23.** $7.00 \text{ lb/in.}^2/\text{s}$
25. 170 km/h **27.** 8.33 ft/s **29.** 1.61 in./s

Exercise 4

1. 1 **3.** $6\dfrac{2}{3}$ and $13\dfrac{1}{3}$ **5.** $18 \text{ m} \times 24 \text{ m}$, the 24 m is the shared side.   **7.** $9 \text{ yd} \times 18 \text{ yd}$
9. $d = 6.51 \text{ cm}$; $h = 3.26 \text{ cm}$ **11.** 25 in.^2 **13.** 6 **15.** 3.46 in. **17.** $(2, 2)$ **19.** $5.00 \text{ in.} \times 4.33 \text{ in.}$
21. $9.90 \text{ units} \times 14.1 \text{ units}$ **23.** $r = 4.08 \text{ cm}$; $h = 5.77 \text{ cm}$ **25.** 12.0 ft
27. $r = 5.24 \text{ m}$; $h = 5.24 \text{ m}$ **29.** $10.4 \text{ in.} \times 14.7 \text{ in.}$ **31.** 59.1 in **33.** 7.5 A
35. 5.0 A **37.** 0.65 **39.** $r = 11.0 \text{ ft}$; $h = 26.3 \text{ ft}$; cost $= \$11,106$

Review Problems

1. 190 ft/s **3.** 4.0 km/h **5.** $3.33 \text{ cm}^2/\text{min}$ **7.** $s/2$ **9.** 6.93 **11.** -2.26 lb/in.^2 per second
13. $t = 4$, $s = 108$; $t = -2$, $s = 0$ **15.** 4 sq. units   **17.** 4 and 6 **19.** $10 \text{ in.} \times 10 \text{ in.} \times 5 \text{ in.}$
21. (a) 142 rad/s; (b) 37.0 rad/s^2 **23.** 36.4 A

◆◆◆◆◆ CHAPTER 26 ◆◆◆

Exercise 1

1. $x + C$ **3.** $6x + C$ **5.** $5x + C$ **7.** $\pi x + C$ **9.** $\dfrac{1}{2}x^2 + C$ **11.** $\dfrac{1}{5}x^5 + C$ **13.** $\dfrac{5}{2}x^2 + C$

15. $\dfrac{8}{5}x^5 + C$ **17.** $\dfrac{x^2}{2} + 4x + C$ **19.** $\dfrac{x^3}{3} + 2x^2 + C$ **21.** $x^3 - 12x^2 + 4x + C$ **23.** $\dfrac{2x^{3/2}}{3} + C$

25. $\dfrac{3x^{7/3}}{7} + C$ **27.** $\dfrac{2}{3}\sqrt{5}\,x^{3/2} + C$ **29.** $\dfrac{15}{2^{4/5}}x^{6/5} + C$ **31.** $-\dfrac{1}{x} + C$ **33.** $-\dfrac{1}{3x^3} + C$

35. $2\sqrt{x} + C$ **37.** $\dfrac{6x^{5/2}}{5} - \dfrac{4x^{3/2}}{3} + C$ **39.** $2x^2 - 4\sqrt{x} + C$ **41.** $x - \dfrac{3x^2}{2} + x^3 - \dfrac{x^4}{4} + C$

Exercise 2

1. $\dfrac{1}{4}(x^4 + 1)^4 + C$ **3.** $(x^3 + 1)^3 + C$ **5.** $\dfrac{1}{2}(x^2 + 2x)^4 + C$ **7.** $\dfrac{1}{1-x} + C$ **9.** $\dfrac{1}{3}(x^2 - 2)^{3/2} + C$

11. $-\dfrac{2}{3}(1 - x^3)^{1/2} + C$ **13.** $t^5 + C$ **15.** $\dfrac{(z^3 + 3z)^4}{12} + C$ **17.** $\dfrac{t^2}{2} + 9\ln|t| + C$ **19.** $3\ln|x| + C$

21. $\dfrac{5}{3}\ln|z^3 - 3| + C$ **23.** $x + \ln|x| + C$ **25.** $-\dfrac{1}{12}\ln\left|\dfrac{3t - 2}{3t + 2}\right| + C$

27. $\dfrac{x}{2}\sqrt{25 - 9x^2} + \dfrac{25}{6}\mathrm{Sin}^{-1}\left(\dfrac{3x}{5}\right) + C$

29. $\dfrac{1}{3}\mathrm{Tan}^{-1}\dfrac{x}{3} + C$ **31.** $\dfrac{1}{12}\mathrm{Tan}^{-1}\left(\dfrac{4x}{3}\right) + C$ **33.** $-\dfrac{1}{12}\ln\left|\dfrac{2x - 3}{2x + 3}\right| + C$ **35.** $\dfrac{1}{4}\ln\left|\dfrac{x - 2}{x + 2}\right| + C$

37. $\dfrac{x}{4}\sqrt{x^2 - 4} - \ln\left|x + \sqrt{x^2 - 4}\right| + C$ **39.** $\dfrac{5}{2}\sin^{-1}x^2 + C$

Exercise 3

1. $y = \dfrac{4x^3}{3} + C$ **3.** $y = -\dfrac{1}{2x^2} + C$ **5.** $s = \dfrac{3t^{1/3}}{2} + C$ **7.** $y = 3x^2/2$ **9.** $2x^{3/2} - 3y + 6.34 = 0$ **11.** $8/3$

13. $y = 2x^2 - 5x + 8$ **15.** $y = x^4 - 3x^2$ **17.** $v = 21.5t + 27.6$ m/s; $s = 10.8t^2 + 27.6t + 44.3$ m

Exercise 4

1. 1.50   **3.** 60.7   **5.** 37.3   **7.** 72.7   **9.** 2.30   **11.** 0.236   **13.** 0.732   **15.** 43,400 in.^{3}

Exercise 5

1. $A \approx 33$ **3.** $A \approx 2.63$ **5.** $A \approx 5.33$ **7.** 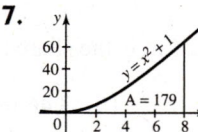   **9.**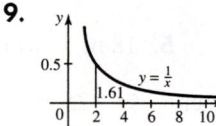

11. 15   **13.** 84   **15.** 40   **17.** 176   **19.** 1.57

Exercise 6

1. 100   **3.** $113\dfrac{1}{3}$ **5.** 64   **7.** 4.24   **9.** 9.83   **11.** 15.75   **13.** 500 ft^2

Review Problems

1. $\dfrac{3x^{2/3}}{2} + C$ **3.** $1.03y^3 + C$ **5.** $\dfrac{4x\sqrt{x}}{3} + C$ **7.** $\dfrac{1}{4}(x^4 - 2x^3)^2 + C$ **9.** $\ln|x + 5| + C$ **11.** $81\dfrac{2}{3}$

13. $\dfrac{a^2}{6}$ **15.** 25.0   **17.** 28.8   **19.** $-\dfrac{1}{12}\ln\left|\dfrac{2x - 3}{2x + 3}\right| + C$ **21.** $\dfrac{1}{6}\mathrm{Tan}^{-1}\left(\dfrac{2x}{3}\right) + C$

23. 0.360   **25.** $N = 100e^{t/2}$ **27.** 7.283   **29.** 16/3 square units

◆◆◆◆◆◆ **CHAPTER 27** ◆◆

Exercise 1

1. $s = 5.80t^2 + 21.4t + 12.6$; 447 cm   3. $s = \dfrac{t^3}{24}$; $2\dfrac{2}{3}$ ft

5. $v = 0.470t^3 + 5.28t + 2.58$ ft/s;   8.33 ft/s   7. $v = \dfrac{t^4}{4.00} - 25.8t + 15.8$ cm/s;   43.1 cm/s

9. $a = 32.2$ ft/s^2; $v = 1.77 + 32.2t$ ft/s; $s = 1.77t + 16.1t^2$ ft; $a = 32.2$ ft/s^2; $v = 98.4$ ft/s; $s = 150$ ft

11. $s = 20t - 16t^2$ ft   13. $v_x = 1670$ cm/s; $v_y = 100$ cm/s   15. $v_x = 344$ cm/s; $v_y = 2250$ cm/s

17. 18,000 rev   19. 15.7 rad/s

Exercise 2

1. 12.1 C   3. 78.4 C   5. 2.25 V   7. 4.25 A   9. 20.2 A

Exercise 3

1. 34   3. 20.4   5. $42\dfrac{2}{3}$ 7. $1\dfrac{1}{6}$ 9. 32.83   11. $2\dfrac{2}{3}$ 13. 2.797   15. 16

17. 26 2/3   19. 2.70   21. 9   23. $\dfrac{4}{3}$ 33. 90.1 ft^2 35. 24.9 ft^3 37. 500 ft^2 39. 34.9 ft^2

Exercise 4

1. 57.4 cubic units   3. π cubic units   5. 0.666 cubic units   7. 0.479 cubic units   9. 60.3 cubic units

11. 32π cubic units   13. 40.2 cubic units   15. $\dfrac{72\sqrt{3}}{5}\pi$ cubic units   17. $\dfrac{8}{9}\pi$ cubic units

19. 3.35 cubic units   21. 102 cubic units   25. 25.1 ft^3 27. 16.9 g

Review Problems

1. 31.4   3. 19.8   5. 25.6   7. 2.13   9. 19,300 rad/s; 15,400 rev

11. $v_x = \dfrac{t^2}{2} + 4$; $v_y = \dfrac{5}{2}t^2 + 15$; $x = \dfrac{t^3}{6} + 4t + 1$; $y = \dfrac{5t^3}{6} + 15t + 1$

13. 151   15. 442   17. 12π 19. 3.30 A

◆◆◆◆◆◆ **CHAPTER 28** ◆◆

Exercise 1

1. 0.704   3. 9.42   5. 5.92   7. 9.29   9. 4.59   11. 1096 ft   13. 223 ft   15. 30.3 ft

Exercise 2

1. 32.1   3. 154   5. 36.2   7. 3.77   9. 1570   11. 36.2   13. $4\pi r^2$ 15. 36.2 ft^2

Exercise 3

1. (5/4, −3/2)   3. 3.22   5. (2.40, 0)   7. 2.7   9. $\overline{x} = 1.067$ 11. $\overline{x} = 9/20$; $\overline{y} = 9/20$ 13. $\overline{x} = 0.300$
15. $\overline{x} = 1.25$ 17. 5   19. 5   21. $h/4$ 23. 4.13 ft   25. 36.1 mm

Exercise 4

1. 88,200 lb   **3.** 41,200 lb   **5.** 62.4 lb   **7.** 2160 lb

Exercise 5

1. 300 in. · lb   **3.** 32 in. · lb   **5.** 50,200 ft · lb   **7.** 5.71×10^5 ft · lb   **9.** 723,000 ft · lb
11. $k/100$ ft · lb   **13.** 13,800 ft · lb

Exercise 6

1. 1/12   **3.** 2.62   **5.** 19.5   **7.** 2/9   **9.** $10.1m$ **11.** $89.4m$ **13.** $\dfrac{3Mr^2}{10}$ **15.** $\dfrac{4Mp^2}{3}$

Review Problems

1. 4.24   **3.** (1.70, 1.27)   **5.** 48,300 ft · lb   **7.** 1120 lb   **9.** 21.80 in. · lb   **11.** 2560 lb   **13.** 51.5
15. 141   **17.** 410

◆◆◆◆◆◆◆ CHAPTER 29 ◆◆

Exercise 1

1. $\cos x$ **3.** $-3 \sin x \cos^2 x$ **5.** $3 \cos 3x$ **7.** $\cos^2 x - \sin^2 x$ **9.** $3.75(\cos x - x \sin x)$
11. $-2 \sin(\pi - x) \cos(\pi - x)$ **13.** $2 \cos 2\theta \cos \theta - \sin 2\theta \sin \theta$ **15.** $2 \sin x \cos^2 x - \sin^3 x$
17. $1.23(2 \cos 3x \sin x \cos x - 3 \sin^2 x \sin 3x)$ **19.** $\dfrac{-\sin 2t}{\sqrt{\cos 2t}}$ **21.** $-\cos x$ **23.** $-2 \sin x - x \cos x$
25. $-\dfrac{\pi^2}{4}$ **27.** $\dfrac{y \cos x + \cos y - y}{x \sin y - \sin x + x}$ **29.** $\sec(x + y) - 1$ **31.** -0.4161 **33.** 1.381
35. $\max(\pi/2, 1)$; $\min(3\pi/2, -1)$; $PI(0, 0)(\pi, 0)$ **37.** $\max(2.50, 5)$; $\min(5.64, -5)$; $PI(0.927, 0)$, $(4.07, 0)$
39. (a) $-735 \sin(2.84t + 0.75) \, \mu A$ (b) $-712 \, \mu A$

Exercise 2

1. $2 \sec^2 2x$ **3.** $-15 \csc 3x \cot 3x$ **5.** $6.50x \sec^2 x^2$ **7.** $-21x^2 \csc x^3 \cot x^3$
9. $x \sec^2 x + \tan x$ **11.** $5 \csc 6x - 30x \csc 6x \cot 6x$ **13.** $2 \sin \theta \sec^2 2\theta + \tan 2\theta \cos \theta$
15. $5 \csc 3t \sec^2 t - 15 \tan t \csc 3t \cot 3t$ **17.** -294 **19.** 853   **21.** $6 \sec^2 x \tan x$ **23.** 18.72
25. $-y \sec x \csc x$ **27.** -1 **29.** $3.43x - y = 1.87$ **31.** $\max(\pi/4, 0.571)$; $\min(3\pi/4, 5.71)$; $PI(0, 0)$, $(\pi, 2\pi)$
33. $v = 23.5$ cm/s; $a = -19.1$ cm/s^2 **35.** -3.58 deg/min   **37.** 15.6 ft   **39.** 31°   **41.** 25 ft

Exercise 3

1. $\operatorname{Sin}^{-1} x + \dfrac{x}{\sqrt{1 - x^2}}$ **3.** $\dfrac{-1}{\sqrt{a^2 - x^2}}$ **5.** $\dfrac{\cos x + \sin x}{\sqrt{1 + \sin 2x}}$ **7.** $-\dfrac{t^2}{\sqrt{1 - t^2}} + 2t \operatorname{Cos}^{-1} t$

9. $\dfrac{1}{1 + 2x + 2x^2}$ **11.** $\dfrac{-a}{a^2 + x^2}$ **13.** $\dfrac{-1}{x\sqrt{4x^2 - 1}}$ **15.** $2t \operatorname{Arcsin} \dfrac{t}{2} + \dfrac{t^2}{\sqrt{4 - t^2}}$

17. $\dfrac{1}{\sqrt{a^2 - x^2}}$ **19.** -0.285 **21.** -0.0537

Exercise 4

1. $\dfrac{1}{x} \log e$ **3.** $\dfrac{3}{x \ln b}$ **5.** $\dfrac{(5 + 9x) \log e}{5x + 6x^2}$ **7.** $\log\left(\dfrac{2}{x}\right) - \log e$ **9.** $\dfrac{1}{x}$ **11.** $\dfrac{2x - 3}{x^2 - 3x}$

13. $8.25 + 2.75 \ln 1.02x^3$ **15.** $\dfrac{1}{x^2(x + 5)} - \dfrac{2 \ln(x + 5)}{x^3}$ **17.** $\dfrac{1}{2t - 10}$ **19.** $\cot x$

21. $\cos x(1 + \ln \sin x)$ **23.** $\dfrac{\sin x}{1 + \ln y}$ **25.** $\dfrac{x + y - 1}{x + y + 1}$ **27.** $-y$ **29.** $-\dfrac{a^2}{x^2\sqrt{a^2 - x^2}}$

31. $x^{\sin x}[(\sin x)/x + \cos x \ln x]$ **33.** $(\text{Arccos } x)^x\left(\ln \text{Arccos } x - \dfrac{x}{\sqrt{1 - x^2}\,\text{Arccos } x}\right)$

35. 1   **37.** 0.3474   **39.** 128°   **41.** min$(1/e, -1/e)$ **43.** min(e, e), $PI/(e^2, \tfrac{1}{2}e^2)$

45. 1.37   **47.** 0.607   **49.** 1.2 per min   **51.** -0.0054 dB/day

Exercise 5

1. $2(3^{2x}) \ln 3$ **3.** $10^{2x+3}(1 + 2x \ln 10)$ **5.** $2x(2^{x^2}) \ln 2$ **7.** $2e^{2x}$ **9.** e^{x+e^x}

11. $-\dfrac{xe^{\sqrt{1-x^2}}}{\sqrt{1 - x^2}}$ **13.** $-\dfrac{2}{e^x}$ **15.** $xe^{3x}(3x + 2)$ **17.** $\dfrac{e^x(x - 1)}{x^2}$ **19.** $\dfrac{(x - 2)e^x + (x + 2)e^{-x}}{x^3}$

21. $2(x + xe^x + e^x + e^{2x})$ **23.** $\dfrac{(1 + e^x)(2xe^x - e^x - 1)}{x^2}$ **25.** $3e^x \sin^2 e^x \cos e^x$

27. $e^\theta(\cos 2\theta - 2 \sin 2\theta)$ **29.** $-e^{(x-y)}$ **31.** $\dfrac{\cos(x + y)}{e^y - \cos(x + y)}$ **33.** 0

35. $e^x\left(\ln x + \dfrac{1}{x}\right)$ **37.** $e^x\left(\dfrac{1}{x} + \ln x\right)$ **39.** $-2e^t \sin t$ **41.** $2e^x \cos x$ **43.** 0.79   **45.** 4.97

47. min$(0.402, 4.472)$ **49.** max$(\pi/4, 3.224)$; min$(5\pi/4, -0.139)$ **51.** \$915.30/yr   **53.** 74.2

55. -563 rev/min^2 **57.** (a) 1000 bacteria/h (b) 22×10^6 bacteria/h   **59.** -245 in. Hg/h

61. 0.436 lb/ft^3/mi   **63.** (a) $679e^{-t/122}\mu$A   (b) 199μA   **65.** (a) $-65.4e^{-t/128}$ (b) -19.5 V

Exercise 6

1. $\dfrac{a^{5x}}{5 \ln a} + C$ **3.** $\dfrac{5^{7x}}{7 \ln 5} + C$ **5.** $\dfrac{a^{3y}}{3 \ln a} + C$ **7.** $4e^x + C$ **9.** $\dfrac{e^{x^2}}{2} + C$ **11.** 27,122   **13.** $2e^{\sqrt{x}} + C$

15. 0.7580   **17.** $2e^{\sqrt{x-2}} + C$ **19.** $ae^{x/a} - ae^{-x/a} + C$ **21.** $x \ln 3x - x + C$ **23.** 1.273   **25.** 6.106

27. 198.0   **29.** 4.51   **31.** 7.21   **33.** 0.930   **35.** 2.12; 0.718   **37.** $V = -2790\,e^{-t/127} + 2790$ V

Exercise 7

1. $-\dfrac{1}{3} \cos 3x + C$ **3.** $-\dfrac{1}{5} \ln|\cos 5\theta| + C$ **5.** $\dfrac{1}{4} \ln|\sec 4x + \tan 4x| + C$ **7.** $-\dfrac{1}{3} \ln|\cos 9\theta| + C$

9. $-\dfrac{1}{2} \cos x^2 + C$ **11.** $-\dfrac{1}{3} \ln|\cos \theta^3| + C$ **13.** $-\cos(x + 1) + C$

15. $\dfrac{1}{5} \ln|\cos(4 - 5\theta)| + C$ **17.** $\dfrac{1}{8} \ln|\sec(4x^2 - 3) + \tan(4x^2 - 3)| + C$ **19.** $\dfrac{\sin x^2}{2} + C$ **21.** 2   **23.** 2

25. 2   **27.** $8/\pi$ **29.** 3   **31.** 14.4   **33.** $(\pi/2, \pi/8)$

35. (a) $q = -7.33 \cos(11.5t + 5.48) + 5.09$ C   (b) 12.3 C

Exercise 8

1. 12   **3.** 4.08   **5.** $\dfrac{1}{2}$ **7.** 7.81   **9.** 19.1   **11.** 1.08

Review Problems

1. $\frac{1}{2}(e^{x/a} + e^{-x/a})$ **3.** $\frac{4}{\sqrt{x}} \sec^2 \sqrt{x}$ **5.** $\frac{4x}{16x^2 + 1} + \text{Arctan } 4x$ **7.** $\frac{\cos 2x}{y}$

9. $x \cos x + \sin x$ **11.** $3x^2 \cos x - x^3 \sin x$ **13.** $\frac{x \cos x - \sin x}{x^2}$ **15.** $\frac{1 + 3x^2}{x^3 + x} \log e$

17. $\frac{1}{\sqrt{x^2 + a^2}}$ **19.** $-3 \csc 3x \cot 3x$ **21.** $\sec^2 x$ **23.** $\frac{6x^2 + 1}{2x^3 + x}$ **25.** $2x \text{ Arccos } x - \frac{x^2}{\sqrt{1 - x^2}}$

27. $(8.96, 0.291)$ **29.** 0.690 **31.** $\frac{3 \sin 2x^2}{2} + C$ **33.** $\frac{3}{5}$ **35.** $\frac{x^2}{2}(\ln x^2 - 1) + C$

37. $\frac{1}{6}(3x^2 - 2) \ln(3x^2 - 2) - \frac{x^2}{2} + C$ **39.** -0.0112 **41.** $\frac{1}{2}$ **43.** $x = \pi/4$

45. 0.0538 **47.** $y - 1/2 = \frac{\sqrt{3}}{2}(x - \frac{\pi}{6})$ **49.** $x = \pi/4$ **51.** 0.517 **53.** $1050°F$ at 5.49 h

55. (a) $-200 \sin(48.3t + 0.95)$ mA (b) 96.4 mA   **57.** (a) $-21.9 \cos(33.6t + 0.73) + 16.3$ C (b) 8.906 C

59. (a) $394e^{-t/335} \mu A$ (b) $217 \mu A$ **61.** $-5.89e^{-t/77.3} + 5.89$

◆◆◆◆◆◆◆ CHAPTER 30 ◆◆◆

Exercise 1

1. First order, first degree, ordinary   **3.** Third order, first degree, ordinary   **5.** Second order, fourth degree, ordinary

7. $y = \frac{7x^2}{2} + C$ **9.** $y = \frac{2x^2}{3} - \frac{5x}{3} + C$ **11.** $y = \frac{x^3}{3} + C$

Exercise 2

1. $y^2 = x^2 + C$ **3.** $y^4 = 4x^3/3 + C$ **5.** $y^3 = x^3 + 1$ **7.** 3 **9.** 1.26

Exercise 3

1. $y = \pm \sqrt{x^2 + C}$ **3.** $\ln|y^3| = x^3 + C$ **5.** $4x^3 - 3y^4 = C$ **7.** $y = \sqrt{x^2 + 1}/C$

9. $\text{Arctan } y + \text{Arctan } x = C$ **11.** $\sqrt{1 + x^2} + \ln y = C$ **13.** $\text{Arctan } x = \text{Arctan } y + C$

15. $2x + xy - 2y = C$ **17.** $x^2 + Cy^2 = C - 1$ **19.** $y = C\sqrt{1 + e^{2x}}$ **21.** $e^{2x} + e^{2y} = C$

23. $(1 + x) \sin y = C$ **25.** $\cos x \cos y = C$ **27.** $2 \sin^2 x - \sin y = C$ **29.** $y^3 - x^3 = 1$

31. $\sin x + \cos y = 2$

Exercise 4

1. $xy - 7x = C$ **3.** $xy - 3x = C$ **5.** $4xy - x^2 = C$ **7.** $\frac{y}{x} - 3x = C$ **9.** $x^3 + 2xy = C$

11. $x^2 + y^2 - 4xy = C$ **13.** $\frac{x}{y} = 2x + C$ **15.** $2y^3 = x + Cy$ **17.** $4xy^2 - 3x^2 = C$ **19.** $2x^2 - xy = 15$

21. $3y^3 - y = 2x$ **23.** $3x^2 + 3y^2 - 4xy = 8$

Exercise 5

1. $x(x - 3y)^2 = C$ **3.** $x \ln y - y = Cx$ **5.** $y^3 = x^3(3 \ln x + C)$ **7.** $x(x - 3y)^2 = 4$ **9.** $y^3 = 3x^3 \ln|x|$

Exercise 6

1. $y = 2x + \dfrac{C}{x}$ **3.** $y = x^3 + \dfrac{C}{x}$ **5.** $y = 1 + Ce^{-x^3/3}$ **7.** $y = \dfrac{C}{\sqrt{x}} - \dfrac{3}{x}$ **9.** $y = x + Cx^2$

11. $(1 + x^2)^2 y = 2\ln x + x^2 + C$ **13.** $y = \dfrac{C}{xe^{x^2/2}}$ **15.** $y = \dfrac{e^x}{2} + \dfrac{C}{e^x}$ **17.** $y = (4x + C)e^{2x}$

19. $x^2 y = 2\ln^2 x + C$ **21.** $y = \dfrac{1}{2}(\sin x - \cos x) + Ce^{-x}$ **23.** $y = \sin x + \cos x + Ce^{-x}$

25. $xy\left(C - \dfrac{3x^2}{2}\right) = 1$ **27.** $y = \dfrac{1}{x^2 + Ce^{-x}}$ **29.** $xy - 2x^2 = 3$ **31.** $y = \dfrac{5x}{2} - \dfrac{1}{2x}$

33. $y = \tan x + \sqrt{2}\,\sin x$

Exercise 7

1. $y = x^2 + 3x - 1$ **3.** $y^2 = x^2 + 7$ **5.** $y = 0.812e^x - x - 1$ **7.** $xy = 4$ **9.** $y = \dfrac{e^x + e^{-x}}{2}$

11. $y = Cx$ **13.** $3xy^2 + x^3 = k$

Exercise 8

3. 628 million bbl/day   **5.** 9.71°F   **7.** 1240°F   **13.** 15.0 ft   **15.** 2.03 s   **17.** 2.36 s

Exercise 9

7. 231 mA; 0.882 V   **9.** 46.4 mA   **11.** 237 mA   **13.** 139 mA

Review Problems

1. $y = \dfrac{e^x}{2x} + \dfrac{C}{xe^x}$ **3.** $y = \sin x + \cos x + Ce^{-x}$ **5.** $x^3 + 2xy = C$ **7.** $x\ln y - y = Cx$

9. $y\ln|1 - x| + Cy = 1$ **11.** $x^2 + \dfrac{x}{y} + \ln y + C$ **13.** $\sin x + \cos y = 1.707$ **15.** $x + 2y = 2xy^2$

17. 14.6 rev/s   **19.** $y = -2\ln|x| + C$ **21.** $y = 2x\ln x + 2x$ **23.** 4740 g

•••••••• **CHAPTER 31** ••

Exercise 1

1. $y = \dfrac{5}{2}x^2 + C_1 x + C_2$ **3.** $y = 3e^x + C_1 x + C_2$ **5.** $y = \dfrac{x^4}{12} + x$

Exercise 2

1. $y = C_1 e^x + C_2 e^{5x}$ **3.** $y = C_1 e^x + C_2 e^{2x}$ **5.** $y = C_1 e^{3x} + C_2 e^{-2x}$ **7.** $y = C_1 + C_2 e^{2x/5}$

9. $y = C_1 e^{2x/3} + C_2 e^{-3x/2}$ **11.** $y = (C_1 + C_2 x)e^{2x}$ **13.** $y = C_1 e^x + C_2 x e^x$ **15.** $y = (C_1 + C_2 x)e^{-2x}$

17. $y = C_1 e^{-x} + C_2 x e^{-x}$ **19.** $y = e^{-2x}(C_1 \cos 3x + C_2 \sin 3x)$ **21.** $y = e^{3x}(C_1 \cos 4x + C_2 \sin 4x)$

23. $y = C_1 \cos 2x + C_2 \sin 2x$ **25.** $y = e^{2x}(C_1 \cos x + C_2 \sin x)$ **27.** $y = 3x e^{-3x}$ **29.** $y = 5e^x - 14x e^x$

31. $y = 1 + e^{2x}$ **33.** $y = (1/4)e^{2x} + (3/4)e^{-2x}$ **35.** $y = e^{-x}\sin x$ **37.** $y = C_1 e^x + C_2 e^{-x} + C_3 e^{2x}$

39. $y = C_1 e^x + C_2 e^{2x} + C_3 e^{3x}$ **41.** $y = C_1 e^x + C_2 e^{-x} + C_3 e^{3x}$ **43.** $y = (C_1 + C_2 x)e^{x/2} + C_3 e^{-x}$

Exercise 3

1. $y = C_1 e^{2x} + C_2 e^{-2x} - 3$ **3.** $y = C_1 e^{2x} + C_2 e^{-x} - 2x + 1$ **5.** $y = C_1 e^{2x} + C_2 e^{-2x} - (1/4)x^3 - (5/8)x$

7. $y = C_1 e^{2x} + C_2 e^{-x} - 3e^x$ **9.** $y = C_1 e^{2x} + C_2 e^{-2x} + e^x - x$ **11.** $y = e^{-2x}(C_1 + C_2 x) + (2e^{2x} + x - 1)/4$

13. $y = C_1 \cos 2x + C_2 \sin 2x - \dfrac{x \cos 2x}{4}$ **15.** $y = (C_1 + C_2 x)e^{-x} + (1/2)\sin x$

17. $y = C_1 \cos x + C_2 \sin x + \cos 2x + x \sin x$ **19.** $y = C_1 \cos x + C_2 \sin x + (e^x/5)(\sin x - 2\cos x)$

21. $y = e^{2x}[C_1 \sin x + C_2 \cos x - (x/2)\cos x]$ **23.** $y = \dfrac{e^{4x}}{2} - 2x - \dfrac{1}{2}$ **25.** $y = \dfrac{1 - \cos 2x}{2}$

27. $y = e^x(x^2 - 1)$ **29.** $y = x \cos x$

Exercise 4

1. (a) 34.5 ms, (b) 3.75 in.   **3.** $x = \sin 15t$ **5.** $x = \cos 6.95t$ **7.** $x = 1.50 \cos \sqrt{2}t$

9.

11. −0.714 in.   **13.** $x = 7.10e^{-t} - 5.60e^{-4t}$

15. (a) Overdamped, (b) 24.1 lb

Exercise 5

13. 9.84 Hz   **15.** 0.496 μF   **17.** 10 A

Review Problems

1. $y = C_1 e^{\sqrt{2}x} + C_2 e^{-\sqrt{2}x} - x^3 - (9/2)x$ **3.** $y = C_1 e^{\sqrt{2}x} + C_2 e^{-\sqrt{2}x} + C_3$ **5.** $y = C_1 e^x + C_2 e^{-3x}$

7. $y = -2e^{-3x} + 2xe^{-3x}$ **9.** $y = C_1 e^{\sqrt{2}x} + C_2 e^{-\sqrt{2}x} - 3e^x(x^3 + 6x^2 + 30x + 72)$

11. $y = C_1 + C_2 e^{-x} - (1/2)\sin 3x - (1/6)\cos 3x$ **13.** $y = C_1 \cos 3x + C_2 \sin 3x$ **15.** $y = 2xe^x$

17. $y = C_1 e^{2x} + C_2 xe^{2x}$ **19.** $y = C_1 e^{-2x} + C_2 xe^{-2x}$ **21.** $y = C_1 e^{-x} + C_2 xe^{-x}$ **23.** $x = 1.5 \cos 7.42t$

25. $i = 98.1 \sin 510t$ mA   **27.** $i = 4.66 e^{-18.75t} \sin 258t$ A

APPLICATIONS INDEX

INDEX TO WRITING QUESTIONS

INDEX TO PROJECTS

GENERAL INDEX

	No.			Formula	Page
ANY TRIANGLE	102	Areas		$\text{Area} = \frac{1}{2}bh$	181
	103			Hero's Formula: $\text{Area} = \sqrt{s(s-a)(s-b)(s-c)}$ where $s = \frac{1}{2}(a+b+c)$	181
	104			Sum of the Angles — $A + B + C = 180°$	182, 212
	105			Law of Sines — $\dfrac{a}{\sin A} = \dfrac{b}{\sin B} = \dfrac{c}{\sin C}$	241
	106			Law of Cosines — $a^2 = b^2 + c^2 - 2bc\cos A$; $b^2 = a^2 + c^2 - 2ac\cos B$; $c^2 = a^2 + b^2 - 2ab\cos C$	247
	107			Exterior Angle — $\theta = A + B$	182
RIGHT TRIANGLES	110			Pythagorean Theorem — $a^2 + b^2 = c^2$	183, 212
	111	Trigonometric Ratios	Sine	$\sin\theta = \dfrac{y}{r} = \dfrac{\text{opposite side}}{\text{hypotenuse}}$	208, 212, 221
	112		Cosine	$\cos\theta = \dfrac{x}{r} = \dfrac{\text{adjacent side}}{\text{hypotenuse}}$	208, 212, 221
	113		Tangent	$\tan\theta = \dfrac{y}{x} = \dfrac{\text{opposite side}}{\text{adjacent side}}$	208, 212, 221
	114		Cotangent	$\cot\theta = \dfrac{x}{y} = \dfrac{\text{adjacent side}}{\text{opposite side}}$	232
	115		Secant	$\sec\theta = \dfrac{r}{x} = \dfrac{\text{hypotenuse}}{\text{adjacent side}}$	232
	116		Cosecant	$\csc\theta = \dfrac{r}{y} = \dfrac{\text{hypotenuse}}{\text{opposite side}}$	232
	117	Reciprocal Relations		(a) $\csc\theta = \dfrac{1}{\sin\theta}$ (b) $\sec\theta = \dfrac{1}{\cos\theta}$ (c) $\cot\theta = \dfrac{1}{\tan\theta}$	234, 462
	118	A and B are Complementary Angles	Cofunctions	(a) $\sin A = \cos B$ (d) $\cot A = \tan B$; (b) $\cos A = \sin B$ (e) $\sec A = \csc B$; (c) $\tan A = \cot B$ (f) $\csc A = \sec B$	235
TRIGONOMETRIC IDENTITIES	123	Quotient Relations		$\tan\theta = \dfrac{\sin\theta}{\cos\theta}$	463
	124			$\cot\theta = \dfrac{\cos\theta}{\sin\theta}$	463
	125	Pythagorean Relations		$\sin^2\theta + \cos^2\theta = 1$	463
	126			$1 + \tan^2\theta = \sec^2\theta$	464
	127			$1 + \cot^2\theta = \csc^2\theta$	464
	128	Sum or Difference of Two Angles		$\sin(\alpha \pm \beta) = \sin\alpha\cos\beta \pm \cos\alpha\sin\beta$	470
	129			$\cos(\alpha \pm \beta) = \cos\alpha\cos\beta \mp \sin\alpha\sin\beta$	471
	130			$\tan(\alpha \pm \beta) = \dfrac{\tan\alpha \pm \tan\beta}{1 \mp \tan\alpha\tan\beta}$	472
	131	Double-Angle Relations		$\sin 2\alpha = 2\sin\alpha\cos\alpha$	474
	132			(a) $\cos 2\alpha = \cos^2\alpha - \sin^2\alpha$ (b) $\cos 2\alpha = 1 - 2\sin^2\alpha$ (c) $\cos 2\alpha = 2\cos^2\alpha - 1$	475
	133			$\tan 2\alpha = \dfrac{2\tan\alpha}{1 - \tan^2\alpha}$	476
	134	Half-Angle Relations		$\sin\dfrac{\alpha}{2} = \pm\sqrt{\dfrac{1 - \cos\alpha}{2}}$	477
	135			$\cos\dfrac{\alpha}{2} = \pm\sqrt{\dfrac{1 + \cos\alpha}{2}}$	478
	136			(a) $\tan\dfrac{\alpha}{2} = \dfrac{1 - \cos\alpha}{\sin\alpha}$ (b) $\tan\dfrac{\alpha}{2} = \dfrac{\sin\alpha}{1 + \cos\alpha}$ (c) $\tan\dfrac{\alpha}{2} = \pm\sqrt{\dfrac{1 - \cos\alpha}{1 + \cos\alpha}}$	479